Condensed Matter Field Theory

The methods of quantum field theory underpin many conceptual advances in contemporary condensed matter physics and neighboring fields. This book provides a praxis-oriented and pedagogical introduction to quantum field theory in many-particle physics, emphasizing the application of theory to real physical systems. This third edition is organized into two parts: the first half of the text presents a streamlined introduction, elevating readers to a level where they can engage with contemporary research literature from the introduction of many-body techniques and functional integration to renormalization group methods, and the second half addresses a range of advanced topics including modern aspects of gauge theory, topological and relativistic quantum matter, and condensed matter physics out of thermal equilibrium. At all stages the text seeks a balance between methodological aspects of quantum field theory and practical applications. Extended problems with worked solutions provide a bridge between formal theory and a research-oriented approach.

Alexander Altland is Professor of Theoretical Physics at the University of Cologne. He is a field theorist who has worked in various areas of condensed matter physics and neighboring fields. His research interests include the physics of disordered and chaotic systems, and the connections between condensed matter and particle physics.

Ben Simons is the Royal Society E. P. Abraham Professor and Herchel Smith Professor of Physics at the University of Cambridge. His research has spanned a broad range of areas in condensed matter and statistical physics, from disordered normal and superconducting compounds to correlated electron and light–matter systems and biological physics.

Condensed Matter Field Theory

Third edition

ALEXANDER ALTLAND
University of Cologne

BEN SIMONS
University of Cambridge

CAMBRIDGE
UNIVERSITY PRESS

CAMBRIDGE
UNIVERSITY PRESS

Shaftesbury Road, Cambridge CB2 8EA, United Kingdom

One Liberty Plaza, 20th Floor, New York, NY 10006, USA

477 Williamstown Road, Port Melbourne, VIC 3207, Australia

314–321, 3rd Floor, Plot 3, Splendor Forum, Jasola District Centre, New Delhi – 110025, India

103 Penang Road, #05–06/07, Visioncrest Commercial, Singapore 238467

Cambridge University Press is part of Cambridge University Press & Assessment, a department of the University of Cambridge.

We share the University's mission to contribute to society through the pursuit of education, learning and research at the highest international levels of excellence.

www.cambridge.org
Information on this title: www.cambridge.org/9781108494601

DOI: 10.1017/9781108781244

First published 2023

A catalogue record for this publication is available from the British Library

A Cataloging-in-Publication data record for this book is available from the Library of Congress

ISBN 978-1-108-49460-1 Hardback

Contents

Preface

Most students who have entered the physics Master's curriculum will have some familiarity with condensed matter physics. But what of "condensed matter *field theory*," the subject of this text? *Fields* are effective continuum degrees of freedom describing macroscopically large numbers of "atomistic" objects. Mundane examples of fields include water waves formed from the interaction of molecules or currents inside a conductor describing the collective motion of electrons. The language of fields reduces the complexity of many-particle systems to a manageable level, the natural degrees of freedom of condensed matter systems.

In condensed matter physics, we neither can, nor want to, trace the dynamics of individual atomistic constituents. Instead, we aim to understand the observable collective properties of matter, their thermal excitations, transport properties, phase behavior and transitions, etc. The art of condensed matter theory is to identify the nature and dynamics of the low-energy degrees of freedom – articulated as fields, and formulated with the framework of effective theories encapsulating universal properties of matter. This program has a long history, and it unfolded in a succession of epochs: in the 1950s and 1960s, the development of high-order perturbation theory; in the 1970s, the advent of renormalization group method; in the 1980s, the development of powerful non-perturbative methods; and, up to the present day, advances in topological field theories. These developments often paralleled, and drew inspiration from, particle physics, where quantum field theory was just as important, if from a slightly different perspective. In the course of its development, field theory has become a lingua franca, providing a unifying framework for the exploration of core concepts of condensed matter physics, as follows.

▷ **Universality:** A comparatively small number of "effective theories" suffices to describe the physics of myriads of different forms of matter. For example, the quantum field theory of vortices in superfluid helium films is the same as that of a plasma of dipoles. Despite different microscopic realizations, these systems fall into the same universality class.

▷ **Emergence:** In condensed matter physics, the conspiracy of large numbers of fundamental degrees of freedom often leads to the emergence of a smaller number of effective ones. For example, in two-dimensional electron systems subject to magnetic fields the emergent degrees of freedom may be effectively pointlike objects carrying *fractional* charge. These quasi-particles are responsible for the observable physics of the system. If one did not know their emergent nature, one might consider them as fractionally charged fermions.

▷ **Broken symmetries and collective fluctuations:** States of matter often show lower symmetry than that of the underlying microscopic theory. For example, a ferromagnetic substance may be magnetized along a specific direction while its Hamiltonian is invariant under global spin rotations. Under such conditions, large collective fluctuations, representing continuous changes between states of different local symmetry, are prevalent. In the vicinity of transition points between different phases, they can induce criticality.

▷ **Criticality**: Fluctuation-induced phenomena are characterized uniquely by just a few dimensionless parameters known as critical exponents. A relatively small number of different critical theories suffices to explain and describe the critical scaling properties of the majority of condensed matter systems close to phase transitions. Yet, the critical theories for some of the most well-known transitions (including, for example, the integer quantum Hall transition) remain unknown, presenting open challenges to future generations of field theorists.

In this third edition, we have separated the text into two major components. In the first part, we introduce core concepts of condensed matter quantum field theory. These chapters will furnish readers with fluency in the language and methodologies of modern condensed matter theory research. No prior knowledge is assumed beyond familiarity with quantum mechanics, statistical mechanics and solid state physics at bachelor's level. We aim to introduce the subject gently, in a language that changes gradually from being prosaic in the beginning to more scientific in later chapters. The subjects covered in the first part reflect developments in condensed matter theory that took place in the second half of the last century. However, in contrast with traditional approaches, the text does not recount these advances in chronological order. Instead, it emphasizes the comparatively modern methods of functional *field integration* – the generalization of the Feynman path integral of quantum point particles to continuum degrees of freedom. We introduce this concept early on and rely on it as an organizational principle throughout the text.

The second part of the text addresses more advanced developments, most of which have come to the fore over the past 30 years. During this period, developments in quantum field theory have proceeded in concert with revolutionary progress in experiment, both in solid state physics and in the neighboring fields of ultracold atom and optical physics. For example, while previous generations of experiments in condensed matter were conducted under close to thermal equilibrium conditions, the micro-fabrication of devices has reached levels such that *nonequilibrium* phenomena can be accessed and controlled. At the same time, we are seeing the advent and impact of *topological* forms of matter, whose physical properties are governed by the mathematical principles of topological order and long-range quantum entanglement. Combined with advances in the ability to manipulate and control quantum states, these developments are beginning to open a window on computational matter, i.e., realizations of condensed matter systems capable of storing and processing quantum information. Indeed, although separate, these new developments are surprisingly interrelated: quantum information may be protected by principles of topology, while nonequilibrium phases of quantum matter may be characterized by principles of topological gauge theory, etc. While it is too early to say where the field may evolve in the next 30 years, concepts from condensed matter field theory will play a key role in shaping new directions of research and in exposing their common themes. The contents and style of this more advanced part of the text reflect these structures.

While the first, introductory, component of the text is arranged in a structured manner, with each new chapter building upon previous chapters, the chapters of the second part of the text can be read independently. Moreover, the writing style of the more advanced chapters is often more succinct, drawing attention to primary, and often contemporary, literature. Perhaps most importantly, a key objective of the second part of the text is to draw readers into modern areas of condensed matter research. Alongside the core material we have also included several forms of supplementary material:

▷ **Info** sections place methodological developments into a given context, contain details on specific applications, or simply provide auxiliary "information" that may enrich the narrative. For example, in chapter 1, an info section is used to describe the concrete realization of "vacuum fluctuations" of fields in condensed matter systems, in the context of Casimir or van der Waals forces.

▷ **Example** sections are used to develop general concepts. For example, the two-sphere is used as an example to illustrate the general concept of differentiable manifolds.

▷ **Remarks** appear as preambles of some sections. They may indicate, for example, whether a section may require knowledge of previous material; this is particularly valuable in the second part of the text, where the chapters are non-sequentially ordered or interlinked. The text also includes sections that, while important, may be safely skipped at a first reading. In such cases, the remarks section provide advice and guidance.

▷ In-text **exercises** (some answered,[1] and some not) provide opportunities for the reader to test their methodological understanding. Alongside these small exercises, each chapter closes with a problem set.

▷ **Problem** set: These problems differ from the in-text exercises both in depth and level, and are chosen to mirror as much as possible the solution of "realistic" condensed matter problems by field-theoretical methods. Their solution requires not just methodological but also conceptual thinking. Many of them reflect the narrative of research papers, some of which are of historical significance. For example, a problem of chapter 2 reviews the construction of the Kondo Hamiltonian as an illustration of the utility of second quantization. Answers are provided for all questions in the problem sets.

---

[1] The reader should not be surprised to find that some of the answers to in-text questions are given ǝpᴉsdn uʍop in footnotes!

▷ Lastly, four short **appendices** introduce or review background material referred to in parts of the main text. They include a review of elements of probability theory, a summary of the Fourier transform conventions used in the text, an introduction to modern concepts of differential geometry, and a concise introduction to conformal symmetry.

This third edition of the text responds in part to the changes that have taken place in the research landscape and emphasis since the first edition was published more than a decade ago. Among these changes, the first and foremost reflects revolutionary developments in topological condensed matter physics. The core chapter on topological field theory has been completely rewritten, and two accompanying chapters – one on gauge theory, and another on relativistic quantum matter – have been added. All other chapters have been substantially revised and brought up to date. In particular, we have taken this opportunity to prune material whose prominence and value to future research may have diminished. At the same time, we have eliminated many "typos" and the occasional embarrassing error, many of which have been drawn to our attention by our friends and colleagues in the community (see below)! We fear that the addition of fresh material will have introduced new errors and will do our best to correct them when notified.

Over the years, many people have contributed to this text, either through constructive remarks and insights, or by spotting typos and errors. In this context, it is a great pleasure to acknowledge with gratitude the substantial input of Sasha Abanov, Piet Brouwer, Christoph Bruder, Chung-Pin Chou, Karin Everschor, Andrej Fischer, Sven Gnutzmann, Colin Kiegel, Jian-Lin Li, Tobias Lück, Jakob Müller-Hill, Julia Meyer, Tobias Micklitz, Jan Müller, Patrick Neven, Sid Parameswaran, Achim Rosch, Max Schäfer, Matthias Sitte, Rodrigo Soto-Garrido, Natalja Strelkova, Nobuhiko Taniguchi, Franjo-Frankopan Velic, Matthias Vojta, Jan von Delft, Andrea Wolff, and Markus Zowislok. We finally thank Martina Markus for contributing hand-drawn portraits of some of the great scientists who pioneered the physics discussed in this book.

PART I

On a fundamental level, all forms of quantum matter can be formulated in terms of a many-body Hamiltonian for a macroscopically large number of constituent particles. However, in contrast with many other areas of physics, the structure of this operator conveys as much information about the properties of the system as, say, the knowledge of the basic chemical constituents tells us about the behavior of a living organism. Rather, it has been a long-standing tenet that the degrees of freedom relevant to the low-energy properties of a system usually are not the microscopic ones. It is a hallmark of many "deep" problems of modern condensed matter physics that the passage between the microscopic and the effective degrees of freedom involves complex and, at times, even controversial mappings. To understand why, it is helpful to review the process of theory building in this field of physics.

The development of early condensed matter physics often hinged on the "unreasonable" success of *non-interacting* theories. The impotency of interactions observed in a wide range of physical systems reflects a **adiabatic continuity**

principle known as **adiabatic continuity**: the quantum numbers characterizing an (interacting) many-body system are determined by fundamental symmetries – translational, rotational, particle exchange, etc. As long as these symmetries are maintained, the system's elementary excitations, or *quasi-particles*, can usually be traced back "adiabatically" to the bare particles of the non-interacting limit. This principle, embodied in Landau's Fermi-liquid theory, has provided a platform for the investigation of a wide range of systems, from conventional metals to ^{3}helium fluids and cold atomic Fermi gases.

However, being contingent on symmetry, it must be abandoned at phase transitions, where interactions effect a rearrangement of the many-body ground state into a state of different, or "broken" symmetry. Symmetry-broken phases generically show excitations different from those of the parent non-interacting phase. They either require classification in terms of new species of quasi-particles, or they may be *collective modes* engaging the cooperative motion of many bare particles. For example, when atoms condense from a liquid into a solid phase, translational symmetry is broken and the elementary excitations (phonons) involve the motion of many individual bare particles.

In this way, each phase of matter is associated with a unique "non-interacting" reference state with its own characteristic quasi-particle excitations – a product of only the relevant symmetries. Within each individual phase, a continuity principle keeps the effects of interactions at bay. This hierarchical picture delivers two profound implications. First, within the quasi-particle framework, the underlying "bare" or elementary particles remain invisible. (To quote from P. W. Anderson's famous article *More is different*, Science **177**, 393 (1972), "The ability to reduce everything to simple fundamental laws does not imply the ability to start from those laws and reconstruct the universe.") Second, while one may conceive an almost unlimited spectrum of interactions, there are comparatively few non-interacting or free theories, constrained by the set of fundamental symmetries. These arguments go a long way in explaining the principle of "universality" observed in condensed matter.

How can these concepts be embedded into a concrete theoretical framework? At first sight, problems with macroscopically many particles seem overwhelmingly daunting. However, our discussion above indicates that representations of manageable complexity may be obtained by focusing on symmetries and restricted sets of excitations. Quantum field theory provides the keys to making this reduction concrete. Starting from an efficient microscopic formulation of the many-body problem, it allows one to systematically develop effective theories for collective degrees of freedom. Such representations afford a classification of interacting systems into a small number of universality classes defined by their fundamental symmetries. This form of complexity reduction has become a potent source of unification in modern theoretical physics. Indeed, several subfields of theoretical physics (such as conformal field theory, random matrix theory, etc.) now define themselves not so much through any specific application as by a certain conceptual or methodological framework.

As mentioned in the preface, the first part of this text is a primer aimed at elevating graduate students to a level where they can engage in independent research. While the discussion of conceptual aspects stands in the foreground, we have endeavored to keep the text firmly rooted in experimental application. As well as routine exercises, it includes extended problems meant to train research-oriented thinking. Some of these answered problems are deliberately designed to challenge. (We all know from experience that the intuitive understanding of formal structures can be acquired only by persistent, and at times even frustrating training.)

1   From Particles to Fields

SYNOPSIS To introduce some basic concepts of field theory, we begin by considering two simple model systems – a one-dimensional "caricature" of a solid and a freely propagating electromagnetic wave. As well as exemplifying the transition from discrete to continuous degrees of freedom, these examples introduce the basic formalism of classical and quantum field theory as well as the notions of elementary excitations, collective modes, symmetries and universality – concepts that will pervade the rest of the text.

One of the appealing facets of condensed matter physics is that phenomenology of remarkable complexity can emerge from a Hamiltonian of comparative simplicity. Indeed, microscopic "condensed matter Hamiltonians" of high generality can be constructed straightforwardly. For example, a prototypical metal or insulator may be described by the **many-particle Hamiltonian**, $H = H_e + H_i + H_{ei}$, where

margin note: many-particle Hamiltonian

$$H_e = \sum_i \frac{\mathbf{p}_i^2}{2m} + \sum_{i<j} V_{ee}(\mathbf{r}_i - \mathbf{r}_j),$$

$$H_i = \sum_I \frac{\mathbf{P}_I^2}{2M} + \sum_{I<J} V_{ii}(\mathbf{R}_I - \mathbf{R}_J), \qquad (1.1)$$

$$H_{ei} = \sum_{iI} V_{ei}(\mathbf{R}_I - \mathbf{r}_i).$$

Here, $\mathbf{r}_i\,(\mathbf{R}_I)$ denotes the coordinates of valence electrons (ion cores), while H_e, H_i, and H_{ei} describe the dynamics of electrons, ions and the interaction of electrons and ions, respectively (see the figure). Of course, the Hamiltonian (1.1) can be made more realistic by, for example, remembering that electrons and ions carry spin, adding potential disorder, or introducing host lattices with multi-atomic unit cells.

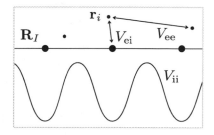

However, for developing our present line of thought, the prototype Hamiltonian H will suffice.

The fact that a seemingly innocuous Hamiltonian like Eq. (1.1) is capable of generating a plethora of phenomenology can be read in reverse: normally, one will not be able to make progress theoretically by approaching the problem in an

ab initio
approach

"**ab initio**" manner, i.e., by an approach that treats all microscopic constituents as equally relevant degrees of freedom. How, then, can successful analytical approaches be developed? The answer lies in several basic principles inherent in generic condensed matter systems.

reduction
principle

1. **Structural reducibility**: Not all components of the Hamiltonian (1.1) need to be treated simultaneously. For example, when our interest is in the vibrational motion of the ion lattice, the dynamics of the electron system can often be neglected or, at least, treated in an "effective" manner. Similarly, the dynamics of the electron system can often be considered independent of the ions, etc.

2. In the majority of condensed matter applications, one is interested not so much in the full profile of a given system but, rather, in its energetically low-lying dynamics. This is motivated partly by practical aspects (in daily life, iron is normally encountered at room temperature and not at its melting point), and partly by the tendency of large systems to behave in a "universal" man-

universality
principle

ner at low temperatures. Here, **universality** implies that systems differing in microscopic detail (i.e., with different types of interaction potentials, ion species, etc.) exhibit common collective behavior at low energy or long length scales. As a physicist, one will normally seek for unifying principles in collective phenomena rather than to describe the peculiarities of individual elements or compounds. However, universality is equally important in the *practice* of condensed matter theory. In particular, it implies that, at low temperatures, system-specific details of microscopic interaction potentials are often of secondary importance, i.e., one may employ *simple* model Hamiltonians.

3. For most systems of interest, the number of degrees of freedom $N = \mathcal{O}(10^{23})$ is formidably large. However, contrary to first impressions, the magnitude of this figure is rather an advantage: in addressing condensed matter problems,

statistical
principles

the **principles of statistics** imply that statistical errors tend to be negligibly small.[1]

symmetries

4. Finally, condensed matter systems typically possess intrinsic **symmetries**. For example, the Hamiltonian (1.1) is invariant under the simultaneous translation and/or rotation of all coordinates, which expresses the global Galilean invariance of the system (these are continuous symmetries). Invariance under spin rotation (continuous) or time reversal (discrete) are other examples of common symmetries. The general importance of symmetries cannot be overemphasized: symmetries support the conservation laws that simplify any problem. Yet, in

---

[1] The importance of this point is illustrated by the empirical observation that the most challenging systems in physical sciences are of *medium*, and not large, scale, e.g., metallic clusters, medium-sized nuclei or large atoms consisting of some $\mathcal{O}(10^1\text{--}10^2)$ fundamental constituents. Such systems are beyond the reach of few-body quantum mechanics while not yet accessible to reliable statistical modeling. The only viable path to approaching systems of this type is often through numerical simulation or the use of phenomenology.

condensed matter, symmetries are even more important. A conserved observable is generally tied to an energetically low-lying excitation. In the universal, low-temperature, regime in which we will typically be interested, it is precisely the dynamics of these excitations that govern the gross behavior of the system. Generally, the identification of fundamental symmetries is the first step in the sequence "symmetry $\mapsto$ conservation law $\mapsto$ low-lying excitations" and one that we will encounter time and again.

To understand how these basic principles can be used to formulate and explore effective low-energy field theories of solid state systems, we will begin by focusing on the **harmonic chain**, a collection of atoms bound by a harmonic potential. In doing so, we will observe that the universal characteristics encapsulated by the low-energy dynamics[2] of large systems relate naturally to concepts of **field theory**.

1.1 Classical Harmonic Chain: Phonons

[Classical Harmonic Chain: Phonons]

Returning to the prototype Hamiltonian (1.1), let us focus on the dynamical properties of the positively charged *core ions* that constitute the host lattice of a crystal. For the moment, we will neglect the fact that atoms are quantum objects and treat the ions as *classical* entities. To further simplify the problem, let us consider a one-dimensional atomic chain rather than a generic *d*-dimensional solid. In this case, the positions of the ions can be specified by a sequence of coordinates with average lattice spacing a. Relying on the structural reducibility principle 1, we will first argue that, to understand the behavior of the ions, the dynamics of the conduction electrons are of secondary importance, i.e., we will set $H_e = H_{ei} = 0$.

At zero temperature, the system freezes into a regularly spaced array of ion cores at coordinates $R_I = \bar{R}_I \equiv Ia$. Any deviation from a perfectly regular configuration incurs a potential energy cost. For low enough temperatures (principle 2), this energy will be

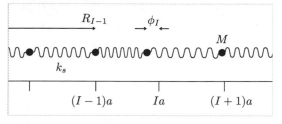

approximately quadratic in the small deviation of the ion from its equilibrium position. The "reduced" low-energy Hamiltonian of the system then reads

$$H = \sum_{I=1}^{N} \left[\frac{P_I^2}{2m} + \frac{k_s}{2}(R_{I+1} - R_I - a)^2 \right],$$ (1.2)

---

[2] In this text, we will focus on the *dynamical* behavior of large systems, as opposed to their *static* structural properties. In particular, we will not address questions related to the formation of definite crystallographic structures in solid state systems.

where the coefficient k_s determines the steepness of the lattice potential. Note that H can be interpreted as the Hamiltonian of N point-like particles of mass m connected by elastic springs with spring constant k_s (see the figure).

1.1.1 Lagrangian formulation and equations of motion

What are the elementary low-energy excitations of the **classical harmonic chain**? To answer this question we might, in principle, attempt to solve Hamilton's equations of motion. Indeed, since H is quadratic in all coordinates, such a program is feasible. However, few of the problems encountered in solid state physics enjoy this property. Further, it seems unlikely that the low-energy dynamics of a macroscopi-

classical
harmonic
chain

> **Joseph-Louis Lagrange**
> **1736–1813**
> was a French mathematician and astronomer (though born in Turin) who excelled in all fields of analysis, number theory and celestial mechanics. In 1788, he published *Mécanique Analytique*, which summarized the field of mechanics since the time of Newton, and is notable for its use of the theory of differential equations. In this text, he transformed mechanics into a branch of mathematical analysis.

cally large chain – which we know from our experience will be governed by *large-scale* wave-like excitations – is adequately described in terms of an "atomistic" language; the relevant degrees of freedom will be of a different type. We should, rather, draw on the basic principles 1–4 set out above. Notably, so far, we have paid attention to neither the intrinsic symmetry of the problem nor the fact that the number of ions, N, is large.

To reduce a microscopic model to an effective low-energy theory, often the Hamiltonian is not a very convenient starting point. Usually, it is more efficient to start from the **classical action**, S. In the present case, $S = \int dt\, L(R, \dot{R})$, where $(R, \dot{R}) \equiv \{R_I, \dot{R}_I\}$ represent the coordinates and their time derivatives. The corresponding **classical Lagrangian** L related to the Hamiltonian (1.2) is given by

classical
action

classical
Lagrangian

$$L = T - U = \sum_{I=1}^{N} \left[\frac{m}{2}\dot{R}_I^2 - \frac{k_\mathrm{s}}{2}(R_{I+1} - R_I - a)^2 \right], \qquad (1.3)$$

where T and U denote, respectively, the kinetic and potential energy.

Since we are interested in the properties of the large-N system, we can expect boundary effects to be negligible. In this case, we may impose periodic boundary conditions, making the identification $R_{N+1} = R_1$. Further, anticipating that the effect of lattice vibrations on the solid is weak (i.e., long-range atomic order is maintained), we may assume that the deviation of ions from their equilibrium position is small (i.e., $|R_I(t) - \bar{R}_I| \ll a$). For $R_I(t) = \bar{R}_I + \phi_I(t)$, with $\phi_{N+1} = \phi_1$, the Lagrangian (1.3) assumes the simplified form

$$L = \sum_{I=1}^{N} \left[\frac{m}{2}\dot{\phi}_I^2 - \frac{k_\mathrm{s}}{2}(\phi_{I+1} - \phi_I)^2 \right].$$

To make further progress, we will now make use of the fact that we are not concerned with behavior on "atomic" scales. For such purposes, our model would, in any case, be much too primitive! Rather, we are interested in experimentally observable behavior that becomes manifest on macroscopic length scales (principle 2). For example, one might wish to study the specific heat of the solid in the limit of infinitely many atoms (or at least a macroscopically large number, $\mathcal{O}(10^{23})$). Under these conditions, microscopic models can usually be simplified substantially (principle 3). In particular, it is often permissible to subject a discrete lattice model **continuum** to a **continuum limit**, i.e., to neglect the discreteness of the microscopic entities **limit** and to describe the system in terms of effective continuum degrees of freedom.

In the present case, taking a continuum limit amounts to describing the lattice fluctuations ϕ_I in terms of smooth functions of a continuous variable x. (See the figure, where the [horizontal] displacement of the point particles is plotted along the vertical axis.)

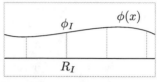

Clearly such a description makes sense only if the relative fluctuations on atomic scales are weak (for otherwise the smoothness condition would be violated). However, if this condition is met – as it will be for sufficiently large values of the stiffness constant k_s – the continuum description is much more powerful than the discrete encoding in terms of the "vector" $\{\phi_I\}$. The steps that we will need to take to go from the Lagrangian to concrete physical predictions will then be much easier to formulate.

Introducing continuum degrees of freedom $\phi(x)$, and applying a first-order Taylor expansion,[3] let us define

$$\phi_I \to a^{1/2}\phi(x)\Big|_{x=Ia}, \quad (\phi_{I+1}-\phi_I) \to a^{3/2}\partial_x\phi(x)\Big|_{x=Ia}, \quad \sum_{I=1}^{N} \to \frac{1}{a}\int_0^L dx,$$

where $L = Na$. Note that, as defined, the functions $\phi(x,t)$ have dimensionality $[\text{length}]^{1/2}$. Expressed in terms of the new degrees of freedom, the continuum limit of the Lagrangian then reads

$$L[\phi] = \int_0^L dx\, \mathcal{L}(\partial_x\phi, \dot\phi), \quad \mathcal{L}(\partial_x\phi, \dot\phi) = \frac{m}{2}\dot\phi^2 - \frac{k_\mathrm{s}a^2}{2}(\partial_x\phi)^2, \qquad (1.4)$$

Lagrangian where the **Lagrangian density** $\mathcal{L}$ has dimensionality [energy]/[length]. Similarly, **density** the classical action assumes the continuum form

$$\boxed{S[\phi] = \int dt\, L[\phi] = \int dt \int_0^L dx\, \mathcal{L}(\partial_x\phi, \dot\phi)} \qquad (1.5)$$

We have thus succeeded in abandoning the N point-particle description in favor of **classical** one involving *continuous* degrees of freedom, a **(classical) field**. The dynamics of **field** the latter are specified by the **functionals** L and S, which represent the continuum generalizations of the discrete classical Lagrangian and action, respectively.

---

[3] Indeed, for reasons that will become clear, higher-order contributions to the Taylor expansion are immaterial in the long-range continuum limit.

field   **INFO** The continuum variable ϕ is our first encounter with a **field**. Before proceeding
with our example, let us pause to make some preliminary remarks on the general definition
of these objects. This will help to place the subsequent discussion of the atomic chain in
a broader context. Formally, a field is a smooth mapping

$$\phi : M \to T$$
$$z \mapsto \phi(z)$$

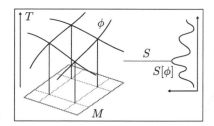

field   from a certain manifold M,[4] often called the "**base**
manifold   **manifold**," into a "**target**" or "**field manifold**"
T (see the figure).[5] In our present example, $M =$
$[0, L] \times [0, t] \subset \mathbb{R}^2$ is the product of intervals in space
and time, and $T = \mathbb{R}$. In fact, the factorization $M \subset$
$\mathcal{R} \times \mathcal{T}$ into a space-like manifold $\mathcal{R}$ multiplied by a
one-dimensional time-like manifold $\mathcal{T}$ is inherent in
most applications of condensed matter physics.[6]

However, the individual factors $\mathcal{R}$ and $\mathcal{T}$ may,
of course, be more complex than in our prototypical problem above. As for the target
manifold, not much can be said in general; depending on the application, the realizations
of T range from real or complex numbers over vector spaces and groups to the "fanciest
objects" of mathematical physics.

functionals   In applied field theory, fields appear not as final objects, but rather as input to **functionals**. Mathematically, a functional $S : \phi \mapsto S[\phi] \in \mathbb{R}$ is a mapping that takes a field
as its argument and maps it into the real numbers. The functional profile $S[\phi]$ essentially
determines the character of a field theory. Notice that the argument of a functional is
commonly indicated in square brackets [ ].

While these formulations may appear unnecessarily abstract, remembering the mathematical backbone of the theory often helps to avoid confusion. At any rate, it takes some
time and practice to get used to the concept of fields and functionals. Conceptual difficulties in handling these objects can be overcome by remembering that any field in condensed
matter physics arises as the limit of a *discrete* mapping. In the present example, the field
$\phi(x)$ is obtained as a continuum limit of the discrete vector $\{\phi_I\} \in \mathbb{R}^N$; the functional
$L[\phi]$ is the continuum limit of the function $L : \mathbb{R}^N \to \mathbb{R}$, etc. While, in practice, fields
are usually easier to handle than their discrete analogs, it is sometimes helpful to think
about problems of field theory in a discrete language. Within the discrete picture, the
mathematical apparatus of field theory reduces to finite-dimensional calculus.

Although the Lagrangian (1.4) contains the full information about the model, we
have not yet learned much about its actual behavior. To extract concrete physical
equations   information from the Lagrangian, we need to derive **equations of motion**. At first
of motion   sight, it may not be entirely clear what is meant by the term "equations of motion"
in the context of an infinite-dimensional model: the equations of motion relevant for

---

[4] If you are unfamiliar with the notion of manifolds (for a crash course, see appendix section
A.1), think of M and T as subsets of some vector space. For the moment, this limitation won't
do any harm.

[5] In some (rare) cases it becomes necessary to define fields in a more general sense (e.g., as
sections of mathematical objects known as fiber bundles). However, in practically all condensed
matter applications, the more restrictive definition above will suffice.

[6] By contrast, the condition of Lorentz invariance implies the absence of such factorizations in
relativistic field theory. In classical statistical field theories, i.e., theories probing the thermodynamic behavior of large systems, M is just space-like.

the present problem are obtained as the generalization of the conventional Lagrange equations of N point-particle classical mechanics to a model with infinitely many degrees of freedom. To derive these equations, we need to generalize Hamilton's extremal principle (i.e., the route from an action to the associated equations of motion) to infinite dimensions. As a warm-up, let us briefly recapitulate how the extremal principle works for a system with one degree of freedom.

Suppose the dynamics of a classical *point* particle with coordinate $x(t)$ is described by the classical Lagrangian $L(x, \dot{x})$ and action $S[x] = \int dt\, L(x, \dot{x})$. **Hamilton's extremal principle** states that the configurations $x(t)$ that are *realized* are those that extremize the action, $\delta S[x] = 0$, i.e., for any smooth curve $t \mapsto y(t)$,

Hamilton's extremal principle

$$\lim_{\epsilon \to 0} \frac{1}{\epsilon} (S[x + \epsilon y] - S[x]) = 0. \tag{1.6}$$

(For a more rigorous discussion, see section 1.2 below.) To first order in ϵ, the action has to remain invariant. Applying this condition, one finds that it is fulfilled if and only if x satisfies **Lagrange's equation of motion**

Lagrange's

$$\boxed{\frac{d}{dt}(\partial_{\dot{x}} L) - \partial_x L = 0} \tag{1.7}$$

EXERCISE Recapitulate the derivation of (1.7) from the classical action.

In Eq. (1.5) we are dealing with a system of infinitely many degrees of freedom, $\phi(x, t)$. Yet Hamilton's principle is general and we may see what happens if Eq. (1.5) is subjected to an extremal principle analogous to Eq. (1.6). In this case, we require the action (1.5) to be invariant under variations $\phi(x, t) \to \phi(x, t) + \epsilon\eta(x, t)$, to first order

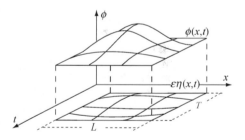

in ϵ. Note that field variations must respect boundary conditions, if present. For example, if $\phi|_{\text{boundary}} = \text{const.}$, then $\eta|_{\text{boundary}} = 0$ (see the figure). When applied to the specific Lagrangian (1.4), substituting the "varied" field leads to

$$S[\phi + \epsilon\eta] = S[\phi] + \epsilon \int dt \int_0^L dx \left(m\dot{\phi}\dot{\eta} - k_s a^2 \partial_x \phi\, \partial_x \eta\right) + \mathcal{O}(\epsilon^2).$$

Integrating by parts and requiring the contribution linear in ϵ to vanish, one obtains

$$\lim_{\epsilon \to 0} \frac{1}{\epsilon} (S[\phi + \epsilon\eta] - S[\phi]) = - \int dt \int_0^L dx \left(m\ddot{\phi} - k_s a^2 \partial_x^2 \phi\right) \eta \overset{!}{=} 0.^7$$

(Notice that the boundary terms vanish identically.) Now, since η was defined to be an arbitrary smooth function, the integral above can vanish only if the factor in

---

[7] Here and throughout $a \overset{!}{=} b$ means "we require a to be equal to b."

parentheses is globally vanishing. Thus the equation of motion takes the form of a **classical wave equation**

$$\left(m\partial_t^2 - k_s a^2 \partial_x^2 \right) \phi = 0 \tag{1.8}$$

The solutions of (1.8) have the form $\phi_+(x-vt)+\phi_-(x+vt)$, where $v = a\sqrt{k_s/m}$, and $\phi_\pm$ are arbitrary smooth functions of

the argument. From this we can deduce that the low-energy **elementary excitations** of our model are lattice vibrations propagating as **classical sound waves** to the left or right at a constant velocity v (see figure).[8] The trivial behavior of the model is, of course, a direct consequence of its simplistic definition – no dissipation, dispersion, or other nontrivial ingredients. Adding these refinements leads to the general classical theory of lattice vibrations (see, e.g., Ashcroft and Mermin[9]). Finally, notice that the elementary excitations of the chain have little in common with its "microscopic" constituents (the atoms). Rather they are **collective excitations**, i.e., elementary excitations comprising a macroscopically large number of microscopic degrees of freedom.

INFO The "relevant" **excitations of a condensed matter system** can, but need not, be of collective type. For example, the interacting electron gas (a system to be discussed in detail below) supports microscopic excitations – charged quasi-particles standing in 1:1 correspondence with the electrons of the original microscopic system – while the collective excitations are plasmon modes of large wavelength that involve many electrons. Typically, the nature of the fundamental excitations cannot be straightforwardly inferred from the microscopic definition of a model. Indeed, the mere *identification* of the relevant excitations often represents the most important step in the solution of a condensed matter problem.

1.1.2 Hamiltonian formulation

An important characteristic of any excitation is its *energy*. How much energy is stored in the sound waves of the harmonic chain? To address this question, we need to switch back to a Hamiltonian formulation. Once again, this is achieved by generalizing standard manipulations from point-particle

Sir William Rowan Hamilton 1805–1865 was an Irish mathematician credited with the discovery of quaternions, the first non-commutative algebra to be studied. He also made seminal contributions to the study of geometric optics and classical mechanics.

mechanics to the continuum. Remembering that, in the Lagrangian formulation of

---

[8] Strictly speaking, the modeling of our system enforces a periodicity constraint $\phi_\pm(x + L) = \phi_\pm(x)$. However, in the limit of large system sizes, this aspect becomes inessential.
[9] N. W. Ashcroft and N. D. Mermin, *Solid State Physics* (Holt–Saunders International, 1983).

point particle mechanics, $p \equiv \partial_{\dot{x}} L$ is the momentum conjugate to the coordinate x, let us consider the Lagrangian *density* and define the field[10]

$$\pi(x) \equiv \frac{\partial \mathcal{L}}{\partial \dot{\phi}} \qquad (1.9)$$

canonical momentum

as the **canonical momentum** associated with ϕ. In common with $\phi(x)$, the momentum $\pi(x)$ is a continuum degree of freedom. At each point in space, it may take an independent value. Notice that $\pi(x)$ is nothing but the continuum generalization of the lattice momentum P_I of Eq. (1.2). (Applied to P_I, a continuum

Hamiltonian density

approximation like $\phi_I \to \phi(x)$ would produce $\pi(x)$.) The **Hamiltonian density** is then defined as usual through the Legendre transformation,

$$\mathcal{H}(\partial_x \phi, \pi) = \left.\left(\pi \dot{\phi} - \mathcal{L}\right)\right|_{\dot{\phi} = \dot{\phi}(\phi, \pi)} \qquad (1.10)$$

from which the full Hamiltonian is obtained as $H = \int_0^L dx\, \mathcal{H}$.

EXERCISE Verify that the transition $L \to H$ is a straightforward continuum generalization of the Legendre transformation of the N-particle Lagrangian $L(\{\phi_I\}, \{\dot{\phi}_I\})$.

Having introduced a Hamiltonian, we are in a position to determine the energy of the sound waves. Application of Eqs. (1.9) and (1.10) to the Lagrangian of the atomic chain yields $\pi(x) = m\dot{\phi}(x)$ and

$$H[\pi, \phi] = \int dx \left[\frac{\pi^2}{2m} + \frac{k_s a^2}{2}(\partial_x \phi)^2 \right]. \qquad (1.11)$$

Considering, say, a right-moving sound-wave excitation, $\phi(x, t) = \phi_+(x - vt)$, we find that $\pi(x, t) = -mv\partial_x \phi_+(x - vt)$ and $H[\pi, \phi] = k_s a^2 \int dx [\partial_x \phi_+(x - vt)]^2 = k_s a^2 \int dx\, [\partial_x \phi_+(x)]^2$, i.e., a positive-definite time-independent expression, as one would expect.

Hamiltonian action

INFO For completeness, we mention that the **Hamiltonian representation of the action** (1.5) is given by $S[\phi, \pi] = \int dt \int_0^L dx(\pi\dot{\phi} - \mathcal{H})$. From here, the Hamiltonian version of the equations of motion can be derived by independent variations in ϕ and π, just as in classical mechanics. As an exercise, carry out this variation for the harmonic chain and verify that you obtain equations equivalent to the wave equation (1.8).[11] Whether one prefers to work in a Hamiltonian or Lagrangian formulation of a field theory depends on the context and is often decided on a case-by-case basis.

Before proceeding further, let us note an interesting feature of the energy functional: in the limit of an infinitely shallow excitation, $\partial_x \phi_+ \to 0$, the energy vanishes. This sets the stage for principles 4, hitherto unconsidered, **symmetry**. The Hamiltonian

---

[10] In field theory literature, it is traditional to denote the momentum by a Greek letter.

[11] Variation of the action in ϕ and π leads to (invert this to check the result) $\dot{\phi} \frac{x}{c}\partial_z v^s y = \pi$, $\frac{uu}{u} = \dot{\phi}$. Differentiation of the first equation in time followed by substitution into the second equation yields the wave equation.

of an atomic chain is invariant under simultaneous translation of all atom coordinates by a fixed increment: $\phi_I \rightarrow \phi_I + \delta$, where δ is constant. This expresses the fact that a global translation of the solid as a whole does not affect the internal energy. Now, the ground state of any specific realization of the solid is defined through a static array of atoms, each located at a fixed coordinate $R_I = Ia \Rightarrow \phi_I = 0$. We say that the above translational symmetry is "spontaneously broken," i.e., the solid has to decide where exactly it wants to rest. However, spontaneous breakdown of a symmetry does not imply that the symmetry has disappeared. On the contrary, infinite-wavelength deviations from the pre-assigned ground state come close to global translations of (macroscopically large portions of) the solid and, therefore, cost a vanishingly small amount of energy. This is the reason for the vanishing of the sound-wave energy in the limit $\partial_x \phi \rightarrow 0$. It is also our first encounter with the aforementioned phenomenon that continuous symmetries lead to the formation of soft, i.e., low-energy, excitations. A much more systematic exposition of these connections will be given in chapter 5.

To conclude our discussion of the classical harmonic chain, let us consider the **specific heat**, a quantity directly accessible in experiment. A rough estimate of this quantity can be obtained from the microscopic Hamiltonian (1.2). According to the principles of statistical mechanics, the thermodynamic energy density is given by

$$u = \frac{1}{L}\frac{\int d\Gamma \; e^{-\beta H} H}{\int d\Gamma \; e^{-\beta H}} = -\frac{1}{L}\partial_\beta \ln \mathcal{Z},$$

where $\beta \equiv 1/k_{\mathrm{B}}T$, $\mathcal{Z} \equiv \int d\Gamma e^{-\beta H}$ is the **Boltzmann partition function**, and the phase space volume element $d\Gamma = \prod_{I=1}^{N} dR_I dP_I$. (Hereafter, for simplicity, we set $k_{\mathrm{B}} = 1$.) The specific heat is then obtained as $c = \partial_T u$. To determine the temperature dependence of c, we can make use of the fact that, upon rescaling of the integration variables, $R_I \rightarrow \beta^{-1/2}X_I$, $P_I \rightarrow \beta^{-1/2}Y_I$, the exponent $\beta H(R, P) \rightarrow H(X, Y)$ becomes independent of temperature (a property that relies on the quadratic dependence of H on both R and P). The integration measure transforms as $d\Gamma \rightarrow \beta^{-N} \prod_{I=1}^{N} dX_I \; dY_I \equiv \beta^{-N} d\Gamma'$. Expressed in terms of the rescaled variables, one obtains the energy density as $u = -L^{-1}\partial_\beta \ln(\beta^{-N}K) = \rho T$, where $\rho = N/L$ is the density of the atoms and we have made use of the fact that the constant $K \equiv \int d\Gamma' \; e^{-H(X,Y)}$ is independent of temperature. We thus find a temperature independent specific heat $c = \rho$. Notice that c is fully universal, i.e., independent of the material constants m and k_{s} determining H. (In fact, we could have anticipated this result from the equipartition theorem of classical mechanics, i.e., the law that in a system with N degrees of freedom, the energy scales as $U = NT$.)

How do these findings compare with experiment? Figure 1.1 shows the specific heat of the insulating compound $EuCoO_3$.[12] At large temperatures, the specific heat approaches a constant, which is consistent with our analysis. However, at

---

[12] Note that, in metals, the specific heat due to lattice vibrations exceeds the specific heat of the free conduction electrons for temperatures larger than a few degrees kelvin.

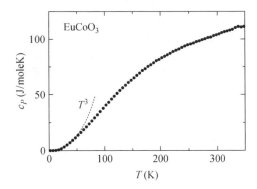

Fig. 1.1   Specific heat c_p of the insulator $EuCoO3$. At large temperatures, c_p approaches a constant value, as predicted by analysis of the classical harmonic chain. However, for small temperatures, deviations from c_p = const. are substantial. Such deviations can be ascribed to quantum effects. (Courtesy of M. Kriener, A. Reichl, T. Lorenz and, A. Freimuth.)

lower temperatures, the specific heat shows deviations from c = const. Yet, this temperature dependence does not reflect a failure of the simplistic microscopic modeling. Rather, the deviation is indicative of a **quantum phenomenon**. Indeed, so far, we have neglected the quantum nature of the atomic variables. In the next chapter we will discuss how an effective low-energy theory of the harmonic chain can be promoted to a quantum field theory. However, before doing so, let us pause to introduce several mathematical concepts that surfaced above, in a way that survives generalization to richer problems.

1.2 Functional Analysis and Variational Principles

Let us revisit the derivation of the equations of motion associated with the harmonic chain, Eq. (1.8). Although straightforward, the calculation was neither efficient, nor did it reveal general structures. In fact, what we did – expanding explicitly to first order in the variational parameter ϵ – has the same status as evaluating derivatives by explicitly taking limits: $f'(x) = \lim_{\epsilon \to 0} \frac{1}{\epsilon}(f(x+\epsilon) - f(x))$. Moreover, the derivation made explicit use of the particular form of the Lagrangian, thereby being of limited use with regard to a general understanding of the construction scheme. Given the importance attached to extremal principles in the whole of field theory, it is worthwhile investing some effort in constructing a more efficient scheme for the general variational analysis of continuum theories. To carry out this program, we first need to introduce the mathematical tool of functional analysis – the concept of **functional differentiation**.

functional
differen-
tiation
    In working with functionals, one is often concerned with how a given functional behaves under (small) variations of its argument function. In particular, given a certain function f that we suspect would make a functional $F[f]$ stationary, one

would like to find out whether the functional remains invariant under variations $f \to f + h$, where h is an infinitely small increment function. In ordinary analysis, questions of this type are commonly addressed by exploring *derivatives*, i.e., we need to generalize the concept of a derivative to functionals. This is achieved by the following definition: a functional F is called differentiable if

$$F[f + \epsilon g] - F[f] = \epsilon\, DF_f[g] + \mathcal{O}(\epsilon^2)$$

differential   where the **differential** DF_f is a linear functional (i.e., one with $DF_f[g_1 + g_2] = DF_f[g_1] + DF_f[g_2]$), ϵ is a small parameter, and g is an arbitrary function. The subscript indicates that the differential depends generally on the "base argument," f. A functional F is said to be **stationary** on f if, and only if, $DF_f = 0$.

In principle, the definition above answers our question concerning a stationarity condition. However, to make use of the definition, we still need to know how to compute the differential DF, and how to relate the differentiability criterion to the concepts of ordinary calculus. To understand how these questions can be systematically addressed, it is helpful to return temporarily to a discrete way of thinking, i.e., to interpret the argument f of a functional $F[f]$ as the limit $N \to \infty$ of a discrete vector $\mathbf{f} = \{f_n \equiv f(x_n), n = 1, \dots, N\}$, where $\{x_n\}$ denotes a discretization of the support of f (cf. the harmonic chain, $\phi \leftrightarrow f$). Prior to taking the continuum limit, $N \to \infty$, $\mathbf{f}$ has the status of an N-dimensional vector and $F(\mathbf{f})$ is a function defined over N-dimensional space. After taking the continuum limit, $\mathbf{f} \to f$ becomes a function itself and $F(\mathbf{f}) \to F[f]$ becomes a functional.

Now, within the discrete picture, it is clear how the variational behavior of functions is to be analyzed. For example, the condition that, for all ϵ and all vectors $\mathbf{g}$, the linear expansion of $F(\mathbf{f} + \epsilon \mathbf{g})$ ought to vanish is simply to say that the ordinary differential, $dF_\mathbf{f}$, defined through $F(\mathbf{f} + \epsilon \mathbf{g}) - F(\mathbf{f}) = \epsilon \cdot dF_\mathbf{f}(\mathbf{g}) + \mathcal{O}(\epsilon^2)$, must be zero. In practice, one often expresses conditions of this type in terms of a certain basis. In a Cartesian basis of N unit vectors, $\mathbf{e}_n$, $n = 1, \dots, N$, $dF_\mathbf{f}(\mathbf{g}) \equiv \langle \nabla F_\mathbf{f}, \mathbf{g} \rangle$, where $\langle \mathbf{f}, \mathbf{g} \rangle \equiv \sum_{n=1}^N f_n g_n$ denotes the standard scalar product, and $\nabla F_\mathbf{f} = \{\partial_{f_n} F\}$ represents the gradient, with the partial derivative defined as

$$\partial_{f_n} F(\mathbf{f}) \equiv \lim_{\epsilon \to 0} \frac{1}{\epsilon} \left[F(\mathbf{f} + \epsilon \mathbf{e}_n) - F(\mathbf{f}) \right]. \tag{1.12}$$

From these identities, the differential is identified as

$$dF_\mathbf{f}(\mathbf{g}) = \sum_n \partial_{f_n} F(\mathbf{f}) g_n. \tag{1.13}$$

The vanishing of the differential amounts to the vanishing of all partial derivatives, $\partial_{f_n} F = 0$.

Equations (1.12) and (1.13) can now be straightforwardly generalized to the continuum limit, whereupon the summation defining the finite-dimensional scalar product translates to an integral,

$$\langle \mathbf{f}, \mathbf{g} \rangle = \sum_{n=1}^N f_n g_n \to \langle f, g \rangle = \int dx\; f(x) g(x).$$

The analog of the nth unit vector is a δ-distribution, $\mathbf{e}_n \to \delta_x$, where $\delta_x(x') \equiv \delta(x - x')$, as can be seen from the following correspondence:

$$f_n \overset{!}{=} \langle \mathbf{f}, \mathbf{e}_n \rangle = \sum_{m=1}^{N} f_m (e_n)_m \to f(x) \overset{!}{=} \langle f, \delta_x \rangle = \int dx' \, f(x') \delta_x(x').$$

Here $(e_n)_m = \delta_{nm}$ denotes the mth component of the nth unit vector. The correspondence (unit vector $\leftrightarrow$ δ-distribution) is easy to memorize: while components of $\mathbf{e}_n$ vanish, save for the nth component, which equals unity, δ_x is a function that vanishes everywhere, save for x where it is *infinite*. That a unit component is replaced by infinity reflects the fact that the support of the δ-distribution is infinitely narrow; to obtain a unit-normalized integral $\int \delta_x$, the function must be singular.

As a result of these identities, (1.13) translates to the continuum differential,

$$DF_f[g] = \int dx \, \frac{\delta F[f]}{\delta f(x)} g(x), \qquad (1.14)$$

where the generalization of the partial derivative,

$$\frac{\delta F[f]}{\delta f(x)} \equiv \lim_{\epsilon \to 0} \frac{1}{\epsilon} \left(F[f + \epsilon \delta_x] - F[f] \right) \qquad (1.15)$$

is commonly denoted by δ instead of ∂. Equations (1.14) and (1.15) establish the connection between ordinary and functional differentiation. Notice that we have not yet learned how to calculate the differential practically, i.e., to evaluate expressions like Eq. (1.15) for concrete functionals. Nevertheless, the identities above are very useful, enabling us to generalize more complex derivative operations of ordinary calculus by straightforward extrapolation. For example, the generalization of the

chain rule   standard **chain rule**, $\partial_{f_n} F(\mathbf{g}(\mathbf{f})) = \sum_m \partial_{g_m} F(\mathbf{g}) \big|_{\mathbf{g}=\mathbf{g}(\mathbf{f})} \partial_{f_n} g_m(\mathbf{f})$ reads

$$\frac{\delta F[g[f]]}{\delta f(x)} = \int dy \, \frac{\delta F[g]}{\delta g(y)} \bigg|_{g=g[f]} \frac{\delta g(y)[f]}{\delta f(x)}.$$

Here $g[f]$ is the continuum generalization of an $\mathbb{R}^m$-valued function, $\mathbf{g} : \mathbb{R}^n \to \mathbb{R}^m$, a function whose components $g(y)[f]$ are functionals by themselves. Furthermore, given some functional $F[f]$, we can construct its **Taylor expansion** as

$$F[f] = F[0] + \int dx_1 \, \frac{\delta F[f]}{\delta f(x_1)} \bigg|_{f=0} f(x_1) + \int \frac{dx_1 \, dx_2}{2} \, \frac{\delta^2 F[f]}{\delta f(x_2) \delta f(x_1)} \bigg|_{f=0} f(x_1) f(x_2) + \cdots,$$

where (exercise)

$$\frac{\delta^2 F[f]}{\delta f(x_2) \delta f(x_1)}$$
$$= \lim_{\epsilon_{1,2} \to 0} \frac{1}{\epsilon_1 \epsilon_2} \left(F[f + \epsilon_1 \delta_{x_1} + \epsilon_2 \delta_{x_2}] - F[f + \epsilon_1 \delta_{x_1}] - F[f + \epsilon_2 \delta_{x_2}] + F[f] \right)$$

generalizes a two-fold partial derivative. The validity of these identities can be made plausible by applying the prescription given in table 1.1 to the corresponding relations of standard calculus. To actually verify the formulae, one has to find the continuum limit of each step taken in the discrete variant of the corresponding

Table 1.1 Summary of basic definitions of discrete and continuum calculus.

Entity	Discrete	Continuum
Argument	vector $\mathbf{f}$	function f
Function(al)	multidimensional function $F(\mathbf{f})$	functional $F[f]$
Differential	$dF_{\mathbf{f}}(\mathbf{g})$	$DF_f[g]$
Cartesian basis	$\mathbf{e}_n$	δ_x
Scalar product $\langle\,,\rangle$	$\sum_n f_n g_n$	$\int dx\, f(x)g(x)$
"Partial derivative"	$\partial_{f_n} F(\mathbf{f})$	$\delta F[f]/\delta f(x)$

proofs. Experience shows that it takes some time to get used to the concept of functional differentiation. However, after some practice, it will become clear that this operation is not only useful but is as easy to handle as conventional partial differentiation.

We finally address the question how to compute functional derivatives in practice. In doing so, we will make use of the fact that, in all but a few cases, the functionals encountered in field theory are of the structure

$$S[\phi] = \int_M dx\, \mathcal{L}(\phi^i, \partial_\mu \phi^i) \tag{1.16}$$

Here, we assume the base manifold M to be parameterized by an m-dimensional coordinate vector $x = \{x_\mu\}$. (In most practical applications $m = d + 1$ and $x = (x_0, x_1, \ldots, x_d)$ contains one time-like component $x_0 = t$ and d space-like components $x_k, k = 1, \ldots, d$.[13]) We further assume that the field manifold has dimensionality n and that ϕ^i, $i = 1, \ldots, n$, are the coordinates of the field. Functionals of this type are called **local functionals**.

local functional

What makes the functional $S[\phi]$ easy to handle is that all of its information is stored in the *function* $\mathcal{L}$. Owing to this simplification, the functional derivative can be related to an ordinary derivative of $\mathcal{L}$. To see this, all we have to do is to evaluate the general definition (1.14) on the functional S:

$$S[\phi + \epsilon\theta] - S[\phi] = \int_M dx\, [\mathcal{L}(\phi + \epsilon\theta, \partial_\mu\phi + \epsilon\partial_\mu\theta) - \mathcal{L}(\phi, \partial_\mu\phi)]$$

$$= \int_M dx\, \left[\frac{\partial\mathcal{L}}{\partial\phi^i}\theta^i + \frac{\partial\mathcal{L}}{\partial(\partial_\mu\phi^i)}\partial_\mu\theta^i \right]\epsilon + O(\epsilon^2)$$

$$= \int_M dx\, \left[\frac{\partial\mathcal{L}}{\partial\phi^i} - \partial_\mu\left(\frac{\partial\mathcal{L}}{\partial(\partial_\mu\phi^i)}\right) \right]\theta^i\epsilon + O(\epsilon^2),$$

where in the last line we have assumed that the field variation vanishes on the boundary of the base manifold, $\theta\,|_{\partial M} = 0$. Comparison with Eq. (1.14) identifies the functional derivative as

$$\frac{\delta S[\phi]}{\delta\phi^i(x)} = \frac{\partial\mathcal{L}}{\partial\phi^i(x)} - \partial_\mu\left(\frac{\partial\mathcal{L}}{\partial(\partial_\mu\phi^i(x))}\right).$$

[13] Following standard convention, we denote space-like components by small Latin indices $k = 1, \ldots, d$. By contrast, space–time indices are denoted by Greek indices $\mu = 0, \ldots, d$.

We conclude that the **stationarity of the functional** (1.16) is equivalent to the condition

$$\forall x, i: \quad \frac{\partial \mathcal{L}}{\partial \phi^i(x)} - \partial_\mu \left(\frac{\partial \mathcal{L}}{\partial(\partial_\mu \phi^i(x))} \right) = 0 \tag{1.17}$$

Euler–
Lagrange
equation

Equation (1.17) is known as the **Euler–Lagrange equation** of field theory. In fact, for $d = 0$ and $x_0 = t$, Eq. (1.17) reduces to the familiar Euler–Lagrange equation for a point particle in n-dimensional space. For $d = 1$ and $(x_0, x_1) = (t, x)$, we get back to the stationarity equations discussed in the previous section. In the next section we will apply the formalism to a higher-dimensional problem.

1.3 Maxwell's Equations as a Variational Principle

REMARK This section requires familiarity with the basic notions of special relativity such as the concepts of 4-vectors, Lorentz transformations, and covariant notation.[14]

classical
electro-
dynamics

As a second example, let us consider *the* archetype of classical field theory, **classical electrodynamics**. As well as exemplifying the application of continuum variational principles for a familiar problem, this example illustrates the unifying potential of the approach: That problems as different as the low-lying vibrational modes of a crystalline solid and electrodynamics can be described by almost identical language indicates that we are dealing with a

James Clerk Maxwell 1831–1879
was a Scottish theoretical physicist and mathematician who made seminal contributions to the study of electricity, magnetism, optics, and the kinetic theory of gases. In particular, he is credited with the formulation of the theory of electromagnetism, synthesizing seemingly unrelated experiments and equations of electricity, magnetism and optics into a consistent theory. He is also known for creating the first true color photograph in 1861!

Maxwell's
equations

useful formalism. Specifically, our aim will be to explore how the equations of motion of electrodynamics, the inhomogeneous **Maxwell's equations**,

$$\nabla \cdot \mathbf{E} = \rho, \qquad \nabla \times \mathbf{B} - \partial_t \mathbf{E} = \mathbf{j}, \tag{1.18}$$

can be obtained from variational principles. For simplicity, we restrict ourselves to a vacuum theory, i.e., $\mathbf{E} = \mathbf{D}$ and $\mathbf{B} = \mathbf{H}$. Further, we have set the velocity of light to unity, $c = 1$. Within the framework of the variational principle, the homogeneous equations,

$$\nabla \times \mathbf{E} + \partial_t \mathbf{B} = 0, \qquad \nabla \cdot \mathbf{B} = 0, \tag{1.19}$$

are regarded as *ab initio* constraints imposed on the degrees of freedom $\mathbf{E}$ and $\mathbf{B}$.

Lorentz
invariance

INFO As preparation for the following discussion, let us briefly recapitulate the notion of **Lorentz invariance**. In this text, we will work mostly in non-relativistic contexts

[14] For a summary of the covariant notation used in this text, see the Info block on 524.

Euclidean field theory

where the time coordinate t and the d space coordinates x_i are bundled into a $(d+1)$-dimensional vector $x = x^\mu = x_\mu = (t, x_i)$ and $\mu = 0, \ldots, d$. In this case, t and x_i may be considered as coordinates of a Euclidean space. Field theories defined in such spaces are called **Euclidean field theories**. By contrast, in relativistic theories we are working in a space–time continuum with a **Minkowski metric**

$$\eta = \{\eta^{\mu\nu}\} = \begin{pmatrix} -1 & & & \\ & +1 & & \\ & & +1 & \\ & & & +1 \end{pmatrix}. \tag{1.20}$$

Here, too, we denote space–time coordinate vectors by $x = x^\mu = (t, x_i)$. However, now the – contravariant or "upstairs" – positioning of the index becomes an essential part of the notation; see Info block on 524 for a summary of the notation conventions of relativity.

Lorentzian field theory

Field theories in space–times with a Minkowski metric are called **Lorentzian field theories**. Recall that a linear coordinate transformation $x^\mu \to x'^\mu \equiv \Lambda^\mu{}_\nu x^\nu$ is a Lorentz transformation if it leaves the Minkowski metric invariant: $x^\mu \eta_{\mu\nu} x^\nu = x'^\mu \eta_{\mu\nu} x'^\nu$. In the covariant notation of relativity, covariant components, x_μ, are obtained from contravariant components, x^μ, by index lowering via the Minkowski metric, $x_\mu = \eta_{\mu\nu} x^\nu$ (this is why the positioning is relevant) and the invariance condition assumes the form $x_\mu x^\mu = x'_\mu x'^\mu$. Expressed as a condition for the Lorentz transformations, this reads $\eta_{\mu\nu} \Lambda^\mu{}_{\mu'} \Lambda^\nu{}_{\nu'} = \eta_{\mu'\nu'}$.

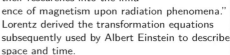

Hendrik Antoon Lorentz
1853–1928
was a Dutch physicist who shared the 1902 Nobel Prize in Physics with Pieter Zeeman "in recognition of the extraordinary service they rendered by their researches into the influence of magnetism upon radiation phenomena." Lorentz derived the transformation equations subsequently used by Albert Einstein to describe space and time.

In cases where we are discussing material which does not depend on the realization of the metric, covariant notation will be used. The Euclidean field theory is then represented by the unit metric $\eta_{\mu\nu} = \delta_{\mu\nu}$.

The representation of Maxwell's theory as a variational principle is best formulated in the language of **relativistically invariant electrodynamics**. As a starting point, we require (1) a field formulated in a set of suitably generalized coordinates and (2) its action. Regarding coordinates, the natural choice will be the coefficients of the electromagnetic (EM) 4-potential, $A^\mu = (\phi, \mathbf{A})$, where ϕ is the scalar potential and $\mathbf{A}$ is the vector potential. The 4-potential A is unconstrained and uniquely determines the fields $\mathbf{E}$ and $\mathbf{B}$ through the standard equations $\mathbf{E} = -\nabla\phi - \partial_t \mathbf{A}$ and $\mathbf{B} = \nabla \times \mathbf{A}$. (In fact, the set of coordinates A_μ is "overly free" in the sense that gauge transformations $A_\mu \to A_\mu + \partial_\mu \Gamma$, where Γ is an arbitrary function, leave the physical fields invariant. Later we will comment explicitly on this point.) The connection between A and the physical fields can be expressed in a more symmetric way by introducing the EM field tensor,[15]

---

[15] Notice that the field tensor (1.21) differs from that in many textbooks on electromagnetism by a sign change, $E_i \leftrightarrow -E_i$. The reason is that in this text we work with a different sign convention for the Minkovski metric, $\eta \leftrightarrow -\eta$; see p.524 for a discussion of this point.

$$F = \{F_{\mu\nu}\} = \begin{pmatrix} 0 & -E_1 & -E_2 & -E_3 \\ E_1 & 0 & -B_3 & B_2 \\ E_2 & B_3 & 0 & -B_1 \\ E_3 & -B_2 & B_1 & 0 \end{pmatrix}. \tag{1.21}$$

The relation between fields and potential now reads $F_{\mu\nu} = \partial_\mu A_\nu - \partial_\nu A_\mu$, where $x_\mu = (-t, \mathbf{x})$ and $\partial_\mu = (\partial_t, \nabla)$.[14]

EXERCISE Confirm that this relation follows from the definition of the vector potential. To verify that the constraint (1.19) is automatically included in the definition (1.21), compute the construct $\partial_\lambda F_{\mu\nu} + \partial_\mu F_{\nu\lambda} + \partial_\nu F_{\lambda\mu}$, where $(\lambda\nu\mu)$ represent arbitrary but *different* indices. This produces four different terms, identified as the left-hand side of Eq. (1.19). Evaluation of the same construct on $F_{\mu\nu} \equiv \partial_\mu A_\nu - \partial_\nu A_\mu$ produces zero, by the symmetry of the right-hand side.

To obtain the structure of the action $S[A]$, we can proceed in different ways. One option would be to regard Maxwell's equations as fundamental, i.e., to construct an action that produces these equations upon variation (by analogy with the situation in classical mechanics where the action functional is designed to reproduce Newton's equations). However, we can also be a little bit more ambitious and ask whether the structure of the action can be motivated independently of Maxwell's equations. In fact, there is just one principle in electrodynamics that is as fundamental as Maxwell's equations: **symmetry**. A theory of electromagnetism must be Lorentz invariant, i.e., invariant under relativistic coordinate transformations.

Aided by the symmetry criterion, we can try to conjecture the structure of the action from three basic assumptions, all independent of Maxwell's equations. The action should be invariant under (i) Lorentz transformations, (ii) gauge transformations, and (iii) it should be simple! The most elementary choice compatible with these conditions is

$$S[A] = \int d^4x \left(c_1 F_{\mu\nu} F^{\mu\nu} + c_2 A_\mu j^\mu \right), \tag{1.22}$$

where $d^4x = dt\, dx_1\, dx_2\, dx_3$ denotes the measure, $j^\mu = (\rho, \mathbf{j})$ the 4-current, and $c_{1,2}$ are undetermined constants. Indeed, up to quadratic order in A, (1.22) defines the only possible structure consistent with gauge and Lorentz invariance.

EXERCISE Using the continuity equation $\partial_\mu j^\mu = 0$, verify that the Aj-coupling is gauge invariant. (Hint: Integrate by parts.) Verify that a contribution like $\int A_\mu A^\mu$ would *not* be gauge invariant.

Having defined a trial action, we can apply the variational principle (1.17) to compute equations of motion. In the present context, the role of the field ϕ is taken by the four components of A. Variation of the action with respect to A_μ gives four equations of motion,

$$\frac{\partial \mathcal{L}}{\partial A_\mu} - \partial_\nu \left(\frac{\partial \mathcal{L}}{\partial(\partial_\nu A_\mu)} \right) = 0, \quad \mu = 0, \ldots, 3, \tag{1.23}$$

where the Lagrangian density is defined by $S = \int d^4x\, \mathcal{L}$. With the specific form of $\mathcal{L}$, it is straightforward to verify that $\partial_{A_\mu}\mathcal{L} = c_2 j^\mu$ and $\partial_{(\partial_\nu A_\mu)}\mathcal{L} = -4c_1 F^{\mu\nu}$. We substitute these building blocks into the equations of motion to obtain $4c_1 \partial_\nu F^{\nu\mu} = c_2 j^\mu$. Comparing this with the definition of the field tensor (1.21), and setting $c_1/c_2 = -1/4$, we arrive at Maxwell's equations (1.18). The overall multiplicative constant c_1 ($= c_2/4$) can be fixed by requiring that the Hamiltonian density associated with the Lagrangian density $\mathcal{L}$ reproduce the known energy density of the EM field (see problem 1.8.2). This leads to $c_1 = -1/4$, so that we have identified

$$\boxed{\mathcal{L}(A_\mu, \partial_\nu A_\mu) = -\frac{1}{4} F_{\mu\nu} F^{\mu\nu} + A_\mu j^\mu} \tag{1.24}$$

electro-
magnetic
Lagrangian

as the **Lagrangian density of the electromagnetic field**. The corresponding action is given by $S[A] = \int d^4x\, \mathcal{L}(A_\mu, \partial_\nu A_\mu)$.

At first sight, this result does not look surprising. After all, Maxwell's equations can be found on the first page of most textbooks on electrodynamics. However, our achievement is actually quite remarkable. By invoking only symmetry, the algebraic structure of Maxwell's equations has been established unambiguously. We have thus proven that Maxwell's equations are relativistically invariant, a fact not obvious from the equations themselves. Further, we have shown that Eqs. (1.18) are the *only* equations of motion linear in the current-density distribution and consistent with the invariance principle. One might object that, in addition to symmetry, we have also imposed an *ad hoc* "simplicity" criterion on the action $S[A]$. However, later we will see that this was motivated by more than mere aesthetics.

Finally, we note that the symmetry-oriented modeling that led to Eq. (1.22) is illustrative of a popular construction scheme in modern field theory. The symmetry-oriented approach stands as complementary to the "microscopic" formulation exemplified in section 1.1. Broadly speaking, these are the two principal approaches to constructing effective low-energy field theories.

▷ **Microscopic analysis:** Starting from a microscopically defined system, one projects onto those degrees of freedom that one believes are relevant for the low-energy dynamics. Ideally, this "belief" is backed up by a small expansion parameter stabilizing the mathematical parts of the analysis. *Advantages:* The method is rigorous and fixes the resulting field theory completely. *Disadvantages:* The method is time consuming and, for complex systems, not even viable.

▷ **Symmetry considerations:** One infers an effective low-energy theory on the basis of only fundamental symmetries of the physical system. *Advantages:* The method is fast and elegant. *Disadvantages:* It is less explicit than the microscopic approach. Most importantly, it does not fix the coefficients of the different contributions to the action.

Thus far, we have introduced some basic concepts of field-theoretical modeling in condensed matter physics. Starting from a microscopic model Hamiltonian, we have illustrated how principles of universality and symmetry can be applied to distill effective continuum field theories, capturing the low-energy content of the system. We have formulated such theories in the language of Lagrangian and Hamiltonian continuum mechanics, and shown how variational principles can be applied to extract concrete physical information. Finally, we have seen that field theory provides a unifying framework whereby analogies between seemingly different physical systems can be uncovered. In the next section we discuss how the formalism of classical field theory can be elevated to the quantum level.

1.4 Quantum Chain

Previously, from measurements of the specific heat, we have seen that at low temperatures the excitation profile of the classical atomic chain differs drastically from that observed experimentally. Generally, in condensed matter physics, low-energy phenomena with pronounced temperature sensitivity are indicative of a quantum mechanism at work. To introduce and exemplify a general procedure whereby quantum mechanics can be incorporated into continuum models, we next consider the low-energy physics of the quantum atomic chain.

The first question to ask is conceptual: how can a model like (1.4) be quantized in general? Indeed, there exists a standard procedure for quantizing continuum theories, which closely resembles the quantization of Hamiltonian point mechanics. Consider the defining equations (1.9) and (1.10) for the canonical momentum and the Hamiltonian, respectively. Classically, the momentum $\pi(x)$ and the coordinate $\phi(x)$ are canonically conjugate variables: $\{\pi(x), \phi(x')\} = -\delta(x - x')$, where $\{\,,\,\}$ is the **Poisson bracket** and the δ-function arises through continuum generalization of the discrete identity $\{P_I, R_{I'}\} = -\delta_{II'}$, $I, I' = 1, \ldots, N$.[16] The theory is quantized by generalization of the canonical quantization procedure for the discrete pair of conjugate coordinates (R_I, P_I) to the continuum: (i) promote $\phi(x)$ and $\pi(x)$ to operators, $\phi \mapsto \hat{\phi}$, $\pi \mapsto \hat{\pi}$, and (ii) generalize the canonical commutation relation $[P_I, R_{I'}] = -i\hbar\delta_{II'}$ to[17]

> **Poisson bracket**

$$\boxed{[\hat{\pi}(x), \hat{\phi}(x')] = -i\hbar\delta(x - x')} \tag{1.25}$$

---

[16] Recall that for conjugate coordinates (R_I, P_I) the Poisson bracket is defined by

$$\{f, g\} = \sum_{I=1}^{N} \left(\frac{\partial f}{\partial R_I} \frac{\partial g}{\partial P_I} - \frac{\partial f}{\partial P_I} \frac{\partial g}{\partial R_I} \right).$$

[17] Note that the dimensionalities of both the quantum and the classical continuum field are compatible with the dimensionality of the Dirac δ-function, $[\delta(x - x')] = [\text{length}]^{-1}$, i.e., $[\phi(x)] = [\phi_I] \times [\text{length}]^{-1/2}$ and similarly for π.

Table 1.2 Relations between discrete and continuum canonically
conjugate variables or operators.

	Classical	Quantum
Discrete	$\{P_I, R_{I'}\} = -\delta_{II'}$	$[\hat{P}_I, \hat{R}_{I'}] = -i\hbar\delta_{II'}$
Continuum	$\{\pi(x), \phi(x')\} = -\delta(x - x')$	$[\hat{\pi}(x), \hat{\phi}(x')] = -i\hbar\delta(x - x')$

**quantum
field**

Operator-valued functions like $\hat{\phi}$ and $\hat{\pi}$ are generally referred to as **quantum fields**.
For clarity, the relevant relations between canonically conjugate classical and quantum fields are summarized in Table 1.2.

INFO By introducing quantum fields, we have departed from the conceptual framework laid out on page 8: being operator-valued, the quantized field no longer represents a mapping into an ordinary differentiable manifold.[18] It is thus legitimate to ask why we bothered to give a lengthy exposition of fields as "ordinary" functions. The reason is that, in the not too distant future, after the framework of functional field integration has been introduced, we will return to the comfortable ground of the definition on page 8.

Employing these definitions, the classical Hamiltonian density (1.10) becomes the quantum operator

$$\hat{\mathcal{H}}(\hat{\phi}, \hat{\pi}) = \frac{1}{2m}\hat{\pi}^2 + \frac{k_{\mathrm{s}}a^2}{2}(\partial_x\hat{\phi})^2. \tag{1.26}$$

The Hamiltonian above represents a quantum field-theoretical *formulation* of the problem, but not yet a solution. In fact, the development of a spectrum of methods for the analysis of quantum field-theoretical models will represent a major part of this text. At this point our objective is merely to exemplify the way in which physical information can be extracted from models like (1.26). As a word of caution, let us mention that the following manipulations, while mathematically straightforward, are conceptually deep. To disentangle different aspects of the problem, we will first concentrate on the plain operational aspects. Later in this section, we will reflect on "what has really happened."

As with any function, operator-valued functions can be represented in a variety of ways. In particular, they can be subjected to Fourier transformation,

$$\begin{Bmatrix} \hat{\phi}_k \\ \hat{\pi}_k \end{Bmatrix} \equiv \frac{1}{L^{1/2}} \int_0^L dx\, e^{\{\mp ikx} \begin{Bmatrix} \hat{\phi}(x) \\ \hat{\pi}(x) \end{Bmatrix}, \quad \begin{Bmatrix} \hat{\phi}(x) \\ \hat{\pi}(x) \end{Bmatrix} = \frac{1}{L^{1/2}} \sum_k e^{\{\pm ikx} \begin{Bmatrix} \hat{\phi}_k \\ \hat{\pi}_k \end{Bmatrix}, \tag{1.27}$$

where $\sum_k$ represents the sum over all Fourier coefficients indexed by the quantized momenta $k = 2\pi m/L$, $m \in \mathbb{Z}$ (not to be confused with the operator momentum $\hat{\pi}$). Note that the *real* classical field $\phi(x)$ quantizes to a *hermitian* quantum field $\hat{\phi}(x)$, implying that $\hat{\phi}_k = \hat{\phi}^\dagger_{-k}$ (and similarly for $\hat{\pi}_k$). The corresponding Fourier representation of the canonical commutation relations reads (exercise)

$$[\hat{\pi}_k, \hat{\phi}_{k'}] = -i\hbar\delta_{kk'}. \tag{1.28}$$

---

[18] At least if we ignore the mathematical subtlety that a linear operator can also be interpreted as an element of a certain manifold.

When expressed in the Fourier representation, making use of the identity,

$$\int_0^L dx\,(\partial_x\hat\phi)^2 = \sum_{k,k'}(-ik\hat\phi_k)(-ik'\hat\phi_{k'})\overbrace{\frac{1}{L}\int_0^L dx\,e^{-i(k+k')x}}^{\delta_{k+k',0}} = \sum_k k^2\hat\phi_k\hat\phi_{-k},$$

together with a similar relation for $\int_0^L dx\,\hat\pi^2$, the Hamiltonian $\hat H = \int_0^L dx\,\mathcal{H}(\hat\phi,\hat\pi)$ assumes the near diagonal form,

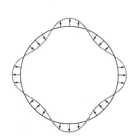

$$\hat H = \sum_k\left[\frac{1}{2m}\hat\pi_k\hat\pi_{-k} + \frac{m\omega_k^2}{2}\hat\phi_k\hat\phi_{-k}\right], \qquad (1.29)$$

where $\omega_k = v|k|$ and $v = a\sqrt{k_s/m}$ denotes the classical sound wave velocity. In this form, the Hamiltonian can be identified as nothing but a superposition of independent **quantum harmonic oscillators**.[19] This result is easy to understand (see the figure). Classically, the system supports a discrete set of wave excitations, each indexed by a wave number $k = 2\pi m/L$. (In fact, we could have performed a Fourier transformation of the classical fields $\phi(x)$ and $\pi(x)$ to represent the Hamiltonian function as a superposition of classical harmonic oscillators.) Within the quantum picture, each of these excitations is described by an oscillator Hamiltonian with a k-dependent frequency. However, it is important not to confuse the atomic constituents, also oscillators (albeit coupled), with the independent *collective* oscillator modes described by $\hat H$.

The description above, albeit perfectly valid, still suffers from a deficiency: the analysis amounts to explicitly describing the effective low-energy excitations of the system (the waves) in terms of their microscopic constituents (the atoms). Indeed the different contributions to $\hat H$ correspond to details of the microscopic oscillator dynamics of individual k-modes. However, it would be much more desirable to develop a picture where the relevant excitations of the system, the waves, appear as fundamental units without an explicit account of the underlying microscopic details. (As with hydrodynamics, information is encoded in terms of collective density variables rather than through individual atoms.) As preparation for the construction of this improved formulation, let us temporarily focus on a single oscillator mode.

1.4.1 Revision of the quantum harmonic oscillator

harmonic oscillator

Consider a standard **harmonic oscillator** (HO) Hamiltonian

$$\hat H = \frac{\hat p^2}{2m} + \frac{m\omega^2}{2}\hat x^2.$$

[19] The only difference between Eq. (1.29) and the canonical form of an oscillator Hamiltonian $\hat H = \hat p^2/2m + m\omega^2\hat x^2/2$ is the presence of the subindices k and $-k$ (a consequence of the relation, $\hat\phi_k^\dagger = \hat\phi_{-k}$). As we will show shortly, this difference is inessential.

The first few energy levels $\epsilon_n = \omega(n + 1/2)$ and the associated Hermite polynomial eigenfunctions are displayed schematically in the figure. (To simplify the notation we henceforth set $\hbar = 1$.) The HO has the status of a single-particle problem. However, the equidistance of its energy levels suggests an alternative interpretation: a given

state ϵ_n may be thought of as an accumulation of n elementary entities, or **quasi-particles**, each having energy ω. What can be said about the features of these new objects? First, they are structureless, i.e., the only "quantum number" identifying the

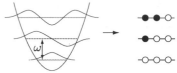

quasi-particles is their energy ω (since otherwise n-particle states formed of the quasi-particles would not be equidistant in energy). This implies that the quasi-particles must be *bosons*. (The same state ω can be occupied by more than one particle.)

quasi-particles

This idea can be formulated in quantitative terms by employing the formalism of ladder operators, in which the operators $\hat{p}$ and $\hat{x}$ are traded for the pair of hermitian adjoint operators $\hat{a} \equiv \sqrt{m\omega/2}(\hat{x} + (i/m\omega)\hat{p})$ and $\hat{a}^\dagger \equiv \sqrt{m\omega/2}(\hat{x} - (i/m\omega)\hat{p})$. Up to a factor of i, the transformation $(\hat{x}, \hat{p}) \to (\hat{a}, \hat{a}^\dagger)$ is canonical, i.e., the new operators obey the canonical commutation relation

$$[\hat{a}, \hat{a}^\dagger] = 1. \tag{1.30}$$

More importantly, the a-representation of the Hamiltonian is very simple, viz.

$$\hat{H} = \omega(\hat{a}^\dagger \hat{a} + 1/2), \tag{1.31}$$

as can be checked by direct substitution. Suppose, now, that we had been given a zero-eigenvalue state $|0\rangle$ of the operator $\hat{a}$: $\hat{a}|0\rangle = 0$. As a consequence, $\hat{H}|0\rangle = (\omega/2)|0\rangle$, i.e., $|0\rangle$ is identified as the ground state of the oscillator.[20] The hierarchy of higher-energy states can then be generated by setting $|n\rangle \equiv (1/\sqrt{n!})\,(\hat{a}^\dagger)^n|0\rangle$.

EXERCISE Using the canonical commutation relation (1.30), verify that $\hat{H}|n\rangle = \omega(n + 1/2)|n\rangle$ and $\langle n|n\rangle = 1$.

Formally, the construction above represents yet another way of constructing eigenstates of the quantum HO. However, its real advantage is that it naturally affords a many-particle interpretation. To this end, let us declare that $|0\rangle$ represents a "vacuum" state, i.e., a state with zero particles. Next, imagine that $\hat{a}^\dagger|0\rangle$ is a state with a single featureless particle (the operator $\hat{a}^\dagger$ does not carry any quantum number labels) of energy ω. Similarly, $(\hat{a}^\dagger)^n|0\rangle$ is considered as a many-body state with n particles; i.e., within the new picture, $\hat{a}^\dagger$ is an operator that "creates" particles. The total energy of these states is given by $\omega \times$ (occupation number). Indeed, it is

---

[20] Switching to a real space representation of the ground state equation, verify that its solution is the familiar ground state wave function $\langle x|0\rangle = \sqrt{m\omega/2\pi}e^{-m\omega x^2/2}$. can be verified by explicit construction. As an exercise, switching to a real space representation of the ground state equation, $0 = \langle 0|x\rangle[(m\omega)/{}^x\varrho + x]$ and verify that its solution is the familiar ground state wave function ${}^{z/_z x m\omega - \partial \mathrm{t} \mathrm{z}/m\omega}\bigwedge = \langle 0|x\rangle$.

straightforward to verify (see the exercise above) that $\hat{a}^\dagger \hat{a}|n\rangle = n|n\rangle$, i.e., the Hamiltonian effectively counts the number of particles in the state. While at first sight, this may look unfamiliar, the new interpretation is internally consistent. Moreover, it achieves what we asked for above: it allows an interpretation of the HO states as a superposition of independent structureless entities.

INFO The representation above shows that we can think about individual quantum problems in **complementary pictures**. This principle finds innumerable applications in modern condensed matter physics. The existence of different interpretations of a given system is by no means heterodox but, rather, reflects a principle of quantum mechanics: there is no "absolute" system that underpins the phenomenology. The only thing that matters is observable phenomena. For example, we will see later that the "fictitious" quasi-particle states of oscillator systems *behave* as "real" particles, i.e., they have dynamics, can interact, can be detected experimentally, etc. From a quantum point of view, these objects can be considered as "real" particles.

1.4.2 Quasiparticle interpretation of the quantum chain

Returning to the oscillator chain, one can transform the Hamiltonian (1.29) to a form analogous to (1.31) by defining the ladder operators[21]

$$\hat{a}_k \equiv \sqrt{\frac{m\omega_k}{2}}\left(\hat{\phi}_k + \frac{i}{m\omega_k}\hat{\pi}_{-k}\right), \quad \hat{a}_k^\dagger \equiv \sqrt{\frac{m\omega_k}{2}}\left(\hat{\phi}_{-k} - \frac{i}{m\omega_k}\hat{\pi}_k\right). \tag{1.32}$$

With this definition, applying the commutation relations (1.28), one finds that the ladder operators obey commutation relations generalizing Eq. (1.30):

$$[\hat{a}_k, \hat{a}_{k'}^\dagger] = \delta_{kk'}, \quad [\hat{a}_k, \hat{a}_{k'}] = [\hat{a}_k^\dagger, \hat{a}_{k'}^\dagger] = 0. \tag{1.33}$$

Expressing the operators $(\hat{\phi}_k, \hat{\pi}_k)$ in terms of $(\hat{a}_k, \hat{a}_k^\dagger)$, it is now straightforward to bring the Hamiltonian into the quasi-particle oscillator form (exercise)

$$\hat{H} = \sum_k \omega_k(\hat{a}_k^\dagger \hat{a}_k + 1/2). \tag{1.34}$$

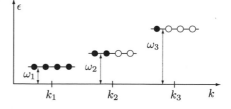

Equations (1.34) and (1.33) represent the final result of our analysis. The Hamiltonian $\hat{H}$ takes the form of a sum of harmonic oscillators with characteristic frequencies ω_k. In the limit $k \to 0$ (i.e., long wavelengths), we have $\omega_k \to 0$; excitations with this property are said to be **massless**.

massless excitation

An excited state of the system is indexed by a set $\{n_k\} = (n_1, n_2, \dots)$ of quasi-particles with energy $\{\omega_k\}$ (see the figure). Physically, the quasi-particles of the

---

[21] As to the consistency of these definitions, recall that $\hat{\phi}_k^\dagger = \hat{\phi}_{-k}$ and $\hat{\pi}_k^\dagger = \hat{\pi}_{-k}$. Under these conditions, the second of the definitions in Eq. (1.32) follows from the first upon taking the hermitian adjoint.

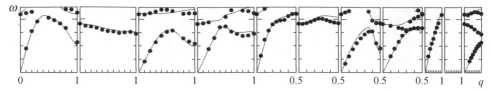

Fig. 1.2   Phonon spectra of the transition metal oxide Sr_2RuO_4 along different axes in momentum space. Notice the approximate linearity of the low-energy branches (**acoustic phonons**) at small momenta q. Superimposed at high frequencies are various branches of **optical phonons**. (Source: Courtesy of M. Braden, University of Cologne.)

phonon   harmonic chain are identified with the **phonon modes** of the solid. A comparison with measured phonon spectra (fig. 1.2) reveals that, at low momenta, we have $\omega_k \sim |k|$ in agreement with our simplistic model (in spite of the fact that the spectrum was recorded for a three-dimensional solid with a nontrivial unit cell – universality!). While the linear dispersion was already a feature of the classical sound wave spectrum, the low-temperature specific heat reflected non-classical behavior. It is left as an exercise (problem 1.8.3) to verify that the quantum nature of the phonons resolves the problem with the low-temperature specific heat discussed in section 1.1.2. (For further discussion of phonon modes in atomic lattices we refer to chapter 2 of the text by Kittel.[22])

EXERCISE Classically, the ground state of the atomic chain comprises a regular array of ions. In the quantum chain, the distance between neighboring ions fluctuates even in the ground state, $|0\rangle$. Using the results above, show that

$$\langle 0|[\phi(x) - \phi(0)]^2|0\rangle = \frac{1}{mL} \sum_k \frac{1 - \cos(kx)}{\omega_k}.$$

In the limit $|x| \gg a$, show that $\langle 0|[\phi(x) - \phi(0)]^2|0\rangle \sim (1/a\sqrt{k_s m}) \ln |x/a|$. What does this imply for the stability of crystalline order in the one-dimensional chain?

1.5 Quantum Electrodynamics

The generality of the procedure outlined above suggests that the quantization of the EM field (1.24) proceeds in a manner analogous to the phonon system. However, there are a number of practical differences that make this task harder (but also more interesting!). First, the vectorial character of the potential, in combination with the condition of relativistic covariance, gives the problem a nontrivial internal geometry. Closely related, the gauge freedom of the vector potential introduces redundant degrees of freedom whose removal on the quantum level is not easily achieved. For example, quantization in a setting where only physical degrees of freedom are kept – i.e., the two polarization directions of the transverse photon field – is technically cumbersome, the reason being that the relevant gauge condition is not relativistically covariant. In contrast, a manifestly covariant

[22] C. Kittel, *Quantum Theory of Solids*, 2nd edition (Wiley, 1987).

scheme, while technically more convenient, introduces spurious "ghost degrees of freedom" that are difficult to remove. To circumvent a discussion of these issues, we will not discuss the problem of EM field quantization in detail.[23] On the other hand, the photon field plays a much too important role in condensed matter physics for us to drop the problem altogether. We will therefore aim at an intermediate exposition, largely insensitive to the problems outlined above, but sufficiently general to illustrate the main principles.

1.5.1 Field quantization

Consider the Lagrangian of the matter-free EM field, $L = -\frac{1}{4} \int d^3x \; F_{\mu\nu} F^{\mu\nu}$. As a first step towards quantization of this system, a gauge choice must be made. In the absence of charge, a particularly convenient choice is the **Coulomb gauge** $\nabla \cdot \mathbf{A} = 0$, with scalar component $\phi = 0$. (Keep in mind that, once a gauge has been set, we cannot expect further results to display "gauge invariance.") Using the gauge conditions, one may verify that the Lagrangian assumes the form

$$L = \frac{1}{2} \int d^3x \left[(\partial_t \mathbf{A})^2 - (\nabla \times \mathbf{A})^2 \right]. \tag{1.35}$$

By analogy with the atomic chain, we would now proceed to "decouple" the theory by expanding the action in terms of eigenfunctions of the Laplace operator. The difference to our previous discussion is that we are dealing (i) with the full three-dimensional Laplacian (instead of a simple second derivative) acting on (ii) the vector quantity $\mathbf{A}$ that is (iii) subject to the constraint $\nabla \cdot \mathbf{A} = 0$. It is these aspects that lead to the complications outlined above.

We can circumvent these difficulties by considering cases where the geometry of the system reduces the complexity of the eigenvalue problem. This restriction is less artificial than it might appear. For example, in anisotropic electromagnetic waveguides, the solutions of the eigenvalue equation can be formulated as[24]

$$-\nabla^2 \mathbf{R}_k(\mathbf{x}) = \lambda_k \mathbf{R}_k(\mathbf{x}), \tag{1.36}$$

where $k \in \mathbb{R}$ is a *one-dimensional* index and the vector-valued functions $\mathbf{R}_k$ are real and orthonormalized: $\int \mathbf{R}_k \cdot \mathbf{R}_{k'} = \delta_{kk'}$. The dependence of the eigenvalues λ_k on k is governed by details of the geometry (see Eq. (1.38) below) and need not be specified for the moment.

waveguide   **INFO** An **electromagnetic waveguide** is a quasi-one-dimensional cavity with metallic boundaries (see fig. 1.3). The practical use of waveguides is that they are good at confining EM waves. At large frequencies, where the wavelengths are of order meters or less, radiation loss in conventional conductors is high. In this frequency domain, hollow conductors provide the only practical way of transmitting radiation. Field propagation inside a waveguide is constrained by boundary conditions. Assuming the walls of the system to be perfectly conducting,

$$\mathbf{E}_\parallel(\mathbf{x}_b) = 0, \qquad \mathbf{B}_\perp(\mathbf{x}_b) = 0, \tag{1.37}$$

where $\mathbf{x}_b$ is a point at the system boundary and $\mathbf{E}_\parallel$ ($\mathbf{B}_\perp$) is the parallel (perpendicular) component of the electric (magnetic) field.

---

[23] Readers interested in learning more about EM field quantization are referred to, e.g., L. H. Ryder, *Quantum Field Theory* (Cambridge University Press, 1996).

[24] More precisely, one should say that Eq. (1.36) defines the set of eigenfunctions relevant for the low-energy dynamics of the waveguide. More-complex eigenfunctions of the Laplace operator exist, but they carry much higher energy.

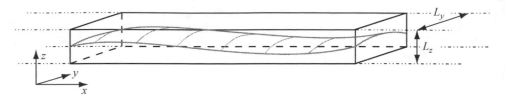

Fig. 1.3   EM waveguide with rectangular cross-section. The structure of the eigenmodes of the field
             is determined by the boundary conditions at the walls of the cavity.

Regarding the problem of field quantization, let us for concreteness consider a cavity
with uniform rectangular cross-section $L_y \times L_z$. To conveniently represent the Lagrangian
of the system, we need to express the vector potential in terms of eigenfunctions $\mathbf{R_k}$ that
are consistent with the boundary conditions (1.37). A complete set of functions fulfilling
this condition is given by

$$\mathbf{R_k} = \mathcal{N}_k \begin{pmatrix} c_1 \cos(k_x x) \sin(k_y y) \sin(k_z z) \\ c_2 \sin(k_x x) \cos(k_y y) \sin(k_z z) \\ c_3 \sin(k_x x) \sin(k_y y) \cos(k_z z) \end{pmatrix}.$$

Here, $k_i = n_i \pi / L_i$ with positive integer n_i, $\mathcal{N}_k$ is a factor normalizing $\mathbf{R_k}$ to unit modulus, and the coefficients c_i are subject to the condition $c_1 k_x + c_2 k_y + c_3 k_z = 0$. Indeed, it is straightforward to verify that a general superposition of the type $\mathbf{A}(\mathbf{x}, t) \equiv$
$\sum_\mathbf{k} \alpha_\mathbf{k}(t) \mathbf{R_k}(\mathbf{x})$, with $\alpha_\mathbf{k}(t) \in \mathbb{R}$, is divergenceless and generates an EM field compatible with (1.37). Substitution of $\mathbf{R_k}$ into Eq. (1.36) identifies the eigenvalues as $\lambda_\mathbf{k} =$
$k_x^2 + k_y^2 + k_z^2$. In the physics and electronic engineering literature, eigenfunctions of the
field modes   Laplace operator in a quasi-one-dimensional geometry are commonly described as **modes**.
As we will see shortly, the energy of a mode (i.e., the Hamiltonian evaluated on a specific
mode configuration) grows with $|\lambda_\mathbf{k}|$. In cases where one is interested in the low-energy
dynamics of the EM field, only configurations with small $|\lambda_\mathbf{k}|$ are relevant. If we consider
a massively anisotropic waveguide with $L_z < L_y \ll L_x$, the modes with smallest $|\lambda_k|$ are
those with $k_z = 0$, $k_y = \pi / L_y$, and $k_x \equiv k \ll L_{z,y}^{-1}$. (Consider why it is not possible to set
both k_y and k_z to zero.) With this choice,

$$\lambda_k = k^2 + (\pi / L_y)^2 \tag{1.38}$$

and a scalar index k suffices to label both eigenvalues and eigenfunctions $\mathbf{R}_k$. A schematic
of the spatial structure of the functions $\mathbf{R}_k$ is shown in fig. 1.3. The dynamical properties
of these configurations will be discussed in the text.

Returning to the problem posed by Eq. (1.35) and (1.36), one can expand the vector
potential in terms of eigenfunctions $\mathbf{R}_k$ as $\mathbf{A}(\mathbf{x}, t) = \sum_k \alpha_k(t) \mathbf{R}_k(\mathbf{x})$, where the sum runs
over all allowed values of the index parameter k. (In a waveguide, $k = \pi n / L$ where $n \in \mathbb{N}$
and L is the length of the guide.) Substituting this expansion into Eq. (1.35) and using
the normalization properties of $\mathbf{R}_k$, we obtain $L = \frac{1}{2} \sum_k \left(\dot{\alpha}_k^2 - \lambda_k \alpha_k^2 \right)$, i.e., a decoupled
representation where the system is described in terms of independent dynamical systems
with coordinates α_k. From this point on, quantization proceeds along the lines of the
standard algorithm, as follows.

First, define momenta through the relation $\pi_k = \partial_{\dot{\alpha}_k} L = \dot{\alpha}_k$. This yields the Hamiltonian $H = \frac{1}{2} \sum_k (\pi_k \pi_k + \lambda_k \alpha_k \alpha_k)$. Next, quantize the theory by promoting fields to
operators, $\alpha_k \to \hat{\alpha}_k$ and $\pi_k \to \hat{\pi}_k$, and declare that $[\hat{\pi}_k, \hat{\alpha}_{k'}] = -i\delta_{kk'}$. The quantum
Hamiltonian operator, again of harmonic oscillator type, then reads

$$\hat{H} = \frac{1}{2} \sum_k \left(\hat{\pi}_k \hat{\pi}_k + \omega_k^2 \hat{\alpha}_k \hat{\alpha}_k \right),$$

where $\omega_k^2 = \lambda_k$. Following the same logic as that marshaled in section 1.4.2, we then define ladder operators

$$a_k \equiv \sqrt{\frac{\omega_k}{2}} \left(\hat{\alpha}_k + \frac{i}{\omega_k} \hat{\pi}_k \right), \qquad a_k^\dagger \equiv \sqrt{\frac{\omega_k}{2}} \left(\hat{\alpha}_k - \frac{i}{\omega_k} \hat{\pi}_k \right),$$

whereupon the Hamiltonian assumes the now familiar form

$$\hat{H} = \sum_k \omega_k (a_k^\dagger a_k + 1/2). \tag{1.39}$$

For the specific problem of the first excited mode in a waveguide of width L_y, $\omega_k = [k^2 + (\pi/L_y)^2]^{1/2}$. Equation (1.39) represents our final result for the quantum Hamiltonian of the EM waveguide. Before concluding this section let us make a few comments on the structure of the result.

▷ The construction above parallels almost perfectly our previous discussion of the harmonic chain.[25] The structural similarity between the two systems finds its origin in the fact that the free field Lagrangian (1.35) is quadratic and, therefore, bound to map onto an oscillator-type Hamiltonian. That we obtained a simple *one-dimensional* superposition of oscillators is due to the boundary conditions specific to a narrow waveguide. For less restrictive geometries, e.g., free space, a more complex superposition of vectorial degrees of freedom in three-dimensional space would have been obtained. However, the principle that the free EM field is mapped onto a superposition of oscillators is independent of geometry.

photon   ▷ Physically, the quantum excitations described by Eq. (1.39) are, of course, the **photons** of the EM field. The unfamiliar appearance of the dispersion relation ω_k is, again, a peculiarity of the waveguide geometry. However, in the limit of large longitudinal wave numbers, $k \gg L_y^{-1}$, the dispersion approaches the form $\omega_k \sim |k|$, i.e., the relativistic dispersion of the photon field. Also, notice that, owing to the equality of the Hamiltonians (1.34) and (1.39), all that has been said about the behavior of the phonon modes of the atomic chain carries over to the photon modes of the waveguide.

▷ As with their phononic analog, the oscillators described by Eq. (1.39) exhibit zero-point fluctuations. It is a fascinating aspect of quantum electrodynamics that these oscillations, caused by quantization of the ultra-relativistic photon field, have various manifestations in non-relativistic physics:

INFO Without going into detail, let us mention some manifestations of **vacuum fluctuations in the phenomenology of condensed matter systems**. One of the most important phenomena induced by vacuum fluctuations is the **Casimir effect**.[26] Two parallel conducting plates embedded into the vacuum exert an attractive force on each other. This phenomenon is not only of conceptual importance – it demonstrates that the vacuum is "alive" – but also of practical relevance. For example, the force balance of hydrophobic

Casimir
effect

---

25 Technically, the only difference is that, instead of index pairs $(k, -k)$, all indices (k, k) are equal and positive. This can be traced to the fact that we have expanded in terms of the real eigenfunctions of the closed waveguide instead of the complex eigenfunctions of the circular oscillator chain.

26 H. B. G. Casimir and D. Polder, *The influence of retardation on the London–van der Waals forces*, Phys. Rev. **73**, 360 (1948); H. B. G. Casimir, *On the attraction between two perfectly conducting plates*, Proc. Kon. Nederland. Akad. Wetensch. **51**, 793 (1948).

suspensions of particles of size $0.1 - 1\,\mu m$ in electrolytes is believed to be strongly influenced by Casimir forces. Qualitatively, the origin of the Casimir force is readily understood. In common with their classical analog, quantum photons exert a certain radiation pressure on macroscopic media. The difference to the classical case is that, due to zero-point oscillations, even the quantum vacuum is capable of creating radiation pressure. For a single conducting body embedded into the infinite vacuum, the net pressure vanishes by symmetry. However, for two parallel plates, the situation is different. Mode quantization arguments similar to those used in the previous section show that the density of quantum modes between the plates is lower than in the semi-infinite outer spaces. Hence, the force (density) created by outer space exceeds the counter-pressure from the inside; the plates "attract" each other.

Van der Waals forces

A second context where vacuum fluctuations play a role is the physics of **van der Waals forces**. Atoms or molecules attract each other by a potential that, at small separation r, scales as r^{-6}. While a detailed discussion of the unusually high power at which this force decays would lead us too far astray, the essence of the argument is as follows. The zero-point fluctuations of the EM field may induce a dipole moment in atoms, which in turn generate a dipole–dipole interaction between close-by atoms, whose detailed evaluation[27] leads to the r^{-6} power–law dependence. Seen in this way, geckos and spiders owe their ability to climb walls to a deeply microscopic principle of quantum field theory.[28]

1.6 Noether's Theorem

It is a basic paradigm of physics that every continuous symmetry entails a conservation law.[29] Conservation laws, in turn, simplify greatly the solution of any problem, which is why one gets acquainted with the correspondence (symmetry $\leftrightarrow$ conservation law) at a very early stage of the physics curriculum, e.g., the connection between rotational symmetry and the conservation of angular momentum. However, it is not trivial to see (at least within the framework of Newtonian mechanics) that the former entails the latter. One needs to know what to look for (viz. angular momentum) to identify the corresponding conserved quantity (rotational invariance). A major advantage of Lagrangian over Newtonian mechanics is that it provides a tool – Noether's theorem – to automatically identify the conservation laws generated by the symmetries of classical mechanics.

What happens when one advances from point to continuum mechanics? Clearly, multi-dimensional continuum theories leave more room for the emergence of complex symmetries but, even more so than in classical mechanics, we are in need of a tool to identify the corresponding conservation laws.

[27] P. W. Milonni, *The Quantum Vacuum* (Academic Press, 1994).

[28] The feet of geckos and spiders are covered with bushels of ultra-fine hair (about three orders of magnitude thinner than human hair). The tips of these hairs come close enough to the atoms of the substrate material to make the van der Waals force sizable. Impressively, this mechanism provides a force of about two orders of magnitude larger than that required to support a spider's full body weight. Both spiders and geckos have to "roll" their feet off the surface to prevent getting stuck by the enormous power of the forces acting on their many body hairs!

[29] Before exploring the ramifications of symmetries and conservation laws for fields, it may be instructive to recapitulate Noether's theorem in the context of classical point-particle mechanics – see, e.g., L. D. Landau and E. M. Lifshitz, *Classical Mechanics* (Pergamon, 1960).

Fortunately, it turns out that Noether's theorem of point mechanics affords a more or less straightforward generalization to higher dimensions. Starting from the general form of the action of a continuum system, Eq. (1.16), the continuum version of Noether's theorem will be derived below. In that we do not refer to a specific physical problem, our discussion will be somewhat dry. This lack of physical context is, however, more than outweighed by the general applicability of the result. The generalized form of Noether's theorem can be – without much further thought – applied to generate the conservation laws of practically any physical symmetry. In this section, we will illustrate the application of the formalism on the simple (yet important) example of space–time translational invariance. A much more intriguing case study will be presented in section 3.6 after some further background of quantum field theory has been introduced.

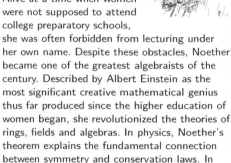

Amalie E. Noether 1882–1935 was a German mathematician known for her ground-breaking contributions to abstract algebra and theoretical physics. Alive at a time when women were not supposed to attend college preparatory schools, she was often forbidden from lecturing under her own name. Despite these obstacles, Noether became one of the greatest algebraists of the century. Described by Albert Einstein as the most significant creative mathematical genius thus far produced since the higher education of women began, she revolutionized the theories of rings, fields and algebras. In physics, Noether's theorem explains the fundamental connection between symmetry and conservation laws. In 1933, she lost her teaching position owing to her being a Jew and a woman, and was forced out of Germany by the Nazis.

1.6.1 Symmetry transformations

The symmetries of a physical system are manifest in the invariance of its action under certain transformations. Mathematically, symmetry transformations are described by two pieces of input data: first, a mapping $f : M \to M$, $x \mapsto f(x) \equiv x'(x)$ that assigns to any point of the base manifold some "transformed" point; second, the field configurations themselves may undergo some change, i.e., there may be a mapping $(\phi : M \to T) \mapsto (\phi' : M \to T)$ that defines a transformed "new field" ϕ' in terms of the "old" ϕ. In principle, there is unlimited freedom in defining such transformations. However, for most applications it is sufficient to consider

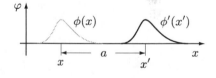

$$\phi'(x') = F(\phi(x)), \qquad (1.40)$$

where F is a function: the new field in the transformed space–time coordinates is obtained as a function of the old field at the original coordinates. With $x' = f(x)$, this correspondence may be equivalently represented as $\phi'(x) = F(\phi(f^{-1}(x)))$. However, irrespective of the representation, it is impor-

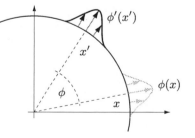

tant to understand that the two operations, $x \mapsto x'$ and $\phi \mapsto \phi'$ may, in general, be independent of each other. The working of such transformations is best illustrated on a few examples:

The invariance properties of a theory under translations in space–time are probed by the mapping $x' = x + a$, $a = \text{const.}$, $\phi'(x') = \phi(x)$. This describes the translation of a field by a fixed offset a in space–time (see the figure). The system is translationally invariant if $S[\phi] = S[\phi']$ for all fields ϕ. As a second example, let us probe the rotational symmetry, $x' = Rx$, where $R \in O(m)$ is a rotation of Euclidean space–time. In this case it would, in general, be unphysical to define $\phi'(x') = \phi(x)$. To illustrate this point, consider the example of a vector field in two dimensions $n = m = 2$ (see the figure). A properly rotated field configuration is defined by $\phi'(x') = R\phi(x)$, i.e., the field amplitude actively participates in the operation. In fact, one does often consider symmetry operations where only the fields are transformed while the base manifold is left untouched.[30] For example, the intrinsic[31] rotational invariance of a magnet is revealed by setting $x' = x$, $m'(x) = R \cdot m(x)$, where the vector field m describes the local magnetization. Conversely, a scalar field $\phi \in \mathbb{R}$ will transform as $\phi(x') = \phi(x)$. These examples show how the extrinsic effects of rotation, $x \mapsto Rx$, and the intrinsic effects, $\phi \mapsto R\phi$, may appear in all sorts of combinations.

To understand the consequences of a symmetry transformation, it is sufficient to consider its infinitesimal version. (Note that any finite transformation can be generated by successive application of infinitesimal transformations.) Consider the two mappings

$$x^\mu \to x^{\mu'} \quad = x^\mu + \partial_{\omega_a} x^\mu|_{\omega=0} \omega_a(x),$$
$$\phi^i(x) \to \phi'^i(x') = \phi^i(x) + \omega_a(x) F_a^i(\phi(x)), \tag{1.41}$$

expressing the change of fields and coordinates to first order in a set of parameter functions ω_a characterizing the transformation. (For a three-dimensional rotation, $(\omega_1, \omega_2, \omega_3) = (\phi, \theta, \psi)$ would be the rotation angles, etc.) The functions F_a^i – which need not depend linearly on the field ϕ, and may explicitly depend on the coordinate x – *define* the incremental change $\phi'(x') - \phi(x)$.

We now ask how the action (1.16) changes under the transformation (1.41), i.e., we wish to compute the difference

$$\Delta S = \int dx' \; \mathcal{L}(\phi'^i(x'), \partial_{x'^{\mu'}}\phi'^i(x')) - \int dx \; \mathcal{L}(\phi^i(x), \partial_{x^\mu}\phi^i(x)),$$

where dx is a shorthand for the integration measure over m coordinates x. Inserting Eq. (1.41), using the identity $\partial_{x'^\nu} x'^\mu = \delta^\mu_{\;\nu} + \partial_{x'^\nu}(\omega_a \partial_{\omega_a} x^\mu)$, together with[32] the Jacobian matrix $\det(\partial x'/\partial x) = 1 + \partial_{x^\mu}(\omega_a \partial_{\omega_a} x^\mu) + \mathcal{O}(\omega^2)$, one obtains

$$\Delta S \simeq \int dx \, (1 + \partial_{x^\mu}(\omega_a \partial_{\omega_a} x^\mu)) \mathcal{L}((\phi^i + \omega)_a F_a^i, (\delta_\mu^{\;\nu} - \partial_{x^\mu}(\omega_a \partial_{\omega_a} x^\nu))\partial_{x_\nu}(\phi^i + \omega_a F_a^i))$$

$$- \int dx \, \mathcal{L}(\phi^i(x), \partial_{x_\mu}\phi^i(x)).$$

So far, we did not use the fact that the transformation was actually meant to be a symmetry transformation. By definition, we are dealing with a symmetry if, for constant parameters ω_a – a uniform rotation or global translation, etc. – the action difference ΔS vanishes. In other words, we may ignore terms in the expansion of ΔS which do not contain derivatives acting on ω_a, as they will not be present in the case where ω_a parameterizes a symmetry. The straightforward expansion of ΔS to leading order in $\partial_\mu \omega_a$ then leads to

$$\Delta S \overset{\text{sym.}}{=} -\int dx \; j^{a\mu}(x)\partial_\mu \omega_a(x), \tag{1.42}$$

---

[30] For example, the standard symmetry transformations of classical mechanics, $q(t) \to q'(t)$, belong to this class: the coordinate vector of a point particle, q (a "field" in $0 + 1$ space–time dimensions) changes while the "base" (time t) does not.

[31] "Intrinsic" means that we rotate just the spins but not the entire system (as we did in our second example, rotational symmetry).

[32] Note that $\det(\partial x'/\partial x) = \exp \operatorname{tr} \ln(\partial x'/\partial x) \simeq \exp[\partial_{x^\mu}(\omega_a \partial_{\omega_a} x^\mu)] \simeq 1 + \partial_{x^\mu}(\omega_a \partial_{\omega_a} x^\mu)$. (Exercise: Show that $\det A = \exp \operatorname{tr} \ln A$, where A is a linear operator.)

Noether
current

where the components of the so-called **Noether currents** j^a are given by

$$j^{a\mu} = \left(\frac{\partial \mathcal{L}}{\partial(\partial_\mu \phi^i)} \partial_\nu \phi^i - \mathcal{L} \, \delta^\mu{}_\nu \right) \frac{\partial x^\nu}{\partial \omega_a}\bigg|_{\omega=0} - \frac{\partial \mathcal{L}}{\partial(\partial_\mu \phi^i)} F^i_a(\phi) \qquad (1.43)$$

For a general field configuration, not much can be said about the Noether current (no matter whether or not the theory possesses a symmetry). However, if the field ϕ obeys the classical equations of motion *and* the theory is symmetric, the Noether current is locally conserved,

$$\partial_\mu j^{a\mu} = 0 \qquad (1.44)$$

This follows from the fact that, for a solution ϕ of the Euler–Lagrange equations, the linear variation of the action in any parameter must vanish. Specifically, integration by parts in Eq. (1.42) leads to $\Delta S = \Delta S[\phi] = \int (\partial_\mu j^{a\mu}) \omega_a$. The vanishing of this expression for arbitrary solutions ϕ and arbitrary ω requires Eq. (1.44). (As an exercise in partial differentiation, try to derive this identity directly from Eq. (1.43). You will need to use the Euler–Lagrange equations Eq. (1.17).) It is very important to keep in mind that the conservation law holds only for solutions of the equations of motion. Therefore, in summary, we have Noether's theorem:

> A continuous symmetry entails a classically conserved current.

We call the current "classically conserved" because, as we will discuss later, in section 9.2, quantum fluctuations around classical solutions may spoil the conservation of currents via

anomaly

the so-called **quantum anomaly**.

The local conservation of a current entails the existence of a globally conserved "charge." For a theory with $d + 1$ space–time coordinates $x = (x^0, x^i) = (t, x^i)$, integration over the space-like directions, and application of Stokes' theorem (exercise), gives $d_t Q^a = 0$, where[33]

$$Q^a(t) \equiv \int d^d x \, j^{a0}(t, x^i) \qquad (1.45)$$

conserved
charge

is the **conserved charge** and we have assumed that the current density vanishes at spatial infinity.

Notice that nowhere in the discussion above have we made any assumption about the internal structure of the Lagrangian. In particular, all results apply equally to the Minkowskian and the Euclidean formulations of the theory.

1.6.2 Examples of symmetries

translational
invariance

Condensed matter systems are often translationally invariant, in space and/or in time. **Translational invariance** may hold down to the microscopic level, where it assumes the form of a discrete symmetry under translation by multiples of the lattice spacing, or it may be emergent only at larger length scales. For example, the fluctuating spin configurations of a paramagnet look locally random, however the system becomes translationally invariant on average over mesoscopic volumes containing many spins. In either case, translational invariance appears as a continuous symmetry of the effective theories relevant to the low-energy physics.

The corresponding symmetry transformation is defined by $x'^\mu = x^\mu + a^\mu$, $\phi'(x') = \phi(x)$. The infinitesimal version of this transformation reads $x'^\mu = x^\mu + \omega^\mu$, where we have

---

[33] Notice that the integral involved in the definition of Q runs only over spatial coordinates.

energy–
momentum
tensor
identified the parameter index a with the space–time index μ. Noether's current, which in the case of translational invariance is called the **energy–momentum tensor** or **stress–energy tensor**, is given by $T^\mu{}_\nu$:

$$T^\mu{}_\nu = \frac{\partial\mathcal{L}}{\partial(\partial_\mu\phi^i)}\partial_\nu\phi^i - \delta^\mu{}_\nu\,\mathcal{L} \qquad (1.46)$$

The conserved "charges" corresponding to this quantity are

$$P_\nu \equiv \int d^d x \left(\frac{\partial\mathcal{L}}{\partial(\partial_0\phi^i)}\partial_\nu\phi^i - \delta^0{}_\nu\,\mathcal{L}\right),$$

where P_0 is the energy and P_i, $i = 1,\dots,d$, the total momentum carried by the system.

EXAMPLE Evaluation of the zeroth component $T^0{}_0$ for the Lagrangian (1.4) of the harmonic chain yields

$$T^0{}_0 = \frac{m}{2}\dot\phi^2 + \frac{k_s a^2}{2}(\partial_x\phi)^2,$$

which is identical to the Hamiltonian density of Eq. (1.11), with $\pi = m\dot\phi$. For a discussion of the momentum density of the chain and of the energy–momentum tensor of the electromagnetic field we refer to problem 1.8.4.

scale
invariance
Systems positioned at the critical point of a second-order phase transition are **scale invariant**. Here, the system looks the same at all length scales, a feature formally expressed as symmetry under *dilatation*, $x \to \lambda x$. The ramifications of this symmetry in field theories will be the central theme of chapter 6. However, at the critical

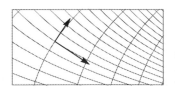

point, systems generally show an even larger set of sym-
conformal
symmetries
metries, known as **conformal symmetries**. By definition, conformal transformations of space–time are angle-preserving in that they map curves intersecting at a certain angle onto image curves intersecting at the same angle. For example, the figure shows the conformal image of a rectangular grid. Besides dilatations, *translations* and *rotations* have this feature. The final, and less obviously angle-preserving, representatives of conformal transformations in general dimensions are the *special conformal transformations* $x^\mu \to (x^\mu - b^\mu x^2)/(1 - 2x_\mu b^\mu + b^2 x^2)$. Geometrically, these are a composition of inversion $x^\mu \to x^\mu/x^2$ followed by translation by b^μ and then by another inversion. The set of
conformal
group
all these transformations defines the **conformal group**, a finite-dimensional symmetry group. (Exercise: How many parameters define the group?[34])

Where they exist, conformal symmetries have far-reaching consequences for the physical properties of a theory. This principle is driven to an extreme in the special and important case of two-dimensional conformal invariance (i.e., the physics of critical two-dimensional systems). The reason is that the two-dimensional conformal group is actually infinite dimensional. Referring to appendix section A.3 for a discussion of two-dimensional conformal invariance, here we note only that the existence of infinitely many symmetries, and as many conserved currents, suffices to almost fully characterize two-dimensional conformal theories. The mechanisms by which this happens are discussed in the appendix, which is perhaps best read at a later stage after more concepts of field theory have been introduced.

Translational and conformal symmetry are examples of space–time symmetries. Later,

[34] $z/(1+p)(z+p) = p + z/(1-p)p + p + 1$ for dilation, translation, rotations, and special transformation, respectively.

when we have introduced field manifolds of richer geometric structure, we will encounter numerous realizations of **internal symmetries**.

1.7 Summary and Outlook

In this chapter we have introduced the general procedure whereby classical continuum theories are quantized. Employing the elementary harmonic oscillator as a example, we have seen that the Hilbert spaces of these theories afford different interpretations. Of particular use is a quasi-particle picture in which the collective excitations of the continuum theories acquired the status of elementary particles. Both examples discussed in this text, the quantum harmonic chain and free quantum electrodynamics, lead to exactly solvable **free field theories**. However, it takes little imagination to foresee that few continuum theories will be as simple. Indeed, the exact solvability of the atomic chain would have been lost had we included higher-order contributions in the expansion in powers of the lattice displacement. Such terms would hinder the free wave-like propagation of the phonon modes. Put differently, phonons would begin to scatter, i.e., interact. Similarly, the free status of electrodynamics is lost once the EM field interacts with a matter field. Needless to say, **interacting field theories** are much more complex, but also more interesting, than the systems considered so far.

Technically, we have seen that the phonon or photon interpretation of the field theories discussed in this chapter could be conveniently formulated in terms of ladder operators. However, the applications discussed so far provide only a glimpse of the advantages of this language. In fact, the formalism of ladder operators, commonly described as "second quantization," represents a central, and historically the oldest, element of quantum field theory. The next chapter is devoted to a more comprehensive discussion of both the formal aspects and applications of this formulation.

1.8 Problems

1.8.1 Electrodynamics from a variational principle

Choosing the Lorentz-gauged components of the vector potential as generalized coordinates, the aim of this problem is to show how the wave equations of electrodynamics can be obtained as a variational principle.

Lorentz gauge

Electrodynamics can be described by Maxwell's equations or, equivalently, by wave-like equations for the vector potential. Working in the **Lorentz gauge**, $\partial_t \phi = -\nabla \cdot \mathbf{A}$, these equations read $(\partial_t^2 - \nabla^2)\phi = \rho$, $(\partial_t^2 - \nabla^2)\mathbf{A} = \mathbf{j}$. Using relativistically covariant notation, the form of the equations can be compressed further to $\partial_\mu \partial^\mu A^\nu = j^\nu$. Starting from the action, $S[A] = -\int d^4 x (\frac{1}{4} F_{\mu\nu} F^{\mu\nu} + A_\mu j^\mu)$, obtain these equations by applying the variational principle. Compare the Lorentz gauge representation of the action with that of the elastic chain. What are the differences and parallels?

Answer:

Substituting the EM field tensor $F_{\mu\nu} = \partial_\mu A_\nu - \partial_\nu A_\mu$ and integrating by parts, the action assumes the form

$$S[A] = -\int d^4 x \left(-\frac{1}{2} A_\nu \left[\partial_\mu \partial^\mu A^\nu - \partial_\mu \partial^\nu A^\mu \right] + j_\mu A^\mu \right).$$

Owing to the Lorentz gauge condition, the second contribution within the square brackets vanishes, and we obtain $S[A] = -\int d^4x (\frac{1}{2}\partial_\mu A_\nu \partial^\mu A^\nu + j_\mu A^\mu)$, where we have again integrated by parts. Applying the general variational equation (1.17), one obtains the wave equation.

1.8.2 Hamiltonian of electromagnetic field

Here, it is shown that the Hamiltonian canonically conjugate to the Lagrangian of the EM field does indeed coincide with the energy density familiar from elementary electrodynamics.

Consider the EM field in the absence of matter, $j = 0$. Verify that the total energy stored in the field is given by $H \equiv \int d^3x\, \mathcal{H}(\mathbf{x})$ where $\mathcal{H}(\mathbf{x}) = \mathbf{E}^2(\mathbf{x}) + \mathbf{B}^2(\mathbf{x})$ is the familiar expression for the EM energy density. (Hint: Use the vacuum form of Maxwell's equations and the fact that, for an infinite system, the energy is defined only up to surface terms.)

Answer:

Following the canonical prescription, let us first consider the Lagrangian density

$$\mathcal{L} = -\frac{1}{4}F_{\mu\nu}F^{\mu\nu} = \frac{1}{2}\sum_{i=1}^{3}(\partial_0 A_i - \partial_i A_0)^2 - \frac{1}{4}\sum_{i,j=1}^{3}(\partial_i A_j - \partial_j A_i)^2.$$

We next determine the components of the canonical momentum through the relation $\pi_\mu = \partial_{\partial_0 A^\mu}\mathcal{L} : \pi_0 = 0$, $\pi_i = \partial_0 A_i - \partial_i A_0 = -E_i$. Using the fact that $\partial_i A_j - \partial_j A_i$ is a component of the magnetic field, the Hamiltonian density can now be written as

$$\mathcal{H} = \pi_\mu \partial_0 A^\mu - \mathcal{L} = \frac{1}{2}(-2\mathbf{E}\cdot\partial_0\mathbf{A} - \mathbf{E}^2 + \mathbf{B}^2) \overset{(1)}{=} \frac{1}{2}(2\mathbf{E}\cdot\nabla\phi + \mathbf{E}^2 + \mathbf{B}^2)$$

$$\overset{(2)}{=} \frac{1}{2}(2\nabla\cdot(\mathbf{E}\phi) + \mathbf{E}^2 + \mathbf{B}^2),$$

where equality (1) is based on addition and subtraction of a term $2\mathbf{E}\cdot\nabla\phi$ and equality (2) on the relation $\nabla\cdot\mathbf{E} = 0$ combined with the identity $\nabla\cdot(\mathbf{a}f) = \nabla\cdot\mathbf{a}f + \mathbf{a}\cdot\nabla f$ (valid for general vector [scalar] functions $\mathbf{a}$ [f]). Substitution of this expression into the definition of the Hamiltonian yields

$$H = \frac{1}{2}\int d^3x\left(2\nabla\cdot(\mathbf{E}\phi) + \mathbf{E}^2 + \mathbf{B}^2\right) = \frac{1}{2}\int d^3x\left(\mathbf{E}^2 + \mathbf{B}^2\right),$$

where we have used the fact that the contribution $\nabla\cdot(\mathbf{E}\phi)$ is a surface term that vanishes upon integration by parts.

1.8.3 Phonon specific heat

Previously, we stated that the mode quantization of elastic media manifests itself in low-temperature anomalies of the specific heat. In this problem, concepts of elementary quantum statistical mechanics are applied to determine the temperature profile of the specific heat.

Compute the energy density $u = -L^{-1}\partial_\beta \ln \mathcal{Z}$ of one-dimensional longitudinal phonons with dispersion $\omega_k = v|k|$, where $\mathcal{Z} = \mathrm{tr}\, e^{-\beta\hat{H}}$ denotes the quantum partition function. First show that the thermal expectation value of the energy density can be represented as

$$u = \frac{1}{L}\sum_k\left[\frac{\omega_k}{2} + \omega_k n_{\mathrm{B}}(\omega_k)\right], \tag{1.47}$$

where $n_\mathrm{B}(\epsilon) = (e^{\beta\epsilon} - 1)^{-1}$ is the Bose–Einstein distribution. Approximate the sum over k by an integral and show that the specific heat $c_v \equiv \partial_T u \sim T$. At what temperature T_cl does the specific heat cross over to the classical result, $c_v = \mathrm{const}$? (Remember that the linear dispersion $\omega_k = v|k|$ is based on a quadratic approximation to the Hamiltonian and, therefore, holds only for $|k| < \Lambda$, where Λ is some cutoff momentum.) Recalling the discussion in section 1.4, for a d-dimensional isotropic solid of volume L^d (with the atomic exchange constants remaining the same in all directions), show that the dispersion generalizes to $\omega_\mathbf{k} = v|\mathbf{k}|$, where $\mathbf{k} = 2\pi(n_1,\dots,n_d)/L$ and $n_i \in \mathbb{Z}$. Show that the specific heat shows the temperature dependence $c_v \sim T^d$.

Answer:

As discussed in the text, the eigenstates of the system are given by $|n_1, n_2, \dots\rangle$, where n_m is the number of phonons of wavenumber $k_m = 2\pi m/L$, $E_{|n_1, n_2, \dots\rangle} = \sum_m \omega_{k_m}(n_m + 1/2) \equiv \sum_m \epsilon_m^{n_m}$ the eigenenergy, and $\omega_m = v|k_m|$. In the energy representation, the quantum partition function then takes the form

$$\mathcal{Z} = \mathrm{tr}\, e^{-\beta \hat{H}} = \sum_{\mathrm{states}} e^{-\beta E_{\mathrm{state}}} = \prod_{m=1,2,\dots} \sum_{n_m=0}^{\infty} e^{-\beta \omega_m(n_m + 1/2)} = \prod_{m=1,2,\dots} \frac{e^{-\beta \omega_m/2}}{1 - e^{-\beta \omega_m}},$$

where n_m is the occupation number of the state with wavenumber k_m. Hence, $\ln \mathcal{Z} = -\sum_m [\beta \omega_m/2 + \ln(1 - e^{-\beta \omega_m})]$. Differentiation with respect to β yields Eq. (1.47) and, making the replacement $\sum_m \to \frac{L}{2\pi} \int dk$, we arrive at $u = C_1 + \frac{1}{2\pi} \int_{|k| < \Lambda} dk\, \frac{v|k|}{e^{\beta v|k|} - 1} = C_1 + \beta^{-2} C_2$, where C_1 is the temperature-independent constant accounting for the "zero-point energies" $\omega_m/2$. In the second equality, we have scaled $k \to \beta k$. This produces a prefactor β^{-2} multiplied by a temperature-independent (up to the temperature dependence of the boundaries $\Lambda \to \beta\Lambda$) integral that we denoted by C_2. Differentiation with respect to T then leads to the relation $c_v = \partial_T u \sim T$. However, for temperatures $T > v\Lambda$ higher than the highest frequencies stored in the phonon modes, the procedure above no longer makes sense (formally, owing to the now non-negligible temperature dependence of the boundaries). Yet, in this regime, we may expand $e^{\beta v|k|} - 1 \simeq \beta v|k|$, which brings us back to the classical result $c_v = \mathrm{const}$.

Consider now a d-dimensional solid with isotropic coupling, $\frac{k_s}{2} \sum_{i=1}^{d} (\phi_{\mathbf{R}+\mathbf{e}_i} - \phi_\mathbf{R})^2$ with $\mathbf{e}_i$ a unit vector in the direction i. Taking the continuum limit leads to a contribution $\frac{k_s a^2}{2} (\nabla \phi(\mathbf{x}))^2$. Proceeding as in the one-dimensional system, the relevant excitations are now waves with wavevector $\mathbf{k} = 2\pi(n_1, \dots, n_d)/L$ and energy $\omega_\mathbf{k} = v|\mathbf{k}|$. Setting $\sum_\mathbf{k} \sim \int d^d k$ and scaling $k_i \to \beta k_i$ then generates a prefactor $\beta^{-(d+1)}$, and we arrive at the relation $c_v \sim T^d$.

1.8.4 Energy–momentum tensor of the harmonic chain

In this problem we analyze the energy–momentum (EM) tensor of the harmonic chain. We discuss its computation and how to make sense of its components.

(a) Show that the two independent components $T^0{}_0$ and $T^1{}_0$ of the EM tensor of the harmonic chain defined via the Lagrangian (1.4) are given by

$$T^0{}_0 = \frac{m}{2}\dot{\phi}^2 + \frac{k_s a^2}{2}(\partial_x \phi)^2, \qquad T^0{}_1 = m\,\partial_t \phi\,\partial_x \phi.$$

(b) In section 1.6.2 we identified $T^0{}_0$ as the energy density of the system. But what is the meaning of the second component? Turning back to the discrete representation of the chain, compute the total momentum carried by weak dynamical fluctuations $\phi_I(t)$ of the mass center coordinates and show that it turns into an integral over $T^0{}_1$ in the continuum limit. This construction identifies $T^0{}_1$ as the momentum density of the chain.

Answer:

(a) This part involves a straightforward application of Eq. (1.46). (b) We can consider the total momentum of the chain as $P = \sum_I a \times \delta(\text{mass density}) \times \dot{\phi}_I$, where $\delta(\text{mass density})$ are the fluctuations in mass density associated with a deviation profile ϕ_I. The local particle density at site I is given by (one particle) / (distorted particle distance), i.e., $1/(a - \phi_{I+1} + \phi_I) \simeq a^{-1} + a^{-2}(\phi_{I+1} - \phi_I) \simeq a^{-1} + a^{-1/2}\partial_x\phi$, where we used the definition of the continuum variable $\phi(x) = a^{-1/2}\phi_I$. This leads to $\delta(\text{mass density}) \simeq ma^{-1/2}\partial_x\phi$. With the particle velocity $\dot{\phi}_I = a^{1/2}\partial_t\phi(x,t)$, we obtain $P = \sum_I a\,m\,\partial_x\phi\,\partial_t\phi \simeq \int dx\,m\,\partial_x\phi\,\partial_t\phi$.

1.8.5 Stress–energy tensor from variation in metric

This problem is for advanced readers. It requires familiarity with integration over manifolds of nontrivial geometry, as reviewed in section A.1, and fluency in variational calculus. Other readers should not tackle this problem just yet. We offer an interpretation of the stress–energy tensor generalizing that given in section 1.6.2: the stress tensor describing how a field theory responds to variations in the underlying geometry.

In section 1.6.2 we derived the stress–energy tensor by investigating how a theory changes under variations $x^\mu \to x^\mu + \omega^\mu(x)$, where the infinitesimal shift may be coordinate dependent. Such deviations describe a local distortion in the geometry of the base manifold. To substantiate this view, consider a situation where the base manifold has a nontrivial geometry, described by a metric tensor $g = \{g_{\mu\nu}\}$. For example, in the field theories of gravity, the base manifold is the universe, and $g_{\mu\nu}$ is its space–time metric. A more mundane example would be a field theory formulated in curvilinear coordinates, where $g_{\mu\nu}$ is the (square of the) Jacobian describing the transformation from Cartesian coordinates. The generalization of the Lagrangian Eq. (1.16) to this case is given by

$$S[\phi] = \int dx\,\sqrt{g}\,\mathcal{L}(\phi, \partial_\mu, \partial^\mu\phi),$$

where $g = |\det(g_{\{\mu\nu\}})|$, and the notation emphasizes that derivatives in the Lagrangian appear in invariant combinations such as $\partial^\mu\phi\partial_\mu\phi$. Their dependence on the metric is hidden in $\partial^\mu\phi = g^{\mu\nu}\partial_\nu\phi$, where $g^{\mu\nu}$ are the coefficients of the inverse of the metric tensor, $g^{\mu\nu}g_{\nu\lambda} = \delta^\mu{}_\lambda$. (We have omitted the internal field index ϕ^i to lighten the notation.)
(a) Prove the auxiliary relations $\partial g^{\rho\sigma}/\partial g_{\mu\nu} = -\delta^{\rho\mu}\delta^{\nu\sigma}$, $\partial_{g_{\mu\nu}}\sqrt{g} = \frac{1}{2}\sqrt{g}g^{\mu\nu}$, and $\partial F/\partial(\partial_\mu\phi) = (\partial F/\partial(\partial^\nu\phi))g^{\nu\mu}$.
(b) Show that the stress tensor is obtained by variation of the action in the metric:

$$T^{\mu\nu}(x) = -\frac{2}{\sqrt{g}}\frac{\delta S}{\delta g_{\mu\nu}(x)}. \tag{1.48}$$

(c) As an example, consider the theory of a free scalar field, $\mathcal{L} = -\frac{1}{2}\partial_\mu\phi\partial^\mu\phi$. Compute the stress tensor via Eq. (1.48) and convince yourself that the result is compatible with that of the example below Eq. (1.46) for the harmonic chain Eq. (1.4) in the case where the differentiation is carried out on the two-dimensional Minkowski metric, $g = \text{diag}(-1, 1)$, and the constants are scaled as $m = k_s a^2 = 1$.

Conceptually, Eq. (1.48) demonstrates that the stress tensor answers the question how a theory responds to variations in the geometry of its base manifold. (The terminology *stress* tensor underpins this interpretation.) Methodologically, it is often convenient to compute the stress tensor via Eq. (1.48), including in cases where the theory is varied at a trivial metric $g_{\mu\nu} = \delta_{\mu\nu}$.

Answer:

(a) Using that $\partial g_{\rho\sigma}/\partial g_{\mu\nu} = \delta^\mu{}_\rho\delta^\sigma{}_\nu,$[35] the first identity is obtained from the matrix relation $0 = \partial_{g_{\mu\nu}}(gg^{-1}) = (\partial_{g_{\mu\nu}}g)g^{-1} + g\partial_{g_{\mu\nu}}g^{-1}$. Written in components, it yields the desired relation. With $g = \pm\det(g),$[36] the second follows from $\partial_{g_{\mu\nu}}\sqrt{g} = 1/(2\sqrt{g})\partial_{g_{\mu\nu}}(\pm\det(g))$. Using $\det g = \exp\operatorname{tr}\ln g$, and $\partial_{g_{\mu\nu}}\operatorname{tr}\ln g = (g^{-1})^{\nu\mu} = g^{\mu\nu}$, we obtain the relation. The final relation follows from the chain rule applied to $\partial_\mu\phi = g_{\mu\nu}\partial^\nu\phi$.

(b) The metric enters the action in two places, the first being the factor $\sqrt{g}$, the second the derivatives $\phi^\mu = g^{\mu\nu}\phi^\nu$. We thus have

$$T^{\mu\nu} = -\frac{2}{\sqrt{g}}\frac{\delta S}{\delta g_{\mu\nu}} = -\frac{2}{\sqrt{g}}\left(\mathcal{L}\frac{\partial\sqrt{g}}{\partial g_{\mu\nu}} + \sqrt{g}\frac{\partial\mathcal{L}}{\partial(\partial^\rho\phi)}\frac{\partial(\partial^\rho\phi)}{\partial g_{\mu\nu}}\right)$$

$$= -\frac{2}{\sqrt{g}}\left(\mathcal{L}\frac{\partial\sqrt{g}}{\partial g_{\mu\nu}} + \sqrt{g}\frac{\partial\mathcal{L}}{\partial(\partial^\rho\phi)}\left(\frac{\partial g^{\rho\sigma}}{\partial g_{\mu\nu}}\right)\partial_\sigma\phi\right) = -\mathcal{L}g^{\mu\nu} + \frac{\partial\mathcal{L}}{\partial(\partial_\mu\phi)}\partial^\nu\phi,$$

where in the final step we used the three relations in (a). Lowering the right index, $T^\mu{}_\nu = T^{\mu\rho}g_{\rho\sigma}$, we get back to Eq. (1.46).

Now considering the relation $\det g = \exp\operatorname{tr}\ln g$, it is varied as $\partial_{g_{\mu\nu}}\sqrt{g} = \frac{1}{2}\sqrt{g}\partial_{g_{\mu\nu}}\operatorname{tr}\ln g = \sqrt{g}(g^{-1})_{\nu\mu} = \sqrt{g}g^{\mu\nu}$, where we have used the symmetry $g_{\mu\nu} = g_{\nu\mu}$ of the metric tensor, and the notation $g^{\mu\nu} = (g^{-1})_{\mu\nu}$ for its inverse.

The differentiation in the second occurrence of the metric, $\partial_\mu\phi = g_{\mu\nu}\partial^\nu\phi$, is done as follows: $\partial_{g_{\mu\nu}}\mathcal{L} = \frac{\partial\mathcal{L}}{\partial(\partial_\mu\phi)}\partial_{g_{\mu\nu}}\partial_\mu\phi = \frac{\mathcal{L}}{\partial(\partial_\mu\phi)}\partial^\nu\phi$. Adding the two terms we get

$$T^{\mu\nu} = \sqrt{g}\left(g^{\mu\nu}\mathcal{L} + \frac{\partial\mathcal{L}}{\partial(\partial_\mu\phi)}\partial^\nu\phi\right).$$

(c) For the free field theory in Minkowski space, we have $\sqrt{g} = 1$ and $\partial_{\partial_\mu\phi}\mathcal{L} = -\partial^\mu\phi$. This gives $T^0{}_0 = \frac{1}{2}(\partial_\mu\phi\partial^\mu\phi) - \partial^0\phi\partial_0\phi = \frac{1}{2}(-\partial^0\phi\partial_0\phi + \partial^1\phi\partial_1\phi) = \frac{1}{2}((\partial_0\phi)^2 + (\partial_1\phi)^2))$. Identifying the zero-coordinate with time, and the one-coordinate with space, this equals the Hamiltonian density (kinetic energy+potential energy density) of the harmonic chain.

---

[35] All derivatives are carried out for a general matrix, and then evaluated at the symmetric configuration $g_{\rho\sigma} = g_{\sigma\rho}$. We are not differentiating within the class of symmetric matrices. Think about this difference.

[36] It is common practice to denote the modulus determinant $g = \pm\det(\{g_{\mu\nu}\})$, and the matrix $g = \{g_{\mu\nu}\}$ by the same symbol g. Which is which should always be clear from the context.

2 Second Quantization

SYNOPSIS The aim of this chapter is to introduce and apply the method of second quantization, a technique that underpins the formulation of quantum many-particle theories. The first part of the chapter focuses on the development of methodology and notation, while the remainder is devoted to applications designed to engender familiarity with, and fluency in, the approach. Indeed, many of these examples will subsequently reappear as applications in our discussion of the methods of quantum field theory.

second
quanti-
zation

In the previous chapter, we encountered two field theories that could conveniently be represented in the language of **second quantization**, i.e., a formulation based on the algebra of ladder operators $\hat{a}_k$.[1] There are two remarkable facts about this formulation. First, second quantization provides a compact way of representing the many-body space of excitations; second, the properties of the ladder operators are encoded in a simple set of commutation relations (cf. Eq. (1.33)) rather than in some explicit Hilbert space representation.

Apart from its aesthetic appeal, these observations would not be of much relevance if it were not for the fact that the formulation can be generalized to a comprehensive and highly efficient formulation of many-body quantum mechanics in general. In fact, second quantization can be considered as the first major cornerstone on which the theoretical framework of quantum field theory was built. This being so, extensive introductions to the concept can be found throughout the literature. We will therefore not develop the formalism in full mathematical rigor but rather will proceed pragmatically by first motivating and introducing its basic elements, followed by a discussion of the second quantized version of standard operations of quantum mechanics (taking matrix elements, changing bases, representing operators, etc.). The second part of the chapter is concerned with developing fluency in the method by addressing several applications. Readers familiar with the formalism may therefore proceed directly to these sections.

---

[1] The term "second quantization" is unfortunate. Historically, this terminology was motivated by the observation that the ladder operator algebra fosters an interpretation of quantum excitations as discrete "quantized" units. However, fundamentally, there is nothing like two superimposed quantization steps in single- or many-particle quantum mechanics. Rather, there is only a particular representation of the "first and only" quantized theory tailored to the problem at hand.

2.1 Introduction to Second Quantization

We begin by recapitulating some basic notions of many-body quantum mechanics, as formulated in the traditional language of symmetrized/antisymmetrized wave functions. Consider the normalized set of wave functions $|\lambda\rangle$ of some single-particle Hamiltonian $\hat{H}$: $\hat{H}|\lambda\rangle = \epsilon_\lambda|\lambda\rangle$, where ϵ_λ are the eigenvalues. With this definition, the normalized two-particle wave function $\psi_F(\psi_B)$ of two fermions (bosons) populating levels λ_1 and λ_2 is given by the anti-symmetrized (symmetrized) product

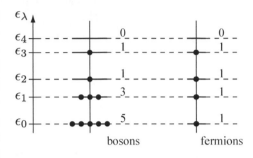

$$\psi_F(x_1, x_2) = \frac{1}{\sqrt{2}}\left(\langle x_1|\lambda_1\rangle\langle x_2|\lambda_2\rangle - \langle x_1|\lambda_2\rangle\langle x_2|\lambda_1\rangle\right),$$

$$\psi_B(x_1, x_2) = \frac{1}{\sqrt{2}}\left(\langle x_1|\lambda_1\rangle\langle x_2|\lambda_2\rangle + \langle x_1|\lambda_2\rangle\langle x_2|\lambda_1\rangle\right).$$

In Dirac notation, the two-body states $|\lambda_1, \lambda_2\rangle_{F(B)}$ corresponding to $\psi_{F(B)}(x_1, x_2) = (\langle x_1|\otimes\langle x_2|)\,|\lambda_1, \lambda_2\rangle_{F(B)}$ above can be represented as

$$|\lambda_1, \lambda_2\rangle_{F(B)} \equiv$$
$$\frac{1}{\sqrt{2}}\left(|\lambda_1\rangle \otimes |\lambda_2\rangle + \zeta|\lambda_2\rangle \otimes |\lambda_1\rangle\right),$$

where $\zeta = -1$ for fermions and $\zeta = +1$ for bosons. Symmetrization is necessitated by quantum **particle indistinguishability**: for fermions (bosons)

particle indistinguishability

Enrico Fermi 1901–1954 was the recipient of the Nobel Prize in Physics in 1938 for "his demonstrations of the existence of new radioactive elements produced by neutron irradiation, and for his related discovery of nuclear reactions brought about by slow neutrons." Born in Rome, Fermi left Italy in 1938 to escape Mussolini's regime. In Chicago, Fermi led the team that designed and built the first nuclear reactor, and he become centrally involved in the Manhattan Project during World War II.

the wave function must be antisymmetric (symmetric) under particle exchange.[2] Generally, a symmetrized N-particle wave function can be expressed as

$$\boxed{|\lambda_1, \lambda_2, \ldots, \lambda_N\rangle \equiv \frac{1}{\sqrt{N! \prod_{\lambda=0}^{\infty} n_\lambda!}} \sum_{\mathcal{P}} \zeta^{(1-\mathrm{sgn}\,\mathcal{P})/2}|\lambda_{\mathcal{P}1}\rangle \otimes |\lambda_{\mathcal{P}2}\rangle \otimes \cdots |\lambda_{\mathcal{P}N}\rangle}$$

(2.1)

where n_λ denotes the number of particles in state λ (for fermions, Pauli exclusion enforces the constraint $n_\lambda \leq 1$) – see the schematic figure above. The summation runs over all $N!$ permutations of the quantum numbers $\{\lambda_1, \ldots, \lambda_N\}$, and sgn $\mathcal{P}$

---

[2] Note, however, that in two dimensions, the standard doctrine of fully symmetric/antisymmetric many-particle wave functions is too narrow, and more general types of exchange statistics can be realized (cf. our discussion on page 42).

denotes the **sign** of the permutation $\mathcal{P}$. (Note that $\mathrm{sgn}\,\mathcal{P} = +1\,[-1]$ if the number of transpositions of two elements that brings the permutation $(\mathcal{P}_1, \mathcal{P}_2, \ldots, \mathcal{P}_N)$ back to its original form $(1, 2, \ldots, N)$ is even [odd].) The prefactor $1/\sqrt{N!\prod_\lambda n_\lambda!}$ normalizes the wave function. In the fermionic case, the wave functions are known

Slater determinant as **Slater determinants**.

Finally, it will be useful to assume that the quantum numbers $\{\lambda_i\}$ defining the state $|\lambda_1, \lambda_2, \ldots, \lambda_N\rangle$ are ordered according to some convention; e.g., for $\lambda_i = x_i$, a one-dimensional coordinate representation, we might order according to the rule $x_1 \leq x_2 \leq \cdots \leq x_N$. Once an ordered sequence of states has been fixed we may – for notational convenience – label our quantum states by integers, $\lambda_i = 1, 2, \ldots$ Any initially non-ordered state (e.g., $|2, 1, 3\rangle$) can be brought to an ordered form ($|1, 2, 3\rangle$) at the cost of, at most, a change of sign.

> **David Hilbert** 1862–1943 was a German who contributed to many branches of mathematics including the theory of algebraic number fields, functional analysis, integral equations, mathematical physics and the calculus of variations. His work in geometry had the greatest influence after Euclid. A systematic study of the axioms of Euclidean geometry led Hilbert to propose 21 such axioms and he analyzed their significance.

permutation group INFO For the sake of completeness, let us spell out the connection between the **permutation group** and **many-body quantum mechanics** in a more mathematical language. The basic arena wherein N-body quantum mechanics takes place is the product space,

$$\mathcal{H}^N \equiv \underbrace{\mathcal{H} \otimes \cdots \otimes \mathcal{H}}_{N\ \text{copies}}$$

of N single-particle Hilbert spaces. In this space, we have a linear representation of the permutation group, S^N,[3] assigning to each $\mathcal{P} \in S^N$ the permutation (with no ordering of the λs implied at this stage),

$$\mathcal{P} : \mathcal{H}^N \to \mathcal{H}^N, \quad |\lambda_1\rangle \otimes \cdots \otimes |\lambda_N\rangle \mapsto |\lambda_{\mathcal{P}1}\rangle \otimes \cdots \otimes |\lambda_{\mathcal{P}N}\rangle.$$

The identification of all irreducible subspaces of this representation is a formidable task, which, thanks to a fundamental axiom of quantum mechanics, we need not address in full. All we need to know is that S^N has two particularly simple one-dimensional irreducible

group representation [3] Recall that a **linear representation of a group** G is a mapping that assigns to each $g \in G$ a linear mapping $\rho_g : V \to V$ for some vector space V. For notational convenience, one usually writes $g : V \to V$ instead of $\rho_g : V \to V$. Conceptually, however, it is often important to distinguish carefully between the abstract group elements g and the *matrices* (also g) assigned to them by a given representation. (Consider, for example, the symmetry group $G = \mathrm{SU}(2)$ of quantum mechanical spin. $\mathrm{SU}(2)$ is the two-dimensional group of unitary matrices with determinant unity. However, when acting in the Hilbert space of a quantum spin $S = 5$, say, the elements of $\mathrm{SU}(2)$ are represented by $(2S+1 = 11)$-dimensional matrices.) Two representations ρ and ρ' that differ only by a unitary transformation, $\forall g \in G : \rho_g = U\rho_g' U^{-1}$, are called unitary equivalent. If a transformation U can be found such that all representation matrices ρ_g assume a block structure, the representation is called **reducible**, otherwise **irreducible**. Notice that the different sub-blocks of a reducible representation by themselves form irreducible representation spaces. The identification of all distinct irreducible representations of a given group is one of the most important objectives of group theory.

representations: one wherein each $\mathcal{P} \in S^N$ acts as the identity transform $\mathcal{P}(\Psi) \equiv \Psi$ and another, the alternating representation $\mathcal{P}(\Psi) = \text{sgn}\,\mathcal{P}\,\Psi$. According to a basic postulate of quantum mechanics, the state vectors $\Psi \in \mathcal{H}^N$ describing bosons/fermions must transform according to the **identity/alternating representation**. The subset $\mathcal{F}^N \subset \mathcal{H}^N$ of all states showing this transformation behavior defines the physical N-body Hilbert space. To construct a basis of $\mathcal{F}^N$, one may apply the symmetrization operator $P^s \equiv \sum_{\mathcal{P}} \mathcal{P}$ (antisymmetrization operator $P^a \equiv \sum_{\mathcal{P}}(\text{sgn}\,\mathcal{P})\mathcal{P}$) to the basis vectors $|\lambda_1\rangle \otimes \cdots, |\lambda_N\rangle$ of $\mathcal{H}^N$. Up to normalization, this operation obtains the states (2.1).

Readers may wonder why we mention these representation-theoretic aspects, since, being pragmatic, all we really need to know is the symmetrization/antisymmetrization postulate and its implementation through Eq. (2.1). Note, however, that one may question what we actually mean when we talk about the permutation exchange of quantum numbers. For example, when we compare wave functions that differ by an exchange of coordinates we should be able to tell by what physical operation we effect this exchange (for, otherwise, we cannot really compare them other than in a formal and in fact ambiguous manner).

Surprisingly, decades passed before this crucial issue in quantum mechanics was critically addressed. In a seminal work by Leinaas and Myrheim,[65] it was shown that the standard paradigm of permutation exchange is far from innocuous. Indeed, its applicability is tied to the dimensionality of space! In two dimensions, a more elaborate scheme is needed. (Nevertheless, one may use representation-theoretic concepts to describe particle exchange. However, the relevant group – the **braid group** – now differs from the permutation group.) Physically, these requirements are manifest in the emergence of quantum particles different from either bosons or fermions. For a further discussion of these so-called **anyons**, we refer to section 8.6.1. (In spite of being included in a later chapter, this section is not difficult to read!)

braid
group

anyons

While representations like (2.1) *can* be used to represent the full Hilbert space of many-body quantum mechanics, a moment's thought shows that this formulation is not at all convenient.

▷ It takes little imagination to anticipate that practical computation in the language of Eq. (2.1) will be cumbersome. For example, to compute the overlap of two wave functions, one needs to form no less than $(N!)^2$ different products.

▷ The representation is tailor-made for problems with fixed particle number N. However, we know from statistical mechanics that, for $N = O(10^{23})$, it is more convenient to work in a grand canonical formulation, where N is allowed to vary.

▷ Similarly, in applications one will often ask the question "what is the amplitude for the injection of a particle into a system at a certain space–time coordinate (x_1, t_1) followed by annihilation at some later time (x_2, t_2)?" Ideally, one would work with a representation that supports the intuition afforded by thinking in terms of such processes, i.e., a representation where the quantum numbers of individual quasi-particles, rather than the entangled set of quantum numbers of all constituents, are fundamental.

The second-quantized formulation of many-body quantum mechanics, as introduced below, will remove all these difficulties in an elegant and efficient manner.

2.1.1 The apparatus of second quantization

Some of the disadvantages of the representation (2.1) can be avoided with relatively little effort. Specifically, it pays to label many-body states in a more efficient manner than Eq. (2.1), and to define a subset of the many-body Hilbert space just large enough to accommodate all states of definite exchange statistics.

Occupation number representation and Fock space

In our present notation, quantum states are represented by "N-letter words" of the form $|1,1,1,1,2,2,3,3,3,4,6,6,\ldots\rangle$. Obviously, this notation contains a lot of redundancy. A more efficient encoding of the state above might read $|4,2,3,1,0,2,\ldots\rangle$, where the ith number signals how many particles occupy state number i; no more information is needed to characterize a symmetrized state. (For fermions, these occupation numbers take a value of either zero or one.) This defines the **occupation number representation**. In the new representation, the basis states of $\mathcal{F}^N$ are specified by $|n_1, n_2, \ldots\rangle$, where $\sum_i n_i = N$. Any state $|\Psi\rangle$ in $\mathcal{F}^N$ can be obtained as a linear superposition,

$$|\Psi\rangle = \sum_{\substack{n_1, n_2, \ldots, \\ \sum_i n_i = N}} c_{n_1, n_2, \ldots} |n_1, n_2, \ldots\rangle.$$

As pointed out above, eventually we will want to emancipate ourselves from the condition of a fixed particle number N. A Hilbert space large enough to accommodate a state with an undetermined number of particles is given by

$$\mathcal{F} \equiv \bigoplus_{N=0}^{\infty} \mathcal{F}^N. \tag{2.2}$$

Notice that the direct sum contains a curious contribution $\mathcal{F}^0$, the **vacuum space**. This is a one-dimensional Hilbert space which describes the sector of the theory with no particles present. Its single normalized basis state, the **vacuum state**, is denoted by $|0\rangle$. We will soon see why it is convenient to add this space to our family of basis states. The spaces $\mathcal{F}$ are called **Fock spaces** and they define

Vladimir Aleksandrovich Fock 1898–1974
was a Soviet physicist who played a key role in the development of the general theory of relativity and many-body theory. His ground-breaking contributions include the introduction of the Fock space and the development of perhaps the most important many-particle approximation scheme, the Hartree–Fock approximation.

the principal arenas of quantum many-body theory. Note that the construction of a Fock space builds on a given single-particle basis defining the number of different labels i.[4] Note that we also need to distinguish between *fermionic* and *bosonic* Fock spaces, depending on the exchange symmetry of their states.

To obtain a basis of $\mathcal{F}$, we need only take the totality of our previous basis states $\{|n_1, n_2, \ldots\rangle\}$, and drop the condition $\sum_i n_i = N$ on the occupation numbers. A general many-body state $|\Psi\rangle$ can then be represented by a linear superposition $|\Psi\rangle = \sum_{n_1, n_2, \ldots} c_{n_1, n_2, \ldots} |n_1, n_2, \ldots\rangle$. Notice that states with a different particle numbers may contribute to the linear superposition forming $|\Psi\rangle$. We shall see that such mixtures play an important role in, for example, the theory of superconductivity.

Foundations of second quantization

The occupation number representation introduced above provides a step in the right direction, but it does not yet solve our main problem: the need for explicit symmetrization or antisymmetrization of a large number of states in each quantum operation. As a first step towards the construction of a more powerful representation, let us recall an elementary fact of linear algebra: a linear map $A : V \to V$ of a vector space into itself is fully determined by defining the images $w_i \equiv A v_i$ of the action of A on a basis $\{v_i\}$. Now let us use this scheme to introduce a set of linear operators acting in Fock space. For every $i = 1, 2, \ldots$, we define operators $a_i^\dagger : \mathcal{F} \to \mathcal{F}$ through

$$a_i^\dagger |n_1, \ldots, n_i, \ldots\rangle \equiv (n_i + 1)^{1/2} \zeta^{s_i} |n_1, \ldots, n_i + 1, \ldots\rangle, \qquad (2.3)$$

where $s_i = \sum_{j=1}^{i-1} n_j$. In the fermionic case, the occupation numbers n_i have to be understood $\mathrm{mod}\, 2$, so that $(1 + 1) = 0 \,\mathrm{mod}\, 2$, i.e., the application of $a_i^\dagger$ to a state with $n_i = 1$ leads to its annihilation.

Notice that, by virtue of this definition, we are able to generate every basis state of $\mathcal{F}$ by repeated application of $a_i^\dagger$s to the vacuum state. (From a formal point of view, this fact alone gives sufficient motivation to add the vacuum space to the definition of Fock space.) Indeed, repeated application of (2.3) leads to the important relation

$$\boxed{|n_1, n_2, \ldots\rangle = \prod_i \frac{1}{(n_i!)^{1/2}} (a_i^\dagger)^{n_i} |0\rangle} \qquad (2.4)$$

Notice that Eq. (2.4) constitutes a strong statement: the complicated permutation "entanglement" implied by the definition (2.1) of the Fock states can be generated by straightforward application of a set of linear operators to a single reference state. Physically, the N-fold application of operators $a^\dagger$ to an empty vacuum state generates an N-particle state, which is why the $a^\dagger$s are commonly called **creation operators**. Of course, the introduction of creation operators might still turn out to be useless, i.e., the requirement of consistency with the properties of the Fock states

creation operators

---

[4] Depending on the application, we use more complex labels than integers i to denote states in occupation number representation. For example, working with spinful fermions on a lattice with sites x_l, $i \to (l, \sigma)$ would be an appropriate notation.

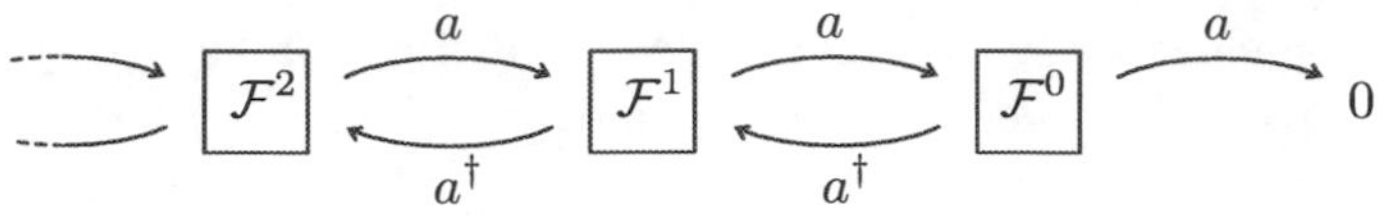

Fig. 2.1   Visualization of the generation of the Fock subspaces $\mathcal{F}^N$ by repeated action of creation operators $a^\dagger$ on the vacuum space $\mathcal{F}^0$.

(such as the fact that, in the fermionic case, the numbers $n_i = 0, 1$ are defined only mod 2) might invalidate the simple relation (2.3) with its (n_i-independent!) operator $a_i^\dagger$. However, as we shall demonstrate below, this is not the case.

Consider two operators $a_i^\dagger$ and $a_j^\dagger$ for $i \neq j$. From the definition (2.3), one may readily verify that $(a_i^\dagger a_j^\dagger - \zeta a_j^\dagger a_i^\dagger)|n_1, n_2, \ldots\rangle = 0$. As it holds for every basis vector, this relation implies that

$$\forall i, j: \qquad [a_i^\dagger, a_j^\dagger]_\zeta = 0, \qquad (2.5)$$

where $[\hat{A}, \hat{B}]_\zeta \equiv \hat{A}\hat{B} - \zeta\hat{B}\hat{A}$, i.e., $[\,,\,]_{\zeta=1} \equiv [\,,\,]$ is the commutator and $[\,,\,]_{\zeta=-1} \equiv \{\,,\,\} \equiv [\,,\,]_+$ is the anticommutator. Turning to the case $i = j$, we note that, for fermions, the two-fold application of $a_i^\dagger$ to any state leads to its annihilation. Thus, $(a_i^\dagger)^2 = 0$ is nilpotent, a fact that can be formulated as $[a_i^\dagger, a_i^\dagger]_+ = 0$. For bosons we have, of course, $[a_i^\dagger, a_i^\dagger] = 0$ (identical operators commute!).

Now, quantum mechanics is a unitary theory so, whenever one meets a new operator $\hat{A}$, one should determine its hermitian adjoint $\hat{A}^\dagger$. To understand the action of the hermitian adjoints $\left(a_i^\dagger\right)^\dagger = a_i$ of the creation operators, we may take the complex conjugates of all basis matrix elements of Eq. (2.3):

$$\langle n_1, \ldots, n_i, \ldots | a_i^\dagger | n_1', \ldots, n_i', \ldots \rangle = (n_i' + 1)^{1/2} \zeta^{s_i'} \delta_{n_1, n_1'} \cdots \delta_{n_i, n_i'+1} \cdots$$

$$\Rightarrow \quad \langle n_1', \ldots, n_i', \ldots | a_i | n_1, \ldots, n_i, \ldots \rangle^* = n_i^{1/2} \zeta^{s_i} \delta_{n_1', n_1} \cdots \delta_{n_i', n_i-1} \cdots$$

As it holds for every bra $\langle n_1', \ldots, n_i', \ldots |$, the last line tells us that

$$a_i|n_1, \ldots, n_i, \ldots\rangle = n_i^{1/2} \zeta^{s_i} |n_1, \ldots, n_i - 1, \ldots\rangle, \qquad (2.6)$$

annihilation operators

a relation that identifies a_i as an operator that "annihilates" particles. The action of creation and **annihilation operators** in Fock space is illustrated in fig. 2.1. Creation operators $a^\dagger: \mathcal{F}^N \to \mathcal{F}^{N+1}$ increase the particle number by one, while annihilation operators $a: \mathcal{F}^N \to \mathcal{F}^{N-1}$ lower it by one. The application of an annihilation operator to the vacuum state, $a_i|0\rangle = 0$, annihilates it. (Do not confuse $|0\rangle$ with the zero vector.)

Taking the hermitian adjoint of Eq. (2.5), we obtain $[a_i, a_j]_\zeta = 0$. Further, a straightforward calculation based on the definitions (2.3) and (2.6) shows that $[a_i, a_j^\dagger]_\zeta = \delta_{ij}$. Altogether, we have shown that the creation and annihilation operators satisfy the algebraic closure relations,

$$\boxed{[a_i, a_j^\dagger]_\zeta = \delta_{ij}, \quad [a_i, a_j]_\zeta = 0, \quad [a_i^\dagger, a_j^\dagger]_\zeta = 0} \qquad (2.7)$$

Given that the full complexity of Fock space is generated by application of $a_i^\dagger$s to a single reference state, the simplicity of the relations obeyed by these operators seems remarkable and surprising.

INFO Perhaps less surprising is that, behind this phenomenon, there lingers some mathematical structure. Suppose that we are given an abstract algebra $\mathcal{A}$ of objects $a_i, a_i^\dagger$ satisfying (2.7). (Recall that an algebra is a vector space whose elements can be multiplied by each other.) Further, suppose that $\mathcal{A}$ is irreducibly represented in some vector space V, i.e., that there is a mapping assigning to each $a_i \in \mathcal{A}$ a linear mapping $a_i : V \to V$ such that every vector $|v\rangle \in V$ can be reached from any other $|w\rangle \in V$ by (possibly iterated) application of operators a_i and $a_i^\dagger$ (due to irreducibility).[5] According to the **Stone–von Neumann theorem**, (i) such a representation is unique (up to unitary equivalence) and (ii) there is a unique state $|0\rangle \in V$ that is annihilated by every a_i. All other states can then be reached by repeated application of $a_i^\dagger$s. The precise formulation of this theorem and its proof – a good practical exercise in working with creation and annihilation operators – is left to problem 2.4.1. From the Stone–von Neumann theorem, we can infer that the Fock space basis could have been constructed in reverse. Not knowing the basis $\{|n_1, n_2, \ldots\rangle\}$, we could have started from a set of operators obeying the commutation relations (2.7) acting in some *a priori* unknown space $\mathcal{F}$. Starting from the unique state $|0\rangle$, the prescription (2.4) would then have yielded an equally unique basis of the entire space $\mathcal{F}$ (up to unitary transformations). In other words, the algebra (2.7) fully characterizes the operator action and provides information equivalent to the definitions (2.3) and (2.6).

Stone–von Neumann theorem (margin note)

Practical aspects of the second quantization

Our next task will be to promote the characterization of Fock space bases to a full reformulation of many-body quantum mechanics. To this end, we need to establish how changes from one single-particle basis $\{|\lambda\rangle\}$ to another, $\{|\tilde{\lambda}\rangle\}$, affect the operator algebra $\{a_\lambda\}$. (In this section, we shall no longer use integers to identify different elements of a given single-particle basis. Rather, we use Greek labels λ; i.e., $a_\lambda^\dagger$ creates a particle in the state λ.) We also need to understand in what way generic operators acting in many-particle Hilbert spaces can be represented in terms of creation and annihilation operators.

▷ **Change of basis:** Using the resolution of identity $\mathrm{id} = \sum_{\lambda=0}^{\infty} |\lambda\rangle\langle\lambda|$, the relations $|\tilde{\lambda}\rangle = \sum_\lambda |\lambda\rangle\langle\lambda|\tilde{\lambda}\rangle$, $|\lambda\rangle \equiv a_\lambda^\dagger|0\rangle$, and $|\tilde{\lambda}\rangle \equiv a_{\tilde{\lambda}}^\dagger|0\rangle$ immediately give rise to the transformation law,

$$\boxed{a_{\tilde{\lambda}}^\dagger = \sum_\lambda \langle\lambda|\tilde{\lambda}\rangle a_\lambda^\dagger, \qquad a_{\tilde{\lambda}} = \sum_\lambda \langle\tilde{\lambda}|\lambda\rangle a_\lambda} \tag{2.8}$$

---

[5] To characterize this representation, we need to be a bit more precise. Within $\mathcal{A}$, a_i and $a_i^\dagger$ are independent objects, i.e., in general, there exists no notion of hermitian adjointness in $\mathcal{A}$. We require, though, that the representation assigns to $a_i^\dagger$ the hermitian adjoint (in V) of the image of a_i. Also, we have to require that $[a_i, a_j^\dagger] \in \mathcal{A}$ be mapped onto $[a_i, a_j^\dagger] : V \to V$ where, in the latter expression, the commutator involves the ordinary product of the matrices $a_i, a_j^\dagger : V \to V$.

In many applications, we will be dealing with continuous sets of quantum numbers (such as position coordinates). In such cases the quantum numbers are commonly denoted as $a(x) = \sum_\lambda \langle x|\lambda\rangle a_\lambda$ and the summations appearing in the transformation formula above translate to integrals: $a_\lambda = \int dx \langle \lambda|x\rangle a(x)$. For example, the transformation from the coordinate represetation to the momentum representation in a system of length L would read

$$a_k = \int_0^L dx \, \langle k|x\rangle a(x), \qquad a(x) = \sum_k \langle x|k\rangle a_k, \qquad (2.9)$$

where $\langle k|x\rangle \equiv \langle x|k\rangle^* = L^{-1/2} e^{-ikx}$ and $k = 2\pi m/L$, $m \in \mathbb{Z}$.

▷ **Representation of operators (one-body):** Single-particle or one-body operators $\hat{\mathcal{O}}_1$ acting in the N-particle Hilbert space $\mathcal{F}^N$ generally take the form $\hat{\mathcal{O}}_1 = \sum_{n=1}^N \hat{o}_n$, where $\hat{o}_n$ is an ordinary single-particle operator acting on the nth particle. A typical example is the kinetic energy operator $\hat{T} = \sum_n \hat{p}_n^2/(2m)$, where $\hat{p}_n$ is the momentum operator acting on the nth particle. Other examples include the one-particle potential operator $\hat{V} = \sum_n V(\hat{x}_n)$, where $V(x)$ is a scalar potential, and the total spin operator $\sum_n \hat{\mathbf{S}}_n$. Since we have seen that, by applying field operators to the vacuum space, we can generate the Fock space in general and any N-particle Hilbert space in particular, it must be possible to represent any operator $\hat{\mathcal{O}}_1$ in an a-representation.

Although the representation of n-body operators is straightforward in principle, it can, at first sight, seem daunting. A convenient way of finding it is to express the operator in terms of the basis in which it is diagonal, and only later transform to an arbitrary basis. For this purpose it is useful to define the **occupation number operator**

occupation
number

$$\boxed{\hat{n}_\lambda = a_\lambda^\dagger a_\lambda} \qquad (2.10)$$

with the property that, for bosons or fermions, $\hat{n}_\lambda \left(a_\lambda^\dagger\right)^n |0\rangle = n \left(a_\lambda^\dagger\right)^n |0\rangle$ (exercise). Since $\hat{n}_\lambda$ commutes with all $a_{\lambda' \neq \lambda}^\dagger$, Eq. (2.4) implies that $\hat{n}_\lambda |n_{\lambda_1}, \ldots\rangle = n_{\lambda_j} |n_{\lambda_1}, \ldots\rangle$, i.e., $\hat{n}_\lambda$ simply counts the number of particles in state λ. Let us now consider a one-body operator $\hat{\mathcal{O}}_1$ which is diagonal in the basis $|\lambda\rangle$, with $\hat{o} = \sum_i o_{\lambda_i} |\lambda_i\rangle\langle\lambda_i|$, $o_{\lambda_i} = \langle\lambda_i|\hat{o}|\lambda_i\rangle$. With this definition, one finds that

$$\langle n_{\lambda_1}', n_{\lambda_2}', \ldots |\hat{\mathcal{O}}_1|n_{\lambda_1}, n_{\lambda_2}, \ldots\rangle = \sum_i o_{\lambda_i} n_{\lambda_i} \langle n_{\lambda_1}', n_{\lambda_2}', \ldots |n_{\lambda_1}, n_{\lambda_2}, \ldots\rangle$$

$$= \langle n_{\lambda_1}', n_{\lambda_2}', \ldots | \sum_i o_{\lambda_i} \hat{n}_{\lambda_i} |n_{\lambda_1}, n_{\lambda_2}, \ldots\rangle.$$

Since this equality holds for any set of states, one can infer the second-quantized representation of the operator,

$$\hat{\mathcal{O}}_1 = \sum_{\lambda=0}^\infty o_\lambda \hat{n}_\lambda = \sum_{\lambda=0}^\infty \langle\lambda|\hat{o}|\lambda\rangle a_\lambda^\dagger a_\lambda.$$

The result is straightforward: a one-body operator engages a single particle at a time – the others are just spectators. In the diagonal representation, one simply counts the number of particles in a state λ and multiplies by the corresponding eigenvalue of the one-body operator. By transforming from the diagonal representation to some arbitrary basis, one obtains the general result,

$$\boxed{\hat{\mathcal{O}}_1 = \sum_{\mu\nu} \langle\mu|\hat{o}|\nu\rangle a_\mu^\dagger a_\nu} \tag{2.11}$$

To consolidate these ideas, let us consider some specific examples. Representing the matrix elements of the single-particle spin operator as $\langle\alpha|\hat{S}_i|\alpha'\rangle \equiv (S_i)_{\alpha\alpha'} = \frac{1}{2}(\sigma_i)_{\alpha\alpha'}$, where α, α' is a two-component spin index and the σ_i are Pauli spin matrices,

$$\sigma_1 = \begin{pmatrix} 0 & 1 \\ 1 & 0 \end{pmatrix}, \quad \sigma_2 = \begin{pmatrix} 0 & -i \\ i & 0 \end{pmatrix}, \quad \sigma_3 = \begin{pmatrix} 1 & 0 \\ 0 & -1 \end{pmatrix}, \tag{2.12}$$

spin operator

the **spin operator** of a many-body system assumes the form

$$\hat{\mathbf{S}} = \sum_{\lambda,\alpha\alpha'} a_{\lambda\alpha'}^\dagger \mathbf{S}_{\alpha'\alpha} a_{\lambda\alpha}. \tag{2.13}$$

Here, λ denotes a set of additional quantum numbers, such as a lattice site index. When second-quantized in the position representation, one can show

one-body Hamiltonian

that the **one-body Hamiltonian** for a free particle is given as the sum of kinetic and potential energy:

> **Wolfgang Ernst Pauli** 1900–1958
>
> was an Austrian physicist who received the Nobel Prize in Physics in 1945 for the all-important "Pauli Principle" according to which two fermions cannot occupy the same quantum state. Pauli also was the first to recognize the existence of the neutrino.

$$\boxed{\hat{H} = \int d^d r \, a^\dagger(\mathbf{r}) \left(\frac{\hat{\mathbf{p}}^2}{2m} + V(\mathbf{r}) \right) a(\mathbf{r})} \tag{2.14}$$

EXERCISE Starting with the momentum representation (in which the kinetic energy is diagonal), transform to the position representation and thereby establish Eq. (2.14).

local density operator

The **local density operator** $\hat{\rho}(\mathbf{r})$, measuring the particle density at coordinate $\mathbf{r}$, is given by

$$\hat{\rho}(\mathbf{r}) = a^\dagger(\mathbf{r})a(\mathbf{r}). \tag{2.15}$$

number

Finally, the **total occupation number operator**, obtained by integrating over the particle density, is defined by $\hat{N} = \int d^d r \, a^\dagger(\mathbf{r})a(\mathbf{r})$. In a theory with discrete quantum numbers, this operator assumes the form $\hat{N} = \sum_\lambda a_\lambda^\dagger a_\lambda$.

▷ **Representation of operators (two-body):** Two-body operators $\hat{\mathcal{O}}_2$ are needed to describe pairwise interactions between particles. Although pair-interaction potentials can be included straightforwardly in classical many-body theories, their embedding into conventional many-body quantum mechanics is made cumbersome by particle indistinguishability. The formulation of interaction processes within the language of second quantization is considerably more straightforward. Initially, let us consider a symmetric two-body potential $V(\mathbf{r}_m, \mathbf{r}_n) \equiv V(\mathbf{r}_n, \mathbf{r}_m)$ between two particles at position $\mathbf{r}_m$ and $\mathbf{r}_n$. Our aim is to find an operator $\hat{V}$ in second-quantized form whose action on a many-body state gives

$$\hat{V}|\mathbf{r}_1, \ldots, \mathbf{r}_N\rangle = \sum_{n<m}^{N} V(\mathbf{r}_m, \mathbf{r}_n)|\mathbf{r}_1, \ldots, \mathbf{r}_N\rangle = \frac{1}{2} \sum_{m \neq n}^{N} V(\mathbf{r}_m, \mathbf{r}_n)|\mathbf{r}_1, \ldots, \mathbf{r}_N\rangle.$$

Here, it is more convenient to use the original representation (2.1) rather than the occupation number representation. When this is compared with the one-point function, one might guess that

$$\hat{V} = \frac{1}{2} \int d^d r \int d^d r' a^\dagger(\mathbf{r}) a^\dagger(\mathbf{r}') V(\mathbf{r}, \mathbf{r}') a(\mathbf{r}') a(\mathbf{r}).$$

That this is the correct answer can be confirmed by applying the operator to a many-body state (exercise):

$$a^\dagger(\mathbf{r}) a^\dagger(\mathbf{r}') a(\mathbf{r}') a(\mathbf{r})|\mathbf{r}_1, \ldots, \mathbf{r}_N\rangle = a^\dagger(\mathbf{r}) a^\dagger(\mathbf{r}') a(\mathbf{r}') a(\mathbf{r}) \, a^\dagger(\mathbf{r}_1) \cdots a^\dagger(\mathbf{r}_N)|0\rangle$$

$$= \sum_{m,n \neq m}^{N} \delta(\mathbf{r} - \mathbf{r}_m) \delta(\mathbf{r}' - \mathbf{r}_n)|\mathbf{r}_1, \ldots, \mathbf{r}_N\rangle.$$

Multiplying by $V(\mathbf{r}, \mathbf{r}')/2$, and integrating over $\mathbf{r}$ and $\mathbf{r}'$, one confirms the validity of the expression. It is left as an exercise to confirm that the naïve expression $\frac{1}{2} \int d^d r \int d^d r' V(\mathbf{r}, \mathbf{r}') \hat{\rho}(\mathbf{r}) \hat{\rho}(\mathbf{r}')$ does not reproduce the two-body operator. More generally, turning to a non-diagonal basis, it is straightforward to confirm that a general **two-body operator** can be expressed in the form

two-body
operator

$$\boxed{\hat{\mathcal{O}}_2 = \frac{1}{2} \sum_{\lambda \lambda' \mu \mu'} \mathcal{O}_{\mu,\mu',\lambda,\lambda'} a_\mu^\dagger a_{\mu'}^\dagger a_{\lambda'} a_\lambda} \tag{2.16}$$

where $\mathcal{O}_{\mu,\mu',\lambda,\lambda'} \equiv \langle \mu, \mu'|\hat{\mathcal{O}}_2|\lambda, \lambda'\rangle$.

Besides the **Coulomb interaction** to be discussed shortly, another important interaction is that between spins. From our discussion of the second-quantized representation of spin $\hat{\mathbf{S}}$, we infer that the general **spin-spin interaction** affords the representation

$$\hat{V} = \frac{1}{2} \int d^d r \int d^d r' \sum_{\alpha \alpha' \beta \beta'} J(\mathbf{r}, \mathbf{r}') \, \mathbf{S}_{\alpha\beta} \cdot \mathbf{S}_{\alpha'\beta'} \, a_\alpha^\dagger(\mathbf{r}) a_{\alpha'}^\dagger(\mathbf{r}') a_{\beta'}(\mathbf{r}') a_\beta(\mathbf{r}),$$

where $J(\mathbf{r}, \mathbf{r}')$ denotes the exchange interaction.

In principle, one may proceed in the same manner and represent general n-body interactions in terms of second-quantized operators. However, as $n > 2$ interactions appear infrequently, we refer to the literature for further discussion.

This completes our formal introduction to the method of second quantization. To develop fluency in the operation of the method, we will continue by addressing several problems chosen from the realm of condensed matter. In doing so, we will see that second quantization often leads to considerable simplification of the analysis of many-particle systems. The effective model Hamiltonians that appear below provide the input for subsequent applications considered in this text. Readers not wishing to become distracted from our main focus – the development of modern methods of quantum field theory in the condensed matter setting – may safely skip the next sections and turn directly to chapter 3. It is worthwhile keeping in mind, however, that the physical motivation for the study of various prototypical model systems considered later in the text is given in section 2.2.

2.2 Applications of Second Quantization

Starting from the prototype Hamiltonian (1.1), we have already explored aspects of lattice dynamics in condensed matter. In much of the remaining text, we will focus on the complementary system, the electron degrees of freedom. Drawing on the first of the principles articulated in chapter 1, we begin our discussion by reducing the Hamiltonian to a form that contains the essential elements of the electron dynamics. As well as the pure electron sub-Hamiltonian H_e, the reduced Hamiltonian involves the interaction between electrons and the ionic background lattice. However, typically lattice distortions due to the motion of the ions and the ion–ion interaction couple only indirectly. (Exercise: Think of a prominent example where the electron sector is strongly influenced by the dynamics of the lattice.) To a first approximation, we may therefore describe the electron system through the Hamiltonian $\hat{H} = \hat{H}_0 + \hat{V}_{\text{ee}}$, where

$$
\hat{H}_0 = \int d^d r \sum_\sigma a_\sigma^\dagger(\mathbf{r}) \left[\frac{\hat{\mathbf{p}}^2}{2m} + V(\mathbf{r}) \right] a_\sigma(\mathbf{r}),
$$

$$
\hat{V}_{\text{ee}} = \frac{1}{2} \int d^d r \int d^d r' \sum_{\sigma\sigma'} V_{\text{ee}}(\mathbf{r} - \mathbf{r}')\, a_\sigma^\dagger(\mathbf{r}) a_{\sigma'}^\dagger(\mathbf{r}') a_{\sigma'}(\mathbf{r}') a_\sigma(\mathbf{r}).
$$

$$(2.17)$$

Here, $V(\mathbf{r}) = \sum_I V_{\text{ei}}(\mathbf{R}_I - \mathbf{r})$ denotes the lattice potential experienced by the electrons, and the coordinates of the lattice ions $\mathbf{R}_I$ are assumed fixed. For completeness, we have also endowed the electrons with a spin index, $\sigma = \uparrow / \downarrow$.

Despite its seemingly innocuous structure, the interacting-electron Hamiltonian (2.17) accommodates a wide variety of electron phases, from metals and magnets to insulators. To classify the phase behavior of the model, it is helpful to divide our considerations, focusing first on the properties of the non-interacting single-particle system $\hat{H}_0$ and then, later, exploring the influence of the electron interaction V_{ee}.

2.2.1 Electrons in a periodic potential

From **Bloch's theorem**, the eigenstates of a periodic Hamiltonian can be presented in the form of Bloch waves[6] $\psi_{\mathbf{k}n}(\mathbf{r}) = e^{i\mathbf{k}\cdot\mathbf{r}}u_{\mathbf{k}n}(\mathbf{r})$, where the components of the crystal momentum $\mathbf{k}$ take values inside the Brillouin zone, $k_i \in [-\pi/a, \pi/a]$. Here, for simplicity, the periodicity of the lattice potential is assumed to be the same in all directions,

> **Felix Bloch** 1905–1983 was a Swiss-American physicist who, in 1952, shared the Nobel Prize in Physics with Edward M. Purcell "for the development of new methods for nuclear magnetic precision measurements and discoveries in connection therewith." Bloch served as the first Director-General of CERN.

i.e., $V(\mathbf{r} + a\mathbf{e}_i) = V(\mathbf{r})$. The index n labels the separate energy bands of the solid, and the functions $u_{\mathbf{k}n}(\mathbf{r} + a\mathbf{e}_i) = u_{\mathbf{k}n}(\mathbf{r})$ are periodic on the lattice. Now, depending on the nature of the bonding, there are two complementary classes of materials where the general structure of the Bloch functions can be simplified significantly.

Nearly free electron systems

For certain materials, notably the elemental metals drawn from groups I–IV of the periodic table, the outermost itinerant conduction electrons behave as if they were "nearly free," i.e., their dynamics is largely oblivious to both the Coulomb potential created by the positively charged ion background and their mutual interaction.

INFO Loosely speaking, Pauli blocking by the bound state inner core electrons prevents the conduction electrons from exploring the region close to the ion core, thereby screening the nuclear charge. In practice, the conduction electrons experience a renormalized **pseudopotential**, which accommodates the effect of the lattice ions and core electrons. Moreover, the high mobility of the conduction electrons provides an efficient method of screening their own mutual Coulomb interaction. In nearly free electron compounds, complete neglect of the lattice potential is usually a good approximation (as long as one considers crystal momenta remote from the boundaries of the Brillouin zone, $k_i - \pm\pi/a$).

In practice, this means that we may set the Bloch function to unity, $u_{\mathbf{k}n} = 1$, and regard the eigenstates of the non-interacting Hamiltonian as plane waves. This motivates the representation of the field operators in momentum space (2.9), whereupon the non-interacting part of the Hamiltonian assumes the free particle form

$$\hat{H}_0 = \sum_{\mathbf{k}\sigma} \frac{\mathbf{k}^2}{2m} a_{\mathbf{k}\sigma}^\dagger a_{\mathbf{k}\sigma}, \tag{2.18}$$

where the sum runs over wave vectors $\mathbf{k}$ (and, once again, we have set $\hbar = 1$). In the Fourier representation, the two-body Coulomb interaction takes the form

---

[6] For a further discussion, we refer to one of the many texts on the elements of solid state physics, e.g., N.D. Ashcroft and N. Mermin, *Solid State Physics* (Holt-Saunders International, 1995).

$$\hat{V}_{ee} = \frac{1}{2L^d} \sum_{\mathbf{k},\mathbf{k'},\mathbf{q},\sigma\sigma'} V_{ee}(\mathbf{q})\, a^\dagger_{\mathbf{k}-\mathbf{q}\sigma} a^\dagger_{\mathbf{k'}+\mathbf{q}\sigma'} a_{\mathbf{k}\sigma'} a_{\mathbf{k'}\sigma}, \qquad (2.19)$$

where (choosing units such that $4\pi\epsilon = 1$), $V_{ee}(\mathbf{q}) = e^2/q^2$ represents the Fourier transform of the potential $V_{ee}(\mathbf{r}) = e^2/|\mathbf{r}|$. Now, as written, this expression neglects the fact that in ionized solids the negative charge density of the electron cloud will be compensated by the charge density of the positively ionized background.

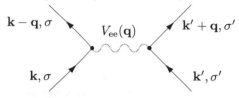

The latter can be incorporated into (2.19) by placing on the sum over $\mathbf{q}$ the restriction that $\mathbf{q} \neq 0$ (exercise). Taken together, the free electron Hamiltonian $\hat{H}_0$ and the Coulomb interaction potential $\hat{V}_{ee}$ are known as the **jellium model**.

jellium model

The interaction described by Eq. (2.19) can be illustrated graphically, as shown in the figure (for a more elaborate discussion of such diagrams, see chapter 4): an electron of momentum $\mathbf{k}$ is scattered into a new momentum state $\mathbf{k} - \mathbf{q}$ while another electron is scattered from $\mathbf{k'} \to \mathbf{k'} + \mathbf{q}$.

In concrete applications of condensed matter physics, one typically considers low excitation energies. The analysis of such systems is naturally organized around the zero-temperature ground state as a reference platform. However, the accurate calculation of the ground state energy of the system

Niels Henrik David Bohr
1885–1962
was a Danish physicist and philosopher who, in 1922, was awarded the Nobel Prize in Physics "for his services in the investigation of the structure of atoms and of the radiation emanating from them."

is a complicated problem of many-body physics that cannot be solved in closed form. Therefore, assuming that interactions will not substantially alter the ground state of the free particle problem (2.18) – which is often not the case! – one uses the ground state of the latter as a reference state.

INFO Deferring a more qualified discussion to later, a preliminary justification for this assumption can be given as follows: suppose that the density of an electron gas is such that each of its N constituent particles occupies an average volume of $O(a^d)$. The average kinetic energy per particle is then estimated to be $T \sim 1/ma^2$, while the Coulomb interaction potential will scale as $V \sim e^2/a$. Thus, for a much smaller than the **Bohr radius** $a_0 = 1/e^2 m$, the interaction contribution is much smaller than the average kinetic energy. In other words, for the dense electron gas, the interaction energy can indeed be treated as a perturbation. Unfortunately, for most metals, one finds that $a \sim a_0$ and neither high- nor low-density approximations are strictly justified.

Bohr radius

The ground state of the system occupied by N non-interacting electrons can be readily inferred from Eq. (2.18). The Pauli principle implies that all energy states $\epsilon_\mathbf{k} = k^2/2m$ will be uniformly occupied up to a cutoff **Fermi energy**, E_F. For a system of size L, the allowed momentum states $\mathbf{k}$ have components $k_i = 2\pi n_i/L$, $n_i \in \mathbb{Z}$. The summation extends up to momenta with $|\mathbf{k}| \leq k_F$, where the **Fermi**

Fermi energy

momentum k_F is defined through the relation $k_F^2/2m = E_F$. The relation between the Fermi momentum and the occupation number can be established by dividing the volume of the **Fermi sphere** $\sim k_F^d$ by the momentum space volume per mode $(2\pi/L)^d$, viz. $N = C(k_F L)^d$, where C denotes a dimensionless geometry-dependent constant (see the figure).

In the language of second quantization, the ground state is represented as

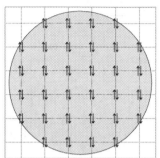

$$|\Omega\rangle \equiv \mathcal{N} \prod_{|\mathbf{k}| \leq k_F, \sigma} a_{\mathbf{k}\sigma}^\dagger |0\rangle, \qquad (2.20)$$

where $|0\rangle$ denotes the state with zero electrons present. When the interaction is weak, one may anticipate that low-temperature properties will be governed by energetically low-lying excitations superimposed upon the state $|\Omega\rangle$. Therefore, remembering the philosophy whereby excitations rather than microscopic constituents play a prime role, one would like to declare the filled Fermi sea, $|\Omega\rangle$ (rather than the empty state $|0\rangle$), to be the "physical vacuum" of the theory. To make this compatible with the language of second quantization, we need to identify a new operator algebra such that the operators $c_{\mathbf{k}\sigma}$ annihilate the Fermi sea. This is easily engineered by defining

$$c_{\mathbf{k}\sigma}^\dagger = \begin{cases} a_{\mathbf{k}\sigma}^\dagger & k > k_F \\ a_{\mathbf{k}\sigma} & k \leq k_F \end{cases}, \qquad c_{\mathbf{k}\sigma} = \begin{cases} a_{\mathbf{k}\sigma} & k > k_F \\ a_{\mathbf{k}\sigma}^\dagger & k \leq k_F \end{cases}. \qquad (2.21)$$

It is then straightforward to verify that $c_{\mathbf{k}\sigma}|\Omega\rangle = 0$ and that the canonical commutation relations are preserved.

The Hamiltonian defined through Eqs. (2.18) and (2.19), represented in terms of the operator algebra (2.21) and the vacuum (2.20), forms the basis of the theory of interactions in highly mobile electron compounds. The investigation of the role of Coulomb interactions in such systems will provide a useful arena in which to apply the methods of quantum field theory formulated in subsequent chapters. Following our classification of electron systems, let us now turn our attention to a complementary class of materials where the lattice potential presents a strong perturbation to the conduction electrons. In such situations realized, for example, in transition metal oxides, a description based on "almost localized" electron states will be used to represent the Hamiltonian (2.17).

Tight–binding systems

Let us consider a "rarefied" lattice in which the constituent ion cores are separated by a distance in excess of the typical Bohr radius a_B of the valence-band electrons. In this "atomic limit," the electron wave functions are tightly bound to the lattice sites. Here, to formulate a microscopic theory of interactions, it is convenient to expand the Hamiltonian in a local basis that reflects the atomic orbital states of

Wannier
states

the isolated ion. Such a representation is presented by the basis of **Wannier states**, defined by

$$|\psi_{\mathbf{R}n}\rangle \equiv \frac{1}{\sqrt{N}} \sum_{\mathbf{k}}^{\text{B.Z.}} e^{-i\mathbf{k}\cdot\mathbf{R}}|\psi_{\mathbf{k}n}\rangle, \qquad |\psi_{\mathbf{k}n}\rangle = \frac{1}{\sqrt{N}} \sum_{\mathbf{R}} e^{i\mathbf{k}\cdot\mathbf{R}}|\psi_{\mathbf{R}n}\rangle, \qquad (2.22)$$

where $\mathbf{R}$ denotes the coordinates of the lattice centers, and $\sum_{\mathbf{k}}^{\text{B.Z.}}$ represents a summation over all momenta $\mathbf{k}$ in the first Brillouin zone. For a system with a vanishingly weak interatomic overlap, i.e., a lattice where V approaches a superposition of independent atomic potentials, the Wannier function $\psi_{\mathbf{R}n}(\mathbf{r}) \equiv \langle\mathbf{r}|\psi_{\mathbf{R}n}\rangle$ converges on the nth orbital of an isolated atom centered at coordinate $\mathbf{R}$. However, when the interatomic coupling is non-zero, i.e., in a "real" solid, the N formerly degenerate states, labeled by n, split to form an energy band (see the figure below). Here, the Wannier functions (which are not eigenfunctions of the Hamiltonian) differ from those of the atomic orbitals through residual oscillations in the region of overlap, to ensure orthogonality of the basis. Significantly, in cases where the Fermi energy lies between two energetically separated bands, the system presents **insulating** behavior. Conversely, when the Fermi energy is located within a band, one may expect **metallic** behavior. Ignoring the complications that arise when bands begin to overlap, we will henceforth focus on metallic systems where the Fermi energy is located within a definite band, n_0.

band
insulator
metal

How can the Wannier basis be exploited to obtain a simplified representation of the general Hamiltonian (2.17)? The first thing to notice is that the Wannier states $\{|\psi_{\mathbf{R}n}\rangle\}$ form an orthonormal basis of the single-particle Hilbert space, i.e., the transformation between the real space and Wannier representation is unitary, $|\mathbf{r}\rangle = \sum_{\mathbf{R}} |\psi_{\mathbf{R}}\rangle\langle\psi_{\mathbf{R}}|\mathbf{r}\rangle = \sum_{\mathbf{R}} \psi_{\mathbf{R}}^*(\mathbf{r})|\psi_{\mathbf{R}}\rangle.$[7] Being unitary, it induces a transformation

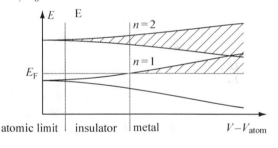

$$a_\sigma^\dagger(\mathbf{r}) = \sum_{\mathbf{R}} \psi_{\mathbf{R}}^*(\mathbf{r}) a_{\mathbf{R}\sigma}^\dagger \equiv \sum_{i} \psi_{\mathbf{R}_i}^*(\mathbf{r}) a_{i\sigma}^\dagger \qquad (2.23)$$

between the real- and Wannier-space operator bases, respectively. In the second representation, following a convention commonly used in the literature, we have labeled the lattice center coordinates $\mathbf{R} \equiv \mathbf{R}_i$ using a counting index $i = 1, \ldots, N$. Similarly, the unitary transformation between Bloch and Wannier states (2.22) induces an operator transformation

---

[7] Here, since we are interested only in contributions arising from the particular "metallic" band n_0 in which the Fermi energy lies, we have dropped the remaining set of bands $n \neq n_0$ and, with them, reference to the specific band index. Exercise: By focusing on just a single band n_0, in what sense is the Wannier basis now complete?

$$a_{\mathbf{k}\sigma}^\dagger = \frac{1}{\sqrt{N}} \sum_i e^{i\mathbf{k}\cdot\mathbf{R}_i} a_{i\sigma}^\dagger, \qquad a_{i\sigma}^\dagger = \frac{1}{\sqrt{N}} \sum_{\mathbf{k}}^{\mathrm{B.Z.}} e^{-i\mathbf{k}\cdot\mathbf{R}_i} a_{\mathbf{k}\sigma}^\dagger. \qquad (2.24)$$

We can now use the transformation formulae (2.23) and (2.24) to formulate a Wannier representation of the Hamiltonian (2.17). Using the fact that the Bloch states diagonalize the single-particle component $\hat{H}_0$, we obtain

$$\hat{H}_0 = \sum_{\mathbf{k}} \epsilon_k a_{\mathbf{k}\sigma}^\dagger a_{\mathbf{k}\sigma} \overset{(2.24)}{=} \frac{1}{N} \sum_{ii'} \sum_{\mathbf{k}} e^{i\mathbf{k}(\mathbf{R}_i - \mathbf{R}_{i'})} \epsilon_k a_{i\sigma}^\dagger a_{i'\sigma} \equiv \sum_{ii'} t_{ii'} a_{i\sigma}^\dagger a_{i'\sigma},$$

where we have set $t_{ii'} = \frac{1}{N} \sum_{\mathbf{k}} e^{i\mathbf{k}\cdot(\mathbf{R}_i - \mathbf{R}_{i'})} \epsilon_k$. The new representation of $\hat{H}_0$ describes electrons hopping from one lattice center i' to another, i. The strength of the hopping matrix element $t_{ii'}$ is controlled by the effective overlap of neighboring atoms. In the extreme atomic limit, where the levels $\epsilon_k = $ const. are degenerate, $t_{ii'} \propto \delta_{ii'}$ and no interatomic transport is possible. The tight-binding representation becomes useful when $t_{i\neq i'}$ is non-vanishing, but the orbital overlap is so weak that only nearest-neighbor hopping effectively contributes.

EXERCISE Taking a square lattice geometry and setting $t_{ii'} = -t$ when i, i' are nearest neighbors and zero otherwise, diagonalize the two-dimensional tight-binding Hamiltonian $\hat{H}_0$. Show that the eigenvalues are given by $\epsilon_{\mathbf{k}} = -2t(\cos(k_x a) + \cos(k_y a))$. Sketch contours of constant energy in the Brillouin zone and note the geometry at half-filling.

To assess the utility of the tight-binding approximation, let us consider its application to *graphene*, a prominent carbon-based lattice system.

graphene   **INFO** **Graphene** is a single layer of graphite, a planar hexagonal lattice of sp^2-hybridized carbon atoms connected by strong covalent bonds of their three planar σ-orbitals (see fig. 2.2 and the schematic overleaf). The remaining p_z orbitals – oriented perpendicular to the lattice plane – overlap weakly to form a band of mobile π-electrons. For a long time, it was thought that graphene sheets in isolation would inevitably be destabilized by thermal fluctuations; only layered stacks of graphene would form a stable compound – graphite. It thus came as a surprise when, in 2004, a team of researchers[8] succeeded in the isolation of large (micron-sized) graphene flakes on an SiO_2 substrate. (Since then, the isolation of even free standing graphene layers has become possible. In fact, our whole conception of the stability of the compound has changed. It is now believed that whenever you draw a line in pencil, a trail of graphene flakes will be left behind!)

Soon after its discovery, it became clear that graphene possesses unconventional conduction properties. Nominally a gapless semiconductor, it has an electron mobility $\sim 2 \times 10^5 \mathrm{cm}^2/\mathrm{Vs}$, far higher than that of even the purest silicon-based semiconductors; it shows manifestations of the integer quantum Hall effect qualitatively different from those of conventional two-dimensional electron compounds (cf. chapter 8 for a general discussion of the quantum Hall effect); etc. Although an in-depth discussion of graphene is beyond the scope of this text, we note that most of its fascinating properties are due to its band structure: electrons in graphene show a linear dispersion and behave like two-dimensional relativistic (Dirac) fermions! By way of an illustration of the concepts discussed above, here we derive this unconventional band dispersion from a tight-binding formulation.

[8] K. S. Novoselov, *et al.*, *Electric field effect in atomically thin carbon films*, Science **306**, 666 (2004).

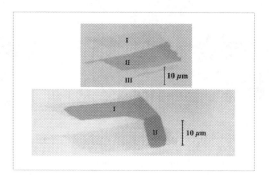 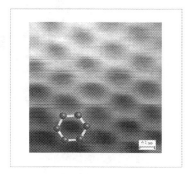

Fig. 2.2   Left: Optical microscopy image of graphene flakes. Regions labeled by 'I' define monolayer graphene sheets of size ca. $10\,\mu$m. Right: STM image of the graphene samples shown in the left part. Images taken from E. Stolyarova *et al.*, *High-resolution scanning tunneling microscopy imaging of mesoscopic graphene sheets on an insulating surface*, PNAS **104**, 9209 (2007). Copyright (2007) National Academy of Sciences.

To a first approximation, graphene's π-electron system can be modeled as a tight-binding Hamiltonian characterized by a single hopping matrix element between neighboring atoms, $-t$ (with t real and positive), and the energy off-set ϵ of the π-electron states. To determine the spectrum of the system, a system of bi-atomic unit cells can be introduced (see the ovals in the schematic) and two (non-orthogonal) unit vectors of the hexagonal lattice, $\mathbf{a}_1 = (\sqrt{3}, 1)a/2$ and 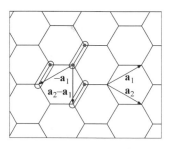 $\mathbf{a}_2 = (\sqrt{3}, -1)a/2$, where $a = |\mathbf{a}_1| = |\mathbf{a}_2| \simeq 2.46$ Å denotes the lattice constant. The tight-binding Hamiltonian is then represented as $\hat{H} = -t \sum_{\langle \mathbf{r}, \mathbf{r}' \rangle} (a_1^\dagger(\mathbf{r}) a_2(\mathbf{r}') + \text{h.c.})$, where h.c. denotes the Hermitian conjugate, the sum runs over all nearest-neighbor pairs of sites and $a_{1(2)}^\dagger(\mathbf{r})$ creates a state in the first (second) atom of the cell at position vector $\mathbf{r}$. Switching to a Fourier representation, the Hamiltonian takes the form

$$\hat{H} = \sum_{\mathbf{k}\sigma} \begin{pmatrix} a_{1\sigma}^\dagger & a_{2\sigma}^\dagger \end{pmatrix} \begin{pmatrix} 0 & -tf(\mathbf{k}) \\ -tf^*(\mathbf{k}) & 0 \end{pmatrix} \begin{pmatrix} a_{1\sigma} \\ a_{2\sigma} \end{pmatrix}, \qquad (2.25)$$

where $f(\mathbf{k}) = 1 + e^{-ik_1 a} + e^{i(-k_1 + k_2)a}$.

EXERCISE Revise the concept of the reciprocal lattice in solid state theory. To derive the Fourier representation above, show that a system of two reciprocal lattice vectors conjugate to the unit vectors above is given by $\mathbf{G}_{1/2} = 2\pi/(\sqrt{3}a)(1, \pm\sqrt{3})$. Next, show that the Fourier decomposition of a field operator reads

$$a_a(\mathbf{r}) = \frac{1}{\sqrt{N}} \sum_{\mathbf{k}} e^{-i\frac{a}{2\pi}(k_1 \mathbf{G}_1 + k_2 \mathbf{G}_2) \cdot \mathbf{r}} a_{a,\mathbf{k}},$$

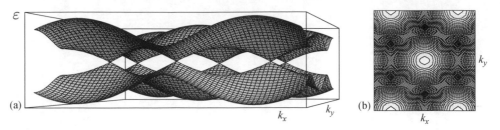

Fig. 2.3   (a) Spectrum of the tight-binding Hamiltonian (2.24) showing the point-like structure of the Fermi surface when $E_F = 0$. (b) A contour plot of the same.

where $k_i \in [0, 2\pi/a]$ is quantized in units $2\pi/L_i$. (L_i is the extension of the system in the direction of $\mathbf{a}_i$ and N its total number of unit cells.) Substitute this decomposition into the real space representation of the Hamiltonian to arrive at the Fourier representation.

Diagonalizing the Hamiltonian (2.25), one obtains the dispersion[9]

$$\epsilon_{\mathbf{k}} = \pm t \left[3 + 2\cos(k_1 a) + 2\cos((k_1 - k_2)a) + 2\cos(k_2 a)\right]^{1/2}. \qquad (2.26)$$

Here, in contrast with the square-lattice tight-binding Hamiltonian, the half-filled system is characterized by a point-like Fermi surface (fig. 2.3). When lightly doped away from half-filling, the spectrum divides into Dirac-like spectra with a linear dispersion. Notice that, of the six Dirac points (fig. 2.3), only two are independent. The complementary four can be reached from those two points by the addition of a reciprocal lattice vector and, therefore, do not represent independent states.

Dirac
Hamil-
tonian
EXERCISE Derive an explicit representation of the **Dirac Hamiltonian** describing the low-energy physics of the system. To this end, choose two inequivalent (i.e., not connected by reciprocal lattice vectors) zero-energy points $\mathbf{k}_{1,2}$ in the Brillouin zone. Expand the Hamiltonian (2.25) around these two points in small momentum deviations $\mathbf{q} \equiv \mathbf{k} - \mathbf{k}_{1,2}$ up to linear order. Show that, in this approximation, $\hat{H}$ reduces to the sum of two two-dimensional Dirac Hamiltonians.

The physics of low-energy Dirac Hamiltonians in condensed matter physics is the subject of chapter 9. There, we will discuss numerous phenomena that owe their existence to the "relativistic invariance" emerging in graphene and related systems at low energy scales.

2.2.2 Interaction effects in the tight-binding system

Although the pseudopotential of the nearly free electron system accommodates the effects of Coulomb interactions between the conduction and valence band electrons, the mutual Coulomb interaction between the conduction electrons themselves may lead to new physical phenomena. These effects can alter substantially the material parameters (e.g., the effective conduction electron mass). However, they change

[9] P. R. Wallace, *The band theory of graphite*, Phys. Rev. **71**, 622 (1947).

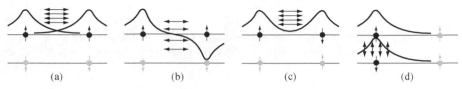

        (a)   (b)   (c)   (d)

Fig. 2.4   Different types of interaction mechanism induced by the tight-binding interaction V_{ee}. The curves symbolically indicate wave function envelopes. (a) Direct Coulomb interaction between neighboring sites. Taking account of the exchange interaction, parallel alignment of spins (b) is preferred since it enforces antisymmetry of the spatial wave function, lowering the electron presence between sites. By contrast, for anti-parallel spin configurations (c), the wave function amplitude in the repulsion zone is enhanced. (d) The Coulomb interaction between electrons of opposite spin occupying the same site.

the nature of the neither ground state nor that of the elementary quasi-particle excitations in any fundamental way; this is the basis of Fermi-liquid theory and a matter to which we will return. By contrast, interactions influence significantly the physics of the tight-binding system: at "half-filling," even weak interactions may drive the system into a correlated magnetic state or insulating phase.

To understand why, let us re-express the interaction in the field operators associated with the Wannier states. Once again, to keep our discussion simple (yet generic in scope), let us focus on a single sub-band and drop any reference to the band index. Then, applied to the Coulomb interaction, the transformation (2.23) leads to the expansion $\hat{V}_{ee} = \sum_{ii'jj'} U_{ii'jj'} a_{i\sigma}^\dagger a_{i'\sigma'}^\dagger a_{j\sigma} a_{j'\sigma'}$, where

$$U_{ii'jj'} = \frac{1}{2} \int d^d r \int d^d r' \, \psi_{\mathbf{R}_i}^*(\mathbf{r}) \psi_{\mathbf{R}_j}(\mathbf{r}) V(\mathbf{r} - \mathbf{r}') \psi_{\mathbf{R}_{i'}}^*(\mathbf{r}') \psi_{\mathbf{R}_{j'}}(\mathbf{r}'). \qquad (2.27)$$

Taken together, the combination of the contributions,

$$\hat{H} = \sum_{ii'\sigma} t_{ii'} a_{i\sigma}^\dagger a_{i'\sigma} + \sum_{ii'jj'\sigma\sigma'} U_{ii'jj'} a_{i\sigma}^\dagger a_{i'\sigma'}^\dagger a_{j'\sigma'} a_{j\sigma},$$

defines the tight-binding representation of the interaction Hamiltonian. Apart from neglect of the neighboring sub-bands, the Hamiltonian is exact. Yet, to assimilate the effects of the interaction, it is useful to assess the relative importance of the different matrix elements, drawing on the nature of the atomic limit that justified the tight-binding description. We will thus focus on contributions to $U_{ii'jj'}$, where the indices are either equal or, at most, those of nearest neighbors. Focusing on the most relevant of these matrix elements, a number of physically different contributions can be identified.

▷   The **direct terms** $U_{ii'ii'} \equiv V_{ii'}$ involve integrals over square moduli of Wannier functions and couple density fluctuations at neighboring sites, $\sum_{i\neq i'} V_{ii'} \hat{n}_i \hat{n}_{i'}$, where $\hat{n}_i = \sum_\sigma a_{i\sigma}^\dagger a_{i\sigma}$. This contribution accounts for the – essentially classical – interaction between charges localized at neighboring sites (see fig. 2.4). In certain materials, interactions of this type may induce **charge-density wave** instabilities.

▷ A second important contribution derives from the **exchange coupling**, which induces magnetic correlations among the electron spins. Setting $J_{ij}^{\mathrm{F}} \equiv U_{ijji}$, and making use of the Pauli matrix identities (see below), one obtains

$$\sum_{i \neq j, \sigma\sigma'} U_{ijji} a_{i\sigma}^\dagger a_{j\sigma'}^\dagger a_{i\sigma'} a_{j\sigma} = -2 \sum_{i \neq j} J_{ij}^{\mathrm{F}} \left(\hat{\mathbf{S}}_i \cdot \hat{\mathbf{S}}_j + \frac{1}{4} \hat{n}_i \hat{n}_j \right).$$

Such contributions tend to induce weak **ferromagnetic coupling** of neighboring spins (i.e., $J^{\mathrm{F}} > 0$). The fact that an effective *magnetic* coupling is born out of the *electrostatic* interaction between quantum particles is easily understood. Consider two electrons inhabiting neighboring sites. The Coulomb repulsion between the particles is minimized if the orbital two-particle wave function is antisymmetric and, therefore, has low amplitude in the interaction zone between the particles. Since the overall wave function must be antisymmetric, the energetically favored real space configuration enforces a symmetric alignment of the spins (fig. 2.4). Such a mechanism is familiar from atomic physics where it

manifests as **Hund's rule**. In general, magnetic interactions in solids are usually generated as an indirect manifestation of the stronger Coulomb interaction.

EXERCISE Making use of the Pauli matrix identity $\boldsymbol{\sigma}_{\alpha\beta} \cdot \boldsymbol{\sigma}_{\gamma\delta} = 2\delta_{\alpha\delta}\delta_{\beta\gamma} - \delta_{\alpha\beta}\delta_{\gamma\delta}$, show that $\hat{\mathbf{S}}_i \cdot \hat{\mathbf{S}}_j = -\frac{1}{2} \sum_{\alpha\beta} a_{i\alpha}^\dagger a_{j\beta}^\dagger a_{i\beta} a_{j\alpha} - \frac{1}{4} \hat{n}_i \hat{n}_j$, where $\hat{\mathbf{S}}_i = \frac{1}{2} \sum_{\alpha\beta} a_{i\alpha}^\dagger \boldsymbol{\sigma}_{\alpha\beta} a_{i\beta}$ denotes the operator for spin $1/2$, and the lattice sites i and j are assumed distinct.

▷ Finally, deep in the atomic limit, where the atoms are well separated and the overlap between neighboring orbitals is weak, the matrix elements t_{ij} and J_{ij}^{F} are exponentially small in the interatomic separation variables. In this limit,

the "on-site" Coulomb or **Hubbard interaction**, $\sum_{i\sigma\sigma'} U_{iiii} a_{i\sigma}^\dagger a_{i\sigma'}^\dagger a_{i\sigma'} a_{i\sigma} = \sum_i U \hat{n}_{i\uparrow} \hat{n}_{i\downarrow}$, where $U_{iiii} \equiv U/2$, dominates (fig. 2.4). Taking only the nearest-neighbor contribution to the hopping matrix elements, and neglecting the energy offset due to the diagonal term, the effective Hamiltonian takes a simplified form

known as the **Hubbard model**,

$$\boxed{\hat{H} = -t \sum_{\langle ij \rangle \sigma} a_{i\sigma}^\dagger a_{j\sigma} + U \sum_i \hat{n}_{i\uparrow} \hat{n}_{i\downarrow}} \qquad (2.28)$$

where $\langle ij \rangle$ denotes neighboring lattice sites. In hindsight, a model of this structure could have been proposed from the outset on purely phenomenological grounds: electrons tunnel between atomic orbitals localized on individual lattice sites, while the double occupancy of a lattice site incurs an energy penalty associated with the mutual Coulomb interaction.

2.2.3 Mott–Hubbard transition and the magnetic state

REMARK In this section we discuss condensed matter phases deriving from those of the Hubbard Hamiltonian Eq. (2.28) at low energies, and their description in terms of effective Hamiltonians. The section illustrates the application of second quantization in condensed matter contexts, and puts some of the effective models discussed later into a wider physical context. However, readers wishing to progress as quickly as possible to the introduction of quantum field-theoretical concepts may skip it at first reading.

Deceptive in its simplicity, the Hubbard model is acknowledged as a paradigm of strong-electron physics. Yet, after more than half a century of intense investigation, the properties of this seemingly simple model are still the subject of debate (at least in dimensions higher than one – see below). Thus, given the importance attached to this system, we will close this section with a brief discussion of some of the remarkable phenomenology that characterizes the Hubbard model.

As well as dimensionality, the phase behavior of the Hubbard Hamiltonian is characterized by three dimensionless parameters: the ratio of the Coulomb interaction scale to the bandwidth U/t, the particle density or filling fraction n (i.e., the average number of electrons per site), and the (dimensionless) temperature, T/t. The symmetry of the Hamiltonian under particle–hole interchange (exercise) allows one to limit consideration to densities in the range $0 \leq n \leq 1$, while densities $1 < n \leq 2$ can be inferred by "reflection."

Focusing first on the low-temperature system, in the dilute limit $n \ll 1$, the typical electron wavelength is greatly in excess of the site separation and the dynamics are free. Here the local interaction presents only a weak perturbation and one can expect the properties of the Hubbard system to mirror those of the weakly interacting nearly free electron system. While the interaction remains weak, one expects metal-

> **Sir Neville Francis Mott**
> **1905–1996**
> was a British physicist who, in 1977, shared with Philip W. Anderson and John H. van Vleck the Nobel Prize in Physics for their "fundamental theoretical investigations of the electronic structure of magnetic and disordered systems." Amongst his contributions to science, Mott provided a theoretical basis to understand the transition of materials from metallic to non-metallic states (the Mott transition).

lic behavior to prevail. By contrast, in the half-filled system, where the average site occupancy is unity, if the interaction is weak, $U/t \ll 1$, one may again expect properties reminiscent of a weakly interacting electron system. If, on the other hand, the interaction is strong, $U/t \gg 1$, site double occupancy is inhibited and electrons in the half-filled system become "jammed": migration of an electron to a neighboring lattice site would necessitate site double occupancy, incurring an energy cost U. Here, in this strongly correlated state, the mutual Coulomb interaction between the electrons drives the system from a metallic to an insulating phase with properties very different from those of a conventional band insulator.

Experimentally, it is often found that the low-temperature phase of the Mott insulator is accompanied by **antiferromagnetic ordering** of the local moments. The origin of these magnetic correlations can be traced to a mechanism known as **super-exchange**[10] and can be understood straightforwardly within the framework

super-
exchange

---

[10] P. W. Anderson, *Antiferromagnetism. Theory of superexchange interaction*, Phys. Rev. **79**, 350 (1950).

of the Hubbard model. To this end, one may consider a simple "two-site" system from which the characteristics of the lattice system can be inferred. At half-filling (i.e., with just two electrons to share between the two sites), one can identify a total of six basis states: two spin polarized states $a_{1\uparrow}^\dagger a_{2\uparrow}^\dagger |\Omega\rangle$, $a_{1\downarrow}^\dagger a_{2\downarrow}^\dagger |\Omega\rangle$, and four states with $S_{\rm total}^z = 0$: $|s_1\rangle = a_{1\uparrow}^\dagger a_{2\downarrow}^\dagger |\Omega\rangle$, $|s_2\rangle = a_{2\uparrow}^\dagger a_{1\downarrow}^\dagger |\Omega\rangle$, $|d_1\rangle = a_{1\uparrow}^\dagger a_{1\downarrow}^\dagger |\Omega\rangle$ and $|d_2\rangle = a_{2\uparrow}^\dagger a_{2\downarrow}^\dagger |\Omega\rangle$. Recalling the constraints imposed by the Pauli principle, it is evident that the fully spin polarized states are eigenstates of the Hubbard Hamiltonian with zero energy, while the remaining eigenstates involve superpositions of the basis states $|s_i\rangle$ and $|d_i\rangle$. In the strong-coupling limit $U/t \gg 1$, the ground state will be composed predominantly of states with no double occupancy, $|s_i\rangle$. To determine the precise structure of the ground state, we could simply diagonalize the 4×4 Hamiltonian – a procedure evidently infeasible in the lattice system. Instead, to gain some intuition for the extended system, we will use a perturbation theory which projects the insulating system onto a low-energy effective spin Hamiltonian. Specifically, we will treat the hopping part of the Hamiltonian $\hat{H}_t$ as a weak perturbation of the Hubbard interaction $\hat{H}_U$.

To implement the perturbation theory, it is helpful to invoke a **canonical transformation** of the Hamiltonian,

canonical
transfor-
mation

$$\hat{H} \mapsto \hat{H}' \equiv e^{-t\hat{O}} \hat{H} e^{t\hat{O}} = e^{-t[\hat{O},\,]} \hat{H} \equiv \hat{H} - t[\hat{O},\hat{H}] + \frac{t^2}{2!}[\hat{O},[\hat{O},\hat{H}]] + \cdots, \quad (2.29)$$

where the exponentiated commutator is defined by the series expansion on the right.

EXERCISE Considering the derivative of $\hat{H}'$ with respect to t, prove the second equality.

By choosing the operator $\hat{O}$ such that $\hat{H}_t + t[\hat{H}_U, \hat{O}] = 0$, all terms of first order in t can be eliminated from the transformed Hamiltonian. As a result, the effective Hamiltonian is brought to the form

$$\hat{H}' = \hat{H}_U + \frac{t}{2}[\hat{H}_t, \hat{O}] + O(t^3). \quad (2.30)$$

Applying the ansatz $t\hat{O} = [\hat{P}_{\rm s}\hat{H}_t\hat{P}_{\rm d} - \hat{P}_{\rm d}\hat{H}_t\hat{P}_{\rm s}]/U$, where $\hat{P}_{\rm s}$ and $\hat{P}_{\rm d}$ are operators that project onto the singly and doubly occupied subspaces respectively, the first-order cancellation is assured.

EXERCISE To verify this statement, take the matrix elements of the first-order equation with respect to the basis states. Alternatively, it can be confirmed by inspection, noting that $\hat{P}_{\rm s}\hat{P}_{\rm d} = 0$, $\hat{H}_U\hat{P}_{\rm s} = 0$ and, in the present case, $\hat{P}_{\rm s}\hat{H}_t\hat{P}_{\rm s} = \hat{P}_{\rm d}\hat{H}_t\hat{P}_{\rm d} = 0$.

Substituting $t\hat{O}$ into Eq. (2.30) and projecting onto the singly occupied subspace one obtains

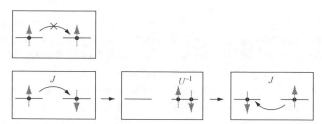

Fig. 2.5   Top: The hybridization of parallel spin polarized states is forbidden by Pauli exclusion. Bottom: The superexchange mechanism by which two antiparallel spins can lower their energy by a virtual process in which the upper Hubbard band is occupied.

$$\hat{P}_{\rm s}\hat{H}'\hat{P}_{\rm s} = -\frac{1}{U}\hat{P}_{\rm s}\hat{H}_t\hat{P}_{\rm d}\hat{H}_t\hat{P}_{\rm s} = -\frac{2t^2}{U}\hat{P}_{\rm s}\left(1 + a_{1\sigma}^\dagger a_{2\sigma'}^\dagger a_{1\sigma'} a_{2\sigma}\right)\hat{P}_{\rm s} = J\left(\hat{\mathbf{S}}_1 \cdot \hat{\mathbf{S}}_2 - \frac{1}{4}\right),$$

exchange
interaction

where $J = 4t^2/U$ denotes the strength of the **antiferromagnetic exchange interaction** coupling the spins on neighboring sites.

EXERCISE   Noting the anticommutation relations of the electron operators, find the matrix elements of the Hubbard Hamiltonian with respect to the four basis states $|s_i\rangle$ and $|d_i\rangle$. Diagonalizing the 4×4 matrix Hamiltonian, obtain the eigenstates of the system. In the strong-coupling system $U/t \gg 1$, determine the spin and energy dependences of the ground state.

The perturbation theory above shows that electrons subject to a strong local repulsive Coulomb interaction have a tendency to adopt an antiparallel spin configuration between neighboring sites. This has a simple physical interpretation: electrons with antiparallel spins can take advantage of the hybridization (however small) and reduce their kinetic energy by hopping to a neighboring site (see fig. 2.5). Parallel spins, on the other hand, are unable to participate in this virtual process due to Pauli exclusion.

The calculation presented above is easily generalized to an extended lattice system. Once again, projecting onto a basis in which all sites are singly occupied, virtual exchange processes favor an antiferromagnetic arrangement of neighboring spins. Such

> **Philip Warren Anderson**
> 1923–2020
> was an American physicist who, in 1977, shared with Sir Neville Mott and John H. van Vleck the Nobel Prize in Physics for their "fundamental theoretical investigations of the electronic structure of magnetic and disordered systems." Anderson made numerous contributions to theoretical physics, from theories of localization and antiferromagnetism to superconductivity and particle physics. He is also credited with promoting the term "condensed matter" in the 1960s as the field expanded beyond studies of the solid state.

a correlated magnetic insulator is described by the quantum spin-1/2 **Heisenberg**

Heisenberg
Hamil-
tonian

Hamiltonian

$$\boxed{\hat{H} = J \sum_{\langle mn \rangle} \hat{\mathbf{S}}_m \cdot \hat{\mathbf{S}}_n}$$ (2.31)

where, as usual, $\langle mn \rangle$ denotes a sum over nearest neighbors and the positive exchange constant $J \sim t^2/U$. While, in the insulating magnetic phase, the charge degrees of freedom remain "quenched," spin fluctuations can freely propagate.

When doped away from half-filling, the behavior of the Hubbard model is notoriously difficult to resolve. The removal of electrons from a half-filled system introduces vacancies into the lower Hubbard band that may propagate through the lattice. For a low concentration of holes, the strong-coupling Hubbard system may be described effectively by the t–J **Hamiltonian**

t–J Hamiltonian

$$\hat{H}_{t-J} = -t \sum_{\langle mn \rangle \sigma} \hat{P}_s a_{m\sigma}^\dagger a_{n\sigma} \hat{P}_s + J \sum_{\langle mn \rangle} \hat{\mathbf{S}}_m \cdot \hat{\mathbf{S}}_n$$

However, the passage of vacancies is frustrated by the antiferromagnetic spin correlations of the background. Here, transport depends sensitively on the competition between the exchange and kinetic energy of the holes. Oddly, at $J = 0$ (i.e., $U = \infty$), the ground state spin configuration is driven ferromagnetic by a *single* hole (exercise: consider why!) while, for $J > 0$, it is generally accepted that a critical concentration of holes is required to destabilize antiferromagnetic order.

cuprate materials

INFO The rich behavior of the Mott–Hubbard system is nowhere more exemplified than in the **ceramic cuprate compounds** – the class of materials that comprise the high-temperature superconductors. Cuprates are built of layers of CuO_2 separated by heavy rare earth ions such as lanthanum. According to band theory, the half-filled system (one electron per Cu site) should be metallic. However, strong electron interactions drive the cuprate system into an insulating antiferromagnetic Mott–Hubbard phase. When doped away from half-filling, charge carriers are introduced into the lower Hubbard band. In this case, the collapse of the Hubbard gap and the loss of antiferromagnetic (AF) order is accompanied by the development of a high-temperature unconventional superconducting (SC) phase, whose mechanism is believed to be rooted in the exchange of antiferromagnetic spin fluctuations. Whether the rich phenomenology of the cuprate system is captured by the Hubbard model remains a subject of great interest and debate. At increasing temperatures, the cuprates pass through a "pseudogap" phase with a partially gapped Fermi surface into that of conventional metallic behavior. (See the figure, where the phase diagram of $La_{2-x}Sr_xCuO_4$ is shown as a function of temperature and the concentration x of Sr atoms replacing La atoms.)

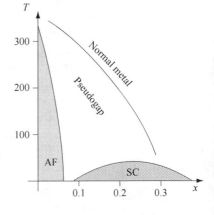

This concludes our preliminary survey of the rich phenomenology of the interacting electron system. Notice that, so far, we have merely discussed ways to distill a reduced model from the original microscopic many-body Hamiltonian (2.17). However, save for the two examples of free field theories analyzed in chapter 1, we have not yet learned how methods of second quantization can be applied to actually solve problems. To this end, in the following section, we will illustrate the application of the method on a prominent *strongly* interacting problem.

2.2.4 Interacting fermions in one dimension

free theory   Within the context of many-body physics, a theory is termed **free** if the Hamiltonian is bilinear in creation and annihilation operators, i.e., $\hat{H} \sim \sum_{\mu\nu} a_\mu^\dagger H_{\mu\nu} a_\nu$, where H may be a finite- or infinite-dimensional matrix.[11] Such models are "solvable" in the sense that the solution of the problem simply amounts to a diagonalization of the matrix $H_{\mu\nu}$ (subject to the preservation of the commutation relations of the operators a and $a^\dagger$). However, only a few models of interest belong to this category. In general, interaction contributions that are typically quartic in the field operators are present, and complete analytical solutions are out of reach.

Yet there are a few precious examples of genuinely interacting systems that are amenable to (nearly) exact solution. In this section, we address an important representative of this class, the one-dimensional interacting electron gas. Not only is its analysis physically interesting but, in addition, it provides an opportunity to practice working with the second-quantized operator formalism on a deeper level. To this end, consider the electron Hamiltonian (2.18) and (2.19) in one dimension. Including the chemical potential E_F into the free part, and neglecting spin degrees of freedom (e.g., one might consider a fully spin-polarized band) we have

$$\hat{H} = \sum_k a_k^\dagger \left(\frac{k^2}{2m} - E_\mathrm{F} \right) a_k + \frac{1}{2L} \sum_{kk',q\neq 0} V(q) a_{k-q}^\dagger a_{k'+q}^\dagger a_{k'} a_k. \qquad (2.32)$$

INFO At first sight, the treatment of a one-dimensional electron system may seem academic. However, effective one-dimensional interacting fermion systems are realized in a surprisingly rich spectrum of materials. For example, a **carbon nanotube** is formed from a graphene layer rolled into a cylindrical geometry. The carbon lattice is surrounded by clouds of mobile (itinerant) electrons (see the upper panel of the figure). Confinement in the circumferential direction divides the system into a series of one-dimensional bands, each classified by a sub-band index and wave number k. At low temperatures, the Fermi surface typically intersects a single sub-band, allowing attention to be concentrated on a strictly one-dimensional system. A similar mechanism renders certain **organic molecules** (such as the Bechgaard salt $(\mathrm{TMTSF})_2\mathrm{PF}_6$, where TMTSF stands for tetramethyl-tetraselenafulvalene) one dimensional.

carbon nanotube

organic conductor

A third realization is presented by artificial low-dimensional structures fabricated from semiconducting devices. The redistribution of charge at the interface of a GaAs/AlGaAs heterostructure results in the formation of a two-dimensional electron gas. By applying external gates, it is possible to fabricate quasi-one-dimensional **semiconductor quantum wires**, in which electron motion in the transverse direction is impeded by a large potential gradient. At sufficiently low Fermi energies, only the lowest eigenstate of the transverse

quantum wire

---

[11] More generally, a free Hamiltonian may also contain contributions $\sim a_\mu a_\nu$ and $a_\mu^\dagger a_\nu^\dagger$.

Schrödinger equation (the lowest "quantum mode") is populated and one is left with a strictly one-dimensional electron system (lower panel). There are other realizations, such as the edge modes in **quantum Hall systems**, "**stripe phases**" in high-temperature superconductors or certain **inorganic crystals**; but we shall not discuss these here.

The one-dimensional fermion system exhibits a number of features not shared by higher-dimensional systems. The origin of these peculiarities can be understood using a simple qualitative picture. Consider an array of interacting fermions confined to a line. To optimize their energy, the electrons can merely "push" each other around, thereby creating density fluctuations. By contrast, in higher-dimensional systems, electrons are free to avoid contact by moving around each other.

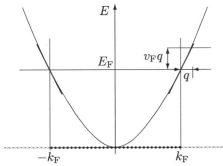

A slightly different formulation of the same picture can be given in momentum space. The Fermi "sphere" of the one-dimensional system is defined through the interval $[-k_F, k_F]$ of filled momentum states. The Fermi "surface" consists of two isolated points, $\{k_F, -k_F\}$ (see the figure). By contrast, higher-dimensional systems typically exhibit extended Fermi surfaces, thus providing more phase space for two-particle interaction processes. The one-dimensional electron system represents a rare exception of an interacting system that can be solved under few, physically weak, simplifying assumptions. This makes it an important test system on which non-perturbative manifestations of many-body interactions can be explored.

We now proceed to develop a quantitative picture of the charge density excitations of the one-dimensional electron system. Anticipating that, at low temperatures, the relevant dynamics takes place in the vicinity of the two Fermi points $\{k_F, -k_F\}$, we will reduce the Hamiltonian (2.32) to an effective model describing the propagation of left- and right-moving excitations. To this end, we first introduce subscripts R/L to indicate that an operator $a^\dagger_{(+/-)k_F+q}$ creates an electron that moves to the right/left with velocity $\simeq v_F \equiv k_F/m$. We next note (see the figure) that,

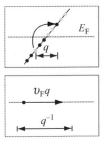

in the vicinity of the Fermi points, the dispersion relation is approximately linear, implying that the non-interacting part of the Hamiltonian assumes the approximate form (exercise)

$$\hat{H}_0 \simeq \sum_{s=\mathrm{R,L}} \sum_q a^\dagger_{sq} \sigma_s v_F q \, a_{sq}, \qquad (2.33)$$

where $\sigma_s = (+/-)$ for $s = \mathrm{R/L}$, and the summation over q is restricted by some momentum cutoff $|q| < \Gamma$ beyond which the linearization of the dispersion is invalid. (Throughout this section, all momentum summations are subject to this constraint.) Turning to the interacting part of the Hamiltonian, let us first define the operator

$$\hat{\rho}_{sq} = \sum_k a^\dagger_{sk+q} a_{sk}. \tag{2.34}$$

These operators afford two interpretations. Application of $\hat{\rho}_{sq}$ to the ground state creates superpositions of particles ($a^\dagger_{sk+q}$) with momentum $k + q$ and energy ϵ_{k+q} and holes (a_{sq}) with momentum k and ϵ_k. This may be interpreted as the excitation of a particle from a filled state k to an empty state $k + q$. Such particle–hole excitations cost energy $\epsilon_{k+q} - \epsilon_k = v_F q$, independent of k (see the upper panel of the figure). Alternatively, thinking of $\hat{\rho}_{sq}$ as the Fourier transform of the real space density operator $\hat{\rho}_s(x)$ (exercise), the particle–hole excitation can be interpreted as a density modulation of characteristic wavelength $\sim q^{-1}$. Since both, particles and holes travel with the same velocity, v_F, these excitations do not disperse and are expected to show a high level of stability. On this basis, we expect the operators $\hat{\rho}_{sq}$ to represent the central degrees of freedom of the theory.

Represented in terms of density operators, the interaction contribution to the Hamiltonian may be recast as

$$\hat{V}_{ee} = \frac{1}{2L} \sum_{kk'q} V_{ee}(q)\, a^\dagger_{k-q} a^\dagger_{k'+q} a_{k'} a_k \equiv \frac{1}{2L} \sum_{qs} [g_4 \hat{\rho}_{sq} \hat{\rho}_{s-q} + g_2 \hat{\rho}_{sq} \hat{\rho}_{\bar{s}-q}], \tag{2.35}$$

where $\bar{s} = \text{L/R}$ denotes the complement of $s = \text{R/L}$, and the constants g_2 and g_4 measure the strength of the interaction in the vicinity of the Fermi points, i.e., where $q \simeq 0$ and $q \simeq 2k_F$. (The notation $g_{2,4}$ follows a common nomenclature.)

EXERCISE Explore the relation between the coupling constants $g_{2,4}$ and the Fourier transform of V_{ee}. Show that, to the summation $\sum_{kk'q}$, not only terms with $(k, k', q) \simeq (\pm k_F, \pm k_F, 0)$, but also terms with $(k, k', q) \simeq (\pm k_F, \mp k_F, 2k_F)$ contribute. When adequately ordered (check it!), these contributions can be arranged into the form of the right-hand side of Eq. (2.35). (For a detailed discussion see, e.g., the books by Giamarchi[12] and Mahan[13]). The only point that matters for our present discussion is that the interaction *can* be represented through density operators with positive constants $g_{2,4}$ determined by the interaction strength.

INFO Working with second-quantized theories, one frequently needs to compute commutators of operators $\hat{A}(a, a^\dagger)$ that are polynomial in the elementary boson/fermion operators of the theory (e.g., $\hat{A} = aa^\dagger$, $\hat{A} = aaa^\dagger a^\dagger$, etc., where we have omitted the quantum number subscripts carried by a and $a^\dagger$). Such types of operation are made easier by elementary

commutator   **commutator algebra**. The most basic identity, from which many other formulae can be derived, reads

$$\boxed{[\hat{A}, \hat{B}\hat{C}]_\pm = [\hat{A}, \hat{B}]_\pm \hat{C} \mp \hat{B}[\hat{A}, \hat{C}]_\pm} \tag{2.36}$$

Iteration of this equation for boson operators $a, a^\dagger$ shows that

$$[a^\dagger, a^n] = -n a^{n-1}. \tag{2.37}$$

[12] T. Giamarchi, *Quantum Physics in One Dimension* (Oxford University Press, 2004).
[13] G. Mahan, *Many Particle Physics* (Plenum Press, 1981).

(Owing to the fact that $a^2 = 0$ in the fermionic case, there is no fermion analog of this equation.) Taylor expansion then shows that, for any analytic function $F(a)$, $[a^\dagger, F(a)] = -F'(a)$. Another useful formula is $a^\dagger F(aa^\dagger) = F(a^\dagger a)a^\dagger$, which is also verified by series expansion.

So far, we have merely rewritten parts of the Hamiltonian in terms of density operators. Ultimately, however, we wish to arrive at a representation whereby these operators, instead of the original electron operators, represent the fundamental degrees of freedom of the theory. Since the definition of the operators $\hat{\rho}$ involves the squares of two Fermi operators, we expect the density operators to resemble bosonic excitations. As a first step towards the construction of the new picture, we therefore explore the commutation relations between the operators $\hat{\rho}_{sq}$.

From the definition (2.34), and the auxiliary identity (2.36), it is straightforward to verify the commutation relation $[\hat{\rho}_{sq}, \hat{\rho}_{s'q'}] = \delta_{ss'} \sum_k (a^\dagger_{sk+q} a_{sk-q'} - a^\dagger_{sk+q+q'} a_{sk})$. As it stands, this relation is not of much practical use. To make further progress, we must resort to a mild approximation. Ultimately we will want to compute observables as expectation values taken in the zero-temperature ground state of the theory, $\langle \Omega | \dots | \Omega \rangle$. To simplify the structure of the theory, we may thus replace the right-hand side of the commutation relation by its ground state expectation: $[\hat{\rho}_{sq}, \hat{\rho}_{s'q'}] \approx \delta_{ss'} \sum_k \langle \Omega | a^\dagger_{sk+q} a_{sk-q'} - a^\dagger_{sk+q+q'} a_{sk} | \Omega \rangle = \delta_{ss'} \delta_{q,-q'} \sum_k \langle \Omega | (\hat{n}_{sk+q} - \hat{n}_{sk}) | \Omega \rangle$, where, as usual, $\hat{n}_{sk} = a^\dagger_{sk} a_{sk}$ and we have made use of the fact that $\langle \Omega | a^\dagger_{sk} a_{sk'} | \Omega \rangle = \delta_{kk'}$. Although this is an uncontrolled approximation, it is expected to become better at low excitation energies.

EXERCISE Critically assess the validity of the above approximation. (For a comprehensive discussion, see Giamarchi's text.[12])

At first sight, it would seem that the right-hand side of our simplified commutator relation vanishes. A simple shift of the summation index, $\sum_k \langle \Omega | \hat{n}_{sk+q} | \Omega \rangle \overset{?}{=} \sum_k \langle \Omega | \hat{n}_{sk} | \Omega \rangle$, indicates that the two terms contributing to the sum cancel. However, this argument is naïve: it ignores the fact that our summation is limited by a cutoff momentum Γ. Since the shift $k \to k - q$ changes the cutoff, the interpretation above is invalid. To obtain a more accurate result, let us consider the case $s = R$ and $q > 0$. We know that, in the ground state, all states with momentum $k < 0$ are occupied, while all states with $k \geq 0$ are empty. This implies that

$$\sum_k \langle \Omega | (\hat{n}_{Rk+q} - \hat{n}_{Rk}) | \Omega \rangle = \left(\sum_{-\Gamma < k \leq -q} + \sum_{-q < k \leq 0} + \sum_{0 < k < \Gamma} \right) \langle \Omega | (\hat{n}_{Rk+q} - \hat{n}_{Rk}) | \Omega \rangle$$

$$= \sum_{-q \leq k \leq 0} \langle \Omega | (\hat{n}_{Rk+q} - \hat{n}_{Rk}) | \Omega \rangle = -\frac{qL}{2\pi},$$

where, in the last equality, we have used the fact that a momentum interval of size q contains $q/(2\pi/L)$ quantized momentum states. Similar reasoning for $s = L$ shows that the effective form of the commutator relation reads

$$[\hat{\rho}_{sq}, \hat{\rho}_{s'q'}] = -\delta_{ss'} \delta_{q,-q'} \sigma_s \frac{qL}{2\pi}. \tag{2.38}$$

If it were not for the q-dependence of the right hand side (r.h.s. throughout), we would indeed have found bosonic commutation relations. To make the connection to bosons explicit, let us define

$$b_q \equiv n_q \hat{\rho}_{\mathrm{L}q}, \qquad b_q^\dagger \equiv n_q \hat{\rho}_{\mathrm{L}-q},$$

$$b_{-q} \equiv n_q \hat{\rho}_{\mathrm{R}-q}, \qquad b_{-q}^\dagger \equiv n_q \hat{\rho}_{\mathrm{R}q}, \tag{2.39}$$

where $q > 0$ and $n_q \equiv (2\pi/Lq)^{1/2}$. The operators $\{b_q, b_q^\dagger\}$ do indeed obey canonical commutation relations (check this). We conclude that, apart from the scaling factors n_q, the quantum density excitations of the system indeed behave as bosonic "particles."

Expressed in terms of the b-operators, the interaction part of the Hamiltonian takes the form (exercise)

$$V_{\mathrm{ee}} = \frac{1}{2\pi} \sum_{q>0} q \begin{pmatrix} b_q & b_{-q}^\dagger \end{pmatrix} \begin{pmatrix} g_4 & g_2 \\ g_2 & g_4 \end{pmatrix} \begin{pmatrix} b_q^\dagger \\ b_{-q} \end{pmatrix}.$$

Notice that we have succeeded in representing a genuine two-body interaction, a contribution that usually renders a model unsolvable, in terms of a quadratic representation. However, this representation of the interaction term is of little use until the kinetic part of the Hamiltonian $\hat{H}_0$ is represented in terms of the b operators.

It turns out that the direct construction of a representation of $\hat{H}_0$ in b's, is cumbersome in practice. However, there exists a more efficient alternative: As follows from the discussion of section 2.1.1, the properties of second quantized operators are fixed by their commutation relations.[14] If we manage to identify an operator $\hat{H}_0'(b, b^\dagger)$ having the same commutation relations with the b-operators as the kinetic energy operator $\hat{H}_0(a, a^\dagger)$, we know that $H_0 = H_0'$, up to an undetermined (and inessential) constant.

Using Eq. (2.33), the definition (2.34), and the auxiliary identity (2.36), it is straightforward to verify that $[\hat{H}_0, \hat{\rho}_{sq}] = q v_{\mathrm{F}} \sigma_s \hat{\rho}_{sq}$. On the other hand, using Eq. (2.38), one finds that the same commutation relations, $[\hat{H}_0', \hat{\rho}_{sq}] = q v_{\mathrm{F}} \sigma_s \hat{\rho}_{sq}$, hold for

$$\hat{H}_0' = \frac{2\pi v_F}{L} \sum_{qs} \hat{\rho}_{sq} \hat{\rho}_{s-q}.$$

On this basis, we may substitute $\hat{H}_0'$ for the non-interacting Hamiltonian.

EXERCISE To gain some confidence in the identification $\hat{H}_0 = \hat{H}_0' + \text{const.}$, and to establish that the undetermined constant actually equals zero, show that the energy expectation values of the state $|\Psi_{sq}\rangle \equiv \hat{\rho}_{sq}|\Omega\rangle$ for both $\langle\Psi_{sq}|\hat{H}_0|\Psi_{sq}\rangle$ and $\langle\Psi_{sq}|\hat{H}_0'|\Psi_{sq}\rangle$ coincide.

---

[14] This argument can be made quantitative by group-theoretic reasoning: Eqs. (2.4) and (2.7) define the irreducible representation of an operator algebra – an *algebra* because [,] defines a product in the space of generators $\{a_\lambda, a_\lambda^\dagger\}$; a *representation* because the operators act in a vector space (namely Fock space $\mathcal{F}$) which is *irreducible* because all states $|\lambda_1, \ldots, \lambda_N\rangle \in \mathcal{F}$ can be reached by the iterative application of operators to a unique reference state (e.g., $|\Omega\rangle$). Under these conditions, Schur's lemma – to be discussed in more detail in section 3.4.1 – states that two operators $\hat{A}_1$ and $\hat{A}_2$, having identical commutation relations with all $\{a_\lambda, a_\lambda^\dagger\}$ are equal up to a constant.

Finally, using Eq. (2.39), and adding the interaction contribution V_{ee}, we arrive at the effective Hamiltonian

$$\hat{H} = \frac{1}{2\pi} \sum_{q>0} q \begin{pmatrix} b_q & b_{-q}^{\dagger} \end{pmatrix} \begin{pmatrix} 2\pi v_{\mathrm{F}} + g_4 & g_2 \\ g_2 & 2\pi v_{\mathrm{F}} + g_4 \end{pmatrix} \begin{pmatrix} b_q^{\dagger} \\ b_{-q} \end{pmatrix}. \qquad (2.40)$$

At this point, we have succeeded in mapping the full interacting problem onto a *free* bosonic theory. The mapping $a \to \hat{\rho} \to b$ is our first example of a concept known as **bosonization**. This technique plays an important role in $2(= 1$ space $+$ 1 time$)$-dimensional field theory in general, and more sophisticated schemes will be discussed in section 3.6. Conversely, it is sometimes useful to represent a boson problem in terms of fermions, via **fermionization**. One may wonder why it is possible to represent the low-lying excitations of a gas of fermions in terms of bosons. **Fermi–Bose transmutability** is indeed a peculiarity of one-dimensional quantum systems. Particles confined to a line cannot pass "around" each other. That means that the whole issue of sign factors arising from the interchange of particle coordinates does not arise, and much of the exclusion-type characteristics of the Fermi system are inactivated.

Now, there is one last problem that needs to be overcome to solve the interacting problem. In chapter 1, we learned how to interpret Hamiltonians with the structure $\sum_q b_q^{\dagger} b_q$ as superpositions of harmonic oscillators. However, in our present problem, terms of the type $b_q b_{-q}$ and $b_{-q}^{\dagger} b_q^{\dagger}$ appear. To return to familiar terrain, we need to eliminate these terms. Before doing so, it is instructive to discuss their physical meaning.

Recall that the number operator of a theory described by operators $\{b_{\lambda}^{\dagger}, b_{\lambda}\}$ is given by $\hat{N} = \sum_{\lambda} b_{\lambda}^{\dagger} b_{\lambda}$. If the Hamiltonian has the form $\hat{H} = \sum_{\mu\nu} b_{\mu}^{\dagger} H_{\mu\nu} b_{\nu}$, this number operator commutes with $\hat{H}$, i.e., $[\hat{N}, \hat{H}] = 0$ (exercise), meaning that the dynamics conserves the total number of particles. Formally, **particle number conservation** implies that $\hat{H}$ and $\hat{N}$ can be simultaneously diagonalized. More generally, any Hamiltonian containing only operators with as many bs as $b^{\dagger}$s (e.g., $b^{\dagger} b^{\dagger} b b$, $b^{\dagger} b^{\dagger} b^{\dagger} b b b$, etc.) creates and annihilates particles in equal numbers and hence is number conserving. Conversely, in situations where the number of particles is not fixed (e.g., a theory of photons or phonons), terms like bb or $b^{\dagger} b^{\dagger}$ can appear. Such a situation is realized in the present problem: the number of density excitations in an electron system is certainly not a conserved quantity, which explains why contributions like $b_q b_{-q}$ appear in $\hat{H}$.

To finally solve the problem, we must find a way to diagonalize the matrix

$$K \equiv \frac{1}{2\pi} \begin{pmatrix} 2\pi v_{\mathrm{F}} + g_4 & g_2 \\ g_2 & 2\pi v_{\mathrm{F}} + g_4 \end{pmatrix}.$$

To this end, let us introduce the shorthand notation $\Psi_q \equiv (b_q^{\dagger}, b_{-q})^T$, and rewrite the Hamiltonian as $H = \sum_{q>0} q \, \Psi_q^{\dagger} K \Psi_q$. If we now define $\Psi_q' \equiv T^{-1} \Psi_q$, where T is a 2×2 matrix acting on Ψ_q,[15] the Hamiltonian transforms as follows:

---

[15] Since K does not depend on q, T can be chosen to be likewise q-independent.

bosonization

fermioni- zation

number conser- vation

$$H = \sum_{q>0} q\, \Psi_q^\dagger K \Psi_q \rightarrow \sum_{q>0} q\, \Psi_q'^\dagger \underbrace{T^\dagger K T}_{K'} \Psi_q' \qquad (2.41)$$

with a new matrix $K' \equiv T^\dagger K T$. We will seek a transformation T that makes K' diagonal. Crucially, however, not all 2×2 matrices T qualify as legitimate transformations. We must ensure that the transformed "vector" again has the structure $\Psi_q' \equiv (b_q'^\dagger, b_{-q}')^T$, with a boson creation/annihilation operator in the first/second component; the bosonic commutation relations of the representation must be conserved by the transformation (think about this point). This invariance condition is expressed in mathematical form as $[\Psi_{qi}, \Psi_{qj}^\dagger] = (-\sigma_3)_{ij} \stackrel{!}{=} [\Psi_{qi}', \Psi_{qj}'^\dagger]$. Substitution of $\Psi' = T^{-1}\Psi$, yields the pseudo-unitarity condition $T^\dagger \sigma_3 T \stackrel{!}{=} \sigma_3$.

With this background, we may identify the transformation bringing K to a diagonal form. To this end, we multiply the definition $T^\dagger K T = K'$ by σ_3 to obtain

$$(\underbrace{\sigma_3 T^\dagger \sigma_3}_{T^{-1}})\sigma_3 K T = \sigma_3 K'.$$

This equation states that the diagonal matrix $\sigma_3 K' \equiv \mathrm{diag}(+v_\rho, -v_\rho) = v_\rho \sigma_3$ is obtained by a *similarity transformation* $T^{-1}(\cdots)T$ from $\sigma_3 K$. The diagonal $\sigma_3 K'$ contains the eigenvalues $\pm v_\rho$ of $\sigma_3 K$, which sum to zero since $\mathrm{tr}(\sigma_3 K) = 0$. These eigenvalues are readily computed as

$$v_\rho = \frac{1}{2\pi}\left[(2\pi v_\mathrm{F} + g_4)^2 - g_2^2\right]^{1/2}. \qquad (2.42)$$

Thus, with $\sigma_3 K' = \sigma_3 v_\rho$, we arrive at $K' = v_\rho \times \mathrm{id.}$, where id. denotes the unit matrix.[16] Substitution of this result into Eq. (2.41) finally leads to the diagonal Hamiltonian $\hat{H} = v_\rho \sum_{q>0} q\, \Psi_q'^\dagger \Psi_q'$ or, equivalently, making use of the identity $\Psi_q'^\dagger \Psi_q' = b_q^\dagger b_q + b_{-q}^\dagger b_{-q} + 1$,

$$\hat{H} = v_\rho \sum_q |q| b_q^\dagger b_q. \qquad (2.43)$$

> **Nicolai Nikolaevich Bogoliubov 1909–1992**
> was a Soviet mathematician and theoretical physicist acclaimed for his works in nonlinear mechanics, statistical physics, the theory of superfluidity and superconductivity, quantum field theory, renormalization group theory, the proof of dispersion relations, and elementary particle theory.

Here we have ignored an overall constant and omitted the prime on the new Bose operators.

In the literature, the transformation procedure outlined above is known as a **Bogoliubov transformation**. Transformations of this type are frequently applied

Bogoliubov transformation

---

[16] Explicit knowledge of the transformation matrix T, i.e., knowledge of the relation between the operators b and b', is not needed for our construction. However, for the sake of completeness, we mention that

$$T = \begin{pmatrix} \cosh\theta_k & \sinh\theta_k \\ \sinh\theta_k & \cosh\theta_k \end{pmatrix},$$

with $\tanh(2\theta) = -g_2/(2\pi v_\mathrm{F} + g_4)$, represents a suitable parameterization.

in quantum magnetism (see below), superconductivity or, more generally, all problems where the particle number is not conserved. Notice that the possibility of transforming to a representation $\sim b^\dagger b$ does not imply that, miraculously, the theory has become particle number conserving. The new "quasi-particle" operators b are related to the original Bose operators through a transformation that mixes b and $b^\dagger$. While the quasi-particle number is conserved, the number of original density excitations is not.

Equations (2.42) and (2.43) represent our final solution of the problem of spinless interacting fermions in one dimension. We have succeeded in mapping the problem onto a form analogous to our previous results (1.34) and (1.39) for the phonon and the photon system, respectively. Indeed, all that has been said about those Hamiltonians applies equally to Eq. (2.43): the basic elementary excitations of the one-dimensional fermion system are waves, i.e., excitations with linear dispersion $\omega = v_\rho |q|$. In the present context, they are termed **charge density waves (CDW)**. The Bose creation operators describing these excitations are, up to the Bogoliubov transformation, and a momentum-dependent scaling factor $(2\pi/Lq)^{1/2}$, equivalent to the density operators of the electron gas. For a non-interacting system, $g_2 = g_4 = 0$, and the CDW propagates with the velocity of the free Fermi particles, v_F. A fictitious interaction that does not couple particles of opposite Fermi momentum, $g_2 = 0$, $g_4 \neq 0$, speeds up the CDW. Heuristically, this can be interpreted as an "acceleration process" whereby a CDW pushes its own charge front. By contrast, interactions between left and right movers, $g_2 \neq 0$, diminish the velocity, i.e., owing to the Coulomb interaction it is difficult for distortions of opposite velocities to penetrate each other. (Notice that for a theory with $g_2 = 0$, no Bogoliubov transformation is needed to diagonalize the Hamiltonian. In this case, undisturbed left- and right-moving waves are the basic excitations of the theory.)

Our discussion above neglected the spin carried by conduction electrons. Had we included spin, the following picture would have emerged (see problem 2.4.6). The long-range dynamics of the electron gas is governed by two independently propagating wave modes, the charge density wave discussed above, and a **spin density wave (SDW)**.[17] The SDW carries a **spin current**, but is electrically neutral. As with the CDW, its dispersion is linear, with an interaction-renormalized velocity, v_s (which, however, is generally larger than the velocity v_ρ of the CDW). To understand the consequences of this phenomenon, imagine that an electron has been thrown into the system (e.g., by attaching a tunnel contact somewhere along the wire). As discussed above, a single electron does not represent a stable excitation of the one-dimensional electron gas. What will happen is that the spectral weight of the particle[18] disintegrates into a collective charge excitation and a spin excitation. The

----

[17] The charge density of the electron gas $\rho = \rho_\uparrow + \rho_\downarrow$ is the sum of the densities of the spin-up and spin-down populations, respectively. The local spin density is given by $\rho_\mathrm{s} \equiv \rho_\uparrow - \rho_\downarrow$. After what has been said above, it is perhaps not too surprising that fluctuations of these two quantities represent the dominant excitations of the electron gas. What *is* surprising, though, is that these two excitations do not interact with each other.

[18] For a precise definition of this term, see chapter 7.

Tomonaga–
Luttinger
liquid

newly excited waves then freely propagate into the bulk of the system at different velocities $\pm v_\rho$ and $\pm v_s$. The collective quantum state defined by the independent and free propagation of CDWs and SDWs is called the **Tomonaga–Luttinger liquid**[19] or just the **Luttinger liquid**. We will return to the discussion of of this phase of quantum matter in chapter 3, from a field-theoretical perspective.

spin–
charge
separation

INFO   The "disintegration" of electrons into collective spin and charge excitations is a phenomenon known as **spin–charge separation**. Such types of effective "fractionalizations" of elementary quantum particles into collective excitations are ubiquitous in modern condensed matter physics. In fractionalization, the quantum numbers carried by elementary particles become absorbed by different excitation channels. One of the most prominent manifestations of this effect is the appearance of fractionally charged excitations in quantum Hall systems, to be discussed in more detail in chapter 8.

Although the theory of spin and charge density waves in one-dimensional conductors has a long history spanning many decades, its experimental verification proved challenging but was eventually achieved.[20]

2.2.5 Quantum spin chains

In section 2.2.1, we discussed how Coulomb interactions may lead to the indirect generation of magnetic interactions. We saw in the previous section how in one dimension this principle manifests itself via the generation of (magnetic) spin density wave excitations. However, to introduce the phenomena brought about by quantum magnetic correlations, it is best to first consider systems where the charge degrees of freedom are frozen and only spin excitations remain. Such systems

Werner Heisenberg 1901–1976 was a German theoretical physicist who, in 1932, received the Nobel Prize in Physics "for the creation of quantum mechanics, the application of which has, *inter alia*, led to the discovery of the allotropic forms of hydrogen." As well as his uncertainty principle, Heisenberg made important contributions to the theories of turbulence, ferromagnetism, cosmic rays, and subatomic particles, and was instrumental in planning the first West German nuclear reactor at Karlsruhe.

are realized, for example, in Mott insulators, where the interaction between the spins of localized electrons is mediated by virtual exchange processes between neighboring electrons. One can describe these correlations through models of localized quantum

---

[19] J. M. Luttinger, *An exactly soluble model of a many-fermion system*, J. Math. Phys. **4**, 1154 (1963); S. Tomonaga, *Remarks on Bloch's method of sound waves applied to many–fermion problems*, Prog. Theor. Phys. **5**, 544 (1950).

[20] C. Kim *et al.*, *Observation of spin-sharge separation in one-dimensional SrCuO2*, Phys. Rev. Lett. **77**, 4054 (1996); B. J. Kim *et al.*, *Distinct spinon and holon dispersions in photoemission spectral functions from one-dimensional SrCuO2*, Nature Phys. **2**, 397 (2006); J. N. Fuchs *et al.*, *Spin waves in a one-dimensional spinor bose gas*, Phys. Rev. Lett. **95**, 150402 (2005); J. Vijayan *et al.*, *Time-resolved observation of spin–charge deconfinement in fermionic Hubbard chains*, Science **367**, 168 (2020).

spins – either in chains or, more generally, in higher-dimensional quantum spin lattices. We begin our discussion with the ferromagnetic spin chain.

Quantum ferromagnet

The quantum **Heisenberg ferromagnet** is specified by the Hamiltonian

$$\hat{H} = -J \sum_{\langle mn \rangle} \hat{\mathbf{S}}_m \cdot \hat{\mathbf{S}}_n \qquad (2.44)$$

where $J > 0$, $\hat{\mathbf{S}}_m$ represents the quantum mechanical spin operator at lattice site m and $\langle mn \rangle$ denotes summation over neighboring sites. In section 2.1.1 (see Eq. (2.13)) the quantum mechanical spin was represented through an electron basis. However, one can conceive of situations where the spin sitting at site m is carried by a different object (e.g., an atom with non-vanishing magnetic moment). For the purposes of our present discussion, we need not specify the microscopic origin of the spin. All we need to know is that (i) the lattice operators $\hat{S}_m^i$ obey the SU(2) commutator algebra ($\hbar = 1$),

$$\left[\hat{S}_m^i, \hat{S}_n^j \right] = i\delta_{mn} \epsilon^{ijk} \hat{S}_n^k, \qquad (2.45)$$

characteristic of quantum spins, and (ii) the total spin at each lattice site is S.[21]

Now, owing to the positivity of the coupling constant J, the Hamiltonian favors configurations where the spins at neighboring sites are aligned in the same direction (see the figure). A ground state of the system 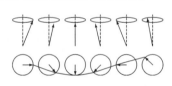 is given by $|\Omega\rangle \equiv \bigotimes_m |S_m\rangle$, where $|S_m\rangle$ represents a state with maximal z-component: $S_m^z |S_m\rangle = S|S_m\rangle$. We have written "a" ground state instead of "the" ground state because the system is highly degenerate: a simultaneous change in the orientation of all spins does not change the ground state energy, i.e., the system possesses a global rotation symmetry.

EXERCISE Compute the energy expectation value of the state $|\Omega\rangle$. Defining *global* spin operators through $\hat{S}^i \equiv \sum_m \hat{S}_m^i$, consider the state $|\alpha\rangle \equiv \exp(i\alpha \cdot \hat{\mathbf{S}})|\Omega\rangle$. Verify that the state α is degenerate with $|\Omega\rangle$. Explicitly compute the state $|(\pi/2, 0, 0)\rangle$. Convince yourself that, for general α, $|\alpha\rangle$ can be interpreted as a state with rotated quantization axis.

As with our previous examples, we expect that a global continuous symmetry entails the presence of energetically low-lying excitations. Indeed, it is obvious that, in the limit of long wavelength λ, a weak distortion of a ground state configuration will cost vanishingly small energy. To explore the physics of these excitations or **spin**
waves quantitatively, we adopt a "semiclassical" picture, where the spin $S \gg 1$ is

---

[21] Remember that the finite-dimensional representations of the spin operator are of dimension $2S + 1$, where S may be integer or half integer. While a single electron has spin $S = 1/2$, the total magnetic moment of electrons bound to an atom may be much larger.

taken to be large. In this limit, the rotation of the spins around the ground state configuration becomes similar to the rotation of a classical magnetic moment.

semiclassical approx- imation

INFO To better understand the mechanism behind the **semiclassical approximation**, consider the Heisenberg uncertainty relation $\Delta S^i \Delta S^j \sim |\langle[\hat{S}^i, \hat{S}^j]\rangle| = \epsilon^{ijk}|\langle\hat{S}^k\rangle|$, where ΔS^i is the root mean square of the quantum uncertainty of spin component i. Using the fact that $|\langle\hat{S}^k\rangle| \leq S$, we obtain for the relative uncertainty, $\Delta S^i/S$, the relation

$$\frac{\Delta S^i}{S}\frac{\Delta S^j}{S} \sim \frac{S}{S^2} \xrightarrow{S \gg 1} 0,$$

i.e., for $S \gg 1$, quantum fluctuations of the spin become increasingly less important.

In the limit of large spin S, and at low excitation energies, it is natural to de- scribe the ordered phase in terms of small fluctuations of the spins around their expectation values (cf. the description of the ordered phase of a crystal in terms of small fluctuations of the atoms around the ordered lattice sites). These fluctuations are conveniently represented in terms of spin raising and lowering operators. With $\hat{S}_m^{\pm} \equiv S_m^1 \pm i S_m^2$, it is straightforward to verify that

$$\boxed{\left[\hat{S}_m^z, \hat{S}_n^{\pm}\right] = \pm\delta_{mn}S_m^{\pm}, \qquad \left[\hat{S}_m^+, \hat{S}_n^-\right] = 2\delta_{mn}S_m^z} \qquad (2.46)$$

Application of $\hat{S}_m^{-(+)}$ lowers (raises) the z-component of the spin at site m by one. To make use of the fact that deviations around $|\Omega\rangle$ are small, it is convenient to represent spins in terms of bosonic creation and annihilation operators $a^{\dagger}$ and a through the **Holstein–Primakoff transformation**:[22]

Holstein– Primakoff transfor- mation

$$\boxed{\hat{S}_m^- = a_m^{\dagger}\left(2S - a_m^{\dagger}a_m\right)^{1/2}, \quad \hat{S}_m^+ = \left(2S - a_m^{\dagger}a_m\right)^{1/2}a_m, \quad \hat{S}_m^z = S - a_m^{\dagger}a_m}$$

EXERCISE Confirm that the spin operators satisfy the commutation relations (2.46).

The utility of this representation is clear. When the spin is large, $S \gg 1$, an expan- sion in powers of $1/S$ gives $\hat{S}_m^z = S - a_m^{\dagger}a_m$, $\hat{S}_m^- \simeq (2S)^{1/2}a_m^{\dagger}$, and $\hat{S}_m^+ \simeq (2S)^{1/2}a_m$. In this approximation, the one-dimensional Heisenberg Hamiltonian takes the form

$$\hat{H} = -J\sum_m \left\{\hat{S}_m^z\hat{S}_{m+1}^z + \frac{1}{2}\left(\hat{S}_m^+\hat{S}_{m+1}^- + \hat{S}_m^-\hat{S}_{m+1}^+\right)\right\}$$

$$= -JNS^2 + JS\sum_m (a_{m+1}^{\dagger} - a_m^{\dagger})(a_{m+1} - a_m) + O(S^0).$$

Keeping fluctuations at leading order in S, the quadratic Hamiltonian can be diagonalized by Fourier transformation. Imposing periodic boundary conditions,

---

[22] T. Holstein and H. Primakoff, *Field dependence of the intrinsic domain magnetization of a ferromagnet*, Phys. Rev. **58**, 1098 (1940).

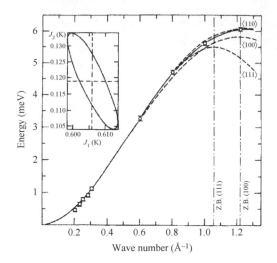

Fig. 2.6   Spin-wave spectrum of EuO as measured by inelastic neutron scattering at a reference
temperature of 5.5 K. Note that, at low values of momentum q, the dispersion is quadratic,
in agreement with the low-energy theory. (Exercise: Close inspection of the data shows the
existence of a small gap in the spectrum at $\mathbf{q} = 0$. To what may this gap be attributed?)
Figure reprinted with permission from L. Passell, O. W. Dietrich and J. Als-Nielser, *Neutron
scattering from the Heisenberg ferromagnets EuO and EuS I: the exchange interaction*,
Phys. Rev. B **14**, 4897 (1976). Copyright (1976) by the American Physical Society.

$\hat{S}_{m+N} = \hat{S}_m$ and $a_{m+N} = a_m$, where N denotes the total number of lattice sites,
and setting $a_m = \frac{1}{\sqrt{N}} \sum_k^{\text{B.Z.}} e^{-ikm} a_k$, the Hamiltonian takes the form (exercise)

$$\hat{H} = -JNS^2 + \sum_k^{\text{B.Z.}} \omega_k a_k^\dagger a_k + O(S^0) \tag{2.47}$$

where $\omega_k = 2JS(1 - \cos k) = 4JS \sin^2(k/2)$. In the limit $k \to 0$, the excitation
energy vanishes as $\omega_k \to JSk^2$. These massless low-energy excitations, known

magnons   as **magnons**, represent the elementary spin-wave excitations of the ferromagnet.
At higher order in S, interactions between the magnon excitations emerge, which
broaden and renormalize the dispersion. Nevertheless, comparison with experiment
(fig. 2.6) confirms that the low-energy spin-wave excitations are quadratic in k.

Quantum antiferromagnet

Having explored the elementary excitation spectrum of the ferromagnet, we now
Heisenberg   turn to the spin S **Heisenberg antiferromagnet**
antiferro-
magnet

$$\hat{H} = J \sum_{\langle mn \rangle} \hat{\mathbf{S}}_m \cdot \hat{\mathbf{S}}_n$$

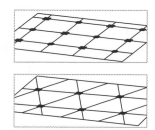

where $J > 0$. As we have seen, such antiferromagnetic systems occur in the arena of correlated electron compounds. Although the Hamiltonian differs from the ferromagnet "only" by a change of sign, the differences in the physics are drastic. First, the phenomenology displayed by the antiferromagnetic Hamiltonian $\hat{H}$ depends sensitively on the geometry of the lattice.

bipartite lattice

For a **bipartite lattice**, i.e., one in which the neighbors of one sublattice A also belong to the other sublattice B (see upper panel of figure), the ground states of the antiferromagnet are close[23] to a staggered spin configuration, known as a **Néel state**, in which all neighboring spins are anti-parallel.

Néel state

Again the ground state is degenerate, i.e., a global rotation of all spins by the same amount does not change the energy. By contrast, on non-bipartite lattices, such as the triangular one shown in the lower panel, there exists no spin configuration wherein each bond is assigned the full exchange energy J. Spin models of this kind are said to be **frustrated**.

frustrated magnet

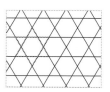

EXERCISE Employing only symmetry considerations, identify a possible classical ground state of the triangular lattice Heisenberg antiferromagnet. (Hint: construct the classical ground state of a three-site plaquette and then develop the periodic continuation.) Show that the classical antiferromagnetic ground state of the Kagomé lattice – a periodic array of corner-sharing Stars of David – has a continuous spin degeneracy generated by *local* spin rotations. How might the degeneracy affect the transition to an ordered phase?

Returning to the one-dimensional system, we first note that a chain is trivially bipartite. As before, our strategy will be to expand the Hamiltonian in terms of bosonic operators. However, before doing so, it is convenient to apply a canonical transformation to the Hamiltonian in which the spins on one

> **Louis Néel 1904–2000**
> was a French physicist who, with Hannes Alfvén shared the 1970 Nobel Prize in Physics for his "fundamental work and discoveries concerning antiferromagnetism and ferrimagnetism that have led to important applications in solid state physics."

sublattice, say B, are rotated through 180° about the x-axis, i.e., $S_B^x \to \widetilde{S}_B^x = S_B^x$, $S_B^y \to \widetilde{S}_B^y = -S_B^y$, and $S_B^z \to \widetilde{S}_B^z = -S_B^z$. When represented in terms of the new operators, the Néel ground state looks like a ferromagnetic state, with all spins aligned. We expect that a gradual distortion of this state will produce the antiferromagnetic analog of the spin waves discussed in the previous section (see the figure).

---

[23] It is straightforward to verify that the classical ground state – the Néel state – is not an exact eigenstate of the quantum Hamiltonian. The true ground state exhibits zero-point fluctuations reminiscent of the quantum harmonic oscillator or atomic chain. However, when $S \gg 1$, the Néel state serves as a useful reference state from which fluctuations can be examined.

In terms of the transformed operators, the Hamiltonian takes the form

$$\hat{H} = -J\sum_m \left[S_m^z \tilde{S}_{m+1}^z - \frac{1}{2}(S_m^+ \tilde{S}_{m+1}^+ + S_m^- \tilde{S}_{m+1}^-) \right].$$

Once again, using the Holstein–Primakoff representation $S_m^- \simeq (2S)^{1/2}a_m^\dagger$, etc.,

$$\hat{H} = -NJS^2 + JS\sum_m \left[a_m^\dagger a_m + a_{m+1}^\dagger a_{m+1} + a_m a_{m+1} + a_m^\dagger a_{m+1}^\dagger \right] + O(S^0).$$

At first sight, the structure of this Hamiltonian, albeit quadratic in the Bose operators, looks awkward. However, after applying the Fourier transformation $a_m = \frac{1}{\sqrt{N}}\sum_k e^{-ikm}a_k$, it assumes a more symmetric form:

$$\hat{H} = -NJS(S+1) + JS\sum_k \begin{pmatrix} a_k^\dagger & a_{-k} \end{pmatrix} \begin{pmatrix} 1 & \gamma_k \\ \gamma_k & 1 \end{pmatrix} \begin{pmatrix} a_k \\ a_{-k}^\dagger \end{pmatrix} + O(S^0),$$

where $\gamma_k = \cos k$. Apart from the definition of the matrix kernel between the Bose operators, $\hat{H}$ is equivalent to the Hamiltonian (2.40) discussed in connection with the charge density wave. Performing the same steps as before, the non-particle-number-conserving contributions $a^\dagger a^\dagger$ can be removed by Bogoliubov transformation. As a result, the transformed Hamiltonian assumes the diagonal form

$$\hat{H} = -NJS(S+1) + 2JS\sum_k |\sin k|(\alpha_k^\dagger \alpha_k + 1/2) \qquad (2.48)$$

Thus, in contrast with the ferromagnet, the spin-wave excitations of the antiferromagnet (fig. 2.7) display a *linear dispersion* in the limit $k \to 0$. Surprisingly, although developed in the limit of large spin, experiment shows that even for $S = 1/2$ spin chains, this linear dispersion is maintained (see fig. 2.7).

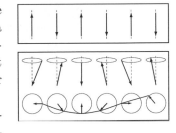

More generally, it turns out that, for chains of arbitrary half-integer spin $S = 1/2, 3/2, 5/2, \ldots$, the low-energy spectrum is linear, in agreement with the results of the harmonic approximation. In contrast, for chains of integer spin $S = 1, 2, 3\ldots$, the low-energy spectrum contains a gap, i.e., these systems do not support long-range excitations. As a rule, the sensitivity of a physical phenomenon to the characteristics of a sequence of *numbers* – such as half integer versus integer – signals the presence of a mechanism of topological origin.[24] At the same time, the formation of a gap (observed for integer chains) represents an interaction effect; at orders beyond the harmonic approximation, spin waves begin to interact nonlinearly with each other, a mechanism that may (for S integer) but need not (for S half integer) destroy the wave-like nature of the low-energy excitations. In section 8.4.6 – in a chapter devoted to a discussion

---

[24] Specifically, the topological signature of a spin field configuration will turn out to be the number of times the classical analog of a spin (a vector on the unit sphere) will wrap around the sphere in $(1 + 1)$-dimensional space-time.

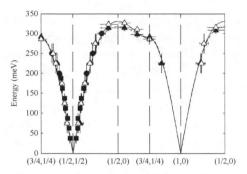

Fig. 2.7   Experimental spin-wave dispersion of the high-T_c parent compound $LaCuO_4$ – a prominent spin-$1/2$ antiferromagnet. The x-axis denotes individual trajectories between distinct points of the two-dimensional Brillouin zone, Γ: $(0,0)$, M: $(1/2, 1/2)$, and X: $(1/2, 0)$. Reprinted with permission from R. Coldea *et al.*, *Spin waves and electronic excitations in* La_2CuO_4, Phys. Rev. Lett. **86**, 5377–(2001). Copyright (2001) by the American Physical Society.

of the intriguing phenomena generated by the conspiracy of global (topological) structures with local interaction mechanisms – we will discuss these phenomena at a deeper level.

2.3 Summary and Outlook

This concludes our discussion of second quantization and some of its applications. Additional examples can be found in the problems below. In this chapter, we have introduced second quantization as a tool whereby problems of many-body quantum mechanics can be cast and addressed more efficiently than by the traditional language of symmetrized many-body wave functions. We have discussed how the two approaches are related to each other, and how the standard operations of quantum mechanics can be performed by second-quantized methods.

Beyond qualitative discussions, the list of concrete applications encountered in this chapter involved problems that either were non-interacting from the outset, or could be reduced to a quadratic operator form by a number of suitable manipulations. However, we carefully avoided dealing with interacting problems where no such reductions are possible – the vast majority of problems encountered in condensed matter physics. What can be done in situations where interactions, i.e., operator contributions of fourth or higher order, are present and no tricks such as bosonization can be performed? Generically, either interacting problems of many-body physics are fundamentally inaccessible to perturbation theory, or they necessitate perturbative analyses of *infinite* order in the interaction contribution. Situations where a satisfactory result can be obtained by first- or second-order perturbation theory are exceptional. Within second quantization, large-order perturbative expansions in interaction operators lead to complex polynomials of creation and annihilation operators. Quantum expectation values taken over such structures can be computed by a reductive algorithm, known as *Wick's theorem*. However, from a modern perspective, the formulation of perturbation theory in this way is not very efficient. More importantly, problems of particular interest are more often non-perturbative in character.

To understand the language of modern condensed matter physics, we need to develop another layer of theory, known as *field integration*. In essence, the latter is a concept generalizing the effective action approach of chapter 1 to the quantum level. However, before discussing quantum field theory, we should understand how the concept works in principle, i.e., on the level of single-particle quantum mechanics. This will be the subject of the next chapter.

2.4 Problems

2.4.1 Stone–von Neumann theorem

In the main text, we introduced creation and annihilation operators in a constructive manner, i.e., by specifying their action on a fixed Fock space state. We saw that this definition implied remarkably simple algebraic relations between the newly-introduced operators – the Heisenberg algebra (2.7). In this problem we explore the mathematical structure behind this observation. (The problem is included for the benefit of the mathematically inclined. Readers primarily interested in practical aspects of second quantization may safely skip it!)

Let us define an abstract algebra of objects a_λ and $\tilde{a}_\lambda$ by

$$[a_\lambda, \tilde{a}_\mu]_\zeta = \delta_{\lambda\mu}, \quad [a_\lambda, a_\mu]_\zeta = [\tilde{a}_\lambda, \tilde{a}_\mu]_\zeta = 0.$$

Further, let us assume that this algebra is unitarily represented in some vector space $\mathcal{F}$. This means that (i) to every a_λ and $\tilde{a}_\lambda$ we assign a linear map $T_{a_\lambda} : \mathcal{F} \to \mathcal{F}$ such that (ii) $T_{[a_\lambda, \tilde{a}_\mu]_\zeta} = [T_{a_\lambda}, T_{\tilde{a}_\mu}]_\zeta$, and (iii) $T_{\tilde{a}_\lambda} = T_{a_\lambda}^\dagger$. To keep the notation simple, we will denote T_{a_λ} by a_λ (now regarded as a linear map $\mathcal{F} \to \mathcal{F}$) and $T_{\tilde{a}_\lambda}$ by $a_\lambda^\dagger$.

Stone–von
Neumann
theorem

The **Stone–von Neumann theorem** states that the representation above is unique, i.e., that, up to unitary basis transformations, there is only one such representation. The statement is proven by explicit construction of a basis on which the operators act in a specific and well-defined way. We will see that this action is given by Eq. (2.6), i.e., the reference basis is simply the Fock space basis used in the text. This proves that the Heisenberg algebra encapsulates the full mathematical structure of the formalism of second quantization.

(a) We begin by noting that the operators $\hat{n}_\lambda \equiv a_\lambda^\dagger a_\lambda$ are hermitian and commute with each other, i.e., they are simultaneously diagonalizable. Let $|n_{\lambda_1}, n_{\lambda_2}, \ldots\rangle$ be an orthonormalized eigenbasis of the operators $\{\hat{n}_\lambda\}$, i.e., $\hat{n}_{\lambda_i}|n_{\lambda_1}, n_{\lambda_2}, \ldots\rangle = n_{\lambda_i}|n_{\lambda_1}, n_{\lambda_2}, \ldots\rangle$. Show that, up to unit-modular factors, this basis is unique. (Hint: Use the irreducibility of the transformation.) **(b)** Show that $a_{\lambda_i}|n_{\lambda_1}, n_{\lambda_2}, \ldots\rangle$ is an eigenstate of $\hat{n}_{\lambda_i}$ with eigenvalue $n_{\lambda_i} - 1$. Use this information to show that all eigenvalues n_{λ_i} are positive integers. (Hint: Note the positivity of the scalar norm.) Show that the explicit representation of the basis is given by

$$|n_{\lambda_1}, n_{\lambda_2}, \ldots\rangle = \prod_i \frac{(a_{\lambda_i}^\dagger)^{n_{\lambda_i}}}{\sqrt{n_{\lambda_i}!}} |0\rangle, \qquad (2.49)$$

where $|0\rangle$ is the unique state which has eigenvalue 0 for all $\hat{n}_i$. Comparison with Eq. (2.4) shows that the basis constructed above indeed coincides with the Fock space basis considered in the text.

Answer:

(a) Suppose we have identified two bases $\{|n_{\lambda_1}, n_{\lambda_2}, \ldots\rangle\}$ and $\{|n_{\lambda_1}, n_{\lambda_2}, \ldots\rangle'\}$ on which all operators $\hat{n}_i$ assume equal eigenvalues. The irreducibility of the representation implies the existence of a polynomial $P(\{a_{\mu_i}, a_{\mu_i}^\dagger\})$ such that $|n_{\lambda_1}, n_{\lambda_2}, \ldots\rangle = P(\{a_{\mu_i}, a_{\mu_i}^\dagger\})|n_{\lambda_1}, n_{\lambda_2}, \ldots\rangle'$. Now, the action of P must not change any of the eigenvalues of $\hat{n}_i$, which means that P contains the operators a_μ and $a_\mu^\dagger$ in equal numbers. Reordering operators, we may thus bring P into the form $P(\{a_{\mu_i}, a_{\mu_i}^\dagger\}) = \tilde{P}(\{\hat{n}_{\mu_i}\})$. However, the action of this latter expression on $|n_{\lambda_1}, n_{\lambda_2}, \ldots\rangle'$ just produces a number, i.e., the bases are equivalent.

(b) For a given state $|n\rangle$ (concentrating on a fixed element of the single-particle basis, we suppress the subscript λ_i throughout), let us choose an integer q such that $\hat{n} a^{q-1}|n\rangle = (n - q + 1)a^{q-1}|n\rangle$ with $n - q + 1 > 0$ while $n - q \leq 0$. We then obtain

$$0 \geq (n - q)\langle n|(a^\dagger)^q a^q|n\rangle = \langle n|(a^\dagger)^q \hat{n} a^q|n\rangle = \langle n|(a^\dagger)^{q+1} a^{q+1}|n\rangle \geq 0.$$

The only way to satisfy these inequalities is to require that $\langle n|(a^\dagger)^{q+1} a^{q+1}|n\rangle = 0$ and $n - q = 0$. The last equation implies the integer-valuedness of n. (In principle, we ought to prove that a zero-eigenvalue state $|0\rangle$ exists. To show this, take any reference state $|n_{\lambda_1}, n_{\lambda_2}, \ldots\rangle$ and apply operators a_{λ_i} until all eigenvalues n_{λ_i} are lowered to zero.) Using the commutation relations, it is then straightforward to verify that the right hand side of Eq. (2.49) is (a) unit-normalized and (b) has eigenvalue n_{λ_i} for each $\hat{n}_{\lambda_i}$.

2.4.2 Semiclassical spin waves

In chapter 1, the development of a theory of lattice vibrations in the harmonic atom chain was motivated by the quantization of the continuum classical theory. The latter provided insight into the nature of the elementary collective excitations. Here we will employ the semiclassical theory of spin dynamics to explore the nature of elementary spin-wave excitations.

(a) Making use of the spin commutation relation, $[\hat{S}_i^\alpha, \hat{S}_j^\beta] = i\delta_{ij}\epsilon^{\alpha\beta\gamma}\hat{S}_i^\gamma$, apply the operator identity $i\dot{\hat{S}}_i = [\hat{S}_i, \hat{H}]$ to express the equation of motion of a spin in a nearest-neighbor spin-S one-dimensional Heisenberg ferromagnet as a difference equation. **(b)** Interpreting the spins as classical vectors, and taking the continuum limit, show that the equation of motion takes the form $\dot{\mathbf{S}} = J\mathbf{S} \times \partial^2\mathbf{S}$, where we

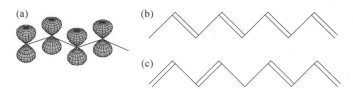

Fig. 2.8   (a) An sp^2-hybridized polymer chain. (b) One of the configurations of the Peierls distorted
chain. The double bonds represent the short links of the lattice. (c) A topological defect
separating two domains of the ordered phase.

have assumed a unit lattice spacing. Find and sketch a wave-like solution describing
small-angle precession around a globally magnetized state $\mathbf{S}_i = S\mathbf{e}_z$.

Answer:

(a) Making use of the equation of motion, and the commutation relation, substitution of the Heisenberg ferromagnetic Hamiltonian gives the difference equation
$\dot{\hat{\mathbf{S}}}_i = J\hat{\mathbf{S}}_i \times (\hat{\mathbf{S}}_{i+1} + \hat{\mathbf{S}}_{i-1})$. (b) Interpreting the spins as classical vectors, and
applying the Taylor expansion $\mathbf{S}_{i+1} \mapsto \mathbf{S}(x+1) = \mathbf{S} + \partial\mathbf{S} + \partial^2\mathbf{S}/2 + \cdots$, one obtains the classical equation of motion shown. Making the ansatz $\mathbf{S} = (c\cos(kx -
\omega t), c\sin(kx - \omega t), \sqrt{S^2 - c^2})$ one may confirm that the equation of motion is sat-
isfied if $\omega = Jk^2\sqrt{S^2 - c^2}$.

2.4.3 Su–Shrieffer–Heeger model of a conducting polymer chain

Polyacetylene consists of bonded CH groups forming an isomeric long-chain polymer. According
to molecular orbital theory, the carbon atoms are sp^2-hybridized, suggesting a planar configuration of the molecule. An unpaired electron is expected to occupy a single π-orbital that is
oriented perpendicular to the plane. The weak overlap of the π-orbitals delocalizes the electrons into a narrow conduction band. According to the nearly-free electron theory, one might
expect the half-filled conduction band of a polyacetylene chain to be metallic. However, the
energy of a half-filled band of a one-dimensional system can always be lowered by imposing a
Peierls
instability
periodic lattice distortion known as a **Peierls instability** (see fig. 2.8). The aim of this problem
is to explore this instability.

(a) At its simplest level, the conduction band of polyacetylene can be modeled as
a simple (arguably over-simplified) microscopic Hamiltonian, due to Su, Shrieffer
and Heeger,[25] in which the hopping matrix elements of the electrons are modulated
by the lattice distortion of the atoms. By taking the displacement of the atomic
sites to be u_n, and treating their dynamics as classical, the effective Hamiltonian
assumes the form

$$\hat{H} = -t \sum_{n=1,\sigma}^{N} (1 + u_n) \left[c_{n\sigma}^\dagger c_{n+1\sigma} + \text{h.c.} \right] + \sum_{n=1}^{N} \frac{k_s}{2} (u_{n+1} - u_n)^2,$$

[25] W. P. Su, J. R. Schrieffer and A. J. Heeger, *Solitons in polyacetylene*, Phys. Rev. Lett. **42**,
1698 (1979).

where, for simplicity, the boundary conditions are taken to be periodic. The first term describes the hopping of electrons between neighboring sites in terms of a matrix element modulated by the periodic distortion of the bond-length, while the second term represents the associated increase in the elastic energy. Taking the lattice distortion to be periodic, $u_n = (-1)^n \alpha$, and the number of sites to be even, bring the Hamiltonian to diagonal form. (Hint: Note that the lattice distortion lowers the symmetry of the lattice. The Hamiltonian is most easily diagonalized by distinguishing the two sites of the sublattice – i.e., doubling the size of the elementary unit cell.) Show that the Peierls distortion of the lattice opens a gap in the spectrum at the Fermi level of the half-filled system.

(b) By estimating the total electronic and elastic energy of the half-filled band (i.e., it has an average of one electron per lattice site), show that the one-dimensional system is always unstable towards the Peierls distortion. To complete this calculation, you will need the approximate formula $\int_{-\pi/2}^{\pi/2} dk \left(1 - \left(1 - \alpha^2\right) \sin^2 k\right)^{1/2} \simeq 2 + (a_1 - b_1 \ln \alpha^2)\alpha^2 + O(\alpha^2 \ln \alpha^2)$, where a_1 and b_1 are (unspecified) numerical constants.

(c) For an even number of sites, the Peierls instability has two degenerate configurations (for one of these, see fig. 2.8 (a)), ABABAB... and BABABA... Comment on the qualitative form of the ground state lattice configuration if the number of sites is odd (see fig. 2.8 (b)). Explain why such configurations give rise to mid-gap states.

Answer:

(a) Since each unit cell has twice the dimension of the original lattice, we begin by recasting the Hamiltonian in a sublattice form,

$$\hat{H} = -t \sum_{m=1,\sigma}^{N/2} \left\{ (1+\alpha) \left[a_{m\sigma}^\dagger b_{m\sigma} + \text{h.c.} \right] + (1-\alpha) \left[b_{m\sigma}^\dagger a_{m+1\sigma} + \text{h.c.} \right] \right\} + 2Nk_s \alpha^2,$$

where the creation operators $a_m^\dagger$ and $b_m^\dagger$ act on the two sites of the elemental unit cell of the distorted lattice. Switching to the Fourier basis, $a_m = \sqrt{2/N} \sum_k e^{2ikm} a_k$ (similarly for b_m), where k takes $N/2$ values uniformly on the interval $[-\pi/2, \pi/2]$ and the lattice spacing of the undistorted system is taken to be unity, the Hamiltonian takes the form

$$\hat{H} = 2Nk_s \alpha^2$$
$$- t \sum_k \begin{pmatrix} a_{k\sigma}^\dagger & b_{k\sigma}^\dagger \end{pmatrix} \begin{pmatrix} 0 & (1+\alpha) + (1-\alpha)e^{2ik} \\ (1+\alpha) + (1-\alpha)e^{-2ik} & 0 \end{pmatrix} \begin{pmatrix} a_{k\sigma} \\ b_{k\sigma} \end{pmatrix}.$$

Diagonalizing the 2×2 matrix, one obtains $\epsilon(k) = \pm 2t \left[1 + (\alpha^2 - 1) \sin^2 k\right]^{1/2}$. In the limit $\alpha \to 0$, one recovers the cosine spectrum characteristic of the undistorted tight-binding problem, while, when $\alpha \to 1$, pairs of monomers become decoupled and we obtain a massively degenerate bonding and antibonding spectrum.

(b) According to the formula given, the total shift in energy is given by $\delta\epsilon = -4t(a_1 - b_1 \ln \alpha^2)\alpha^2 + 2k_s\alpha^2$. Maximizing the energy gain with respect to α, a stable configuration is found when $\alpha^2 = \exp\left((a_1/b_1) - 1 - k_s/(2tb_1)\right)$.

(c) If the number of sites is odd, the Peierls distortion is inevitably frustrated. The result is that the polymer chain must accommodate a *topological excitation*. The excitation is said to be topological since the defect cannot be removed by a smooth continuous deformation. Its effect on the spectrum of the model is to introduce a state that lies within the band gap of the material. (We will return to the discussion of the topology of this system later, in section 8.1.1.) The consideration of an odd number of sites forces a topological defect into the system. However, even if the number of sites is even, one can create low-energy topological excitations of the system either by doping (see fig. 2.8(b)) or by the creation of **excitons**, particle–hole excitations. Indeed, such topological excitations can dominate the transport properties of the system.

excitons

2.4.4 Schwinger boson representation

As with the Holstein–Primakoff representation, the Schwinger boson provides another representation of quantum spin. The aim here is to confirm the validity of this representation. For practical purposes, the value of the particular representation depends on its application.

Schwinger
boson

In the **Schwinger boson representation**, the quantum mechanical spin is expressed in terms of two bosonic operators a and b:

$$\hat{S}^+ = a^\dagger b, \qquad \hat{S}^- = (\hat{S}^+)^\dagger, \qquad \hat{S}^z = \frac{1}{2}\left(a^\dagger a - b^\dagger b\right).$$

(a) Show that this definition is consistent with the commutation relations for spin: $[\hat{S}^+, \hat{S}^-] = 2\hat{S}^z$. **(b)** Using the bosonic commutation relations, show that

$$|S, m\rangle = \frac{(a^\dagger)^{S+m}}{\sqrt{(S+m)!}} \frac{(b^\dagger)^{S-m}}{\sqrt{(S-m)!}} |\Omega\rangle$$

is compatible with the definition of an eigenstate of the total spin operator $\mathbf{S}^2$ and S^z. Here $|\Omega\rangle$ denotes the vacuum of the Schwinger bosons, and the total spin S defines the physical subspace $\{|n_a, n_b\rangle | n_a + n_b = 2S\}$.

Answer:

(a) Using the commutation relation for bosons, one finds $[\hat{S}^+, \hat{S}^-] = a^\dagger b \, b^\dagger a - b^\dagger a \, a^\dagger b = a^\dagger a - b^\dagger b = 2\hat{S}^z$, as required. **(b)** Using the identity $\hat{\mathbf{S}}^2 = (\hat{S}^z)^2 + \frac{1}{2}(\hat{S}^+ \hat{S}^- + \hat{S}^- \hat{S}^+) = \frac{1}{4}(\hat{n}_a - \hat{n}_b)^2 + \hat{n}_a \hat{n}_b + \frac{1}{2}(\hat{n}_a + \hat{n}_b)$ one finds that $\hat{\mathbf{S}}^2 |S, m\rangle = [m^2 + (S+m)(S-m) + S]|S, m\rangle = S(S+1)|S, m\rangle$, as required. Similarly, one finds $\hat{S}^z |S, m\rangle = \frac{1}{2}(n_a - n_b)|n_a = S+m, n_b = S-m\rangle = m|S, m\rangle$, showing $|S, m\rangle$ to be an eigenstate of the operator $\hat{S}^z$ with eigenvalue m.

2.4.5 Jordan–Wigner transformation

So far we have shown how the algebra of quantum mechanical spin can be expressed using boson operators – cf. the Holstein–Primakoff transformation and the Schwinger boson representation. Here, we show that a representation for spin $1/2$ can be obtained in terms of fermion operators.

In the **Jordan–Wigner transformation**, spin-$1/2$ degrees of freedom are represented in terms of a single structureless fermion. Consider an up spin as a particle and a down spin as the vacuum, i.e., $|\uparrow\rangle \equiv |1\rangle = f^\dagger|0\rangle, |\downarrow\rangle \equiv |0\rangle = f|1\rangle$. In this representation the spin raising and lowering operators are expressed in the form $\hat{S}^+ = f^\dagger$ and $\hat{S}^- = f$, while $\hat{S}^z = f^\dagger f - 1/2$.

(a) With this definition, confirm that the spins obey the algebra $[\hat{S}^+, \hat{S}^-] = 2\hat{S}^z$. However, there is a problem: spin operators on different sites commute, while fermion operators anticommute, e.g., $S_i^+ S_j^+ = S_j^+ S_i^+$, but $f_i^\dagger f_j^\dagger = -f_j^\dagger f_i^\dagger$. To obtain a faithful spin representation, it is necessary to cancel this unwanted sign. Although a general procedure is hard to formulate, in one dimension this can be achieved by a nonlinear transformation,

$$\hat{S}_l^+ = f_l^\dagger \exp\left(i\pi \sum_{j<l} \hat{n}_j\right), \qquad \hat{S}_l^- = \exp\left(-i\pi \sum_{j<l} \hat{n}_j\right) f_l, \qquad \hat{S}_l^z = f_l^\dagger f_l - \frac{1}{2}.$$

Operationally, this seemingly complicated transformation is straightforward: in one dimension, the particles can be ordered on the line. By counting the number of particles "to the left," we assign an overall sign of $+1$ or -1 to a given configuration and thereby "transmute" the particles into fermions. (Put differently, the exchange of two fermions induces a sign change that is compensated
by a factor arising from the phase – the **Jordan–Wigner string**.) (b) Using the Jordan–Wigner representation, show that $\hat{S}_m^+ \hat{S}_{m+1}^- = f_m^\dagger f_{m+1}$. (c) For the spin-$1/2$ anisotropic quantum Heisenberg spin chain, the spin Hamiltonian assumes the form $\hat{H} = -\sum_n (J_z \hat{S}_n^z \hat{S}_{n+1}^z + \frac{J_\perp}{2}(\hat{S}_n^+ \hat{S}_{n+1}^- + \hat{S}_n^- \hat{S}_{n+1}^+))$. Turning to the Jordan–Wigner representation, show that the Hamiltonian can be cast in the form

$$\hat{H} = -\sum_n \left(\frac{J_\perp}{2}\left(f_n^\dagger f_{n+1} + \text{h.c.}\right) + J_z \left(\frac{1}{4} - f_n^\dagger f_n + f_n^\dagger f_n f_{n+1}^\dagger f_{n+1}\right)\right).$$

(d) The construction shows the equivalence between the one-dimensional spin-$1/2$ XY-model, defined as the spin chain with one coupling absent, $J_z = 0$, and a non-interacting theory of spinless fermions. In this case, show that the spectrum assumes the form $\epsilon(k) = -J_\perp \cos ka$.

Answer:

(a) From the fermionic anticommutation relations, $[\hat{S}^+, \hat{S}^-]_- = [f^\dagger, f]_- = f^\dagger f - ff^\dagger = 2f^\dagger f - 1 = 2\hat{S}^z$. (b) Using the commutativity of number operators on different sites, we obtain $\hat{S}_m^+ \hat{S}_{m+1}^- = f_m^\dagger \exp(i\pi \sum_{j<m} n_j) \exp(-i\pi \sum_{l<m+1} n_l) f_{m+1} =$

$f_m^\dagger e^{-i\pi n_m} f_{m+1} = f_m^\dagger f_{m+1}$ where we have used the relation $f_m^\dagger e^{-i\pi n_m} = f_m^\dagger$. **(c)** The fermion representation is simply obtained by substitution of the above relations into the spin Hamiltonian. **(d)** With $J_z = 0$, the spin Hamiltonian assumes the form of a non-interacting tight-binding Hamiltonian $\hat{H} = \frac{J_\perp}{2} \sum_n (f_n^\dagger f_{n+1} + \text{h.c.})$. This Hamiltonian, which has been encountered previously, is diagonalized in Fourier space, after which one obtains the cosine band dispersion.

2.4.6 Spin–charge separation in one-dimension

In section 2.2.4, a free theory of interacting spinless fermions was developed in one dimension making use of the bosonization formalism. This analysis showed that the low-energy degrees of freedom were described by hydrodynamic charge (i.e., density) fluctuations propagating with a linear dispersion. However, as well as charge, the electron degrees of freedom carry spin. The aim of this problem is to explore the fate of the spin degrees of freedom in a one-dimensional environment.

As a first step, we introduce operators (cf. Eq. (2.34)) $\hat{\rho}_{sq\alpha} = \sum_k a_{s(k+q)\alpha}^\dagger a_{sk\alpha}$, $\alpha = \uparrow, \downarrow$, generalizing the previously introduced density operators for the presence of spin. Similarly, the bosonic degrees of freedom of the theory (cf. Eq. (2.39)) now carry a spin index, so that $b_q \to b_{q\alpha}$. One aspect that makes the problem more difficult to tackle than the previously explored spinless case is that the $2k_F$-momentum transfer interaction $|k_F + q + q_1, \uparrow; k_F + q - q_1, \downarrow\rangle \to |-k_F + q + q_2, \uparrow; -k_F + q - q_2, \downarrow\rangle$, in which a right-moving spin-up electron is scattered to a left-moving spin-up electron, cannot be expressed in terms of slowly fluctuating density operators. (If you don't believe this, try!) However, using the renormalization group methods to be introduced in chapter 6, it can be shown that this type of interaction is largely irrelevant physically and can be neglected from the outset.

Concentrating on the low-momentum-transfer interaction, the effective bosonic Hamiltonian assumes the form (verify)

$$\hat{H} = \sum_{q>0,s,\alpha} v_F q b_{sq\alpha}^\dagger b_{sq\alpha} + \sum_{q>0,s,\alpha\alpha'} |q| \left[\frac{g_2}{2\pi} \left(b_{sq\alpha}^\dagger b_{\bar{s}q\alpha'}^\dagger + \text{h.c.} \right) + \frac{g_4}{2\pi} b_{sq\alpha}^\dagger b_{sq\alpha'} \right].$$

Introducing operators that create charge (ρ) and spin (σ) fluctuations, $b_{sq\rho} = \frac{1}{\sqrt{2}}(b_{sq\uparrow} + b_{sq\downarrow})$, $b_{sq\sigma} = \frac{1}{\sqrt{2}}(b_{sq\uparrow} - b_{sq\downarrow})$, rearrange the Hamiltonian, and thereby show that it assumes a diagonal form with the spin and charge degrees of freedom exhibiting different velocities. This is a manifestation of **spin–charge separation**: even without the introduction of spin-dependent forces, the spin and charge degrees of freedom of the electron in the metallic conductor separate and propagate at different velocities. In this sense, there is no way to adiabatically continue from non-interacting electrons to the collective charge and spin excitations of the system.

Answer:

Motivated by the separation into spin and charge degrees of freedom, a rearrangement of the Hamiltonian gives

$$\hat{H} = \sum_{q>0,s} \left\{ v_F q \left(b^\dagger_{sq\rho} b_{sq\rho} + b^\dagger_{sq\sigma} b_{sq\sigma} \right) + |q| \left[\frac{g_2}{\pi} \left(b^\dagger_{sq\rho} b^\dagger_{\bar{s}q\rho} + \text{h.c.} \right) + \frac{g_4}{\pi} b^\dagger_{sq\rho} b_{sq\rho} \right] \right\}.$$

Applying a Bogoliubov transformation, the Hamiltonian is brought to the diagonal form

$$\hat{H} = \sum_{q>0,s} \left[|q| \sqrt{(v_F + g_4/\pi)^2 - (g_2/\pi)^2} \, \alpha^\dagger_{sq\rho} \alpha_{sq\rho} + |q| v_F \alpha^\dagger_{sq\sigma} \alpha_{sq\sigma} \right] + \text{const.}$$

2.4.7 The Kondo problem

Historically, the Kondo problem has assumed great significance in the field of strongly correlated quantum systems. It represents perhaps the simplest example of a phenomenon driven by strong electron interaction and, unusually for this arena of physics, admits a detailed theoretical understanding. Further, in respect of the principles established in chapter 1, it exemplifies a number of important ideas from the concept of reducibility – the collective properties of the system may be captured by a simplified effective Hamiltonian that includes only the relevant low-energy degrees of freedom – and the renormalization group. In the following problem, we will seek to develop the low-energy theory of the "Kondo impurity" system, leaving the discussion of its phenomenology to problems 4.6.3 and 6.7.3 in subsequent chapters.

The Kondo effect is rooted in the observation that, when small amounts of magnetic ion impurities are embedded in a metallic host (such as manganese in copper or iron in CuAu alloys), a pronounced minimum develops in the temperature dependence of the resistivity. Although the phenomenon was discovered experimentally in 1934,[26] it was not until 1964 that it was understood by Kondo.[27] Historically, the first step towards the solution of the problem came with a suggestion by Anderson that the system could be modeled as an itinerant band of electron states interacting with dilute magnetic moments associated with the ion impurities.[28] Anderson proposed that the integrity of the local moment was protected by a large local Coulomb repulsion which inhibited multiple occupancy of the orbital state – a relative of the Hubbard U-interaction. Such a system is described by the **Anderson impurity Hamiltonian**,

Anderson
impurity
Hamiltonian

$$\hat{H} = \sum_{\mathbf{k}\sigma} \left(\epsilon_{\mathbf{k}} c^\dagger_{\mathbf{k}\sigma} c_{\mathbf{k}\sigma} + \left(V_{\mathbf{k}} d^\dagger_\sigma c_{\mathbf{k}\sigma} + \text{h.c.} \right) \right) + \sum_\sigma \epsilon_d n_{d\sigma} + U n_{d\uparrow} n_{d\downarrow},$$

where the operators $c^\dagger_{\mathbf{k}\sigma}$ create an itinerant electron of spin σ and energy $\epsilon_{\mathbf{k}}$ in the metallic host while $d^\dagger_\sigma$ creates an electron of spin σ on the local impurity at position $\mathbf{d}$. Here, $V_{\mathbf{k}} = e^{i\mathbf{k}\cdot\mathbf{d}} \langle \phi_d | \hat{H} | \psi_{\mathbf{d}} \rangle$ denotes the coupling between these states, where ϕ_d is the atomic d level and $\psi_{\mathbf{d}}$ is the Wannier state of the conduction electrons at site

[26] W.J. de Haas *et al.*, *The electrical resistance of gold, copper, and lead at low temperatures*, Physica **1**, 1115 (1934).

[27] J. Kondo, *Resistance minimum in dilute magnetic alloys*, Prog. Theor. Phys. **32**, 37 (1964).

[28] P. W. Anderson, *Localized magnetic states in metals*, Phys. Rev. **124**, 41 (1961).

d. Here, we have used $n_{d\sigma} = d_\sigma^\dagger d_\sigma$ to denote the number operator. While electrons in the band are described as free fermions, those associated with the impurity state experience an on-site Coulomb interaction of strength U. In the Kondo regime, the Fermi level ϵ_F lies between the impurity level ϵ_d and $\epsilon_d + U$, so that the average impurity site occupancy is unity. Nevertheless, the coupling of the impurity to the itinerant electron states admits virtual processes in which the site occupancy can fluctuate between zero and two. These virtual fluctuations allow the spin on the impurity site to flip through exchange.

On the basis of our discussion of the half-filled Hubbard model in section 2.2.2 it makes sense to transform the Anderson impurity Hamiltonian into an effective low-energy theory. To this end, we can express the total wave function of the Hamiltonian $|\psi\rangle$ as the sum of terms $|\psi_0\rangle$, $|\psi_1\rangle$, and $|\psi_2\rangle$, where the subscript denotes the occupancy of the impurity site. With this decomposition, the Schrödinger equation for the Hamiltonian can be cast in matrix form, $\sum_{n=0}^{2} \hat{H}_{mn}|\psi_n\rangle = E|\psi_m\rangle$, where $\hat{H}_{mn} = \hat{P}_m \hat{H} \hat{P}_n$ and the operators $\hat{P}_m$ project onto the subspace with m electrons on the impurity (i.e., $\hat{P}_0 = \prod_\sigma (1 - n_{d\sigma})$, $\hat{P}_1 = (1 + n_{d\uparrow} + n_{d\downarrow} - 2n_{d\uparrow}n_{d\downarrow})$ and $\hat{P}_2 = n_{d\uparrow}n_{d\downarrow})$).

(a) Construct the operators $\hat{H}_{mn}$ explicitly and explain why $\hat{H}_{20} = \hat{H}_{02} = 0$. Since we are interested in the effect of virtual excitations from the $|\psi_1\rangle$ subspace, we may proceed by formally eliminating $|\psi_0\rangle$ and $|\psi_2\rangle$ from the Schrödinger equation. By doing so, show that the equation for $|\psi_1\rangle$ can be written as

$$\left[\hat{H}_{10} \frac{1}{E - \hat{H}_{00}} \hat{H}_{01} + \hat{H}_{11} + \hat{H}_{12} \frac{1}{E - \hat{H}_{22}} \hat{H}_{21} \right] |\psi_1\rangle = E|\psi_1\rangle.$$

(b) At this stage, the equation for $|\psi_1\rangle$ is exact. Show that an expansion to leading order in $1/U$ and $1/\epsilon_d$ leads to the expression

$$\hat{H}_{12} \frac{1}{E - \hat{H}_{22}} \hat{H}_{21} + \hat{H}_{10} \frac{1}{E - \hat{H}_{00}} \hat{H}_{01}$$

$$\simeq - \sum_{\mathbf{k}\mathbf{k}'\sigma\sigma'} V_{\mathbf{k}} V_{\mathbf{k}'}^* \left(\frac{c_{\mathbf{k}\sigma}^\dagger c_{\mathbf{k}'\sigma'} d_\sigma d_{\sigma'}^\dagger}{U + \epsilon_d - \epsilon_{\mathbf{k}'}} + \frac{c_{\mathbf{k}'\sigma'} c_{\mathbf{k}\sigma}^\dagger d_{\sigma'}^\dagger d_\sigma}{\epsilon_{\mathbf{k}} - \epsilon_d} \right).$$

To obtain the first term in the expression, consider the commutation of $(E - \hat{H}_{22})^{-1}$ with $\hat{H}_{21}$ and make use of the fact that the total operator acts upon the singly occupied subspace. A similar line of reasoning will lead to the second term in the expression. Here, $U + \epsilon_d - \epsilon_{\mathbf{k}'}$ and $\epsilon_d - \epsilon_{\mathbf{k}}$ denote the respective excitation energies of the virtual states.

Making use of the Pauli matrix identity, $\boldsymbol{\sigma}_{\alpha\beta} \boldsymbol{\sigma}_{\gamma\delta} = 2\delta_{\alpha\delta}\delta_{\beta\gamma} - \delta_{\alpha\beta}\delta_{\gamma\delta}$, it follows that (exercise)

$$\sum_{\sigma\sigma'} c_{\mathbf{k}\sigma}^\dagger c_{\mathbf{k}'\sigma'} d_{\sigma'}^\dagger d_\sigma = 2\hat{\mathbf{s}}_{\mathbf{k}\mathbf{k}'} \cdot \hat{\mathbf{S}}_d + \frac{1}{2} \sum_{\sigma\sigma'} c_{\mathbf{k}\sigma}^\dagger c_{\mathbf{k}'\sigma} n_{d\sigma'},$$

where $\hat{\mathbf{S}}_d = \sum_{\alpha\beta} d_\alpha^\dagger \boldsymbol{\sigma}_{\alpha\beta} d_\beta / 2$ denotes the spin-1/2 degree of freedom associated with the impurity and $\hat{\mathbf{s}}_{\mathbf{k}\mathbf{k}'} = \sum_{\alpha\beta} c_{\mathbf{k}\alpha}^\dagger \boldsymbol{\sigma}_{\alpha\beta} c_{\mathbf{k}'\beta} / 2$. Combining this result with that obtained above, up to an irrelevant constant the total effective Hamiltonian (including $\hat{H}_{11}$) acting in the projected subspace $|\psi_1\rangle$ is given by

$$\hat{H}_{\text{sd}} = \sum_{\mathbf{k}\sigma} \epsilon_{\mathbf{k}} c_{\mathbf{k}\sigma}^{\dagger} c_{\mathbf{k}\sigma} + \sum_{\mathbf{k}\mathbf{k}'} \left[2 J_{\mathbf{k},\mathbf{k}'} \, \hat{\mathbf{s}}_{\mathbf{k}\mathbf{k}'} \cdot \hat{\mathbf{S}}_d + K_{\mathbf{k},\mathbf{k}'} \sum_{\sigma} c_{\mathbf{k}\sigma}^{\dagger} c_{\mathbf{k}'\sigma} \right],$$

where

$$J_{\mathbf{k},\mathbf{k}'} = V_{\mathbf{k}'}^{*} V_{\mathbf{k}} \left(\frac{1}{U + \epsilon_d - \epsilon_{\mathbf{k}'}} + \frac{1}{\epsilon_{\mathbf{k}} - \epsilon_d} \right),$$

$$K_{\mathbf{k},\mathbf{k}'} = \frac{V_{\mathbf{k}'}^{*} V_{\mathbf{k}}}{2} \left(\frac{1}{\epsilon_{\mathbf{k}} - \epsilon_d} - \frac{1}{U + \epsilon_d - \epsilon_{\mathbf{k}'}} \right).$$

With both $U + \epsilon_d$ and ϵ_d greatly in excess of the typical excitation energy scales, one may safely neglect the particular energy dependence of the parameters $J_{\mathbf{k},\mathbf{k}'}$ and $K_{\mathbf{k},\mathbf{k}'}$. In this case, the exchange interaction $J_{\mathbf{k},\mathbf{k}'}$ can be treated as local, the scattering term $K_{\mathbf{k},\mathbf{k}'}$ can be absorbed into a shift of the single-particle energy of the itinerant band, and the positive (i.e., antiferromagnetic) exchange coupling can be accommodated through the effective **sd-Hamiltonian**

sd-
Hamiltonian

$$\boxed{\hat{H}_{\text{sd}} = \sum_{\mathbf{k}\sigma} \epsilon_{\mathbf{k}} c_{\mathbf{k}\sigma}^{\dagger} c_{\mathbf{k}\sigma} + 2 J \hat{\mathbf{S}}_d \cdot \hat{\mathbf{s}}(\mathbf{0})} \tag{2.50}$$

where $\hat{\mathbf{s}}(\mathbf{0}) = \sum_{\mathbf{k}\mathbf{k}'\sigma\sigma'} c_{\mathbf{k}\sigma}^{\dagger} \boldsymbol{\sigma}_{\sigma\sigma'} c_{\mathbf{k}'\sigma'}/2$ denotes the local spin density of the itinerant electron band at the impurity site, $\mathbf{d} = \mathbf{0}$. To understand how the magnetic impurity affects the low-temperature transport, we refer to problem 4.6.3, where the sd–Hamiltonian is explored in the framework of a diagrammatic perturbation theory in the spin interaction.

Answer:

(a) Since the diagonal elements $\hat{H}_{mm}$ leave the occupation number fixed, they may be identified with the diagonal elements of the Hamiltonian,

$$\hat{H}_{00} = \sum_{\mathbf{k}} \epsilon_{\mathbf{k}} c_{\mathbf{k}\sigma}^{\dagger} c_{\mathbf{k}\sigma}, \quad \hat{H}_{11} = \sum_{\mathbf{k}} \epsilon_{\mathbf{k}} c_{\mathbf{k}\sigma}^{\dagger} c_{\mathbf{k}\sigma} + \epsilon_d, \quad \hat{H}_{22} = \sum_{\mathbf{k}} \epsilon_{\mathbf{k}} c_{\mathbf{k}\sigma}^{\dagger} c_{\mathbf{k}\sigma} + 2\epsilon_d + U.$$

The off-diagonal terms arise from the hybridization between the free electron states and the impurity. Since the coupling involves only the transfer of single electrons, $\hat{H}_{02} = \hat{H}_{20} = 0$ and

$$\hat{H}_{10} = \sum_{\mathbf{k}\sigma} V_{\mathbf{k}} d_{\sigma}^{\dagger} (1 - n_{d\bar{\sigma}}) c_{\mathbf{k}\sigma}, \qquad \hat{H}_{12} = \sum_{\mathbf{k}\sigma} V_{\mathbf{k}} d_{\sigma}^{\dagger} n_{d\bar{\sigma}} c_{\mathbf{k}\sigma},$$

where $\bar{\sigma} = \uparrow$ for $\sigma = \downarrow$ and vice versa, $\hat{H}_{01} = \hat{H}_{10}^{\dagger}$ and $\hat{H}_{12} = \hat{H}_{21}^{\dagger}$. Since $\hat{H}_{00}|\psi_0\rangle + \hat{H}_{01}|\psi_1\rangle = E|\psi_0\rangle$, one may set $|\psi_0\rangle = (E - \hat{H}_{00})^{-1}\hat{H}_{01}|\psi_1\rangle$ and, similarly, $|\psi_2\rangle = (E - \hat{H}_{22})^{-1}\hat{H}_{21}|\psi_1\rangle$. Then, substituting into the equation for $|\psi_1\rangle$, one obtains the given expression.

(b) Making use of the expressions from part (a), we obtain

$$\hat{H}_{12} \frac{1}{E - \hat{H}_{22}} \hat{H}_{21} = \sum_{\mathbf{k}\mathbf{k}'\sigma\sigma'} V_{\mathbf{k}} V_{\mathbf{k}'}^{*} c_{\mathbf{k}\sigma}^{\dagger} n_{d\bar{\sigma}} d_{\sigma} \frac{1}{E - \hat{H}_{22}} d_{\sigma'}^{\dagger} n_{d\bar{\sigma}'} c_{\mathbf{k}'\sigma'}$$

$$\hat{H}_{10}\frac{1}{E-\hat{H}_{00}}\hat{H}_{01} = \sum_{\mathbf{k}\mathbf{k}'\sigma\sigma'} V_{\mathbf{k}}V_{\mathbf{k}'}^* \, d_\sigma^\dagger(1-n_{d\bar{\sigma}})c_{\mathbf{k}\sigma}\frac{1}{E-\hat{H}_{00}}c_{\mathbf{k}'\sigma'}^\dagger(1-n_{d\bar{\sigma}'})d_{\sigma'}.$$

Then, substituting for $\hat{H}_{22}$ and $\hat{H}_{00}$ from (a), and commuting operators, we have

$$\frac{1}{E-\hat{H}_{22}}d_{\sigma'}^\dagger n_{d\bar{\sigma}'}c_{\mathbf{k}'\sigma'} = -\frac{d_{\sigma'}^\dagger n_{d\bar{\sigma}'}c_{\mathbf{k}'\sigma'}}{U+\epsilon_d-\epsilon_{\mathbf{k}'}}\left[1-\frac{E-\epsilon_d-\hat{H}_{00}}{U+\epsilon_d-\epsilon_{\mathbf{k}'}}\right]^{-1},$$

$$\frac{1}{E-\hat{H}_{00}}c_{\mathbf{k}'\sigma'}^\dagger(1-n_{d\bar{\sigma}'})d_{\sigma'} = -\frac{c_{\mathbf{k}'\sigma'}^\dagger(1-n_{d\bar{\sigma}'})d_{\sigma'}}{\epsilon_{\mathbf{k}'}-\epsilon_d}\left[1-\frac{E-\epsilon_d-\hat{H}_{00}}{\epsilon_{\mathbf{k}'}-\epsilon_d}\right]^{-1}.$$

Expanding in large U and ϵ_d, to leading order we obtain

$$\hat{H}_{12}\frac{1}{E-\hat{H}_{22}}\hat{H}_{21} + \hat{H}_{10}\frac{1}{E-\hat{H}_{00}}\hat{H}_{01}$$

$$\simeq -\sum_{\mathbf{k}\mathbf{k}'\sigma\sigma'} V_{\mathbf{k}}V_{\mathbf{k}'}^* \left(\frac{c_{\mathbf{k}\sigma}^\dagger n_{d,\bar{\sigma}}d_\sigma d_{\sigma'}^\dagger n_{d\bar{\sigma}'}c_{\mathbf{k}'\sigma'}}{U+\epsilon_d-\epsilon_{\mathbf{k}'}} + \frac{d_\sigma^\dagger(1-n_{d,\bar{\sigma}})c_{\mathbf{k}\sigma}c_{\mathbf{k}'\sigma'}^\dagger(1-n_{d\bar{\sigma}'})d_{\sigma'}}{\epsilon_{\mathbf{k}'}-\epsilon_d}\right).$$

Finally, noting that this operator acts upon the singly occupied subspace spanned by $|\psi_1\rangle$, we see that the factors involving $n_{d\sigma}$ are redundant and can be dropped. As a result, swapping the momentum and spin indices in the second part of the expression, we obtain the required expression.

3   Path Integral

SYNOPSIS The aim of this chapter is to introduce the concept and methodology of the path integral, starting with single-particle quantum mechanics and then generalizing the approach to many-particle systems. Emphasis is placed on establishing the interconnections between the quantum mechanical path integral, classical Hamiltonian mechanics, and statistical mechanics. The practice of Feynman path integration is discussed in the context of several pedagogical applications. As well as the canonical examples of a quantum particle in a single- or double-well potential, we discuss the generalization of the path integral scheme to the tunneling of extended objects (quantum fields), and dissipative and thermally assisted quantum tunneling. In the final part of the chapter, the concept of path integration is extended to that of field integration for many-body systems.

To introduce the path integral formalism, we leave temporarily the arena of many-body physics and return to single-particle quantum mechanics. By establishing the path integral approach for ordinary quantum mechanics, we set the stage for the introduction of field integral methods for many-body theories. We will see that the path integral not only represents a gateway to higher-dimensional functional integral methods but, when viewed from an appropriate perspective, already represents a field-theoretical approach in its own right. Exploiting this connection, various concepts of field theory, including stationary phase analysis, the Euclidean formulation of field theory, and instanton techniques will be introduced.

3.1 The Path Integral: General Formalism

Broadly speaking, there are two approaches to the formulation of quantum mechanics: the "operator approach," based on the canonical quantization of physical observables and the associated operator algebra and the Feynman path integral.[1] While canonical quantization is usually encountered first in elementary courses on quan-

---

[1] For a more extensive introduction to the Feynman path integral, we refer to one of the many standard texts including R. P. Feynman and A. R. Hibbs, *Quantum Mechanics and Path Integrals* (McGraw-Hill, 1965); J. W. Negele and H. Orland, *Quantum Many Particle Systems* (Addison-Wesley, 1988); and L. S. Schulman, *Techniques and Applications of Path Integration* (Wiley, 1981). Alternatively, one may turn to the original literature, R. P. Feynman, *Space–time approach to non-relativistic quantum mechanics*, Rev. Mod. Phys. **20**, 362 (1948). Historically, Feynman's development of the path integral was motivated by earlier work by Dirac on the connection between classical and quantum mechanics, P. A. M. Dirac, *On the analogy between classical and quantum mechanics*, Rev. Mod. Phys. **17**, 195 (1945).

tum mechanics, path integrals have acquired the reputation of being a sophisticated concept that is better reserved for advanced courses. Yet this reputation is hardly justified! In fact, the path integral formulation has many advantages, most of which explicitly support an intuitive understanding of quantum mechanics. Moreover, integrals – even the infinite-dimensional ones encountered below – are hardly more abstract than infinite-dimensional linear operators. Further merits of the path integral include the following.

Richard P. Feynman 1918–1988 was an American physicist who, with Sin-Itiro Tomonaga, and Julian Schwinger, shared the 1965 Nobel Prize in Physics for "fundamental work in quantum electrodynamics, with far-reaching consequences for the physics of elementary particles." He was well known for his unusual life-style, as well as his popular books and lectures on mathematics and physics.

▷ Although the classical limit is not always easy to retrieve within the canonical formulation of quantum mechanics, it constantly remains visible in the path integral approach. The latter makes explicit use of classical mechanics as a "platform" on which to build a theory of quantum fluctuations. The classical solutions of Hamilton's equation of motion always remain central to the formalism.

▷ Path integrals allow for an efficient formulation of *non-perturbative* approaches in quantum mechanics. Examples include the "instanton" formulation of quantum tunneling, discussed below. The extension of such methods to continuum theories has led to some of the most powerful concepts of quantum field theory.

▷ The Feynman path integral represents a prototype of higher-dimensional field integrals. Yet, even the "zero-dimensional" path integral is of relevance to applications in many-body physics. Very often, one encounters environments, such as the superconductor or correlated electron devices, where a macroscopically large number of degrees of freedom "lock" to form a single collective variable. (For example, to a first approximation, the phase information carried by the order parameter in moderately large superconducting grains can often be described by a *single* phase degree of freedom, i.e., a "quantum particle" on a unit circle.) ideally suited to the analysis of such systems.

What then is the basic idea of the path integral approach? More than any other formulation of quantum mechanics, the path integral formalism is based on connections to classical mechanics. The variational approach employed in chapter 1 relied on the fact that classically allowed trajectories in configuration space extremize an action functional. A principal constraint to be imposed on any such trajectory is energy conservation. By contrast, quantum particles have a little more freedom than their classical counterparts. In particular, by the uncertainty principle, energy conservation can be violated by an amount ΔE over a time $\sim \hbar/\Delta E$ (here, and throughout this chapter, we will make $\hbar$ explicit for clarity). The connection to the action principles of classical mechanics becomes particularly apparent in problems of quantum tunneling: a particle of energy E may tunnel through a potential barrier of height $V > E$. However, this process is penalized by a damping factor $\sim \exp(i \int_{\text{barrier}} dx \, p/\hbar)$, where $p = \sqrt{2m(E - V)}$, i.e., the exponent of the (imaginary) action associated with the classically forbidden path.

These observations motivate a new formulation of quantum propagation. Could it be that, as in classical mechanics, the quantum amplitude A for propagation between any two points in coordinate space is again controlled by the action functional– controlled in a relaxed sense, where not just a single extremal path $x_{\text{cl}}(t)$ but an entire manifold of neighboring paths contribute? More specifically, one might speculate that the quantum amplitude is obtained as $A \sim \sum_{x(t)} \exp(iS[x]/\hbar)$, where $\sum_{x(t)}$ symbolically stands for a summation over all paths compatible with the initial conditions of the problem, and S denotes the *classical* action. Although at this stage no formal justification for the path integral has been presented, with this ansatz some features of quantum mechanics would obviously be borne out correctly. Specifically, in the classical limit ($\hbar \to 0$), the quantum mechanical amplitude would become increasingly dominated by the contribution to the sum from the classical path $x_{\text{cl}}(t)$. This is so because non-extremal configurations would be weighted by a rapidly oscillating amplitude associated with the large phase $S/\hbar$ and would, therefore, average to zero.[2] Second, quantum mechanical tunneling would be a natural element of the theory; non-classical paths do contribute to the net amplitude, but at the cost of a damping factor specified by the imaginary action (as in the traditional formulation).

Fortunately, no fundamentally novel picture of quantum mechanics needs to be declared in order to promote the idea of the path "integral" $\sum_{x(t)} \exp(iS[x]/\hbar)$ to a working theory. As we will see in the next section, the new formulation can be developed from the established principles of canonical quantization.

3.2 Construction of the Path Integral

time
evolution
operator

All information about an autonomous[3] quantum system is contained in its **time evolution operator**. A formal integration of the time-dependent Schrödinger equation $i\hbar\partial_t|\Psi\rangle = \hat{H}|\Psi\rangle$ gives the time evolution operator

$$|\Psi(t')\rangle = \hat{U}(t', t)|\Psi(t)\rangle, \quad \hat{U}(t', t) = e^{-\frac{i}{\hbar}\hat{H}(t'-t)}\Theta(t' - t). \tag{3.1}$$

The operator $\hat{U}(t', t)$ describes dynamical evolution under the influence of the Hamiltonian from a time t to time t'. Causality implies that $t' > t$, as indicated by the step or Heaviside Θ-function. In the real space representation, we can write

---

[2] More precisely, in the limit of small $\hbar$, the path sum can be evaluated by saddle-point methods, as detailed below.

[3] A system is classified as **autonomous** if its Hamiltonian does not explicitly depend on time. Actually, the construction of the path integral can be straightforwardly extended to include time-dependent problems. However, in order to keep the introductory discussion as simple as possible, here we assume time independence.

$$\Psi(q', t') = \langle q'|\Psi(t')\rangle = \langle q'|\hat{U}(t', t)\Psi(t)\rangle = \int dq\, U(q', t'; q, t)\Psi(q, t),$$

propagator

where $U(q', t'; q, t) = \langle q'|e^{-\frac{i}{\hbar}\hat{H}(t'-t)}|q\rangle\Theta(t' - t)$ defines the (q', q)-component of the time evolution operator. As the matrix element expresses the probability amplitude for a particle to propagate between points q and q' in a time $t' - t$, it is known as the **propagator** of the theory.

The idea behind Feynman's approach is easy to formulate. Rather than taking on the Schrödinger equation for general times t, one may first solve the simpler problem of time evolution for infinitesimally small times Δt. We thus divide the time evolution into $N \gg 1$ time steps,

$$e^{-i\hat{H}t/\hbar} = \left(e^{-i\hat{H}\Delta t/\hbar}\right)^N, \tag{3.2}$$

where $\Delta t = t/N$. Although nothing more than a formal rewriting of (3.1), the representation (3.2) has the advantage that the factors $e^{-i\hat{H}\Delta t/\hbar}$ are close to the unit operator. (More precisely, if Δt is much smaller than the [reciprocal of the] eigenvalues of the Hamiltonian in the regime of physical interest, the exponents are small in comparison with unity and, as such, can be treated perturbatively.) A first simplification arising from this fact is that the exponentials can be factorized into two pieces, each of which can be readily diagonalized. To achieve this factorization, we make use of the identity[4]

$$e^{-i\hat{H}\Delta t/\hbar} = e^{-i\hat{T}\Delta t/\hbar}e^{-i\hat{V}\Delta t/\hbar} + O(\Delta t^2),$$

where the Hamiltonian $\hat{H} = \hat{T} + \hat{V}$ is the sum of the kinetic energy $\hat{T} = \hat{p}^2/2m$ and potential energy $\hat{V}$.[5] (The following analysis, restricted for simplicity to a one-dimensional Hamiltonian, is easily generalized to arbitrary spatial dimension.) The advantage of this factorization is that the eigenstates of each factor $e^{-i\hat{T}\Delta t/\hbar}$ and $e^{-i\hat{V}\Delta t/\hbar}$ are known independently. To exploit this fact we consider the time evolution operator factorized as

$$\langle q_{\text{f}}|\left(e^{-i\hat{H}\Delta t/\hbar}\right)^N|q_{\text{i}}\rangle \simeq \langle q_{\text{f}}|\wedge e^{-i\hat{T}\Delta t/\hbar}e^{-i\hat{V}\Delta t/\hbar}\wedge\cdots\wedge e^{-i\hat{T}\Delta t/\hbar}e^{-i\hat{V}\Delta t/\hbar}|q_{\text{i}}\rangle \tag{3.3}$$

and insert at positions indicated by "$\wedge$" the resolution of identity

$$\text{id.} = \int dq_n \int dp_n\, |q_n\rangle\langle q_n|p_n\rangle\langle p_n|. \tag{3.4}$$

Here $|q_n\rangle$ and $|p_n\rangle$ represent complete sets of position and momentum eigenstates respectively, and $n = 1, \ldots, N$ serves as an index keeping track of the time steps at which the unit operator is inserted. The rationale behind the particular choice (3.4) is clear. The unit operator is arranged in such a way that both $\hat{T}$ and $\hat{V}$ act on

---

[4] Note that, by the Baker–Campbell–Hausdorff formula, for operators $\hat{A}$ and $\hat{B}$, we have $e^{\hat{A}+\hat{B}} = e^{\hat{A}}e^{\hat{B}}(1 - \frac{1}{2}[\hat{A}, \hat{B}] + \cdots)$.

[5] Although this ansatz covers a wide class of quantum problems, many applications (e.g., Hamiltonians involving spin or magnetic fields) do not fit into this framework. For a detailed exposition covering its realm of applicability, we refer to the specialist literature.[1]

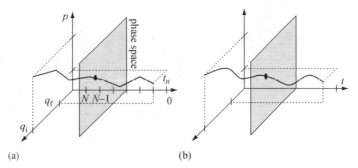

Fig. 3.1   (a) Visualization of a set of phase space points contributing to the discrete-time configuration integral (3.5). (b) In the continuum limit, the set of points becomes a smooth
curve.

the corresponding eigenstates. Inserting Eq. (3.4) into (3.3), and making use of the
identity $\langle q|p\rangle = \langle p|q\rangle^* = e^{iqp/\hbar}/(2\pi\hbar)^{1/2}$, one obtains

$$\langle q_{\mathrm{f}}|e^{-i\hat{H}t/\hbar}|q_{\mathrm{i}}\rangle \simeq \int_{q_N=q_{\mathrm{f}},q_0=q_{\mathrm{i}}} \prod_{n=1}^{N-1} dq_n \prod_{n=1}^{N} \frac{dp_n}{2\pi\hbar}$$

$$\times \exp\left(-i\frac{\Delta t}{\hbar}\sum_{n=0}^{N-1}\left(V(q_n) + T(p_{n+1}) - p_{n+1}\frac{q_{n+1}-q_n}{\Delta t}\right)\right). \quad (3.5)$$

Thus, the matrix element of the time evolution operator has been expressed as a
$(2N-1)$-dimensional integral over eigenvalues. Up to corrections of higher order
in $V\Delta t/\hbar$ and $T\Delta t/\hbar$, the expression (3.5) is exact. At each "time step" $t_n =
n\Delta t$, $n = 1, \ldots, N$, we are integrating over a pair of coordinates $x_n \equiv (q_n, p_n)$
parameterizing the **classical phase space**. Taken together, the points $\{x_n\}$ form
an N-point discretization of a path in this space (see fig. 3.1).

**classical
phase
space**

To make further progress, we need to develop intuition for the behavior of the integral (3.5). We first notice that rapid fluctuations of the integration arguments x_n
as a function of the index n are inhibited by the structure of the integrand. When
taken together, contributions for which $(q_{n+1} - q_n)p_{n+1} > O(\hbar)$ (i.e., when the
phase of the exponential exceeds 2π) lead to "random phase cancellations." In the
language of wave mechanics, the superposition of partial waves of erratically different phases leads to destructive interference. The *smooth* variation of the paths that
contribute significantly motivates the application of a continuum limit analogous
to that employed in chapter 1.

Therefore, sending $N \to \infty$ at fixed $t = N\Delta t$, the formerly discrete set $t_n =
n\Delta t, n = 1, \ldots, N$, becomes dense on the time interval $[0, t]$ and the set of phase
space points $\{x_n\}$ becomes a continuous curve $x(t)$. In the same limit,

$$\Delta t \sum_{n=0}^{N-1} \mapsto \int_0^t dt', \qquad \frac{q_{n+1}-q_n}{\Delta t} \mapsto \partial_{t'} q\,|_{t'=t_n} \equiv \dot{q}|_{t'=t_n},$$

while $[V(q_n) + T(p_{n+1})] \mapsto [T(p|_{t'=t_n}) + V(q|_{t'=t_n})] \equiv H(x|_{t'=t_n})$ denotes the classical Hamiltonian. In the limit $N \to \infty$, the fact that kinetic and potential energies are evaluated at neighboring time slices n and $n+1$, becomes irrelevant.[6] Finally,

$$\lim_{N \to \infty} \int_{q_N = q_f, q_0 = q_i} \prod_{n=1}^{N-1} dq_n \prod_{n=1}^{N} \frac{dp_n}{2\pi\hbar} \equiv \int_{\substack{q(t)=q_f \\ q(0)=q_i}} Dx,$$

where $Dx \equiv D(q, p)$ defines the integration measure.

functional integrals **INFO** Integrals extending over infinite-dimensional measures, such as $D(q, p)$, are generally called **functional integrals** (recall our discussion of functionals in chapter 1). The question how functional integration can be defined rigorously is not altogether straitghtforward and represents a subject of ongoing mathematical research. However, in this book – as in most applications in physics – we take a pragmatic view and deal with the infinite-dimensional integration naïvely unless mathematical problems arise (which actually will not be the case!).

Then, applying these conventions to Eq. (3.5), one finally obtains

$$\boxed{\langle q_f | e^{-i\hat{H}t/\hbar} | q_i \rangle = \int_{\substack{q(t)=q_f \\ q(0)=q_i}} Dx \exp \left(\frac{i}{\hbar} \int_0^t dt' \, (p\dot{q} - H(p, q)) \right)} \qquad (3.6)$$

Hamiltonian path integral Equation (3.6) represents the **Hamiltonian formulation of the path integral**. The integration extends over all possible paths through the classical phase space, which begin and end at the same *configuration* points q_i and q_f respectively (cf. fig. 3.1). The contribution of each path is weighted by its Hamiltonian action.

classical action **INFO** Remembering the connection of the Hamiltonian to the Lagrangian through the Legendre transform, $H(p, q) = p\dot{q} - L(p, q)$, the **classical action** of a trajectory $t \mapsto q(t)$ is given by $S[p, q] = \int_0^t dt' \, L(q, \dot{q}) = \int_0^t dt' \, [p\dot{q} - H(p, q)]$.

Before we turn to the discussion of the path integral (3.6), it is useful to recast the integral in an alternative form which will be both convenient and instructive. The search for an alternative formulation is motivated by the resemblance of Eq. (3.6) to the Hamiltonian formulation of classical mechanics. Given that Hamiltonian and Lagrangian mechanics can be equally employed to describe dynamical evolution, it

---

[6] To see this formally, one may Taylor expand $T(p_{n+1}) = T(p(t' + \Delta t))|_{t'=n\Delta t}$ around $p(t')$. For smooth $p(t')$, all but the zeroth-order contribution $T(p(t'))$ scale with powers of Δt, thereby becoming irrelevant. Note, however, that all such arguments are based on the assertion that the dominant contributions to the path integral are smooth in the sense $q_{n+1} - q_n \sim O(\Delta t)$. A closer inspection, however, shows that in fact $q_{n+1} - q_n \sim O(\Delta t)^{1/2}$.[1] In some cases, the most prominent one being the quantum mechanics of a particle in a magnetic field, the lowered power of Δt spoils the naïve form of the continuity argument above and so more care must be applied in taking the continuum limit. In cases where a "new" path integral description of a quantum mechanical problem is developed, it is imperative to delay taking the continuum limit until the fluctuation behavior of the discrete integral across individual time slices has been thoroughly examined.

is natural to seek a Lagrangian analog of Eq. (3.6). Focusing on Hamiltonians for which the kinetic energy $T(p)$ is quadratic in p, the Lagrangian form of the path integral can indeed be inferred from (3.6) by straightforward Gaussian integration.

To make this point clear, let us rewrite the integral in a way that emphasizes its dependence on the momentum variable p:

$$\langle q_\mathrm{f}|e^{-i\hat{H}t/\hbar}|q_\mathrm{i}\rangle = \int_{\substack{q(t)=q_\mathrm{f}\\q(0)=q_\mathrm{i}}} Dq\ e^{-\frac{i}{\hbar}\int_0^t dt' V(q)} \int Dp\ e^{-\frac{i}{\hbar}\int_0^t dt'(\frac{p^2}{2m}-p\dot{q})}. \tag{3.7}$$

The exponent is quadratic in p, a continuum generalization of a Gaussian integral. Carrying out the integration (for the details, see Eq. (3.13) below), one obtains

$$\boxed{\langle q_\mathrm{f}|e^{-i\hat{H}t/\hbar}|q_\mathrm{i}\rangle = \int_{\substack{q(t)=q_\mathrm{f}\\q(0)=q_\mathrm{i}}} Dq\ \exp\left(\frac{i}{\hbar}\int_0^t dt' L(q,\dot{q})\right)} \tag{3.8}$$

where $Dq = \lim_{N\to\infty}\left(\frac{Nm}{2\pi i\hbar t}\right)^{N/2}\prod_{n=1}^{N-1} dq_n$ denotes the functional measure of the remaining q-integration, and $L(q,\dot{q}) = m\dot{q}^2/2 - V(q)$ represents the classical Lagrangian. Strictly speaking, the finite-dimensional Gaussian integral (see the Info section below) is not directly applicable to the infinite-dimensional integral (3.7). This, however, does not represent a substantial problem as we can always rediscretize the integral (3.7), and reinstate the continuum limit after integration (exercise).

Together, Eqs. (3.6) and (3.8) represent the central results of this section. A quantum mechanical transition amplitude has been expressed in terms of an infinite-dimensional integral extending over paths through phase space (3.6) or coordinate space (3.8). All paths begin (end) at the initial (final) coordinate of the matrix element. Each path is weighted by its *classical* action. Notice in particular that the quantum transition amplitude is represented without reference

Johann Carl Friedrich Gauss
1777–1855
was a German mathematician and physicist who worked in a wide variety of fields including number theory, analysis, differential geometry, geodesy, magnetism, astronomy and optics. As well as several books, Gauss published a number of memoirs (reports of his experiences), mainly in the journal of the Royal Society of Göttingen. However, in general, he was unwilling to publish anything that could be regarded as controversial and, as a result, some of his most brilliant work was found only after his death.

to Hilbert-space operators. Nonetheless, quantum mechanics is still fully present! The point is that the integration extends over all paths and not just the subset of solutions of the classical equations of motion. (The distinct role played by classical paths in the path integral will be discussed below in section 3.2.2.) The two forms of the path integral, (3.6) and (3.8), represent the formal implementation of the "alternative picture" of quantum mechanics proposed heuristically at the beginning of the chapter.

Gaussian
integration

INFO **Gaussian integration**: With a few exceptions, all integrals encountered in this book are of Gaussian form. In most cases, the dimension of the integrals will be large, if not infinite. Yet, after some practice, it will become clear that high-dimensional Gaussian integrals are no more difficult to handle than their one-dimensional counterparts. Therefore, considering the important role played by Gaussian integration in field theory, here we derive the principal formulae once and for all. Our starting point is the one-dimensional integral. The proofs of the one-dimensional formulae provide the key to more complex functional identities that will be used throughout the text.

Gaussian integration: The ancestor of all Gaussian integrals is the identity

$$\int_{-\infty}^{\infty} dx\ e^{-\frac{a}{2}x^2} = \sqrt{\frac{2\pi}{a}}, \qquad \text{Re } a > 0 \tag{3.9}$$

In the following we will need various generalizations of Eq. (3.9). First, $\int_{-\infty}^{\infty} dx\ e^{-ax^2/2}x^2 = (2\pi/a^3)^{1/2}$, a result established by differentiating (3.9). Often one encounters integrals where the exponent is not purely quadratic from the outset but rather contains both quadratic and linear components. The generalization of Eq. (3.9) to this case reads

$$\int_{-\infty}^{\infty} dx\ e^{-\frac{a}{2}x^2+bx} = \sqrt{\frac{2\pi}{a}}\,e^{\frac{b^2}{2a}}. \tag{3.10}$$

To prove this identity, one simply eliminates the linear term by the change of variables $x \to x + b/a$, which transforms the exponent $-ax^2/2 + bx \to -ax^2/2 + b^2/2a$. The constant factor scales out and we are left with Eq. (3.9). Note that Eq. (3.10) holds even for complex b. The reason is that, as a result of shifting the integration contour into the complex plane, no singularities are encountered, i.e., the integral remains invariant.

Later, we will be concerned with the generalization of the Gaussian integral to complex arguments. The extension of Eq. (3.9) to this case reads

$$\int d(\bar{z}, z)\ e^{-\bar{z}wz} = \frac{\pi}{w}, \qquad \text{Re } w > 0,$$

where $\bar{z}$ denotes the complex conjugate of z. Here, $\int d(\bar{z}, z) \equiv \int_{-\infty}^{\infty} dx\ dy$ represents the independent integration over the real and imaginary parts of $z = x + iy$. The identity is easy to prove: owing to the fact that $\bar{z}z = x^2 + y^2$, the integral factorizes into two pieces, each of which is equivalent to Eq. (3.9) with $a = w$. Similarly, it may be checked that the complex generalization of Eq. (3.10) is given by

$$\int d(\bar{z}, z)\ e^{-\bar{z}wz+\bar{u}z+\bar{z}v} = \frac{\pi}{w}\,e^{\frac{\bar{u}v}{w}}, \qquad \text{Re } w > 0. \tag{3.11}$$

More importantly $\bar{u}$ and v may be independent complex numbers; they need not be related to each other by complex conjugation (exercise).

Gaussian integration in more than one dimension: All the integrals above have higher-dimensional counterparts. Although the real and complex versions of the N-dimensional integral formulae can be derived in a perfectly analogous manner, it is better to discuss them separately in order not to confuse the notation.

(a) **Real case**: The multi-dimensional generalization of the prototype integral (3.9) reads

$$\int d\mathbf{v}\ e^{-\frac{1}{2}\mathbf{v}^T \mathbf{A}\mathbf{v}} = (2\pi)^{N/2}(\det \mathbf{A})^{-1/2}, \tag{3.12}$$

where $\mathbf{A}$ is a positive-definite real symmetric N-dimensional matrix and $\mathbf{v}$ is an N-component real vector. The proof makes use of the fact that $\mathbf{A}$ (by virtue of being symmetric) can be diagonalized by orthogonal transformation, $\mathbf{A} = \mathbf{O}^T \mathbf{D}\mathbf{O}$, where the matrix $\mathbf{O}$ is orthogonal and all elements of the diagonal matrix $\mathbf{D}$ are positive.

The matrix $\mathbf{O}$ can be absorbed into the integration vector by means of the variable transformation $\mathbf{v} \mapsto \mathbf{O}\mathbf{v}$, which has unit Jacobian, $|\det \mathbf{O}| = 1$. As a result, we are left with a Gaussian integral with exponent $-\mathbf{v}^T \mathbf{D} \mathbf{v}/2$. Owing to the diagonality of $\mathbf{D}$, the integral factorizes into N independent Gaussian integrals, each of which contributes a factor $\sqrt{2\pi/d_i}$, where $d_i, i = 1, \ldots, N$, is the ith entry of the matrix $\mathbf{D}$. Noting that $\prod_{i=1}^{N} d_i = \det \mathbf{D} = \det \mathbf{A}$, Eq. (3.12) is obtained.

The multi-dimensional generalization of Eq. (3.10) reads

$$\boxed{\int d\mathbf{v} \; e^{-\frac{1}{2}\mathbf{v}^T \mathbf{A} \mathbf{v} + \mathbf{j}^T \cdot \mathbf{v}} = (2\pi)^{N/2}(\det \mathbf{A})^{-1/2} e^{\frac{1}{2}\mathbf{j}^T \mathbf{A}^{-1}\mathbf{j}}} \tag{3.13}$$

where $\mathbf{j}$ is an arbitrary N-component vector. Equation (3.13) is proven by analogy with (3.10), i.e., by shifting the integration vector according to $\mathbf{v} \to \mathbf{v} + \mathbf{A}^{-1}\mathbf{j}$, which does not change the value of the integral but removes the linear term from the exponent: $-\frac{1}{2}\mathbf{v}^T \mathbf{A}\mathbf{v} + \mathbf{j}^T \cdot \mathbf{v} \to -\frac{1}{2}\mathbf{v}^T \mathbf{A}\mathbf{v} + \frac{1}{2}\mathbf{j}^T \mathbf{A}^{-1}\mathbf{j}$. The resulting integral is of the type (3.12), and we arrive at Eq. (3.13).

The integral (3.13) is not only of importance in its own right, but also serves as a "generator" of other useful integral identities. Applying the differentiation operation $\partial^2_{j_m j_n}|_{\mathbf{j}=0}$ to the left- and the right-hand side of Eq. (3.13), one obtains the identity[7] $\int d\mathbf{v} \; e^{-\frac{1}{2}\mathbf{v}^T \mathbf{A}\mathbf{v}} v_m v_n = (2\pi)^{N/2}(\det \mathbf{A})^{-1/2} A_{mn}^{-1}$. This result can be formulated as

$$\langle v_m v_n \rangle = A_{mn}^{-1}, \tag{3.14}$$

where we have introduced the shorthand notation

$$\langle \cdots \rangle \equiv (2\pi)^{-N/2} \det \mathbf{A}^{1/2} \int d\mathbf{v} \; e^{-\frac{1}{2}\mathbf{v}^T \mathbf{A}\mathbf{v}}(\cdots), \tag{3.15}$$

suggesting an interpretation of the Gaussian weight as a probability distribution.

Indeed, the differentiation operation leading to Eq. (3.14) can be iterated. Differentiating four times, one obtains $\langle v_m v_n v_q v_p \rangle = A_{mn}^{-1} A_{qp}^{-1} + A_{mq}^{-1} A_{np}^{-1} + A_{mp}^{-1} A_{nq}^{-1}$. One way of memorizing the structure of this important identity is that the Gaussian "expectation" value $\langle v_m v_n v_p v_q \rangle$ is given by all "pairings" of the type (3.14) that can be formed from the four components v_m. This rule generalizes to expectation values of arbitrary order: $2n$-fold differentiation of Eq. (3.13) yields

$$\boxed{\langle v_{i_1} v_{i_2} \cdots v_{i_{2n}} \rangle = \sum_{\substack{\text{pairings of} \\ \{i_1, \ldots, i_{2n}\}}} A_{i_{k_1} i_{k_2}}^{-1} \cdots A_{i_{k_{2n-1}} i_{k_{2n}}}^{-1}} \tag{3.16}$$

Wick's theorem

This result is the mathematical identity underlying **Wick's theorem** (for real bosonic fields), to be discussed in more physical terms below.

(b) **Complex case**: The results above can be extended straightforwardly to multi-dimensional complex Gaussian integrals. The complex version of Eq. (3.12) is given by

$$\boxed{\int d(\mathbf{v}^\dagger, \mathbf{v}) \; e^{-\mathbf{v}^\dagger \mathbf{A}\mathbf{v}} = \pi^N (\det \mathbf{A})^{-1},} \tag{3.17}$$

where $\mathbf{v}$ is a complex N-component vector, $d(\mathbf{v}^\dagger, \mathbf{v}) \equiv \prod_{i=1}^{N} d(\operatorname{Re} v_i) \, d(\operatorname{Im} v_i)$, and $\mathbf{A}$ is a complex matrix with positive definite hermitian part. (Remember that every matrix can be decomposed into a hermitian and an anti-hermitian component, $\mathbf{A} = \frac{1}{2}(\mathbf{A} + \mathbf{A}^\dagger) + \frac{1}{2}(\mathbf{A} - \mathbf{A}^\dagger)$.) For hermitian $\mathbf{A}$, the proof of Eq. (3.17) is analogous to

---

[7] Note that the notation A_{mn}^{-1} refers to the mn-element of the matrix $\mathbf{A}^{-1}$.

that of Eq. (3.12), i.e., $\mathbf{A}$ is unitarily diagonalizable, $\mathbf{A} = \mathbf{U}^\dagger \mathbf{A} \mathbf{U}$, the matrices $\mathbf{U}$ can be transformed into $\mathbf{v}$, the resulting integral factorizes, etc. For non-hermitian $\mathbf{A}$ the proof is more elaborate, if unedifying, and we refer to the literature for details. The generalization of Eq. (3.17) to exponents with linear contributions reads

$$\boxed{\int d(\mathbf{v}^\dagger, \mathbf{v}) \ \exp\left(-\mathbf{v}^\dagger \mathbf{A} \mathbf{v} + \mathbf{w}^\dagger \mathbf{v} + \mathbf{v}^\dagger \mathbf{w}'\right) = \pi^N (\det \mathbf{A})^{-1} \exp\left(\mathbf{w}^\dagger \mathbf{A}^{-1} \mathbf{w}'\right)} \qquad (3.18)$$

Note that $\mathbf{w}$ and $\mathbf{w}'$ may be independent complex vectors. The proof of this identity mirrors that of Eq. (3.13), i.e., by first effecting the shift $\mathbf{v}^\dagger \to \mathbf{v}^\dagger + \mathbf{w}^\dagger A^{-1}$, $\mathbf{v} \to \mathbf{v} + A^{-1}\mathbf{w}'$.[8] As with (3.13), Eq. (3.18) may also serve as a generator of integral identities. Differentiating the integral twice according to $\partial^2_{w'_m, \bar{w}_n}|_{\mathbf{w}=\mathbf{w}'=0}$ gives

$$\langle \bar{v}_m v_n \rangle = A_{nm}^{-1},$$

where $\langle \cdots \rangle \equiv \pi^{-N} (\det \mathbf{A}) \int d(\mathbf{v}^\dagger, \mathbf{v}) \, e^{-\mathbf{v}^\dagger \mathbf{A} \mathbf{v}} (\cdots)$. Iteration to more than two derivatives gives $\langle \bar{v}_n \bar{v}_m v_p v_q \rangle = A_{pm}^{-1} A_{qn}^{-1} + A_{pn}^{-1} A_{qm}^{-1}$ and, eventually,

$$\boxed{\langle \bar{v}_{i_1} \bar{v}_{i_2} \cdots \bar{v}_{i_n} v_{j_1} v_{j_2} \cdots v_{j_n} \rangle = \sum_P A_{j_1 i_{P1}}^{-1} \cdots A_{j_n i_{Pn}}^{-1}}$$

where $\sum_P$ represents a sum over all permutations of n integers.

Gaussian functional integration: With this preparation, we are in a position to define the main practice of field theory – the method of Gaussian functional integration. Turning to Eq. (3.13), let us suppose that the components of the vector $\mathbf{v}$ parameterize the weight of a real scalar field on the sites of a one-dimensional lattice. In the continuum limit, the set $\{v_i\}$ translates to a function $v(x)$, and the matrix A_{ij} is replaced by an operator kernel $A(x, x')$. In applications, this kernel often assumes the role of the inverse of the effective **propagator** of a theory, and we will use this denotation. The generalization of Eq. (3.13) to the infinite-dimensional case reads

$$\int Dv(x) \ \exp\left(-\frac{1}{2} \int dx \, dx' \ v(x) A(x, x') v(x') + \int dx \ j(x) v(x)\right)$$
$$\propto (\det A)^{-1/2} \exp\left(\frac{1}{2} \int dx \, dx' \ j(x) A^{-1}(x, x') j(x')\right), \qquad (3.19)$$

where the inverse kernel $A^{-1}(x, x')$ satisfies the equation

$$\boxed{\int dx' \ A(x, x') A^{-1}(x', x'') = \delta(x - x'')} \qquad (3.20)$$

The notation $Dv(x)$ is used to denote the measure of the functional integral. Although the constant of proportionality, $(2\pi)^N$, left out of Eq. (3.19), is formally divergent in the thermodynamic limit $N \to \infty$, it does not affect averages that are obtained from derivatives of such integrals. For example, for Gaussian-distributed functions, Eq. (3.14) has the generalization

$$\boxed{\langle v(x) v(x') \rangle = A^{-1}(x, x')}$$

Gaussian functional integration

propagator

---

[8] For an explanation of why $\mathbf{v}$ and $\mathbf{v}^\dagger$ may be shifted independently of each other, cf. the analyticity remarks made in connection with Eq. (3.11).

where the r.h.s. features the propagator. Accordingly, Eq. (3.16) assumes the form

$$\langle v(x_1)v(x_2)\cdots v(x_{2n})\rangle = \sum_{\substack{\text{pairings of}\\ \{x_1,\dots,x_{2n}\}}} A^{-1}(x_{k_1},x_{k_2})\cdots A^{-1}(x_{k_{2n-1}},x_{k_{2n}}) \qquad (3.21)$$

The generalizations of the other Gaussian averaging formulae above should be obvious.

To make sense of Eq. (3.19), one must interpret the meaning of the determinant, $\det A$. When the variables entering the Gaussian integral are discrete, the integral simply represents the determinant of the (real symmetric) matrix. In the present case, however, one must interpret A as a hermitian operator having an infinite set of eigenvalues. The determinant simply represents the product over this infinite set (see, e.g., section 3.2.4).

Before turning to specific applications of the Feynman path integral, let us stay with the general structure of the formalism and identify two fundamental connections, to classical point mechanics and to classical and quantum statistical mechanics.

3.2.1 Path integral and statistical mechanics

The path integral reveals a connection between quantum mechanics and statistical mechanics whose importance can hardly be exaggerated. To reveal this link, let us for a moment forget about quantum mechanics and consider, by way of an example, a classical one-dimensional model of a "flexible string" held under constant tension and confined to a "gutter-like" potential (see the figure). For simplicity, let us assume that the mass density of the string is high,

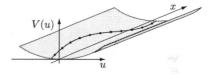

so that its fluctuations are "asymptotically slow" (i.e., the kinetic contribution to its energy is negligible). Transverse fluctuations of the string are then penalized by its line tension and the external potential. Assuming that the transverse displacement $u(x)$ is small, the potential energy stored in the string separates into two parts: the first arises from the line tension and the second from the external potential. Starting with the former, a transverse fluctuation of a line segment of length dx by an amount du leads to a potential energy of magnitude $dV_{\text{tension}} = \sigma((dx^2 + du^2)^{1/2} - dx) \simeq \sigma dx\,(\partial_x u)^2/2$, where σ denotes the tension. Integrating over the length of the string, one obtains $V_{\text{tension}}[\partial_x u] \equiv \int dV_{\text{tension}} = \frac{1}{2}\int_0^L dx\,\sigma(\partial_x u)^2$. The second contribution, arising from the external potential, is given by $V_{\text{external}}[u] \equiv \int_0^L dx\,V(u)$. Adding the two contributions, the total energy is given by $V = V_{\text{tension}} + V_{\text{external}} = \int_0^L dx\,[\frac{\sigma}{2}(\partial_x u)^2 + V(u)]$.

EXERCISE Find an expression for the kinetic energy contribution assuming that the string has mass per unit length m. How does this model compare with the continuum model of lattice vibrations discussed in chapter 1? Convince yourself that, in the limit $m \to \infty$, the kinetic contribution to the partition function $\mathcal{Z} = \text{tr}(e^{-\beta H})$ is negligible.

According to the general principles of statistical mechanics, the equilibrium properties of a system are encoded in the partition function $\mathcal{Z} = \text{tr}[e^{-\beta V}]$, where "tr" denotes a summation over all possible configurations of the system and V is the total potential energy functional. Applied to the present case, $\text{tr} \to \int Du$, where $\int Du$ stands for functional integration over all configurations of the string $u(x)$, $x \in [0, L]$. Thus, the partition function of the string is given by

$$\mathcal{Z} = \int Du \, \exp\left(-\beta \int_0^L dx \left(\frac{\sigma}{2}(\partial_x u)^2 + V(u)\right)\right). \tag{3.22}$$

A comparison of this result with Eq. (3.8) shows that the partition function of the *classical* system coincides with the *quantum mechanical* amplitude

$$\boxed{\mathcal{Z} = \int dq \, \langle q | e^{-it\hat{H}/\hbar} | q \rangle \Big|_{\substack{\hbar = 1/\beta, \\ t = -iL}}}$$

evaluated at an imaginary "time" $t \to -i\tau \equiv -iL$, where $\hat{H} = \hat{p}^2/2\sigma + V(q)$, and Planck's constant is identified with the "temperature" $\hbar = 1/\beta$. (Here, we have assumed that our string is subject to periodic boundary conditions.)

To see this explicitly, let us consider quantum mechanics in a time formally made imaginary as $e^{-it\hat{H}/\hbar} \to e^{-\tau\hat{H}/\hbar}$, or $t \to -i\tau$. Assuming convergence (i.e., positivity of the eigenvalues of $\hat{H}$), a construction scheme perfectly analogous to the one outlined in section 3.1 would lead to a path integral of the structure (3.8). Formally, the only differences would be (i) that the Lagrangian would be integrated along the imaginary time axis $t' \to -i\tau' \in [0, -i\tau]$ and (ii) that there would be a change of the sign of the kinetic energy term, i.e., $(\partial_{t'}q)^2 \to -(\partial_{\tau'}q)^2$. After a suitable exchange of variables, $\tau \to L, \hbar \to 1/\beta$, the coincidence of the resulting expression with the partition function (3.22) is clear.

The connection between quantum mechanics and classical statistical mechanics outlined above generalizes to higher dimensions. There are close analogies between quantum field theories in d dimensions and classical statistical mechanics in $d + 1$ dimensions. (The equality of the path integral above with the one-dimensional statistical model is merely the $d = 0$ version of this connection.) In fact, the connection turned out to be one of the major driving forces behind the success of path integral techniques in modern field theory and statistical mechanics. It offered, for the first time, the possibility of drawing connections between systems that had seemed unrelated.

However, the concept of imaginary time not only provides a bridge between quantum and classical statistical mechanics but also plays a role within a purely quantum mechanical context. Consider the partition function of a *single-particle* quantum system,

$$\boxed{\mathcal{Z} = \text{tr}(e^{-\beta\hat{H}}) = \int dq \, \langle q | e^{-\beta\hat{H}} | q \rangle}$$

The partition function can be interpreted as a trace over the transition amplitude $\langle q|e^{-i\hat{H}t/\hbar}|q\rangle$ evaluated at an imaginary time $t = -i\hbar\beta$. Thus, real-time dynamics and quantum statistical mechanics can be treated on the same footing, provided that we allow for the appearance of imaginary times.

Later we will see that the concept of imaginary, or even generalized complex, time plays an important role in field theory. There is even some nomenclature regarding imaginary times. The transformation $t \to -i\tau$ is described as a **Wick rotation** (alluding to the fact that multiplication by i can be interpreted as a $(\pi/2)$-rotation in the complex plane). Imaginary-time representations of Lagrangian actions are termed **Euclidean**, whereas the real-time forms are called **Lorentzian**.

Wick rotation

Euclidean action

INFO The origin of this terminology can be understood by considering the structure of the action of, say, the phonon model (1.4). Forgetting for a moment about the magnitude of the coupling constants, we see that the action has the bilinear structure $\sim x_\mu g^{\mu\nu} x_\nu$, where $\mu = 0, 1$, the vector $x_\mu = \partial_\mu \phi$, and the diagonal matrix $g = \mathrm{diag}(-1, 1)$ is the two-dimensional version of a Minkowski metric. (In three spatial dimensions g would take the form of the standard Minkowski metric of special relativity; see the discussion in section 9.1.) On Wick rotation of the time variable, the factor -1 in the metric changes sign to $+1$, and g becomes a positive definite Euclidean metric. The nature of this transformation motivates the notation above.

Once one has grown accustomed to the idea that the interpretation of time as an imaginary quantity can be useful, yet more general concepts can be conceived. For example, one may contemplate propagation along temporal contours that are neither purely real nor purely imaginary but are generally complex. Indeed, it turns out that path integrals with curvilinear integration contours in the complex time plane find numerous applications in statistical and quantum field theory.

3.2.2 Semiclassical analysis of the path integral

In deriving the two path integral representations (3.6) and (3.8), no approximations were made. Yet the majority of quantum mechanical problems are unsolvable in closed form, and the situation regarding the path integral approach is no different. In fact, only the path integrals of problems with a quadratic Hamiltonian – corresponding to the quantum mechanical harmonic oscillator and generalizations thereof – can be carried out in closed form. Yet, what counts more than the availability of exact solutions is the flexibility with which approximation schemes can be developed. As to the path integral formulation, it is particularly strong in cases where **semiclassical limits** of quantum theories are explored. Here, by "semiclassical" we mean the limit $\hbar \to 0$, i.e., the case where the theory is governed by classical structures with small quantum fluctuations superimposed.

To see how classical structures enter the path integral, consider Eqs. (3.6) and (3.8) at small $\hbar$. In this case, the path integrals are dominated by configurations with stationary action. (Nonstationary contributions to the integral imply phase fluctuations which largely average to zero.) Since the exponents of the two path integrals (3.6) and (3.8) involve the Hamiltonian or Lagrangian action, the extremal path configurations are just the solutions of the corresponding equations of motion,

$$\text{Hamiltonian}: \quad \delta S[x] = 0 \quad \Rightarrow \quad d_t x = \{H, x\} \equiv \partial_p H \ \partial_q x - \partial_q H \ \partial_p x,$$

$$\text{Lagrangian}: \quad \delta S[q] = 0 \quad \Rightarrow \quad (d_t \partial_{\dot{q}} - \partial_q) \, L = 0.$$

These equations must be solved subject to the boundary conditions $q(0) = q_\mathrm{i}$ and $q(t) = q_\mathrm{f}$. (Note that these boundary conditions do not uniquely specify a solution, i.e., in general there may be many solutions – try to invent examples!)

Now, although the stationary phase configurations are classical, quantum mechanics is still present. Technically, we are evaluating an integral in a stationary phase approximation. In such cases, *fluctuations* around stationary points are an essential part of the integral. At the very least it is necessary to integrate out Gaussian (quadratic) fluctuations around the stationary point. In the case of the path integral, fluctuations of the action around the stationary phase configurations involve non-classical (in that they do not solve the classical equations of motion) trajectories through phase or coordinate space. Before exploring how this mechanism works, let us consider the stationary phase analysis of functional integrals in general.

INFO Consider a functional integral $\int Dx \, e^{-F[x]}$ where $Dx = \lim_{N \to \infty} \prod_{n=1}^{N} dx_n$ represents a functional measure resulting from taking the continuum limit of some finite-dimensional integration space, and the "action" $F[x]$ may be an arbitrary complex functional of x (leading to convergence of the integral). The function resulting from the limit of infinitely many discretization points, $\{x_n\}$, is denoted by $x : t \mapsto x(t)$ (where t plays the role of the discrete index n). Evaluating the integral above within a **stationary phase approximation** amounts to performing the following steps:

stationary
phase
approx-
imation

1. Find the "points" of stationary phase, i.e., configurations $\bar{x}$ qualified by the condition of vanishing functional derivative (see section 1.2),

$$DF_x = 0 \quad \Leftrightarrow \quad \forall t : \left. \frac{\delta F[x]}{\delta x(t)} \right|_{x=\bar{x}} = 0.$$

Although there may be more than one solution, we first discuss the case in which the configuration $\bar{x}$ is unique.

2. Taylor expand the functional to second order around $\bar{x}$, i.e.,

$$F[x] = F[\bar{x} + y] = F[\bar{x}] + \frac{1}{2} \int dt \, dt' \, y(t') A(t, t') y(t) + \cdots \tag{3.23}$$

where $A(t, t') = \left. \frac{\delta^2 F[x]}{\delta x(t) \, \delta x(t')} \right|_{x=\bar{x}}$ denotes the second functional derivative. Owing to the stationarity of $\bar{x}$, no first-order contribution can appear.

3. Check that the operator $\hat{A} \equiv \{A(t, t')\}$ is positive definite. If it is not, there is a problem – the integration over Gaussian fluctuations y diverges. (In practice, where the analysis is rooted in a physical context, such eventualities arise only rarely. The resolution can usually be found in a judicious rotation of the integration contour.) For positive definite $\hat{A}$, the functional integral over y is doable and one obtains

$\int Dx\, e^{-F[x]} \simeq e^{-F[\bar{x}]} \det(\frac{\hat{A}}{2\pi})^{-1/2}$ (see the above discussion of Gaussian integrals and, in particular, (3.19)).

4.   Finally, if there are many stationary phase configurations, $\bar{x}_i$, the individual contributions must be added:

$$\int Dx\, e^{-F[x]} \simeq \sum_i e^{-F[\bar{x}_i]} \det\left(\frac{\hat{A}_i}{2\pi}\right)^{-1/2}. \qquad (3.24)$$

Equation (3.24) represents the most general form of the stationary phase evaluation of a (real) functional integral.

EXERCISE As applied to the Gamma function $\Gamma(z+1) = \int_0^\infty dx\, x^z e^{-x}$, with z complex, show that the stationary phase approximation is consistent with Stirling's approximation, $\Gamma(s+1) = \sqrt{2\pi s}\, e^{s(\ln s - 1)}$.

When applied to the Lagrangian form of the Feynman path integral, this programme can be implemented directly. In this case, the extremal field configuration $\bar{q}(t)$ is identified as the classical solution, $\bar{q}(t) \equiv q_{\mathrm{cl}}(t)$. Defining $r(t) = q(t) - q_{\mathrm{cl}}(t)$ as the deviation of a general path $q(t)$ from a nearby classical path $q_{\mathrm{cl}}(t)$ (see

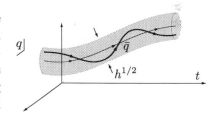

the figure), and assuming that there exists only one classical solution connecting q_{i} with q_{f} in time t, a stationary phase analysis yields the approximation

$$\langle q_{\mathrm{f}}|e^{-i\hat{H}t/\hbar}|q_{\mathrm{i}}\rangle \simeq e^{iS[q_{\mathrm{cl}}]/\hbar} \int_{r(0)=r(t)=0} Dr \exp\left(\frac{i}{2\hbar}\int_0^t dt_1 dt_2\, r(t_1) \left.\frac{\delta^2 S[q]}{\delta q(t_1)\,\delta q(t_2)}\right|_{q=q_{\mathrm{cl}}} r(t_2)\right),$$

$$(3.25)$$

cf. Eq. (3.23). For free Lagrangians of the form $L = \frac{m}{2}\dot{q}^2 - V(q)$, the second functional derivative of the action can be computed by rules of functional differentiation formulated in chapter 1. Alternatively, one can obtain this result by expanding the action to second order in fluctuations $r(t)$ (exercise):

$$\frac{1}{2}\int_0^t dt dt'\, r(t) \left.\frac{\delta^2 S[q]}{\delta q(t)\,\delta q(t')}\right|_{q=q_{\mathrm{cl}}} r(t') = -\frac{1}{2}\int dt\, r(t)\left(m\partial_t^2 + V''(q_{\mathrm{cl}}(t))\right) r(t), \quad (3.26)$$

where $V''(q_{\mathrm{cl}}(t)) \equiv \partial_q^2 V(q)|_{q=q_{\mathrm{cl}}}$ represents an ordinary derivative of the potential. Thus, the Gaussian integration over r yields the square root of the determinant of the operator $m\partial_t^2 + V''(q_{\mathrm{cl}}(t))$ – interpreted as an operator acting in the space of functions $r(t)$ with boundary conditions $r(0) = r(t) = 0$. (Note that, as we are dealing with a differential operator, the issue of boundary conditions is crucial.)

This concludes our conceptual discussion of the path integral. Before turning to its applications, let us briefly summarize the main steps in its construction.

3.2.3 Construction recipe for the path integral

Consider a general quantum transition amplitude $\langle\psi|e^{-i\hat{H}t/\hbar}|\psi'\rangle$, where t may be real, purely imaginary, or in general complex. To construct a functional integral representation of the amplitude:

1. Partition the time interval into $N \gg 1$ steps, $e^{-i\hat{H}t/\hbar} = [e^{-i\hat{H}\Delta t/\hbar}]^N$, $\Delta t = t/N$.

2. Regroup the operator content appearing in the expansion of each factor $e^{-i\hat{H}\Delta t/\hbar}$ according to the relation

$$e^{-i\hat{H}\Delta t/\hbar} = 1 + \Delta t \sum_{mn} c_{mn}\hat{A}^m\hat{B}^n + O(\Delta t^2),$$

where the eigenstates $|a\rangle, |b\rangle$ of $\hat{A}, \hat{B}$ are known and the coefficients c_{mn} are c-numbers. (In the quantum mechanical application above, $\hat{A} = \hat{p}, \hat{B} = \hat{q}$.) This "normal ordering" procedure emphasizes that distinct quantum mechanical systems may be associated with the same classical action.

3. Insert resolutions of identity as follows:

$$e^{-i\hat{H}\Delta t/\hbar} = \sum_{a,b} |a\rangle\langle a| \left(1 + \Delta t \sum_{mn} c_{mn}\hat{A}^m\hat{B}^n + O(\Delta t^2) \right) |b\rangle\langle b|$$
$$= \sum_{a,b} |a\rangle\langle a|e^{-iH(a,b)\Delta t/\hbar}|b\rangle\langle b| + O(\Delta t^2),$$

where $H(a,b)$ is the Hamiltonian evaluated at the eigenvalues of $\hat{A}$ and $\hat{B}$.

4. Regroup terms in the exponent: owing to the "mismatch" of the eigenstates at neighboring time slices, n and $n+1$, not only the Hamiltonians $H(a,b)$, but also sums over differences of eigenvalues appear (the last factor in (3.5)).

5. Take the continuum limit.

3.2.4 Example: quantum particle in a well

Free particle propagator

The simplest example of a quantum mechanical problem is a **free particle**, $\hat{H} = \hat{p}^2/2m$. The corresponding free propagator is given by

$$\boxed{G_{\text{free}}(q_{\text{f}}, q_{\text{i}}; t) \equiv \left\langle q_{\text{f}} \left| \exp\left(-\frac{i}{\hbar}\frac{\hat{p}^2 t}{2m} \right) \right| q_{\text{i}} \right\rangle \Theta(t) = \left(\frac{m}{2\pi i\hbar t} \right)^{\frac{1}{2}} \exp\left(\frac{i}{\hbar}\frac{m(q_{\text{f}}-q_{\text{i}})^2}{2t} \right) \Theta(t)}$$

$$(3.27)$$

where the Heaviside Θ-function is a reflection of causality.[9]

---

[9] Motivated by its interpretation as a Green function, in the following we refer to the quantum transition probability amplitude by the symbol G (as opposed to the symbol U used above).

EXERCISE Verify this result by the standard methodology of quantum mechanics. (Hint: Insert a resolution of identity and perform a Gaussian integral.)

Irritatingly, the derivation of Eq. (3.27) by the methods of path integration is not trivial; in a way, the problem is "too free." In the path integral construction, the absence of a confining potential shows up as divergences that must be regularized by re-discretization. Instead of discussing such regularization methods in detail, we proceed pragmatically and simply postulate that whenever a path integral over a zero-potential action is encountered, it may be formally replaced by the propagator Eq. (3.27). This substitution will be applied in the discussion of various physically more interesting path integrals below.

EXERCISE Starting from a formal path integral representation, obtain a perturbative expansion of the amplitude $\langle \mathbf{p}'|U(t \to \infty, t' \to -\infty)|\mathbf{p}\rangle$ for the scattering of a free particle from a short-range central potential $V(r)$. In particular, show that application of the above substitution rule to the first-order term in the expansion recovers the Born scattering amplitude, $-i\hbar e^{-i(t-t')E(p)/\hbar}\delta(E(p) - E(p'))\langle \mathbf{p}'|V|\mathbf{p}\rangle$.

As a first application of the path integral, let us consider a quantum particle in a one-dimensional potential well (see the figure), $\hat{H} = \hat{p}^2/2m + V(\hat{q})$. A discussion of this example illustrates how the semiclassical evaluation scheme discussed above works in practice. For simplicity we assume the potential to be symmetric, $V(q) = V(-q)$, with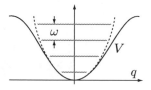
$V(0) = 0$. Consider then the amplitude for a particle initialized at $q_I = 0$ to return after a time t, $G(0, 0; t) \equiv \langle q_f = 0|e^{-i\hat{H}t/\hbar}|q_i = 0\rangle\Theta(t)$. Drawing on our previous discussion, the path integral representation of the transition amplitude for positive time $t > 0$ is given by

$$G(0, 0; t) = \int_{q(t)=q(0)=0} Dq \, \exp\left(\frac{i}{\hbar}\int_0^t dt' \, L(q, \dot{q})\right)$$

with Lagrangian $L = m\dot{q}^2/2 - V(q)$.

Now, for a generic potential $V(q)$, the path integral cannot be evaluated exactly. Instead, we wish to invoke the semiclassical analysis outlined above. Accordingly, we must first find solutions to the classical equation of motion. Minimizing the action with respect to variations of $q(t)$, one obtains the Euler–Lagrange equation of motion $m\ddot{q} = -V'(q)$. This equation must be solved subject to the boundary conditions $q(t) = q(0) = 0$. One solution is obvious: $q_{cl}(t) = 0$. Assuming that this is the only solution,[10] we obtain (cf. Eqs. (3.25) and (3.26))

[10] In general, this assumption is wrong. For smooth potentials $V(q)$, a Taylor expansion of V at small q gives the harmonic oscillator potential, $V(q) = V_0 + m\omega^2 q^2/2 + \cdots$. For times t commensurate with π/ω, one has multiple periodic solutions, $q_{cl}(t) \propto \sin(\omega t)$, starting out from the origin at time $t = 0$ and returning to it at time t, as required by the boundary conditions. In the next section, we will see why the restriction to the trivial solution was nonetheless legitimate.

$$G(0,0;t) \simeq \int_{r(0)=r(t)=0} Dr \, \exp\left(-\frac{i}{\hbar}\int_0^t dt' r(t')\frac{m}{2}\left(\partial_{t'}^2 + \omega^2\right)r(t')\right),$$

where $m\omega^2 \equiv V''(0)$ is the second derivative of the potential at the origin.[11] Note that, in this case, the contribution to the action from the stationary phase field configuration vanishes: $S[q_{cl}] = 0$. Following the discussion in section 3.2, Gaussian functional integration over r then leads to the semiclassical expansion

$$G(0,0;t) \simeq J \det\left(-m(\partial_t^2 + \omega^2)/2\right)^{-1/2}, \tag{3.28}$$

where J absorbs various constant prefactors.

It is often most convenient to represent operator determinants such as Eq. (3.28) as infinite products over eigenvalues. In the present case, the eigenvalues ϵ_n are determined by the equation $-(m/2)\left(\partial_t^2 + \omega^2\right)r_n = \epsilon_n r_n$, which must be solved subject to the boundary condition $r_n(t) = r_n(0) = 0$. A complete set of solutions to this equation is given by[12] $r_n(t') = \sin(n\pi t'/t)$, $n = 1, 2, \ldots$, with eigenvalues $\epsilon_n = m[(n\pi/t)^2 - \omega^2]/2$. Applied to the determinant,

$$\det\left(-\frac{m}{2}(\partial_t^2 + \omega^2)\right)^{-1/2} = \prod_{n=1}^{\infty}\left(\frac{m}{2}\left(\left(\frac{n\pi}{t}\right)^2 - \omega^2\right)\right)^{-1/2}.$$

To interpret this result, one must make sense of the infinite product (which even seems divergent for times commensurate with π/ω). Moreover the value of the constant J has yet to be determined.

To resolve these difficulties, one may exploit the fact that (i) we know the transition amplitude (3.27) of the *free* particle system, and (ii) the latter coincides with the transition amplitude G in the special case where the potential $V \equiv 0$. In other words, had we computed G_{free} via the path integral, we would have obtained the same constant J and the same infinite product, but with $\omega = 0$. This allows the transition amplitude to be "regularized" as

$$G(0,0;t) = \frac{G(0,0;t)}{G_{\text{free}}(0,0;t)}G_{\text{free}}(0,0;t) = \prod_{n=1}^{\infty}\left(1 - \left(\frac{\omega t}{n\pi}\right)^2\right)^{-1/2}\left(\frac{m}{2\pi i\hbar t}\right)^{1/2}\Theta(t).$$

Then, with the identity $\prod_{n=1}^{\infty}(1 - (x/n\pi)^2)^{-1} = x/\sin x$, one obtains

$$G(0,0;t) \simeq \left(\frac{m\omega}{2\pi i\hbar \sin(\omega t)}\right)^{1/2}\Theta(t). \tag{3.29}$$

In the case of the harmonic oscillator, the expansion of the potential truncates at quadratic order and our result is exact. (For a more wide-ranging discussion of the path integral for the quantum harmonic oscillator, see problem 3.8.1.) For a general

[11] Those who are uncomfortable with functional differentiation can arrive at the same expression by substituting $q(t) = q_{cl}(t) + r(t)$ into the action and expanding in $r(t)$.

[12] To find the solutions of this equation, recall the structure of the Schrödinger equation for a particle in a one-dimensional box of width $L = t$.

potential, the semiclassical approximation involves the replacement of $V(q)$ by a quadratic potential with the same curvature. The calculation above also illustrates how coordinate-space fluctuations around a completely static solution may reinstate the zero-point fluctuations characteristic of quantum mechanical bound states.

EXERCISE Using the expression for the free particle propagator, use the Feynman path integral to show that, in an **infinite square well potential** between $q = 0$ and $q = L$, $\langle q_F | \exp(-\frac{i}{\hbar}\frac{\hat{p}^2}{2m}t)|q_I\rangle = \frac{2}{L}\sum_{n=1}^{\infty} e^{-\frac{i}{\hbar}E_n t}\sin(k_n q_I)\sin(k_n q_F)$, where $E_n = (\hbar k_n)^2/2m$ and $k_n = \pi n/L$. (Hint: When considering contributions from different paths, note that each reflection from an infinite potential barrier imparts an additional phase factor of -1. Note also the **Poisson summation formula**, $\sum_{m=-\infty}^{\infty} f(m) = \sum_{n=-\infty}^{\infty}\int_{-\infty}^{\infty} d\phi\, f(\phi)\, e^{2\pi i n\phi}$.) Compare the result with that for a quantum particle on a ring of circumference L.

Poisson summation formula

3.3 Advanced Applications of the Feynman Path Integral

SYNOPSIS In this section we discuss applications of path integration in the description of quantum mechanical tunneling, decay and dissipation. Readers on a fast track may skip this section at first reading and return to it at a later stage when reference to concepts introduced in this section is made.

The path integral was introduced roughly half a century after the advent of quantum mechanics. Since then it has not replaced the operator formalism but developed into an alternative formulation of quantum mechanics. Depending on the context, operator or path integral techniques may be superior in the description of quantum mechanical problems. Here, we discuss the application of the path integral in fields where it is particularly strong (or even indispensable): quantum mechanical tunneling, decay, and dissipation.

3.3.1 Double well potential: tunneling and instantons

Consider a quantum particle confined to a double well potential (see figure). Our aim will be to estimate the quantum probability amplitude for the particle either to stay at the bottom of one of the local minima or to go from one minimum to the other. In doing so, it is understood that the energy range accessible to the particle ($\Delta E \sim \hbar/t$) is well below the potential barrier height, i.e., quantum mechanical transfer between minima is by **tunneling**. Here, in contrast with the single well, it is far from clear what kind of classical stationary-phase solutions may

quantum tunneling

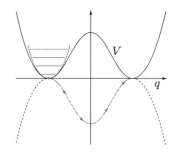

serve as a basis for a description of the quantum dynamics; there appear to be no classical paths connecting the two minima! Of course one may think of particles "rolling" over the potential hill. Yet, these are singular and energetically inaccessible by assumption.

The key to resolving these difficulties is an observation already made above, that the time argument appearing in the path integral should be considered as a general complex quantity that can (according to convenience) be set to any value in the complex plane. In the present case, a Wick rotation to imaginary times will reveal a stationary point of the action. At the end of the calculation, the real-time amplitudes that we seek can be obtained by analytic continuation.

glasses   **INFO** The mechanism of quantum tunneling plays a role in a number of problems of condensed matter. A prominent example is the physics of amorphous solids such as **glasses**.

A schematic of a glass is shown in the figure. The absence of long-range order implies that individual chemical bonds cannot assume their optimal binding lengths. For under-stretched bonds, this leads to the formation of two metastable minima around the ideal binding axis (see inset). The energetically lowest excitations of the system are transitions of individual atoms between nearly degenerate minima 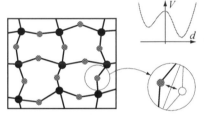 of this type, i.e., flips of atoms around the binding axis. A prominent phenomenological model[13] describes the system by an ensemble of quantum double wells of random center height and width. This model explains the existence of a vast system of metastable points in the landscape of low-energy configurations of glassy systems.

Consider the imaginary-time transition amplitudes

$$G_E(a, \pm a; \tau) \equiv \langle a | e^{-\frac{\tau}{\hbar} \hat{H}} | \pm a \rangle = G_E(-a, \mp a; \tau) \tag{3.30}$$

for the double well, where the coordinates $\pm a$ coincide with the two minima of the potential. From Eq. (3.30), the real-time amplitudes $G(a, \pm a; t) = G_E(a, \pm a; \tau \to it)$ can be recovered by the analytic continuation, $\tau \to it$. According to section 3.2.1, the **Euclidean-time path integral** of the transition amplitudes is given by

$$\boxed{G_E(a, \pm a; \tau) = \int_{q(0) = \pm a, q(\tau) = a} Dq \, \exp\left(-\frac{1}{\hbar} \int_0^\tau d\tau' \left(\frac{m}{2} \dot{q}^2 + V(q) \right) \right)} \tag{3.31}$$

where the function q now depends on imaginary time. From Eq. (3.31) we obtain the stationary-phase (or saddle-point) equations

$$- m\ddot{q} + V'(q) = 0. \tag{3.32}$$

This result indicates that the Wick rotation amounts to an effective *inversion* of the potential, $V \to -V$ (shown in the figure above). Crucially, in the inverted potential,

[13] P. W. Anderson, B. I. Halperin and C. M. Varma, *Anomalous low-temperature thermal properties of glasses and spin glasses*, Phil. Mag. **25**, 1 (1972).

the barrier has become a trough. Within this new formulation, there *are* classical solutions connecting the two points, $\pm a$. More precisely, there are three different types of classical solution fulfilling the condition for the particle to be at coordinates $\pm a$ at times 0 and τ: (a) a solution where the particle rests permanently at a;[14] (b) a corresponding solution where the particle stays at $-a$; and, most importantly, (c) a solution in which the particle leaves its initial position at $\pm a$, accelerates through the minimum at $q = 0$, and reaches the final position $\mp a$ at time τ. In computing the transition amplitudes, all three types of path must be taken into account. As to (a) and (b), by computing quantum fluctuations around these solutions, one can recover the physics of the zero-point motion described in section 3.2.4 for each well individually. (Exercise: Convince yourself that this is true!) Now let us see what happens if the paths connecting the two coordinates are added to this picture.

The instanton gas

instanton

The classical solution of the Euclidean equation of motion that connects the two potential maxima is called an **instanton**, while a solution traversing the same path but in the opposite direction is called an anti-instanton. The name was conceived by 't Hooft with the idea that these objects are similar in their structure to "solitons," particle-like solutions of classical field theories. However, unlike solitons, they

> **Gerardus 't Hooft** 1946–
> is a Dutch theoretical physicist who, with Martinus J. G. Veltman, received the 1999 Nobel Prize in Physics "for elucidating the quantum structure of electroweak interactions in physics." Together, they were able to identify the properties of the W and Z particles. The 't Hooft–Veltman model allowed scientists to calculate the physical properties of other particles, including the mass of the top quark, which was directly observed in 1995.

are structures in (Euclidean) time; hence the term "instant-." Moreover, the syllable "-on" hints at a particle-like interpretation of the solution. The reasoning is that, as a function of time, instantons are almost everywhere stationary save for a short region of variation (see below). Considering time as akin to a spatial dimension, these states can be interpreted as a localized excitation or, according to standard field-theoretical practice, a *particle*.[15]

---

[14] Note that the potential inversion answers a question that arose above, i.e., whether or not the classical solution staying at the bottom of the single well was actually the only one that could be considered. As with the double well, we could have treated the single well within an imaginary time representation, whereupon the well would have become a hill. Clearly, the boundary condition requires the particle to start and finish at the top of the hill, i.e., the solution that stays there for ever. By formulating the semiclassical expansion around that path, we would have obtained Eq. (3.29) with $t \to -i\tau$ which, upon analytic continuation, would have led back to the real-time result.

[15] The instanton method has inspired a variety of excellent and pedagogical reviews including A. M. Polyakov, *Quark confinement and topology of gauge theories*, Nucl. Phys. **B120**, 429 (1977). See also A. M. Polyakov, *Gauge Fields and Strings* (Harwood, 1987); S. Coleman, *Aspects of Symmetry – Selected Erice Lectures* (Cambridge University Press, 1985), chapter 7; and A. I. Vainshtein et al., *ABC of instantons*, Sov. Phys. Usp. **25**, 195 (1982).

To proceed, we must first compute the action of
the instanton solution. Multiplying (3.32) by $\dot{q}_{\mathrm{cl}}$, integrating over time (i.e., performing the first integral
of the equation of motion), and using the fact that, at
$q_{\mathrm{cl}} = \pm a$, $\partial_\tau q_{\mathrm{cl}} = V = 0$, one finds

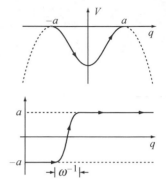

$$\frac{m}{2}\dot{q}_{\mathrm{cl}}^2 = V(q_{\mathrm{cl}}). \tag{3.33}$$

With this result, one obtains the instanton action

$$S_{\mathrm{inst}} = \int_0^\tau d\tau' \left(\frac{m}{2}\dot{q}_{\mathrm{cl}}^2 + V(q_{\mathrm{cl}}) \right) = \int_0^\tau d\tau' \frac{dq_{\mathrm{cl}}}{d\tau'}(m\dot{q}_{\mathrm{cl}})$$

$$= \int_{-a}^a dq\,(2mV(q))^{1/2}. \tag{3.34}$$

Note that S_{inst} is determined solely by the functional profile of the potential V
(i.e., it does not depend on the structure of the solution q_{cl}). Second, let us explore
the structure of the instanton as a function of time. Defining the second derivative
of the potential at $\pm a$ by $V''(\pm a) = m\omega^2$, Eq. (3.33) implies that, for large times
(when the particle is close to the right-hand maximum), $\dot{q}_{\mathrm{cl}} = -\omega(q_{\mathrm{cl}} - a)$, which
integrates to $q_{\mathrm{cl}}(\tau) \overset{\tau \to \infty}{\longrightarrow} a - e^{-\tau\omega}$. Thus the temporal extension of the instanton is
set by the oscillator frequencies of the local potential minima (the maxima of the
inverted potential) and, in cases where tunneling takes place on time scales much
larger than that, it can be regarded as short (see the figure).

The confinement of the instanton to a narrow interval of time has an important
implication – there must exist *approximate* solutions of the stationary equation
involving further anti-instanton–instanton pairs (physically, solutions with the particle repeatedly bouncing to-and-fro in the inverted potential). According to the
general philosophy of the saddle-point scheme, the path integral is obtained by
summing over all solutions of the saddle-point equations and hence over all instanton configurations. The summation over multi-instanton configurations – termed
instanton   the **instanton gas** – is simplified by the fact that individual instantons have short
temporal support (events of overlapping configurations are rare) and that not too
many instantons can be accommodated in a finite time interval (the instanton gas
is dilute). The actual density is dictated by the competition between the configurational "entropy" (favoring high density) and the "energetics," the exponential
weight implied by the action (favoring low density) – see the estimate below.

In practice, multi-instanton configurations imply a transition amplitude

$$G(a, \pm a; \tau) \simeq \sum_{n \text{ even/odd}} K^n \int_0^\tau d\tau_1 \int_0^{\tau_1} d\tau_2 \cdots \int_0^{\tau_{n-1}} d\tau_n\, A_n(\tau_1, \cdots, \tau_n), \tag{3.35}$$

where A_n denotes the amplitude associated with n instantons, and we
have noted that, in order to connect
a with $\pm a$, the number of instantons
must be even or odd. The n instantons

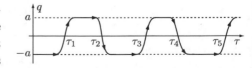

contributing to each A_n can take place at arbitrary times $\tau_i \in [0, \tau], i = 1, \ldots, n$, and all these possibilities have to be summed (i.e., integrated). Here, K denotes a constant absorbing the temporal dimension $[\text{time}]^n$ introduced by the time integrations, and $A_n(\tau_1, \ldots, \tau_n)$ is the transition amplitude, evaluated within the semiclassical approximation around a configuration of n instantons at times $0 \leq \tau_n \leq \tau_{n-1} \leq \cdots \leq \tau_1 \leq \tau$ (see the figure). In the following, we first focus on the transition amplitude A_n, which controls the exponential dependence of the tunneling amplitude, and will return later to consider the prefactor K.

According to the general semiclassical principle, each amplitude $A_n = A_{n,\text{cl}} \times A_{n,\text{qu}}$ factorizes into two parts: a classical contribution $A_{n,\text{cl}}$ accounting for the action of the instanton configuration; and a quantum contribution $A_{n,\text{qu}}$ resulting from quadratic fluctuations around the classical path. Focusing initially on $A_{n,\text{cl}}$ we note that, at intermediate times, $\tau_i \ll \tau' \ll \tau_{i+1}$, where the particle rests on top of either of the maxima at $\pm a$, no action accumulates (see the previous section). However, each instanton has a finite action S_{inst} (3.34), and these contributions sum to give the full classical action,

$$A_{n,\text{cl}}(\tau_1, \ldots, \tau_n) = e^{-nS_{\text{inst}}/\hbar}, \tag{3.36}$$

which is independent of the time coordinates τ_i, i.e., instantons are "non-interacting."

Regarding $A_{n,\text{qu}}$, there are two contributions. First, while the particle rests on either of the hills (corresponding to the straight segments in the figure above), quadratic fluctuations around the classical (i.e., spatially constant) configuration play the same role as the quantum fluctuations considered in the previous section, the only difference being that we are working in a Wick-rotated picture. There, it was found that quantum fluctuations around a classical configuration, which stays for a (real) time t at the bottom of a well, result in a factor $\sqrt{1/\sin(\omega t)}$ (the remaining constants being absorbed into the prefactor K^n). Rotating to imaginary time, $t \to -i\tau$, it follows that the quantum fluctuation accumulated during the stationary time $\tau_{i+1} - \tau_i$ is given by

$$\left(\frac{1}{\sin(-i\omega(\tau_{i+1} - \tau_i))} \right)^{1/2} \sim e^{-\omega(\tau_{i+1} - \tau_i)/2},$$

where we have used the fact that the typical separation times between instantons are much larger than the inverse of the characteristic oscillator scales of each minimum. (It takes the particle much longer to tunnel through a high barrier than to oscillate in either of the wells of the *real* potential.)

Second, there are also fluctuations around the instanton segments of the path. However, owing to the fact that an instanton takes a time $O(\omega^{-1}) \ll \Delta\tau$, where $\Delta\tau$ represents the typical time between instantons, one can neglect these contributions (which is to say that they can be absorbed into the prefactor K without explicit calculation). Within this approximation, setting $\tau_0 \equiv 0, \tau_{n+1} \equiv \tau$, the overall quantum fluctuation correction is given by

$$A_{n,\text{qu}}(\tau_1, \ldots, \tau_n) = \prod_{i=0}^{n} e^{-\omega(\tau_{i+1} - \tau_i)/2} = e^{-\omega\tau/2}, \tag{3.37}$$

which again is independent of the particular spacing configuration $\{\tau_i\}$. Combining Eq. (3.36) and (3.37), one finds that

$$G(a,\pm a;\tau) \simeq \sum_{\substack{n=0 \\ n\,\mathrm{even/odd}}}^{\infty} K^n e^{-nS_{\mathrm{inst}}/\hbar} e^{-\omega\tau/2} \overbrace{\int_0^\tau d\tau_1 \int_0^{\tau_1} d\tau_2 \cdots \int_0^{\tau_{n-1}} d\tau_n}^{\tau^n/n!} \qquad (3.38)$$

$$= e^{-\omega\tau/2} \sum_{\substack{n \\ n\,\mathrm{even/odd}}} \frac{1}{n!}\left(\tau K e^{-S_{\mathrm{inst}}/\hbar}\right)^n.$$

Finally, performing the sum, one obtains the transition amplitude

$$G(a,\pm a;\tau) \simeq C e^{-\omega\tau/2} \left\{ \begin{array}{l} \cosh(\tau K e^{-S_{\mathrm{inst}}/\hbar}), \\ \sinh(\tau K e^{-S_{\mathrm{inst}}/\hbar}), \end{array} \right. \qquad (3.39)$$

where the factor C depends in a non-exponential way on the transition time.

Before we turn to a discussion of the physical content of this result, let us check the self-consistency of our central working hypothesis – the diluteness of the instanton gas. To this end, consider the representation of G in terms of the partial amplitudes (3.38). To determine the typical number of instantons contributing to the sum, note that for a general sum $\sum_n c_n$ of positive quantities $c_n > 0$, the "typical" value of the summation index can be estimated as $\langle n \rangle \equiv \sum_n c_n n / \sum_n c_n$. With the abbreviation $X \equiv \tau K e^{-S_{\mathrm{inst}}/\hbar}$, it follows that

$$\langle n \rangle \equiv \frac{\sum_n n X^n/n!}{\sum_n X^n/n!} = X,$$

where, as long as $\langle n \rangle \gg 1$, the even/odd distinction in the sum is irrelevant. Thus, we can infer that the average instanton density, $\langle n \rangle/\tau = K e^{-S_{\mathrm{inst}}/\hbar}$, is both exponentially small in the instanton action S_{inst} and also independent of τ, confirming the validity of our diluteness assumptions above.

Finally, let us consider how the form of the transition amplitude (3.39) is understood in physical terms. To this end, consider the basic quantum mechanics of the problem (see the figure). Provided that there is no coupling across the barrier, the Hamiltonian has two independent, oscillator-like, sets of low-lying eigenstates sitting in the two local minima. Allowing for a weak inter–barrier coupling, these states individually split into a doublet of symmetric and antisymmetric eigenstates, $|S\rangle$ and $|A\rangle$, with energies ϵ_A and ϵ_S, respectively. Focusing on the low-energy sector formed by the ground state doublet, the transition amplitudes (3.30) can be expressed as

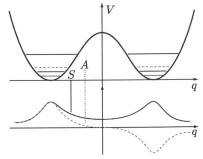

$$G(a,\pm a;\tau) \simeq \langle a| \left(|S\rangle e^{-\epsilon_S \tau/\hbar}\langle S| + |A\rangle e^{-\epsilon_A \tau/\hbar}\langle A| \right) |\pm a\rangle.$$

Setting $\epsilon_{A/S} = (\hbar\omega \pm \Delta\epsilon)/2$, the symmetry properties $|\langle a|S\rangle|^2 = |\langle -a|S\rangle|^2 = C/2$ and $\langle a|A\rangle\langle A|-a\rangle = -|\langle a|A\rangle|^2 = -C/2$ imply

$$G(a, \pm a; \tau) \simeq \frac{C}{2} \left(e^{-(\hbar\omega - \Delta\epsilon)\tau/2\hbar} \pm e^{-(\hbar\omega + \Delta\epsilon)\tau/2\hbar} \right) = Ce^{-\omega\tau/2} \begin{cases} \cosh(\Delta\epsilon\tau/\hbar), \\ \sinh(\Delta\epsilon\tau/\hbar), \end{cases}$$

Comparing this expression with Eq. (3.39), the interpretation of the instanton calculation becomes clear: at long times, the transition amplitude is obtained from the two lowest states – the symmetric and anti-symmetric combinations of the two oscillator ground states. The energy splitting $\Delta\epsilon$ accommodates the energy shift due to the tunneling between the two wells. Remarkably, the effect of tunneling has been obtained from a purely classical picture formulated in imaginary time! The instanton calculation also produces a prediction for the splitting of the energies due to tunneling,

$$\Delta\epsilon = \hbar K \exp(-S_{\text{inst}}/\hbar),$$

which, up to the prefactor, agrees with the result of a WKB-type analysis.

Before leaving this section, some general remarks on instantons are in order:

▷ In hindsight, was the approximation scheme used above consistent? In particular, terms at second order in $\hbar$ were neglected while terms non-perturbative in $\hbar$ (the instanton) were kept. Yet the former typically give rise to a larger correction to the energy than the latter. However, the large perturbative shift affects the energies of the symmetric and antisymmetric states equally. The instanton contribution gives the *leading* correction to the splitting of the levels. It is the latter that is likely to be of more physical significance.

▷ Second, it may appear that the machinery above was a bit of an "overkill" for describing a simple tunneling process. Indeed, the basic result (3.39) could have been obtained by more elementary means such as the WKB method. Why then have we discussed instantons at such length? One reason is that, even within a purely quantum mechanical framework, the instanton formulation is much stronger than WKB. The latter represents, by and large, an uncontrolled approximation: in general it is hard to tell whether WKB results are accurate or not. In contrast, the instanton approximation is controlled by well-defined expansion parameters. For example, by going beyond the semiclassical approximation and/or softening the diluteness assumption, the calculation of the transition amplitudes can, in principle, be driven to arbitrary accuracy.

▷ A second and, for our purposes, more important motivation is that instanton techniques are of crucial importance within higher-dimensional field theories (here we regard the path integral formulation of quantum mechanics as a (0 space + 1 time) one-dimensional field theory). The reason is that instantons are intrinsically non-perturbative, i.e., instanton solutions to stationary-phase equations describe physics inaccessible to perturbative expansions around a non-instanton sector of the theory. (For example, the instanton orbits in the example are not reachable by perturbative expansion around a trivial orbit.)

▷ A related feature of the instanton analysis above is that the *number* of instantons involved is a stable quantity; "stable" in the sense that, by including perturbative fluctuations around the n-instanton sector, say, one does not connect with the $n+2$ sector. Although no rigorous proof of this statement has been given, it should be clear heuristically: a trajectory involving n instantons between the hills of the inverted potential cannot be smoothly connected with one of a different number. Attempts to perturbatively interpolate between such configurations inevitably "cost" large action, greatly exceeding any stationary phase-like value. In this way, the individual instanton sectors are stabilized by large intermediate energy barriers.

INFO The analysis above provides a method to extract the tunneling rate between the quantum wells to a level of exponential accuracy. However, it is sometimes necessary to compute the exponential prefactor K. Although such a computation follows the general principles outlined for the single well, there are some idiosyncrasies in the tunneling system that warrant discussion. According to the general principles outlined in section 3.2.2, after integrating over Gaussian fluctuations around the saddle-point field configurations, the contribution to the transition amplitude from the n-instanton sector is given by

$$G_n = J \det \left(-m\partial_\tau^2 + V''(q_{\mathrm{cl},n}) \right) e^{-nS_{\mathrm{inst}}},$$

fluctuation determinant

where $q_{\mathrm{cl},n}(\tau)$ represents an n-instanton configuration and J the normalization. In the zero-instanton sector, the evaluation of the functional determinant recovers the familiar harmonic oscillator result, $G(a, a, \tau) = (m\omega/\pi\hbar)^{1/2} \exp[-\omega\tau/2]$. Let us now consider the one-instanton sector of the theory. To evaluate the **fluctuation determinant**, one must consider the spectrum of the operator $-m\partial_\tau^2 + V''(q_{\mathrm{cl},1})$. Differentiating the defining equation for $q_{\mathrm{cl},1}$ (3.32), one may confirm that

$$\left(-m\partial_\tau^2 + V''(q_{\mathrm{cl},1}) \right) \partial_\tau q_{\mathrm{cl},1} = 0,$$

i.e., the function $\partial_\tau q_{\mathrm{cl},1}$ presents a "zero mode" of the operator! The origin of the zero mode is elucidated by noting that a translation of the instanton along the time axis, $q_{\mathrm{cl},1}(\tau) \to q_{\mathrm{cl},1}(\tau + \delta\tau)$, leaves the action approximately invariant. However, for small $\delta\tau$, we have $q_{\mathrm{cl},1}(\tau + \delta\tau) \simeq q_{\mathrm{cl},1}(\tau) + \delta\tau \partial_\tau q_{\mathrm{cl},1}$, showing that, to linear order in $\delta\tau$, the function $\partial_\tau q_{\mathrm{cl},1}$ describes a zero action shift. For the same reason, $\delta\tau$ is a "zero-mode coordinate."

With this interpretation, it becomes clear how to repair the formula for the fluctuation determinant. While the Gaussian integral over fluctuations is controlled for the non-zero eigenvalues, its execution for the zero mode must be rethought. Indeed, by integrating over the coordinate of the instanton, $\int_0^\tau d\tau_0 = \tau$, one finds that the contribution to the transition amplitude in the one-instanton sector is given by

$$J\tau \sqrt{\frac{S_{\mathrm{inst}}}{2\pi\hbar}} \det{}' \left(-m\partial_\tau^2 + V''(q_{\mathrm{cl},1}) \right)^{-1/2} e^{-S_{\mathrm{inst}}},$$

where the prime indicates the exclusion of the zero mode from the determinant and the factor $\sqrt{S_{\mathrm{inst}}/2\pi\hbar}$ reflects the Jacobian associated with the change to a new set of integration variables which contains the zero-mode coordinate τ as one of its elements.[16] To fix the overall constant J, we normalize by the fluctuation determinant of the imaginary-time harmonic oscillator, i.e., we use the fact that (see section 3.2.4) the return amplitude of the latter evaluates to $G(a, a, \tau) = J \det(m(-\partial_\tau^2 + \omega^2)/2)^{-1/2} = (m\omega/\pi\hbar)^{1/2} e^{-\omega\tau/2}$, where the first/second representation is the imaginary time variant of Eq. (3.28)/Eq.(3.29). Using

[16] See, e.g., J. Zinn-Justin, *Quantum Field Theory and Critical Phenomena* (Oxford University Press, 1993) for an explicit calculation of this Jacobian.

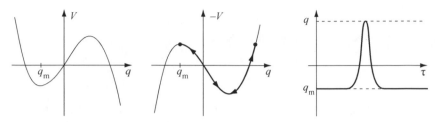

Fig. 3.2   Effective potential showing a metastable minimum together with the inverted potential and a sketch of a bounce solution. To obtain the tunneling rate, it is necessary to sum over a dilute gas of bounce trajectories.

this result, and noting that the zero-mode analysis above generalizes to the n-instanton sector, we find that the pre-exponential constant K used in our analysis of the double well affords the explicit representation

$$K = \omega \sqrt{\frac{S_{\text{inst}}}{2\pi\hbar}} \left(\frac{m\omega^2 \det' \left(-m\partial_\tau^2 + V''(q_{\text{cl},1}) \right)}{\det \left(-m\partial_\tau^2 + m\omega^2 \right)} \right)^{-1/2}.$$

Note that the instanton determinant depends sensitively on the particular nature of the potential $V(q)$. For the quartic potential, $V(q) = m\omega^2(x^2 - a^2)^2/8a^2$, the term in parentheses above is given by $1/12$ while $S_{\text{inst}} = 2m\omega a^2/3$. For further details, we refer to, e.g., Zinn-Justin (1993).[16]

EXERCISE A quantum particle moves in a periodic potential V with period a. Taking the Euclidean action for the instanton connecting two neighboring minima to be S_{inst}, express the propagator $G(ma, na; \tau)$, with m and n integer, as a sum over instantons and anti-instantons. Using the identity $\delta(q - q') = \int_0^{2\pi} \frac{d\theta}{2\pi} e^{i(q-q')\theta}$, show that $G(ma, na; \tau) \sim e^{-\omega\tau/2} \int_0^{2\pi} \frac{d\theta}{2\pi} e^{-i(n-m)\theta} \exp[2\Delta\epsilon\,\tau\cos\theta]$, where the notation follows section 3.3.1. Noting that, in the periodic system, the eigenfunctions are Bloch states $\psi_{p\alpha}(q) = e^{ipq} u_{p\alpha}(q)$, the Bloch function, show that the propagator is consistent with the spectrum $\epsilon_p = \omega/2 - 2\Delta\epsilon\cos(pa)$.

Escape from a metastable minimum: "bounces"

The instanton gas formulation can be adapted easily to describe quantum tunneling from a metastable state, such as that of an unstable nucleus. Consider the "survival probability" $|G(q_{\text{m}}, q_{\text{m}}; t)|^2$ of a particle captured in a metastable minimum q_{m} of a one-dimensional potential, such as that shown in fig. 3.2. As with the double well, in the Euclidean-time formulation of the path integral, the dominant contribution to $G(q_{\text{m}}, q_{\text{m}}; \tau)$ arises from the classical paths minimizing the action corresponding to the inverted potential (see fig. 3.2). However, in contrast with the double well, **bounce configuration** the classical solution takes the form of a **"bounce,"** i.e., the particle spends only a short time away from the potential minimum as there is only a single (metastable) minimum. As with the double well, one can expect multiple bounce trajectories to present a significant contribution. Summing over all bounce trajectories (note that, in this case, we have an exponential series and no even–odd parity effect), one obtains the survival probability

$$G(q_{\mathrm{m}}, q_{\mathrm{m}}; \tau) = Ce^{-\omega\tau/2} \exp\left(\tau K e^{-S_{\mathrm{bounce}}/\hbar}\right).$$

Analytically continuing to real time, $G(q_m, q_m; t) = Ce^{-i\omega t/2}e^{-\Gamma t/2}$, where the decay rate is given by $\Gamma/2 = |K|e^{-S_{\mathrm{bounce}}/\hbar}$. (Note that, on physical grounds, K must be imaginary.[17]).

EXERCISE Consider a heavy nucleus having a finite rate of α-decay. The nuclear forces are short range so that the rate of α-emission is controlled by the tunneling of α-particles under a Coulomb barrier. Taking the effective potential to be spherically symmetric, with a deep well of radius r_0 beyond which it decays as $U(r) = 2(Z-1)e^2/r$, where Z is the nuclear charge, find the temperature T of the nuclei above which α-decay is thermally assisted if the energy of the emitted particles is E_0. Estimate the mean energy of the α particles as a function of T.

EXERCISE A uniform electric field E is applied perpendicularly to the surface of a metal with work function W. Assuming that the electrons in the metal describe a Fermi gas of density n, with exponential accuracy, find the tunneling current at zero temperature ("cold emission"). Show that, effectively, only electrons with energy near the Fermi level are able to tunnel. With the same accuracy, find the current at non-zero temperature ("hot emission"). What is the most probable energy of tunneling electrons as a function of temperature?

3.3.2 Tunneling of quantum fields: "fate of the false vacuum"

Hitherto, we have focused on applications of the path integral to the quantum mechanics of point-like particles. However, the formalism can be straightforwardly extended to richer physical contexts. As an illustration, here we consider a setting where the tunneling object is not a point particle but an elastic continuum with infinitely many degrees of freedom.

Consider a situation where a continuous classical field can assume two equilibrium states with different energy densities. To be concrete, one may consider a harmonic chain confined to one or other minimum of an asymmetric quasi-one-dimensional "gutter-like" double well potential defined on an interval of length L (see the figure). When quantized, the state of higher energy density becomes unstable through barrier penetration – it is said to be a **false vacuum**.[18] Specifically, drawing on our discussion of the harmonic

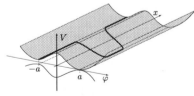

false
vacuum

---

[17] In fact, a more careful analysis shows that this estimate of the decay rate is too large by a factor 2 (for further details see, e.g., the discussion in Coleman[15]).

[18] For a detailed discussion of the history and ramifications of this idea, we refer to the original insightful paper by S. Coleman, *Fate of the false vacuum: Semiclassical theory*, Phys. Rev.

chain in chapter 1, let us consider a quantum system specified by the Hamiltonian density

$$\hat{\mathcal{H}} = \frac{\hat{\pi}^2}{2m} + \frac{k_s a^2}{2}(\partial_x \hat{\phi})^2 + V(\hat{\phi}), \qquad (3.40)$$

where $[\hat{\pi}(x), \hat{\phi}(x')] = -i\hbar\delta(x-x')$ and $V(\phi)$ represents a double well potential. The inclusion of a weak bias $-f\phi$ in $V(\phi)$ identifies a stable and a metastable potential minimum. Previously, we have seen that, in the absence of a confining potential, the quantum string exhibits low-energy collective wave-like excitations – phonons. In a confining potential, these harmonic fluctuations are rendered massive. However, drawing on the quantum mechanical principles established in the single-particle system, one might assume that the string tunnels freely between the two minima. To explore the capacity of the system to tunnel, let us suppose that, at time $t = 0$, the string is located in the (metastable) minimum of the potential at $\phi = -a$. What is the probability that the *entire* string will tunnel across the barrier into the potential minimum at $\phi = a$ in time t?

INFO The tunneling of fields between nearly degenerate ground states plays a role in numerous physical contexts. By way of example, consider a **superheated liquid**. In this context, the "false" vacuum is the liquid state, the true one the gaseous phase. The role of the field is taken by the local density distribution in the liquid. Thermodynamic fluctuations trigger the continuous appearance of vapor bubbles in the liquid. For small bubbles, the gain in volume energy is outweighed by the surface energy cost – the bubble will collapse. However, for bubbles beyond a certain critical size the energy balance is positive. The bubble will grow and, eventually, swallow the entire mass density of the system; the liquid has vaporized or, more formally, the density field has tunneled[19] from the false ground state into the true ground state.

More speculative (but more dramatic) manifestations of the phenomenon have been suggested in the context of **cosmology**: what if the Big Bang released our universe not into its true vacuum configuration, but into a state separated by a huge barrier from a more favorable sector of the energy landscape? In this case, the fate depends on the tunneling rate:

> *If this time scale is of the order of milliseconds, the universe is still hot when the false vacuum decays... if this time is of the order of years, the decay will lead to a sort of secondary Big Bang with interesting cosmological consequences. If this time is of the order of 10^9 years, we have occasion for anxiety.* (S. Coleman)

Previously, for the point-particle system, we saw that the transition probability between the minima of the double well is most easily accessed in the Euclidean-time framework. In the present case, anticipating our discussion of the quantum

D **15**, 2929 (1977). In fact, many ideas developed in this work were anticipated in an earlier analysis of metastability in the context of classical field theories by J. S. Langer, *Theory of the condensation point*, Ann. Phys. (NY) **41**, 108 (1967).

[19] At this point, readers should no longer be confused regarding the concept of "tunneling" in the context of a classical system. Within the framework of the path integral, the classical partition sum maps onto the path integral of a fictitious quantum system. It is this tunneling that we have in mind.

field integral later in the chapter, the Euclidean-time action associated with the
Hamiltonian density (3.40) assumes the form[20]

$$S[\phi] = \int_0^T d\tau \int_0^L dx \left(\frac{m}{2}(\partial_\tau \phi)^2 + \frac{k_s a^2}{2}(\partial_x \phi)^2 + V(\phi) \right),$$

where the time integral runs over the interval $[0, T = it]$. Here, for simplicity,
let us assume that the string obeys periodic boundary conditions in space, $\phi(x +$
$L, \tau) \equiv \phi(x, \tau)$. To estimate the tunneling amplitude, we will explore the survival
probability of the metastable state, imposing the boundary conditions $\phi(x, \tau = 0) =$
$\phi(x, \tau = T) = -a$ on the path integral. Once again, when the potential barrier is
high, and the time T is long, one may assume that the path integral is dominated
by the saddle-point field configuration of the Euclidean action. In this case, varying
the action with respect to the field $\phi(x, \tau)$, one obtains the classical equation of
motion

$$m\partial_\tau^2 \phi + k_s a^2 \partial_x^2 \phi = \partial_\phi V(\phi),$$

with the boundary conditions above.

Motivated by our consideration of the point-
particle problem, we might seek a solution in
which the string tunnels as a single rigid entity
without "flexing." However, it is evident from
the spatial translational invariance of the system that the instanton action would scale with
the system size L. In the infinite system $L \to \infty$,
such field configurations cannot contribute to
the tunneling amplitude. Instead, one must consider different ones, in which the transfer of the
chain occurs by degree. In this case, elements
of the string cross the barrier in a consecutive

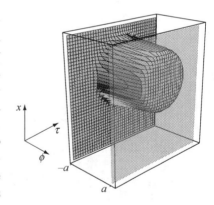

sequence as two outwardly propagating "domain walls" (see the figure, where the
emergence of such a "double-kink" configuration is shown as a function of space
and time). Such a field configuration is motivated by symmetry. After the rescaling
$x \mapsto v_s x$, where $v_s = \sqrt{k_s a^2/m}$ denotes the classical sound wave velocity, the saddle-
point equation assumes the isotropic form $m\partial^2 \phi = \partial_\phi V(\phi)$, where $\partial^2 = \partial_\tau^2 + \partial_x^2$.
Then, setting $r = \sqrt{x^2 + (\tau - T/2)^2}$, and sending $(T, L) \to \infty$, the space–time
rotational symmetry suggests a solution of the form $\phi = \phi(r)$, where $\phi(r)$ obeys
the radial diffusion equation

$$m\partial_r^2 \phi + \frac{m}{r}\partial_r \phi = \partial_\phi V$$

[20] Those readers who wish to verify this formula right away may (re-)discretize the harmonic chain,
present the transition amplitude as a product of Feynman path integrals for each element of
the string and, finally, take the continuum limit.

with the boundary condition $\lim_{r\to\infty}\phi(r) = -a$. This equation describes the one-dimensional motion of a particle in a potential $-V$ and subject to an apparent "friction force" $-mr^{-1}\partial_r\phi$ whose strength is inversely proportional to "time" r.

To understand the profile of the **bounce solution** suppose that, at time $r = 0$, the particle has been released from rest at a position slightly to the left of the (inverted) potential maximum at a. After rolling through the potential minimum it will climb the potential hill at $-a$. Now, the initial position may be fine-tuned in such a way that the viscous damping of the particle compensates for the excess potential energy (which would otherwise make the particle overshoot and disappear to infinity): there exists a solution where the particle starts close to $\phi = a$ and eventually ends up at $\phi = -a$, in accord with the imposed boundary conditions. In general, the analytical solution for the bounce depends sensitively on the form of the confining potential. However, if we assume that the well asymmetry imposed by external potential $-f\phi$ is small, the radial equation may be considerably simplified. In this limit, we may invoke a "thin-wall" approximation in which we assume that the bounce configuration is described by a domain wall of thickness Δr, at a radius $r_0 \gg \Delta r$, separating an inner region where $\phi(r < r_0) = a$ from the outer region where $\phi(r > r_0) = -a$. In this case, and to lowest order in an expansion in f, the action of the friction force is immaterial, i.e., we may set $m\partial_r^2\phi = \partial_\phi V$ – the very instanton equation formulated earlier for the point-particle system!

Substituting the solution back into S, one finds that the bounce (or kink-like) solution is characterized by the Euclidean action

$$S = v_{\mathrm{s}}\left(2\pi r_0 S_{\mathrm{inst}} - \pi r_0^2 2af\right),$$

where S_{inst} denotes the action of the instanton of the point-particle system (3.34), and the second term accommodates the effect of the potential bias on the field configuration. Crucially, the instanton contribution to the action scales with the circumference of the domain wall in space–time, while the contribution of the potential bias scales with the area of the domain. From this scaling dependence it is evident that, however small the external force f, at large enough r_0 the contribution of the second term will always outweigh the first and the string will tunnel from the metastable to the global minimum. More precisely, the optimal size of domain is found by minimizing the action with respect to r_0. In doing so, one finds that $r_0 = S_{\mathrm{inst}}/2af$. Substituting back into the action, one obtains the tunneling rate

$$\Gamma \sim \exp\left(-\frac{1}{\hbar}\frac{\pi v_{\mathrm{s}} S_{\mathrm{inst}}^2}{2af}\right).$$

It follows that, in the absence of an external force f, tunneling of the string across the barrier is *completely inhibited*! In the zero-temperature unbiased system, the symmetry of the quantum Hamiltonian is spontaneously broken: the ground state exhibits a two-fold degeneracy in which the string is confined to one potential minimum or another.

3.3.3 Tunneling in a dissipative environment

In the condensed matter context, it is infeasible to completely divorce a system from its environment. For example, the tunneling of an atom from one interstitial site in a crystal to another is influenced by its coupling to the phonon degrees of freedom that characterize the crystal lattice. By exchanging energy with the phonons, which act in the system as an external bath, a quantum particle can lose its phase coherence and, with it, its quantum mechanical character. Beginning with the seminal work of Caldeira and Leggett,[21] there have been numerous theoretical investigations of the effect of its environment on the quantum properties of a system. Such effects are particularly acute in systems where the quantum mechanical degree of freedom is *macroscopic*, such as the magnetic flux trapped in a superconducting quantum interference device (SQUID). In the following, we show that the Feynman path integral provides a natural (and almost unique) setting in which the effects of the environment on a microscopic or macroscopic quantum degree of freedom can be explored. For further discussion of the response of quantum wave coherence to environmental coupling, we refer to chapter 12.

macroscopic quantum tunneling

Before we begin, let us note that the phenomenon of **macroscopic quantum tunneling** is an active area of research with applications in atomic, molecular and optical (AMO) physics, and other fields. By contrast, our discussion here will target a particular illustrative application, and highlight only the guiding principles. For an in-depth discussion, we refer the reader to one of the many comprehensive reviews.[22]

Caldeira–Leggett model

Previously, we applied the path integral to study quantum tunneling of a particle q across a potential barrier $V(q)$. Here, we consider the influence of an external environment on tunneling. Following Caldeira and Leggett, we represent the environment by a **"bath" of N quantum harmonic oscillators** characterized by a set of frequencies $\{\omega_\alpha\}$,

oscillator bath

$$\hat{H}_{\text{bath}}[q_\alpha] = \sum_\alpha^N \left(\frac{\hat{p}_\alpha^2}{2m_\alpha} + \frac{m_\alpha}{2} \omega_\alpha^2 q_\alpha^2 \right).$$

For simplicity, let us suppose that the particle–bath coupling is linear in the bath coordinates, $\hat{H}_c[q, q_\alpha] = -\sum_\alpha^N f_\alpha[q] q_\alpha$, where $f_\alpha[q]$ represents some function of the particle coordinate q. Expressed as a path integral, the survival probability of a particle confined to a metastable minimum at a position $q = a$ can then be expressed as (taking $\hbar = 1$)

[21] A. O. Caldeira and A. J. Leggett, *Influence of dissipation on quantum tunneling in macroscopic systems*, Phys. Rev. Lett. **46**, 211 (1981).

[22] See, e.g., A. J. Leggett *et al.*, *Dynamics of the dissipative two-state system*, Rev. Mod. Phys. **59**, 1 (1976), U. Weiss, *Quantum Dissipative Systems* (World Scientific Publishing, 1993).

$$\langle a|e^{-i\hat{H}t/\hbar}|a\rangle = \int_{q(0)=q(t)=a} Dq\, e^{iS_{\mathrm{part}}[q]} \int Dq_\alpha\, e^{iS_{\mathrm{bath}}[q_\alpha]+iS_{\mathrm{c}}[q,q_\alpha]},$$

where $\hat{H} = \hat{H}_{\mathrm{part}} + \hat{H}_{\mathrm{bath}} + \hat{H}_{\mathrm{c}}$ denotes the total Hamiltonian of the system,

$$S_{\mathrm{part}}[q] = \int_0^t dt'\left(\frac{m}{2}\dot{q}^2 - V(q)\right), \quad S_{\mathrm{bath}}[q_\alpha] = \int_0^t dt' \sum_\alpha \frac{m_\alpha}{2}\left(\dot{q}_\alpha^2 - \omega_\alpha^2 q_\alpha^2\right)$$

denote, respectively, the actions of the particle and bath, while

$$S_{\mathrm{c}}[q,q_\alpha] = \int_0^t dt' \sum_\alpha \left(f_\alpha[q]q_\alpha + \frac{f_\alpha^2[q]}{2m_\alpha\omega_\alpha^2}\right)$$

represents their coupling.[23] Here, we assume that the functional integral over $q_\alpha(t)$ is taken over all field configurations of the bath, while the path integral over $q(t)$ is subject to the boundary conditions $q(0) = q(t) = a$. Since we are addressing a tunneling problem, it will again be useful to transfer to the Euclidean-time representation. For convenience, we assume the boundary conditions on the fields $q_\alpha(\tau)$ to be periodic on the interval $[0, T^{-1} \equiv \beta]$.[24]

To reveal the effect of the bath, we can integrate out the fluctuations q_α and thereby obtain an effective action for q. Being Gaussian in the coordinates q_α, the integration can be performed straightforwardly, and it induces a time-nonlocal interaction of the particle (exercise) $\langle a|e^{-i\hat{H}t/\hbar}|a\rangle = \int Dq\, e^{-S_{\mathrm{eff}}[q]}$, where the constant of integration has been absorbed into the measure and

$$S_{\mathrm{eff}}[q] = S_{\mathrm{part}}[q] + \frac{1}{2T} \sum_{\omega_n,\alpha} \frac{\omega_n^2 f_\alpha[q(\omega_n)]f_\alpha[q(-\omega_n)]}{m_\alpha\omega_\alpha^2(\omega_\alpha^2 + \omega_n^2)}.$$

Here, the sum $\sum_{\omega_n}$ runs over the discrete set of Fourier frequencies $\omega_n = 2\pi n T$ with n integer.[25] Then, if the coupling to the bath is linear, $f_\alpha[q(\tau)] = c_\alpha q(\tau)$, the effective action assumes the form (exercise)

$$\boxed{S_{\mathrm{eff}}[q] = S_{\mathrm{part}}[q] - T\int_0^\beta d\tau\, d\tau'\, q(\tau)K(\tau-\tau')q(\tau')}$$

where the kernel $K(\tau) = \int_0^\infty \frac{d\omega}{\pi} J(\omega)D_\omega(\tau)$, with

$$J(\omega) = \frac{\pi}{2}\sum_\alpha \frac{c_\alpha^2}{m_\alpha\omega_\alpha}\delta(\omega-\omega_\alpha), \qquad D_\omega(\tau) = -\sum_{\omega_n} \frac{\omega_n^2}{\omega(\omega^2+\omega_n^2)}e^{i\omega_n\tau}.$$

[23] The second term in the action of the coupling has been introduced to keep the effect of the environment minimally invasive (purely dissipative). If it were not present, the coupling to the oscillator degrees of freedom would effectively shift the extremum of the particle potential, i.e., change its potential landscape. (Exercise: Substitute the solutions of the Euler–Lagrange equations $\delta_{q_\alpha}S[q,q_\alpha] = 0$, computed for a fixed realization of q, into the action to obtain the said shift.)

[24] In section 3.4, we will see that these boundary conditions emerge naturally in the derivation of the integral from a many-body Hamiltonian.

[25] More precisely, anticipating our discussion of the Matsubara frequency representation below, we have defined the Fourier decomposition on the Euclidean-time interval β, setting $q(\tau) = \sum_m q_m e^{-i\omega_m\tau}, q_m = T\int_0^\beta d\tau\, q(\tau)e^{i\omega_m\tau}$, where $\omega_m = 2\pi m/\beta$ with m integer.

Physically, the time non-locality of the action is easily understood. Taken as a whole, the particle and the bath maintain quantum phase coherence. However, by exchanging fluctuations with the external bath, the particle experiences a self-interaction, retarded in time. The integration over the bath degrees of freedom involved in the generation of this interaction implies a "loss of information," which we expect to generate quantum mechanical phase decoherence. However, before developing this point, we first need to take a closer look at the dissipation kernel K itself.

In the representation above, which is standard in the field, the kernel K separates

bath spectral function

into a **bath spectral function** $J(\omega)$ and a time-dependent factor $D_\omega(\tau)$. While the latter describes the temporal retardation, the former describes the bath. Its job is to bundle the information contained in the oscillator masses, frequencies, and coupling constants into a single frequency-dependent function $J(\omega)$. For the small frequencies relevant to the description of the macroscopic degree of freedom, q, we expect the "density of bath modes" (i.e., the number of oscillators per frequency interval) to be a power law, implying that $J(\omega) \sim \omega^\alpha$ will be a power law too. In principle, α may take an arbitrary value. However, the most frequently encountered and physically important is the case $\alpha = 1$.

ohmic dissipation

INFO Consider, the coupling of a particle to a continuum of bosonic modes whose spectral density $J(\omega) = \eta\omega$ grows linearly with frequency. In this case of **ohmic dissipation**,

$$K(\omega_n) = \frac{\omega_n^2}{\pi} \int_0^\infty d\omega \, \frac{J(\omega)}{\omega(\omega^2 + \omega_n^2)} = \frac{\eta}{2}|\omega_n|.$$

Fourier transforming this expression, we obtain

$$K(\tau) = \frac{\pi T \eta}{2} \frac{1}{\sin^2(\pi T \tau)} \overset{\tau \ll T^{-1}}{\simeq} \frac{\eta}{2\pi T} \frac{1}{\tau^2}, \tag{3.41}$$

i.e., a strongly time-non-local "self-interaction" of the particle. To understand why $J(\omega) \sim \omega$ is termed *ohmic*, note that the induced linearity $K(\omega) \sim \omega$ corresponds to a single derivative in the time representation.[26] Thinking of q as a mechanical degree of freedom, this time derivative represents a friction term in the equations of motion. If q assumes the role of a fluctuating charge, the time derivative describes the presence of a resistor, hence the denotation "ohmic".

To explore the properties of the dissipative action, it is helpful to separate the non-local interaction according to the identity $q(\tau)q(\tau') = \frac{1}{2}(q(\tau)^2 + q(\tau')^2) - (q(\tau) - q(\tau'))^2/2$. The first contribution effectively renormalizes the potential $V(q)$ and presents an inessential perturbation, which can be absorbed in a redefined $V(q)$. By contrast, the remaining contribution, which is always positive, plays an important role.

---

[26] Here, we are sweeping the modulus in $|\omega_n|$ under the rug. The proper formulation of the argument requires the techniques of non equilibrium path integration to be introduced in chapter 12. However, the conclusion remains the same.

Dissipative quantum tunneling

Previously we have seen that the tunneling rate of a particle from a metastable potential minimum can be inferred from the extremal field configurations of the Euclidean action: the bounce trajectory. To explore the effect of dissipative coupling, it is necessary to understand how it revises the structure of the bounce solution. Now, in general, the non-local character of the interaction prohibits an exact solution of the classical equation of motion. In such cases, the effect of the dissipative coupling can be explored perturbatively or with the assistance of the renormalization group (see the discussion in section 6.1.2). However, by tailoring our choice of potential $V(q)$, we can gain some intuition about the more general situation.

To this end, let us consider a particle of mass m confined in a metastable minimum by a (semi-infinite) harmonic potential trap (see the figure),

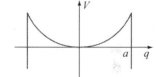

$$V(q) = \begin{cases} \frac{1}{2}m\omega_c^2 q^2, & 0 < |q| \le a, \\ -\infty, & |q| > a. \end{cases}$$

Further, let us assume that the environment imparts an ohmic dissipation with damping or "viscosity" η. To keep our discussion general, let us consider the combined impact of dissipation and temperature on the rate of tunneling from the potential trap. To do so, following Langer,[27] it is natural to investigate the "quasi-equilibrium" quantum partition function $\mathcal{Z}$ of the combined system. In this case, the tunneling rate appears as an imaginary contribution to the free energy $F = -T \ln \mathcal{Z}$, i.e., $\Gamma = -2 \operatorname{Im} F$.

Expressed as a path integral, the quantum partition function of the system takes the form $\mathcal{Z} = \int_{q(\beta)=q(0)} Dq \, e^{-S_{\text{eff}}/\hbar}$, where the ohmic dissipation kernel (3.41) enters the effective action via the contribution $-\frac{\eta}{2\pi} \int d\tau d\tau' \, q(\tau) K(\tau - \tau') q(\tau')$. Setting $q(\tau)q(\tau') = (q(\tau)^2 + q(\tau')^2)/2 - (q(\tau) - q(\tau'))^2/2$, and absorbing the first term into the potential $V(q)$, the Euclidean action assumes the form

$$S_{\text{eff}}[q] = \int_0^\beta d\tau \left(\frac{m}{2}\dot{q}^2 + V(q) \right) + \frac{\eta}{4\pi} \int_0^\beta d\tau \, d\tau' \left(\frac{q(\tau) - q(\tau')}{\tau - \tau'} \right)^2.$$

Once again, to estimate the tunneling rate, we will suppose that the barrier is high and the temperature is low, so that the path integral is dominated by stationary configurations of the action. In this case, one may identify three distinct solutions. First, the particle may remain at $q = 0$, poised precariously on the maximum of the inverted harmonic potential. Contributions from this solution and the associated harmonic fluctuations reproduce terms in the quantum partition function associated with states of the infintely extended harmonic potential trap. Second, there exists a singular solution in which the particle remains at the minimum of the inverted potential, i.e., perched on the potential barrier. The latter provides a negligible contribution to the quantum partition function and can be neglected. Finally, there

[27] J. S. Langer, *Theory of the condensation point*, Ann. Phys. (NY) **41**, 108 (1967).

exists a bounce solution in which a particle injected at a position q inside the well accelerates down the inverted potential gradient, is reflected from the potential barrier, and returns to the initial position q in a time β. While, in the limit $\beta \to \infty$, the path integral singles out the boundary condition $q(0) = q(\beta) \to 0$, at finite β the boundary condition will depart from 0 in a manner that depends nontrivially on the temperature. It is this general bounce solution that governs the decay rate.

Since, in the inverted potential, the classical bounce trajectory stays within the interval over which the potential is quadratic, a variation of the Euclidean action with respect to $q(\tau)$ yields the classical equation of motion

$$-m\ddot{q} + m\omega_c^2 q + \frac{\eta}{\pi}\int_0^\beta d\tau' \, \frac{q(\tau) - q(\tau')}{(\tau - \tau')^2} = A\delta(\tau - \beta/2),$$

where the term on the right-hand side of the equation imparts an impulse that changes the velocity of the particle discontinuously, while the coefficient A is chosen to ensure symmetry of the bounce solution on the Euclidean-time interval. Turning to the Fourier representation, the solution of the saddle-point equation then assumes the form

$$q_n = ATe^{-i\omega_n\beta/2}g(\omega_n), \quad g(\omega_n) \equiv [m(\omega_n^2 + \omega_c^2) + \eta|\omega_n|]^{-1}. \tag{3.42}$$

Imposing the condition that $q(\tau = \beta/2) = a$, one finds that $A = a/f$ where $f \equiv T\sum_{\omega_n} g(\omega_n)$. Finally, the action of the bounce is given by

$$S_{\text{bounce}} = \frac{1}{2T}\sum_{\omega_n}\left(m(\omega_n^2 + \omega_c^2) + \eta|\omega_m|\right)|q_n|^2 = \frac{a^2}{2f}. \tag{3.43}$$

In the following, we discuss the meaning of this expression in a number of limiting cases.

1. Let us first determine the **zero-temperature tunneling rate in the absence of dissipation** as a point of reference: $\eta \to 0$ and $\beta \to \infty$. In this case, the frequency summation translates to a continuous integral, $f = \int_{-\infty}^\infty \frac{d\omega}{2\pi}g(\omega) = (2m\omega_c)^{-1}$. Using this result, the bounce action (3.43) takes the form $S_{\text{bounce}} = m\omega_c a^2$. As one would expect, the tunneling rate $\Gamma \sim e^{-S_{\text{bounce}}}$ is controlled by the ratio of the potential barrier height $m\omega_c^2 a^2/2$ to the attempt frequency ω_c. Also notice that the bounce trajectory is given by

    $$q(\tau) = \frac{a}{f}\int_{-\infty}^\infty \frac{d\omega}{2\pi}\, e^{i\omega(\tau - \beta/2)}g(\omega) = a\, e^{-\omega_c|\tau - \beta/2|},$$

    i.e., as expected from our discussion in section 3.3.1, the particle spends only a time $1/\omega_c$ in the under-barrier region.

2. Staying at zero temperature, we next consider the **influence of dissipation** on the capacity for tunneling. We focus on the limit where the dynamics of the particle is overdamped, $\eta \gg m\omega_c$, with $f = \int_{-\infty}^\infty \frac{d\omega}{2\pi}g(\omega) \simeq (2/\pi\eta)\ln(\eta/m\omega_c)$, which implies that $S_{\text{bounce}} = \pi\eta a^2/(4\ln(\eta/m\omega_c))$. This result shows that, the coupling of the particle to the ohmic bath leads to an *exponential* suppression of the tunneling rate, while only a weak dependence on the jump frequency

persists. Physically, this result is easy to rationalize: under-barrier tunneling is a feature of the quantum mechanical system. In the transfer of energy to and from the external bath, the phase coherence of the particle is lost. At zero temperature, the tunneling rate becomes suppressed and the particle becomes confined.

3. Let us now consider the **influence of temperature on the tunneling rate when the dissipative coupling is inactive**, $\eta \to 0$. In this case, the discrete frequency summation takes the form[28] $f = T \sum_{\omega_n} g(\omega_n) = \coth(\beta\omega_c/2)/2\omega_c m$. Using this result, one obtains $S_{\text{bounce}} = m\omega_c a^2 \tanh(\beta\omega_c/2)$. In the low-temperature limit $\beta \to \infty$, we have $S_{\text{bounce}} = m\omega_c a^2$, as discussed above. At high temperatures $\beta \to 0$, one recovers a classical thermal dependence of the escape rate, $S_{\text{bounce}} \simeq \beta m\omega_c^2 a^2/2$, as expected from statistical mechanics.

4. We conclude with a brief remark on the **interplay of thermal activation with ohmic dissipation**. Applying the the Euler–Maclaurin formula $\sum_{m=0}^{\infty} f(m) = \int_0^{\infty} dx\, f(x) + \frac{f(0)}{2} - \frac{f'(0)}{12} + \cdots$ to relate discrete sums over Matsubara frequencies to their zero-temperature integral limits, one finds that $S_{\text{bounce}}(T) - S_{\text{bounce}}(T = 0) \propto \eta T^2$. This shows that, in the dissipative regime, an increase in temperature diminishes the tunneling rate with a scale proportion to the damping.

This concludes our introductory discussion of the application of path integration methods to dissipative quantum tunneling. Thanks to recent progress in atomic, molecular, and optical physics, large varieties of quantum mechanical few-body systems in and out of equilibrium have come under experimental control. In this line of research the concepts of path integration introduced in this chapter (and further advanced in chapter 12 to the many-body context) are crucially important. Readers interested in learning more are encouraged to study the classic references[22] and stay tuned to ongoing developments!

3.4 Construction of the Many-Body Field Integral

Having developed the single-particle path integral, we now consider its extension to many-particle systems: quantum field theory. Our starting point is analogous to that outlined at the beginning of the chapter. Just as there are two different approaches to quantum mechanics, quantum field theory can also be formulated in two different ways: the formalism of canonically quantized field operators and functional integration. As to the former, although much of the technology needed to implement this framework – essentially Feynman diagrams – originated in high-energy physics, it was with the development of condensed matter physics through

---

[28] For details on how to implement the discrete frequency summation, see the Info block on page 141.

the 1950s to 1970s that this approach was driven to unprecedented sophistication. The reason is that, almost as a rule, problems in condensed matter investigated at that time necessitated perturbative summations to infinite order in the nontrivial content of the theory (typically interactions). This requirement led to the development of advanced techniques to sum perturbation series in many-body interaction operators.

However, in the 1970s, non-perturbative problems began to attract increasing attention – a still prevailing trend – and it turned out that the formalism of canonically quantized operators was not tailored to this type of physics. By contrast, the alternative approach to many-body problems, functional integration, is ideally suited to it! The situation is similar to the one described earlier, where we saw that the Feynman path integral provided a spectrum of novel routes to approaching quantum mechanical problems (parametrically controlled semiclassical limits, analogies to classical mechanics, statistical mechanics, etc.). Similarly, the introduction of field integration in many-body physics spawned new theoretical developments. In fact, the advantage of the path integral approach in many-body physics is more pronounced than in single-particle quantum mechanics: higher dimensionality introduces more complex fields, and along with them concepts of geometry and topology enter the stage. The ensuing structures are conveniently exposed within the field integral framework. Moreover, the connections to classical statistical mechanics play a more important role than in single-particle quantum mechanics. These concepts will be addressed in subsequent chapters when applications of the field integral are discussed.

Before turning to the quantitative construction of the field integral, it is instructive to anticipate the structures we should expect to be emerging. In quantum mechanics, we were starting from a point particle characterized by a coordinate q (or other quantum numbers for that matter). Path integration then meant integration over all time-dependent configurations $q(t)$, i.e., a set of curves $t \mapsto q(t)$ (see fig. 3.3, upper panel). By contrast, the degrees of freedom of field theory are continuous objects $\phi(x)$ in themselves: x parameterizes a d-dimensional base manifold and ϕ takes values in some target manifold (fig. 3.3, lower panel). The natural generalization of a "path" integral then implies integration over a single copy of these objects at each instant of time, i.e., we must integrate over generalized surfaces, mappings from $(d + 1)$-dimensional space–time into the field manifold, $(x, t) \mapsto \phi(x, t)$. While this notion may sound daunting, it is important to realize that, conceptually, nothing much changes in comparison with the path integral: instead of a one-dimensional manifold – a curve – our object of integration will be a $(d+1)$-dimensional manifold.

3.4.1 Construction of the field integral

The construction of the many-body path integral (henceforth *field integral* for brevity) follows the general scheme outlined at the end of section 3.2.3. As before, we start with the segmentation of the time evolution of a quantum many-body Hamiltonian into infinitesimal time slices. The goal then is to absorb as much as

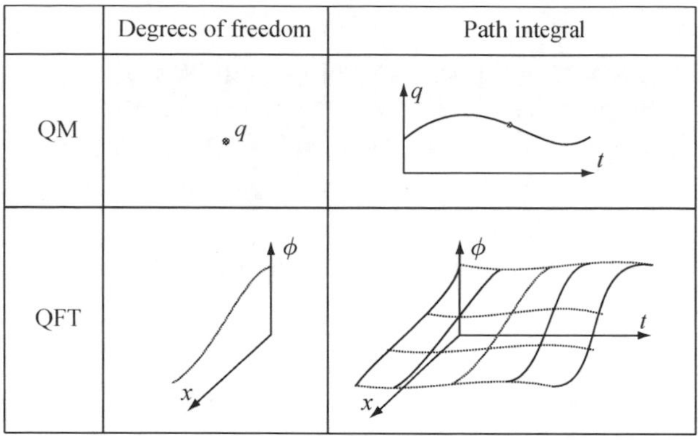

	Degrees of freedom	Path integral
QM	q	
QFT		

Fig. 3.3   The concept of field integration. Upper panels: Path integral of quantum mechanics – integration over all time-dependent configurations of a point particle degree of freedom leads to integrals over curves. Lower panels: Field integral – integration over time-dependent configurations of d-dimensional continuum mappings (fields) leads to integrals over generalized $(d+1)$-dimensional surfaces.

possible of the quantum dynamical phase accumulated during the short-time propagation into a set of suitably chosen eigenstates. But how should these eigenstates be chosen? In the context of single-particle quantum mechanics, the structure of the Hamiltonian suggested a representation in terms of coordinate and momentum eigenstates. Remembering that many-particle Hamiltonians are conveniently expressed in terms of creation and annihilation operators, an obvious idea would be to search for eigenstates of *these* operators. Such states indeed exist and are called **coherent states**.

Coherent states (bosons)

Our goal is to find eigenstates of the (non-hermitian) Fock space operators $a^\dagger$ and a. Although the general form of these states will turn out to be the same for bosons and fermions, there are differences regarding their algebraic structure. The point is that the anticommutation relations of fermions require that the eigenvalues of an annihilation operator themselves anticommute, i.e., they *cannot* be ordinary numbers. Postponing the introduction of the unfamiliar concept of anticommuting "numbers" to the next section, we first concentrate on the bosonic case, where problems of this kind do not arise.

So, what form do the eigenstates $|\phi\rangle$ of the bosonic Fock space operators a and $a^\dagger$ take? Being a state in the Fock space, an eigenstate $|\phi\rangle$ can be expanded as

$$|\phi\rangle = \sum_{n_1, n_2, \dots} C_{n_1, n_2, \dots} |n_1, n_2, \dots\rangle, \qquad |n_1, n_2, \dots\rangle = \frac{(a_1^\dagger)^{n_1}}{\sqrt{n_1!}} \frac{(a_2^\dagger)^{n_2}}{\sqrt{n_2!}} \dots |0\rangle,$$

where $a_i^\dagger$ creates a boson in state i, $C_{n_1, n_2, \dots}$ are expansion coefficients, and $|0\rangle$ is the vacuum. (Here, it is convenient to adopt this convention for the vacuum as opposed to the notation $|\Omega\rangle$ used previously.) Furthermore, the many-body state $|n_1, n_2, \ldots\rangle$ is indexed by a set of occupation numbers: n_1 in state $|1\rangle$, n_2 in state $|2\rangle$, and so on. Importantly, the state $|\phi\rangle$ may contain superpositions of basis states containing different numbers of particles. Now, if the minimum number of particles in state $|\phi\rangle$ is n_0, the minimum of $a_i^\dagger|\phi\rangle$ must be $n_0 + 1$. Clearly, the creation operators $a_i^\dagger$ themselves cannot possess eigenstates.

However, with annihilation operators this problem does not arise. Indeed, the annihilation operators do possess eigenstates, known as **bosonic coherent states**,

bosonic
coherent
states

$$\boxed{|\phi\rangle \equiv \exp\Big(\sum_i \phi_i a_i^\dagger\Big)|0\rangle} \tag{3.44}$$

where the elements of $\phi = \{\phi_i\}$ represent a set of *complex* numbers. The states $|\phi\rangle$ are eigenstates in the sense that, for all i,

$$\boxed{a_i|\phi\rangle = \phi_i|\phi\rangle} \tag{3.45}$$

i.e., they simultaneously diagonalize all annihilation operators. Noting that a_i and $a_j^\dagger$, with $j \neq i$, commute, Eq. (3.45) can be verified by showing that $a \exp(\phi a^\dagger)|0\rangle = \phi \exp(\phi a^\dagger)|0\rangle$.[29] Although not crucial to the practice of field integration, in the construction of the path integral it will be useful to assimilate some further properties of coherent states.

▷ By taking the hermitian conjugate of (3.45), we find that the "bra" associated with the "ket" $|\phi\rangle$ is a left eigenstate of the creation operators, i.e., for all i,

$$\langle\phi|a_i^\dagger = \langle\phi|\bar{\phi}_i, \tag{3.46}$$

where $\bar{\phi}_i$ is the complex conjugate of ϕ_i, and $\langle\phi| = \langle0|\exp(\sum_i \bar{\phi}_i a_i)$.

▷ It is a straightforward matter – e.g., by a Taylor expansion of (3.44) – to show that the action of a creation operator on a coherent state yields the identity

$$a_i^\dagger|\phi\rangle = \partial_{\phi_i}|\phi\rangle. \tag{3.47}$$

Reassuringly, it may be confirmed that Eq. (3.47) and (3.45) are consistent with the commutation relations $[a_i, a_j^\dagger] = \delta_{ij}$: we have $[a_i, a_j^\dagger]|\phi\rangle = (\partial_{\phi_j}\phi_i - \phi_i\partial_{\phi_j})|\phi\rangle = \delta_{ij}|\phi\rangle$.

▷ Making use of the relation $\langle\theta|\phi\rangle = \langle0|\exp(\sum_i \bar{\theta}_i a_i)|\phi\rangle = \exp(\sum_i \bar{\theta}_i\phi_i)\langle0|\phi\rangle$, one finds that the overlap between two coherent states is given by

$$\langle\theta|\phi\rangle = \exp\Big(\sum_i \bar{\theta}_i\phi_i\Big). \tag{3.48}$$

▷ From this result, it follows that the norm of a coherent state is given by

$$\langle\phi|\phi\rangle = \exp\Big(\sum_i \bar{\phi}_i\phi_i\Big). \tag{3.49}$$

---

[29] Using the result $[a, (a^\dagger)^n] = n(a^\dagger)^{n-1}$ (cf. Eq. (2.37)), a Taylor expansion shows that $ae^{\phi a^\dagger}|0\rangle = [a, e^{\phi a^\dagger}]|0\rangle = \sum_{n=0}^{\infty} \frac{\phi^n}{n!}[a, (a^\dagger)^n]|0\rangle = \phi\sum_{n=1}^{\infty} \frac{\phi^{n-1}}{(n-1)!}(a^\dagger)^{n-1}|0\rangle = \phi e^{\phi a^\dagger}|0\rangle$.

▷ Most importantly, the coherent states form a complete – in fact an over-complete – set of states in Fock space:

$$\boxed{\int \prod_i \frac{d\phi_i}{\pi} \exp\left(-\sum_i \bar\phi_i \phi_i\right) |\phi\rangle\langle\phi| = \mathbb{1}_{\mathcal{F}}} \tag{3.50}$$

where $d\phi_i = d\,\mathrm{Re}\,\phi_i\, d\,\mathrm{Im}\,\phi_i$, and $\mathbb{1}_{\mathcal{F}}$ represents the unit operator or identity in the Fock space.

Schur's lemma

INFO The proof of Eq. (3.50) proceeds by application of **Schur's lemma**. The action of the creation and annihilation operators in Fock space is *irreducible* in the sense that any state can be represented by the action of these operators on a reference state (such as the vacuum). Under these circumstances, Schur's lemma states that if an operator commutes with all $\{a_i, a_i^\dagger\}$ it must be proportional to the unit operator. (Refer for a comprehensive discussion of Schur's lemma to the mathematical literature, the essence of the statement is that only multiples of the unit matrix commute with all linear transformations of a vector space.) Specifically, the commutation property of the l.h.s. of Eq. (3.50) is verified as follows

$$\begin{aligned}
a_i \int d\phi\, e^{-\bar\phi\phi} |\phi\rangle\langle\phi| &= \int d\phi\, e^{-\bar\phi\phi} \phi_i |\phi\rangle\langle\phi| \\
&= -\int d\phi\, \left(\partial_{\bar\phi_i} e^{-\bar\phi\phi}\right) |\phi\rangle\langle\phi| \stackrel{\text{by parts}}{=} \int d\phi\, e^{-\bar\phi\phi} |\phi\rangle \left(\partial_{\bar\phi_i}\langle\phi|\right) \\
&= \int d\phi\, e^{-\bar\phi\phi} |\phi\rangle\langle\phi| a_i,
\end{aligned} \tag{3.51}$$

where we have set $d\phi \equiv \prod_i d\phi_i/\pi$ for brevity, and used the shorthand notation $\sum_i \bar\phi_i \phi_i \equiv \bar\phi\phi$. Taking the adjoint of Eq. (3.51), one verifies the commutativity with the creation operators, and hence the above proportionality statement. To fix the constant of proportionality, one can simply take the overlap with the vacuum:

$$\int d\phi\, e^{-\bar\phi\phi} \langle 0|\phi\rangle\langle\phi|0\rangle = \int d\phi\, e^{-\bar\phi\phi} = 1. \tag{3.52}$$

Taken together, Eqs. (3.51) and (3.52) prove (3.50). Note that the coherent states are over-complete in the sense that they are not pairwise orthogonal (see Eq. (3.48)). The exponential weight $\exp(-\bar\phi\phi)$ appearing in the resolution of the identity compensates for the overcounting achieved by integrating over the whole set of coherent states.

With these definitions we have the basis to construct the path integral for bosonic systems. However, before doing so, we will first introduce the fermionic version of the coherent state. This will allow us to construct the path integrals for bosons and fermions simultaneously, thereby emphasizing the similarity of their structure.

Coherent states (fermions)

Much of the formalism above generalizes to the fermionic case. As before, it is evident that creation operators cannot possess eigenstates. Following the bosonic case, let us suppose that the annihilation operators are characterized by a set of coherent states such that, for all i,

$$\boxed{a_i|\eta\rangle = \eta_i|\eta\rangle} \tag{3.53}$$

where η_i is the eigenvalue. Although the structure of this equation appears to be equivalent to its bosonic counterpart (3.45) it has one frustrating feature: the anticommutativity of the fermionic operators, $[a_i, a_j]_+ = 0$, where $i \neq j$, implies that the eigenvalues η_i also have to anticommute:

$$\boxed{\eta_i\eta_j = -\eta_j\eta_i} \tag{3.54}$$

Clearly, these objects cannot be ordinary numbers. In order to define a fermionic version of coherent states, we have two choices: we may (a) accept (3.54) as a working definition and pragmatically explore its consequences, or (b) try to remove any mystery from the definitions (3.53) and (3.54). This latter task is tackled in the Info block below, where objects $\{\eta_i\}$ with the desired properties are defined in a mathematically clean manner. Readers wishing to proceed more rapidly may skip this exposition and turn directly to the more praxis-oriented discussion below.

Hermann Günther Grassmann 1809–77 was a German linguist, publisher, physicist, and mathematician. Decades before the formal definition of linear spaces around 1920, he invented foundations of linear algebra. Curiously, his best remembered single contribution, exterior algebra, appeared as a byproduct of his thinking about tidal waves.

INFO   There is a mathematical structure ideally suited to generalize the concept of ordinary number (fields): **algebras**. An algebra $\mathcal{A}$ is a vector space endowed with a multiplication rule $\mathcal{A} \times \mathcal{A} \to \mathcal{A}$. We can construct an algebra $\mathcal{A}$ tailored to our needs by starting out from a set of elements, or generators, $\eta_i \in \mathcal{A}, i = 1, \ldots, N$, and imposing the following rules:

(i)   The elements η_i can be added and multiplied by complex numbers, e.g.,

$$c_0 + c_i\eta_i + c_j\eta_j \in \mathcal{A}, \qquad c_0, c_i, c_j \in \mathbb{C}, \tag{3.55}$$

such that $\mathcal{A}$ is a complex vector space.

(ii)   The product, $\mathcal{A} \times \mathcal{A} \to \mathcal{A}, (\eta_i, \eta_j) \mapsto \eta_i\eta_j$, is defined to be associative and anticommutative, viz. (3.54). Owing to the associativity of this operation, there is no ambiguity when it comes to forming products of higher order, i.e., $(\eta_i\eta_j)\eta_k = \eta_i(\eta_j\eta_k) \equiv \eta_i\eta_j\eta_k$. The definition requires that products of odd order in the number of generators anticommute, while (even, even) and (even, odd) combinations commute (exercise).

By virtue of (i) and (ii), the set $\mathcal{A}$ of all linear combinations

$$c_0 + \sum_{n=1}^{\infty} \sum_{i_1, \ldots i_n = 1}^{N} c_{i_1, \ldots, i_n} \eta_{i_1} \ldots \eta_{i_n}, \qquad c_0, c_{i_1, \ldots, i_n} \in \mathbb{C},$$

spans a finite-dimensional associative algebra $\mathcal{A}$,[30] known as the **Grassmann algebra** or **exterior algebra**. For completeness we mention that Grassmann algebras find other realizations, a particularly important one being the algebra of alternating differential forms, which will play a role later in the text (see appendix section A.1).

[30]   Whose dimension can be shown to be 2^N (exercise).

Apart from their anomalous commutation properties, the generators $\{\eta_i\}$, and their product generalizations $\{\eta_i\eta_j, \eta_i\eta_j\eta_k, \ldots\}$, resemble ordinary, if anticommutative, numbers. (In practice, the algebraic structure underlying their definition can safely be ignored. All we will need to remember is the rule (3.54) and the property (3.55).) We emphasize that $\mathcal{A}$ contains not only anticommuting but also commuting elements, i.e., linear combinations of an even number of Grassmann numbers η_i are overall commutative. (This mimics the behavior of the Fock space algebra: products of an even number of annihilation operators $a_i a_j \ldots$ commute with all other linear combinations of operators a_i. In spite of this similarity, the "numbers" η_i must not be confused with the Fock space operators; there is nothing on which they act.)

To make practical use of the new concept, we need to go beyond the level of pure arithmetic. Specifically, we need to introduce functions of anticommuting numbers and also elements of calculus. Remarkably, most of the concepts of calculus – differentiation, integration, etc. – generalize to anticommuting number fields, and in fact turn out to be *simpler* than in ordinary calculus!

▷ Functions of Grassmann numbers are defined via their Taylor expansion:

$$f(\xi_1, \ldots, \xi_k) = \sum_{n=0}^{\infty} \sum_{i_1, \ldots, i_n=1}^{k} \frac{1}{n!} \frac{\partial^n f}{\partial \xi_{i_1} \cdots \partial \xi_{i_n}}\bigg|_{\xi=0} \xi_{i_n} \cdots \xi_{i_1}, \qquad (3.56)$$

where $\xi_1, \ldots, \xi_k \in \mathcal{A}$ and f is an analytic function. Note that the anticommutation properties of the algebra imply that the series terminates after a finite number of terms. For example, in the simple case where η is first order in the generators of the algebra, $f(\eta) = f(0) + f'(0)\eta$ (since $\eta^2 = 0$) – functions of Grassmann variables are fully characterized by a finite number of Taylor coefficients!

▷ Differentiation with respect to Grassmann numbers is defined by

$$\boxed{\partial_{\eta_i} \eta_j = \delta_{ij}} \qquad (3.57)$$

Note that, in order to be consistent with the commutation relations, the differential operator ∂_{η_i} must itself be anticommutative. In particular, $\partial_{\eta_i} \eta_j \eta_i \overset{i \neq j}{=} -\eta_j$.

▷ Integration over Grassmann variables is defined by

$$\boxed{\int d\eta_i = 0, \qquad \int d\eta_i \, \eta_i = 1} \qquad (3.58)$$

Note that the definitions (3.56), (3.57), and (3.58) imply that the actions of Grassmann differentiation and integration are effectively identical, i.e.,

$$\int d\eta \, f(\eta) = \int d\eta \, (f(0) + f'(0)\eta) = f'(0) = \partial_\eta f(\eta).$$

With this background, let us now apply the Grassmann algebra to the construction of fermion coherent states. To this end, we must enlarge the algebra so as to allow for the multiplication of Grassmann numbers by fermion operators. In order to be consistent with the anticommutation relations, we require that fermion operators and Grassmann generators anticommute,

$$[\eta_i, a_j]_+ = 0. \tag{3.59}$$

It is then straightforward to demonstrate that

$$\boxed{|\eta\rangle = \exp\Big(-\sum_i \eta_i a_i^\dagger \Big)|0\rangle} \tag{3.60}$$

fermionic coherent states

for the **fermionic coherent states**, i.e., by a structure perfectly analogous to the bosonic states (3.44).[31] It is a straightforward and useful exercise to demonstrate that the properties (3.46), (3.47), (3.48), (3.49) and, most importantly, (3.50) carry over to the fermionic case. One merely has to identify a_i with a fermionic operator and replace the complex variables ϕ_i by $\eta_i \in \mathcal{A}$. Apart from a few sign changes and the $\mathcal{A}$-valued nature of the arguments, the fermionic coherent states differ only in two respects from their bosonic counterpart: first, the Grassmann variables $\bar{\eta}_i$ appearing in the adjoint of a fermion coherent state,

$$\boxed{\langle \eta| = \langle 0| \exp\Big(-\sum_i a_i \bar{\eta}_i \Big) = \langle 0| \exp\Big(\sum_i \bar{\eta}_i a_i \Big)}$$

are not related to the η_i of the states $|\eta\rangle$ via some kind of complex conjugation. Rather η_i and $\bar{\eta}_i$ are independent variables.[32] Second, the Grassmann version of a Gaussian integral (see below), $\int d\eta\, e^{-\bar{\eta}\eta} = 1$, does not contain the factors of π characteristic of standard Gaussian integrals. Thus, the measure of the fermionic analog of (3.50) does not contain a π in the denominator.

For the sake of future reference, the most important properties of Fock space coherent states are summarized in Table 3.1.

Grassmann Gaussian integration

Grassmann Gaussian integral

Before turning to the field integral, we need to consider the generalization of **Gaussian integration for Grassmann variables**. The prototype of all Grassmann Gaussian integration formulae reads

---

[31] To prove that the states (3.60) indeed fulfill the defining relation (3.53), we note that
$$a_i e^{-\eta_i a_i^\dagger}|0\rangle \overset{(3.56)}{=} a_i(1 - \eta_i a_i^\dagger)|0\rangle \overset{(3.59)}{=} \eta_i a_i a_i^\dagger|0\rangle = \eta_i|0\rangle = \eta_i(1 - \eta_i a_i^\dagger)|0\rangle = \eta_i e^{-\eta_i a_i^\dagger}|0\rangle.$$
This, in combination with the fact that a_i and $\eta_j a_j^\dagger$ ($i \neq j$) commute, proves Eq. (3.53). Note that the proof is simpler than in the bosonic case: the fermionic Taylor series terminates after the first contribution. This observation is representative of a general rule: Grassmann calculus is simpler than standard calculus – all series are finite, integrals always converge, etc.

[32] In the literature, a complex conjugation of Grassmann variables is sometimes defined. Although appealing from an aesthetic point of view – symmetry between bosons and fermions – this **supersymmetry** concept is problematic. The difficulties become apparent when **supersymmetric theories** are considered, i.e., theories where operator algebras contain both bosons and fermions (so-called superalgebras). It is not possible to introduce a complex conjugation consistent with the super-algebra commutation relations and therefore the idea had better be abandoned altogether. (Unlike the bosonic case, where complex conjugation is required to define convergent Gaussian integrals, no such need arises in the fermionic case.)

Table 3.1 Properties of coherent states for bosons ($\zeta = 1$, $\psi_i \in \mathbb{C}$) and fermions ($\zeta = -1$, $\psi_i \in \mathcal{A}$). In the last line, the integration measure is defined as $d(\bar{\psi}, \psi) \equiv \prod_i \frac{d\bar{\psi}_i \, d\psi_i}{\pi^{(1+\zeta)/2}}$.

Definition	$\lvert\psi\rangle = \exp\left(\zeta \sum_i \psi_i a_i^\dagger\right)\lvert 0\rangle$
Action of a_i	$a_i\lvert\psi\rangle = \psi_i\lvert\psi\rangle, \quad \langle\psi\rvert a_i = \partial_{\bar{\psi}_i}\langle\psi\rvert$
Action of $a_i^\dagger$	$a_i^\dagger\lvert\psi\rangle = \zeta\partial_{\psi_i}\lvert\psi\rangle, \quad \langle\psi\rvert a_i^\dagger = \langle\psi\rvert\bar{\psi}_i$
Overlap	$\langle\psi'\lvert\psi\rangle = \exp\left(\sum_i \bar{\psi}_i'\psi_i\right)$
Completeness	$\int d(\bar{\psi}, \psi)\, e^{-\sum_i \bar{\psi}_i\psi_i}\lvert\psi\rangle\langle\psi\rvert = \mathbb{1}_F$

$$\boxed{\int d\bar{\eta}d\eta\; e^{-\bar{\eta}a\eta} = a} \tag{3.61}$$

This equation is derived by a first-order Taylor expansion of the exponential and application of Eq. (3.58). Its multi-dimensional generalization to matrix and vector structures is given by

$$\int d\bar{\phi}d\phi\, e^{-\bar{\phi}^T \mathbf{A}\phi} = \det \mathbf{A}, \tag{3.62}$$

where $\bar{\phi}$ and ϕ are N-component vectors of Grassmann variables, the measure $d\bar{\phi}d\phi \equiv \prod_{i=1}^N d\bar{\phi}_i d\phi_i$, and $\mathbf{A}$ is an *arbitrary* complex matrix. For matrices that are unitarily diagonalizable, $\mathbf{A} = \mathbf{U}^\dagger\mathbf{D}\mathbf{U}$, Eq. (3.62) is proven in the same way as its complex counterpart (3.17): through the change of variables $\phi \to \mathbf{U}^\dagger\phi$, $\bar{\phi} \to \mathbf{U}^T\bar{\phi}$. Since $\det \mathbf{U} = 1$, the measure remains invariant (see the Info block below) and leaves us with N decoupled integrals of the type (3.61). The resulting product of N eigenvalues is just the determinant of $\mathbf{A}$ (see the later discussion of the partition function of non-interacting gas). For non-unitarily diagonalizable $\mathbf{A}$, the identity is established by a straightforward expansion of the exponent. The expansion terminates at Nth order and, by commuting through integration variables, it may be shown that the resulting Nth-order polynomial of matrix elements of $\mathbf{A}$ is the determinant.

INFO As with ordinary integrals, Grassmann integrals can be subjected to **variable transforms**. Suppose we are given an integral $\int d\bar{\phi}d\phi\, f(\bar{\phi}, \phi)$ and wish to change variables according to

$$\bar{\nu} = \mathbf{M}\bar{\phi}, \qquad \nu = \mathbf{M}'\phi, \tag{3.63}$$

where, for simplicity, $\mathbf{M}$ and $\mathbf{M}'$ are complex matrices (i.e., we here restrict ourselves to linear transforms). One can show that[33]

$$\bar{\nu}_1 \cdots \bar{\nu}_N = (\det \mathbf{M})\bar{\phi}_1 \cdots \bar{\phi}_N, \quad \nu_1 \cdots \nu_N = (\det \mathbf{M}')\phi_1 \cdots \phi_N. \tag{3.64}$$

---

[33] There are different ways to prove this identity. The most straightforward is by explicitly expanding Eq. (3.63) in components and commuting all Grassmann variables to the right. A more elegant way is to argue that the coefficient relating the right- and left-hand sides of Eq. (3.64) must be an Nth-order polynomial of matrix elements of $\mathbf{M}$. In order to be consistent with the anticommutation behavior of Grassmann variables, the polynomial must obey commutation relations, which uniquely characterize a determinant. Exercise: Check the relation for $N = 2$.

On the other hand, the integral of the new variables must obey the defining relation $\int d\bar{\nu}\,\bar{\nu}_1 \cdots \bar{\nu}_N = \int d\nu\,\nu_1 \cdots \nu_N = (-)^{N(N-1)/2}$, where $d\bar{\nu} = \prod_{i=1}^{N} d\bar{\nu}_i$ and the sign on the r.h.s. is attributed to ordering of the integrand, i.e., $\int d\nu_1 d\nu_2\,\nu_1\nu_2 = -\int d\nu_1\,\nu_1 \int d\nu_2\,\nu_2 = -1$. Together Eqs. (3.64) and (3.63) enforce the identities $d\bar{\nu} = (\det \mathbf{M})^{-1} d\bar{\phi}$, $d\nu = (\det \mathbf{M}')^{-1} d\phi$, which combine to give

$$\int d\bar{\phi} d\phi\, f(\bar{\phi}, \phi) = \det(\mathbf{M}\mathbf{M}') \int d\bar{\nu} d\nu\, f(\bar{\phi}(\bar{\nu}), \phi(\nu)).$$

Keeping the analogy with ordinary commuting variables, the Grassmann version of Eq. (3.18) reads

$$\int d\bar{\phi} d\phi\, \exp\left(-\bar{\phi}^T \mathbf{A} \phi + \bar{\nu}^T \cdot \phi + \bar{\phi}^T \cdot \nu\right) = \det \mathbf{A}\, \exp\left(\bar{\nu}^T \mathbf{A}^{-1} \nu\right). \qquad (3.65)$$

To prove the latter, we note that $\int d\eta\, f(\eta) = \int d\eta\, f(\eta + \nu)$, i.e., in Grassmann integration, one can shift variables as in the ordinary case. The proof of the Gaussian relation above thus proceeds in complete analogy to the complex case. As with Eq. (3.18), (3.65) can also be employed to generate further integration formulae. Defining $\langle \cdots \rangle \equiv \det \mathbf{A}^{-1} \int d\bar{\phi} d\phi\, e^{-\bar{\phi}^T \mathbf{A} \phi}(\cdots)$, and expanding both the left- and the right-hand sides of Eq. (3.65) to leading order in the "monomial" $\bar{\nu}_j \nu_i$, one obtains $\langle \phi_j \bar{\phi}_i \rangle = A_{ji}^{-1}$. The N-fold iteration of this procedure gives

$$\langle \phi_{j_1} \phi_{j_2} \cdots \phi_{j_n} \bar{\phi}_{i_n} \cdots \bar{\phi}_{i_2} \bar{\phi}_{i_1} \rangle = \sum_P (\operatorname{sgn} P) A_{j_1 i_{P1}}^{-1} \cdots A_{j_n i_{Pn}}^{-1}$$

where the sign of the permutation accounts for the sign changes accompanying the interchange of Grassmann variables. Finally, as with Gaussian integration over commuting variables, by taking $N \to \infty$ the Grassmann integration can be translated to a Gaussian functional integral.

3.5 Field Integral for the Quantum Partition Function

Having introduced the coherent states, the construction of path integrals for many-body systems is now straightforward. However, before proceeding, we should address the question of what a "path integral for many-body systems" actually means. In the next chapter, we will see that much of the information

Josiah Willard Gibbs 1839–1903 was an American scientist who is credited with the development of chemical thermodynamics, and introduced the concepts of free energy and chemical potential.

on quantum many-particle systems is encoded in expectation values of products of creation and annihilation operators, i.e., expressions of the structure $\langle a^\dagger a \cdots \rangle$. Objects of this type are generally called *correlation functions*. At any non-zero

**Gibbs dis-
tribution**

temperature, the average $\langle \cdots \rangle$ entering the definition of the correlation function runs over the quantum **Gibbs distribution** $\hat{\rho} \equiv e^{-\beta \hat{H}}/\mathcal{Z}$, where

$$\mathcal{Z} = \operatorname{tr} e^{-\beta \hat{H}} = \sum_n \langle n | e^{-\beta \hat{H}} | n \rangle \tag{3.66}$$

is the quantum partition function. Here, the sum extends over a complete set of Fock space states $\{|n\rangle\}$, and we have included a chemical potential, μ, in the definition of $\hat{H} = \hat{H}' - \mu \hat{N}$ for notational simplicity. (For the present, we specify neither the statistics of the system – bosonic or fermionic – nor the structure of the Hamiltonian.)

Ultimately, we will want to construct the path integral representations of correlation functions. Later, we will see that they can be derived from a path integral for $\mathcal{Z}$ itself. The latter is actually of importance in its own right, as it contains the information needed to characterize the thermodynamic properties of a many-body quantum system.[34] We thus begin our journey into many-body field theory with a construction of the path integral for $\mathcal{Z}$.

To prepare the representation of the partition function (3.66) in terms of coherent states, one must insert the resolution of identity (see table 3.1)

$$\mathcal{Z} = \int d(\bar{\psi}, \psi) \, e^{-\sum_i \bar{\psi}_i \psi_i} \sum_n \langle n | \psi \rangle \langle \psi | e^{-\beta \hat{H}} | n \rangle. \tag{3.67}$$

We now wish to get rid of the – now redundant – Fock space summation over $|n\rangle$ (another resolution of identity). To bring the summation to the form $\sum_n |n\rangle\langle n| = \mathbb{1}_{\mathcal{F}}$, we must commute the factor $\langle n|\psi\rangle$ to the right-hand side. However, in performing this operation, we must be careful not to miss a potential sign change whose presence will have important consequences for the structure of the fermionic path integral: whilst for bosons $\langle n|\psi\rangle\langle\psi|n\rangle = \langle\psi|n\rangle\langle n|\psi\rangle$, the fermionic coherent states change sign upon permutation, $\langle n|\psi\rangle\langle\psi|n\rangle = \langle-\psi|n\rangle\langle n|\psi\rangle$. (Exercise: With $\langle-\psi| \equiv \langle 0| \exp[-\sum_i \bar{\psi}_i a_i]$, verify that this sign follows as a direct consequence of the anticommutation of Grassmann variables and Fock space operators.) Note that, as both $\hat{H}$ and $\hat{N}$ contain elements that are even in the creation and annihilation operators, the sign is insensitive to the presence of the Boltzmann factor in (3.67). Making use of the sign factor ζ, the result of the interchange can be formulated as

**thermody-
namic pote-
ntial**

[34] In fact, the statement above is not entirely correct. Thermodynamic properties involve the **thermodynamic potential** $\Omega = -T \ln \mathcal{Z}$ rather than the partition function itself. At first sight, it seems that the difference between the two is artificial – one might first calculate $\mathcal{Z}$ and then take the logarithm. However, typically, one is unable to determine $\mathcal{Z}$ in closed form, but rather one has to perform a perturbative expansion, i.e., the result of a calculation of $\mathcal{Z}$ will take the form of a series in some small parameter ϵ. Now a problem arises when the logarithm of the series is taken. In particular, the Taylor series expansion of $\mathcal{Z}$ to a given order in ϵ does not automatically determine the expansion of Ω to the same order. Fortunately, the situation is not all that bad. As we will see in the next chapter, the logarithm essentially rearranges the perturbation series in an order known as a *cumulant expansion*.

$$\mathcal{Z} = \int d\psi \, e^{-\sum_i \bar{\psi}_i \psi_i} \sum_n \langle \zeta \psi | e^{-\beta \hat{H}} | n \rangle \langle n | \psi \rangle = \int d\psi \, e^{-\sum_i \bar{\psi}_i \psi_i} \langle \zeta \psi | e^{-\beta \hat{H}} | \psi \rangle. \quad (3.68)$$

INFO For **notational brevity** we will denote integration measures as $d(\bar{\psi}, \psi) \to d\psi$, i.e., we indicate only one representative of a group of variables or fields over which we are integrating. In the same spirit, we set $S[\bar{\psi}, \psi] \to S[\psi]$, etc.

Equation (3.68) now becomes the starting point for the construction scheme of a path integral. To be concrete, let us assume that the Hamiltonian is limited to a maximum of two-body interactions (see Eqs. (2.11) and (2.16)),

$$\hat{H}(a^\dagger, a) = \sum_{ij} h_{ij} a_i^\dagger a_j + \sum_{ijkl} V_{ijkl} a_i^\dagger a_j^\dagger a_k a_l. \quad (3.69)$$

normal ordering

Note that we have arranged for all the annihilation operators to stand to the right of the creation operators. Fock space operators of this structure are said to be **normal ordered**.[35] The reason for emphasizing normal ordering is that such an operator can be readily diagonalized by means of coherent states. Dividing the "time interval" β into N segments and inserting coherent state resolutions of identity (steps 1, 2, and 3 of the general scheme), Eq. (3.68) assumes the form

$$\mathcal{Z} = \int_{\substack{\bar{\psi}^0 = \zeta \bar{\psi}^N \\ \psi^0 = \zeta \psi^N}} \prod_{n=1}^{N} d\psi^n \exp\left(-\delta \sum_{n=0}^{N-1} \left(\frac{(\bar{\psi}^n - \bar{\psi}^{n+1})}{\delta} \psi^n + H(\bar{\psi}^{n+1}, \psi^n) \right) \right), \quad (3.70)$$

where $\delta = \beta/N$ and $\frac{\langle \psi | \hat{H}(a^\dagger, a) | \psi' \rangle}{\langle \psi | \psi' \rangle} = \sum_{ij} h_{ij} \bar{\psi}_i \psi'_j + \sum_{ijkl} V_{ijkl} \bar{\psi}_i \bar{\psi}_j \psi'_k \psi'_l \equiv H(\bar{\psi}, \psi')$ and we have adopted the shorthand $\psi^n = \{\psi_i^n\}$, etc. Finally, sending $N \to \infty$, and taking limits analogous to those leading from Eq. (3.5) to (3.6), we obtain the

field integral

continuum version of the **field integral**,[36]

$$\boxed{\mathcal{Z} = \int D\psi \, e^{-S[\psi]}, \quad S[\psi] = \int_0^\beta d\tau \left(\bar{\psi} \partial_\tau \psi + H(\bar{\psi}, \psi) \right)} \quad (3.71)$$

where $D\psi = \lim_{N \to \infty} \prod_{n=1}^{N} d\psi^n$ and the fields satisfy the condition

$$\bar{\psi}(0) = \zeta \bar{\psi}(\beta), \quad \psi(0) = \zeta \psi(\beta). \quad (3.72)$$

---

[35] More generally, an operator is defined to be "normal ordered" with respect to a given vacuum state $|0\rangle$ if, and only if, it annihilates $|0\rangle$. Note that the vacuum need not necessarily be defined as a zero-particle state. If the vacuum contains particles, normal ordering will not lead to a representation where all annihilators stand to the right. If, for whatever reason, one is given a Hamiltonian whose structure differs from Eq. (3.69), one can always effect a normal ordered form at the expense of introducing commutator terms. For example, normal ordering the quartic term leads to the appearance of a quadratic contribution that can be absorbed into h_{ij}.

[36] Whereas the bosonic continuum limit is indeed perfectly equivalent to that taken in constructing the quantum mechanical path integral ($\lim_{\delta \to 0} \delta^{-1}(\bar{\psi}^{n+1} - \bar{\psi}^n) = \partial_\tau |_{\tau=n\delta} \bar{\psi}(\tau)$ gives an ordinary derivative, etc.), a novelty arises in the fermionic case. The notion of replacing differences by derivatives is purely symbolic for Grassmann variables. There is no sense in which $\bar{\psi}^{n+1} - \bar{\psi}^n$ is small. The symbol $\partial_\tau \bar{\psi}$ rather denotes the formal (and well-defined) expression $\lim_{\delta \to 0} \delta^{-1}(\bar{\psi}^{n+1} - \bar{\psi}^n)$.

Written in a more explicit form, the action associated with the general pair-interaction Hamiltonian (3.69) can be cast in the form

$$S[\psi] = \int_0^\beta d\tau \Big(\sum_{ij} \bar\psi_i(\tau)\left(\partial_\tau \delta_{ij} + h_{ij}\right)\psi_j(\tau) + \sum_{ijkl} V_{ijkl}\bar\psi_i(\tau)\bar\psi_j(\tau)\psi_k(\tau)\psi_l(\tau) \Big).$$

(3.73)

Notice that the structure of the action fits nicely into the general scheme discussed in the previous section. By analogy, one would expect that the exponent of the many-body path integral would carry the significance of the Hamiltonian action, $S \sim \int d(p\dot q - H)$, where (q, p) symbolically stands for a set of generalized coordinates and momenta. In the present case the natural pair of canonically conjugate operators is $(a, a^\dagger)$. One would then interpret the eigenvalues $(\psi, \bar\psi)$ as "coordinates" (much as (q, p) are the eigenvalues of the operators $(\hat q, \hat p)$). Adopting this interpretation, we see that the exponent of the path integral indeed has the canonical form of a Hamiltonian action and is, therefore, easy to memorize.

Equations (3.71) and (3.73) define the **field integral in the time representation** (in the sense that the fields are functions of a time variable). In practice it is often useful to represent the action in an alternative, Fourier conjugate, representation. To this end, note that, owing to the boundary conditions (3.72), the functions $\psi(\tau)$ can be interpreted as functions on the entire Euclidean-time axis that are periodic or antiperiodic on the interval $[0, \beta]$. As such, they can be represented in terms of a Fourier series,[37]

$$\psi(\tau) = T \sum_{\omega_n} \psi_{\omega_n} e^{-i\omega_n \tau}, \qquad \psi_n = \int_0^\beta d\tau\, \psi(\tau) e^{i\omega_n \tau},$$

$$\omega_n = \begin{cases} 2n\pi T & \text{bosons} \\ (2n+1)\pi T & \text{fermions} \end{cases} \qquad n \in \mathbb{Z}$$

(3.74)

Matsubara
frequencies

where the ω_n are known as **Matsubara frequencies**. Substituting this representation into (3.71) and (3.73), we obtain $\mathcal{Z} = \int D\psi\, e^{-S[\bar\psi,\psi]}$, where $D\psi = \prod_n d\psi_n$ defines the measure (for each Matsubara frequency index n we have an integration over a coherent state basis $\{|\psi_n\rangle\}$), and the action takes the form

$$S[\psi] = T \sum \bar\psi_{in}\left(-i\omega_n \delta_{ij} + h_{ij}\right)\psi_{jn} + T^3 \sum V_{ijkl}\bar\psi_{in_1}\bar\psi_{jn_2}\psi_{kn_3}\psi_{ln_4}\delta_{n_1+n_2,n_3+n_4},$$

(3.75)

where the summations run over all Hilbert space and Matsubara indices and we used the identity $\int_0^\beta d\tau\, e^{-i\omega_n \tau} = \beta\delta_{\omega_n 0}$. Equation (3.75) defines the **frequency representation of the action**.

INFO In performing calculations in the Matsubara frequency representation, one sometimes runs into convergence problems (which will manifest themselves in the form of

---

[37] As always, the concrete definition of a Fourier transform leaves freedom for different conventions. However, in imaginary-time quantum field theory, it is often convenient if a factor of T multiplies each Matsubara sum.

ill-convergent Matsubara frequency summations). In such cases, it is important to remember that Eq. (3.75) does not actually represent the precise form of the action. What is missing is a convergence-generating factor whose presence follows from the way in which the integral was constructed and which regularizes otherwise non-convergent sums (except, of course, in cases where the divergences have a physical origin). Since the fields $\bar\psi$ are evaluated infinitesimally later than the fields ψ (cf. Eq. (3.70)), the h- (including μ)-dependent contributions to the action acquire a factor $\exp(-i\omega_n\delta)$, with δ a positive infinitesimal. Similarly, the V contribution acquires a factor $\exp(-i(\omega_{n_1}+\omega_{n2})\delta)$. In cases where the convergence is not critical, we will omit these contributions. However, once in a while it is necessary to remember their presence.

Partition function of non-interacting gas

As a first exercise, let us consider the quantum partition function of non-interacting gas. In some sense, the field integral of the non-interacting partition function has a status similar to that of the path integral for the harmonic oscillator: the direct quantum mechanical solution of the problem is straightforward and application of the full artillery of the field integral seems somewhat ludicrous. From a pedagogical point of view, however, the free partition function is a useful problem; it provides us with a welcome opportunity to introduce concepts of field integration within a comparatively simple setting. Also, the field integral of the free partition function will be an important building block for our subsequent analysis of interacting problems.

Consider, then, the partition function (3.71) with $H_0(\psi) = \sum \bar\psi_i H_{0,ij}\psi_j$. Diagonalizing H_0 by a unitary transformation U, $H_0 = UDU^\dagger$, and transforming the integration variables, $U^\dagger\psi \equiv \phi$, the action assumes the form $S = T\sum \bar\phi_{an}(-i\omega_n + \xi_a)\phi_{an}$, where ξ_a are the single-particle eigenvalues.[38] Remembering that the fields $\phi_a(\tau)$ are independent integration variables (exercise: why does the transformation $\psi \to \phi$ have a Jacobian unity?), we find that the partition function decouples, $\mathcal{Z} = \prod_a \mathcal{Z}_a$, where

$$\mathcal{Z}_a = \int D\phi_a \exp\left(-T\sum_{\omega_n}\bar\phi_{an}(-i\omega_n + \xi_a)\phi_{an}\right) = \prod_n (\beta(-i\omega_n + \xi_a))^{-\zeta} \quad (3.76)$$

where the last equality follows from the fact that the integrals over ϕ_{an} are of a one-dimensional complex or Grassmann Gaussian type. In performing these integrals, we recalled that, as per the definition (3.74), $[\phi_n] = (\text{energy})^{-1}$. This dimension is accounted for by a factor $\beta^{-\zeta}$ accompanying each integration variable in the measure, where $\zeta = 1 \ (-1)$ for bosonic (fermionic) fields.[39]

At this stage, we have left all aspects of field integration behind and reduced the problem to one of computing an infinite product over factors $i\omega_n - \xi_a$. Since

---

[38] We use the standard symbol $\xi_a \equiv \epsilon_a - \mu$ to denote single-particle eigenvalues including a chemical potential term.

[39] In the later analysis of physical observables, we will never need to worry about such scaling conventions – as usual with path integrals, the normalization factor of the measure will cancel against a matching one in the denominator.

products are usually more difficult to bring under control than sums, we take the logarithm of $\mathcal{Z}$ to obtain the free energy

$$F = -T \ln \mathcal{Z} = T\zeta \sum_{an} \ln(\beta(-i\omega_n + \xi_a)). \tag{3.77}$$

INFO Before proceeding with this expression, let us take a second look at the intermediate identity (3.76). Our calculation showed the partition function to be the product over all eigenvalues of the operator $-i\hat{\omega} + \hat{H}$ defining the action of the non-interacting system (here, $\hat{\omega} = \{\omega_n \delta_{nn'}\}$). As such, it can be written compactly as:

$$\boxed{\mathcal{Z} = \det\left(\beta(-i\hat{\omega} + \hat{H})\right)^{-\zeta}} \tag{3.78}$$

This result was derived by first converting to an eigenvalue integration and then performing one-dimensional integrals over "eigencomponents" ϕ_{an}. While technically straightforward, this – explicitly representation-dependent – procedure is not well suited to generalization to more complex problems. (Keep in mind that later on we will want to embed the action of the non-interacting problem into the more general framework of an interacting theory.)

Indeed, it is not necessary to refer to an eigenbasis at all. In the bosonic case, Eq. (3.17) tells us that Gaussian integration over a bilinear form $\sim \bar{\phi}\hat{X}\phi$ generates the inverse determinant of $\hat{X}$. Similarly, as we have seen, Gaussian integration extends to the Grassmann case with the determinants appearing in the numerator rather than in the denominator (as exemplified by Eq. (3.78)). (As a matter of fact, (3.76) is already a proof of this relation.)

We now have to confront a technical problem: how do we compute Matsubara sums of the form $\sum_n \ln(-i\omega_n + x)$? Indeed, it takes little imagination to foresee that sums of the type $\sum_{n_1, n_2, \dots} X(\omega_{n_1}, \omega_{n_2}, \dots)$, where X stands symbolically for some function, will be a recurrent structure in the analysis of functional integrals. A good ansatz would be to argue that, for sufficiently low temperatures (i.e., temperatures smaller than any other characteristic energy scale in the problem), the sum can be traded for an integral, $T\sum_{\omega_n} \to \int d\omega/2\pi$. However, this approximation is too crude to capture much of the characteristic temperature dependence in which one is usually interested. Yet, there exists an alternative, and much more accurate, way of computing sums over Matsubara frequencies.

Matsubara summation

INFO Consider a single **Matsubara frequency summation**,

$$S \equiv \sum_n h(\omega_n), \tag{3.79}$$

where h is some function and ω_n may be either bosonic or fermionic (cf. Eq. (3.74)). The basic idea behind the standard scheme of evaluating such sums is to introduce a complex auxiliary function $g(z)$ that has simple poles at $z = i\omega_n$. The sum S then emerges as the sum of residues obtained by integrating the product gh along a suitably chosen path in the complex plane. Typical choices of g include

$$g(z) = \begin{cases} \frac{\beta}{\exp(\beta z)-1} & \text{bosons,} \\ \frac{\beta}{\exp(\beta z)+1} & \text{fermions,} \end{cases} \quad \text{or } g(z) = \begin{cases} (\beta/2)\coth(\beta z/2) & \text{bosons,} \\ (\beta/2)\tanh(\beta z/2) & \text{fermions,} \end{cases} \tag{3.80}$$

where, in much of this section, we will
employ the former. (Notice the similarity between these functions and
the familiar Fermi and Bose distribution functions.) In practice, the
choice of counting function is mostly
a matter of taste, save for some cases
where one of the two options is dictated by convergence criteria.

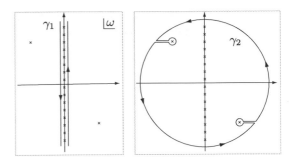

Integration over the path shown in
the left panel of the figure then produces

$$\frac{\zeta}{2\pi i} \oint dz \; g(z)h(-iz) = \zeta \sum_n \text{Res} \left(g(z)h(-iz)\right)_{z=i\omega_n} = \sum_n h(\omega_n) = S,$$

where, in the third identity, we have used the fact that the "counting functions" g are
chosen to have residue ζ and it is assumed that the integration contour closes at $z \to \pm i\infty$.
Now, the integral along a contour in the immediate vicinity of the poles of g is usually
intractable. However, as long as we are careful not to cross any singularities of g or the
function $h(-iz)$ (symbolically indicated by the isolated crosses in the figure[40]) we are
free to distort the integration path, ideally to a contour along which the integral can be
performed. Finding a suitable contour is not always straightforward. If the product gh
decays sufficiently fast as $|z| \to \infty$ (i.e., faster than z^{-1}), usually one tries to "inflate" the
original contour to an infinitely large circle (see the right panel of figure).[41] The integral
along the outer perimeter of the contour then vanishes and one is left with the integral
around the singularities of the function h. In the simple case where $h(-iz)$ possesses a
number of isolated singularities at $\{z_k\}$ (i.e., the situation indicated in the figure) we thus
obtain

$$S = \frac{\zeta}{2\pi i} \oint dz \; g(z)h(-iz) = -\zeta \sum_k \text{Res} \left(g(z)h(-iz)\right|_{z=z_k}, \tag{3.81}$$

where the contour integral encircles the singularities of $h(-iz)$ in a clockwise direction.
The computation of the infinite sum S has been now been reduced to the evaluation of a
finite number of residues – a task that is always possible!

To illustrate the procedure for a simple example, let us consider the function

$$h(\omega_n) = -\frac{\zeta T}{i\omega_n e^{-i\omega_n \delta} - \xi},$$

where δ is a positive infinitesimal.[42] To evaluate the sum $S = \sum_{\omega_n} h(\omega_n)$, we first observe
that the product $g(z)h(-iz)$ has benign convergence properties. Further, the function

---

[40] Remember that a function that is bounded and analytic in the entire complex plane is constant,
i.e., every "interesting" function will have singularities.

[41] Notice that the condition $\lim_{|z|\to\infty} |gh| < z^{-1}$ is not as restrictive as it may seem. The reason
is that the function h will be mostly related to physical observables approaching some limit
(or vanishing) for large excitation energies. This implies vanishing in at least portions of the
complex plane. The convergence properties of g depend on the concrete choice of the counting
function. (Exercise: Explore the convergence properties of the functions shown in Eq. (3.80).)

[42] In fact, this choice of h is not as artificial as it may seem. The expectation value of the **number
of particles** in the grand canonical ensemble is defined through the identity $N \equiv -\partial F/\partial \mu$,
where F is the free energy. In the non-interacting case, F is given by Eq. (3.77) and, remembering that $\xi_a = \epsilon_a - \mu$, one obtains $N \approx \zeta T \sum_{an} 1/(-i\omega_n + \xi_a)$. Now, why did we write
"$\approx$" instead of "$=$"? The reason is that the right-hand side, obtained by naïve differentiation

$h(-iz)$ has a simple pole that, in the limit $\delta \to 0$, lies on the real axis at $z = \xi$. This leads to the result

$$\sum_n h(\omega_n) = -\zeta \operatorname{Res}\left(g(z)h(-iz)\right)_{z=\xi} = \frac{1}{e^{\beta\xi} - \zeta}.$$

We have thus arrived at the important identity

$$-\zeta T \sum_{\omega_n} \frac{1}{i\omega_n - \xi_a} = \left\{ \begin{array}{ll} n_{\rm B}(\epsilon_a) & \text{bosons} \\ n_{\rm F}(\epsilon_a) & \text{fermions} \end{array} \right. \tag{3.82}$$

where

$$n_{\rm F}(\epsilon) = \frac{1}{\exp(\epsilon - \mu) + 1}, \qquad n_{\rm B}(\epsilon) = \frac{1}{\exp(\epsilon - \mu) - 1} \tag{3.83}$$

Fermi/Bose distribution are the **Fermi and Bose distribution functions**. As a corollary, we note that the expectation value for the number of particles in a non-interacting quantum gas assumes the familiar form $N = \sum_a n_{\rm F/B}(\epsilon_a)$.

Before returning to our discussion of the partition function, let us note that life is not always as simple as the example above. More often than not, the function h contains not only isolated singularities but also cuts or, worse, singularities. In such circumstances, finding a good choice of the integration contour can be far from straightforward!

Returning to the problem of computing the sum (3.77), consider for a moment a fixed eigenvalue $\xi_a \equiv \epsilon_a - \mu$. In this case, we need to evaluate the sum $S \equiv \sum_n h(\omega_n)$, where $h(\omega_n) \equiv \zeta T \ln[\beta(-i\omega_n + \xi)] = \zeta T \ln[\beta(i\omega_n - \xi)] + C$ and C is an inessential constant. As discussed before, the sum can be represented as $S = \frac{\zeta}{2\pi i} \oint dz\, g(z)h(-iz)$, where $g(z) = \beta(e^{\beta z} - \zeta)^{-1}$ is (β times) the distribution function and the contour encircles the poles of g. Now, there is an essential difference from the example discussed previously: the function $h(-iz) = \zeta T \ln(z - \xi) + C$ has a branch cut along the real axis, $z \in (-\infty, \xi)$ (see figure). To avoid contact with this singularity one must distort the integration contour as shown in the figure. Noticing that the (suitably regularized, see our previous discussion) integral along the perimeter vanishes, we conclude that

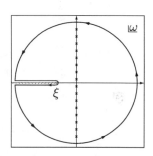

of Eq. (3.77), is ill-convergent. (The sum $\sum_{n=-\infty}^{\infty} 1/(n + x)$, x arbitrary, does not exist!) At this point we have to remember the remark made on page 139: had we carefully treated the discretization of the field integral, both the logarithm of the free energy and $\partial_\mu F$ would have acquired infinitesimal phases $\exp(-i\omega_n\delta)$. As an exercise, try to keep track of the discretization of the field integral from its definition to Eq. (3.77) to show that the accurate expression for N reads

$$N = \zeta T \sum_{an} \frac{1}{-i\omega_n e^{-i\omega_n\delta} + \xi_a} = \sum_{an} h(\omega_n)|_{\xi=\xi_a},$$

where h is the function introduced above. (Note that the necessity to keep track of the lifebuoy $e^{-i\delta\omega_n}$ does not arise too often. Most Matsubara sums of physical interest relate to functions f that decay faster than z^{-1}.)

$$S = \frac{T}{2\pi i} \int_{-\infty}^{\infty} d\epsilon \, g(\epsilon) \left(\ln(\epsilon^+ - \xi) - \ln(\epsilon^- - \xi) \right),$$

where $\epsilon^\pm = \epsilon \pm i\eta$, η is a positive infinitesimal, and we have used the fact that $g(\epsilon^\pm) \simeq g(\epsilon)$ is continuous across the cut. Also, without changing the value of the integral (exercise: why?), we have enlarged the integration interval from $(-\infty, \xi]$ to $(-\infty, \infty)$. To evaluate the integral, we observe that $g(\epsilon) = \zeta \partial_\epsilon \ln(1 - \zeta e^{-\beta\epsilon})$ and integrate by parts:

$$S = -\frac{\zeta T}{2\pi i} \int d\epsilon \ln \left(1 - \zeta e^{-\beta\epsilon} \right) \left(\frac{1}{\epsilon^+ - \xi} - \frac{1}{\epsilon^- - \xi} \right) = \zeta T \ln \left(1 - \zeta e^{-\beta\xi} \right).$$

Dirac identity

Here, the second equality is based on the **Dirac identity**

$$\lim_{\delta \searrow 0} \frac{1}{x + i\delta} = -i\pi\delta(x) + \mathcal{P}\frac{1}{x}, \tag{3.84}$$

where $\mathcal{P}$ denotes the principal value. Insertion of this result into Eq. (3.77) finally gives the familiar expression

$$F = \zeta T \sum_a \ln \left(1 - \zeta e^{-\beta\xi_a} \right) \tag{3.85}$$

free energy of the Fermi/Bose gas

for the **free energy of the Fermi/Bose gas**. While this result could have been obtained more straightforwardly by methods of quantum statistical mechanics, we will see shortly the power of Matsubara frequency summations in the analysis of less elementary problems.

3.6 Field-Theoretical Bosonization: A Case Study

The field integral (3.71) provides an exact representation of the quantum partition function; it contains full information about the microscopic Hamiltonian. However, our main interest lies in the universal large-scale behavior of a quantum system. To extract this information from the field integral, we must identify the relevant long-range degrees of freedom and transition from the microscopic field theory to some effective theory defined in terms of those variables.

In chapter 1 we saw that there are two principal strategies to execute this program: explicit construction – the subject of the next two chapters – and more phenomenological approaches based on consistency and symmetry arguments. Besides its low level of rigor, a disadvantage of the second route is the lack of quantitative control of the results (which implies susceptibility to mistakes). On the other hand, the phenomenological approach is far less laborious and involves a minimal amount of technical preparation. Often, the phenomenological deduction of a low-energy field theory precedes its rigorous construction (sometimes by decades). Indeed, there are cases where phenomenology is the only viable route.

Below we will illustrate the phenomenological approach for an interacting one-dimensional electron gas. We will map the microscopic partition function of the system onto a free (and thus exactly solvable) low-energy field theory.[43] Here, emphasis will be placed on purely methodological aspects of the analysis, i.e., we will derive an effective theory but will not explore it. (Nonetheless, the derivation is instructive and will help to understand the essential physics of the system!) In later chapters, the theory will serve as the starting point for the discussion of various interesting applications.

3.6.1 One-dimensional electron gas (fermionic theory)

Our starting point is the action of a non-interacting one-dimensional electron, gas

$$S_0[\psi] = \sum_{s=\pm 1} \int dx\, d\tau\, \psi_s^\dagger \left(-isv_F\partial_x + \partial_\tau\right)\psi_s,$$

where $\psi_{+/-}$ are right- and left-moving fermions and we have denoted the Grassmann conjugate to ψ by $\psi^\dagger$.[44] Recall that the right- and left-moving fermion operators are projections of the global momentum-dependent fermion operator to the vicinites of the right and left Fermi points, i.e., $\psi_+^\dagger(q) = \psi_{k_F+q}^\dagger$, $\psi_-^\dagger(q) = \psi_{-k_F+q}^\dagger$, where $|q| \ll k_F$. Fourier transforming this expression, we therefore obtain the approximate decomposition $\psi(x) = e^{ik_Fx}\psi_+(x) + e^{-ik_Fx}\psi_-(x)$. Before proceeding, let us rewrite the action in a form that emphasizes the symmetries of the problem:

$$S_0[\psi] = \int d^2x\, \psi^\dagger \left(\sigma_0\partial_{x_0} - i\sigma_3\partial_{x_1}\right)\psi = \int d^2x\, \bar\psi \left(\sigma_1\partial_{x_0} - \sigma_2\partial_{x_1}\right)\psi, \qquad (3.86)$$

where we have set $v_F = 1$ for notational simplicity. Here, $\psi = (\psi_+, \psi_-)^T$ is a two-component field comprising right- and left-moving fermions, $x = (x_0, x_1) = (\tau, x)$ parameterizes $(1+1)$-dimensional Euclidean space–time, and $\bar\psi \equiv \psi^\dagger\sigma_1$. The second equality identifies the action of the free one-dimensional fermion gas with that of the $(1+1)$-dimensional Dirac field. We can make this connection to Dirac theory more visual by defining the two-dimensional **Euclidean γ-matrices** $\gamma^0 = \sigma_1, \gamma^1 = -\sigma_2, \gamma^5 = -\sigma_3$, to represent the action in the form

margin note: **Euclidean γ-matrices**

$$S_0[\psi] = \int d^2x\, \bar\psi\partial_\mu\gamma^\mu\psi, \qquad (3.87)$$

which may be familiar from particle physics textbooks. For later reference, we will use the γ-matrix notation throughout parts of this section. However, keep in mind that these are just ordinary Pauli matrices, with the historical convention that $\sigma_3 = \gamma^5$. (Chapter 9 contains an introduction to Dirac field theory in condensed

---

[43] A preliminary account of the ideas underlying this mapping has already been given in section 2.2.4.

[44] Following the remarks earlier, this is a formal notation; there is no Grassmann analog of complex conjugation. However, within the context of relativistic fermions, our standard symbol $\bar\psi$ is reserved for another object (see below).

matter physics. However, this material will not be needed now.) Independently of the concrete realization, the feature that will be essential is the relation

$$\gamma^\mu \gamma^\nu + \gamma^\nu \gamma^\mu = 2\delta^{\mu\nu}, \tag{3.88}$$

i.e., different γ-matrices anticommute, and they square to unity.

Symmetries

We next turn to a discussion of the **symmetries** of the problem. First, the action is clearly invariant under the transformation, $\psi \to e^{i\phi_v}\psi$, where $\phi_v = $ const. What is the resulting conserved current? The infinitesimal variant of this transformation is described by $\psi \to \psi + (i\delta\phi_v)\psi$ or, in a notation adapted to (1.41), $\psi \leftrightarrow \phi^i$, $\omega^a \leftrightarrow i\delta\phi_v$, $\psi \leftrightarrow F_a^i$. Equation (1.43) then gives the conserved current $j_v^\mu = \frac{\partial \mathcal{L}}{\partial(\partial_\mu \psi)}\psi = \bar{\psi}\gamma^\mu\psi$. For later reference, we mention that, under a rotation of space–time, $x^\mu \to (R \cdot x)^\mu$, the components of j_v transform like a vector, $j^\mu \to (R \cdot j)^\mu$. In relativistic field theory, j_v is therefore usually called a **vector current**.

vector
current

Notice that the two components of the vector, $j^0 = \psi^\dagger\psi = \psi_+^\dagger\psi_+ + \psi_-^\dagger\psi_- \equiv \rho$ and $j^1 = -i\psi^\dagger\sigma_3\psi = -i(\psi_+^\dagger\psi_+ - \psi_-^\dagger\psi_-) \equiv -ij$, are the charge density, ρ, of the system and ($-i$ times) the current density, j, respectively.[45] Thus, the equation $-i\partial_\mu j^\mu = i\partial_\tau\rho + \partial_x j = 0$ expresses the conservation of particle current in imaginary time.

INFO This is a manifestation of the general result that the U(1)-symmetry of quantum mechanics (the freedom to multiply wave functions – or operators, in a second-quantized approach – by a constant phase $e^{i\phi}$) implies the **conservation of particle current**. We will encounter various other realizations of this symmetry later in the text.

current
conser-
vation

EXERCISE Subject the action of the general field integral (3.71) to the transformation $\psi \to e^{i\phi}\psi$, $\bar{\psi} \to \bar{\psi}e^{-i\phi}$ and compute the resulting Noether current. Convince yourself that the components of the current are the coherent state representation of the standard density/current operator of quantum mechanics.

Now the action (3.86) possesses a less obvious second symmetry: it remains invariant under the transformation

$$\psi \to e^{i\phi_a\gamma^5}\psi, \qquad \bar{\psi} \to \bar{\psi}e^{i\phi_a\gamma^5}. \tag{3.89}$$

Using the anticommutativity relation (3.88), it follows that $[\gamma^\mu, \gamma^5]_+ = 0$ for $\mu = 0, 1$, from which one obtains $(\bar{\psi}e^{i\phi_a\gamma^5})\partial_\mu\gamma^\mu(e^{i\phi_a\gamma^5}\psi) = \bar{\psi}\partial_\mu\gamma^\mu\psi$. A straightforward application of Noether's theorem to the infinitesimal transformation $\psi \to \psi + (i\phi_a)\gamma^5\psi$ gives the conserved current $j_a^\mu = i\bar{\psi}\gamma^\mu\gamma^5\psi = \epsilon^{\mu\nu}\bar{\psi}\gamma^\nu\psi$, where $\epsilon^{\mu\nu}$ is the antisymmetric tensor. Introducing a unit vector e_2 pointing into a fictitious third dimension perpendicular to the space–time plane, the current can be written as

---

[45] Notice that, for a one-dimensional Fermi system with velocity $v_F = 1$, the current density equals the density of right movers minus that of the left movers.

$j_a = e_2 \times j_v$. This representation shows that j_a transforms like an axial vector under rotations (as, say, a magnetic field). For this reason, j_a is called an **axial current**.

axial current

EXERCISE In the transformation (3.89), unusually for a quantum symmetry, the two phase factors enter with equal signs. What is the **physical meaning of this strange looking symmetry**? To find out, pass back to the original first representation of the action in Eq. (3.86) and show that, in that language, $\psi \to e^{i\phi_a\sigma_3}\psi$ and $\psi^\dagger \to \psi^\dagger e^{-i\phi_a\sigma_3}$ assume the form of a conventional unitary symmetry. In combination, the two transformations $\phi_{v,a}$ reflect the freedom to independently change the phase of the left- and right-moving states. This means that, in the low-energy approximation, the numbers of left- and right-moving fermions, $n_{L,R}$ are individually conserved. (Formally, the two sectors are described via independent Hilbert spaces.) Specifically, the axial symmetry represents the conservation of the difference $n_L - n_R$.

INFO The axial symmetry of the relativistic electron gas is an example of symmetry that does not survive at the quantum level. The conservation of the axial current breaks down once quantum fluctuations are taken into account, a phenomenon known as the **chiral anomaly** or **axial anomaly**. In field theory, "anomaly" is the general terminology for the breaking of symmetries of classical actions by fluctuations in the field integral. This principle plays an important role in various areas of condensed matter physics and will be discussed in section 9.2. However, it is easy to understand why axial symmetry does not survive extension to the full quantum theory: the left- and right-moving sectors are only seemingly independent. In fact, they represent small sections of a band of one-dimensional fermions with a cosine dispersion; see fig. 9.2 for a visualization. Changes applied to the large-scale dispersion (i.e., via application of an electric field shifting all momentum states) may alter the number of left- or right moving fermions (violating the axial symmetry) for a conserved total fermion number (vectorial symmetry).

chiral anomaly

Given that we are dealing with the simplest possible one-dimensional theory, a formal discussion of symmetries may seem to be something of an overkill. However, we shall see shortly that the effort was well invested: as soon as we switch on interactions, the fermionic theory ceases to be exactly solvable. It turns out, however, that our symmetry discussion above provides the key to a bosonic reformulation of the problem which *does* enjoy exact solvability. Yet, before turning to the bosonic approach, let us briefly recapitulate how interactions couple to the model.

Interacting case

As in section 2.2.4, we assume a short-range interaction between the left- and right-moving densities. Quantitatively, this is described by the coherent-state representation of the second quantized Hamiltonian (2.35), i.e.,

$$S_{\text{int}}[\psi] = \frac{1}{2}\sum_s \int dx d\tau \ (g_2\hat{\rho}_s\hat{\rho}_{\bar{s}} + g_4\hat{\rho}_s\hat{\rho}_s) \tag{3.90}$$

where $\hat{\rho}_s \equiv \psi_s^\dagger\psi_s$. Notice that this interaction term leaves the vectorial or axial symmetry of the system intact (exercise: why?). But what else can we say about the interacting system? In fact, we saw in section 2.2.4 that it is difficult to understand the physics of the system in the language of microscopic fermion states. Rather, one should turn to a formulation in terms of the effective long-range degrees of freedom of the model – non-dispersive charge density fluctuations. In the next section, we will formulate the dynamics of these excitations in a field-theoretical language. Remarkably, it will turn out that it takes only a minimal investment of phenomenological input plus symmetry considerations to extract this formulation from the microscopic model. (For a more rigorous, but also more laborious, construction of the theory, see the article[46])

3.6.2 One-dimensional electron gas (bosonic theory)

The native degrees of freedom of the one-dimensional electron gas are fermions, here assumed spinless for simplicity. On the other hand, we saw in section 2.2.4 that, at low energies, the system is governed by *bosonic* excitations representing charge density waves. As a first step we thus aim to identify a set of operator representations, which in the field integral framework turn into integration variables, representing the different realizations of degrees of freedom of the system. We begin by considering an algebra of boson operators $b(x), b^\dagger(x)$ with commutation relations $[b(x), b^\dagger(x)] = \delta(x - x')$. Anticipating the emergence of *charge* fluctuations, the boson operator b is factored into the **number–phase representation**

number–
phase rep-
resentation

$$b(x) \equiv \hat{\rho}(x)^{1/2} e^{i\hat{\phi}(x)}, \qquad b^\dagger(x) \equiv e^{-i\hat{\phi}(x)} \hat{\rho}(x)^{1/2} \qquad (3.91)$$

where the hermitian operators $\hat{\rho}$ and $\hat{\phi}$ represent charge and phase of the complex boson field, respectively.

INFO Representations of this type are frequently employed in bosonic theories. Importantly, $(b, b^\dagger) \to (\hat{\rho}, \hat{\phi})$ is a **canonical transformation**, i.e., the density and the phase of a bosonic excitation form a canonically conjugate pair, $[\hat{\phi}(x), \hat{\rho}(y)] = -i\delta(x - y)$.[47] To check this assertion, let us temporarily shift the density operator $\hat{\rho} \to \hat{\rho} + k_F/\pi$, so that $\hat{\rho}$ now describes the total density. (A shift by a number does not alter the commutation relations.) We thus write $b = \hat{\rho}^{1/2}\exp(i\hat{\phi})$ and $b^\dagger = \exp(-i\hat{\phi})\,\hat{\rho}^{1/2}$, where the position index has been dropped for notational transparency, while $[b, b^\dagger] = 1$ requires that $[\hat{\phi}, \hat{\rho}] = -i$:

canonical
transfor-
mation

$$[b, b^\dagger] = [\hat{\rho}^{1/2} e^{i\hat{\phi}} e^{-i\hat{\phi}} \hat{\rho}^{1/2}] = \hat{\rho} - e^{-i\hat{\phi}} \hat{\rho} e^{i\hat{\phi}}$$

$$= \hat{\rho} - e^{-i[\hat{\phi},]} \hat{\rho} = \hat{\rho} - \hat{\rho} + i[\hat{\phi}, \hat{\rho}] - \frac{1}{2}[\hat{\phi}, [\hat{\phi}, \hat{\rho}]] + \cdots \overset{[\hat{\rho}, \hat{\phi}]=i}{=} 1.$$

(In the last line, we have used the general operator identity $e^{\hat{A}} \hat{B} e^{-\hat{A}} = e^{[\hat{A},]} \hat{B}$.)

[46] H. Schöller and J. von Delft, *Bosonization for beginners – Refermionization for experts*, Ann. Phys. **7**, 225 (1998).
[47] This implies, in particular, that the corresponding field integral transformation $(\psi, \bar{\psi}) \to (\rho, \phi)$ from complex integration variables to the two real variables (ρ, ϕ) has a unit Jacobian (exercise).

We next aim to represent our actual degrees of freedom, spinless fermions, in terms of the bosonic operators $\hat{\rho}$ and $\hat{\phi}$. To this end, we can think of the fermion as a structureless charge endowed with fermionic exchange statistics. In the above representation, a charge is created by application of $\exp(i\hat{\phi})$ to the vacuum. To see this, notice that $\hat{\phi}$ is the "momentum" conjugate to $\hat{\rho}$, i.e., the "translation operator" $\exp(i\hat{\phi}(x))$ increases the charge at x by unity. We next complement this operator with a second one responsible for the bookkeeping of the fermionic exchange statistics. Applying the **Jordan–Wigner transformation**, previously applied in problem 2.4.5 to represent bosonic spin operators as fermions, we first define $\hat{\theta}(x) \equiv \pi \int_{-\infty}^{x} dx' \, \hat{\rho}(x')$ as a *string operator* satisfying the commutation relations

$$[\hat{\phi}(x), \hat{\theta}(y)] = -i\pi\Theta(y-x) \tag{3.92}$$

With this definition, it is straightforward to verify that for arbitrary integers s and s', we have

$$[e^{is\hat{\theta}(x)}e^{i\hat{\phi}(x)}, e^{is'\hat{\theta}(x')}e^{i\hat{\phi}(x')}]_+ \propto e^{i\pi\Theta(x'-x)} + e^{i\pi\Theta(x-x')} = 0. \tag{3.93}$$

Eq. (3.93) implies that the operators $\Gamma \exp(is\hat{\theta}) \exp(i\hat{\phi})$ are candidates for fermion operators. To identify the constant Γ, we need to consider the anticommutators of these fermions and their adjoint, as follows.

EXERCISE Verify the anticommutator identity above. (Hint: Use the general identities $e^{\hat{A}}\hat{B}e^{-\hat{A}} = e^{[\hat{A},\,]}\hat{B}$ and $e^{\hat{A}}e^{\hat{B}}e^{-\hat{A}} = \exp[e^{\hat{A}}\hat{B}e^{-\hat{A}}]$.) To investigate the anticommutator of creation and annihilation operators, we temporarily turn back to a lattice representation, $\hat{\phi}(x) \to \hat{\phi}_i$ and $\hat{\theta}(x) \to \hat{\theta}_i \equiv \pi \sum_{j<i} \hat{\rho}_j$. Verify that, with these definitions, $c_i = \frac{1}{\sqrt{2}} \exp(is\hat{\theta}_i) \exp(i\hat{\phi}_i)$ and its adjoint satisfy canonical anticommutation relations $[c_i, c_j^\dagger]_+ = \delta_{ij}$ (Hint: compare with problem 2.4.5). In the continuum limit, the Kronecker δ must turn into a singular δ-function, which explains the need for a formally divergent prefactor Γ (see also problem 3.8.10).

In practice, the value of Γ is not of particular relevance, and we can consider

$$c(x) = \Gamma \sum_{s=\pm 1} e^{is(k_F x + \hat{\theta}(x))} e^{i\hat{\phi}(x)} \tag{3.94}$$

as a bosonic representation of the fermion operator. This ansatz checks all the boxes required of a proper one-dimensional fermion field operator: the freedom to choose s in Eq. (3.93) is exploited to split the fermion into right- and left-moving contributions, as before. We have absorbed the "rapidly oscillating" part of the fermion operator in the c-number-valued factors $\exp(\pm ik_F x)$, so that $\hat{\theta}, \hat{\phi}$ may be considered as "slowly fluctuating" operator fields. In the language of Eq. (3.91), the shift $\hat{\theta} \to \hat{\theta} + k_F x$ means that $\hat{\rho} = \frac{1}{\pi}\partial_x\hat{\theta} \to \hat{\rho} + k_F/\pi$. The shifted operator $\hat{\rho}$ thus describes *fluctuations* of the density around the average value k_F/π.

Eq. (3.94) is an example of a **bosonization** identity, a transformation representing fermions in terms of boson operators. At first sight, it is not obvious what one

Jordan–
Wigner
transfor-
mation

bosoni-
zation

might gain from this representation: we have traded our simple fermion operator for a nonlinear expression in terms of bosonic degrees of freedom – but why? The point is that we are usually interested not in expressions linear in the fermion operators but in fermion bilinears (currents, densities, etc.). In contrast with the case of a single fermion, fermion bilinears have very simple expressions in terms of bosons. In particular, the Hamiltonian operator of the interacting system becomes quadratic (manifestly solvable) when expressed in terms of bosons. This makes bosonization a singularly powerful approach to the physics of one-dimensional fermion systems, as we now explore.

Non-interacting system

We now apply a combination of symmetry and dynamical arguments to identify the Hamiltonian of the non-interacting system. The fermionic prototype action is invariant under global rotations of space–time, $x \to x' = R \cdot x$, $\psi'(x') = \psi(x)$. This symmetry must pertain to the bosonic description of the theory. Turning to the "intrinsic" **symmetries** of the system, we note that the vectorial and axial symmetry operations considered in the previous section act on the left- and right-moving fermion states as $\psi_s \to e^{i\phi_v}\psi_s$ and $\psi_s \to e^{is\phi_a}\psi_s$, respectively. Of course, these transformations must continue to be symmetries no matter which representation of the theory (bosonic, fermionic or whatever) is chosen. A glance at Eq. (3.94) shows that the symmetry transformation acts on the bosonic variables by a simple shift operation: vectorial, $(\phi, \theta) \to (\phi + \phi_v, \theta)$; axial, $(\phi, \theta) \to (\phi, \theta + \phi_a)$. For $\phi_{a,v} = \text{const.}$, these transformations must not change the action, which excludes the presence of non-derivative terms. (For example, a contribution such as $\sim \int \theta^2$ would not be invariant under a uniform axial transformation, etc.)

Of course, symmetries alone do not suffice to fix the action of the system. What we need, in addition, is a minimal amount of **dynamical input**. Specifically, we will use the fact that the creation of a density distortion $\rho = \partial_x \theta / \pi$ costs a certain amount of energy U. Assuming that $U \sim \rho^2$ (i.e., that screening has rendered the Coulomb interaction effectively short-range), the Lagrangian action of the charge displacement field θ will contain a term $\sim (\partial_x \theta)^2$. However, this expression lacks rotational invariance. Its unique rotationally invariant extension reads $(\partial_x \theta)^2 + (\partial_\tau \theta)^2$. Thus, up to second order in derivatives, the Lagrangian action is given by $S_0[\theta] = (c/2) \int dx\, d\tau\, [(\partial_x \theta)^2 + (\partial_\tau \theta)^2]$, where the coupling constant c needs to be specified. This expression tells us that the field θ has a linear dispersion and propagates at constant velocity (recall our discussion of, e.g., the phonon action in chapter 1). Recalling that $\partial_x \theta \sim \rho$, this confirms our results regarding the behavior of density distortions in one-dimensional electron systems, derived in section 2.2.4.

To fix the value of the coupling constant c, we must compute the correlation function[48]

$$C(x, \tau) \equiv \langle (\psi_+^\dagger \psi_-)(x, \tau)(\psi_-^\dagger \psi_+)(0, 0) \rangle_\psi \qquad (3.95)$$

[48] The function C describes the correlation of the bilinear $\bar{\psi}_+ \psi_-$ with itself measured at different values of space and time. We have more to say on the subject of correlation functions in the next chapter.

first in the fermionic, then in the bosonic, language and require coinciding answers. Referring for a detailed discussion to problem 3.8.10, we merely note that the result obtained for the free fermion action (3.86) reads $C(x, \tau) = (4\pi^2(x^2 + \tau^2))^{-1}$. The bosonic variant $C(x, \tau) = \Gamma^4 \left\langle e^{2i\theta(x,\tau)} e^{-2i\theta(0,0)} \right\rangle_\theta$ leads to the same expression provided that we set $c = 1/\pi$ and $\Gamma = 1/(2\pi a)^{1/2}$. Thus, our final result for the Lagrangian form of the action of the non-interacting system reads

$$S[\theta] = \frac{1}{2\pi} \int d\tau dx \left((\partial_\tau \theta)^2 + (\partial_x \theta)^2\right).$$

Before proceeding to include interactions, let us turn from the Lagrangian to the Hamiltonian formulation. The transcription between the two languages is most naturally formulated in a **real- (Lorentzian-) time framework**, $t = -i\tau$, where the action reads[49]

$$S_{\mathrm{M}}[\theta] \equiv \int dtdx \, \mathcal{L}(\partial_x \theta, \partial_t \theta) = \frac{1}{2\pi} \int dtdx \left((\partial_t \theta)^2 - (\partial_x \theta)^2\right).$$

From here, we obtain the canonical momentum associated with the field, $\pi_\theta = \partial_{\partial_t \theta} \mathcal{L} = \frac{1}{\pi} \partial_t \theta$, and the Hamiltonian density as $\mathcal{H} = \frac{1}{2\pi}((\pi\pi_\theta)^2) + (\partial_x \theta)^2)$. On the other hand, we know from the commutator relation Eq. (3.92) that $[\partial_x \hat{\phi}(x), \hat{\theta}(y)] = i\pi\delta(x - y)$. Comparison with the structure of canonical commutation relations between coordinate and momenta $[\hat{\pi}_\theta(x), \hat{\theta}(y)] = -i\delta(x - y)$ leads to the identification $\pi_\theta = -\frac{1}{\pi} \partial_x \phi$. In this way, the Hamiltonian action $S_{\mathrm{M}} = \int dtdx(\pi_\theta \partial_t \theta - \mathcal{H})$ is identified as

$$S_{\mathrm{M}}[\theta, \phi] = \frac{1}{2\pi} \int dtdx \left(-2\partial_x \phi \, \partial_t \theta - (\partial_x \phi)^2 - (\partial_x \theta)^2\right).$$

We finally pass back to Euclidean-time to obtain

$$S[\theta, \phi] = \frac{1}{2\pi} \int dx \, d\tau \left((\partial_x \theta)^2 + (\partial_x \phi)^2 + 2i\partial_\tau \theta \partial_x \phi\right). \tag{3.96}$$

EXERCISE As a quick and dirty alternative to the Hamiltonian construction, decouple the $\partial_\tau \theta^2$ term in the Lagrangian in terms of a Hubbard–Stratonovich field $\partial_x \phi$ to obtain the representation Eq. (3.96).

It is instructive to inspect the Noether current corresponding to the **vectorial symmetry** in the bosonic language. A straightforward application of Noether's theorem to the transformation $\phi \to \phi + \phi_{\mathrm{v}}$ obtains $\partial_\mu j_{\mathrm{v}}^\mu = 0$, where $j_{\mathrm{v}}^0 \overset{\mathrm{p.146}}{=} \rho = \partial_x \theta/\pi$ and $j_{\mathrm{v}}^1 \overset{\mathrm{p.146}}{=} -ij = -i\partial_x \phi/\pi$.[50] While the first of these relations merely reiterates the definition of the displacement field, the second contains new information. Remembering that $\rho = \rho_+ + \rho_-$ and $j = \rho_+ - \rho_-$, we obtain the identification

$$\boxed{\rho_\pm = \frac{1}{2\pi} \left(\partial_x \theta \pm \partial_x \phi\right)} \tag{3.97}$$

of the left- and right-moving densities in terms of the Bose field.

[49] Recall that the overall sign of the action is determined by the equality $\exp(-S) = \exp(iS_{\mathrm{M}})$.
[50] The (arbitrary) normalization constant π^{-1} has been chosen to obtain consistency with previous definitions; see below.

EXERCISE Compute the bosonic representation of the axial Noether current.

Interacting system

We are now in a position to turn to the interacting case. In fact, all the hard work necessary to bring the interacting problem under control has already been done! We simply substitute the bosonic representation (3.97) into the Coulomb action (3.90) and obtain

$$
\begin{aligned}
S_{\text{int}} &= \frac{1}{8\pi^2} \sum_s \int dx\, d\tau \; (g_2 \partial_x(\theta - s\phi)\partial_x(\theta + s\phi) + g_4 \partial_x(\theta - s\phi)\partial_x(\theta - s\phi)) \\
&= \frac{1}{4\pi^2} \int dx\, d\tau \; \left((g_2 + g_4)(\partial_x\theta)^2 + (g_4 - g_2)(\partial_x\phi)^2\right),
\end{aligned}
$$

i.e., an action that is still quadratic and thus exactly solvable.

INFO In the field-theoretical literature (especially the literature on conformal field theory), interactions of this type are commonly referred to as **current–current interactions**. The reaseon is that S_{int} can be expressed as a bilinear form in the Noether currents generated by the symmetries of the system.

current–current interactions

Adding S_{int} to the action of the non-interacting theory (3.96), we arrive at the final expression

$$
\boxed{S[\theta, \phi] = \frac{1}{2\pi} \int dx\, d\tau \; \left(\frac{v}{g}(\partial_x\theta)^2 + gv(\partial_x\phi)^2 + 2i\,\partial_\tau\theta\partial_x\phi\right)} \qquad (3.98)
$$

where we have introduced the parameters

$$
v \equiv \frac{1}{2\pi}\left((2\pi v_{\text{F}} + g_4)^2 - g_2^2\right)^{1/2}, \qquad g \equiv \left(\frac{2\pi v_{\text{F}} + g_4 - g_2}{2\pi v_{\text{F}} + g_4 + g_2}\right)^{1/2} \qquad (3.99)
$$

and have reinstated v_{F}. Comparison with the results of section 2.2.4 identifies v as the effective velocity of the charge-density wave excitations of the system.[51] Equation (3.98) represents the main result of our analysis. We have succeeded in mapping a non-linear interacting problem onto a linear bosonic field theory. Critical readers may object that this result does not contain much new information. After all, a second-quantized representation of the theory in terms of free bosons was derived previously in section 2.2.4. Nonetheless, the analysis of this section is valuable from both a methodological and a conceptual point of view. Methodologically, it turns out that the field integral formulation $\int D(\theta, \phi) \exp(-S[\theta, \phi])$ is the most convenient starting point to address the many intriguing phenomena displayed by one-dimensional electron systems (we will meet some examples). From a conceptual point of view, it is remarkable that no "microscopic" calculations had to be

---

[51] In fact, the structure of Eq. (3.98) by itself determines the interpretation of v as an effective velocity. To see this (exercise), integrate over ϕ to arrive at a wave-like Lagrangian for θ whose characteristic velocity is set by v.

performed to determine the effective field theory. All we had to invest was symmetry arguments, some phenomenological input, and consistency checks. In chapter 8 we will meet a related problem – the low-energy field theory of quantum spin chains – where no microscopic construction for an effective field theory is known. In this case, symmetry analysis represents the *only* viable route towards a solution of the problem.

3.7 Summary and Outlook

In this chapter we introduced the path integral formulation of quantum mechanics, an approach independent of, yet equivalent to, the standard route of canonical operator quantization (modulo certain mathematical imponderabilities related to continuum functional integration). While a few exactly solvable quantum problems are more efficiently formulated by the standard approach, a spectrum of features makes the path integral an indispensable tool of modern quantum mechanics: it is intuitive, powerful in the treatment of non-perturbative problems, and tailor-made to the formulation of semiclassical limits. We have also seen that it provides a unifying link relating quantum problems to classical statistical mechanics.

Building on this approach, we learned how to represent the partition function of a quantum many-body system in terms of a path integral. The field integral representation of the partition function will be the platform on which further developments will be based. In fact, we are now in a position to face the main problem addressed in this text: practically none of the "nontrivial" field integrals relevant to applications are tractable in an exact way. This reflects the absence of closed solutions for the majority of interacting many-body problems. Before employing the field integral to solve "serious" problems, we need to develop a spectrum of approximation strategies – perturbation theory, linear response theory, mean field methods, instanton techniques, and the like. The construction and application of such methods will be the subject of the following chapters.

3.8 Problems

3.8.1 Quantum harmonic oscillator

The quantum harmonic oscillator provides a valuable setting in which to explore the Feynman path integral and methods of functional integration. Along with a few other precious examples, the path integral may be computed exactly, and the propagator explored rigorously.

(a) Starting with the Feynman path integral, show that the propagator for the quantum harmonic oscillator, $\hat{H} = \hat{p}^2/2m + m\omega^2\hat{q}^2/2$, takes the form

$$\langle q_{\mathrm{f}}|e^{-i\hat{H}t/\hbar}|q_{\mathrm{i}}\rangle = \left(\frac{m\omega}{2\pi i\hbar\sin\omega t}\right)^{1/2}\exp\left(\frac{i}{2\hbar}m\omega\left((q_{\mathrm{i}}^2+q_{\mathrm{f}}^2)\cot\omega t - \frac{2q_{\mathrm{i}}q_{\mathrm{f}}}{\sin\omega t}\right)\right).$$

Suggest why the propagator varies periodically in time, and explain the origin of singularities at $t = n\pi/\omega$, $n = 1, 2, \ldots$ Taking the frequency $\omega \to 0$, show that the propagator for the free particle is recovered.

(b) Show that the wave packet $\psi(q, t = 0) = (2\pi a)^{-1/4}\exp[-q^2/4a]$ remains Gaussian at all subsequent times. Obtain the width $a(t)$ as a function of time.

(c) Semiclassical limit: Taking the initial wave packet to be of the form $\psi(q, t = 0) = (2\pi a)^{-1/4}\exp[\frac{i}{\hbar}mvq - \frac{1}{4a}q^2]$ (corresponding to a wave packet centered at $q = 0$ with a velocity v), find the wave packet at times $t > 0$, and determine its mean position, mean velocity and mean width as functions of time.

Answer:

(a) Making use of the Feynman path integral, the propagator can be expressed as the functional integral,

$$\langle q_{\mathrm{f}}|e^{-i\hat{H}t/\hbar}|q_{\mathrm{i}}\rangle = \int_{q(0)=q_{\mathrm{i}}}^{q(t)=q_{\mathrm{f}}} Dq\, e^{iS[q]/\hbar}, \quad S[q] = \int_0^t dt'\,\frac{m}{2}\left(\dot{q}^2 - \omega^2 q^2\right).$$

The evaluation of the integral over field configurations $q(t')$ is facilitated by parameterizing paths in terms of fluctuations around the classical trajectory. Setting $q(t') = q_{\mathrm{cl}}(t') + r(t')$, where $q_{\mathrm{cl}}(t')$ satisfies the classical equation of motion, we have $m\ddot{q}_{\mathrm{cl}} = -m\omega^2 q_{\mathrm{cl}}$, and, applying the boundary conditions, one obtains $q_{\mathrm{cl}}(t') = A\sin(\omega t') + B\cos(\omega t')$ with $B = q_{\mathrm{i}}$ and $A = q_{\mathrm{f}}/\sin(\omega t) - q_{\mathrm{i}}\cot(\omega t)$. Gaussian in q, the action separates as $S[q] = S[q_{\mathrm{cl}}] + S[r]$, where $S[q_{\mathrm{cl}}] = \frac{m\omega^2}{2}[(A^2 - B^2)\frac{\sin(2\omega t)}{2\omega} + 2AB\frac{\cos(2\omega t)-1}{2\omega}] = \frac{m\omega}{2}[(q_{\mathrm{i}}^2 + q_{\mathrm{f}}^2)\cot(\omega t) - \frac{2q_{\mathrm{i}}q_{\mathrm{f}}}{\sin(\omega t)}]$. Finally, integrating over the fluctuations and applying the identity $z/\sin z = \prod_{n=1}^{\infty}(1 - z^2/\pi^2 n^2)^{-1}$, one obtains the required result, periodic in t with frequency ω and singular at $t = n\pi/\omega$. In particular, a careful regularization of the expression for the path integral shows that

$$\langle q_{\mathrm{f}}|e^{-i\hat{H}t/\hbar}|q_{\mathrm{i}}\rangle \mapsto \begin{cases} \delta(q_{\mathrm{f}} - q_{\mathrm{i}}), & t = 2\pi n/\omega, \\ \delta(q_{\mathrm{f}} + q_{\mathrm{i}}), & t = \pi(2n+1)/\omega. \end{cases}$$

Physically, the origin of the singularity is clear. The harmonic oscillator is peculiar in having a spectrum with energies uniformly spaced in units of $\hbar\omega$. Noting the eigenfunction expansion $\langle q_{\mathrm{f}}|e^{-i\hat{H}t/\hbar}|q_{\mathrm{i}}\rangle = \sum_n\langle q_{\mathrm{f}}|n\rangle\langle n|q_{\mathrm{i}}\rangle e^{-i\omega nt}$, this means that when $\hbar\omega \times t/\hbar = 2\pi \times$ integer there is a coherent superposition of the states and the initial state is recovered. Furthermore, since the ground state and its even integer descendants are symmetric while the odd states are antisymmetric, it is straightforward to prove the identity for the odd periods.

(b) Given the initial condition $\psi(q, t = 0)$, the time evolution of the wave packet can be determined from the propagator as $\psi(q, t) = \int_{-\infty}^{\infty} dq'\,\langle q|e^{-i\hat{H}t/\hbar}|q'\rangle\psi(q', 0)$, from which one obtains

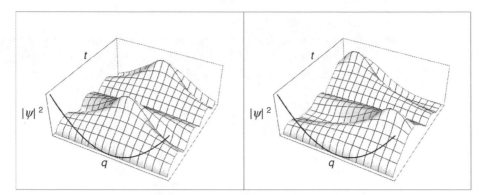

Fig. 3.4   (a) Variation of a "stationary" Gaussian wave packet in a harmonic oscillator, taken from the solution, and (b) variation of the moving wave packet.

$$\psi(q,t) = J(t) \int_{-\infty}^{\infty} dq' \, \frac{e^{-\frac{q'^2}{4a}}}{(2\pi a)^{1/4}} \exp\left(\frac{i}{\hbar} \frac{m\omega}{2} \left([q^2 + q'^2] \cot(\omega t) - \frac{2qq'}{\sin(\omega t)} \right) \right),$$

where $J(t)$ represents the time-dependent contribution arising from the fluctuations around the classical trajectory. Being Gaussian in q', the integral can be performed explicitly. Setting $\alpha = 1/2a - im\omega \cot(\omega t)/\hbar, \beta = im\omega q/(\hbar \sin(\omega t))$, and performing the Gaussian integral over q', one obtains

$$\psi(q,t) = J(t) \frac{1}{(2\pi a)^{1/4}} \sqrt{\frac{2\pi}{\alpha}} e^{\beta^2/2\alpha} \exp\left(\frac{i}{2\hbar} m\omega q^2 \cot(\omega t) \right),$$

where $\beta^2/2\alpha = -(1 + i\kappa \cot(\omega t))q^2/4a(t)$. Rearranging terms, it is straightforward to show that $\psi(q,t) = (2\pi a(t))^{-1/4} \exp[-q^2/(4a(t))]e^{i\varphi(q,t)}$, where $a(t) = a(\cos^2(\omega t) + \kappa^{-2} \sin^2(\omega t))$, $\kappa = 2am\omega/\hbar$, and $\varphi(q,t)$ represents a pure phase.[52] As required, under the action of the propagator, the normalization of the wave packet is preserved. (A graphical representation of the time evolution is shown in fig. 3.4a.) Note that, if $a = \hbar/2m\omega$ (i.e., $\kappa = 1$), $a(t) = a$ for all times – i.e., it is a pure eigenstate.

(c) Still in Gaussian form, the integration can again be performed explicitly for the new initial condition. In this case, we obtain an expression of the form above but with $\beta = \frac{i}{\hbar} \frac{m\omega}{\sin(\omega t)} \times (q - \frac{v}{\omega} \sin(\omega t))$. Reading off the coefficients, we find that the position and velocity of the wave packet have the forms $q_0(t) = (v/\omega) \sin(\omega t)$, $v(t) = v \cos(\omega t)$, coinciding with those of the classical oscillator. Note that, as above, the width $a(t)$ of the wave packet oscillates at frequency ω. (A graphical representation of the time evolution is shown in fig. 3.4 (b).)

3.8.2 Density matrix

Using the results derived in the previous example, this problem explores how real-time dynamical information can be converted into quantum statistical information.

---

[52] For completeness, we note that $\varphi(q,t) = -\frac{1}{2} \tan^{-1}(\frac{1}{\kappa} \cot(\omega t)) - \frac{\kappa q^2}{4a} \cot(\omega t)(\frac{a}{a(t)} - 1)$.

Using the results above, obtain the density matrix $\rho(q, q') = \langle q|e^{-\beta \hat{H}}|q'\rangle$ for the harmonic oscillator at finite temperature, $\beta = 1/T$ ($k_B = 1$). Obtain and comment on the asymptotics: (i) $T \ll \hbar\omega$ and (ii) $T \gg \hbar\omega$. (Hint: In the high-temperature case, carry out the expansion in $\hbar\omega/T$ to second order.)

Answer:

The density matrix can be deduced from the general solution above. In the Euclidean-time formulation,

$$\rho(q, q') = \langle q|e^{-\beta H}|q'\rangle = \langle q|e^{-(i/\hbar)H(\hbar\beta/i)}|q'\rangle$$
$$= \left(\frac{m\omega}{2\pi\hbar\sinh(\beta\hbar\omega)}\right)^{1/2} \exp\left(-\frac{m\omega}{2\hbar}\left((q^2 + q'^2)\coth(\beta\hbar\omega) - \frac{2qq'}{\sinh(\beta\hbar\omega)}\right)\right).$$

(i)   For $T \ll \hbar\omega$ $(\beta\hbar\omega \gg 1)$, $\coth(\beta\hbar\omega) \to 1$, $\sinh(\beta\hbar\omega) \to e^{\beta\hbar\omega}/2$, and

$$\rho(q, q') \simeq \left(\frac{m\omega}{\pi\hbar e^{\beta\hbar\omega}}\right)^{1/2} \exp\left(-\frac{m\omega}{2\hbar}(q^2 + q'^2)\right) = \langle q|n = 0\rangle \, e^{-\beta E_0} \, \langle n = 0|q'\rangle.$$

(ii)   Using the relations $\coth x \overset{x \leqslant 1}{=} 1/x + x/3 + \cdots$ and $1/\sinh x \overset{x \leqslant 1}{=} 1/x - x/6 + \cdots$, the high-temperature expansion $(T \gg \hbar\omega)$ gives

$$\rho(q, q') \simeq \left(\frac{m}{2\pi\beta\hbar^2}\right)^{1/2} e^{-m(q-q')^2/2\beta\hbar^2} e^{-\beta m\omega^2(q^2 + q'^2 + qq')/6} \simeq \delta(q - q')e^{-\beta m\omega^2 q^2/2},$$

i.e., the classical Maxwell–Boltzmann distribution.

3.8.3 Depinning transition and bubble nucleation

In section 3.3.2 we explored the capability for a quantum field to tunnel from the metastable minimum of a potential, the "false vacuum." Yet, prior to the early work of Coleman on the quantum mechanical problem, similar ideas had been developed by Langer[18] in the context of classical bubble nucleation. The following problem draws connections between the classical and the quantum problem. As posed, the quantum formulation describes the depinning of a flux line in a superconductor from a columnar defect.

Consider a quantum elastic string embedded in a three-dimensional space and "pinned" by a columnar defect potential V oriented parallel to the z-axis. The corresponding Euclidean-time action is given by

$$S[\mathbf{u}] = \int dz \, d\tau \left(\frac{1}{2}\rho\dot{\mathbf{u}}^2 + \frac{1}{2}\sigma(\partial_z \mathbf{u})^2 + V(|\mathbf{u}|)\right),$$

where the two-dimensional vector field $\mathbf{u}(z,\tau)$ denotes the
string displacement within the xy-plane, ρ represents the density
per unit length, and σ defines the tension in the string. On this
system (see the figure), let us suppose that an external in-plane
field f is imposed along the x-direction, $S_{\text{ext}} = -f \int dz\, d\tau\, \mathbf{u} \cdot \mathbf{e_x}$.
Following the steps below, determine the probability (per unit
time and per unit length) for the string to detach from the defect.

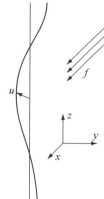

(a) Derive a saddle-point equation in the two-dimensional $z\tau$-
space. Rescaling the coordinates, transform the equation of motion to a problem with circular symmetry.

(b) If the field is weak, one can invoke a "thin-wall" or "bubble"
approximation to describe the saddle-point solution $\mathbf{u}(z,t)$ by
specifying two regions of space–time, where the string is either free or is completely
locked to the defect, respectively. In this approximation, find $\mathbf{u}(z,t)$. (Hint: use the
fact that, in either case, of complete locking or complete freedom, the potential
does not exert a net force on the string.)

(c) With exponential accuracy, determine the detaching probability. You may as-
sume that, for all values of u_x obtained in (b), $V(|u|) \simeq V_0 = \text{const}$.

(Exercise: Finally, consider how the quantum model can be related to the classical
system.)

Answer:

(a) Varying the action with respect to u_x, the saddle-point equation assumes the
form $\rho \ddot{u}_x + \sigma \partial^2 u_x = -f + V'(u)(u_x/u)$, where $u = |\mathbf{u}|$. Applying the rescaling $\tau = (\rho/\sigma)^{1/4}\tilde{\tau}$ and $z = (\sigma/\rho)^{1/4}\tilde{z}$, the equation takes the symmetrized form $\sqrt{\sigma\rho}\partial^2 u_x = -f + V'(u)(u_x/u)$, where $\partial^2 = \partial_{\tilde{\tau}}^2 + \partial_{\tilde{z}}^2$, and the following boundary conditions on
the radial coordinate $(\tilde{\tau}, \tilde{z}) \mapsto (r, \phi)$ are imposed,

$$
u_x(r) = \begin{cases} 0, & r > R, \\ g(r), & r < R. \end{cases}
$$

(b) In the "thin-wall" approximation, the potential gradient can be neglected. In
this case, the saddle-point equation assumes the form $\partial^2 g = -f/\sqrt{\sigma\rho}$, with the
solution $g = (R^2 - r^2)f/\sqrt{4\sigma\rho}$.

(c) With this result, the tunneling rate can be estimated from the saddle-point
action

$$
S_{\text{bubble}} = \int^R d^2 r \left(\frac{\sqrt{\sigma\rho}}{2}(\tilde{\partial}g)^2 + V_0 - fg \right) = -\pi R^2 \left(\frac{3f^2 R^2}{16(\sigma\rho)^{1/2}} - V_0 \right).
$$

Minimizing over R, one obtains the optimal radius $R_*^2 = 8V_0\sqrt{\sigma\rho}/(3f^2)$. As a result,
we obtain the tunneling rate $W \propto e^{-S(R_*)} = \exp(-4\pi V_0^2 \sqrt{\sigma\rho}/(3f^2))$.

3.8.4 Tunneling in a dissipative environment

In section 3.3.3 we considered the influence of dissipation on the action of a point-particle in a quantum well. There, a model was chosen in which the degrees of freedom of the environment were represented phenomenologically by a bath of harmonic oscillators. In the following, we will explore a model in which the particle is coupled to the fluctuations of a quantum mechanical "string." We will discuss how this model is relevant to the physics of impurity scattering in interacting quantum wires.

(a) A quantum particle of mass m is confined by a sinusoidal potential $U(q) = 2g \sin^2(\pi q/q_0)$. Employing the Euclidean-time Feynman path integral ($\hbar = 1$),

$$\mathcal{Z} = \int Dq \; e^{-S_{\text{part}}[q]}, \quad S_{\text{part}}[q] = \int_{-\infty}^{\infty} d\tau \left(\frac{m}{2} \dot{q}^2 + U(q) \right),$$

confirm by direct substitution that the extremal contribution to the propagator connecting two neighboring degenerate minima ($q(\tau = -\infty) = 0$ and $q(\tau = \infty) = q_0$) is given by the instanton trajectory $q_{\text{cl}}(\tau) = (2q_0/\pi) \arctan(e^{\omega_0 \tau})$, where $\omega_0 = (2\pi/q_0)\sqrt{g/m}$. Show that $S[q_{\text{cl}}] = (2/\pi^2)mq_0^2\omega_0$. (Note: Although the equation of motion associated with the minimum of the path integral is nonlinear, the solution above is exact.)

(b) If the quantum particle is coupled at one point $x = 0$ to an infinite "string" $u(x, t)$, the path integral is given by

$$\mathcal{Z} = \int Du \int Dq \; \delta\left(q(\tau) - u(\tau, x = 0) \right) e^{-S_{\text{string}}[u] - S_{\text{part}}[q]},$$

where the classical action of the string is given by (cf. the action functional for phonons discussed in section 1.1)

$$S_{\text{string}}[u] = \frac{1}{2} \int_{-\infty}^{\infty} d\tau \int_{-\infty}^{\infty} dx \left(\rho \dot{u}^2 + \sigma(\partial_x u)^2 \right).$$

Here $\delta(q(\tau) - u(\tau, x = 0))$ represents a functional δ-function, which enforces the condition $q(\tau) = u(\tau, x = 0)$ for all times τ. Operationally, it can be understood from the discretized form $\prod_n \delta(q(\tau_n) - u(\tau_n, x = 0))$. By representing the functional δ-function as the functional integral

$$\delta(q(\tau) - u(\tau, 0)) = \int Df \; \exp\left(i \int_{-\infty}^{\infty} d\tau \; f(\tau)(q(\tau) - u(\tau, 0)) \right),$$

and integrating over string fluctuations, show that the dynamics of the particle is governed by the effective action $S_{\text{eff}}[q] = S_{\text{part}}[q] + \frac{\eta}{2} \int \frac{d\omega}{2\pi} |\omega| |q(\omega)|^2$, where $\eta = \sqrt{4\rho\sigma}$. Compare this result with the dissipative action discussed in section 3.3.3.

(c) Treating the correction to the particle action as a perturbation, use your result from (a) to show that the effective action for an instanton–anti-instanton pair $q(\tau) = q_{\text{cl}}(\tau + \bar{\tau}/2) - q_{\text{cl}}(\tau - \bar{\tau}/2)$, where $\omega_0\bar{\tau} \gg 1$, is given approximately by

$$S_{\text{eff}}[q] = 2S_{\text{part}}[q_{\text{cl}}] - \frac{\eta q_0^2}{\pi} \ln(\omega_0 \bar{\tau}).$$

(Hint: Note that, in finding the Fourier decomposition of $q_{\rm cl}(\tau)$, a crude estimate is sufficient.)

(d) Using this result, estimate the typical separation of the pair (i.e., interpret the overall action as an effective probability distribution function for $\bar\tau$, and evaluate $\langle\bar\tau\rangle = \int d\bar\tau\,\bar\tau e^{-S_{\rm eff}}$). Comment on the implications of your result for the nature of the tunneling probability.

INFO The model above is directly applicable to the problem of **impurity scattering in interacting quantum wires**. Consider a clean interacting wire, as described by Eq. (3.98) in bosonized representation. Now assume the presence of an impurity potential $\hat V = u c^\dagger(0)c(0) \to \gamma\cos(\hat\theta(0))$, where γ is a constant; here, we have used the bosonization identity (3.94) and anticipated that the most important physical effect of the impurity will be scattering from left to right movers, i.e., we neglect the left–left and right–right contributions to the scattering operator. Adding this term to the action, the latter is Gaussian everywhere, except for the point $x = 0$ where the impurity sits. In particular, the ϕ-action is Gaussian everywhere, and we may integrate over this field to obtain the effective action

$$
S[\theta] = \frac{1}{2\pi g}\int dx d\tau \left(v(\partial_x\theta)^2 + \frac{1}{v}(\partial_\tau\theta)^2\right) + \gamma\int d\tau\cos(2\theta(0)).
$$

At this point, the connection to the above problem becomes apparent: $q = \theta(0)$ is considered as the coordinate of the quantum particle and $u(x) = \theta(x)$, $x \neq 0$ are the coordinates of the attached harmonic string with stiffness $\sigma = v/(\pi g)$ and mass density $1/(\pi vg)$. At $x = 0$, these variables are rigidly locked to each other.

Considering the path integration over $\theta(x \neq 0)$ as integration over the string, we then obtain the effective action for the $\theta(0) \equiv \theta$ coordinate:

$$
S_{\rm eff}[\theta] = \frac{1}{\pi g}\int \frac{d\omega}{2\pi}|\theta(\omega)|^2 + \gamma\int d\tau\cos(2\theta). \tag{3.100}
$$

In section 6.1.2 we will investigate this theory by renormalization group methods and study how dissipation affects the scattering properties off the cosine potential.

Answer:

(a) Varying the Euclidean-time path integral with respect to $q(\tau)$, the extremal field configuration obeys the classical **sine–Gordon equation**

$$
m\ddot q - \frac{2\pi g}{q_0}\sin\left(\frac{2\pi q}{q_0}\right) = 0.
$$

sine–Gordon
equation

Applying the trial solution, the equation of motion is satisfied if $\omega_0 = (2\pi/q_0)\sqrt{g/m}$. From this result one obtains the classical action $S[q_{\rm cl}] = \int_0^\infty d\tau(\frac{m}{2}\dot q_{\rm cl}^2 + U(q_{\rm cl})) = \int_0^\infty d\tau\, m\dot q_{\rm cl}^2 = m\int_0^{q_0} dq\,\dot q_{\rm cl} = 2\frac{mq_0^2}{\pi^2}\omega_0$.

(b) In Fourier space, the action of the classical string takes the form $S_{\rm string} = \frac{1}{2}\int_{-\infty}^\infty \frac{d\omega}{2\pi}\int_{-\infty}^\infty \frac{dk}{2\pi}(\rho\omega^2 + \sigma k^2)|u(\omega,k)|^2$. Representing the functional δ-function as the functional integral

$$
\int Df \exp\left(i\int_{-\infty}^\infty \frac{d\omega}{2\pi} f(\omega)\left(q(-\omega) - \int_{-\infty}^\infty \frac{dk}{2\pi}u(-\omega,-k)\right)\right),
$$

and performing the integral over the degrees of freedom of the string, one obtains

$$\int Du \; e^{-S_{\text{string}} - i \int d\tau \, f(\tau) u(\tau,0)} \propto \exp\left(-\frac{1}{2} \int_{-\infty}^{\infty} \frac{d\omega}{2\pi} \int_{-\infty}^{\infty} \frac{dk}{2\pi} \frac{1}{\rho\omega^2 + \sigma k^2} |f(\omega)|^2 \right).$$

Integrating over k, and performing the Gaussian functional integral over the Lagrange multiplier field $f(\omega)$, one obtains the given effective action.

(c) Approximating the instanton–anti-instanton pair $q(\tau) = q_{\text{cl}}(\tau + \bar\tau) - q_{\text{cl}}(\tau - \bar\tau)$ by a "top-hat" function, we obtain $q(\omega) = \int_{-\bar\tau/2}^{\bar\tau/2} d\tau \, q_0 e^{i\omega\tau} = q_0 \bar\tau \sin(\omega\bar\tau/2)/(\omega\bar\tau/2)$. Treating the dissipative term as a perturbation, the action then takes the form

$$S_{\text{eff}} - 2S_{\text{part}} = \frac{\eta}{2} \int_0^{\omega_0} \frac{d\omega}{2\pi} |\omega| (q_0\bar\tau)^2 \frac{\sin^2(\omega\bar\tau/2)}{(\omega\bar\tau/2)^2} \simeq \frac{q_0^2}{\pi} \eta \ln(\omega_0\bar\tau),$$

where ω_0 serves as a high-frequency cutoff.

(d) Interpreted as a probability distribution for the instanton separation, one finds

$$\langle \bar\tau \rangle = \int d\bar\tau \; \bar\tau \, \exp\left(-\frac{q_0^2}{\pi} \eta \ln(\omega_0\bar\tau) \right) \sim \int^{\infty} d\bar\tau \; \bar\tau^{1 - q_0^2 \eta/\pi}.$$

The divergence of the integral shows that, for $\eta > 2\pi/q_0^2$, instanton–anti-instanton pairs are confined and particle tunneling no longer occurs. Later, in chapter 6, we will revisit the dissipative phase transition from the standpoint of the renormalization group.

3.8.5 Exercises on fermion coherent states

To practice the coherent state method, we include a few technical exercises on the fermionic coherent state which complement the structures discussed in the main text.

Considering a fermionic coherent state $|\eta\rangle$, verify the following identities: (a) $\langle\eta|a_i^\dagger = \langle\eta|\bar\eta_i$, (b) $a_i^\dagger|\eta\rangle = -\partial_{\eta_i}|\eta\rangle$ and $\langle\eta|a_i = \partial_{\bar\eta_i}\langle\eta|$, (c) $\langle\eta|\nu\rangle = \exp\left(\sum_i \bar\eta_i \nu_i\right)$, and (d) $\int d\eta \, e^{-\bar\eta\eta}|\eta\rangle\langle\eta| = \mathbb{1}_{\mathcal{F}}$, where $d\eta = \prod_i d\bar\eta_i \, d\eta_i$, and $\sum_i \bar\eta_i \eta_i = \bar\eta\eta$. Finally, (e) show that $\langle n|\psi\rangle\langle\psi|n\rangle = \langle\zeta\psi|n\rangle\langle n|\psi\rangle$, where $|n\rangle$ is an n-particle state in Fock space while $|\psi\rangle$ is a coherent state.

Answer:

Making use of the rules of Grassmann algebra,

(a) $\langle\eta|a_i^\dagger = \langle 0| \exp[-\sum_j a_j \bar\eta_j] a_i^\dagger = \langle 0| \prod_j (1 - a_j\bar\eta_j) a_i^\dagger = \langle 0|(1 - a_i\bar\eta_i) a_i^\dagger \prod_{j\neq i}(1 - a_j\bar\eta_j) =$

$\underbrace{\langle 0| a_i a_i^\dagger}_{\langle 0|[a_i, a_i^\dagger]_+ = \langle 0|} \bar\eta_i \prod_{j\neq i}(1 - a_j\bar\eta_j) = \langle 0| \prod_j (1 - a_j\bar\eta_j) \bar\eta_i = \langle\eta|\bar\eta_i.$

(b) $a_i^\dagger|\eta\rangle = \underbrace{a_i^\dagger(1 - \eta_i a_i^\dagger)}_{a_i^\dagger = \partial_{\eta_i} \eta_i a_i^\dagger = -\partial_{\eta_i}(1 - \eta_i a_i^\dagger)} \prod_{j\neq i}(1 - \eta_j a_j^\dagger)|0\rangle = -\partial_{\eta_i} \prod_j (1 - \eta_j a_j^\dagger)|0\rangle = -\partial_{\eta_i}|\eta\rangle,$

$\langle\eta|a_i = \langle 0| \prod_{j\neq i}(1 - a_j\bar\eta_j) \underbrace{(1 - a_i\bar\eta_i) a_i}_{a_i = -\partial_{\bar\eta_i} a_i\bar\eta_i = \partial_{\bar\eta_i}(1 - a_i\bar\eta_i)} = \partial_{\bar\eta_i} \langle 0| \prod_j (1 - a_j\bar\eta_j) = \partial_{\bar\eta_i}\langle\eta|.$

(c) $\langle\eta|\nu\rangle = \langle\eta| \prod_i \underbrace{(1 - \nu_i a_i^\dagger)}_{(1+a_i^\dagger\nu_i)} |0\rangle = \langle\eta| \prod_i \underbrace{(1 + \bar\eta_i\nu_i)}_{\exp(\bar\eta_i\nu_i)} |0\rangle = \exp(\sum_i \bar\eta_i\nu_i).$

(d) To prove the completeness of fermion coherent states, we apply Schur's lemma, i.e., we need to show that $[a_j^{(\dagger)}, \int d\eta e^{-\bar\eta\eta}|\eta\rangle\langle\eta|] = 0$:

$$a_j^\dagger \int d\eta\, e^{-\bar\eta\eta}|\eta\rangle\langle\eta| = -\int d\eta\, e^{-\bar\eta\eta}\partial_{\eta_j}|\eta\rangle\langle\eta| = \int d\eta\, \underbrace{\partial_{\eta_j} e^{-\bar\eta\eta}}_{\bar\eta_j e^{-\bar\eta\eta}} |\eta\rangle\langle\eta| = \int d\eta\, e^{-\bar\eta\eta}|\eta\rangle\langle\eta|a_j^\dagger,$$

$$a_j \int d\eta\, e^{-\bar\eta\eta}|\eta\rangle\langle\eta| = \int d\eta\, \underbrace{e^{-\bar\eta\eta}\eta_j}_{-\partial_{\bar\eta_j} e^{-\bar\eta\eta}} |\eta\rangle\langle\eta| = \int d\eta\, e^{-\bar\eta\eta}|\eta\rangle\partial_{\bar\eta_j}\langle\eta| = \int d\eta\, e^{-\bar\eta\eta}|\eta\rangle\langle\eta|a_j.$$

The constant of proportionality is fixed by taking the expectation value with the vacuum:

$$\langle 0| \int d\eta e^{-\bar\eta\eta}|\eta\rangle\langle\eta|0\rangle = \int d\eta\, e^{-\bar\eta\eta} = 1.$$

(e) Representing a general n-particle state by $|n\rangle = a_1^\dagger \ldots a_n^\dagger|0\rangle$, $\langle n| = \langle 0|a_n \ldots a_1$, the matrix element $\langle n|\psi\rangle$ reads

$$\langle n|\psi\rangle = \langle 0|a_n \ldots a_1|\psi\rangle = \langle 0|\psi_n \ldots \psi_1|\psi\rangle = \psi_n \ldots \psi_1.$$

Similarly, we obtain $\langle\psi|n\rangle = \bar\psi_1 \ldots \bar\psi_n$. Using these results,

$$\langle n|\psi\rangle\langle\psi|n\rangle = \psi_n \ldots \psi_1\bar\psi_1 \ldots \bar\psi_n = \psi_1\bar\psi_1 \ldots \psi_n\bar\psi_n$$
$$= (\zeta\bar\psi_1\psi_1) \ldots (\zeta\bar\psi_n\psi_n) = (\zeta\bar\psi_1) \ldots (\zeta\bar\psi_n)\psi_n \ldots \psi_1 = \langle\zeta\psi|n\rangle\langle n|\psi\rangle.$$

3.8.6 Feynman path integral from the functional field integral

The abstract nature of the coherent-state representation conceals the close similarity between the Feynman and coherent-state path integrals. To help elucidate the connection, the goal of this problem is to confirm that the Feynman path integral of the quantum harmonic oscillator follows from the coherent state path integral.

Consider the simplest bosonic many-body Hamiltonian, $\hat H = \omega(a^\dagger a + \frac{1}{2})$, where $a^\dagger$ creates "structureless" particles, i.e., states in a one-dimensional Hilbert space. Note that $\hat H$ can be interpreted as the Hamiltonian of a single oscillator degree of freedom. Show that the field integral for the partition function $\mathcal{Z} = \text{tr}\exp(-\beta\hat H)$ can be mapped onto the (imaginary-time) path integral of a harmonic oscillator by a suitable variable transformation.

Answer:

In the coherent-state representation, the quantum partition function of the oscillator Hamiltonian is expressed in terms of the path integral,

$$\mathcal{Z} = \int D\phi \exp\left(-\int_0^\beta d\tau \left(\bar\phi\partial_\tau\phi + \omega\bar\phi\phi\right)\right), \tag{3.101}$$

where $\phi(\tau)$ denotes a complex scalar field, the constant factor $e^{-\beta\omega/2}$ has been absorbed into the measure of the functional integral $D(\bar{\phi}, \phi)$, and we have set the chemical potential $\mu = 0$. The connection between the coherent-state and Feynman integral is established by the change of field variables, $\phi(\tau) = (\frac{m\omega}{2})^{1/2}(q(\tau) + \frac{i}{m\omega}p(\tau))$, where $p(\tau)$ and $q(\tau)$ represent real fields. Substituting this representation into Eq. (3.101), and rearranging some terms by integrating by parts, the connection is established: $\mathcal{Z} = \int D(p, q) \exp[-\int_0^\beta d\tau(-ip\dot{q} + \frac{p^2}{2m} + \frac{m\omega^2}{2}q^2)]$. (Of course, the "absorption" of the constant $\hbar\omega/2$ in the Hamiltonian into the measure constitutes a "slight of hand." In operator quantum mechanics, the correspondence between the (q, p) and the $(a, a^\dagger)$ representations of the Hamiltonian includes that constant. Keeping in mind that the constant reflects the non-commutativity of operators, think how it might be recovered within the path integral formalism. Hint: It is best to consider the mapping between representations within a time-slice discretized setting.)

3.8.7 Quantum partition function of the harmonic oscillator

The following is an exercise on elementary field integration and infinite products.

Compute the partition function of the harmonic oscillator Hamiltonian in the field integral formulation. To evaluate the resulting infinite product over Matsubara frequencies, apply the formula $x/\sin x = \prod_{n=1}^\infty (1 - x^2/(\pi n)^2)^{-1}$. (Hint: The normalization of the result can be fixed by requiring that, in the zero-temperature limit, the oscillator occupies its ground state.) Finally, compute the partition function by elementary means and check your result. As an additional exercise, repeat the same steps for the "fermionic oscillator," i.e., with $a, a^\dagger$ fermion operators. Here you will need the auxiliary identity $\cos x = \prod_{n=1}^\infty (1 - x^2(\pi(n + 1/2))^{-2})$.

Answer:

Making use of the Gaussian functional integral for complex fields, from Eq. (3.101) one obtains $\mathcal{Z}_\mathrm{B} \sim \det(\partial_\tau + \omega)^{-1} \sim \prod_n(-i\omega_n + \omega)^{-1} \sim \prod_{n=1}^\infty(1 + (\frac{\beta\omega}{2\pi n})^2)^{-1} \sim (\sinh(\beta\omega/2))^{-1}$. In the limit of low temperatures, the partition function is dominated by the ground state, $\lim_{\beta\to\infty} \mathcal{Z}_\mathrm{B} = \exp(-\beta\omega/2)$, which fixes the constant of proportionality. Thus, $\mathcal{Z}_\mathrm{B} = (2\sinh(\beta\hbar\omega/2))^{-1}$.

In the fermionic case, Gaussian integration gives a product over eigenvalues in the numerator and we have to use fermionic Matsubara frequencies, $\omega_n = (2n + 1)\pi/\beta$: $\mathcal{Z}_\mathrm{F} \sim \det(\partial_\tau + \omega) \sim \prod_n(-i\omega_n + \omega) \sim \prod_{n=1}^\infty(1 + (\frac{\beta\omega}{(2n+1)\pi})^2) \sim \cosh(\beta\omega/2)$. Fixing the normalization, one obtains $\mathcal{Z}_\mathrm{F} = 2e^{-\beta\omega}\cosh(\beta\omega/2)$. Taken together, these results are easily confirmed by direct computation: $\mathcal{Z}_\mathrm{B} = e^{-\beta\omega/2}\sum_{n=0}^\infty e^{-n\beta\omega} = (2\sinh(\beta\omega/2))^{-1}$, $\mathcal{Z}_\mathrm{F} = e^{-\beta\omega/2}\sum_{n=0}^1 e^{-n\beta\omega} = 2e^{-\beta\omega}\cosh(\beta\omega/2)$.

3.8.8 Frequency summations

Using the frequency summation techniques developed in the text, this problem involves the computation of two basic correlation functions central to the theory of the interacting Fermi gas.

pair cor-
relation
function

(a) The **pair correlation function** $\chi^{\mathrm{c}}_{\omega_n,\mathbf{q}}$ is an important building block entering the calculation of the Cooper pair propagator in superconductors (see section 5.3.2). It is defined by

$$\chi^{\mathrm{c}}_{\omega_n,\mathbf{q}} \equiv -\frac{T}{L^d} \sum_{\omega_m,\mathbf{p}} G_0(i\omega_m,\mathbf{p})G_0(i\omega_n - i\omega_m,\mathbf{q}-\mathbf{p}) = \frac{1}{L^d} \sum_{\mathbf{p}} \frac{1 - n_{\mathrm{F}}(\xi_{\mathbf{p}}) - n_{\mathrm{F}}(\xi_{\mathbf{q}-\mathbf{p}})}{i\omega_n - \xi_{\mathbf{p}} - \xi_{\mathbf{q}-\mathbf{p}}},$$

where $G_0(\mathbf{p},i\omega_m) = (i\omega_m - \xi_{\mathbf{p}})^{-1}$. Verify the second equality. (Note that $\omega_m = (2m+1)\pi T$ are fermionic Matsubara frequencies, while $\omega_n = 2\pi nT$ are bosonic.)

density–
density
response
function

(b) Similarly, verify that the **density–density correlation function**, central to the theory of the Fermi gas (see section 4.2), is given by

$$\chi^{\mathrm{d}}_{\omega_n,\mathbf{q}} \equiv -\frac{T}{L^d} \sum_{\omega_m,\mathbf{p}} G_0(i\omega_m,\mathbf{p})G_0(i\omega_m + i\omega_n,\mathbf{p}+\mathbf{q}) = -\frac{1}{L^d} \sum_{\mathbf{p}} \frac{n_{\mathrm{F}}(\xi_{\mathbf{p}}) - n_{\mathrm{F}}(\xi_{\mathbf{p}+\mathbf{q}})}{i\omega_n + \xi_{\mathbf{p}} - \xi_{\mathbf{p}+\mathbf{q}}}.$$

Answer:

(a) To evaluate the sum over fermionic frequencies ω_m, we employ the Fermi function $\beta n_{\mathrm{F}}(z) = \beta(e^{\beta z} + 1)^{-1}$ defined in the left column of (3.80). Noting that the function $G_0(z,\mathbf{p}_1)G_0(z+i\omega_n,\mathbf{p}_2)$ has simple poles at $z = \xi_{\mathbf{p}_1}$ and $z = \xi_{\mathbf{p}_2} - i\omega_n$, and applying (3.81) (with the identification $S = \sum h$ and $h = G_0 G_0$), we obtain $S = \frac{-n_{\mathrm{F}}(\xi_{\mathbf{p}_1}) + n_{\mathrm{F}}(\xi_{-\mathbf{p}_2} + i\omega_n)}{i\omega_n - \xi_{\mathbf{p}_1} - \xi_{\mathbf{p}_2}}$. Using the fact that $n_{\mathrm{F}}(x + i\beta\omega_n) = n_{\mathrm{F}}(x)$ and $n_{\mathrm{F}}(-x) = 1 - n_{\mathrm{F}}(x)$ we arrive at the given result.
(b) One proceeds as in part (a).

3.8.9 Pauli paramagnetism

There are several mechanisms whereby a Fermi gas subject to an external magnetic field responds to such a perturbation. One of these, the phenomenon of Pauli paramagnetism, is purely quantum mechanical in nature. Its origin lies in the energy balance of spinful fermions rearranging at the Fermi surface in response to the field. In this problem, we explore the resulting contribution to the magnetic susceptibility of the electron gas.

Fermions couple to a magnetic field by their orbital momentum as well as by their spin. Concentrating on the latter mechanism, consider the Hamiltonian

$$\hat{H}_z = -\mu_0 \mathbf{B} \cdot \hat{\mathbf{S}}, \qquad \hat{\mathbf{S}} = \frac{1}{2} a^{\dagger}_{\alpha\sigma} \boldsymbol{\sigma}_{\sigma\sigma'} a_{\alpha\sigma'},$$

where $\boldsymbol{\sigma}$ is the vector of Pauli matrices, α is an orbital quantum number, and $\mu_0 = e/2m$ is the Bohr magneton.

It turns out that the presence of $\hat{H}_z$ in the energy balance leads to the generation of a net paramagnetic response of purely quantum mechanical origin. Consider a two-fold (i.e., spin) degenerate single-particle band of free electrons states (see the figure). Both bands are filled up to a certain chemical potential μ. Upon the switching on of an external field, the degeneracy is lifted and the two bands shift in opposite directions by an amount $\sim \mu_0 B$. While, deep in the bands, the Pauli principle forbids a rearrangement of spin configurations, at the Fermi energy $\downarrow$ states can turn to energetically more favorable $\uparrow$ states. More precisely, for bands shifted by an amount $\sim \mu_0 B$, a number $\sim \mu_0 B \rho(\mu)$ of states may change their spin orientation, which leads to a total energy change of $\Delta E \sim -\mu_0^2 B^2 \rho(\mu)$. Differentiating twice with respect to the magnetic field gives a positive contribution to the magnetic susceptibility, $\chi \sim -\partial_B^2 \Delta E \sim \mu_0^2 \rho(\mu)$.

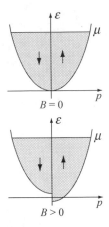

(a) To convert the qualitative estimate above into a quantitative result, construct the coherent state action of the full Hamiltonian $\hat{H} = \hat{H}_0 + \hat{H}_z$, where $\hat{H}_0 = \sum_{\alpha,\sigma} a_{\alpha\sigma}^\dagger \epsilon_\alpha a_{\alpha\sigma}$ is the non-magnetic part of the Hamiltonian. Integrate out the Grassmann fields to obtain the free energy F as a sum over Matsubara frequencies.

(b) Show that, at low temperatures, the spin contribution to the magnetic susceptibility $\chi \equiv -\partial_B^2\big|_{B=0} F$ is given by

$$\boxed{\chi \xrightarrow{T\to 0} \frac{\mu_0^2}{2}\rho(\mu)} \tag{3.102}$$

where $\rho(\epsilon) = \sum_\alpha \delta(\epsilon - \epsilon_\alpha)$ denotes the single-particle density of states. (Hint: Perform the field derivatives prior to the frequency summation.)

Answer:

(a) Choosing the quantization axis parallel to the magnetic field, the Hamiltonian assumes a diagonal form $\hat{H} = \sum_{\alpha\sigma} a_{\alpha\sigma}^\dagger [\epsilon_\alpha - (\mu_0 B/2)(\sigma_z)_{\sigma\sigma}] a_{\alpha\sigma}$ and the (frequency representation of the) action reads $S[\psi] = \sum_{\alpha\sigma n} \bar{\psi}_{\alpha\sigma n}(-i\omega_n + \xi_\alpha - (\mu_0 B/2)(\sigma_z)_{\sigma\sigma})\psi_{\alpha\sigma n}$. Integrating over ψ, we obtain the partition function $\mathcal{Z} = \prod_{\alpha n} \beta^2 [(-i\omega_n + \xi_\alpha)^2 - \frac{1}{4}(\mu_0 B)^2]$ and

$$F = -T \ln \mathcal{Z} = -T \sum_{\alpha n} \ln\left(\beta^2\left((-i\omega_n + \xi_\alpha)^2 - \frac{1}{4}(\mu_0 B)^2\right)\right).$$

(b) Differentiating the free energy, we obtain $\chi = -\frac{1}{2}\mu_0^2 T \sum_{\alpha\omega_n}(-i\omega_n + \xi_\alpha)^{-2}$. Defining $\chi = \sum_{\alpha n} h_\alpha(\omega_n)$, where $h_\alpha(\omega_n) = \frac{1}{2}\mu_0^2 T(-i\omega_n + \xi_\alpha)^{-2}$, Eq. (3.81) can be applied to perform the frequency sum. Noting that the function $h(-iz)$ has poles of *second* order at $z = \xi_\alpha$, i.e., $\text{Res}[g(z)h(-iz)]_{z=\xi_\alpha} = g'(\xi_\alpha)$, we obtain

$$\chi = -\frac{\mu_0^2}{2}\sum_\alpha n_F'(\xi_\alpha) = -\frac{\mu_0^2}{2}\int_{-\infty}^\infty d\epsilon\, \rho(\epsilon) n_F'(\epsilon - \mu).$$

At low temperatures, $T \to 0$, the Fermi distribution function approaches a step function, $n_F(\epsilon) \to \theta(-\epsilon)$, i.e., $n_F'(\epsilon) = -\delta(\epsilon)$, and our result reduces to (3.102).

3.8.10 Boson–fermion duality

The equivalence of the bosonic and fermionic representations of the one-dimensional electron gas is exemplified by computation of the correlation function (3.95) considered in the text.

(a) Employ the free fermion field integral with action (3.86) to compute the zero-temperature limit of the correlation function (3.95). (Assume $x > 0$ and, for simplicity, work in the limit of low temperatures, evaluating Matsubara sums as integrals.) (b) For a free scalar bosonic field θ with action $S[\theta] = (1/2c) \int dx\, d\tau\, [(\partial_\tau \theta)^2 + (\partial_x \theta)^2]$, compute the correlation function $K(x, \tau) \equiv \langle \theta(x, \tau)\theta(0, 0) - \theta(0, 0)\theta(0, 0) \rangle$ for $x > 0$. (c) Compute the correlation function $C(x, \tau) = \Gamma^2 \left\langle e^{2i\theta(x,\tau)}\, e^{-2i\theta(0,0)} \right\rangle$ and find for which values of Γ equivalence to the fermionic representation is obtained.

Answer:

(a) We set $v_F = 1$ and defined the partition sum $\mathcal{Z}_\pm \equiv \int D\psi\, \exp(-S_\pm[\psi])$, where $S_\pm[\psi] = \int dx\, d\tau\, \bar\psi(\partial_\tau \mp i\partial_x)\psi$. Next consider the Green function $G_\pm(x, \tau) = \mathcal{Z}_\pm^{-1} \int D\psi\, \bar\psi(x,\tau)\psi(0,0)\, e^{-S_\pm[\psi]}$, which we compute by Gaussian integration as $G_\pm(x,\tau) = -(\partial_{\tau'} \mp i\partial_{x'})^{-1}_{(0,0;x,\tau)} = -\frac{T}{L}\sum_{p,\omega_n} \frac{1}{i\omega_n \mp p} e^{-ipx + i\omega_n \tau}$. Assuming for definiteness that $x > 0$, and integrating over momenta, we arrive at $G_\pm(x, \tau) = \mp iT \sum_n \Theta(\mp n) e^{\omega_n(\pm x + i\tau)} \simeq \frac{1}{2\pi} \frac{1}{\pm ix - \tau}$, where, in the last equality, we have approximated the frequency sum by an integral. Thus, the correlation function (3.95) is given by $C(x, \tau) = G_+(x, \tau)G_-(-x, \tau) = \frac{1}{(2\pi)^2} \frac{1}{x^2 + \tau^2}$. (b) Expressed in a frequency/momentum Fourier representation, $S[\theta] = \frac{L}{2cT}\sum_{q,n} |\theta_{qn}|^2 (q^2 + \omega_n^2)$. Performing the Gaussian integral over θ, we obtain

$$K(x, \tau) = \frac{cT}{L}\sum_{qn} \frac{e^{iqx - i\omega_n \tau} - 1}{q^2 + \omega_n^2} \simeq \frac{cT}{2}\sum_n \frac{e^{-|\omega_n||x| - i\omega_n \tau} - 1}{|\omega_n|}$$

$$\simeq \frac{c}{4\pi}\int_0^{a^{-1}} d\omega\, \frac{e^{-\omega(x+i\tau)} - 1}{\omega} + \text{c.c.} \overset{x,\tau \gg a}{\simeq} -\frac{c}{4\pi}\ln\left(\frac{x^2 + \tau^2}{a^2}\right),$$

where we have approximated the momentum and frequency sums by integrals, the latter cutoff at large frequencies by $E_F \simeq v_F a^{-1} \overset{v_F=1}{=} a^{-1}$. (c) Using the results derived in (b),

$$C(x, \tau) = \Gamma^2 \left\langle e^{2i(\theta(x,\tau) - \theta(0,0))} \right\rangle = \Gamma^2 e^{-2\langle(\theta(x,\tau) - \theta(0,0))^2\rangle}$$

$$= \Gamma^2 \exp\left(-\frac{c}{\pi}\ln\left(\frac{x^2 + \tau^2}{a^2}\right)\right) = \Gamma^2 \left(\frac{a^2}{x^2 + \tau^2}\right)^{\frac{c}{\pi}}.$$

Setting $c = \pi$ and $\Gamma = 1/(2\pi a)$, we obtain equivalence to the fermionic representation of the correlation function considered in (a).

3.8.11 Electron–phonon coupling

As follows from the structure of the "master" Hamiltonian of condensed matter (1.1), mobile electrons in solids are influenced by the vibrations of the host ions, the phonons. This coupling mechanism may generate a net attractive interaction between electrons. Referring to the info on page 258 for a qualitative discussion of this interaction mechanism, the purpose of this problem is to explore quantitatively the profile of the phonon mediated electron–electron interaction. In section 5.3 we will see that this interaction lies at the root of conventional BCS superconductivity.

Consider the three-dimensional variant of the phonon Hamiltonian (1.34),

$$\hat{H}_{\mathrm{ph}} = \sum_{\mathbf{q},j} \omega_q a^\dagger_{\mathbf{q},j} a_{\mathbf{q},j} + \mathrm{const.},$$

where ω_q is the phonon dispersion (here assumed to depend only on the modulus of the momentum, $|\mathbf{q}| = q$), and the index $j = 1, 2, 3$ accounts for the fact that the lattice ions can oscillate in three directions in space (i.e., there are three linearly independent oscillator modes.[53]) Electrons in the medium sense the induced charge $\rho_{\mathrm{ind}} \sim \nabla \cdot \mathbf{P}$, where $\mathbf{P} \sim \mathbf{u}$ is the polarization generated by the local distortion $\mathbf{u}$ of the lattice ($\mathbf{u}(\mathbf{r})$ is the three-dimensional generalization of the displacement field $\phi(r)$ considered in chapter 1). Expressed in terms of phonon creation and annihilation operators (cf. Eq. (1.32)), $\mathbf{u_q} = \mathbf{e}_j(a_{\mathbf{q},j} + a^\dagger_{-\mathbf{q},j})/(2m\omega_q)^{1/2}$, where $\mathbf{e}_j$ is the unit vector in the j-direction,[53] from which follows the **electron–phonon Hamiltonian**

electron–
phonon
Hamil-
tonian

$$\hat{H}_{\mathrm{el-ph}} = \gamma \int d^d r\, \hat{n}(\mathbf{r}) \nabla \cdot \mathbf{u}(\mathbf{r}) = \gamma \sum_{\mathbf{q},j} \frac{iq_j}{(2m\omega_q)^{1/2}} \hat{n}_\mathbf{q}(a_{\mathbf{q},j} + a^\dagger_{-\mathbf{q},j}).$$

Here, $\hat{n}_\mathbf{q} \equiv \sum_\mathbf{k} c^\dagger_{\mathbf{k+q}} c_\mathbf{k}$ denotes the electron density and, for simplicity, the electron spin has been suppressed.
(a) Formulate the coherent state action of the electron–phonon system.
(b) Integrate out the phonon fields and show that an attractive interaction between electrons is generated.

Answer:

(a) Introducing a Grassmann field ψ (and a complex field ϕ) to represent the electron (phonon) operators, one obtains the coherent state field integral, $\mathcal{Z} = \int D\psi D\phi \exp\left(-S_{\mathrm{el}}[\psi] - S_{\mathrm{ph}}[\phi] - S_{\mathrm{el-ph}}[\psi, \phi]\right)$, where

$$S_{\mathrm{ph}}[\phi] = \sum_{q,j}(-i\omega_n + \omega_q)|\phi_{qj}|^2, \quad S_{\mathrm{el-ph}}[\psi, \phi] = \gamma \sum_{q,j} \frac{iq_j}{(2m\omega_q)^{1/2}} \rho_q(\phi_{qj} + \bar{\phi}_{-qj}),$$

$\rho_q = \sum_k \bar\psi_{k+q} \psi_k$, and the electron action need not be specified explicitly. Here we

[53] For details, see N. W. Ashcroft and N. D. Mermin, *Solid State Physics* (Holt-Saunders International, 1983).

have adopted a short-hand convention setting $q = (\omega_n, \mathbf{q})$; this is not to be confused with the modulus $|\mathbf{q}|$.

(b) Performing the Gaussian integration over the phonon fields, we obtain the effective electron action

$$S_{\text{eff}}[\psi] = S_{\text{el}}[\psi] - \ln \int D\phi \, e^{-S_{\text{ph}}[\phi] - S_{\text{el-ph}}[\psi,\phi]} = S_{\text{el}}[\psi] - \frac{\gamma^2}{2m} \sum_q \frac{|\mathbf{q}|^2}{\omega_n^2 + \omega_q^2} \rho_q \rho_{-q},$$

where we have omitted a ψ-independent contribution from the integration over the uncoupled ϕ-action. Transforming from Matsubara to real frequencies, $\omega_n \to -i\omega$, one may notice that, for every momentum mode $\mathbf{q}$, the interaction is attractive at low frequencies, $\omega < \omega_q$.

3.8.12 Disordered quantum wires

In this problem, we consider a one-dimensional interacting Fermi system – a "quantum wire" – in the presence of impurities. Building on the results obtained in section 3.6, we derive an effective low-energy action. (Analysis of the large-scale behavior of the disordered quantum wire requires renormalization group methods and is postponed to chapter 6.)

In section 3.6 we discussed the physics of interacting fermions in one dimension. We saw that, unlike in a Fermi liquid, the fundamental excitations of the system are charge and spin density waves – collective excitations describing the wave-like propagation of the charge and spin degrees of freedom, respectively. Going beyond the level of an idealized translationally invariant environment, the question we wish to address currently is to what extent the propagation of these modes will be hampered by the presence of spatially localized imperfections. This problem is of considerable practical relevance. All physical realizations of one-dimensional conductive systems – semiconductor quantum wires, conducting polymers, carbon nanotubes, quantum Hall edges, etc. – generally contain imperfections. Further, and unlike systems of higher dimensionality, a spin or charge degree of freedom propagating down a one-dimensional channel will inevitably hit any impurity blocking its way. We thus expect that impurity scattering has a stronger impact on the transport coefficients than in higher dimensions.

However, there is a second and less obvi-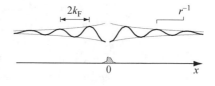
ous mechanism behind the strong impact of
disorder scattering on the conduction behavior of one-dimensional quantum wires. Imagine
a wave packet of characteristic momentum k_F
colliding with an impurity at position $x = 0$
(see the figure). The total wave amplitude to the left of the impurity, $\psi(x) \sim \exp(ik_F x) + r \exp(-ik_F x)$, will be a linear superposition of the incoming amplitude $\sim \exp(ik_F x)$ and the reflected outgoing amplitude $\sim r \exp(-ik_F x)$, where r is the reflection coefficient. Thus, the electronic density profile is given by $\rho(x) = |\psi(x)|^2 \sim 1 + |r|^2 + 2 \operatorname{Re}(re^{-2ik_F x})$, which contains an oscillatory con-

tribution known as a **Friedel oscillation**. Moreover, a closer analysis (see the exercise below) shows that, in one dimension, the amplitude of these oscillations decays rather slowly, varying as $\sim |x|^{-1}$. The key point is that, in the presence of electron–electron interactions, other particles approaching the impurity will notice not only the impurity itself but also the charge density pattern of the Friedel oscillation. The additional scattering potential then creates a secondary Friedel oscillation, etc. We thus expect that even a weak imperfection in a Luttinger liquid acts as a "catalyst" for the recursive accumulation of a *strong* potential. In this problem, we will derive the effective low-energy action describing the interplay of interaction and impurity scattering. The actual catalytic amplification mechanism outlined above is then explored in chapter 6 by renormalization group methods.

EXERCISE To explore the **Friedel oscillatory response of the one-dimensional electron gas** to a local perturbation, consider the density–density correlation function

$$\Pi(x,t) = \langle \hat{\rho}(x,t)\hat{\rho}(0,0) \rangle - \langle \hat{\rho}(x,t) \rangle \langle \hat{\rho}(0,0) \rangle,$$

where $\langle \cdots \rangle$ denotes the ground state expectation value, $\hat{\rho} = a^\dagger a$, and $a(x) = e^{ik_F x}a_+(x) + e^{-ik_F x}a_-(x)$ splits into a left- and a right-moving part as usual. Using the fact that $\hat{H} = \sum_{q,s} v_F(p_F + sq)a_{sq}^\dagger a_{sq}$ and the von Neumann equation $\dot{a}_{sq} = i[\hat{H}, a_{sq}]$, show that the time dependence of the annihilation operators is given by $a_{sq}(t) = e^{-iv_F(p_F + sq)t}a_{sq}$. Use this result, the canonical operator commutation relations, and the ground state property $a_{\pm,q}|\Omega\rangle = 0$ for $\pm q > 0$, to show that

$$\Pi(x,t) = \frac{1}{4\pi^2}\left(\frac{1}{(x - v_F t)^2} + \frac{1}{(x + v_F t)^2} + \frac{2\cos(2p_F x)}{x^2 - (v_F t)^2} \right).$$

Use this result to argue why the *static* response to an impurity potential decays as $\sim |x|^{-1}$.

Consider the one-dimensional quantum wire, as described by the actions (3.86) and (3.90). Further, assume that, at $x = 0$, the system contains an imperfection or impurity. Within the effective action approach, this is described by

$$S_{\text{imp}}[\psi] = \int d\tau \left(v_+ \psi_+^\dagger \psi_+ + v_- \psi_-^\dagger \psi_+ + v\psi_+^\dagger \psi_- + \bar{v}\psi_-^\dagger \psi_+ \right),$$

where all field amplitudes are evaluated at $x = 0$ and the constants $v_\pm \in \mathbb{R}$ and $v \in \mathbb{C}$ describe the amplitudes of forward and backward scattering, respectively.

(a) Show that the forward scattering contributions can be removed by a gauge transformation. This demonstrates that forward scattering is inessential as long as only gauge invariant observables are considered. What is the reason for the insignificance of forward scattering?

We next reformulate the problem in bosonic language. While the clean system is described by Eq. (3.98), substitution of (3.94) into the impurity action gives $S_{\text{imp}}[\theta] = \gamma \int d\tau \cos(2\theta(\tau, x = 0))$, where $\gamma = 2v\Gamma^2$ and we have assumed the backward scattering amplitude to be real. (Consider how any phase carried by the scattering amplitude can be removed by a global gauge transformation of the fields $\psi_\pm$. Notice also the independence of S_{imp} of the field ϕ.)

(b) Integrate out the Gaussian field ϕ to obtain the Lagrangian form of the action,

$$S[\theta] = \frac{1}{2\pi g} \int dx\, d\tau \left(v(\partial_x \theta)^2 + v^{-1}(\partial_\tau \theta)^2\right) + S_{\text{imp}}[\theta].$$

This formulation still contains redundancy: everywhere, except at $x = 0$, the action is Gaussian. This observation suggests that one may integrate out all field degrees of freedom $\theta(x \neq 0)$, thus reducing the problem to one that is local in space (though, as we shall see, non-local in time). To this end, we may reformulate the field integral as $\mathcal{Z} = \int D\tilde\theta\, e^{-S[\tilde\theta]}$, where $e^{-S[\tilde\theta]} = \int D\theta\, e^{-S[\theta]} \prod_\tau \delta(\tilde\theta(\tau) - \theta(0,\tau))$ is the action integrated over all field amplitudes save for $\theta(0,\tau)$, and $\prod_\tau \delta(\tilde\theta(\tau) - \theta(0,\tau))$ is a product of δ-functions (one for each time slice) imposing the constraints $\theta(0,\tau) = \tilde\theta(\tau)$. Representing the δ-functions as $\delta(\tilde\theta - \theta(0,\tau)) = \frac{1}{2\pi} \int dk(\tau) \exp(ik(\tau)(\tilde\theta(\tau) - \theta(0,\tau)))$, one obtains

$$\exp(-S[\tilde\theta]) = \int D\theta Dk \exp\left(-S[\theta] + i\int d\tau\, k(\tau)(\tilde\theta(\tau) - \theta(0,\tau))\right)$$

$$= e^{-S_{\text{imp}}[\tilde\theta]} \int D\theta Dk \exp\left(-\int dx d\tau \left(\frac{1}{2\pi g}\left(v(\partial_x\theta)^2 + \frac{1}{v}(\partial_\tau\theta)^2\right) + ik(\tilde\theta - \theta)\delta(x)\right)\right).$$

The advantage gained with this representation is that it permits us to replace $\cos(2\theta(0,\tau))$ in S_{imp} by $\cos(2\tilde\theta(\tau))$, whereupon the θ-dependence of the action becomes purely quadratic.

(c) Integrate out the field $\theta(x,\tau)$ and, redefining $\tilde\theta(\tau) \to \theta(\tau)$, obtain the representation $\mathcal{Z} = \int D\theta\, e^{-S_{\text{eff}}[\theta]}$, where

$$\boxed{S_{\text{eff}}[\theta] = \frac{1}{\pi T g} \sum_{\omega_n} |\omega_n||\theta_n|^2 + \gamma \int d\tau \cos(2\theta(\tau))} \qquad (3.103)$$

Notice that the entire effect of the bulk electron gas at $x \neq 0$ has gone into the first, dissipative term. Thus we have reduced the problem to one involving a single time-dependent degree of freedom

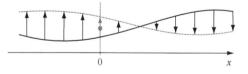

subject to a dissipative damping mechanism and a periodic potential (see our earlier discussion of this problem in problem 3.8.4 above).

INFO To understand the **physical origin of the dissipative damping mechanism**, notice that, in the absence of the impurity, the system is described by a set of harmonic oscillators. We can thus think of the degree of freedom $\theta(0,\tau)$ as the coordinate of a "bead" embedded into an infinitely extended harmonic chain. From the point of view of this bead, the neighboring degrees of freedom hamper its free kinematic motion; i.e., in order to move, the bead has to drag an entire "string" of oscillators behind it. In other words, a local excitation of the $x = 0$ oscillator will lead to the dissipation of kinetic energy into the continuum of neighboring oscillators. Clearly, the rate of dissipation will increase with both the stiffness of the oscillator chain (g^{-1}) and the frequency of the excitation (ω_n), as described by the first term in Eq. (3.103).

Answer:

(a) Consider the gauge transformation: $\psi_+(x, \tau) \to \exp(-iv_{\mathrm{F}}^{-1} v_+ \theta(x))$. While S_{int} and S_{imp} are gauge invariant and do not change, substitution of the transformed field into the non-interacting action leads to $S_0[\psi^\dagger, \psi] \to S_0[\psi^\dagger, \psi] - v_+ \int d\tau \psi_+^\dagger \psi_+$. The induced term cancels the v_+ contribution to S_{imp}. A similar transformation removes the v_- contribution. The physical reason for the insignificance of the forward scattering operators is that they describe the scattering of states $|\pm k_{\mathrm{F}}\rangle$ into the same states $|\pm k_{\mathrm{F}}\rangle$. The optional phase shift picked up in these processes is removed by the transformation above.

(b) This step involves an elementary Gaussian integral.

(c) Expressed in momentum space, the effective action assumes the form

$$e^{-S[\tilde{\theta}]} = e^{-S_{\mathrm{imp}}[\tilde{\theta}]} \int Dk\, e^{i \sum_n k_n \tilde{\theta}_{-n}}$$

$$\times \int D\theta \exp\left\{ -\frac{T}{L} \sum_{qn} \left(\frac{1}{2\pi g} \left(vq^2 + v^{-1}\omega_n^2 \right) |\theta_{q,n}|^2 + ik_n \theta_{q,-n} \right) \right\}$$

$$= e^{-S_{\mathrm{imp}}[\tilde{\theta}]} \int Dk\, e^{i \sum_n k_n \tilde{\theta}_{-n}} \exp\left(-\frac{\pi gT}{2L} \sum_{qn} (vq^2 + v^{-1}\omega_n^2)^{-1} |k_n|^2 \right)$$

$$= \mathcal{N} e^{-S_{\mathrm{imp}}[\tilde{\theta}]} \int Dk \exp\left(-\sum_n \left(|\omega_n|^{-1} |k_n|^2 + i \sum_{\omega_n} k_n \tilde{\theta}_{-n} \right) \right).$$

Finally, integrating over k, and denoting $\tilde{\theta}(\tau)$ by $\theta(\tau)$, we obtain (3.103).

4   Perturbation Theory

SYNOPSIS In this chapter, we introduce the analytical machinery to investigate the properties of many-body systems perturbatively. Employing the paradigmatic "ϕ^4-theory" as an example, we learn here how to describe systems that are not too far from a known reference state. Diagrammatic methods are introduced as a tool to implement efficiently perturbation theory at large orders. These new concepts are then applied to the analysis of various properties of the weakly interacting or disordered electron gas.

Previously, we have emphasized that the majority of many-particle problems cannot be solved in closed form; in general, one is compelled to think about approximation strategies. One ansatz leans on the fact that, when approaching the low-temperature physics of a many-particle system, we often have some preliminary ideas of its preferred states and excitations. One may then try to explore the system by using these configurations as a platform. For example, one might expand the Hamiltonian in the vicinity of a reference state and check that, indeed, residual "perturbations" are weak and can be dealt with by some kind of approximation scheme. Consider, for example, the quantum Heisenberg magnet. In dimensions higher than one, an exact solution of this system is out of reach. However, we expect that at low temperatures the spins will be frozen into configurations aligned along some (domain-wise) constant magnetization axis. Residual fluctuations around these configurations, spin waves, can be described in terms of a controlled expansion scheme. Similar programs work for countless other physical systems.

These considerations dictate much of our further strategy. We will need to construct methods to identify and describe the lowest-energy configurations of many-particle systems – often called "mean fields" – and learn how to implement perturbation theory around them. In essence, the first part of that program amounts to solving a variational problem, a relatively straightforward task. However, the formulation of perturbation strategies requires some preparation and a good deal of critical caution (since many systems notoriously defy perturbative assaults – a situation easily overlooked or misjudged!). We thus turn the logical sequence of the two steps upside down and devote this chapter to an introduction to many-body perturbation theory. This will include several applications to problems where the mean field is trivial and perturbation theory on its own suffices to produce meaningful results. Perturbation theory superimposed on nontrivial mean fields will then form the subject of the next chapter.

4.1 General Concept and Low-Order Expansions

As with any perturbative approach, many-body perturbation theory amounts to an expansion of observables in powers of some parameter – typically the coupling strength of an interaction operator. However, before discussing how this program is implemented in practice, it is important to develop some understanding of the mathematical status of such "series expansions." (To motivate our discussion, it may, and often *does*, happen that infinite-order expansion in the "small parameter" of the problem does not exist.)

4.1.1 An instructive integral

Consider the integral

$$I(g) = \int_{-\infty}^{\infty} \frac{dx}{(2\pi)^{1/2}} \exp\left(-\frac{x^2}{2} - gx^4\right). \tag{4.1}$$

This can be regarded as a caricature of a particle subject to a harmonic potential (x^2) and an "interaction" (x^4). For $g \ll 1$, it seems natural to treat the interaction perturbatively, i.e., to develop the expansion $I(g) \approx \sum_n g^n I_n$ where, applying **Stirling's approximation**, $n! \stackrel{n \gg 1}{\approx} n^n e^{-n}$,

$$g^n I_n = \frac{(-g)^n}{n!} \int_{-\infty}^{\infty} \frac{dx}{(2\pi)^{1/2}} e^{-x^2/2} x^{4n} = (-g)^n \frac{(4n-1)!!}{n!} \stackrel{n \gg 1}{\approx} \left(-\frac{16gn}{e}\right)^n.$$

This estimate should alarm us: strictly speaking, it states that a series expansion in the "small parameter" g does not exist. No matter how small g, at roughly the $(1/g)$th order in the perturbative expansion, the series begins to diverge. In fact, it is easy to predict this breakdown on qualitative grounds: for $g > 0$ $(g < 0)$, the integral (4.1) is convergent (divergent). This implies that the series expansion of the function $I(g)$ around $g = 0$ must have zero radius of convergence.

However, there is also a more "physical" way of understanding the phenomenon. Consider a one-dimensional version of (3.16), where the "Gaussian average" is given by Eq. (3.15):

$$\int_{-\infty}^{\infty} \frac{dx}{(2\pi)^{1/2}} e^{-x^2/2} x^{4n} = \sum_{\substack{\text{all possible} \\ \text{pairings of } 4n \text{ objects}}} 1 = (4n-1)!!.$$

The factor $(4n-1)!!$ measures the combinatorial freedom to pair up $4n$ objects. This suggests an interpretation of the breakdown of the perturbative expansion as the result of a competition between the smallness of the expansion parameter g and the combinatorial proliferation of equivalent contributions, or "pairings," to the Gaussian integral. Physically, the combinatorial factor can be interpreted as the number of different "partial amplitudes" contributing to the net result at any given order of perturbation theory. Eventually, the exponential growth of this figure overpowers the smallness of the expansion parameter, which is when perturbation theory breaks down. (Oddly, the existence of this rather general mechanism is usually not highlighted in textbook treatments of quantum perturbation theory!)

Does the ill-convergence of the series imply that perturbative approaches to problems of the structure of (4.1) are doomed to fail? Fortunately, this is not the case. While the infinite series $\sum_{n=0}^{\infty} g^n I_n$ is divergent, a partial resummation $\sum_{n=0}^{n_{\max}} g^n I_n$ can yield excellent approximations to the exact result. To see this, let us use the fact that $|e^{-gx^4} - \sum_{n=0}^{n_{\max}} \frac{(-gx^4)^n}{n!}| \leq \frac{(gx^4)^{n_{\max}+1}}{(n_{\max}+1)!}$ to estimate the error as follows

$$\left| I(g) - \sum_{n=0}^{n_{\max}} g^n I_n \right| \leq g^{n_{\max}+1} |I_{n_{\max}+1}| \overset{n_{\max} \gg 1}{\sim} \left(\frac{16g n_{\max}}{e} \right)^{n_{\max}}.$$

Variation with respect to $n_{\max}$ shows that the error reaches its minimum when $n_{\max} \approx (16g)^{-1}$, where it scales as $e^{-1/16g}$. (Note the exponential dependence of the error on g. For example, for a small coupling $g \approx 0.005$, 12th order perturbation theory would lead to an approximation of high precision, $\sim 10^{-5}$.) By contrast, for $g \approx 0.02$, perturbation theory becomes poor after the third order!

Summarizing, the moral to be taken from the analysis of the integral (4.1) is that perturbative expansions should not be confused with rigorous Taylor expansions. **asymptotic expansions** Rather they represent **asymptotic expansions** in the sense that, for weaker and weaker coupling, a *partial* resummation of the perturbation series leads to an ever more precise approximation to the exact result. For weak enough coupling, the distinction between Taylor expansion and asymptotic expansion becomes academic (at least for physicists). However, for intermediate or strong coupling theories, the asymptotic character of perturbation theory must be kept in mind.

4.1.2 ϕ^4-theory

While the ordinary integral $I(g)$ conveys something of the general status of perturbation theory, we need to proceed to the level of the functional integral to learn about the practical implementation of perturbative methods. The simplest interacting field theory displaying all relevant structures is defined by

$$\mathcal{Z} \equiv \int D\phi \, e^{-S[\phi]}, \qquad S[\phi] \equiv \int d^d x \left(\frac{1}{2}(\partial \phi)^2 + \frac{r}{2}\phi^2 + g\phi^4 \right) \qquad (4.2)$$

ϕ^4-theory where ϕ is a scalar bosonic field. Owing to the structure of the interaction, this model is often referred to as the ϕ^4-**theory**. The ϕ^4-model not only provides a prototypical setting in which features of interacting field theories can be explored, but also appears in numerous applications. For example, close to its critical point, the d-dimensional Ising model is described by the ϕ^4-action (see the Info block below). More generally, it can be shown that the long-range behavior of classical statistical systems with a single order parameter (e.g., the density of a fluid, the uniaxial magnetization of a solid, etc.) is described by the ϕ^4-action.[1] Within the

---

[1] Heuristically, this is explained by the fact that $S[\phi]$ is the simplest interacting (i.e., non-Gaussian) model action invariant under inversion $\phi \longleftrightarrow -\phi$. (The action of a uniaxial magnet

Ginzburg–
Landau
theory

context of statistical mechanics, $S[\phi]$ is known as the **Ginzburg–Landau free energy functional** (and less frequently as the **Landau–Wilson model**).

Ising
model

INFO The d-dimensional classical **Ising model** describes the magnetism of a lattice of moments S_i that can take one of two values, ± 1. It is defined through the Hamiltonian

$$H_{\text{Ising}} = \sum_{ij} S_i C_{ij} S_j - H \sum_i S_i, \tag{4.3}$$

where $C_{ij} = C(|i-j|)$ is a (translationally invariant) correlation matrix describing the mutual interaction of the spins, H is an external magnetic field, and the sums run over the sites of a d-dimensional lattice (assumed hypercubic, for simplicity). The Ising model represents the simplest Hamiltonian describing classical magnetism. In low dimensions, $d = 1, 2$, it can be solved exactly, i.e., the partition function and all observables depending on it can be computed rigorously (see chapter 6). However, for higher dimensions, no closed solutions exist and one has to resort to approximation strategies to analyze the properties of the partition function. Below, we will show that the long-range physics of the system is described by ϕ^4-theory. Notice that, save for the exceptional case of $d = 1$ discussed in section 6.1.1, the system is expected to display a magnetic phase transition. As a corollary, this implies that the ϕ^4-model must exhibit much more interesting behavior than its innocuous appearance suggests!

Consider then the classical partition function

$$\mathcal{Z} = \operatorname{tr} e^{-\beta H_{\text{Ising}}} = \sum_{\{S_i\}} \exp\left(\sum_{ij} S_i K_{ij} S_j + \sum_i h_i S_i \right), \tag{4.4}$$

where $K \equiv -\beta C$, and we have generalized Eq. (4.3) to the case of a spatially varying magnetic field, $h_i \equiv \beta H_i$. The feature that prevents us from computing the configurational sum is, of course, the interaction between spins. However, at a price, the interaction can be removed. Consider the identity for the "fat unity," $1 = \mathcal{N} \int D\psi \, \exp(-\frac{1}{4} \sum_{ij} \psi_i (K^{-1})_{ij} \psi_j)$, where $D\psi \equiv \prod_i d\psi_i$, K^{-1} is the inverse of the correlation matrix, and $\mathcal{N} = 1/\det(4\pi K)^{1/2}$ is the normalization factor. A shift of the integration variables, $\psi_i \to \psi_i - 2\sum_j K_{ij} S_j$, brings the integral into the form

$$1 = \mathcal{N} \int D\psi \, \exp\left(-\frac{1}{4} \sum_{ij} \psi_i (K^{-1})_{ij} \psi_j + \sum_i S_i \psi_i - \sum_{ij} S_i K_{ij} S_j \right).$$

Incorporating the fat unity under the spin sum in the partition function, one obtains

$$\mathcal{Z} = \mathcal{N} \int D\psi \sum_{\{S_i\}} \exp\left(-\frac{1}{4} \sum_{ij} \psi_i (K^{-1})_{ij} \psi_j + \sum_i S_i(\psi_i + h_i) \right). \tag{4.5}$$

Thus, we have removed the interaction between spin variables at the expense of introducing a new continuous field $\{\psi_i\}$. Why should we do this? First, a multidimensional integral $\int D\psi$ is usually easier to work with than a multidimensional sum $\sum_{\{S_i\}}$ over

should depend on the value of the local magnetization, but not on its sign.) A purely Gaussian theory might describe wave-like fluctuations of the magnetization, but not the "critical" phenomenon of a magnetic transition. One thus needs a ϕ^4-interaction term at least. Later on we will see that more complex monomials of ϕ, such as ϕ^6 or $(\partial\phi)^4$, are inessential in the long-range limit.

Hubbard–Stratonovich transformation

discrete objects. Moreover, the new representation may provide a more convenient platform for approximation strategies. The transformation leading from Eq. (4.4) to (4.5) is our first example of a **Hubbard–Stratonovich transformation**. The interaction of one field is decoupled at the expense of the introduction of another. Notice that, in spite of the somewhat high-minded terminology, the transformation is tantamount to a simple shift of a Gaussian integration variable, a feature shared by all Hubbard–Stratonovich transformations!

The summation $\sum_{\{S_i\}} = \prod_i \sum_{S_i}$ can now be performed trivially:

$$\mathcal{Z} = \mathcal{N} \int D\psi \, \exp\left(-\frac{1}{4}\sum_{ij}\psi_i(K^{-1})_{ij}\psi_j\right)\prod_i(2\cosh(\psi_i + h_i))$$

$$= \mathcal{N} \int D\psi \, \exp\left(-\frac{1}{4}\sum_{ij}(\psi_i - h_i)(K^{-1})_{ij}(\psi_j - h_j) + \sum_i \ln(\cosh\psi_i)\right),$$

where we have absorbed the inessential factor $\prod_i 2$ into a redefinition of $\mathcal{N}$. Finally, changing integration variables from ψ_i to $\phi_i \equiv \frac{1}{2}\sum_j(K^{-1})_{ij}\psi_j$, one obtains

$$\mathcal{Z} = \mathcal{N} \int D\phi \, \exp\left(-\sum_{ij}\phi_i K_{ij}\phi_j + \sum_i \phi_i h_i + \sum_i \ln\cosh\left(2\sum_j K_{ij}\phi_j\right)\right).$$

This representation still does not look very inviting. To bring it into a form amenable to further analytical evaluation, we need to make the simplifying assumption that we are working at temperatures low enough that the exponential weight $K_{ij} = -\beta C(|i - j|)$ inhibits strong fluctuations of the field ϕ. More precisely, we assume that $|\phi_i| \ll 1$ and that the spatial profile of the field is smooth. To make use of these conditions, we switch to a Fourier representation, $\phi_i = \frac{1}{\sqrt{N}}\sum_{\mathbf{k}}e^{-i\mathbf{k}\cdot\mathbf{r}_i}\phi(\mathbf{k})$, $K_{ij} = \frac{1}{N}\sum_{\mathbf{k}}e^{-i\mathbf{k}\cdot(\mathbf{r}_i - \mathbf{r}_j)}K(\mathbf{k})$, and expand $\ln\cosh(x) = x^2/2 - x^4/12 + \cdots$. Noting that $(K\phi)(\mathbf{k}) = K(\mathbf{k})\phi(\mathbf{k}) = K(0)\phi(\mathbf{k}) + \frac{1}{2}\mathbf{k}^2 K''(0)\phi(\mathbf{k}) + \mathcal{O}(\mathbf{k}^4)$, we conclude that the low-temperature expansion of the action has the general structure

$$S[\phi] = \sum_{\mathbf{k}}\left(\phi_{\mathbf{k}}(c_1 + c_2\mathbf{k}\cdot\mathbf{k})\phi_{-\mathbf{k}} + c_3\phi_{\mathbf{k}}h_{-\mathbf{k}}\right)$$

$$+ \frac{c_4}{N}\sum_{\mathbf{k}_1,\ldots,\mathbf{k}_4}\phi_{\mathbf{k}_1}\phi_{\mathbf{k}_2}\phi_{\mathbf{k}_3}\phi_{\mathbf{k}_4}\delta_{\mathbf{k}_1+\mathbf{k}_2+\mathbf{k}_3+\mathbf{k}_4,0} + \mathcal{O}(\mathbf{k}^4, h^2, \phi^6).$$

EXERCISE Identify the coefficients $c_{1\ldots4}$ of the expansion.[2]

Switching back to a real space representation and taking the continuum limit, $S[\phi]$ assumes the form of a prototypical ϕ^4-action:

$$S[\phi] = \int d^dx \left(c_2(\partial\phi)^2 + c_1\phi^2 + c_3\phi h + c_4\phi^4\right).$$

A rescaling of variables $\phi \to \phi/\sqrt{2c_2}$ finally brings the action into the form of Eq. (4.2) with coefficients $r = c_1/c_2$ and $g = c_4/2c_2$.[3]

---

[2] You should find $c_1 = K(0)(1 - 2K(0))$, $c_2 = \frac{1}{2}K''(0)(1 - 4K(0))$, $c_3 = -1$, $c_4 = \frac{K(0)}{3}$.

[3] The only difference is that the magnetic ϕ^4-action contains a term linear in ϕ and h. The reason is that, in the presence of a magnetic field, the action is no longer invariant under inversion $\phi \to -\phi$.

We have thus succeeded in describing the low-temperature phase of the Ising model in terms of a ϕ^4-model. While the structure of the action could have been guessed on symmetry grounds, the "microscopic" derivation has the advantage that it yields explicit expressions for the coupling constants. There is actually one interesting aspect of the dependence of these constants on the parameters of the microscopic model. Consider the constant c_1 controlling the **k**-independent contribution to the Gaussian action: $c_1 \propto K(0)(1 - 2K(0)) \propto (1 - 2\beta C(0))$. Since $C(0)$ must be positive to ensure the overall stability of the model (exercise: why?) the constant c_1 will change sign at a certain "critical temperature" β^*. For temperatures lower than β^*, the Gaussian action is unstable (i.e., fluctuations with low wavevector become unbound) and the anharmonic term ϕ^4 alone controls the stability of the model. Clearly, the behavior of the system will change drastically at this point. Indeed, the critical temperature $c_1(\beta^*) = 0$ marks the position of the magnetic phase transition, a point to be discussed in more detail below.

Let us begin our primer of perturbation theory by introducing some nomenclature.[4] For simplicity, let us define the functional integral

$$\langle \cdots \rangle \equiv \frac{\int D\phi \, e^{-S[\phi]}(\cdots)}{\int D\phi \, e^{-S[\phi]}}. \tag{4.6}$$

Owing to the structural similarity to a thermal average, $\langle \cdots \rangle$ is sometimes called

functional
expecta-
tion value a **functional average** or **functional expectation value**. Similarly, let us define

$$\langle \cdots \rangle_0 \equiv \frac{\int D\phi \, e^{-S_0[\phi]}(\cdots)}{\int D\phi \, e^{-S_0[\phi]}} \tag{4.7}$$

as the functional average over the Gaussian action $S_0 \equiv S|_{g=0}$. The average over a product of field variables,

$$C_n(\mathbf{x}_1, \mathbf{x}_2, \ldots, \mathbf{x}_n) \equiv \langle \phi(\mathbf{x}_1)\phi(\mathbf{x}_2) \cdots \phi(\mathbf{x}_n) \rangle, \tag{4.8}$$

n-point
correlation
function is known as an n-**point correlation function** or **n-point function**.[5]

The one-point function $C_1(\mathbf{x}) = \langle \phi(\mathbf{x}) \rangle$ simply measures the expectation value of the field amplitude. For the particular case of the ϕ^4-problem above the phase transition and, more generally, the majority of field theories with an action that is even in the field amplitudes, $C_1 = 0$ and the first non-vanishing correlation function is the two-point function

$$G(\mathbf{x}_1 - \mathbf{x}_2) \equiv C_2(\mathbf{x}_1, \mathbf{x}_2). \tag{4.9}$$

> **George Green** 1793–1841
> With just four terms of formal schooling, his early life was spent running a windmill in Nottingham, UK. The inventor of Green functions, he made major contributions to potential theory, although where he learned his mathematical skills is a mystery. Green published only ten mathematical works, the first and most important being published at his own expense in 1828, "An essay on the application of mathematical analysis to the theories of electricity and magnetism." He left his mill and became an undergraduate at Cambridge in 1833 at the age of 40, and then a Fellow of Gonville and Caius College in 1839.

---

[4] Needless to say, the jargon introduced below is not restricted to the ϕ^4 example!

[5] Notice that, depending on the context and/or scientific community, the phrase "n-point function" sometimes refers to C_{2n} instead of C_n.

propagator

(Exercise: Why does C_2 depend only on the *difference* of its arguments?) The two-point function is sometimes also called the **propagator** of the theory, the **Green function** or, especially in the more formal literature, the **resolvent operator**.

The existence of different names suggests that we have met with an important object. Indeed, we will shortly see that the Green function is both, a central building block of the theory and a carrier of profound physical information.

INFO To understand the **physical meaning of the correlation function**, let us recall that the average of a linear field amplitude, $\langle \phi(\mathbf{0}) \rangle$, vanishes. (See the figure, where a few "typical" field configurations are sketched as functions of a coordinate.) However, the average of the squared amplitude $(\phi(\mathbf{0}))^2$ is certainly non-vanishing, simply because we are integrating over a positive object. Now, what happens if we split our single observation point into two, $\langle \phi^2(\mathbf{0}) \rangle \to \langle \phi(\mathbf{0})\phi(\mathbf{x}) \rangle = G(\mathbf{x})$? For asymptotically large values of $\mathbf{x}$, it is likely that the two amplitudes fluctuate independently, i.e., $G(\mathbf{x}) \to 0$ asymptotically. However, this decoupling will not happen locally. The reason is that the field amplitudes are correlated over a certain range in space. For example, if $\phi(\mathbf{0}) > 0$ the field amplitude will, on average, stay positive in an entire neighborhood of $\mathbf{0}$ since rapid fluctuations of the field are energetically costly (owing to the gradient term in the action). The spatial correlation profile of the field is described by the function $G(\mathbf{x})$.

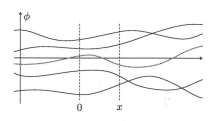

How does the correlation behavior of the field relate to the basic parameters of the action? A quick answer can be given by **dimensional analysis**. The action of the theory must be dimensionless (because it appears as the argument of an exponential). Denoting the dimension of any quantity X by $[X]$, and using the fact that $\left(\int d^d x \right) = L^d$, $[\partial] = L^{-1}$, inspection of Eq. (4.2) yields the set of relations

$$L^{d-2}[\phi]^2 = 1, \quad L^d[r][\phi]^2 = 1, \quad L^d[g][\phi]^4 = 1,$$

from which it follows that $[\phi] = L^{-(d-2)/2}$, $[r] = L^{-2}$, $[g] = L^{d-4}$. In general, both parameters, g and r, carry a non-zero length dimension. However, temporarily concentrating on the non-interacting sector, $g = 0$, the only parameter of the theory, r, has dimensionality L^{-2}. Arguing in reverse, we conclude that any intrinsic length scale produced by the theory (e.g., the range over which the fields are correlated), must scale as $\sim r^{-1/2}$.

A more quantitative description can be obtained by considering the **free propagator** of the theory,

$$G_0(\mathbf{x}) \equiv \langle \phi(\mathbf{0})\phi(\mathbf{x}) \rangle_0. \tag{4.10}$$

Since the momentum representation of the Gaussian action is simply given by $S_0[\phi] = \frac{1}{2}\sum_{\mathbf{p}} \phi_{\mathbf{p}}(p^2 + r)\phi_{-\mathbf{p}}$, it is convenient to first compute G_0 in reciprocal space: $G_{0,\mathbf{p}} \equiv \int d^d x e^{i\mathbf{p}\cdot\mathbf{x}} G_0(\mathbf{x}) = \sum_{\mathbf{p}'} \langle \phi_{\mathbf{p}}\phi_{\mathbf{p}'} \rangle_0$. Using the Gaussian contraction rule (3.14), the free functional average takes the form $\langle \phi_{\mathbf{p}}\phi_{\mathbf{p}'} \rangle_0 = \delta_{\mathbf{p}+\mathbf{p}',0}(p^2 + r)^{-1}$, i.e.,[6]

$$G_{0,\mathbf{p}} = \langle \phi_{\mathbf{p}}\phi_{-\mathbf{p}} \rangle_0 = \frac{1}{p^2 + r}. \tag{4.11}$$

---

[6] This result clarifies why G is referred to as a Green function. Indeed, $G_{0,\mathbf{p}}$ is (the Fourier representation of the) Green function of the differential equation $(-\partial_{\mathbf{r}}^2 + r)G(\mathbf{r},\mathbf{r}') = \delta(\mathbf{r} - \mathbf{r}')$.

dimensional
analysis

free
propagator

To obtain $G(\mathbf{x})$, we need to compute the inverse transform

$$G_0(\mathbf{x}) = \frac{1}{L^d}\sum_{\mathbf{p}} e^{-i\mathbf{p}\cdot\mathbf{x}} G_{0,\mathbf{p}} \approx \int \frac{d^d p}{(2\pi)^d} \frac{e^{-i\mathbf{p}\cdot\mathbf{x}}}{p^2 + r}, \qquad (4.12)$$

where we have assumed that the system is large, i.e., the sum over momenta can be exchanged for an integral.

For simplicity, let us compute the integral for a one-dimensional system. (For the two- and three-dimensional cases, see the exercise below.) Setting $p^2 + r = (p + ir^{1/2})(p - ir^{1/2})$, we note that the (complex extension of the) p-integral has simple poles at $\pm ir^{1/2}$. For x smaller (larger) than zero, the integrand is analytic in the upper (lower) complex p-plane, and closure of the integration contour to a semicircle of infinite radius gives

$$G_0(x) = \int \frac{dp}{2\pi} \frac{e^{-ipx}}{(p + ir^{1/2})(p - ir^{1/2})} = \frac{e^{-r^{1/2}|x|}}{2r^{1/2}}. \qquad (4.13)$$

correlation
length This result conveys an interesting observation: typically, correlations decay exponentially, at a rate set by the **correlation length**, $\xi \equiv r^{-1/2}$. However, as r approaches 0, the system becomes long-range correlated. The origin of this phenomenon can be understood by inspecting the structure of the Gaussian contribution to the action (4.2). For $r \to 0$ (and still neglecting the ϕ^4 contribution), nothing prevents the constant field mode $\phi(\mathbf{x}) = \phi_0 = $ const. from becoming infinitely large, i.e., the fluctuating contribution to the field becomes relatively less important than the constant offset. The increasing "stiffness" of the field in turn manifests itself in a growth of spatial correlations. Notice that this dovetails with our previous statement that $r = 0$ marks the position of a phase transition. Indeed, the build-up of infinitely long-range spatial correlations is a hallmark of second-order phase transitions (see chapter 6).

EXERCISE Referring to Eq. (4.12), show that, in dimensions $d = 2$ and $d = 3$,

$$G_0(\mathbf{x}) \overset{d=2}{=} \int \frac{d^2 k}{(2\pi)^2} \frac{e^{-i\mathbf{k}\cdot\mathbf{x}}}{k^2 + r} = \frac{1}{2\pi} K_0(\sqrt{r}|\mathbf{x}|) = \begin{cases} -\frac{1}{2\pi}\ln\frac{\sqrt{r}|\mathbf{x}|}{2} & |\mathbf{x}| \ll 1/\sqrt{r}, \\ \frac{1}{2}(2\pi\sqrt{r}|\mathbf{x}|)^{-\frac{1}{2}} e^{-\sqrt{r}|\mathbf{x}|} & |\mathbf{x}| \gg 1/\sqrt{r}, \end{cases}$$

$$G_0(\mathbf{x}) \overset{d=3}{=} \int \frac{d^3 k}{(2\pi)^3} \frac{e^{-i\mathbf{k}\cdot\mathbf{x}}}{k^2 + r} = \frac{e^{-\sqrt{r}|\mathbf{x}|}}{4\pi|\mathbf{x}|}.$$

Notice that, in both cases, the Green function diverges in the limit $|\mathbf{x}| \to 0$ and decays exponentially (at a rate $\sim r^{-1/2}$) for $|\mathbf{x}| \gg r^{-1/2}$.

4.1.3 Perturbation theory at low orders

REMARK To navigate the following section, it is helpful to first recapitulate section 3.2 on continuum Gaussian integration.

Having discussed the general structure of the theory and of its free propagator, let us turn our attention to the role of the interaction contribution to the action,

$$S_{\text{int}}[\phi] \equiv g \int d^d x \, \phi^4.$$

vertex
operator Within the jargon of field theory, an integrated monomial of a field variable (such as ϕ^4) is commonly called an **(interaction) operator** or a **vertex (operator)**. Keeping in mind the words of caution given in section 4.1.1, we wish to explore perturbatively how the interaction vertex affects the functional expectation value of any given field observable, i.e., we wish to analyze expansions of the type

$$\langle X[\phi]\rangle \approx \frac{\sum_{n=0}^{\cdots} \frac{(-g)^n}{n!}\langle X[\phi](\int d^dx\,\phi^4)^n\rangle_0}{\sum_{n=0}^{\cdots} \frac{(-g)^n}{n!}\langle(\int d^dx\,\phi^4)^n\rangle_0} \approx \sum_{n=0}^{n_{\max}} X^{(n)}, \tag{4.14}$$

where X may be any observable and $X^{(n)}$ denotes the contribution of nth order to the expansion in g. The limits on the summation in the numerator and denominator are symbolic because, as explained above, we will need to terminate the total perturbative expansion at a certain finite order $n_{\max}$.

To keep the discussion concrete, let us focus on the perturbative expansion of the propagator in the coupling constant, g. (A physical application relating to this expansion will be discussed below.) The zeroth-order contribution $G^{(0)} = G_0$ has been discussed before; so the first nontrivial term we have to explore is (exercise)

$$G^{(1)}(\mathbf{x},\mathbf{x}') = -g\left(\left\langle \phi(\mathbf{x})\int d^dy\,\phi(\mathbf{y})^4\phi(\mathbf{x})\right\rangle_0 - \left\langle \phi(\mathbf{x})\phi(\mathbf{x}')\right\rangle_0 \left\langle \int d^dy\,\phi(\mathbf{y})^4\right\rangle_0\right). \tag{4.15}$$

Since the functional average is now over a Gaussian action, this expression can be evaluated by **Wick's theorem** (3.21). For example, the functional average of the first of the two terms leads to (integral signs and constants having been stripped off for clarity)

$$\begin{aligned}
\left\langle \phi(\mathbf{x})\phi(\mathbf{y})^4\phi(\mathbf{x}')\right\rangle_0 &= 3\langle\phi(\mathbf{x})\phi(\mathbf{x}')\rangle_0\left(\langle\phi(\mathbf{y})\phi(\mathbf{y})\rangle_0\right)^2 \\
&\quad + 12\langle\phi(\mathbf{x})\phi(\mathbf{y})\rangle_0\langle\phi(\mathbf{y})\phi(\mathbf{y})\rangle_0\langle\phi(\mathbf{y})\phi(\mathbf{x}')\rangle_0 \\
&= 3G_0(\mathbf{x}-\mathbf{x}')G_0(\mathbf{0})^2 + 12G_0(\mathbf{x}-\mathbf{y})G_0(\mathbf{0})G_0(\mathbf{y}-\mathbf{x}'), \tag{4.16}
\end{aligned}$$

where we have used the fact that the operator inverse of the Gaussian action is, by definition, the free Green function (see Eq. (4.10)). Further, notice that the total number of terms appearing on the right-hand side is equal to $15 = (6-1)!!$, which is simply the number of distinct pairings of six objects (see Eq. (3.21) and our discussion in section 4.1.1). Similarly, the second contribution to $G^{(1)}$ leads to

$$\langle\phi(\mathbf{x})\phi(\mathbf{x}')\rangle_0\langle\phi(\mathbf{y})^4\rangle_0 = 3\langle\phi(\mathbf{x})\phi(\mathbf{x}')\rangle_0[\langle\phi(\mathbf{y})^2\rangle_0]^2 = 3G_0(\mathbf{x}-\mathbf{x}')G_0(\mathbf{0})^2.$$

Before analyzing these structures in more detail, let us make some general observations. The first-order expansion of G contains a number of factors of $G_0(\mathbf{0})$, the free Green function evaluated at coinciding points. This bears disturbing consequences. To see this, consider $G_0(\mathbf{0})$ evaluated in momentum space,

$$G_0(\mathbf{0}) = \int \frac{d^dp}{(2\pi)^d}\frac{1}{p^2+r}. \tag{4.17}$$

For dimensions $d > 1$, the integral is divergent at large momenta or short wavelengths; we have met with an **ultraviolet (UV) divergence**. Physically, the divergence implies that, already at first order, our expansion runs into a difficulty that is obviously related to the short-distance structure of the system. How can this problem be overcome? One way out is to remember that field theories such as the ϕ^4-model represent effective low-temperature, or long-wavelength, approximations to more microscopic models. The range of applicability of the action must be limited to wavelengths in excess of some microscopic lattice cutoff a (e.g., the **UV cutoff** lattice spacing), or momenta $k < a^{-1}$. It seems that, once that **ultraviolet cutoff** has been built in, the convergence problem is solved. However, there is something unsatisfactory in this argument. All our perturbative corrections, and therefore the final result of the analysis, exhibit sensitivity to the microscopic cutoff parameter. But this is not what we expect of a sensible low-energy theory (see the discussion in chapter 1)! The UV problem signals that something more interesting is going on than a naïve cutoff regularization has the capacity to describe. We discuss this point extensively in chapter 6.

However, even if we temporarily close our eyes to the UV phenomenon, there is another problem. For dimensions $d \leq 2$, and in the limit $r \to 0$, $G_0(\mathbf{0})$ also diverges **infrared (IR) di- vergence** at *small momenta*, an **infrared (IR) divergence**. Being related to structures at large wavelengths, this type of singularity should attract our attention even more than the UV divergence mentioned above. Indeed, it is intimately related to the accumulation of long-range correlations in the limit $r \to 0$ (cf. the structure of the integral (4.12)). We will come back to the discussion of the IR singularities, and their connection to the UV phenomenon, in chapter 6.

The considerations above show that the perturbative analysis of functional integrals will be accompanied by all sorts of divergences. Moreover, there is another, less fundamental, but also important, point: referring to Eq. (4.16), we have to concede that the expression does not look particularly inviting. To emphasize the point, let us consider the core contribution to the expansion at second order in g.

EXERCISE Show that the 10th-order contraction leads to the $945 = (10 - 1)!!$ terms

$$
\begin{aligned}
\left\langle \phi(\mathbf{x})\phi(\mathbf{y})^4\phi(\mathbf{y}')^4\phi(\mathbf{x}') \right\rangle_0 &= 9G_0(\mathbf{x} - \mathbf{x}')G_0(\mathbf{0})^4 + 72G_0(\mathbf{x} - \mathbf{x}')G_0(\mathbf{y} - \mathbf{y}')^2G_0(\mathbf{0})^2 \\
&+ 24G_0(\mathbf{x} - \mathbf{x}')G_0(\mathbf{y} - \mathbf{y}')^4 + \big(36G_0(\mathbf{x} - \mathbf{y})G_0(\mathbf{x}' - \mathbf{y})G_0(\mathbf{0})^3 \\
&+ 144(G_0(\mathbf{x} - \mathbf{y})G_0(\mathbf{x}' - \mathbf{y})G_0(\mathbf{y} - \mathbf{y}')^2G_0(\mathbf{0}) + G_0(\mathbf{x} - \mathbf{y})G_0(\mathbf{x}' - \mathbf{y}')G_0(\mathbf{0})^2G_0(\mathbf{y} - \mathbf{y}')) \\
&+ 96G_0(\mathbf{x} - \mathbf{y})G_0(\mathbf{x}' - \mathbf{y}')G_0(\mathbf{y}' - \mathbf{y})^3 + (\mathbf{y} \leftrightarrow \mathbf{y}')\big].
\end{aligned}
\tag{4.18}
$$

Note: Our further discussion will not rely on this result, which is only illustrative.

Eq. (4.18) is not an illuminating expression. There are eight groups of different terms; but it is not obvious how to attribute any meaning to these contributions. Further, if we considered the full second-order Green function $G^{(2)}$, i.e., took account of the expansion of both numerator and denominator in Eq. (4.14), we would find that some contributions canceled (see problem 4.6.1). Clearly, the situation will not improve at third and higher orders in g.

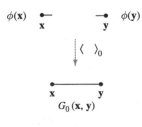

To apply perturbative concepts beyond lowest orders, a more efficient formulation of the expansion is needed. The key to the construction of a better language lies

Fig. 4.1 Graphical representation of a first-order-in-g contraction contributing to the expansion of the Green function.

in the observation that our previous notation is full of redundancy. For example, in the contraction of a perturbative contribution, we represent our fields by $\phi(\mathbf{x})$. A more compact way of keeping track of the presence of that field is shown in the upper portion of the unnumbered figure. Draw a point (with an optional "$\mathbf{x}$" labeling its position) and attach a little leg to it. The leg indicates that the fields are sociable objects, i.e., they need to find a partner with which to pair. After the contraction, a pair $\langle \phi(\mathbf{x})\phi(\mathbf{y}) \rangle \to G_0(\mathbf{x} - \mathbf{y})$ becomes a free Green function. Graphically, this information can be represented by a pairwise connection of the legs of the field symbols to lines, where each line is identified with a Green function connecting the two terminating points. The full contraction of a free correlation **diagrams**   function $\langle \phi(\mathbf{x}_1)\phi(\mathbf{x}_2) \cdots \phi(\mathbf{x}_{2n}) \rangle_0$ is represented by the set of all distinct **diagrams** formed by pairwise connection of the field vertices.

Feynman diagrams   **INFO** Graphical codes as a means for the representation of complex perturbation series were introduced by Feynman in 1948[7] and have since been called **Feynman diagrams**. After an initial phase of skepticism, they were soon recognized as immensely powerful and became an indispensable element of almost all areas of theoretical physics. Importantly, Feynman diagrams are not just computational tools but also a means of communication: when discussing in front of a blackboard, it takes seconds to draw the diagrams represented in, say, fig. 4.2, but much longer to write down the corresponding formulae. Not to mention that formulae are often difficult to read while diagrams – with a bit of practice – afford intuitive interpretations. Their status in many-body physics will become more tangible as we proceed and work with numerous different classes of diagrams.

Figure 4.1 shows the graphical representation of the contraction of Eq. (4.16). (The cross appearing on the left-hand side represents four field operators sitting at the same point $\mathbf{y}$.) According to our rule formulated above, each of the two diagrams on the right-hand side represents the product of three Green functions, taken between the specified coordinates. Further, each contribution is weighted by a combinatorial factor, i.e., the number of identical diagrams of that structure. Consider, for example, the second contribution on the right-hand side. It is formed by connecting the "external" field vertex at $\mathbf{x}$ to any of the legs of the internal vertex at $\mathbf{y}$: four possibilities. Next, the vertex at $\mathbf{x}'$ is connected with one of the remaining three unsaturated vertices at $\mathbf{y}$: three possibilities. The last contraction $\mathbf{y} \leftrightarrow \mathbf{y}$ is fixed, i.e., we obtain altogether $3 \times 4 = 12$ equivalent diagrams – "equivalent" in that each of these represents the same configuration of Green functions.

EXERCISE Verify that the graphical representation of the second-order contraction (4.18) is as shown in fig. 4.2.[8] Associate the diagrams with individual contributions appearing

[7] R. Feynman, *The theory of positrons*, Phys. Rev. **76**, 749 (1949).
[8] In the figure, the coordinates carried by the field vertices have been dropped for notational

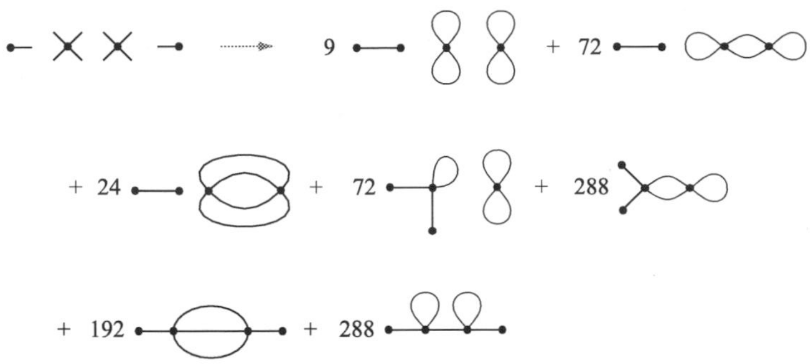

Fig. 4.2   Graphical representation of the second-order correction to the Green function. In the main
text, the seven types of diagram contributing to the contraction will be referred to (in the
order they appear above) as diagrams 1 to 7.

in Eq. (4.18) and try to reproduce the combinatorial factors. (For more details, see problem 4.6.1.)

The graphical representation of the contractions shown in figs. 4.1 and 4.2 provides us with sufficient background to list some general aspects of the diagrammatic approach.

▷ First, diagrammatic methods help to efficiently *represent* the perturbative expansion. However, we are still left with the problem (see the discussion above) of *computing* the analytical expressions corresponding to individual diagrams. To go back from an nth-order graph to its analytical representation, one (a) attaches coordinates to all field vertices, (b) identifies lines between points with Green functions, (c) multiplies the graph by the overall constant $g^n/n!$, and (d) integrates over all the internal coordinates. When one encounters expressions like $G^{(n)} = $ "sum of graphs," the operations (a)–(d) are implicit.

▷ As should be clear from the formulation of our basic rules, there is no fixed rule as to how to represent a diagram. As long as no lines are cut, any kind of reshaping, twisting, rotating, etc. of the diagram leaves its content invariant. (At large orders of perturbation theory, it often takes a second look to identify two diagrams as equivalent.)

▷ From the assembly of diagrams contributing to any given order, a number of internal structures common to the series expansion become apparent. For example, looking at the diagrams shown in fig. 4.2, we notice that some are connected, and some are not. Among the set of **connected diagrams** (nos. 5, 6, 7), there are some whose "core portion," i.e., the content of the diagram after the legs connecting to the external vertices have been removed, can be cut into two pieces just by cutting one more line (no. 7). Diagrams of this type are called

connected diagrams

simplicity. To restore the full information carried by any of these "naked" graphs, one attaches coordinates $\mathbf{x}$ and $\mathbf{x}'$ to the external field vertices and integration coordinates $\mathbf{y}_i$ to each of the i nodes that do not connect to an external field vertex. Since no information is lost, diagrams are often represented without explicit reference to coordinates.

one-particle reducible while the others are termed one-particle irreducible. More generally, a diagram whose core region can be cut by severing n lines is called n-**particle reducible**. (For example, no. 6 is three-particle reducible, no. 7 is one-particle reducible, etc.) One can also attach a **loop order** to a diagram, i.e., the number of inequivalent loops formed by segments of Green functions (for fig. 4.2, 4, 3, 3, 3, 2, 2, 2, in that order). One (correctly) expects that these structures, which are difficult to discern from the equivalent analytical representation, will reflect themselves in the mathematics of the perturbative expansion. We return to the discussion of this point below.

▷ Then there is the issue of **combinatorics**. The diagrammatic representation simplifies the determination of the combinatorial factors appearing in the expansion. However, the problem of getting the combinatorics right remains nontrivial. (If you are not impressed with the factors entering the second-order expansion, consider the $(14 - 1)!! = 135,135$ terms contributing at third order!) In some sub-disciplines of theoretical physics, the art of identifying the full set of combinatorial coefficients at large orders of perturbation theory has been developed to a high degree of sophistication. Indeed, one can set up refined sets of diagrammatic construction rules that automate the combinatorics to a considerable extent. Pedagogical discussions of these rules can be found, for example, in the textbooks by Negele and Orland, and by Ryder.[9] However, as we will see shortly, the need to explicitly carry out a large-order expansion taking account of all diagrammatic sub-processes rarely arises in modern condensed matter physics; mostly one is interested in subclasses of diagrams, for which the combinatorics is less problematic. For this reason, the present text does not contain a state-of-the-art exposition of all diagrammatic tools, and interested readers are referred to the literature.

▷ Finally, and perhaps most importantly, the diagrammatic representation of a given contribution to the perturbative expansion often suggests a **physical interpretation** of the corresponding physical process. (After all, any term contributing to the expansion of a physical observable must correspond to some "real" physical process.) Unfortunately, the ϕ^4-theory is not well suited to illustrate this aspect – void of any dynamical content, it is a little bit too simple. However, the possibility of "reading" individual diagrams will become evident in the next section when we discuss an application to the interacting electron gas.

Above, we have introduced the diagrammatic approach for the example of field expectation values $\langle \phi(\mathbf{x})(\phi(\mathbf{y})^4)^n \phi(\mathbf{x}') \rangle_0$. However, to obtain the Green function to any given order in perturbation theory, we need to add to these expressions the contributions emanating from the expansion of the denominator of the functional average (see Eqs. (4.14) and (4.15)). While, at first sight, the need to keep track of even more terms seems to complicate matters, we will see that, in fact, quite the

---

[9] J. W. Negele and H. Orland, *Quantum Many Particle Systems* (Addison-Wesley, 1988); L. H. Ryder, *Quantum Field Theory* (Cambridge University Press, 1996).

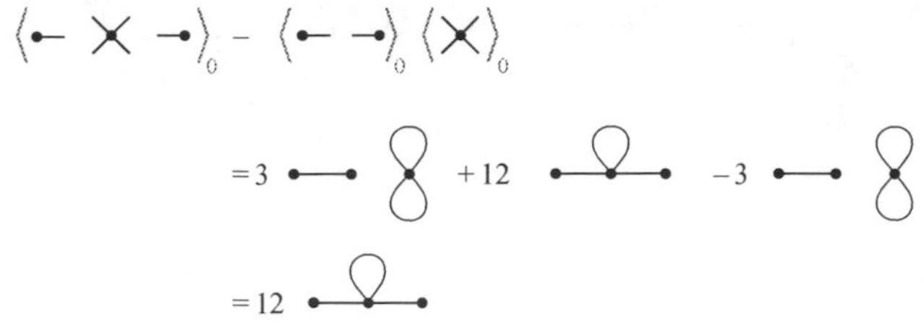

$$\left\langle \leftarrow \times \rightarrow \right\rangle_0 - \left\langle \leftarrow \rightarrow \right\rangle_0 \left\langle \times \right\rangle_0$$

$$= 3 \;\bullet\!\!\!-\!\!\!\bullet\; \, 8 \; + 12 \; \bullet\!\!\!-\!\!\!\overset{\displaystyle \varphi}{\bullet}\!\!\!-\!\!\!\bullet \; - 3 \; \bullet\!\!\!-\!\!\!\bullet \; 8$$

$$= 12 \; \bullet\!\!\!-\!\!\!\overset{\displaystyle \varphi}{\bullet}\!\!\!-\!\!\!\bullet$$

Fig. 4.3   Graphical representation of the first-order correction to the Green function: Vacuum graphs cancel out.

opposite is true! The combined expansion of numerator and denominator leads to a miraculous "cancellation mechanism" that greatly simplifies the analysis.

Let us exemplify the mechanism of cancellation on $G^{(1)}$. The three diagrams corresponding to the contractions of Eq. (4.15) are shown in fig. 4.3, where integral signs and coordinates have been dropped for simplicity. On the left-hand side of the equation, the brackets $\langle \cdots \rangle_0$ indicate that the second contribution comes from the expansion of the denominator. The point to be noticed is that the graph produced by the contraction of that term cancels against a contribution arising from the numerator. One further observes that the cancelled graph is of a special type: it contains an interaction vertex that does not connect to any of the external vertices. Diagrams with that property are commonly termed **vacuum diagrams**.[10]

vacuum diagrams

EXERCISE Construct the diagrammatic representation of $G^{(2)}$, verify that vacuum graphs in the denominator cancel against the numerator, and show that the the result is obtained as the sum of connected diagrams in fig. 4.4. (For more details, see problem 4.6.1.)

Indeed, the cancellation of vacuum graphs pertains to higher-order correlation functions and to all orders of the expansion:

> The contribution to a correlation function $C^{(2n)}(\mathbf{x}_1, \ldots, \mathbf{x}_{2n})$ at lth order of perturbation theory is given by the sum of all graphs, excluding vacuum graphs.

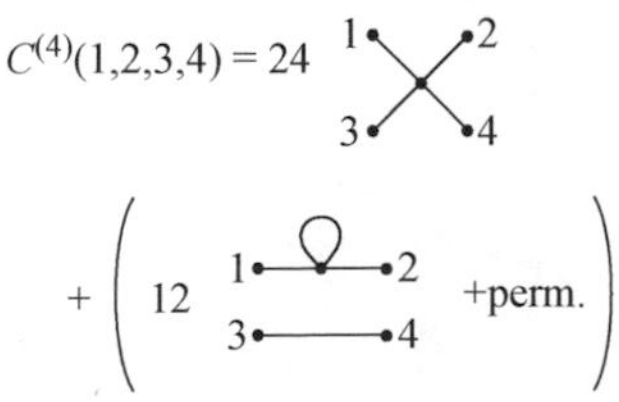

For example, the first-order expansion of the four–point function $C^{(4)}(\mathbf{x}_1, \ldots \mathbf{x}_4)$ is shown in the figure, where coordinates $\mathbf{x}_i \leftrightarrow i$ are abbreviated by their indices and "+ perm." stands for the six permutations obtained by interchanging arguments. In the literature, the statement of vacuum graph cancellation is sometimes referred to as the **linked cluster theorem**. Notice that the linked cluster feature takes care of two problems: first,

linked cluster theorem

$$C^{(4)}(1,2,3,4) = 24 \begin{matrix} 1\bullet & & \bullet 2 \\ & \times & \\ 3\bullet & & \bullet 4 \end{matrix}$$

$$+ \left(12 \; \begin{matrix} 1\bullet\!\!-\!\!\overset{\varphi}{\bullet}\!\!-\!\!\bullet 2 \\ 3\bullet\!\!-\!\!-\!\!-\!\!\bullet 4 \end{matrix} \; + \text{perm.} \right)$$

---

[10] The term "vacuum graph" has its origin in the diagrammatic methods invented in the 1950s in the context of particle theory. Instead of thermal averages $\langle \cdots \rangle_0$, one considered matrix elements $\langle \Omega | \cdots | \Omega \rangle$ taken on the ground state or "vacuum" of the field theory. This caused matrix elements $\langle \Omega | (S_{\mathrm{int}}[\phi])^n | \Omega \rangle$ not containing an external field vertex to be dubbed "vacuum diagrams."

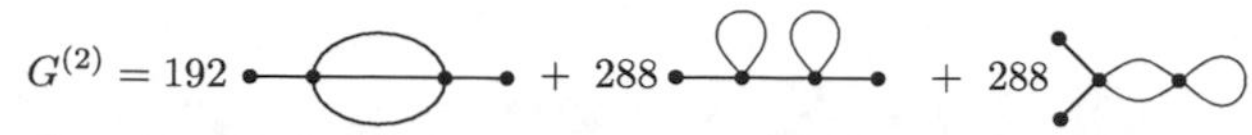

$$G^{(2)} = 192 \quad\text{}\quad + \ 288 \quad\text{}\quad + \ 288 \quad\text{}$$

Fig. 4.4   Graphical representation of the second-order contribution to the Green function.

we are relieved of the burden of a double expansion of numerator and denominator and, second, only non-vacuum contributions to the expansion of the former need to be kept.

INFO The **proof of the linked cluster theorem** is straightforward. Consider a contribution of nth order to the expansion of the numerator of (4.14), $\frac{(-g)^n}{n!}\langle X[\phi](\int d_x^d \phi^4)^n\rangle_0$. The contraction of this expression will lead to a sum of vacuum graphs of pth order and non-vacuum graphs of $(n-p)$th order, where p runs from 0 to n. The pth-order contribution is given by

$$\frac{1}{n!}\binom{n}{p}\left\langle X[\phi]\left(\int \phi^4\right)^{n-p}\right\rangle_0^{\mathrm{n.v.}}\left\langle\left(\int \phi^4\right)^p\right\rangle_0,$$

where the superscript n.v. indicates that the contraction excludes vacuum graphs, and the combinatorial coefficient counts the ways of choosing p vertices ϕ^4 of a total of n vertices to form a vacuum graph. Summing over p, we find that the expansion of the numerator, split into vacuum and non-vacuum contributions, reads

$$\sum_{n=0}^{\infty}\sum_{p=0}^{n}\frac{(-g)^n}{(n-p)!\,p!}\left\langle X[\phi]\left(\int \phi^4\right)^{n-p}\right\rangle_0^{\mathrm{n.v.}}\left\langle\left(\int \phi^4\right)^p\right\rangle_0.$$

By a straightforward rearrangement of the summations, this can be rewritten as

$$\sum_{n=0}^{\infty}\frac{(-g)^n}{n!}\left\langle X[\phi]\left(\int \phi^4\right)^n\right\rangle_0^{\mathrm{n.v.}}\sum_{p=0}^{\infty}\frac{(-g)^p}{p!}\left\langle\left(\int \phi^4\right)^p\right\rangle_0.$$

The p-summation recovers exactly the expansion of the denominator; so we are left with the sum over all non-vacuum contractions.

Before concluding this section, let us discuss one last technical point. The translational invariance of the ϕ^4-action suggests a representation of the theory in reciprocal space. Indeed, the **momentum space representation** of the propagator (4.11) is much simpler than the real space form, and the subsequent analytical evaluation of diagrams will be formulated in momentum space anyway (cf. the prototypical expression (4.17)).

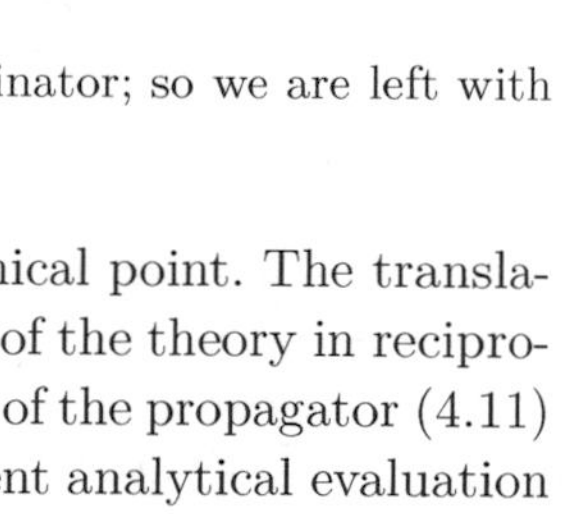

The diagrammatic formulation of the theory in momentum space is straightforward. All we need to do is to slightly adjust the graphical code. Inspection of Eq. (4.11) shows that the elementary contraction should now be formulated as indicated in the figure. Only fields with opposite momentum can be contracted; the line carries this momentum as a label. Notice that the momentum representation of the field vertex $\phi^4(\mathbf{x})$ is **not** given by $\phi_{\mathbf{p}}^4$.
Rather, Fourier transformation of the vertex leads to the three-fold convolution

Fig. 4.5   Momentum-space representation of a first-order contribution to the Green function. Internal momenta $\mathbf{p}_i$ are integrated over.

$$\int d^d x\, \phi^4(\mathbf{x}) \rightarrow \frac{1}{L^{3d}} \sum_{\mathbf{p}_1,\dots\mathbf{p}_4} \phi_{\mathbf{p}_1} \phi_{\mathbf{p}_2} \phi_{\mathbf{p}_3} \phi_{\mathbf{p}_4} \delta_{\mathbf{p}_1+\mathbf{p}_2+\mathbf{p}_3+\mathbf{p}_4,\mathbf{0}}.$$

The graphical representation of the first-order correction to the Green function (i.e., the momentum space analog of fig. 4.3) is shown in fig. 4.5. It is useful to think about the vertices of the momentum space diagrammatic language in the spirit of "Kirchhoff laws": the sum of all momenta flowing into a vertex is equal to zero. Consequently (exercise) the total sum of all momenta "flowing" into a diagram from external field vertices must also equal zero: $\langle \phi_{\mathbf{p}_1} \phi_{\mathbf{p}_2} \cdots \phi_{\mathbf{p}_n} \rangle_0 \rightarrow (\cdots) \delta_{\mathbf{p}_1+\mathbf{p}_2\cdots+\mathbf{p}_n,\mathbf{0}}.$ This fact expresses the conservation of total momentum characteristic of theories with translational invariance.

EXERCISE Represent the diagrams of the second-order contraction shown in fig. 4.2. Convince yourself that the "Kirchhoff law" suffices to fix the result. Note that the number of summations over internal momenta is equal to the number of loops.

This concludes the first part of our introduction to perturbation theory. Critical readers will object that, while we have undertaken some effort to efficiently represent the perturbative expansion, we have not discussed how interactions actually modify the results of the free theory. Indeed, we are not yet in a position to address this problem, the reason being that we first need to better understand the origin and remedy of the UV and IR divergences observed above.

However, temporarily ignoring the presence of this roadblock, let us try to outline what kind of information can be obtained from perturbative analyses. We first note that, in condensed matter physics,[11] low-order perturbation theory is usually not enough to obtain quantitative results. The fact that the "perturbation" couples to a macroscopic number of degrees of freedom[12] usually necessitates the summation of infinite (sub)series of a perturbative expansion or even the application of non-perturbative methods. This, however, does not mean that the tools developed above are useless: given a system subject to unfamiliar interactions, low-order perturbation theory will usually be applied as a first step to explore the situation. For example, a malign divergence of the expansion in the interaction operator may signal the

---

[11] There are sub-disciplines of physics where the situation is different. For example, consider the high-precision scattering experiments of atomic and sub-atomic physics. In these areas, the power of a theory to quantitatively predict the dependence of scattering rates on the strength of the projectile–target interaction (the "perturbation") is a measure of its quality. Such tests involve large-order expansions in the physical coupling parameters.

[12] In contrast, low-order expansions in the *external* perturbation (e.g., experimentally applied electric or magnetic fields) are usually secure; see chapter 7.

presence of an instability towards the formation of a different phase. Or it may turn out that certain contributions to the expansion are "physically more relevant" than others. Technically, such contributions usually correspond to diagrams of a regular graphical structure. If so, a summation over all "relevant processes" may be in reach. In either case, low-order expansions provide vital hints, in the planning of a more complete analysis. In the following, we discuss two examples that may help to make these considerations more transparent.

4.2 Ground State Energy of the Interacting Electron Gas

In section 2.2.1, we began to consider the physics of itinerant electron compounds. We argued that such systems can be described in terms of the free particle Hamiltonian (2.18) together with the interaction operator (2.19). While we have reviewed the physics of non-interacting systems, the role of electron–electron interactions has not been addressed. Yet, by now, we have developed enough analytical machinery to consider this problem. Below we will apply concepts of perturbation theory to estimate the contribution of electron correlations to the ground state energy of a Fermi system. However, before plunging into the technicalities of this analysis, it is useful to discuss some qualitative aspects of the problem.

4.2.1 Qualitative aspects

A principal question that we will need to address is under what physical conditions are interactions "weak" (in comparison with the kinetic energy), i.e., when does a perturbative approach with the interacting electron system make any sense? To estimate the relative magnitude of the two contributions to the energy, let us assume that each electron occupies an average volume r_0^3. According to the uncertainty relation, the minimum kinetic energy per particle will be of order $\mathcal{O}(\hbar^2/mr_0^2)$. On the other hand, assuming that each particle interacts predominantly with its nearest neighbors, the Coulomb energy is of order $\mathcal{O}(e^2/r_0)$. The ratio of the two energy scales defines the **dimensionless density parameter** (see the figure)

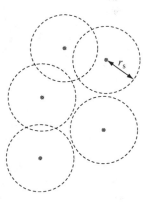

dimensionless
density
parameter

$$\boxed{\frac{e^2}{r_0}\frac{mr_0^2}{\hbar^2} = \frac{r_0}{a_0} \equiv r_{\mathrm{s}}}$$

where $a_0 = \hbar^2/e^2 m$ denotes the Bohr radius.[13] Physically, r_s is the radius of the sphere containing one electron on average; for the Coulomb interaction, the denser the electron gas, the smaller r_s. We thus identify the electron density as the relevant parameter controlling the strength of electron–electron interactions. Below, we will be concerned with the regime of high density, $r_s \ll 1$, or weak Coulomb interaction, accessible to perturbation theory controlled by the smallness of this parameter.

Wigner crystal

INFO In the opposite limit, $r_s \gg 1$, properties become increasingly dominated by electron correlations. It is believed that, for sufficiently large r_s (low density), the electron gas undergoes a (first-order) transition to a condensed phase known as a **Wigner crystal**. Although Wigner crystals have never been unambiguously observed, several experiments performed on low-density electron gases are consistent with a Wigner crystal ground state. Monte Carlo simulation suggests that Wigner crystallization may occur for densities with $r_s > 31$ in the two-dimensional electron gas and for $r_s > 106$ in three dimensions. (Note that this scenario relies on the system's being at low temperature and on the long-range nature of the Coulomb interaction. If the Coulomb interaction is screened, $V(r) \sim e^{-r/\lambda}$, then $r_s \sim (r_0/a_0)e^{-r_0/\lambda}$ and the influence of the Coulomb interaction at low densities becomes diminished.)

Fermi liquid theory

INFO Most metals lie in a regime $2 < r_s < 6$ of intermediate coupling, difficult to access in terms of controlled perturbation theory. Its description is the subject of Landau's **Fermi liquid Theory**,[14] an ingenious theoretical framework relying on the principle of **adiabatic continuity**.

In the absence of an electronic phase transition (such as Wigner crystallization), a non-interacting ground state evolves smoothly or adiabatically into the interacting ground state as the strength of the interaction is increased.[15] An elementary excitation of the non-interacting system represents an "approximate excitation" of the interacting system (i.e., its "lifetime" is long). The excitations are quasi-particles (and quasi-holes) above a sharply defined Fermi surface.

The starting point of Fermi liquid theory is a few phenomenological assumptions, all rooted in the adiabaticity principle. For example, it is postulated that the density of quasi-particles can be described in terms of a momentum-dependent density distribution $n(\mathbf{p})$, which, in the limit of zero

> **Lev Davidovich Landau**
> **1908–1968**
> Nobel Laureate in Physics in 1962 "for his pioneering theories for condensed matter, especially liquid helium." Landau's work covered all branches of theoretical physics, ranging from fluid mechanics to quantum field theory. Starting in 1936, a large portion of his papers refer to the theory of the condensed state, and eventually led to the construction of a complete theory of "quantum liquids" at very low temperatures. In 1938, Landau was arrested for comparing Stalinism to Nazism, and was released a year later only after Kapitsa petitioned Stalin, vouching for Landau's behavior.

---

[13] Notice that the estimate of the relative magnitude of energy scales mimics Bohr's famous qualitative discussion of the average size of the hydrogen atom.

[14] L. D. Landau, *The theory of a Fermi liquid*, Sov. Phys. JETP **3**, 920 (1956).

[15] As a simple example, consider the evolution of the bound states of a quantum particle as the confining potential is changed from a box to a harmonic potential well. While the wave functions and energies evolve, the topological characteristics of the wave functions, i.e., the number of nodes, and therefore the assignment of the corresponding quantum numbers, remain unchanged.

interaction, evolves into the familiar Fermi distribution. From this assumption (and a few more postulates) a broad spectrum of observables can be analyzed without further "microscopic" calculation. Its remarkable success (as well as few notorious failures) has made Landau's Fermi liquid theory a powerful tool in the development of modern condensed matter physics, but one which we are not going to explore in detail.[16]

4.2.2 Perturbative approach

The starting point of the perturbative analysis is the functional representation of the free energy $F = -T \ln \mathcal{Z}$ through the quantum partition function. (Here, as usual, we set $k_{\mathrm{B}} = 1$.) Expressed as a field integral, we have $\mathcal{Z} = \int D\psi \, e^{-S[\psi]}$, where

$$
S[\psi] = T \sum \bar{\psi}_{p\sigma} \left(-i\omega_n + \frac{\mathbf{p}^2}{2m} - \mu \right) \psi_{p\sigma} + \frac{T^3}{2L^3} \sum \bar{\psi}_{p+q\sigma} \bar{\psi}_{p'-q\sigma'} V(\mathbf{q}) \psi_{p'\sigma'} \psi_{p\sigma}.
$$

Here, we have introduced the **4-momentum** $p \equiv (\mathbf{p}, \omega_n)$, comprising both frequency and momentum.[17] As usual, the sums extend over all momenta (and spin variables, σ).

Zeroth-order perturbation theory

As with the Green function discussed previously, the free energy can be expanded in terms of an interaction parameter. To fix a reference scale against which to compare the correlation energies, let us begin by computing the free energy Eq. (3.85) of the non-interacting electron gas:

$$
F^{(0)} = -T \sum_{\mathbf{p}\sigma} \ln \left(1 + e^{-\beta \left(\frac{\mathbf{p}^2}{2m} - \mu \right)} \right) \xrightarrow{T \to 0} \sum_{\mathbf{p}^2/2m < \mu, \sigma} \left(\frac{\mathbf{p}^2}{2m} - \mu \right) \simeq -\frac{2}{5} N\mu, \quad (4.19)
$$

where $\mu \equiv p_{\mathrm{F}}^2/2m$, $N = (2mL^2\mu)^{3/2}/3\pi^2$ is the number of particles, and the estimate on the right is obtained by replacing the sum over momenta by an integral (exercise). According to Eq. (4.19), the average kinetic energy per particle is equal to $3/5$ of the Fermi energy. To relate this scale to the density parameter r_{s}, we choose to measure all energies in units of the Rydberg energy (viz. the ionization energy of hydrogen), $E_{\mathrm{Ry}} = me^4/2\hbar^2 = 13.6 \, \mathrm{eV}$,

$$
\frac{F^{(0)}}{E_{\mathrm{Ry}}} \sim \frac{N}{r_{\mathrm{s}}^2}. \quad (4.20)
$$

---

[16] Interested readers are referred to one of several excellent reviews, e.g., P. W. Anderson, *Basic Notions in Condensed Matter Physics* (Benjamin, 1984).

[17] Be careful not to confuse the 4-momentum p with the modulus of the 3-momentum, $p = |\mathbf{p}|$.

First-order perturbation theory

If we turn now to a discussion of interactions, a formal expansion of F to first order in V gives

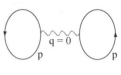

$$F^{(1)} = \frac{T^4}{2L^3} \left\langle \sum_{pp'q,\sigma\sigma'} \bar{\psi}_{p+q\sigma}\bar{\psi}_{p'-q\sigma'}V(\mathbf{q})\psi_{p'\sigma'}\psi_{p\sigma} \right\rangle_0 , \quad (4.21)$$

where $\langle \cdots \rangle_0$ denotes the functional average with respect to the non-interacting action. The two[18] diagrams contributing to this expression are shown in the figure. To account for the specifics of the electron gas, we are using a diagrammatic code slightly different from that of the previous section:

▷ The Coulomb interaction is represented by a wavy line labeled by the momentum argument $\mathbf{q}$.

free Green function of the electron gas

▷ A contraction $\langle \psi_{p\sigma}\bar{\psi}_{p\sigma}\rangle_0$ is indicated by a solid arrow representing the **free Green function of the electron gas**,

$$\boxed{G_p \equiv \frac{1}{i\omega_n - \frac{\mathbf{p}^2}{2m} + \mu}} \quad (4.22)$$

i.e., the inverse of the free action. The labeling of the contraction by an arrow (instead of a unidirectional line as in ϕ^4-theory) is motivated by two considerations. First, it indicates that a contraction $\langle \psi_\eta \bar{\psi}_\lambda \rangle_0$ describing the creation of an electron with quantum numbers λ followed by the annihilation of an electron with quantum numbers η is a directed process; second, there are situations (e.g., when a magnetic field is present) where $\langle \psi_{n\sigma}(\mathbf{r})\bar{\psi}_{n\sigma}(\mathbf{r}')\rangle_0 \neq \langle \psi_{n\sigma}(\mathbf{r}')\bar{\psi}_{n\sigma}(\mathbf{r})\rangle_0$.

▷ The sum of all 4-momenta emanating from an interaction vertex formed by a wavy line and two electron field lines is equal to zero – the "Kirchhoff law."

▷ Finally, we have to be careful about **sign factors** arising when Grassmann variables are interchanged. However, the anticommutativity of the fields merely leads to an overall factor $(-)^{N_l}$, where N_l is the number of loops of a diagram.[19]

Hartree diagram

Turning to the discussion of the two individual contributions in the figure, we notice that the first, generally known as a **Hartree contribution**, vanishes. Technically, this is a consequence of the fact that the interaction line connecting the two loops carries zero momentum.

---

[18] Remember that, in a theory with complex or Grassmann fields, only contractions $\sim \langle \bar{\psi}\psi \rangle_0$ exist, i.e., there is a total of $n!$ distinct contributions to a contraction $\langle \bar{\psi}\psi \ldots \psi \rangle_0$ of $2n$ field operators.

[19] To verify this claim, one may notice that a loop is formed by a "ring-wise contraction" of $\langle \bar{\psi}_1\psi_2\bar{\psi}_3 \ldots \psi_N \rangle_0$, i.e., $(2 \rightarrow 3)(4 \rightarrow 5)\ldots((N-2) \rightarrow (N-1))(N \rightarrow 1)$. The last contraction introduces the minus sign.

However, as discussed in section 2.2.1, $V(\mathbf{q} = 0) = 0$: physically, the vanishing of the Hartree contribution is a consequence of charge neutrality. Indeed, the two Green function loops $\sum_p G_p$ measure the local particle density of the electron gas (see the discussion on page 142). Global charge neutrality requires that the electron density cancels against that of the ionic background.

Douglas R. Hartree 1897– 1958 was a British applied mathematician who became one of the first "computational physicists." He developed methods for numerically solving the Schrödinger equation, among them the celebrated "Hartree approximation."

INFO However, notice that this cancellation mechanism relies on our assumption of overall spatial homogeneity. Only in a spatially uniform system does the density of the electron gas locally compensate the positive counter-density. In the context of real metals, the inevitable presence of impurities breaks translational invariance and there is no reason for the Hartree contribution to vanish. Indeed, the analysis of Hartree-type contributions to the correlation energy in disordered electronic media is a subject of ongoing research.

Fock diagram

While the Hartree term describes the classical interaction of charge densities through the Coulomb potential, the second diagram shown in the figure, known as a **Fock diagram**, is quantum. Translating the diagrammatic language back into Green functions, we obtain

$$F^{(1)} = -\frac{T^2}{L^3} \sum_{p,p'} G_p V(\mathbf{p} - \mathbf{p}') G_{p'} = -\frac{1}{L^3} \sum_{\mathbf{p},\mathbf{p}'} n_F(\epsilon_\mathbf{p}) \frac{e^2}{4\pi |\mathbf{p} - \mathbf{p}'|^2} n_F(\epsilon_{\mathbf{p}'})$$

$$\overset{T \to 0}{=} -\frac{1}{L^3} \sum_{\epsilon_\mathbf{p}, \epsilon_{\mathbf{p}'} < \mu} \frac{e^2}{|\mathbf{p} - \mathbf{p}'|^2} \sim -e^2 L^3 p_F^4. \tag{4.23}$$

Here, the sign factor in the first equality arises because there is an odd number of fermion loops, and the parametric scaling of the sum follows from dimensional considerations: we are integrating the inverse square of the distance in momentum space ($[\text{momentum}]^{-2}$) over two Fermi spheres ($[\text{momentum}]^{6}$). Since the integral is convergent at low momenta, it must scale as the fourth power of the upper cutoff $\sim p_F^4$.[20] Division by the Rydberg energy leads to the scaling

$$\frac{F^{(1)}}{E_\text{Ry}} = \text{const.} \times \frac{N}{r_\text{s}}, \tag{4.24}$$

where the constant is of order unity. This result conforms with our previous estimate of the density dependence of correlation energies.

---

[20] For a more detailed computation, see C. Kittel, *Quantum Theory of Solids* (Wiley, 1963).

Second-order perturbation theory

Let us consider the second-order contribution

$$F^{(2)} = -\frac{T}{2}\left(\frac{T^3}{2L^3}\right)^2 \left\langle \left(\sum_{pp'q} \bar\psi_{p+q\sigma}\bar\psi_{p'-q\sigma'} V(\mathbf{q})\psi_{p'\sigma'}\psi_{p\sigma} \right)^2 \right\rangle_0^{\mathrm{c}}$$

where the superscript indicates that only connected diagrams contribute.

EXERCISE Using the field integral representation of $\mathcal{Z}$, show that second-order expansion of $F = -T \ln \mathcal{Z}$ in the interaction operator S_{int} leads to the identity $F^{(2)} = -\frac{T}{2}[\langle S_{\mathrm{int}}^2\rangle_0 - \langle S_{\mathrm{int}}\rangle_0^2]$. Convince yourself that the second term cancels disconnected diagrams. Apply arguments similar to those involved in the proof of the linked cluster theorem to verify that the cancelation of disconnected graphs pertains to all orders in the expansion of F, i.e., the free energy can be obtained by expanding the partition function $\mathcal{Z}$ (not its logarithm) and keeping only connected diagrams; dropping disconnected contributions is equivalent to taking the logarithm.

Connected contraction of the eight field operators leads to four distinct types of diagram (exercise) of which two are of Hartree type (i.e., contain a zero-momentum interaction line $V(\mathbf{q} = \mathbf{0})$). The non-vanishing diagrams $F^{(2),1}$ and $F^{(2),2}$ are shown in the figure.[21] Translating these diagrams into momentum summations over Green functions, one obtains (exercise)

$$F^{(2),1} = -\frac{T^3}{L^6} \sum_{p_1,p_2,q} G_{p_1}G_{p_1+q}G_{p_2}G_{p_2+q}V^2(\mathbf{q}),$$

$$F^{(2),2} = \frac{1}{2}\frac{T^3}{L^6} \sum_{p,q_1,q_2} G_pG_{p-q_1}G_{p-q_1-q_2}G_{p-q_2}V(\mathbf{q}_1)V(\mathbf{q}_2).$$

(4.25)

While, at first sight, these expressions do not look very illuminating, closer inspection reveals some structure. Reflecting the fact that electron transport in solids is carried by excitations at the Fermi energy, the electron Green function (4.22) assumes large values for momenta $\mathbf{p} \simeq p_{\mathrm{F}}$. This implies that only configurations where all momentum arguments carried by the Green function are close to the Fermi surface contributes significantly to the sums in (4.25). Considering the first sum, we see that, for small $|\mathbf{q}|$ and $|\mathbf{p}_i| \simeq p_{\mathrm{F}}$, this condition is met, i.e., there are two unbound summations over momentum shells around the Fermi surface. However, for the second sum, the

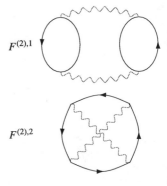

$F^{(2),1}$

$F^{(2),2}$

---

[21] In fact, one more non-vanishing contribution is obtained by drawing a single "ring" of Green functions containing two non-crossing interaction lines. This diagram, and its obvious generalization to higher-order processes containing sequential "self-interactions" of a single Green function do not play an essential role (in the present context). The reasons why will become clear in section 4.3.1 when we introduce the notion of self-energies.

situation is less favourable. For fixed $|\mathbf{p}_1| \simeq p_F$, fine-tuning of both $\mathbf{q}_1$ and $\mathbf{q}_2$ is necessary to bring all momenta close to p_F, i.e., effectively one momentum summation is frozen out. There is no need to enter into detailed calculations to predict that $F^{(2),1} \gg F^{(2),2}$ as a consequence. The ratio of the two terms will be proportional to the area of the Fermi surface, which, in turn, is proportional to the density of the electron gas. For large densities, the second Fock diagram can thus be neglected in comparison with the first.

INFO Of course, there must be a more physical way of understanding this observation. The Green function lines in the diagrams $F^{(2),i}$ describe the propagation of quasi-particles and quasi-holes[22] on the background of the interacting medium. Now, the diagram

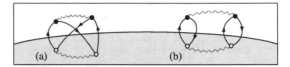

$F^{(2),2}$ contains a simply connected propagator line: a single particle–hole excitation at the Fermi surface undergoes a second-order interaction process with itself (see part (a) of the figure). By contrast, the first diagram $F^{(2),1}$ involves two independent electron–hole excitations, as shown in (b). Since, in a dense electron gas, a second-order interaction process will more likely involve different particles, this type of contribution is more important. Notice that the process shown in (b) can be interpreted as a "polarization" of the medium due to the excitation of electron–hole pairs.

Higher orders in perturbation theory: RPA approximation

The picture above readily generalizes to interaction processes of higher order. In the high-density limit, dominant contributions to the free energy should contain one free integration over the Fermi momentum per interaction process. A moment's thought shows that only diagrams of "ring graph" structure (see the figure) meet this condition. Expanding the free energy functional to nth order in the interaction operator, and retaining only diagrams having that structure, one obtains

$$F_{\mathrm{RPA}}^{(n)} = -\frac{T}{2n} \sum_q \left(\frac{2T}{L^3} V(\mathbf{q}) \sum_p G_p G_{p+q} \right)^n. \tag{4.26}$$

(To understand the origin of the multiplicative factor $1/n$, notice that $F^{(n)}$ results from the connected contraction of an operator $\sim (n!)^{-1} \langle S_{\mathrm{int}}^n \rangle_0^c$. There are $(n-1)!$ different ways of arranging the interaction operators S_{int} in a ring-shaped structure, i.e., the diagram carries a global factor $(n-1)!/n! = 1/n$.)

---

[22] In principle, the system consists of physical electrons immersed into a globally positive background. However (cf. the discussion in section 2.2.1), keeping in mind that at low temperatures dynamical processes take place in the immediate vicinity of the Fermi surface, a more problem-oriented way of thinking about states is in terms of quasi-particles and quasi-holes, i.e., electron states immediately above and below the Fermi surface.

random
phase
approx-
imation
In Eq. (4.26), the subscript RPA stands for **"random phase approxima-tion."**[23] However, more important than the designation are the facts: (i) we have managed to identify a particularly relevant subclass of diagrams contributing to the plethora of interaction processes; (ii) there is a physical parameter controlling the dominance of these diagrams; and (iii) we are apparently able to sum up the entire series of nth-order RPA-interaction contributions. Indeed, summation over n leads to the simple result,

$$F_{\mathrm{RPA}} \equiv \sum_n F_{\mathrm{RPA}}^{(n)} = \frac{T}{2} \sum_q \ln\left(1 - V(\mathbf{q})\Pi_q\right), \qquad (4.27)$$

**polarization
operator**
where we have introduced the **polarization operator**,[24]

$$\Pi_q \equiv \frac{2T}{L^3} \sum_p G_p G_{p+q} \qquad (4.28)$$

Equation (4.27) represents our first example of an infinite-order expansion. However, before turning to the discussion of further aspects of infinite-order perturbation theory, let us stay for a moment with the RPA approximation to the weakly interacting electron gas.

The last unknown we need to compute before understanding its physics is the polarization operator. Drawing on the frequency summation in problem 3.8.8, the polarization operator can be written explicitly as

$$\Pi_q = \frac{2T}{L^3} \sum_p \frac{1}{i\omega_n - \xi_{\mathbf{p}}} \frac{1}{i\omega_{n+m} - \epsilon_{\mathbf{p+q}}} = \frac{2}{L^3} \sum_{\mathbf{p}} \frac{n_{\mathrm{F}}(\epsilon_{\mathbf{p}}) - n_{\mathrm{F}}(\epsilon_{\mathbf{p+q}})}{i\omega_m - \epsilon_{\mathbf{p+q}} + \epsilon_{\mathbf{p}}}. \qquad (4.29)$$

In this intermediate result, the polarization operator is expressed as a function of imaginary Matsubara frequencies. Referring to chapter 7 for a detailed discussion, the extraction of real-time information from it requires an analytic continuation from imaginary to real frequencies ($i\omega_m \to \omega$). Specifically, the above function is analytic in the complex half-planes of positive and negative Matsubara frequencies (singularities are confined to the real frequency axis), so that the analytic continuation amounts to a straightforward substitution, and the frequency representation of the **real-time polarization operator** reads

$$\Pi_{\mathbf{q},\omega} = \frac{2}{L^3} \sum_{\mathbf{p}} \frac{n_{\mathrm{F}}(\epsilon_{\mathbf{p}}) - n_{\mathrm{F}}(\epsilon_{\mathbf{p+q}})}{\omega^+ - \epsilon_{\mathbf{p+q}} + \epsilon_{\mathbf{p}}}. \qquad (4.30)$$

---

[23] The attribute "random phase" seems to allude to the fact that the quantum mechanical phase carried by the particle–hole excitations stirred up by interactions gets lost after each elementary polarization process. This contrasts with more generic contributions to F, where quantum phases may survive more complex interaction processes. Also note that more than one approximation scheme in statistical physics has been dubbed "random phase."

[24] The definition (4.28) applies to the specific case of a three-dimensional translationally invariant system. More generally, the polarization operator is defined as the frequency/momentum Fourier transform of the connected average $\langle \bar{\psi}(\mathbf{x}, \tau)\psi(\mathbf{x}, \tau)\bar{\psi}(\mathbf{x}', \tau')\psi(\mathbf{x}', \tau')\rangle_c$.

(Here, we have kept an infinitesimal real part in $\omega^+ = \omega + i0$ to stay clear of the poles lying on the real axis; cf. an analogous frequency shift entering the definition of the retarded Green function, $G = 1/(\omega^+ - H)$ of quantum mechanics.) Eq. (4.30) is an intuitive result. The Fermi factors in the numerator tell us that the polarization of the medium requires empty states $\epsilon_{\mathbf{p+q}}$ and occupied states $\epsilon_{\mathbf{p}}$, or vice versa – polarization is confined to the neighborhood of the Fermi surface. The energy denominator indicates that polarization for AC frequencies ω dominantly engages particle–hole configurations with energy difference $\omega = \epsilon_{\mathbf{p+q}} - \epsilon_{\mathbf{p}}$.

Lindhard function

INFO The summation over $\mathbf{p}$ in Eq. (4.30) defines the **Lindhard function** $\Pi_{\mathbf{q},\omega}$. Referring to the literature[13] for a detailed discussion of $\Pi_{\mathbf{q},\omega}$, we note that it depends on two scales with dimension energy, ω and $v_F q$. For later reference, let us consider the Lindhard function in the limiting cases of small and large values of the ratio $\omega/v_F q$.

Low frequencies – For $\omega \ll v_F q$, we may neglect the frequency dependence of the denominator of (4.30) and make the approximation $(n_F(\epsilon_{\mathbf{p+q}}) - n_F(\epsilon_{\mathbf{p}}))(\epsilon_{\mathbf{p+q}} - \epsilon_{\mathbf{p}})^{-1} \simeq d_\epsilon n_F(\epsilon_{\mathbf{p}}) \simeq -\delta(\epsilon_{\mathbf{p}} - \mu)$. In this case,

$$\Pi_{\mathbf{q},\omega} \simeq -\nu_0, \tag{4.31}$$

where

$$\nu_0 \equiv \frac{2}{L^3} \sum_{\mathbf{p}} \delta(\mu - \epsilon_p) = 2 \int \frac{d^3p}{(2\pi)^3} \delta(\epsilon_p - \mu) = \frac{m p_F}{\pi^2} \tag{4.32}$$

density of states

denotes the **density of states** per volume of spinful non-interacting electrons at the Fermi surface. Occasionally – for example, in the context of magnetic phase transitions addressed in problems 5.6.7 and 6.7.2 – one needs to push the expansion to higher orders. As a result of a somewhat tedious calculation (you can try; the integrals are elementary) one finds, e.g., that for low frequencies, $|\omega_n| < v_F|\mathbf{q}|$,

$$\Pi_{\mathbf{q},\omega} = \nu_0 \left(1 - k_F^{-2}\mathbf{q}^2 - \frac{|\omega_n|}{c v_F|\mathbf{q}|} + \dots \right), \tag{4.33}$$

where c is a numerical constant.

High frequencies – In the opposite limit, $\omega \gg v_F q$, we can expand to first order in $\epsilon_{\mathbf{p+q}} - \epsilon_{\mathbf{p}} \simeq \mathbf{v}_F \cdot \mathbf{q}$. Noting that the sum over a single scalar product of this type vanishes by rotational symmetry, this leads to (exercise)

$$\Pi_{\mathbf{q},\omega} \simeq -\frac{2}{L^3\omega^2} \sum_{\mathbf{p}} n_F'(\epsilon_{\mathbf{p}})(\mathbf{v}_F \cdot \mathbf{q})^2 \simeq \frac{\nu_0(v_F q)^2}{3\omega^2}, \tag{4.34}$$

where $n_F'(\epsilon) \equiv d_\epsilon n_F(\epsilon)$. We will provide physical interpretations of these two limits shortly when we discuss the polarization operator in the context of screening.

Eq. (4.30) is an important building block in the theory of the weakly interacting electron gas. For example, one may ask what contribution to the **expansion of the free energy** ensues when Π_q is substituted into Eq. (4.27). While this calculation is straightforward in principle, the final summation over q turns out to be nontrivial in practice. Here, we simply quote the result of a famous study of Gell-Mann and Brueckner[25] where they obtained the free energy per particle

$$\frac{F_{\text{RPA}}}{E_{\text{Ry}} N} = -0.142 + 0.0622 \ln r_{\text{s}}. \tag{4.35}$$

When combined with the kinetic energy (4.20) and the first-order correlation energy (4.24), the structure of the density expansion of the free energy becomes clear: the sum over all RPA diagrams yields the coefficient of $\mathcal{O}(r_{\text{s}}^0)$ in the expansion in r_{s}.[26] We conclude that the sum of the RPA diagrams provides the next term in the expansion of the free energy in the dimensionless density parameter.

However, more important for our present discussion is the conceptual meaning of the RPA – notably its role in the **physics of screening**. To this end, let us temporarily consider the expectation value of the particle number $N = -\partial_\mu F$, rather than the free energy itself. Specifically, we wish to compare the first-order correction (to the non-interacting result), $N^{(1)} = -\partial_\mu F$, with the RPA, $N_{\text{RPA}} = -\partial_\mu F_{\text{RPA}}$. Noting that $\partial_\mu G_p = -(G_p)^2$ we readily find that (see Eq. (4.23))

$$N^{(1)} = -\frac{2T^2}{L^3} \sum_{p,q} (G_p)^2 G_{p+q} V(\mathbf{q}).$$

See fig. 4.6(a) for a diagrammatic visualization. Now, consider the μ-derivative of F_{RPA}, Eq. (4.27):

$$N_{\text{RPA}} = \frac{T}{2} \sum_q \frac{V(\mathbf{q}) \partial_\mu \Pi_q}{1 - V(\mathbf{q}) \Pi_q} = -\frac{T^2}{L^3} \sum_q \frac{V(\mathbf{q})}{1 - V(\mathbf{q}) \Pi_q} \left[\sum_p G_{p+q} (G_p)^2 + (q \leftrightarrow -q) \right]$$

$$= -\frac{2T^2}{L^3} \sum_q \frac{V(\mathbf{q})}{1 - V(\mathbf{q}) \Pi_q} \sum_p G_{p+q} (G_p)^2 = -\frac{2T^2}{L^3} \sum_q V_{\text{eff}}(q) \sum_p G_{p+q} (G_p)^2, \tag{4.36}$$

where we have defined the effective interaction

$$\boxed{V_{\text{eff}}(q) \equiv \frac{1}{V(\mathbf{q})^{-1} - \Pi_q} \equiv \frac{V(\mathbf{q})}{\epsilon(q)}} \tag{4.37}$$

dielectric function

with the generalized **dielectric function**

$$\boxed{\epsilon(q) \equiv 1 - V(\mathbf{q}) \Pi_q} \tag{4.38}$$

Structurally, the expression for N_{RPA} resembles the first-order expression $N^{(1)}$, but with the "bare" Coulomb interaction replaced by the effective interaction V_{eff}. From its definition, it is clear that V_{eff} represents a geometric series over polarization "bubbles," augmented by bare interaction lines. This is visualized in fig. 4.6(b),

---

[25] M. Gell-Mann and K. Brueckner, *Correlation energy of an electron gas at high density*, Phys. Rev. **106**, 364 (1957); see also reference [13].

[26] Here we follow a convention (used mostly in the older literature) where the RPA starts from the *second*-order ring diagram $F^{(2),1}$. However, henceforth we will refer to the RPA as the sum over all ring diagrams, including the first, $F^{(1)}$.

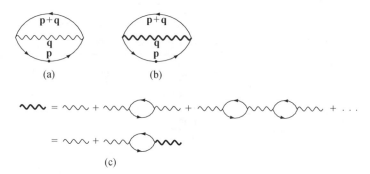

Fig. 4.6   Diagrammatic visualization of expectation value of the particle number. (a) First-order Fock correction and (b) the RPA approximation, where the definition of the RPA interaction line is shown in the lower part of the figure.

where the bold wavy line is defined in the bottom part. In fact, we do not need to look hard at the analytical expression (4.36) to understand its origin. The μ-differentiation acting on F_{RPA} may pick any of the n rings contributing to $F_{\mathrm{RPA}}^{(n)}$ in Eq. (4.26). The "differentiated ring" becomes the bubble in fig. 4.6, while all other rings conspire to form the $(n-1)$th-order contribution to the effective interaction line.

At this stage, the connection between RPA and the collective electromagnetic response of the charged system becomes discernible. We remember that the electric field $\mathbf{E}$ in a medium is related to the vacuum field $\mathbf{D}$ through the relation $\mathbf{D}(q) = \epsilon(q)\mathbf{E}(q)$, where the dielectric function $\epsilon(q) = 1 + 4\pi\chi(q)$ is determined by the **electric susceptibility** and where $q = (\mathbf{q}, \omega)$. The function χ measures the tendency of the medium to "respond" or adjust to an external electromagnetic perturbation. Identifying $\mathbf{E}/\mathbf{D}$ with the ratio of the "dressed" potential V_{eff} and the "bare" potential V, we conclude that, on the level of the microscopic theory, $4\pi\chi(q) = -V(\mathbf{q})\Pi_q$, i.e., the susceptibility is proportional to the polarization operator Π_q. These connections motivate the introduction of the dielectric function, as in Eq. (4.38) above.

It is instructive to consider the form of the screened interaction in the two limiting cases of low and high frequencies discussed on p. 195. At low frequencies, $\omega \ll v_F q$, the electron gas has enough time to adjust to the spatial variation $\sim q^{-1}$ of the potential, thus screening out electroneutrality-violating potential fluctuations. Substitution of Eq. (4.31) shows that, in this **static limit**, the electron gas interacts through the effective potential,

$$V_{\mathrm{eff}}(q) \overset{\omega \ll q v_F}{=} \frac{1}{V(\mathbf{q})^{-1} + \nu_0} = \frac{4\pi e^2}{q^2 + 4\pi e^2 \nu_0} \equiv \frac{4\pi e^2}{q^2 + \lambda^{-2}}, \qquad (4.39)$$

where $\lambda \equiv \left(4\pi e^2 \nu_0\right)^{-1/2}$ denotes the **Thomas–Fermi screening** length, and $V(\mathbf{q}) = 4\pi e^2/q^2$ is the bare Coulomb potential. Indeed, it is straightforward to verify that the inverse Fourier transform leads to an effective interaction potential (exercise)

$$V_{\text{eff}}(\mathbf{r}) = e^2 \frac{e^{-|\mathbf{r}|/\lambda}}{|\mathbf{r}|} \qquad (4.40)$$

that is exponentially suppressed, or screened, on length scales $|\mathbf{r}| > \lambda$.

INFO Let us briefly recapitulate the **heuristic interpretation of Thomas–Fermi screening**. Imagine a test charge e has been immersed in an electron gas. The host system will respond to this perturbation by a local distortion of its density. To compute the distortion, we note that the effective potential $V_{\text{eff}}(\mathbf{r})$ created by the test charge changes the electronic energy levels according to $\epsilon_p \to \epsilon_p - V_{\text{eff}}$. (Here, we assume that the external perturbation changes so slowly that the simultaneous usage of momentum and coordinate quantum numbers is not in conflict with the uncertainty relation.) The induced charge density thus reads as

$$\rho_{\text{ind}} = -2e \int \frac{d^3p}{(2\pi)^3} \left(n_{\text{F}}(\epsilon_p - V_{\text{eff}}) - n_{\text{F}}(\epsilon_p) \right) \approx 2e V_{\text{eff}} \int \frac{d^3p}{(2\pi)^3} n_{\text{F}}'(\epsilon_p) \simeq -e V_{\text{eff}} \nu_0.$$

Substitution of this result into the Fourier transform of the Poisson equation $\nabla^2 V_{\text{eff}}(\mathbf{r}) = -4\pi e \rho(\mathbf{r}) = -4\pi e (e\delta(\mathbf{r}) + \rho_{\text{ind}}(\mathbf{r}))$ leads to Eq. (4.39).

dynamic screening

Next, we consider the case of **dynamic screening**, $\omega \gg v_{\text{F}} q$. In this case, the substitution of Eq. (4.34) into the effective interaction gives

$$V_{\text{eff}}(\mathbf{q}, \omega) = \frac{4\pi e^2}{q^2} \frac{1}{1 - (\omega_{\text{p}}/\omega)^2},$$

plasma frequency

where $\omega_{\text{p}} \equiv (4\pi n e^2/m)^{1/2}$ denotes the **plasma frequency** and we have used the fact that (exercise) the **particle density** $n \equiv N/L^3$ is related to the density of states by $n = m v_{\text{F}}^2 \nu_0/3$.

INFO The form of the denominator hints at a collective instability of the electron gas at frequencies $\sim \omega_{\text{p}}$. Its quadratic dependence in frequency further suggests an instability akin to those in undamped oscillating systems. The "mode" responsible for this divergence is known as the **plasmon mode**, and its origin can be understood as follows. Imagine that the electron gas is uniformly displaced by a distance x against the positively charged background (see the figure). This will lead to the formation of oppositely charged surface layers at the two ends of the

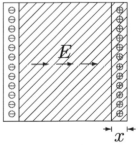

plasmon mode

system. The surface charge densities $\rho_{\pm} = \pm exn$ lead to an electric field $E = 4\pi e n x$ directed opposite to the displacement vector. Mobile charge carriers inside the system are thus subject to a force $-4\pi e^2 n x$. The solution of the equation of motion $m\ddot{x} = -4\pi e^2 n x$ oscillates at a frequency $\omega_{\text{p}} = (4\pi e^2 n/m)^{1/2}$, the plasma frequency. Since the motion of the charge carriers is in turn responsible for the accumulation of the charged surface layers, we conclude that the system performs a collective oscillatory motion, known as the plasmon excitation.

At this point, we conclude our preliminary discussion of the electron gas. We have seen that large-order perturbation theory can be applied to successfully explain various features of the interacting system: energetic lowering due to quantum correlation, screening, and even collective instabilities.

The interacting electron gas is but one example of the many applications of the diagrammatic perturbation theory. After the full potential of the approach had become evident – in the late 1950s and early 1960s – diagrammatic techniques of great sophistication were developed and applied to a plethora of many-body problems. Indeed, more than two decades passed before large-order perturbation theory eventually ceased to be *the* most important tool of theoretical condensed matter physics. Reflecting the practical relevance of the approach, there is a large body of textbook literature concentrating on perturbative methods.[27] Although it would make little sense to develop the field in its full depth once again, a few generally important concepts of diagrammatic perturbation theory are summarized in the next section.

4.3 Infinite-Order Expansions

Turning back to the prototypical ϕ^4-model, it is the purpose of the present section to introduce a number of general concepts relating to infinite-order perturbative summations. As should be clear from the discussion above, a meaningful summation over an infinite set of diagrams necessitates the existence of a class of perturbative corrections that is "more important" than others. In practice, what we need is a small parameter distinguishing between diagrams of different structure. In our example above, this parameter was the effective density r_s of the electron gas. However, in other settings, the control parameter N may be defined quite differently: large values of a spin, S, the number of colors N_c in QCD, the number of spatial dimensions, d, the number of modes of an optical wave guide, etc. Unfortunately, in most physical contexts, these parameters are far from large: $S = 1/2$, $d = N_c = 3$, etc. So we have to resort to a "poor man's" strategy where we develop a controlled and self-consistent theory in the limit of asymptotically large control parameters and hope that some fragments of truth survive in the limit down to more mundane values of N. Perhaps unexpectedly, this strategy often works astonishingly well down to values $N = \mathcal{O}(1)$.

So, let us, then, begin by introducing a large control parameter into a ϕ^4-type theory. To this end, we promote ϕ from a scalar to an N-component vector field $\phi = \{\phi^a\}$, $a = 1, \ldots, N$. The self-interaction of the field is modeled as $g \int d^d x \, \phi^a \phi^a \phi^b \phi^b$, i.e., an expression that is "rotationally" invariant in ϕ-space. The action of our modified theory is thus given by

$$S[\phi] \equiv \int d^d x \left(\frac{1}{2} \partial\phi \cdot \partial\phi + \frac{r}{2} \phi \cdot \phi + \frac{g}{4N} (\phi \cdot \phi)^2 \right), \qquad (4.41)$$

---

[27] See, e.g., A. A. Abrikosov, L. P. Gorkov, and I. E. Dzyaloshinkii, *Methods of Quantum Field Theory in Statistical Physics* (Dover Publications, 1975), A. Fetter and J. D. Walecka, *Quantum Theory of Many-Particle Systems* (McGraw-Hill, 1971), D. Pines and P. Nozières, *The Theory of Quantum Liquids – Normal Fermi Liquids* (Addison-Wesley, 1989), and S. Doniach and E. H. Sondheimer, *Green Functions for Solid State Physicists* (Benjamin Cummings, 1974).

where the factor $1/N$ in front of the interaction constant has been introduced for later convenience.

As before, we shall concentrate on the Green function $G^{ab}(\mathbf{x} - \mathbf{y}) = \langle \phi^a(\mathbf{x})\phi^b(\mathbf{y}) \rangle$ as a "test observable." Denoting the Green function by a bold line, the free Green function $G_0 \equiv \langle \phi^a(\mathbf{x})\phi^b(\mathbf{y}) \rangle_0 \propto \delta^{ab}$ by a thin line, and the interaction operator by a wavy line,[28] the structure of the first- and second-order expansion of the Green function are shown in the upper portion of fig. 4.7. For simplicity, the combinatorial factors weighting individual diagrams have been omitted.

4.3.1 Self-energy operator

Even without resorting to the large-N structure of the theory, it is possible to bring some order to the spaghetti of diagrams contributing to the expansion. Indeed, there are two distinct sub-classes of diagrams: those that are one-particle reducible (i.e., can be cut into two halves by cutting a single internal line; see the classification on page 183) and those that are not. This observation motivates the collection of all one-particle irreducible sub-portions of the diagrammatic expansion into a structural unit. In fig. 4.7, this entity, which is commonly called the **self-energy** and sometimes also the **effective mass operator**, is denoted by a hatched circle. The first- and second-order expansion of the self-energy are shown in the bottom part of the figure.

Freeman Dyson 1923–2020 was a British scientist who, trained as a mathematician, turned to physics in the 1940s. His work in condensed matter physics, statistical mechanics, and several other areas has had a lasting influence on the development of modern theoretical physics. Beyond his professional work in physics, Dyson has written several books on the social implications of modern science.

self-energy

With that definition, the Green function becomes a "chain" of self-energy operators, separated by free Green function lines, as shown in the second identity of the figure. A convenient representation of the expansion is shown in the third identity. An insertion of the full Green function after the first self-energy correction recursively generates the full series. Let us translate these statements into the language of formulae. Denoting the set of all self-energy diagrams by $\hat{\Sigma} = \{\Sigma^{ab}(\mathbf{x} - \mathbf{y})\}$,[29] the expansion of the Green function assumes the form

$$\hat{G} = \hat{G}_0 + \hat{G}_0 \hat{\Sigma} \hat{G}_0 + \hat{G}_0 \hat{\Sigma} \hat{G}_0 \hat{\Sigma} \hat{G}_0 + \cdots = \hat{G}_0 + \hat{G}_0 \hat{\Sigma} \hat{G}. \qquad (4.42)$$

Dyson equation

Here, the operator products involve summation over coordinates and internal indices, i.e., $(\hat{A}\hat{B})^{ab}(\mathbf{x} - \mathbf{y}) = \int d^d z \, A^{ac}(\mathbf{x} - \mathbf{z})B^{cb}(\mathbf{z} - \mathbf{y})$. Recursion relations of this type are commonly referred to as **Dyson equations**. The Dyson equation states that the problem of calculating $\hat{G}$ is essentially tantamount to understanding the

---

[28] Since the four field vertices entering the interaction are no longer indiscriminate, the interaction "point" representation of section 4.1.3 is no longer suitable.

[29] The conservation of global momentum in the theory implies (exercise: think about it!) that, like the Green function, the self-energy depends only on the difference of its coordinate arguments.

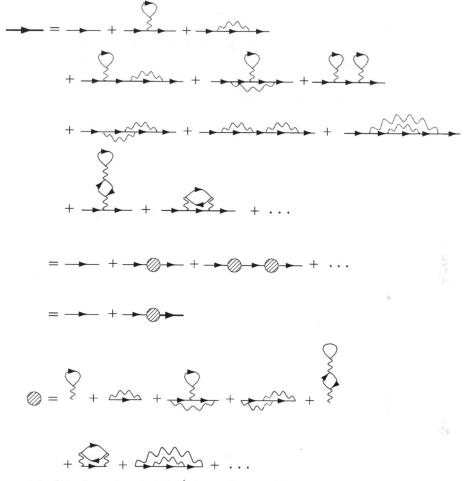

Fig. 4.7   Expansion of the Green function of ϕ^4-theory. Bottom: Expansion of the self-energy operator.

self-energy. To make this point more explicit, let us reformulate the Dyson equation in momentum space:

$$\hat{G}_{\mathbf{p}} = \hat{G}_{0,\mathbf{p}} + \hat{G}_{0,\mathbf{p}}\hat{\Sigma}_{\mathbf{p}}\hat{G}_{\mathbf{p}} \quad \Leftrightarrow \quad [\mathbf{1} - \hat{G}_{0,\mathbf{p}}\hat{\Sigma}_{\mathbf{p}}]\hat{G}_{\mathbf{p}} = \hat{G}_{0,\mathbf{p}}.$$

Here we have used the convolution theorem or, more physically, the fact that all scattering processes lumped into the self-energy conserve momentum separately. Matrix multiplication of this identity from the left by $[\mathbf{1} - \hat{G}_{0,\mathbf{p}}\hat{\Sigma}_{\mathbf{p}}]^{-1}$ leads to the expression

$$\hat{G}_{\mathbf{p}} = [\mathbf{1} - \hat{G}_{0,\mathbf{p}}\hat{\Sigma}_{\mathbf{p}}]^{-1}\hat{G}_{0,\mathbf{p}} = \left(\hat{G}_{0,\mathbf{p}}^{-1} - \hat{\Sigma}_{\mathbf{p}}\right)^{-1}.$$

Finally, using the fact that $(\hat{G}_{0,\mathbf{p}}^{-1})^{ab} = (p^2 + r)\delta^{ab}$, we arrive at the formal solution

$$\boxed{G_p^{ab} = \left((p^2 + r - \hat{\Sigma}_p)^{-1}\right)^{ab}} \tag{4.43}$$

This equation provides two lessons: first, the full information about the Green function is indeed stored in the self-energy; second, the self-energy somehow "adds" to the arguments p^2 and r entering the quadratic action, a point to be discussed in more detail below.

But how then do we compute the self-energy operator? In fact, the construction recipe follows from what has been said above. By definition, the nth-order contributions to the self-energy are generated by the connected and one-particle irreducible contraction of n interaction operators (weighted with the appropriate combinatorial factor $1/n!$). Two field vertices stay uncontracted as connectors to the free Green function lines contacting the self-energy. For example, the first-order contribution is given by (exercise)

$$(\Sigma_{\mathbf{p}}^{(1)})^{ab} = -\delta^{ab} \frac{g}{L^d} \Big(\frac{1}{N} \sum_{\mathbf{p}'} G_{0,\mathbf{p}'} + \sum_{\mathbf{p}'} G_{0,\mathbf{p}-\mathbf{p}'} \Big),$$

where the first (second) contribution corresponds to the first (second) diagram in the self-energy expansion in fig. 4.7.

EXERCISE Represent the second-order contribution $\Sigma^{(2)}$ in terms of Green functions.

Once the self-energy has been computed to any desired order, the result is substituted into Eq. (4.43) and one obtains the Green function.

INFO A critical reader will note that there are some problems with the line of argument above. First, we have tacitly ignored the **issue of combinatorics**. (How do we know that, once we have plugged the expansion of the self-energy into the Dyson equation, we get the same result as a brute-force direct expansion of the Green function?) To understand that the two-step program – "first compute the self-energy, and then substitute into the Dyson series" – indeed produces correct results, let us consider the nth-order contribution to the expansion of the Green function, with its overall combinatorial factor $1/n!$. Now imagine that we want to distribute those diagrams that contain, say, one free internal Green function over two self-energy operators according to the expression $G_0 \Sigma G_0 \Sigma G_0$. Assuming that the first self-energy operator is of order $m < n$ and the second is of order $n - m$, we notice that there are $\binom{n}{m}$ possibilities to distribute the interaction vertices over the two self-energies. That means that we obtain an overall combinatorial factor of $\frac{1}{n!}\binom{n}{m} = \frac{1}{m! \, (n-m)!}$. But $1/m!$ and $1/(n-m)!$ are precisely the combinatorial factors that appear in the definition of an mth-order and an $(n-m)$th-order self-energy operator, respectively. Arguing in reverse, we conclude that the prescription above indeed produces the correct combinatorics.

A second objection concerns the **consistency of the expansion**, i.e., the nth-order expansion of the self-energy is, of course, by no means equivalent to nth-order expansion of the Green function, nor to any specific order of the expansion. Indeed, when working with the concept of a self-energy, structuring the expansion according to its order in the interaction operator does not make much sense. We should rather focus on the summation of specific infinite-order diagram classes, as exemplified in the previous section and discussed in more general terms below.

4.3.2 Large-N expansion

So far we have not made reference to the N-component structure of the theory. However, let us now assume that N is very large, i.e., that we may be content with an expansion of the Green function to leading order in $1/N$. This condition can be made explicit by sending $N \to \infty$ and declaring $\lim_{N \to \infty} G_{\mathbf{p}}^{aa}$ to be our observable of interest.

The limit of large N entails a drastic simplification of the diagrammatic expansion. Each interaction vertex brings an overall factor of $1/N$, which must be compensated by a summation over field components to produce a contribution that survives the limit $N \to \infty$. This condition removes numerous diagrams contributing to the series. For example, in the Green function expansion of fig. 4.7, only the first, third, eighth, and ninth diagrams survive the limit. In all other contributions, interaction and Green function lines are interwoven in a way that does not leave room for one field-index summation per interaction vertex.

Inspection of the series shows that only diagrams void of crossing interaction lines (cf. the figure) survive the limit of large N. The approximation – indeed in the limit of infinite N it becomes exact – that retains only these contributions is commonly called the **non-crossing approximation (NCA)**. More poetically, the diagrams contributing to the reduced expansion are sometimes called "rainbow diagrams."

Importantly, the NCA self-energy can be computed in closed form. All one has to realize is that the summation over all rainbow diagrams amounts to substitution of the full NCA Green function under a single interaction line (exercise). Since the NCA self-energy is proportional to unity in the field-index space, we can express this fact through the formula

> **Max Born** 1882–1970
> was a German theorist who worked on the mathematical basis of quantum mechanics, and in particular on its probabilistic interpretation. Born shared the 1954 Nobel Prize in Physics with Walter Bothe for "for his fundamental research in quantum mechanics, especially for his statistical interpretation of the wave function."

$$\Sigma_{\mathbf{p}} \stackrel{\text{NCA}}{=} -\frac{g}{L^d} \sum_{\mathbf{p}'} G_{\mathbf{p}'} = -\frac{g}{L^d} \sum_{\mathbf{p}'} \left(p'^2 + r - \Sigma_{\mathbf{p}'} \right)^{-1}. \qquad (4.44)$$

In the literature, this equation goes under the name **self-consistent Born approximation** (SCBA). It is a "Born approximation" because, formally, it resembles a first-order perturbative correction (due to the overall factor of g). The approximation is "self-consistent" because the self-energy recursively appears on the right-hand side of the equation again, i.e., the equation is in fact not of first but of infinite order.

INFO Although the objective of the present section is to expose general structures, let us briefly review the **solution of a Born equation**. To keep things simple, let us assume that we are dealing with an effective low-energy model, so that the momentum summations must be terminated at an upper cutoff Λ. We further make the assumption (to be checked self-consistently) that the solution for the self-energy will turn out to be momentum independent: $\Sigma_{\mathbf{p}} = \Sigma$. This leads to the expression

$$\Sigma \approx -g \int^{\Lambda} \frac{d^d p}{(2\pi)^d} \frac{1}{p^2 + r - \Sigma}.$$

The evaluation of the integral depends on dimensionality and on the analytical structure of the energy denominator. For example, taking $d = 2$ and assuming that the parameter r is much smaller than the self-energy induced by scattering – an assumption also to be checked self-consistently – we obtain

$$\Sigma \approx -\frac{g}{4\pi} \int_0^{\Lambda^2} d(p^2) \frac{1}{p^2 - \Sigma} \simeq -\frac{g}{4\pi} \ln\left(-\frac{\Lambda^2}{\Sigma}\right).$$

A solution $\Sigma(g, \Lambda)$ can now be sought either by approximate analytical methods or graphically. (One plots both sides of the equation as functions of Σ and seeks a crossing point.)

However, for our present discussion, more important than the detailed dependence of Σ on the input parameters g and Λ is the principal meaning of the self-energy: apparently, Σ adds to the parameter r of the naked Green function. (Notice that the solution of the equation determining Σ will be negative.) Remembering that $r \sim \xi^{-2}$, one concludes that the interaction operator lowers the spatial correlation of the system. This is indeed what one should expect intuitively: scattering due to interactions acts as a source of "disorder" inside the system.

At this stage, it is worth taking a step back and seeing what we have achieved. We have managed to compute the Green function to infinite order in an expansion in the set of "relevant" diagrams. How does that fit with what has been said in section 4.1.1 about the "asymptotic" nature of perturbative series? In fact, the exponential proliferation of the number of diagrams, i.e., the mechanism that led to the eventual breakdown of the perturbative expansion, is blocked by the limit $N \to \infty$. Only subclasses of diagrams, with far fewer members, contribute and the series remains summable.

The large-N principle is actually not limited to the expansion of the Green function. To illustrate the point, let us briefly consider the expansion of the **four-point** **four-point function** **correlation function**,

$$C_{\mathbf{q}}^{(4)} = \frac{1}{N} \sum_{ab} \frac{1}{L^{2d}} \sum_{\mathbf{p},\mathbf{p}'} \langle \phi_{\mathbf{p}}^a \phi_{-\mathbf{p}-\mathbf{q}}^a \phi_{-\mathbf{p}'}^b \phi_{\mathbf{p}'+\mathbf{q}}^b \rangle. \tag{4.45}$$

In the next chapter we will see that objects of this structure often represent the most important information carriers of the theory. Unlike the Green function discussed previously, they relate directly to observable quantities. In the many-body **two-** literature, the *four*-point function is described as a **two-particle propagator**, in-**particle** dicating that it describes the joint propagation of two particles. Leaving a more **propagator** substantial discussion to the next chapter, we concentrate here on the formal aspects of its perturbative expansion.

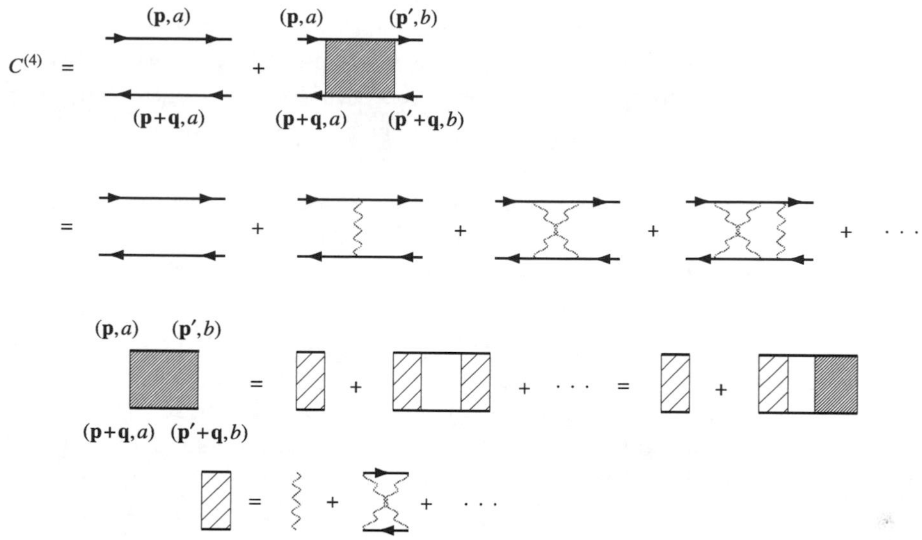

$$C^{(4)} = $$

$$= $$

Fig. 4.8   Expansion of the four-point function. Notice that the arrows represent the full Green function, i.e., all diagrams "renormalizing" the two-particle subunits of the diagram are automatically included.

The structure of the expansion of the four-point function is shown in fig. 4.8 where, for simplicity, momentum and component–indices are indicated only once. The simplest diagram contributing to the expansion consists of just two Green functions. It encapsulates all disconnected contractions, i.e., $\langle \phi^a_{\mathbf{p}} \phi^b_{-\mathbf{p}'} \rangle \langle \phi^a_{-\mathbf{p}-\mathbf{q}} \phi^b_{\mathbf{p}'+\mathbf{q}} \rangle \sim \delta^{ab} \delta_{\mathbf{p}\mathbf{p}'}$ contributing to $C^{(4)}$. All other contractions simultaneously involve all four field operators, i.e., they contain interaction lines between the Green functions. The sum of all these contributions is represented by the diagram containing the hatched surface. A few low-order contributions to the expansion are explicitly shown in the second line. Notice that all arrows appearing in these diagrams are fat. This indicates that diagrams "dressing" the two-particle sub-units of the expansion are automatically included. For example, the second contribution, containing just a single interaction line between the two-particle propagators, in fact represents an entire series of diagrams obtained by substituting the expansion of fig. 4.7 for the full Green function. (In an analytical calculation, one takes account of these contributions simply by substitution of the self-energy renormalized Green function for each arrow.)

EXERCISE Write down analytical expressions contributing to the low-order diagrams shown in the first and second lines of the figure.

All diagrams involving interactions between the two Green functions have in common that they contain four external legs, the Green function connectors to the external field operators. Considering these legs as removed, one ends up with a core contribution often called the **vertex**. (The designation is motivated by the fact that the first contribution to the vertex is indeed the contracted interaction vertex of the action; see diagram no. 2. in the second line,) In fig. 4.8, the vertex is denoted by a tightly hatched area.

As with the Green function, the expansion of the vertex can be given some structure. To this end, notice that some of the diagrams contributing to the vertex (e.g., the fourth diagram in the second line with external legs removed) can be cut into two simply by cutting two internal Green function lines. Vertex diagrams of this type are called **two-particle reducible**, by analogy with the "one-particle reducible" contributions to the expansion of the Green function. As with the expansion of the latter, we lump all irreducible contributions to the vertex (e.g., the second and third diagram in the second line, external legs removed) into a substructure called the **irreducible vertex**. In the shaded box of fig. 4.8, the irreducible vertex is denoted by a lightly hatched area. The first two diagrams contributing to it are shown in the bottom row of the box. Here, one can see that the irreducible vertex plays a role similar to the self-energy of the Green function. Expressed in terms of the irreducible vertex, the expansion of the vertex assumes the regular form shown in the first row of the box.

To represent these graphical relations in analytical form, we denote the full vertex by the symbol $\hat{\Gamma} = \{\Gamma^{aa',bb'}_{\mathbf{p},\mathbf{p}',\mathbf{q}}\}$ and the irreducible vertex by $\hat{\Gamma}_0 = \{\Gamma^{aa',bb'}_{0,\mathbf{p},\mathbf{p}',\mathbf{q}}\}$, where the indices a,a',b,b' keep track of the index labels carried by the four Green functions entering the vertex. (Although we have defined our correlation function in such a way that $a = a'$, $b = b'$, a generalization to

> **Hans Albrecht Bethe 1906–2005**
> was a German-American Nobel Laureate in physics in 1967 for his "contributions to the theory of nuclear reactions, especially his discoveries concerning the energy production in stars." As well as nuclear matter, he has also contributed substantially to atomic and condensed matter physics.

a four-fold index label is necessary to formulate the recursion Eq. (4.46).) The three momentum arguments represent the momenta of the Green functions connecting to the vertex operators, as indicated explicitly in the first line of the figure. (Remember that the theory has overall momentum conservation, i.e., three momentum arguments suffice to fix unambiguously the momentum dependences of Γ and $C^{(4)}$.) The content of the third line of the figure can then be expressed in terms of a closed recursion relation:

$$\Gamma^{aa',bb'}_{\mathbf{p},\mathbf{p}',\mathbf{q}} = \Gamma^{aa',bb'}_{0,\mathbf{q},\mathbf{p}',\mathbf{p}} + \frac{1}{L^d} \sum_{c,c',\mathbf{p}''} \Gamma^{aa',cc'}_{0,\mathbf{p},\mathbf{p}'',\mathbf{q}} G^c_{\mathbf{p}''} G^{c'}_{\mathbf{p}''+\mathbf{q}} \Gamma^{cc',bb'}_{\mathbf{p}'',\mathbf{p}',\mathbf{q}}. \tag{4.46}$$

Expressions of this type are (often) called **Bethe–Salpeter equations**. Comparison with Eq. (4.42) shows that this equation appears to be conceptually similar to the Dyson equation for the one-particle Green function.[30] Indeed, the principle behind most recursion relations of perturbation theory is a structure like

$$\boxed{\hat{X} = \hat{X}_0 + \hat{X}_0 * \hat{Z} * \hat{X},}$$

(4.47)

where $\hat{X}$ is our object of interest (e.g., $\hat{\Gamma}$), $\hat{X}_0$ is its free version, $\hat{Z}$ is a subunit that is, in some sense, irreducible ($\hat{\Sigma}$ or $\hat{\Gamma}_0 GG$), and $*$ some generalized matrix convolution.

Owing to the importance of the two-particle propagator, the solution of Bethe–Salpeter equations is a central issue in many areas of many-body physics. In most cases, only approximate solutions can be obtained. With our present example, "approximate" means that one sends N to large values and seeks a solution to leading order in N^{-1}. In that limit, the only surviving contribution to the irreducible vertex is the first, a plain interaction line (see the figure). As with the self-energy operator discussed in the previous section, all diagrams with entangled interaction lines are frustrated in the sense that we do not have as many index summations (each producing a factor N) as interaction constants (each proportional to $1/N$). Such contributions vanish in the limit of large N. We thus conclude that the Bethe–Salpeter equation assumes the simple form

$$\Gamma^{ab}_{\mathbf{p},\mathbf{p}',\mathbf{q}} = -\frac{g}{N} - \frac{g}{N}\frac{1}{L^d}\sum_{c,\mathbf{p}''} G^a_{\mathbf{p}''} G^c_{\mathbf{p}''+\mathbf{q}}\, \Gamma^{cb}_{\mathbf{p}'',\mathbf{p}',\mathbf{q}},$$

where $\Gamma^{ab} \equiv \Gamma^{aa',bb'}\delta^{aa'}\delta^{bb'}$.[31] This equation can be simplified even further by making the ansatz $\Gamma^{ab}_{\mathbf{p},\mathbf{p}',\mathbf{q}} = \Gamma_{\mathbf{q}}$, where $\Gamma_{\mathbf{q}}$ is independent of discrete indices and input momenta $\mathbf{p}$ and $\mathbf{p}'$. (When solving a perturbative recursion relation, it is always a good idea to try an ansatz of maximal simplicity, i.e., one that is no more complex than the constituting elements of the equation.) Then,

$$\Gamma_{\mathbf{q}} = -\frac{g}{N} - gP_{\mathbf{q}}\Gamma_{\mathbf{q}} \quad \Rightarrow \quad \Gamma_{\mathbf{q}} = -\frac{g}{N}\frac{1}{1 + gP_{\mathbf{q}}},$$

(4.48)

where we have introduced the abbreviation $P_{\mathbf{q}} = \frac{1}{L^d}\sum_{\mathbf{p}} G_{\mathbf{p}} G_{\mathbf{p}+\mathbf{q}}$. In principle, one may now proceed by substituting the large-N expansion of the Green

---

[30] We can make the analogy perfect by defining a "one-particle vertex" $\hat{\Gamma}^{(1)} = \hat{G}^{(0)-1}[\hat{G} - \hat{G}^{(0)}]\hat{G}^{(0)-1}$. Inspection of the second identity of fig. 4.7 shows that the expansion of $\Gamma^{(1)}$ starts and ends with a self-energy operator, i.e., the first free Green function line G_0 is removed, and so are the two external G_0 lines connecting to the self-energy operator. In direct analogy to Eq. (4.46), the analytical formula for $\Gamma^{(1)}$ then reads $\hat{\Gamma}^{(1)} = \Sigma + \Sigma G_0 \hat{\Gamma}^{(1)}$.

[31] Here we have used the fact that the large-N approximation of the irreducible vertex forces the two input indices a, a' to be equal (and the same for the output indices). As a word of caution our large-N approximation of the irreducible vertex explicitly uses the fact that the input/output indices entering our definition of the vertex are equal, i.e., should we compute a correlation function where the two input/output indices are different (a rare occurrence in realistic applications), the large-N approximation of the four-point functions would no longer assume the simple form of a regular ladder.

function (4.43) and computing the function $P_{\mathbf{q}}$ by integration over $\mathbf{p}$. This would produce a closed expression for Γ and, by virtue of the relation

$$C_{\mathbf{q}}^{(4)} = L^d P_{\mathbf{q}} \left(1 + N\Gamma_{\mathbf{q}} P_{\mathbf{q}} \right],$$

our correlation function. Since the emphasis in this section is on conceptual aspects of perturbation theory, we will not pursue the analysis to its very end. (For an analysis of the Bethe–Salpeter equation in a context more interesting than ϕ^4-theory, see the discussion of the quantum disorder problem in section 4.4.) Yet there is one aspect of the expression for $\Gamma_{\mathbf{q}}$ worth noticing here. Consider $P_{\mathbf{q}}$ expanded as a Taylor series in $\mathbf{q}$ and focus on the zeroth-order contribution, P_0. From the definition of the Green function (4.43) we have

$$P_0 = \frac{1}{L^d} \sum_{\mathbf{p}} G_{\mathbf{p}}^2 = \partial_\Sigma \frac{1}{L^d} \sum_{\mathbf{p}} G_{\mathbf{p}} \overset{(4.44)}{=} -g^{-1}\partial_\Sigma \Sigma = -g^{-1}. \tag{4.49}$$

When substituted into the formula for Γ, this shows that for small momenta the expansion of the numerator of the vertex starts with a power of $\mathbf{q}$. (By symmetry, the first non-vanishing contribution will be of $\mathcal{O}(q^2)$.) This means that both the vertex and the four-point correlation function are long-range objects, i.e., unlike the Green function, they do not decay exponentially, but as a power law. The long-range character of the four-point function has observable consequences, as discussed in various different contexts below.

Summarizing, our discussion of the two- and four-point functions has shown that if a large parameter is present, relevant subclasses of perturbative contributions can be identified and summed to infinite order. As a matter of fact, there are not too many of these summable diagram classes: ring-diagrams, rainbow-diagrams and ladder-diagrams nearly exhaust the set of "friendly" corrections, amenable to analytical summation. Notice that so far we have largely considered abstract summation schemes; i.e., we still need to learn more about the way in which intermediate results like Eqs. (4.43) or (4.48) can be translated into concrete physical information. These aspects are discussed – on applications more rewarding than plain ϕ^4-theory – a little later. However, at this point we leave the discussion of formal perturbation theory. While a state-of-the-art exposition of the subject would require much more space – for condensed-matter-oriented texts on perturbative methods, we refer to the references given in footnote[27] The material introduced in this section suffices for nearly all purposes of the present text.

4.4 Perturbation Theory of the Disordered Electron Gas

REMARK In this section we consider the disordered electron gas as a case study for application of the general concepts developed earlier in this chapter. Readers on a fast track may skip this section and return to it, for example, when they encounter a problem subject to quantum disorder in their own work (which sooner or later will be the case). The perturbation theory applied in this section largely parallels that developed in section 4.3.

We take this as an incentive to present some of the material in a question and answer style, giving readers an opportunity to test their understanding of diagrammatic perturbation theory on the fly.

Having introduced the formal building blocks of infinite order perturbation theory, we now illustrate their application to the important example of the disordered electron gas. There is perhaps no better case study to exemplify the rationale of perturbative approaches – identification of relevant physical processes in low-order perturbation theory, followed by infinite order summation of parametrically distinct diagram classes – than the physics of this system. In this problem, we take a perspective complementary to that of section 4.2: Coulomb interactions are assumed to be irrelevant. Instead, we take the presence of translational symmetry breaking impurities seriously. Of course, "real" problems are subject to both interactions and disorder. However, there is a surprisingly large class of applications that can be addressed neglecting either one or the other. In this section, we emphasize the disorder perspective.

4.4.1 Field theory representation of the disordered electron gas

No semiconductor, or metal of macroscopic[32] extent, is ever free of imperfections and impurities. Indeed, the effect of disorder on the phenomenology of metals or semiconductors could not be more varied. In some cases, disorder plays an essential role (for example, conventional light bulbs would not function without impurity scattering!), in others the effect is parasitic (imparting only a "blurring" of otherwise structured experimental data) or it conspires to give rise to completely unexpected types of electron dynamics (as is the case in the quantum Hall effect to be discussed in chapter 8).

Some, but not all of these phenomena can be addressed in terms of the perturbative concepts introduced in this chapter. In this section, we will start our venture into the physics of the disordered electron gas by formulating a functional integral framework capable of describing perturbative and non-perturbative effects alike. Building on this platform, we will then develop a diagrammatic framework suitable to describe perturbative manifestations of impurity scattering: notably the physics of diffusion, and quantum processes heralding the eventual breakdown of perturbation theory in low-dimensional disordered materials.

Replica trick

In realistic materials, the inevitable presence of defect atoms or other types of lattice imperfections leads to the presence of static "random" potentials $V(\mathbf{r})$ adding to the translationally invariant crystalline background potential (set to zero throughout,

---

[32] In ultraclean semiconducting devices, electrons may travel up to distances of several microns without experiencing impurity scattering. Even so, the "chaotic" scattering from the typically irregular boundaries of the system has an equally invasive effect on charge carrier dynamics.

assuming that we may adopt a free electron approximation describing a good metal). In many applications, one assumes that V is drawn from some random distribution $P[V]$. One may then compute physical observables averaged over realizations of V – describing those aspects of the physics that are effectively self-averaging – or statistical fluctuations of observables in cases where sample-to-sample variations are a focus of interest. To further simplify matters, it is often assumed that V is Gaussian distribute, with

$$P[V]DV = \exp\left(-\frac{1}{2\gamma^2}\int d^d r\; V^2(\mathbf{r})\right) DV, \qquad (4.50)$$

where the measure includes a factor unit-normalizing the distribution. In this case, fluctuations of $V(\mathbf{r})$ are short-range correlated, as described by the second moment

$$\langle V(\mathbf{r})V(\mathbf{r}')\rangle_{\mathrm{dis}} = \gamma^2 \delta(\mathbf{r}-\mathbf{r}'). \qquad (4.51)$$

Here, $\langle\ldots\rangle_{\mathrm{dis}} \equiv \int DV\, P[V](\ldots)$, with the coefficient γ^2, of dimension energy^2/volume, parameterizing the effective strength of the randomness.

As discussed above, our aim is to average the quantum expectation value of a certain observable $\mathcal{O}$ over the disorder ensemble. Let us assume that $\mathcal{O}$ is represented as a derivative of the functional free energy, $\mathcal{O} = -\frac{\delta}{\delta J}|_{J=0} \ln \mathcal{Z}$, where J is a parameter function (also known as a *source field* in the parlance of field theory),[33]

$$\langle\mathcal{O}\rangle_{\mathrm{dis}} = -\frac{\delta}{\delta J}\Big|_{J=0}\langle\ln\mathcal{Z}\rangle_{\mathrm{dis}} = -\int DV\, P[V]\frac{1}{\mathcal{Z}[V,0]}\frac{\delta}{\delta J}\Big|_{J=0}\mathcal{Z}[V,J]. \qquad (4.52)$$

This fundamental relation presents a technical challenge: the appearance of the function V in both the numerator and denominator makes such integrals largely intractable.

There exist three different approaches circumventing the denominator problem: the **supersymmetry approach**,[34] the **Keldysh technique**,[35] and the **replica trick**.[36] All three share the feature that they alter the definition of the functional partition function in such a way that (i) $\mathcal{Z}[J=0] = 1$ (i.e., the disorder dependence of the denominator disappears), while (ii) Eq. (4.52) remains valid, and (iii) the algebraic structure of $\mathcal{Z}[J]$ is left largely unchanged. Since the disorder appears linearly in the Hamiltonian, point (iii) implies that we need to average functionals with actions linear in the potential V, an enterprise that turns out to be quite feasible.

INFO All three approaches have different strength and weaknesses. Supersymmetry is technically the most demanding and tailored to problems non-perturbative in the disorder strength. A serious weakness is that (with few exceptions), it can not be applied to systems with particle interactions. The Keldysh technique is the method of choice for problems

[33] For example, the parameter μ is a source generating expectation values of the particle number: $-\partial_\mu F = \langle\hat{N}\rangle$.

[34] K. B. Efetov, *Supersymmetry method in localisation theory*, Sov. Phys. JETP **55**, 514 (1982).

[35] For a review see, e.g., A. Kamenev, *Many body theory of non-equilibrium systems* in *Nanophysics: Coherence and transport*, eds. H. Bouchiat *et al.* (Elsevier, Amsterdam, 2005).

[36] S. F. Edwards and P. W. Anderson, *Theory of spin glasses*, J. Phys. F **5**, 965 (1975).

Fig. 4.9   The idea behind the replica trick; for an explanation, see the main text.

with or without quantum disorder out of thermal equilibrium. It will be discussed in depth in chapter 12. Finally, the replica trick is the oldest and most widely applied applied formulation. Its applicability extends beyond disordered electron systems to numerous other disordered classical or quantum system. For example, the method has proven most fruitful in the theory of conventional and spin glasses.[37] However, its application to non-perturbative problems is not innocuous and has occasionally been criticized (see below).

replica
trick
The basis of the **replica trick** is to consider the Rth power of the partition function, $\mathcal{Z}^R$. For integer R one may think of $\mathcal{Z}^R$ as the partition function of R identical copies of the original system (see fig. 4.9), hence the name "replica" trick. To appreciate the merit of this procedure, note the formal relations

$$\langle \mathcal{O} \rangle = -\frac{\delta}{\delta J} \ln \mathcal{Z}[J] = -\frac{\delta}{\delta J} \lim_{R \to 0} \frac{1}{R} \left(e^{R \ln \mathcal{Z}} - 1 \right) = -\frac{\delta}{\delta J} \lim_{R \to 0} \frac{1}{R} (\mathcal{Z}^R - 1).$$

The crucial last equality states that expectation values of observables can be obtained by performing computations with the Rth power of $\mathcal{Z}$ (instead of its logarithm). In the coherent state representation, the expression for the replicated partition function involves an effective action which is still linear in the disorder, $(e^{\int dV})^R = e^{\sum \int dV}$, and hence comparatively easy to average. However, the replica-averaging procedure involves one unusual feature – at the end of the calculation, one must implement the analytic continuation $R \to 0$. More precisely, we will have to compute the function $f(R) \equiv \delta_J R^{-1} \langle \mathcal{Z}^R \rangle_{\text{dis}}$ for every integer R and then analytically continue $R \to 0$. As long as we are doing perturbation theory, $f(R)$ will be *polynomial* in R (think why) and hence analytic. However, in non-perturbative problems, analyticity is no longer guaranteed and the application of the formalism becomes a tricky affair.

Specifically, for the disordered electron system, the replica partition function assumes the form

$$\mathcal{Z}^R[J] = \int D\psi \, \exp \left(-\sum_{a=1}^{R} S[\psi^a, J] \right), \tag{4.53}$$

where $\psi^a, a = 1, \ldots, R$, denotes the Grassmann field representing partition function number a, $D\psi \equiv \prod_{a=1}^{R} D\psi^a$, and the action of the individual replicas in the absence of sources,

$$S[\psi^a] = \sum_n \int d^d r \, \bar{\psi}_n^a(\mathbf{r}) \left(-i\omega_n - \frac{\nabla^2}{2m} - E_{\mathrm{F}} + V(\mathbf{r}) \right) \psi_n^a(\mathbf{r}), \tag{4.54}$$

[37] For a review of replica-based theoretical approaches in these fields we refer to G. Parisi, *Glasses, replicas and all that*, in *Proc. Les Houches – Ecole d'Eté de Physique Théorique*, vol. 77, ed. J.-L. Barrat *et al.* (Elsevier, 2004).

describes non-interacting fermions subject to the disorder potential. Since the source couples to all replicas identically, $S_{\text{source}}[\psi, J] = \sum_a \int J\mathcal{O}(\psi^a)$ in symbolic notation, our our expectation value assumes the form

$$\langle \mathcal{O} \rangle = -\lim_{R \to 0} \frac{1}{R} \frac{\delta \mathcal{Z}^R[J]}{\delta J} = \lim_{R \to 0} \frac{1}{R} \sum_{a=1}^{R} \langle \mathcal{O}(\psi^a) \rangle_\psi \qquad (4.55)$$

Assuming that all observables are evaluated as in Eq. (4.55), we no longer need to keep an explicit reference to the source field J.

Now, let us average the functional (4.53) over the distribution (4.50). A straightforward application of the Gaussian integral formula (3.19) leads us to the result

$$\langle \mathcal{Z}^R \rangle_{\text{dis}} = \int D\psi \, \exp \left(-\sum_{a=1}^{R} S_{\text{cl}}[\psi^a] - \sum_{a,b=1}^{R} S_{\text{dis}}[\psi^a, \psi^b] \right), \qquad (4.56)$$

where $S_{\text{cl}} = S\big|_{V=0}$ denotes the action of the clean (non-disordered) system, and

$$S_{\text{dis}}[\psi^a, \psi^b] \equiv -\frac{\gamma^2}{2} \sum_{mn} \int d^d r \; \bar{\psi}^a_m(\mathbf{r}) \psi^a_m(\mathbf{r}) \bar{\psi}^b_n(\mathbf{r}) \psi^b_n(\mathbf{r}) \qquad (4.57)$$

represents an effective quartic interaction generated by the disorder average. Notice the superficial similarity between S_{dis} and an attractive short-range "interaction" term. However, in contrast with a dynamically generated interaction, (a) S_{dis} does not involve frequency-exchanging processes (the reason being that scattering off static impurities is energy conserving), and (b) it describes interactions between particles with different replica indices.

To understand the physics behind the attractive inter-replica interaction, consider the potential landscape of a given impurity configuration (see the figure). Irrespective of their replica indices, all Feynman amplitudes will try to trace out those regions in configuration space where the potential energy is low, i.e., there will be a tendency to propagate through the same regions in the potential landscape. (Recall that 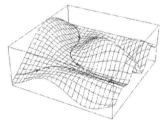 all replica fields are presented with the same potential profile.) On average, this looks as if the replica fields are subject to an attractive interaction mechanism.

In summary, one may account for the presence of quenched or static disorder by (a) replicating the formalism, (b) representing observables as in Eq. (4.55), and (c) adding the replica non-diagonal contribution Eq. (4.57) to the action. This results in a theory wherein the disorder no longer appears explicitly. (Technically, the effective action has become translationally invariant.) The price to be paid is that the action now contains the non-linearity Eq. (4.57).

4.4.2 Propagator of the disordered electron gas

Following the rationale of this chapter, we start our analysis of the theory with a discussion of the two-point function or propagator, $G(\mathbf{x}, \mathbf{y}; \tau) \equiv \langle \psi(\mathbf{x}, \tau)\, \bar{\psi}(\mathbf{y}, 0) \rangle_\psi$. The propagator is key to the understanding of the problem, both conceptually and methodologically, when it becomes a building block in the computation of observables in the next section.

Scattering time

Before turning to the computation of the propagators, we need to introduce the concept of the **elastic scattering time**, τ.

To this end, consider the quantum amplitude $U(\mathbf{y}, \mathbf{x}; t) = \langle \mathbf{y} | \exp(-i\hat{H}t) | \mathbf{x} \rangle$ for a particle to propagate from a point $\mathbf{x}$ to a point $\mathbf{y}$ in a time t for a particular realization of the disorder potential. One may think of this amplitude as the sum of all Feynman paths connecting the points $\mathbf{x}$ and $\mathbf{y}$. On its journey along each path, the particle scatters (see the figure), implying that the action of the path depends sensitively on the particular realization of impurities. For large separations, $|\mathbf{x} - \mathbf{y}|$, the scattering phase becomes a "quasi-random" function of the impurity configuration. The same applies, of course, to the linear superposition of all paths, the net amplitude U.

Let us now consider the impurity-averaged value of the transition amplitude $\langle U(\mathbf{x}, \mathbf{y}; t) \rangle_{\text{dis}}$. As we are averaging over a superposition of random phases, one may expect that the disorder average will be translationally invariant and, as a result of the random phase cancelation, rapidly decaying, $\langle U(\mathbf{x}, \mathbf{y}; t) \rangle_{\text{dis}} \sim \exp(-|\mathbf{x} - \mathbf{y}|/2\ell)$. The decay constant ℓ of the averaged transition amplitude defines the **elastic mean free path** while the related time $\tau \equiv \ell/v_{\text{F}}$ denotes the elastic scattering time. In the following, we develop a quantitative description of this "damping" process.

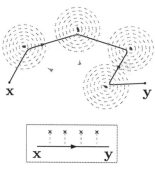

Within the framework of the coherent-state functional the transition amplitude (continued to imaginary time) is represented by the correlation function $G(\mathbf{x}, \mathbf{y}; \tilde{\tau}) \equiv \langle \psi(\mathbf{x}, \tilde{\tau})\, \bar{\psi}(\mathbf{y}, 0) \rangle_\psi$, where the averaging is over the Grassmann action (4.54), and we write $\tilde{\tau}$ for the imaginary-time argument to avoid confusion with the scattering time, τ. As usual, it will be convenient to perform the intermediate steps of the computation in frequency–momentum space. We thus represent the correlation function as

$$G(\mathbf{x}, \mathbf{y}; \tilde{\tau}) = \frac{T}{L^d} \sum_{\omega_n, \mathbf{p}, \mathbf{p}'} e^{-i\omega_n \tau + i\mathbf{p} \cdot \mathbf{x} - i\mathbf{p}' \cdot \mathbf{y}} G_{\mathbf{p}, \mathbf{p}'; \omega_n}, \qquad (4.58)$$

where ω_n is a fermionic Matsubara frequency and $G(\mathbf{p}, \mathbf{p}'; \omega_n) = \langle \psi_{n, \mathbf{p}} \bar{\psi}_{n\mathbf{p}'} \rangle_\psi$. (Keep in mind that, prior to the impurity average, the system lacks translational

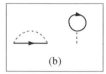 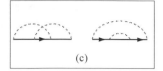

Fig. 4.10   In the literature the scattering off impurities is frequently denoted by dashed lines. Building blocks of the ensuing diagrammatic code include (a) the impurity scattering vertex,
(b) the first-order self-energy diagrams, (c) the second-order self-energy diagrams, and (d)
SCBA self-energy. In (d), the bold line represents the full Green function and the diagram
implies that the self-energy is computed neglecting all crossed lines (see the discussion in
section 4.3.2).

invariance, i.e., the Green function depends on two independent momentum arguments.) Following the general prescription developed in the previous section, the
correlation function averaged over a Gaussian disorder distribution is then given by
(see Eq. (4.55))

$$\langle G_{\mathbf{p},\mathbf{p}',n}\rangle_{\mathrm{dis}} = \lim_{R\to 0}\frac{1}{R}\sum_{a=1}^{R}\langle\psi_{n,\mathbf{p}}^{a}\,\bar{\psi}_{n,\mathbf{p}}^{a}\rangle_{\psi}\delta_{\mathbf{p},\mathbf{p}'}, \qquad (4.59)$$

where ψ^{a} is the ath component of the R-fold replicated field and $\langle\cdots\rangle_{\psi}$ now stands
for the functional average, with an action including the interaction term (4.57) and
the free fermion action $S_{\mathrm{cl}}[\psi^{a}] = \sum_{n,\mathbf{p}}\bar{\psi}_{n,\mathbf{p}}^{a}(-i\omega_n + (\mathbf{p}^2/2m) - E_{\mathrm{F}})\psi_{n,\mathbf{p}}^{a}$.

In the following, we discuss the perturbative analysis of the propagator as a
sequence of assignments whose detailed solution is given below. Readers are invited
to test their understanding of perturbation theory by answering as many as possible
of the questions below without consulting of the solutions.

Following the general arguments of section 4.3.1, the principal object of interest
is the impurity-generated self-energy operator Σ. Let us prepare the analysis of
this object by introducing some diagrammatic notation. We depict the impurity
scattering vertex defining the action (4.57) as in fig. 4.10(a). As usual, setting
$p = (\omega_n, \mathbf{p})$, the free particle Green function $G_{0,p} \equiv (i\omega_n - \xi_{\mathbf{p}})^{-1}$ is denoted by a
thin (directed) line. Using this notation, and following the rules of diagrammatic
perturbation theory developed in chapter 4:

Q1: Consider $\Sigma \approx \Sigma^{(1)}$ at first order in the scattering amplitude (fig. 4.10(b)). Show
that the "Hartree-type" diagram (on the right) does not contribute (in the replica
limit!). Identify the real and imaginary parts of the "Fock" contribution (left) to
the self-energy, and show that

$$\mathrm{Im}\,\Sigma^{(1)}(\omega_n) \equiv -\mathrm{sgn}(\omega_n)\frac{1}{2\tau} = -\mathrm{sgn}(\omega_n)\pi\nu\gamma^2, \qquad (4.60)$$

where $\nu = \nu(E_{\mathrm{F}})$ is the density of states at the Fermi energy. (Hint: Use the identity
$\sum_{\mathbf{p}} F(\epsilon_{\mathbf{p}}) = \int d\epsilon\,\nu(\epsilon)F(\epsilon)$ and make use of the Dirac identity (3.84).) Show that, in
dimensions $d \geq 2$, $\mathrm{Re}\,\Sigma^{(1)}$ actually diverges. Convince yourself (both formally and
heuristically) that this divergence is an artefact of our modeling of the impurity

potential by a δ-correlated function (see Eq. (4.50) and the related discussion). Consider what could be the reason why the real part of the self-energy does not play a crucial role.

Q2: Turning to the second-order contribution $\Sigma^{(2)}$ (see fig. 4.10(c)), convince yourself that, in dimensions $d > 1$, the diagram with crossed impurity lines is parametrically smaller than the contribution on the right. (What is the small parameter of the expansion?)

Q3: This motivates computation of the self-energy in the self-consistent Born approximation (SCBA) (see fig. 4.10(d)). Show that the SCBA equation is solved by a self-energy whose imaginary part continues to be given by Eq. (4.60).

Q4: Putting everything together, we obtain the important result

$$\langle G_p \rangle_{\mathrm{dis}} = \frac{1}{i\omega_n + E_{\mathrm{F}} - \frac{p^2}{2m} + \frac{i}{2\tau}\mathrm{sgn}(\omega_n)} \tag{4.61}$$

for the averaged propagator of the weakly disordered ($k_{\mathrm{F}}\ell \sim E_{\mathrm{F}}\tau \gg 1$) electron gas.

We next aim to justify the identification of the self-energy with (one half of) the inverse scattering time. To this end, perform the Matsubara frequency summation by contour methods (noting the singularities of the Green function (4.61) and its decay at infinity). Then, taking $\tilde{\tau} \gg E_{\mathrm{F}}^{-1}$, consider the momentum integral to conclude that

$$\langle G(\mathbf{x}, \mathbf{y}; \tilde{\tau}) \rangle_{\mathrm{dis}} = G_{\mathrm{cl}}(\mathbf{x}, \mathbf{y}; \tilde{\tau})e^{-|\mathbf{x}-\mathbf{y}|/2\ell}. \tag{4.62}$$

The decay at scales $|\mathbf{x} - \mathbf{y}| \sim 2\ell = v_{\mathrm{F}}(2\tau)$ underlies the identification above.

Q5: Why is the replica method exact in perturbation theory?

A1: Unlike the Fock diagram, where all replica indices are locked to the index of the incoming Green functions, the Hartree diagram contains one free replica summation. This summation yields an excess factor R that, in the limit $R \to 0$, vanishes. For the same reason, all diagrams with closed fermion loops (loops connected to the external field amplitudes only by impurity lines or not at all) do not contribute to the expansion. Technically, the excluded contributions represent vacuum diagrams,[38] i.e., on the level of perturbation theory, the only[39] effect of the replica limit is the

---

[38] To understand this assertion, consider the non-replicated theory prior to the impurity average. Owing to the absence of "real" interactions, any closed fermion loop appearing in the expansion must be a vacuum diagram. After taking the impurity average, the loop may become connected to the external amplitudes by an impurity line. However, it remains a vacuum loop and would cancel against the expansion of the normalization denominator, if we were crazy enough to formulate the numerator–denominator expansion of the theory explicitly.

[39] In a connected diagram, all replica indices are locked to the index a of the external field vertices. We thus obtain (symbolic notation) $\langle G \rangle_{\mathrm{dis}} \sim \lim_{R\to 0} \frac{1}{R}\sum_a \langle \psi^a \bar{\psi}^a \rangle \propto \lim_{R\to 0} \frac{R}{R} \times \mathrm{const.} = \mathrm{const.}$ where the factor R results from the summation over a and we have used the fact that the correlation function $\langle \psi^a \bar{\psi}^a \rangle$ is independent of a (replica symmetry).

elimination of all vacuum processes. We have thus shown that the replica theory exactly simulates the effect of the normalizing partition function present in the denominator of the unreplicated theory (see the discussion of the linked cluster theorem in section 4.1.3). This proves – all on the level of perturbation theory – the equivalence of the representations.

The representation of the disorder-generated interaction (4.57) in momentum space emphasizes the fact that impurity scattering exchanges arbitrary momentum, but not frequency. A straightforward Wick contraction along the lines of our discussion in section 4.3.1 then yields the first-order contribution

$$\Sigma_p^{(1)} = \gamma^2 \sum_{\mathbf{p}'} G_{\mathbf{p}',n}^{(0)} \simeq \gamma^2 \int d\epsilon \, \frac{\nu(\epsilon)}{i\omega_n + E_{\mathrm{F}} - \epsilon} \simeq \gamma^2 \mathrm{P} \int d\epsilon \, \frac{\nu(\epsilon)}{E_{\mathrm{F}} - \epsilon} - i\pi\gamma^2\nu \, \mathrm{sgn}\,(\omega_n),$$

where $\mathrm{P}\int$ stands for the principal value integral. For $d \geq 2$, the increase in the DoS $\nu(\epsilon)$ as a function of ϵ makes the real part of the self-energy formally divergent. This divergence is an immediate consequence of the unbounded summation over $\mathbf{p}'$ – which is an artefact of the model.[40] In any case, the real part of the self-energy is not of prime interest to us: all that $\mathrm{Re}\,\Sigma_{\mathbf{p}n} = \mathrm{const.}$ describes is a frequency- and momentum-independent shift of the energy. By contrast, the imaginary part $\mathrm{Im}\,\Sigma_{\mathbf{p},n}^{(1)} = -\pi\gamma^2\nu \, \mathrm{sgn}\,(\omega_n)$ describes the attenuation of the quasi-particle amplitude due to impurity scattering.

A2: The analysis of the second-order contribution $\Sigma^{(2)}$ parallels our discussion of the RPA in chapter 4: the Green functions $G_{\mathbf{p}}$ are sharply peaked around the Fermi surface $|\mathbf{p}| = p_{\mathrm{F}}$. (Since the Matsubara index n in $p = (\mathbf{p}, n)$ is conserved in impurity scattering, we will not always write it out explicitly.) Representing the diagram with non-crossing lines in momentum space, one obtains $\Sigma_{\mathrm{n.c.}}^{(2)} \sim \sum_{\mathbf{p}_1,\mathbf{p}_2} (G_{\mathbf{p}_1}^{(0)})^2 G_{\mathbf{p}_2}^{(0)}$, restricting both momenta to the Fermi surface, i.e., $|\mathbf{p}_1|, |\mathbf{p}_2| \simeq p_{\mathrm{F}}$. By contrast, the contribution with crossed lines takes the form $\Sigma_{\mathrm{c.}}^{(2)} \sim \sum_{\mathbf{p}_1,\mathbf{p}_2} G_{\mathbf{p}_1}^{(0)} G_{\mathbf{p}_2}^{(0)} G_{\mathbf{p}_2+\mathbf{p}-\mathbf{p}_1}^{(0)}$. Since all three momentum arguments have to be tuned to values close to p_{F}, only one summation runs freely over the Fermi surface. To estimate the relative weight of the two contributions, we need to know the width of the "shell" centered around the Fermi surface in which the Green functions assume sizeable values. Since $|G| = [(E_{\mathrm{F}} - p^2/2m)^2 + (\mathrm{Im}\,G^{-1})^2]^{-1}$, the width of the Lorenzian profile is set by $\mathrm{Im}\,G^{-1}$. As long as we are working with the bare Green function, $\mathrm{Im}\,[G^{(0)}]^{-1} = \omega_n \propto T$ is proportional to the temperature. However, a more physical approach is to anticipate that impurity scattering will broaden the width $\mathrm{Im}\,[G^{(0)}]^{-1} \sim \tau^{-1}$ to a constant value (to be identified shortly as the inverse scattering time). Then, the relative weight of the two diagrams can be estimated as $p_{\mathrm{F}}^{2(d-1)}/(p_{\mathrm{F}}(v_{\mathrm{F}}\tau)^{-1})^{d-1} = (p_{\mathrm{F}}\ell)^{d-1}$, where $\ell = v_{\mathrm{F}}\tau$ is the elastic mean free path and the numerator and denominator estimate the volume in momentum space accessible to the $\mathbf{p}_{1,2}$ summations in the

[40] In reality, the summation will be finite because either (a) there is an underlying lattice structure (i.e., the $\mathbf{p}'$-summation is limited to the Brillouin zone), or (b) the kernel $K(\mathbf{r})$ describing the profile of the impurity potential varies on scales large in comparison with the Fermi wavelength. In this latter case, its Fourier transform has to be added to the definition of the scattering vertex above. The presence of this function then limits the $\mathbf{p}'$-summation to values $|\mathbf{p}' - \mathbf{p}| < \xi^{-1} \ll p_{\mathrm{F}}$.

two diagrams. The important message to be taken away from this discussion is that, for weak disorder $p_F \gg \ell^{-1}$ (which we have assumed throughout), and in dimensions $d > 1$, scattering processes with crossed impurity lines are negligible. Under these conditions, we are entitled to evaluate the self-energy (and, for that matter, all other observables) within the **self-consistent Born approximation** approximation. Since this approximation neglects crossing impurity lines, it is sometimes called the **non-crossing approximation (NCA)** to the weakly disordered electron gas.

non-crossing approx-imation

A3: Drawing on the analogous discussion in section 4.3.2, the SCBA for the self-energy is given by the diagram shown in fig. 4.10(d). The corresponding analytical expression takes the form (cf. Eq. (4.44))

$$\Sigma_{\mathbf{p},n} = \frac{\gamma^2}{L^d} \sum_{\mathbf{p}'} \frac{1}{i\omega_n + E_F - p'^2/2m - \Sigma_{\mathbf{p}',n}}. \tag{4.63}$$

Guided by our results obtained at the first order of perturbation theory, we may seek a solution of (the imaginary part of) this equation by the ansatz $\operatorname{Im}\Sigma_{\mathbf{p},n} = -\mathrm{sgn}(\omega_n)/2\tau$. Substitution of this expression into the SCBA equation gives

$$-\frac{1}{2\tau}\mathrm{sgn}(\omega_n) \simeq \gamma^2 \operatorname{Im}\int d\epsilon \, \frac{\nu(\epsilon)}{i\omega_n + E_F - \epsilon + \frac{i}{2\tau}\mathrm{sgn}(\omega_n)} \simeq -\pi\gamma^2\nu\,\mathrm{sgn}(\omega_n),$$

where we have assumed that $E_F\tau \gg 1$. We have thus arrived at the identification of $\tau^{-1} = 2\pi\nu\gamma^2$ and obtain the important result Eq. (4.61).

A4: The sign function contained in the self-energy of Eq. (4.61) implies that, in the contour integral representation of the Matsubara sum in terms of a complex variable z, we have an extended singularity (a cut) $\mathrm{sgn}(\omega_n) \to \mathrm{sgn}(\operatorname{Im}z)$ at $\operatorname{Re}z = 0$. We thus integrate along the contour indicated in the figure to obtain

$$\langle G(\mathbf{x},\mathbf{y};\tilde{\tau})\rangle_{\mathrm{dis}} = \sum_{\mathbf{p}}\int \frac{d\epsilon}{2\pi} e^{-\epsilon\tilde{\tau}}(1 - n_F(\epsilon))e^{i\mathbf{p}\cdot(\mathbf{x}-\mathbf{y})} \operatorname{Im}\left(\frac{1}{E_F + \epsilon - \frac{p^2}{2m} + \frac{i}{2\tau}}\right).$$

Here, we are using $1 - n_F$ instead of n_F as a pole function because $e^{-z\tau}(1 - n_F(z))$ is finite for large $|z|$, whereas $e^{-z\tau}n_F(z)$ is not. Turning to the momentum integral, we note that the essential difference from the Green function of the clean electron gas is an upgrade of the infinitesimal causality parameter in $\epsilon + i\delta$ to the finite $\delta \to 1/2\tau$. We need not do the momentum integral explicitly to conclude that, for $\epsilon \sim \tilde{\tau}^{-1} \ll E_F$, its poles are located at $p \simeq p_F \pm \frac{i}{2\ell}$. In the evaluation of the integral by the method of residues, the exponen-

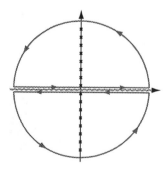

tials $\exp(ip|\mathbf{x} - \mathbf{y}|)$ thus need to be evaluated at $\exp(ip_F|\mathbf{x} - \mathbf{y}| - |\mathbf{x} - \mathbf{y}|/(2\ell))$. Consequently, the disorder-averaged Green function is related to the Green function of the clean system G_{cl} by the exponential damping factor factored out in Eq. (4.62).

A5: We refer to the discussion in A1.

4.4.3 Diffusion

How do local fluctuations in the electron density $\delta\hat{\rho}(x) \equiv \hat{\rho}(x) - \langle\hat{\rho}(x)\rangle$, $x = (\mathbf{r}, \tau)$ relax in a given realization of a random potential? In the following, we consider the correlation function $D(x) \equiv \langle\langle\delta\hat{\rho}(x)\delta\hat{\rho}(0)\rangle\rangle_{\text{dis}}$ to answer this question. A good way to think about it is as a "response function" probing how $\delta\hat{\rho}(x)$ changes in response to a density fluctuation at $0 = (\mathbf{0}, 0)$. Notice that, since $\rho \sim \bar{\psi}\psi$, D represents a four-point function.

INFO Classically, we expect D to show signatures of *diffusion*, a relaxation process that takes place over length scales much larger than the microscopic mean free path, ℓ. To identify the underlying mechanism of "stability" think of $D(x) = \langle\langle\bar{\psi}(x)\psi(0)\rangle \langle\psi(x)\bar{\psi}(0)\rangle\rangle_{\text{dis}}$ as the product of two quantum mechanical amplitudes. The first, $\langle\psi(x)\bar{\psi}(0)\rangle$, describes the propagation of a particle created at 0 to x, and the second, $\langle\bar{\psi}(x)\psi(0)\rangle$, the propagation of a *hole* between the same space–time points. (Formally, the hole amplitude is the complex conjugate of the particle amplitude.)

particle–
hole
propagator

To develop intuition for the behavior of this **particle–hole propagator** in the disordered medium, we temporarily switch to a real-time description $\tau \to it$ and imagine the individual amplitudes represented as sums over Feynman paths weighted by their classical action. Individual paths, α, connect $0 = (\mathbf{0}, 0) \to x = (\mathbf{r}, t)$ and are traversed at energy ϵ. This leads to the symbolic representation

$$D \sim \int d\epsilon\, d\epsilon' \sum_{\alpha\beta} A_\alpha A_\beta^* \exp\left(\frac{i}{\hbar}(S_\alpha(\epsilon) - S_\beta(\epsilon'))\right),$$

where the prefactors A_α are inessential to our discussion. As with the single-particle propagator, the strong sensitivity of the actions $S_{\alpha,\beta}$ on the impurity potential implies that generic contributions (α, β) to the path double sum (fig. 4.11 (a)) vanish upon impurity averaging. By contrast, the "diagonal" contribution $D_{\text{diag}} = \int d\epsilon\, d\epsilon' \sum_\alpha |A_\alpha|^2 \exp(\frac{i}{\hbar}(S_\alpha(\epsilon) - S_\alpha(\epsilon')))$, is positive definite and will survive averaging. A glance at (fig. 4.11 (b)) indicates that it describes classical diffusion in the system.

However, the path double sum provides more than an elaborate quantum mechanical description of classical diffusion: we may expect the presence of select doublets (α, β) formed by topologically distinct paths that even so provide a stable contribution to the pair propagator. For example, the two amplitudes depicted in fig. 4.11 (c) are different but locally propagate along the same path. This indicates that their actions should be approximately identical and hence cancel out in the exponent. Configurations of this loop topology are candidates for stable yet essentially non-classical contributions to the pair propagator. One may easily come up with other loop topologies (try it!) sharing the same qualitative signatures.

The merit of the above representation is that it anticipates the topologies of Feynman diagrams describing classical and non-classical aspects of the disordered electron gas. In the following, we will show how these contributions enter the quantitative description of the particle–hole propagator.

Below, we apply concepts very similar to those developed in section 4.3.2 to understand the spatial long-range character of the four-point function. Specifically,

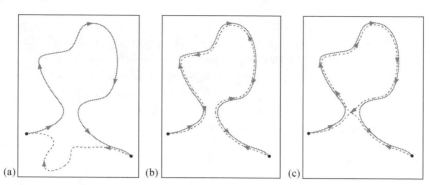

Fig. 4.11 (a) A generic pair of Feynman paths contributing to the density–density correlation function prior to averaging. (b) Particle and hole propagating along the same path in configuration space – the "diffuson." (c) A non-classical contribution to the path sum.

we will elucidate the diffusive character of this correlation function, and identify quantum contributions beyond diffusion.

Q1: Before turning to explicit computations, let us derive two exact relations obeyed by the Fourier transform $D_{\mathbf{q},\omega_m}$: show that $\lim_{\mathbf{q}\to 0} D_{\mathbf{q},0} = L^{-d}\partial_\mu N$ is determined by the **thermodynamic density of states**, $\partial_\mu N$, and $D_{0,\omega_m} = 0$, where $N = \langle \hat{N} \rangle_\psi$ denotes the number of particles in the system and, as usual, $\mu \leftrightarrow E_F$ represents the chemical potential. Explain the origin of these two sum rules. (Hint: Consider particle number conservation.)

thermo-dynamic density of states

Q2: Represent the correlation function as $D(x) = \delta^2_{\mu(x)\mu(0)} \langle \ln \mathcal{Z} \rangle_{\mathrm{dis}}|_{\mu(x)=\mu}$, where $\mu(x)$ is the generalization of the chemical potential to a space–time-dependent source field. As in the previous subsection, express the logarithm in terms of a replica limit and show that the momentum–frequency representation of D resembles that of the correlation function $C^{(4)}$ introduced in section 4.3.2. Using a Fourier convention somewhat different from our standard choice, $\psi(x)=(T/L^d)^{1/2}\sum_p e^{-ip\cdot x}\psi_p$, show that the frequency–momentum representation of D is similar to that of Eq. (4.45).

Q3: To compute the two-particle correlation function, one may apply concepts similar to those introduced in section 4.3.2. In doing so, we will benefit from two major simplifications. First, the large parameter $p_F\ell \gg 1$ plays a role similar to that of N in section 4.3.2. Second, we may make use of the fact that the momentum $\mathbf{q}$ Fourier conjugate to the argument $|\mathbf{r}| \gg \ell$ is much smaller than the Fermi momentum. Show that, under these conditions, the irreducible vertex $\Gamma_{0,q,p,p'} = (2\pi\nu\tau)^{-1}\delta_{\omega_n,\omega'_n}$ collapses to (the Fourier representation of) a single impurity vertex. (Since all field amplitudes that contribute in the replica limit carry the same replica index a, one may drop the replica structure from the notation.) Write down the Bethe–Salpeter equation for the full vertex.

Q4: Denoting the two individually conserved Matsubara frequencies of the two Green functions entering the particle–hole propagators by ω_n and ω_{n+m}, respectively, argue – formally and physically – why the cases of equal and opposite $\text{sgn}(n)$ and $\text{sgn}(n+m)$ must be considered separately. Why is $n(n+m) < 0$ the more interesting case? Assuming opposite signs, we aim to solve the Bethe–Salpeter equation for the vertex $\Gamma_{p,p',q}$ by expansion to leading order in the small parameters $|\mathbf{q}|\ell < 1$, and $|\omega_n|\tau \sim |\omega_{n+m}|\tau \ll 1$. To begin, reason why, for these frequency/momentum values, the vertex will not depend on the "fast" momenta $\mathbf{p}, \mathbf{p}'$. To formulate the expansion, we will need two auxiliary identities: first, apply symmetry considerations to verify $\int d^d p \, f(|\mathbf{p}|)(\mathbf{v} \cdot \mathbf{p})^2 = (v^2/d) \int d^d p \, f(|\mathbf{p}|)\mathbf{p}^2$. Second, apply the theorem of residues to show that

$$L^{-d} \sum_{\mathbf{p}} \left(G_{\mathbf{p}}^+\right)^{n_+ +1} \left(G_{\mathbf{p}}^-\right)^{n_- +1} = 2\pi i^{n_- -n_+} \frac{(n_+ + n_-)!}{n_+! \, n_-!} \nu \tau^{n_+ + n_- +1},$$

where $G_{\mathbf{p}}^{\pm} \equiv (E_{\mathrm{F}} - p^2/2m \pm i/2\tau)^{-1}$. Armed with these identities, expand the Bethe–Salpeter equation and show that $\Gamma_q = (2\pi\nu\tau^2 L^d(|\omega_m| + Dq^2))^{-1}$, where $D = v_{\mathrm{F}}\ell/d$

diffusion constant defines the **diffusion constant** of a disordered metal and $q = |\mathbf{q}|$.

A few remarks about this result are in order. First, note the absence of $(\mathbf{q}, \omega_m)$-independent constants in the denominator, which technically results from the cancellation of two terms in the vertex equation. (We have met with a similar cancellation in our discussion of the vertex of the generalized ϕ^4-theory.) Thanks to this cancellation, $\Gamma(\mathbf{r}, \tau)$ becomes a long-range object. Second, a Fourier transformation to real space shows that Γ is a solution of the diffusion equation $(\partial_\tau - D\nabla^2)\Gamma(\tau, \mathbf{r}) = (2\pi L^d \tau^2)^{-1}\delta^d(\mathbf{r})\delta(\tau)$ (exercise), describing the manner in which a distribution initially centered at $\mathbf{x} = 0$ spreads out in time.

Having the vertex for $\omega_n\omega_{n+m} < 0$ under control, we are now a few steps away from the final result for the density correlation function $D(q)$. What remains to be done is (a) an attachment of Green function legs to the vertex, and (b) the addition of the "empty bubble" (see the second and first term in the first row of fig. 4.8), (c) summation over ω_n, and (d) the addition of the non-singular contribution of Green functions with Matsubara frequencies of equal sign. These calculations do not add much further insight (but are recommended as instructive technical exercises), and we just quote their result. Attachment of external Green functions to the vertex, as in (a) above, brings a multiplicative factor $-L^d(2\pi\nu\tau)^2$, so that we obtain[41]

$$D_{n,n+m}(\mathbf{q}) = -\frac{2\pi\nu}{Dq^2 + |\omega_m|}, \qquad \omega_n\omega_{n+m} < 0, \qquad (4.64)$$

for the (singular contribution) to the impurity-averaged product of two Green functions of different causality. This mode plays an important role as a building block in the analysis of the disordered electron gas and in the literature is called the dif-

diffusion mode fusion mode, or the **diffuson**. The empty bubble, (b), contributes a factor $L^d 2\pi\nu\tau$, which we neglect in comparison to the singular vertex. There are m terms with $\omega_n\omega_{n+m} < 0$ in the sum $T\sum_n(\ldots)$, and so the diffuson enters the final result multiplied by a factor $\omega_m/2\pi$. Finally, one can show that the infinitely many equal-sign

---

[41] The overall minus sign comes from the reordering of Grassmann variables involved in the Wick contraction of (4.65); symbolically $\langle \bar{\psi}\psi\bar{\psi}\psi \rangle \to -GG$.

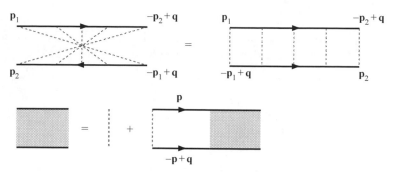

Fig. 4.12   Maximally crossed contribution to the irreducible vertex. Second line: the corresponding Bethe–Salpeter equation.

configurations $\omega_n \omega_{n+m} > 0$ sum to the density of states, ν.[42] Adding everything up, we thus obtain $D(q) = -\nu D q^2 (|\omega_m| + D q^2)^{-1}$ as the final result for our correlation function. Note how this result conforms with the two limiting conditions discussed in (A1) below. (For a non-interacting fermion system, the single particle density of states, $L^d \nu$, equals the thermodynamic one, $\nu = L^{-d} \partial_\mu N$.)

Q5: Can we extend the above analysis to observe signatures of quantum modifications of classical diffusion? A key hint follows from inspection of fig. 4.11(c). While the diffusion-like processes describe the propagation of particles and holes with nearly identical momenta, $\mathbf{p} + \mathbf{q}$ and $\mathbf{p}$, respectively, the situation if the loop is traversed in the opposite direction is different. Here, the momenta are nearly *opposite*, $\mathbf{p} + \mathbf{q}$ and $-\mathbf{p}$. This observation suggests that there must exist a long-range (singular for small $\mathbf{q}$ and ω_m) contribution, Γ^C, to the irreducible vertex representing this type of correlated motion. The diagrammatic representation of Γ^C is shown in fig. 4.12: a sum over diagrams whose "maximally crossed" impurity lines reflect the fact that the particle and hole traverse a sequence of scattering events in reverse order. Formally, this diagram class is significant because it, too, contains one Fermi sphere summation per impurity line. The most economical way to see this is to imagine the lower of the two fermion propagators twisted by 180° (the second diagram in the figure). Superficially, it now resembles the previously explored vertex. The important difference, however, is that the arrows marking the fermion propagators point in the same direction. Proceed as in Q4 to compute Γ^C and show that it equals Eq. (4.64).

Cooperon   In the literature, Γ^C is called the *Cooperon mode* or **Cooperon**.[43] The fact that the Cooperon mode has the same algebraic structure as the diffuson mode indicates that we have identified an important distinction between classical and

[42] If you want to check this assertion, you will need to do the frequency summation before the momentum summation.

[43] The denotation alludes to a similarity with the Cooper pair modes to be discussed in connection with superconductivity in the next chapter. They, too, represent pair correlations of quasi-particles with nearly opposite momenta.

quantum dynamics. However, to fully explore its consequences, we need the more powerful field-theoretical machinery introduced in the next chapter.

Here are the answers to the above five questions.

A1: The functional expectation value of the particle number is given by $\langle N \rangle = \int d\tau \, d^d r \, \langle \hat{\rho}(\mathbf{r}, \tau) \rangle_\psi = \int d^d r \, \langle \hat{\rho}_{\omega_m = 0}(\mathbf{r}) \rangle_\psi$. Differentiating this expression with respect to μ, and noting that the chemical potential couples to the action through the term $\mu \int d^d r \int d\tau \hat{\rho}(\mathbf{r}, \tau)$, we obtain[44]

$$\partial_\mu \langle N \rangle = \int d^d r \, d^d r' \, (\langle \hat{\rho}_0(\mathbf{r}) \hat{\rho}_0(\mathbf{r}') \rangle_\psi - \langle \hat{\rho}_0(\mathbf{r}) \rangle_\psi \langle \hat{\rho}_0(\mathbf{r}') \rangle_\psi]$$

$$= \int d^d r \, d^d r' \, D_{\omega_n = 0}(\mathbf{r} - \mathbf{r}') = L^d \lim_{\mathbf{q} \to 0} D_{\mathbf{q}, 0}.$$

Particle number conservation demands that $\int d^d r \, \langle \delta \hat{\rho}(\mathbf{r}, \tau) \rangle_\psi = 0$ at all times. Consequently, $\int d^d r \, D(\mathbf{r}, \tau) = 0$ or, equivalently, $D_{\mathbf{q} = 0, \omega_m} = 0$.

A2: It is straightforward to verify that the two-fold μ-differentiation above yields the correlation function D. Now let us employ the replica formulation $\langle \ln \mathcal{Z} \rangle_{\mathrm{dis}} = \lim_{R \to 0} \frac{1}{R} \langle \mathcal{Z}^R - 1 \rangle$. Differentiating the right-hand side of this equation, one obtains

$$D(x) = \lim_{R \to 0} \frac{1}{R} \langle \bar{\psi}^a(x) \psi^a(x) \bar{\psi}^b(0) \psi^b(0) \rangle_\psi.$$

To avoid the vanishing of this expression in the limit $R \to 0$, we need to connect operators $\bar{\psi}^a$ and ψ^b ($\bar{\psi}^b$ and ψ^a) by fermion lines (thus enforcing $a = b$ – otherwise the two-fold summation over a and b would produce an excessive factor R which would result in the vanishing of the expression in the replica limit). We thus obtain a structure similar to that discussed in section 4.3.2 (see fig. 4.13): two propagators connecting the points x and 0, where the role of the wavy interaction line of section 4.3.2 is now played by the "interaction" generated by the impurity correlator $\langle VV \rangle$.

Finally, substituting the Fourier representation of $\psi^a(x)$ into the definition of D, using the fact that $\langle \bar{\psi}^a_{\mathbf{p}_1} \psi^a_{\mathbf{p}_1 + q} \bar{\psi}^a_{\mathbf{p}_2 + q'} \psi^a_{\mathbf{p}_2} \rangle \propto \delta_{qq'}$ (momentum conservation in the averaged theory), and that the impurity lines do not exchange frequency, we obtain

$$D_q = \frac{T}{L^d} \lim_{R \to 0} \frac{1}{R} \sum_{\omega_n} \sum_{\mathbf{p}_1 \mathbf{p}_2} \langle \bar{\psi}^a_{\mathbf{p}_1, \omega_n} \psi^a_{\mathbf{p}_1 + \mathbf{q}, \omega_{n+m}} \bar{\psi}^a_{\mathbf{p}_2 + \mathbf{q}, \omega_{n+m}} \psi^a_{\mathbf{p}_2, \omega_n} \rangle_\psi. \tag{4.65}$$

A3: As in our discussion of the self-energy, the condition that only diagrams containing one free summation over the Fermi surface per impurity line contribute to leading order in $(p_F \ell)^{-1}$ eliminates all contributions with crossed impurity lines. Specifically, the only remaining contribution to the irreducible vertex is the single impurity line. As a consequence, the diagrammatic expansion of the correlation

---

[44] From the expressions above it is, in fact, not quite clear why first we set $\omega_n = 0$ and only then $\mathbf{q} = 0$. That this is the correct order of limits can be seen by generalizing $\mu \to \mu(\mathbf{r})$ to a smoothly varying static field, evaluating the corresponding functional derivatives $\delta/\delta\mu(\mathbf{r})$, and setting $\mu = \text{const.}$ at the end of the calculation.

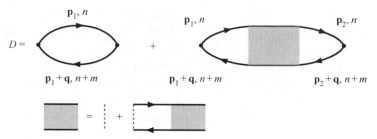

Fig. 4.13   Diagrammatic expansion of the diffuson mode.

function assumes the form shown in fig. 4.13. The Bethe–Salpeter equation for the impurity vertex (shaded in the figure) reads

$$\Gamma_{p_1,p_2,q} = \frac{1}{2\pi L^d \nu \tau} + \frac{1}{2\pi L^d \nu \tau} \sum_{\mathbf{p}} G_{\mathbf{p+q},n+m} G_{\mathbf{p},n} \Gamma_{p,p_2,q}, \qquad (4.66)$$

where the Gs denote the impurity-averaged single-particle Green functions evaluated in the SCBA and discussed in the previous section.

A4: In the solution of the Bethe–Salpeter equation, we encounter momentum integrals over Green functions with Matsubara frequencies ω_n, ω_{n+m}. Comparing with Eq. (4.61), we notice that for $\omega_n \omega_{n+m} > 0$ these Green functions have their poles on the same side of the real axis, and the evaluation of the integrals by the theorem of residues vanishes. Physically, the continuation from positive (negative) Matsubara frequencies to the real axis defines a retarded (advanced) Green function. Products of two propagators of the same causality are rapidly oscillating, which is the physical principle behind the vanishing of the integral.

The expansion of the product of Green functions in the Bethe–Salpeter equation to leading order in the small energies ω_n, ω_{n+m} and in $\mathbf{q} \cdot \mathbf{p}/m$ reads as

$$G_{p+q}G_p = G_{\mathbf{p}}^+ G_{\mathbf{p}}^- \left(1 - iG_{\mathbf{p}}^+ \omega_{n+m} - iG_{\mathbf{p}}^- \omega_n + (G_{\mathbf{p}}^+ \mathbf{q} \cdot \mathbf{p}/m)^2 + \cdots \right), \qquad (4.67)$$

where a term linear in $\mathbf{q}$ has been omitted as it will vanish upon integration over the angular coordinates of $\mathbf{p}$.

We next assume $\Gamma_{p,p',q} \simeq \Gamma_q$. Physically, this is in the anticipation that, for the long time scales corresponding to $|\omega_m|\tau \ll 1$, the diffusion process "forgets" about the direction of the incoming and outgoing microscopic electron trajectories. The integration over $\mathbf{p}$ – now decoupled from the vertex – is then carried out with the help of the two auxiliary identities, and we arrive at the relation

$$\Gamma_q = \frac{1}{2\pi \nu \tau L^d} + \left(1 - \tau \omega_m - q^2 \frac{\tau^2 v_F^2}{d}\right) \Gamma_q.$$

Solving this equation for Γ_q, we obtain the final result for Γ_q stated in the question.

A5: Let us now consider the irreducible vertex in a sector in momentum space where the sum of the upper incoming momentum, $\mathbf{p}_1$, and lower incoming momentum, $-\mathbf{p}_1 + \mathbf{q}$, is small. We can convince ourselves that, in this case, $\mathbf{q}$ is also conserved and each impurity insertion comes with an integration over a fast momentum. Now, consider the lower line turned around in such a way that the diagram assumes the form of a ladder structure. (Note that it still remains "irreducible" because the notion of irreducibility has been defined for fermion lines of opposite orientation.) The corresponding Bethe–Salpeter equation takes the form

$$\Gamma_q^{\mathrm{C}} = \frac{1}{2\pi\nu\tau L^d} + \frac{1}{2\pi\nu\tau L^d} \sum_{\mathbf{p}} G_{-\mathbf{p}+\mathbf{q},n_1+m} G_{-\mathbf{p},n_1} \Gamma_q^{\mathrm{C}}.$$

Again, we assume that the Matsubara frequencies ω_n and ω_{n+m} carried by the upper and lower propagators are of opposite signs. An expansion of the Green functions followed by application of the two auxiliary identities of Q4 then readily leads to the result.

4.5 Summary and Outlook

This concludes our introduction to the concepts of perturbation theory. We have seen that general perturbative expansions mostly have the status of "asymptotic" rather than convergent series. We have learned how to efficiently encode perturbative series by graphical methods and how to assess the "importance" of individual contributions. Further, we have seen how the presence of a large parameter can be utilized to firmly establish infinite-order expansions. A number of recursive techniques have been introduced to sum diagram sequences of infinite order.

However, a second look at the discussion in the previous sections shows that the central tool, the functional integral, did not play much of a role. All it did was to provide the combinatorial framework of the perturbative expansion of correlation functions. However, for that we hardly need the full machinery of functional integration. Indeed, the foundations of the perturbative approach were laid down in the 1950s, long before people even began to think about the conventional path integral. (For a pure operator construction of the perturbative expansion, see the problem set.)

More importantly, the analysis so far has a serious methodological weakness: all subclasses of relevant diagrams have the common feature that they contain certain sub-units, more structured than the elementary propagator or the interaction line. For example, the RPA diagrams are organized in terms of polarization bubbles, the NCA diagrams have their rainbows, and the ladder diagrams their rungs. Within the diagrammatic approach, in each diagram these units are reconstructed from scratch, i.e., in terms of elementary propagators and interaction lines. However, taking seriously the general philosophy on information reduction declared at the beginning of the text, we should strive to make the "important" structural elements of an expansion our *elementary* degrees of freedom. This program is hardly feasible within purely diagrammatic theory. However, the functional integral is ideally suited to introducing degrees of freedom hierarchically, i.e., trading microscopic objects for entities of higher complexity. The combination of functional integral techniques with perturbation theory presents a powerful theoretical machinery, which is the subject of the next chapter.

4.6 Problems

4.6.1 Technical aspects of diagrammatic perturbation theory

Taking the second-order expansion of the ϕ^4 Green function as an example, the aim of this problem is to discuss a number of technical aspects relating to the classification and the combinatorics of diagrams.

(a) In the ϕ^4-theory, show that the full contraction of $\langle \phi(\mathbf{x})\phi^4(\mathbf{y})\phi^4(\mathbf{y}')\phi(\mathbf{x}') \rangle$ generates the 945 terms

$$\underbrace{9G_0(\mathbf{x}-\mathbf{x}')G_0^4(\mathbf{0})}_{1} + \underbrace{72G_0(\mathbf{x}-\mathbf{x}')G_0^2(\mathbf{y}-\mathbf{y}')G_0^2(\mathbf{0})}_{2} + \underbrace{24G_0(\mathbf{x}-\mathbf{x}')G_0^4(\mathbf{y}-\mathbf{y}')}_{3}$$

$$+\Bigg(\underbrace{36G_0(\mathbf{x}-\mathbf{y})G_0(\mathbf{x}'-\mathbf{y})G_0^3(\mathbf{0})}_{4} + \underbrace{144G_0(\mathbf{x}-\mathbf{y})G_0(\mathbf{x}'-\mathbf{y})G_0^2(\mathbf{y}-\mathbf{y}')G_0(\mathbf{0})}_{5}$$

$$+\underbrace{96G_0(\mathbf{x}-\mathbf{y})G_0(\mathbf{x}'-\mathbf{y}')G_0^3(\mathbf{y}-\mathbf{y}')}_{6}$$

$$+\underbrace{144G_0(\mathbf{x}-\mathbf{y})G_0(\mathbf{x}'-\mathbf{y}')G_0(\mathbf{y}-\mathbf{y}')G_0^2(\mathbf{0})}_{7} +(\mathbf{y} \longleftrightarrow \mathbf{y}'). \Bigg) \qquad (4.68)$$

Try to reproduce the combinatorial prefactors. (b) Check that the disconnected terms cancel the "vacuum loops" from the expansion of the denominator. (c) Represent the corresponding diagrams in momentum space. Convince yourself that the "Kirchhoff law" discussed in the main text suffices to unambiguously fix the result.

Answer:

(a) The seven distinct pairings of the ten ϕ-fields are shown in fig. 4.2 (and referred to here as diagrams 1–7 in the order in which they appear in the figure). By way of example, let us consider the combinatorial factor of diagram 2. There are $3\cdot2 = 6$ ways to pair two fields $\phi(\mathbf{y})$ and the same for $\phi(\mathbf{y}')$. From these two contractions, we obtain a factor $(6G(\mathbf{0}))^2$. Then, there are two possible ways of pairing the two remaining $\phi(\mathbf{y})$ with the two remaining $\phi(\mathbf{y}')$ leading to a factor $2G_0^2(\mathbf{y} - \mathbf{y}')$. Finally, the pairing of $\phi(\mathbf{x})$ with $\phi(\mathbf{x}')$ gives a single factor $G_0(\mathbf{x} - \mathbf{x}')$. Multiplying all contributions, we obtain term 2. (b) To obtain all terms contributing to

$G^{(2)}$, both numerator and denominator of Eq. (4.14) have to be expanded to second order (where the identification $X[\phi] = \phi(\mathbf{x})\phi(\mathbf{x}')$ is understood). In symbolic notation,

$$\frac{N_0 + N_1 + \frac{1}{2}N_2 + \cdots}{1 + D_1 + \frac{1}{2}D_2 + \cdots} \simeq N_0 + \underbrace{N_1 - N_0 D_1}_{G^{(1)}} + \underbrace{\frac{1}{2}N_2 - N_1 D_1 + N_0 D_1^2 - \frac{1}{2}N_0 D_2}_{G^{(2)}}.$$

The expression for N_2 is given by formula (4.68). From the main text, we know that $N_1 = 3G_0(\mathbf{x} - \mathbf{x}')G_0^2(\mathbf{0}) + 12G_0(\mathbf{x} - \mathbf{y})G_0(\mathbf{0})G_0(\mathbf{y} - \mathbf{x}')$ and $D_1 = 3G_0^2(\mathbf{0})$. Further, $N_0 = G_0(\mathbf{x} - \mathbf{x}')$. Finally, D_2 reads

$$\langle \phi^4(\mathbf{y})\phi^4(\mathbf{y}') \rangle = 9G_0^4(\mathbf{0}) + 72G_0^2(\mathbf{0})G_0^2(\mathbf{y} - \mathbf{y}') + 24G_0^4(\mathbf{y} - \mathbf{y}').$$

Collecting all the terms, we obtain

$$N_2 - 18G_0(\mathbf{x} - \mathbf{x}')G_0^4(\mathbf{0}) \underbrace{-72G_0(\mathbf{x} - \mathbf{y})G_0^3(\mathbf{0})G_0(\mathbf{y} - \mathbf{x}')}_{-(4 + \mathbf{y} \leftrightarrow \mathbf{y}')} + 18G_0(\mathbf{x} - \mathbf{x}')G_0^4(\mathbf{0})$$

$$\underbrace{-9G_0(\mathbf{x} - \mathbf{x}')G_0^4(\mathbf{0})}_{-1} \underbrace{-72G_0(\mathbf{x} - \mathbf{x}')G_0^2(\mathbf{0})G_0^2(\mathbf{y} - \mathbf{y}')}_{-2}$$

$$\underbrace{-24G_0(\mathbf{x} - \mathbf{x}')G_0^4(\mathbf{y} - \mathbf{y}')}_{-3}$$

$$= \underbrace{144G_0(\mathbf{x} - \mathbf{y})G_0(\mathbf{x}' - \mathbf{y})G_0^2(\mathbf{y} - \mathbf{y}')G_0(\mathbf{0})}_{5}$$

$$+ \underbrace{144G_0(\mathbf{x} - \mathbf{y})G_0(\mathbf{x}' - \mathbf{y}')G_0(\mathbf{y} - \mathbf{y}')G_0^2(\mathbf{0})}_{6}$$

$$+ \underbrace{96G_0(\mathbf{x} - \mathbf{y})G_0(\mathbf{x}' - \mathbf{y}')G_0^3(\mathbf{y} - \mathbf{y}')}_{7}$$

$$+ (\mathbf{y} \longleftrightarrow \mathbf{y}'),$$

i.e., the set of connected diagrams shown in fig. 4.4. (c) This translates to a straightforward exercise in Fourier transformation.

4.6.2 Self-consistent T-matrix approximation

In section 4.4, we described the effects of static impurities on the electron gas within an effective model with Gaussian-distributed random potential. Here, we discuss an alternative approach, modeling impurities as dilute strong scattering centers in real space. Although many of the results obtained from these formulations are similar, it is sometimes preferable, or even necessary, to employ one or the other. The concrete aim of the present problem is to determine the scattering rate imposed by a collection of isolated impurities using diagrammatic perturbation theory.

Consider a system of spinless electrons subject to a random pattern of $N_{\rm imp}$ non-magnetic scattering impurities and described by the Hamiltonian $\hat{H} = \hat{H}_0 + \hat{H}_{\rm imp}$, where $\hat{H}_0 = \sum_{\mathbf{k}} \epsilon_{\mathbf{k}} c_{\mathbf{k}}^\dagger c_{\mathbf{k}}$,

$$\hat{H}_{\rm imp} = \gamma \int d^d r \sum_{i=1}^{N_{\rm imp}} \delta(\mathbf{r} - \mathbf{R}_i) c^\dagger(\mathbf{r}) c(\mathbf{r}) = \gamma \sum_{i=1}^{N_{\rm imp}} c^\dagger(\mathbf{R}_i) c(\mathbf{R}_i),$$

and we have assumed that an impurity at position $\mathbf{R}_i$ creates a local potential $\gamma \delta(\mathbf{r} - \mathbf{R}_i)$ whose strength is determined by the parameter γ (of dimensionality energy/volume). Our aim is to compute the Green function $G_n(\mathbf{r} - \mathbf{r}') \equiv \langle\langle c_n^\dagger(\mathbf{r}) c_n(\mathbf{r}') \rangle\rangle_{\rm imp}$, where n is a Matsubara frequency index and the configurational average $\langle \ldots \rangle_{\rm imp} \equiv L^{-dN_{\rm imp}} \prod_i \int d^d R_i$ is defined by integration over all impurity coordinates.

(a) To begin, let us consider scattering from a single impurity, i.e., $N_{\rm imp} = 1$. By developing a perturbative expansion in the impurity potential, show that the Green function can be written as $G_n = G_{0,n} + G_{0,n} T_n G_{0,n}$, where

$$T_n = \langle H_{\rm imp} + H_{\rm imp} G_{0,n}(\mathbf{0}) H_{\rm imp} + H_{\rm imp} G_{0,n}(\mathbf{0}) H_{\rm imp} G_{0,n}(\mathbf{0}) H_{\rm imp} + \cdots \rangle_{\rm imp} \quad (4.69)$$

denotes the T-**matrix** and $H_{\rm imp} = \gamma \sum_i \delta(\mathbf{r} - \mathbf{R}_i)$ is the first-quantized representation of the impurity Hamiltonian. Show that the T-matrix equation is solved by $T_n = L^{-d} \left(\gamma^{-1} - G_{0,n}(\mathbf{0}) \right)^{-1}$.

T-matrix | **INFO** In the present problem, the T-**matrix** actually comes out as a c-number. More generally, for a Hamiltonian $H = H_0 + V$ split into a free part and an "interaction," the T-matrix is implicitly defined through $G = G + G_0 T G$, where $G_0 = G(z) = (z - H)^{-1}$ and $G_0 = (z - H_0)^{-1}$. Comparison with the geometric series expansion of G in V shows that this is equivalent to the recursion relation

$$T = V + V G_0 T. \qquad (4.70)$$

For a general interaction operator, T is an operator (possibly of infinite range), hence the denotation T-*matrix*. The denotation also hints at a close relation to the scattering S-matrix; the two objects carry essentially the same information. In the physics of scattering or that of quantum impurity problems, the T-matrix is a popular object to work with because it describes the "essence" of the scattering process via the recursion relation (4.70), which involves only V in the numerator and G_0 in the denominator. This architecture is convenient for, e.g, the splitting of Hilbert space into physically distinct sectors in the solution of T-matrix equations (see the later problem 6.7.3 for an illustration).

For an in-depth discussion of the T-matrix formalism, we refer to textbooks on scattering theory.

(b) For a collection of random impurities, the bookkeeping in terms of a series expansion as in Eq. (4.69) becomes cumbersome. Instead, we turn to a diagrammatic representation as in fig. 4.14, where different crosses represent different impurities. In this series, we identify two types of diagrams: those with crossing impurity lines, and those without. Argue why the former class becomes irrelevant upon averaging and in the limit of weak disorder (work in momentum space). What is the physical reason for their relative smallness (best argued in real space)? Convince

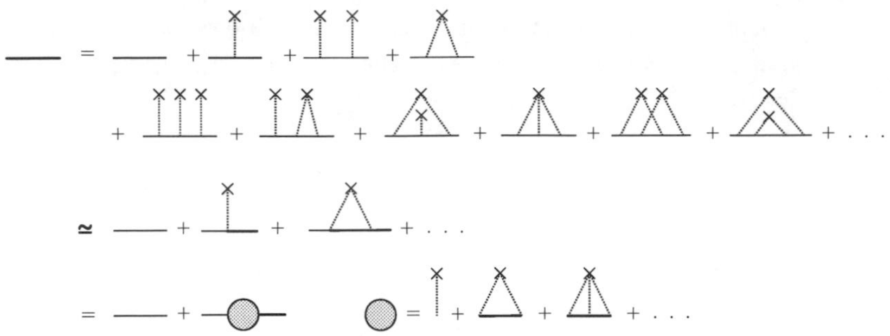

Fig. 4.14   Diagrammatic series expansion of the impurity-averaged Green function (for a discussion, see the main text).

yourself that the series is self-consistently summed by the representation in the final line. Identifying the self-energy with the circle, the algebraic representation of this expression reads

$$G_k = \frac{1}{G_{0,k}^{-1} - \Sigma_n}, \qquad \Sigma_n = T_n \equiv n_{\text{imp}} \frac{1}{\gamma^{-1} - G_n(\mathbf{0})}, \qquad (4.71)$$

T-matrix approximation

where $n_{\text{imp}} = N_{\text{imp}}/L^d$ is the impurity concentration. In the literature, this result is known as the **self-consistent T-matrix approximation (SCTA)**. Comparing with the T-matrix of the single impurity, the difference is that the T-matrix of the dilute impurity system makes reference to the full Green function, which in turn contains the T-matrix. Hence the attribute self-consistent.

(c) As in the case of the Gaussian-distributed potential, the impurity self-energy defines the scattering time, τ. Using the operator identity $(1-\hat{O}_2)^{-1}-(1-\hat{O}_1)^{-1} = (1-\hat{O}_2)^{-1}(\hat{O}_2-\hat{O}_1)(1-\hat{O}_1)^{-1}$, and defining the real-frequency analytical continuation $\hat{T}^{\pm}(\epsilon) \equiv \hat{T}(i\omega_n \to \epsilon \pm i\delta)$ (and the same for $\hat{G}$), show that

$$T^+(\epsilon) - T^-(\epsilon) = T^-(\epsilon)(\hat{G}^+(\epsilon,\mathbf{0}) - \hat{G}^-(\epsilon,\mathbf{0}))T^+(\epsilon) = -2\pi i |T^+(\epsilon)|^2 \nu,$$

where, as usual, $\nu = \nu(\epsilon)$ denotes the density of states per unit volume. This leads to the "Golden Rule" result $1/2\tau \equiv -\text{Im}\,\Sigma^+ = \pi|T^+|^2 n_{\text{imp}}\nu$, expressing the scattering rate as a product of the transition rate $\sim |T|^2$ and density of states.

Answer:

(a) A formal expansion of the Green function $G_n = (i\omega_n - H_0 - H_{\text{imp}})^{-1}$ in H_{imp} leads immediately to the series (4.69). Substitution of $H_{\text{imp}} = \gamma\delta(\mathbf{r} - \mathbf{R})$, where $\mathbf{R}$ is the impurity position, we obtain $T = \langle \delta(\mathbf{r} - \mathbf{R})\delta(\mathbf{r}' - \mathbf{R})\rangle_{\text{imp}}\gamma(1 - \gamma G_{0,n}(\mathbf{0},\mathbf{0}))^{-1}$. Performing the average over $\mathbf{R}$ we arrive at the result. (b) The diagrams in fig. 4.14 are in one-to-one relation to the different terms appearing when the expansion from (a) is generalized to multiple impurity Hamiltonians. Among them, there appear diagrams with crossing lines, such as number 5 in the second line. When averaged over the independent positions of the crosses, constraints in momentum space appear which violate the "one integration per impurity line rule." In real space, the crossing fourth-order diagram corresponds to a scattering path $\mathbf{R}_1 \to \mathbf{R}_2 \to \mathbf{R}_1 \to \mathbf{R}_2$, which for large impurity separation has less phase space than the competing sequence $\mathbf{R}_1 \to \mathbf{R}_1 \to \mathbf{R}_2 \to \mathbf{R}_2$. We arrive at the self-consistent T-matrix relation by straightforward diagrammatic manipulations, as indicated in the bottom part of fig. 4.14. (c) is obtained straightforwardly.

4.6.3 Kondo effect: perturbation theory

In problem 2.4.7 we introduced the Kondo effect, describing the interaction of a local impurity with an itinerant band of carriers. (Those unfamiliar with the physical context and background to the problem are referred back to that section.) There we determined an effective Hamiltonian for the coupled system, describing the spin exchange interaction that acts between the local moment of the impurity state and the itinerant band. In the following, motivated by the seminal work of Kondo,[45] we employ methods of perturbation theory to explore the impact of magnetic fluctuations on transport. In doing so, we elucidate the mechanism responsible for the observed temperature dependence of the electrical resistance found in magnetic impurity systems.

The perturbation theory of the Kondo effect is one of those problems where the traditional formalism of second-quantized operators is superior to the field integral (the reason being that perturbative manipulations on quantum spins are difficult to formulate in a field integral language). More specifically, the method of choice would be second-quantized perturbation theory formulated in the interaction picture, a concept discussed in great detail in practically any textbook on many-body perturbation theory but not in this text. For this reason, the solution scheme discussed below is somewhat awkward and not very efficient (yet fully sufficient for understanding the essence of the phenomenon).

The phenomenology of the Kondo system can be explored in several different ways, ranging from the exact analytical solution of the Kondo Hamiltonian system[46] to a variational analysis of the Anderson impurity Hamiltonian.[47] In the following, we focus on the perturbative scheme developed in the original study of Kondo. Later, in problem 6.7.3, we will introduce a more advanced approach based on the renormalization group. The starting point of the analysis is the effective sd–Hamiltonian (2.50) introduced in problem 2.4.7. Setting $\hat{H}_{\mathrm{sd}} = \hat{H}_0 + \hat{H}_{\mathrm{imp}}$, where

$$\hat{H}_0 = \sum_{\mathbf{k}\sigma} \epsilon_{\mathbf{k}} c_{\mathbf{k}\sigma}^\dagger c_{\mathbf{k}\sigma}, \quad \hat{H}_{\mathrm{imp}} = 2J\hat{\mathbf{S}} \cdot \hat{\mathbf{s}}(\mathbf{r} = 0),$$

and $\hat{\mathbf{s}}(0) = \frac{1}{2} \sum_{\mathbf{k}\mathbf{k}'\sigma\sigma'} c_{\mathbf{k}\sigma}^\dagger \boldsymbol{\sigma}_{\sigma\sigma'} c_{\mathbf{k}'\sigma'}$, our aim is to develop a perturbative expansion in J to explore the scattering properties of the model. Here we have assumed that

[45] J. Kondo, *Resistance minimum in dilute magnetic alloys*, Progr. Theor. Phys. **32**, 37 (1964).

[46] N. Andrei, *Diagonalization of the Kondo Hamiltonian*, Phys. Rev. Lett. **45**, 379 (1980), P. B. Wiegmann, *Exact solution of the s–d exchange model (Kondo problem)*, J. Phys. C **14**, 1463 (1981).

[47] C. M. Varma and Y. Yafet, *Magnetic susceptibility of mixed-valence rare-earth compounds*, Phys. Rev. B **13**, 2950 (1976).

the exchange constant J is characterized by a single parameter, positive in sign (i.e., antiferromagnetic).

In problem 4.6.2 (c), we saw that the scattering rate associated with a system of impurities is specified by the T-matrix. Now, we are dealing with an impurity at a fixed position (no averaging). In this case, the impurity scattering does not conserve momentum, so that the scattering rate is given by

$$\frac{1}{2\tau} = \pi \frac{n_{\text{imp}}}{L^d} \sum_{\mathbf{k}',\sigma'} \langle |\langle \mathbf{k}, \sigma | T^+(\epsilon) | \mathbf{k}', \sigma' \rangle|^2 \rangle_{\text{S}} \, \delta(\epsilon_{\mathbf{k}} - \epsilon_{\mathbf{k}'}),$$

where the symbol $\langle \cdots \rangle_{\text{S}} \equiv \text{tr}_{\text{S}}(\cdots)/\text{tr}_{\text{S}}(\mathbf{1})$ indicates that the calculation of the electron self-energy implies an average over all configurations of the impurity spin. (Indeed, it is the presence of an internal impurity degree of freedom which makes the problem distinct from conventional disorder scattering.) The scattering time is proportional to the electric conductivity, and in this way experimentally observable.

(a) Show that to leading order in the exchange constant J the scattering rate at $\epsilon = \epsilon_F$ is given by

$$\frac{1}{2\tau} = \pi c_{\text{imp}} \nu J^2 S(S+1),$$

where ν denotes the density of states at the Fermi level. From this result, one can infer a temperature independent resistivity; hence, we have not yet explained the essence of the effect.

(b) At second order in the expansion in $\hat{H}_{\text{imp}}$, the T-matrix assumes the form $\hat{T}^{(2)} = \hat{H}_{\text{imp}}(\epsilon^+ - \hat{H}_0)^{-1}\hat{H}_{\text{imp}}$. Assuming for simplicity that we are working at zero temperature, $|\mathbf{k}, \sigma\rangle = c^\dagger_{\mathbf{k}\sigma}|\Omega\rangle$ is a configuration where a single particle with momentum $\mathbf{k}$, $|\mathbf{k}| > p_{\text{F}}$, and spin σ is superimposed on the filled Fermi sphere. Convince yourself that the state $\hat{H}_{\text{imp}}|\mathbf{k}, \sigma\rangle$ is a linear combination of a single-particle state and a two-particle–one-hole state (see fig. 4.15). Determine the excitation energies of the two configurations. Use this result to compute $(\epsilon_{\mathbf{k}}^+ - \hat{H}_0)^{-1}\hat{H}_{\text{imp}}|\mathbf{k}, \sigma\rangle$. Then show that the real part of the second-order matrix element is given by

$$\text{Re}\, \langle \mathbf{k}', \sigma' | \hat{T}^{(2)} | \mathbf{k}, \sigma \rangle = J^2 \sum_{\mathbf{p}}' \frac{1}{\xi_{\mathbf{k}} - \xi_{\mathbf{p}}} \left(S(S+1) - \mathbf{S} \cdot \sigma_{\sigma\sigma'}(\Theta(\xi_{\mathbf{p}}) - \Theta(-\xi_{\mathbf{p}})) \right).$$

$$(4.72)$$

Hint: (i) In deriving this result, all vacuum contributions have to be discarded (since they cancel the partition function denominator of the properly normalized perturbation series). The defining property of a vacuum contribution is that it factorizes into two independent ground state matrix elements. For example, $\langle \Omega | c_{\mathbf{k}\sigma} \hat{H}_{\text{imp}}(\epsilon^+ - \hat{H}_0)^{-1} c^\dagger_{\mathbf{k}'\sigma'} |\Omega\rangle \langle \Omega | \hat{H}_{\text{imp}} |\Omega\rangle$ is of this type. (ii) Note that $(\hat{\mathbf{S}} \cdot \sigma)^2 = \hat{\mathbf{S}}^2 - \sigma \cdot \hat{\mathbf{S}}$. (iii) Taking the real part of the T-matrix amounts to omitting the infinitesimal imaginary increment in the energy denominators.

At finite temperature, the Pauli blocking factors generalize to Fermi functions, $\Theta(\epsilon) \to (1 - n_{\text{F}}(\epsilon))$ and $\Theta(-\epsilon) \to n_{\text{F}}(\epsilon)$, and we obtain

$$\text{Re}\langle \mathbf{k}', \sigma' | \hat{T}^{(2)} | \mathbf{k}, \sigma \rangle = J^2 \sum_{\mathbf{p}} \frac{1}{\xi_{\mathbf{k}} - \xi_{\mathbf{p}}} \left(S(S+1)\delta_{\sigma'\sigma} + (2n_{\text{F}}(\xi_{\mathbf{p}}) - 1)\hat{\mathbf{S}} \cdot \sigma_{\sigma'\sigma} \right).$$

$$(4.73)$$

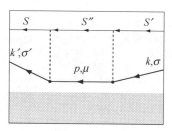 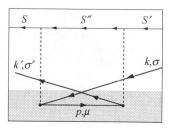

Fig. 4.15   The two spin-scattering processes contributing to the electron T-matrix at second-order perturbation theory. The second process involves a two-particle–one-hole configuration as an intermediate state.

Neglecting the first (non-singular, why?) contribution, one may note that the second can be absorbed into a renormalization of the term first order in J derived above, $1/(2\tau) = \pi\nu c_{\text{imp}} J_{\text{eff}}^2 S(S+1)$, where $J_{\text{eff}} = J(1 + 2Jg(\epsilon, T))$ and the function $g(\epsilon, T) = \frac{\nu}{2}\mathcal{P}\int_{-D}^{D} d\xi\, \frac{\tanh(\beta\xi/2)}{\xi - \epsilon}$ depends sensitively on the bandwidth $2D$ of the itinerant electrons and on the energy $\epsilon \equiv \epsilon_{\mathbf{k}}$ of the reference state. Noting that

$$\lim_{\epsilon \to 0} g(\epsilon, 0) = \nu \ln\left|\frac{D}{\epsilon}\right|, \qquad \lim_{T \to 0} g(0, T) = \nu \ln\left|\frac{D}{k_B T}\right|,$$

the effective exchange constant can be written as

$$J_{\text{eff}} = J\left(1 + 2\nu J \ln\left|\frac{D}{\max(\epsilon, T)}\right|\right).$$

On substituting into the expression for the scattering rate, one finds that the resistivity diverges logarithmically with temperature,

$$\frac{1}{\tau(T)} \simeq \frac{1}{\tau_0}\left(1 - 4\nu J \ln\left(\frac{T}{D}\right)\right),$$

which is the prime perturbative signature of the Kondo effect.

Although the perturbation theory suggests a divergence of the resistivity with temperature, the result remains valid only up to the characteristic **Kondo temperature** scale,

Kondo temperature

$$T_K = D \exp\left(-\frac{1}{2\nu J}\right).$$

At this temperature, the logarithmic correction becomes comparable to the first-order term, signaling a breakdown of perturbation theory. Experimentally, it is found that, at temperatures $T \ll T_K$, the resistance saturates. The origin of this saturation is that electrons in the itinerant band combine with the electron on the impurity site to form a singlet and in this way effectively screen the magnetic impurity.

INFO   Although it was originally conceived for the problem of magnetic impurities in metals, effects of **Kondo resonance formation** have been observed in artificial

quantum dot structures.[48] Here a microscopic quantum dot (with dimensions of about $1\,\mu\text{m}$) is sandwiched between two metallic leads. In the so-called Coulomb blockade regime (see problem 5.6.4), the charging energy of the dot plays the role of the local Hubbard interaction while the leads act as the Fermi sea. The development of a Kondo resonance below T_K appears as a signature in the quantum transport through the dot. In particular, for temperatures $T < T_K$, the differential conductance dI/dV shows a peak corresponding to the suppression of scattering off the impurity state.

Answer:

(a) To leading order in $\hat{H}_{\text{imp}}$, the T-matrix is given by $\hat{T} \simeq \hat{T}^{(1)} = \hat{H}_{\text{imp}}$. Using the facts that $\langle \mathbf{k}', \sigma' | \hat{H}_{\text{imp}} | \mathbf{k}, \sigma \rangle = J L^{-d} \langle \sigma | \mathbf{S} \cdot \sigma | \sigma' \rangle$, we then obtain

$$\frac{1}{2\tau} \simeq \pi N_{\text{imp}} \sum_{\mathbf{k}', \sigma'} \left\langle |\langle \mathbf{k}, \sigma | \hat{H}_{\text{imp}} | \mathbf{k}', \sigma' \rangle|^2 \right\rangle_S \delta(\epsilon_{\mathbf{k}} - \epsilon_{\mathbf{k}'})$$

$$= \pi c_{\text{imp}} J^2 L^{-d} \sum_{\sigma'} \langle\langle \sigma | \mathbf{S} \cdot \sigma | \sigma' \rangle \langle \sigma' | \mathbf{S} \cdot \sigma | \sigma \rangle\rangle_S \sum_{\mathbf{k}} \delta(\epsilon_{\mathbf{k}} - \epsilon_{\mathbf{k}'})$$

$$= \pi c_{\text{imp}} \nu J^2 \langle\langle \sigma | \mathbf{S} \cdot \sigma \mathbf{S} \cdot \sigma | \sigma \rangle\rangle_S = \pi J^2 S(S+1) c_{\text{imp}} \nu,$$

where in the last line we have used the fact that $\sum_{\mathbf{k}} \delta(\epsilon_{\mathbf{k}} - \epsilon_{\mathbf{k}'}) = L^d \nu$ and $\langle(\mathbf{S} \cdot \sigma)(\mathbf{S} \cdot \sigma)\rangle_S = S(S+1) \cdot \mathbf{1}$.

(b) Substituting the explicit form of $\hat{H}_{\text{imp}}$ and using the anticommutation relations of fermions, we obtain (summation convention)

$$\hat{H}_{\text{imp}} | \mathbf{k}', \sigma' \rangle = J \sum_{\mathbf{p}_1 \mathbf{p}_2} c^\dagger_{\mathbf{p}_1 \mu_1} (\mathbf{S} \cdot \sigma_{\mu_1 \mu_2}) c_{\mathbf{p}_2 \mu_2} c^\dagger_{\mathbf{k}', \sigma'} | \Omega \rangle$$

$$= J \sum_{\mathbf{p}_1} (\mathbf{S} \cdot \sigma_{\mu_1 \sigma'}) c^\dagger_{\mathbf{p}_1 \mu_1} | \Omega \rangle - J \sum_{\mathbf{p}_1 \mathbf{p}_2} (\mathbf{S} \cdot \sigma_{\mu_1 \mu_2}) c^\dagger_{\mathbf{p}_1 \mu_1} c^\dagger_{\mathbf{k}' \sigma'} c_{\mathbf{p}_2 \mu_2} | \Omega \rangle,$$

i.e., a linear combination of a one-particle state and a two-particle–one-hole state. Noting that the energies of the two contributions are given by $\epsilon_{\mathbf{p}_1}$ and $\epsilon_{\mathbf{p}_1} + \epsilon_{\mathbf{k}'} - \epsilon_{\mathbf{p}_2}$, respectively, multiplying by the "bra" $\langle \mathbf{k}, \sigma | \hat{H}_{\text{imp}}$, and observing that the overlap between a one-particle state and a two-particle–one-hole state vanishes, we obtain

$$\text{Re}\langle \mathbf{k}, \sigma | \hat{T}^{(2)} | \mathbf{k}', \sigma' \rangle = J^2 (\mathbf{S} \cdot \sigma_{\sigma \mu_1'})(\mathbf{S} \cdot \sigma_{\mu_1 \sigma'}) \sum_{\mathbf{p}_1 \mathbf{p}_1'} \frac{\langle \Omega | c_{\mathbf{p}_1' \mu_1'} c^\dagger_{\mathbf{p}_1 \mu_1} | \Omega \rangle}{\epsilon_{\mathbf{k}} - \epsilon_{\mathbf{p}_1}}$$

$$+ J^2 (\mathbf{S} \cdot \sigma_{\mu_2' \mu_1'})(\mathbf{S} \cdot \sigma_{\mu_1 \mu_2}) \sum_{\mathbf{p}_1 \mathbf{p}_2 \mathbf{p}_1' \mathbf{p}_2'} \frac{\langle \Omega | c^\dagger_{\mathbf{p}_2' \mu_2'} c_{\mathbf{k}\sigma} c_{\mathbf{p}_1' \mu_1'} c^\dagger_{\mathbf{p}_1 \mu_1} c^\dagger_{\mathbf{k}' \sigma'} c_{\mathbf{p}_2 \mu_2} | \Omega \rangle}{\epsilon_{\mathbf{k}} - \epsilon_{\mathbf{p}_1} - \epsilon_{\mathbf{k}'} + \epsilon_{\mathbf{p}_2}}$$

$$\to J^2 (\mathbf{S} \cdot \sigma_{\sigma \mu})(\mathbf{S} \cdot \sigma_{\mu \sigma'}) \sum_{\mathbf{p}} \frac{\Theta(\xi_{\mathbf{p}})}{\epsilon_{\mathbf{k}} - \epsilon_{\mathbf{p}}} + J^2 (\mathbf{S} \cdot \sigma_{\mu \sigma'})(\mathbf{S} \cdot \sigma_{\sigma \mu}) \sum_{\mathbf{p}} \frac{\Theta(-\xi_{\mathbf{p}})}{\epsilon_{\mathbf{k}} - \epsilon_{\mathbf{p}}},$$

where the arrow indicates that vacuum contributions have been discarded. Application of the spin identity then leads directly to Eq. (4.72).

---

[48] D. Goldhaber-Gordon, H. Shtrikman, D. Mahalu *et al.*, *Kondo effect in a single-electron transistor*, Nature **391**, 156 (1998).

5 Broken Symmetry and Collective Phenomena

SYNOPSIS Previously, we have seen how perturbative descriptions of weakly interacting theories may be formulated in a manner essentially detached from the field integral. (The same formalism could have been, and has been, developed within the framework of the "old" quantum field theory and second quantization.) In this chapter, we will learn how this approach becomes stronger, both conceptually and methodologically, when integrated more tightly into the field integral framework. In doing so, we will see how the field integral provides a method for identifying and exploring nontrivial reference ground states – "mean fields." A fusion of perturbative and mean field methods will provide us with analytical machinery powerful enough to address a rich variety of applications including superfluidity and superconductivity, metallic magnetism, and the interacting and disordered electron gas.

As mentioned in chapter 4, the perturbative machinery is but one part of a larger framework. In fact, the diagrammatic series already contained hints indicating that a straightforward expansion of a theory in the interaction operator might not always be an optimal strategy: all previous examples that contained a large parameter N – and usually it is only problems of this type that are amenable to controlled analytical analysis – shared the property that the diagrammatic analysis bore structures of higher complexity. (For example, series of polarization operators appeared rather than series of the elementary Green functions, etc.) This phenomenon suggests that large-N problems should qualify for a more efficient and, indeed, a more physical formulation.

While these remarks might appear to be largely methodological, the rationale behind searching for an improved theoretical formulation is much deeper. With our previous examples, the perturbative expansion was benign. However, we already saw some glimpses indicating that more drastic things may happen. For example, for frequencies approaching the plasma frequency, the polarization operator of the weakly interacting electron gas developed an instability. The appearance of such instabilities usually indicates that one is formulating a theory around the wrong reference state (in that case, the uniformly filled Fermi sphere of the non-interacting electron gas). Thus, what we would like to develop is a theoretical framework that is capable of detecting the right reference states, or mean fields, of a system. We would like to efficiently apply perturbative methods around these states and do so in a language drawing upon the physical rather than the plain microscopic degrees of freedom.

In this chapter, we develop a functional-integral-based approach that meets these criteria. In contrast with the previous chapters, the discussion here is decidedly biased towards concrete application to physically motivated problems. After the formulation of the general strategy of mean field methods, the following section addresses a problem that we have encountered before, the weakly interacting electron gas. The exemplification of the new concepts on a known problem enables us to understand the connection between the mean field approach and straightforward perturbation theory. In subsequent sections we then turn to the discussion of problems that lie firmly outside the range of direct perturbative summation, superfluidity and superconductivity.

Roughly speaking, the functional integral approach to problems with a large parameter proceeds according to the following program.

1. First, one must identify the relevant structural units of the theory. (This part of the program *can* be carried out efficiently using the methods discussed previously.)

2. Second, it is necessary to introduce a new field – let us call it ϕ for concreteness – that encapsulates the relevant degrees of freedom of the low-energy theory.

3. With this in hand, one can then trade integration over the microscopic fields for integration over ϕ, a step often effected by an operation known as the Hubbard–Stratonovich transformation.

4. The low-energy content of the theory is then usually explored by subjecting the resulting action $S[\phi]$ to a stationary phase analysis. (The justification for applying stationary phase methods is provided by the existence of a large parameter $N \gg 1$.) Solutions to the stationary phase equations are generically called *mean fields*. Mean fields can either be uniquely defined or come in discrete or continuous families. In the latter case, summation or integration over the mean field manifold is required. Often, at this stage, instabilities in the theory show up – an indication of a physically interesting problem!

5. Finally, the nature of the excitations above the ground state is described by expanding the functional integral around the mean fields. From the low-energy effective (mean field + fluctuations) theory one can then compute physical observables.

In the next section, we will illustrate this program on a specific example studied earlier by diagrammatic methods.

5.1 Case Study: Plasma Theory of the Electron Gas

REMARK In this section, the electron spin does not play a significant role, and we assume spinless fermions for simplicity. (Alternatively, one may assume the electron gas to be spin-polarized by application of a strong magnetic field.)

Let us return to the field theory of the interacting electron gas (see section 4.2.2),

$$S[\psi] = T \sum_p \bar{\psi}_p \left(-i\omega_n + \frac{\mathbf{p}^2}{2m} - \mu \right) \psi_p + \frac{T^3}{2L^3} \sum_{pp'q} \bar{\psi}_{p+q} \bar{\psi}_{p'-q} V(\mathbf{q}) \psi_{p'} \psi_p,$$

where $V(\mathbf{q}) = 4\pi e^2 / |\mathbf{q}|^2$, and $p = (\omega_n, \mathbf{p})$ are four-momenta. Being quartic in the fields ψ, the Coulomb interaction prevents explicit computation of the ψ-integral. However, it is actually a straightforward matter to reduce, or "decouple," the interaction operator, bringing it to a form quadratic in the fields ψ. This is achieved by a technical trick involving a tailor-made Gaussian integral:

5.1.1 Hubbard–Stratonovich transformation

Let us multiply the functional integral by the "fat unity"

$$1 \equiv \int D\phi \exp \left(-\frac{T}{2L^3} \sum_q \phi_q V^{-1}(\mathbf{q}) \phi_{-q} \right),$$

where ϕ represents a real bosonic field variable, and a normalization constant has been absorbed into the definition of the functional measure $D\phi$. Notice that, here, the sum runs over a four-momentum $q = (\omega_m, \mathbf{q})$ containing a *bosonic* Matsubara frequency. Employing the variable shift $\phi_q \to \phi_q + iV(\mathbf{q})\rho_q$, where $\rho_q \equiv T \sum_p \bar{\psi}_p \psi_{p+q}$, one obtains

$$1 = \int D\phi \exp \left(-\frac{T}{2L^3} \sum_q \left(\phi_q V^{-1}(\mathbf{q}) \phi_{-q} - 2i\rho_{-q}\phi_q - \rho_q V(\mathbf{q})\rho_{-q} \right) \right).$$

The rationale behind this procedure can be seen in the last contribution to the exponent: this term is equivalent to the quartic interaction contribution to the fermionic path integral, albeit with opposite sign. Therefore, multiplication of our unity by $\mathcal{Z}$ leads to the field integral $\mathcal{Z} = \int D\phi D\psi \, e^{-S[\phi,\psi]}$, where

$$S[\phi, \psi] = \frac{T}{2L^3} \sum_q \phi_q V^{-1}(\mathbf{q}) \phi_{-q} + T \sum_{p,q} \bar{\psi}_p \left(\left(-i\omega_n + \frac{\mathbf{p}^2}{2m} - \mu \right) \delta_{q,0} + \frac{iT}{L^3}\phi_q \right) \psi_{p-q}$$

$$\tag{5.1}$$

denotes the action, i.e., an expression that is free of quartic field interactions of ψ_σ.

To gain intuition for the nature of the action, it is helpful to rewrite S in a **real space representation**. With $\phi_q = \int_0^\beta d\tau \int d^d r \, e^{-i\mathbf{q}\cdot\mathbf{r}+i\omega\tau} \phi(\mathbf{r}, \tau)$, one may confirm that (exercise)

$$S[\phi, \psi] = \int d\tau \int d^3r \left\{ \frac{1}{8\pi e^2}(\partial\phi)^2 + \bar{\psi} \left(\partial_\tau - \frac{\partial^2}{2m} - \mu + i\phi \right) \psi \right\}.$$

Physically, ϕ couples to the electron degrees of freedom as a space–time-dependent (imaginary) potential, while the first term reflects the Lagrangian energy density associated with the electric component of the electromagnetic field. Before proceeding, let us step back and discuss the general philosophy of the manipulations that led from the original partition function to the two-field representation above.

INFO The "decoupling" of quartic interactions by Gaussian integration over an auxiliary field, is generally known as a **Hubbard–Stratonovich transformation**. (For a previous example, see our discussion of the Ising model on page 174.) To make the working of the transformation more transparent, let us reformulate it in a notation that is not burdened by the presence of model-specific constants. Consider an interaction operator of the form $S_{\text{int}} = V_{\alpha\beta\gamma\delta}\bar{\psi}_\alpha\psi_\beta\bar{\psi}_\gamma\psi_\delta$ (summation convention), where $\bar{\psi}$ and ψ may be either bosonic or fermionic field variables, the indices $\alpha, \beta, \dots$ refer to an unspecified set of quantum numbers, Matsubara frequencies, etc., and $V_{\alpha\beta\gamma\delta}$ is an interaction matrix element. Now, let us introduce composite fields $\rho_{\alpha\beta} \equiv \bar{\psi}_\alpha\psi_\beta$ in order to rewrite the interaction as $S_{\text{int}} = V_{\alpha\beta\gamma\delta}\,\rho_{\alpha\beta}\rho_{\gamma\delta}$. The notation can be simplified still further by introducing composite indices $m \equiv (\alpha\beta)$, $n \equiv (\gamma\delta)$, whereupon the action $S_{\text{int}} = \hat{\rho}_m V_{mn}\hat{\rho}_n$ acquires the structure of a generalized bilinear form. To reduce the action to a form quadratic in the ψs, one may simply multiply the exponentiated action by unity, i.e.,

$$\exp\left(-\rho_m V_{mn}\rho_n\right) = \underbrace{\int D\phi \, \exp\left(-\frac{1}{4}\phi_m V_{mn}^{-1}\phi_n\right)}_{1} \exp\left(-\rho_m V_{mn}\rho_n\right),$$

where ϕ is bosonic. (Note that here V_{mn}^{-1} represents the matrix elements of the inverse and not the inverse $(V_{mn})^{-1}$ of individual matrix elements.) Finally, applying the variable change $\phi_m \to \phi_m + 2i(V\rho)_m$, where the notation $(V\rho)$ is shorthand for $V_{mn}\rho_n$, one obtains

$$\boxed{\exp\left(-\rho_m V_{mn}\rho_n\right) = \int D\phi \exp\left(-\frac{1}{4}\phi_m V_{mn}^{-1}\phi_n - i\phi_m\rho_m\right)}$$

i.e., the term quadratic in $\hat{\rho}$ has been cancelled.[1] This completes the formulation of the Hubbard–Stratonovich transformation. The interaction operator has been traded for integration over an auxiliary field coupled to a ψ-bilinear (the field $\phi_m\rho_m$).

▷ In essence, the Hubbard–Stratonovich transformation is tantamount to the Gaussian integral identity (3.13) but read in reverse. An exponentiated square is removed in exchange for a linear coupling. (In (3.13) we showed how terms linear in the integration variable can be removed.)

▷ To make the skeleton outlined above into a well-defined prescription, one has to be specific about the meaning of the Gaussian integration over the kernel $\phi_m V_{mn}^{-1}\phi_n$: one needs to decide whether the integration variables are real or complex, and safeguard convergence by making sure that V is a positive matrix (which is usually the case on physical grounds).

▷ There is some freedom as to the choice of the integration variable. For example, the factor $1/4$ in front of the Gaussian weight $\phi_m V_{mn}^{-1}\phi_n$ has been introduced for mere convenience (i.e., to generate a coupling $\phi_m\rho_m$ free of numerical factors). If one does not like to invert the matrix kernel V_{mn}, one can scale $\phi_m \to (V\phi)_m$, whereupon the key formula reads

$$\exp\left(-\rho_m V_{mn}\rho_n\right) = \int D\phi \, \exp\left(-\frac{1}{4}\phi_m V_{mn}\phi_n - i\phi_m V_{mn}\rho_n\right).$$

EXERCISE Show that the passage from the Lagrangian to the Hamiltonian formulation of the Feynman path integral can be interpreted as a Hubbard–Stratonovich transformation.

---

[1] Here, we have assumed that the matrix V is symmetric. If it is not, we can apply the relation $\rho_m V_{mn}\rho_n \equiv \rho^T V \rho = \frac{1}{2}\left(\rho^T (V + V^T)\rho\right)$ to symmetrize the interaction.

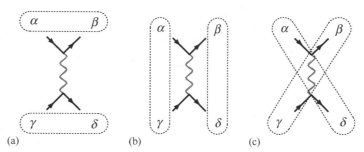

(a)   (b)   (c)

Fig. 5.1   The different channels of decoupling an interaction by Hubbard–Stratonovich transformation. (a) Decoupling in the "density" channel; (b) decoupling in the "pairing" or "Cooper" channel; and (c) decoupling in the "exchange" channel.

As defined, the Hubbard–Stratonovich transformation is exact. However, to make it a meaningful operation, it must be motivated by physical considerations. In our discussion above, we split up the interaction by choosing $\rho_{\alpha\beta}$ as a composite field. However, there is clearly some arbitrariness with this choice. Why not, for example, pair the fermion-bilinears according to $(\bar{\psi}_\alpha\psi_\beta)(\bar{\psi}_\gamma\psi_\delta)$, or otherwise? The three inequivalent choices of pairing up fields are shown in fig. 5.1 where, as usual, the wavy line with attached field vertices represents the interaction, and the dashed ovals indicate how the fields are paired.

The version of the transformation discussed above corresponds to fig. 5.1(a). This type of pairing is sometimes referred to as decoupling in the **direct channel**. The designation becomes more transparent if we consider the example of the spinful electron–electron interaction,

$$S_{\mathrm{int}} = \frac{1}{2}\int d\tau \int d^d r\, d^d r'\; \bar{\psi}_\sigma(\mathbf{r},\tau)\bar{\psi}_{\sigma'}(\mathbf{r}',\tau)V(\mathbf{r}-\mathbf{r}')\psi_{\sigma'}(\mathbf{r}',\tau)\psi_\sigma(\mathbf{r},\tau),$$

i.e., here, $\alpha = \beta = (\mathbf{r},\tau,\sigma)$, $\gamma = \delta = (\mathbf{r}',\tau,\sigma')$, and $V_{\alpha\beta\gamma\delta} = V(\mathbf{r}-\mathbf{r}')$. The "direct" decoupling proceeds via the most obvious choice, i.e., the density $\rho(\mathbf{r},\tau) = \bar{\psi}_\sigma(\mathbf{r},\tau)\psi_\sigma(\mathbf{r},\tau)$. One speaks about decoupling in a "channel" because, as will be elucidated below, the propagator of the decoupling field can be interpreted in terms of two Green function lines tied together by multiple interactions, a sequential object reminiscent of a channel.

However, there are other choices for ρ. Decoupling in the **exchange channel** (fig. 5.1(c)) is generated by the choice $\rho_{\alpha\delta} \sim \bar{\psi}_\alpha\psi_\delta$, where, in the context of the Coulomb interaction, the reversed pairing of fields is reminiscent of the exchange contraction generating Fock-type contributions. Finally, one may decouple in the **Cooper channel** (fig. 5.1(b)) $\rho_{\alpha\gamma} \sim \bar{\psi}_\alpha\bar{\psi}_\gamma$, $\rho_{\beta\gamma} = \rho^\dagger_{\gamma\beta}$. Here, the pairing field is conjugate to two creation operators. Below, we will see that this type of decoupling is tailored to problems involving superconductivity.

The remarks above may convey the impression of arbitrariness. Indeed, the "correct" choice of decoupling can be motivated only by physical reasoning. Put differently, the transformation as such is exact, no matter what channel we choose. However, later on we will want to derive an effective low-energy theory based on the decoupling field. In cases where one has accidentally decoupled in an "unphysical" channel, it will be difficult, if not impossible, to distill a meaningful low-energy theory for the field ϕ conjugate to ρ. Although the initial model still contains the full microscopic information (by virtue of the exactness of the transformation) it is not amenable to further approximation schemes.

In fact, one is frequently confronted with situations where more than one Hubbard–Stratonovich field is needed to capture the full physics of the problem. To appreciate this point, consider the Coulomb interaction of spinful fermions in momentum space:

direct channel

exchange channel

Cooper channel

$$S_{\text{int}}[\psi] = \frac{T^3}{2L^3} \sum_{p_1,\ldots,p_4} \bar{\psi}_{\sigma p_1} \bar{\psi}_{\sigma' p_3} V(\mathbf{p}_1 - \mathbf{p}_2) \psi_{\sigma' p_4} \psi_{\sigma p_2} \delta_{p_1 - p_2 + p_3 - p_4}. \tag{5.2}$$

We can decouple this interaction in any one of the three channels discussed above. However, "interesting" physics is usually generated by processes where one of the three unbounded momenta entering the interaction vertex is small. Only these interaction processes have a chance to accumulate an overall collective excitation of low energy (see our previous discussion of the RPA, the interacting electron gas, and many of the examples to follow). Geometrically, the three-dimensional Cartesian space of free momentum coordinates (p_1, p_2, p_3) entering the vertex, contains three thin layers, where one of the momenta is small, (q, p_2, p_3), (p_1, q, p_3), (p_1, p_2, q), $|q| \ll |p_i|$. (Why not make all momenta small? Because that would be in conflict with the condition that the Green functions connecting to the vertex be close to the Fermi surface.) One may thus break down the full momentum summation to a restricted summation over the small-momentum sub-layers:

$$S_{\text{int}}[\psi] \simeq \frac{T^3}{2L^3} \sum_{p,p',q} \left(\bar{\psi}_{\sigma p} \psi_{\sigma p+q} V(\mathbf{q}) \bar{\psi}_{\sigma' p'} \psi_{\sigma' p'-q} - \bar{\psi}_{\sigma p} \psi_{\sigma' p+q} V(\mathbf{p}' - \mathbf{p}) \bar{\psi}_{\sigma' p'+q} \psi_{\sigma p'} \right.$$
$$\left. - \bar{\psi}_{\sigma p} \bar{\psi}_{\sigma' -p+q} V(\mathbf{p}' - \mathbf{p}) \psi_{\sigma p'} \psi_{\sigma' -p'+q} \right).$$

Now, each of these three contributions defines its own slow decoupling field. The first term should be decoupled in the direct channel $\rho_{\text{d},q} \sim \sum_p \bar{\psi}_{\sigma p} \psi_{\sigma p+q}$, the second in the exchange channel $\rho_{\text{x},\sigma\sigma' q} \sim \sum_p \bar{\psi}_{\sigma p} \psi_{\sigma' p+q}$, and the third in the Cooper channel $\rho_{\text{c},\sigma\sigma' q} \sim \sum_p \bar{\psi}_{\sigma p} \bar{\psi}_{\sigma' -p+q}$. One thus winds up with an effective theory that contains three independent slow Hubbard–Stratonovich fields. (Notice that the decoupling fields in the exchange and in the Cooper channel explicitly carry a spin structure.)

In our discussion of the high-density limit of the electron gas above, we decoupled in the direct channel. That choice was made since, drawing on our previous discussion, we knew that the relevant contributions are generated by an RPA-type contraction of fields $\rho_{\text{d},q} \sim \sum_p \bar{\psi}_{\sigma p} \psi_{\sigma p+q}$, where q is small. If we had not known this, an analysis of the three-fold Hubbard–Stratonovich decoupled action would have led to the same conclusion. Generally, if in doubt, one should decouple in all available channels and let the mean field analysis outlined below discriminate between the relevant fields.

At the expense of introducing a second field, the Hubbard–Stratonovich transformation provides an action quadratic in the fermion fields. The advantage of this representation is that the fermion integration can be carried out by the Gaussian integral formula (3.62), with the result

$$\mathcal{Z} = \int D\phi \, \exp\left(-\frac{T}{2L^3} \sum_q \phi_q V^{-1}(\mathbf{q}) \phi_{-q} \right) \det(\hat{G}^{-1}), \quad \hat{G}^{-1} = -i\hat{\omega} + \frac{\hat{\mathbf{p}}^2}{2m} - \mu + i\hat{\phi},$$

where, as usual, the circumflexes appearing in the argument of the determinant indicate that symbols have to be interpreted as operators (acting in the space of Matsubara and Hilbert space components), and the notation $\hat{G}^{-1}$ is motivated by the structural similarity to an inverse Green function.

The standard procedure to deal with the determinants generated at intermediate stages of the manipulation of a field integral is to simply re-exponentiate them. This is achieved by virtue of the identity

$$\boxed{\ln \det \hat{A} = \operatorname{tr} \ln \hat{A}} \tag{5.3}$$

valid for arbitrary (non-singular) operators $\hat{A}$.[2] Thus, the quantum partition function takes the form $\mathcal{Z} = \int D\phi \, e^{-S[\phi]}$, where

$$S[\phi] = \frac{T}{2L^3} \sum_q \phi_q V^{-1}(\mathbf{q}) \phi_{-q} - \operatorname{tr} \ln(\hat{G}^{-1}). \tag{5.4}$$

At this point, steps 1, 2, and 3 of the general program outlined above have been accomplished, and the problem has been mapped to an integral over the auxiliary field. No approximations have been made so far.

5.1.2 Stationary phase analysis

Since the remaining integral over ϕ cannot be made exactly (but we expect the emergence of a large control parameter), we turn to the next best approach, a **stationary phase analysis**. We thus seek solutions of the saddle-point equations

$$\frac{\delta S[\phi]}{\delta \phi_q} \overset{!}{=} 0$$

for all $q = (\mathbf{q}, \omega_n)$ with $\mathbf{q} \neq 0$. (Why is $\mathbf{q} = 0$ excluded? Recall our discussion of the Hartree zero-momentum mode and charge neutrality in section 4.2.2.) Solutions **mean fields** of such equations are commonly referred to as **mean fields**. The origin of this denotation is that, from the perspective of the fermions, $\hat{\mathbf{p}}^2/2m - \mu + i\phi$ resembles the effective Hamiltonian in the presence of a background potential, or "mean" field.

The concrete evaluation of the functional derivative $\delta S/\delta \phi$ leads us to question how one differentiates the trace of the logarithm of the *operator* $\hat{G}^{-1}$, with respect to ϕ. Owing to the presence of the trace, the differentiation can be carried out as if $\hat{G}$ were a function:[3] $\delta_{\phi_q} \operatorname{tr} \ln(\hat{G}^{-1}) = \operatorname{tr}(\hat{G} \delta_{\phi_q} \hat{G}^{-1})$. Comparing with the momentum representation of Eq. (5.1), we find that $(\delta_{\phi_q} G^{-1})_{p,p'} = iTL^{-3}\delta_{p',p-q}$ and from there follows $\operatorname{tr}(\hat{G}\delta_{\phi_q}\hat{G}^{-1}) = iTL^{-3}\sum_p G_{p-q,p}$. The resulting saddle-point equation assumes the form

---

[2] Equation (5.3) is readily established by switching to an eigenbasis, whereupon one obtains $\ln \det \hat{A} = \sum_a \ln \epsilon_a = \operatorname{tr} \ln \hat{A}$, where ϵ_a are the eigenvalues of $\hat{A}$ and we have used the fact that the eigenvalues of $\ln \hat{A}$ are $\ln \epsilon_a$.

[3] Consider an operator $\hat{A}(x)$ depending on some parameter x. Let $f(\hat{A})$ be an arbitrary function $(f(\hat{A}) = \ln \hat{A}$ in the present application). Then

$$\partial_x \operatorname{tr}(f(\hat{A})) = \partial_x \sum_n \frac{f^{(n)}(0)}{n!} \operatorname{tr}(\hat{A}^n) = \sum_n \frac{f^{(n)}(0)}{n!} \operatorname{tr}((\partial_x \hat{A})\hat{A}^{n-1} + \hat{A}(\partial_x \hat{A})\hat{A}^{n-2} + \cdots)$$

$$= \sum_n \frac{n}{n!} f^{(n)}(0) \operatorname{tr}(\hat{A}^{n-1}(\partial_x \hat{A})) = \operatorname{tr}(f'(\hat{A})\partial_x \hat{A}),$$

where, in the third equality, we have used the cyclic invariance of the trace.

$$\frac{\delta}{\delta \phi_q} S[\phi] = V^{-1}(\mathbf{q}) \phi_{-q} - i \sum_p G[\phi]_{p-q,p} \overset{!}{=} 0, \qquad (5.5)$$

where the notation emphasizes the dependence of G on ϕ.

INFO Almost every functional stationary-phase program involves the **differentiation of trace logs** of operators containing auxiliary fields. While the details of the program vary from case to case, the general strategy is always the same: use the presence of the trace to differentiate the logarithm to an inverse, and then represent the operator as a *matrix* containing the matrix elements of the auxiliary field. Since the auxiliary field enters that matrix linearly, the differentiation is easy. Finally, perform the trace. The procedure needs some familiarization but after a while becomes routine.

Equations of the above form are generally solved by making a physically motivated *ansatz*, i.e., by guessing the solution. This ansatz should take the symmetries of the equation (both space–time symmetries and internal symmetries of the fields) into account. At present, all we have to work with is the space–time translational invariance of the problem, and so the first guess is a homogeneous solution, $\phi(\mathbf{r}, t) \equiv \bar{\phi} = \text{const.}$ One thus relies on the picture that a spatially and temporally varying field configuration will be energetically more costly than a constant one, and therefore could not provide a stable extremal point.

INFO Be aware that there exist translationally invariant problems with inhomogeneous mean fields; or a homogeneous solution exists, but it is energetically inferior to a textured field configuration. Indeed, there may be sets of degenerate solutions, etc. Often, when new theories describing an unknown territory have been developed, the search for the "correct" mean field turns out to be a matter of long, and sometimes controversial, research.

In the present context, spatiotemporal homogeneity and charge neutrality translate to the trivial solution $\phi_q = 0$. To see this, recall that ϕ_0 is not defined (a consequence of the charge neutrality condition). With all components $\phi_{q\neq 0} = 0$, $G_{p,p'} \propto \delta_{p,p'}$ becomes diagonal in momentum space. As a consequence, both terms in the stationary phase equation vanish. We have thus found an (admittedly trivial) solution, and thus completed step 4 of the general program outlined at the start of the chapter.

5.1.3 Fluctuations

We now proceed to expand the functional in **fluctuation**s around $\phi = 0$. Since the mean field solution vanishes, it makes no sense to introduce new notation, i.e., we will also denote the fluctuations by the symbol ϕ. As regards the first term in the action (5.4), it already has a quadratic form. The logarithmic contribution can be expanded as if we were dealing with a function (again, a consequence of the trace), i.e.,

$$\text{tr} \ln \hat{G}^{-1} = \text{tr} \ln \hat{G}_0^{-1} + i\, \text{tr}(\hat{G}_0 \phi) + \frac{1}{2} \text{tr}(\hat{G}_0 \phi \hat{G}_0 \phi) + \cdots,$$

where $\hat{G}_0^{-1} \equiv i\hat{\omega} - \hat{\mathbf{p}}^2/(2m) + \mu$ is the momentum and frequency diagonal operator whose matrix elements give the free Green function of the electron gas. Now, let us discuss the terms appearing on the right-hand side in turn. Being ϕ-independent, the first term is a constant describing the non-interacting content of the theory. Indeed, $\exp(\text{tr}\ln\hat{G}_0^{-1}) = \exp(-\text{tr}\ln\hat{G}_0) = \det\hat{G}_0^{-1} \equiv \mathcal{Z}_0$ is just the partition function of the non-interacting electron gas. As it is linear in ϕ, the second term of the expansion vanishes by virtue of the stationary phase condition. (Perhaps pause for a moment to think about this statement.) The third term is the interesting one. Remembering that ϕ represents an effective voltage, since this term describes the way in which potential fluctuations couple to the electron gas it must encode screening.

To resolve this connection, let us make the momentum dependence of the second-order term explicit (exercise):

$$\frac{1}{2}\text{tr}(\hat{G}_0\phi\hat{G}_0\phi) = \frac{1}{2}\left(\frac{T}{L^3}\right)^2 \sum_{p,q} G_{0,p}\phi_q G_{0,p+q}\phi_{-q} = \frac{T}{2L^3}\sum_q \Pi_q \phi_q \phi_{-q},$$

where, once again, we encounter the polarization operator (4.28) (excluding a factor of 2 as we are presently working with spinless fermions.) Combining with the first term in the action, one finally obtains

$$S_{\text{eff}}[\phi] = \frac{T}{2L^3}\sum_q \phi_q \left(\frac{\mathbf{q}^2}{4\pi e^2} - \Pi_q\right)\phi_{-q} + \mathcal{O}(\phi^4), \qquad (5.6)$$

where we note that odd powers of ϕ vanish by symmetry (exercise).

Physically, the coupling to the medium modifies the long-range correlation $\sim \mathbf{q}^{-2}$ of the effective potential ϕ. This screening effect, via the $\text{tr}(G\phi G\phi)$ polarization bubble, coincides with that of the effective RPA interaction (4.37) derived diagrammatically in the previous chapter. From here it is a one-line calculation to reproduce the result for the RPA free energy of the electron gas discussed previously. Gaussian integration over the field ϕ (step 5 of the program) leads to the expression $\mathcal{Z}_{\text{RPA}} = \mathcal{Z}_0\prod_q(1 - 4\pi e^2\mathbf{q}^{-2}\Pi_q)^{-1/2}$, where we note that the ϕ-integration is normalized to unity, i.e., for $\Pi = 0$ the integral collapses to unity. Taking the logarithm, we obtain the free energy

$$F_{\text{RPA}} = -T(\ln\mathcal{Z} - \ln\mathcal{Z}_0) = \frac{T}{2}\sum_q \ln\left(1 - \frac{4\pi e^2}{\mathbf{q}^2}\Pi_q\right),$$

in agreement with Eq. (4.27).

At this point, it is instructive to compare the two approaches to the problem: diagrammatics and field integration. We first note that the latter indeed leads to the "reduced" description we sought: an effective degree of freedom, ϕ, coupling to the single physically relevant building block of the theory, the polarization bubble, Π_q. The downside is that we had to go through some preparatory analysis to arrive at this representation. However, this turned out to be an effort well invested. After the identification of ϕ as the appropriate Hubbard–Stratonovich field, the construction proceeded along the lines of a largely automated program (seeking saddle-points, expanding, etc.). In particular, there was no need to do battle with combinatorial problems. Further, the risk of missing relevant contributions, or diagrams, in the expansion of the theory is less pronounced than in diagrammatic approaches. But, undoubtedly, the most important advantage of the functional integral is its extensibility. For example, an expansion of the theory to higher orders in ϕ would have generated an interacting theory of voltage fluctuations. The direct and error-free summation of such correlations beyond the RPA level by diagrammatic methods would require more refined skills.

The mean field optimizing the problem above is particularly simple, with $\phi = 0$. More interesting situations arise when one encounters non-vanishing mean field configurations and the perturbation theory has to be organized around a nontrivial reference state. In the next section we discuss an important case study where this situation is realized. (See problem 5.6.7 for an example closer to the current problem of the electron plasma.)

5.2 Bose–Einstein Condensation and Superfluidity

Previously, we considered the influence of weak Coulomb interaction on the properties of the electron gas. In the following, our goal will be to consider the phases realized by a weakly interacting Bose gas. To this end, let us introduce the quantum partition function $\mathcal{Z} = \int D\psi \, e^{-S[\psi]}$, where

$$S[\psi] = \int d\tau \int d^d r \left(\bar\psi(\mathbf{r}, \tau)(\partial_\tau + \hat H_0 - \mu)\psi(\mathbf{r}, \tau) + \frac{g}{2}(\bar\psi(\mathbf{r}, \tau)\psi(\mathbf{r}, \tau))^2 \right). \quad (5.7)$$

Here ψ represents a complex field subject to the periodic boundary condition $\psi(\mathbf{r}, \beta) = \psi(\mathbf{r}, 0)$. The functional integral $\mathcal{Z}$ describes the physics of a system of bosonic particles in d dimensions subject to a repulsive contact interaction of strength $g > 0$. For the moment the specific structure of the one-body operator $\hat H_0$ need not be specified. The most remarkable phenomena displayed by systems of this type are Bose–Einstein condensation (BEC) and superfluidity. However, contrary to a widespread belief, these two effects do not depend on each other: superfluidity can arise without condensation and vice versa. We begin our discussion with the more elementary of the two phenomena.

5.2.1 Bose–Einstein condensation

At sufficiently low temperatures the ground state of a bosonic system can involve the condensation of a macroscopic fraction of particles into a single state. This phenomenon, predicted in a celebrated work by Einstein, is known as **Bose–Einstein condensation**. To understand it within the framework of the present formalism, let us temporarily switch off the interaction and assume the one-particle Hamiltonian to be diagonalized. In the frequency representation, the partition function is then given by

$$\mathcal{Z}_0 \equiv \mathcal{Z}|_{g=0} = \int D\psi \exp\left(-T\sum_{an} \bar{\psi}_{an}\left(-i\omega_n + \xi_a\right)\psi_{an}\right),$$

where $\xi_a = \epsilon_a - \mu$. Without loss of generality, we assume that the eigenvalues $\epsilon_a \geq 0$ are positive with a ground state $\epsilon_0 = 0.$[4] (In contrast with the fermionic systems discussed above, we should not have in mind low-energy excitations superimposed on high-energy microscopic degrees of freedom. Here, everything will take place in the vicinity of the ground state of the microscopic single-particle Hamiltonian.)

Satyendranath Bose 1894–1974
was an Indian mathematician and physicist who undertook important work in quantum theory, in particular on Planck's black body radiation law. His work was enthusiastically endorsed by Einstein. He also published on statistical mechanics, leading most famously to the concept of Bose–Einstein statistics. Dirac coined the term "boson" for particles obeying such statistics.

Further, we note that the chemical potential determining the number of particles in the system must be negative for, otherwise, the Gaussian weight corresponding to the low-lying states $\epsilon_a < -\mu$ would change sign, resulting in an ill-defined theory.

From our discussion of section 3.5 we recall that the number of particles in the system is given by

$$N(\mu) = -\partial_\mu F = T\sum_{na} \frac{1}{i\omega_n - \xi_a} = \sum_a n_{\mathrm{B}}(\epsilon_a),$$

where, $n_{\mathrm{B}}(\epsilon) = (e^{\beta(\epsilon-\mu)} - 1)^{-1}$ denotes the Bose distribution. For a given number of particles, this equation determines the temperature dependence of the chemical potential, $\mu(T)$. As the temperature is reduced, the distribution function controlling the population of individual states decreases. Since the number of particles must be kept constant, this scaling

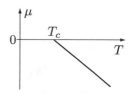

must be counter-balanced by a corresponding increase in the chemical potential.

Below a certain critical temperature T_c, even the maximum value of the chemical potential, $\mu = 0$, will not suffice to keep the distribution function $n_{\mathrm{B}}(\epsilon_{a\neq0})$ large enough to accommodate all particles in states of non-vanishing energy, i.e.,

$\sum_{a>0} n_{\mathrm{B}}(\epsilon_a)|_{\mu=0} \overset{T<T_c}{\equiv} N_1 < N$. Below this **critical temperature**, the chemical potential stays constant at $\mu = 0$ (see the figure). As a result, a macroscopic number of particles, $N - N_1$, must accumulate in the single-particle ground state: **Bose–Einstein condensation**.

---

[4] The chemical potential μ can always be adjusted so as to meet this condition.

EXERCISE For a three-dimensional free particle spectrum, $\epsilon_k = \hbar^2 k^2/2m$, show that the critical temperature is set by $T_c = c_0 \hbar^2/ma^2$, where $a = \rho^{-1/3}$ is the average inter-particle spacing, and c_0 is a constant of order unity. Show that for temperatures $T < T_c$, the density of particles in the condensate ($\mathbf{k} = 0$) is given by $\rho_0(T) = \rho[1 - (T/T_c)^{3/2}]$.

INFO After the theoretical prediction of Bose–Einstein condensation in the 1920s, it took about seven decades before the phenomenon was directly[5] observed in experiment. The reason for this delay is that the critical condensation temperature of particles (atoms) that are comfortably accessible to experiment is extremely low.

However, in 1995, the groups of Cornell and Wieman at Colorado University and, soon after, Ketterle at MIT succeeded in cooling a system of rubidium atoms down to temperatures of 20 billionths(!) of a kelvin.[6] To reach these temperatures, a gas of rubidium atoms was caught in a magnetic trap, i.e., a configuration of magnetic field gradients that coupled to the magnetic moments of the atoms so as to keep the system spatially localized.

Eric A. Cornell 1961– (left)
Wolfgang Ketterle 1957– (center)
Carl E. Wieman 1951– (right)
were joint recipients of the 2001 Nobel Prize in Physics "for the achievement of Bose–Einstein condensation in dilute gases of alkali atoms, and for early fundamental studies of the properties of the condensates."

The temperature of the gas of atoms was then lowered in a two-stage cooling process: laser cooling down to temperatures of $\mathcal{O}(10^{-5})$ K, combined with *evaporative* cooling whereby a fraction of particles of comparatively high thermal energy was allowed to escape and one was left with a residual system of particles of ultra low energy. This procedure, which took years of experimental preparation, eventually led to temperatures low enough to achieve condensation.

The preparation of a Bose–Einstein condensed state of matter was recognized with the award of the 2001 Nobel Prize in physics. Since 1995, research on atomic condensates has blossomed into a broad arena of research. It is now possible to prepare complex states of Bose condensed matter such as atomic vortices in rotating Bose–Einstein condensates, condensates in different dimensionalities, or even synthetic crystalline states of matter. Regrettably, a detailed discussion of these interesting developments is beyond the scope of the present text. Those interested in learning more about this area are referred to the many reviews of the field.

With this background, let us now discuss how Bose–Einstein condensation is described by the functional integral. Evidently, the characteristics of the condensate will be described by the zero field component $\psi_0(\tau)$. The problem with this zero mode is that, below the condensation transition, both the chemical potential and the eigenvalue are zero. This means that the action of the zero Matsubara component $\psi_{0,0}$ vanishes, and we have an ill-defined integral. We will deal with this difficulty in a pragmatic way. That is, we will treat $\psi_0(\tau)$ not as an integration field

---

[5] Here, by "direct," we refer to the controlled preparation of a state of condensed massive bosonic particles. There are numerous indirect manifestations of condensed states, e.g., the anomalous properties of the helium liquids at low temperatures, or of Cooper-pair condensates in superconductors.

[6] M. H. Anderson, J. R. Ensher, M. R. Matthews, C. E. Wieman and E. A. Cornell, *Observation of Bose–Einstein condensation in a dilute atomic vapor*, Science **269**, 198 (1995).

but rather as a time-independent Lagrange multiplier to be used to fix the number of particles below the transition. More precisely, we introduce a reduced action of the form

$$S_0[\psi] = -\bar\psi_0 \mu \psi_0 + T \sum_{a\neq 0, n} \bar\psi_{an}(-i\omega_n + \xi_a)\psi_{an},$$

where we have not yet set $\mu = 0$ (since we still need μ as a differentiation variable). To understand the rationale behind this simplification, note that

$$N = -\partial_\mu F_0|_{\mu=0^-} = T\partial_\mu \ln \mathcal{Z}_0|_{\mu=0^-} = T\bar\psi_0\psi_0 + T\sum_{a\neq 0,n} \frac{1}{i\omega_n - \epsilon_a} = T\bar\psi_0\psi_0 + N_1 \tag{5.8}$$

determines the number of particles. According to this expression, $T\bar\psi_0\psi_0 = N_0$ sets the number of particles in the condensate. Now, what enables us to regard ψ_0 as a time-independent field? Remembering the construction of the path integral, time slicing and the introduction of time-dependent integration variables were required because quantum Hamiltonians generically contain non-commuting operators. (Otherwise we could have decoupled the expression $\mathrm{tr}(e^{-\beta \hat H(a^\dagger, a)}) \simeq \int d\psi \, e^{-\beta \hat H(\bar\psi, \psi)}$ in terms of a single coherent state resolution, i.e., a "static" configuration.) Reading this observation in reverse, we conclude that the dynamic content of the field integral represents the quantum character of a theory.

In view of this fact, the temporal fluctuations of field variables are often referred to as **quantum fluctuations**. Conversely:

> A static approximation in a field integral $\psi(\tau) = \psi_0 = \text{const.}$ amounts to replacing a quantum degree of freedom by its classical approximation.

quantum fluctu- ations

In order to distinguish them from quantum fluctuations in the "classical" static sector of the theory are called **thermal fluctuations** (see the following Info block). To justify the approximation of $a_0 \leftrightarrow \psi_0$ by a classical object, notice that, upon condensation, $N_0 = \langle a_0^\dagger a_0 \rangle$ will assume "macroscopically large" values. On the other hand, the commutator $[a_0, a_0^\dagger] = 1$ continues to be of $\mathcal{O}(1)$. It thus seems to be legitimate to neglect all commutators of the zero operator a_0 in comparison with its expectation value – a classical approximation.[7]

Now, we are still left with the problem that the ψ_0-integration appears to be undefined. The way out is to remember that the partition function should extend over those states that contain an average number N of particles. That is, Eq. (5.8) has to be interpreted as a relation that fixes the modulus $\bar\psi_0\psi_0$ in such a way as to adjust the appropriate value of N.

---

[7] Unfortunately, the actual state of affairs regarding the classical treatment of the condensate is somewhat more complex than the simple argument above suggests. (For a good discussion, see A. A. Abrikosov, L. P. Gorkov, and I. E. Dzyaloshinskii, *Methods of Quantum Field Theory in Statistical Physics*, Dover Publications, 1975.) However, the net result of a more thorough analysis, i.e., an integration over all dynamically fluctuating components $\psi_{0,n\neq0}$, shows that the treatment of ψ_0 as classical represents a legitimate approximation.

INFO The identification "**time-independent** = **classical** = **thermal**" is suggestive, but can be misleading. To see why, recall that, to leading order in an expansion in $\hbar$, $[\hat{O}(\hat{q},\hat{p}),\hat{O}'(\hat{q},\hat{p})] = i\{O(q,p),O'(q,p)\} + \ldots$, where $\{\ ,\ \}$ are the classical Poisson brackets and (q,p) on the r.h.s. are c-numbers (see section A.4.3 for more details). To this order, the time evolution of operators according to the von Neumann equation $\hbar\partial_t\hat{O} = -i[\hat{H},\hat{O}]$ is approximated as $\partial_t O = \{H,O\}$, which is the Hamilton equation for the classical variable O represented by the quantum operator $\hat{O}$. The construction shows that classical dynamics is entirely due to a quantum commutator, and that **the identification "commutator = quantum" is premature**. Similarly, "classical = time independent" cannot be generally correct: the saddle-point equations of the Feynman path integral have the classical, yet generically time-dependent, trajectories of classical mechanics as solutions.

A safer reading is that commutators (and Poisson brackets) make a theory dynamical. In many-body theories, we often encounter states ψ_0 of macroscopic occupation $\mathcal{O}(N)$. Both, classically and quantum mechanically, such states show a high degree of dynamical inertia, the formal statement being that expectation values of commutators, which are $\mathcal{O}(1)$, are much smaller than the occupation numbers themselves. In this interpretation, the time independence of ψ_0 simply reflects the dynamical inertia of a macroscopically occupied state.

It is instructive to discuss the tendency to temporal stationarity from the perspective of the field integral. Consider the action $S[\theta] = N \int_0^\beta d\tau \int d^dr\, (\partial_\tau\theta)^2 + \ldots$, where $N \gg 1$ is a large coupling constant representing the macroscopic occupation of a state with field degree of freedom $\theta = \theta(\tau,\mathbf{r})$, and the ellipses stand for other contributions to the action. Contributions of the "soft-est" non-vanishing time variation have derivatives $\partial_\tau\theta \sim \beta^{-1}\theta$ of the order of the inverse of the extension of the system in the time direction, β (see the figure). (If you wish to formulate this estimate in more precise terms, inspect the Matsubara series expansion of θ.) Such fluctuations have actions scaling as $S[\theta] \sim N\beta \int d^dr\, (\beta^{-1}\theta)^2 = NT \int d^dr\, \theta^2$. This estimate shows that, above temperatures scaling as $\sim N^{-1}$, temporal fluctuations are blocked, and we are left with the "classical" configurations $\theta = \text{const}$. The action of these scale as $S[\theta] = T^{-1} \int d^dr\,(\ldots)$, resembling a thermal Arrhenius factor; hence the denotation "**thermal fluctuations.**"

thermal
fluctu-
ations

5.2.2 The weakly interacting Bose gas

Now, with this background, let us restore the interactions and consider a small but nonzero coupling constant g in the action (5.7). To keep the discussion concrete, we specialize to the case of a free single-particle system, $\hat{H}_0 = \hat{\mathbf{p}}^2/2m$. In this case, variation of the action yields

$$\left(\partial_\tau + \frac{\hat{\mathbf{p}}^2}{2m} - \mu\right)\psi + \frac{g}{L^d}\bar{\psi}\psi\psi = 0, \tag{5.9}$$

Gross–
Pitaevskii
equation

a result known as the (time-dependent) **Gross–Pitaevskii equation**.[8] Following the paradigm to seek solutions reflecting the symmetries of variational equations, we note that translational invariance in space and time suggests a constant solution, $\psi_0 = \text{const}$. To better understand the meaning of the reduced equation

---

[8] E. P. Gross, *Structure of a quantized vortex in boson systems*, Il Nuovo Cimento **20**, 454 (1961); L. P. Pitaevskii, *Vortex lines in an imperfect Bose gas*, Sov. Phys. JETP **13**, 451 (1961).

$$\left(-\mu + \frac{g}{L^d}|\psi_0|^2\right)\psi = 0, \tag{5.10}$$

it is instructive to consider the action evaluated on constant field configurations

$$S[\psi] \stackrel{\psi=\text{const.}}{=} \frac{L^d}{T}\left(-\mu|\psi|^2 + \frac{g}{2L^d}|\psi|^4\right). \tag{5.11}$$

(Notice the similarity of the action to the integrand of the toy problem discussed in section 4.1.1.) Crucially, the stability of the action is now guaranteed by the interaction vertex, no matter how small $g > 0$ is (see the figure.) As a consequence, ψ_0 may be treated as an ordinary integration variable;

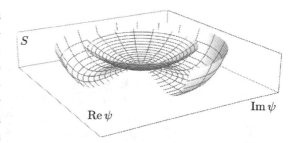

there are no longer any convergence issues. Integration over all field components will produce a partition function $\mathcal{Z}(\mu)$ parametrically dependent on the chemical potential. As is usual in statistical physics, this function is then employed to fix the particle number. (Notice that, with regard to thermodynamics, the interacting system behaves more "naturally" than its non-interacting approximation. This reflects a general feature of bosonic systems: interactions "regularize" a number of pathological features of the ideal gas.)

For negative μ, the action has a single minimum at $\psi = 0$, corresponding to the unique solution of the stationary phase equation in this case. However, for positive μ, the equation is solved by any constant ψ_0 with modulus $|\psi_0| = \sqrt{\mu L^d/g} \equiv \gamma$. There is quite a bit of physics hiding in this innocent-looking result.

▷ For $\mu < 0$ (i.e., above the condensation threshold of the non-interacting system), $\psi_0 = 0$ means that no stable condensate amplitude exists.

▷ However, below the condensation threshold, $\mu > 0$, the proportionality $\bar\psi_0\psi_0 \propto L^d$, reflects the principle of adiabatic continuity: even in the weakly interacting system, the ground state remains extensively occupied.

▷ The equation couples only to the modulus of ψ_0. That is, the solution of the stationary phase equation is continuously degenerate: each configuration $\psi_0 = \gamma\exp(i\phi)$, $\phi \in [0, 2\pi]$, is a solution.

For our present discussion, the last point is the most important. It raises the question which configuration $\psi_0 = \gamma\exp(i\phi)$ is the "right" one?

Without loss of generality, we may choose $\psi_0 = \gamma \in \mathbb{R}$ as a reference solution. This choice amounts to selecting a particular minimum lying in the "Mexican hat" profile of the action shown above. However, it is clear that an expansion of the action around that minimum will be singular: fluctuations $\psi_0 \to \psi_0 + \delta\psi$ that do not leave the azimuthally symmetric well of degenerate minima do not change the action and, therefore, have vanishing expansion coefficients. As a consequence, we will not be able to implement a simple scheme "saddle-point plus quadratic fluctuations;" there is nothing that constrains the deviations $\delta\psi$ to be small. In the next section, we discuss the general principles behind this phenomenon. In section 5.2.4, we then continue to explore its ramifications in the physics of the Bose system.

5.2.3 Spontaneous symmetry breaking

The mechanism encountered here is one of "spontaneous symmetry breaking." To understand the general principle, consider an action $S[\psi]$ with a global continuous symmetry under some transformation g (not to be confused with the aforementioned coupling constant of the Bose gas). The action then remains invariant under a global transformation of the fields such that $\forall i \in M \colon \psi_i \to g\psi_i$, where M is the base manifold, i.e., $S[\psi] = S[g\psi]$. The transformation is "continuous" in the sense that g takes values in some manifold. Since symmetry operations may be applied in succession and reversed, the set of all g defines a symmetry *group*.

Lie group   **INFO** Recall that a **Lie group** is a group that is also a differentiable manifold (see section A.1.1). In other words, it is a group which is smooth and to which concepts of geometry and calculus can be applied. Lie groups such as O(3) or SU(2) are often represented to act in vector spaces via the **fundamental representation**, $\mathbf{v} \to g\mathbf{v}$, or in an **adjoint representation** $\hat{O} \to g\hat{O}g^{-1}$, where the representing matrices, g, are usually labeled by the same symbol as the abstract group elements. (A frequent source of confusion is the premature identification of the former with the latter.) Also recall that, close to the identity, the group may be parameterized as $g = \exp(\phi_a T_a) = \mathbb{1} + \phi_A T_a + \ldots$, where the ϕ_a are real coordinates, and $\{T_a\}$ is a basis of the **Lie algebra**. In the present textbook, not much more information is required. However, in some areas of quantum field theory, high-powered group theory is crucially involved, and we refer interested readers to the large number of textbooks on the subject.

Examples: The action of a Heisenberg ferromagnet is invariant under **rotation** of all spins simultaneously by the same amount, $\mathbf{S}_i \to g\mathbf{S}_i$. In this case, $g \in G = $ O(3), the three-dimensional group of rotations. The action of the displacement fields $\mathbf{u}$ describing the elastic deformations of a solid (phonons) is invariant under the simultaneous **translation** of all displacements $\mathbf{u}_i \to \mathbf{u}_i + \mathbf{a}$, i.e., the symmetry manifold is the d-dimensional translation group $G \cong \mathbb{R}^d$. In the previous section, we encountered a U(1) symmetry under the phase multiplication $\psi_0 \to e^{i\phi}\psi_0$. This phase freedom expresses the global symmetry of quantum mechanics under **transformations by a phase**, a point we discuss in more detail below.

The breaking of continuous symmetries

Given a theory with an action that is invariant under a global symmetry G, two scenarios are conceivable: either the ground states share the invariance properties of the action or they do not. The two alternatives are illustrated in the figure on p. 247 for the Bose system. For $\mu < 0$, the action $S[\bar{\psi}_0, \psi_0]$ has a single ground state at $\psi_0 = 0$. This state is trivially symmetric under the action of $G = \mathrm{U}(1)$. However, for positive μ, i.e., in the situation discussed above, there is an entire manifold of degenerate ground states, defined through the relation $|\psi_0| = \gamma$. These ground states transform into each other under the action of the gauge group. However, none of them is individually invariant. For the other examples mentioned above, the situation is similar.

In general, the ground states of a theory will be invariant under transformation only by the elements of a certain subgroup $H \subseteq G$ (this includes the two extremes $H = \{1\}$ and $H = G$). For example, below the transition temperature, the ground state of the Heisenberg magnet is defined by aligned configurations of spins. Assuming that the spins are oriented along the z-direction, the state is invariant under the abelian subgroup $H \subset \mathrm{O}(3)$ containing all rotations around the z-axis. However, invariance under the full rotation group is broken. Solids represent states where the translation symmetry is fully broken, i.e., all atoms collectively occupy a fixed pattern of spatial positions in space, $H = \{1\}$, etc. Whenever $H \subset G$, we say that the symmetry **spontaneously broken**.

spontaneous
symmetry
breaking

In spite of the undeniable existence of solids, magnets, and Bose–Einstein condensates of definite phase, the notion of ground states not sharing the full symmetry of a theory may appear paradoxical, or at least unnatural. For example, even if any particular ground state of the Mexican hat potential breaks the symmetry, should not all these states enter the partition sum with equal statistical weight, such that the net outcome of the theory is again fully symmetric?

To understand why symmetry breaking is a natural and observable phenomenon, it is instructive to perform a gedanken experiment. Consider the partition function of a classical[9] ferromagnet,

$$\mathcal{Z} = \mathrm{tr}\left(e^{-\beta(H - \mathbf{h}\cdot\sum_i \mathbf{S}_i)}\right),$$

where H is the rotationally invariant part of the energy functional and $\mathbf{h}$ represents a weak external field. (Alternatively, we can think of $\mathbf{h}$ as an internal field, caused by a slight structural imperfection of the system.) In the limit of vanishing field strength, the theory becomes manifestly symmetric,

$$\lim_{N\to\infty} \lim_{h\to 0} \mathcal{Z} \longrightarrow \text{rot. sym.},$$

where the limit $N \to \infty$ serves as a mnemonic indicating that we are considering systems of macroscopic size. However, keeping in mind that the model ought to describe a physical magnet, the order of limits taken above appears questionable. Since the external perturbation couples to a macroscopic number of spins, a more natural description of an "almost" symmetric situation would be

---

[9] The same argument can be formulated for the quantum magnet.

$$\lim_{h \to 0} \lim_{N \to \infty} \mathcal{Z} \longrightarrow ?$$

The point is that the two orders of limits lead to different results. In the latter case, for any $\mathbf{h}$, the $N \to \infty$ system is described by an explicitly symmetry-broken action. No matter how small the magnetic field, the energetic cost to rotate $N \to \infty$ spins against the field is too high, and the ground state $|\mathbf{S}\rangle$ will be uniquely aligned, $\mathbf{S}_i \parallel \mathbf{h}$. When we then send $\mathbf{h} \to 0$ in a subsequent step, that particular state will remain the observable ground state. Although, formally, a spontaneous thermal fluctuation rotating all spins by the same amount $|\mathbf{S}\rangle \to |g\mathbf{S}\rangle$ would not cost energy, that fluctuation can be excluded by entropic reasoning.[10] (By analogy, one rarely observes macroscopic quantities of heated water hopping out of a kettle as a consequence of a concerted thermal fluctuation of the molecules.)

Goldstone modes

The appearance of nontrivial ground states is just one manifestation of spontaneous symmetry breaking. Equally important, residual fluctuations around the ground state lead to the formation of **soft fluctuation modes**, i.e., field configurations $\phi_{\mathbf{q}}$ whose action has the general structure

$$S[\phi] = c_2 \sum_{\mathbf{q}} q^2 |\phi_{\mathbf{q}}|^2 + \mathcal{O}(\phi^4, q^3), \tag{5.12}$$

Goldstone modes

which vanishes in the limit of long wavelengths, $\mathbf{q} \to 0$.[11] Specifically, the soft modes formed on top of a symmetry-broken ground state are called **Goldstone modes**. We already know from the discussion in section 4.1.2 that the presence of soft modes in a theory has striking consequences, among them the buildup of long-range power-law correlations.

EXERCISE Explore the structure of the propagator $G(\mathbf{q}) \equiv \langle \phi_{\mathbf{q}} \phi_{-\mathbf{q}} \rangle$ associated with $S[\phi]$ and convince yourself that the arguments formulated for the specific case of the ϕ^4-theory are of general validity. To this end, notice that, for small $\mathbf{q}$, $G(\mathbf{q}) \sim |\mathbf{q}|^{-2}$, as determined by the smallest power of $\mathbf{q}$ appearing in the action. The power-law behavior of the correlation function implies a homogeneity relation $G(\mathbf{q}/\lambda) = \lambda^2 G(\mathbf{q})$. Show that this scaling relation implies that the Fourier transform $G(r \equiv |\mathbf{r}|) = \langle \phi(\mathbf{r})\phi(\mathbf{0}) \rangle$ obeys the "scaling law" $G(\lambda \mathbf{r}) = \lambda^{-d+2} G(\mathbf{r})$. This, in turn, implies that the real space correlation function also decays as a power law, $G(\mathbf{r}) \sim |\mathbf{r}|^{-d+2}$, i.e., in a "long-range" manner. Explore the breakdown of the argument for an action with a finite mass term. Convince yourself that, in this case, the decay would be exponential, i.e., short-range. We also notice that something strange is happening in dimensions $d \leq 2$. For a discussion of what is going on below this "lower critical dimension," see section 5.2.3.

---

[10] In chapter 6, we will show that this simple picture in fact breaks down in dimensions $d \leq 2$.

[11] Linear derivative terms $q_j |\phi|^2$ would violate spatial reflection symmetry, and non-analytic terms $|q|^\alpha |\phi|^2$ represent spatially nonlocal correlations. (Fourier transform back to real space to investigate this point.) We ignore these cases as exceptional.

Below, we will see that Goldstone mode fluctuations dominate practically all observable properties of symmetry-broken systems. However, before exploring the consequences of their presence, let us in- 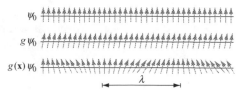 vestigate their cause. To this end, con-
sider the action of a symmetry group element g on a (symmetry broken) ground state ψ_0 (see the middle row of the figure). By definition, $S[g\psi_0] = S[\psi_0]$ still assumes its extremal value. Assuming a fixed group element parameterized as, e.g., $g = \exp(\sum_a \phi_a T_a)$, the action represented as a function of the coordinates vanishes, $S[\phi] = 0$. However, for a weakly fluctuating spatial profile, $\psi_0 \to g(\mathbf{r})\psi_0$ (the bottom row of the figure), some price must be paid, $S[\phi] \neq 0$, where the energy cost depends inversely on the fluctuation rate λ of the coordinate fields $\{\phi_a(\mathbf{r})\}$. The expansion of S in terms of gradients of ϕ is thus bound to lead to a soft mode action of the type (5.12).

In view of their physical significance, it is important to ask how many independent soft modes exist. The answer follows from the geometric picture developed above. Suppose our symmetry group G has dimension r, i.e., its Lie algebra is spanned by r linearly independent generators $T_a, a = 1, \ldots, r$. If the subgroup $H \subset G$ has dimension $s < r$, s of these generators can be chosen in such a way as to leave the ground state invariant. The remaining $p \equiv r - s$ generators create Goldstone modes. In the language of group theory, these generators span the coset space G/H. For example, for the ferromagnet, $H = O(2)$ is the one-dimensional subgroup of rotations around the quantization axis (e.g., the z-axis). Since the rotation group has dimension 3, there must be two independent Goldstone modes. These are generated by the rotation, or angular momentum generators $J_{x,y}$ acting on the z-aligned ground state. The coset space $O(3)/O(2)$ is isomorphic to the 2-sphere, i.e., the sphere traced out by the spins as they fluctuate around the ground state.

Finally, why bother to formulate these concepts in the language of Lie groups and generators? The reason is that the connection between the coordinates parameterizing the Goldstone modes ϕ_a and the original coordinates ψ_i, of the problem, respectively, is often nonlinear and not very transparent. With problems more complex than those mentioned above, it is best to develop a good understanding of the geometry of the problem before turning to specific coordinate choices.

Lower critical dimension

The above discussion indicates that symmetry breaking is to be expected in cases where stationary phase solutions of reduced symmetry are more stable (have lower action) than configurations of higher symmetry. But is this always the case? Or are there situations where a more powerful principle prevents symmetry breaking?

classical
XY-model   To understand that this is indeed the case, consider the **classical XY-model**, a model of classical planar spins of magnitude S defined on a d-dimensional lattice. This system possesses a global $O(2)$ symmetry under simultaneous rotation of all

spins by the same angle θ. Now assume a symmetry-broken configuration where all spins point in the same direction, say, $\mathbf{S}(\mathbf{r}) = S\mathbf{e}_1$. The canonical Goldstone mode action associated with departures from this configuration reads

$$S[\mathbf{S}] = \frac{c}{2} \int d^d r \, (\partial\theta)^2,$$

where $\theta(\mathbf{r})$ is the angular variable relative to $\mathbf{e}_1$ and c is a constant. Now consider the average value $\langle S_1(\mathbf{r}) \rangle = S\langle \cos(\theta(\mathbf{r})) \rangle$ of the local one-component, which we assume is close to unity. An expansion in small angular deviations gives $\langle S_1(\mathbf{r}) \rangle = S - (S/2)\langle \theta^2(\mathbf{r}) \rangle + \cdots$. Performing the Gaussian integral over θ in a momentum space representation, we obtain

$$\langle S_1(\mathbf{r}) \rangle \approx S \left(1 - \frac{1}{2cL^d} \sum_{\mathbf{q}} \frac{1}{q^2} \right) \approx S \left(1 - \frac{1}{2c} \int \frac{d^d q}{(2\pi)^d} \frac{1}{q^2} \right).$$

The crucial observation now is that, in dimensions $d \leq 2$, the integral is divergent. In the marginal case $d = 2$, $\int_{L^{-1}}^{a^{-1}} d^2 q q^{-2} = \pi \ln(L/a)$, where we have used the fact that the momentum integral should be limited by a short- (long-)wavelength cutoff of the order of the inverse lattice spacing (the system size). In the thermodynamic limit, $L \to \infty$, the integral grows without bound, implying that the assumption of an ordered state was ill-founded. We observe that

> The phenomenon of spontaneous symmetry breaking occurs only in systems of sufficiently large dimensionality.

lower critical dimension

The threshold dimension below which entropic mechanisms exclude spontaneous symmetry breaking is called the **lower critical dimension**, d_c.

While derived for a specific model, a moment's thought shows that the conclusion of no symmetry breaking below $d_c = 2$ is of general validity: a general continuous symmetry entails a Goldstone mode action (5.12), which in turn leads to fluctuations diverging in $d \leq d_c = 2$. The consequence, that of no symmetry breaking below $d_c = 2$, is known as the **Mermin–Wagner theorem**:[12]

Mermin–Wagner theorem

> The lower critical dimension of systems with broken continuous symmetries is $d_c = 2$.

Notice that the divergence of the fluctuation integral reflects a competition between the volume element of the integration $\sim q^{d-1} dq$ and the dispersion of the Goldstone modes $\sim q^2$. The former measures the phase space, or "entropy" available to Goldstone mode fluctuations, and the latter their action, or "energy" cost. From this perspective, the Mermin–Wagner theorem is a statement about energy versus entropy, a competition frequently encountered in statistical mechanics and field

---

[12] N. D. Mermin and H. Wagner, *Absence of ferromagnetism or antiferromagnetism in one- or two-dimensional isotropic Heisenberg models*, Phys. Rev. Lett. **17**, 1133 (1966).

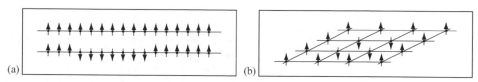

Fig. 5.2   (a) Illustrating the absence of spontaneous symmetry breaking in the one-dimensional Ising model. No matter how low the temperature, the energy cost associated with the creation of a segment of flipped spins is outweighed by the entropy gain. (b) In higher dimensions, entropic factors no longer have the capacity to overpower the extensive growth of energy associated with the formation of mismatched regions.

theory.[13] Its role in symmetry breaking becomes even more evident when we turn to the case of *discrete* symmetries.

Discrete symmetries

While the existence of Goldstone modes relies crucially on the continuity of symmetries, the phenomenon of symmetry breaking does not. As an example, consider the Hamiltonian of the classical **one-dimensional Ising model** $H = -J \sum_{i=1}^{N} S_i S_{i+1} - B \sum_{i=1}^{N} S_i$, where $S_i \in \{1, -1\}$ and $J > 0$ is a ferromagnetic exchange constant. In the absence of an external field B, the model has a $\mathbb{Z}_2$ symmetry under simultaneous exchange of all spins $S_i \to -S_i$, and there are two candidates for symmetry-breaking ground states, $\dots 1, 1, \dots$ and $\dots (-1), (-1), \dots$ (fig. 5.2 (a)).

one-dimensional Ising model

Now, let us imagine that a segment of M consecutive spins were to flip. In doing so, they would incur an energetic cost of $U = 2J$ associated with the alignment mismatch at the domain boundaries. However, there are $\simeq N$ different choices for placing the domain, i.e., the energy loss is counteracted by an entropic factor of $S \sim \ln N$ (exercise). No matter how small T is, for large systems the *free* energy balance $F = U - TS$ for domain creation is positive, implying that the system will favor a disordered phase.

EXERCISE By enumerating the number of spin configurations with the same energy, obtain a formal expression for the classical partition function $\mathcal{Z} = \sum_{\{S\}} \exp(-\beta H)$ of an Ising chain of length N with periodic boundary conditions. (Hint: Consider the number of domain wall configurations.) Making use of Stirling's approximation $\ln n! \simeq n(\ln n - 1)$, determine the temperature dependence of the correlation length (domain size) at low temperatures. Confirm that the system does not order at any finite temperature.

How do these arguments carry over to **Ising systems of higher dimensionality**? For example, in two dimensions (fig.5.2 (b)) a large connected region of M mismatched spins has a circumference of $\mathcal{O}(M^{1/2})$, and therefore incurs an energy cost $U \sim M^{1/2} J$. However, the entropic gain still scales as $\sim \text{const.} \times \ln N$. Thus, for the two-dimensional system, $F \sim J M^{1/2} - T \ln N$. We conclude that it

---

[13] See our previous analysis of instantons, where a large instanton action could be overcompensated by the entropic freedom to place the instanton anywhere in time.

is energetically unfavorable to flip a thermodynamic $M \sim N^{x>0}$ number of spins; the $(d \geq 2)$-dimensional Ising model *does* possess an ordered symmetry-broken low-temperature phase, and we conclude that

> The lower critical dimension of systems with discrete symmetries is $d_c = 1$.

In the next section, we return to the Bose gas and discuss how the general concepts developed here manifest themselves in that context.

5.2.4 Superfluidity

The theory of the weakly interacting superfluid to be discussed below was originally conceived by Bogoliubov, then in the language of second quantization.[14] We will reformulate the theory in the language of the field integral, starting with the action of the weakly interacting Bose gas (5.7). Focusing on temperatures below T_c ($\mu > 0$), we begin by expanding the theory around an arbitrary mean field, say $\bar{\psi}_0 = \psi_0 = (\mu L^d/g)^{1/2} = \gamma$.

INFO Notice that the quantum ground state corresponding to the configuration ψ_0 is unconventional in the sense that it cannot have a definite particle number. The reason is that, according to the correspondence $\psi \leftrightarrow a$ between coherent states and operators, respectively, a non-vanishing functional expectation value of ψ_0 is equivalent to a non-vanishing quantum expectation value $\langle a_0 \rangle$. Assuming that, at low temperatures, the thermal average $\langle \ldots \rangle$ will project onto the ground state $|\Omega\rangle$, we conclude that $\langle \Omega | a_0 | \Omega \rangle \neq 0$, i.e., $|\Omega\rangle$ cannot be a state with a definite number of particles.[15]

The symmetry group acts on this state by multiplication, $\psi_0 \to e^{i\phi}\psi_0$ and $\bar{\psi}_0 \to e^{-i\phi}\bar{\psi}_0$. Knowing that the action of a weakly space-time modulated phase $\phi(r)$, $r = (\mathbf{r}, \tau)$ will be massless, let us introduce coordinates

$$\psi(r) = \rho^{1/2}(r)e^{i\phi(r)}, \qquad \bar{\psi}(r) = \rho^{1/2}(r)e^{-i\phi(r)},$$

where $\rho(r) = \rho_0 + \delta\rho(r)$ and $\rho_0 = \bar{\psi}_0\psi_0/L^d$ is the condensate density. Evidently, the variable $\delta\rho$ parameterizes deviations of the field $\psi(r)$ from the extremum. These excursions are energetically costly, i.e., $\delta\rho$ will turn out to be a massive mode. Also notice that the transformation of coordinates $(\bar{\psi}, \psi) \to (\rho, \phi)$, viewed as a change of integration variables, has a Jacobian unity.

INFO As we are dealing with a (functional) integral, there is freedom as to the choice of integration parameters. (In contrast with the operator formulation, there is no *a priori* constraint for a transform to be canonical.) However, physically meaningful changes of **canonical transformation** representation usually involve **canonical transformations** in the sense that the corresponding transformations of operators would conserve the commutation relations. Indeed,

[14] N. N. Bogoliubov, *On the theory of superfluidity*, J. Phys. (USSR) **11**, 23 (1947) (reprinted in D. Pines, *The Many-Body Problem*, Benjamin, 1961).

[15] However, as usual with grand canonical descriptions, in the thermodynamic limit the relative uncertainty in the number of particles, $(\langle \hat{N}^2 \rangle - \langle \hat{N} \rangle^2)/\langle \hat{N} \rangle^2$, will become vanishingly small.

we saw in the Info block starting on page 148 that the operator transformation $a(\mathbf{r}) \equiv \hat{\rho}(\mathbf{r})^{1/2} e^{i\hat{\phi}(\mathbf{r})}, a^\dagger(\mathbf{r}) \equiv e^{-i\hat{\phi}(\mathbf{r})} \hat{\rho}(\mathbf{r})^{1/2}$ fulfills this criterion. Notice that, as in Lagrangian mechanics, a canonical transformation leaves the $\int$ (variable)$\times \partial_\tau$(momentum) part of the action form-invariant.

We next substitute the density–phase relation into the action and expand to second order around the mean field. Ignoring gradients acting on the density field (in comparison with the "potential" cost of these fluctuations), we obtain

$$S[\rho, \phi] \approx \int d\tau \int d^d r \left(i\delta\rho\partial_\tau\phi + \frac{\rho_0}{2m}(\nabla\phi)^2 + \frac{g\rho^2}{2} \right). \tag{5.13}$$

Here, the second term measures the energy cost of spatially varying phase fluctuations. Notice that fluctuations with $\phi(r) = $ const. do not incur an energy cost – ϕ is a Goldstone mode. Finally, the third term records the energy cost of massive fluctuations from the potential minimum. Equation (5.13) represents the Hamiltonian version of the action, i.e., an action comprising coordinates ϕ and momenta ρ. Gaussian integration over the field $\delta\rho$ leads us to the Lagrangian form of the action (exercise):

$$S[\phi] \approx \frac{1}{2} \int d\tau \int d^d r \left(\frac{1}{g}(\partial_\tau\phi)^2 + \frac{\rho_0}{m}(\nabla\phi)^2 \right). \tag{5.14}$$

Comparison with Eq. (1.4) identifies this action as that for the familiar d-dimensional oscillator. Drawing on the results of chapter 1 (see, e.g., Eq. (1.29)), we find that the energy $\omega_\mathbf{k}$ carried by elementary excitations of the system scales linearly with momentum, $\omega_\mathbf{k} = (g\rho_0/m)^{1/2}|\mathbf{k}|$.

phenomenon of super-fluidity

INFO **Superfluidity** is one of the most counterintuitive and fascinating phenomena displayed by condensed matter systems. Experimentally, the most straightforward access to superfluid states of matter is provided by the helium liquids. As representatives of the many unusual effects displayed by superfluid states of helium, we mention the capability of thin films to flow up the walls of a vessel (if the reward is that on the outer side of the container, a low-lying basin can be reached – the fountain experiment), or to effortlessly propagate through porous media in a way that a normal fluid cannot. Readers interested in learning more about the phenomenology of superfluid states of matter may refer to the classic text by Pines and Nozières.[16]

The actions (5.13) and (5.14) describe the phenomenon of **superfluidity**. To make the connection between the fundamental observable of a superfluid system, the **supercurrent**, and the phase field explicit, let us consider the quantum mechanical current operator

supercurrent

$$\hat{\mathbf{j}}(r) = \frac{i}{2m} \left((\nabla a^\dagger(r))a(r) - a^\dagger(\mathbf{r}, \tau)\nabla a(r) \right)$$

$$\xrightarrow{\text{fun. int.}} \frac{i}{2m} \left((\nabla\bar{\psi}(r))\psi(r) - \bar{\psi}(r)\nabla\psi(r) \right) \approx \frac{\rho_0}{m}\nabla\phi(r), \tag{5.15}$$

[16] D. Pines and P. Nozières, *The Theory of Quantum Liquids: Superfluid Bose Liquids* (Addison-Wesley, 1989).

where the arrow indicates the functional integral correspondence of the operator description and we have neglected all contributions arising from spatial fluctuations of the density profile. (Indeed, these – massive – fluctuations describe the "normal" contribution to the current flow.)

The gradient of the phase variable is therefore a measure of the (super)current flow in the system. The behavior of that degree of freedom can be understood by inspection of the stationary phase equations – i.e., the Hamilton or Lagrange equations of motion – associated with the action (5.13) or (5.14). Turning to the Hamiltonian formulation, one obtains (exercise)

$$i\partial_\tau \phi = -g\rho + \mu, \quad i\partial_\tau \rho = \frac{\rho_0}{m}\nabla^2\phi = \nabla \cdot \mathbf{j}.$$

The second of these equations represents (the Euclidean-time version of) a continuity equation. A current flow with non-vanishing divergence is accompanied by dynamic distortions in the density profile. The first equation tells us that the system adjusts to spatial density fluctuations by a dynamic phase fluctuation. The most remarkable feature of these equations is that they possess steady state solutions with non-vanishing current flow. Setting $\partial_\tau \phi = \partial_\tau \rho = 0$, we obtain the conditions $\delta\rho = 0$ and $\nabla \cdot \mathbf{j} = 0$, i.e., below the condensation temperature, a configuration with a uniform density profile can support a steady state divergenceless (super)current. Notice that a "mass term" in the ϕ-action would spoil this property, i.e.,

> Supercurrent flow is intimately linked to the condensate phase being a Goldstone mode.

EXERCISE Add a fictitious mass term to the ϕ-action and explore its consequences. How do the features discussed above present themselves in the Lagrange picture?

INFO It is instructive to interpret the phenomenology of supercurrent flow from a more microscopic perspective. Steady state current flow in normal environments is prevented by the mechanism of **energy dissipation**: particles constituting the current scatter off imperfections inside the system, thereby converting part of their kinetic energy into elementary excitations, which is observed macroscopically as heat production. How does the superfluid state of matter avoid this mechanism of dissipative loss? Trivially, no energy can be exchanged if there are no elementary excitations to create. This happens, e.g., when the excitations of a system are so high–lying that the kinetic energy stored in the current-carrying particles is insufficient to create them. But this is not the situation that we encounter in the superfluid. As we saw above, there is no energy gap separating the quasi-particle excitations of the system from the ground state. Rather, the dispersion $\omega(\mathbf{k})$ vanishes linearly as $\mathbf{k} \to 0$. However, an ingenious argument due to Landau shows that this very linearity indeed suffices to stabilize dissipationless transport as follows.

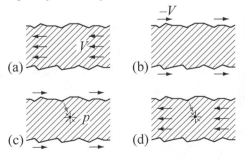

Consider a fluid flowing at uniform velocity $\mathbf{V}$ through a pipe (see figure (a)). Considering a fluid volume of mass M, the current carries a kinetic energy $E_1 = M\mathbf{V}^2/2$. Now, suppose we view the situation from the point of view of the fluid, i.e., we perform a Galilean transformation into its own rest frame, (b). From the perspective of the fluid, the walls of the pipe appear as though they were moving with velocity $-\mathbf{V}$. Now, suppose that frictional forces between fluid and the wall lead to the creation of an elementary excitation of momentum $\mathbf{p}$ and energy $\epsilon(\mathbf{p})$, i.e., the fluid is no longer at rest but carries kinetic energy, (c). After a Galilean transformation back to the laboratory frame (d), one finds that the energy of the fluid after the creation of the excitation is given by (exercise)

$$E_2 = \frac{M\mathbf{V}^2}{2} + \mathbf{p} \cdot \mathbf{V} + \epsilon(\mathbf{p}).$$

Now, since all the energy needed to manufacture the excitation must have been provided by the liquid itself, energy conservation requires that $E_1 = E_2$, or $-\mathbf{p} \cdot \mathbf{V} = \epsilon(\mathbf{p})$. Since $\mathbf{p} \cdot \mathbf{V} > -|\mathbf{p}||\mathbf{V}|$, this condition can only be met if $|\mathbf{p}||\mathbf{V}| > \epsilon(\mathbf{p})$. While systems with a "normal" gapless dispersion, $\epsilon(\mathbf{p}) \sim \mathbf{p}^2$, are compatible with this energy-balance relation (i.e., no matter how small $|\mathbf{V}|$ is, quasi-particles of low momentum can always be excited),

critical velocity both gapped dispersions $\epsilon(\mathbf{p}) \xrightarrow{\mathbf{p}\to 0}$ const. and linear dispersions are disallowed if $\mathbf{V}$ becomes smaller than a certain **critical velocity** V_*. Specifically, for a linear dispersion $\epsilon(\mathbf{p}) = v|\mathbf{p}|$, the critical velocity is given by $V_* = v$. For currents slower than that, the flow is necessarily dissipationless.

We conclude our discussion of the interacting Bose gas with an important remark. Superficially, Eqs. (5.13) and (5.14) suggest that we have managed to describe the superfluid phase in terms of a Gaussian theory. However, one must recall that ϕ is a phase field, defined only modulo 2π. (In (5.13) and (5.14) this condition is understood implicitly.) This phase nature reflects the fact that the Goldstone mode manifold is the group $U(1)$, which has the topology of a circle. It turns out that phase configurations that wind by multiples of 2π as one moves around a center point define the most interesting excitations of the superfluid: **superfluid**

superfluid vortices **vortices.** Coexisting with the harmonic phonon-like excitations discussed above, these excitations lead to a wealth of observable phenomena, to be discussed in chapter 6.

The anticipation of phase vortices exemplifies an important aspect of working with Goldstone modes in general: one should always keep an eye on their "global" geometric structure. Straightforward perturbative expansions such as those leading to Eq. (5.13) are prone to miss topological features. Where present, these usually impact strongly on the physics of symmetry-broken phases and it is crucial to keep them on board.

5.3 Superconductivity

In this section, we discuss the phenomenon of superconductivity from the perspective of the field integral. As in the previous section, the *phase* of a macroscopically occupied state will be the main protagonist. However, unlike the neutral superfluid, this phase is now coupled to the electromagnetic field, and this causes striking phenomenological differences. This section will introduce the conceptual foundations of these differences, cast into the language of the field integral. Although the presentation is self-contained, our focus is on theoretical aspects and some readers may find it useful to first consult an elementary introduction to superconductivity for further motivation.

5.3.1 Basic concepts of BCS theory

The electrical resistivity of many metals and alloys drops abruptly to zero when the material is cooled to a sufficiently low temperature. This phenomenon, which goes by the name of **superconductivity**, was first observed by Kammerlingh Onnes in Leiden in 1911, three years after he first liquefied helium.

Kammerlingh Onnes 1853–1926 was a Dutch physicist awarded the Nobel Prize in Physics in 1913 "for his investigations on the properties of matter at low temperatures which led, inter alia, to the production of liquid helium."

Superconductivity involves an ordered state of conduction electrons in a metal, caused by the presence of a residual attractive interaction at the Fermi surface. The nature and origin of the ordering were elucidated in a seminal work by Bardeen, Cooper and Schrieffer – BCS theory[17] – some 50 years after its discovery: at low temperatures, an attractive pairwise interaction can induce an instability of the electron gas towards the formation of bound pairs of time-reversed states $\mathbf{k} \uparrow$ and $-\mathbf{k} \downarrow$ in the vicinity of the Fermi surface.

INFO From where does an **attractive interaction** between charged particles appear? In conventional[18] (BCS) superconductors, attractive correlations between electrons are due to the exchange of lattice vibrations, or phonons: the motion of an electron through a metal causes a dynamic local distortion of the ionic crystal. Crucially, this process is governed by two totally different time scales. For an electron, it takes a time $\sim E_{\mathrm{F}}^{-1}$ to traverse the immediate vicinity of a lattice ion and to trigger a distortion out of its equilibrium position into a configuration that both particles find energetically beneficial (see the second panel of fig. 5.3).

However, once the ion has been excited it needs a time of $\mathcal{O}(\omega_{\mathrm{D}}^{-1} \gg E_{\mathrm{F}}^{-1})$ to relax back into its equilibrium position. Here, ω_{D} denotes the Debye frequency, the characteristic scale for phonon excitations. This means that, long after the first electron has passed, a second electron may benefit from the distorted ion potential (third panel). Only after the ion has been left alone for a time $> \omega_{\mathrm{D}}^{-1}$ does it relax back into its equilibrium configuration (fourth panel). The net effect of this retardation mechanism is an attractive interaction

---

[17] J. Bardeen, L. N. Cooper, and J. R. Schrieffer, *Microscopic theory of superconductivity*, Phys. Rev. **106**, 162 (1957); *Theory of superconductivity*, Phys. Rev. **108**, 1175 (1957).

[18] Since the discovery of the class of high-temperature cuprate superconductors in 1986, it has become increasingly evident that the physical mechanisms responsible for high-temperature and "conventional" superconductivity are likely to be different. The pairing mechanism in the cuprates remains enigmatic, although the consensus is that its origin is rooted in spin fluctuations.

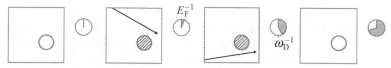

Fig. 5.3   On the dynamical origin of an attractive fermion interaction caused by phonon scattering. For a discussion, see the Info block below.

between the two electrons. Since the maximum energy scale of ionic excitations is given by the Debye frequency, the range of the interaction is limited to energies $\sim \omega_D$ around the Fermi surface. (For a more quantitative formulation, see problem 3.8.11.)

Cooper pairs

Comprising two fermions, the electron–electron bound states, known as **Cooper pairs**, mimic the behavior of bosonic composite particles.[19] At low temperatures, these quasi-bosonic degrees of freedom form a condensate which is responsible for the remarkable properties of superconductors.

To understand the tendency to pair formation, consider the diagram, where the region of attractive correlation is indicated as a shaded ring of width $\sim \omega_D/v_F$ centered around the Fermi surface. Now, consider a two-electron state $|\mathbf{k}\uparrow, -\mathbf{k}\downarrow\rangle$ formed by two particles of (near) opposite momentum and spin.[20] Momentum conserving scattering will lead to the formation of a new state $|(\mathbf{k}+\mathbf{p})\uparrow, -(\mathbf{k}+\mathbf{p})\downarrow\rangle \equiv |\mathbf{k}'\uparrow, -\mathbf{k}'\downarrow\rangle$ of the same opposite-momentum structure. Crucially, the momentum transfer $\mathbf{p}$ may trace out a large set of values of $\mathcal{O}(k_F^{d-1}\omega_D/v_F)$ without violating the condition that the final states be close to the Fermi momentum. Remembering our previous discussion of the RPA approximation, we recognize a familiar mechanism: an *a priori* weak interaction may amplify its effect by conspiring with a large phase-space volume.

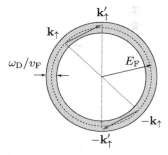

BCS Hamiltonian

Following Bardeen, Cooper, and Schrieffer, we describe the ensuing phenomenology in terms of the **BCS Hamiltonian**

$$\hat{H} = \sum_{\mathbf{k}\sigma} \epsilon_{\mathbf{k}} \hat{n}_{\mathbf{k}\sigma} - \frac{g}{L^d} \sum_{\mathbf{k},\mathbf{k}',\mathbf{q}} c^\dagger_{\mathbf{k}+\mathbf{q}\uparrow} c^\dagger_{-\mathbf{k}\downarrow} c_{-\mathbf{k}'+\mathbf{q}\downarrow} c_{\mathbf{k}'\uparrow} \tag{5.16}$$

[19] However, Cooper pairs typically have a length scale (the coherence length to be introduced below) exceeding the average particle spacing of the electron gas, typically by three orders of magnitude. In this sense, it can be misleading to equate a pair with a single composite particle. However, the crossover between a BEC phase of tightly-bound Fermi pairs and the BCS phase of weakly bound pairs has been explored in the context of atomic condensates. For details of the theory of the BEC–BCS crossover, we refer to the literature, e.g., C. A. R. Sá de Melo, M. Randeria, and J. R. Engletbrecht, *Crossover from BCS to Bose superconductivity: Transition temperature and time-dependent Ginzburg–Landau theory*, Phys. Rev. Lett. **71**, 3202 (1993); J. R. Engletbrecht, M. Randeria and C. A. R. Sá de Melo, *BCS to Bose crossover: Broken-symmetry state*, Phys. Rev. B **55**, 15153 (1997).

[20] We here discuss spin-singlet superconductors. In **spin triplet superconductors**, electrons of equal spin are paired.

where g represents a (positive) constant. The BCS Hamiltonian describes the physics of a thin shell of width $\mathcal{O}(\omega_D)$ centered around the Fermi surface region where a net attractive interaction prevails. It simplifies the realistic momentum-dependent interaction induced by phonon exchange (see problem 3.8.11) to a simple point interaction.[21] As we will see, these simplifications keep the essential physics on board and make the problem amenable to analytical solution.

So far, our discussion has not yet explained why an attractive interaction is so special. Nor have we elucidated the consequences of repeated pair scattering at the Fermi surface. In the following, we will explore these points, first by perturbative methods. A breakdown of perturbation theory will then indicate the formation of a new mean field which we analyze in section 5.3.3 by field-theoretical methods.

5.3.2 Cooper instability

To explore the fate of a Cooper pair under multiple scattering, let us consider the four-point correlation function

$$C(q) = \left\langle \frac{T}{L^d} \sum_k \bar{\psi}_{k+q\uparrow} \bar{\psi}_{-k\downarrow} \frac{T}{L^d} \sum_{k'} \psi_{k'+q\downarrow} \psi_{-k'\uparrow} \right\rangle, \qquad (5.17)$$

describing the propagation of Cooper pairs at characteristic momentum $q = (\mathbf{q}, \omega_m)$. In principle, a Fourier transformation may be applied to obtain information on the space-time correlation function of Cooper pairs; however, this will not be necessary for our purposes.

To calculate $C(q)$, we draw on the perturbative methods introduced in section 4.3. As in the random phase approximation, the density of the electron gas plays the role of a large parameter. We thus expand $C(q)$ in the interaction constant g, retaining only those terms that contain one momentum summation per interaction. Summation over these contributions leads to the ladder diagram series shown

in the figure, where the momentum labels of the Green functions are omitted for clarity. According to the definition of the correlation function, the two Green functions entering the ladder carry momenta $k+q$ and $-k$, respectively. Momentum conservation implies that the consecutive rungs of the ladder also carry near-opposite momenta $p + q$ and $-p$, where p is a summation variable.

EXERCISE Convince yourself that the ladder diagrams shown in the figure are the only diagrams that contain one free momentum summation per interaction vertex.

[21] For a discussion of the important influence of the *repulsive* electron–electron interaction on the physics of the superconductor, see A. I. Larkin and A. A. Varlamov, *Fluctuation phenomena in superconductors*, in *Handbook on Superconductivity: Conventional and Unconventional Superconductors*, eds. K.-H. Bennemann and J. B. Ketterson (Springer-Verlag, 2002).

The central part of the correlation function is the vertex Γ shown in the bottom part of the figure. Translating to an algebraic formulation, one obtains the Cooper version of a Bethe–Salpeter equation $\Gamma_q = g + gTL^{-d}\sum_p G_{p+q}G_{-p}\Gamma_q$, where we have anticipated a solution independent of the intermediate momenta. Solving this equation for Γ_q, we arrive at an equation structurally similar to (4.48):

$$\Gamma_q = \frac{g}{1 - gTL^{-d}\sum_p G_{p+q}G_{-p}}. \tag{5.18}$$

The frequency part of the summation over p gives

$$\frac{T}{L^d}\sum_p G_{p+q}G_{-p} = \frac{1}{L^d}\sum_{\mathbf{p}} \frac{1 - n_F(\xi_{\mathbf{p+q}}) - n_F(\xi_{-\mathbf{p}})}{i\omega_m + \xi_{\mathbf{p+q}} + \xi_{-\mathbf{p}}}$$

$$= \frac{1}{L^d}\sum_{\mathbf{p}} \left(\frac{1}{2} - n_F(\xi_{\mathbf{p}})\right)\left(\frac{1}{i\omega_m + \xi_{\mathbf{p+q}} + \xi_{-\mathbf{p}}} + (\mathbf{q} \leftrightarrow -\mathbf{q})\right),$$

where we have used results from problem 3.8.8 for the frequency summation and, in the last line, noted that $\xi_{\mathbf{p}} = \mathbf{p}^2/2m - \mu = \xi_{-\mathbf{p}}$. For the sake of our present argument, it is sufficient to consider the case of zero external momentum, $q = (\mathbf{0}, 0)$. (The summation for nonzero $\mathbf{q}$ is left as an instructive exercise in Fermi-surface integration.) Once more using the identity $L^{-d}\sum_{\mathbf{p}} F(\epsilon_{\mathbf{p}}) = \int d\epsilon\, \nu(\epsilon) F(\epsilon)$, and remembering that the interaction is limited to a ω_{D}-shell around the Fermi surface, we then obtain

$$\frac{T}{L^d}\sum_p G_p G_{-p} = \int_{-\omega_{\mathrm{D}}}^{\omega_{\mathrm{D}}} d\epsilon\, \nu(\epsilon)\frac{1 - 2n_F(\epsilon)}{2\epsilon} \simeq \nu\int_T^{\omega_{\mathrm{D}}} \frac{d\epsilon}{\epsilon} = \nu\ln\left(\frac{\omega_{\mathrm{D}}}{T}\right), \tag{5.19}$$

where we have used the fact that, at energies $\epsilon \sim T$, the $1/\epsilon$ singularity of the integrand is cutoff by the Fermi distribution function. Substitution of this result back into the expression for the vertex leads to the result

$$\Gamma_{(0,0)} \simeq \frac{g}{1 - g\nu\ln\left(\frac{\omega_{\mathrm{D}}}{T}\right)}.$$

From this, one can read off essential features of the transition to the superconducting phase.

▷ The interaction constant appears in combination with the density of states, where the latter factor measures the number of final states accessible to the pair scattering in figure 5.3. As a consequence, even a weak interaction can lead to sizable effects if ν is large enough.

▷ The effective strength of the Cooper pair correlation grows upon increasing the energetic range ω_{D} or, equivalently, on lowering the temperature. Obviously, something drastic happens when $g\nu\ln(\omega_{\mathrm{D}}/T) = 1$, i.e., when $T = T_{\mathrm{c}}$ with

$$\boxed{T_{\mathrm{c}} \equiv \omega_{\mathrm{D}}\exp\left(-\frac{1}{g\nu}\right)} \tag{5.20}$$

critical
temper-
ature

At this **critical temperature**, the vertex develops a singularity. Since the vertex and the correlation function are related through multiplication by a number of (non-singular) Green functions, the same is true for the correlation function itself.

▷ As we will soon see, T_c marks the transition temperature to the superconducting state. Below T_c, a perturbative expansion around the Fermi sea of the non-interacting system as a reference state breaks down. The Cooper instability signals that we need to look for an alternative ground state or mean field, one that accounts for the strong binding of Cooper pairs.

In the next section, we explore the nature of the superconducting state from a complementary perspective, namely one that gives the identification of that mean field the highest priority.

5.3.3 Mean-field theory of superconductivity

The discussion in the previous section suggests that close to T_c the system develops an instability towards an accumulation of Cooper pairs. The ensuing ground state must combine aspects of the filled Fermi sea with those of macroscopic condensation. Continuing to follow the classic work[17], we now develop a simple mean field analysis revealing the principal characteristics of this state. In section 5.3.4 we then combine the lessons learned in the perturbative and the mean field analyses, respectively, to construct a powerful field integral representation of the superconductor.

John Bardeen 1908–1991
Leon N. Cooper 1930–
John R. Schrieffer 1931–2019
Nobel Laureates in Physics in 1972 for their theory of superconductivity. (Bardeen was also awarded the 1956 Nobel Prize in Physics for his research on semiconductors and discovery of the transistor effect.)

Our starting point here is the assumption that, below T_c, the ground state, $|\Omega_s\rangle$, contains a macroscopic number of Cooper pairs. If so, the creation of another Cooper pair will leave the ground state essentially unchanged, $c^\dagger_{-\mathbf{k}\downarrow}c^\dagger_{\mathbf{k}\uparrow}|\Omega_s\rangle \simeq |\Omega_s\rangle$, and the expectation values

$$\Delta \equiv \frac{g}{L^d}\sum_{\mathbf{k}}\langle\Omega_s|c_{-\mathbf{k}\downarrow}c_{\mathbf{k}\uparrow}|\Omega_s\rangle, \quad \bar{\Delta} \equiv \frac{g}{L^d}\sum_{\mathbf{k}}\langle\Omega_s|c^\dagger_{\mathbf{k}\uparrow}c^\dagger_{-\mathbf{k}\downarrow}|\Omega_s\rangle, \tag{5.21}$$

will assume nonzero values. The anticipation that $\Delta > 0$ assumes non-zero values below the transition temperature T_c and that $\Delta = 0$ above it makes this quantity a candidate **order parameter** of the superconducting (BCS) transition. However, this statement remains a mere presumption until it has been checked self-consistently.

At first sight, the non-vanishing expectation value of Δ indeed looks strange: it implies that the state $|\Omega_s\rangle$ cannot contain a definite number of particles. To make this feature look more natural, remember the bosonic nature of the two-fermion pair state $|\mathbf{k}\uparrow, -\mathbf{k}\downarrow\rangle$. From this perspective, the operator $c^\dagger_{\mathbf{k}\uparrow}c^\dagger_{-\mathbf{k}\downarrow}$ creates a bosonic excitation, and the non-vanishing of its expectation value implies a condensation phenomenon akin to that discussed in section 5.2.

To take this analogy further, we substitute

$$\sum_{\mathbf{k}} c_{-\mathbf{k}+\mathbf{q}\downarrow}c_{\mathbf{k}\uparrow} = \frac{L^d\Delta}{g} + \underbrace{\sum_{\mathbf{k}} c_{-\mathbf{k}+\mathbf{q}\downarrow}c_{\mathbf{k}\uparrow} - \frac{\Delta L^d}{g}}_{\text{small}}$$

into the Hamiltonian (5.16) and retain only terms up to second order in the electron operators. Adding the chemical potential, and setting $\xi_{\mathbf{k}} = \epsilon_{\mathbf{k}} - \mu$, we obtain the mean field Hamiltonian

$$\hat{H} - \mu\hat{N} \simeq \sum_{\mathbf{k}} \left(\xi_{\mathbf{k}} c^\dagger_{\mathbf{k}\sigma}c_{\mathbf{k}\sigma} - \left(\bar{\Delta}c_{-\mathbf{k}\downarrow}c_{\mathbf{k}\uparrow} + \Delta c^\dagger_{\mathbf{k}\uparrow}c^\dagger_{-\mathbf{k}\downarrow} \right) \right) + \frac{L^d|\Delta|^2}{g}. \qquad (5.22)$$

BdG Hamiltonian

While the linearization around a macroscopically occupied state was formulated by Gorkov[22] (inspired by earlier work of Bogoliubov[14] on superfluidity), the linearized Hamiltonian itself is commonly known as the **Bogoliubov–de Gennes (BdG) Hamiltonian**, reflecting the promotion of the mean field description by de Gennes.

Consistently with the anticipation of a ground state of indefinite particle number, the BCS Hamiltonian is not number conserving. To understand its ground state, we proceed in a manner analogous to that of section 2.2.4 (where there appeared a Hamiltonian of similar structure, namely $a^\dagger a + aa + a^\dagger a^\dagger$). Specifically, let us recast the fermion operators in a two-component **Nambu spinor** representation,

Nambu spinor

$$\Psi^\dagger_{\mathbf{k}} = \left(c^\dagger_{\mathbf{k}\uparrow}, c_{-\mathbf{k}\downarrow} \right), \quad \Psi_{\mathbf{k}} = \begin{pmatrix} c_{\mathbf{k}\uparrow} \\ c^\dagger_{-\mathbf{k}\downarrow} \end{pmatrix},$$

comprising $\uparrow$-creation and $\downarrow$-annihilation operators in a single object. It is then straightforward to show that the Hamiltonian assumes the bilinear form

$$\hat{H} - \mu\hat{N} = \sum_{\mathbf{k}} \Psi^\dagger_{\mathbf{k}} \begin{pmatrix} \xi_{\mathbf{k}} & -\Delta \\ -\bar{\Delta} & -\xi_{\mathbf{k}} \end{pmatrix} \Psi_{\mathbf{k}} + \sum_{\mathbf{k}} \xi_{\mathbf{k}} + \frac{L^d|\Delta|^2}{g}.$$

---

[22] L. P. Gor'kov, *About the energy spectrum of superconductors*, Soviet Phys. JETP **7**, 505 (1958).

Being bilinear in operators, the mean field Hamiltonian can be diagonalized by a unitary transformation (a fermionic version of the **Bogoliubov transformation** previously applied to a similarly structured bosonic problem in section 2.2.4)

$$\chi_{\mathbf{k}} \equiv \begin{pmatrix} \alpha_{\mathbf{k}\uparrow} \\ \alpha^{\dagger}_{-\mathbf{k}\downarrow} \end{pmatrix} = \begin{pmatrix} \cos\theta_{\mathbf{k}} & \sin\theta_{\mathbf{k}} \\ \sin\theta_{\mathbf{k}} & -\cos\theta_{\mathbf{k}} \end{pmatrix} \begin{pmatrix} c_{\mathbf{k}\uparrow} \\ c^{\dagger}_{-\mathbf{k}\downarrow} \end{pmatrix} \equiv U_{\mathbf{k}}\Psi_{\mathbf{k}},$$

(under which the anticommutation relations of the new electron operators $\alpha_{\mathbf{k}\sigma}$ are maintained – exercise). Note that the operators $\alpha^{\dagger}_{\mathbf{k}\uparrow}$ are superpositions of $c^{\dagger}_{\mathbf{k}\uparrow}$ and $c_{-\mathbf{k}\downarrow}$, i.e., the quasi-particle states created by these operators contain linear combinations of particle and hole states. Choosing Δ to be real,[23] and setting $\tan(2\theta_{\mathbf{k}}) = -\Delta/\xi_{\mathbf{k}}$, i.e., $\cos(2\theta_{\mathbf{k}}) = \xi_{\mathbf{k}}/\lambda_{\mathbf{k}}$, $\sin(2\theta_{\mathbf{k}}) = -\Delta/\lambda_{\mathbf{k}}$, where

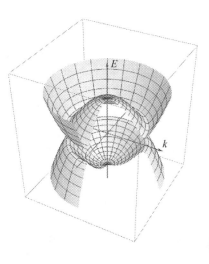

$$\lambda_{\mathbf{k}} = (\Delta^2 + \xi_{\mathbf{k}}^2)^{1/2}, \qquad (5.23)$$

the transformed Hamiltonian takes the form (exercise)

$$\hat{H} - \mu\hat{N} = \sum_{\mathbf{k}\sigma} \lambda_{\mathbf{k}} \alpha^{\dagger}_{\mathbf{k}\sigma} \alpha_{\mathbf{k}\sigma} + \sum_{\mathbf{k}} (\xi_{\mathbf{k}} - \lambda_{\mathbf{k}}) + \frac{\Delta^2 L^d}{g}.$$
$$(5.24)$$

This result shows that the elementary excitations, the **Bogoliubov quasi-particles**, created by $\alpha^{\dagger}_{\mathbf{k}\sigma}$, have a minimum energy Δ known as the **energy gap**. The full dispersion $\pm\lambda_{\mathbf{k}}$ is shown in the figure above. Owing to the energy gap separating filled and empty quasi-particle states, elementary excitations are difficult to excite at low temperatures, implying a rigidity of the ground state.

To determine the ground state wave function, one needs to identify the vacuum state of the algebra $\{\alpha_{\mathbf{k}}, \alpha^{\dagger}_{\mathbf{k}}\}$, i.e., the state that is annihilated by all the quasi-particle annihilation operators $\alpha_{\mathbf{k}\sigma}$. This condition is met uniquely by the state

$$\boxed{|\Omega_s\rangle \equiv \prod_{\mathbf{k}} \alpha_{\mathbf{k}\uparrow}\alpha_{-\mathbf{k}\downarrow}|\Omega\rangle \sim \prod_{\mathbf{k}} \left(\cos\theta_{\mathbf{k}} - \sin\theta_{\mathbf{k}} c^{\dagger}_{\mathbf{k}\uparrow} c^{\dagger}_{-\mathbf{k}\downarrow}\right)|\Omega\rangle}$$

where $|\Omega\rangle$ represents the vacuum state of the fermion operator algebra $\{c_{\mathbf{k}}, c^{\dagger}_{\mathbf{k}}\}$, and $\sin\theta_{\mathbf{k}} = \sqrt{(1 - \xi_{\mathbf{k}}/\lambda_{\mathbf{k}})/2}$. Since the vacuum state of any algebra of canonically conjugate operators is unique, the state $|\Omega_s\rangle$ must, up to normalization, be *the* vacuum state. From the representation above, it is straightforward to verify that the normalization is unity (exercise).

[23] If $\Delta \equiv |\Delta|e^{i\phi}$ is not real, it can be made so by the global gauge transformation $c_a \to e^{i\phi/2}c_a, c^{\dagger}_a \to e^{-i\phi/2}c^{\dagger}_a$. Notice the similarity to the gauge freedom that led to Goldstone mode formation in the previous section! Indeed, we will see shortly that the gauge structure of the superconductor has equally far-reaching consequences.

Finally, we need to solve Eq. (5.21) self-consistently for the input parameter Δ:

$$\Delta = \frac{g}{L^d} \sum_{\mathbf{k}} \langle \Omega_{\mathrm{s}} | c_{-\mathbf{k}\downarrow} c_{\mathbf{k}\uparrow} | \Omega_{\mathrm{s}} \rangle = -\frac{g}{L^d} \sum_{\mathbf{k}} \sin \theta_{\mathbf{k}} \cos \theta_{\mathbf{k}} = \frac{g}{2L^d} \sum_{\mathbf{k}} \frac{\Delta}{(\Delta^2 + \xi_{\mathbf{k}}^2)^{1/2}}$$

$$\simeq \frac{g\Delta}{2} \int_{-\omega_{\mathrm{D}}}^{\omega_{\mathrm{D}}} \frac{\nu(\xi)d\xi}{(\Delta^2 + \xi^2)^{1/2}} = g\Delta\nu \sinh^{-1}(\omega_{\mathrm{D}}/\Delta), \tag{5.25}$$

where we have assumed that the pairing interaction g extends uniformly over an energy scale ω_{D} (over which the density of states ν is roughly constant). Rearranging this equation for Δ, we obtain the important relation

$$\boxed{\Delta = \frac{\omega_{\mathrm{D}}}{\sinh(1/g\nu)} \overset{g\nu \ll 1}{\simeq} 2\omega_{\mathrm{D}} \exp\left(-\frac{1}{g\nu}\right)} \tag{5.26}$$

This is the second time that have we encountered the combination of energy scales on the right-hand side of the equation. Previously, we identified $T_{\mathrm{c}} = \omega_{\mathrm{D}} \exp[-(g\nu)^{-1}]$ as the transition temperature at which the Cooper instability takes place. Our current discussion indicates that T_{c} and the quasi-particle energy gap Δ at $T = 0$ coincide. In fact, this identification might have been anticipated from the discussion above. At temperatures $T < \Delta$, thermal fluctuations are not capable of exciting quasi-particle states above the ground state. One thus expects that $T_{\mathrm{c}} \sim \Delta$ separates a low-temperature phase, characterized by the features of the anomalous pairing ground state, from a Fermi-liquid-like high-temperature phase, where free quasi-particle excitations prevail.

In the mean field approximation, the ground state $|\Omega_{\mathrm{s}}\rangle$ and its quasi-particle excitations formally diagonalize the BCS Hamiltonian. Before proceeding with the further development of the theory, let us pause to discuss a number of important properties of these states.

Ground state

In the limit $\Delta \to 0$, $\sin^2 \theta_{\mathbf{k}} \to \theta(\mu - \epsilon_{\mathbf{k}})$, and the ground state collapses to the filled Fermi sea with chemical potential μ. As Δ becomes nonzero, states in the vicinity of the Fermi surface rearrange themselves into a condensate of paired states. The latter

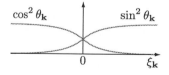

involves the population of single-particle states with energy $\epsilon_{\mathbf{k}} > \mu$. (This follows simply from the energy dependence of the weight function $\sin \theta_{\mathbf{k}}$ entering the definition of the ground state – see the figure.) However, it is straightforward to show that, for any value $g > 0$, the total energy of the ground state, $E_{|\Omega_{\mathrm{s}}\rangle} \equiv \langle \Omega_{\mathrm{s}} | \hat{H} - \mu \hat{N} | \Omega_{\mathrm{s}} \rangle = \sum_{\mathbf{k}} (\xi_{\mathbf{k}} - \lambda_{\mathbf{k}}) + \Delta^2 L^d / g$, is lower than the energy $E_0 \equiv 2 \sum_{|\mathbf{k}| < p_{\mathrm{F}}} \xi_{\mathbf{k}}$ of the Fermi sphere when $g = 0$.

EXERCISE To show that $E_{|\Omega_{\mathrm{s}}\rangle} < E_0$, it is convenient to represent the ground state energy of the Fermi sea as $E_0 = \lim_{\Delta \to 0} E_{|\Omega_{\mathrm{s}}\rangle}$. Use this representation (and the solution of the mean field equation) to verify that the superconductor ground state energy lies below that of the uncorrelated Fermi sea. It is also instructive to ask for the minimum value that $E_{|\Omega_s\rangle}$ may assume upon variation of Δ for fixed g. Show that the solution of the variational equation $\partial_\Delta E_{|\Omega_s\rangle} = 0$ leads back to the mean field equation for Δ discussed above.

Excitations

It is important to distinguish between quasi-particle states and "excitations." Quasi-particle states are the eigenstates of the BCS Hamiltonian. Their energy–momentum relation is shown in the figure on page 264. Notice that there is a positive- and a negative-energy branch of quasi-particles. In the limit $\Delta \to 0$, the quasi-particles evolve into ordinary electrons. By contrast, the energy of *excitations* (as created by the operators $\alpha_{\mathbf{k}}^\dagger$) is always positive. An excitation can be either the creation of a quasi-particle at positive energy or the elimination of a quasi-particle (the creation of a quasi-hole) at negative energy. (In the ground state, all negative-energy quasi-particle states are filled.)

As $\Delta \to 0$, the excitation operators evolve into the operator algebra introduced in Eq. (2.21). Notice that the total number of excitations is equal to the number of quasi-particle states. However, their density of states (non-vanishing for positive energies only) is twice as large. This is so because the dispersion of the excitations is obtained by superimposing the positive branch of the quasi-particle spectrum onto the sign-inverted negative branch. For a particle–hole symmetric system, the quasi-particle spectrum is invariant under sign inversion, which implies that the two branches contribute equally to the density of excitations.

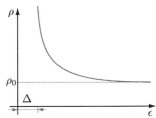

After these general remarks we turn to the specific discussion of the excitations of the BCS superconductor. According to Eq. (5.24), the excitation spectrum is gapped, i.e., it takes a minimum energy Δ to create an excitation above the BCS ground state, compare the continuous spectrum at $g = 0$. To better understand the profile of the spectrum, let us compute the density of excitations $\rho(\epsilon)$,[24] in the vicinity of the Fermi surface, $\epsilon \approx \mu$:

$$\rho(\epsilon) = \frac{1}{L^d} \sum_{\mathbf{k}\sigma} \delta(\epsilon - \lambda_{\mathbf{k}}) = \int d\xi \underbrace{\frac{1}{L^d} \sum_{\mathbf{k}\sigma} \delta(\xi - \xi_{\mathbf{k}})}_{\nu(\xi)} \delta(\epsilon - \lambda(\xi))$$

$$\approx \nu \sum_{s=\pm 1} \int_0^\infty d\xi \, \frac{\delta\left(\xi - s[\epsilon^2 - \Delta^2]^{1/2}\right)}{\left|\partial_\xi[\xi^2 + \Delta^2]^{1/2}\right|} = 2\nu\Theta(\epsilon - \Delta)\frac{\epsilon}{(\epsilon^2 - \Delta^2)^{1/2}}.$$

A schematic plot of the BCS quasi-particle density of states is shown in the figure above. It is apparent that the spectral weight of the quasi-particles has been transferred from the Fermi surface to the interval $[\Delta, \infty]$. The divergence at Δ signals that the majority of quasi-particle states populate the spectral region just above the gap.

---

[24] To distinguish the density of states of the quasi-particles (ν) from that of the excitations, we denote the latter by ρ.

EXERCISE Integrate the result above to confirm that $\int_\Delta^E d\epsilon\, \rho(\epsilon) \xrightarrow{E \gg \Delta} 2E\nu$. This demonstrates that the excitation density of states is indeed twice as large as that of the quasiparticles.

The analysis above explains important aspects of the physics of the BCS superconductor: the instability towards condensation, and the presence of an excitation gap. Indeed, it would be tempting to make the latter responsible for the absence of electrical resistivity: should current flow, the absence of low-lying quasi-particle excitations would require it to be dissipationless. Yet that picture neglects the most important excitation of the system, the collective phase mode. Previously, we made the *ad hoc* decision to set the phase of the order parameter to zero. However, as with the superfluid, the phase represents a Goldstone mode and its presence must have important consequences. Indeed, it will turn out that the phase mode is responsible for most of the electromagnetic properties of the superconductor.

d-wave supercon- ductor

INFO The mean field ansatz considered in this section assumes the pairing of particles of opposite spin into a condensate state that is spatially structureless. Reflecting the trivial rotational symmetry of the condensate, such superconductors are called *spin singlet s-wave* superconductors. Most conventional superconductors fall into this category. However, there are others with more interesting pairing properties. For example, high-temperature superconductors generically exhibit *d*-**wave pairing**, which means that the condensate wave function changes sign under $\pi/2$ rotations around a symmetry axis. Importantly, this behavior implies the existence of zeros of the order parameter function on the Fermi surface. In the vicinity of these regions, the system admits quasi-particle excitations with linearly vanishing, effectively relativistic dispersion, see the exercise below. (The physical properties of such forms of two-dimensional *Dirac quantum matter* are discussed in chapter 9.) Other types of superconductors – such as strontium ruthenate – exhibit *p*-**wave pairing**. In such cases, the orbital wave function of the Cooper pairs is spatially antisymmetric, implying that **spin triplets** with aligned spins are paired (to make the overall pair wave function antisymmetric under particle exchange). The symmetries of the condensate wave function are responsible for numerous fascinating properties of *p*-wave superconductors, notably the presence of *topologically twisted* ground state wave functions, to be discussed in detail in chapter 8.

spin-triplet supercon- ductor

EXERCISE The order parameter of a lattice *d*-wave superconductor is given by $\Delta_\mathbf{k} = \Delta_0(\cos(k_x a) - \cos(k_y a))$, where $\mathbf{k} = (k_x, k_y)^T$ is a two-dimensional lattice momentum and a is the lattice spacing. At two lines in momentum space, $k_x = \pm k_y$, the order parameter vanishes and low-energy quasi-particles persist. Assuming that the kinetic energy of the lattice problem is given by $t_\mathbf{k} = -t(\cos(k_x a) + \cos(k_y a))$, compute the quasi-particle energies of the Bogoliubov Hamiltonian $\hat{H} = \sum_\mathbf{k} \Psi_\mathbf{k}^\dagger((t_\mathbf{k} - \mu)\sigma_3 + \Delta_\mathbf{k}\sigma_1)\Psi_\mathbf{k}$. Show that, at four points in the Brillouin zone, $\mathbf{k} = (\pm 1, \pm 1)\pi/2a$, the quasi-particle energy vanishes. Linearize the Hamiltonian in momentum space around one of these hot spots, say $(+1, +1)\pi/2a$, to show that it assumes the form

$$\hat{H}_{++} = \begin{pmatrix} ta(k_x + k_y) & \Delta_0 a(k_x - k_y) \\ \Delta_0 a(k_x - k_y) & -ta(k_x + k_y) \end{pmatrix}.$$

Apply a unitary transformation to bring $\hat{H}_{++}$ to the form of a **two-dimensional Dirac Hamiltonian** $\hat{H}_{++} \to k_1\sigma_1 + k_2\sigma_2$, where $k_{1/2} = k_x \mp k_y$, we have rescaled $k_1 \to (\Delta_0/t)k_1$, and set $v_F = 1$. Show that the quasi-particle density of states of this Hamiltonian is given by $\rho(\epsilon) \sim |\epsilon|$.

5.3.4 Superconductivity from the field integral

REMARK For notational simplicity, natural units $\hbar = e = 1$ are used throughout.

In the previous section we discussed the physics of the BCS superconductor from the complementary perspectives of perturbation and mean field theory, respectively; the elephant in the room, the phase Goldstone mode, was ignored. We now turn to the field integral formulation, which combines all these facets into the most powerful, and arguably also the most physical, approach to the description of superconductor physics. Our starting point is the coordinate representation of the BCS Hamiltonian (5.16),

$$\hat{H}_{BCS} = \int d^d r \left(c_\sigma^\dagger \left(\frac{1}{2m}(-i\nabla - \mathbf{A})^2 + \phi - \mu \right) c_\sigma - g c_\uparrow^\dagger c_\downarrow^\dagger c_\downarrow c_\uparrow \right).$$

Anticipating the emergence of nontrivial electromagnetic phenomena, we have coupled the single-particle Hamiltonian to a vector potential $\mathbf{A}$ and a scalar potential ϕ. The origin and physical consequences of these fields are discussed somewhat later. Expressed in the form of the coherent state path integral, the corresponding quantum partition function takes the form $\mathcal{Z} = \int D\psi \, e^{-S[\psi]}$, where

$$S[\psi] = \int d\tau d^d r \left(\bar{\psi}_\sigma \left(\partial_\tau + i\phi + \frac{1}{2m}(-i\nabla - \mathbf{A})^2 - \mu \right) \psi_\sigma - g\bar{\psi}_\uparrow \bar{\psi}_\downarrow \psi_\downarrow \psi_\uparrow \right). \quad (5.27)$$

INFO The substitution

$$\partial_\tau \to \partial_\tau + i\phi, \quad -i\nabla \to -i\nabla - \mathbf{A}, \tag{5.28}$$

minimal coupling

is often called the **minimal coupling** of an electromagnetic field. It is minimal in the sense that only orbital and potential coupling of the field are taken into account. (For example, the field–spin interaction is neglected.) At the same time, it is the minimally required framework to endow the theory with **local gauge invariance** under U(1) transformations

$$\psi \to e^{i\theta}\psi, \quad \bar{\psi} \to e^{-i\theta}\bar{\psi}, \quad \phi \to \phi - \partial_\tau \theta, \quad \mathbf{A} \to \mathbf{A} + \nabla\theta, \tag{5.29}$$

where $\theta = \theta(\tau, \mathbf{x})$ is an arbitrary space–time-dependent phase configuration. Minimal coupling introduces the quantum gauge principle into the theory. This should be compared with the discussion of the neutral superfluid, where only invariance under global U(1) transformations was required. In field-theoretical parlance, one sometimes says that a global U(1) symmetry has been *gauged* to become a local one. Importantly, however, this does not mean that the symmetry has become richer. As we will discuss in detail in chapter 10, the different gauges (different choices of θ) of a theory simply represent different ways of describing identical physical states.

In addition, some care must be exercised in interpreting the status of the zeroth component, ϕ, of the potential. Introduced to establish gauge invariance in an imaginary-time formalism, this potential couples as an imaginary contribution to the action (5.27). It thus differs from the true potential, ϕ_M, of real-time electromagnetism by a factor i, where the subscript M stands for the real-time (Minkowski) formalism, in distinction from the present imaginary-time (Euclidean) framework with potential $\phi \equiv \phi_E$. As with the analytic continuation of $i\omega_m \to E$ to real energies, a continuation $i\phi = i\phi_E \to \phi_M$ must be applied before interpreting the potential in a real-time context. To keep the notation slim, we will avoid subscripts E, M unless they are required to avoid confusion.

We know that the field integral cannot be done in closed form, that a perturbative expansion in g will fail below T_c, and that the relevant degrees of freedom of the problem are Cooper pairs. All this suggests a Hubbard–Stratonovich transformation in the Cooper channel (see the Info block on p.236) as a natural approach. We thus start from the identity

$$\exp\left(g \int d\tau\, d^d r\, \bar{\psi}_\uparrow \bar{\psi}_\downarrow \psi_\downarrow \psi_\uparrow \right)$$
$$= \int D\Delta \exp\left(-\int d\tau\, d^d r \left(\frac{1}{g}|\Delta|^2 - \left(\bar{\Delta}\psi_\downarrow \psi_\uparrow + \Delta \bar{\psi}_\uparrow \bar{\psi}_\downarrow \right) \right) \right),$$

where $\Delta(\mathbf{r}, \tau)$ is a complex field subject to periodic boundary conditions, $\Delta(0) = \Delta(\beta)$ (why?). Notice that, if Δ were constant, it would couple to the theory as the mean field order parameter of the BCS Hamiltonian (5.22). This analogy suggests to introduce once again a Nambu representation,

$$\bar{\Psi} = \begin{pmatrix} \bar{\psi}_\uparrow & \psi_\downarrow \end{pmatrix}, \qquad \Psi = \begin{pmatrix} \psi_\uparrow \\ \bar{\psi}_\downarrow \end{pmatrix},$$

comprising particle and hole degrees of freedom in a single object. Expressed in terms of the Nambu spinor, the partition function takes the form

$$\mathcal{Z} = \int D\psi D\Delta \exp\left(-\int d\tau d^d r \left[\frac{1}{g}|\Delta|^2 - \bar{\Psi}\hat{\mathcal{G}}^{-1}\Psi \right] \right),$$

where

$$\hat{\mathcal{G}}^{-1} = \begin{pmatrix} \hat{G}_0^{(p)-1} & \Delta \\ \bar{\Delta} & \hat{G}_0^{(h)-1} \end{pmatrix} \tag{5.30}$$

Gorkov Green function

is the **Gorkov Green function**, and $\hat{G}_0^{(p)-1} = -\partial_\tau - i\phi - (-i\nabla - \mathbf{A})^2/(2m) + \mu$, $\hat{G}_0^{(h)-1} = -\partial_\tau + i\phi + (+i\nabla - \mathbf{A})^2/(2m) - \mu$ represent the non-interacting Green functions of the particle and hole, respectively.

INFO Computing the Ψ-representation of the action for a general single-particle Hamiltonian $\hat{H}$, one finds that $\hat{G}_0^{(p)-1} = -\partial_\tau - \hat{H} + \mu$, and $\hat{G}_0^{(h)-1} = -\partial_\tau + \hat{H}^T - \mu$. (With $\nabla^T = -\nabla$, the expression above is identified as a special case of the more general form.) This representation is actually very revealing: it tells us that the Green function of a hole is obtained from that of the electron by a sign change $\hat{H} \to -\hat{H}$ (the energy of a hole is the negative of the corresponding particle energy) followed by transposition $-\hat{H} \to -\hat{H}^T$, i.e., a quantum time reversal operation (a hole can be imagined as a particle propagating backwards in time). Noting that $\hat{\mathbf{p}}^T = -\hat{\mathbf{p}}$ and $\hat{\mathbf{x}}^T = \hat{\mathbf{x}}$, the pair of Green functions can equivalently be represented as

$$\hat{G}_0^{(\text{p})-1} = -\partial_\tau - \hat{H}(\hat{\mathbf{x}}, \hat{\mathbf{p}}) + \mu, \quad \hat{G}_0^{(\text{h})-1} = -\partial_\tau + \hat{H}(\hat{\mathbf{x}}, -\hat{\mathbf{p}}) - \mu. \qquad (5.31)$$

The Gaussian integration over Grassmann fields can now be performed and yields the formal expression (see the analogous formula (5.4) for the normal electron system)

$$\mathcal{Z} = \int D\Delta \, \exp\left(-\frac{1}{g} \int d\tau d^d r \, |\Delta|^2 + \ln \det \hat{\mathcal{G}}^{-1} \right). \qquad (5.32)$$

This exact representation of the problem will now be our starting point for the mean field analysis of the superconducting phase.

Mean field theory

A variation of the action in Δ generates the mean field equation[25] $g^{-1}\bar{\Delta}(x) - \text{tr}(\hat{\mathcal{G}}(x,x)E_{12}^{\text{ph}}) = 0$, where $x = (\tau \times \mathbf{x})$, the 2×2 matrix E_{ij}^{ph} in Nambu space takes the value of unity at position (i,j) and is zero otherwise. Temporarily ignoring the presence of the fields $(\phi, \mathbf{A})$, and employing a uniform solution ansatz, $\Delta(\tau, \mathbf{x}) \equiv \Delta_0 = \text{const.}$, the equation simplifies to

$$\frac{1}{g}\bar{\Delta}_0 = \text{tr}\left(\begin{pmatrix} -\partial_\tau + \frac{\nabla^2}{2m} + \mu & \Delta_0 \\ \bar{\Delta}_0 & -\partial_\tau - \frac{\nabla^2}{2m} - \mu \end{pmatrix}^{-1} (x, x) \begin{pmatrix} & 1 \\ 0 & \end{pmatrix} \right)$$

$$= \frac{T}{L^d} \sum_p \begin{pmatrix} i\omega_n - \xi_{\mathbf{p}} & \Delta_0 \\ \bar{\Delta}_0 & i\omega_n + \xi_{\mathbf{p}} \end{pmatrix}_{21}^{-1} = \frac{T}{L^d} \sum_p \frac{\bar{\Delta}_0}{\omega_n^2 + \xi_{\mathbf{p}}^2 + |\Delta_0|^2},$$

where, in the second line, we have switched to a frequency–momentum representation and $p = (\omega_n, \mathbf{p})$. Rearranging the equation, we arrive at $g^{-1} = TL^{-d}\sum_p(\omega_n^2 + \lambda_{\mathbf{p}})^{-2}$, where $\lambda_{\mathbf{p}} = (\xi_{\mathbf{p}}^2 + \Delta_0^2)^{1/2} > 0$ (see Eq. (5.23)). Performing the Matsubara sum by the usual contour integration (see section 3.5), and remembering that the range of the interaction is limited to values $|\xi_{\mathbf{p}}| \lesssim \omega_{\text{D}}$, the equation reduces to

$$\frac{1}{g} = \frac{1}{L^d} \sum_{\mathbf{p}} \frac{1 - 2n_{\text{F}}(\lambda_{\mathbf{p}})}{2\lambda_{\mathbf{p}}} = \int_{-\omega_{\text{D}}}^{\omega_{\text{D}}} d\xi \, \nu(\xi) \frac{1 - 2n_{\text{F}}(\lambda(\xi))}{2\lambda(\xi)}.$$

BCS gap equation

Noting that the integrand is even in ξ, and using the identity $1 - 2n_{\text{F}}(\epsilon) = \tanh(\epsilon/2T)$, we arrive at the celebrated **BCS gap equation**

$$\frac{1}{g\nu} = \int_0^{\omega_{\text{D}}} d\xi \, \frac{\tanh\left(\frac{\lambda(\xi)}{2T} \right)}{\lambda(\xi)}, \qquad (5.33)$$

where $\nu(\xi) \simeq \nu$ is assumed approximately constant at the Fermi surface.

[25] For the differentiation of the tr ln term with respect to Δ, we refer to the analogous calculation in Eq. (5.5).

For temperatures $T \ll \Delta_0$, the approximation $\tanh(\lambda/2T) \simeq 1$ gets us back to the $T = 0$ gap equation (5.25) analyzed above. But what happens as the temperature is increased? Since the onset of superconductivity has the character of a second-order phase transition (see section 6.2.3), we expect a singular dependence of its *order parameter* $\Delta_0(T)$ on temperature. Indeed, it turns out (see problem 5.6.2) that the gap $\Delta(T_c) = 0$ vanishes at a **critical temperature** T_c given by, up to numerical constants, Eq. (5.20). In other words, the order parameter begins to form at the same temperature that we previously saw marked the destabilization of the metallic Fermi surface. For temperatures slightly smaller than T_c (see problem 5.6.2),

$$\Delta_0 = \text{const.} \times \sqrt{T_c(T_c - T)} \tag{5.34}$$

shows square root behavior, similarly to the magnetization order parameter of a ferromagnet. This temperature dependence has been accurately confirmed for a large class of superconductors. Also, again up to numerical factors, the critical temperature T_c coincides with the zero-temperature value of the gap Δ, Eq. (5.26).

5.3.5 Ginzburg–Landau theory

Having explored the large-scale profile of the gap function, we next turn our attention to the vicinity of the superconductor transition. For temperatures $\delta T = T_c - T \ll T$, the gap parameter $\Delta \ll T$ is small, providing a small parameter for the expansion of the action (5.32). The result of this expansion will be an effective theory of the superconducting phase transition, elucidating the similarity to the neutral superfluid (as well as important differences).

Our task is to expand $\text{tr} \ln \hat{\mathcal{G}}^{-1}$ in powers of Δ. For simplicity, we continue to ignore the coupling to the external field. (It will be straightforward to put it back in later.) We also define $\hat{\mathcal{G}}_0^{-1} \equiv \hat{\mathcal{G}}^{-1}|_{\Delta=0}$, and $\hat{\Delta} \equiv \left(\begin{smallmatrix} & \Delta \\ \bar{\Delta} & \end{smallmatrix} \right)$, so that

$$\text{tr} \ln \hat{\mathcal{G}}^{-1} = \text{tr} \ln \left[\hat{\mathcal{G}}_0^{-1}(1 + \hat{\mathcal{G}}_0 \hat{\Delta}) \right] = \underbrace{\text{tr} \ln \hat{\mathcal{G}}_0^{-1}}_{\text{const.}} + \underbrace{\text{tr} \ln [1 + \hat{\mathcal{G}}_0 \hat{\Delta}]}_{-\sum_{n=0}^{\infty} \frac{1}{2n} \text{tr}(\hat{\mathcal{G}}_0 \hat{\Delta})^{2n}}.$$

Here we have used the relation $\text{tr} \ln [\hat{A}\hat{B}] = \text{tr} \ln \hat{A} + \text{tr} \ln \hat{B}$.[26] Further, note that only even contributions in $\hat{\Delta}$ survive. The free energy of the non-interacting electron gas, $\text{tr} \ln \hat{\mathcal{G}}_0^{-1}$, provides an inessential contribution to the action, which we will ignore.

To give this formal expansion some meaning, let us consider the second-order term in more detail. By substituting the explicit form of $\hat{\mathcal{G}}_0^{-1}$ it is straightforward to verify that

$$-\frac{1}{2}\text{tr}\,(\hat{\mathcal{G}}_0\hat{\Delta})^2 = -\text{tr}\left([\hat{\mathcal{G}}_0]_{11}\Delta[\hat{\mathcal{G}}_0]_{22}\bar{\Delta}\right) = -\sum_q \underbrace{\frac{T}{L^d}\sum_p [\hat{\mathcal{G}}_{0,p}]_{11}[\hat{\mathcal{G}}_{0,p-q}]_{22}}_{-\frac{T}{L^d}\sum_p G_p G_{-p+q}} \Delta(q)\bar{\Delta}(q),$$

$$\tag{5.35}$$

---

[26] Notice that the relation $\text{tr} \ln [\hat{A}\hat{B}] = \text{tr} \ln \hat{A} + \text{tr} \ln \hat{B}$ applies to non-commutative matrices.

where we have made use of the representation of the composite Green function $\hat{\mathcal{G}}_0$ in terms of the single-particle Green function $G_p^{(\mathrm{p})} = G_p$ and the hole Green function $G_p^{(\mathrm{h})} = -G_{-p}^{(\mathrm{p})} = -G_{-p}$ (see Eq. (5.31)). On combining with the first term in Eq. (5.32), we arrive at the quadratic action for the order parameter field,

$$S^{(2)}[\Delta] = \sum_q \Gamma_q^{-1} |\Delta(q)|^2, \quad \Gamma_q^{-1} = \frac{1}{g} - \frac{T}{L^d} \sum_p G_p G_{-p+q}. \qquad (5.36)$$

This is our second encounter with the vertex function Γ_q^{-1}: in our perturbative analysis of the Cooper channel (see Eq. (5.18)) we identified the same expression. To appreciate the connection, we should revise the general philosophy of the Hubbard–Stratonovich scheme. The field Δ was introduced to decouple an attractive interaction in the Cooper channel. By analogy with the field ϕ used in the development of the RPA approximation to the direct channel, the action of the field $\Delta \sim \bar\psi_\uparrow \bar\psi_\downarrow$ can be interpreted as the "propagator" of the composite object $\bar\psi_\uparrow \bar\psi_\downarrow$, i.e., a quadratic contraction $\sim \langle \bar\Delta \Delta \rangle$ describes propagation in the Cooper channel $\sim \langle \bar\psi_\uparrow \bar\psi_\downarrow \psi_\downarrow \psi_\uparrow \rangle$, as described by a four-point correlation function. This connection is made explicit by comparison of the quadratic action with the direct calculation of the Cooper four-point function given above.

However, in contrast with our discussion in section 5.3.2 (where all we could do was diagnose an instability as $\Gamma_{q=0}^{-1} \to 0$), we are now in a position to comprehensively explore the consequences of the symmetry breaking. Indeed, $\Gamma_{q=0}^{-1} \to 0$ corresponds to a sign change of the quadratic action of the constant order parameter mode $\Delta(q = 0)$. In the vicinity of this point, the constant contribution to the action must scale as $\sim (T - T_c)$, from which we conclude that the action assumes the form

$$S^{(2)}[\Delta] = \int d\tau d^d r \, \frac{r(T)}{2} |\Delta|^2 + \mathcal{O}(\partial \Delta, \partial_\tau \Delta),$$

where $r(T) \sim T - T_c$ and $\mathcal{O}(\partial \Delta, \partial_\tau \Delta)$ denotes temporal and spatial gradients to be discussed shortly.

EXERCISE Use Eq. (5.19) and the expansion

$$n_{\mathrm{F}}(\epsilon, T) - n_{\mathrm{F}}(\epsilon, T_c) \simeq (T - T_c) \partial_T \big|_{T_c} n_{\mathrm{F}}(\epsilon, T) = -\partial_\epsilon n_{\mathrm{F}}(\epsilon, T_c)(T - T_c) \frac{\epsilon}{T_c}$$

to show that $r(T) = \nu t$, where $t = (T - T_c)/T_c$ defines the reduced temperature.

For temperatures below T_c, the quadratic action becomes unstable and – in direct analogy with our previous discussion of the superfluid condensate action – we have to turn to the fourth-order contribution, $S^{(4)}$, to ensure stability of the functional integral. At orders $n > 2$ of the expansion, spatial and temporal gradients can be safely neglected (since $\Delta \ll T$ and they have a low weight compared with the leading gradient contributions to $S^{(2)}$). For $\Delta = $ const., it is straightforward to verify that an expansion analogous to that in Eq. (5.35) gives

$$S^{(2n)} = \frac{1}{2n}\mathrm{tr}\,(\hat{\mathcal{G}}_0\hat{\Delta})^{2n} = \frac{(-)^n}{2n}\sum_p (G_p G_{-p})^n |\Delta|^{2n} \simeq \frac{\nu|\Delta|^{2n}}{2n}\sum_{\omega_l}\int_{-\omega_D}^{\omega_D}\frac{d\xi}{(\omega_l^2 + \xi^2)^n}$$

$$= \text{const.} \times \nu|\Delta|^{2n}\sum_{\omega_l}\frac{1}{\omega_l^{2n-1}} = \text{const.} \times \nu T\left(\frac{|\Delta|}{T}\right)^{2n},$$

where "const." denotes numerical constant factors. Here, in the fourth equality we noted that, for $\omega_D \gg T$, the integral is dominated by its infrared divergence, $\int_0^{\omega_D} d\xi\,(\omega_l^2 + \xi^2)^{-n} \simeq \int_{\omega_l}^{\infty} d\xi\,\xi^{-2n}$. We conclude that contributions of higher order in the expansion are (i) positive and (ii) small in the parameter $|\Delta|/T \ll 1$. This being so, it is sufficient to retain only the fourth-order term (to counterbalance the unstable second-order term). We thus arrive at the effective action

$$S[\Delta] = \int d\tau\,d^d r\left(\frac{r(T)}{2}|\Delta|^2 + u|\Delta|^4 + \mathcal{O}(\partial\Delta, \partial_\tau\Delta, |\Delta|^6)\right),\qquad (5.37)$$

with $u \sim \nu/T_c$, valid in the vicinity of the transition.

INFO Referring to problem 5.6.3 for details, we note that the straightforward inclusion of spatial gradients in the action leads to

$$\boxed{S_{\mathrm{GL}}[\Delta] = \beta\int d^d r\left(\frac{r}{2}|\Delta|^2 + \frac{c}{2}|\partial\Delta|^2 + u|\Delta|^4\right)}\qquad (5.38)$$

Ginzburg–
Landau
action

where $c \sim \rho_0(v_F/T)^2$. This action is known as the **(classical) Ginzburg–Landau action**. It is termed "classical" because (cf. our remarks on page 245) temporal fluctuations of Δ are ignored. Notice that the form of the action might have been anticipated on symmetry grounds alone. Indeed, Eq. (5.38) was proposed by Ginzburg and Landau as an effective action for superconductivity years before the advent of the microscopic theory.[27] A generalization of the action to include temporal fluctuations leads to the time-dependent Ginzburg–Landau theory, to be discussed below.

Equation (5.37) makes the connection between the superconductor and superfluid explicit. In the limit of a constant order parameter, the action does indeed reduce to that of the superfluid condensate amplitude (5.11). Above T_c, $r > 0$ and the unique mean field configuration is given by $\Delta = 0$. However, below the critical temperature, $r < 0$, and a configuration with non-vanishing Cooper pair amplitude Δ_0 becomes favorable:

$$\frac{\delta S[\Delta_0]}{\delta\Delta} = 0 \;\Rightarrow\; \bar{\Delta}_0\left(\frac{r}{2} + 2u|\Delta_0|^2\right) = 0 \;\Rightarrow\; |\Delta_0| = \sqrt{\frac{-r}{4u}} \sim \sqrt{T_c(T_c - T)},$$

[27] V. L. Ginzburg and L. D. Landau, *On the theory of superconductivity*, Zh. Eksp. Teor. Fiz. **20**, 1064 (1950).

cf. the previous estimate (5.34). As with the superfluid, the mean field equation leaves the **phase of the order parameter** $\Delta = e^{2i\theta}\Delta_0$ unspecified.[28] Below the transition, the symmetry of the action is broken by the formation of a ground state with definite phase, e.g., $\theta = 0$. This entails the formation of a soft phase mode exploring deviations from the reference ground state. Pursuing the parallels with the superfluid, one might conjecture that this is a *Goldstone mode* with linear dispersion, and that the system supports dissipationless supercurrents of charged particles: superconductivity.

However, at this point, we have overstretched the analogies to our previous discussion. The argument has ignored the fact that the symmetry broken by the ground state of the superfluid is a global phase U(1). However, as explained on page 268, the microscopic action of the superconductor possesses a gauged U(1) symmetry. As we will discuss below, this difference implies drastic consequences.

5.3.6 Action of the phase mode

The ramifications of the local gauge structure can be explored only in conjunction with the electromagnetic field $(\phi, \mathbf{A})$. We therefore return to the microscopic action (5.32), where $\mathcal{G}$ now represents the full Gorkov Green function (5.30). At this point, we need not specify the origin of the electromagnetic field – whether it is an external probe field, or a fluctuating ambient field. However, we will assume that it is weak enough not to destroy the superconductivity, so that the modulus of the order parameter is still given by the value Δ_0, as described in the previous section.

What is the form of the action describing the interplay of the phase degree of freedom and the electromagnetic field? As usual in field theory, the question may be answered in two ways: by phenomenology or microscopic construction. Here, both approaches are instructive and beautifully complement each other, and so we present them in turn.

Phenomenology

We begin the phenomenological construction of the phase mode action by listing the general criteria fixing its structure.

▷ The phase θ is a Goldstone mode, i.e., the action cannot contain terms that do not vanish in the limit $\theta \to$ const.

▷ We will assume that gradients acting on the phase θ (but not necessarily the magnitude of the phase) and the electromagnetic potentials are small. That is, we will be content with determining the structure of the action to lowest order in these quantities.

---

[28] We denote the phase by 2θ since, under a gauge transformation $\bar{\psi} \to e^{i\theta}\bar{\psi}$, the composite field $\Delta \sim \bar{\psi}\bar{\psi}$ picks up twice that phase. However, in the functional integral, θ is an integration variable, and the factor of 2 a matter of convention.

▷ By symmetry, the action must not contain terms with an odd number of derivatives or mixed gradients of the type $\partial_\tau \theta \nabla \theta$. Respecting the character of the microscopic model, the action must be rotationally invariant.

▷ The action must be invariant under the local gauge transformation (5.29).

The first three criteria are manifestly satisfied by the trial action

$$S[\theta] = \int d\tau d^d r \ \left(c_1(\partial_\tau \theta)^2 + c_2(\nabla \theta)^2\right),$$

where c_1 and c_2 are constants. However, this action is not invariant under a gauge shift of the phase, $\theta(\tau, \mathbf{r}) \to \theta(\tau, \mathbf{r}) + \varphi(\tau, \mathbf{r})$. To endow it with that quality, we introduce a minimal coupling to the electromagnetic potential,

$$\boxed{S[\theta, A] = \int d\tau \, d^d r \ \left(c_1(\partial_\tau \theta + iA_0)^2 + c_2(\nabla \theta - \mathbf{A})^2\right)} \qquad (5.39)$$

To second order in gradients, this action describes uniquely the energy cost associated with phase fluctuations.[29] Notice, however, that the present line of argument does not fix the coupling constants $c_{1,2}$. To determine their values, we need to derive the action microscopically, as in the next section, or invoke further phenomenological input (see info below). Either way one obtains $c_2 = n_s/2m$ and $c_1 = \nu$, where we have defined n_s as the density of the Cooper pair condensate. (For a precise definition, see below.)

INFO Beginning with coefficient c_2, let us briefly discuss the **phenomenological derivation** of the coupling constants. Our starting point is the representation of the current as a functional derivative, $\langle \delta_{\mathbf{A}(\tau, \mathbf{r})} S \rangle = \langle \mathbf{j}(\tau, \mathbf{r}) \rangle$, of the action. (Readers not familiar with this relation will find it explained in the Info block on page 408.) Indeed, the differentiation of the microscopic action Eq. (5.27) yields

$$\left\langle \frac{\delta S}{\delta \mathbf{A}} \right\rangle = -\frac{1}{2m} \left\langle \bar{\psi}_\sigma (-i\nabla - \mathbf{A})\psi_\sigma + ((i\nabla - \mathbf{A})\bar{\psi}_\sigma)\psi_\sigma \right\rangle \equiv \langle \mathbf{j} \rangle,$$

where $\mathbf{j}$ is the quantum current density operator. Now let us assume that a fraction of the electronic states participate in the condensate, such that $\mathbf{j} = \mathbf{j}_n + \mathbf{j}_s$, where the current carried by the normal states of the system, $\mathbf{j}_n$, will not be of concern, while $\mathbf{j}_s$ is the supercurrent carried by the condensate. Let us further assume that those states ψ^s participating in the condensate carry a collective phase θ with a non-vanishing average, $\psi^s = e^{i\theta}\tilde{\psi}^s$, where the residual phase carried by the local amplitude $\tilde{\psi}^s$ averages to zero. Then, concentrating on the phase information carried by the condensate, and neglecting density fluctuations,

---

[29] Actually, this statement must be taken with a pinch of salt: to the list of criteria determining the low-energy action, we should have added symmetry under Galilean transformations, $\mathbf{x} \to \mathbf{x} + \mathbf{v}t$, describing the change to a moving reference frame. The microscopic BCS action has this symmetry, but Eq. (5.39) does not. One may patch up this problem by adding to the action combinations of full derivatives and higher order derivatives in space and time (see I. J. R. Aitchinson *et al.*, *Effective Lagrangians for BCS superconductors at $T = 0$*, Phys. Rev. B **51**, 6531 (1995)). However, here we do not discuss this extension but work with the lowest order approximation (5.39). (We thank Sid Parameswaran for drawing attention to this issue.)

$$\left\langle \frac{\delta S}{\delta \mathbf{A}} \right\rangle \simeq \langle \mathbf{j}^s \rangle \simeq -\frac{n_s}{m} \left\langle \nabla \theta - \mathbf{A} \right\rangle,$$

where $n_s \equiv \bar{\psi}^s \psi^s$ is the density of the condensate.

Now, let us evaluate the relation $\langle \delta_{\mathbf{A}} S \rangle = \langle \mathbf{j} \rangle$ on our trial action (5.39),

$$\langle \mathbf{j} \rangle \stackrel{!}{=} \left\langle \frac{\delta S[A]}{\delta \mathbf{A}} \right\rangle = -2c_2 \langle \nabla \theta - \mathbf{A} \rangle.$$

Comparison with the phenomenological expectation value of the supercurrent operator leads to the identification of the so far unspecified coupling constant $c_2 = n_s/2m$.

Turning to the coupling constant c_1, let us assume that the electron system has been subjected to a weak external potential perturbation $\phi(\tau, \mathbf{r})$. Assuming that the potential fluctuates slowly, we expect screening by a readjustment of the charge density, similar to that generating Thomas–Fermi screening (see the Info block on p. 198). This suggests a contribution to the effective Hamiltonian $\sim (-)\nu \int d^d r \, \phi^2$, where the minus sign reflects a lowering of the energy by screening. Comparing this expression with our trial action – which contains the imaginary time "Euclidean" potential, $A_0 = i\phi$ – we suspect an identification $c_1 \sim \nu$. However, the argument must be taken with caution: the presence of the order parameter implies a modification of the charge density close to the Fermi energy, where the charge redistribution takes place; we need a more microscopic calculation to gain certainty.

Microscopic derivation

REMARK This section is more technical than most in this text. It introduces various tricks routinely applied in the construction of low energy effective actions, and the authors thought it instructive to document such a derivation for a case study of a complexity matching that of research applications. However, readers on a fast track may skip this section, or return to it at a later stage. (To keep the notation concise, we write $\phi = A_0$ for the temporal component of the vector potential. Note, however, that we are in an Euclidean framework, $\phi_{\text{Euclidean}} = i\phi_{\text{Lorentzian}}$.)

The starting point for the derivation of the phase action from first principles is the Gorkov Green function appearing under the tr ln of Eq. (5.32),

$$\hat{\mathcal{G}}_\theta^{-1} = \begin{pmatrix} -\partial_\tau - i\phi - \frac{1}{2m}(-i\nabla - \mathbf{A})^2 + \mu & \Delta_0 e^{2i\theta} \\ \Delta_0 e^{-2i\theta} & -\partial_\tau + i\phi + \frac{1}{2m}(+i\nabla - \mathbf{A})^2 - \mu \end{pmatrix},$$

now minimally coupled to the electromagnetic potential. To simplify the analysis, we assume a constant modulus, Δ_0, of the order parameter, thus neglecting massive fluctuations around the mean field amplitude. We next make use of the gauge freedom inherent in the theory to remove the phase dependence of the order parameter field. To do so, we introduce the unitary matrix $\hat{U} \equiv \text{diag}\left(e^{-i\theta}, e^{i\theta}\right)$ and transform the Green function as $\hat{\mathcal{G}}_\theta^{-1} \to \hat{U}\hat{\mathcal{G}}_\theta^{-1}\hat{U}^\dagger \equiv \hat{\mathcal{G}}^{-1}$, where (exercise)

$$\hat{\mathcal{G}}^{-1} \equiv \begin{pmatrix} -\partial_\tau - i\tilde{\phi} - \frac{1}{2m}(-i\nabla - \tilde{\mathbf{A}})^2 + \mu & \Delta_0 \\ \Delta_0 & -\partial_\tau + i\tilde{\phi} + \frac{1}{2m}(+i\nabla - \tilde{\mathbf{A}})^2 - \mu \end{pmatrix}$$

contains the phase field through a gauge transformed potential $\tilde{\phi} = \phi + \partial_\tau \theta$, $\tilde{\mathbf{A}} = \mathbf{A} - \nabla \theta$. Since $\text{tr} \ln \hat{\mathcal{G}}^{-1} = \text{tr} \ln \left(\hat{U}\hat{\mathcal{G}}^{-1}\hat{U}^\dagger\right)$, the two Green functions represent the theory equivalently.

Josephson
junction

INFO The transformation above conveys an important message – under gauge transformations, the order parameter $\Delta \sim \bar{\psi}_\uparrow \bar{\psi}_\downarrow$ changes as $\Delta \to e^{2i\theta}\Delta$. Lacking gauge invariance, it cannot be an experimentally accessible observable nor serve as a conventional order parameter such as, e.g., the magnetization of a ferromagnet. Perhaps unexpectedly, the **conceptual status of the order parameter field** is a subject of ongoing research, and there are different proposed interpretations (see Ref.[30] for a pedagogical discussion). A pragmatic, and experimentally relevant one, is to notice that, while Δ may be unobservable, overlaps $\langle \Delta\bar{\Delta}' \rangle$ between the order parameters of different superconductors are gauge invariant; the complex conjugate $\bar{\Delta}' \to e^{-i2\theta}$ changes by the opposite phase, and hence may leave direct signatures in physical observables. This principle is used in the **Josephson junction** , which is essentially a system of two tunnel-coupled superconductors (see problem 5.6.6), and represents one of the most important probes into the physics of superconductors. A more radical view[31] abandons the concept of a local order parameter for the superconductor altogether and proposes a topological phase, which by definition is a phase whose ordering principle evades classification by local order parameters (cf. section 8.1). However, in the context of the present discussion, Δ has the status of a complex Hubbard–Stratonovich field. As long as we do not over-interpret its conceptual status, we may proceed pragmatically and describe the low energy physics of the superconductor in terms of this effective field, which, despite these complications, we will continue to call an order parameter, as most people do.

To make further progress, we assume that both spatio-temporal gradients of θ and the electromagnetic potential are small, and expand in powers of $(\tilde{\phi}, \tilde{\mathbf{A}})$.

To facilitate the expansion, it will be useful to represent the 2×2 matrix structure of the Green function through a Pauli matrix expansion:

$$
\hat{\mathcal{G}}^{-1} = -\sigma_0 \partial_\tau - \sigma_3 \left(i\tilde{\phi} + \frac{1}{2m}(-i\nabla - \tilde{\mathbf{A}}\sigma_3)^2 - \mu \right) + \sigma_1 \Delta_0
$$

$$
= \underbrace{-\sigma_0 \partial_\tau - \sigma_3 \left(-\frac{1}{2m}\nabla^2 - \mu \right) + \sigma_1 \Delta_0}_{\hat{\mathcal{G}}_0^{-1}} \underbrace{- i\sigma_3\tilde{\phi} + \frac{i}{2m}\sigma_0[\nabla, \tilde{\mathbf{A}}]_+}_{\hat{\mathcal{X}}_1} \underbrace{- \sigma_3 \frac{1}{2m}\tilde{\mathbf{A}}^2}_{\hat{\mathcal{X}}_2},
$$

where we define $\sigma_0 \equiv \mathbb{1}$ as the unit matrix. Noting that $\hat{\mathcal{X}}_i$ are of first and second order in $\tilde{A}$, respectively, the expansion takes the form

$$
S[\tilde{A}] = -\mathrm{tr}\ln\left(\hat{\mathcal{G}}_0^{-1} - \hat{\mathcal{X}}_1 - \hat{\mathcal{X}}_2 \right) = \mathrm{const.} - \mathrm{tr}\ln\left(1 - \hat{\mathcal{G}}_0[\hat{\mathcal{X}}_1 + \hat{\mathcal{X}}_2] \right)
$$

$$
= \mathrm{const.} + \underbrace{\mathrm{tr}\left(\hat{\mathcal{G}}_0 \hat{\mathcal{X}}_1 \right)}_{S^{(1)}[\tilde{A}]} + \underbrace{\mathrm{tr}\left(\hat{\mathcal{G}}_0 \hat{\mathcal{X}}_2 + \frac{1}{2}\hat{\mathcal{G}}_0 \hat{\mathcal{X}}_1 \mathcal{G}_0 \hat{\mathcal{X}}_1 \right)}_{S^{(2)}[\tilde{A}]} + \cdots, \qquad (5.40)
$$

where we have used the fact that $\hat{\mathcal{X}}_{1,2}$ are of first and second order in the field, respectively.

[30] M. Greiter, *Is electromagnetic gauge invariance spontaneously violated in superconductors?*, Ann. Phys. **319**, 217 (2005).

[31] T. H. Hansson, Vadim Oganesyan, and S. L. Sondhi, *Superconductors are topologically ordered*, Ann. Phys. **313**, 497 (2004).

INFO Expansions of this sort, which in the literature are known as **gradient expansions**, appear frequently in the construction of low energy theories. After the introduction of a Hubbard–Stratonovich field, ϕ, and integration over the microscopic degrees of freedom, one obtains $\operatorname{tr}\ln(\hat{\mathcal{G}}_0^{-1} + \hat{X}[\phi])$. Specifically, in cases where ϕ parameterizes the Goldstone mode of a symmetry-breaking phase, the operator $\hat{X}[\phi]$ coupling to the new field contains only gradients of it (spatially uniform fluctuations cost no action). Assuming that these gradients are small compared with the native action scales hidden in the free Green function $\hat{\mathcal{G}}_0^{-1}$, one then performs a gradient expansion.

Writing $\hat{\mathcal{G}}_l \to \hat{\mathcal{G}}$ for notational simplicity, the first-order action $S^{(1)}$ takes the form (exercise)

$$S^{(1)}[\tilde{A}] = \frac{T}{L^d} \sum_p \operatorname{tr}\left(\mathcal{G}_p \mathcal{X}_1(p,p)\right) = \frac{T}{L^d} \sum_p \operatorname{tr}\left(\mathcal{G}_p\left(i\sigma_3\tilde{\phi}_0 + \frac{1}{m}\sigma_0 \mathbf{p} \cdot \tilde{\mathbf{A}}_0\right)\right),$$

where the subscripts 0 refer to the zero-momentum components of the fields $\tilde{\phi}$ and $\tilde{\mathbf{A}}$. Since the Green function $\hat{\mathcal{G}}$ is even in the momentum, the second contribution $\propto \mathbf{p}$ vanishes by symmetry. Further $(\partial_\tau \theta)_0 = 0 \cdot \theta_0 = 0$, i.e., $\tilde{\phi}_0 = \phi_0$ and $S^{(1)}[\tilde{A}] = iTL^{-d}\sum_p (\mathcal{G}_{p,11} - \mathcal{G}_{p,22})\phi_0$, where the indices refer to particle–hole space. To understand the meaning of this expression, notice that $\mathcal{G}_{p,11} = \langle \bar{\psi}_{\uparrow,p}\psi_{\uparrow,p}\rangle_0$ gives the expectation value of the spin-up electron density operator on the background of a fixed-order parameter background. Similarly, $-\mathcal{G}_{p,22} = -\langle \psi_{\downarrow,p}\bar{\psi}_{\downarrow,p}\rangle_0 = +\langle \bar{\psi}_{\downarrow,p}\psi_{\downarrow,p}\rangle_0$ gives the spin-down density. Summation over frequencies and momenta recovers the full electron density: $TL^{-d}\sum_p(\mathcal{G}_{p,11} - \mathcal{G}_{p,22}) = NL^{-d} \equiv n$, or

$$S^{(1)}[\tilde{A}] = iN\phi_0 = in\int d\tau d^d r\, \phi(\tau,\mathbf{r}).$$

Thus, the first contribution to our action simply describes the electrostatic coupling of the scalar potential to the total charge of the electron system. However, as with the electron plasma discussed earlier, the "correct" interpretation of this expression is $S^{(1)} = 0$: the coupling of the potential to the electron density cancels an equally strong coupling to a positive background, whose presence we leave implicit as always.

We thus turn to the discussion of the second-order contribution $S^{(2)}$. The term containing $\hat{\mathcal{X}}_2$ is reminiscent in structure to the $S^{(1)}$ contribution discussed before. Thus, replacing $\hat{\mathcal{X}}_1$ by $\hat{\mathcal{X}}_2$, we immediately infer that

$$\operatorname{tr}(\hat{\mathcal{G}}\hat{\mathcal{X}}_2) = \frac{n}{2m}\int d\tau\, d^d r\, \mathbf{A}^2(\tau,\mathbf{r}). \tag{5.41}$$

This contribution derives from the diamagnetic contribution $\mathbf{A}^2/2m$ to the electron Hamiltonian and is hence called a **diamagnetic term**. If we had only the diamagnetic contribution, an external field would lead to an increase of the energy. However, we have not yet included the operator $\hat{\mathcal{X}}_1$ in our analysis.

Substituting for $\hat{\mathcal{X}}_1$, and noting that cross-terms $\sim \tilde{\phi}\,\mathbf{p}\cdot\tilde{\mathbf{A}}$ vanish on integration due to their oddness in momenta, we obtain

$$\frac{1}{2}\mathrm{tr}\left(\hat{\mathcal{G}}_0\hat{\mathcal{X}}_1\hat{\mathcal{G}}\hat{\mathcal{X}}_1\right) = \frac{T}{2L^d}\sum_{p,q}\mathrm{tr}\left(-\mathcal{G}_p\sigma_3\tilde{\phi}_q\mathcal{G}_p\sigma_3\tilde{\phi}_{-q} + \frac{1}{m^2}\mathcal{G}_p\sigma_0\mathbf{p}\cdot\tilde{\mathbf{A}}_q\mathcal{G}_p\sigma_0\mathbf{p}\cdot\tilde{\mathbf{A}}_{-q}\right).$$

Here, noting that we are already working at the second order of the expansion, the residual dependence of the Green functions $\hat{\mathcal{G}}$ on the small momentum variable q has been neglected.[32] Originating in the paramagnetic operator $\sim [\mathbf{p},\tilde{\mathbf{A}}]_+$ in the electron Hamiltonian, the magnetic contribution to this expression is called a **paramagnetic term**. Paramagnetic contributions to the action lead to a lowering of the energy in response to external magnetic fields, i.e., the diamagnetic and the paramagnetic term act in competition.

To proceed, we need the explicit form of the Green function, which follows from a bit of Pauli matrix algebra[33] as

$$\mathcal{G}_p = (i\sigma_0\omega_n - \sigma_3\xi_{\mathbf{p}} + \sigma_1\Delta_0)^{-1} = \frac{1}{\omega_n^2 + \xi_{\mathbf{p}}^2 + \Delta_0^2}\left(-i\sigma_0\omega_n - \sigma_3\xi_{\mathbf{p}} + \sigma_1\Delta_0\right). \quad (5.42)$$

Using this representation, and noting that for rotationally invariant functions $F(\mathbf{p}^2)$ (exercise), $\sum_{\mathbf{p}}(\mathbf{p}\cdot\mathbf{v})(\mathbf{p}\cdot\mathbf{v}')F(\mathbf{p}^2) = \frac{\mathbf{v}\cdot\mathbf{v}'}{d}\sum_{\mathbf{p}}\mathbf{p}^2 F(\mathbf{p}^2)$, one obtains

$$\frac{1}{2}\mathrm{tr}\left(\hat{\mathcal{G}}\hat{\mathcal{X}}_1\hat{\mathcal{G}}\hat{\mathcal{X}}_1\right) = \frac{T}{L^d}\sum_{p,q}\frac{\tilde{\phi}_q\tilde{\phi}_{-q}(\omega_n^2 - \lambda_{\mathbf{p}}^2 + 2\Delta_0^2) - \frac{1}{3m^2}\mathbf{p}^2\tilde{\mathbf{A}}_q\cdot\tilde{\mathbf{A}}_{-q}(\omega_n^2 - \lambda_{\mathbf{p}}^2)}{(\omega_n^2 + \lambda_{\mathbf{p}}^2)^2},$$

where $\lambda_{\mathbf{p}}^2 = \xi_{\mathbf{p}}^2 + \Delta_0^2$. We now substitute this result together with the diamagnetic contribution (5.41) back into the expansion (5.40), partially transform back to real space $\sum_q f_q f_{-q} = \int d\tau\, d^d r\, f^2(\tau,\mathbf{r})$, and arrive at the action

$$S[\tilde{A}] = \int d\tau\, d^d r\left(\underbrace{\frac{T}{L^d}\sum_p\frac{\omega_n^2 - \lambda_{\mathbf{p}}^2 + 2\Delta_0^2}{(\omega_n^2 + \lambda_{\mathbf{p}}^2)^2}}_{c_1}\tilde{\phi}^2(\tau,\mathbf{r})\right.$$

$$\left.+ \underbrace{\left(\frac{n}{2m} - \frac{1}{dm^2}\frac{T}{L^d}\sum_p\frac{\mathbf{p}^2(\omega_n^2 - \lambda_{\mathbf{p}}^2)}{(\omega_n^2 + \lambda_{\mathbf{p}}^2)^2}\right)}_{c_2}\tilde{\mathbf{A}}^2(\tau,\mathbf{r})\right).$$

This intermediate result identifies the coupling constants $c_{1,2}$. The last step of the derivation, i.e., the sum over the "fast" momenta p, is now a relatively straightforward exercise. Beginning with the frequency summations, one may note that the denominator has two isolated poles of second order at $\omega_n = \pm i\lambda_{\mathbf{p}}$. Applying

---

[32] i.e., we have set $\sum_{pq}(\mathcal{G}_p\tilde{\phi}_q\mathcal{G}_{p+q}\tilde{\phi}_{-q}) \approx \sum_{pq}(\mathcal{G}_p\tilde{\phi}_q\mathcal{G}_p\tilde{\phi}_{-q})$.

[33] Here, we use the general matrix identity $(v_0\sigma_0 + \mathbf{v}\cdot\sigma)^{-1} = (v_0\sigma_0 - \mathbf{v}\cdot\sigma)/(v_0^2 - \mathbf{v}^2)$. Other useful relations include $(i,j = 1,2,3; \mu,\nu = 1,2,3,4)$ $\sigma_i^2 = 1, i \neq j$, $[\sigma_i,\sigma_j]_+ = 0$, $\sigma_i\sigma_j = i\epsilon^{ijk}\sigma_k$, $\mathrm{tr}\,\sigma_\mu = 2\delta_{\mu,0}$.

the standard summation rules, it is then straightforward to evaluate the coupling constant of the **potential contribution** as (exercise)

$$
c_1 = \frac{1}{L^d} \sum_{\mathbf{p}} \frac{1}{2\lambda_{\mathbf{p}}} \left(n_F(-\lambda_{\mathbf{p}}) \left(\frac{\Delta_0}{\lambda_{\mathbf{p}}}\right)^2 + n_F'(-\lambda_{\mathbf{p}}) \frac{\xi_{\mathbf{p}}^2}{\lambda_{\mathbf{p}}} \right) + (\lambda_{\mathbf{p}} \leftrightarrow -\lambda_{\mathbf{p}})
$$

$$
\approx \frac{1}{L^d} \sum_{\mathbf{p}} \frac{\Delta_0^2}{2\lambda_{\mathbf{p}}^3} \approx \frac{\nu}{4} \int_{-\infty}^{\infty} d\xi \, \frac{\Delta_0^2}{(\xi^2 + \Delta_0^2)^{3/2}} = \frac{\nu}{2}, \tag{5.43}
$$

in accord with the expectation $c_1 \sim \nu$. With the **magnetic contribution**, the situation is more interesting. Here, an analogous construction yields

$$
c_2 = \frac{n}{2m} - \frac{\beta}{dm^2 L^d} \sum_{\mathbf{p}} (n_F(\lambda_{\mathbf{p}})(1 - n_F(\lambda_{\mathbf{p}}))) \simeq \frac{n_s}{2m},
$$

$$
n_s = n \left(1 - \int_{-\infty}^{\infty} d\xi \, \beta(n_F(\lambda)(1 - n_F(\lambda))) \right), \tag{5.44}
$$

where we note that the integrand is strongly peaked at the Fermi surface, i.e., the factor $\mathbf{p}^2 \approx 2m\mu$ can be removed from under the integral, and that[34] $n = 2\nu\mu/d$.

superfluid density

The parameter n_s is called the **superfluid density**. Its definition reflects a competition between dia- and paramagnetism in the magnetic response of the system. At *low temperatures*, $T \ll \Delta_0$, the positivity of $\lambda_{\mathbf{p}} \geq \Delta_0$ implies that $n_F(\lambda_{\mathbf{p}}) \approx 0$, and hence the vanishing of the integral. In this limit, $c_2 \approx n/(2m)$ and $n_s = n$. The diamagnetic contribution to the action dominates which, as we will discuss below, leads to the expulsion of magnetic fields by the superconductor. In the opposite regime of *high temperatures*, $T \gg \Delta_0$, the integral extends over energy domains much larger than Δ_0 and we can approximate $-\int d\xi \, \beta(n_F(\lambda)(1 - n_F(\lambda))) \approx -\int d\xi \, \beta(n_F(\xi)(1 - n_F(\xi)) = 1$ to obtain $c_2 \approx 0$. The near[35] cancellation of dia- and paramagnetic contributions is typical of the response of normal conductors to external magnetic fields.

INFO At intermediate temperatures, the superfluid density n_s lies between zero and the full density. Historically, the concept of a "superfluid density" was introduced prior to the

---

[34] This follows straightforwardly from the two definitions (the factor 2 accounts for the electron spin)

$$
\left.\begin{array}{c}\nu\\n\end{array}\right\} = \frac{2}{L^d} \sum_{\mathbf{p}} \left\{\begin{array}{c}\delta(\mu - \xi_{\mathbf{p}})\\\Theta(\mu - \xi_{\mathbf{p}})\end{array}\right\} = \frac{2}{(2\pi)^d} \int d^d p \left\{\begin{array}{c}\delta(\mu - \xi)\\\Theta(\mu - \xi)\end{array}\right\}.
$$

[35] In reality, the magnetic response of conducting materials is more nuanced. Going beyond second lowest order of perturbation theory in $\mathbf{A}$, a careful analysis of the coupling of a (small) magnetic field to the orbital degrees of freedom of the Fermi gas shows that the cancellation of diamagnetic and paramagnetic contributions is not perfect. The total response is described by a weak diamagnetic contribution, χ_d, a phenomenon known as **Landau diamagnetism**. The diamagnetic orbital response is overcompensated by **Pauli paramagnetism**, a three times larger coupling of opposite sign $\chi_p = -3\chi_d$. For large magnetic fields, the situation changes totally, and more pronounced effects such as **Shubnikov–de Haas oscillations** or even the **quantum Hall effect** are observed.

two-fluid
model

BCS theory, when a phenomenological model known as the **two-fluid model** defined the state-of-the-art understanding of superconductivity. The model assumes that, below the transition, a fraction of the electron system condenses into a dissipationless superfluid of density n_s, while the rest of the electrons remain in the state of a "normal" Fermi liquid of density $n_n = n - n_s$. On this basis, various properties characteristic of superconductivity could be successfully explained.

However, BCS theory showed that the bulk of electrons are oblivious to the pairing interaction and that only a small region around the Fermi surface undergoes restructuring. On this basis, the idea of an independent superposition of two independent "fluids" has been superseded by the microscopic (and not much more complicated) theory of superconductivity.

Before leaving this section let us discuss one last technical point, the **validity of the gradient expansion**. Previously, we expanded the phase action up to leading order in the gradients $\partial_\tau \theta$ and $\nabla\theta$. Indeed, why is such a truncation permissible? This question arises in most derivations of low energy theories, and it is worthwhile to address it in generality. To this end, consider the action of a Hubbard–Stratonovich field $S[\phi]$ and assume its independence under constant displacements $\phi(x) \to \phi(x) + \phi_0$. In this case, the action depends only on derivatives $\partial\phi$, and expansion leads to a series with symbolic momentum space representation

$$S \sim N \sum_q \left((l_0 q)^2 \phi_q \phi_{-q} + (l_0 q)^4 \phi_q \phi_{-q} + \cdots \right),$$

where l_0 is some microscopic reference scale of dimension [length] or [time], the ellipses stand for terms of higher order in q and/or ϕ_q, and N is the large parameter of the theory (which usually appears as a global multiplicative factor). Since only configurations with $S \sim 1$ contribute significantly to the field integral, we obtain the estimate $\phi_q \sim N^{-1/2}(l_0 q)^{-1}$ from the leading order term of the action. This means that terms of higher order in the field variable, $N(l_0 q \phi_q)^{n>2} \sim N^{1-n/2}$, are small in inverse powers of N and can be neglected. Similarly, terms like $N(l_0 q)^{n>2}\phi_q \phi_{-q} \sim (l_0 q)^{n-2}$. As long as we are interested in large-scale fluctuations on scales $q^{-1} \gg l_0$, these terms can be neglected, too.

Notice that our justification for neglecting terms of higher order relies on *two* independent parameters; large N and the smallness of the scaling factor $q l_0$. If $N = 1$ but still $q l_0 \ll 1$, terms involving two gradients but large powers of the field $\sim q^2 \phi_q^{n>2}$ are no longer negligible. Conversely, if $N \gg 1$ but one is interested in scales $q l_0 \simeq 1$, terms of second order in the field weighted by a large number of gradients $\sim q^{n>2}\phi_q^2$ must be taken into account. An incorrect treatment of this point has been the source of numerous errors, including in the published literature.

5.3.7 Meissner effect and Anderson–Higgs mechanism

"Vanishing resistivity" is often the first thing that comes to mind in association with superconductors. However, from a conceptual perspective, the collective ordering of a macroscopically large numbers of quantum degrees of freedom is an equally striking phenomenon. Unlike in a metal, where the phases of individual fermion wave functions tend to cancel out, here they get locked to define one collective degree of freedom, mathematically represented by the order parameter amplitude. Its macroscopic occupation makes the condensate a **macroscopic quantum state** possessing a level of stability against local decoherence that conventional wave functions do not have. The presence of this state is responsible for most of the phenomena of superconductivity, including vanishing resistivity.

We have already emphasized that superconductors share the phase ordering principle with superfluids. However, we now turn to a discussion of the principal difference between the two: the fact that the particles entering the superconducting condensate are charged. To appreciate the consequences, recall that in superfluids long range phase fluctuations could build up at vanishing action cost – the nature of the Goldstone mode. What happens in the superconductor? We have already seen that phase and charge are a canonically conjugate pair and that their fluctuations are inseparably linked. However, in the presence of electromagnetic fields, long range fluctuations of the charge do cost energy, and hence action. We must therefore expect that the phase mode "talks" to the electromagnetic field in a manner that leaves room for the buildup of large (massive) action contributions destroying the Goldstone mode character of the dispersion. This is the principle behind the Anderson–Higgs mechanism.

To describe this phenomenon in more detail, we consider the case where temperature is high enough to inhibit quantum fluctuations of the phase (cf. the remarks on page 245), $\theta(\tau, \mathbf{r}) = \theta(\mathbf{r})$. We also assume the absence of electric fields and work in a gauge $\phi = \partial_\tau \mathbf{A} = 0$. Under these conditions, the action simplifies to

$$S[\mathbf{A}, \theta] = \frac{\beta}{2} \int d^d r \left(\frac{n_s}{m} (\nabla \theta - \mathbf{A})^2 + (\nabla \times \mathbf{A})^2 \right),$$

where we included the action $\frac{1}{4} \int d\tau \, d^d r \, F_{\mu\nu} F^{\mu\nu} \longrightarrow \frac{\beta}{2} \int d^d r \, (\nabla \times \mathbf{A})^2 = \frac{\beta}{2} \int d^d r \, \mathbf{B}^2$ of the static magnetic field. Since the action is gauge invariant under transformations $\mathbf{A} \to \mathbf{A} + \nabla \phi$, $\theta \to \theta + \phi$, we expect that integration over all realizations of θ – a feasible task since $S[\mathbf{A}, \theta]$ is quadratic — will produce a likewise gauge invariant effective action $e^{-S[\mathbf{A}]} \equiv \int D\theta \, e^{-S[\mathbf{A},\theta]}$. As usual with translationally invariant systems, the problem is best formulated in momentum space, where

$$S[\mathbf{A}, \theta] = \frac{\beta}{2} \sum_{\mathbf{q}} \left(\frac{n_s}{m} (i\mathbf{q}\theta_{\mathbf{q}} - \mathbf{A}_{\mathbf{q}}) \cdot (-i\mathbf{q}\theta_{-\mathbf{q}} - \mathbf{A}_{-\mathbf{q}}) + (\mathbf{q} \times \mathbf{A}_{\mathbf{q}}) \cdot (\mathbf{q} \times \mathbf{A}_{-\mathbf{q}}) \right)$$

$$= \frac{\beta}{2} \sum_{\mathbf{q}} \left(\frac{n_s}{m} (\theta_{\mathbf{q}} q^2 \theta_{-\mathbf{q}} - 2i\theta_{\mathbf{q}} \mathbf{q} \cdot \mathbf{A}_{-\mathbf{q}} + \mathbf{A}_{\mathbf{q}} \cdot \mathbf{A}_{-\mathbf{q}}) + (\mathbf{q} \times \mathbf{A}_{\mathbf{q}}) \cdot (\mathbf{q} \times \mathbf{A}_{-\mathbf{q}}) \right).$$

Integration over the components $\theta_{\mathbf{q}}$ is now straightforward and leads to

$$S[\mathbf{A}] = \frac{\beta}{2} \sum_{\mathbf{q}} \left(\frac{n_s}{m} \left(\mathbf{A}_{\mathbf{q}} \cdot \mathbf{A}_{-\mathbf{q}} - \frac{(\mathbf{q} \cdot \mathbf{A}_{\mathbf{q}})(\mathbf{q} \cdot \mathbf{A}_{-\mathbf{q}})}{q^2} \right) + (\mathbf{q} \times \mathbf{A}_{\mathbf{q}}) \cdot (\mathbf{q} \times \mathbf{A}_{-\mathbf{q}}) \right).$$

To bring this result into a more transparent form, we split the vector potential into a longitudinal and a transverse component:[36]

$$\mathbf{A_q} = \underbrace{\mathbf{A_q} - \frac{\mathbf{q}(\mathbf{q} \cdot \mathbf{A_q})}{q^2}}_{\mathbf{A_q^\perp}} + \underbrace{\frac{\mathbf{q}(\mathbf{q} \cdot \mathbf{A_q})}{q^2}}_{\mathbf{A_q^\parallel}} . \tag{5.45}$$

To motivate this decomposition, notice that the transverse component determines the physical magnetic field: $\mathbf{B_q} = i\mathbf{q} \times \mathbf{A_q}$ and $\mathbf{q} \times \mathbf{q} = 0$ imply $\mathbf{B_q} = i\mathbf{q} \times \mathbf{A_q^\perp}$. Also note that the transverse component is properly gauge invariant under transformations $\mathbf{A_q} \to \mathbf{A_q} + i\mathbf{q}\phi_\mathbf{q}$ (since $(\mathbf{q}\phi_\mathbf{q})^\perp = 0$).

Substituting the above decomposition, a straightforward calculation shows that

$$S[\mathbf{A}] = \frac{\beta}{2} \sum_\mathbf{q} \left(\frac{n_s}{m} + q^2 \right) \mathbf{A_q^\perp} \cdot \mathbf{A_{-q}^\perp} \tag{5.46}$$

Anderson–Higgs mechanism

This action describes the essence of the **Anderson–Higgs mechanism**:[37] integration over all realizations of the phase mode has generated a mass term for the gauge field. As we will see, the mass term, proportional to the superfluid density, is responsible for the curious magnetic phenomena displayed by superconductors in the Higgs phase.

INFO In the early 1960s, the discovery of the Anderson–Higgs mechanism had a disruptive influence both in the physics of superconductivity and that of the standard model of particle physics. The latter described the fundamental forces of nature in terms of the exchange of gauge fields, which upon quantization assumed the identity of gauge particles. The problem was that, to explain the short range nature of the weak interaction, the corresponding bosonic particles – the W and Z vector bosons – had to be quite massive, about two orders of magnitude heavier than the proton. Peter Higgs,[37] and independently several other researchers, proposed a solution by coupling the gauge bosons to a complex scalar particle with a symmetry breaking ϕ^4-type action similar to that of the superconducting order parameter. Via the Anderson–Higgs mechanism, the fluctuations of this field introduced the otherwise inexplicable mass generation. Upon quantization, the Higgs field assumed the identity of a scalar boson,[38] with accurately predictable mass, as it had to determine the known masses of the vector bosons. In 2012 signatures matching these expectations where seen at the LHC facility at CERN. Subsequent research corroborated

Higgs particle

the discovery of the **Higgs particle**, whose existence is now considered established. The theoretical research leading to the experimental discovery five decades later was awarded with the 2013 Nobel Prize.

---

[36] The terminology "longitudinal component" emphasizes the fact that $\mathbf{F_q^\parallel}$ is the projection of a vector field $\mathbf{F_q}$ onto the argument vector $\mathbf{q}$. Correspondingly, the "transverse component" is the orthogonal complement of the longitudinal component.

[37] P. W. Anderson, *Plasmons, gauge invariance, and mass*, Phys. Rev. **130**, 439 (1962); P. W. Higgs, *Broken symmetries and the masses of gauge bosons*, Phys. Rev. Lett. **13**, 508 (1964).

[38] The same is true for the order parameter. For example, the zero and one occupation number states of the order parameter "phase boson" define the states of a superconducting qubit in quantum computation.

To understand the phenomenological consequences of gauge field mass generation, let us vary the action (5.46) in $\mathbf{A}$ (keeping in mind the transversality condition $\mathbf{q} \cdot \mathbf{A}_{\mathbf{q}}^{\perp} = 0$, we henceforth drop the superscript $\perp$) to obtain $(n_{\mathrm{s}}/m + q^2)A_{\mathbf{q}} = 0$, or

$$\left(\frac{n_{\mathrm{s}}}{m} - \nabla^2\right) A(\mathbf{r}) = 0. \tag{5.47}$$

first London equation

Remembering that $\mathbf{B} = \nabla \times \mathbf{A}$, multiplication of this equation by $\nabla\times$ produces the **first London equation**

$$\boxed{\left(\frac{n_{\mathrm{s}}}{m} - \nabla^2\right) \mathbf{B}(\mathbf{r}) = 0} \tag{5.48}$$

For $n_{\mathrm{s}} \neq 0$, this equation does not have a non-vanishing constant solution and we conclude that:

> A bulk superconductor cannot accommodate a magnetic field.

Meissner effect

This phenomenon is known as the **Meissner effect**. To understand what happens at the interface between vacuum threaded by a constant magnetic field $\mathbf{B}_0$ and a superconductor, we can solve the London equation to obtain $\mathbf{B}(x) \sim \mathbf{B}_0 \exp(-x/\lambda)$, where

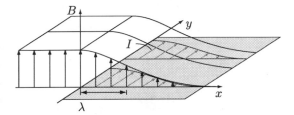

$$\lambda = \sqrt{\frac{m}{n_{\mathrm{s}}}}$$

penetration depth

is known as the **penetration depth** and x is the direction perpendicular to the interface (see the figure). The physical mechanism behind the Meissner phenomenon is as follows: previously, we saw that the magnetic response of a superconductor is fully diamagnetic. That is, in response to an external field, diamagnetic screening currents will be generated. The magnetic field generated by these currents counteracts the unwanted external field. To see this explicitly, we obtain the current density induced by the field by differentiating the first term[39] of the action in $\mathbf{A}$:

$$\mathbf{j}(\mathbf{r}) = \frac{\delta}{\delta \mathbf{A}(\mathbf{r})} \int d^d r \, \frac{n_{\mathrm{s}}}{2m} \mathbf{A}^2 = \frac{n_{\mathrm{s}}}{m} \mathbf{A}(\mathbf{r}), \tag{5.49}$$

second London equation

i.e., the current density is directly proportional to $\mathbf{A}$. This is the **second London equation**. Since the vector potential and the magnetic field show the same decay, Eqs. (5.47) and (5.48), the current density also decays exponentially inside the superconductor. However, in doing so, it annihilates the external field.

INFO For increasing external field strength, eventually a point must be reached where the superconductor is no longer able to sustain zero resistance. It is instructive to explore this

[39] See remarks on page 275. Only the first term, generated by the field–matter current, contributes to the current. It is instructive to think about this point.

breakdown from the perspective of the effective Ginzburg Landau field theory (5.38). To this end, we consider the minimal coupling of the latter to the external field. Remembering that under gauge transformations $\Delta \to \Delta e^{2i\phi}$, i.e., $\nabla\Delta\,\nabla\bar{\Delta} \to (\nabla + 2i\nabla\phi)\Delta(\nabla - 2i\nabla\phi)\bar{\Delta}$, the gauge invariant extension of the action reads as

$$S_{\mathrm{GL}}[\Delta,\bar{\Delta}] = \beta \int d^d r \left(\frac{r}{2}|\Delta|^2 + \frac{c}{2}|(\nabla - 2i\mathbf{A})\Delta|^2 + g|\Delta|^4 \right),$$

where $\mathbf{A}$ transforms as $\mathbf{A} \to \mathbf{A} + \nabla\phi$. To monitor what happens as $|\mathbf{A}| \propto |\mathbf{B}|$ increases, we consider the mean field equation (exercise)

$$\left(r + c(-i\nabla - 2A)^2 + 4g|\Delta|^2 \right) \Delta = 0,$$

where $r < 0$, since we are below the superconducting transition temperature. Due to the positivity of the third term, a non-vanishing solution can exist only if the first two terms add to a negative contribution. This in turn requires that the eigenvalues of the kinetic operator satisfy

$$\mathrm{EV}(-i\nabla - 2\mathbf{A})^2 \overset{!}{<} \frac{|r|}{c}.$$

Formally, $(-i\nabla - 2\mathbf{A})^2$ is the kinetic energy operator of a particle of mass $1/2$ and charge $q = 2$ in a uniform magnetic field. Its eigenvalues are the Landau levels, $\omega_c (n + 1/2)$, $n = 0, 1, \ldots$, familiar from elementary quantum mechanics (see section 8.4.7). Here, ω_c is the cyclotron frequency, $\omega_c = qB/m = 4B$. Thus, a finite pairing amplitude can be obtained only if $|r|/c$ is larger than the energy of the lowest Landau level, or

$$B < B_{c2} \equiv \frac{|r|}{2c}.$$

For magnetic fields larger than this, the energy needed to expel the field is larger than the maximum gain of condensation energy $S[\Delta]$, and superconductivity breaks down. For a detailed discussion of the ways by which strong fields force their entry into superconductors – which include the formation of flux tube lattices in type II superconductors[40] – we refer to the literature.[41]

To conclude this section, let us discuss the most prominent superconducting phenomenon, **absence of electrical resistivity**. Assume a gauge with $\phi = 0$, where an external electric field $\mathbf{E}$ is represented by $\mathbf{E} = -i\partial_\tau \mathbf{A}$. In this case, a time derivative of the second London equation (5.49) gives $-i\partial_\tau \mathbf{j} = -in_s/m\,\partial_\tau \mathbf{A} = n_s/m\,\mathbf{E}$. Continuing back to real times we conclude that

$$\boxed{\partial_t \mathbf{j} = \frac{n_s}{m}\mathbf{E}}$$

i.e., in the presence of an electric field the current increases linearly at a rate inversely proportional to the carrier mass and proportional to the carrier density. The unbound increase of current is indicative of uniform and dissipationless motion of the condensate particles inside the superconductor. Now, an unbound increasing

[40] A. A. Abrikosov, *On the magnetic properties of superconductors of the second group*, Sov. Phys. JETP **5**, 1174 (1957).

[41] For example, L. D. Landau and E. M. Lifshitz, *Course of Theoretical Physics, Vol. 9 – Statistical Physics 2* (Butterworth–Heinemann, 1981).

current is clearly unphysical, i.e., what the relation above really tells us is that a superconductor cannot maintain non-vanishing field gradients.

EXERCISE Assuming that each particle is subject to Newton's equation of motion $m\ddot{\mathbf{r}} = \mathbf{E}$, obtain the current–field relation above. How would the relation between field and current change if the equation of motion contained a friction term (modeling dissipation) so that $m\ddot{\mathbf{r}} = -\frac{m}{\tau}\dot{\mathbf{r}} + \mathbf{E}$?

5.4 Field Theory of the Disordered Electron Gas

REMARK Building on the perturbative analysis of section 4.4, we here construct a powerful field-theoretical approach to the physics of the disordered electron gas. As with its predecessor, this section is optional. Readers on a fast track may advance directly to the next chapter.

In the perturbative analysis of section 4.4 we identified diffusion modes as long-range excitations of the disordered electron gas (see the top panel of the figure for a schematic). As with the Cooper channel mode of the superconductor, or the RPA mode of the electron plasma, we expect these diffusion modes to be the excitations of an effective low energy theory. However, the perturbative analysis also hinted at a structures beyond diffusion. Without becoming quantitative, we reasoned that it supports excitations which are locally classically diffusive, but globally of non-classical nature.

A semiclassical illustration of such a process was shown in the right panel of fig. 4.11, and is redrawn in a more stripped down way in the middle panel of the figure shown here. Here, the propagators swap roles at a vertex to traverse a scattering path in reversed order. Owing to this reversal of amplitudes, such quantum contributions to transition probabilities, known as a **weak localization** corrections, require time reversal invariance as a microscopic symmetry. (For example, an external magnetic field would make the classical action picked up along a path dependent on its direction of traversal, and in this way kill this weak localization correction.)

weak lo-
calization

However, there exist other processes which do not require this symmetry, e.g., that shown in the bottom panel. Looking at such representations from afar, they resemble "interaction vertices" of effective degrees of freedom whose elementary propagators are the diffusion modes. On this basis, we suspect the existence of an effective quantum field theory that, unlike the theories discussed so far, must be strongly nonlinear and possess interaction vertices of higher complexity. We would like to derive this theory and understand what it has to say about the physics of the disordered electron gas at large distance and low temperature scales.

5.4.1 Mean-field theory and spontaneous symmetry breaking

As usual, our starting point is a microscopic description of the system, which in this case would be defined by the replicated partition sum (4.56). However, instead of working with the full partition sum, we simplify matters by exploiting the fact that impurity scattering does not transfer energy. In concrete terms, this means that the frequencies ω_n and ω_m carried by the solid and dashed line amplitudes in the diagrams above remain conserved throughout the entire scattering process. We have also seen that long range contributions arise only if the two frequencies have different signs. Finally, we will eventually perform an analytic continuation to real frequencies, say, $\omega_n \to E_F + \frac{1}{2}(\omega + i\delta)$ and $\omega_m \to E_F - \frac{1}{2}(\omega + i\delta)$, with $\delta > 0$. On this basis, it is sufficient to consider the stripped down functional $\mathcal{Z} = \int D\psi \, \exp(-S[\psi])$, where

$$S[\psi] = \int d^d r \, \bar{\psi} \left(H_0 + V - E_F + \frac{1}{2}(\omega + i\delta)\tau_3 \right) \psi \tag{5.50}$$

is the action of a doublet of fields $\psi = (\psi^+, \psi^-)$ representing the two quasi-particle amplitudes of retarded and advanced signature $\pm i\delta$, and τ_3 acts in this causal subspace. As before, $\psi^\pm = \{\psi^{\pm a}\}$ carry a replica index $a = 1, \ldots, R$, and $H_0 = p^2/2m$ is the free electron Hamiltonian. Correlation functions may be generated by adding suitable source terms to the action. For example, we may define

$$J = (a_+ \tau^+ \delta(\mathbf{r} - \mathbf{r}_1) + a_- \tau^- \delta(\mathbf{r} - \mathbf{r}_2)) \otimes P^1, \tag{5.51}$$

where $\tau^\pm = \frac{1}{2}(\tau_1 \mp \tau_2)$ and P^1 is a projector onto the first replica subspace. Adding the source contribution $\int d^d r \, \bar{\psi} J \psi$ to the action and differentiating, we have

$$\frac{\partial^2 \mathcal{Z}}{\partial a^+ \partial a^-} = \langle \bar{\psi}^{+1}(\mathbf{r}_1)\psi^{-1}(\mathbf{r}_1)(\bar{\psi}^{-1}(\mathbf{r}_2)\psi^{+1}(\mathbf{r}_2)) \rangle = -G^+_{E_F + \frac{1}{2}\omega}(\mathbf{r}_2, \mathbf{r}_1) G^-_{E_F - \frac{1}{2}\omega}(\mathbf{r}_1, \mathbf{r}_2),$$

where $G^\pm_E = (E - H_0 - V)^{-1}$, and on the r.h.s. the implicit replica limit $R \to 0$ leads to the vanishing of the determinant factor $\det(G^+ G^-)^R$. In this way, the correlation function diagrammatically depicted above is extracted from the functional integral. However, to keep the notation simple, we will neglect source contributions in the notation until we need them.

As in our previous analysis, we start by averaging the action over the Gaussian distributed potential with second moment $\langle V(\mathbf{r})V(\mathbf{r}') \rangle = \delta(\mathbf{r} - \mathbf{r}')/(2\pi\nu\tau)$, where we represent the variance in terms of the scattering time via Eq. (4.60). Performing the Gaussian average, we obtain the fixed-energy restriction of the action governing the partition sum (4.56)

$$S[\psi] = \int d^d r \left(\bar{\psi} \left(H_0 - E_F + \frac{1}{2}(\omega + i\delta)\tau_3 \right) \psi - \frac{1}{4\pi\nu\tau} (\bar{\psi}\psi)^2 \right). \tag{5.52}$$

We now subject the quartic term to a somewhat strange looking operation: $(\bar{\psi}\psi)^2 = (\bar{\psi}^\alpha \psi^\alpha)(\bar{\psi}^\beta \psi^\beta) = -(\psi^\beta \bar{\psi}^\alpha)(\psi^\alpha \bar{\psi}^\beta)$, where $\alpha = (\sigma, a)$, $\sigma = \pm$ is a composite index comprising the causal signature and the replica index, and the minus sign accounts for the fact 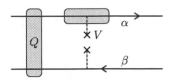 that in the last step we exchange an odd number of Grassmann variables. The motivation behind this reordering is indicated in the figure, where the vertical shaded area represents the bilinears $\bar{\psi}^\alpha \psi^\alpha$ coupling to the scattering potential V. However, we suspect the "slow" variables of the theory to be pairs of amplitudes correlated by scattering off the same potential vertices, bilinears $\psi^\alpha \bar{\psi}^\beta$, as indicated by the vertical shaded area. The reordering of the averaged scattering vertex groups two of these vertical composite objects in such a way that they can be subjected to a subsequent Hubbard–Stratonovich decoupling.

INFO The concrete formulation of the Hubbard–Stratonovich decoupling depends on how ambitious one is. For time reversal invariant systems, the inclusion of both processes, with and without partially reversed sense of traversal (exemplified in the second and first diagram in the figure above), requires Hubbard–Stratonovich decoupling in the direct and in the Cooper channel (see Info block on page 236). However, to keep things simple, here we assume time reversal symmetry breaking by, e.g., a magnetic field whose presence in the action we keep suppressed for notational simplicity. In this case, the assumption that $\psi^\alpha \bar{\psi}^\beta$ are slowly fluctuating bilinears is equivalent to direct channel decoupling in terms of a single field.

Q1: In the limit $\omega, \delta \to 0$, what are the symmetries of the action? And how do the symmetries reduce for finite ω, δ?

Q2: Formulate a Hubbard–Stratonovich transformation to decouple the quartic term of an *anti*-hermitian matrix field $A = \{A^{\alpha\beta}\}$. Then integrate out the fermions to obtain the A-field action,

$$S[A] = -\pi\nu\tau \int d^d r \operatorname{tr}(A^2) - \operatorname{tr}\ln G^{-1}[A],$$

$$G^{-1}[A] = E_F + \frac{1}{2}(\omega + i\delta)\tau_3 - H_0 + A. \tag{5.53}$$

Q3: Neglecting the symmetry breaking shift ω, vary the action in A to obtain a mean field equation. Comparing with the previous discussion of the SCBA in section 4.4.2, discuss why the *non-hermitian* configuration $\bar{A} = x + iy\tau_3$ with real x, y defines a good ansatz. Show that $y = 1/2\tau$ is a solution for the imaginary part and argue why we may anticipate that the real part (which we neglect throughout) does not play an interesting role.

Q4: Explain why the finiteness of y represents a symmetry breaking phenomenon. Show that $A = iQ/(2\tau)$, where $Q = T\tau_3 T^{-1}$ defines a natural representation of the Goldstone modes. Notice that transformations of the unbroken symmetry group $T = \mathrm{bdiag}(T^+, T^-) \in \mathrm{U}(R) \times \mathrm{U}(R)$ (where bdiag means block diagonal) commute with τ_3 and hence do not change the saddle point. A concrete parameterization of the Goldstone mode manifold $\mathrm{U}(2R)/\mathrm{U}(R) \times \mathrm{U}(R)$ is given by

$$T = \exp(W), \qquad W = \begin{pmatrix} & B \\ -B^\dagger & \end{pmatrix}, \tag{5.54}$$

where the generator matrices are anti-hermitian for unitarity and off-diagonal to act non-redundantly on the diagonal mean field, τ_3.

Notice that the fields on the Goldstone mode configurations satisfy the equation $Q^2 = \mathbb{1}$. We interpret this as a nonlinear constraint imposed on the degrees of freedom of the ensuing low energy field theory. In general, field theories whose degrees of freedom are subject to nonlinear constraints are called **nonlinear σ-**

nonlinear
σ-models

models. For example, the field theory describing Goldstone mode fluctuations of the directional configurations $\mathbf{n}(\mathbf{r})$, $\mathbf{n}^2 = 1$ of a Heisenberg ferromagnet is another example of a nonlinear σ-model. The present one is a matrix model, as opposed to the vector nature of the magnetic model. One may expect the nonlinearity of the field theory to lead to a complex hierarchy of "interaction vertices" when the theory is formulated in terms of linear coordinates (the B-generator matrices) of the non-linear field manifold. As outlined above, this is what we expect for the disordered electron gas, so we are on the right track in identifying its low energy effective description. In the next section, we derive the corresponding action in explicit terms.

A1: For $\omega, \delta = 0$ the action possesses a continuous global symmetry under rotations, $\psi \to T\psi$, $\bar\psi \to \bar\psi T^{-1}$, $T \in \mathrm{U}(2R)$. The symmetry is broken by ω, δ to $\mathrm{U}(R) \times \mathrm{U}(R)$ where the two factors act in the replica subspaces of definite causality $\pm$.

A2: The Hubbard–Stratonovich decoupling follows standard protocol, except that we need to account for the α-indices. We thus consider the functional integral multiplied by the unit normalized Gaussian integral

$$1 = \int DA \, \exp\left(\pi\nu\tau \int d^dr \, \mathrm{tr}(A^2)\right),$$

where the integral is over anti-hermitian matrices $A = \{A^{\alpha\beta}\}$ for convergence. (Why does anti-hermiticity safeguard convergence?) Next shift $A \to A + \Psi\bar\Psi/(2\pi\nu\tau)$. The generated quartic term cancels against the impurity vertex, and the mixed term is rewritten as $-\mathrm{tr}(A\Psi\bar\Psi) = +\bar\Psi A\Psi$, where we again gain a sign from Grassmann exchange. Doing the integral over the Ψ-fields, we obtain the A-field action.

A3: Variation of the action $\delta/\delta A^{\beta\alpha}(\mathbf{r})$ leads to the equation

$$2\pi\nu\tau A^{\alpha\beta}(\mathbf{r}) + G[A]^{\alpha\beta}(\mathbf{r}, \mathbf{r}) = 0. \tag{5.55}$$

We now seek for an ansatz reflecting the symmetries of the problem: translational invariance of the averaged action, and rotational symmetry in causal space, infinitesimally broken by $i\delta$. This motivates the quoted configuration where, for $y > 0$, the imaginary contribution $iy\tau_3$ guarantees that we do not hit the real cut line of the Green function (something to be categorically avoided). An imaginary part in the mean field solution will upgrade the symmetry breaking in causal space $i\delta \rightarrow iy$ from infinitesimal δ to a finite value, y. In other words, it will represent a mechanism of spontaneous symmetry breaking in the wake of which we expect Goldstone modes. By contrast, a real part only shifts the energy, and fluctuations around it are predictably massive. We therefore focus on the imaginary part throughout, and obtain the simplified equation

$$2\pi\nu\tau y\,\tau_3 = -\mathrm{Im} \int \frac{d^d p}{(2\pi)^d} \frac{1}{-\xi_p + i\tau_3 y} = -\pi\nu\tau_3, \tag{5.56}$$

where $\xi_p = p^2/2m - E_{\mathrm{F}}$ and we note that the integral on the r.h.s. defines the density of states. This equation leads to $y = 1/2\tau$, showing that the imaginary part of the mean field is defined by the inverse of the scattering time. Indeed, the above mean field equation is identical to the SCBA equation (4.63), and hence the imaginary part of the mean field equals that of the scattering self-energy, $1/2\tau$.

A4: Thinking of $i\delta$ as the analog of an infinitesimal magnetic field, the solution $A = i\tau_3/(2\tau)$ has a status analogous to a finite magnetization. Much as the Goldstone modes of a magnet are obtained by acting on the mean field magnetization vector with the symmetry group, we note here that, under symmetry transformations, $\bar{\psi}A\psi \rightarrow \bar{\psi}T^{-1}AT\psi$. This leads to the representation of the Goldstone modes quoted in the question.

5.4.2 Low-energy field theory

In the previous section, we showed that the disordered electron gas is subject to a symmetry breaking principle, and identified the corresponding Goldstone mode degrees of freedom, $Q = T\tau_3 T^{-1}$. We next identify their effective action, which we expect to combine aspects of classical diffusion with non-classical quantum interference. Again, we have the choice between an explicit derivation – substitution $A \rightarrow iQ/(2\tau)$ into the Hubbard–Stratonovich action Eq. (5.53) followed by a leading order expansion in gradients acting on the Q-fields, and in the symmetry breaking parameter ω[42] – and a symmetry based approach. Referring to Ref.[43] for numerous examples of explicit derivations of σ-model actions for different realizations of disordered electron systems, here we proceed by symmetry reasoning.

[42] A more cautious approach would decompose the full field manifold $A = A_{\mathrm{gs}} + A_{\mathrm{m}}$ into the Goldstone mode fluctuations, A_{gs}, and a massive contribution, A_{m}. However, it can be shown (and is physically expected) that the integration over massive fluctuations does not alter the Goldstone mode action in a significant way.

[43] K. B. Efetov, *Supersymmetry in Disorder and Chaos* (Cambridge University Press, 1999).

Q1: What symmetries must an action of the Goldstone mode matrices $Q = Q^{\alpha\beta}(\mathbf{r})$ satisfy? Justify why, to leading order in gradients and in the symmetry breaking parameter ω, the action assumes the form $S[Q] = \int d^d r \left(c_1 \mathrm{tr}(\nabla Q)^2 + c_2 \frac{\omega}{2} \mathrm{tr}(\tau_3 Q)\right)$, with as yet unspecified constants $c_{1,2}$.

Q2: The action above looks deceptively simple, but it is not. Substitution of the parameterization (5.54) followed by expansion in the generators will lead to an action of infinite order in the matrices B. We tentatively identify the vertices of $\mathcal{O}(B^{2n})$ with the effective interactions indicated in the diagrams on p.286. For $\omega \to 0$, there is no obvious small parameter suppressing the contribution of these nonlinearities. However, for finite ω, we are probing time scales of $\mathcal{O}(\omega^{-1})$. In this case, classical diffusion must emerge as a dominant principle, and nonlinearities will play a lesser role.

To substantiate this intuition, expand the action to leading order in B and show that it reduces to

$$S^{(2)}[B] = -\int d^d r \, \mathrm{tr}\left(B(8c_1 \nabla^2 + 2c_2 \omega)B^\dagger\right). \tag{5.57}$$

Q3: To make sense of this expression, and fix the coupling constants, we need to compute from it an observable. At this point, we remember that our functional integral contains a source term (5.51). Show that the inclusion of the source and expansion to leading order in B generates an additional contribution $2c_2[a_+\mathrm{tr}(B^{11}(\mathbf{r}_1)) + a_-\mathrm{tr}(B^{\dagger 11}(\mathbf{r}_2))]$ in the action. Here, B^{11} is a complex-valued matrix field. The trace in $S^{(2)}$ implies that this field couples to the action as $-\int B^{11}(8c_1\nabla^2 + 2c_2\omega)\overline{B^{11}}$. With this action, differentiation in the source parameters leads to

$$\frac{\partial^2 \mathcal{Z}}{\partial a^+ \partial a^-} = 4c_2^2 \langle B^{11}(\mathbf{r}_1)\overline{B^{11}}(\mathbf{r}_2)\rangle = -4c_2^2(8c_1\nabla^2 + 2c_2\omega)^{-1}(\mathbf{r}_2, \mathbf{r}_1). \tag{5.58}$$

We identify this expression with the diffusion mode, previously computed by perturbative methods, see Eq. (4.64). More precisely, the previous result was for the imaginary-time diffusion mode expressed in momentum space. Referring to chapter 7 for precise translation rules between imaginary-time and real-time frequencies, suggest a meaningful substitution of the Matsubara frequency ω_m by the real frequency $\omega + i\delta$ (Hint: As with all propagators, the denominator of the diffusion mode must never vanish), and on this basis identify the action parameters as $c_1 = D\pi\nu/4$, $c_2 = \pi i\nu$. We thus arrive at

$$\boxed{S[Q] = \frac{\pi\nu}{4}\int d^d r \left(D\,\mathrm{tr}(\nabla Q)^2 + 2i\omega\,\mathrm{tr}(\tau_3 Q)\right)} \tag{5.59}$$

for the action of the **nonlinear σ-model of the disordered electron gas**.

The field theory (5.59) provides the most advanced tool for the description of quantum state propagation in disordered media. We fixed its structure by demanding that a quadratic expansion around the trivial mean field $Q = \tau_3$ in Goldstone mode generators should yield a diffusive propagator. However, the principal advantage of the field theory is that it allows us to go beyond quadratic order and

**Hikami
box**

study the modifications of classical diffusion by quantum fluctuations. For example, the third diagram shown in the figure at the beginning of the section contains two structures resembling fourth-order "interaction vertices" (known as **Hikami boxes** or **encounter regions**). Indeed one can show (this is a tedious exercise, yet much simpler than a first principle computation by perturbative methods) that these corrections arise from a fourth-order expansion of Q in B, followed by a contraction of the structure $B\,(BB^\dagger BB^\dagger)^2\,B^\dagger$. Each Wick contraction of the B-fields brings down a factor $1/(Dq^2 - i\omega)$, which presents us with two questions: under which circumstances are these modifications of classical diffusion relevant? And if they are, how do we get them under control? The first of these is comparatively easy to answer. At large length scales, $q \to 0$, and long time scales, $\omega \to 0$, the Goldstone mode propagators become singular. In the diagrammatic analysis of the field theory,[44] we will encounter loop integrals such as $\int d^d q\, 1/(Dq^2 - i\omega)$. These become infrared singular for small ω and $d \le 2$, indicating that something happens below the critical dimension $d = 2$. However, at this point we are not yet in a position to explore this fluctuation-dominated regime. We will return to this question in chapter 6 after the required methodology of the renormalization group has been introduced.

Here are the answers to the above questions.

A1: In the absence of explicit symmetry breaking, the action must be invariant under uniform transformations $Q \to TQT^{-1}$. It must also be rotationally invariant in the spatial coordinates, $\mathbf{r}$ (which excludes terms of first order in gradients, or rotationally non-invariant second-order gradients). The gradient term in the action is the unique choice satisfying these criteria. Notice that higher order field polynomials, such as $\mathrm{tr}(\partial QQ\partial QQQQ) = -\mathrm{tr}(\partial Q\partial Q)$, all collapse to the stated term because of $Q^2 = \mathbb{1}$ (exercise). The parameter ω reduces the symmetry down to that of the unbroken subgroup $\mathrm{U}(R) \times \mathrm{U}(R)$, and the unique non-gradient operator having this symmetry is $\mathrm{tr}(\tau_3 Q)$. In this way we arrive at the second term, which naturally must be proportional to ω.

A2: This follows from straightforward expansion and an integration by parts.

A3: The source term defined in Eq. (5.51) adds to the symmetry breaking term $\frac{\omega}{2}\tau_3$, and so will appear in the action as $c_2 \int d^d r\, \mathrm{tr}(JQ) = 2c_2(a_+ \,\mathrm{tr}(Q(\mathbf{r}_1)\tau^+ \otimes P^1) + a_- \,\mathrm{tr}(Q(\mathbf{r}_2)\tau^- \otimes P^1))$. Expansion to leading order in the generators then readily leads to the stated result.

A4: The above real space expression is transformed to momentum space by substitution $\nabla^2 \to -q^2$. As for the energy arguments, we substitute the previous energy difference of Green functions, $i\omega_m$, by the present one, $\omega + i\delta$. The finite imaginary part ensures that the denominator $Dq^2 - i\omega + \delta$ will never vanish, even if $q^2 = \omega = 0$. With

---

[44] Compared with the microscopic diagrams in terms of Green functions, we are now on a meta-level where the primitive building blocks of the diagrammatic code are diffusion modes.

these rules, we obtain the correspondence $-2\pi\nu/(Dq^2 - i\omega) = 4c_2^2/(8c_1q^2 - 2c_2\omega)$, and from there the constants as stated in the question.

5.5 Summary and Outlook

In this introduction to functional mean field methods, we have learned how integrals over microscopic quantum fields can be traded for integrals over effective degrees of freedom adjusted to the low-energy characteristics of a system. We found that the functional dependence of the action on the new coordinates is usually nonlinear – the notorious "trace logarithms" – and has to be dealt with by stationary phase analysis. While a first principles solution of the mean field equations is often not possible, all applications discussed in this chapter have shared the feature that solutions could be found by an educated guess. These solutions could display the full symmetry of an action or – the more interesting scenario – spontaneously break them down to a reduced symmetry.

INFO The mean fields discussed in this chapter shared the property spatial uniformity. However, there are numerous problem classes with **spatially non-uniform mean fields**. These include lattice systems with discrete translational symmetry, where the optimal

staggering solutions show **staggering**, i.e., they change sign under translation by one lattice spacing. Deceptively, these configurations usually exist in parallel with a uniform solution of higher energy (which therefore is unphysical); see problem 5.6.1 for an example.[45]

frustration In other systems the existence of meaningful stationary phase configurations is excluded by mechanisms of **frustration**. The principle of frustration is illustrated in the figure for an Ising system with antiferromagnetic exchange interaction on a triangular lattice. In this case, the non-bipartite structure of the latter prevents the spins from aligning in a uniformly alternating manner.

glasses **Glasses** (see the remarks on page 110) possess a macroscopic number of irregular *metastable* extremal configurations that are close in energy but differ from each other in a restructuring of the atomic configuration at remote places. The dynamics of such

aging systems is governed by the process of **aging**, i.e., transitions between different extremal configurations on very long time scales (witness the apparent rigidity of window glass!).

Finally, even systems possessing spatially homogeneous mean field configurations may host other solutions of physical significance. We have encountered realizations of this scenario already in chapter 3. The imaginary-time Euler–Lagrange equations of a particle in a double well (its "mean field equations") had a metastable constant solution (a constant mean field). However, in addition to that, we found **instantons** (non-uniform mean fields). Superficially, it seemed that the non-vanishing action of instantons would make them irrelevant. However, for sufficiently long times, this action cost was overcompensated by the freedom to place instantons at arbitrary times. In the language of statistical mechanics, the energy cost associated with instanton production could be overwhelmed by the *free* energy gain in configurational entropy. Later we will see that entropy–energy competitions of this type are realized in many other contexts in quantum and statistical field theory.

---

[45] In problems where lattice structures matter, it is advisable to test various trial solutions transforming differently under translation.

For example, the phase actions of superfluids possess extremal configurations wherein the phase winds around a fixed reference point in space to define a *vortex*. While the energies of individual vortices are high, the positional entropy may dominate and lead to a proliferation of vortices at high temperatures (section 6.5).

Building on mean field solutions, we considered various applications and described how effective actions for fluctuations are derived and evaluated. In cases where the mean field broke a continuous symmetry, these actions were soft, reflecting the presence of Goldstone modes. We saw that Goldstone mode fluctuations have a dominant effect on the observable physical properties of a system.

The theoretical machinery developed thus far already enables us to tackle complex problems. However, two gaps need to be filled to make the functional integral a universally applicable tool. First, our discussion of field fluctuations has been limited so far: we expanded actions to second order and discussed the corresponding fluctuation determinants. However, interesting problems usually contain significant anharmoticities. For example, the quartic term in ϕ^4-theory led to singularities whose nature remained largely obscure. To understand the physics of generic field theories, we thus need to (i) develop criteria indicating in what circumstances they can be limited to second order and (ii), in cases where they cannot, learn how to deal with anharmonic content. Second, the emphasis so far has been on the conceptual description of different phases of condensed matter systems. However, in applications, one is typically interested in comparison with experimentally accessible data. We thus need to develop ways to extract concrete observables from the functional integral. Theses two topics (which are not as unrelated as one might think) are discussed in chapters 6 and 7, respectively, and they conclude our introduction to the basics of condensed matter field theory.

5.6 Problems

5.6.1 Peierls instability

In problem 2.4.3 we saw that the half-filled one-dimensional lattice system is unstable to the formation of a commensurate periodic lattice distortion. It turns out that this is a general phenomenon: at zero temperature, one-dimensional crystals comprising electrons and ions tend to spontaneously develop lattice distortions transforming them into dielectrics. This lattice instability is known as a Peierls distortion, and in this problem we study its nature.

The starting point of the theory is the Euclidean action for a one-dimensional gas of electrons,

$$S_{\text{el}}[\psi] = \int d\tau \int dx \, \bar{\psi} \left(\partial_\tau - \frac{\partial_x^2}{2m} - \mu \right) \psi,$$

where, for simplicity, we assume spinless electrons. Anticipating that the detailed nature of lattice vibrations may not be essential, we describe the lattice in terms of a harmonic Euclidean action (section 1.1),

$$S_{\mathrm{ph}}[u] = \frac{\rho}{2} \int d\tau \int dx \left((\partial_\tau u)^2 + c^2 (\partial_x u)^2 \right),$$

where $u(\tau, x)$ denotes the scalar bosonic displacement field. We next need to couple the electron system to lattice distortions. While one might try to derive a "realistic" model of the coupling, here we are interested in general stability issues, which show a high degree of universality. We therefore consider the phenomenological "Fröhlich's deformation potential approximation" (for more details, see problem 3.8.11),

Rudolph E. Peierls 1907–1995
Born in Germany, Peierls did fundamental work during the early years of quantum mechanics. He also studied the physics of lattice vibrations (phonons) and is credited with the development of the concept of the "hole" in condensed matter. In collaboration with Otto Frisch, Peierls was the first to realize that an atom bomb based on $^{235}\mathrm{U}$ would be feasible, and was engaged in the Manhattan Project.

wherein the coupling is proportional to the gradient of the deformation $\partial_x u$,

$$S_{\mathrm{el-ph}}[\psi, u] = g \int d\tau \int dx \, \bar{\psi}\psi \, \partial_x u.$$

The full Euclidean action $S = S_{\mathrm{el}} + S_{\mathrm{ph}} + S_{\mathrm{el-ph}}$ represents an interacting theory, which is not exactly solvable. Here, we analyze it perturbatively in the coupling parameter g and then explore the stability of the ensuing effective theory by mean field methods.

(a) As a first step, integrate out the fermion field ψ to obtain an effective action for the displacement field u. Taking g to be small, expand the action up to second order in u to obtain the effective action

$$S_{\mathrm{eff}}[u] \simeq \sum_{q, \omega_m} \left(\frac{\rho}{2}(\omega_m^2 + c^2 q^2) - \frac{g^2}{2} q^2 \chi(q, \omega_m) \right) |u_{\omega_m, q}|^2,$$

where $\chi(q, \omega_m) = -TL^{-1} \sum_{p, \omega_n} G_{0, p, \omega_n} G_{0, p+q, \omega_n + \omega_m}$ contains the Green functions of the free one-dimensional electron gas.

(b) When seeking saddle-points of effective actions, the first guess is usually homogeneous, such as $u_0(x, \tau) \equiv u_0$. However, such solutions are not necessarily the best; it may be favorable to spontaneously break the translational symmetry of an action. Show that the static solution $u_0 \cos(2k_{\mathrm{F}}x + \varphi)$ has this feature, with $S[u_0 \cos(2k_{\mathrm{F}}x + \varphi)] < S[u_0]$ below a critical temperature T_{c}. At low temperatures, the system thus is unstable towards the formation of a static sinusoidal lattice distortion. Use the approximation $\chi(2k_{\mathrm{F}}, 0) \simeq \ln(\beta\omega_{\mathrm{D}})/4\pi v_{\mathrm{F}}$, where ω_{D} is the Debye frequency, to determine the transition temperature T_{c}.

Referring for detailed discussions of the instability to solid state textbooks, we note that a distortion of periodicity $2k_{\mathrm{F}}$ effectively doubles the periodicity of the crystal, $a \to 2a$. The corresponding electron dispersion relation contains band gaps at the new effective reciprocal lattice vector π/a, which lower the energy of the fermion system by an amount outweighing the energy required to distort the ion centers.

Answer:

(a) Integrating out the electron degrees of freedom ψ, one obtains

$$\int D\psi\, e^{-S_{\text{el}} - S_{\text{el-ph}}} = \exp\left(\operatorname{tr}\ln\left(\partial_\tau - \frac{\partial_x^2}{2m} - \mu + g\partial_x u\right)\right).$$

With the bare electron Green function $G_0^{-1} = \partial_\tau - \partial_x^2/2m - \mu$, a second-order expansion in g gives $\operatorname{tr}\ln[G_0^{-1} + g(\partial_x u)] \simeq \operatorname{tr}\ln G_0^{-1} - (g^2/2)\sum_{\omega_m, q} q^2\chi(q, \omega_m)|u_{\omega_m, q}|^2$, where we note that the first-order term vanishes by symmetry. Adding the coupling term to the free action of the displacement field, we obtain the required result.

(b) In the presence of the static periodic modulation, $u_{\omega_m, \mathbf{q}} = \pi u_0[e^{i\varphi}\delta(q - 2k_F) + e^{-i\varphi}\delta(q + 2k_F)]\delta(\omega_m)$ the action takes the value $S_{\text{eff}} = 4\pi^2 u_0^2 k_F^2[\rho c^2 - g^2\chi(2k_F, 0)]$. This must be compared with $S_{\text{eff}}[u_0] = 0$. The transition is realized when $\rho c^2 < 4\pi^2 g^2\chi(2k_F, 0)$. Substituting the expression for χ, one obtains the critical temperature $T_c = \omega_D \exp\left(-4\pi^2 v_F \rho c^2/g^2\right)$.

5.6.2 Temperature profile of the BCS gap

This problem addresses the dependence of the BCS gap on temperature. In particular, we wish to understand its nonanalytic vanishing at the transition temperature (this being a hallmark of a second-order phase transition).

Consider the BCS gap equation (5.33)

$$\frac{1}{g\nu} = \int_0^{\omega_D/2T} dx\, \frac{\tanh(x^2 + \kappa^2)^{1/2}}{(x^2 + \kappa^2)^{1/2}},$$

where we have introduced a dimensionless integration variable and $\kappa \equiv \Delta/2T$. In spite of its innocuous appearance, the temperature dependence of this equation is not straightforward to infer. Referring for a quantitative discussion to Abrikosov et al.,[46] we here restrict ourselves to exploring the gap profile in the vicinity of the transition temperature.

(a) To determine the value of T_c we proceed somewhat indirectly, assuming that this is determined through the condition $\Delta(T_c) = 0$. Use this criterion to obtain Eq. (5.20) for the critical temperature. (You may assume the hierarchy of energy scales $\omega_D \gg T_c \sim \Delta_0 \gg \Delta(T \simeq T_c)$, where $\Delta_0 = \Delta(T = 0)$.)

(b) Now, let us derive the approximate profile Eq. (5.34) of the gap for temperatures T slightly smaller than T_c. To this end, add to and subtract from the right-hand side of the gap equation the integral $\int dx\, x^{-1}\tanh x$. Then expand to leading order in the small parameters $\delta T/T_c$ and Δ/T_c, where $\delta T = T_c - T > 0$.

[46] A. A. Abrikosov, L. P. Gorkov, and I. E. Dzyaloshinskii, *Methods of Quantum Field Theory in Statistical Physics* (Dover Publications, 1975).

Answer:

(a) For $\Delta = 0$, $\lambda(\xi) = (\xi^2 + \Delta^2)^{1/2} = |\xi|$ and (5.33) assumes the form

$$\frac{1}{g\nu} = \int_0^{\omega_D/2T_c} dx \, \frac{\tanh x}{x},$$

where we have introduced $x \equiv \xi/2T_c$ as a dimensionless integration variable. The dominant contribution to the integral comes from the region $x \gg 1$, where $\tanh x \simeq 1$. As a result, one obtains $1/g\nu \simeq \ln(\omega_D/2T_c)$. Solving for T_c, we arrive at Eq. (5.20).

(b) Adding and subtracting the integral given above, we have

$$\frac{1}{g\nu} = \int_0^{\omega_D/2T} dx \left(\frac{\tanh(x^2 + \kappa^2)^{1/2}}{(x^2 + \kappa^2)^{1/2}} - \frac{\tanh x}{x} \right) + \int_0^{\omega_D/2T} dx \, \frac{\tanh x}{x}.$$

Arguing as in (a), the second integral can be estimated as $\ln(\omega_D/2T) \approx \ln(\omega_D/2T_c) + (\delta T/T_c) = (1/g\nu) + (\delta T/T_c)$, where we have expanded to linear order in δT. Thus,

$$-\frac{\delta T}{T_c} \approx \int_0^{\omega_D/2T} dx \left(\frac{\tanh(x^2 + \kappa^2)^{1/2}}{(x^2 + \kappa^2)^{1/2}} - \frac{\tanh x}{x} \right).$$

Now, the remaining integral can be split into a "low-energy region" $0 \le x \le 1$, and a "high-energy region" $1 < x < \omega_D/2T$. Using the small-x expansion $\tanh x \simeq x - x^3/3$, we find that the first region gives a contribution $\sim \kappa^2$. With $\tanh x \overset{x>1}{\approx} 1$, the second region contributes a term $\mathcal{O}(\kappa^2)$ that is, however, approximately independent of the large-energy cutoff $\omega_D/2T$. Altogether, we obtain $\delta T/T_c \approx \text{const.} \times \kappa^2 \approx \text{const.} \times (\Delta_0^2/T_c^2)$ from which one obtains Eq. (5.34).

5.6.3 Fluctuation contribution to the superconductor action

Here, as a technical exercise, we derive the energy cost corresponding to large-scale spatial fluctuations of the order parameter of a BCS superconductor.

Consider the second-order contribution to the Ginzburg–Landau action of the BCS superconductor (5.36). In problem 3.8.8 we saw that the frequency summation involved in the definition of the integral kernel evaluates to

$$\chi_q^c \equiv -\frac{T}{L^d} \sum_p G_p G_{-p+q} = \frac{1}{L^d} \sum_{\mathbf{P}} \frac{1 - n_F(\xi_{\mathbf{p}}) - n_F(\xi_{-\mathbf{p}+\mathbf{q}})}{i\omega_n - \xi_{\mathbf{p}} - \xi_{-\mathbf{p}+\mathbf{q}}},$$

with the four-momenta $q = (\omega_n, \mathbf{q})$ and $p = (\omega_m, \mathbf{p})$. Expand $\chi^c(\mathbf{q}) \equiv \chi_{(0,\mathbf{q})}^c$ to second order in $\mathbf{q}$. (Hint: You may trade the momentum summation for an integral, and linearize the dispersion as $\xi_{\mathbf{p}+\mathbf{q}} \simeq \xi_{\mathbf{p}} + \mathbf{p} \cdot \mathbf{q}/m$. Also, use the identity $\int d\epsilon \, \epsilon^{-1} \partial_\epsilon^2 n_F(\epsilon) = cT^{-2}$, where the numerical constant $c = 7\zeta(3)/2\pi^2$ and $\zeta(x) = \sum_{n=1}^{\infty} n^x$ defines the ζ-function.)

Answer:

Using the fact that $\xi_{\mathbf{p}} = \xi_{-\mathbf{p}}$,

$$
\begin{aligned}
\chi^c(\mathbf{q}) &= -\int \frac{d^d p}{(2\pi)^d} \frac{1 - n_F(\xi_{\mathbf{p}+\mathbf{q}/2}) - n_F(\xi_{\mathbf{p}-\mathbf{q}/2})}{\xi_{\mathbf{p}+\mathbf{q}/2} + \xi_{\mathbf{p}-\mathbf{q}/2}} \\
&\simeq -\int \frac{d^d p}{(2\pi)^d} \frac{1 - 2n_F(\xi_{\mathbf{p}}) - \partial_\xi^2 n_F(\xi_{\mathbf{p}})(\mathbf{q}\cdot\mathbf{p})^2/(4m^2)}{2\xi_{\mathbf{p}}} \\
&= \chi^c(\mathbf{0}) - \frac{\nu\mu\mathbf{q}^2}{12m} \int d\epsilon\; \epsilon^{-1}\partial_\epsilon^2 n_F(\epsilon) = \chi^c(\mathbf{0}) - \frac{c}{24}\frac{\nu v_F^2}{T^2}\mathbf{q}^2,
\end{aligned}
$$

where c is a numerical constant and in the second equality we have used the fact that, for $\xi = \mathcal{O}(T)$, $\mathbf{p}^2/2m \simeq \mu$. Substituting this expansion into the quadratic action, we obtain the gradient term in Eq. (5.38).

5.6.4 Coulomb blockade

quantum
dots

The low temperature physics of **quantum dots** – metallic or semiconducting devices of extension $1\,\mu$m and less – is predominantly influenced by charging effects. In this problem, we explore the impact of charging on the most basic characteristic of a quantum dot, the tunneling density of states (DoS).

Quantum dots are of increasing technological importance as building blocks of quantum electronic devices. Consider one such dot weakly[47] connected to an environment representing the external components of the device (see fig. 5.4 for a realization of such a setup). Due to its small size, the removal or addition of an electron incurs a significant energy cost $E_C = e^2/2C$, where C is the capacitance of the system. The

charging
energy

discreteness of this **charging energy** leads to a plethora of observable effects, the most basic of which is a strong suppression of the DoS at the Fermi surface.

For a non-interacting system, the **single-particle DoS** is defined as $\rho(\epsilon) = \mathrm{tr}\,\delta(\epsilon - \hat{H}) = -\pi^{-1}\,\mathrm{Im}\,\mathrm{tr}\,\hat{G}(\epsilon + i0) = -\pi^{-1}\,\mathrm{Im}\,\mathrm{tr}\,\hat{G}_n|_{i\omega_n\to\epsilon+i0}$, where $\hat{G}(z) = (z - $

tunneling
DoS

$\hat{H})^{-1}$ is the Green function. The **tunneling DoS**[48] generalizes this definition to the interacting case: $\nu(\epsilon) = -\pi^{-1}\,\mathrm{Im}\,\mathrm{tr}\,\hat{G}_n|_{i\omega_n\to\epsilon+i0}$, where the coherent state path integral representation of the Green function is given by

$$
G_{\alpha\beta}(\tau) = \mathcal{Z}^{-1}\int D\psi\, e^{-S[\psi]}\bar{\psi}_\beta(\tau)\psi_\alpha(0), \tag{5.60}
$$

and the indices α, β label the eigenstates of the single-particle contribution to the Hamiltonian. Having an irregular structure, these states are unknown. However, this is not an issue for understanding the nature of the Coulomb blockade. A model Hamiltonian describing the joint effects of single particle physics and charging effects reads as $\hat{H} = \hat{H}_0 + E_C(\hat{N} - N_0)^2$, where $\hat{H}_0 = \sum_\alpha \epsilon_\alpha a_\alpha^\dagger a_\alpha$, $\hat{N} = \sum_\alpha a_\alpha^\dagger a_\alpha$ is

---

[47] By "weak," we mean that the conductance of all external leads attached to the system is such that $g < g_0$, where $g_0 = e^2/h \simeq (25.8\,\mathrm{k}\Omega)^{-1}$ is the quantum unit of conductance.

[48] The terminology *tunneling* DoS is motivated by the fact that $\nu(\epsilon)$ is an important building block in the calculation of tunneling currents. (Recall the Golden Rule: tunneling rates are obtained by the multiplication of transition probabilities with state densities.)

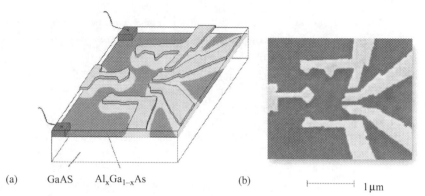

(a)   GaAS   Al$_x$Ga$_{1-x}$As   (b)   ⊢---------⊣ 1 μm

Fig. 5.4   (a) Schematic picture of a confined two-dimensional electron gas (a quantum dot) formed
at the interface between a GaAs and an AlGaAs layer. (b) Electron microscopic image of
the "real" device. (Source: Courtesy of C. M. Marcus.)

the number operator and N_0 represents the preferred number of particles (as set
by capacitively coupled gate electrodes; see the wiring indicated in fig. 5.4). The
action controlling the behavior of the Green function (5.60) is thus given by

$$S[\psi] = \int_0^\beta d\tau \left(\sum_\alpha \bar{\psi}_\alpha (\partial_\tau + \epsilon_\alpha - \mu)\psi_\alpha + E_C \left(\sum_\alpha \bar{\psi}_\alpha \psi_\alpha - N_0 \right)^2 \right). \qquad (5.61)$$

(a) Introducing a bosonic field variable $V(\tau)$, decouple the interaction by a
Hubbard–Stratonovich transformation. Bring the functional representation of the
Green function to the form $G_\alpha(\tau) = \mathcal{Z}^{-1} \int DV\, e^{-S[V]} \mathcal{Z}_\alpha[V] G_\alpha[V](\tau)$, where G_α is
the diagonal element of the Green function (the representation above implies that
all off-diagonal elements vanish), $G_\alpha[V]$ represents the Green function of the non-
interacting system subject to an imaginary time-dependent potential $iV(\tau)$, and
$\mathcal{Z}_\alpha[V]$ is the corresponding partition function.
(b) Turning to a Matsubara Fourier representation $V(\tau) = T \sum_m e^{-i\omega_m \tau} V_m$, split
the static zeroth component $V_0 = 2\pi k + \tilde{V}/T$ into an integer multiple of 2π and a
residual term $\tilde{V} \in [0, 2\pi T]$, and show that all but the static component $\tilde{V}$ can be
removed from the action by a gauge transformation. Why cannot $\tilde{V}$ be gauged as
well? Explore the transformation behavior of the Green function and integrate over
the non-zero mode components $V_{m \neq 0}$.
(c) Making use of the relation $\sum_{k=1}^\infty \frac{\cos(kx)}{k^2} = \frac{\pi^2}{6} - \frac{\pi|x|}{2} + \frac{x^2}{4} + \cdots$, perform the
Matsubara summation, $\sum_{\omega_m \neq 0}$. Show that the Green function can be expressed as
$G(\tau) = F(\tau)\tilde{G}(\tau)$, where the function $F(\tau) = \exp(-E_C(\tau - \beta^{-1}\tau^2))$ is obtained
by integration over the dynamical components of V, while $\tilde{G}$ is a non-interacting
Green function averaged over the static component V_0.
(d) The remaining integration over the static component $\tilde{V}_0$ is achieved by the
stationary phase method. Neglecting the weak dependence of the non-interacting
Green function on $\tilde{V}_0$, derive and interpret the saddle-point equation. Approximate
the functional by its value at the saddle-point (i.e., neglect quadratic fluctuations

around the saddle-point value for $\tilde{V}_0$). As a result, obtain a representation $G(\tau) = F(\tau)G_0(\tau)$ where G_0 is a non-interacting Green function evaluated at a renormalized chemical potential.

(e) Assuming $E_C \gg T$, compute an approximation of the Fourier transform of $F(\tau)$. (You may approximate Matsubara sums by integrals.) Use your result to obtain the zero-temperature DoS[49]

$$\nu(\epsilon) = \nu_0(\epsilon - E_C \, \mathrm{sgn}(\epsilon))\Theta(|\epsilon| - E_C),$$

where ν_0 is the DoS of the non-interacting systems. Owing to the large charging energy, the single-particle DoS vanishes in a window of width $2E_C$ centered around the Fermi energy: this is the **Coulomb blockade**. Particles of energy $\epsilon > E_C$ larger than the charging threshold are free to enter the dot. However, in doing so they lose an amount E_C of (charging) energy, which explains the energy shift in the factor ν_0.

Coulomb blockade

Answer:

(a) Using the identity (summation over α is implied)

$$\exp\left(-E_C\int\left(\bar{\psi}_\alpha\psi_\alpha - N_0\right)^2\right) = \int DV \exp\left(-\int\left(\frac{V^2}{4E_C} - iN_0V + i\bar{\psi}_\alpha V\psi_\alpha\right)\right),$$

the quantum partition function takes the form $\mathcal{Z} = \int D\psi DV\, e^{-S[V]-S[\psi,V]}$, where $S[\psi,V] = \int d\tau \sum_\alpha \bar{\psi}_\alpha(\partial_\tau + \epsilon_\alpha - \mu + iV)\psi_\alpha$ and $S[V] = \int d\tau\,(\frac{V^2}{4E_C} - iN_0V)$. Thus,

$$G_\alpha(\tau) = \frac{1}{\mathcal{Z}}\int D\psi DV\, e^{-S[V]-S[\psi,V]} = \frac{1}{\mathcal{Z}}\int DV\, e^{-S[V]}\mathcal{Z}_\alpha[V]G_\alpha[V](\tau),$$

where $G[V]$ is obtained from a non-interacting theory with action $S[V] \equiv S|_{\mu \to \mu - iV}$.

(b) The gauge transformation $\psi_\alpha(\tau) \to \exp\left(-i\int_0^\tau d\tau'\,(V(\tau') - \tilde{V})\right)\psi_\alpha(\tau)$ (with $\bar{\psi}_\alpha$ transforming under the complex conjugate phase) removes much of the time-dependent potential from the tr ln. The zero mode offset $\tilde{V}$ has to be excluded from the transformation to preserve the time periodicity of the gauge factor. Substitution of the transformed field leads to removal of the dynamic V-components from the action, $S[V] \to S(\tilde{V})$, and the appearance of a gauge factor multiplying the pre-exponential terms. We thus obtain

$$G(\tau) = \frac{1}{\mathcal{Z}}\int D\psi DV\, e^{-S[V]-S[\psi,\tilde{V}_0]}e^{i\int_0^\tau d\tau'\,(V(\tau')-\tilde{V}_0)}\bar{\psi}_\alpha(\tau)\psi_\alpha(0)$$

$$= \frac{1}{\mathcal{Z}}\int DV\, e^{-S[V]}e^{i\int_0^\tau d\tau'(V(\tau')-\tilde{V})}\mathcal{Z}(\tilde{V})G_\alpha(\tilde{V})(\tau)$$

$$= \frac{F(\tau)}{\mathcal{Z}}\int d\tilde{V}\, e^{-\frac{\beta}{4E_C}\tilde{V}^2 + i\beta N_0\tilde{V}}\mathcal{Z}(\tilde{V})G_\alpha(\tilde{V})(\tau),$$

---

[49] For a more elaborate analysis of finite-temperature corrections, we refer to A. Kamenev and Y. Gefen, *Zero bias anomaly in finite-size systems*, Phys. Rev. B **54**, 5428 (1996).

where we have omitted the $2\pi kT$ contribution to the zero mode (as it does not play much of a role in the context of this problem – for the physical meaning of the integers k see section 12.6.1),

$$F(\tau) = \prod_{n \neq 0} \int dV_n \, e^{-\frac{T}{4E_C} V_n V_{-n} + \frac{TV_n}{\omega_n}(\exp(i\omega_n\tau)-1)} = \prod_{n \neq 0} e^{-\frac{2E_C T}{\omega_n^2}(1-\exp(-i\omega_n\tau))},$$

and in the second equality we have performed the Gaussian integral over $V_{n \neq 0}$.
(c) Using the formulae given, the Matsubara summation gives

$$-2E_C T \sum_{n \neq 0} \frac{1}{\omega_n^2}(1 - \exp(-i\omega_n\tau)) = -E_C(|\tau| - \beta^{-1}\tau^2),$$

i.e., $F(\tau) = \exp[-E_C(|\tau| - \beta^{-1}\tau^2)]$, and $G_\alpha(\tau) = F(\tau)\tilde{G}_\alpha(\tau)$, where

$$\tilde{G}_\alpha(\tau) = \mathcal{Z}^{-1} \int d\tilde{V} \, e^{-\frac{\beta}{4E_C}\tilde{V}^2 + i\beta N_0\tilde{V}} \mathcal{Z}(\tilde{V}) G_\alpha(\tilde{V})(\tau).$$

(d) Defining a $\tilde{V}$-dependent free energy by $\mathcal{Z}(\tilde{V}) = \exp(-\beta F(\tilde{V}))$, noting that $\tilde{V}$ shifts the chemical potential, $F(\tilde{V}, \mu) = F(\mu - i\tilde{V})$, and neglecting the $\tilde{V}$-dependence of $G(\tilde{V}) \simeq G$, we obtain the saddle-point equation

$$0 = \frac{\partial}{\partial\tilde{V}}\left(\frac{1}{4E_C}\tilde{V}^2 - iN_0\tilde{V} - F(\mu - i\tilde{V})\right) = \frac{1}{2E_C}\tilde{V} - iN_0 + i\langle\hat{N}\rangle_{\mu - i\tilde{V}},$$

where, in the second equality, we have used the fact that $\partial_{\tilde{V}} F(\mu - i\tilde{V}) = -i\partial_\mu F(\mu - i\tilde{V}) = -i\langle\hat{N}\rangle_{\mu - i\tilde{V}}$. Substituting the solution of the saddle-point equation $\bar{V} = 2iE_C(N_0 - \langle\hat{N}\rangle_{\mu - i\bar{V}})$ amounts to replacing the chemical potential μ by an effective chemical potential $\bar{\mu} = \mu + 2E_C(N_0 - \langle\hat{N}\rangle_{\bar{\mu}})$. As a preliminary result we thus obtain $G_\alpha(\tau) = F(\tau)G_{0\alpha}$, where the non-interacting Green function is evaluated at $\bar{\mu}$. In passing, we note that the condition for the applicability of the saddle-point approximation reads as (exercise: why?) $1/(2E_C) - \partial_\mu\langle\hat{N}\rangle_{\bar{\mu}} \gg \beta$.
(e) For $E_C \gg T$, the dominant contribution to the Fourier transform of $F(\tau)$ comes from the boundary regions of the imaginary-time interval, $\tau \ll \beta$ and $\beta - \tau \ll \beta$. Linearizing the exponent of F in these regions we obtain

$$F_m = \int_0^\beta d\tau \, e^{i\omega_m\tau} F(\tau) \simeq \int_0^\infty d\tau \left(e^{i\omega_m\tau} + e^{-i\omega_m\tau}\right) e^{-E_C\tau} = \frac{2E_C}{E_C^2 + \omega_m^2}.$$

Using the fact that $G_{0n\alpha} = (i\omega_n - \xi_\alpha)^{-1}$, where $\xi_\alpha = \epsilon_\alpha - \bar{\mu}$, we then obtain

$$G_{n\alpha} = T\sum_m F_m G_{0(n-m)\alpha} \simeq \int \frac{d\omega}{2\pi} \frac{2E_C}{(i\omega_n - i\omega - \xi_\alpha)(E_C^2 + \omega^2)}$$

$$= \frac{1}{i\omega_n - E_C\,\mathrm{sgn}(\xi_\alpha) - \xi_\alpha}.$$

Turning back to real energies, $i\omega_n \to \omega + i\delta$, and taking the imaginary part, we arrive at

$$\nu(\epsilon) = \sum_\alpha \delta(\epsilon - E_C \, \mathrm{sgn}(\xi_\alpha) - \xi_\alpha) = \int d\omega \, \nu_0(\omega) \delta(\epsilon - E_C \, \mathrm{sgn}(\omega) - \omega),$$

and performing the final integral, we obtain the required result.

5.6.5 Action of a tunnel junction

In the previous problem, we considered the physics of a perfectly isolated quantum dot. However, in practice (cf. the dot depicted in fig. 5.4) the system is usually connected to an external environment by some leads. It is the purpose of this problem to derive an effective action accounting for the joint effect of charging and the coupling to an environment.

Consider a quantum dot connected to an external lead (it is straightforward to generalize to the presence of several leads). We model the latter as a wave guide with eigenstates ψ_a whose detailed structure we need not specify. The full system is described by an action $S[\psi_\alpha, \psi_a] = S_{\mathrm{dot}}[\psi_\alpha] + S_{\mathrm{lead}}[\psi_a] + S_{\mathrm{T}}[\psi_\alpha, \psi_a]$, where $S_{\mathrm{dot}}[\psi_\alpha]$ is given by Eq. (5.61),

$$S_{\mathrm{lead}}[\bar{\psi}_a, \psi_a] = \sum_a \int_0^\beta d\tau \, \bar{\psi}_a (\partial_\tau + \epsilon_a - \mu) \psi_a,$$

and the coupling between dot and lead is described by

$$S_{\mathrm{T}}[\psi_\alpha, \bar{\psi}_\alpha, \psi_a, \bar{\psi}_a] = \sum_{a\alpha} \int d\tau \, \bar{\psi}_\alpha T_{\alpha a} \psi_a + \mathrm{h.c.}.$$

Throughout we will assume that the coupling is sufficiently weak that contributions of $\mathcal{O}(T^4)$ to the effective action are negligibly small – the "tunneling approximation."
(a) Proceeding as in the previous problem, decouple the charging interaction by a Hubbard–Stratonovich transformation. Integrate out the fermions and subject the problem to the same gauge transformation as used above to remove the dynamical contents of the Hubbard–Stratonovich field V. You will observe that the gauge phase transforms the coupling matrices T.
(b) Expand the action to leading (i.e., second) order in the coupling matrix elements $T_{\alpha a}$. (You may ignore the integration over the static component of the Hubbard–Stratonovich field; as discussed above, it leads merely to a shift of the chemical potential.) Assuming that the single-particle DoS of dot and lead do not vary significantly on the energy scales at which the field V fluctuates, determine the dependence of the tunneling term on the gauge phase $\phi(\tau) \equiv \int^\tau d\tau'(V(\tau') - \tilde{V}_0)$ and identify its coupling constant as the Golden Rule tunneling rate $4g_T$.[50] (To obtain a finite result, subtract from the tunneling action $S_{\mathrm{tun}}[\phi]$ the constant $S_{\mathrm{tun}}[0]$.

---

[50] In problem 7.6.3, we show that the coupling constant equals four times the tunneling conductance, g_T, of the barrier; hence the denotation $4g_T$.

dissipative
tunnel-
ing action

Note that the Fourier transform of $|\omega_m|$ is given by (see Eq. (3.41)) $\pi T \sin^{-2}(\pi T \tau)$. Expressing the charging action $S_c[V]$ in terms of the gauge phase $V = \dot{\phi} + \tilde{V}$ (and neglecting the constant offset $\tilde{V}$), the complete **dissipative tunneling action** takes the form

$$S[\phi] = \int d\tau \left(\frac{\dot{\phi}^2}{4E_C} - iN_0\dot{\phi} \right) - g_T \int d\tau d\tau' \frac{\sin^2(\frac{1}{2}(\phi(\tau) - \phi(\tau')))}{\sin^2(\pi T(\tau - \tau'))} \tag{5.62}$$

INFO First derived by Ambegaokar, Eckern, and Schön,[51] the action (5.62) is often called the **AES action**. Identifying the phase ϕ with a circular coordinate, it describes the quantum dynamics of a particle on a ring with kinetic energy $\sim \dot{\phi}^2/E_C$ and subject to a dissipative damping mechanism of strength $\sim g_T$. The latter describes the dissipation of the energy stored in dynamical voltage fluctuations $V \sim \dot{\phi}$ into the microscopic degrees of freedom of the quasi-particle continuum.

In the absence of dissipation, the action describes ballistic motion, the ring topology reflecting the 2π-periodicity of the quantum phase, which in turn relates to the quantization of charge (recall that charge and phase are canonically conjugate). This periodicity is the main source of charge quantization phenomena in the AES approach. For strong ($g_T > 1$) dissipation, full traversals of the ring get increasingly less likely, and the particle begins to forget that it lives on a ring. This damping manifests itself in an exponential suppression of charge quantization phenomena. Indeed, for increasing coupling between lead and dot, the charge on the latter fluctuates and is no longer effectively quantized. For a detailed account of the physical crossover phenomena associated with the strength of g_T, we refer to the review Ref.[52]

Answer:

(a) Decoupling the action and integrating over the fermionic degrees of freedom, one obtains the functional $\mathcal{Z} = \exp(-S_{\text{eff}}[V])$, where

$$S_{\text{eff}}[V] = S_c[V] - \text{tr} \ln \begin{pmatrix} \partial_\tau - \mu + iV + \hat{\epsilon}_d & T \\ T^\dagger & \partial_\tau - \mu + \hat{\epsilon}_l \end{pmatrix}$$

$$= S_c[V] - \text{tr} \ln \begin{pmatrix} \partial_\tau - \mu + i\tilde{V} + \hat{\epsilon}_d & Te^{i\phi} \\ e^{-i\phi}T^\dagger & \partial_\tau - \mu + \hat{\epsilon}_l \end{pmatrix}. \tag{5.63}$$

Here, as in the previous problem, $S_c[V] = \int d\tau(V^2/(4E_C) - iN_0V)$, $\hat{\epsilon}_d = \{\epsilon_\alpha \delta_{\alpha\alpha'}\}$ and $\hat{\epsilon}_l = \{\epsilon_a \delta_{aa'}\}$ contain the single-particle energies of dot and lead, respectively, the matrix structure is in dot–lead space, and $\phi(\tau) = \int d\tau' (V(\tau') - \tilde{V})$. In passing from the first to the second equality, we have subjected the argument of the tr ln to the unitary transformation (gauge transformation) described by the matrix $\text{diag}(e^{i\phi}, 1)$ (with the block structure in dot–lead space).

(b) Expanding the tr ln to second order in T, and regularizing by subtracting the constant $S_{\text{tun}}[0]$, we obtain

51 U. Eckern, G. Schön and V. Ambegaokar, *Quantum dynamics of a superconducting tunnel junction*, Phys. Rev. B **30**, 6419 (1984).
52 I. L. Aleiner, P. W. Brouwer and L. I. Glazman, *Quantum effects in Coulomb blockade*, Phys. Rep. **358**, 309 (2002).

$$S_{\text{tun}}[\phi] = |T|^2 \sum_{\alpha a} \sum_n \left(\sum_m G_{\alpha,n}(e^{i\phi})_m G_{a,n+m}(e^{-i\phi})_{-m} - G_{\alpha,n} G_{a,n} \right) + S_{\text{tun}}[0],$$

(5.64)

where $G_{\alpha/a,n} = (i\omega_n - \hat{\epsilon}_{\alpha/a} + \mu)^{-1}$ are the Green functions of dot and lead, respectively, and the constant $S_{\text{tun}}[0]$ will be omitted throughout. Approximating the Green functions as

$$\sum_{\alpha/a} G_{\alpha/a,n} = -\int d\epsilon \, \rho_{\text{d/l}}(\epsilon + \mu) \frac{i\omega_n + \epsilon}{\omega_n^2 + \epsilon^2} \simeq -\pi i \rho_{\text{d,l}} \operatorname{sgn}(\omega_n),$$

(5.65)

where $\rho_{\text{d,l}} \equiv \rho_{\text{d,l}}(\mu)$ is the density of states at energy μ, we arrive at the result

$$S_{\text{tun}}[\phi] = \frac{g_T}{2} \sum_{n,m} (-\operatorname{sgn}(\omega_n) \operatorname{sgn}(\omega_{n+m}) + 1)(e^{i\phi})_m (e^{-i\phi})_{-m}$$

$$= \frac{g_T}{2\pi T} \sum_m |\omega_m| (e^{i\phi})_m (e^{-i\phi})_{-m},$$

(5.66)

where we note that $\sum_m (e^{i\phi})_m (e^{-i\phi})_{-m} = 1$, $g_T = 2\pi^2 \rho_l \rho_d |T|^2$ is proportional to the Golden Rule tunneling rate between dot and lead, and the appearance of a term $\sim |\omega_m|$ is a signature of a dissipative damping mechanism (for a discussion of this point, see section 12.6.4). Using the given formula for the Fourier transform of $|\omega_m|$, we obtain the time representation of the action as

$$S_{\text{tun}}[\phi] = -\frac{g_T}{2} \int d\tau d\tau' \frac{e^{i\phi(\tau) - i\phi(\tau')}}{\sin^2(\pi T(\tau - \tau'))}$$

$$= g_T \int d\tau \, d\tau' \frac{\sin^2((\phi(\tau) - \phi(\tau'))/2)}{\sin^2(\pi T(\tau - \tau'))} + \text{const.}$$

We finally add the charging action to obtain the result (5.62).

5.6.6 Josephson junction

Building on the results obtained in the previous problem, here we derive an effective action of a Josephson junction – a system comprising two superconductors separated by an insulating or normal conducting interface region. The problem includes a preliminary discussion of the physics of the Josephson junction, notably its current–voltage characteristics. In chapter 6, renormalization group methods will be applied to explore in detail the phenomenology of the system.

Consider two superconducting quantum dots separated by a tunneling barrier. Generalizing the model discussed in the last two problems, we describe each dot by an action

$$S^i[\psi_\alpha^i, \phi_i] = \int_0^\beta d\tau \sum_\alpha \bar{\psi}_\alpha^i (\partial_\tau + \xi_{i\alpha}\sigma_3 + e^{i\phi_i(\tau)\sigma_3}\Delta\sigma_1)\psi_\alpha^i, \qquad i = 1, 2,$$

where $\psi_\alpha^i = (\psi_{\alpha\uparrow}^i, \bar{\psi}_{\alpha\downarrow}^i)^T$ are Nambu spinors, σ_i Pauli matrices in particle–hole space, and ϕ_i the phase of the order parameter on dot i. Noting that two dots form a capacitor, we assume the presence of a "capacitive interaction"

$$S_{\text{int}} = \frac{E_C}{4} \int d\tau \, (\hat{N}_1 - \hat{N}_2)^2,$$

where $\hat{N}_i = \sum_\alpha \bar{\psi}_\alpha^i \sigma_3 \psi_\alpha^i$ is the charge operator on dot i, and $1/2E_C$ the capacitance of the system. Finally, the tunneling between the two dots is described by the action

$$S_{\text{T}}[\psi_\alpha] = \sum_{\alpha\beta} \int d\tau \, \bar{\psi}_\alpha^1 (T_{\alpha\beta}\sigma_3)\psi_\beta^2 + \text{h.c.},$$

where $T_{\alpha\beta} = \langle \alpha | \hat{T} | \beta \rangle$ denotes the tunneling matrix elements between the single-particle states $|\alpha\rangle$ and $|\beta\rangle$. Now, were it not for the presence of the superconducting order parameter, the low-energy physics of the system would again be described by the effective action (5.62).

EXERCISE Convince yourself of the validity of this statement, i.e., check that the dot–dot system can be treated along the same lines as the dot–lead system considered above and trace the phase dependence of the various contributions to the action.

(a) Turning to the superconducting case, show that the Hubbard–Stratonovich action is given by

$$S_{\text{eff}}[V, \phi] = S_{\text{c}}[V]$$
$$- \text{tr}\ln \begin{pmatrix} \partial_\tau + (\hat{\xi}_1 + \frac{i}{2}(\dot{\phi}_1 + V))\sigma_3 + \Delta\sigma_1 & e^{-\frac{i}{2}(\phi_1 - \phi_2)\sigma_3}T \\ T^\dagger e^{\frac{i}{2}(\phi_1 - \phi_2)\sigma_3} & \partial_\tau + (\hat{\xi}_2 + \frac{i}{2}(\dot{\phi}_2 - V))\sigma_3 + \Delta\sigma_1 \end{pmatrix},$$

where $\hat{\xi}_i, i = 1, 2$, comprise the single-particle energies of the system and $S_{\text{c}}[V] = (1/4E_C) \int d\tau \, V^2$ is the charging action. Previously we have seen that the functional integrals over Hubbard–Stratonovich fields decoupling charging interactions are concentrated around vanishing mean field solutions. Presently, the $\text{tr}\ln$ is coupled to two such field combinations, $V + \dot{\phi}_1$ and $-V + \dot{\phi}_2$, respectively. We assume that these combinations remain close to their vanishing stationary phase values, and neglect fluctuations around them. In other words, we impose the *Josephson conditions* $\dot{\phi}_1 = -\dot{\phi}_2 \equiv -\dot{\phi}$ and $V = \dot{\phi}$. (If you are ambitious, explore the stability of these conditions by second-order expansion of the $\text{tr}\ln$ in the field combinations above.)

(b) Expanding the action to second order in T (i.e., the leading order) and using the Josephson conditions, show that $S = S_{\text{c}} + S_{\text{tun}}$, where

$$S_{\text{tun}}[\phi] = |T|^2 \sum_{\alpha\alpha'\omega_m\omega_n} \text{tr}\left(G_{1,\alpha\omega_n}(e^{i\sigma_3\phi})_m G_{2,\alpha'\omega_n + \omega_m}(e^{-i\sigma_3\phi})_{-m} - (\phi \leftrightarrow 0)\right),$$

$G_{i\omega_n} = (i\omega_n - \xi_i\sigma_3 - \Delta\sigma_1)^{-1}$ is the bare Gorkov Green function, and we again regularize the action by subtracting $S_{\text{tun}}[\phi = 0]$. Denoting the block diagonal and off-diagonal contributions to the Green function by $G_{i,\text{d}/\text{o}}$, respectively, the action splits into two terms, $\text{tr}(G_{1\text{d}}e^{i\phi\sigma_3}G_{2\text{d}}e^{-i\phi\sigma_3}+G_{1\text{o}}e^{i\phi\sigma_3}G_{2\text{o}}e^{-i\phi\sigma_3})$. Show that, up to small corrections of $\mathcal{O}(\omega_n/\Delta)$, the diagonal terms vanish, and interpret this result. (Hint: Compare with the discussion of the previous problem.)

(c) Turning to the particle-hole off-diagonal sector, show that

$$S_{\text{tun}}[\Delta,\phi] = \gamma\int_0^\beta d\tau\ \cos(2\phi(\tau)), \quad \gamma = |T|^2(\pi\rho)^2\Delta.$$

Josephson
action

Combining everything, one obtains the **action of the Josephson junction**,

$$\boxed{S[\phi] = \frac{1}{4E_C}\int d\tau\ \dot\phi^2 + \gamma\int d\tau\ \cos(2\phi(\tau)) + \Gamma S_{\text{diss}}[\phi]} \tag{5.67}$$

While our analysis above suggests that the coefficient of the dissipative term should be zero, voltage fluctuations in "real" Josephson junctions *do* seem to be dissipatively damped, even at low fluctuation frequencies. Although there is no obvious explanation of this phenomenon, it is common to account for the empirically observed loss of energy by adding a dissipation term to the action.

(d) Finally, explore the current–voltage characteristics of the non-dissipative junction. To this end, perform a Hubbard–Stratonovich transformation on the quadratic charging interaction. What is the physical meaning of the Hubbard–Stratonovich auxiliary field? Interpret your result as the Hamiltonian action of a conjugate variable pair and compute the equations of motion. Show that the **Josephson current** flowing between the superconductors is given by

Josephson
current

$$I = -2\gamma\sin(2\phi). \tag{5.68}$$

According to this equation, a finite-order parameter phase difference causes the flow of a static current carried by Cooper pairs tunneling coherently across the barrier: the **DC Josephson effect**. Application of a finite voltage difference or, equivalently, the presence of a finite charging energy, renders the phase $\dot\phi = V$ dynamical. For a static voltage difference, ϕ increases uniformly in time and the current across the barrier behaves as time-oscillatory: the **AC Josephson effect**. Finally, if the voltage difference becomes very large, $V \gg \Delta$, the Fourier spectrum of ϕ contains frequencies $|\omega_m| > \Delta$ (think about it). At these frequencies, phase variations have the capacity to create quasi-particle excitations, which in turn may tunnel incoherently across the barrier (thereby paying a price in condensate energy but benefiting from the voltage drop). The tunneling of independent quasi-particles is described by the dissipative term in the action (which, we recall, is negligible only at frequencies $|\omega_m| < \Delta$).

DC
Josephson
effect

AC
Josephson
effect

Answer:

(a) The given result is proven by decoupling the capacitive interaction by a field V and integrating out the fermions. This leads to the action

$$S[V,\phi] = S_{\text{c}}[V] - \text{tr}\ln\begin{pmatrix} G_1^{-1}[V,\phi_1] & -T\sigma_3 \\ -T^\dagger\sigma_3 & G_2^{-1}[-V,\phi_2] \end{pmatrix},$$

where the Gorkov Green functions of the two superconductors, $G_i^{-1}[V, \phi] = (-\partial_\tau - (\xi_i + iV\sigma_3/2 - \Delta e^{i\phi_i\sigma_3}\sigma_1)$, appear coupled to the Hubbard–Stratonovich voltage field, and $\xi_i = \{\xi_{i,\alpha}\}$ is shorthand notation for the vector of energy eigenvalues. Noting that $e^{i\phi\sigma_3}\sigma_1 = D(\phi)\sigma_1 D^{-1}(\phi)$, where $D(\phi) = \exp(\frac{i}{2}\phi\sigma_3)$, we factor out the phase dependence as $G_i^{-1}[V, \phi] = D(\phi_i)G_i^{-1}[V + \dot\phi_i]D^{-1}(\phi_i)$, where $G[V] = G[V, 0]$ is the Gorkov Green function with real order parameter. Substituting these representations into the tr ln and using the cyclic invariance of the trace, we conclude that the tr ln is equivalent to one with $G^{-1}[V, \phi_i] \to G^{-1}[V + \dot\phi_i]$ and $T \to TD^{-1}(\phi_1)D(\phi_2)$, which is the stated form of the action.

(b) The first part of the problem is shown by straightforward expansion of the action derived in part (a), with Josephson conditions in place. Substituting the diagonal contribution to the Green function $G_{d,n} = (-i\omega_n - \xi\sigma_3)/(\omega_n^2 + \xi^2 + \Delta^2)$ into the tunneling action and comparing with Eq. (5.64), we find that the sum Eq. (5.65) becomes replaced by

$$\sum_a (G_{i,d})_{a,\omega_n} = \int d\epsilon\, \rho_i(\epsilon + \mu)\frac{-i\omega_n - \epsilon\sigma_3}{\omega_n^2 + \epsilon^2 + \Delta^2} \simeq -i\pi\rho_i\frac{\omega_n}{(\omega_n^2 + \Delta^2)^{1/2}}.$$

Comparing with Eq. (5.66), we obtain

$$\sum_n \left(-\frac{\omega_n}{(\omega_n^2 + \Delta^2)^{1/2}}\frac{\omega_{n+m}}{(\omega_{n+m}^2 + \Delta^2)^{1/2}} + 1\right) \simeq \begin{cases} |\omega_m|/\pi T, & |\omega_m| \gg \Delta, \\ 0 + \mathcal{O}(\omega_m/\Delta), & |\omega_m| \ll \Delta, \end{cases}$$

instead of a global factor $|\omega_m|/2\pi T$. The physical interpretation of this result is that only high-frequency ($\omega > \Delta$) fluctuations of the voltage field $V = \dot\phi$ have the capacity to overcome the superconductor gap and dissipate their energy by creating quasi-particle excitations. In contrast, low-frequency fluctuations do not suffer from dissipative damping.

(c) Substituting the off-diagonal term $G_{o,n} = -\Delta\sigma_1/(\omega_n^2 + \xi^2 + \Delta^2)$ into the tunneling action and neglecting contributions of $\mathcal{O}(|\omega_m|/\Delta)$, we find

$$S_{\text{tun}}[\phi] \simeq |T|^2 \sum_{nm}\sum_{\alpha\alpha'}\frac{\Delta}{\omega_n^2 + \xi_\alpha^2 + \Delta^2}\frac{\Delta}{\omega_n^2 + \xi_{\alpha'}^2 + \Delta^2}\text{tr}\left(\sigma_1(e^{i\sigma_3\phi})_m\sigma_1(e^{-i\sigma_3\phi})_{-m}\right)$$

$$= \frac{|T|^2(\pi\rho)^2\Delta}{2T}\sum_m \text{tr}\left(\sigma_1(e^{i\sigma_3\phi})_m\sigma_1(e^{-i\sigma_3\phi})_{-m}\right)$$

$$= |T|^2(\pi\rho)^2\Delta\int d\tau\, \cos(2\phi(\tau)).$$

(d) Think of the non-dissipative Josephson action as the action of a point particle with kinetic energy $\sim E_C^{-1}\dot\phi^2$ and potential energy $\cos(2\phi)$. In this language, passage to the Hubbard–Stratonovich decoupled action

$$S[\phi, N] = E_C \int d\tau\, N^2 + \gamma \int d\tau\, \cos(2\phi(\tau)) + i\int d\tau\, N\dot\phi,$$

amounts to a transition from the Lagrangian to the Hamiltonian picture. The notation emphasizes that the momentum conjugate to the phase variable is the number operator of the system. (More precisely, N measures the difference in the charges carried by the two superconductors; the total charge of the system is conserved.) Varying the action, we obtain the Hamilton equations

$$\dot{\phi} = i2E_C N, \quad \dot{N} = 2i\gamma \sin(2\phi).$$

Now, $I = i\partial_\tau N$ is the current flowing from one dot to the other, i.e., the second relation gives the Josephson current (5.68). The first relation states that, for a finite charging energy, mismatches in the charge induce time variations in the phase. By virtue of the Josephson relation, such time variations give a finite voltage drop.

5.6.7 Metallic magnetism

Previously we considered the Hubbard–Stratonovich transformed functional integral as an efficient tool in the perturbation theory of the interacting electron gas. However, we have also seen that interactions may have non-perturbative effects and cause phase transitions, the Mott transition of section 2.2.3 being one such example. In this problem, we consider another interaction–driven transition of the electron gas, the Stoner transition[53] into a magnetic phase, and explore it by methods of functional integration.

Historically, the Stoner transition has assumed a special place in the theoretical literature. Developments in statistical mechanics through the 1950s and 1960s highlighted the importance of fluctuation phenomena in the classification and phenomenology of classical phase transitions (see chapter 6). The universal properties of classical finite temperature transitions are characterized by universal sets of critical exponents. In a quantum mechanical system, a phase transition can be tuned by a change in an external parameter even at zero temperature – a **quantum phase transition**. In a seminal work by John Hertz,[54] it was proposed by that the region surrounding a quantum critical point was itself characterized by **quantum critical phenomena**.

quantum
phase
transition

    In this context, the problem of metallic magnetism presents a useful prototype. It is also one for which the class of heavy fermion materials provides a rich arena for experimental observation. In the following, we develop a low-energy theory of the interacting electron system and discuss the nature of the mean field transition to the itinerant ferromagnetic phase. Later, following our discussion of the renormalization group methods in chapter 6, we will use the low-energy theory as a platform to discuss the general phenomenology of quantum criticality (see problem 6.7.2).

    Our starting point is the lattice Hamiltonian for a non-interacting electron gas perturbed by a local "on-site" Hubbard interaction, $\hat{H} = \hat{H}_0 + \hat{H}_U$ where

$$\hat{H}_0 = \sum_{\mathbf{p}\sigma} \epsilon_{\mathbf{p}} c_{\mathbf{p}\sigma}^\dagger c_{\mathbf{p}\sigma}, \quad \hat{H}_U = U \sum_i^N \hat{n}_{i\uparrow} \hat{n}_{i\downarrow}.$$

[53] E. C. Stoner, *Ferromagnetism*, Rep. Prog. Phys. **11**, 43 (1947).
[54] J. A. Hertz, *Quantum critical phenomena*, Phys. Rev. B **14**, 1165 (1976).

Here, the sum runs over the N lattice sites i and $\hat{n}_{i\sigma} = c_{i\sigma}^{\dagger} c_{i\sigma}$. The electron dispersion relation $\epsilon_{\mathbf{p}}$ (a function of the lattice geometry) as well as the dimensionality are left unspecified at this stage.

As we saw in chapter 2, the phase diagram of the lattice Hubbard Hamiltonian is rich, exhibiting a range of correlated ground states depending on the density and strength of interaction. In the lattice system, close to half filling, strong interactions may induce a transition into an insulating antiferromagnetic Mott–Hubbard state. Here, we will show that in the opposite case of low densities, the system may assume a spin-polarized magnetic phase. In this **Stoner ferromagnet**, the charge carriers remain in a mobile state. Magnetic systems with this property are called **itinerant magnets**.

As usual, the formation of this magnetic phase relies on a competition between kinetic and interaction energies: since electrons of the same spin cannot occupy the same state, they escape the Hubbard interaction. However, the same exclusion principle requires them to occupy different single-particle states, which implies a less favorable kinetic energy. When the total reduction in potential energy outweighs the increase in kinetic energy, a transition to a spin-polarized or ferromagnetic phase is induced.

Once again, our first step in the quantitative exploration of the phenomenon is a Hubbard–Stratonovich decoupling. We begin by separating the interaction into contributions coupling to charge and spin densities,

$$\hat{H}_U = \frac{U}{4} \sum_i (\hat{n}_{i\uparrow} + \hat{n}_{i\downarrow})^2 - \frac{U}{4} \sum_i (\hat{n}_{i\uparrow} - \hat{n}_{i\downarrow})^2.$$

We expect fluctuations in the charge density to have little influence on the physics of the low-density system, and therefore neglect their contribution to the interaction, so that $\hat{H}_U \simeq -U \sum_i (\hat{S}_i^z)^2$, where $\hat{S}_i^z = (\hat{n}_{i\uparrow} - \hat{n}_{i\downarrow})/2$.

EXERCISE Here, for simplicity, we have isolated a component of the Hubbard interaction that couples to the spin degrees of freedom but violates the spin symmetry of the original interaction. How could the local interaction be recast in a manner which makes the spin symmetry explicit while isolating the coupling to the spin degrees of freedom? (Hint: Recall our discussion of the exchange interaction in section 2.2.2.)

(a) Express the quantum partition function of system as a functional field integral. Decouple the interaction by a scalar magnetization field $\hat{m} = \{m_i(\tau)\}$ and show that, after integration over the fermions, the partition function takes the form

$$\mathcal{Z} = \mathcal{Z}_0 \int Dm \exp\left(-\frac{U}{4} \int_0^\beta d\tau \sum_i m_i^2(\tau) + \mathrm{tr}\ln\left(1 - \frac{U}{2}\sigma_3 \hat{m}\hat{G}\right) \right),$$

where $\mathcal{Z}_0$ is the partition function of the non-interacting system, the free Green function in momentum representation is given by $G_p = (i\epsilon_n - \xi_{\mathbf{p}})^{-1}$, and σ_i are Pauli matrices in spin space. For simplicity, we assume a nearly free electron dispersion $\xi_{\mathbf{p}} = p^2/2m$ throughout, which in physical terms means that we are working in a low density limit (why?).

We suspect that there is a phase transition to a magnetic phase at some critical value of the interaction, $U = U_c$, signaled by the appearance of a non-zero magnetization field m_i. Assuming this transition to be of second order (which means that at U_c a magnetization profile develops continuously, without jumps), we aim to identify a field theory that captures its critical behavior. As with the Ginzburg–Landau theory of the superconductor, we proceed by self-consistent expansion of the tr ln in a regime where m is small.

(b) Drawing on the RPA expansion of the weakly interacting electron gas (section 5.1), expand the action to fourth order in the magnetization field and show that it assumes the form $S[m] = S^{(2)}[m] + S^{(4)}[m]$, where

$$S^{(2)}[m] = \frac{1}{2} \sum_q \left(r + \xi^2 \mathbf{q}^2 + \frac{|\omega_n|}{v|\mathbf{q}|} \right) |m_q|^2, \quad S^{(4)}[m] = \frac{u}{4} \int dx\, m(x)^4. \quad (5.69)$$

Here $\int dx \equiv \int d\tau\, d^d x$, $r = 2/U\nu - 1$, the velocity $v = cv_F$ with a numerical constant c, ν is the fermion density of states (including the spin multiplicity 2), and the magnetization field has been rescaled so that, in the limit $U \to \infty$, the coefficient $r \to -1$. (Hint: In the term of quartic order, the momentum carried by the magnetization field is assumed to be negligibly small compared with that of the electron Green function. The term thus assumes the form $\int m^4$ times a prefactor, which you need not evaluate in detail. However, convince yourself of its positivity.) For the "polarization bubble" multiplying the second-order expansion, use the results discussed following Eq. (4.28), and especially Eq. (4.33).

In problem 6.7.2, we will use this action as the starting point for the analysis of quantum criticality in itinerant magnetism. However, for the time being let us consider the problem on the simple mean field level, and for a moment assume constancy of the magnetization field. In this limit, the action collapses to

$$S[m] = \int dx \left(\frac{r}{2} m^2 + u m^4 \right).$$

This action indicates that the system has a transition into a magnetized phase at an interaction strength set by the **Stoner criterion** $U_c \nu/2 = 1$. The density of states $\nu \propto \Delta n/\Delta E$ tells us how much (non-interacting) energy ΔE is required to populate Δn energy levels. If the gain in interaction energy, $\propto U \Delta n$, exceeds that energy, we have a transition.

INFO Disclaimer: The gradient expansion leading to the action (5.69) is far from innocuous. Its validity relies on the benign nature of fluctuations in the magnetization density. Subsequent work, however, has cast doubt on the validity of this approximation.[55] Moreover, in real lattice systems, the transition generically occurs at parameter values where interaction effects are strong, and the free fermion approximation used above becomes unreliable. Readers interested in the detailed theory of itinerant ferromagnetism are advised to study the specialized literature.

[55] D. Belitz, T. R. Kirkpatrick, and T. Vojta, *Non-analytic behavior of the spin susceptibility in clean Fermi systems*, Phys. Rev. B **55**, 9453 (1997).

Answer:

(a) Starting with the functional partition function

$$
\mathcal{Z} = \int D\psi \, \exp\left\{ -\int_0^\beta d\tau \left(\sum_{\mathbf{p}} \bar\psi_{\mathbf{p}}(\partial_\tau + \xi_{\mathbf{p}})\psi_{\mathbf{p}} - \frac{U}{4} \sum_i (\bar\psi_i \sigma_3 \psi_i)^2 \right) \right\},
$$

the interaction can be decoupled by a scalar field conjugate to the local magnetization density,

$$
\mathcal{Z} = \int Dm D\psi \, \exp\left\{ -\int d\tau \left(\frac{U}{4} \sum_i m_i^2 + \sum_{\mathbf{p}} \bar\psi_{\mathbf{p}}(\partial_\tau + \xi_{\mathbf{p}})\psi_{\mathbf{p}} + \frac{U}{2} \sum_i \bar\psi_i \sigma_3 m_i \psi_i \right) \right\}.
$$

Integrating over ψ, one obtains the required partition function.

(b) The expansion of the action mirrors the RPA of the weakly interacting electron gas. Terms odd in powers of m vanish identically (a property compatible with the symmetry $m \to -m$). To leading (second) order we obtain

$$
S[m] = \frac{1}{2} \sum_q \frac{U}{2} \left(1 - \frac{U}{2} \Pi_q \right) |m_q|^2,
$$

with the polarization operator defined by Eq. (4.28). Substitution of the expansion (4.33) leads to

$$
S[m] = \frac{U^2 \nu}{8} \sum_q \left(\frac{2}{U\nu} - 1 + \xi^2 \mathbf{q}^2 + \frac{|\omega_n|}{v|\mathbf{q}|} \right) |m_q|^2. \tag{5.70}
$$

A final rescaling $m_q \to 4m_q/\sqrt{\nu}U$ brings the action into the given form. Here, the coupling constant $v_2 = U(1 + U\Pi_0)/2 = U(1 - U\nu/2)/2$ contains the polarization operator (4.28), and we have used Eq. (4.31) for its static limit.

The quartic coupling constant contains the product of four Green functions, $v_4 = (TU^4/8N) \sum_p G_p^4$, which we need not evaluate in detail for the purposes of the present problem. Transforming back to a real-space representation, we obtain the quartic action as given.

6   Renormalization Group

SYNOPSIS The renormalization group (RG) is a powerful tool for the exploration of interacting theories in regimes where perturbation theory fails. In this chapter, we introduce the concept on the basis of two examples, one classical, the other quantum. With this background, we then discuss RG methods in more general terms, introducing the notion of scaling, dimensional analysis, and the connection to the general theory of phase transitions and critical phenomena. In the final parts of the chapter, we discussion various examples including the RG analysis of the ferromagnetic transition, phase transitions in models with vortex formation, and the critical physics of nonlinear sigma models.

In this chapter we will frequently integrate over frequency or momentum variables. For improved readability, we use the notation $\int (d\omega) \equiv \int \frac{d\omega}{2\pi}$ and $\int (dp) \equiv \int \frac{d^d p}{(2\pi)^d}$.

In chapter 4 we introduced ϕ^4-theory as a prototype of interacting theories, and developed a perturbative framework to describe its nonlinearities in the language of diagrams. However, being critical, one might argue that both the derivation and the analysis of the theory presented only limited understanding: the ϕ^4-continuum description was obtained as a gradient expansion of a d-dimensional Ising model. But what controls the validity of the low-order expansion? (The same question could be asked for most of the continuum approximations performed throughout previous chapters.) Individual terms contributing to the expansion of the model (see Eq. (4.17)) contained divergences at both large and small momenta. We had to concede that we had no clue about how to overcome these problems. It also became evident that such difficulties are not specific to the ϕ^4-model but endemic in field theory.

To better address these problems, new ideas must be developed. First, let us remember that our principal objective is the understanding of long-range characteristics. On the other hand, models such as ϕ^4 exhibit fluctuations on all length scales, and it were the short-scale fluctuations that were responsible for the majority of difficulties. In principle, we already know how to deal with situations of this kind: we need a theory of slow fluctuations obtained by integrating over all rapid fluctuations. For example, in section 3.3.3 we derived the effective action of a quantum particle by integration over the rapid fluctuations of an external environment.

The problem with our current application is that in the ϕ^4-theory there is no clear-cut separation into "fast" and "slow" degrees of freedom. Rather, all fluctuations, ranging from the shortest scales (of the order of some microscopic cutoff, a, limiting the applicability of the theory) to the longest scales (of the order of, say, the system size, L) are treated on the same footing.

To implement the scheme of integrating over fast modes to generate an effective action of the slow degrees of freedom, one must *declare* artificially a certain length scale $a^{(1)} \equiv ba > a$ as the scale separating "short-wavelength fluctuations" on scales $[a, a^{(1)}]$ from "long-wavelength fluctuations" on scales $[a^{(1)}, L]$. Integration over the former will modify the action of the latter. Since the short-range action is no simpler than the long-range action, this step generically involves approximations. Indeed, various scenarios are possible. For example, the integration may corrupt the structure of the long-wavelength action, leaving us with a theory different from the one with which we began. Alternatively, the post-integration long-range action may be structurally identical to the original, in which case the effect of the integration can be absorbed in a modified set of coupling constants.

If the latter, the procedure is well motivated: we have arrived at a theory identical to the original but for (a) a different, or **renormalized**, set of coupling constants, and (b) an increased short-distance cutoff $a \to a^{(1)} = ba$. Evidently, one may then iterate this procedure; that is, declare a new cutoff $a^{(2)} = ba^{(1)} = b^2 a$ and integrate out fluctuations on length scales $[a^{(1)}, a^{(2)}]$, etc. Along with the recursive integration of more layers of fluctuations, the coupling constants of the theory change, or *flow*, until the cutoff $a^{(n)} \sim L$ has become comparable with the length scales in which we are interested. We will see that this flow of coupling constants encodes much of the long-range behavior of the theory.

The general line of reasoning above summarizes the essence of the renormalization group. Of course, a step by step algorithm would be useless had each reduction $a^{(n-1)} \to a^{(n)}$ to be performed explicitly. However, the program is recursive. Since the model reproduces itself at each step, a single step already encodes the full information about its renormalization.

history of
the RG

INFO The **formulation of renormalization group concepts** has a long history, reflecting the versatility of the method.[1] The advent of these ideas in the 1960s and early 1970s marked the transition between two different epochs. While, hitherto, the focus in many-body theory had been on the development of ever more-sophisticated perturbative techniques, the seventies stood under the spell of the renormalization program.

In the second half of the 1960s, ideas to recursively generate flows of coupling constants arose – apparently in independent developments – both in condensed matter and in particle physics. However, it took the insight of Kenneth Wilson to realize the full potential of the approach and to develop it into a widely applicable tool[2]. Wilson's renormalization program, and its later extension by others, led to revolutionary progress in condensed

---

[1] A perspective on the development of the renormalization group can be found in the review article by M. E. Fisher, *Renormalization group theory: Its basis and formulation in statistical physics*, Rev. Mod. Phys. **70**, 653 (1998), or in the text by J. Cardy, *Scaling and Renormalization in Statistical Physics* (Cambridge University Press, 1996).

[2] See K. Wilson, *The renormalization group: critical phenomena and the Kondo problem*, Rev. Mod. Phys. **47**, 773 (1975), still one of the best introductions to the approach!

matter physics, particle physics, and general statistical mechanics. In this way the RG concept became a major driving force behind the partial unification of these fields.

In fact, the (unfortunate) denotation "renormalization group" reflects the historical origins of the approach: by the late 60th, breakthroughs in the development of the standard model of elementary particles had cemented the belief that the fundamental structure of matter could be understood in terms of symmetries and their implementation through groups. In an attempt to absorb the newly developed RG approach into this general framework, it became dubbed the renormalization *group*. Of course, this link would not have been drawn had it not been justified:

Kenneth G. Wilson 1936–2013 an American physicist who was awarded the 1982 Nobel Prize in Physics, awarded for "discoveries he made in understanding how bulk matter undergoes phase transitions, i.e., sudden and profound structural changes resulting from variations in environmental conditions." Wilson's background ranged from elementary particle theory and condensed matter physics (critical phenomena and the Kondo problem) to quantum chemistry and computer science.

formally the sequence of RG transformations outlined above defines a semigroup.[3] However, this analogy to groups has never been essential and is little more than a historical name tag.

6.1 Renormalization: Two Examples

While the procedures outlined above may at first sight look mysterious, their working becomes quite natural when formulated in the context of concrete models. We therefore start our discussion of the renormalization group with two case studies, the renormalization of the classical one-dimensional Ising model, and of dissipative quantum tunneling. Along the way, we will introduce a number of concepts to be defined in more generality in later sections of the chapter.

6.1.1 One-dimensional Ising model

one-dimensional Ising model

Consider the classical **one-dimensional Ising model** defined through the Hamiltonian

$$H_{\mathrm{I}} = -J \sum_{i=1}^{N} S_i S_{i+1} - H \sum_{i=1}^{N} S_i,$$

where $S_i = \pm 1$ denotes the uniaxial magnetization of site i (periodic boundary conditions, $S_{N+1} = S_1$, imposed), and H represents an external field. Thinking of

---

[3] We interpret individual RG transformations as abstract mappings of actions, $R : S \mapsto S'$. The concatenation of such transformations, $R \circ R'$, satisfies two group axioms: there is a unit transformation (nothing is integrated out) and the iteration is associative. However, since there is no inverse to R, we only have a semigroup rather than a full group.

$-\beta H_{\mathrm{I}}[S]$ as a functional of the discrete field $\{S_i\}$, we will solve the model in two different ways: the first is exact, the second makes recourse to an RG algorithm.

Exact solution

The model defined by the Hamiltonian above is exactly solvable, and all of its correlation functions can be computed in closed form. The starting point of this solution is the observation that the Boltzmann weight can be factorized as

$$e^{-\beta H_{\mathrm{I}}} = \exp\left(\sum_{i=1}^{N}(KS_iS_{i+1} + hS_i)\right) = \prod_{i=1}^{N} T(S_i, S_{i+1}),$$

where we have introduced the dimensionless parameters $K \equiv \beta J > 0$ and $h = \beta H$, and the weight is defined as $T(S, S') = \exp(KSS' + h(S + S')/2)$. Defining a two-dimensional matrix T with elements $T_{nm} = T((-1)^{n+1}, (-1)^{m+1})$, or

$$T = \begin{pmatrix} e^{K+h} & e^{-K} \\ e^{-K} & e^{K-h} \end{pmatrix},$$

the partition function assumes the form

$$\mathcal{Z} = \sum_{\{S_i\}} e^{-\beta H_{\mathrm{I}}} = \sum_{\{S_i\}} \prod_{i=1}^{N} T(S_i, S_{i+1}) = \sum_{\{n_i\}} \prod_{i=1}^{N} T_{n_i n_{i+1}} = \operatorname{tr} T^N.$$

transfer
matrix We have thus managed to represent the partition function as a trace of the Nth power of the so-called **transfer matrix** T.[4] The advantage of this representation is that it describes the partition sum in terms of the two eigenvalues of T, which are readily computed as $\lambda_{\pm} = e^K(\cosh(h) \pm (\sinh^2(h) + e^{-4K})^{1/2})$. With $\mathcal{Z} = \operatorname{tr} T^N = \lambda_+^N + \lambda_-^N$, and noting that $\lambda_+ > \lambda_-$, we conclude that in the thermodynamic limit $N \to \infty$ we have $\mathcal{Z} \overset{N \to \infty}{\longrightarrow} \lambda_+^N$. Restoring the original microscopic parameters, we obtain the free energy as

$$F \equiv -\frac{1}{\beta}\ln\mathcal{Z} = -N\left(J + T\ln\left(\cosh(\beta H) + \sqrt{\sinh^2(\beta H) + e^{-4\beta J}}\right)\right). \quad (6.1)$$

From this result, we obtain the magnetization density $m = M/N$ by differentiation with respect to H,

$$m = \frac{\sinh(\beta H)}{\sqrt{\sinh^2(\beta H) + e^{-4\beta J}}}. \quad (6.2)$$

---

[4] The terminology *transfer matrix* originates in an interpretation of the Ising model as a fictitious dynamical process in which a state S_i is transferred to a state S_{i+1}, where the transition amplitude is given by $T(S_i, S_{i+1})$.

We observe that, with decreasing temperature, the magnetization changes ever more steeply to adjust to an external field (see the figure). However, the system does not magnetize at any non-zero temperature in the absence of a field, $H = 0$. This finding is in accordance with our

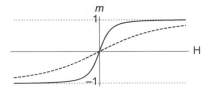

earlier observation (section 5.2.3) that discrete symmetries do not get spontaneously broken in one dimension.

The concept of scaling

Eq. (6.2) represents a full solution of the one-dimensional Ising model. From the magnetization, other thermodynamic characteristics, such as the magnetic susceptibility $\chi = -\partial_H^2 F$, are obtained by further differentiation with respect to H and/or T. However, this is an exceptional situation; in most problems of interest, we are not in possession of a closed solution. This means that, before comparing the exact solution with that produced by the RG program, we should reformulate the former in a universally applicable language, a code that can be used to describe model solutions irrespective of the particular method by which these have been obtained.

A universal theme in the analysis of all many-body systems is their fluctuation behavior at large distance scales, as probed by correlation functions. Specifically, for the Ising model, the two-point correlation function assumes the form

$$C(r_1 - r_2) \equiv \langle S(r_1)S(r_2)\rangle - \langle S(r_1)\rangle\langle S(r_2)\rangle \sim \exp\left(-\frac{|r_1 - r_2|}{\xi}\right),$$

correlation length

where we have switched to a continuum notation $S_i \to S(r)$, and ξ defines the **correlation length** of the system. Such correlation functions are generically related to quantities of thermodynamic significance. For example, the magnetic susceptibility of the Ising model follows from its partition sum as

$$\chi = -\partial_H^2 F\big|_{H=0} = T\partial_H^2 \ln \mathcal{Z}\big|_{H=0} = \beta \int dr dr' \overbrace{(\langle S(r)S(r')\rangle - \langle S(r)\rangle\langle S(r')\rangle)}^{C(r-r')}.$$
(6.3)

Performing the integral, we obtain $\chi \sim \xi$, i.e., a proportionality of the susceptibility to the correlation length of fluctuations in the system.

This result exemplifies the role played by the correlation length in general: Close to zero temperature, or in the vicinity of a (second-order) phase transition, we expect fluctuations to become long-range, and hence a divergence of the correlation length. In these regimes, the correlation length exceeds all microscopically defined length scales; it is the *only* relevant length scale. Accordingly, we expect observables X of dimensionality $[\text{length}]^{D_X}$ to obey the **scaling form**

$$X \sim \xi^{D_X} g_X,$$

where g_X is a dimensionless function of other relevant system parameters.

Let us explore how scaling manifests itself in the Ising model. Computing the susceptibility from the exact solution (6.2), we obtain

$$\xi \sim \chi \sim \partial_H|_{H=0}\, m \sim e^{2\beta J}, \tag{6.4}$$

i.e., the expected divergence of the correlation length at low temperatures. As a scaling observable, consider the so-called **reduced free energy**,

$$f(T) \equiv \frac{F}{TL}, \tag{6.5}$$

a function with dimensionality L^{-1}. Noting that $N \sim L$, a straightforward low-temperature expansion of Eq. (6.1) indeed gives

$$f(T) - f(0) = -\xi^{-1}\left(1 + \frac{1}{2}\xi^2 h^2\right) \equiv \xi^{-1}g(\xi h), \tag{6.6}$$

where we have subtracted the infinite but inessential constant $f(0)$ and assumed that $1 \gg \xi^{-1} \gg h$. (The scaling form above actually suggests that the magnetic field has dimension L^{-1}, a prediction substantiated below.)

Referring for a more in-depth discussion to section 6.2.4, we note that scaling forms provide us with the sought after language for the description of systems in regimes with long-range fluctuations. They contain the "essential" information on physical observables, they can be measured – both experimentally and numerically – and they are obtained from the "natural" objects of study of field theory, correlation functions. In the next section, we will pretend that we did not know the exact solution of the one-dimensional Ising model and will derive its scaling properties from an RG analysis.

Kadanoff's block spin RG

According to the general scheme outlined at the beginning of the chapter, the idea of the RG program is to recursively trace out short-scale fluctuations of a system and assess their influence on the remaining degrees of freedom. Several years before the RG procedure was formulated in generality, in a seminal study Kadanoff elucidated the conceptual power of such an approach for Ising type systems.[5]

His idea was to subdivide a spin chain into regular clusters of b neighboring spins (see the figure for $b = 2$). One then proceeds to sum over the 2^b sub-configurations of each cluster, thereby generating an effective functional describing the  inter-cluster energy balance. While it is clear that this energy functional exists, a less obvious question is whether it will again have the form of an Ising functional. Remarkably, the answer is affirmative: the Ising model is *renormalizable*. Its structural reproduction suggests an interpretation of each cluster as a meta-Ising spin, **block spin** or **block spin**. In this way, the renormalization step qualifies for iteration: in a

---

[5] Leo P. Kadanoff, *Scaling laws for Ising models near T_c*, Physics Physique Fizika **2**, 263 (1966).

second RG step, b block spins are grouped to form a new cluster (now comprising b^2 microscopic spins), which are then traced out, etc. We next discuss how this algorithm is implemented in concrete terms.

Within the transfer matrix approach, a cluster of b spins is represented through b transfer matrices T. Taking the partial trace over its degrees of freedom amounts to passing from these b matrices to the product $T' = T^b$. (The internal index summation involved in taking the product amounts to a trace over intra-cluster degrees of freedom.) The transition from the original partition function $\mathcal{Z}$ to the new partition function $\mathcal{Z}'$ is defined through the relation

$$\mathcal{Z}_N(K,h) = \operatorname{tr} T^N = \operatorname{tr}(T^b)^{N/b} = \operatorname{tr}(T')^{N/b} = \mathcal{Z}_{N/b}(K',h'), \qquad (6.7)$$

where the notation makes explicit the parametric dependence of the partition function on the size of the system, N, and on the coupling constants K, h. Notice that the equation relies on the condition that the reduced trace, $\operatorname{tr}(T')^{N/b}$, again has the form of an Ising partition function or, equivalently, that T' has the same algebraic structure as T.

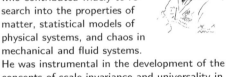

Leo P. Kadanoff 1937–2015 an American theoretical physicist and applied mathematician who contributed widely to research into the properties of matter, statistical models of physical systems, and chaos in mechanical and fluid systems. He was instrumental in the development of the concepts of scale invariance and universality in the physics of phase transitions.

To establish this structure, we compute T' for the case of $b = 2$ block spins. Defining $u \equiv e^{-K}$, $v \equiv e^{-h}$, we have

$$T = \begin{pmatrix} e^{K+h} & e^{-K} \\ e^{-K} & e^{K-h} \end{pmatrix} = \begin{pmatrix} u^{-1}v^{-1} & u \\ u & u^{-1}v \end{pmatrix},$$

and

$$T' \equiv T^2 = \begin{pmatrix} u^2 + u^{-2}v^{-2} & v + v^{-1} \\ v + v^{-1} & u^2 + u^{-2}v^2 \end{pmatrix} \overset{!}{=} C\begin{pmatrix} u'^{-1}v'^{-1} & u' \\ u' & u'^{-1}v' \end{pmatrix}.$$

In the last equality, we require that the new transfer matrix be of the same structure as the original. However, noting that this requirement will introduce three conditions (for the three independent entries of the symmetric matrices T and T'), we are willing to tolerate the appearance of an overall multiplicative constant C.[6] Having introduced this new parameter, we have enough freedom to solve the three equations, from which one finds (exercise)

$$u' = \frac{\sqrt{v + v^{-1}}}{(u^4 + u^{-4} + v^2 + v^{-2})^{1/4}}, \quad v' = \frac{\sqrt{u^4 + v^2}}{\sqrt{u^4 + v^{-2}}}, \qquad (6.8)$$

and the factor $C = \sqrt{v + v^{-1}}(u^4 + u^{-4} + v^2 + v^{-2})^{1/4}$, which will be of lesser importance. The new transfer matrix describes an Ising spin system,

[6] Taking the product of the new transfer matrices, we see that this constant appears in the partition function as $\mathcal{Z}' \sim C^{N/b}$, i.e., the free energy acquires an overall additive constant $F' \sim -NT \ln C/b$, which will be of no further significance.

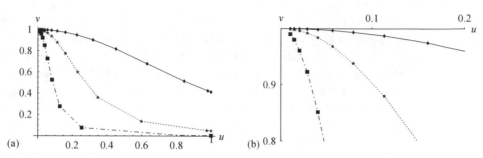

Fig. 6.1   (a) The flow of coupling constants of the one-dimensional Ising model generated by iteration of the RG transformation. The three curves shown are for starting values $(u, v) = (0.01, 0.9999)(\blacklozenge)$, $(0.01, 0.999)(\star)$, and $(0.01, 0.99)(\blacksquare)$. (b) Magnification of the zero-temperature fixed point region.

▷ whose Hamiltonian is defined at a different temperature, magnetic field, and exchange constant (as described by the new values of the coupling constants (u', v')) and

▷ which is defined for a lattice length scale twice as large as in the original system.

To make further progress, we note that the result of the block spin transformation can be represented as the discrete map

$$\begin{pmatrix} u' \\ v' \end{pmatrix} = \begin{pmatrix} f_1(u, v) \\ f_2(u, v) \end{pmatrix},$$

where the functions $f_{1,2}$ are defined through Eq. (6.8). In fig. 6.1 sequences of points generated by an iterative application of the map f are shown for different values of starting values (u_0, v_0). It is evident from these **RG trajectories** that the map f possesses two sets of **fixed points**, i.e., points (u^*, v^*) that remain invariant under the application of the map f:

RG fixed points

$$\begin{pmatrix} u^* \\ v^* \end{pmatrix} = \begin{pmatrix} f_1(u^*, v^*) \\ f_2(u^*, v^*) \end{pmatrix}.$$

Inspection of Eq. (6.8) shows that this is the case for the point $(u^*, v^*) = (0, 1)$, and the line $(u^*, v^*) = (1, v)$.

The set of its fixed points defines the most fundamental element of an RG analysis. Fixed points organize the space of "flowing" coupling constants into regions of different behavior. At the fixed points themselves, all characteristics of the model, including its correlation length ξ, remain invariant. On the other hand, individual RG steps increase the fundamental length scale of the system. Consistency requires that either $\xi = 0$ or $\xi = \infty$.

In the present case, the line of fixed points is identified with $u = \exp[-\beta J] = 1$, i.e., $\beta = 0$. This is the limit of infinitely large temperatures, for which we expect a state of maximal thermal disorder, i.e., $\xi = 0$. Besides the high-temperature fixed line, there is a zero-temperature fixed point $(u, v) = (\exp[-\beta J], \exp[-\beta h]) = (0, 1)$

implying that $T \to 0$ and $h = 0$. Upon approaching zero temperature, the system is expected to form larger and larger segments of aligned spins, $\xi \to \infty$.

Another important difference between the high- and low-temperature fixed points is that the former is **attractive** and the latter **repulsive**: recursion relations started from a low-temperature initial configuration will iterate towards high-temperature configurations. Physically, this means that, even if the system looks ordered at microscopic length scales (low temperature), at larger length scales we will notice the presence of domain walls – the absence of spontaneous symmetry breaking again; large block spins are in a less ordered (high temperature) state.

Beyond the fixed points themselves, their attractive or repulsive behavior is the second most fundamental signature of an RG analysis. This information is required in order to understand the actual flow patterns of a system under changes of scale. Further, in the vicinity of fixed points, flows are "slow" and recursion relations afford linearization. It is in this regime that their solutions simplify and derived physical observables assume scaling forms.

To illustrate these features on the present example, consider our RG map in the vicinity of the $T = 0$ fixed point. Defining a two-component vector, $\mathbf{x}^* \equiv (u^*, v^*)^T = (0, 1)^T$, and with $\Delta \mathbf{x}$ a small deviation from the fixed point, we can write $\mathbf{x}^* + \Delta \mathbf{x}' = \mathbf{f}(\mathbf{x}^* + \Delta \mathbf{x}) \approx \mathbf{f}(\mathbf{x}^*) + \partial_{\mathbf{x}} \mathbf{f} \cdot \Delta \mathbf{x} + \mathcal{O}(\Delta \mathbf{x}^2)$. Since $\mathbf{f}(\mathbf{x}^*) = \mathbf{x}^*$, we obtain the linearized map $\Delta \mathbf{x}' = \partial_{\mathbf{x}} \mathbf{f} \cdot \Delta \mathbf{x} + \mathcal{O}(\Delta \mathbf{x}^2)$. To explore this linearization in more detail, we introduce another pair of variables, $r \equiv u^4$, $s \equiv v^2$, whereupon the RG transformation becomes rational.[7] Differentiating this map at $(r, s) = (0, 1)$, it is straightforward to show that

$$\begin{pmatrix} \Delta r' \\ \Delta s' \end{pmatrix} = \begin{pmatrix} 4 & \\ & 2 \end{pmatrix} \begin{pmatrix} \Delta r \\ \Delta s \end{pmatrix}.$$

Noting that a transformation with $b = 4$, say, is equivalent to a two-fold application of a $b = 2$ transformation, we can recast the relation above in the form

$$\begin{pmatrix} \Delta r' \\ \Delta s' \end{pmatrix} = \begin{pmatrix} b^2 & \\ & b \end{pmatrix} \begin{pmatrix} \Delta r \\ \Delta s \end{pmatrix}, \tag{6.9}$$

applicable to arbitrary b.

To make use of this linearized flow equation, we consider the free energy (6.5), $f(\Delta r, \Delta s) \equiv -N^{-1} \ln \mathcal{Z}_N(K(\Delta r), h(\Delta s)) \equiv -N^{-1} \ln \mathcal{Z}_N(\Delta r, \Delta s)$, and reformulate Eq. (6.7) according to

$$f(\Delta r, \Delta s) = -\frac{1}{N} \ln \mathcal{Z}_N(\Delta r, \Delta s) = -\frac{1}{N'b} \ln \mathcal{Z}_{N'}(\Delta r', \Delta s') = \frac{1}{b} f(b^2 \Delta r, b \Delta s).$$

This equation describes the scaling of the free energy density under block spin transformations or, equivalently, changes in the fundamental length scale at which we consider our model (all in the linearizable low-temperature regime). The right-hand side describes how it would look from a "blurred" perspective, where all degrees of freedom on scales $< b$ have been combined into a single structural unit.

---

[7] Show that $(_{\text{I}} - s + s + {}_{\text{I}} - \imath + \imath)/(_{\text{I}} - s + s + \mathsf{z}) = {}_{\prime}\imath$ and $(_{\text{I}} - s + \imath)/(s + \imath) = {}_{\prime}s$.

Now, b is a free parameter without intrinsic significance; it can be set to any desired value. For example, we may find it convenient to consider the model at scales where $b^2 \Delta r = 1$. With this choice one obtains $f(\Delta r, \Delta s) = \Delta r^{1/2} f(1, \Delta s / \Delta r^{1/2}) \equiv \Delta r^{1/2} g(\Delta s / \Delta r^{1/2})$, where the dimensionless one-parameter function g is defined through the second equality. Finally, we relate back to the physical parameters,

$$\Delta r = r - 0 = r = u^4 = e^{-4K}, \quad \Delta s = s - 1 = v^2 - 1 = e^{-2h} - 1 \simeq -2h,$$

which brings us to the scaling relation

$$f = e^{-2K} g(e^{2K} h). \tag{6.10}$$

This is the scaling form predicted by the RG analysis. Notice that the dependence of the free energy on two independent parameters K, h is reduced to a one-parameter function, multiplied by an overall prefactor. This finding is consistent with the assumed scaling form $f \sim \xi^{-1} g$. Noting that there is no reason for the rescaled free energy $g(x) = f(1, x)$ to be singular; the divergence of ξ is driven by the prefactor, i.e., $\xi \sim e^{2K}$. On this basis, we have $g = g(\xi h)$, i.e., the magnetic field appears in conjunction with the correlation length and we have reproduced the exact asymptotic, Eq. (6.6).

INFO Do not be concerned about the accumulation of vague proportionalities "$\sim$" in these constructions. Scaling laws describe the behavior of observables in the vicinity of phase transition points where power laws prevail. For example, different systems in the Ising universality class may show a common power law dependence of magnetization versus temperature, $M \sim T^\beta$. However, the numerical prefactors in such proportionalities vary between materials, and depend on the chosen system of units. They are non-universal and likely to be of lesser importance both theoretically and experimentally.

EXERCISE Apply the block spin RG procedure to the one-dimensional q-state Potts spin model $\beta H = -K \sum_{i=1}^{N} \delta_{s_i, s_{i+1}}$, where $s_i = (1, 2, \ldots, q)$. Identify all fixed points and note their stability.

6.1.2 Dissipative quantum tunneling

In a second case study, we apply the RG concept in a Fourier space representation, where the distinction between fast and slow modes is met on the basis of their momentum (or frequency). This is the standard procedure in field theory applications, where one is working with continuum fields that have "forgotten" about underlying lattice structures.

In section 3.3.3, we explored the influence of external environments in quantum mechanical tunneling. We saw that an environment affects the tunneling action of a system with coordinate θ via an ohmic contribution to the action $S_{\text{diss}}[\theta] = (\pi T g)^{-1} \sum_{\omega_n} |\omega_n| |\theta_n|^2$, where g^{-1} is a coupling constant. We also noticed that such dissipative structures appear in a number of different settings, for example in the problem of impurity scattering in a quantum wire (see problem 3.8.12).

In the following, we apply an RG program to understand the effect of dissipation on quantum tunneling. For concreteness, let us assume that a quantum mechanical particle inhabits a periodic potential $U(\theta) = c\cos\theta$. In this case, the quantum partition function $\mathcal{Z} = \int D\theta \exp(-S[\theta])$ is governed by the action[8]

$$S[\theta] = \frac{1}{4\pi g}\int (d\omega)\,|\omega||\theta(\omega)|^2 + c\int d\tau\,\cos(\theta(\tau)). \tag{6.11}$$

Momentum shell renormalization

Following the general philosophy outlined at the beginning of the chapter, we begin by arbitrarily subdividing the set of field modes θ_n into short- and long-wavelength degrees of freedom. For example, assuming that the effective bosonic action applies up to a cutoff frequency Λ, we might say that fluctuations on scales $\Lambda/b < |\omega| < \Lambda$ are fast $(b > 1)$ while those with $|\omega| < \Lambda/b$ are slow.

momentum shell renor-malization

INFO Referring to frequency as the 0-component of a generalized momentum, the strategy outlined above is called **momentum shell renormalization**. It is particularly popular in condensed matter field theory, where effective theories always come with an upper cutoff defined by the lattice spacing. (The assumption of an upper cutoff is less natural in particle physics, which generically prefers to implement RG steps without reference to such a scale; see the remarks in the Info block on page 332.) While momentum shell renormalization is a method of choice in continuum field theory, numerical approaches or lattice statistical mechanics often favor **real-space renormalization**, as in the previous chapter. In cases where different renormalization programs are applied to the same problem, the results must agree, and such comparisons can provide valuable consistency checks.

We thus begin by decomposing a general field amplitude $\theta(\tau) \equiv \theta_s(\tau) + \theta_f(\tau)$ into a slow contribution $\theta_s(\tau)$ and its fast complementary $\theta_f(\tau)$ part, where

$$\theta_{s,f}(\tau) \equiv \int_{s,f} (d\omega)e^{-i\omega\tau}\theta(\omega), \tag{6.12}$$

with $\int_s \equiv \int_{|\omega|<\Lambda/b}$ and $\int_f \equiv \int_{\Lambda/b\leq|\omega|<\Lambda}$. Substituting this split into the action (6.11), one obtains $S[\theta_s, \theta_f] = S_s[\theta_s] + S_f[\theta_f] + S_U[\theta_s, \theta_f]$, where

$$S_{s,f}[\theta_{s,f}] = \frac{1}{4\pi g}\int_{s,f}(d\omega)|\theta(\omega)|^2|\omega|, \quad S_U[\theta_s,\theta_f] = c\int d\tau\,\cos(\theta_s(\tau)+\theta_f(\tau)).$$

Our goal now is to derive an effective action $S_{\text{eff}}[\theta_s]$ of the slow fields, after the fast ones have been integrated out, i.e., $e^{-S_{\text{eff}}[\theta_s]} \equiv e^{-S_s[\theta_s]}\langle e^{-S_U[\theta_s,\theta_f]}\rangle_f$, where $\langle\cdots\rangle_f \equiv \int D\theta_f\, e^{-S_f[\theta_f]}(\cdots)$. In view of the nonlinearity of the action, this step must

---

[8] Here, we assume temperatures sufficiently low that Matsubara sums can be replaced by frequency integration, $\sum|\omega_m||\theta_m|^2 \to T\int\frac{d\omega}{2\pi}|\omega|\,|\theta(\omega)|^2$, with $\theta(\omega) = \theta_n/T$. We also assume the coupling $1/g$ to be large so that, for low frequencies, the dissipative term dominates over the kinetic contribution $m\dot\theta^2/2$. In the language of mechanics, this is the overdamped limit of an oscillator degree of freedom.

necessarily be perturbative. Assuming that the coupling constant c is small, we approximate

$$e^{-S_{\text{eff}}[\theta_s]} = e^{-S_s[\theta_s]} \langle 1 - S_U[\theta_s, \theta_f] + \cdots \rangle_f \approx e^{-S_s[\theta_s]} e^{-\langle S_U[\theta_s, \theta_f] \rangle_f}. \qquad (6.13)$$

In words, we have expanded the action to first order in c, averaged the resulting expression over fast fluctuations, and in the final step re-exponentiated the result. This linearization followed by re-exponentiation is reminiscent of the steps involved in the construction of the path integral by Trotter decomposition. While at first sight it may seem to be limited to asymptotically weak coupling, c, we will see later in the chapter that the approach is much stronger than that.

The influence of the fast fluctuations on the effective action is now contained in the average, which we compute as

$$\langle S_U[\theta_s, \theta_f] \rangle_f = c \int D\theta_f\, e^{-S_f[\theta_f]} \int d\tau \cos(\theta_f(\tau) + \theta_s(\tau))$$

$$= \frac{c}{2} \int d\tau e^{i\theta_s(\tau)} \int D\theta_f \exp\left(-\frac{1}{4\pi g} \int_f (d\omega)\theta(\omega)|\omega|\theta(-\omega) + i \int_f (d\omega)e^{i\omega\tau}\theta(\omega) \right) + \text{c.c.}$$

$$= \frac{c}{2} \int d\tau \exp\left(i\theta_s(\tau) - \pi g \int_f (d\omega)|\omega|^{-1} \right) + \text{c.c.} = c \int d\tau \cos(\theta_s) e^{-2\pi g \int_{\Lambda/b}^{\Lambda} (d\omega) \frac{1}{\omega}}$$

$$= c \int d\tau \cos(\theta_s) e^{-g\ln b} = cb^{-g} \int d\tau \cos(\theta_s).$$

We arrive at the remarkable conclusion that the effective action for the slow field,

$$S_{\text{eff}}[\theta_s] = \frac{1}{4\pi g} \int_s (d\omega)\, |\theta(\omega)|^2 |\omega| + cb^{-g} \int d\tau \cos(\theta_s),$$

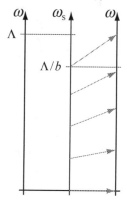

is structurally identical to the action from which we started; the action is renormalizable. However, S_{eff} differs from the bare action S in two respects: first, the fast field integration induces a change in the coupling constant c and, second, the new action is defined for field configurations fluctuating in a reduced range $|\omega| < \Lambda/b$.

The essential step of the RG program is a comparison of the model before and after the integration over the fast fields. However, model actions defined for different sets of field configurations, θ vs. θ_s, cannot be sensibly related to each other. To facilitate the comparison, we first **rescale** frequency/time,

$$(\tau, \omega) \to (\tau', \omega') \equiv (\tau b^{-1}, \omega b),$$

in such a way that the product $\omega\tau = \omega'\tau'$ remains invariant but the new frequency range $|\omega'| < \Lambda$ (see the figure).

Recalling that $\theta_s(\omega)$ are just integration variables, the question remains as to how to choose a meaningful set of variables $\theta'(\omega')$ defined on the full frequency range. In principle, there is ample of freedom for further rescaling – which we will turn into a creative resource when we discuss the RG program in generality. However, in the present context, $\theta(\tau)$ is an angular variable (the cos term) defined on a fixed interval $[0, 2\pi]$, and so we define $\theta'(\tau') \equiv \theta_s(\tau)$. Since $\theta'(\omega')$ is related to $\theta'(\tau')$ by Fourier transform, Eq. (6.12), this relation implies

$$\theta'(\omega') = b^{-1}\theta(\omega).$$

Substitution of the new variables $\omega', \tau', \theta'(\tau')$, and $\theta'(\omega')$ into the effective action then gives

$$S_{\text{eff}}[\theta_s] = S'[\theta'] \equiv \frac{1}{4\pi g} \int_{|\omega'|<\Lambda} (d\omega')\, |\theta'(\omega')|^2 |\omega'| + cb^{1-g} \int d\tau' \cos(\theta'(\tau')).$$

INFO The transformation of the action $S_{\text{eff}}[\theta_s]$ under rescaling could have been anticipated without calculation from **dimensional analysis**: the definition $\omega' = b\omega$ implies that all contributions to the action of dimension [frequency]^d change by a factor b^{-d}. Since $\theta(\tau)$ is a dimensionless phase, we have $[\theta(\tau)] = 1$ while $[\theta(\omega)] = [\text{frequency}]^{-1}$. Thus, the first term of the action has dimension 1 and remains invariant. The second operator carries the dimension $[d\tau] = [\text{frequency}]^{-1}$ and, therefore, changes by a factor b.

Flow equations

Comparing the effective actions $S[\theta]$ and $S'[\theta']$ before and after the integration over the fast modes, we note that the obvious difference is a change in the coupling constant,

$$c \to c(b) \equiv c\, b^{1-g}. \tag{6.14}$$

However, in addition to that, we have an implicit change: the new action describes fluctuations on slower frequency scales or larger temporal scales. This difference remains implicit because we have chosen to measure the "new" frequency continuum in rescaled variables $\omega' = b\omega$. The combination of these two changes implies that the effective strength of the potential changes when looked at from a somewhat larger length scales.

INFO To understand this mechanism in physical terms, consider the case where the action Eq. (6.11) describes scattering off an isolated impurity in a quantum wire, see Eq. (3.100) and the Info section on page 159. The scattering of electrons off this impurity creates a $2k_{\text{F}}$-oscillatory density modulation in the electron gas known as a *Friedel*

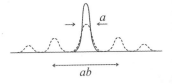

oscillation. In the presence of repulsive electron–electron interactions, $g < 1$, further incoming electrons witness a stronger potential, realized as the superposition of the native potential and the density modulation. (Conversely, attractive interactions, $g > 1$, weaken the scattering potential.) The support of this effective potential exceeds that of the impurity, which is consistent with the scale increase under renormalization.

At this point, we have found out how the coupling constant changes after one RG step, Eq. (6.14). Notice that this result depends on how we choose to dissect the frequency spectrum via the parameter b. However, what we ultimately want to know is the coupling constant after all fluctuations down to a given infrared scale $\omega_{\min}$ have been integrated out. To obtain this information, the RG step needs to be iterated: from $c \equiv c^{(0)}$ and $c' \equiv c^{(1)}$, on to a sequence of coupling constants $c^{(0)} \to c^{(1)} \to c^{(2)} \to \cdots$. As in the previous discussion of the Ising model, we interpret this sequence as a dynamical system, where individual time steps involves an update $c \to c' = c\,b^{1-g} = c\,e^{\ln b(1-g)}$. Assuming that in each step only a thin layer in frequency space is shaven off, $b = 1 + \epsilon$ is close to unity and the coupling constant update small. We may thus turn to a continuum description as $\epsilon^{-1}(c' - c) \approx d_\epsilon c \approx d_{\ln b}c = c(1 - g)$. The evolution equation describing the continuous change of the coupling constant,

$$\boxed{\frac{dc}{d\ln b} \equiv \beta(c) = c(1 - g)} \qquad (6.15)$$

RG flow equation

is an example of an **RG flow equation**. For historical reasons, the right-hand side of the flow equation is called the β-**function**.

Interpreting $t \equiv \ln b$ as a time-like coordinate, we can now integrate the evolution equation to obtain[9] $c(t) = c(0)e^{(1-g)t}$, where $c(t = 0) = c(b = 1)$ is the bare coupling constant of the theory. (Remember that, for $b = 1$, $\Lambda_{\text{eff}} = \Lambda/b = \Lambda$, which is the unrenormalized theory.)

At what time scale $t = \ln b$ should the RG flow be stopped? The answer depends on the realization of the cutoff scale, $\omega_{\min}$, which can be temperature, T, the oscillation frequency ω_0 of an external perturbation, the inverse time of flight through a system of finite extension, v_F/L, or some other natural scale. In either case, the final cutoff value equals $\Lambda_{\text{eff}} = \omega_{\min}$, or $t = \ln(\Lambda/\omega_{\min})$. The effective theory at these scales looks structurally identical to the microscopic model but with a renormalized coupling constant $c = c(0)(\Lambda/\omega_{\min})^{1-g} \sim \omega_{\min}^{g-1}$.

Notice that both the bare constant $c(0)$ and the cutoff Λ depend non-universally on microscopic parameters of the model. However, following the standard scaling paradigm, we will focus on the dependence of the coupling constant on the low-energy scale $\omega_{\min}$. The dependence of the coupling constant on the time scales $\sim \omega_{\min}^{-1}$ is illustrated in the figure. For interaction parameters $g > 1$ ($g < 1$), it decreases (increases). The

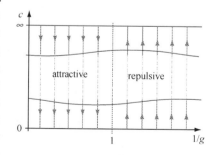

"non-interacting" case $g = 1$ defines a fixed line where the coupling strength does not change.

However, at this stage we remember that the analysis has been based on an expansion of the action to first order in the impurity operator. In the repulsive case, even if the initial value of the coupling constant is small it will soon flow

---

[9] Alert readers will notice that this result coincides with the change of the coupling constant obtained after the first RG step. However, this is a coincidence due to the simple structure of the present model.

weak–
strong
coupling
duality

into a region where this approach loses its meaning (indicated by a wavy line in the figure). How do we know what happens after the coupling constant has disappeared into the forbidden zone? It turns out that, for that for this particular model, there exists a **duality** mapping the model at large c and $g < 1$ to a structurally identical model at small $\tilde{c} \sim c^{-1}$ and $\tilde{g} > 1$ (see problem 6.7.1). We therefore conjecture – a hypothesis backed up by conformal field theory – that the coupling constant will grow in a regime of strong coupling, become equivalent to one with $\tilde{c} < 1$, which then continues to diminish ($\tilde{g} > 1$). This leads to the global flow diagram indicated in the figure, and to interesting physical conclusions: Dissipation strengths exceeding the critical value $g = 1$ block quantum tunneling at large time scales; repulsive interactions effectively enhance impurity scattering in Luttinger liquids and eventually block quantum transport, etc.

duality **INFO** The term **duality** is used by different communities in physics, although there does not seem to be a universal definition. Duality may refer to a situation where identical physics is described by different theoretical descriptions. The classical wave–particle duality of quantum mechanics is an example of this type. In this case, the two formulations of quantum mechanics in momentum or position space representations, respectively, describe identical physics. However, the terminology is also used in the reverse situation where different physics is described by identical theories. For example, the vacuum Maxwell equations, $\nabla \cdot \mathbf{E} = 0$, $\nabla \cdot \mathbf{B} = 0$, $\nabla \times \mathbf{E} + \partial_t \mathbf{B} = 0$, $\nabla \times \mathbf{B} - \partial_t \mathbf{E} = 0$ map onto each other under exchange $\mathbf{E} \to \mathbf{B}$, $\mathbf{B} \to -\mathbf{E}$: the magnetic and electric sectors of vacuum electrodynamics are described identically, a parallelism driving the search for fundamental magnetic monopoles and extension to the charged theory.

In condensed matter physics with its many different effective theories, dualities can be powerful assets. For example, **Kramers–Wannier duality**[10] maps Ising models in low temperature phases onto other Ising models at high temperatures. In the language of the old models, the variables of the new models describe different observables, and in this way crucial information on both partners of the duality is obtained. Duality becomes particularly powerful in the cases where a duality transformation maps a model onto a structurally identical one, albeit with different coupling constants, and different physical interpretations of its variables. An example of such **self-duality** is the two-dimensional Ising model on a square lattice, which is Kramers–Wannier self-dual to itself under an exchange of high and low temperatures. (The Ising variables of the dual theory describe the degree of "disorder" present in the original one, and hence are in a low expectation value/high-temperature phase if the former are in a large expectation value/low temperature phase.)

self-duality

Our discussion above refers to another self dual theory, namely that described by Eq. (6.11). For a discussion of the duality transformation and its physical interpretation we refer to problem 6.7.1.

Operator relevance

The above derivation of flow equations was based on a seemingly ill-controlled first-order expansion followed by re-exponentiation, Eq. (6.13). However, contrary to what one might suspect, the expansion is not just stabilized by the presumed initial

[10] H. A. Kramers and G. H. Wannier, *Statistics of the two-dimensional ferromagnet*, Phys. Rev. **60**, 252 (1941).

weakness of c. More important is the fact that the model action does not admit "important" contributions other than $\cos\theta$; even if they are generated at higher orders in perturbation theory in c via correlations induced by the f-fluctuations, we need not consider them. This is the rationale behind the expansion (symbolic notation) $\langle\exp(-\int\cos\theta)\rangle = \sum\frac{1}{n!}\langle(-\int\cos\theta)^n\rangle \to \sum\frac{1}{n!}(-\int\langle\cos\theta\rangle)^n = \exp(-\int\langle\cos\theta\rangle)$. Let us explore this point in more detail.

To understand the legitimacy of the reduction above, we need to assess the relevance of corrections to it. For example, expansion of the exponent to second order in the potential operator leads to

$$\left\langle c^2 \int d\tau\, d\tau'\, \cos(\theta_s(\tau) + \theta_f(\tau)) \cos(\theta_s(\tau') + \theta_f(\tau')) \right\rangle_c,$$

where we consider the connected average $\langle \hat{A}\hat{B}\rangle_c \equiv \langle\hat{A}\hat{B}\rangle - \langle\hat{A}\rangle\langle\hat{B}\rangle$ since the square of the averaged action $\sim (\langle c \int \cos\theta\rangle)^2$ is already accounted for by the previous scheme. Now consider the double time integral $\int = \int_{|\tau-\tau'|>b/\Lambda} + \int_{|\tau-\tau'|<b/\Lambda}$ to be decomposed into an off-diagonal contribution and one with nearly coinciding time arguments. In the first, the cosine terms are averaged at very different time coordinates τ and τ' over the rapidly fluctuating field, θ_s. We thus expect correlations to be weak, and the connected average to be close to zero.

INFO Readers finding the above argument too vague, may consider the following estimate:

$$\left\langle \exp\left(\pm i(\theta_s(\tau) + \theta_f(\tau))\right) \exp\left(\pm i(\theta_s(\tau') + \theta_f(\tau'))\right)\right\rangle_c$$

$$\propto \left\langle \exp\left(i\int_f (d\omega)\left(\pm e^{i\omega\tau} \pm e^{i\omega\tau'}\right)\theta(\omega)\right)\right\rangle - \left(\left\langle\exp\left(\pm i\int_f (d\omega)e^{i\omega\tau}\theta(\omega)\right)\right\rangle\right)^2$$

$$= \exp\left(-2\pi g\int_f (d\omega)\,|\omega|^{-1}\left(1 \pm \cos(\omega(\tau-\tau'))\right)\right) - \left(\exp\left(-\pi g\int_f (d\omega)\,|\omega|^{-1}\right)\right)^2$$

$$= b^{-2g}\left(\exp\left(\mp 4\pi g\int_{\Lambda/b}^{\Lambda}(d\omega)\,\frac{\cos(\omega(\tau-\tau'))}{\omega}\right) - 1\right) \approx 0.$$

Here, the $\pm$ signs in the first line indicate the four different possible combinations of signs in the cos functions; in the second equality we have used our previous results on the integrals over the high-lying frequency shell; and in the crucial third equality, we noticed that, typically, $(\tau - \tau') > b/\Lambda, 1/\Lambda$ so that the oscillatory term integrates to something close to zero and can be neglected in comparison with the constant.

However, from the complementary regime of nearly coinciding time arguments, we obtain

$$\left\langle c^2 \int_{|\tau-\tau'|<b/\Lambda} d\tau\, d\tau'\, \cos(\theta_s(\tau) + \theta_f(\tau)) \cos(\theta_s(\tau') + \theta_f(\tau'))\right\rangle_c$$

$$\propto b\int d\tau\, \left\langle \cos^2(\theta_s(\tau)) + \theta_f(\tau)\right\rangle_c \propto b^{-4g+1}\int d\tau\, \cos^2(\theta_s(\tau)) + \text{const.},$$

where the prefactor b is the width of the integration domain, $\int_{|\tau-\tau'|<b/\Lambda} d\tau\, d\tau' \sim b\int d\tau$, and we noted that, for such narrow time windows, the field integration will

be oblivious to the difference between $\theta(\tau)$ and $\theta(\tau')$. The second proportionality is obtained by averaging the integrand along the lines of our previous calculations.

The above computations conveys two important messages. First, new operators may be generated by the RG procedure – here, a term $\int d\tau \cos(2\theta)$, not present in the original action. One may anticipate that the appearance of new operators in RG programs is the rule rather than the exception. However, second, one needs to ask if these new contributions are important or *relevant* contributions to the theory. The computation above shows that fast field integration in consecutive RG steps will lead to a factor $\langle\cos(2\theta)\rangle \to \cos(2\theta)b^{-4g}$, where the factor 4 relative to the b^{-g} of the original cosine term is due to the doubling of the angular argument. This suppression by fast field fluctuations is much stronger than the factor b^{-g} of the native $\cos\theta$ term, leading to the conclusion that the addition to the action is

irrelevant
operator

an **irrelevant operator**.

Obviously, the classification of operators according to their relevance is an important element of renormalization group theory, and we will discuss it from a general perspective in the next section. For now, we conclude that the principle backing our realization of the RG step as in Eq. (6.13) is the operator irrelevance of higher order correlations missed by it. (Convince yourself that the argument extends to correlation contributions of higher than second order.)

Before leaving this section, let us make a few **general observations about the renormalization procedure**. We first notice that it would have been futile to attack the problem by the perturbative methods developed in chapter 4. The reason is that the propagator of the $(0+1)$-dimensional effective theory, $|\omega|^{-1}$, leads to logarithmic divergences when integrated over frequency; the present theory shows the UV and IR divergences problematic in perturbation theory. In section 4.1, we argued that the ensuing divergences may be removed by introduction of a UV and, if needed, an IR cutoff. However, this solution did not look attractive, as it would lead to non-universal cutoff dependences in all results.

The present approach solves this problem by introducing not one, but an entire hierarchy of cutoffs, $\Lambda, \Lambda/b, \Lambda/b^2, \dots$ In each RG step, the effect of UV fluctuations in one hierarchical cutoff interval on the effective IR theory at the next lower level is studied. While there remains some memory of the non-universal starting value in the solution of the flow equations, e.g., $c = c(0)(\Lambda/\omega_{\min})^{1-g}$, their dependence on the IR scales of the theory $c \sim \omega_{\min}^{1-g}$ is universal. Since these proportionalities are the sole object of interest in scaling theory, the RG program has effectively overcome the issue of singularities. In fact, a stronger statement can be made:

> RG programs extract nontrivial information on the IR scale
> dependence of a theory from its UV singularities.

Previously, we argued that in scaling theory a physical length dimension is attributed to observables of interest. This dimension then determines how the observables scale in the vicinity of fixed points. For example, an observable with dimension $[X] = (\text{length})^{D_X}$ should change as $X \to X b^{D_X}$ under an RG step changing length by a factor of b. Presently, our "length" is "time," and θ is a dimensionless field.

(Appearing as an argument of a cos function, it has to be.) On this basis, we conclude that, e.g., the nonlinear contribution to the action $X \equiv \int d\tau \cos\theta$ carries dimension (length)^1. However, our observation was that in a single RG step, this operator picks up a scaling factor, Eq. (6.14), different from b^1. What is going on?

anomalous dimension

We first note that dimensions attributed on the basis of plain dimensional analysis are called **engineering dimensions**. Deviations in scaling from the engineering dimensions are **anomalous dimensions**. For example, the operator above has engineering dimension 1 and anomalous dimension $-g$. Tracing the origin of the anomalous dimension, we realize that it has, once again, to do with the UV cutoff Λ. The latter is a dimensionful scale, and it affects the length-scaling of X through the backdoor, via the cutoff dependence of fluctuations. In relations such as $c = c(0)(\Lambda/\omega_{\min})^{1-g}$, we see that the scale dependence of the operator is contained in dimensionless ratios (UV cutoff)/(IR scales). In this way, the actual dimension of the operator may deviate from its engineering dimension. Technically, determining the anomalous dimensions of all relevant operators is the goal of all RG programs.

6.2 Renormalization Group: General Theory

Having looked at two case studies, we are now in a position to discuss RG theory in its generality. Suppose we consider a field theory with action

$$S[\phi] \equiv \sum_{a=1}^{N} g_a \mathcal{O}_a[\phi],$$

where $\phi = \{\phi(\mathbf{x})\}$ is a (generally multi-component) field, g_a are coupling constants, and $\mathcal{O}_a[\phi]$ operators,[11] often realized as $\mathcal{O}_a = \int d^d x \, (\nabla\phi)^n \phi^m$ – products of field amplitudes and their derivatives.[12] The goal of the renormalization program is the derivation of flow equations describing the change of the coupling constants $\{g_a\}$ as fast field fluctuations are successively integrated out.

6.2.1 Renormalization group flow

Irrespective of their concrete technical implementation, the derivation of RG flow equations in statistical field theory involves three canonical steps, outlined in the following. (For the somewhat different rationale underlying renormalization in high energy physics – where the assumption of a UV cutoff as an anchor point for the procedure is unnatural – we refer to the literature; see also the Info block below.)

---

[11] In RG theory, it is customary to call individual contributions to an action "operators."
[12] In our previous example of the Luttinger liquid, there appeared an operator $\int (d\omega)\theta(\omega)|\omega|\theta(-\omega)$. When represented in space–time, this operator is highly non-local.

I: Subdivision of the field manifold

We begin with a separation of the field manifold $\{\phi\}$ into a sector to be integrated out, $\{\phi_\mathrm{f}\}$, and a complementary set, $\{\phi_\mathrm{s}\}$. For example:

▷ In the real space renormalization of lattice problems, where the base manifold $\mathbf{x} \to \mathbf{x}_i$ is realized through a set of lattice points, one may adopt a **block spin scheme**. In this case, the slow degrees of freedom live on a coarse-grained lattice with enlarged unit cell and are defined as suitable averages taken over the lattice degrees of freedom inside those cells.

momentum shell renormalization

▷ In **momentum shell renormalization**, frequently applied in continuum field theory, one integrates over high lying regions in momentum space, $\Lambda/b \leq |\mathbf{p}| < \Lambda$. The explicit cutoff dependence introduced in momentum shell renormalization is avoided by an alternative scheme known as **dimensional regularization**. The idea of this approach is a formal modification of the physical dimension as $d \to d \pm \epsilon$. In this way, integrals that would be UV singular in the native dimension are made finite. For example, the logarithmically singular $\int dp/p$ becomes convergent in $1 - \epsilon$ dimensions, $\int d^{1-\epsilon} p/p$. (For the definition of integrals in non-integer dimensions, and the application of dimensional regularization in field theory, we refer to, e.g., Ref.[13])

dimensional regularization

▷ For a discussion of alternative schemes, such as the introduction of short-distance real space cutoffs underlying the so-called **operator product expansion**, we also refer to the literature (see, e.g., Ref.[1]).

II: RG step

In the second step, one integrates over the short-range fluctuations singled out in the first. These integrals generally require approximation schemes, several of which will be introduced below. However, after that fast field integration has been carried out, one obtains an action

$$S'[\phi_s] \equiv \sum_a g_a' \mathcal{O}_a'[\phi_s],$$

with a new set of coupling constants, and potentially newly generated operators. If new operators form, one needs to find out whether they are *relevant* (see below) in their scaling behavior. If so, they have to be included in subsequent RG steps, with initial coupling constants whose specific values are generally not of much importance. One then needs to find out whether the augmented set of operators defines a complete system, i.e., one that does not generate further relevant operators under renormalization. If not, back to square one and repeat.

[13] L. H. Ryder, *Quantum Field Theory* (Cambridge University Press, 1996).

III: Flow equations

In the third step, one rescales frequency/momentum in such a way that the field amplitude ϕ' fluctuates on the same scales as the original field ϕ. This is achieved by the transformation

$$q \to bq, \quad \omega \to b^z \omega,$$

dynamical exponent

where the **frequency renormalization exponent** or **dynamical exponent**, z, may take arbitrary values. (Depending on the dispersion $k^z \sim \omega$ of the theory, common realizations include $z = 1, 2$, but sometimes also $z = 3$ (see problem 6.7.2), fractional values, or even $z = \infty$.) Finally, ϕ is an integration variable, which may be rescaled arbitrarily in a transformation

$$\phi \to b^{-d_\phi} \phi,$$

field renormalization

called **field renormalization**. This freedom is often used to select one or several terms in the action as representatives of the free theory – a canonical candidate being the gradient operator $\sim \int d^d r (\partial \phi)^2$ – and require that they remain fully invariant under the RG step. This is achieved by choosing b^{d_ϕ} so as to cancel the factor b^x arising after the renormalization of the operator. (Exercise: Which field renormalization keeps the above gradient operator invariant?[14])

INFO Notation such as $X \to b^{d_X} X$, which is ubiquitous in the literature, can be quite confusing. For example, when we write $p \to bp$, what we mean is "introduce a new momentum variable, $p' = bp$, related to the old one by multiplication by b." However, in an integral, the change of variables implies $\int d^d p f(p) = b^{-d} \int d^d p' f(p' b^{-1})$. We then call the new variables p again, to write this as $\int d^d p f(pb^{-1})$. So, the change $p \to bp$ means that in integrals we are supposed to replace all $p \to b^{-1} p$ – confusing! The situation with field renormalization $\phi \to b^{d_\phi} \phi$ is similar.

A safe way not to get lost is to handle the "$\ldots \to \ldots$" notation with care and emphasize the integral structure: all that we are doing in RG theory is change integration variables in integrals and remembering that, after an RG step, one changes to (momentum) variables reinstating the old cutoff via a stretching by b.

As a result of these manipulations, we obtain a renormalized action,

$$S[\phi] = \sum_a g'_a \mathcal{O}_a[\phi],$$

differing from the original one only in the set of coupling constants. Introducing the vector $g = \{g_a\}$, we represent this change in terms of a function $\tilde{R}$ as

$$g' = \tilde{R}(g).$$

For small values of the control parameter, $\ell \equiv \ln b$, this function is close to the identity map, and we may represent the difference $g' - g = \tilde{R}(g) - g$ in terms of the

RG flow equation

RG flow equation

---

[14] Under rescaling, $\int d^d r (\partial \phi)^2$ picks up a scaling factor $b^{d_\phi p_\phi + z - p + 2 d_\phi} q$. So, $z/(p - z) = {}^\phi p$ does the job.

$$\boxed{\frac{dg}{d\ell} = R(g)} \qquad\qquad (6.16)$$

β-function

where the functions on the r.h.s., $R(g) = \lim_{\ell \to 0} \ell^{-1}(\tilde{R}(g) - g)$, define the β-**functions** of the theory.

INFO It is interesting to compare the RG strategy above with the **alternative renormalization schemes** common in high-energy physics. In particle physics, the bare, unrenormalized theory has no physical identity. Unlike in condensed matter, where field theories are derived from underlying microscopic theories, it is inaccessible, or maybe even not fully known. However, one may legitimately require that, after integration over UV-divergent fluctuations, the "renormalized" coupling constants of the theory – which *are* accessible in terms of physical observables such as the physical electron mass – are finite. On this basis, the rationale of the procedure is upside down: without introducing a physical cutoff, infinities arising in the theory are removed either by postulating cancelling infinities in the bare coupling constants and/or the introduction of **counter terms**. The latter are operators added ad hoc to the action for the purpose of cancelling infinities generated by other operators in the RG process.

Fortunately, the different RG schemes yield identical results in the form of β-functions governing the flow of an effective set of couplings.

6.2.2 Analysis of the flow equation

The flow equations (6.16) contain the full information on the renormalization of a theory. They predict the flow of its coupling constants under changes in the effective length scale, l. As exemplified in section 6.1, one generally starts the analysis of these equations with an identification of their **fixed points**, the set $\{g^*\}$ of coupling constants for which the flow remains stationary, $R(g^*) = 0$. A theory fine-tuned to a fixed point configuration does not change under renormalization.

RG fixed
points

self-
similarity

Self-similarity is the defining property of *fractals* and is approximately realized in many systems in nature. (The picture shows a tendency towards self-similarity in romanesco broccoli.) We have already mentioned that self-similarity is incompatible with the existence of any fixed length scales in a system, including that of a finite correlation length:

> At an RG fixed point, the correlation length, ξ, is either infinity or zero.

second-
order

Specifically, a diverging correlation length, $\xi \to \infty$, is a hallmark of a **second-order phase transition**. This correspondence leads to the tentative identification of RG fixed points as signatures of phase transitions of the physical system. (For a review of phase transitions and the critical phenomena accompanying them, see the following section.) The flow of the coupling constants in the immediate vicinity of a fixed point configuration must then describe the **critical phenomena** accompanying a transition.

For couplings g sufficiently close to a fixed point it is sufficient to consider the linearized mapping

$$R(g) \equiv R(g^* + (g - g^*)) \simeq W(g - g^*), \quad W_{ab} = \left.\frac{\partial R_a}{\partial g_b}\right|_{g=g^*},$$

as described by a matrix W.

To explore the properties of flow, assume that W has been diagonalized with eigenvalues $\lambda_\alpha, \alpha = 1, \ldots, N$ and *left*-eigenvectors ϕ_α,

$$\phi_\alpha^T W = \phi_\alpha^T \lambda_\alpha.$$

(In general, W does not have any symmetries besides being real, so its left- and right-eigenvectors may be different.) The advantage of using left-eigenvectors is that they allow us to conveniently describe the flow of the physical coupling constants under renormalization: with $v_\alpha = \phi_\alpha^T(g-g^*)$ the αth component of the vector $g-g^*$ expanded in the basis $\{\phi_\alpha\}$, we have

$$\frac{dv_\alpha}{d\ell} = \phi_\alpha^T \frac{d}{d\ell}(g - g^*) = \phi_\alpha^T W(g - g^*) = \lambda_\alpha \phi_\alpha^T(g - g^*) = \lambda_\alpha v_\alpha.$$

scaling fields
Under renormalization, the coefficients v_α change by a mere scaling factor λ_α, and hence are called **scaling fields** – an unfortunate nomenclature, the coefficients v_α are ℓ-dependent functions, not fields. These equations are trivially integrated to obtain

$$\boxed{v_\alpha(\ell) \sim \exp(\ell \lambda_\alpha)}$$

This result suggests a distinction between three types of scaling fields:

(ir)relevance
▷ For $\lambda_\alpha > 0$ the associated scaling field is said to be **relevant**, in the sense that it drives the system away from the critical region. In fig. 6.2, v_2, with associated vector ϕ_2, is relevant.

▷ Scaling fields with negative eigenvalues (v_1, v_3) are **irrelevant**; they vanish under renormalization, and the corresponding (linear combinations of) coupling constants generally do not play much of a role.[15]

marginality
▷ Finally, stationary scaling fields, where $\lambda_\alpha = 0$, are called **marginal**. A marginal scaling field corresponds to a direction in coupling constant space with vanishing partial derivative, $\partial_{\phi_\alpha} R|_{g^*=0} = 0$. In this case, one often considers the second-order derivative, $\partial_{\phi_\alpha}^2 R|_{g^*=0} \equiv 2x$, and this leads to scaling, $d_\ell v_\alpha = x v_\alpha^2$. For **marginally (ir)relevant** $x < 0$ ($x > 0$), the field is **marginally (ir)relevant**: in the vicinity of the fixed point, its flow becomes vanishingly slow. However it is still directed, and in this sense (ir)relevant. In cases where no truly relevant fields exist, marginally relevant ones are the next most important objects to consider.

---

[15] However, it may happen that physical observables depend in a singular manner on irrelevant scaling variables, in which case the latter are called **dangerously irrelevant**. For example, we will see that the coefficient of the quartic term in a ϕ^4-theory of magnetism is irrelevant in dimensions $d > 4$. However, the magnetization depends on this coefficient in a non-analytical manner, so cannot be ignored.

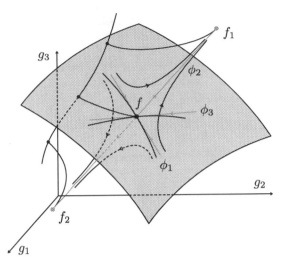

Fig. 6.2   RG flow in the vicinity of a fixed point with two irrelevant (ϕ_1, ϕ_3) and one relevant (ϕ_2) scaling fields. The critical surface (shaded) is defined through the vanishing of the relevant field, $\phi_2 = 0$. On it, the RG flow is directed towards the fixed point f. Deviations away from criticality make the system approach one of the stable fixed points f_1 and f_2.

The distinction of relevant/irrelevant/marginal scaling fields in turn implies a classification of different types of fixed points:

fixed points

▷ **Stable fixed points** have only irrelevant, or perhaps marginal, scaling fields. They define "stable phases of matter." A system initiated at small length scales in the vicinity of such an attractor will scale towards it and eventually resemble the self-similar fixed-point configuration. (Recall the example of the high-temperature fixed line of the one-dimensional Ising model.)

▷ **Unstable fixed points** realize the opposite extreme, with only relevant scaling fields. These fixed points are Platonic ideals: they can never be realized fully, and no matter how closely a model system resembles the fixed point limit at small distance scales, it will be driven away from it at large scales (cf. the $T = 0$ fixed point of the 1D Ising model).

▷ However, **generic fixed points** have both relevant and irrelevant scaling fields. Such fixed points are associated with the physics of *phase transitions*. To understand this correspondence, we note that the eigenvectors ϕ_α associated with irrelevant scaling fields are tangent to a generalized surface, S, in coupling constant space, the **critical surface** (see fig. 6.2 for a two-dimensional realization.) This surface defines the **basin of attraction** of the fixed point; a system with couplings $g \in S$ will be attracted towards it. Unlike unstable fixed points, there is a finite, if fine-tuned, set of material configurations realizing the critical physics of the phase transition.

critical surface

However, even small deviations from S make the system flow away from the critical surface (see fig. 6.2 for one such relevant direction). For example, in the case of the ferromagnetic phase transition – discussed in more detail in the next section – deviations from the critical temperature T_c are relevant. If we consider a system only slightly above or below T_c, it may initially (on intermediate length scales) appear to be critical. However, the relevant deviation will grow, and at large scales the system looks as if it were at a higher or lower temperature, depending on the sign of the initial deviation. Eventually, it ends up in either the ferromagnetic low temperature or the paramagnetic high temperature phase. The scenario outlined above suggests that systems with generic fixed points typically possess complementary stable fixed points towards which the flow is directed after it has left the critical region. We also notice that a scaling direction that is relevant at one fixed point (e.g., ϕ_2 at the critical fixed point) may be irrelevant at others (ϕ_2 at the high- and low-temperature fixed points).

6.2.3 Review of critical phenomena

REMARK In this section we review elements of the theory of critical phenomena often taught in advanced courses in statistical mechanics. Readers familiar with the concept of criticality and critical exponents may skip this section, or use it for reference purposes.

The discussion above revealed a close connection between the concept of renormalization and that of phase transitions and critical phenomena. This section provides a quick review of the concepts of criticality required to put the machinery of field-theoretical renormalization into a physical context.

Second-order phase transitions

order pa-
rameter

The most fundamental signature of a phase transition is its **order parameter**, M. The order parameter is a measurable quantity whose value identifies the phase of a system. Order parameters often reflect the breaking of locally defined symmetries at a phase transition – the breaking of rotational symmetry at the ferromagnetic transition, as evidenced by the magnetization order parameter, being a classic example. However, they need not necessarily be locally defined, nor represent the breaking of a symmetry. For example, at the quantum Hall transition (to be discussed in detail in section 8.4.7), an integer-valued order parameter signifies the change of non-local topological order in a quantum ground state.

In statistical mechanics, we learn that transitions between different phases of matter fall into two major categories: In **first-order transitions**, the order parameter changes discontinuously; at

second-
order
phase
transition

a **second-order phase transition** it changes in a continuous yet non-analytic manner. The distinction between the two classes is illustrated in the figure for the case of the ferromagnet. For temperatures $T < T_c$, a variation in the external magnetic field H leads to a discontinuous jump in the order parameter at

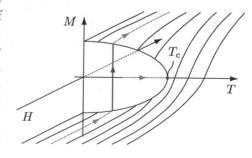

$H = 0$ – first order. However, increasing T at zero H leads to a continuous vanishing of the magnetization at T_c – second order. The figure also illustrates that lines of first-order phase transitions in parameter space ($H = 0, T < T_c$) often end at a second-order transition point ($H = 0, T = T_c$).

The phenomenology of second-order transitions is generally richer than that of first-order transitions: being a thermodynamic state variable, the order parameter is coupled to a **conjugate field**, $H : M = -\partial_H F$, defined here as a "source field" of the order parameter in terms of the free energy. At a second order transition M changes non-analytically, which implies that the second-order derivative, a **thermodynamic susceptibility**, $\chi = -\partial_H^2 F$, develops a singularity. However, the susceptibility is intimately linked to the field fluctuation behavior of a system (see the discussion around Eq. (6.3)). In particular, a divergent susceptibility requires a similarly divergent correlation length, an indication of self-similarity.

thermody-
namic sus-
ceptibility

Through this mechanism, the divergence of a susceptibility implies (potentially singular) power law scaling of various other physical quantities. Suppose that $X(t)$ is one such observable, and t a dimensionless control variable proportional to a relevant scaling field measuring the distance to the transition. Under a change of length scale, $x \to x/b$ and $t \to tb^{-D_t}$. The observable X can change only by a factor reflecting its own scaling dimension, $X(t) = b^{D_X} X(tb^{-D_t})$. (A more far-reaching change would be in conflict with asymptotic self-similarity.) The *homogeneity* of the function $X(t)$ is equivalently expressed as $X \sim t^{D_X/D_t}$, a power law dependence.

Critical exponents

The exponents characterizing the power laws of relevant thermodynamic observables are called the **critical exponents** of a phase transition. For various reasons, they represent data of outstanding physical importance:

critical
exponents

1. Critical exponents are universal in that the complete set of them – usually just a handful – uniquely identifies a transition.

2. The spectrum of critical exponents carries information identical to that contained in the spectrum of relevant scaling dimensions. In fact, it contains excessive information: for example, of the six critical exponents of the magnetic transition, only two are independent. The others are interrelated by[16] by **scaling laws** or **exponent identities**, to be discussed momentarily.

scaling
laws

3. Critical exponents are universal; they are pure numbers depending on characteristics such as space–time dimensionality, the identity of a system's Goldstone mode manifold, or the number of components of an order parameter.

4. Critical exponents can be measured, both in real and in numerical experiments probing power law scaling. Their universality and phenomenological relevance make them quantities of singular experimental interest.

---

[16] Unfortunately, the parlance of critical phenomena makes excessive use of the prefix "scaling."

The following is a brief list of the **most prominent exponents**, α, β, γ, δ, η, ν, and z:

α: In the vicinity of the critical temperature, the **specific heat** $C = -T\partial_T^2 F$ scales as $C \sim |t|^{-\alpha}$, where $t = (T - T_c)/T_c$ measures the distance to the critical point. Note that two physical statements are being made here: first, the reduced temperature is a relevant scaling field and, second, the scaling exponents controlling the behavior of C above and below the transition are identical. The same symmetry applies to most other exponents.

β: Approaching the transition temperature from below, the **order parameter** vanishes as $M \equiv -\partial_H F\big|_{H\searrow 0} \sim (-t)^{\beta}$.

γ: The **susceptibility** behaves as $\chi \equiv \partial_h M|_{h\searrow 0} \sim |t|^{-\gamma}$.

δ: At the critical temperature, $t = 0$, the **field dependence of the order parameter** is given by $M \sim |h|^{1/\delta}$.

ν: Upon approaching the transition point, the **correlation length** diverges as $\xi \sim |t|^{-\nu}$.

η: This implies that the correlation function,

$$C(r) \sim \begin{cases} r^{2-d-\eta}, & r \ll \xi, \\ \exp(-r/\xi), & r \gg \xi, \end{cases}$$

crosses over from exponential to a power law scaling behavior at the length scale ξ. Notice that $C \sim \langle \phi\phi \rangle$ carries twice the dimension of the field ϕ. The engineering dimension of the latter follows from the dimensionlessness of the gradient operator $\int d^d r \, (\nabla\phi)^2$: $[\phi] = L^{(2-d)/2}$, according to which $C(\mathbf{r})$ has canonical dimension L^{2-d}. The exponent **anomalous dimension** η, commonly referred to as the **anomalous dimension** of the correlation function, measures the mismatch between the observed and the canonical dimension.

z: The exponent z is special in that it applies to the theory of quantum phase transitions. A quantum theory in d space dimensions can be conceptualized as a classical theory in **quantum phase transition** $d+1$ space–time dimensions. It possesses a **quantum phase transition** if that effective classical theory has a phase transition in the ordinary sense. In the vicinity of the phase transition, the correlation length in both the space and time directions diverges. However, the scaling of these scales need not be identical. Denoting the correlation length in the temporal direction by τ, we define $\tau \sim \xi^z$, where deviations $z \neq 1$ in the **dynamical exponent** **dynamical exponent** measure the degree of anisotropy.

Now, a moment's thought shows that not all the six classical exponents can be independent. Previously we have noted that the flow in the vicinity of a transition point is governed by the relevant scaling fields. We anticipate that the field conjugate to the order parameter is relevant. For example, a magnetic field has a relevant influence on a magnetic transition. Deviations from the critical temperature, $t \neq 0$, are also relevant.[17]

However, for the majority of classical phase transitions, there are no further relevant scales, and the flow away from criticality is governed by a two-dimensional dynamical system. We therefore expect that only two of the six exponents are **scaling laws** independent. Historically, the four **scaling laws** constraining the system of six

---

[17] Recall that, in the derivation of the ϕ^4-model of the magnetic transition, the coupling constant of the "mass operator" $r \int d^d r \, \phi^2$ was proportional to the reduced temperature $t = |T - T_c|/T_c$. Deviations away from $r \sim t = 0$ are relevant, and on this basis we consider t to be a (relevant) coupling.

constants down to $6 - 4 = 2$ independent constants were discovered at a time when the universal concept of "scaling fields" had not been fully understood. The table below lists these equations along with their discoverers, after whom they are named. In section 6.2.4, we will demonstrate with an example how these laws naturally emerge as a consequence of scaling theory.

Fisher	$\nu(2 - \eta) = \gamma$	Rushbrooke	$\alpha + 2\beta + \gamma = 2$
Widom	$\beta(\delta - 1) = \gamma$	Josephson	$2 - \alpha = \nu d$

The consequence of this discussion is that:

> Only two arbitrarily chosen exponents need be specified to characterize comprehensively a classical phase transition.

Universality of phase transitions

We conclude this section with the discussion of a feature following from the geometric picture of scaling: universality. For example, there is a plethora of very different physical two-dimensional systems – from confined classical Coulomb gases and $(1 + 1)$ dimensional disordered interacting quantum wires, to classical models of planar magnetism – that are in the universality class of the Kosterlitz–Thouless transition (to be discussed in section 6.5). The concept of universality implies that all these systems will behave identically when described in the language of critical flows.

universality classes

More generally, there is only a small number – tens – of different **universality classes** of phase transitions that are physically relevant, a number that should be compared with the infinity of microscopically distinct many-particle systems. Consider then an experimentalist investigating a system that is known to exhibit a phase transition. Searching

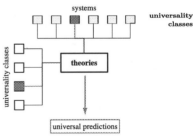

for the critical phenomena accompanying the transition, she will vary control parameters, X_i (e.g., temperature, pressure, magnetic field, etc.), until the system begins to exhibit large fluctuations. Variation of the X_i changes the couplings of the effective theory, i.e., it parameterizes a curve in the space of coupling constants. At some stage the curve may intersect the critical surface of the system (see fig. 6.2). For this particular set of coupling constants, the system is critical. In the vicinity of the crossing point, when looked at on larger and larger length scales, the system will display the universal behavior characteristic of the phase transition point. This is the origin of universality: variation of the system parameters in a different manner (or considering a system with different material constants) will generate another coupling trajectory. However, as it intersects S, identical critical flows of observables will ensue.

The principle assigning to different systems common universality classes through their effective theories (see the figure for a schematic) is the most powerful source of universality we have in physics. It is not tied to specific physical realizations, nor does it respect boundaries between fields (for example, the Kosterlitz–Thouless universality class is realized in condensed matter, particle physics, quantum optics, and beyond).

6.2.4 Scaling theory

The dynamical system of scaling fields fixes the power law behaviors of all observables described by the theory, including those that are accessible in experiment. To see how, suppose that we had represented an observable of experimental interest, X, as

$$X = C(p_i, g_\alpha)$$

in the language of the functional integral. Here, C is an n-point correlation function $C(p_i, g_\alpha) = (\cdots)\langle \phi\phi \cdots \phi \rangle_\phi$, and the notation indicates that C may depend on the momentum scale at which it is evaluated (e.g., through the momentum dependence of current operators in the Fourier transform of a conductance correlation function). The ellipses indicate the dependence of the correlation function on coupling constants and on other possible structures such as derivatives acting on the fields.

Assuming renormalizability, C can be evaluated before or after an RG step; the result must be the same. However, the RG transformation will modify the individual constituents entering the definition of C: it will change coupling constants, $g_\alpha \to g_\alpha b^{\lambda_\alpha}$, momenta $p_i \to b p_i$, and field amplitudes $\phi \to b^{d_\phi} \phi$ so that, after the RG-step, the correlation function assumes the form

$$\boxed{C(p_i, g_\alpha) = b^{n d_\phi} C(p_i b, g_\alpha b^{\lambda_\alpha})} \tag{6.17}$$

where we have simplified the notation by assuming that the coupling constants themselves scale. (Otherwise, the rescaling assumes the form of a matrix relation, a complication that would make this section harder to read without offering further insight.) For notational convenience, let us also assume that the coupling constants are measured relative to a fixed point, i.e., we apply a shift $g \to g - g^*$.

The essential statement made by Eq. (6.17) is that multiple scale changes combine to give a net result that is scale invariant. In the next section, we demonstrate how the presence of the dummy parameter b can be used as a vehicle to obtain nontrivial results for the behavior of observables near phase transitions.

Scaling functions

In this section, we will formulate the scaling principle, previously applied to the Ising model, in a more general form. To begin with, let us assume that there is just a single relevant scaling field g, while all remaining g_α are irrelevant or marginally irrelevant. We can then write

$$
\begin{aligned}
C(p_i, g, g_\alpha) &= b^{nd_\phi} C(p_i b, g b^\lambda, g_\alpha b^{\lambda_\alpha}) = g^{-nd_\phi/\lambda} C(p_i g^{-1/\lambda}, 1, g_\alpha g^{-\lambda_\alpha/\lambda}) \\
&\stackrel{g \ll 1}{\approx} g^{-nd_\phi/\lambda} C(p_i g^{-1/\lambda}, 1, 0) \equiv g^{-nd_\phi/\lambda} F(p_i g^{-1/\lambda}),
\end{aligned}
$$

where we have used the arbitrariness of b to set $g b^\lambda = 1$ and in the third equality assumed that we are close enough to the transition that the dependence of C on irrelevant scaling fields is negligible. The effective single parameter correlation function C, defined as

$$
C(p_i, g) = g^{-nd_\phi/\lambda} F(p_i g^{-1/\lambda}), \tag{6.18}
$$

scaling function

is an example of a **scaling function**.

In other applications (for example, if C represents a thermodynamic observable or a global transport coefficient), one might be interested in correlation functions $C(g, g_\alpha)$ which do not depend on specifically chosen momenta. The freedom of scaling can then be used to describe the dependence of correlations on the most relevant *and* the second most relevant control parameter g' (which may be relevant, marginal, or irrelevant). Following the same logic as above, one obtains

$$
C(g, g') = g^{-nd_\phi/\lambda} G(g' g^{-\lambda'/\lambda}).
$$

INFO While the results of analytical theories are often interpreted in thermodynamic limits, $L \to \infty$, numerical simulations are carried out for systems of limited size. To compare numerical data obtained for different system sizes L with analytical computations, one considers L as one of the coupling parameters of the theory. (For example, L might enter an analytical computation as an IR momentum cutoff $p_{\min} \sim L^{-1}$.) Since $[L] = [\text{length}]$, this parameter changes as $L \to L/b$ under RG rescaling, and with the choice $b \sim L$ one

finite size scaling

obtains the **finite size scaling** relation

$$
G(g_1, g_2, \ldots, L) = L^{nd_\phi} G_{\text{fs}}(g_1^{1/\lambda_1} L, g_2^{1/\lambda_2} L, \ldots).
$$

In a numerical experiment carried out for couplings $\{g_\alpha\}$, one may then check whether data obtained for different system sizes can be collapsed onto a single curve when plotted as a function of the scaling variables $g_\alpha L^{\lambda_\alpha}$. One may even proceed in reverse: determine the dimension of coupling constants by numerical identification of a scaling variable that leads to data collapse.

Figure 6.3 illustrates the principle on the example of the percolation transition (section 11.7.1), which is a transition with a single relevant coupling, $g = p - p_c$. With $\xi \sim g^{-\nu}$, we know that $\lambda = 1/\nu$ and $g \to g b^{1/\nu}$. The figure illustrates how raw data plotted as a function of p collapses onto a single curve when plotted as a function of $g L^{1/\nu}$.

While the form of specific scaling functions is context dependent, the construction principle is always the same. The free scaling parameter b is chosen to reduce the number of independent variables by one. The resulting scaling functions define powerful interfaces between theory and experiment. As illustrated in the Info block above, data collapse under rescaling is a prime signature of criticality near phase

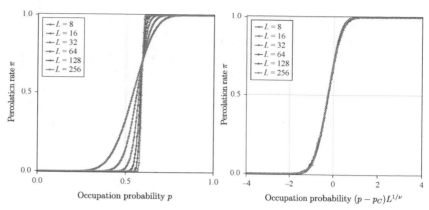

Fig. 6.3   Finite-size scaling of the observable *percolation rate* in the vicinity of the percolation transition (data courtesy of Simon Trebst). Left: Raw data as a function of the scaling variable
"occupation probability," p, for different system sizes. Right: Data collapsing onto a single
curve when represented as a function of $(p - p_c)L^{1/\nu}$.

transitions. The principle may be applied to confirm or to predict scaling dimensions
of relevant system parameters.[18]

INFO In the particle physics and the field theory communities, the information contained
in scale dependent correlation functions is often represented in a somewhat different manner. Consider the relation

$$C(g_\alpha) = e^{n\ell d_\phi} C(g_\alpha(\ell)), \qquad (6.19)$$

where we have set $b = e^\ell$, and do not yet require *power law* scaling, i.e., the coupling constants may change in an unspecified manner as $g_\alpha \to g_\alpha(\ell)$ under an RG transformation.
Using the ℓ-independence as $0 = d_\ell$(r.h.s.), we obtain

$$\boxed{(nd_\phi + \partial_\ell + \beta_\alpha(g)\partial_{g_\alpha}) C(g_\alpha(\ell)) = 0} \qquad (6.20)$$

renormal-
ization
group
equation

where $\beta_\alpha(g_\alpha) = \partial_\ell g_\alpha$ are the β-functions of the theory, and the partial derivative ∂_l acts on
optional scale dependent parameters in C (such as external momenta, as in Eq. (6.18)).
Equation (6.20) is called a **renormalization group equation**. The RG equation and
the scaling form (6.19) used to derive it express equivalently the scaling behavior of the
correlation function.

Scaling functions and critical exponents

The critical exponents describe the scaling of very different physical observables
in the vicinity of a phase transition. The fact that they are all determined by the

---

[18] However, the collapsing of experimental data onto scaling functions requires considerable skill.
If the data set consists of patches of only limited overlap, errors are easily made; the cost of
numerical finite-size scaling increases exponentially in L. For example, numerical research of
the quantum Hall correlation function exponent (section 8.4.7), which is analytically unknown,
has a history spanning four decades. With results hovering around 2.5 ± 0.2, there is still no
conclusive answer.

smaller set of relevant couplings is another manifestation of universality in critical phenomena. To see how these correlations arise in concrete terms, we consider the case of a classical transition, with six exponents $\alpha, \beta, \gamma, \delta, \nu, \eta$ (see page 337), and two relevant scaling fields, reduced temperature t, and the likewise reduced conjugate field $h \equiv H/T_c$. Under a renormalization group transformation, the dimensionless reduced free energy $f = F/TL^d$ will behave as[19] $f(t, h) = b^{-d} f(tb^{y_t}, hb^{y_h})$. We next fix $tb^{y_t} = 1$ to reduce the number of independent variables to one:

$$\boxed{f(t, h) = t^{d/y_t} f(h/t^{y_h/y_t})} \tag{6.21}$$

Modulo irrelevant perturbations, this function contains the full information on scaling near the transition, and in particular on all critical exponents. Comparing with the definitions summarized on page 337, it is straightforward to show that

$$\alpha = 2 - \frac{d}{y_t}, \qquad \beta = \frac{d - y_h}{y_t}, \qquad \gamma = \frac{2y_h - d}{y_t},$$

$$\delta = \frac{y_h}{d - y_h}, \qquad \nu = \frac{1}{y_t}, \qquad \eta = 2 + d - 2y_h, \tag{6.22}$$

from which follow the scaling laws summarized in the table on page 338 by elimination of the dimensions $y_{h,t}$.

hyperscaling relation

EXERCISE Verify the equations above. To obtain the fifth, the **hyperscaling relation**, $\nu = 1/y_t$, notice that $\xi \to b\xi$. On the other hand, we know that $t \sim \xi^{-1/\nu}$. The sixth relation is obtained from Eq. (6.3) by a substitution of the η-dependent correlation function into the integral to obtain a relation between the critical exponents γ and η (Fisher's scaling law).

To summarize, Eqs. (6.22) underpin the assertion that:

> The dimensions of the relevant scaling fields have a more fundamental status than the critical exponents.

6.3 RG Analysis of the Ferromagnetic Transition

In this section we will illustrate the general concepts introduced above for the classical theory of the uniaxial ferromagnetic transition – a paradigmatic transition whose universality class not only includes magnetism but also the transition between liquid and gaseous phases of matter. While the technical elements of our discussion are specific to the above transition, the solution strategy is generic.

---

[19] We here note that f does not carry an anomalous dimension. The reason ist that, by construction, $F = -T \ln \mathcal{Z}$ does not change under renormalization (renormalization merely changes the resolution at which the integral $\mathcal{Z}$ is evaluated). The scaling of the reduced free energy is therefore carried by the prefactor L^{-d}.

In section 4.1.2, we identified ϕ^4-theory as an effective low-energy model of the ferromagnetic system. We saw how, within this theory, the transition showed up at the level of mean field theory: Above the transition temperature, $\bar{\phi} = 0$, and below $\bar{\phi}$ acquires a non-zero expectation value $\bar{\phi} = \pm$const., where the sign ambiguity reflects the $\mathbb{Z}_2$ spin up/down symmetry breaking in the ferromagnetic phase. However, beyond these statements, not much could be said, reflecting the absence of powerful analytical methods to handle the fluctuation singularities of the model. In this section, we will see that RG methods, and only these, are the appropriate tool to understand the critical physics of the system.

6.3.1 Engineering dimensions

The first question that we wish to address has a somewhat technical status: what is the justification for representing the Ising model as[20]

$$S[\phi] = \int d^d r \left(\frac{r}{2} \phi^2 + \frac{1}{2} (\partial \phi)^2 + \frac{\lambda}{4!} \phi^4 - h\phi \right), \qquad (6.23)$$

in terms of the ϕ^4 action? Why is it legitimate to neglect the higher powers and field gradients present in an exact reformulation of the Ising problem in terms of ϕ-variables?

To rationalize the neglect of these terms, we proceed by dimensional analysis. We assume that the actual dimensions of the different operators in Eq. (6.23) will not differ strongly from their engineering dimensions (an assumption to be checked self-consistently), and we begin by identifying the latter. We first use the freedom of field, or integration variable, rescaling to give ϕ a dimension $[\phi] = L^{(2-d)/2}$. In this way, the leading gradient term is made dimensionless, $\left(\int d^d r (\partial \phi)^2 \right) = L^{d-2+2\times(2-d)/2} = L^0$, and the dimension of all other operators is fixed as

$$\left(\int \phi^2 \right) = L^2, \left(\int \phi^4 \right) = L^{-d+4}, \left(\int \phi^n \right) = L^{d+(2-d)\frac{n}{2}}, \left(\int (\partial^m \phi)^2 \right) = L^{2(1-m)}.$$

According to these relations:

▷ The engineering dimension of the non-gradient operator $\sim \phi^2$ is positive in all dimensions, indicating its general relevance.

▷ The ϕ^4 operator is irrelevant in dimensions $d > 4$; in these high dimensions, a harmonic approximation ($\lambda = 0$) should be justified. At lower dimensions, it is relevant, and it is not obvious how to deal with it. (Unlike for the problem studied in section 6.1.2, the fluctuation integrals of ϕ^4-theory cannot be dealt with in straightforward first-order perturbation theory.) However, we also anticipate that the watershed $d = 4$ plays an interesting role: in this case, the interaction operator is no longer irrelevant, but not yet relevant as in $d < 4$. One may thus

---

[20] Generalizing the result of section 4.1.2, we have incorporated a coupling to an external field. (Exercise: Recapitulate the construction of section 4.1.2 to convince yourself how it is that coupling the Ising system to an external field generates the fourth term of Eq. (6.23). A lazy alternative: invoke symmetry arguments to explain the structure of this term.)

hope that renormalized perturbation theory will get the situation under control, and perhaps even allow one to extrapolate and draw conclusions as to what is happening below four dimensions.

▷ Operators $\phi^{n>4}$ become relevant in dimensions $d < (-1/n + 1/2)^{-1} < 4$. However, even then, they are less relevant than the dominant ϕ^4-operator. This is the justification for the neglect of $\phi^{n>4}$ operators in the effective theory.

▷ Operators with more than two gradients are generally irrelevant and negligible in all dimensions.

▷ However, the operator $\int \phi$ coupling to the external field carries dimension $1+d/2$ and is strongly relevant.

Guided by the orientation provided by dimensional analysis, we next analyse the model in a succession of steps: mean field analysis, analysis of quadratic fluctuations, and finally a renormalized analysis of nonlinear fluctuations.

6.3.2 Landau mean field theory

As usual, we begin our analysis of the functional integral by considering spatially homogeneous configurations, $\bar\phi$. Inspection of the potential part of the field-free Lagrangian, $r\phi^2/2 + \lambda\phi^4/4!$, shows that, depending on the sign of r, the system possesses two different stationary points.

For $r > 0$, the action has a global minimum at $\phi = 0$, with $\bar\phi = 0$ the unique mean field. Remembering that ϕ is a measure of the system's magnetization and that r is proportional to the reduced temperature, we identify this regime with the high temperature **paramagnetic phase**. Conversely, for $r < 0$, the action has two $\mathbb{Z}_2$ symmetry breaking minima at $\bar\phi = \pm\phi_0 \equiv \pm(6|r|/\lambda)^{1/2}$ – the low temperature

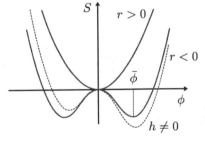

ferromagnetic phase. Finally, in the presence of an external field, the extrema shift, as described by the solution of the variational equation $\delta_{\bar\phi}S = 0$, i.e.,

$$r\bar\phi + \frac{\lambda}{6}\bar\phi^3 - h = 0. \tag{6.24}$$

The explicit breaking of $\mathbb{Z}_2$ symmetry manifests itself in a unique stationary configuration of least action whose sign is determined by that of h.

In this way, the stationary phase analysis shows how the sign change of the parameter r triggers a phase transition. This observation is consistent with the general relevance of the operator $r\int \phi^2$: the vanishing of its coupling constant defines a critical surface, and deviations away from it initiate a flow in either one of two different directions. Identifying the critical point with a critical temperature, we have the proportionality $r \sim (T - T_c)$, a relation that may be just postulated, or

Table 6.1 Critical exponents of the ferromagnetic transition. The experimental exponents represent cumulative data from various materials. Values taken from K. Huang, *Statistical Mechanics* (Wiley, 1987).

exponent	experiment	mean field	Gaussian	ϵ^1	ϵ^5
α	0–0.14	0	1/2	1/6	0.109
β	0.32–0.39	1/2	1/4	1/3	0.327
γ	1.3–1.4	1	1	7/6	1.238
δ	4–5	3	5	4	4.786
ν	0.6–0.7	–	1/2	7/12	0.631
η	0.05	–	0	0	0.037

confirmed via microscopic derivations of the model (recall the derivation on page 174, where we found that r exhibits a sign change as a function of temperature).

mean field critical exponents　We next ask the obvious question, what does mean field theory say about the **critical exponents** of the transition? Identifying $\bar{\phi} \sim M$ with the magnetization-order parameter of the transition, and referring back to the list on page 337, the low-temperature profile is given by $|\bar{\phi}| = (12|r|/\lambda)^{1/2} \sim |t|^{1/2}$, implying that $\beta = 1/2$. The exponent γ is obtained by differentiating Eq. (6.24) with respect to h. With $\chi \sim \partial_h \phi$, it is then straightforward to verify that $\chi \sim |t|^{-1}$, or $\gamma = 1$. The action evaluated on the mean field configuration takes the form

$$\frac{S[\bar{\phi}]}{L^d} = \frac{r}{2}\bar{\phi}^2 + \frac{\lambda}{4!}\bar{\phi}^4 \sim \begin{cases} \lambda^{-1}t^2, & t < 0, \\ 0, & t > 0. \end{cases} \tag{6.25}$$

Given the free energy $F = TS[\bar{\phi}]$, the specific heat $C = -T^2\partial_T^2 F \sim \partial_t^2 S$ behaves as a step function at the transition point, implying $\alpha = 0$. At criticality, $r = 0$, the magnetization depends on h as $\bar{\phi} \sim h^{1/3}$, implying that $\delta = 3$. Finally, the correlation length exponents ν, η cannot be computed from mean field theory as they probe the spatial profile of *fluctuating* field configurations.

In table 6.1, the mean field exponents are compared with experimentally obtained values. At first sight, the differences do not seem dramatic, which in view of the total neglect of field fluctuations is surprising. On the other hand, we must keep in mind that observables depend sensitively on the values of the exponents via singular power laws, where a difference between 1.3 and 1, say, makes a huge difference. We will therefore aim to improve our understanding of the transition by including fluctuations.

6.3.3 Gaussian model

Following the standard doctrine (mean field + fluctuations), we will start by including quadratic fluctuations around the constant value $\bar{\phi}$. We approach the transition from high temperatures, $r > 0$ and $\bar{\phi} = 0$, and consider the quadratic expansion,

$$S[\phi] \approx \int d^d r \left(\frac{r}{2} \phi^2 + \frac{1}{2} (\nabla \phi)^2 - h\phi \right), \tag{6.26}$$

where the presence of the linear term is due to the fact that the external field h has been neglected at the mean field level. In principle, we may now do the Gaussian integral to obtain an effective free energy. However, in view of the fact that this strategy will not work beyond the quadratic level, we pursue a different approach, via a baby version of momentum shell renormalization.

Following the algorithm of section 6.2, we split the field as $\phi = \phi_s + \phi_f$ into fast and slow components. Substitution into Eq. (6.26) then leads to a breakup $S[\phi_s, \phi_f] = S_s[\phi_s] + S_f[\phi_f]$ into a fast and a slow action. Here, the simplicity of the Gaussian model is reflected in the absence of a coupling action $S_c[\phi_s, \phi_f]$. As a consequence, the integration over ϕ_f merely leads to an inessential constant. The effect of the RG transformation is reduced to its final step, the rescaling of the slow action. According to our previous discussion, the ensuing scaling factors are determined by the engineering dimensions of the operators appearing in the action, i.e., $r \to b^2 r$ and $h \to b^{d/2+1} h$. Using the fact that $r \sim t$, this rescaling defines two relevant scaling dimensions, $y_t = 2$ and $y_h = d/2 + 1$. In a final step, we compare with Eq. (6.22) to obtain the list of exponents,

$$\alpha = 2 - \frac{d}{2}, \quad \beta = \frac{d}{4} - \frac{1}{2}, \quad \gamma = 1, \quad \delta = \frac{d+2}{d-2}, \quad \nu = \frac{1}{2}, \quad \eta = 0.$$

These results now do depend on the dimensionality of the system, which is natural, since they were obtained from spatial fluctuations. Table 6.1 contains their values for a three-dimensional system. We cannot really say that the results improve those of the mean field analysis. Some exponents (e.g., δ) agree better with the experimental data, others (e.g., α) are worse.

As a corollary, we note that the Gaussian model possesses only one fixed point, $r = h = 0$. This fixed point is present in all free field theories and is generally called the **Gaussian fixed point**.

Gaussian
fixed point

6.3.4 Renormalization group analysis

We can now no longer postpone facing the main challenge in the problem, the manifestations of the nonlinear ϕ^4 operator in its fluctuation behavior. In section 6.3.1, we observed that this operator becomes relevant in dimension $d = 4$, a finding corroborated from a different perspective in the Info section below. It thus seems that this operator is strongly relevant, and therefore difficult to handle, in the dimensions $2, 3$ in which one is usually interested.

However, there is a technical trick one can play to turn its marginality in $d = 4$ into an asset: the functional integral is just an integral, and there is no reason not to consider it in *non-integer dimensions*. Specifically, one may focus on dimensions slightly below the threshold, $d = 4 - \epsilon$, and hope that small ϵ provides an expansion parameter in which the effective "strength" of the nonlinearity can be controlled. (We *do* know that it becomes irrelevant for negative ϵ.) At the end of the calculation, one will then need to "analytically continue" to the physical dimension $d = 3 = 4-1$ and trust in a mantra of faith of theoretical physics:

> Good theories usually work outside their formal confines of justification.

ϵ-expansion This is the idea behind the ϵ-**expansion** of renormalization group theory.

INFO Besides dimensional analysis, the stability of a theory to **fluctuations around mean fields** provides another indication of the relevance of its operator contents. In the following, we outline such a stability analysis for the case of ϕ^4-theory. While the line of reasoning may be somewhat contrived, it provides us with more insight into the nature of fluctuations than formal dimensional analysis. For concreteness, consider the magnetic susceptibility,

$$\chi = -\partial_H^2 F \sim \int d^d r \, \langle \phi(\mathbf{r})\phi(0)\rangle_c \sim G(\mathbf{k} = 0),$$

where we note that $\langle \phi(\mathbf{r})\phi(\mathbf{r}')\rangle = G(\mathbf{r} - \mathbf{r}')$ is the Green function of the model. With this identification, we know that a divergence of the susceptibility indicating the transition is equivalent to a singularity of the zero-momentum Green function.

In the Gaussian theory, Eq. (6.26) implies that $G(\mathbf{k}) = (r + k^2)^{-1}$. Thus, $\chi \sim r^{-1}$, consistent with the identification of the mean field transition through $r \sim t = 0$. Now let us consider corrections to this result in perturbation theory in the ϕ^4 interaction. From our discussion of ϕ^4 perturbation theory in section 4.3.1, we recall that the Green function then picks up a self-energy, $G^{-1} \to r + k^2 - \Sigma$, which to leading one-loop order is given by $\Sigma = -\frac{\lambda}{2}\int^\Lambda (dk)(r + k^2)^{-1}$,[21] where we note the presence of a UV cutoff limiting the integration. As a result, the susceptibility becomes

$$\chi^{-1} \sim G(0)^{-1} = r - \Sigma = r + \frac{\lambda}{2}\int^\Lambda (dk)\frac{1}{r + k^2}.$$

Approaching the transition from above, $r > 0$, the fluctuation-induced contribution is positive. With $r \sim T - T_c$, we consider this as a *lowering* of the transition temperature. Since interactions between fluctuation modes are likely to disorder the system, this looks like a natural interpretation. However, we also observe that the cutoff Λ is needed to prevent the fluctuation contribution from diverging in dimensions $d \geq 2$.

Let us try to make sense of this divergence without invoking the fully fledged RG machinery. To this end, we argue that the parameter r of the bare theory has no intrinsic physical meaning; at any rate, we do know that it will be strongly modified by the integration over fluctuations. This ansatz suggests that we should absorb the leading singularity into a redefinition $r \to \tilde{r}$:

$$\chi^{-1} = \tilde{r} + \frac{\lambda}{2}\int^\Lambda (dk)\left(\frac{1}{r + k^2} - \frac{1}{k^2}\right) \approx \tilde{r} - \frac{\lambda\tilde{r}}{2}\int^\Lambda (dk)\frac{1}{(\tilde{r} + k^2)k^2}. \qquad (6.27)$$

In this equation $\tilde{r}$ is *formally* defined as

$$\tilde{r} \equiv r + \frac{\lambda}{2}\int^\Lambda (dk)\frac{1}{k^2},$$

involving a strongly UV singular integral. However, *physically*, we define the leading contribution to the inverse susceptibility, $\tilde{r}$, as the effective temperature of the transition.

---

[21] Compare the first two diagrams in fig. 4.7, which are identical in the present $N = 1$ version of ϕ^4-theory.

This is an observable quantity and therefore it must be finite. Also note that, in the second equality of Eq. (6.27), we replaced $r \to \tilde{r}$ in the fluctuation contribution, which is permissible to lowest order in perturbation theory in λ.

Via this "bootstrap" construction, we have shuffled the UV divergence of the theory into a reinterpreted coupling $\tilde{r}$. However, we still need to deal with the second fluctuation contribution in Eq. (6.27). On dimensional grounds, we estimate the parameter dependence of this integral as $\sim \lambda L^d \tilde{r}^{(d-4)/2}$. For dimensions lower than $d \equiv d_c = 4$, this term becomes large for small $\tilde{r}$, i.e., upon approaching the transition. This observation is the

Ginzburg criterion

essence of the **Ginzburg criterion**. The latter defines the **upper critical dimension**, d_c, of a theory as the dimension below which fluctuation contributions acquire a singular dependence on the effective control parameter, $\tilde{r}$, of a phase transition. The upper critical dimension defines the threshold between a (mean field + quadratic fluctuations) scenario in $d > d_c$ from strongly fluctuation dominated physics below d_c. However, what the Ginzburg criterion does not do is instruct us on how to bring these fluctuations under control; this is the job of the RG program to be discussed next.

We finally note that the above – admittedly contrived – reasoning parallels the rationale of renormalization in particle physics: the bare couplings (r) of a theory are denied any physical meaning; it is permissible to add formally divergent quantities to them if this is required to make observable predictions of a theory ($\tilde{r}$) finite. For an illustration of the predictive power of this approach, we suggest the renormalization of quantum electrodynamics as discussed in Ref.[13].

Turning to the concrete formulation of the RG analysis, we have to realize that ϕ^4 theory is sufficiently complex that we need to make recourse to all available approximation schemes. Besides the dimensional ϵ-expansion, this will include an

loop expansion

expansion in the loop number of momentum integrals. In fact, loop expansions are engaged in the majority of RG analyses of field theories. To understand the rationale behind them, consider a fluctuation action multiplied by some large parameter (which could be $\hbar^{-1}$ in a semiclassical expansion, or some large N parameter). The expansion in the number of loops is then equivalent to an expansion in the inverse of that parameter.

EXERCISE To verify this statement, assume the action $S[\phi]$ to be multiplied by a parameter, a. Contributions of nth order in perturbation theory in the ϕ^4 vertex are then weighted by a factor a^n. At the same time, each of the $4n/2$ propagator lines involved in their contraction comes with a factor a^{-1}, so that the overall power is $a^{n-4n/2} = a^{-n}$. Next express this parameter in terms of the number L of independent momentum integrals. Notice that the momentum space representation of each vertex contains three free momentum indices. (The fourth is locked by momentum conservation.) Contractions further reduce the number of free momentum indices. Do the bookkeeping to show that the overall power of the graph is a^{-L+1}.

The advantage of using L, rather than n, as a counting parameter is that no explicit reference to the vertex order in perturbation theory is made. In cases where more than one effective vertex is engaged (as will be the case when we introduce the slow field/fast field decomposition of the action), this way of fixing the parameter order of diagrams is simpler.

However, even if a theory does not provide us the favor of a large parameter – as is the case here – one often pretends that one is present and performs a loop expansion nonetheless. The idea behind this formally uncontrolled expansion is then that fluctuations must somehow be controllably small to make the fluctuation integral meaningful.

In the following, we formulate the RG analysis of ϕ^4-theory to lowest, one-loop order.

Step I

The first, straightforward, step of the RG analysis involves the introduction of slow and fast fields, $\phi = \phi_\mathrm{s} + \phi_\mathrm{f}$. Substitution into the action leads to $S[\phi_\mathrm{s}, \phi_\mathrm{f}] = S_\mathrm{f}[\phi_\mathrm{f}] + S_\mathrm{s}[\phi_\mathrm{s}] + S_\mathrm{c}[\phi_\mathrm{s}, \phi_\mathrm{f}]$, where

$$S_\mathrm{f}[\phi_\mathrm{f}] = \int d^d r \left(\frac{r}{2} \phi_\mathrm{f}^2 + \frac{1}{2} (\nabla \phi_\mathrm{f})^2 \right) + \dots,$$

$$S_\mathrm{s}[\phi_\mathrm{s}] = \int d^d r \left(\frac{r}{2} \phi_\mathrm{s}^2 + \frac{1}{2} (\nabla \phi_\mathrm{s})^2 + \frac{\lambda}{4!} \phi_\mathrm{s}^4 - h \phi_\mathrm{s} \right),$$

$$S_\mathrm{c}[\phi_\mathrm{s}, \phi_\mathrm{f}] = \frac{\lambda}{4} \int d^d r \, \phi_\mathrm{s}^2 \phi_\mathrm{f}^2 + \dots$$

Here, we have neglected terms of $\mathcal{O}(\phi_\mathrm{f}^4)$ since their contraction leads to two-loop diagrams. The same applies to terms of $\mathcal{O}(\phi_\mathrm{s} \phi_\mathrm{f}^3)$ (exercise). Terms of $\mathcal{O}(\phi_\mathrm{s}^3 \phi_\mathrm{f})$ do not arise because the addition of a fast momentum and three slow momenta is incompatible with momentum conservation.

Step II

To simplify the notation, let us rescale the momentum according to $\mathbf{q} \to \mathbf{q}/\Lambda$, implying that coordinates are measured in units of the inverse cutoff $\mathbf{r} \to \mathbf{r}\Lambda$. With the coupling constants rescaled according to their engineering dimensions, $r \to r\Lambda^2$, $\lambda \to \lambda\Lambda^{4-d}$, the action remains unchanged, while the fast and slow momenta are now integrated over the dimensionless intervals $|\mathbf{q}_\mathrm{s}| \in [0, b^{-1}]$ and $|\mathbf{q}_\mathrm{f}| \in [b^{-1}, 1]$, respectively. We next construct an effective action by integration over the fast field: $e^{-S_\mathrm{eff}[\phi_\mathrm{s}]} = e^{-S_\mathrm{s}[\phi_\mathrm{s}]} \left\langle e^{-S_\mathrm{c}[\phi_\mathrm{s}, \phi_\mathrm{f}]} \right\rangle_\mathrm{f}$. In performing the average over fast fluctuations, $\langle \cdots \rangle_\mathrm{f}$, we retain only contributions of one-loop order. We also neglect terms that lead to the appearance of $\phi_\mathrm{s}^{n>4}$ contributions in the action. (For example, the contraction $\langle (\int \phi_\mathrm{s}^2 \phi_\mathrm{f}^2)^3 \rangle$ generates such a term.) To this level of approximation, one obtains

$$e^{-S_\mathrm{eff}[\phi_\mathrm{s}]} = e^{-S_\mathrm{s}[\phi_\mathrm{s}]} \exp \left(-\langle S_\mathrm{c}[\phi_\mathrm{s}, \phi_\mathrm{f}] \rangle_\mathrm{f} + \frac{1}{2} \langle S_\mathrm{c}[\phi_\mathrm{s}, \phi_\mathrm{f}]^2 \rangle_\mathrm{f}^\mathrm{c} \right),$$

where the superscript c denotes a connected average. (Exercise: Check the consistency of this expansion.) The skeleton structure of the two diagrams corresponding to the contractions $\langle S_\mathrm{c}[\phi_\mathrm{s}, \phi_\mathrm{f}] \rangle_\mathrm{f}$ and $\langle S_\mathrm{c}[\phi_\mathrm{s}, \phi_\mathrm{f}]^2 \rangle_\mathrm{f}^\mathrm{c}$ is shown in parts (a) and (b) of the figure below, respectively, where the external line segments indicate the passive ϕ_s amplitudes. The first of the two diagrams, (a), evaluates to

$$\langle S_c[\phi_s, \phi_f]\rangle_f = \frac{\lambda}{4} \int_f (dq') \frac{1}{r + q'^2} \int_s (dq)\phi_s(\mathbf{q})\phi_s(-\mathbf{q}).$$

We now consider the summation over fast momenta appearing in this expression. Assuming that we are in the vicinity of the critical point, where r is small, we expand to first order, $\int_f (dq)(r + q^2)^{-1} = I_1 - rI_2$, where

$$I_\alpha \equiv \int_f (dq) \frac{1}{q^{2\alpha}}. \tag{6.28}$$

These integrals are conveniently computed in polar coordinates,

$$I_\alpha = \Omega_d \int_{b^{-1}}^1 dq\, q^{d-2\alpha-1} = \frac{\Omega_d}{d - 2\alpha}(1 - b^{2\alpha-d}), \tag{6.29}$$

where $\Omega_d = (2\pi)^{-d} \times 2\pi^{d/2}/\Gamma(d/2)$ is the area of the d-dimensional unit sphere expressed in terms of the Γ-function and measured in units of $(2\pi)^d$. After rescaling $\mathbf{q} \to b\mathbf{q}, \phi \to b^{(d-2)/2}\phi$, the quadratic part of the action takes the form

$$S^{(2)}[\phi] = \frac{b^2}{2}\left(r + \frac{\lambda\Omega_d}{2(d-2)}(1 - b^{2-d}) - \frac{r\lambda\Omega_d}{2(d-4)}(1 - b^{4-d})\right)\int d^d r\, \phi^2. \tag{6.30}$$

Turning to the second diagram, (b), the presence of four external legs means that it is proportional to ϕ_s^4. Owing to momentum conservation, the internal propagator lines (a) depend on both a fast internal momentum variable and the slow external momenta carried by the fields ϕ_s. Conveniently, however, the dependence on the latter is negligible. To see why, note that the integral with all momenta taken into account yields a result with structure (b) $F(\mathbf{q}_1, \mathbf{q}_2, \mathbf{q}_3)\phi(\mathbf{q}_1)\phi(\mathbf{q}_2)\phi(\mathbf{q}_3)\phi(-\mathbf{q}_1 - \mathbf{q}_2 - \mathbf{q}_3)$, where $\mathbf{q}_{1,2,3}$ represent slow momenta and F is a function. Taylor expansion of the latter would generate powers of $\mathbf{q}_i$, i.e., *derivatives*, acting on the slow field; we have seen that these are irrelevant, and we may thus neglect the $\mathbf{q}$ dependence of F from the outset.

Simplified in this way, diagram (b) leads to the result

$$\frac{1}{2}\langle S_c[\phi_s, \phi_f]^2\rangle_f \simeq \frac{\lambda^2}{16}\int d^d r\, \phi_s^4 \int_f (dq)\frac{1}{(r + q^2)^2} = \frac{\lambda^2 I_2}{16}\int d^d r\, \phi_s^4 + \mathcal{O}(\lambda^2 r).$$

We evaluate the integral with Eq. (6.29) and rescale to obtain the quartic contribution to the renormalized action

$$S^{(4)}[\phi] = b^{4-d}\left(\frac{\lambda}{4!} - \frac{\lambda^2\Omega_d}{16}\frac{1 - b^{4-d}}{d - 4}\right)\int d^d r\, \phi^4.$$

Finally, there are no one-loop diagrams affecting the linear part of the action, i.e.,

$$S^{(1)}[\phi] = h b^{d/2+1}\int d^d r\, \phi$$

rescales according to its engineering dimension.

Step III

Combining everything, we find that the coupling constants transform as

$$r \to b^2 \left(r + \frac{\lambda \Omega_d}{2(d-2)}(1 - b^{2-d}) - \frac{r\lambda \Omega_d}{2(d-4)}(1 - b^{4-d}) \right),$$

$$\lambda \to b^{4-d} \left(\lambda - \frac{3}{2}\lambda^2 \Omega_d \frac{1 - b^{4-d}}{d-4} \right),$$

$$h \to h b^{d/2+1}.$$

We next set $d = 4 - \epsilon$ and evaluate the right-hand sides of these expressions to leading order in ϵ. With $\Omega_{4-\epsilon} \approx \Omega_4 = 1/(8\pi^2)$, we thus obtain

$$r \to b^2 \left(r + \frac{\lambda}{32\pi^2}(1 - b^{-2}) - \frac{r\lambda}{16\pi^2} \ln b \right),$$

$$\lambda \to (1 + \epsilon \ln b) \left(\lambda - \frac{3\lambda^2}{16\pi^2} \ln b \right),$$

$$h \to h b^{3-\epsilon/2},$$

flow equations which, setting $b = e^\ell$, lead to the **flow equations**:

$$
\boxed{
\begin{aligned}
\frac{dr}{d\ell} &= 2r + \frac{\lambda}{16\pi^2} - \frac{r\lambda}{16\pi^2} \\
\frac{d\lambda}{d\ell} &= \epsilon\lambda - \frac{3\lambda^2}{16\pi^2} \\
\frac{dh}{d\ell} &= \frac{6-\epsilon}{2}h
\end{aligned}
}
\tag{6.31}
$$

These equations demonstrate the meaning of the ϵ-expansion. According to the second equation, a perturbation away from the Gaussian fixed point will initially grow at a rate set by the engineering dimension ϵ. However, this runaway flow of the nonlinearity is countered by the one-loop contribution $\sim \lambda^2$, and is eventually stopped at the value $\lambda \sim \epsilon$. A similar competition governs the flow of the coupling r in the first equation.

Temporarily ignoring the magnetic field and equating the r.h.s. of Eq. (6.31) to zero, we indeed find that, for $\epsilon > 0$, besides the Gaussian point $(r_1^*, \lambda_1^*) = (0,0)$, a **nontrivial fixed point** $(r_2^*, \lambda_2^*) = (-\epsilon/6, 16\pi^2\epsilon/3)$ has appeared. This second fixed point is $\mathcal{O}(\epsilon)$ away from the Gaussian one and coalesces with it as ϵ is sent to zero. Plotting the β-function for the coupling constant λ, we find that λ is relevant around the Gaussian fixed point but irrelevant at the nontrivial fixed point. In the opposite case of dimensions larger than four, $\epsilon < 0$, the nontrivial fixed point disappears and the Gaussian one remains as the only, now stable, fixed point.

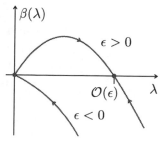

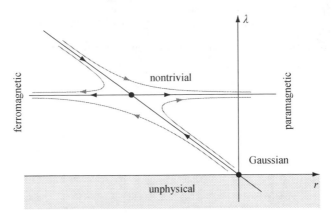

Fig. 6.4   Phase diagram of the ϕ^4-model as obtained from the ϵ-expansion.

To understand the full flow diagram of the system, we linearize the β-function around the two fixed points. Denoting the respective linearized mappings by $W_{1,2}$, we find

$$W_1 = \begin{pmatrix} 2 & 1/(16\pi^2) \\ 0 & \epsilon \end{pmatrix}, \quad W_2 = \begin{pmatrix} 2 - \epsilon/3 & (1 + \epsilon/6)/(16\pi^2) \\ 0 & -\epsilon \end{pmatrix}.$$

Figure 6.4 shows the flow in the vicinity of the two fixed points, as described by the matrices $W_{1,2}$, as well as the extrapolation to a global flow chart. Notice that the critical surface of the system – the straight line interpolating between the two fixed points – is tilted with respect to the r ($\sim$ temperature) axis of the phase diagram. This implies that it is not the physical temperature alone that dictates whether the system will eventually wind up in the paramagnetic ($r \gg 0$) or ferromagnetic ($r \ll 0$) sector of the phase diagram. Instead one has to relate temperature ($\sim r$) to the strength of the nonlinearity ($\sim \lambda$) to decide on which side of the critical surface one is positioned. For example, for strong enough λ, even a system with r initially negative may eventually flow towards the disordered phase. This type of behavior cannot be predicted from the mean field analysis of the model (which would generally predict a ferromagnetic state for $r < 0$). Rather, it represents a nontrivial effect of fluctuations. Finally, notice that, while we can formally extend the flow into the lower portion of the diagram, $\lambda < 0$, this region is unphysical. The reason is that, for $\lambda < 0$, the action is unstable and, in the absence of a sixth-order contribution, does not describe a physical system.

We now proceed to discuss the **critical exponents** associated with the one-loop approximation. Of the two eigenvalues of W_2, $2 - \epsilon/3$ and $-\epsilon$, only the former is relevant. As with the Gaussian fixed point, it is tied to the scaling of the coupling constant, $r \sim t$, i.e., we have $y_t = 2 - \epsilon/3$ and, as before, $y_h = (d+2)/2 = (6-\epsilon)/2$. An expansion of the exponents summarized in Eq. (6.22) to first order in ϵ then yields the list

$$\alpha = \frac{\epsilon}{6}, \quad \beta = \frac{1}{2} - \frac{\epsilon}{6}, \quad \gamma = 1 + \frac{\epsilon}{6}, \quad \delta = 3 + \epsilon, \quad \nu = \frac{1}{2} + \frac{\epsilon}{12}, \quad \eta = 0.$$

If we are now cavalier enough to extend the radius of the expansion to $\epsilon = 1$, i.e., $d = 3$, we obtain the fifth column of table 6.1. Apparently, agreement with the experimental results has improved – in spite of the fact that the ϵ-expansion has been pushed beyond its range of applicability! (For $\epsilon = 1$, terms of $\mathcal{O}(\epsilon^2)$ are no longer negligible.)

How can one rationalize the **success of the ϵ-expansion**? One might simply speculate that nature seems to be sympathetic to the concept of renormalization and the loop expansion. However, a more qualified approach is to explore what happens at higher orders in the expansion. Although this extension comes at the price of a much more laborious analysis, the first-order expansion prompted researchers to drive the ϵ-expansion up to fifth order. The results of this analysis are summarized in the last column of table 6.1. In view of the fact that we are still extending a series beyond its radius of convergence,[22] the level of agreement with the experimental data is striking. In fact, the exponents obtained by the ϵ-expansion even agree – to an accuracy better than one percent – with the exponents of the *two*-dimensional model,[23] where the "small" parameter $\epsilon = 2$.

6.4 RG Analysis of the Nonlinear σ-Model

Conceptually, the merit of the ϵ-expansion is that it enables one to explore the phase diagrams of nonlinear theories in a more or less controlled manner. Expansions of this type are not only applied close to the upper critical dimension (the mean field threshold) but also in the vicinity of the lower one (the collapse of spontaneous symmetry breaking, section 5.2.3). In this section, we consider a problem in this category: an ϵ-expansion around $d = 2$ to detect the onset of global thermal disorder in models with continuous symmetries.

Real-valued scalar fields describe phase transitions with single-component order parameters. However, throughout the text, we have encountered problems where the order parameter involves more than one component: complex fields describing condensation phenomena, matrix fields associated with the disordered electron gas, or the field theories of spin. In such cases, the field degrees of freedom are often subject to nonlinear constraints. For example, we saw that, below the condensation point, amplitude fluctuations of the complex superfluid order parameter are negligible, and field fluctuations are confined to a circle in the complex plane – the phase mode.

nonlinear
σ-model

Multi-component field theories subject to nonlinear constraints are generally known as **nonlinear σ-models**. In many cases, these constraints reflect the breaking

---

[22] It is believed that the ϵ-series is an asymptotic one and that the agreement with the "true" exponents will become worse at some hypothetical order.

[23] The latter are known from the exact solution of the two-dimensional model, see L. Onsager, *Crystal statistics I. A two-dimensional model with an order–disorder transition*, Phys. Rev. **65**, 117 (1944).

of continuous symmetries at phase transition points. The aim of the present section is to apply RG methods to describe critical phenomena in the vicinity of the lower critical dimension $d = 2$, where symmetry breaking can occur.

We consider a paradigmatic nonlinear σ-model defined by the partition function $\mathcal{Z} = \int Dg\, e^{-S[g]}$, where

$$S[g] = \frac{1}{\lambda} \int d^d r \, \mathrm{tr}\left(\nabla g \nabla g^{-1}\right), \qquad (6.32)$$

and $g \in G$ takes values in a compact Lie group G such as $G = \mathrm{O}(3)$, relevant to three-component spin models. (In this case, the nonlinear constraint would be realized through $g^T g = \mathbb{1}$.) The integration $\int Dg = \prod_x \int d_\mu g(x)$ extends over the *Haar measure* $d_\mu g$ of G.

Haar measure

INFO The **Haar measure** $d_\mu g$ of a compact Lie group G is a (uniquely specified) integration measure defined by the condition that it is unit normalized, $\int_G d_\mu g \cdot 1 = 1$, and invariant under left and right multiplication by a fixed group element:

$$\int_G d_\mu g\, f(g) = \int_G d_\mu g\, f(gh^{-1}) = \int_G d_\mu g\, f(h^{-1}g)$$

for any $h \in G$. Upon translation of the integration variable $g \to hg$ or $g \to gh$, these equations assume the form $\int_G d_\mu(gh)\, f(g) = \int_G d_\mu g\, f(g) = \int_G d_\mu(hg)\, f(g)$. Holding for any f, this implies that $d_\mu(hg) = d_\mu g = d_\mu(gh)$, which means that the Haar measure assigns equal volume density to any point on the group manifold. Owing to this homogeneity property, integrations over groups are almost always performed with respect to the Haar measure. Details on the explicit construction of this measure can be found in textbooks on Lie group theory.[24]

Given that nonlinear σ-models describe Goldstone mode fluctuations, we presume that the physics of the latter is reflected in the functional integral $\mathcal{Z}$. In particular, we know that Goldstone mode fluctuations become uncontrollably strong in dimensions $d \leq d_c = 2$, while in larger dimensions they are weak.

In the present context, the strength of fluctuations is controlled by the parameter λ, $\lambda \to 0$ representing the so-called **weak coupling** limit with suppressed fluctuations, and $\lambda \to \infty$ the **strong coupling** limit, where the field

weak/strong coupling

Alexander M. Polyakov
1945–
is a Russian theoretical physicist and mathematician who has made important contributions to quantum field theory, from non-abelian gauge theory to conformal field theory. His path integral formulation of string theory had profound and lasting impacts in the conceptual and mathematical understanding of the theory. He also played an important role in elucidating the conceptual framework behind renormalization independently of Kenneth Wilson's Nobel Prize winning work. He formulated pioneering ideas in gauge–string duality which later became key in the formulation of the holographic principle.

g is uniformly distributed over the group manifold, representing fluctuations of maximal strength. We wish to understand the phase transition separating weak from

---

[24] See, e.g., J. Fuchs and C. Schweigert, *Symmetries, Lie Algebras and Representations, a Graduate Course for Physicists* (Cambridge University Press, 1997).

strong coupling at $d_c = 2$. To this end, following a seminal work by Polyakov,[25] we will again apply an ϵ-expansion, this time formulated in $d = d_c + \epsilon$ where fluctuations remain controllable.

However, before turning to the actual formulation of the RG analysis, we need to do some preparatory work concerning the perturbative evaluation of functional integrals with group target spaces.

6.4.1 Field integrals over groups

As with any other integral over nonlinear manifolds, group integrals are performed in coordinates. Specifically, in perturbative applications where group field fluctuations are weak, it is convenient to work in an **exponential representation** where $g = \exp(W)$ and $W \in \mathcal{G}$ lives in the Lie algebra (see the Info block on page 248) of the group. With $W = i \sum^a \pi^a T_a$, and T_a the hermitian matrix generators (see info below) of the group, the integration reduces to one over a set of real coordinates, π_a.

In the exponential representation, an expansion in W describes fluctuations of g around the identity. However, owing to to the Haar invariance of the measure, the "anchor point" of the expansion is arbitrary, and the generalization $g \to h \exp(W)$ describes fluctuations around arbitrary group elements, h. (In our RG application, W will parameterize fast field fluctuations around a slow field $h = g_s$.)

INFO Recall that the Lie algebra $\mathcal{G}$ of a matrix Lie group is an M-dimensional vector space whose elements are $N \times N$ matrices. (For example, the Lie algebra su(2) is the three-dimensional vector space of traceless hermitian 2×2 matrices.) A system of **generators,** T^a, **of a Lie group** $\{T^a | a = 1, \ldots, M\}$ is a basis of this algebra. (For example, $T^a = \sigma^a$, the Pauli matrices in the case of su(2).) The commutators between these generator matrices $[T^a, T^b] = -if^{abc}T^c$ define the **structure constants**, f^{abc}, of the group. (For su(2), $f^{abc} = \epsilon^{abc}$ are defined by the antisymmetric tensor.)

For later reference, we note that the generators of the group families $U(N), SU(N)$ and $O(N)$ obey the **completeness relations**

$$\sum_{a=1}^{N^2} T_{ij}^a T_{kl}^a = \frac{1}{2}\delta_{il}\delta_{jk}, \qquad\qquad U(N), \qquad\qquad (6.33)$$

$$\sum_{a=1}^{N^2-1} T_{ij}^a T_{kl}^a = \frac{1}{2}\delta_{il}\delta_{jk} - \frac{1}{2N}\delta_{ij}\delta_{kl}, \qquad SU(N), \qquad\qquad (6.34)$$

$$\sum_{a=1}^{N(N-1)/2} T_{ij}^a T_{kl}^a = \delta_{il}\delta_{jk} - \delta_{ik}\delta_{jl}, \qquad\qquad O(N), \qquad\qquad (6.35)$$

where the upper limits are the respective group dimensions. (Exercise: Verify these relations by inspection of the conditions that Lie algebra elements have to satisfy.)

[25] A. M. Polyakov, *Interactions of Goldstone particles in two dimensions. Applications to ferromagnets and massive Yang–Mills fields*, Phys. Lett. B **59**, 79 (1975).

The generators are often chosen normalized as $\text{tr}(T^a T^b) = c\delta^{ab}$, where c is a constant. Interpreting $\langle W|W'\rangle \equiv \text{tr}(WW')$ as a scalar product on the Lie algebra (exercise: convince yourself that $\langle\,|\,\rangle$ meets all the criteria required of a scalar product), this normalization makes $\{T^a\}$ an orthonormal basis. In the cases $\text{U}(N), \text{SU}(N)$, and $\text{O}(N)$, consistency with the completeness relation requires $c = 1/2, 1/2$, and 1, i.e.,

$$\text{tr}(T^a T^b) = \frac{\delta^{ab}}{2}, \qquad \text{U}(N), \text{SU}(N), \tag{6.36}$$

$$\text{tr}(T^a T^b) = \delta^{ab}, \qquad \text{O}(N). \tag{6.37}$$

The upside of the exponential parameterization is that W is a linear object, integrated over the *vector space* $\mathcal{G}$ (see the Info block below for details). The downside is that the representation of the Haar measure in W-coordinates may involve a nontrivial Jacobian. However, in many applications, like the present one, this Jacobian does not interfere significantly with a perturbative integration.[26]

We next discuss how integrals over fluctuations on the group manifold are performed in practice. Substitution of the expansion $g = e^W = 1 + W + W^2/2 + \cdots$ into Eq. (6.32) produces a series $S = \sum_n S^{(n)}$, where $S^{(n)}$ is of nth order in W. As always, we will organize our analysis around the tractable quadratic action, which is given by

$$\begin{aligned}
S^{(2)}[W] &= -\frac{1}{\lambda}\int d^d r\, \text{tr}\left(\nabla W\, \nabla W\right) = \frac{1}{\lambda}\int d^d r\, \nabla\pi^a \nabla\pi^b\, \text{tr}\left(T^a T^b\right) \\
&\overset{(6.37)}{=} \frac{1}{\lambda}\int d^d r\, \nabla\pi^a \nabla\pi^a = \frac{1}{2}\int (dq)\, \pi_p^a \Pi_p^{-1} \pi_{-p}^a.
\end{aligned}$$

Here we have used the normalization condition (6.37) and in the final step switched to a momentum representation (for a lighter notation, we are avoiding boldface notation in this section, $\mathbf{p} \to p$), where $\Pi_p \equiv \lambda/2p^2$ is the "propagator" of the fields π^a.

At higher orders in the expansion, we encounter integrals $\langle \text{tr}(F(W))\text{tr}(G(W))\cdots\rangle$, where $F, G, \ldots$ are functions of W and $\langle\cdots\rangle \equiv \mathcal{N}\int D\pi \exp(-S^{(2)}[\pi])(\cdots)$ denotes the average over the quadratic action. These integrals may be evaluated with the help of Wick's theorem as the sum over all possible pairings of π-variables or, equivalently, W-matrices. Each pairing is of the form either $\langle \text{tr}(AW)\text{tr}(A'W)\rangle$ or $\langle \text{tr}(AWA'W)\rangle$, where the matrices A and A' may contain W-matrices themselves. (However, when computing an individual pairing, these matrices are temporarily kept fixed.) Specifically, for $G = \text{O}(N)$, the Gaussian integrals give

$$\begin{aligned}
\langle \text{tr}(AW_p)\text{tr}(A'W_{p'})\rangle &= -\langle \pi_p^a \pi_{p'}^{a'}\rangle\, \text{tr}(AT^a)\text{tr}(A'T^{a'}) \\
&= -\delta_{p,-p'}\Pi_p\, \text{tr}(AT^a)\, \text{tr}(A'T^a) \overset{(6.35)}{=} -\delta_{p,-p'}\Pi_p\left(\text{tr}(AA') - \text{tr}(AA'^T)\right),
\end{aligned}$$

---

[26] In the exponential representation, the Haar measure assumes the form $d_\mu g \to \prod_a d\pi^a\, J(\pi)$, where $J(\pi)$ encapsulates both the geometry of the Haar measure and the Jacobian associated with the transformation $g \to \exp(\pi^a T^a)$. It can be shown that its Taylor expansion starts as $J(\pi) = 1 + \mathcal{O}(\pi^4)$. Since terms $\mathcal{O}(\pi^4)$ do not enter the RG transformation at one-loop order, we may ignore J and work with the flat measure. However, at higher loop orders, a more careful analysis becomes necessary.

$$\langle \mathrm{tr}(AW_p A'W_{p'})\rangle = -\langle \pi_p^a \pi_{p'}^{a'}\rangle \ \mathrm{tr}\ (AT^a A'T^{a'})$$

$$= -\delta_{p,-p'}\Pi_p \ \mathrm{tr}(AT^a A'T^a) \overset{(6.35)}{=} -\delta_{p,-p'}\Pi_p \left(\mathrm{tr}(A)\mathrm{tr}(A') - \mathrm{tr}(AA'^{T})\right).$$

General integrals over W-matrices are computed by applying these contraction rules until all possible pairings have been exhausted.

EXERCISE As an example, consider the expression

$$X \equiv \langle \mathrm{tr}(A_{q,p}W_{p+q}W_{-p})$$
$$\mathrm{tr}(A'_{q',p'}W_{p'+q'}W_{-p'})\rangle,$$

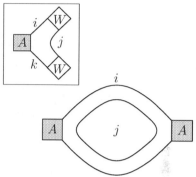

where $A, A' \in \mathrm{o}(N)$ are matrix fields in the Lie algebra of $\mathrm{O}(N)$ (and therefore traceless). Show that the pair contraction of the first W-matrix with the second vanishes owing to the tracelessness of A and A'. Next, perform the contractions $(1-4)(2-3)$ and $(1-3)(2-4)$ to obtain the result

$$X = (N-2)\Pi_p \Pi_{p+q}\mathrm{tr}\left(A_{p,q}\left(A'_{-q,p+q} - A'_{-q,-p}\right)\right). \tag{6.38}$$

Referring to the inset box in the figure for the matrix index structure of the contracted operator, the origin of the factor N in the result lies in the emergence of a free running index sum in the contraction represented diagrammatically in the lower part of the figure.

6.4.2 One-loop expansion

With this background, let us now return to the actual RG program for the $\mathrm{O}(N)$-model. As always, we proceed in stages, beginning with:

Step I

We start with a multiplicative decomposition of the matrix fields g into slow and fast components. This is achieved by defining $g(r) = g_\mathrm{s}(r)g_\mathrm{f}(r)$, where the generators of $g_\mathrm{s}(r)$ and $g_\mathrm{f}(r)$ have momentum components in the ranges $[0, \Lambda b^{-1}]$ and $[\Lambda b^{-1}, \Lambda]$, respectively. Substituting this decomposition into the action, one obtains $S[g_\mathrm{s}g_\mathrm{f}] = S[g_\mathrm{s}] + S[g_\mathrm{f}] + S[g_\mathrm{s}, g_\mathrm{f}]$, with the coupling action

$$S[g_\mathrm{s}, g_\mathrm{f}] = \frac{2}{\lambda}\int d^d r \ \mathrm{tr}\left(g_\mathrm{s}^{-1}\nabla g_\mathrm{s}g_\mathrm{f}\nabla g_\mathrm{f}^{-1}\right). \tag{6.39}$$

Step II

Turning to the integration over the fast field g_f, we will again apply a first-order loop expansion. The strategy is an expansion of $g_\mathrm{f} = \exp(W)$ followed by the identification of all contributions that lead to no more than one momentum integration over the fast field propagator Π_p. Representing a general trace $\int \mathrm{tr}(XW^n)$ with a slow X in momentum space, one may note that terms with $n > 2$ will inevitably lead to more than one integral. On the other hand, terms with $n = 1$ vanish owing to momentum conservation (consider why!). Thus, all we need to do is to consider the action $S[g_s, W]$ expanded to second order in W. Substitution of $g_f = 1 + W + W^2/2$ into the action (6.39) yields the effective coupling

$$S[g_s, W] = \frac{1}{\lambda} \int d^d r \, \mathrm{tr} \left(\Phi_\mu [\nabla_\mu W, W] \right) \simeq -\frac{2i}{\lambda} \int (dq) \int (dp) p_\mu \, \mathrm{tr} \left(\Phi_{\mu, -q} W_p W_{-p} \right),$$

where we have introduced the abbreviation $\Phi_\mu = g_s^{-1} \partial_\mu g_s$ and, in the last representation, neglected the small momentum in comparison with the fast, $p \pm q \simeq p$. To obtain all one-loop corrections, we have to expand the functional in powers of $S[g_s, W]$ and integrate over W. However, since each power of S comes with one derivative acting on a slow field, and terms of more than two such derivatives are irrelevant, it suffices to consider terms of order $\mathcal{O}(S^2)$. To one-loop order, the RG step thus amounts to the replacement

$$S[g] \rightarrow S[g_s] - \ln \left(1 + \frac{1}{2} \langle S[g_s, W]^2 \rangle_W \right) \simeq S[g_s] - \frac{1}{2} \langle S[g_s, W]^2 \rangle_W.$$

Written more explicitly,

$$\langle S[g_s, W]^2 \rangle_W = \int (dq)(dq') \int (dp)(dp') \, \langle \mathrm{tr}(A_{q,p} W_p W_{-p}) \, \mathrm{tr}(A_{q',p'} W_{p'} W_{-p'}) \rangle_W,$$

where $A_{p,q} = -2i L^d \lambda^{-1} p_\mu \Phi_{\mu,-q}$. This is an expression of the type considered in the exercise above. Using Eq. (6.38) with the explicit form of the propagator, $\Pi_p = \lambda/2p^2$, one obtains

$$\langle S[g_s, W^2] \rangle_W \simeq -2(N-2) \int (dp) \frac{p_\mu p_\nu}{p^4} \int (dq) \, \mathrm{tr}(\Phi_{\mu,q} \Phi_{\nu,-q})$$

$$= -C \int d^d r \, \mathrm{tr} \left(\Phi_\mu \Phi_\mu \right) = C \int d^d r \, \mathrm{tr} \left(\nabla g_s \nabla g_s^{-1} \right) = C \lambda S[g_s], \quad (6.40)$$

where the constant

$$C = \frac{2(N-2)}{d} \int (dp) \frac{1}{p^2} = \frac{2(N-2)\Omega^d}{d(2\pi)^d} \int_{\Lambda/b}^{\Lambda} dp \, p^{d-3}$$

$$= \frac{2(N-2)\Omega^d (\Lambda^{d-2} - (\Lambda/b)^{d-2})}{(2\pi)^d d(d-2)} \overset{d=2+\epsilon}{\simeq} \frac{(N-2)\ln b}{2\pi}.$$

Step III

Substituting Eq. (6.40) back into the action, one obtains

$$S[g] \rightarrow \left(1 - \frac{(N-2)\ln b \lambda}{4\pi} \right) S[g_s] \mapsto \left(1 - \frac{(N-2)\ln b \lambda}{4\pi} \right) b^\epsilon S[g],$$

where, in the second step, we performed a rescaling of the momenta, $q \rightarrow bq$, noting that in $2 + \epsilon$ dimensions the engineering dimension of the action is $(2 + \epsilon) - 2 = \epsilon$.

As expected, the RG reproduces the action up to a global scaling factor. Absorbing this factor in a renormalized coupling constant, λ_r, we obtain

$$\lambda_r = \left(1 - \frac{(N-2)\ln b\lambda}{4\pi}\right)^{-1} b^{-\epsilon}\lambda \simeq \lambda\left(\frac{(N-2)\lambda}{4\pi} - \epsilon\right)\ln b + \lambda,$$

where, in the last equality, we assumed both λ and ϵ to be small.[27] Finally, differentiating this result, one obtains the **RG equation of the O(N) nonlinear σ-model**

RG equation

$$\boxed{\frac{d\lambda}{d\ln b} \simeq -\epsilon\lambda + \frac{(N-2)\lambda^2}{4\pi} + \mathcal{O}(\lambda^3, \epsilon^2, \lambda^2\epsilon)} \tag{6.41}$$

The figure shows that, in dimensions $d < 2 \to \epsilon < 0$ and generic group dimensions, the β function is positive, and hence the flow is directed towards large values of λ. In other words, the model inevitably drifts towards a strong coupling phase. While this phase is inaccessible to the perturbative loop expansion applied here, there are no indications of an intermittent competing phase transition. We thus conclude that a disordered phase is the fate of the σ-model, which is consistent with the expected restoration of broken symmetries by large fluctuations in low dimensions. However, for dimensions $d > 2$, the model exhibits a fixed point at finite coupling strength $\lambda_c = 4\pi\epsilon/(N-2)$. Linearizing the flow in the vicinity of the fixed point, we obtain the exponent $y_\lambda = \epsilon$, assuming the role of the "thermal exponent" of ϕ^4-theory. Using the scaling relations (exercise: interpret the scaling relations in the context of the present model), we obtain the correlation length exponent $\nu \approx 1/\epsilon$, and the heat capacity exponent $\alpha \approx 2 - 2/\epsilon$.

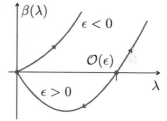

EXERCISE To complete the analysis of the O(3) model, one may explore the magnetic exponent. Introducing a magnetic field perturbation into the action, i.e., $h \int d^d r \operatorname{tr}[g+g^{-1}]$, show that, under renormalization we have $h' = b^{y_h}h$, where $y_h = 2 + \epsilon(N-3)/(2(N-2)) + \mathcal{O}(\epsilon^2)$. From this result, use the exponent identities to show that the anomalous dimension $\eta \approx \epsilon/(N-2)$.

We finally note that, for $d = 2$ and $N = 2$, the coupling constant does not renormalize. This is so because, for O(2) – the abelian group of planar rotations – the action of the σ-model simplifies to $S = \frac{1}{\lambda}\int d^2r\,(\nabla\phi)^2$, where $\phi(\mathbf{r})$ is the field of local rotation angles. This is a free action that is not renormalized by small fluctuations. (However, in the following section, we see that the situation is not quite as simple as it seems. The O(2) model admits large fluctuations – vortex configurations of the angular field – which *do* have a nontrivial impact on its behavior. However,

---

[27] The smallness of λ is needed to justify the perturbative loop expansion. In the opposite regime $\lambda \gg 1$, the fields g fluctuate wildly and perturbative expansion is not an option.

being topological in nature, these excitations are beyond the scope of our present analysis.)

EXERCISE Repeat the analysis above for the SU(N)-model. First derive the intermediate identities,

$$\langle \mathrm{tr}(AW_p)\,\mathrm{tr}(A'W_{p'}) \rangle_W = -\delta_{p,-p'}\frac{\Pi_p}{2}\left(\mathrm{tr}(AA') - \frac{1}{N}\mathrm{tr}(A)\,\mathrm{tr}(A') \right),$$

$$\langle \mathrm{tr}(AW_p A'W_{p'}) \rangle_W = -\delta_{p,-p'}\frac{\Pi_p}{2}\left(\mathrm{tr}(A)\,\mathrm{tr}(A') - \frac{1}{N}\mathrm{tr}(AA') \right),$$

where $\Pi_q = \lambda/L^d p^2$, and

$$p_\mu p_{\mu'}\langle \mathrm{tr}(\Phi_{\mu,q}W_{p+q}W_{-p})\mathrm{tr}(\Phi_{\mu',q'}W_{p'+q'}W_{-p'}) \rangle_W = N\frac{\Pi_p \Pi_{p+q}}{4}p_\mu p_{\mu'}\mathrm{tr}(\Phi_{\mu,q}\Phi_{\mu',-q}).$$

Use these results to obtain the RG equation

$$\frac{d\ln\lambda}{d\ln b} \simeq \frac{N\lambda}{8\pi} - \epsilon + \mathcal{O}(\lambda^2,\epsilon^2,\lambda\epsilon). \tag{6.42}$$

Notice that the right-hand side of the SU($N = 2$) equation coincides with the right-hand side of the O($N = 3$) equation. This reflects the (local) isomorphism of the groups O(3) and SU(2).

6.5 Berezinskii–Kosterlitz–Thouless Transition

Previously, we have discussed systems whose critical physics was encoded in local field fluctuations. However, it often happens that the relevant physics is that of global fluctuations, the latter colloquially defined as configurations that cannot be continuously deformed into flat configurations. As an example, consider the phase mode of a superfluid in two dimensions. This field takes values in O(2), and we saw in the previous section that its effective action is not renormalized by short range fluctuations. However, it may happen that the phase degree of freedom winds around a rotation center in space to form a vortex. Once created, the vortex cannot be annihilated by local deformation; it is a topological excitation. One can imagine that a regime with a finite density of free vortices will be physically distinct from one without, and that there may be a phase transition between them.

In this section, we will apply RG methods to explore the physics of the anticipated vortex transition. The discussion will illustrate how topological structures in field theory – the subject of chapter 8 – require the introduction of concepts beyond the local fluctuation paradigm emphasized thus far.

two-dimensional
XY-model

We consider the **two-dimensional XY-model** defined by the action

$$\boxed{S[\theta] = -J \sum_{\langle ij \rangle} \cos(\theta_i - \theta_j)} \tag{6.43}$$

where the sum $\sum_{\langle ij \rangle}$ extends over nearest neighbors on a square lattice, and the θ_i are locally defined phase variables. The continuum limit of this action is the two-dimensional nonlinear σ-model with target $O(2) \simeq U(1)$ considered in the previous section. (Assuming $g \equiv \exp(i\theta)$ to be smooth, perform a lowest order gradient expansion of Eq. (6.43) to verify this statement.) However, at present, we are interested in the formation of vortices. At the center of a vortex, its phase winds rapidly (is non-differentiable), so we will not take the continuum limit just yet.

INFO Interpreting θ as the angular variable describing the orientation of planar spins, a natural realization of the model above is two-dimensional magnetism and another is superconductivity:

Consider a system of small superconducting islands connected by tunneling barriers, see the figure schematic. (Such arrays mimic the mesoscopic structure of strongly disordered superconducting materials. However, they can also be realized synthetically, the most prominent realization being superconducting qubit arrays for quantum computing. In that context, each cell realizes a single qubit, and externally controllable coupling between them is used to perform qubit operations.)

Microscopically, each tunnel junction defines a Josephson junction, i.e., a tunnel coupling between two superconducting structures (see problem 5.6.6). If these couplings are not too small, the local number of charges, N_i, will fluctuate strongly. However, this implies that fluctuations of the canonically conjugate phase variable, θ_i are weak (recall that $[\theta_i, \hat{N}_j] = -i\delta_{ij}$ form a conjugate pair). To a zeroth order approximation, one may then neglect dynamical fluctuations (i.e, set $\theta_i = \text{const.}$), and describe the full **Josephson junction array** by the action (6.43).

Josephson junction array

However, two-dimensional magnetism and superconductivity are but two of many realizations of the XY-model in- and outside condensed matter physics.

6.5.1 High- and low-temperature phase

Let us begin by looking at the model in the complementary limits of low and high temperatures, respectively. By **high temperatures**, we mean that the dimensionless parameter J – physically the ratio of some magnetic exchange coupling and temperature – approaches zero. In this limit, the partition function may be represented as a series expansion in J,

$$\mathcal{Z} = \int D\theta \, e^{-S[\theta_i]} = \int D\theta \prod_{\langle ij \rangle} \left(1 + J\cos(\theta_i - \theta_j) + \mathcal{O}(J^2)\right),$$

where $D\Theta = \prod_i d\theta_i$. We may interpret the individual factors in the product as lattice links connecting neighboring sites i and j. Each of these contributes a factor of either unity or, with much lower probability, $J\cos(\theta_i - \theta_j)$. Since $\int_0^{2\pi} d\theta_1 \cos(\theta_1 - \theta_2) = 0$, any graph with a single bond emanating from a site vanishes. On the other hand, a site at which two bonds meet yields a factor $\int_0^{2\pi} (d\theta_2/2\pi) \cos(\theta_1 - \theta_2) \cos(\theta_2 - \theta_3) = (1/2)\cos(\theta_1 - \theta_3)$. The argument shows that the partition function can be presented as a closed "loop" expansion, where individual loops contribute factors $\sim J^{\text{loop length}}$. In a similar manner, we may consider the correlation function between phases at different lattice sites,

$$C(r) \equiv \text{Re}\langle e^{i\theta_i} e^{-i\theta_j} \rangle \sim \left(\frac{J}{2} \right)^r = \exp \left(-\frac{r}{\xi} \right),$$

where r is the lattice distance between sites i and j (the minimal number of links connecting them), and we introduced the correlation length $\xi^{-1} = \ln(2/J)$. The exponential decay is due to the fact that a minimum of r link factors $\sim J$ need to be invested to cancel the otherwise independent phase fluctuations in the correlation function.

Conversely, in the **low-temperature** phase, fluctuations between neighboring phases θ_i are penalized, and we may pass to a continuum representation $\theta_i \to \theta(x)$ governed by the gradient action $(J/2) \int d^2x \, (\nabla \phi)^2$ – the O(2) nonlinear σ-model. Neglecting the fact that the phase θ is defined only mod(2π), it becomes an ordinary real field, and the correlation function may be evaluated as $C(r) = \text{Re}\langle e^{i(\theta(0)-\theta(x))} \rangle = e^{-\langle(\theta(0)-\theta(r))^2\rangle/2}$ (exercise). In two dimensions, Gaussian fluctuations grow logarithmically, $\langle(\theta(0) - \theta(x))^2\rangle/2 = \ln(r/a)/2\pi J$, where a denotes a short-distance cutoff.[28] We thus obtain a power law $C(r) \simeq (a/r)^{\frac{1}{2\pi J}}$, indicating a **quasi-long-range order** in the system.

The qualitatively different decay behavior of the correlation allows for two options: a gradual destruction of order as temperature is increased or a nonzero temperature phase transition. Here, it is important to notice that the argument above was not specific to the O(2) model. For example, with a few technical modifications (think how), it would likewise apply to the O(3) or SU(2) model. The only difference is that, in these cases, the continuum action would not be that of a single real variable, but that of the nonlinear σ-model studied in the previous section. And the RG analysis showed that the former scenario is realized. No matter what the value of J, fluctuations will gradually increase and make the system end up in a disordered phase when looked at from sufficiently large distance scales.

However, in $d = 2$, the O(2) theory is non-committal; its β-function (6.41) vanishes and short-range fluctuations do not alter the result of the continuum estimate. The construction above thus does not tell us how to interpolate between the two physically different regimes. However, there is an "elephant in the room," which we have hitherto neglected: the option of vortex formation or, in other words, the phase periodicity in 2π of the field θ. In the next section, we discuss how this aspect is included in the analysis, and how it leads to the prediction of a phase transition with unique properties.

---

[28] The same correlation function was considered in problem 3.8.10 in a slightly different notation. Consult that problem for a derivation of the logarithm and/or argue on dimensional grounds why no other length dependence is possible.

6.5.2 Vortices and the topological phase transition

In separate and seminal works,[29] Berezinskii, and Kosterlitz and Thouless, proposed that the disordering of the two-dimensional O(2) model is driven by a proliferation of phase vortices. (In recognition of this pioneering work on topological phase transitions, Kosterlitz and Thouless were awarded the 2016 Nobel Prize, shared with Haldane.)

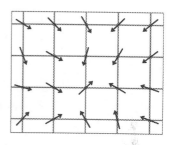

vortex

We thus consider a field configuration $\{\theta_i\}$ for which the phase winds an integer multiple n of a full 2π-rotation as one follows a closed path $\{i\}$ around some anchor point. Such configurations are called **vortices** and the winding number n defines their **topological charge**. The discreteness of the winding number n makes it impossible to find a continuous deformation returning the vortex to a non-winding configuration. Another observation to the same effect is that, at the vortex center, the variable θ_i must vary strongly from site to site: the center of the vortex is not accessible to continuum approximations in terms of a globally continuous variable θ. (Owing to the discontinuous variation of θ_i, a vortex may not be seen in the blink of an eye. Exercise: Find the vortex center of the configuration shown in the figure.) Finally, the sign of n distinguishes between clockwise and anticlockwise winding configurations.

topological defects

INFO Phase vortices are representatives of a wider class of **topological defects**. Other defect structures include domain walls in Ising-like models, or hedgehog-like configurations in three-dimensional Heisenberg models with free rotating spin variables. Besides these, there are many topological defects which are not easy to visualize; for example, winding configurations in field theories with an SU(2) target space. Chapter 8 is devoted to the discussion of topological structures in continuum systems and their description by field-theoretical methods.

Far from a vortex center, θ turns into a smooth variable – strong variations of θ incur an action cost – and continuum methods can be applied to describe the presence of a winding configuration. Specifically, the integral $\oint \nabla\theta \cdot d\mathbf{l} = 2\pi n$ along a path encircling the vortex yields its charge. A model phase configuration producing this winding number upon integration is defined by $\nabla\theta = (n/r)\mathbf{e}_\phi$, where $\mathbf{e}_\phi$ is the azimuthal

> **David J. Thouless 1934–2019**
> was a British theorist known for breakthrough contributions to various areas of condensed matter including the physics of glasses, localization theory and, in particular, topological phenomena. Thouless was among the first to appreciate the importance of topology in solid state critical phenomena, and was awarded the 2016 Nobel Prize (shared with Kosterlitz and Haldane) for his work on topological phase transitions.

[29] V. L. Berezinskii, *Violation of long-range order in one-dimensional and two-dimensional systems with a continuous symmetry group. I. Classical systems*, Sov. Phys. JETP **32**, 493 (1971); J. M. Kosterlitz and D. J. Thouless, *Ordering, metastability, and phase transitions in two-dimensional systems*, J. Phys. C **6**, 1181 (1973).

vector relative to the vortex center. (Notice the inevitable divergence of the derivative at the center.) All other phase configurations representing the vortex may be obtained by adding globally continuous fields to this rotationally symmetric configuration. Without loss of generality, we may thus describe the vortex in terms of the conveniently symmetric configuration. We also note that the vortex contents of a field can be visualized by plotting the "field lines," $\nabla\theta$, away from the centers (see fig. 6.5 for the example of a two-vortex system).

Before approaching the physics of vortices in terms of field theory, it is instructive to sketch an argument due to Kosterlitz and Thouless indicating the presence of a phase transition in the system. The action cost of a single vortex of charge n is the sum of a core contribution, S_n^c, and a contribution from the homogeneous distortions away from the center. Assuming that the boundary between the two regions is defined by a circle of radius a (whose detailed choice is arbitrary), the vortex action is given by

$$S_n = S_n^c + \frac{J}{2}\int_a d^2r\,(\nabla\theta)^2 = S_n^c + \pi Jn^2 \ln\left(\frac{L}{a}\right).$$

The dominant contribution to the action arises from the region outside the core and diverges logarithmically with the system size L.

EXERCISE It is interesting to observe that **defect structures in higher dimensional target spaces** do not lead to divergences. As an example, consider the generalization of the XY-model to three-dimensional spin degrees of freedom, $\mathbf{n} = \mathbf{n}(\theta,\phi)$ with $\mathbf{n}^2 = 1$. Formulate the nonlinear σ-model action $S = \int d^2x\,(\partial\mathbf{n}\cdot\partial\mathbf{n})$ in spherical coordinates. Next invent a spherical topological defect configuration $\mathbf{n}(\mathbf{x})$ winding once around the sphere as a function of the two-dimensional argument $\mathbf{x}$. Show that its action remains finite, even in the case of an infinitely large system size.[30]

This observation is consistent with the absence of a phase transition in two-dimensional nonlinear σ-models with higher dimensional target spaces; defect structures of arbitrary complexity can be created at finite action cost and do not interfere with the flow of these systems into a regime of strong structural disorder.

The large energy cost associated with the defects inhibits their spontaneous formation at low temperatures and protects the integrity of the phase with quasi-long-range order. To find out over what range vortices are suppressed, we consider the partition function of just a single $n = 1$ vortex configuration,

$$\mathcal{Z}_1 \approx \left(\frac{L}{a}\right)^2 \exp\left(-S_1^c - \pi J \ln\left(\frac{L}{a}\right)\right) = \exp\left(-S_1^c - (\pi J - 2)\ln\left(\frac{L}{a}\right)\right). \tag{6.44}$$

Here the factor of $(L/a)^2$ counts the different options for placing the vortex center into the system. Exponentiating that factor as in the second equality produces an entropic factor, logarithmic in system size as is the vortex energy. For temperatures

---

[30] The spherical coordinate representation of the action is given by (ɹoɟ ʎɐ...)... [footnote printed upside-down]

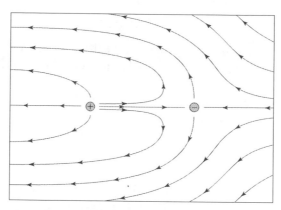

Fig. 6.5   Spin configurations of the two-dimensional XY-model showing vortices of charge ± 1.

exceeding a critical temperature implicitly defined by the condition $J_c = 2/\pi$, that entropy outweighs the energy cost and the creation of a vortex becomes statistically probable.

In fact, the estimate above for the onset of instability is even too conservative: *Pairs* of vortices may appear at lower J, before isolated vortices become affordable. To understand why, consider two oppositely charged ± 1 vortices separated by a distance d (see fig. 6.5). In this case, the individual phase windings tend to cancel out at distances $\gtrsim d$. (Exercise: Describe the pair as the superposition of two oppositely winding configurations and show that $\partial\theta \sim r^{-2}$ far from the center. Discuss the analogy with the field created by an electric dipole.) The integral over this phase profile yields a finite energy, implying that vortex dipoles (whose entropy is even higher than that of individual vortices) will be present at any non-zero temperature.

The same argument also shows that multiply charged vortices are statistically irrelevant. Constructing an $n = 2$ vortex by the superposition of two $n = 1$ vortices, it is evident that its energy cost is $S_2^c \simeq 2 \times S_1^c$. However, by forcing the two centers to sit on top of each other, we loose a factor $2\ln(L/a)$ in entropy. The 2×1 vortex configuration is therefore more likely than 1×2.

The argument above predicts the essential **phenomenology of the Berezinskii–Kosterlitz–Thouless (BKT) transition**. At low temperatures, the system will be in a phase containing tightly bound vortex dipoles, whose number diminishes with lowering temperature. However, beyond a critical temperature, $T_c \sim J_c^{-1}$, entropy favors the unbinding of vortices and a transition into a vortex plasma phase takes place. (For this reason, the BKT transition is sometimes called vortex unbinding transition.)

BKT transition

6.5.3 RG analysis of the BKT transition

The phenomenological model above predicts a phase transition but is too coarse to describes its critical physics. For this purpose, we now engage the more refined machinery of an RG analysis.

Coulomb plasma representation

To prepare for application of the RG, we need to introduce an efficient syntax describing both the vortex contents and the local fluctuations of the phase field. We first note that the gradient form $u_i \equiv \partial_i \theta$ does not capture the vorticity of field configurations (for otherwise, the loop integrals $\oint dl\, u$ could not assume finite values). However, any two-dimensional vector field can be described by the more general representation $u_i = \partial_i \theta + \epsilon_{ij} \partial_j \psi$, where the two scalar fields (θ, ψ) implicitly define the component fields (u_1, u_2). Integrating this ansatz around a vortex center, and using Stokes theorem, we obtain $\pm 2\pi = \oint dl\, u = \int dS \Delta \psi$, where the second integral is over the area bounded by the surrounding curve. Since the choice of the latter is arbitrary (as long as it stays away from the vortex center), we must require $\Delta \psi = \pm 2\pi \delta(r)$, where we assume the center to sit at $r = (r_1, r_2) = (0, 0)$, and the δ-function is smeared over scales comparable to the extension of the core. This two-dimensional Poisson equation is solved by

$$\psi(r) = 2\pi C(r), \qquad C(r) \equiv \frac{1}{2\pi} \ln(|r|), \tag{6.45}$$

where $C(r)$ is the two-dimensional Coulomb potential.

EXERCISE Verifiy this identity either by using the results of problem 3.8.10, or by direct computation. It is useful to regularize the logarithm as $\ln(|r|) = \lim_{\epsilon \to 0} \frac{1}{2} \ln(r^2 + \epsilon^2)$.

On this basis, we may represent the action of a general field configuration away from a set of vortices with charges $\{n_i\}$ at coordiantes $\{r_i\}$ as

$$S[u] = \frac{J}{2} \int d^2 r\, u^2, \qquad u_i = \partial_i \theta_i + \epsilon_{ij} \partial_j \psi, \qquad \psi(r) = \sum_i n_i \ln(r - r_i).$$

An integration by parts shows that

$$S[u] = \frac{J}{2} \int d^2 r (\partial_i \theta \partial_i \theta - \psi \Delta \psi) = \frac{J}{2} \int d^2 r\, (\partial \theta)^2 - 2\pi^2 J \sum_{i,j} n_i n_j C(r_i - r_j).$$

Here, the second term contains a singularity for $i = j$ which, however, need not concern us: associated with the short-distance physics of individual vortices, it can be absorbed into a redefined core energy, $S_c \to S_c + 2\pi^2 J C(0)$. The first term describes the contribution of pure gradient (non-winding) phase fluctuations. We have seen in the previous chapter that these do not lead to interesting physics in the RG sense. We therefore focus on the second term throughout, and describe the system by the partition sum

$$\boxed{Z = \sum_{N=0}^{\infty} \frac{y_0^{2N}}{(N!)^2} \int Dr \exp\left(4\pi^2 J \sum_{i<j} C(r_i - r_j) \right)} \tag{6.46}$$

Here, the sum is over N pairs of oppositely charged vortices,[31] and the integration $\int Dr = \int \prod_i dr_i$ is over the $2N$ center coordinates r_i, where the combinatorial factor $(N!)^{-2}$ prevents overcounting. (Permutations between positive or negative vortices, respectively, lead to identical configurations.) Finally, the **vortex fugacity**, $y_0 = \exp(-S_c)$, contains the core energies.

Notice that Eq. (6.46) implies a remarkable statement: we have reduced the partition function of the XY-model to that of a two-dimensional Coulomb plasma, i.e., a classical system of logarithmically interacting point charges. This mapping underpins the universality of BKT criticality – it applies to a charge plasma in a plane as much as to the XY–model.

EXERCISE In fact, the list of systems in the BKT universality class does not end here. The **sine–Gordon model** defined by the action

$$S[\theta] = \frac{c}{2} \int d^2r \, (\nabla\theta)^2 + g \int d^2r \, \cos\theta \qquad (6.47)$$

also belongs to this class. This constitutes one of the simplest models with a non-polynomial action. It frequently arises in the context of the $(1 + 1)$-dimensional electron gas, where the cosine operator describes potential scattering in the bosonized language of section 3.6.

To see the correspondence with the Coulomb plasma (and hence the vortex system), expand the partition function in powers of g and show that

$$\mathcal{Z} = \sum_{N=0}^{\infty} \frac{y_0^{2N}}{(N!)^2} \prod_{i=1}^{2N} \int d^d r_i \left\langle \exp\left(i \sum_{i=1}^{2N} (-)^i \theta(r_i) \right) \right\rangle,$$

where $y_0 = g/2$ and the angle brackets denote averaging over the free action. (To prove that positive and negative phases appear in equal numbers, consider the role of the integration over the zero mode $\theta(r) = \text{const.}$) Using the fact that $\langle \theta(r)\theta(r') \rangle = C(|r - r'|)/c$, and neglecting the infinite self-interaction of the fields at coinciding points, show that the partition function becomes identical to Eq. (6.46) when $c = 1/(8\pi^2 J)$.

In the continuum representation of the sine-Gordon model, the BKT transition becomes accessible to momentum shell RG methods quite similar to those applied in section 6.1.2. For a detailed discussion of this particular RG analysis, we refer to the text by Gogolin, Nersesyan, and Tsvelik referenced in footnote [34] of chapter 8.

Perturbation theory

We have already mentioned that, in the language of the Coulomb plasma, the BKT transition describes the unbinding of tightly bound dipoles into a gas of charges. However, to fully understand this phenomenon, we must take into account that the "bare" logarithmic interaction between charges is subject to *screening*. The quantitative description of this mechanism is a job for an RG analysis. However, unlike the momentum shell RG discussed in previous sections, this time it will be more natural to work in real space.

---

[31] In the thermodynamic limit, configurations violating overall charge neutrality would cost divergent energy and hence may be neglected.

To begin, let us compute the effective interaction between two external charges at r and r' perturbatively in the fugacity y_0. To lowest order, the bare interaction is modified by the presence of two internal charges, positioned at s and s', and the effective interaction assumes the form

$$e^{-S_{\text{eff}}(r-r')} \equiv \langle e^{-4\pi^2 JC(r-r')} \rangle_t \simeq$$

$$\frac{(r \oplus \!\!-\!\!-\!\! \ominus r') + y_0^2 \int d^2s\, d^2s' \left(\begin{array}{ccc} s \oplus & \!\!-\!\!-\!\! & \ominus s' \\ | & \times & | \\ r \oplus & \!\!-\!\!-\!\! & \ominus r' \end{array} \right)}{1 + y_0^2 \int d^2s\, d^2s'\, (s \oplus \!\!-\!\!-\!\! \ominus s')},$$

where we consider only terms up to $\mathcal{O}(y_0^2)$ and the dashes represent the interaction between charges. Susbsitution of the corresponding inteaction terms leads to

$$e^{-S_{\text{eff}}(r-r')+4\pi^2 JC(r-r')} \simeq \frac{1 + y_0^2 \int d^2s\, d^2s'\, e^{-4\pi^2 JC(s-s')+4\pi^2 JD(r,r',s,s')}}{1 + y_0^2 \int d^2s\, d^2s'\, e^{-4\pi^2 JC(s-s')}}$$

$$\simeq 1 + y_0^2 \int d^2s\, d^2s'\, e^{-4\pi^2 JC(s-s')} \left(e^{4\pi^2 JD(r,r',s,s')} - 1 \right), \qquad (6.48)$$

where $D(r,r',s,s') = C(r-s) - C(r-s') - C(r'-s) + C(r'-s')$ denotes the interaction between the internal and external dipoles, while the direct interaction $C(s-s')$ tends to keep the separation $x = s' - s$ small. Defining the center of mass $X = (s+s')/2$, we can change variables to $s = X - x/2$ and $s' = X + x/2$, and expand the dipole–dipole interaction in small x as $D(r,r',s,s') \simeq -x \cdot \nabla_X C(r - X) + x \cdot \nabla_X C(r' - X) + \mathcal{O}(x^3)$. To the same order

$$e^{4\pi^2 JD(r,r',s,s')} - 1 \simeq -4\pi^2 Jx \cdot \nabla_X (C(r - X) - C(r' - X))$$
$$+ 8\pi^4 J^2[x \cdot \nabla_X(C(r - X) - C(r' - X))]^2 + \mathcal{O}(x^3).$$

Substituting this expression into Eq. (6.48), and changing integration variables, one finds that the term linear in x integrates to zero while the angular average of $(x \cdot \nabla_X C)^2$ leads to $x^2 (\nabla_X C)^2/2$. Thus, to $\mathcal{O}(r^4)$, one obtains

$$e^{-S_{\text{eff}}(r-r')} \simeq e^{-4\pi^2 JC(r-r')}$$
$$\times \left(1 + y_0^2 \int d^2x\, e^{-4\pi^2 JC(x)} 8\pi J^2 \frac{x^2}{2} \int d^2X\, (\nabla_X(C(r - X) - C(r' - X)))^2 \right).$$

Using the identity $\nabla^2 C(r) = \delta^2(r)$, integration by parts shows that $\int d^2X\, [\nabla_X (C(r-X)-C(r'-X))]^2 = 2(C(r-r')-C(0))$. Finally, absorbing the short-distance divergence into an appropriate cutoff with $C(x) \to \ln(|x|/a)/2\pi$, one arrives at the expression

$$e^{-S_{\text{eff}}(r-r')} = e^{-4\pi^2 JC(r-r')} \left(1 + 16\pi^5 J^2 y_0^2 C(r - r') \int_1^\infty ds\, s^3\, e^{-2\pi J \ln s} + \mathcal{O}(y_0^4) \right),$$

where $s = |x|/a$. Exponentiating the second term, one obtains the effective interaction $S_{\text{eff}}(r - r') \simeq 4\pi^2 J_{\text{eff}} C(r - r')$, where

$$J_{\text{eff}} = J - 4\pi^3 J^2 y_0^2 \int_1^\infty ds\, s^{3-2\pi J} + \mathcal{O}(y_0^4). \tag{6.49}$$

effective dielectric constant

According to this result, the perturbative inclusion of screening charge–anti-charge pairs leads to a gradual suppression of their interaction, or a deviation of the **effective dielectric constant** of the medium $\varepsilon = J/J_{\text{eff}}$ away from the vacuum value of unity. However, as long as the integral over the screening charge separation, sa, remains at large values, that correction is small. The breakdown of perturbation theory at $J \sim J_c = 2/\pi$ occurs at the point where the free energy of an isolated charge changes sign.

Renormalization

real space renormalization

The difficulty associated with the divergence at small J is overcome by a **real space renormalization** procedure, introduced by José *et al.*[32] The idea is to break the integral in Eq. (6.49) into two parts, $\int_1^\infty \to \int_1^b + \int_b^\infty$, and absorb the non-singular short-distance contribution into a redefined $J \to \tilde{J}$. This procedure is carried out order by order in y_0 even though the full integral multiplying y_0^2 remains formally divergent. As a result, one obtain a new equation $J_{\text{eff}}^{-1} = \tilde{J}^{-1} + 4\pi^3 y_0^2 \int_b^\infty ds\, s^{3-2\pi J}$, where $\tilde{J}^{-1} = J^{-1} + 4\pi^3 y_0^2 \int_1^b ds\, s^{3-2\pi J}$. The variable in the remaining integral is now rescaled $(s \to s/b)$ to yield an equation for J_{eff}^{-1} equivalent to Eq. (6.49) but with renormalized parameters J and y_0:

$$J_{\text{eff}}^{-1} = \tilde{J}^{-1} + 4\pi^3 \tilde{y}_0^2 \int_1^\infty dx\, x^{3-2\pi \tilde{J}},$$

where $\tilde{y}_0 = b^{2-\pi J} y$. For an infinitesimal renormalization step, $b = e^\ell \approx 1 + \ell$, one obtains the differential recursion relations

$$\frac{dJ^{-1}}{d\ell} = 4\pi^3 y_0^2 + \mathcal{O}(y_0^4), \tag{6.50}$$

$$\frac{dy_0}{d\ell} = (2 - \pi J)y_0 + \mathcal{O}(y_0^3). \tag{6.51}$$

These equations predict a monotonic increase of the (inverse) coupling J^{-1} in ℓ, while the recursion relation for y_0 changes sign at $J_c^{-1} = \pi/2$. At high temperatures, when J is small, y_0 is relevant, while at lower temperatures it becomes irrelevant. The RG flows, shown in fig. 6.6(a), separate the parameter space into two regions. At low temperatures, and small y_0, flows terminate on a fixed line at $y_0 = 0$ and $J_{\text{eff}} \geq 2/\pi$. This is the insulating phase, in which only dipoles of finite size occur (hence the vanishing of y_0 under coarse-graining). The strength of the effective interaction is given by the point on the fixed line at which the flow terminates. Flows that do not terminate on the fixed line asymptote to larger values of J^{-1} and y_0, where perturbation theory eventually breaks down. This is the signal of the high temperature phase, where charges dissociate.

[32] J. V. José, L. P. Kadanoff, S. Kirkpatrick, and D. R. Nelson, *Renormalization, vortices, and symmetry-breaking perturbations in the two-dimensional planar model*, Phys. Rev. B **16**, 1217 (1977).

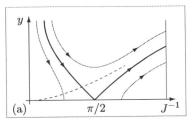 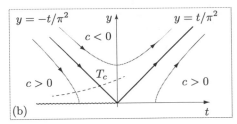

Fig. 6.6   RG flow diagram of the XY-model (a) far from the critical point and (b) close to the critical point.

The critical trajectory separating the two regions of the phase diagram flows to a fixed point ($J_c^{-1} = \pi/2$, $y_0 = 0$). To explore the critical behavior at the transition, we expand the recursion relations in the vicinity of this point. Setting $t = J^{-1} - \pi/2$, Eq. (6.51) simplifies to

$$\frac{dt}{d\ell} = 4\pi^3 y^2 + \mathcal{O}(ty^2, y^4),$$

$$\frac{dy}{d\ell} = \frac{4}{\pi}ty + \mathcal{O}(t^2 y, y^3). \tag{6.52}$$

RG flow

In contrast with the flow equations discussed in previous sections, Eqs. (6.52) remain nonlinear in the vicinity of the fixed point (see fig. 6.6b). To understand these nonlinearities, we first note that $t^2 - \pi^4 y^2$ is a conserved quantity. As a consequence, the flow proceeds along hyperbolae characterized by different values of $c \equiv t^2 - \pi^4 y^2$. For $c < 0$, the focus[33] of the hyperbola is on the y-axis, and the flows proceed to $(t, y) \to \infty$. Conversely, hyperbolae with $c > 0$ have foci on the t-axes, and have two branches in the half-plane $y \geq 0$: the branches for $t < 0$ flow to the fixed line, while those in the $t > 0$ quadrant flow to infinity. The critical trajectory separating flows to zero and to infinite y corresponds to $c = 0$, or $t_c = -\pi^2 y_c$. Therefore, a small but non-zero fugacity y_0 reduces the critical temperature to $J_c^{-1} = \pi/2 - \pi^2 y_0 + \mathcal{O}(y_0^2)$.

low-temperature phase   In terms of the original XY-model, the **low-temperature phase** is characterized by a line of fixed points with $J_{\text{eff}} = \lim_{\ell \to \infty} J(\ell) \geq 2/\pi$. Here the phase correlations decay as a power law, i.e., $\langle \cos(\theta(r) - \theta(0)) \rangle \sim |r|^{-\eta}$, with $\eta = 1/2\pi J_{\text{eff}} \leq 1/4$. Since the parameter c is negative and vanishes at the critical point, we can define $c = -b^2(T_c - T)$ close to the transition with a constant b. When described in this way, the trajectories of initial points track hyperbolic lines $(t(T), y(T))$ whose foci $c = t^2 - \pi^4 y^2 \propto (T_c - T)$ define a linear measure of the proximity of the phase transition. In the thermodynamic limit, $\ell \to \infty$, they end up at fixed

---

[33] Recall that the foci of hyperbolae are the coordinates where they intersect the coordinate axes. Depending on the sign of c, the hyperbolae are oriented in the horizontal or vertical directions, as indicated in fig. 6.6.

points $(y, t) = (0, -b\sqrt{T_c - T})$. Thus, in the vicinity of the transition, the effective interaction parameter,

$$J_{\text{eff}} = \frac{2}{\pi} - \frac{4}{\pi^2} \lim_{\ell \to \infty} t(\ell) = \frac{2}{\pi} + \frac{4b}{\pi^2} \sqrt{T_c - T},$$

exhibits a square root singularity.

INFO The stiffness J_{eff} can be measured directly in **experiments on superfluid films**. We saw in section 5.2.4 that the neutral superfluid is described by a phase action whose stiffness, $J = T^{-1} \rho_s / m^2$, contains the superfluid density, ρ_s. This quantity is measurable by examining the changes in the inertia of a torsional oscillator; the superfluid fraction, ρ_s, experiences no friction and does not oscillate.

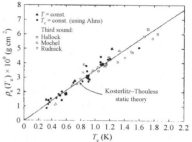

The figure[34] shows data for the superfluid density at the critical temperature taken for different superfluid ^{4}He films. These experiments confirm a universal jump in J of $2/\pi$ at the critical point, and scaling consistent with a square root singularity for $T < T_c$. The universality of the ratio $\rho_s(T_c)/T_c = J_c m^2 = 2m^2/\pi$ visible in the figure demonstrates that $J_c = 2/\pi$ is independent of any material parameters.

high-temperature phase

In the **high-temperature phase**, correlations decay exponentially. The parameter $c = t^2 - \pi^4 y^2 = b^2(T - T_c)$ characterizing individual hyperbolic trajectories is now positive. The recursion relation $d_\ell t = 4\pi^3 y^2 = (4/\pi)(t^2 + b^2(T - T_c))$ can be integrated to give

$$\frac{4}{\pi} \ell \simeq \frac{1}{b\sqrt{T - T_c}} \arctan\left(\frac{t}{b\sqrt{T - T_c}} \right).$$

The integration must be terminated at values $t(\ell) \sim y(\ell) \sim 1$ beyond which the near fixed point theory loses its validity. This occurs for a value $\ell^* \approx \pi^2/(8b\sqrt{T - T_c})$, where we have approximated $\arctan(1/b\sqrt{T - T_c}) \approx \pi/2$. The resulting correlation length is then given by

$$\xi \approx a e^{l^*} \approx a \exp\left(\frac{\pi^2}{8b\sqrt{T - T_c}} \right). \tag{6.53}$$

Notably, the correlation length diverges but, unusually, not as a power law – again a consequence of the nonlinearity of the flow equations. From here, we obtain a reduced free energy

$$f_{\text{sing}} \propto \xi^{-2} \propto \exp\left(-\frac{\pi^2}{4b\sqrt{T - T_c}} \right), \tag{6.54}$$

whose derivatives with respect to temperature at the transition are all finite. In particular, the second derivative – the heat capacity – remains non-singular, which is again unusual for a second-order phase transition.

[34] Reprinted with permission from D. J. Bishop and J. D. Reppy, *Study of the superfluid transition in two-dimensional ^{4}He films*, Phys. Rev. Lett. **40**, 1727 (1978). Copyright (1978) by the American Physical Society.

6.6 Summary and Outlook

In this chapter we have introduced renormalization group methods as a powerful option for the understanding of theories beyond perturbation theory. The focus here has been on applications in quantum statistical mechanics: we introduced real space renormalization, tailored to spin models or local defect structures, and different versions of momentum space renormalization. We discussed concepts of renormalized perturbation theory, such as loop and ϵ-expansion, and how the flow equations obtained by RG methods characterize the critical physics of phase transitions. Differently geared introductions to the RG might have put more emphasis on the question of renormalizability, alternative regularization schemes, renormalization through counter-terms, or the triumphs that renormalization group methods have had in QED and QCD. However, we hope that the present discussion hs been substantial enough to convince the reader of the power and versatility of RG methods and that it will motivate further study.

At this point, we have introduced most of the basic concepts of field theory in condensed matter physics. The final missing piece is the discussion of concepts interfacing between field integrals and observable quantities via correlation functions. This will be the subject of the next chapter, which will conclude the introductory part of this text.

INFO Before leaving this chapter one final comment on the role of symmetries in critical phenomena is due. We have seen the importance of the scale invariance of theories close to phase transitions – both methodologically in the execution of the RG program and conceptually in the manifestations of critical phenomena. However, it can be shown that, under quite mild conditions, critical theories enjoy an even richer set of symmetries, namely **conformal** invariance under **conformal transformations**. Conformal transformations are mappings
symmetry of space(–time) which are locally angle preserving. Besides scale transformations, they include translation, rotation, or angle preserving inversions of space. In the particular case of two-dimensional theories, the group of conformal transformations is even infinite dimensional, and conformal symmetry is a principle so rich that it almost fully determines the contents of a theory.

Readers wishing to learn more about conformal invariance are invited to take a look at the nutshell introduction in appendix section A.3, or to consult one of the many available texts on the subect.[6]

6.7 Problems

6.7.1 Dissipative quantum tunneling: strong potential limit

In section 6.1.2 we considered a quantum particle subject to a periodic potential and dissipative damping. We applied a combination of perturbation theory and renormalization to show that the potential will grow if the disispation is sufficiently strong. Eventually, however, it will become so large that the perturbative treatment is no longer valid. In this problem we consider the situation in this high-potential limit. As a result, we obtain the upper portion of the flow diagram discussed in section 6.1.2.

Let us consider the action (6.11) in the limit where the potential strength c is large in comparison to the high-frequency cutoff Λ. To implement the condition $\omega < \Lambda$, we add a kinetic energy term $S_{\text{kin}}[\theta] = (l^2 m/2) \int d\tau\, \dot{\theta}^2$ to the action, where l is a parameter of dimensionality [length]. The mass parameter is chosen such that, at $\omega_n = \Lambda$, the kinetic energy term becomes of the same order as the dissipative term, and for larger frequencies dominates. Its strong frequency dependence, $\dot{\theta}^2 \to \omega^2 \theta(\omega)$ then effectively regularizes all frequency integrals. (Consider the fast field integrals of the RG step to convince yourself that this is true.)

We thus start out from the generalization of Eq. (6.11),

$$S[\theta] = \frac{1}{4\pi g} \int (d\omega)\, |\omega|\, |\theta(\omega)|^2 + \int d\tau \left(\frac{ml^2}{2} \dot{\theta}^2 + \cos(\theta(\tau)) \right).$$

Assuming the dissipation term to be weak, let us consider the variational equations in the $g \to \infty$ limit: $ml^2 d_\tau^2 \bar{\theta} + \sin \bar{\theta} = 0$, i.e., the equation of the mathematical pendulum in imaginary time. Recalling that these equations describe the real-time problem in an effectively inverted potential (section 3.3.1), the relevant stationary phase configurations contain extended periods where the particle rests at the maxima of the potential. These resting periods are interspersed by instanton events, where it quickly rolls from one maximum to another.

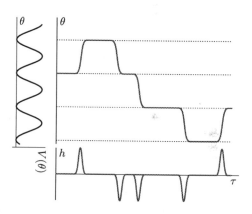

Throughout, it will be convenient to change variables to $h \equiv d_\tau \bar{\theta}$ (see the bottom part of the figure). For a configuration containing n_+ instanton events and n_- anti-instanton events, $h(\tau) = \sum_{i=1}^{n} e_i f(\tau - \tau_i)$, where τ_i is the time of the event, $e_i = +1\,(-1)$ for an instanton (anti-instanton), $n = n_- + n_+$, and $f(\tau)$ is a function that is peaked around zero, has a width $\sim c$, and integrates to $2\pi = \int_{-\infty}^{\infty} d\tau\, f(\tau) = h(\infty) - h(-\infty)$.

(a) Show that the action of a general instanton trajectory is given by

$$S[\bar{\theta}] = \frac{1}{4\pi g} \int (d\omega) \sum_{i,j=1}^{n} e_i e_j e^{-i\omega(\tau_i - \tau_j)} \frac{|f(\omega)|^2}{|\omega|} + nS_{\text{inst}},$$

where S_{inst} is the action of a single instanton event and we neglect fluctuation determinants. Then, apply a Hubbard–Stratonovich transformation to bring the partition function to the form

$$\mathcal{Z} = \sum_n \frac{1}{n!} \prod_{i=1}^n \int d\tau_i \sum_{e_i = \pm 1} e^{-nS_{\mathrm{inst}}} \int Dq \, e^{-\frac{g}{4\pi} \int (d\omega)|q(\omega)|^2|\omega|} \, e^{i \sum_i e_i q(\tau_i)}.$$

You may use the fact that, in the frequency range of interest, $f(\omega) \simeq f(0) = 2\pi$.
(c) Sum over all configurations $\{e_i\}$ to obtain

$$\mathcal{Z} = \int Dq \exp\left(-\frac{g}{4\pi T} \sum_{\omega_m} |q_m|^2 |\omega_m| + \gamma \int d\tau \cos(q(\tau)) \right), \qquad (6.55)$$

where $\gamma = 2e^{-S_{\mathrm{inst}}}$.

Intriguingly, we have arrived at a functional integral whose action is structurally equivalent to the starting point (6.11). The only difference is the changes in the coupling constants, $g \to g_{\mathrm{dual}} \equiv g^{-1}$ and $c \to \gamma = 2e^{-S_{\mathrm{inst}}}$. This is an example of **self-duality** of a field theory. In the present realization, large values of the potential c map onto a small potential in the dual representation; dissipation strengths $g > 1$ map into $g_{\mathrm{dual}} < 1$. Use this information to convince yourself of the validity of the upper portion of the flow diagram discussed in section 6.1.2.

self-duality

Answer:

(a) This result is obtained by substitution of $\bar{\theta}_m = -if_m\omega_m^{-1}\sum_i e_i e^{i\omega_m \tau_i}$ into the action. The decoupling of the $e_i e_j$-term by an auxiliary field $q(\omega)$ is achieved by standard Gaussian integration. In the partition function, we need to integrate over all intermediate time coordinates τ_i, and sum over all sign configurations $\{e_i\}$.
(c) Reorganizing terms we obtain

$$\mathcal{Z} = \int Dq \, e^{-\frac{g}{4\pi} \int (d\omega)|q(\omega)|^2|(\omega)|} \sum_n \frac{1}{n!} \left(e^{-S_{\mathrm{inst}}} \sum_{e=\pm 1} \int d\tau \, e^{ieq(\tau)} \right)^n$$

$$= \int Dq \, e^{-\frac{g}{4\pi} \int (d\omega)|q(\omega)|^2|(\omega)|} \sum_n \frac{1}{n!} \left(\gamma \int d\tau \cos(q(\tau)) \right)^n,$$

and, upon resummation, the result is Eq. (6.55).

6.7.2 Quantum criticality

quantum phase transition

A **quantum phase transition** is a qualitative change of the ground state of a quantum system in response to the variation of an external parameter. If the order parameter characterizing the transition remains continuous, power law singularities and critical phenomena are to be expected. In this problem, we discuss what sets these zero temperature quantum phase transitions apart from their non-zero temperature classical cousins.

We have frequently seen how d-dimensional quantum systems bear similarity with classical $(d + 1)$-dimensional systems – the imaginary-time coordinate adds one dimension, limited to a finite range $[0, \beta]$ at finte temperatures. However, in most applications, time and space enter the action in an anisotropic manner, and their scaling properties must be discussed separately. This intrinsic anisotropy is at the origin of the phenomena characterizing *quantum criticality*. The theoretical phenomenology of the quantum critical system was formulated by Hertz[35] soon after the development of the renormalization group method. However, it took much longer before experimentalists were able to probe quantum criticality in, for example, the context of metallic magnetism in heavy fermion compounds.

heavy fermion materials

INFO **Heavy fermion materials** contain rare-earth elements such as Ce or Yb, or actinide elements such as U (examples including UBe_{13}, $CeCu_2Si_2$, and many more). The effective masses of their inner shell conduction electrons often exceed the bare electron mass by two orders of magnitude and, as a consequence, the Fermi energy is unusually small (exercise). At low temperatures, many of these materials are magnetically ordered, others show strong paramagnetic behavior, and some unconventional superconductivity.

In this problem, we will consider quantum criticality in itinerant magnetism. Our starting point is the effective action of a magnetization field, m, in a Fermi liquid, assumed uniaxial, or scalar for simplicity. In problem 5.6.7 we found that, for low frequencies, $|\omega_n| \lesssim \Gamma_q$ (see Eq. (5.69)),

$$S[m] = \frac{T}{2} \sum_n \int (dq) \left(\delta + \mathbf{q}^2 + \frac{|\omega_n|}{\Gamma_q} \right) |m_q|^2 + \frac{u}{4} \int dx\, m^4(x).$$

In this expression, the lattice spacing is set to unity and hence $\mathbf{q}$ is dimensionless, with the q-integration cut off at $\Lambda = 1$. In problem 5.6.7 we considered ferromagnetic correlations, for which $\Gamma_q \sim v_F |\mathbf{q}|$. However, here it will be instructive to compare with the antiferromagnetic case, for which one can show that $\Gamma_q = \text{const}$. Finally, the parameter r measures the distance to the quantum critical point.

In the limit of high temperature, fluctuations are dominantly static, $m(\tau) = \text{const.}$, and the problem reduces to that of ϕ^4 theory. Recall that in dimensions dimensions $1 > d > d_u$, and for $\delta < 0$, we observed the formation of a fixed point with finite magnetization and nontrivial critical fluctuations around it. Drawing on the seminal works of Hertz[35] and Millis[36] our aim here is to explore the role played by quantum fluctuations in this setting.

(a) As a warm-up exercise, show that the free energy of the **Gaussian theory**, $\delta > 0$, $u = 0$, takes the form

$$F_{\text{Gauss}} = 2L^d \int (dq) \int_0^{\Gamma_q} (d\omega) \coth\left(\frac{\omega\beta}{2}\right) \tan^{-1}\left(\frac{\omega/\Gamma_q}{\delta + \mathbf{q}^2}\right)$$

[35] J. A. Hertz, *Quantum critical phenomena*, Phys. Rev. B **14**, 1165 (1976).
[36] A. J. Millis, *Effect of a non-zero temperature on quantum critical points in itinerant fermion systems*, Phys. Rev. B **48**, 7183 (1993).

up to a formally divergent temperature-independent constant. (Hints: To extract the δ-dependent contribution above, first compute the derivative $\partial_\delta F$, to later regain the δ-sensitive part of F by integration, $\int d\delta\, \partial_\delta F$.)

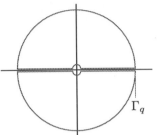

For the Matsubara summation, work with the integration contours shown in the figure. The integral along the small semicircles cancels against the contribution of the zero Matsubara frequency, ω_0, to the sum (why?). Neglect the contribution from the closing semicircles at $|\omega| \sim \Gamma_q$.

(b) To guide our analysis of the non-Gaussian theory, it is helpful to develop the RG on the Gaussian model first. As usual, we start with the introduction of a fast/slow field separation scheme. Here, the presence of independent frequency and coordinate axes complicates matters somewhat compared with earlier studies. Where momentum is concerned, the fast layer will be defined as $\Lambda/b < |\mathbf{q}| < \Lambda$ as usual. Frequencies, on the other hand, are cut off at $\omega \simeq \Gamma_q = \Gamma_q(\mathbf{q})$. We thus declare that frequencies $\Gamma_q/b^{-z} < |\omega| < \Gamma_q$ are fast and all others slow. Here, the presence of the dynamical exponent z reminds us that frequencies and momenta may be scaled differently. The anisotropic dissection of the coordinate domain defines the fast wedge indicated in the figure.

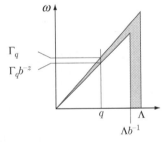

Perform the RG step for the Gaussian theory, and then a rescaling such that the coefficients of the $\mathbf{q}^2$ and the $|\omega|/\Gamma_q$ terms remain constant. To facilitate the bookkeeping, represent the Matsubara sum as an integral $T \sum_n \to \int (d\omega)$. Then rescale as $\mathbf{q}' = z\mathbf{q}, \omega' = b^z\omega, m'_q = b^{-d_m}m_q$, and determine the dynamical exponent z and the field renormalization d_m needed to achieve the above constancy. Determine the scaling dimension y_δ of the δ-term.

Finally show that the quartic interaction scales with $y_u = 4 - d - z$, so that $d_u = 4 - z$ is the **upper critical dimension** of the model.

upper critical dimension

(c) The inclusion of the interaction term generates fast-slow field coupling $S_c[m_s, m_f] = (3u/2) \int dx\, m_s^2 m_f^2$. We represent the RG step as

$$\mathcal{Z} = \int Dm_s\, e^{-S[m_s]} \langle e^{-U[m_s, m_f]} \rangle_f,$$

where $\langle \cdots \rangle_f$ denotes the fast field integration over the quadratic action $S_f[m_f]$. Perform this integral to first order in an expansion in u.[37] For the ensuing loop integrals over the fast field propagator, use the results obtained in part (a) and take the particular form of the fast momentum/frequency integration domain into account. Finally rescale as in (b) to obtain the the RG equations

---

[37] Note that this is different from the one-loop expansion performed earlier for the ϕ^4-model. The latter included terms of $\mathcal{O}(u^2)$.

$$\frac{dT}{d\ln b} = zT,$$

$$\frac{du}{d\ln b} = (4 - d - z)u,$$

$$\frac{d\delta}{d\ln b} = 2\delta + uf(T), \tag{6.56}$$

where

$$f(T) = c\int_0^1 \frac{ds}{\pi}\left(\coth\left(\frac{s}{2T}\right)\frac{s}{1+s^2} + z\coth\left(\frac{s^{z-2}}{2T}\right)\frac{s^{d-3+z}}{s^4+1}\right)$$

and c is a numerical constant. The first and second of these equations say that temperature $T = T(b)$ and interaction constant, $u = u(b)$ scale according to their engineering dimension at lowest order in perturbation theory. The third equation describes the change in the parameter $\delta = \delta(b)$ measuring the distance from the quantum critical point in terms of a function $f(T)$. The complicated-looking structure of that function reflects the joint influence of fluctuations in spatial and in temporal directions on the flow of the coupling constant.

(d) Solve Eqs. (6.56) to obtain

$$T(b) = Tb^z,$$

$$u(b) = ub^{4-d-z},$$

$$\delta(b) = \delta b^2 + b^2 u \int_0^{\ln b} dx\, e^{(2-d-z)x} f(Te^{xz}), \tag{6.57}$$

where (T, u, δ) defines the starting configuration of the flow.

The function f governing the flow equations (6.56) contains a wealth of information on the nature of the quantum critical point. Referring for their full analysis to Ref.[36], we summarize here a few conclusions.

quantum phase transition

At $T = 0$, a **quantum phase transition** occurs at a critical value $\delta = \delta_c$, where

$$\delta_c = \frac{uf(0)}{z + d - 2} \tag{6.58}$$

is proportional to the interaction coefficient (and hence close to zero in the present linearized theory). The critical point separates an ordered phase with finite magnetization ($\delta < \delta_c$) from a disordered one ($\delta > \delta_c$).

Focusing on the disordered side of the transition (the analysis of the ordered side is more compli-

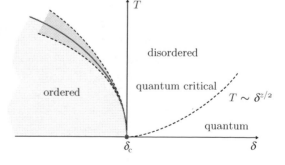

cated and will not be addressed here), $\delta > \delta_c$, both δ and T increase under renormalization. Depending on which one becomes large first, $\delta \sim 1$ or $T \sim 1$, we expect different behaviors.

To find the demarcation line between these two different regimes, we substitute $1 \sim \delta(b) \sim \delta b^2$ into $T(b) \sim Tb^z \sim T\delta^{-z/2}$. If $T(b) \ll 1$ when $\delta(b)$ reaches unity, i.e., the starting value $T \ll \mathrm{const.} \times \delta^{z/2}$, the critical properties of the transition are largely identical to those of the $T = 0$ quantum theory. This is the **quantum regime**, or $T = 0$ regime.

We may identify the scale $\xi \sim b \sim \delta^{-1/2}$ at which $\delta(b) \sim 1$ becomes large as the correlation length of the model and $\xi_\tau \sim \delta^{-z/2}$ as the corresponding correlation time. For temperatures larger than the crossover scale above, the extension of the system in the imaginary-time direction, we have $T^{-1} < \xi_\tau$. The system looks strongly correlated at all scales, as if it were at the critical point. This temperature range defines the regime of **quantum criticality**. Notice the counterintuitive correspondence (large temperatures) $\leftrightarrow$ (quantum criticality).

Specifically, for temperatures $T \gg 1$, the function $f(T) \simeq \tilde{c}$, with $\tilde{c}$ a numerical constant. We may then introduce the product $v = uT$ and convince ourselves (do it) that the scaling equations reduce to

$$\frac{dv}{d\ln b} = (4 - d)v,$$

$$\frac{d\delta}{d\ln b} = 2\delta + \tilde{c}v.$$

These equations are structurally identical to the (linearization of the) flow equations of ϕ^4-theory, Eq.(6.31). The interpretation of this finding is that, at high temperatures, the criticality of our system reduces to that of the classical ferromagnet.

The discussion of the ensuing critical phenomena requires matching the flow in the $T \gg 1$ regime with that at lower temperatures $T \ll 1$. Referring for a detailed discussion to Ref.[36], we note that the system supports a line of classical (finite temperature) phase transitions in the ϕ^4-class at a critical temperature

$$T_c \propto \left(\frac{\delta_c - \delta}{u}\right)^{z(d+z-2)^{-1}}.$$

This line terminates in the quantum critical point at $T_c = 0$. It also turns out that the critical line lies inside a narrow region (see the dashed line) in which the theory becomes essentially non-Gaussian, i.e., the interaction u cannot be treated in lowest order perturbation theory. Inside that region, the previous analysis of the classical ferromagnetic transition applies.

The above characterization of the phase diagram may be translated to measurable predictions of, e.g., the specific heat of quantum magnetic materials. For a discussion of this and other experimental signatures we refer to the huge body of literature on quantum magnetism.

Answer:

(a) For the Gaussian theory, integration over $m(q)$ produces a determinant $\mathcal{Z}$ assuming the form of a product over values $q = (\mathbf{q}, \omega_n)$. Taking the logarithm to obtain the free energy $F = -T \ln \mathcal{Z}$, and then the δ-derivative, one finds

$$\partial_\delta F = L^d \int (dq) T \sum_{\omega_n} \frac{1}{\delta + q^2 + |\omega_n|/\Gamma_q} = L^d \int (dq) P \int_{-\Gamma_q}^{\Gamma_q} \frac{d\omega}{2\pi i} \frac{\coth(\beta\omega/2)}{\delta + q^2 - i\omega/\Gamma_q}$$

$$= -L^d \int (dq) \int_0^{\Gamma_q} \frac{d\omega}{\pi} \frac{\coth(\beta\omega/2)\, \omega/\Gamma_q}{(\delta + q^2)^2 + (\omega/\Gamma_q)^2}.$$

To arrive at the first integral, we have integrated over the contour described in the problem. Using the identity $\partial_\delta \tan^{-1}\left(\frac{\omega/\Gamma_q}{\delta + q^2}\right) = -\frac{\omega/\Gamma_q}{(\delta+q^2)^2+(\omega/\Gamma_q)^2}$, we obtain the required result.

(b) In the Gaussian theory, fast and slow fluctuations decouple, and so the RG step is trivial. The remaining slow field action is given by

$$S[m_{\rm s}] = \frac{1}{2} \int^{\Lambda/b} (dq) \int (d\omega) \left(\delta + \mathbf{q}^2 + \frac{|\omega|}{\Gamma_q}\right) |m_{\rm s}(q)|^2,$$

where the cutoff dependence of the frequency integral is left implicit. Now rescale to obtain

$$S[m] = \frac{b^{-d-z}}{2} \int^\Lambda (dq) \int (d\omega) \left(\delta + \mathbf{q}^2 b^{-2} + \frac{|\omega|}{\Gamma_q} b^{-z+1}\right) |m(q)|^2 b^{2d_m}. \qquad (6.59)$$

In the ferromagnetic case, $\Gamma_q \sim |\mathbf{q}|$, the $\mathbf{q}^2$ term and the $|\omega|/\Gamma_q$ term remain invariant if and only if $d + 2 + z - 2d_m = 0$ and $d + 2z - 1 - 2d_m = 0$, respectively. This requires $d_m = (d + 2 + z)/2$ with $z = 3$. For the antiferromagnetic case, we obtain $z = 2$. The δ-term picks up a scaling factor b^2, so $y_\delta = 2$. Similarly, the dimension of the operator $u \int dx\, m^4$ is given by $y_u = -3(d+z) + 4d_m = -d - z + 4$. Here, the factor 3 represents the three frequency/momentum integrals that are equivalent to a single coordinate integral.

(c) The first term in the u-expansion, $(3u/2) \int dx\, m_{\rm s}^2 \langle m_{\rm f}^2 \rangle$, gives rise to a renormalization of the coefficient δ. Evaluating this contribution, and expressing the frequency summation by a contour integral as in (a), we obtain

$$\langle m_{\rm f}^2 \rangle = 2 \int_{\partial\Lambda} (dq)(d\omega) \coth\left(\frac{\omega\beta}{2}\right) \frac{\omega/\Gamma_q}{(\delta + \mathbf{q}^2)^2 + (\omega/\Gamma_q)^2},$$

where the integral runs over the momentum shell $\partial\Lambda$ (see the figure on page 376). Taking each component of the integral in turn, one obtains

$$\langle m_{\rm f}^2 \rangle = 2\left(\Lambda - \frac{\Lambda}{b}\right) \Lambda^{d-1} \Omega_d \int_0^{\Gamma_\Lambda} (d\omega) \coth\left(\frac{\omega\beta}{2}\right) \frac{\omega/\Gamma_\Lambda}{(\delta + \Lambda^2)^2 + (\omega/\Gamma_\Lambda)^2}$$

$$+ \frac{1}{\pi} \int^\Lambda (dq) \left(\Gamma_q - \frac{\Gamma_q}{b^z}\right) \coth\left(\frac{\Gamma_q \beta}{2}\right) \frac{1}{(\delta + \mathbf{q}^2)^2 + 1}.$$

Noting that $\Lambda(1 - e^{-\ln b}) \simeq \Lambda \ln b$ and $\Lambda(1 - e^{-z\ln b}) \simeq \Lambda z \ln b$, this result assumes the form $\langle m_{\rm f}^2 \rangle = f(T, \delta) \ln b$. Finally, applying the rescaling $s = \omega/\Gamma_\Lambda$ or $s = |\mathbf{q}|/\Lambda$ as appropriate, setting $\Lambda = \Lambda_q = 1$, and neglecting $\delta \ll 1$, we obtain the required expression for $f(T) \simeq f(T, 0)$, and the update equation $\delta' = \delta(1 + 2\ln b) + 12u f(T) \ln b$. At lowest order in perturbation theory, the coupling constant u changes according to its engineering dimension, $u' = b^{4-d-z}u \simeq (1 + (4 - d - z) \ln b)u$. Similarly,

$T' = b^z T \simeq (1 + z \ln b)T$. Differentiating these expressions, we arrive at the required differential recursion relations.

(d) This is a straightforward exercise in solving linear differential equations.

6.7.3 Kondo effect: poor man's scaling

In problem 2.4.7, we saw that the low-energy properties of a magnetic impurity immersed into a metal are described by the sd-Hamiltonian, Eq. (2.50): a system of free fermions coupled to the impurity spin by an exchange coupling J. In problem 4.6.3, we applied perturbation theory to find that the impurity scattering rate diverged at decreasing temperatures – the principle behind the increase in low-temperature resistivity in the Kondo problem. However, in perturbation theory, nothing could be said about the extrapolation of that temperature dependence to the low temperature/strong coupling regime. In the following, we fill this gap by renormalization group methods.

Our starting point is the sd-Hamiltonian

$$\hat{H}_{sd} = \sum_{\mathbf{k}\sigma} \epsilon_{\mathbf{k}} c_{\mathbf{k}\sigma}^{\dagger} c_{\mathbf{k}\sigma} + \sum_{\mathbf{kk'}} \left(J_z \hat{S}^z \left(c_{\mathbf{k}\uparrow}^{\dagger} c_{\mathbf{k'}\uparrow} - c_{\mathbf{k}\downarrow}^{\dagger} c_{\mathbf{k'}\downarrow} \right) + J_+ \hat{S}^+ c_{\mathbf{k}\downarrow}^{\dagger} c_{\mathbf{k'}\uparrow} + J_- \hat{S}^- c_{\mathbf{k}\uparrow}^{\dagger} c_{\mathbf{k'}\downarrow} \right),$$

where $\hat{\mathbf{S}}$ denotes the impurity spin and, anticipating anisotropy, we allow for independent components of the exchange coupling.

In the following, we will implement a version of the RG procedure taken from the original paper by Anderson.[38] To this end, let us divide the conduction band of width D into high-lying electron–hole states within an energy shell $D/b < |\epsilon_{\mathbf{k}}| < D$ and the remaining states $0 < |\epsilon_{\mathbf{k}}| < D/b$. To eliminate the high-lying excitations, we apply a procedure similar to that leading to the sd-Hamiltonian in problem 2.4.7: we write the eigenstates of the Hamiltonian as a sum $|\psi\rangle = |\psi_0\rangle + |\psi_1\rangle + |\psi_2\rangle$, where $|\psi_1\rangle$ contains no conduction electrons or holes close to the band edge, $|\psi_0\rangle$ has at least one high energy hole, and $|\psi_2\rangle$ has at least one electron. Following the discussion in problem 2.4.7, we eliminate $|\psi_0\rangle$ and $|\psi_2\rangle$ from the Schrödinger equation $\sum_{n=0}^{2} \hat{H}_{mn}|\psi_n\rangle = E|\psi_m\rangle$ to obtain

$$\left(\hat{H}_{10} \frac{1}{E - \hat{H}_{00}} \hat{H}_{01} + \hat{H}_{11} + \hat{H}_{12} \frac{1}{E - \hat{H}_{22}} \hat{H}_{21} \right) |\psi_1\rangle = E|\psi_1\rangle. \tag{6.60}$$

Here, the Hamiltonians $\hat{H}_{00}$ and $\hat{H}_{22}$ appearing in the high-energy shell propagators can be approximated by the free conduction electron Hamiltonian $\hat{H}_0 = \sum_{\mathbf{k}\sigma} \epsilon_{\mathbf{k}} c_{\mathbf{k}\sigma}^{\dagger} c_{\mathbf{k}\sigma}$. Corrections to this approximations would include higher powers of the shell thickness and may be neglected in the limit $b \searrow 1$.

---

[38] P. W. Anderson, *A poor man's derivation of scaling laws for the Kondo problem*, J. Phys. C **3**, 2436 (1970). For an extended and much more detailed discussion, see A. C. Hewson, *The Kondo Problem to Heavy Fermions* (Cambridge University Press, 1993).

(a) Identify the concrete form of the third term on the l.h.s. of Eq. (6.60). In particular, show that there exists a contribution

$$J_+ J_- \sum_{\mathbf{k}'_s \mathbf{k}_s \mathbf{k}_f} \hat{S}^- c^\dagger_{\mathbf{k}'_s \uparrow} c_{\mathbf{k}_f \downarrow} \frac{1}{E - \hat{H}_{22}} \hat{S}^+ c^\dagger_{\mathbf{k}_f \downarrow} c_{\mathbf{k}_s \uparrow} |\psi_1\rangle,$$

where the wave vectors $\mathbf{k}_f (\mathbf{k}_s)$ lie within (outside) the band edge.

(b) Focusing on this contribution alone (for now), perform the summation over f-states to show that the operator assumes the form

$$J_+ J_- \sum_{\mathbf{k}'_s \mathbf{k}_s} \nu_0 D(1 - b^{-1}) \hat{S}^- \hat{S}^+ c^\dagger_{\mathbf{k}'_s \uparrow} c_{\mathbf{k}_s \uparrow} \frac{1}{E - D + \epsilon_{\mathbf{k}_s} - \hat{H}_0},$$

where ν_0 denotes the density of states at the band edge. Assuming that we interested in states close to the ground state, and E is close to the ground state energy, all three contributions to the energy denominator, $\hat{H}_0, E, \epsilon_{\mathbf{k}_s}$ may be neglected compared to the high energy scale D. Using the fact that for spin $S = 1/2$, $\hat{S}^- \hat{S}^+ = 1/2 - \hat{S}^z$, the contribution thus takes the form

$$-J_+ J_- \sum_{\mathbf{k}'_s \mathbf{k}_s} \nu_0 (1 - b^{-1}) \left(\frac{1}{2} - \hat{S}^z \right) c^\dagger_{\mathbf{k}'_s \uparrow} c_{\mathbf{k}_s \uparrow}.$$

We leave it as an unanswered part of the exercise to confirm that the parallel contribution from the process in which a hole is created in the lower band edge leads to the expression

$$-J_+ J_- \sum_{\mathbf{k}'_s \mathbf{k}_s} \nu_0 (1 - b^{-1}) \left(\frac{1}{2} + \hat{S}^z \right) c_{\mathbf{k}'_s \uparrow} c^\dagger_{\mathbf{k}_s \uparrow}.$$

(Try to use arguments based on particle–hole symmetry rather than developing a first principles analysis.) Following a similar procedure, one may confirm that the second class of spin-conserving terms lead to the contributions

$$-\frac{J_z^2}{4} \sum_{\mathbf{k}'_s \mathbf{k}_s \sigma} \nu_0 D(1 - b^{-1}) \begin{cases} c^\dagger_{\mathbf{k}'_s \sigma} c_{\mathbf{k}_s \sigma}, \\ c_{\mathbf{k}'_s \sigma} c^\dagger_{\mathbf{k}_s \sigma}. \end{cases}$$

This completes the analysis of contributions to the effective Hamiltonian for $|\psi_1\rangle$ which leave the electron and impurity spin unchanged. The four remaining contributions involve a spin flip. Following a similar procedure to the one outlined above, one may identify two further contributions,

$$\frac{J_z J_+}{2} \sum_{\mathbf{k}'_s \mathbf{k}_s} \nu_0 (1 - b^{-1}) \hat{S}^+ \begin{cases} -c^\dagger_{\mathbf{k}'_s \downarrow} c_{\mathbf{k}_s \uparrow}, \\ c_{\mathbf{k}'_s \uparrow} c^\dagger_{\mathbf{k}_s \downarrow} \end{cases}$$

where we have used the identity $\hat{S}^z \hat{S}^+ = \hat{S}^+ / 2$. One may confirm that the corresponding terms with the order reversed generate an equal contribution.

When combined with terms from a spin-reversed process, altogether one finds that the components of the exchange constant become renormalized according to

$$J_\pm(b) = J_\pm - 2J_z J_\pm \nu_0 (1 - b^{-1}),$$
$$J_z(b) = J_z - 2J_+ J_- \nu_0 (1 - b^{-1}).$$

Setting $b = e^\ell$ as usual, we arrive at the differential recursion relations

$$\frac{dJ_\pm}{d\ell} = 2\nu_0 J_z J_\pm, \qquad \frac{dJ_z}{d\ell} = 2\nu_0 J_+ J_-.$$

A crucial observation now is that these equations are mathematically identical to those of the BKT transition (6.52) – another manifestation of universality in critical phenomena. To understand the flow, we just need to import our earlier analysis. As a result, we obtain the flow of coupling constants shown in the figure. In particular, the combination of coupling constants is conserved,

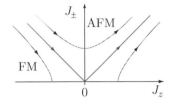

$$J_z^2 - J_\pm^2 = \text{const.},$$

and the constant J_z always increases. The lines $J_z^2 = J_\pm^2$ separate the phase diagram into a ferromagnetic region (J_z terminating at a non-zero negative value) from a more interesting antiferromagnetic domain, where J_z and $J_\pm$ flow to large values.

Setting $J_z = J_\pm \equiv J > 0$, integration of the scaling equations $dJ/d\ell = 2\nu_0 J^2$ gives

$$\frac{1}{J} - \frac{1}{J(\ell)} = 2\nu_0 \ell = 2\nu_0 \ln\left(\frac{D(\ell)}{D}\right).$$

Kondo
temper-
ature
This equation motivate the definition of the **Kondo temperature** through

$$D \exp\left(-\frac{1}{\nu_0 J}\right) = D(\ell) \exp\left(-\frac{1}{\nu_0 J(\ell)}\right) \equiv T_K.$$

When the running cutoff $D(\ell) \sim T_K$ reaches the Kondo temperature, $J(\ell) \sim \nu_0^{-1}$ has become large and perturbation theory breaks down. Physically, the large antiferromagnetic coupling corresponds to a complete screening of the impurity spin by the spins of itinerant electrons. At temperatures far below the Kondo temperature, the effective spin singlet formed in this way behaves as an ordinary impurity, and transport coefficients such as the impurity resistivity cease to show strong temperature dependence.

Answer:

(a) The operator transferring conduction electrons into the band edge is given by

$$\hat{H}_{21} = \sum_{\mathbf{k}_s, \mathbf{k}_f} \left(J_z \hat{S}^z \left(c^\dagger_{\mathbf{k}_f\uparrow} c_{\mathbf{k}_s\uparrow} - c^\dagger_{\mathbf{k}_f\downarrow} c_{\mathbf{k}_s\downarrow} \right) + J_+ \hat{S}^+ c^\dagger_{\mathbf{k}_f\downarrow} c_{\mathbf{k}_s\uparrow} + J_- \hat{S}^- c^\dagger_{\mathbf{k}_f\uparrow} c_{\mathbf{k}_s\downarrow} \right).$$

Since $\hat{H}_{22} \simeq \hat{H}_0$ does not change spin states, the operator $\hat{H}_{12}(E - \hat{H}_{22})^{-1}\hat{H}_{21}$ involves the following combinations:

(1) $J_- \hat{S}^- c^\dagger_{\mathbf{k}'_s\uparrow} c_{\mathbf{k}_f\downarrow} \times J_+ \hat{S}^+ c^\dagger_{\mathbf{k}_f\downarrow} c_{\mathbf{k}_s\uparrow},$

(2) $- J_z \hat{S}^z c^\dagger_{\mathbf{k}'_s\downarrow} c_{\mathbf{k}_f\downarrow} \times J_+ \hat{S}^+ c^\dagger_{\mathbf{k}_f\downarrow} c_{\mathbf{k}_s\uparrow},$

(3) $J_z \hat{S}^z c^\dagger_{\mathbf{k}'_s\uparrow} c_{\mathbf{k}_f\uparrow} \times J_- \hat{S}^- c^\dagger_{\mathbf{k}_f\uparrow} c_{\mathbf{k}_s\downarrow},$

(4) $J_z \hat{S}^z c^\dagger_{\mathbf{k}'_s\uparrow} c_{\mathbf{k}_f\uparrow} \times J_z \hat{S}^z c^\dagger_{\mathbf{k}_f\uparrow} c_{\mathbf{k}_s\uparrow}.$

Of the four processes, terms (2) and (3) involve a spin-flip process while terms (1) and (4) preserve the electron spin orientation. When combined with the energy denominator, the particular contribution specified in the question corresponds to the first process, (1).

(b) Since the band edge occupancy of the reference state $|\psi_1\rangle$ is zero, one can set

$$\sum_{\mathbf{k}_f} c^\dagger_{\mathbf{k}'_s\uparrow} c_{\mathbf{k}_f\downarrow} \frac{1}{E - \hat{H}_{22}} c^\dagger_{\mathbf{k}_f\downarrow} c_{\mathbf{k}_s\uparrow} |\psi_1\rangle = \nu_0 D(1 - b^{-1}) c^\dagger_{\mathbf{k}'_s\uparrow} c_{\mathbf{k}_s\uparrow} \frac{1}{E - D + \epsilon_{\mathbf{k}_s} - \hat{H}_0},$$

where $\hat{H}_0$ denotes the single-particle Hamiltonian of the band electrons, $\sum_{\mathbf{k}_f} = \int_{D/b}^{D} d\epsilon \nu(\epsilon) \simeq \nu_0 D(1 - b^{-1})$, and we have set $\epsilon_{\mathbf{k}_f} \simeq D$. As a result, one obtains the required formula.

7   Response Functions

SYNOPSIS   The chapter begins with a brief survey of the concepts and techniques of *experimental* condensed matter physics. We will show how correlation functions provide a bridge between concrete experimental data and the theoretical formalism developed in previous chapters. Specifically, we discuss an example of outstanding practical importance: how the response of many-body systems to various types of electromagnetic perturbation is described by correlation functions, and how these functions can be computed by field-theoretical means. We will exploit the fact that, in most cases, the external probes are weak compared with the intrinsic electromagnetic forces of a system, and hence can be treated in a linear approximation. The ensuing linear response theory defines the structure of a large group of probe functions routinely considered in field theory. Physically, these functions define transport coefficients, the electric conductance being the best known representative.

In the previous chapters we learned how microscopic representations of many-body systems are mapped onto effective low-energy models. However, to test these theories, we need to connect them to experiment. Modern condensed matter physics benefits from a plethora of powerful techniques, including electric and thermal transport, neutron, electron, and light scattering, calorimetric measurements, induction experiments, scanning tunneling microscopy, etc. We begin this chapter with a synopsis of structures common to most of these approaches. This will anticipate the links between experiment and theory to be addressed in later sections.

7.1   Experimental Approaches to Condensed Matter

7.1.1   Basic concepts

Broadly speaking, experimental condensed matter physics can be subdivided into three[1] categories:

▷   experiments probing thermodynamic coefficients,

▷   transport experiments,

▷   spectroscopy.

---

[1]   Some classes of experiment (such as scanning tunneling microscopy) do not really fit into this scheme. Moreover, the physics of atomic condensates and optics – which on the theoretical side has an overlap with the concepts discussed previously – has its own portfolio of techniques.

Most references to experimental data made in previous chapters were to thermodynamic properties. Thermodynamic coefficients are extracted from the partition sum by differentiation with respect to globally defined parameters (temperature, homogeneous magnetic field, etc.). This simplicity has advantages but also limitations: thermodynamic data is highly universal[2] and therefore represents an important signature of any system. On the other hand, it is blind with regard to microscopic structures and dynamical features, and therefore does not suffice for fully understanding the physics of a system.

By contrast, transport and spectroscopic measurements – the focus of the present chapter – probe more fine-grained structures of a system. In spite of their large variety, these experimental approaches share a few common features:

▷ Many-body systems usually interact with their environments via *electromagnetic* forces.[3] Experiments use this principle to subject systems under consideration to electromagnetic perturbations (voltage gradients, influx of spin magnetic moments carried by a beam of neutrons, local electric fields formed at the tips of scanning tunneling microscopes, etc.), and detect their response by measuring devices.

Formally, the perturbation is described by an addition

$$\hat{H}_{\mathrm{F}} = \int d^d r \, F'_i(\mathbf{r}, t) \hat{X}'_i(\mathbf{r}) \qquad (7.1)$$

to the system Hamiltonian, where the generalized forces F'_i are (time-dependent) coefficients, coupled to the system through operators $\hat{X}'$. For example, $F'_i(\mathbf{r}, t) = \phi(\mathbf{r}, t)$ for a dynamical voltage coupling to the electronic charge carriers via the density operator $\hat{X}'_i = \hat{\rho}$.

▷ The use of the term "perturbation" is appropriate because the forces $\{F'_i\}$ are generically weak.

▷ The forces perturb the system out of its $F'_i = 0$ reference state. The measurable consequence is that certain operators, $\hat{X}_i$, build up non-vanishing expectation values $X_i(\mathbf{r}, t) = \langle \hat{X}_i(\mathbf{r}, t) \rangle$. (For example, a current $\hat{X}_i = \hat{j}_i$ may begin to flow in response to an applied voltage $F'_i = \phi$.) The goal of theory is to understand the dependence of the measured values of X_i on the forces F'_j.

▷ For a general external influence, $X_i[F'_j]$ will be some functional of the forces. The situation simplifies under the assumed condition of weakness. In this case, the functional relation is approximately linear, i.e., it is of the form

$$X_i(\mathbf{r}, t) = \int d^d r' \int dt' \, \chi_{ij}(\mathbf{r}, t; \mathbf{r}', t') F'_j(\mathbf{r}', t') + \mathcal{O}(F'^2). \qquad (7.2)$$

---

[2] Remember that a few thermodynamic variables are sufficient to characterize the state of a homogeneous system in equilibrium.
[3] An important exception is heat conduction.

While the quantities $\{F_j'\}$ and $\{X_i\}$ are externally adjustable and observable, respectively, the integral kernel χ represents an intrinsic property of the system. These **response functions** or **generalized susceptibilities** describe how the system responds to the application of $\{F_i'\}$, and they are the prime objects of interest of **linear response theory**.

▷ We are generally interested in response functions probing a system's energetically low lying excitations. For this reason, condensed matter experiments are mostly carried out at low temperatures: from $\mathcal{O}(1\,\mathrm{K})$ in the physics of correlated quantum matter, down to $\mathcal{O}(1\,\mathrm{mK})$ in transport experiments.

These considerations show that response functions are a principal interface between experiment and theory; they are measurable, while theory attempts to predict them, or at least understand the experimental observation. However, before turning to the more specific discussion of response functions in field-theoretical frameworks, it may be illustrative to introduce a few concrete types of experimental probe and the information they yield.

INFO   The linearity criterion may break down if either the perturbation is strong, or the system under consideration is small. Both scenarios are realized in modern physics. For instance, **nonlinear optics** probes the response of systems to strong electromagnetic forces, usually exerted via *lasers*. Phenomena such as frequency doubling (see below) then indicate a nonlinear response. Conversely, for the small-sized systems realized in **nanoelectronics** or atomic condensates, small perturbations may by themselves induce a shift out of thermal equilibrium (chapter 12). Finally, the measurement itself may play a non-trivial role: quantum measurement is an invasive process changing the state of a system. In some system classes, notably those considered in connection with **quantum computing**, these changes become crucially relevant. However, in this chapter, we restrict ourselves to the class of linear probe experiments.

7.1.2 Experimental methods

REMARK   This section provides background material and can be skipped at first reading. Alternatively, it can be used as a (non-alphabetically ordered) "glossary" of experimental techniques in condensed matter physics.

In this section, we illustrate the connection above (perturbation $\leadsto$ response) for a few examples. Theorists interested in applied aspects of condensed matter field theory will need a more substantial background in experimentation (see Kuzmany's text[4] for the important class of spectroscopic experiments); the following is no more than a quick overview.

[4] H. Kuzmany, *Solid State Spectroscopy* (Springer-Verlag, 1998).

Thermodynamic experiments

thermo-dynamic suscept-ibilities

Thermodynamic experiments usually probe **susceptibilities**, derivatives of extensive variables with respect to intensive ones.[5] Examples include the **specific heat** $c_v = \partial U/\partial T$, the rate of change of internal energy under a change of temperature, the **magnetic susceptibility** $\chi = \partial M/\partial H$, the change of magnetization in response to a static magnetic field, and the **(isothermal) compressibility** $\kappa = -V^{-1}\partial V/\partial p$, the volume change in response to external pressure, etc. Note that the magnetic susceptibility and the isothermal compressibility are tensor quantities. However, as with other tensorial observables, that complication is usually suppressed in the notation unless it becomes relevant.

Thermodynamic response functions are highly universal. (Remember that a few thermodynamic state variables suffice to characterize unambiguously the state of a given system.) For given values of chemical potential, magnetic field, pressure, etc., a calorimetric experiment will produce a one-dimensional function $c_v(T)$. The low-temperature profile of that function generally contains hints as to the nature of the low-energy excitations of a system.[6] However, the universality of thermodynamic data also implies a limitation: thermodynamic coefficients do not contain information about the spatial fluctuations of a given system or about its dynamics.

Transport experiments

The application of a generalized "voltage" U may trigger current flow through a device (see the figure). That voltage may be electrical, $U = V$, or represent a temperature drop $U = \Delta T$, or even a difference in magnetization between two attached reservoirs, $\mathbf{U} = \Delta \mathbf{M}$. Accordingly, the induced current can be the **electrical current** I carried by the charge of mobile carriers, the "**thermal current**" I_T carried by their energy, or the "**spin current**" I_S carried by their magnetic moments. These currents need not be parallel to the voltage gradient. For example, in a perpendicular magnetic field, a voltage gradient will give rise to a transverse **Hall current** $I_\perp$. The ratio of a current and a *static* voltage difference defines a **direct current (DC) conductance** coefficient, $g = \frac{I}{U}$. (Quantum transport in response to time-varying voltages is described by alternating current (AC) transport coefficients, and will be addressed below.)

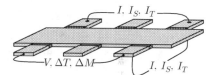

Conductance measurements probe the electrical transport behavior of metals or the thermal conduction properties of insulators and superconductors. The required attachment of contacts makes transport experiments invasive, which can be a disadvantage. For example, there are situations where injection processes at the contacts,

---

[5] Recall that a thermodynamic variable is *extensive* (*intensive*) if it is proportional (independent of) system size.

[6] For example, the specific heat of a Fermi liquid, $c_{v,\text{Fermi liquid}} \sim T$, is linear in temperature and that of phonons, $c_{v,\text{phonon}} \sim T^3$, is cubic, while in a system without low-lying excitations, it vanishes exponentially (exercise: consider why).

rather than the bulk transport in which one is interested, determine the measurement outcome. (For a further discussion of this point, we refer to problem 7.6.1.)

Spectroscopic experiments

spectro-
scopy

The general setup of a **spectroscopic experiment** is shown in the figure. A beam of particles p – either massive (electrons, neutrons, muons, atoms, etc.), or quanta of electromagnetic radiation – is generated at a source and then directed

onto a sample. The kinematic information about the source beam is contained in the dispersion relation $(\mathbf{k}, \omega(\mathbf{k}))$.[7] The particles of the source beam then interact with constituents X of the sample to generate a secondary beam of scattered particles p'. Symbolically,

$$
\begin{array}{ccccccc}
p & + & X & \longrightarrow & p' & + & X' \\
\updownarrow & & \updownarrow & & \updownarrow & & \updownarrow \\
\mathbf{k}, \omega(\mathbf{k}) & & \mathbf{K}, \Omega(\mathbf{K}) & & \mathbf{k}', \omega(\mathbf{k}') & & \mathbf{K}', \Omega(\mathbf{K}'),
\end{array}
$$

where X' represents the final state of the process inside the sample. Notice that the particles p' leaving the sample need not be identical to those incident on the sample. (For example, in photoemission spectroscopy, X-ray quanta displace electrons from the core levels of atoms in a solid.) The dominant scattering process may be elastic (e.g., X-rays scattering off the static lattice structure) or inelastic (e.g., neutrons scattering off phononic excitations). In either case, the accessible information about the scattering process is contained in the frequency–momentum distribution $P(\omega(\mathbf{k}'), \mathbf{k}')$ of the scattered particles, as monitored by a detector.

From these data, one would like to obtain the dispersion $(\Omega(\mathbf{K}), \mathbf{K})$ of the states inside the solid, and this is where the detective work of spectroscopy begins. We know that $(\mathbf{k}, \omega(\mathbf{k}))$ and $(\mathbf{K}, \Omega(\mathbf{k}))$ are related through an energy–momentum conservation law, i.e.,

$$
\mathbf{k} + \mathbf{K} = \mathbf{k}' + \mathbf{K}',
$$
$$
\omega(\mathbf{k}) + \Omega(\mathbf{K}) = \omega(\mathbf{k}') + \Omega(\mathbf{K}').
$$

According to this relation, a peak in the recorded distribution $P(\mathbf{k}', \omega(\mathbf{k}'))$ signals the existence of an internal physical structure (for example, an excitation, or lattice structure) of momentum $\Delta\mathbf{K} \equiv \mathbf{K}' - \mathbf{K} = \mathbf{k} - \mathbf{k}'$ and frequency $\Delta\Omega \equiv \Omega' - \Omega = \omega - \omega'$. However, what sounds straightforward in principle may be complex in practice: solid state components interact almost exclusively through electromagnetic forces. When charged particles are used as scattering probes, the resulting interactions may actually be too strong. This happens when scattering electrons strongly interact with surface states (rather than probing a bulk), or when complicated

---

[7] For some sources, e.g., a laser, the preparation of a near-monochromatic source beam is (by now) standard. For others, such as neutrons, it requires considerable experimental skills (and a lot of money!).

Table 7.1 Different types of condensed matter spectroscopy.

Type	Incident	Outgoing	Application
Raman	visible light	same	dispersion of optical phon-ons and other collective excitations
Infrared	infrared light	same	spectrosocopy of bandgaps, chemical composition
X-ray	X-ray	same	crystalline structure
X-ray absorption	X-ray	same	electronic structure via solid state binding energies
X-ray fluorescence	X-ray	same	chemical analysis, detection of impurities
Photoemission spectroscopy (PES)	X-ray	electrons	photoemission of electrons probing binding energies
Angle-resolved photoemission spectroscopy (ARPES)	X-ray	electrons	angle-resolved detection of photoemission electrons probing band structures
Neutron scattering	neutrons	same	low-lying collective excitations (phonons, magnons), crystallographic structure
Nuclear magnetic resonance (NMR)	magnetic field	same	magnetic properties of conduction electrons

processes of large order in the interaction parameters make the interpretation of scattering amplitudes difficult. For this reason, many spectroscopic techniques employ neutrons or electromagnetic radiation as scattering agents.

A few of the most important types of spectroscopy applied in condensed matter physics are summarized in table 7.1. The list is not exhaustive; other classes of spectroscopy include Auger, Mössbauer, positron–electron annihilation, electron energy loss spectroscopy. The large number of specialized spectroscopic probes emphasizes the importance of these approaches in condensed matter.

Other experimental techniques

STM microscopy

Among the few experimental probes of condensed matter that do not fit comfortably into the three-fold "transport–thermodynamics–spectroscopy" classification, **scanning tunneling microscopy (STM)** is particularly important. STM provides visual images of quasi two-dimensional condensed matter systems; their invention by Binnig and Rohrer in the 1980s (for which they were awarded the Nobel Prize in 1986) triggered a revolution in the area of nanotechnology.

The figure (courtesy of T. Michely and T. Knispel) illustrates the principle for the example of an $(8.5\,\mathrm{nm})^2$ quasi two-dimensional sample of NbS_2 – one layer of niobium atoms sandwiched between two layers of sulfur atoms – resting on graphene.

A small tip is brought into proximity to the surface. When the tip–surface separation becomes comparable with atomic scales, electrons begin to tunnel from the top sulfur layer. The resulting tunnel current is fed into a piezoelectric crystal that in turn levels the height of the tip. Through this mechanism, the surface–tip separation is kept constant, with an accuracy of fractions of atomic separations. A horizontal sweep then generates an accurate image of the crystalline structure of the top layer of atoms. In this particular case, the system supports a charge density wave (CDW) of wavelength three

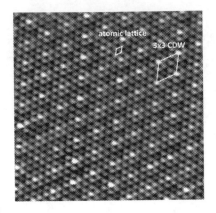

times the atomic separation. The image illustrates how such surface excitations are also resolved with microscopic precision.

7.2 Linear Response Theory

REMARK Throughout this section we will alternately use operator and functional integral representations. If appropriate, we will use a caret notation, $\hat{X}$, to distinguish second-quantized operators from their field integral represenation, X.

In the previous section, we argued that experiments often probe the (linear) response of a system to the application of weak perturbations $\{F_j'\}$. Such linear responses can be cast in terms of a generalized susceptibility χ; see Eq. (7.2). In the following, we give this formal expression (7.2) a more concrete meaning by relating it to the theory familiar from previous chapters. However, before entering this discussion, let us list a few properties of χ that follow from commonsense reasoning:

causality   ▷ *Causality* – The generalized forces $F_j'(t')$ cannot cause an effect prior to their action, i.e., $\chi_{ij}(\mathbf{r}, \mathbf{r}'; t, t') = 0$ for $t < t'$. Formally, we say that the response is **retarded**.

   ▷ If the system Hamiltonian does not depend on time, the response depends only on the difference in the time coordinates, $\chi_{ij}(\mathbf{r}, \mathbf{r}'; t, t') = \chi_{ij}(\mathbf{r}, \mathbf{r}'; t - t')$. In this case it is convenient to Fourier transform the temporal convolution (7.2),

$$X_i(\mathbf{r}, \omega) = \int d^d r' \, \chi_{ij}(\mathbf{r}, \mathbf{r}'; \omega) F_j'(\mathbf{r}', \omega) + \mathcal{O}(F'^2). \qquad (7.3)$$

The important statement made by Eq. (7.3) is that perturbations acting at a characteristic frequency ω cause a linear response at the same frequency. For example, an AC voltage with frequency ω will drive an AC current of the same frequency, etc. We can read this statement in reverse to say that, if the system responds at frequencies $\neq \omega$, we have triggered a nonlinear response. Indeed, it is straightforward to verify (exercise) that the formal extension of the functional observable–force relation (7.2) to nth order generates a response with frequency $n\omega$. According to Eq. (7.3), a peak in the response $X_i(\omega)$ at a characteristic frequency ω indicates a maximum of the response function χ. Such peaks, in turn, reflect the presence of intrinsic excitations and represent structures of interest.

▷ For translationally invariant systems, the response function depends only on the differences between spatial coordinates: $\chi_{ij}(\mathbf{r}, \mathbf{r}'; t - t') = \chi_{ij}(\mathbf{r} - \mathbf{r}'; t - t')$. Spatial Fourier transformation then leads to the relation

$$X_i(\mathbf{q}, \omega) = \chi_{ij}(\mathbf{q}; \omega) F_j'(\mathbf{q}, \omega) + \mathcal{O}(F'^2). \tag{7.4}$$

By analogy with what was said above about the frequency response, we conclude that a peak of the function $X_i(\mathbf{q}, \omega)$ signals the presence of an excitation with frequency ω and momentum $\mathbf{q}$. We thus see that, at least in principle, linear response measurements probe the full dispersion of a system's excitations.

This is as much as one can say on general grounds. We next turn to a more concrete level and relate the response function to the microscopic constituents characterizing a system.

EXERCISE Consider X-rays or neutron radiation probing a crystalline medium whose unit cells are spanned by vectors $\mathbf{a}_i$, $i = 1, \ldots, d$. Show that the response function χ reflects this periodicity through the condition $\chi(\mathbf{k}, \mathbf{k}'; \omega) \propto \delta_{\mathbf{k}-\mathbf{k}'-\mathbf{G}}$, where $\mathbf{G}$ belongs to the reciprocal lattice of the system. In this way, angle-resolved scattering in **spectroscopic crystallography** probes the periodicity of the reciprocal lattice and, therefore, of the original lattice.

spectro-
scopic crys-
tallography

7.2.1 Microscopic response theory

We now set out to relate the response function to the microscopic elements of the theory. Our starting point is the representation of the response signal $X(t)$ as the expectation value of some (single-particle) operator[8] $\hat{X} = \sum_{aa'} c_a^\dagger X_{aa'} c_{a'} \equiv c^\dagger X c$, where c_a may be a bosonic or fermionic operator. Within the framework of the field integral this expectation value assumes the form

$$X(\tau) = \langle \bar{\psi}(\tau) X \psi(\tau) \rangle \equiv \frac{1}{\mathcal{Z}} \int D\psi \, e^{-S[F', \psi]} \, \bar{\psi}(\tau) X \psi(\tau), \tag{7.5}$$

where $S[F', \psi] = S_0[\psi] + \delta S'[F', \psi]$ contains the action, S_0, of the unperturbed system and a contribution

---

[8] For notational transparency, we drop operator indices in this section. For example, $\int d\tau \, F \bar{\psi} X \psi$ might represent $\int d\tau \int d^d r \, F(\mathbf{r}, \tau) \bar{\psi}_\sigma(\tau, \mathbf{r}) X(\mathbf{r})_{\sigma\sigma'} \psi_{\sigma'}(\tau, \mathbf{r})$.

$$\delta S'[F', \psi] = \int d\tau \hat{H}_{F'} = \int d\tau F'(\tau)\bar{\psi}(\tau)X'\psi(\tau),$$

added by the generalized force (see Eq. (7.1)).

In practice, it is often convenient to represent $X(\tau)$ as a derivative of the free energy functional. To this end, let us formally couple our operator $\hat{X}$ to a second generalized force and define

$$\delta S[F, \psi] \equiv \int d\tau\, F(\tau)\hat{X}(\tau) = \int d\tau\, F(\tau)\bar{\psi}(\tau)X\psi(\tau)$$

as a new element of our action. With $S[F, F', \psi] = S_0[\psi] + \delta S[F, \psi] + \delta S'[F', \psi]$, we then have

$$X(\tau) = -\left.\frac{\delta}{\delta F(\tau)}\right|_{F=0} \ln \mathcal{Z}[F, F'],$$

where $\mathcal{Z}[F, F'] = \int \mathcal{D}\psi\, e^{-S[F,F',\psi]}$ depends on the two generalized forces.

In the absence of the driving force F' the expectation value X would vanish. We also assume that F' is weak in the sense that a linear approximation in F' satisfactorily describes the measured value of X. Noting that the first-order expansion of a general functional is given by[9]

$$G[F'] \simeq G[0] + \int d\tau'\, \frac{\delta G[F']}{\delta F'(\tau')} F'(\tau'),$$

we can write

$$X(\tau) \simeq -\int d\tau' \left(\frac{\delta^2}{\delta F(\tau)\, \delta F'(\tau')} \ln \mathcal{Z}[F, F']\right) F'(\tau').$$

Comparison with (7.2) leads to the identification $\chi(\tau, \tau') = -\frac{\delta^2}{\delta F(\tau)\,\delta F'(\tau')} \ln \mathcal{Z}[F, F']$ of the response kernel. Carrying out the derivatives, we find

$$\chi(\tau, \tau') = -\frac{1}{\mathcal{Z}}\frac{\delta^2}{\delta F(\tau)\delta F'(\tau')}\mathcal{Z}[F, F'] + \frac{1}{\mathcal{Z}}\frac{\delta}{\delta F'(\tau')}\mathcal{Z}[0, F'] \times \frac{1}{\mathcal{Z}}\frac{\delta}{\delta F(\tau)}\mathcal{Z}[F, 0],$$

where $\mathcal{Z} \equiv \mathcal{Z}[0, 0]$. The last term in this expression is the functional expectation value $\langle \hat{X}(\tau)\rangle$ taken over the unperturbed action. We have assumed that this average vanishes, so that we arrive at the representation

$$\boxed{\chi(\tau, \tau') = -\frac{1}{\mathcal{Z}}\left.\frac{\delta^2}{\delta F(\tau)\,\delta F'(\tau')}\right|_{F=F'=0}\mathcal{Z}[F, F']} \qquad (7.6)$$

Performing the two derivatives, we obtain an alternative representation of the response function as a four-point correlation function:

$$\chi(\tau, \tau') = -\langle X(\tau)X'(\tau')\rangle = -\langle \bar{\psi}(\tau)X\psi(\tau)\,\bar{\psi}(\tau')X'\psi(\tau')\rangle. \qquad (7.7)$$

While this result is obtained straightforwardly by first-order expansion of (7.5) in F', the utility of the representation (7.7) will shortly become evident.

---

[9] Here, and throughout, all derivatives are taken at zero, e.g., $\frac{\delta G[F']}{\delta F'(\tau')} \equiv \frac{\delta G[F']}{\delta F'(\tau')}\big|_{F'=0}$.

INFO Equation (7.7) indicates a connection between two seemingly different physical mechanisms. Consider the case where the observed and the driving operator are equal: $\hat{X}' = \hat{X}$. Using the vanishing of the equilibrium expectation values, $\langle X(\tau) \rangle = 0$, we can rewrite (7.7) as

$$\chi(\tau, \tau') = -\left\langle (X(\tau) - \langle X(\tau) \rangle)(X(\tau') - \langle X(\tau') \rangle) \right\rangle. \tag{7.8}$$

fluctuation–dissipation theorem

This relation is called the **fluctuation–dissipation theorem (FDT)**. On its right-hand side we have a correlation function probing the dynamical *fluctuation* behavior of the observable represented by $\hat{X}$. By contrast, the response function χ on the left-hand side describes the ways in which externally imposed fluctuations *dissipate* into the microscopic excitations of the system. For example, for $\hat{X} = \mathbf{j} = $ current density (see below), χ describes the dissipative conductance of the system. The manifold interpretations of the FDT in the context of statistical field theory will be addressed in the "nonequilibrium" chapters 11 and 12.

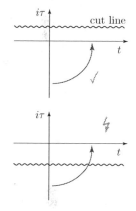

While the function $\chi(\tau, \tau')$ is formulated in **imaginary time**, the quantity in which we are actually interested is the **real time** response of $X(t)$ to driving by $F'(t')$. This tension of real-time questions addressed in imaginary-time frameworks appeared frequently in previous chapters, and we generally resolved it by the "analytical continuation" $t \to -i\tau$. However, this was just high-minded terminology for a naïve substitution of time variables, and now the time has come to address the issue properly.

In the majority of cases, substitution does indeed lead to correct results. It may happen, however, that a function $f(\tau)$ contains singularities in the complex τ-plane.[10] If these are extended singularities blocking interpolation to the real axis (see figure), the simple substitution prescription is problematic. In the following, we discuss the situation for the important case where f is a linear response function, and show how mathematically controlled continuation procedures are developed.

7.3 Analytic Structure of Correlation Functions

REMARK Throughout much of this section we will work in an operator representation in which expressions with circumflexes, $\hat{X}$, represent canonically quantized operators and $\langle \cdots \rangle = \mathcal{Z}^{-1}\mathrm{tr}(\cdots \exp\{-\beta[\hat{H} - \mu\hat{N}]\})$ is the quantum–thermal expectation value.

Discussion of the analytic structure of correlation functions does not require the overhead of the field integral and is best formulated in the language of Fock space

---

[10] It is best to think of $f(\tau)$ as a function on the imaginary axis of a complex-time domain. Analytical continuation defines a generalization, $F(z)$, on a domain around the imaginary axis. When we make the substitution $f(\tau) \to f(\tau \to it)$, we imply that the analyticity domain includes the real axis.

operators. We are interested in the dynamical correlations of two operators $\hat{X}_1$ and $\hat{X}_2$, where

$$\hat{X}_i(\tau) \equiv e^{\tau(\hat{H}-\mu\hat{N})}\hat{X}_i e^{-\tau(\hat{H}-\mu\hat{N})}, \tag{7.9}$$

is the imaginary-time Heisenberg representation,[11] $\hat{H}$ is the full system operator, and $\hat{X}_i = \hat{X}_i(c^\dagger, c)$ are constructed from the fundamental fermion or boson operators. (Think of them as current or spin operators, for definiteness.)

imaginary–
time–
ordered
correlation
function

The **imaginary–time–ordered correlation function** of the two operators is defined as

$$C^\tau(\tau_1 - \tau_2) \equiv -\langle T_\tau \hat{X}_1(\tau_1)\hat{X}_2(\tau_2)\rangle \equiv - \begin{cases} \langle \hat{X}_1(\tau_1)\hat{X}_2(\tau_2)\rangle, & \tau_1 \geq \tau_2 \\ \zeta_{\hat{X}}\langle \hat{X}_2(\tau_2)\hat{X}_1(\tau_1)\rangle, & \tau_2 > \tau_1 \end{cases} \tag{7.10}$$

time-
ordering
operator

Here, the action of the **time-ordering operator** T_τ is defined through the second relation, where the sign factor $\zeta_{\hat{X}} = \pm 1$ is determined by the statistics of $\hat{X}_i$: $\zeta_{\hat{X}} = 1$ if $\hat{X}_i$ is bosonic and -1 if it is fermionic.[12] The action of T_τ is to chronologically order the two operators under the expectation value.

INFO It is instructive to compare the definition (7.10) with the field integral correlation function (7.7) defined above. Within the field integral formalism, operators are replaced by coherent state functions, $\hat{X}(c^\dagger, c) \to X(\bar{\psi}, \psi)$. These functions commute or anticommute, depending on the statistics of the coherent state variables, $X_1 X_2 = \zeta_X X_2 X_1$, so that the time-ordering operation becomes redundant.[13] We conclude that the correlation function (7.7) *is* the field integral representation of (7.10).

In a manner that is difficult to motivate in advance, we next introduce three different *real-time* correlation functions. Substituting in Eq. (7.10) real times for imaginary times, $\tau \to it$, we obtain the **real time ordered correlation function**

real time–
ordered
correlation
function

$$C^T(t_1 - t_2) = -i\langle T_t \hat{X}_1(t_1)\hat{X}_2(t_2)\rangle \tag{7.11}$$

where the factor of i has been introduced for later convenience, T_t chronologically orders real times, and $\hat{X}(t) \equiv e^{it(\hat{H}-\mu\hat{N})}\hat{X}e^{-it(\hat{H}-\mu\hat{N})}$ is a real-time Heisenberg operator. While this expression appears to be the "natural" generalization of (7.10), it is not our prime object of interest. More important is the **retarded correlation function**

retarded
correlation
function

$$C^+(t_1 - t_2) = -i\Theta(t_1 - t_2)\langle [\hat{X}_1(t_1), \hat{X}_2(t_2)]_{\zeta_{\hat{X}}}\rangle \tag{7.12}$$

---

[11] If you are unfamiliar with the (imaginary-time) Heisenberg representation, either look it up in a textbook on second quantization or revisit the construction of the field integral to convince yourself that $\hat{X}_i(\tau)$ is the operator representation of coherent state "operators" $X_i(\tau)$ inserted into the functional integral at time slice τ.

[12] The operator $\hat{X}_i$ is bosonic if $\{c^\dagger, c\}$ are Bose operators or if it is of even order in fermion operators.

[13] In second quantization, explicit time–ordering is required by the presence of nontrivial commutators between the field operators.

This function is "retarded" in that it is non-vanishing only for $t_1 > t_2$. The complementary time domain, $t_1 < t_2$, is described by the **advanced correlation function**

$$C^-(t_1 - t_2) = +i\Theta(t_2 - t_1)\langle[\hat{X}_1(t_1), \hat{X}_2(t_2)]_{\zeta_{\hat{X}}}\rangle \qquad (7.13)$$

INFO To appreciate the **meaning of the retarded response function**, we need to reformulate the construction above in operator language: We want to find the expectation value $X(t) = \langle\hat{U}^{-1}(t)\hat{X}\hat{U}(t)\rangle$, where $\hat{U}^t \equiv \hat{U}(t, -\infty)$ is the quantum mechanical time-evolution operator, computed for a Hamiltonian $\hat{H} = \hat{H}_0 + F'(t)\hat{X}'$ containing a weak time-dependent perturbation $F'(t)\hat{X}'$. Here, the thermal average $\langle\cdots\rangle = \mathcal{Z}^{-1}\text{tr}[(\cdots)\exp\{-\beta(\hat{H}_0 - \mu\hat{N})\}]$ is taken with reference to $\hat{H}_0$ only. This prescription corresponds to a dynamical protocol where, in the distant past $t \to -\infty$, the system was prepared in a thermal equilibrium state of $\hat{H}_0$. A perturbation $\propto F'(t)$ was then gradually switched on to effect the dynamical evolution of $\hat{X}$. The essence of this **"switching on procedure"** is the assumption that $\hat{X}'$ is remains weak enough not to disturb the system out of thermal equilibrium.

To compute the expectation value, it is convenient to switch to a representation in which the evolutionary change due to the perturbation is isolated:

$$X(t) = \langle(\hat{U}^{F'})^{-1}(t)\hat{X}(t)\hat{U}^{F'}(t)\rangle, \qquad (7.14)$$

where $\hat{U}^{F'}(t) = \hat{U}_0^{-1}(t)\hat{U}(t)$, and $\hat{X}(t) = \hat{U}_0^{-1}(t)\hat{X}\hat{U}_0(t)$ evolves according to the $\hat{H}_0$-Heisenberg representation. With these definitions, it is straightforward to verify that $\hat{U}^{F'}$ obeys the differential equation, $d_t\hat{U}^{F'}(t) = -iF'(t)\hat{X}'(t)\hat{U}^{F'}(t)$, i.e., the time-evolution of $\hat{U}^{F'}$ is controlled by the (Heisenberg representation of the) perturbation $\hat{X}'$. The solution of this differential equation (with boundary condition $\hat{U}^{F'}(t \to -\infty) \to \mathbb{1}$) is given by

$$\hat{U}^{F'}(t) = T_t \exp\left(-i\int_{-\infty}^t dt'\, F'(t')\hat{X}'(t')\right) \simeq \mathbb{1} - i\int_{-\infty}^t dt'\, F'(t')\hat{X}'(t') + \cdots.$$

Substituting this expression into Eq. (7.14) we obtain

$$X(t) = -i\int dt'\, \theta(t - t')F'(t')\langle[\hat{X}(t), \hat{X}'(t')]\rangle = \int dt'\, C^+(t - t')F'(t').$$

According to this result, the retarded response function describes the linear response of $\hat{X}$ to the perturbation:

> The retarded response function Eq. (7.12) is the prime object of interest of linear response theory.

7.3.1 Lehmann representation

We next investigate the connections between the different correlation functions defined above. Specifically, we want to understand how to obtain the retarded response function (7.12) (the function of interest) from the imaginary-time-ordered one (7.10) (the function computed from the theory). These connections are revealed by a formal expansion of the correlation functions $C^{T/\tau,+/-}$ in exact eigenfunctions known as the **Lehmann representation**: consider the Hamiltonian diagonalized in terms of eigenfunctions $|\Psi_\alpha\rangle$ and eigenvalues E_α. With $\text{tr}(\cdots) = \sum_\alpha\langle\Psi_\alpha|\cdots|\Psi_\alpha\rangle$,

Lehmann
represen-
tation

and inserting a resolution of unity $\mathbb{1} = \sum_\beta |\Psi_\beta\rangle\langle\Psi_\beta|$ between the two operators appearing in the correlation functions, it is straightforward to show that, e.g.,

$$C^T(t) = -\frac{i}{\mathcal{Z}} \sum_{\alpha\beta} X_{1\alpha\beta} X_{2\beta\alpha} e^{it\Xi_{\alpha\beta}} \left(\Theta(t)e^{-\beta\Xi_\alpha} + \zeta_X \Theta(-t)e^{-\beta\Xi_\beta}\right), \qquad (7.15)$$

where $\Xi_\alpha \equiv E_\alpha - \mu N_\alpha$, $\Xi_{\alpha\beta} \equiv \Xi_\alpha - \Xi_\beta$, and $X_{\alpha\beta} \equiv \langle\Psi_\alpha|\hat{X}|\Psi_\beta\rangle$. We next Fourier transform to find

$$
\begin{aligned}
C^T(\omega) &= \int_{-\infty}^{\infty} dt\, C^T(t) e^{i\omega t - \eta|t|} \\
&= \frac{1}{\mathcal{Z}} \sum_{\alpha\beta} X_{1\alpha\beta} X_{2\beta\alpha} \left[\frac{e^{-\beta\Xi_\alpha}}{\omega + \Xi_{\alpha\beta} + i\eta} - \zeta_{\hat{X}}\frac{e^{-\beta\Xi_\beta}}{\omega + \Xi_{\alpha\beta} - i\eta}\right],
\end{aligned}
$$

where the convergence-generating factor η – which will play a key role throughout – has been introduced to make the Fourier representation well-defined.[14]

With reference to a full solution of the problem, Eq. (7.15) is not of help for practical purposes. However, it is the key to understanding exact connections between the correlation functions that are essential, including from a practical perspective. To see this, consider the Lehmann representation of the other correlation functions. Proceeding as with the real-time function above, it is straightforward to show that

$$
\left.\begin{aligned} C^T(\omega) \\ C^+(\omega) \\ C^-(\omega) \end{aligned}\right\} = \frac{1}{\mathcal{Z}} \sum_{\alpha\beta} X_{1\alpha\beta} X_{2\beta\alpha} \left[\frac{e^{-\beta\Xi_\alpha}}{\omega + \Xi_{\alpha\beta} + \left.\begin{aligned}+\\-\end{aligned}\right\} i\eta} - \zeta_{\hat{X}}\frac{e^{-\beta\Xi_\beta}}{\omega + \Xi_{\alpha\beta} + \left.\begin{aligned}-\\-\end{aligned}\right\} i\eta}\right], \qquad (7.16)
$$

where the horizontal lines on the right-hand side indicate division by various denominators. From this result, a number of important features of the correlation functions follow. Think of $C^{T,+/-}(z)$ as functions of a generalized complex variable z, from which $C^{T,+/-}(\omega)$ are obtained by restricting $z = \omega$ to the real axis. Within the extended framework, $C^{T,+/-}$ are complex functions with singularities in the close neighborhood of the real axis. More specifically:

▷ The retarded correlation function C^+ has singularities for $z = -\Xi_{\alpha\beta} - i\eta$, slightly below the real axis. It is, however, analytic in the entire upper complex half-plane, $\mathrm{Im}(z) \geq 0$.

---

[14] Indeed, we can attach physical significance to this factor. The switching-on procedure outlined above can be implemented by attaching a small damping term $\exp(-|t|\eta)$ to an otherwise purely oscillatory force. If we absorb this factor into the definition of all Fourier integrals, $\int dt\, (F(t)e^{-t|\eta|})e^{i\omega t}(\cdots) \to \int dt\, F(t)\,(e^{-t|\eta|}e^{i\omega t})(\cdots)$, we arrive at the Fourier regularization above.

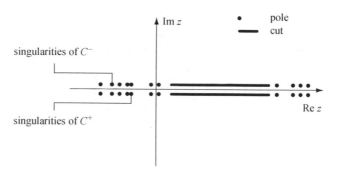

Fig. 7.1   Illustration of the singularities of advanced and retarded correlation functions in the complex plane. The points denote poles and the lines branch cuts.

▷ Conversely, the advanced correlation function C^- has singularities above the real axis. It is analytic in the lower half-plane, $\mathrm{Im}(z) \leq 0$. Notice that C^+ and C^- are connected through complex conjugation,

$$C^+(\omega) = \left[C^-(\omega)\right]^* . \tag{7.17}$$

▷ The time-ordered correlation function has singularities on either side of the real axis (which makes it harder to analyze).

▷ The positions of the singularities contain information about the excitations of the system (see fig. 7.1). To see how, consider the case where $\hat{X}_1 = c_a^\dagger$ and $\hat{X}_2 = c_b$ are creation and annihilation operators corresponding to some single-particle basis $\{|a\rangle\}$. In this case, $N_\alpha - N_\beta \equiv \Delta N = 1$ (independently of the state indices α, β) and $E_\alpha - E_\beta$ are the single-particle energies for a non-interacting system or are of the order of these energies for an interacting system (why?). We observe that the singularities of $C^{T,+/-}$ map out the single-particle spectrum of the problem described by the correlation functions. This correspondence follows in more direct terms from the interpretation of the one-particle correlation function as the amplitude for the creation of a state $|a\rangle$ followed by the annihilation of a state $|b\rangle$ at later time. The time Fourier transform of the amplitude, $|a\rangle \xrightarrow{t} |b\rangle$, becomes "large" when the phase ($\sim \omega t$) of the Fourier argument is in resonance with an eigenphase $\sim (E_\alpha - E_\beta)t$ supported by the system. (Think of the simple example of a plane wave Hamiltonian for illustration.)

   The correspondence between singularities and excitations extends to cases of more complex structure. For example, for a two-particle correlation function, $\hat{X}_1 \sim c_a^\dagger c_b$, the energies $E_\alpha - E_\beta$ describe the spectrum (the "energy cost") of two-particle excitations, etc. Also notice that the single-particle spectrum can be continuous, in which case the functions $C^{T,+/-}$ have *cuts* parallel to the real axis rather than isolated singularities (fig. 7.1).

▷ Once one of the correlation functions is known, the others follow straightforwardly from a simple recipe: using the Dirac identity (3.84) it is straightforward to show that (Exercise)

$$\boxed{\operatorname{Re} C^T(\omega) = \operatorname{Re} C^+(\omega) = \operatorname{Re} C^-(\omega)} \qquad (7.18)$$

and

$$\boxed{\operatorname{Im} C^T(\omega) = \pm \operatorname{Im} C^\pm(\omega) \times \begin{cases} \coth(\beta\omega/2), & \text{bosons} \\ \tanh(\beta\omega/2), & \text{fermions} \end{cases}} \qquad (7.19)$$

i.e., the three different functions store identical information and can be computed from one another.

We next include the **imaginary-time correlation function** C^τ in the general framework. Starting from the imaginary-time analog of Eq. (7.15),

$$C^\tau(\tau) = -\frac{1}{\mathcal{Z}} \sum_{\alpha\beta} X_{1\alpha\beta} X_{2\beta\alpha} e^{\Xi_{\alpha\beta}\tau} \left(\Theta(\tau) e^{-\beta\Xi_\alpha} + \zeta_{\hat{X}} \Theta(-\tau) e^{-\beta\Xi_\beta} \right), \qquad (7.20)$$

we observe that C^τ acquires periodicity properties,

$$C^\tau(\tau) = \zeta_{\hat{X}} C^\tau(\tau + \beta), \quad \tau < 0, \qquad (7.21)$$

consistent with the periodicity properties of bosonic or fermionic correlation functions in the functional integral formalism. Consequently, C^τ can be expanded in a Fourier representation $C^\tau(i\omega_n) = \int_0^\beta d\tau\, C^\tau(\tau) e^{i\omega_n\tau}$ with bosonic or fermionic Matsubara frequencies. Applying this transformation to the Lehmann representation (7.20), we obtain

$$C^\tau(i\omega_n) = \frac{1}{\mathcal{Z}} \sum_{\alpha\beta} \frac{X_{1\alpha\beta} X_{2\beta\alpha}}{i\omega_n + \Xi_{\alpha\beta}} \left[e^{-\beta\Xi_\alpha} - \zeta_X e^{-\beta\Xi_\beta} \right]. \qquad (7.22)$$

Our final task is to relate the four correlation functions defined through Eq. (7.16) and (7.22) to each other. To this end, we define the "master function"

$$C(z) = \frac{1}{\mathcal{Z}} \sum_{\alpha\beta} \frac{X_{1\alpha\beta} X_{2\beta\alpha}}{z + \Xi_{\alpha\beta}} \left[e^{-\beta\Xi_\alpha} - \zeta_X e^{-\beta\Xi_\beta} \right], \qquad (7.23)$$

depending on a complex argument z. When evaluated for $z = \omega^+, \omega^-, i\omega_n$, respectively, the function $C(z)$ coincides with C^+, C^-, C^τ. Further, $C(z)$ is analytic everywhere except for the real axis. This knowledge suffices to construct the relation between the different correlation functions that was sought. Suppose we have succeeded in computing $C^\tau(i\omega_n) = C(z = i\omega_n)$ for all positive Matsubara frequencies.[15]

Further, let us assume that we have managed to find an analytic extension of $C(z = i\omega_n) \to C(z)$ into the entire upper complex half-plane, $\operatorname{Im} z > 0$. The evaluation of this extension on the infinitesimally shifted real axis $z = \omega + i0$ then coincides with the retarded Green function $C^+(\omega)$ (see figure). In other words,

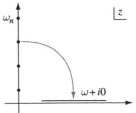

---

[15] Keep in mind that in practical computations we will not proceed through the Lehmann representation.

> To find $C^+(\omega)$, we need to compute $C^\tau(i\omega_n)$ for all positive Matsubara frequencies and then continue the result down to the real axis, $i\omega_n \to \omega + i0$.

(The advanced Green function C^- is obtained analogously, by analytic extension of the thermal correlation function $C^\tau(i\omega_n < 0)$ to frequencies with a negative offset, $\omega - i0$.)

INFO These statements follow from a theorem of complex function theory stating that two analytic functions $F_1(z)$ and $F_2(z)$ coincide if $F_1(z_n) = F_2(z_n)$ on a sequence $\{z_i\}$ with a limit point in the domain of analyticity. (In our case, $i\omega_n \to i\infty$ is the limit point.) From inspection of (7.23), we already know that $F_1(i\omega_n) \equiv C^+(\omega \to i\omega_n)$ coincides with $F_2(i\omega_n) = C^\tau(i\omega_n)$. Thus, any analytic extension of C^τ must coincide with C^+ everywhere in the upper complex half-plane, including the infinitesimally shifted real axis.

EXERCISE Writing $z = \omega \pm i\eta$, transform the spectral representation (7.23) back to the time domain: $C(t) = \frac{1}{2\pi} \int d\omega\, e^{-i\omega t} C(\omega \pm i\eta)$. Convince yourself that, for $\mathrm{Im}(z)$ positive (negative), the temporal correlation function $C(t)$ contains a Θ-function $\Theta(t)$ ($\Theta(-t)$). (Hint: Make use of Cauchy's theorem.) The presence of this constraint does not hinge on η being infinitesimal. It even survives generalization to a frequency-dependent function $\eta(\omega) > 0$. (For the physical relevance of this statement, see below.) All that matters is that, for $\eta > 0$, the function $C(\omega \pm i\eta)$ is analytic in the upper (lower) complex half-plane. This observation implies an important **connection between analyticity and causality**: correlation functions whose frequency representation is analytic in the upper (lower) complex half-plane are causal (anticausal). (A time-dependent function is called "(anti)causal" if it vanishes for (positive) negative times.)

How is the continuation process carried out in practice? Basically, the answer follows from what was said above. If we know the correlation function $C^\tau(i\omega_n)$ for all positive Matsubara frequencies, and if that function remains analytic upon the substitution $C^\tau(i\omega_n \to z)$ of a general argument in the positive half-plane, we merely have to make the replacement $i\omega_n \to \omega + i0$ to obtain $C^+(\omega)$. Often, however, we do not know $C^\tau(i\omega_n)$ for all positive frequencies. For example, we may be working within an effective low-energy theory whose regime of validity is restricted to frequencies $\omega_n < \omega^*$ smaller than some cutoff frequency. Everything then hinges on finding a meaningful model function that can be extended to infinity and whose evaluation for small frequencies $\omega_n < \omega^*$ coincides with our result. While, there are no generally applicable recipes for how to deal with such situations, physically well-defined theories usually admit analytic continuation.

INFO As an example, let us consider the **non-interacting single-particle Green function**, i.e., the choice $\hat{X}_1 = c_a$, $\hat{X}_2 = c_a^\dagger$ in the eigenbasis of a free particle system. With the many-body eigenstates $|\alpha\rangle = |a_1, a_2, \ldots\rangle$ defined as symmetric or antisymmetric combinations of single-particle states $|a_i\rangle$, we obtain $E_\alpha = \epsilon_{a_1} + \epsilon_{a_2} + \cdots$ as sums of one-body energies. Using the fact that $E_\beta = E_\alpha + \epsilon_a$ (exercise: why?), one may verify that the correlation function assumes the simple form

$$C(z) \equiv G_a(z) = \frac{1}{z - \xi_a}.$$

The notable thing here is that the dependence of the correlation function on the partition sum cancels out (check!). The thermal version of this Green function, $G_a(i\omega_n) = (i\omega_n - \xi_a)^{-1}$, appeared previously as a building block of perturbation theory. This is, of course, no coincidence: the Green function featured as the functional expectation value $\langle \bar{\psi}_{a,n} \psi_{a,n} \rangle_0$ in the Gaussian theory. But this is just the functional representation of the operator correlation function considered above.

Building on this representation, it is customary to define a **Green function operator**

$$\hat{G}(z) \equiv \frac{1}{z + \mu - \hat{H}}.$$

By design, the eigenvalues of this operator – which are still functions of z – are given by the correlation function $G_a(z)$ above. Numerous physical observables can be represented compactly in terms of the operator Green function. For example, using the Dirac identity (3.84), the single-particle density of states of the non-interacting system is obtained as the trace of the retarded Green function (operator),

$$\boxed{\rho(\epsilon) = -\frac{1}{\pi} \operatorname{Im} \operatorname{tr} \hat{G}^+(\epsilon)} \tag{7.24}$$

To illustrate the procedure of analytic continuation, let us consider a few examples.

1. For the single-particle Green function ($\hat{X}_1 = c_a, \hat{X}_2 = c_a^\dagger$) of a free system considered in the info block above, the continuation amounts to a mere substitution,

    $$G_a^+(\omega) = \frac{1}{\omega + i0 - \xi_a}. \tag{7.25}$$

2. We have seen that quasi-particle interactions lead to the appearance of a – generally complex – self-energy $\Sigma(z)$: $G_a(\omega_n) \to (i\omega_n - \xi_a - \Sigma(i\omega_n))^{-1}$, where we suppress the potential dependence of the self-energy on the Hilbert space index a. Extension down to the real axis leads to the relation

    $$G_a^+(\omega) = \frac{1}{\omega^+ - \xi_a - \Sigma(\omega^+)}, \tag{7.26}$$

    where $\omega^+ \equiv \omega + i0$ and $\Sigma(\omega^+)$ is the analytic continuation of the function $\Sigma(z)$ to the real axis. Although the concrete form of the self-energy depends on the problem under consideration, a few statements can be made in general. Specifically, decomposing the self-energy into real and imaginary parts, $\Sigma = \Sigma' + i\Sigma''$, we have

    $$\Sigma'(\omega^+) = +\Sigma'(\omega^-), \qquad \Sigma''(\omega^+) = -\Sigma''(\omega^-) < 0. \tag{7.27}$$

    If the imaginary part is finite, the real axis becomes a cut line at which $\operatorname{Im}\Sigma(z)$ changes sign. Formally, the cut structure follows from Eq. (7.17) relating the retarded and advanced Green functions through complex conjugation. To understand it in physical terms, suppose we start from a non-interacting limit and gradually switch on interactions. The (Landau) principle of adiabatic continuity implies that nowhere in this process must the Green function – alias the propagator of the theory – become singular. Thus, the combination

$i(\omega_n - \Sigma''_{i\omega_n})$ can never approach zero. The safeguard preventing the vanishing of the energy denominator is that $-\Sigma''$ and ω_n have opposite signs. This principle can be checked order by order in perturbation theory, but its validity extends beyond perturbation theory.

Transforming back to the time domain we obtain

$$G^+(t) = \int \frac{d\omega}{2\pi} e^{-i\omega t} G^+(\omega) \approx e^{-it(\xi_a + \Sigma') + t\Sigma''} \Theta(t),$$

where we have somewhat oversimplified matters by assuming that the dependence of the self-energy operator on ω is negligible: $\Sigma(\omega) \approx \Sigma$.

EXERCISE Check the second equality above.

Interpreting $G^+(t)$ as the amplitude for propagation in the state $|a\rangle$ during a time interval t, and $|G^+|^2$ as the associated probability density, we observe that the probability of staying in state $|a\rangle$ decays exponentially, $|G^+|^2 \propto e^{2t\Sigma''}$, i.e., $-2\Sigma'' \equiv \frac{1}{\tau}$ defines the inverse of the **effective lifetime** τ of state $|a\rangle$. The appearance of a finite lifetime expresses the fact that, in the presence of interactions, single-particle states decay into a continuum of many-body states. This picture will be substantiated in section 7.3.2 below.

3.   As an example of a situation where the analytic continuation is a little more involved, we can apply Eq. (7.24) to compute the BCS quasi-particle density of states (DoS) of a superconductor. In section 5.3.4 we found that the Gorkov Green function of a superconductor as $\hat{G}(i\omega_n) = [i\omega_n - (\hat{H} - \mu)\sigma_3 - \Delta\sigma_1]^{-1}$. Switching to an eigenrepresentation and inverting the Pauli matrix structure,

$$-\frac{1}{\pi} \mathrm{tr}\, \hat{G}(i\omega_n) = \frac{1}{\pi} \sum_a \mathrm{tr} \left(\frac{i\omega_n + \xi_a \sigma_3 + \Delta\sigma_1}{\omega_n^2 + \xi_a^2 + \Delta^2} \right) = \frac{2i\omega_n}{\pi} \sum_a \frac{1}{\omega_n^2 + \xi_a^2 + \Delta^2}.$$

Next, performing our standard change from a summation over eigenenergies to an integral, we arrive at

$$-\frac{1}{\pi} \mathrm{tr}\, \hat{G}(i\omega_n) \simeq \frac{2i\omega_n}{\pi} \nu \int d\xi \frac{1}{\omega_n^2 + \xi^2 + \Delta^2} = \frac{2i\omega_n \nu}{\sqrt{\omega_n^2 + \Delta^2}},$$

where ν is the normal density of states at the Fermi level. This is the quantity that we need to continue to real frequencies. To this end, we adopt the standard convention whereby the cut of the square root function is on the negative real axis, i.e., $\sqrt{-r + i0} = -\sqrt{-r - i0} = i\sqrt{|r|}$ for r positive real. Then,

$$-\frac{1}{\pi} \mathrm{tr}\, \hat{G}(i\omega_n \to \epsilon^+) = \frac{2\epsilon^+ \nu}{\sqrt{-\epsilon^{+2} + \Delta^2}} \simeq \frac{2\epsilon\nu}{\sqrt{-\epsilon^2 - i0\,\mathrm{sgn}\,(\epsilon) + \Delta^2}},$$

where we anticipate that the infinitesimal offset of ϵ in the numerator is irrelevant (trace its fate!) and, making use of the fact that, for ϵ approaching the real axis, only the sign of the imaginary offset matters: $(\epsilon + i0)^2 \simeq \epsilon^2 + 2i0\epsilon \simeq$

$\epsilon^2 + 2i0 \operatorname{sgn} \epsilon$. Taking the imaginary part of that expression, we arrive at the standard BCS form

$$\rho(\epsilon) = \operatorname{Im} \frac{2\epsilon\nu}{\sqrt{-\epsilon^2 - i0 \operatorname{sgn}(\epsilon) + \Delta^2}} = \begin{cases} 0, & |\epsilon| < \Delta, \\ \dfrac{2|\epsilon|\nu}{\sqrt{\epsilon^2 - \Delta^2}}, & |\epsilon| > \Delta. \end{cases}$$

7.3.2 Sum rules and other exact identities

REMARK This section introduces a number of useful exact identities obeyed by correlation functions. It is not directly related to the development of linear response theory and can be skipped at first reading.

Besides the connection between correlation functions, the Lehmann representation implies a number of additional identities. These formulae are not specific to any particular context. Based only on analytical structures, they are exact and enjoy general applicability. They can be used to obtain full knowledge of a correlation function from fragmented information – e.g., we saw in Eq. (7.18) and (7.19) how all three real-time correlation functions can be deduced once any one of them is known – or to gauge the validity of approximate calculations. The violation of an exact identity within an approximate analysis usually indicates that something has gone seriously wrong.

The spectral (density) function

spectral function We begin by defining the **spectral (density) function** as (minus twice) the imaginary part of the retarded correlation function,

$$\boxed{A(\omega) \equiv -2 \operatorname{Im} C^+(\omega)} \tag{7.28}$$

Using Eq. (3.84) and the Lehmann representation (7.16), it is straightforward to verify that

$$A(\omega) = \frac{2\pi}{\mathcal{Z}} \sum_{\alpha\beta} X_{1\alpha\beta} X_{2\beta\alpha} \left[e^{-\beta\Xi_\alpha} - \zeta_{\hat{X}} e^{-\beta\Xi_\beta} \right] \delta(\omega + \Xi_{\alpha\beta}). \tag{7.29}$$

INFO To understand the **physical meaning of the spectral function**, we again consider the case $\hat{X}_1 = c_a$, $\hat{X}_2 = c_a^\dagger$, where $\{|a\rangle\}$ is the eigenbasis of the non-interacting part, $\hat{H}_0$, of some many-body Hamiltonian. In the absence of interactions, $A(\omega) = 2\pi\delta(\omega - \xi_a)$ follows from Eq. (7.25); the spectral function is singularly peaked at the single-particle energy $\xi_a = \epsilon_a - \mu$.

This singularity reflects the fact that, in the free system, the many-body state $c_a^\dagger|\alpha\rangle$ is again an eigenstate. Since it is orthogonal to all other eigenstates, the summation over states $|\beta\rangle$ contains only a single non-vanishing term. We say that the "spectral weight" carried by the (unit-normalized) state $c_a^\dagger|\alpha\rangle$ is concentrated on a single state. However, in the presence of interactions, $c_a^\dagger|\alpha\rangle$ is no longer an eigenstate and we expect the spectral weight carried by this state to be distributed over a potentially large number of states $|\beta\rangle$.

It is instructive to explore this distribution of weight in the representation (7.26), where the effect of interactions has been lumped into a self-energy operator Σ. Taking the imaginary part of this expression, we find

$$A_a(\omega) = -2\frac{\Sigma''(\omega)}{(\omega - \xi_a - \Sigma'(\omega))^2 + (\Sigma''(\omega))^2}, \tag{7.30}$$

where we have neglected the infinitesimal imaginary increment ω^+ in comparison with the finite imaginary contribution $i\Sigma''$. This result indicates that interactions shift single-particle energies, $\epsilon_a \to \epsilon_a + \Sigma'$, and smear the previously singular distribution of weights into a Lorentzian of width Σ'' (see figure). This scales in proportion to the inverse of the lifetime τ discussed in the previous section, indicating that the resolution in energy (the width of the broadened δ-function) is limited by the inverse of the longest time scale (the lifetime of the quasi-particle state) by Heisenberg reciprocity.

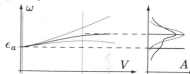

spectral function as probability

The spectral density function obeys the general normalization condition, $\int \frac{d\omega}{2\pi} A_a(\omega) = 1$, and it is positive (see the exercise below). These features suggest an interpretation in terms of a **probability measure** describing in what way the state $c_a^\dagger |\alpha\rangle$ is spread out over the continuum of many-body states. The signatures of this broadening are schematically illustrated in the figure: for vanishing interaction strength, A is a δ-function at a free particle eigenvalue. With increasing interaction, we observe a shift of the center and the broadening to a Lorentzian, as described by the self-energy representation (7.30). For still stronger interactions, the spectral function need no longer look Gaussian, and the analysis of its energy dependence reveals valuable information on the system's excitations. Indeed, the spectral function is an important object of study in numerical analyses of strongly correlated systems.

EXERCISE Apply analyticity arguments to argue that Eq. (7.30) obeys the unit normalization condition $\int \frac{d\omega}{2\pi} A_a(\omega) = 1$, including for frequency-dependent $\Sigma(\omega)$. Show the exactness of this statement on the basis of Eq. (7.29). Verify that all terms contributing to the spectral expansion are positive.

The spectral expansion may be applied to demonstrate other useful, and exact features of the spectral measure. As an example, show that the integral of the spectral function weighted by the Fermi or Bose distribution function yields the average state occupation:

$$\int \frac{d\omega}{2\pi} n_{F/B}(\omega) A_a(\omega) = \langle \hat{n}_a \rangle. \tag{7.31}$$

A number of exact identities involving correlation functions are formulated in terms of the spectral function. For example, the spectral function carries the same information as the correlation function itself. (In view of the fact that A is obtained by removing the real part of C, this statement might be surprising at first sight.) To see this, consider the integral

$$I = \int_{-\infty}^{\infty} \frac{d\omega}{2\pi} \frac{A(\omega)}{z - \omega}$$

for $\mathrm{Im}\, z > 0$, for definiteness. With $A(\omega) = i(C^+(\omega) - C^-(\omega))$, the second term in the integral is analytic in the entire lower half-plane, and $C^-(\omega)$ decays sufficiently fast to apply the Cauchy theorem and conclude that it does not contribute. By contrast, the first term is analytic in the upper half-plane, except for a simple pole at $\omega = z$. Application of the theorem of residues to an infinite semicircular contour surrounding the pole yields $I = C^+(z) = C(z)$. The same reasoning applied to $\mathrm{Im}\, z < 0$ shows that $I = C(z)$ in general, or

$$C(z) = \int_{-\infty}^{\infty} \frac{d\omega}{2\pi} \frac{A(\omega)}{z - \omega} \tag{7.32}$$

Notice that this is a highly nonlocal relation: yes, we can obtain the full expression for $C(z)$ from just its imaginary part. However, for this we need to know the imaginary part along the entire real axis.

Equation (7.32) is one variant of a few other identities obtained from the above reasoning. Specifically, setting $z = \omega^+$, we have

$$C^+(\omega) = -\frac{1}{2\pi i} \int d\omega' \frac{C^+(\omega')}{\omega - \omega' + i0}.$$

Representing the denominator under the integral in terms of the Dirac identity and collecting terms, this yields

$$C^+(\omega) = \frac{1}{\pi i} \int d\omega' C^+(\omega') \mathrm{P}\frac{1}{\omega' - \omega}$$

where P denotes the principal part. It is customary to consider the real and imaginary parts of this relation separately, whence one arrives at the **Kramers–Kronig relations** or **dispersion relations**:

Kramers–
Kronig
relations

$$\begin{aligned}
\mathrm{Re}\, C^+(\omega) &= \frac{1}{\pi} \int d\omega' \, \mathrm{Im}\, C^+(\omega') \mathrm{P}\frac{1}{\omega' - \omega} \\
\mathrm{Im}\, C^+(\omega) &= -\frac{1}{\pi} \int d\omega' \, \mathrm{Re}\, C^+(\omega') \mathrm{P}\frac{1}{\omega' - \omega}
\end{aligned} \tag{7.33}$$

INFO To point out the **physical content of the Kramers–Kronig relations**, we anticipate a little and note that the scattering amplitude of particles of incident energy ω is proportional to the retarded Green function $C^+(\omega)$ of the scattering target. The Kramers–Kronig relations imply that the real part of the scattering amplitude, the index of refraction, is proportional to the imaginary part, the index of absorption, integrated over all energies. In this way, these relations establish a connection between two seemingly unrelated physical mechanisms.

The dielectric function: a case study

As an example of the application of the Kramers–Kronig relations, we consider the **dielectric function** $\epsilon(\mathbf{q}, \omega)$ describing the polarization properties of a medium in the presence of an electromagnetic field. In section 4.2, we computed the dielectric function within the framework of the random phase approximation (RPA), cf. Eq. (4.38). However, here we wish to stay on a more general footing. Indeed, it is a straightforward exercise in linear response theory to show that (see problem 7.6.2)

$$\epsilon(\mathbf{q},\omega)^{-1} = 1 - \frac{V_0(\mathbf{q})}{L^d} \int d\tau \, e^{i\omega_m \tau} \langle \hat{n}(\mathbf{q},\tau)\hat{n}(-\mathbf{q},0)\rangle_c \bigg|_{i\omega_m \to \omega^+}$$

$$= \epsilon(\mathbf{q},\omega)^{-1} = 1 - \frac{V_0(\mathbf{q})}{L^d}C^+(\mathbf{q},\omega), \qquad (7.34)$$

where $V_0(\mathbf{q}) = 4\pi e^2/q^2$ is the bare Coulomb potential, $\langle \hat{n}\hat{n}\rangle_c$ denotes the connected thermal average of two density operators $\hat{n}(\mathbf{q},\tau) = c_\mathbf{q}^\dagger(\tau)c_\mathbf{q}(\tau)$, and $i\omega_m \to \omega^+$ indicates the analytic continuation to real frequencies. Heuristically, Eq. (7.34) can be understood by noting that $1/\epsilon = V_{\mathrm{eff}}/V_0$ measures the ratio between the effective potential felt by a test charge in a medium and the vacuum potential. The difference between these two quantities is due to the polarization of the medium, which, in turn, is a measure of its inclination to build up charge distortions $\delta\langle\hat{n}\rangle$ in response to the action of the potential operator $\sim \int dV_0\hat{n}$. In linear response theory,[16] $\delta\langle\hat{n}\rangle$ is given by the kernel $\int d\langle\hat{n}\hat{n}\rangle V_0$, i.e., the second term in Eq. (7.34). We thus observe that the dielectric function is determined by the retarded correlation function $C^+(\mathbf{q},\omega)$ with $\hat{X}_1 = \hat{X}_2 = \hat{n}$. This particular response function is called the retarded **density–density response function**, and it appears as an important building block in various contexts of many-body theory.

density–
density
response
function

We now focus on the response function, $\epsilon^{-1} - 1 \propto C^+$, and investigate what we can learn from analyticity criteria. Application of the Kramers–Kronig relation gives

$$\mathrm{Re}\,\epsilon(\mathbf{q},\omega)^{-1} - 1 = \frac{1}{\pi}\int_{-\infty}^{\infty} d\omega' \,\mathrm{Im}\,\epsilon(\mathbf{q},\omega')^{-1}\frac{1}{\omega'-\omega}.$$

Using the fact that $\mathrm{Im}\,\epsilon(\mathbf{q},\omega')^{-1} = -\mathrm{Im}\,\epsilon(-\mathbf{q},-\omega')^{-1} = -\mathrm{Im}\,\epsilon(\mathbf{q},-\omega')^{-1}$, where the first identity holds for the Fourier transform of arbitrary real-valued functions and the second follows from real-space symmetry, the integral is brought to the form

$$\mathrm{Re}\,\epsilon(\mathbf{q},\omega)^{-1} = 1 + \frac{2}{\pi}\int_0^{\infty} d\omega' \,\mathrm{Im}\,\epsilon(\mathbf{q},\omega')^{-1}\frac{\omega'}{\omega'^2-\omega^2}. \qquad (7.35)$$

Now consider this relation in the limit $\omega = 0$ and $|\mathbf{q}| \to 0$, describing the response to a quasi-static electromagnetic perturbation. Previously, we saw that in this limit the dielectric function behaves as (cf. Eq. (4.39))

$$\epsilon(\mathbf{q},0)^{-1} = (1 + 4\pi\nu|\mathbf{q}|^{-2})^{-1} \overset{|\mathbf{q}|\to 0}{\longrightarrow} 0. \qquad (7.36)$$

While this result was derived in RPA, it is based on the principles of Thomas–Fermi screening and should therefore be of general validity. Substitution of it into

sum rule Eq. (7.35) leads to the **sum rule**[17]

$$\boxed{\lim_{\mathbf{q}\to 0}\int_0^{\infty} d\omega \,\frac{\mathrm{Im}\,\epsilon(\mathbf{q},\omega)^{-1}}{\omega} = -\frac{\pi}{2}} \qquad (7.37)$$

---

[16] The linear response approximation is appropriate here since the standard definition of the dielectric function $\epsilon = \lim_{V_0\to 0}(V_0/V_{\mathrm{eff}})$ implies an infinitesimally weak external perturbation.

[17] Various other sum rules obeyed by the dielectric function are derived in Ref. [13].

Exact sum rules play an important role as consistency checks for approximate physical theories. In the present case, we learn that the full integral over the frequency-dependent absorption properties of a medium (which are complicated) must integrate to a value determined by its static screening properties (which is simple). The derivation above exemplifies the standard principle for the construction of other sum rules. (a) A quantity of physical interest is represented in terms of a retarded response function, which then (b) is substituted into a Kramers–Kronig-type relation. This produces a frequency-non-local connection between the response function at a given frequency and an integral ("sum") over all other frequencies. (c) The integral is evaluated at a specific frequency, where the reference quantity is known, and so we obtain a "known = $\int$ unknown" relation.

Experimental access to the spectral density function

Previously, we showed how the structure of the spectral function reflects the dispersion of the fundamental excitations of a system. In this section, we emphasize a complementary aspect: the spectral function is measurable. In this way, it defines one of the most direct interfaces between theory and experiment.

For concreteness, consider a scattering setup, as indicated in the figure on page 388. To describe the scattering rate from an incoming state $(\epsilon, \mathbf{k})$ into an outgoing state $(\epsilon', \mathbf{k}')$, we first note that the full Hilbert space $\mathcal{H}$ is the direct product of the Fock space $\mathcal{F}$ of the target system and the single-particle space $\mathcal{H}_1$ of the incoming particle: $\mathcal{H} = \mathcal{F} \otimes \mathcal{H}_1$. For simplicity, we describe the system–particle interaction by a point-like interaction

$$\hat{H}_{\text{int}} = C \int d^d r \, \delta(\hat{\mathbf{r}} - \mathbf{r}) c^\dagger(\mathbf{r}) c(\mathbf{r}) = C \int \frac{d^d q}{(2\pi)^d} e^{i\mathbf{q}\cdot\hat{\mathbf{r}}} \hat{\rho}(\mathbf{q}), \tag{7.38}$$

where $\hat{\mathbf{r}}$ is the position operator in $\mathcal{H}_1$ and, in the second equation, we have transformed to momentum space, with $\hat{\rho}(\mathbf{q}) \equiv \int dr \, e^{-i\mathbf{q}\cdot\mathbf{r}} c^\dagger(\mathbf{r}) c(\mathbf{r})$ the particle density operator. Scattering processes of first order in the interaction Hamiltonian are described by transition matrix elements $\mathcal{A}(\mathbf{q}) = \langle \beta, \mathbf{k} - \mathbf{q} | \hat{H}_{\text{int}} | 0, \mathbf{k} \rangle$ in the system eigenbasis $\{|\alpha\rangle\}$. Substitution of the above interaction Hamiltonian leads to

$$\mathcal{A}(\mathbf{q}) = \langle \beta, \mathbf{k} - \mathbf{q} | \hat{H}_{\text{int}} | 0, \mathbf{k} \rangle \propto \langle \beta | \hat{\rho}_\mathbf{q} | 0 \rangle. \tag{7.39}$$

We conclude that the scattering amplitude probes density modulations in the bulk.

Turning to observable consequences, we note that the transition *rate* between incoming and outgoing states differing in momentum $\mathbf{q} = \mathbf{k}' - \mathbf{k}$ and frequency $\omega = \epsilon' - \epsilon$ is computed from the scattering amplitude as (consult a quantum mechanics textbook if you feel unfamiliar with this statement)

$$\mathcal{P}(q) = \mathcal{P}(\omega, \mathbf{q}) = 2\pi \sum_\beta |\langle \beta | \hat{\rho}(\mathbf{q}) | 0 \rangle|^2 \delta(\omega - \Xi_{\beta 0}), \tag{7.40}$$

where the δ-function enforces energy conservation. The summation over β reflects the fact that only scattering states are monitored, while the final state of the target remains unobserved. To relate this to response functions, we abbreviate $\hat{\rho}(\mathbf{q})_{\alpha\beta}$ as $\langle\alpha|\hat{\rho}(\mathbf{q})|\beta\rangle$ and reformulate (7.40) as

$$
\mathcal{P}(q) = -2\,\mathrm{Im}\sum_{\beta}\frac{\rho(\mathbf{q})_{\beta 0}\rho(-\mathbf{q})_{0\beta}}{\omega^+ + \Xi_{0\beta}} = -2\lim_{T\to 0}\mathrm{Im}\frac{1}{\mathcal{Z}}\sum_{\alpha\beta}\frac{\rho(\mathbf{q})_{\beta\alpha}\rho(-\mathbf{q})_{\alpha\beta}e^{-\beta\Xi_\alpha}}{\omega^+ + \Xi_{\alpha\beta}}
$$

$$
= -2\lim_{T\to 0}\mathrm{Im}\frac{1}{\mathcal{Z}}\sum_{\alpha\beta}\frac{\rho(\mathbf{q})_{\beta\alpha}\rho(-\mathbf{q})_{\alpha\beta}(e^{-\beta\Xi_\alpha} - e^{-\beta\Xi_\beta})}{\omega^+ + \Xi_{\alpha\beta}}
$$

$$
= -2\lim_{T\to 0}\mathrm{Im}\,C^+(\omega) = A(\mathbf{q},\omega),
$$

where $A(\mathbf{q},\omega)$ is the spectral density function of the operators $\hat{X}_1 = \hat{\rho}(\mathbf{q})$ and $\hat{X}_2 = \hat{\rho}(-\mathbf{q})$. Here, we have made use of the fact that, for $T\to 0$, the Boltzmann weight $\exp(-\beta\Xi_\alpha)$ projects onto the ground state. Similarly, for $\omega > 0$, the contribution $\exp(-\beta\Xi_\beta)$ vanishes (exercise: why?).

We conclude that:

> The inelastic scattering cross-section for momentum transfer $\mathbf{q}$ and energy exchange ω is a direct probe of the spectral density function $A(\mathbf{q},\omega)$.

Building on a first-order perturbative representation of the scattering amplitude, this result belongs to the general framework of linear response relations.

7.4 Electromagnetic Linear Response

In this section we apply the general formalism to the particular case of the response to an electromagnetic field. In numerous applications discussed in this text, we "minimally couple" the vector potential representing a field to a theory and study the consequences, usually in some weak-field approximation. Here, we highlight the general features that all these coupling schemes have in common: they represent the retarded linear response of a system.

Consider a system of charged particles in the presence of an electromagnetic field with potential $A_\mu(x) = (\phi(x), \mathbf{A}(x))$, where $x^\mu = (t, \mathbf{x})$ and we are working in a real-time ("Lorentzian") framework in d space dimensions. The system will respond to this perturbation by the build-up of charge and/or current expectation values, $j^\mu = \langle\hat{j}^\mu\rangle$, where $j^\mu = (\rho, \mathbf{j})$. We wish to identify the linear functional $j = K[A] + \mathcal{O}(A^2)$ relating the current to its driving potential. Written more explicitly,

$$
j^\mu(x) = \int_{t'<t}dx'\,K^{\mu\nu}(x, x')A_\nu(x'), \tag{7.41}
$$

where the condition $t' < t$ indicates that the response is retarded.

In equilibrium quantum field theory (for the discussion of out-of-equilibrium electromagnetic responses, see chapter 12), we first compute the **response tensor in imaginary time** as $K(\mathbf{r}, \mathbf{r}'; i\omega_n)$ and then continue to real frequencies,

$i\omega_n \to \omega + i0$. In this Euclidean framework, the Lorentzian coupling $j^\mu A_\mu = -j_0 A_0 + \mathbf{j} \cdot \mathbf{A}$, with $j_0 = \rho$ and $A_0 = \phi$, becomes $j^\mu A_\mu = +j_0 A_0 + \mathbf{j} \cdot b\mathbf{A}$, with $j_0 = i\rho$ and $A_0 = i\phi$. The additional i mirrors the i from the temporal component of the imaginary-time vector $(t, \mathbf{r}) \to (-i\tau, \mathbf{r})$. Note that, in either framework, $j^\mu A_\mu = -\rho\phi + \mathbf{j} \cdot \mathbf{A}$.

Before turning to a more concrete approach, we note **two constraints obeyed by the response kernel** K on general grounds. First, a change of gauge $A_\mu \to A_\mu + \partial_\mu f$ cannot alter the response current. Substituting this condition into Eq. (7.41), we find

$$0 \overset{!}{=} \int_{t' < t} dx' \, K^{\mu\nu}(x, x') \partial_\nu f(x') = -\int_{t' < t} dx' \, (\partial_{x'_\nu} K^{\mu\nu}(x, x')) f(x').$$

Since f is arbitrary, this implies $0 = K^{\mu\nu} \overset{\leftarrow}{\partial}_\nu$ where the arrow indicates that the derivative acts to the left. Second, current conservation demands $\partial_\mu j^\mu = 0$, or $0 = \int dx' \, \partial_{x_\mu} K^{\mu\nu}(x, x') A_\nu(x')$ in the language of Eq. (7.41). As it holds for general A, this relation implies that $\partial_\mu K^{\mu\nu} = 0$. Summarizing,

> Gauge invariance and particle number conservation demand that
> $$\partial_\mu K^{\mu\nu} = K^{\mu\nu} \overset{\leftarrow}{\partial}_\nu = 0.$$

Our starting point in the derivation of a more explicit representation of the response tensor is the definition

$$j^\mu = \frac{\delta S_c[A]}{\delta A_\mu}, \tag{7.42}$$

for the current in a system with action $S = S_{\text{EM}} + S_c$. Here, the subscript c indicates that only the action, S_c, describing field–matter coupling is differentiated, while the Maxwell action S_{EM} of the electromagnetic field itself remains unaffected.[18]

INFO For readers not familiar with Eq. (7.42), we note that the vector potential $\mathbf{A}$ couples to a system of charged particles at coordinates $\mathbf{r}_i$ through the term

$$S_c[\mathbf{A}] \equiv \int d\tau \sum_i \dot{\mathbf{r}}_i \cdot \mathbf{A}(\mathbf{r}_i) = \int d\tau \int d^d r \sum_i \delta(\mathbf{r} - \mathbf{r}_i) \dot{\mathbf{r}}_i \cdot \mathbf{A}(\tau, \mathbf{r}).$$

However, $\mathbf{j} = \delta_{\mathbf{A}} S_c = \sum_i \delta(\mathbf{r} - \mathbf{r}_i) \dot{\mathbf{r}}_i$ is just the definition of the total particle current density. Similarly, the coupling to the scalar potential (in imaginary time) is given by

$$S_c[A_0] = -\int d\tau \sum_i \phi(\mathbf{r}_i) = \int d\tau \int d^d r \sum_i \delta(\mathbf{r} - \mathbf{r}_i) i A_0(\mathbf{r}),$$

so that $\delta_{A_0} S_c = \sum_i \delta(\mathbf{r} - \mathbf{r}_i) = i\rho = j_0$.

Equation (7.42) is all we need to formulate a general expression for the linear response kernel. Since the electromagnetic potential A_μ both drives a current as a

---

[18] Why do we not start from a more explicit representation such as $S_c[A] = \int j^\mu A_\mu$? The reason is that, in some cases – the microscopic theory of non-relativistic charged particles (section 7.4) being the most prominent example – the current $j^\mu = j^\mu[A]$ depends on A itself, so that $S_c[A]$ is nonlinear in A.

generalized force and generates its expectation value via (7.42), Eq. (7.6) assumes the form

$$\boxed{K^{\mu\nu}(x, x') = \frac{1}{\mathcal{Z}} \frac{\delta^2}{\delta A_\mu(x)\, \delta A_\nu(x')} \mathcal{Z}[A]} \tag{7.43}$$

where, as usual, the derivative is taken at zero. We note that the symmetry of this relation implies that $K^{\mu\nu}(x, x') = K^{\nu\mu}(x', x)$. As a consequence, $\partial_\mu K^{\mu\nu} = K^{\mu\nu} \overleftarrow{\partial}_\nu$, and we conclude that, in linear response, gauge invariance (right-hand side) and particle number conservation (left-hand side) represent flip sides of the same coin.

This is about as much as can be said in general. In the following we use two examples to illustrate how an electromagnetic linear response manifests itself in concrete terms in microscopic and in effective theories.

INFO Equation (7.46) describes the general electromagnetic response of many particle systems. For example, working in a gauge with vanishing scalar component A_0, we have $E_i(q) = \omega_m A_i(q)$ for the electric field. On this basis, the **longitudinal conductivity** $j_i(\omega) = \sigma(\omega) E_i(\omega)$ is obtained from the response tensor as

$$\sigma(\omega) = -\lim_{\mathbf{q}\to 0} \frac{1}{\omega_m} K_{ii}(q) \bigg|_{i\omega_m \to \omega + i0}. \tag{7.44}$$

In the same manner, the off-diagonal components $K_{i\neq j}$ yield the **transverse or Hall conductivity**. The temporal components K_{00} describe the **density response**, an observable that, as we saw, is essential in the analysis of scattering data. Notice that gauge invariance/current conservation imply some redundancy in this scheme. For example, in a gauge with non-vanishing scalar component, information on the conductivity is contained in the temporal components of the response tensor, etc. This freedom may be, by an approximate calculation, used to focus on the components that one finds easiest to compute, and/or to check that gauge invariance is respected.

Electromagnetic response of the microscopic theory

Consider the microscopic action of a system of spinless[19] fermions or bosons in the presence of an electromagnetic field,

$$S[\psi, A] = \int dx\, \bar{\psi} \left(\partial_\tau + iA_0 + \frac{1}{2m}(-i\nabla - \mathbf{A})^2 - \mu + V_0 \right) \psi + S_{\text{int}}[\psi],$$

where V_0 denotes a (field-independent) potential. In this case, the derivative (7.42) leads to

$$j_0 = i\bar{\psi}\psi, \quad j_i = \frac{1}{m} \bar{\psi} \left(-\frac{i}{2} \overleftrightarrow{\partial}_i + A_i \right) \psi, \tag{7.45}$$

where $f\overleftrightarrow{\partial}g \equiv f\partial g - (\partial f)g$. Note that the spatial components of the current depend on the vector potential. Application of Eq. (7.43) then gives

---

[19] To keep the discussion simple, we do not consider the electromagnetic coupling of spin degrees of freedom, such as the Zeeman or spin–orbit coupling.

$$K^{\mu\nu}(x,x') = \frac{1}{Z} \frac{\delta^2}{\delta A_\mu(x)\,\delta A_\nu(x')} \int D\psi\, e^{-S[\psi,A]}$$

$$= -\frac{1}{Z} \frac{\delta}{\delta A_\nu(x')} \int D\psi\, j^\mu(x)\, e^{-S[\psi,A]}$$

$$= \left\langle -\delta(x-x')\delta^{\mu\nu}(1-\delta^{\mu,0}) \frac{1}{m}(\bar\psi\psi)(x) + j^\mu(x)j^\nu(x') \right\rangle,$$

where the first term comes from the A-derivative of the current operator, and the current operators in the second term are defined at $A = 0$. We thus obtain

$$\boxed{K^{\mu\nu}(x,x') = -\frac{\langle \rho(x)\rangle}{m}\delta(x-x')\delta^{\mu\nu}(1-\delta^{\mu 0}) + \langle j^\mu(x)j^\nu(x')\rangle} \qquad (7.46)$$

for the response tensor of the electromagnetic field. The first term, whose presence originates in the $\mathbf{A}^2$ contribution to the microscopic Hamiltonian, describes the *diamagnetic* contribution to the response, and the second term gives the *paramagnetic* contribution. The diamagnetic and paramagnetic contributions tend to oppose each other in electromagnetic response phenomena. We studied such a competition in the context of the response of the BCS superconductor in section 5.3.6.

Electromagnetic response of effective theories

The tensor K describing the response of low-energy effective theories may look different from Eq. (7.46). To compute K, we first need to identify the coupling of the electromagnetic potential to the effective action. (It is often sufficient to minimally couple the field, i.e., to introduce A^μ in order to give the action local U(1)-invariance.) In a second step, we need to compute the derivative (7.43).

As an example, consider the **phase action of the BCS superconductor** minimally coupled to the potential, Eq. (5.39), where the coupling constants $c_1 = \nu_2$ and $c_2 = n_s/2m$ are defined through microscopic material parameters. Differentiating with respect to the components $\mathbf{A}_i$, one obtains, e.g.,

$$K_{ij}(x,x') = -\frac{n_s}{m}\left[\delta_{ij}\delta(x-x') - \frac{n_s}{m}\langle \partial_i\theta(x)\partial_j\theta(x')\rangle\right] \qquad (7.47)$$

for the spatial components of the response tensor.

EXERCISE Evaluate the correlator $\langle \partial_i\theta\,\partial_j\theta\rangle$ to show that the conductivity (7.44) diverges in the limit $\omega \to 0$. This **divergence of the conductivity** can be traced to the fact that, in a superconductor, there is no cancellation between the diamagnetic and paramagnetic responses. Compute the response of the system to a static magnetic field. Show that, in the London gauge $\nabla \cdot \mathbf{A} = 0$, the current and vector potential are proportional to each other, $\mathbf{j} \propto \mathbf{A}$. Recall (see section 5.3.7) that this result, in combination with the Maxwell equation $\nabla \times \mathbf{H} = 4\pi\mathbf{j}$, implies the **Meissner effect**.

7.4.1 Longitudinal conductivity of the disordered electron gas

REMARK In this intentionally technical section, we consider the example of a disordered electron gas to illustrate how response functions such as Eq. (7.46) are computed in concrete terms. We will introduce a few tricks related to the analytical structure of correlations functions and their continuation to real times. Although formulated for a specific example, the strategy parallels that applied in other applications of response theory. Since most steps of the computation are routine repetitions of what we have done in previous chapters, we will not formulate every step in ultimate detail.

longitudinal AC conductivity Our aim in this section is to compute the **longitudinal AC conductivity of the disordered electron gas**, $\mathbf{j}(\omega) = \sigma(\omega)\mathbf{E}(\omega)$, from the response tensor (7.46). We thus consider a situation with no current flow in directions transverse to the electric field (such as would be the case in the presence of strong magnetic fields outside the regime of linear response), and where the system is spatially isotropic, $K^{ij} \equiv K\delta^{ij}$.

Drude theory **EXERCISE** Before entering into the microscopic calculation, it is instructive to formulate an expectation of the result based on classical **Drude transport theory**. Consider a classical electron at coordinate $\mathbf{r}$ subject to the force of an electric field, $-\mathbf{E}(t)$ and to a friction force $-\frac{m}{\tau}\dot{\mathbf{r}}$ inhibiting its free acceleration due to the presence of disorder, where τ is the scattering time. Solve the equation of motion $m\ddot{\mathbf{r}} = -\mathbf{E} - \frac{m}{\tau}\dot{\mathbf{r}}$ in a real-time Fourier frequency representation for the velocity $\mathbf{v}(\omega)$, and from there show that the current $\mathbf{j} = -n\mathbf{v}$ (n is the particle density) is given by $\mathbf{j}(\omega) = \sigma(\omega)\mathbf{E}(\omega)$ with

$$\boxed{\sigma(\omega) = \frac{n}{m}\frac{1}{\tau^{-1} - i\omega}} \qquad (7.48)$$

In spite of the oversimplifying nature of this derivation – neglecting the presence of a Fermi surface, etc. – it underpins the expectation that the presence of disorder is an essential factor in electronic transport, at least at frequency scales $\omega \lesssim \tau^{-1}$ lower than the scattering rates.

We consider an electric field $\mathbf{E}$ of negligible spatial variation but finite temporal variation. Our object of interest is the induced current density, $\mathbf{j}(\mathbf{q} \to 0, \omega)$. Representing the vector potential as $\partial_t \mathbf{A} = -\mathbf{E}$, it is computed from the spatial components $K^{ii} \equiv K$ of the response tensor (7.46). Although the presence of impurities is crucial, we expect "self averaging," so that the response $K(x, x') = K(x - x')$ will depend only on coordinate differences. In this case, our starting point is the Fourier transformation

$$K(q) = \frac{1}{L^d}\left(-\frac{1}{m}\langle\rho\rangle_V + \langle j_{i,q}j_{i,-q}\rangle_V\right), \qquad (7.49)$$

at external momentum $q = (i\omega_m, \mathbf{0})$, where the index i is arbitrary (no summation). We are working in an imaginary-time framework and, at the end of the calculation, we need to perform the analytic continuation (7.44). With the current operators given by Eq. (7.45) (with $A_i = 0$), we have the Fourier representation

$$\rho = \sum_p \bar{\psi}_p \psi_p, \quad j_{i,q} = \frac{1}{m}\sum_p p_i \bar{\psi}_p \psi_{p+q}.$$

We substitute these expressions into (7.49) and apply Wick's theorem to compute the functional expectation value as

$$K(q) = -\frac{T}{L^d} \sum_p \left(\frac{1}{m} G_p + \frac{p_i^2}{m^2} G_{p+q} G_p \right), \tag{7.50}$$

where $G_p \equiv (i\omega_n - \xi_\mathbf{p} + \frac{i}{2\tau} \operatorname{sgn} \omega_n)^{-1}$ are the free electron Green functions averaged over realizations of the disorder potential (cf. Eq. (4.61)),[20] and τ defines the mean impurity scattering time.

Turning to the summation over frequencies by contour integration, the key point is that the extrapolation of the Green functions to the complex plane, $G_p = G_\mathbf{p}(z) = (z - \xi_\mathbf{p} + \frac{i}{2\tau} \operatorname{sgn} \operatorname{Im} z)^{-1}$, has a cut along the real axis; this, in turn, means that the product $G_{\mathbf{p}+\mathbf{q}}(z + i\omega_m) G_\mathbf{p}(z)$ has cuts at $\operatorname{Im}(z) = 0$ and $\operatorname{Im}(z + i\omega_m) = 0$ (see figure). We need to integrate over a contour encompassing all Matsubara frequencies while avoiding the cuts. While

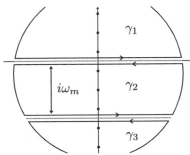

there is no connected integration path, the union of the three contours $\gamma_{1,2,3}$ shown in the figure suffices:

$$T \sum_{\omega_n} G_p G_{p+q} = -\frac{1}{2\pi i} \oint_{\gamma_1 \cup \gamma_2 \cup \gamma_3} dz \, n_\mathrm{F}(z) \, G_\mathbf{p}(z) G_\mathbf{p}(z + i\omega_m).$$

In these integrals, the imaginary parts of the Green function either have the same ($\gamma_{1,3}$) or opposite signs (γ_2). We know from experience that products containing propagators of opposite causality are "more interesting" than those without, and the present case is no exception. Indeed, one can show (this is a useful exercise in Green function algebra) that the two integrals over the contours γ_1 and γ_3 cancel the first (diamagnetic) term in Eq. (7.50). In this way, their role realizes the well-known diamagnetic–paramagnetic competition in the present electromagnetic response realization.

Turning to the integral over the central contour, γ_2:

$$\frac{1}{2\pi i} \oint_{\gamma_2} dz \, n_\mathrm{F}(z) G_\mathbf{p}(z) G_\mathbf{p}(z + i\omega_m)$$

$$= \frac{1}{2\pi i} \int_{-\infty}^{\infty} d\epsilon \, n_\mathrm{F}(\epsilon) \left[-G_\mathbf{p}(\epsilon - i0) G_\mathbf{p}(\epsilon + i\omega_m) + G_\mathbf{p}(\epsilon - i\omega_m) G_\mathbf{p}(\epsilon + i0) \right]$$

---

[20] Critical readers may wonder why we ignore "vertex" contributions, diagrammatically described by impurity scattering lines *between* the two Green functions, as in section 4.4.3. This is a somewhat subtle issue: owing to the strong momentum dependence of the current operators $j \propto p\bar\psi\psi$, such contributions tend to cancel. Physically, $\langle jj \rangle$ represents the expectation value of two velocity operators at different points in space. Impurity scattering leads to a rapid randomization of these velocities, which is why we do not expect "long-range" diffusion contributions to this particular correlation function.

$$\xrightarrow{i\omega_m \to \omega + i0} \frac{1}{2\pi i} \int_{-\infty}^{\infty} d\epsilon\, n_F(\epsilon) \left[-G_{\mathbf{p}}(\epsilon^-) G_{\mathbf{p}}(\epsilon^+ + \omega) + G_{\mathbf{p}}(\epsilon^- - \omega) G_{\mathbf{p}}(\epsilon^+) \right]$$

$$= -\frac{1}{2\pi i} \int_{-\infty}^{\infty} d\epsilon\, [n_F(\epsilon) - n_F(\epsilon + \omega)] G_{\mathbf{p}}(\epsilon^-) G_{\mathbf{p}}(\epsilon^+ + \omega),$$

where, in the first equality, we have used the symmetry $n_F(\epsilon + i\omega_m) = n_F(\epsilon)$ of the distribution function under translation by bosonic frequencies. Substituting this result into (7.50) and (7.44), we arrive at the intermediate result

$$\sigma(\omega) = \int_{-\infty}^{\infty} \frac{d\epsilon}{2\pi} \frac{n_F(\epsilon) - n_F(\epsilon + \omega)}{\omega} \frac{2}{L^d m^2} \sum_{\mathbf{p}} p_i^2 G_{\mathbf{p}}^-(\epsilon) G_{\mathbf{p}}^+(\epsilon + \omega) \qquad (7.51)$$

where $G_{\mathbf{p}}^{\pm}(\epsilon) = \left(\epsilon - \xi_{\mathbf{p}} \pm \frac{i}{2\tau} \right)^{-1}$. Equation (7.51) is an intuitive result: application of an AC field creates electron–hole pairs with excitation energy ω. The difference of Fermi functions demands that the energy of the excited electron lie within a shell $[\mu - \omega, \mu]$ at the Fermi surface. Its dynamics is described by a retarded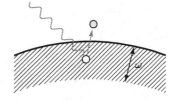
Green function of energy $\epsilon + \omega$, and that of the hole left behind by the advanced Green function. (Hole propagation can be understood as particle propagation backwards in time.) The weighting product p_i^2/m^2 probes velocity correlation as the observable for the conduction process.

The remaining two integrals are elementary. For low temperatures, $n_F(\epsilon) - n_F(\epsilon + \omega) \simeq -\omega d_\epsilon n_F(\epsilon) \simeq \omega \delta(\epsilon)$, so that the energy integral collapses and $G_{\mathbf{p}}^{\pm}(0) = (-\xi_{\mathbf{p}} \pm i/2\tau)^{-1}$ depends on momentum only through the energy variable $\xi_{\mathbf{p}}$. Owing to rotational invariance, we can make the replacement $p_i \to \mathbf{p}^2/d$, and the straightforward final integral over momentum confirms the result (7.48).

7.5 Summary and Outlook

In this chapter we discussed how real-time retarded response functions, represent a principal interface between experiment and theory. The connection was established by linear response, where the expectation values of observables modified by external influences are weak compared with the intrinsic energy scales of many-body systems. These retarded response functions have to be related to the imaginary-time correlation functions produced by finite-temperature equilibrium quantum field theory. We saw how these connections could be established in general on the basis of formal spectral expansions. We finally illustrated the working of this machinery on the important example of an electromagnetic linear response, considering the longitudinal conductivity of an electron gas as a concrete case study.

This chapter concludes the introductory part of the book. We have learned how to formulate quantum field theories, analyze them in terms of a spectrum of approximate methods, and read their contents by correlation functions. In the second part of the book, we discuss the application of field-theoretical concepts in various areas of condensed matter research, among them the theory of out-of-equilibrium systems, and the theory of topological and relativistic quantum matter.

7.6 Problems

7.6.1 Orthogonality catastrophe

This problem describes a scenario often realized in condensed matter physics: a localized quasi-particle is immersed in a large host system. Assuming the latter to be described by its N-body ground state wavefunction, the ensuing "single-particle state $\otimes$ N-particle ground state," turns out to be largely orthogonal to the true ground state of the now $(N + 1)$-particle system. As a consequence, it takes a very long time for the intruding particle to become accommodated to its new environment, a retardation phenomenon known as the orthogonality catastrophe. Following Anderson's original line of argument, here we explore the physical mechanism underlying this phenomenon.

Anderson showed[21] that the ground state $|\Psi'\rangle$ of a system in the presence of a local perturbation $V(\mathbf{r})$ is generically orthogonal to the ground state $|\Psi\rangle$ of the unperturbed system: $|\langle\Psi|\Psi'\rangle|^2 \xrightarrow{N\to\infty} 0$. Here, the local perturbation serves as a model of the interaction potential created by a new particle in the system. The principle applies, for example, to samples connected to external leads at local contacts. In this case, the injection of external charge carriers gets blocked, and the resulting "contact resistance" can become the essential bottleneck in electron transport.

(a) Consider a system of non-interacting fermions. Let the single-particle eigenstates in the absence or presence of the perturbation be given by $\{|m\rangle\}$ and $\{|m'\rangle\}$, respectively, and the corresponding many-body ground states by $|\Psi\rangle$ and $|\Psi'\rangle$. These states are Slater determinants formed from all single-particle states with energies below a Fermi energy E_F. Their overlap is given by $\chi \equiv \langle\Psi|\Psi'\rangle \equiv \det(A)$, where $A_{nm} = \langle n|m'\rangle$.

To understand the magnitude of this determinant, use the nontrivial statement that the determinant of a matrix $\tilde{A} = (v_1, v_2, \ldots)$ formed by *unit normalized* vectors $|v_i| = 1$ obeys the condition $\det(\tilde{A}) \leq 1$.[22] Construct a determinant normalized in this way, and use the bound to reduce the estimate of $\det A$ to one of the norms of its row vectors. Make use of the completeness of the single-particle states $\{|m\rangle\}$ and assume that the overlap $\sum_{\epsilon_{m'}>E_F} |\langle n|m'\rangle|^2 \ll 1$ between occupied states $|n\rangle$ and empty $|m'\rangle$ states is small. On this basis, show that

[21] P. W. Anderson, *Infrared catastrophe in Fermi gases with local scattering potentials*, Phys. Rev. Lett. **18**, 1049 (1967).

[22] This bound can be understood from the interpretation of a determinant as the geometric volume of the parallelepiped spanned by its row vectors. For vectors of length unity, this volume is equal to or less than unity.

$$\tilde{\chi} \equiv |\langle \Psi | \Psi' \rangle|^2 < \chi = \exp\left[-\frac{1}{2} \sum_{\epsilon_n \le E_F, \epsilon_{m'} > E_F} |\langle n | m' \rangle|^2 \right] \equiv \exp[-\mathcal{I}].$$

(b) We next compute the exponent $\mathcal{I}$ for the particularly simple case of a spherically symmetric system of radius R. Focusing on the sector of lowest angular momentum $l = 0$, the unperturbed states are given by $\phi_n(r) = N_n \sin(k_n r)/k_n r$, where $k_n = \pi n/R$, and the normalization $N_n = k_n/\sqrt{2\pi R}$. The asymptotic profile of the perturbed wavefunctions can be approximated as $\phi'_n(r) = N_n \sin(k_n r + \delta_m(1 - r/R))/k_n r$, where δ_m is the s-wave scattering phase shift. Use this ansatz to show that

$$\langle n | m' \rangle = \frac{\sin \delta_m}{\pi(n - m) + \delta_m}.$$

Assuming that the scattering phase shift $\delta_m \to 0$ for states $|m'\rangle$ with energy $\epsilon_m \sim k_m^2/2m_0 \gtrsim \epsilon_0$ (m_0 is the particle mass), argue that $I \propto c \ln N$ and hence $\chi \sim N^{-c}$ is inversely proportional in the system size, where c is a constant.

Answer:

(a) We define a matrix with unit-normalized rows through $\tilde{A}_{nm} \equiv A_{nm}/\mathcal{N}_n$ with $\mathcal{N}_n^2 = \sum_m |\langle n | m' \rangle|^2$. Since $\det \tilde{A} < 1$, $\chi = \det \tilde{A} \prod_n \mathcal{N}_n \le \prod_n \mathcal{N}_n$. Taking the logarithm of this relation, we obtain

$$\ln \chi \le \sum_{\epsilon_n < E_F} \ln(\mathcal{N}_n) = \frac{1}{2} \sum_{\epsilon_n < E_F} \ln \left(\sum_{\epsilon_{m'} < E_F} |\langle n | m' \rangle|^2 \right).$$

Using the closure relation $\sum_{m'} |m'\rangle\langle m'| = 1$ and expanding the logarithm in $\sum_{\epsilon_{m'} > E_F} |\langle n | m' \rangle|^2 \ll 1$, we arrive at the approximation

$$\ln \left(\sum_{\epsilon_{m'} < E_F} |\langle n | m' \rangle|^2 \right) = \ln \left(1 - \sum_{\epsilon_{m'} > E_F} |\langle n | m' \rangle|^2 \right) \simeq - \sum_{\epsilon_{m'} > E_F} |\langle n | m' \rangle|^2 = -2\mathcal{I}.$$

Substituting this result into the formula for $\ln \chi$, we arrive at the stated result.
(b) The overlap matrix A is given by

$$A_{nm} = \langle n | m' \rangle = 4\pi N_n N_m \int_0^R dr \, r^2 \frac{\sin(k_n r)}{k_n r} \frac{\sin\left(k_m r + \delta_m(1 - \frac{r}{R})\right)}{k_m r}$$

$$\simeq \frac{2\pi N_n N_m}{k_n k_m} \frac{\sin \delta_m}{k_n - k_m + \frac{\delta_m}{R}}.$$

Using the above relation between wave numbers and momenta, we arrive at the result. Under the stated assumptions, the double sum over n and m is cutoff at an upper limit determined by the condition $(\pi m/R)^2/2m_0 \sim \epsilon_0$, or $m \sim R$. Since the "integration" over n and m is weighted by a term $\propto (n - m)^{-2}$, we obtain a result logarithmic in the cutoff, $I \sim \ln R \sim \ln N$, where $N \sim R^3$ is the particle number. From here, we obtain the scaling $\chi = \exp(-I) \sim N^{-c}$, proportional to an inverse power of the particle number.

7.6.2 RPA dielectric function

On a fundamental level, the response of a charged fermion system to an external electro-magnetic perturbation is described by the dielectric function ϵ_q. In this problem, we derive the connection between ϵ_q and the density–density response function (7.34). The problem is physically instructive and a good exercise in diagrammatic calculus.

(a) Derive Eq. (7.34). To this end, consider two infinitesimally weak (yet generally time-dependent) test charges immersed in a system of charged fermions. Omitting the self-interaction of the particles, determine the interaction correction to the free energy and use your result to compute the ratio between the vacuum and the actual interaction potential. Show that

$$\epsilon_q = \left(1 - \frac{TV_0(\mathbf{q})}{L^d} \langle \hat{n}_q \hat{n}_{-q} \rangle_c \right)^{-1}, \tag{7.52}$$

where the subscript c indicates that only connected diagrams contribute. Analytic continuation to real frequencies yields Eq. (7.34).

(b) Equation (7.52) represents the exact density–density correlation function, which is generally unknown. As a first step towards a more manageable expression, apply diagrammatic methods to show that Eq. (7.34) is equivalent to

$$\epsilon_q = 1 + \frac{TV_0(\mathbf{q})}{L^d} \langle \hat{n}_q \hat{n}_{-q} \rangle_{\mathrm{irr}}, \tag{7.53}$$

where $\langle \hat{n}\hat{n} \rangle_{\mathrm{irr}}$ is the interaction-irreducible density–density response function, the sum of all diagrams that cannot be cut by cutting one interaction line. (The response function $\langle \hat{n}\hat{n} \rangle_{\mathrm{irr}}$ is to the effective interaction what the self-energy is to the Green function.) In order to show that (7.52) and (7.53) are equivalent, find a series expansion of $C(\mathbf{q}, i\omega_m)$ in terms of V_0 and $\Pi(\mathbf{q}, i\omega_m)$. While this series resembles that of the RPA, it is more general in that the bubble $\langle \hat{n}\hat{n} \rangle_{\mathrm{irr}}$ still contains interaction lines.

Answer:

(a) If we represent the two test charges by two charge distributions $\rho_{1,2}(q)$, the Coulomb interaction becomes $S_{\mathrm{int}} = \frac{T}{2L^d} \sum_q [\hat{n}_q + \rho_{1,q} + \rho_{2,q}] V_0(\mathbf{q}) [\hat{n}_{-q} + \rho_{1,-q} + \rho_{2,-q}]$. An expansion of the field integral (or, equivalently, the partition function) to lowest order in $\rho_{1,2}$ then produces the interaction contribution to the free energy,

$$F_{\mathrm{int}}[\rho_1, \rho_2] = -\frac{T^2}{L^d} \sum_q \rho_{1,q} \left[V_0(\mathbf{q}) - \frac{TV_0^2(\mathbf{q})}{L^d} \langle \hat{n}_{-q} \hat{n}_q \rangle_c \right] \rho_{2,-q}.$$

The term in square brackets determines the effective interaction potential $V_{\mathrm{eff}}(q)$. Remembering that $\epsilon_q = V_0(\mathbf{q})/V_{\mathrm{eff}}(q)$, we obtain Eq. (7.52).

(b) For a diagrammatic representation of the solution, see fig. 7.2.

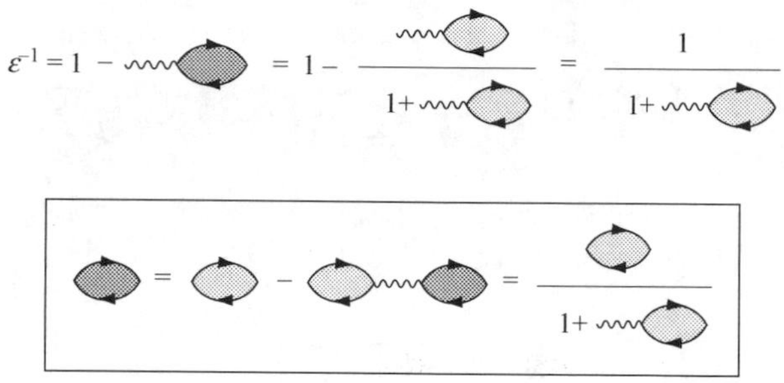

Fig. 7.2   Representation of the dielectric function in terms of the full density–density correlation func-
            tion (the darker-shaded bubbles) and the interaction-irreducible density–density correlation
            function (lighter-shaded bubbles).

7.6.3 Electromagnetic response of a quantum dot

In problem 5.6.5, we derived the effective action of a quantum dot tunnel coupled to external
leads. On this basis, here we derive a formula describing the response of the system to an
applied time-dependent voltage at high temperatures, where it behaves nearly classically.

(a) Consider Eq. (5.62) for a quantum dot coupled to a lead. Assuming that a bias
voltage $iU(\tau)$ between lead and dot has been applied, show that the argument of
the dissipative action generalizes to $\phi(\tau) \to \phi(\tau) + \int_0^\tau d\tau'\, U(\tau')$. (Hint: Apply gauge
arguments at an early stage in the construction of the effective action.) Focusing
on the regime of high temperatures, expand the action to second order in ϕ (argue
why, for high temperatures, anharmonic fluctuations of ϕ are small) and compute
the 00-component of the linear response kernel, K_{00}. Analytically continue back
to real frequencies and show that the current flowing across the tunnel barrier in
response to a voltage $U(\omega)$ is given by

$$I(\omega) = \frac{U(\omega)}{2\pi/g_T - 2E_C/i\omega}. \tag{7.54}$$

(b) To understand the meaning of this result, notice that, classically, the dot is
equivalent to a capacitor connected via a classical resistor to a voltage source.
Apply Kirchhoff's laws to obtain the current–voltage characteristics of the classical
system and compare with the result above.

Answer:

(a) Consider the action of the system at an intermediate stage, immediately after
the fermions have been integrated out (cf. Eq. (5.63)). In the presence of a bias
voltage, the "tr ln" operator contains a term $i(V + U)$, where V is the Hubbard–
Stratonovich field decoupling the interaction. Removing the sum of these two fields
by the same gauge transformation as used in the derivation of the dissipative action,
the tunneling matrix elements acquire the phase factor specified above.

At second order in the expansion in ϕ, the action assumes the form

$$S^{(2)}[\phi, V] = \frac{1}{4E_C T} \sum_m \omega_m^2 |\phi_m|^2 + \frac{g_T}{4\pi T} \sum_m |\omega_m| |\phi_m + U_m/i\omega_m|^2,$$

where the Fourier identity $|\omega_m| \leftrightarrow -\pi T \sin^{-2}(\pi T \tau)$ has been used. Differentiating twice with respect to U, we obtain

$$(K_{00})_m = \frac{1}{Z} T \frac{\delta^2}{\delta U_m \delta U_{-m}} Z[U] = -\frac{g_T}{2\pi |\omega_m|} \left(1 - \frac{g_T |\omega_m|}{2\pi T} \langle \phi_m \phi_{-m} \rangle \right)$$

$$= -\frac{g_T}{2\pi |\omega_m|} \left(1 - \frac{g_T |\omega_m|}{2\pi T} \frac{1}{\frac{\omega_m^2}{2E_C T} + \frac{g_T |\omega_m|}{2\pi T}}\right) = -\frac{1}{\frac{2\pi |\omega_m|}{g_T} + 2E_C}.$$

The 00-element of the linear response tensor describes changes in the particle number[23] in response to an applied potential, $\delta N_m = (K_{00})_m U_m$. Noting that $i\partial_\tau \delta N = I$ is the current through the barrier, and substituting our result for K_{00}, we obtain

$$I_m = U_m \left(\frac{2\pi}{g_T} + \frac{2E_C}{|\omega_m|}\right)^{-1}.$$

Analytical continuation from positive imaginary frequencies, $|\omega_m| \to -i\omega + 0$, to real retarded frequencies gives Eq. (7.54).

(b) According to Kirchhoff's laws, the sum of all voltage drops in the system must vanish, $0 = U(\omega) + C^{-1} \delta N(\omega) - RI(\omega)$, where C and R denote the capacitance and resistance, respectively. Using the fact that $\delta N = I/i\omega$, and solving for I, we readily obtain

$$I(\omega) = \frac{U(\omega)}{R - 1/i\omega C},$$

i.e., the current–voltage characteristics for a classical RC circuit. The identifications $R = 2\pi/G_T$ and $C = 1/2E_C$ bring us back to the formula derived microscopically above.

---

[23] Here, we note that the volume factor distinguishing between particle number and density is contained in our definition of the source variables.

PART II

Readers who have advanced thus far should find that they can fluently read the majority of condensed matter theory research papers. Building on this level of attainment, the second part of the text discusses directions of contemporary condensed matter research – as always from a field-theoretical perspective. Readers will find that the second part is somewhat different from the first, both in content and style. For one thing, its chapters no longer build sequentially on each other; they can be studied in no particular order. At the same time, they are strongly interlinked among each other. To explain this structure, we first note a topical division between the first three chapters of the second part and the final two chapters. The former have in common that they gravitate towards the topic of *topology*. Referring for an introductory discussion to section 8.1, topology has become a dominant theme in the condensed matter physics of the last two decades. Three chapters discuss this development from different perspectives, starting with chapter 8, which introduces field-theoretical concepts for the description of topological phases in condensed matter physics. Transitions between phases of different topological signature are often described by effectively relativistic (Dirac) theories. Chapter 9 introduces relativistic quantum field theory with an emphasis on this application field. Finally, the phenomenology of topological phases of matter is often described in the language of gauge fields. Chapter 10 discusses gauge theory with this application in mind, i.e., a perspective generalizing "minimal substitution schemes," $p \to p - A$, to a more geometrical and global interpretation of gauge theory.

Reflecting their interdependent contents, the three chapters make frequent references to each other. If a particular section essentially relies on material from another chapter, we say so at the beginning. We also note that all three of these chapters use the language of modern differential geometry, and notably that of differential forms. Familiarity with it is not required (in all but a few sections, we offer shortcuts formulated in the traditional language of calculus) but it definitely helps and makes for a more satisfactory reading. The essential background material is provided in appendix section A.1. First-time readers who are not in a rush may find it useful to read these pages before venturing into the topological chapters of this text.

The final two chapters of the text address the physics of *nonequilibrium systems*, first from a classical, then from a quantum perspective. In the last two to three decades, research on condensed matter out of equilibrium has developed from a relatively small sub-field to mainstream. Factors driving this development include experimental device miniaturization (the smaller a quantum system, the easier it is to push it out of equilibrium), interdisciplinary overlap with the fields of atomic condensates and quantum optics (which routinely operate under out-of-equilibrium conditions), and the increase in numerical computing power, which now makes the meaningful simulation of nonequilibrium phenomena an option.

Compared with the relatively few equilibrium phases, matter out of equilibrium realizes many more universality classes – to a degree that the universality principle of many-body physics may reach its limits – and a matching number of different theoretical approaches for investigating them. A principal advantage of the field-theoretical approach to nonequilibrium physics is that it combines a multitude of theoretical formulations under one conceptual roof. In chapter 11 we lay the groundwork for this unified formulation by first reviewing various traditional approaches to *classical nonequilibrium* physics. We then show how these can all be derived from a common field integral representation. This formulation then serves as a platform for an investigation of nonequilibrium physics which otherwise would stay out of reach.

Finally, chapter 12 addresses *quantum nonequilibrium* phenomena via field integrals which have those of chapters 11 as their semiclassical limits. Readers familiar with the foundations of nonequilibrium theory should find it possible to read chapter 12 without previous study of the preceding chapter. However, there are numerous cross references, and a combined reading of the two chapters may be more rewarding overall.

8   Topological Field Theory

SYNOPSIS In the context of condensed matter physics, the attribute "topological" refers to forms of matter whose microscopic physics is characterized by integer-valued indices, or "invariants." In this chapter, we approach the physics of topological quantum matter from the perspective of field theory. Much as continuous deformations do not change the presence or absence of a knot in a string, local perturbations cannot alter topological invariants. This implies a powerful principle of universality with far-reaching physical consequences. Here, we discuss how topological signatures are diagnosed and described by methods of field theory. The chapter starts with a brief overview of topological quantum matter. We then illustrate for a simple example how the topological master principle – robustness against local deformations – manifests itself in field theory via topological terms. This will be followed by a more systematic discussion of three principal families of topological terms (θ-terms, Wess–Zumino terms, and Chern–Simons terms) and their application in the physics of spin systems, the integer and fractional quantum Hall effects, and topological insulators.

The first two decades of the century have witnessed an explosion of research on topological quantum matter. The driving forces behind this development include both advances in our fundamental understanding of matter, and the promise of applications in quantum information technology or materials design. These days the field has become vast, and it is impossible to provide a comprehensive overview in this text. Instead, we will focus on the introduction of basic concepts in the physics of topological quantum matter, as always emphasizing the field theory perspective. However, the field-theoretical approach taken here is only one of several avenues to the physics of topological matter. Others include descriptions in terms of Dirac matter (chapter 9), gauge theory (chapter 10), mathematical classification theory, and more. These concepts are intertwined, and this shows up in the frequent appearance of topological concepts in the later chapters of this text.

We start with a short synopsis of topological quantum matter. Turning to the field integral framework, we then introduce salient features of topological actions for a toy model, that of a quantum particle on a ring. In the main part of the chapter, we discuss three large families of topological terms and their application in condensed matter physics.

8.1 Topological Quantum Matter

The first thing to appreciate when entering the field of topological quantum matter is that we are far from having established a complete understanding. Unlike materials distinguished by, say, different types of local magnetization, topological matter is defined by *non-local* forms of order. These ordering principles are interwoven into macroscopically large numbers of microscopic degrees of freedom and are often hard to identify. While the understanding of two-dimensional topological order is relatively advanced, the extension to higher dimensions is still in its infancy. Besides dimensionality, features entering the classification include quantum statistics (fermionic or bosonic), the presence or absence of excitation gaps distinguishing between insulators and metals, and the degree of particle correlations or *entanglement*.

Of these, the third criterion is the most essential. At the broadest level, we distinguish between topological matter with and without long-range entanglement. There are various equivalent definitions of this distinction. One definition says that a

long-range
entan-
glement

system is **long-range entangled (LRE)** if its ground states cannot be transformed into a disentangled product state by a sequence of local unitary transformations. (For alternative criteria, see below.) It should be evident that the complementary class of short-range entangled (SRE) matter is easier to understand. For example, we now have a more or less complete classification of fermionic gapped SRE topological quantum matter – the class of topological insulators and superconductors – in all dimensions. Numerous such systems have become experimental reality. Compared with that, both the classification and the experimental realization of LRE quantum matter is less developed.

In this text, we will discuss realizations of both SRE and LRE matter. To provide some orientation for beginners in the field, the following two sections contain a brief synopsis of essential features of these two material classes.

8.1.1 Short-range entangled topological matter

short-range
entangled
topological
matter

At first sight, the definition of **short-range entangled topological matter** above may seem a contradiction in terms. How can a system deformable to a trivial product state be topologically ordered? To understand why we have nontrivial forms of SRE topological matter, note that the above criterion requires deformability into a trivial state under the most general class of unitary transformations. However, many systems are protected by symmetries, such as the time-reversal or charge conjuga-

symmetry-
protected
topological
order

tion symmetries of superconductor Hamiltonians. Under these circumstances, only unitary transformations respecting those symmetries are physical. This restriction defines the concept of **symmetry-protected topological (SPT) order**:

▷ Different classes of SPT order cannot be deformed into each other by symmetry-respecting unitary transformations without passing through a phase transition.

▷ However, all SPT phases can be transformed into the same product state under the most general class of local unitary operations.

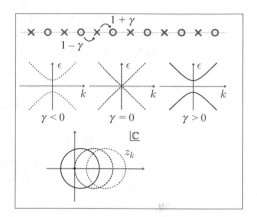

SSH model

To make the meaning of these statements more tangible, let us discuss an example: The **Su–Schrieffer–Heeger (SSH) model** was introduced in 1979 as a minimal model for the one-dimensional organic compound trans-polyacetylene.[1] It is defined as the fermion chain,

$$\hat{H} = \sum_i c_{i+1}^\dagger (1 + \gamma(-1)^i) c_i + \text{h.c.},$$

where γ is a real parameter defining a staggered hopping amplitude, and we have set the average hopping strength to unity. Introducing a two-site unit cell containing even ($\circ$) and consecutive odd ($\times$) sites, and switching to a momentum space representation, the Hamiltonian assumes the form (exercise)

$$\hat{H} = \sum_k C_k^\dagger \begin{pmatrix} & z_k \\ \bar{z}_k & \end{pmatrix} C_k, \qquad z_k = (1 - \gamma) + (1 + \gamma)e^{ik}, \tag{8.1}$$

where $C_k = (c_{\circ,k}, c_{\times,k})^T$, and we have set the lattice spacing to unity. From here the eigenvalues are readily computed as $\epsilon_{\pm,k} = \pm|z_k| = \pm\sqrt{2}((1+\gamma^2)+(1-\gamma^2)\cos k)^{1/2}$. For generic values of γ, this dispersion has a gap with minimal value $4|\gamma|$ at $k = \pi$. Only at $\gamma = 0$ does the gap close and $k = \pi$ becomes the center of a region of approximately linear dispersion (see the figure). To see how $\gamma = 0$ marks a transition between different phases, we need to take a look at the ground state wave function.

The quantum ground state of the insulating phases is defined by the Slater determinant built from all single-particle states with negative energy $\epsilon_-(k)$. Representing these (prior to normalization) as $\Psi_k = (-z_k, |z_k|)^T$, the information on the topological nature of the ground state resides in the upper component z_k: As $k \in [0, 2\pi]$ runs through the Brillouin zone, z_k traces out a circle of radius $1 + \gamma$, centered at $1 - \gamma$ in the complex plane (see bottom of figure). For $\gamma > 0$ ($\gamma < 0$) it does (does not) enclose the origin $z = 0$. It is not possible to change the control parameter γ adiabatically from one option to the other without passing through the point $\gamma = 0$. At this phase transition point, the ground state contains a zero-energy state. The gap to the valence band closes and the system is no longer an insulator. We conclude that the Hamiltonian of the SSH chain defines topologically distinct ground states distinguished by the "winding" of their wave functions. This structure implements the first of the two criteria mentioned above.

[1] W. P. Su, J. R. Schrieffer, and A. J. Heeger, *Solitons in polyacetylene*, Phys. Rev. Lett. **42**, 1698 (1979).

chiral
symmetry

To understand the role of the second criterion, we note that the winding number construction above is stabilized by a symmetry: the Hamiltonian anticommutes with the third Pauli matrix, $[\hat{H}, \sigma_3]_+ = 0$, which is an example of a **chiral symmetry**. The consequences of this symmetry are best discussed in a Pauli spin language, where we define $z_k = x_k + iy_k$, and represent the Hamiltonian as $\hat{H} = \sum_k C_k^\dagger(z_k \cdot \sigma)C_k$, with $z \cdot \sigma \equiv x_k\sigma_x - y_k\sigma_y$. In this representation, $k \mapsto z_k \cdot \sigma$ defines a closed curve in the equatorial plane of the class of spin Hamiltonians $H = v_i\sigma_i$, and the topological index distinguishes between curves that may or may not enclose the origin $z = 0$. Inside the plane, it is not possible to continuously deform one type of curve into the other without crossing the origin (the point of the gap-closing phase transition). However, the situation changes if we allow for unitary transformations $\hat{H} \to \hat{U}\hat{H}\hat{U}^{-1}$ violating the chiral symmetry in that they generate contributions proportional to σ_z. In this case, the freedom to enter the third dimension makes it possible to deform configurations with and without winding into the same trivial insulating state, without ever closing a gap.

EXERCISE Try to come up with a family of unitary transformations $U_{k,s}$ such that $U_{k,0} = \mathbb{1}$ is the identity and $U_{k,1}$ transforms the Hamiltonians H_k into the non-chiral form $H_k = |z_k|\sigma_3$ with trivial ground state $\Psi_k = (0,1)^T$. (Hint: Think geometrically. The curves previously confined to the xy-plane must be contracted to points on the z-axis.[2]) Since $|z_k|$ does not depend on the sign of γ, this is an example of a chiral symmetry-violating **unitary transformation connecting different topological sectors**: Curves with and without winding number around the origin of the plane are deformed into identical, point-like curves in three-dimensional space without compromising the ground state gap.

Do the two states separated by a phase transition differ in any characteristic besides a formal winding number? For an infinite or closed chain the answer is negative. This follows from the observation that a physically inconsequential relabeling $\circ \leftrightarrow \times$ maps the chain with parameter γ to one with $-\gamma$. However, the situation changes if a system with boundaries is considered. Assume the chain has been cut in such a way that a $\circ/\times$-site defines the left/right boundary. We now claim that the configuration $\gamma > 0$ supports a pair of zero-energy states exponentially localized at the boundaries, while there are no such states for $\gamma < 0$. The most straightforward way to verify this assertion is by direct computation in real space: the zero-energy Schrödinger equation $\hat{H}\psi = 0$ defined by the nearest-neighbor hopping Hamiltonian assumes the form $0 = \hat{H}\psi_{2l+1} = t(1 + \gamma)\psi_{2l} + t(1 - \gamma)\psi_{2l+2}$ and $0 = \hat{H}\psi_{2l} = t(1 - \gamma)\psi_{2l-1} + t(1 + \gamma)\psi_{2l+1}$, with boundary conditions $0 = \psi_1 = t(1 - \gamma)\psi_2$ and $0 = \psi_{2N} = t(1 + \gamma)\psi_{2N-1}$. Focusing on the left edge, the boundary condition requires $\psi_2 = 0$, which extends to $\psi_{2l} = 0$ upon iteration into the bulk. On the odd-numbered sites we have the recursion relation $\psi_{2l+1} = c\psi_{2l-1}$, with $c \equiv \frac{1-\gamma}{1+\gamma}$. For positive γ or $c < 1$, this equation has an exponentially decaying solution $\psi_{2l+1} = c^l\psi_1$ living on the odd-numbered sites. No solution is found for $\gamma < 0$. An analogous

[2] The transformation matrices $U_{k,s} = \exp\left(-\frac{\pi s}{2}\frac{|z_k|}{4}(\hat{x}_k\sigma_x + \hat{y}_k\sigma_y)\right)$ do the job. They transform the Hamiltonian to $U_{k,s}H_kU_{k,s}^{-1} = \cos(\pi s/2)(z_k \cdot \sigma) + \sin(\pi s/2)|z_k|\sigma_3$.

construction at the right-hand boundary demonstrates the existence of a second zero-energy state for $\gamma > 0$. The presence or absence of surface states in the finite chain makes the two phases physically distinguishable.

bulk–boundary correspondences

The stabilization above of a pair of zero-energy states by a topologically nontrivial bulk is an example of the **bulk–boundary correspondence** that is ubiquitous in the physics of topological quantum matter: gapped bulk states with topologically nontrivial ground states support gapless surface states. For one-dimensional systems, the surfaces are points, and "gapless" means zero-energy states sitting in the middle of the bulk spectral gap. However, for systems in higher dimensions, the extended surfaces are generally described by surface Hamiltonians with gapless dispersion relations. In practice, this means that we have *conducting* states of matter, protected by topological principles. We will see more examples of this correspondence later in the text.

topological insulators and superconductors

The SSH chain is representative of the general class of **topological insulators and superconductors**, a form of fermionic short-range entangled (SRE) gapped quantum matter. The field of topological insulator physics[3] has exploded in the first and second decades of the century. Compared with other forms of topological matter, the physics of this material class is relatively well understood. Referring for an in-depth discussion to one of a growing number of textbooks, here we emphasize that the question of whether an insulator in d dimensions may support topological ground states depends on symmetries. On the broadest level, there are ten symmetry classes distinguished by their behavior under the fundamental symmetries (see info block below): time-reversal, charge conjugation and combinations thereof.[4] The 10-row table defined by symmetries and physical dimension defines

periodic table

the **periodic table of topological insulators** (see table 8.1).[5] The subsequent inclusion of additional (unitary) crystalline symmetries[6] has led to the discovery

crystalline topological insulators

of the more diversified class of **crystalline topological insulators**. At the time of writing of this text, numerous representatives of these system classes have been realized experimentally, and there are certainly more to come. For a discussion of the complementary class of *bosonic* SRE phases, we refer to the literature, and to the example of the anti-ferromagnetic spin chain to be discussed in section 8.4.6.

INFO In this text, various realizations of topological insulators are discussed as examples. Although the understanding of these does not require familiarity with the systematic classification of topological insulators, here we provide a quick overview, showing how the symmetries and dimensionality of an insulating material determine whether it may carry topological structure or not. The upshot of this discussion is the **periodic table of topological insulators and superconductors** mentioned above.

---

[3] Throughout, we will consider superconductors as materials with a quasiparticle excitation gap, and refer to topological insulators and superconductors as topological insulators.

[4] A. Altland and M. Zirnbauer, *Nonstandard symmetry classes in mesoscopic normal–superconducting hybrid structures*, Phys. Rev. B **55**, 1142 (1997).

[5] A. Kitaev, *Periodic table for topological insulators and superconductors*, in AIP Conference Proceedings **1134**, 22 (2009); S. Ryu *et al.*, *Topological insulators and superconductors: Tenfold way and dimensional hierarchy*, New Journal of Physics **12** (2010).

[6] L. Fu, *Topological crystalline Insulators*, Phys. Rev. Lett. **106**, 106802 (2011).

According to a famous theorem by Wigner,[7] symmetries in quantum mechanics are realized as unitary or anti-unitary transformations of Hilbert space. Specifically, the classification of topological insulators builds on two transformations T and C representing the symmetry of a quantum system under **time-reversal** or **charge conjugation**, respectively.[8] When represented in the language of first quantized free electron Hamiltonians, they

Eugene Paul Wigner 1902–1995
was an American physicist and Nobel Laureate in Physics in 1963 "for his contributions to the theory of the atomic nucleus and the elementary particles, particularly through the discovery and application of fundamental symmetry principles."

time-reversal, charge conjugation

assume the form of anti-unitary transformations (showing up via the presence of a matrix transposition), $T : H \to U_T^\dagger H^T U_T$ and similarly $C : H \to U_C^\dagger H^T U_C$, where $U_{T,C}$ are fixed unitary matrices. We call a system T or C invariant if

$$\text{T-symmetry: } H = U_T^\dagger H^T U_T, \qquad \text{C-symmetry: } H = -U_C^\dagger H^T U_C. \qquad (8.2)$$

(Note the minus sign entering the definition of C-symmetry.) The matrices U_X, $X = C, T$ satisfy a constraint putting our symmetries into one of two categories. This is seen by two-fold application of the symmetry operation, which for an X-symmetric Hamiltonian sends H back to itself: $H \to U_X^\dagger (U_X^\dagger H^T U_X)^T U_X = (\overline{U}_X U_X)^\dagger H (\overline{U}_X U_X) = H$. This relation requires the commutativity of $\overline{U}_X U_X$ with arbitrary Hamiltonians in the symmetry class, which is possible only if

$$\overline{U}_X U_X = \pm \mathbb{1}, \qquad X = C, T.$$

We denote the case of the positive (negative) sign as the symmetry $X = +$ ($X = -$). Cases with absent X-symmetry are called $X = 0$. Note that Eqs. (8.2) describe the symmetry for all Hamiltonians of a given symmetry in a given basis. This is reflected in a universal (simple) form of the matrices U_X. Frequent realizations include $U_T = \mathbb{1}$ (the time reversal of spinless particles in the position basis), $U_C = \sigma_x$ (charge conjugation symmetry of spin-triplet superconductors), and $U_T = \sigma_y$ (the time reversal of spin-1/2 particles in the position basis), etc.

The different choices $T, C = +, -, 0$ yield nine different symmetry classes.[9] To see that the actual number of distinct anti-unitary symmetry classes equals $10 = 9 + 1$, we need

[7] E. P. Wigner, *Gruppentheorie und ihre Anwendung auf die Quantenmechanik der Atomspektren* (Friedrich Vieweg und Sohn, Braunschweig, 1931).

[8] The denotations "time-reversal" and "charge conjugation" symmetry are a notorious source of confusion. The physical origin of Eqs. (8.2) are two symmetry operations that change the sense of traversal of world lines (T) and or that exchange particles with anti-particles (C), respectively. First-principle definitions of C, T are formulated within Dirac theory, which is the fundamental description of fermion matter (see problem 9.4.1). Equations (8.2) represent the non-relativistic descent of these symmetries, formulated in first-quantized language. The confusion starts when symmetries having different physical origin assume the form of C or T and are referred to as charge conjugation, time-reversal, particle–hole symmetry or something else. It can be avoided by being concrete and not paying too much attention to the name tag. For example, if a mean field superconductor Hamiltonian assumes the Nambu form $H = \begin{pmatrix} h & \Delta \\ \Delta^\dagger & -h^T \end{pmatrix}$, with anti-symmetric $\Delta = -\Delta^T$, it satisfies a C-symmetry, $H = -\sigma_x H^T \sigma_x$ with $U_C = \sigma_x$, and there is no need to become philosophical about whether to call it charge conjugation or particle–hole symmetry.

[9] There cannot be additional independent anti-unitary symmetries X'. The reason is that the combination $XX' \equiv U$ is a unitary symmetry. Hence, $X' = XU$, meaning that X' equals X modulo a unitary transformation U.

to inspect the case $(T, C) = (0, 0)$. A system lacking both time-reversal and charge conjugation symmetry may nevertheless be symmetric under the combined operation $CT \equiv S$. Under it, $H \to U_S^\dagger H U_S$, with $U_S = \bar{U}_T U_C$, and we say that a Hamiltonian possesses **chiral symmetry**, S, if

chiral symmetry

$$S\text{-symmetry: } H = -U_S^\dagger H U_S, \tag{8.3}$$

i.e., if it is symmetric under $S = CT$, in which case we write $S = +$.[10] This is neither an anti-unitary symmetry, nor a conventional unitary symmetry. The minus sign, inherited from C-symmetry, puts it in a different category. The chiral symmetry option means that $(T, C) = (0, 0)$ splits into $(T, C, S) = (0, 0, 0)$ and $(T, C, S) = (0, 0, +)$.

We have thus arrived at the conclusion that:

> Free fermion Hamiltonians belong to one of ten different symmetry classes.

It now turns out that the membership in a given symmetry class decides the topological contents of insulating free fermion matter. Referring to Ref.[5] for a comprehensive explanation, we note that the free fermion ground states of d-dimensional gapped Hamiltonians of a given symmetry may carry $\mathbb{Z}$-valued integer or $\mathbb{Z}_2$-valued binary topological invariants. The different options are summarized in the **periodic table of topological insulators** (table 8.1), where the first column labels sym-

Élie Cartan 1869–1951
was a French mathematician famed for work in the theory of Lie groups and differential geometry. His work in these fields has had profound influence in various areas of physics, including particle physics and the theory of general relativity.

metry classes in a notation introduced by the mathematician Cartan in the 1920s.[11] The connection to mathematics originates in the observation that the time-evolution operator $U = \exp(iH)$ of a first quantized Hamiltonian takes values in a class of manifolds known as **symmetric spaces**. For example, if H is just hermitian, $(T, C, S) = (0, 0, 0)$, and U takes values in the N-dimensional unitary group, $U(N)$, a.k.a. the compact symmetric space A_N of rank N. If we impose the additional condition of time-reversal invariance $(T, C, S) = (+, 0, 0)$, and consider the case $U_T = \mathbb{1}$, such that $H = H^T$, the time-evolution operator takes values in $U(N)/O(N)$, which is to say that anti-symmetric generators $iH = -iH^T$ are excluded. This is the symmetric space AI_N, etc. In the theory of topological quantum matter, the seemingly unintuitive, but also unambiguous, Cartan denotation of symmetries has stuck – to say that a superconductor obeys time-reversal invariance and particle–hole symmetry leaves room for confusion; to say that it is in class CI does not.

The labels $\mathbb{Z}, \mathbb{Z}_2, 0$ in the table denote the possible topological invariants. For example, the SSH Hamiltonian considered above is defined in $d = 1$ and symmetry class $(T, C, S) = (0, 0, +)$, or AIII. Its chiral symmetry is expressed as $\sigma_3 H \sigma_3 = -H$, where H is the off-diagonal first-quantized Hamiltonian in Eq. (8.1). This leaves room for a $\mathbb{Z}$-valued topological invariant. In the discussion above, we saw that some SSH chain Hamiltonians were trivial, while others had winding number 1. Such Hamiltonians of higher winding number involve multiple coupled chains. This example demonstrates that Hamiltonians belonging to topologically nontrivial classes may, but need not, have topological ground states. However, the example also illustrates why class A Hamiltonians in $d = 1$ cannot support topological ground states.

---

[10] Invariance under two-fold application of S leads to the condition $U_S = \pm\mathbb{1}$. However, these are not independent choices, since every time that $U_S^2 = \mathbb{1}$, this can be changed into a $-\mathbb{1}$ by the transformation $U_S \to iU_S$. (Why does this reasoning not apply to C, T?)

[11] É. Cartan, *Sur une classe remarquable d'espaces de Riemann, I*, Bulletin de la Société Mathématique de France, **54**, 214 (1926).

Table 8.1 Periodic table of topological
insulators. The first two rows contain the
classes without anti-unitary symmetries.

Class	T	C	S	1	2	3	4
A	0	0	0	0	$\mathbb{Z}$	0	$\mathbb{Z}$
AIII	0	0	+	$\mathbb{Z}$	0	$\mathbb{Z}$	0
AI	+	0	0	0	0	0	$\mathbb{Z}$
BDI	+	+	+	$\mathbb{Z}$	0	0	0
D	0	+	0	$\mathbb{Z}_2$	$\mathbb{Z}$	0	0
DIII	−	+	+	$\mathbb{Z}_2$	$\mathbb{Z}_2$	$\mathbb{Z}$	0
AII	−	0	0	0	$\mathbb{Z}_2$	$\mathbb{Z}_2$	$\mathbb{Z}$
CII	−	−	+	$\mathbb{Z}$	0	$\mathbb{Z}_2$	$\mathbb{Z}_2$
C	0	−	0	0	$\mathbb{Z}$	0	$\mathbb{Z}_2$
CI	+	−	+	0	0	$\mathbb{Z}$	0

Also note the diagonal stripe pattern in the table, known as **Bott periodicity**: the table repeats itself with period $d \to d + 2$ and $d \to d + 8$ in the classes without and with anti-unitary symmetries, respectively. This repetitive pattern affords a physical explanation discussed in Ref.[5] and represents a powerful organizing principle in the theory of quantum matter. For example, if unitary symmetries, such as crystalline symmetries, enter the stage, the classification of invariants becomes much more complicated, but it would be even more complicated if it were not for Bott periodicity. Finally, much of the discussion above survives the inclusion of **particle interactions**. However, it turns out that, in the presence of (Mott) gaps generated by interactions, ground states carrying integer invariants can be deformed to those with different invariants without closure of the gap, i.e., without a phase transition. This amounts to a partial collapse of the classification. For example, in class BDI in $d = 1$, this mechanism causes the reduction $\mathbb{Z} \to \mathbb{Z}_8$.[12]

8.1.2 Long-range entangled topological matter

topological
order

We call matter **topologically ordered** if the ground state cannot be transformed into a product state by local unitary transformations. The terminology "ordered" makes reference to the complementary class of systems where a local order parameter is defined by the breaking of some symmetry. Here, the role of the order parameter is taken by non-local topological signatures of the ground state. Prominent classes of matter included in this category are fractional quantum Hall systems (section 8.6.5), and various forms of spin liquids (section 10.4).

A key signature of *topological* order is the presence of long-range *entanglement*. This correspondence establishes a link between two of the most profound concepts of mathematics and physics. However, showing it in concrete cases is not straightforward, the reason being that entanglement is not an easily accessibe observable in many-body theory. There are essentially two ways to describe entangled states

[12] L. Fidkowski and A. Kitaev, *Topological phases of fermions in one dimension*, Phys. Rev. B **83**, 075103 (2011).

of matter theoretically. The first is constructive, in that long-range entanglement is hardwired into quantum ground states by design, e.g., via the inclusion of spatially extended "string" order (for an example, see section 10.4.1). The second way detects entanglement in given systems via the computation of entropic measures known as **entanglement entropies** (see info block below). This approach has become industrialized in the computational investigation of topological matter. However, except for the rare case of conformally invariant theories, entanglement entropies are usually not analytically computable.

entanglement entropy

entanglement

INFO Composite quantum systems are entangled if their states differ from the product states of their constituents. That **entanglement** need not be easily visible can be illustrated even for a two-spin system: $| \uparrow\uparrow \rangle \equiv | \uparrow \rangle \otimes | \uparrow \rangle$ is a product state and therefore not entangled, the **Bell state** $\frac{1}{\sqrt{2}}(| \uparrow\downarrow \rangle + | \downarrow\uparrow \rangle)$ is maximally entangled, and the *seemingly* entangled state $\frac{1}{2}(| \uparrow\uparrow \rangle + | \uparrow\downarrow \rangle + | \downarrow\uparrow \rangle + | \downarrow\downarrow \rangle)$ is actually not entangled because it is the product state, $| \rightarrow\rightarrow \rangle$, of two spin-$x$ eigenstates $| \rightarrow \rangle \equiv \frac{1}{\sqrt{2}}(| \uparrow \rangle + | \downarrow \rangle)$ in disguise. The example shows that the choice of basis is an important element in the detection of entanglement. This becomes a serious issue for systems with large numbers of degrees of freedom and competing bases, i.e., the standard situation in many-body physics.

In practice, entanglement is often probed via the computation of entropic measures. One starts by applying an **entanglement cut**, separating a given system into two constituents whose relative entanglement is of interest. Such cuts are often applied in real space, for example the division of a quantum wire into two halves. Calling the two subsystems A and B, one then picks one of them and computes the **reduced density matrix** $\rho_A \equiv \text{tr}_B(\rho)$ by tracing over the Hilbert space of the other. The **entanglement entropy** is the von Neumann entropy of ρ_A,

$$\boxed{S_A \equiv -\text{tr}_A[\rho_A \ln(\rho_A)]} \tag{8.4}$$

Of particular interest are the entanglement entropies of **pure states**, i.e., projectors onto single wave functions $\rho = |\Psi\rangle\langle\Psi|$. If $|\Psi\rangle = |\Psi_A\rangle \otimes |\Psi_B\rangle$ is a product state – think of a fully polarized spin system – the entanglement entropy vanishes. (As an exercise, compute the entanglement entropy of the two-spin system with A equal to the first spin for the three states mentioned above.) By contrast, eigenfunctions of many-body Hamiltonians at highly excited energies are generically strongly entangled with $S_A \sim \mathcal{V}_A$ extensive in the subsystem volume, $\mathcal{V}_A$. This is called **volume law entanglement**. (Notice the similarity to the extensive entropy in classical statistical mechanics.)

However, the most interesting forms of entanglement entropies are often observed for states close to the ground state. They usually show **area law entanglement**, $S_A = c\mathcal{A}_A$, where $\mathcal{A}_A$ is the geometric surface area of A and c is a measure of the spatial extension over which microscopic degrees of freedom are entangled across the interface.[13] However, in systems with LRE topological order, $S_A = c\mathcal{A}_S - \gamma$ picks up a correction known as **topological entanglement entropy**.[14] Reflecting the topological background, γ is a universal number, related (but not identical) to the ground state degeneracy. For example, the fractional quantum Hall systems to be discussed in section 8.6.5 have m-fold degenerate ground states and a topological entanglement entropy $\gamma = (\ln m)/2$. The topological $\mathbb{Z}_2$ spin liquid to be discussed in section 10.3.2 has $\gamma = \ln 2$, etc. However, as with entropies

topological entanglement entropy

---

[13] An interesting exception is provided by states at a quantum critical point, where fixed length scales do not exist. For example, for one-dimensional systems at a $(1+1)$-dimensional quantum critical point, $S_A = \text{const.} \times \ln L_A$, where L_A is the length of the domain A.

[14] A. Kitaev and J. Preskill, *Topological entanglement entropy*, Phys. Rev. Lett. **96**, 110404 (2006).

in general, analytical calculations of γ for unknown theories are often impossible and we have to rely on computational methods.

Topological order shows not only in entanglement measures but also in physical observables, which are easier to access by analytical methods. First, the many-body wave functions of LRE topological matter contain robust ground state degeneracies. The ensuing ground *spaces* have very interesting physical properties. For example, in conventional systems, adiabatic (meaning intra-ground space) interchanges of quasiparticles lead to sign factors distinguishing between bosonic or fermionic exchange statistics. However, in systems with topological degeneracies, such "braiding" protocols may generate more complex transformations of ground state wave functions. These include multiplication by phases different from ± 1, or even non-abelian unitary transformations. The braiding statistics of topological matter is the subject of a huge research field, which even includes a future applied perspective. In **topological quantum computation**,[33] engineered braiding protocols are investigated as a resource for computing that is protected against the local influences of decoherence by long-range entanglement.

Second, excitations forming on top of topologically degenerate ground states are *emergent* quasiparticles, different from the microscopic constituents of the system. The emergent nature of these particles shows in unusual properties such as fractional charge and/or exchange statistics.

The above signatures of topological order have the advantage that they show in analytically tractable physical observables. Both, ground state properties and quasiparticle excitations are described via topological field theories, which in the case of two-dimensional matter are the Chern–Simons theories to be discussed in section 8.6 below. (However, compared to the situation in two dimensions, the understanding of topological order in higher dimensions is still in its infancy.)

A natural question to ask is whether the two criteria *long-range entanglement* and *ground state degeneracy* condition each other. The answer is that generically they do, but not always. A case in point is the **integer quantum Hall effect** (section 8.4.7). This archetypal form of topological matter is long-range entangled: it does not show ground state degeneracy, and defines an entry in the periodic table of symmetry-protected topological insulators. In this way, it sits between all stools and connects to various concepts of topological matter at the same time.

After this short overview, we now turn to the introduction of quantum field-theoretical methods in topological matter. In quantum field theory, the presence of topological structures shows up via *topological terms* contributing to effective actions. We will introduce the three largest families of topological terms, the θ-, WZW- and Chern–Simons terms. All three cases will be discussed in the context of concrete examples (the integer quantum Hall effect, antiferromagnetic spin chains, and fractional quantum Hall effect, respectively). This approach is not only illustrative but also underpins an important practical aspect: like the standard contributions to effective theories, topological terms emerge from an underlying microscopic theory.

However, unlike these, they probe non-local characteristics and are therefore easily overlooked in standard gradient expansion approaches. The examples below highlight this aspect, and show how we can avoid missing topological terms. However, as a warm-up to the general discussion, let us begin by introducing a few common characteristics of topological terms in field theory for a simple example.

8.2 Example: Particle on a Ring

Consider a free quantum particle of unit charge confined to one dimension and subject to periodic boundary conditions – a particle on a ring. To make the problem more rich, we assume the ring to be threaded by a magnetic flux Φ. Describing the particle position by an angular variable $\phi \in [0, 2\pi]$, the free Hamiltonian of the system reads as ($\hbar = e = c = 1$),

$$\hat{H} = \frac{1}{2}(-i\partial_\phi - A)^2, \tag{8.5}$$

where $A = \Phi/\Phi_0$ is the vector potential corresponding to the magnetic field (think why?) and $\Phi_0 = h/e = 2\pi$ is the magnetic flux quantum. (For simplicity, we set the radius of the ring and the particle mass to unity.) Periodicity implies that we are working on a Hilbert space of wave functions ψ subject to the condition $\psi(0) = \psi(2\pi)$.

The eigenfunctions and the spectrum of the Hamiltonian are given by

$$\psi_n(\phi) = \frac{1}{\sqrt{2\pi}}\exp(in\phi), \quad \epsilon_n = \frac{1}{2}\left(n - \frac{\Phi}{\Phi_0}\right)^2, \quad n \in \mathbb{Z}. \tag{8.6}$$

Although the above setup is very simple, many of the concepts of topological quantum field theory can be illustrated on it. To explore these connections, let us reformulate the system in the language of the imaginary-time path integral (see problem 8.8.1 for details):

$$\mathcal{Z} = \int_{\phi(\beta)-\phi(0)\in 2\pi\mathbb{Z}} D\phi\, e^{-\int d\tau\, L(\phi,\dot{\phi})}, \tag{8.7}$$

where the boundary condition $\phi(\beta) - \phi(0) \in 2\pi\mathbb{Z}$ allows for multiple ring traversal, and the Lagrangian is given by

$$L(\phi,\dot{\phi}) = \frac{1}{2}\dot{\phi}^2 - iA\dot{\phi}. \tag{8.8}$$

EXERCISE Verify by Legendre transformation that the Hamiltonian corresponding to this Lagrangian is given by Eq. (8.5). Obtain the spectrum (8.6) from the path integral, i.e., represent the partition function in the form $\mathcal{Z} = \sum_n \exp(-\beta\epsilon_n)$. (Hint: Apply a Hubbard–Stratonovich decoupling of the quadratic term.)

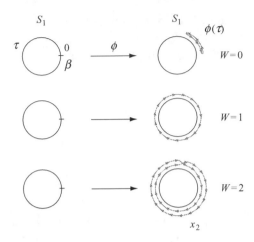

Fig. 8.1   Maps $\phi : S_1 \to S_1$ of different winding numbers on the ring.

Suppose that we did not know the exact solution of the problem. Our routine approach to evaluating the integral would then be a stationary phase analysis. The Euler–Lagrange equations of the action $S[\phi] = \int_0^\beta d\tau\, L(\phi, \dot\phi)$,

$$\frac{\delta S[\phi]}{\delta \phi(\tau)} = 0 \Leftrightarrow \ddot\phi = 0,$$

have two interesting properties. (i) The vector potential does not enter the equations. On the other hand, we saw above that it *does* have a physical effect (the spectrum depends on A). We need to understand how these two seemingly contradictory observations can be reconciled with each other. (ii) There exists a whole family of solutions, $\phi_W(\tau) \equiv W 2\pi\tau/\beta$, and the action of these configurations, $S[\phi_W]|_{A=0} = \frac{1}{2\beta}(2\pi W)^2$, varies discontinuously with W. We have seen other systems supporting discrete families of saddle-point configurations separated by action barriers – for example, the instanton solutions in the quantum double well. However, the present problem is different in that the saddle-point sectors, W, are separated by a topological barrier that is not only energetic but also more profound.

   To understand this point, we note that imaginary time with periodic boundary conditions is topologically a circle, S^1, and the field ϕ is topologically a map $\phi : S^1 \to S^1$ from that circle to the one defined by the phase mod 2π.[15] Each such map is indexed by a **winding number**, W, counting the number of times the phase winds around the target circle: $\phi(\beta) - \phi(0) = 2\pi W$ (see fig. 8.1).

   It is not possible to change W by continuous deformation of ϕ. Since continuity is assumed in field integration (which physical principle stands behind this?), $\int D\phi$ becomes a sum of integrals over functions $\phi(\tau)$ of definite winding numbers, or different **topological sectors**:

winding
number

15 Imaginary time is equivalent to a circle of circumference β. However, in arguments of a topological nature, it is customary to ignore "non-topological" complications such as the finite circumference of a circle.

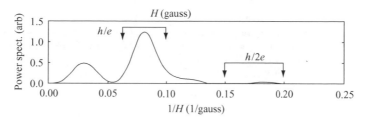

Fig. 8.2   Power spectrum of the persistent current carried by a single gold loop of diameter $\mathcal{O}(1\,\mu\mathrm{m})$, and circumference about ten times larger than the mean free path. (Reprinted from V. Chandrasekhar *et al.*, *Magnetic response of a single, isolated gold loop*, Phys. Rev. Lett. **67**, 3578 (1991). Copyright by the American Physical Society.)

$$\mathcal{Z} = \sum_{W} \int_{\phi(\beta)-\phi(0)=2\pi W} D\phi\, e^{-\int d\tau\, L(\phi,\dot{\phi})} = \sum_{W} e^{2\pi i W A} \int_{\phi(\beta)-\phi(0)=2\pi W} D\phi\, e^{-\frac{1}{2}\int d\tau\, \dot{\phi}^2}, \quad (8.9)$$

where we noted that the A-dependent term in the action,

$$S_{\text{top}}[\phi] \equiv iA \int_0^\beta d\tau\, \dot{\phi} = iA(\phi(\beta) - \phi(0)) = i2\pi W A,$$

is sensitive only to the index W carried by ϕ. This representation makes the topological contents of the problem visible:

▷ The functional integral is an **integral over disjoint topological sectors**.

▷ The contribution to the action, S_{top}, is our first example of a topological term. (It belongs to the class of **θ-terms**, to be discussed in section 8.4.)

▷ Variational equations probe how the action responds to small deformations of a field. These do not alter the topological sector, and for this reason the topological term does not show up in **equations of motion**.

▷ Nevertheless, it *does* affect the result of the functional integration as a W-dependent phase, weighting the contribution of different sectors.

▷ Since S_{top} knows only about topological classes, it is oblivious to the metrics of both the base and the target manifold. Specifically, rescaling $\tau \to i\tau \equiv t$, from imaginary to real time leaves $S_{\text{top}}[\phi] = 2\pi i A W$ invariant; in either case it assumes the form of an imaginary phase. We will see later that this reflects a more general principle: both in in Euclidean and Minkovski space–time, topological terms enter as imaginary phases.

All these features generalize to other realizations of topological field theories. However, before introducing these generalizations, we need to provide a little more mathematical background on the topology of fields. This will be the subject of the next section.

EXERCISE   Reflecting its simplicity, the above particle-on-a-ring paradigm frequently appears in condensed matter physics. Examples encountered previously in this text include the Josephson junction (problem 5.6.6) between superconductors, where ϕ is a phase

persistent
currents

difference of condensates, and the physics of normal metal granules subject to strong charging, where the time-derivative of the phase is electric voltage. Another interesting realization is the physics of normal metallic **persistent currents**:

Consider a ring-shaped conductor subject to a magnetic flux. According to a prediction by Byers and Yang,[16] the magnetic field induces an equilibrium current

$$I(\Phi) = -\frac{\partial F(\Phi)}{\partial \Phi},$$

periodic in Φ with period Φ_0. Remembering that a vector potential enters the free energy as $\sim \int d\mathbf{A} \cdot \mathbf{j}$, derive this formula. Show that, at zero temperature, the persistent current flowing in a perfectly clean one-dimensional metal of non-interacting fermions assumes the form of a Φ_0-periodic sawtooth function, $I(\Phi) = \frac{2\pi v_F}{L}[\Phi/\Phi_0]$, where $[x] = x - n$ and n is the largest integer smaller than x. (Hint: For zero temperature, the free energy of a non-interacting system of particles is equal to the sum of all single-particle energies up to the Fermi energy. Notice that the current is carried by the last occupied state, and the currents $-\partial_\Phi \epsilon_n$ carried by all levels below the Fermi energy cancel.)

For a long time, the prediction of this current was believed to be of only academic relevance: the presence of impurities in realistic materials disrupts the phase of wave functions, so that a phenomenon relying on the phase of the highest occupied single-particle level should be highly fragile. On the other hand, the topological nature of the flux coupling indicates that the effect might be more robust.

To substantiate this second view, show that a gauge transformation $\psi(\phi) \to e^{iA\phi}\psi(\phi)$ removes A from the Hamiltonian while changing the boundary conditions to $\psi(0) = e^{2\pi iA}\psi(2\pi)$. For a clean system, obtain the spectrum in the twisted representation and convince yourself that the information on the field now sits in the boundary conditions. While mathematically equivalent, the second formulation provides a new perspective: Even in the presence of a moderate amount of impurities,[17] wave functions remain sensitive to changes in the boundary conditions, suggesting that a persistent current should be more robust than previously thought. In a series of beautiful experiments (see fig. 8.2 and Refs.[18]), persistent current flow in normal metallic rings was indeed observed for rings exceeding the scattering mean free path by orders of magnitude.

8.3 Homotopy

In this section, we generalize the classification in terms of winding numbers previously exemplified for the particle-on-a-ring system to higher-dimensional settings. This leads to the concept of homotopy in field theory, which will be fundamental to the applications discussed later in the chapter.

[16] B. Byers and C. N. Yang, *Theoretical considerations concerning quantized magnetic flux in superconducting cylinders*, Phys. Rev. Lett. **7**, 46 (1961).

[17] We do not want the impurity potential to be so strong as to localize wave function on length scales shorter than the ring circumference.

[18] A. C. Bleszynski-Jayich *et al.*, *Persistent currents in normal metal rings*, Science **326**, 272 (2009); H. Bluhm *et al.*, *Persistent currents in normal metal rings*, Phys. Rev. Lett. **102**, 136802 (2009).

8.3.1 Homotopy group

We begin by introducing a few mathematical concepts. In chapter 1, fields were defined as maps $\phi : M \to T, z \mapsto \phi(z)$, from a base manifold M into a target manifold T. In practice, we often work with theories with continuous symmetries described by the action of a group G. These symmetries may or may not be spontaneously broken down to a smaller group H. (Think of the canonical example of a magnet with rotation symmetry $G = \mathrm{O}(3)$, broken down to $H = \mathrm{O}(2)$ in the ferromagnetic case.) In all these cases, $T = G/H$ is a coset space, where G is one of the compact classical groups $\mathrm{U}(N)$, $\mathrm{O}(N)$, and $\mathrm{Sp}(N)$, and H is some subgroup thereof. Although this realization of T does not exhaust the repertoire of field-theoretical applications, it is general enough for our present purposes.

Turning to the base manifold, we frequently have situations where fields at spatial infinity decay. The constancy of the field at the boundary, $\phi\big|_{\partial M} = \text{const.}$, can then be used to compactify to a sphere of large radius (see the figure). Again, this setting does not cover every possible situation, but will be general enough for the moment. Topologically, a large sphere is equivalent to a unit sphere, S^d, which means that the topology of a large family of applications is described by fields

$$\phi : S^d \to G/H,$$
$$z \mapsto \phi(z),$$

mapping unit spheres into coset spaces.

topology   **INFO** A few comments on the term **topology** in the present context. The phenomena discussed in this chapter rely on continuity: disruptive changes in field configurations come with large action costs and are beyond the scope of effective theories. They have a status similar to the topology-altering breaking of a circle. However, topological data is oblivious to continuous deformations of fields; it does not depend on metric structures. The minimal structure describing continuity modulo metric is that of a **topological space**.[19] The target and base manifolds introduced above are of higher structure, admitting the definition of metrics probing distances in M or T. However, in this chapter, no reference to these will be made. This comment is more than a formality. For example, the metric independence of a term in an effective action is often the most straightforward way to identify it as a topological term.

We proceed to discuss the **topological contents of fields**: Two fields ϕ_1 and ϕ_2 are topologically equivalent if they can be continuously deformed into each other.

---

[19] Let X be a set and $\mathcal{J} = \{Y_i \subset X | i \in I\}$ a collection of its subsets. The pair $(X, \mathcal{J})$ is called a topological space if and only if, (a) $\{\}, X \in \mathcal{J}$, (b) for $J \subset I$, $\bigcup_{i \in J} Y_i \in \mathcal{J}$, and (c) for any finite subset $J \subset I$, $\bigcap_{i \in J} Y_i \in \mathcal{J}$. The elements of $\mathcal{J}$ define the open subsets of X. A map $\phi : X \to Y$ between two topological spaces is a **continuous map** if for any open set $U \subset Y$, the pre-image $\phi^{-1}(U) \subset X$ is open in X.

Fig. 8.3   Concatenation of two two-dimensional fields into a single field.

homotopy   Technically, this condition requires the existence of a continuous deformation map,
               a **homotopy** in the language of mathematics,

$$\hat{\phi} : S^d \times [0,1] \to T,$$
$$(z,t) \mapsto \hat{\phi}(z,t),$$

such that $\hat{\phi}(.,0) = \phi_1$ and $\hat{\phi}(.,1) = \phi_2$. We denote the *equivalence class* of all
fields topologically equivalent to a given representative ϕ by $[\phi]$. For example, in
the case $\phi : S^1 \to S^1$ discussed in the previous section, individual classes contain all
fields of a specified winding number. The set of all topological equivalence classes
homotopy $\{[\phi]\}$ of maps $\phi : S^d \to T$ is called the **dth homotopy group**, $\pi_d(T)$.
group

INFO In what sense is $\pi_d(T)$ a **group** rather than just a set? To understand this point,
consider the base manifold deformed from a sphere to a d-dimensional unit cube $I^d =$
$[0,1]^d$. This operation is topologically empty provided that the boundary ∂I^d of the cube
is identified with a single point on the sphere, $\phi|_{\partial I^d} = \text{const.}$ (Think of ∂I^d as the infinitely
large boundary of the original base manifold. Without loss of generality, we can assume
the constant to be the same for all fields.)

Two fields ϕ_1 and ϕ_2 described in the cube representation may now be glued together
to form a new field $\phi_3 \equiv \phi_1 * \phi_2$ (fig. 8.3). For example, we might define

$$\phi_3(x_1, x_2, \ldots, x_d) = \begin{cases} \phi_1(2x_1, x_2, \ldots, x_d), & x_1 \in [0, 1/2], \\ \phi_2(2x_1 - 1, x_2, \ldots, x_d), & x_1 \in [1/2, 1]. \end{cases}$$

(Think why the presence of a common reference configuration ϕ^* is essential here.) Each
field is in a homotopy class, and we define the group operation in $\pi_d(T)$ as

$$[\phi_1] * [\phi_2] \equiv [\phi_3]. \tag{8.10}$$

(Think why this is a valid definition and why it does not depend on the choice of coordinates in which the gluing operation is formulated.) It is straightforward to verify that "$*$"
satisfies the criteria required by a group operation:

> The concatenation of fields by gluing defines a group structure of homotopy classes.

8.3.2 Examples of homotopies

Only in simple cases can homotopy group structures be identified by straightforward
inspection. For example, maps $S^1 \to S^1$ are classified by winding numbers $W \in \mathbb{Z}$

Table 8.2 Homotopy groups of a number of common maps. A missing entry means that no general results are available.

	S^1	S^2	$S^{k>2}$	T^k	SU(2)	SU(N)
$d=1$	$\mathbb{Z}$	$\emptyset$	$\emptyset$	$\underbrace{\mathbb{Z} \oplus \cdots \oplus \mathbb{Z}}_{k}$	$\emptyset$	$\emptyset$
$d=2$	$\emptyset$	$\mathbb{Z}$	$\emptyset$	$\emptyset$	$\emptyset$	$\emptyset$
$d>2$	$\emptyset$	$\emptyset^a$	$\begin{matrix} d<k: & \emptyset \\ d=k: & \mathbb{Z} \\ d>k: \end{matrix}$	$\emptyset$	$\pi_3(\mathrm{SU}(2))=\mathbb{Z}$	

^{a} But $\pi_3(S^2) = \mathbb{Z}$.

and concatenation yields fields of added winding numbers: $\pi_1(S^1) = \mathbb{Z}$ with addition as a group operation. On the other hand, it is evident that maps $S^1 \to S^2$ – closed curves on the 2-sphere – can be shrunk to trivial maps: $\pi_1(S^2) = \emptyset$. However, in general, the identification of homotopy groups for the targets $T = G/H$ of field theory is a hard problem and few things can be said in general.

For example, a topological space, T, is called *simply connected* if any closed curve on it is contractible. By the definition of homotopy, this is equivalent to $\pi_1(T) = \emptyset$. Examples include all higher-dimensional spheres $S^{d>1}$ and SU(N): $\pi_1(S^{d>1}) = \pi_1(\mathrm{SU}(N)) = \emptyset$. Conversely, the first homotopy groups of non-simply connected spaces are nontrivial. For example, curves on the d-dimensional torus T^d are classified by (think why!)

$$\pi_1(T^k) = \underbrace{\mathbb{Z} \times \cdots \times \mathbb{Z}}_{k}.$$

The higher-dimensional groups, $\pi_{d>1}(T)$, are less easy to imagine. For example, maps of S^2 into itself are classified according to how often they wrap around the sphere: $\pi_2(S^2) = \mathbb{Z}$. However, the generalization to higher dimensions, $\pi_d(S^k) = \mathbb{Z}$, with $\pi_k(S^{d>k}) = \emptyset$, is less intuitive. Interestingly, maps $S^d \to S^{k<d}$ into lower-dimensional spheres can be nontrivial. For example, Hopf has shown that $\pi_3(S^2) = \mathbb{Z}$. For a summary of these, and a few more results, see table 8.2.

However, even basic applications of physics are not covered by these known results, and in practice the identification of homotopies is often addressed on a case-by-case basis (see Ref.[21] for further discussion).

8.4 θ-terms

We now turn to the question of how field homotopy shows up in field theory. Each field $\phi : M \to T$ is labeled by a homotopy class, which in nontrivial cases assumes the form of an integer or multi-integer index, W, sometimes denoted the

[21] M. Nakahara, *Geometry, Topology and Physics* (IOP Publishing, 2003).

topological
charge

topological charge. Accordingly, the functional integral defining the theory is organized as

$$\mathcal{Z} = \sum_W \int D\phi_W \, e^{-S[\phi_W]},$$

where $\int D\phi_W$ denotes integration over all fields in the **topological sector** labeled by W. It may happen that the action of the theory,

$$S[\phi] = S_0[\phi] + S_{\text{top}}[\phi],$$

topological
action

contains a **topological action**, $S_{\text{top}}[\phi] \equiv F(W)$, defined to be a contribution which depends on the charge, W, but on nothing else. Since S_{top} is constant within each sector, it may be pulled upfront to obtain

$$\mathcal{Z} = \sum_W e^{-F(W)} \int D\phi_W \, e^{-S_0[\phi_W]}. \tag{8.11}$$

At first sight, it may look as if $F(W)$ can be an arbitrary function. However, it turns out that topology almost uniquely fixes the form of allowed $F(W)$. To see how this comes about, consider two fields ϕ_1 and ϕ_2, assumed to be constant everywhere, save for two localized regions somewhere in space–time (see figure). We now glue these fields to obtain a new one, $\phi_1 * \phi_2$, with charge $W_1 + W_2$. Assuming the respective regions of variation to be well sep-

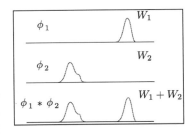

arated, the action must be additive in the sense $S[\phi_1 * \phi_2] = S[\phi_1] + S[\phi_2]$. In particular, $F(W_1 + W_2) = S_{\text{top}}[\phi_1 * \phi_2] = S_{\text{top}}[\phi_1] + S_{\text{top}}[\phi_2] = F(W_1) + F(W_2)$.

This construction tells us that the topological action is linear, $F(W_1 + W_2) = F(W_1) + F(W_2)$. If W is just an integer, it must be of the form,

$$F(W) = i\theta W,$$

where we ignore additive constants, and anticipate that the constant of proportionality, θ, is generally imaginary.

The term $i\theta W$ is called the **topological phase**. Since W is integer, the parameter θ is defined only mod 2π, and therefore represents an angular parameter. This

topological
angle

topological angle is commonly called θ,[22] hence the denotation θ-term for the topological action.

A disadvantage of the representation above is that it makes reference to the index W as an abstract quantity. It would be better to work with representations

$$S_{\text{top}}[\phi] = \int d^d x \, \mathcal{L}_{\text{top}}[\phi, \partial_\mu \phi],$$

[22] The concept of θ-terms was pioneered in 't Hooft's analysis of SU(2) gauge field instantons in four-dimensional compactified space–time (G. 'tHooft, *Magnetic monopoles in unified gauge theories*, Nucl. Phys. **B79**, 276 (1974).) In this paper, he labeled the topological angle by θ, and the denotation has stuck.

directly in terms of the field, where the **topological Lagrangian (density)**, $\mathcal{L}_{\text{top}}$, is implicitly defined by the condition $S_{\text{top}}[\phi] = i\theta W$. In this way, the winding number is computed from a *topological action*, generally referred to as a θ-**term**,[23] and there is no need to explicitly split the field integration into a sum over distinct sectors:

θ-term

$$\mathcal{Z} = \sum_W \int D\phi_W \, e^{-F(W) - S_0[\phi_W]} = \int D\phi \, e^{-S_{\text{top}}[\phi] - S_0[\phi]}.$$

In the next section, we discuss ways to construct topological actions using geometric principles. Knowing this construction can be a useful asset in deciding whether an unfamiliar field theory admits a topological action. However, for standard realizations of topological field theories – with fields defined on circles, or spheres, or other homotopically nontrivial manifolds – the form of the topological action is known, and what remains to check is whether a theory has a finite θ-angle. For example, the magnetic action appearing in our previous discussion of the particle on a ring,

$$S_{\text{top}}[\phi] \equiv i\theta \int_0^\beta \frac{d\tau}{2\pi} \dot{\phi} = iW\theta,$$

is a θ-term with $\theta = 2\pi A$, non-vanishing if there is magnetic flux through the ring. Before turning to the discussion of more advanced realizations in sections 8.4.2 and 8.4.7, let us summarize the defining **criteria satisfied by θ-terms** in general:

▷ θ-terms are defined by integer-valued functionals of fields, and

▷ they couple to the action as imaginary phases.

▷ Their coupling constants, θ, are defined only mod 2π.

▷ Since θ-terms are invariant under local deformations of fields, they do not affect equations of motion.

8.4.1 Geometry of θ-terms

REMARK In this section, we discuss the universal definition of topological θ-actions from a geometric perspective, using the language of differential forms introduced in appendix section A.1. Why coonsider this seemingly abstract perspective before turning to examples? The reason is that various features of θ-terms are hard to understand in the coordinate-heavy representations customary in physics. In fact, it is often not straightforward to decide whether a contribution to an action is a θ-term. Similarly, one would like to know whether a theory specified by a base and a target manifold admits a θ-term and, if so, how it will look. This question, too, is easier to answer if the underlying geometric structures are understood. Readers wishing to see some examples first are invited to skip this section at first reading. However, it will be required for our subsequent discussion of Wess–Zumino terms in section 8.5.

---

[23] Note that it is appropriate to denote topological actions as "terms" and not "operators." Unlike other contributions to functional integral actions, they are not obtained as path integral representations of operators acting in Hilbert spaces.

A θ-term is a functional $S_{\text{top}}[\phi]$ assigning to a field $\phi : M \to T, z \mapsto \phi(z)$ a number, namely the number of times $\phi(z)$ covers T as z runs through M. The systematic construction of this functional starts with the observation that almost all target manifolds relevant to field-theoretical applications come with a natural integral measure. The precise statement is that on the n-dimensional target manifold, T, there exists a canonical volume n-form ω such that $\text{Vol}_T = \int_T \omega$ is the volume of T.[24] Without loss of generality, we may define ω such that $\text{Vol}_T = 1$. For example, for $T = S^2$, a two-sphere, $\omega = \frac{1}{4\pi} \sin\theta d\theta \wedge d\phi$ in a local system of spherical coordinates. First consider the case where $\phi : M \to T$ is a diffeomorphism, i.e., a smooth one-to-one coverage of T. In this case, we may think of ϕ as a coordinate representation of T and M as the coordinate space. The integral $1 = \int_T \omega$ may be computed by a "change of variables" as an integral over M: in differential-form language, this reads, $1 = \int_T \omega = \int_M \phi^*\omega$, i.e., the integral is calculated as the integral of the pullback of the form ω under ϕ over M (see Eq. (A.16)). For example, if $(x, y) \in M$ are base manifold coordinates and the field has spherical coordinate representation $\phi(x, y) = (\theta(x, y), \phi(x, y))$, the integral over the pullback assumes the form

$$1 = \frac{1}{4\pi} \int_M \sin(x, y) \left(\frac{\partial\theta}{\partial x} \frac{\partial\phi}{\partial y} - \frac{\partial\theta}{\partial y} \frac{\partial\phi}{\partial x} \right) dxdy = \frac{1}{4\pi} \int_M \mathbf{n} \cdot (\partial_x \mathbf{n} \times \partial_y \mathbf{n}) \, dxdy,$$

where $\mathbf{n}$ is the unit vector on the sphere defined by the spherical coordinates. The last two expressions already do what we are looking for: they assign to the field $\mathbf{n} = \mathbf{n}(x, y)$ an integer winding number equal to unity, provided that it covers the sphere in a diffeomorphic manner. However, a valid θ-term must do more: it must yield the winding number for fields including those that are not bijective, and it must yield higher-order winding numbers (or zero) for fields winding multiple times (or not at all) around M. In the following, we show that the topological action

$$\boxed{S_{\text{top}}[\phi] = i\theta \int_M \phi^*\omega} \tag{8.12}$$

θ-term   satisfies these criteria and defines the invariant representation of a θ-**term**.

INFO θ-terms usually appear in field theories with equal dimension, $\dim(M) = \text{T}$, of the base and target manifolds. However, there are exceptions to this rule. As an example, we consider the targets $\text{U}(n)$ or $\text{SU}(n)$ on base manifolds of odd dimension $d = 1, 3, \cdots$. In these cases, the θ-terms are built from the forms $\text{tr}(g^{-1}dg)$ for $d = 1$, $\text{tr}(g^{-1}dg \wedge g^{-1}dg \wedge g^{-1}dg)$ for $d = 3$, or $\text{tr}(g^{-1}dg)^{\wedge d}$ for general odd dimensions. The terms themselves are then obtained by pullback under fields $x \mapsto g(x)$ to the base, which in $d = 1$ and $d = 3$ assume the form;

$$d = 1, \qquad S_{\text{top}}[g] = \frac{\theta}{2\pi} \int dx \, \text{tr}(g^{-1}d_x g), \tag{8.13}$$

$$d = 3, \qquad S_{\text{top}}[g] = \frac{i\theta}{24\pi^2} \int d^3x \, \epsilon^{ijk} \text{tr}((g^{-1}\partial_i g)(g^{-1}\partial_j g)(g^{-1}\partial_k g)). \tag{8.14}$$

---

[24] Field manifolds of indefinite volume, for example, hyperboloids, are usually topologically trivial and the present discussion does not apply.

The topological quantization of these integrals is proven in problem 8.8.3. In the second line, we encounter the integral for the **winding number invariant** in three dimensions, where the numerical factor $i/24\pi^2$ is obtained by explicit computation (as an exercise, you could perform the integral using the Euler angle coordinates (8.21) for the group SU(2)).

How can the existence of a winding number of fields from low-dimensional bases, such as $d = 3$, into higher-dimensional targets, such as U(n), $n \geq 2$, be understood? It turns out that the topological properties of the unitary groups are contained in SU(2) $\subset$ U(n). Windings in U(n) reduce to windings around these three-dimensional submanifolds. To illustrate the idea of such a reduction on a simpler example, consider the punctured plane $\mathbb{R} \setminus \{0, 0\}$. Embedded in this two-dimensional manifold, we have the one-dimensional circle around the origin, $S^1 \subset T$. Closed curves in the punctured planes may be continuously deformed to closed curves in the circle, showing that this lower-dimensional manifold classifies the topology of the higher-dimensional manifold. A similar principle applies to the above group manifolds.

We finally note the equivalence SU(2) $\simeq S^3$ of the two-dimensional special unitary group to the three-sphere.[25] This identification, and the various one- and two-dimensional examples discussed in this section, indicate that, where the topology of θ-terms is concerned, targets with spherical topology, $S^{1,2,3}$, are the most important ones in condensed matter physics.

In the info block below, we demonstrate that $S_{\text{top}}[\phi]$ does not vary under continuous deformations of ϕ, and in this way show that S_{top} is sensitive only to the homotopy class $[\phi]$ of the field. For fields in the zero-homotopy class, the topological action vanishes. (To see this, consider the representative $\phi = \text{const.}$, for which $\phi^*\omega = 0$.) For fields covering T once, the integral yields unity, as discussed above. Finally, for fields with W coverages, the base M can be partitioned into sectors M_i, such that $\phi|_{M_i}$ performs a single coverage and $\int_{M_i} \phi^*\omega = 1$. The full integral then yields $S_{\text{top}}[\phi] = W$.

EXERCISE Construct a field $\phi : M \to S^2, (x, y) \mapsto \mathbf{n}(x, y)$ wrapping W times around the unit sphere and show that the θ-term constructed above yields this winding number.

To summarize, Eq. (8.12) defines the invariant form of the θ-action, and concrete representations are obtained by substitution of ω and ϕ for a given set (M, T) and coordinate system. In the following sections, we illustrate the principle for various examples.

INFO Let us demonstrate the invariance of Eq. (8.12) under continuous field deformations. To this end, consider two fields ϕ and ϕ' continuously deformable into each other. More specifically, we set $\phi' = \psi \circ \phi$, where $\psi : T \to T$ is different from unity only inside a local domain $U \subset T$ (see the figure).

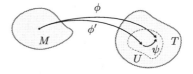

25 To see this explicitly, consider the representation $\begin{pmatrix} a & b \\ -\bar{b} & \bar{a} \end{pmatrix}$ of U(2) matrices for which the unit-determinant condition of SU(2) reads as $|a|^2 + |b|^2 = 1$. With $a = x^0 + ix^1$ and $b = x^2 + ix^3$, this assumes the form $\sum (x^i)^2 = 1$, defining a three-sphere.

We then have

$$S_{\text{top}}[\phi] - S_{\text{top}}[\phi'] = i\theta \int_M (\phi^*\omega - (\psi \circ \phi)^*\omega) = i\theta \int_M \phi^*(\omega - \psi^*\omega),$$

where we have used the fact that the pullback under two consecutive maps equals the succession of pullbacks, $(\psi \circ \phi)^*\omega = \phi^*(\psi^*\omega)$. Now $\phi^*(\omega - \psi^*\omega) \equiv \xi$ is a n-form on M, different from zero only locally. Any form of degree n can locally be represented as the derivative of an $(n-1)$-form.[26] We thus have a representation $\xi = d\kappa$ with $(n-1)$-form κ and from there obtain

$$S_{\text{top}}[\phi] - S_{\text{top}}[\phi'] = i\theta \int_M d\kappa = i\theta \int_{\partial M} \kappa = 0,$$

where we have used Stokes' theorem (A.17); thus, by construction, the field difference on the boundary of M vanishes, $\kappa|_{\partial M} = 0$.

8.4.2 θ-Terms in two-dimensional field theories

Field theories with θ-terms can have arbitrary dimensionality. However, some of the most interesting condensed matter applications are two-dimensional, and we will focus on them here. The target spaces of these theories have the topology of two-spheres, and the question we are addressing is how the homotopy of fields from two-dimensional space into the sphere affects their physics. More specifically, we will consider quantum spin chains (see section 8.4.6), the two-dimensional classical Heisenberg model, and the field theory of the integer quantum Hall effect (see section 8.4.7) as examples. In either case, assuming decaying boundary conditions at spatial infinity, and compactifying space as above, the fields assume the form of maps,

$$\mathbf{n} : S^2 \to S^2,$$

$$\mathbf{x} \mapsto \mathbf{n}(\mathbf{x}), \quad |\mathbf{n}| = 1,$$

and the relevant homotopy group is $\pi_2(S^2) \simeq \mathbb{Z}$. We now reason that the topological action describing this setting is given by

$$\boxed{S_{\text{top}}[\mathbf{n}] = \frac{i\theta}{4\pi} \int d^2x \, \mathbf{n} \cdot (\partial_1 \mathbf{n} \times \partial_2 \mathbf{n})} \tag{8.15}$$

To understand this point, we first verify the insensitivity of S_{top} to small variations $\mathbf{n}(\mathbf{x}) \to \mathbf{n}(\mathbf{x}) + \epsilon^a(\mathbf{x})R^a\mathbf{n}$, where the functions $\epsilon^a, a = 1, 2$, are infinitesimal and

---

[26] The reason is that, locally, an n-form affords the coordinate representation $\xi = f dx^1 \wedge \cdots \wedge dx^n$. With the *ansatz* $\kappa = g dx^2 \wedge \cdots \wedge dx^n$, the equation $d\kappa = \xi$ reduces to the differential equation $\partial_1 g = f$ (of course, one may pick any other of the n coordinates for the definition of this local representation), defined on some open interval of x_1. This is an ordinary differential equation that can be solved. For example, $\sin\theta \, d\theta \wedge d\phi = -d(\cos\theta \, d\phi)$ holds locally on the sphere, but there can be no such representation of the area form on the whole sphere. If it existed, $\omega = d\kappa$, the area integral $\int_{S^2} \omega = \int_{S^2} d\kappa = \int_{\partial S^2} \kappa = 0$ would vanish by Stokes' theorem (A.17) and the absence of a boundary.

R^a are the generators of rotations around two axes perpendicular to $\mathbf{n}$. A few integrations by parts show that the variation of S_{top} assumes the form

$$\delta S_{\text{top}}[\mathbf{n}] = \frac{3i\theta}{4\pi} \int d^2x \, \epsilon^a R^a \mathbf{n} \cdot (\partial_1 \mathbf{n} \times \partial_2 \mathbf{n}).$$

However, $R^a\mathbf{n}$ is perpendicular to $\mathbf{n}$ while (exercise) $(\partial_1\mathbf{n} \times \partial_2\mathbf{n})$ lies parallel, i.e., $\delta S_{\text{top}} = 0$. This construction demonstrates that S_{top} does not affect the equations of motion.

The invariance of S_{top} also implies that the computation $S_{\text{top}}[\mathbf{n}_0]$ for any test configuration determines the value of S_{top} for all fields $\mathbf{n}$ connected to $\mathbf{n}_0$ by continuous interpolation. Specifically, we consider the family

$$\mathbf{n}^{(W)} : \mathbb{R}^2 \to S^2,$$

$$(x_1, x_2) \mapsto \left(\phi = -W \tan^{-1}\left(\frac{x_2}{x_1}\right), \vartheta = 2\tan^{-1}\left(\frac{a^2}{x_1^2 + x_2^2}\right) \right),$$

where the sphere is parameterized in polar coordinates, $\mathbf{n}^{(W)} = \mathbf{n}^{(W)}(\vartheta, \phi)$. In honor of their inventor, Tony Skyrme, [27] these field configurations are called (magnetic) **skyrmions**. As usual, the suffix "on" hints at a stable, particle-like, excitation. The reason for this stability is that skyrmions of nonvanishing winding number are twists of the spin vector around the sphere (see the figure for a visualization with $W = 1$), and such twists cannot

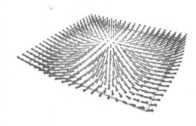

be undone by continuous deformation. Substitution of $\mathbf{n}^{(W)}(\vartheta, \phi)$ into the integral (8.15) shows that the winding number is extracted by the topological action, $S_{\text{top}}[\phi^{(W)}] = i\theta W$.

skyrmion

EXERCISE Verify these statements. Show that the topological charge is insensitive to coordinate changes on the target and the base manifolds. Try to invent other simple field configurations of non-vanishing topological charge.

INFO A word on **semantics**: depending on the context, topologically nontrivial field configurations are referred to as **solitons, instantons, skyrmions**, etc. There is no rigorous systematics in this. However, a rule of thumb is that topological excitations in dynamical quantum theories are called "instantons," while the term "solitons" is preferred for static configurations.

---

[27] See T. H. R. Skyrme, *A nonlinear field theory*, Proc. R. Soc. Lond. A **260**, 127 (1961), where the skyrmion was introduced in an effective model of nuclear matter. The terminology "magnetic skyrmion" is reserved for fields with target space S^2 ($\to$ magnetic moments) in $d = 2$, in distinction to the higher-dimensional skyrmions of particle physics.

8.4.3 Functional integration over topologically charged fields

As discussed above, the integration over topologically nontrivial fields may be organized into an integration over sectors of definite index W. This means that the standard program of field theory – finding stationary configurations, integrating over fluctuations, renormalization group, etc. – needs to be carried out for each sector separately. While this sounds straightforward in principle, there are various practical aspects to consider.

For one, the solution of the extremal equations $\delta_\phi S[\phi]|_{\phi \in \phi_W} = \delta_\phi S_0[\phi]|_{\phi \in \phi_W}$ in topologically charged sectors can be difficult, even though the topological term does not enter. The reason is that solutions with topological charge generally lack translational invariance, so that we are looking for inhomogeneous configurations of extremal action. (For a trick that solves the problem in the present context of fields on the sphere, see the following info block.)

INFO In some cases, the solution of **stationary equations in topologically nontrivial sectors** can be side-stepped by energetic arguments. To see how, consider the positive definite integral

$$0 \le \frac{1}{2} \int d^2x \, (\partial_\mu \mathbf{n} + \epsilon_{\mu\nu} \mathbf{n} \times \partial_\nu \mathbf{n}) \cdot (\partial_\mu \mathbf{n} + \epsilon_{\mu\nu'} \mathbf{n} \times \partial_{\nu'} \mathbf{n})$$
$$= \int d^2x \, (\partial_\mu \mathbf{n} \cdot \partial_\mu \mathbf{n} + \epsilon_{\nu\mu} \mathbf{n} \cdot (\partial_\mu \mathbf{n} \times \partial_\nu \mathbf{n})) \, .$$

The last term in the second line equals -8π times the topological charge, i.e.,

$$W \le \frac{1}{8\pi} \int d^2x \, \partial_\mu \mathbf{n} \cdot \partial_\mu \mathbf{n} = S_0[\phi].$$

This construction shows that W is a lower bound for the action of field configurations of topological charge W. The limit is reached for configurations

$$\partial_\mu \mathbf{n} + \epsilon_{\mu\nu} \mathbf{n} \times \partial_\nu \mathbf{n} = 0, \tag{8.16}$$

on which the integral vanishes. Any deformation of these fields leads to a positive integral, showing that Eq. (8.16) defines a stationary (even minimal) configuration. In this way, stationary phase configurations are obtained by solution of the *first-order* differential equation (8.16).

Equation (8.16) is best solved in complex coordinates, $z \equiv x_1 + ix_2$, and using a stereographic representation of the target vector,

$$n_1 + in_2 = \frac{2w}{1 + |w|^2}, \quad n_3 = \frac{1 - |w|^2}{1 + |w|^2}, \tag{8.17}$$

where $w \in \mathbb{C}$. (Recall the stereographic representation projecting points on the sphere to the infinite two-dimensional plane and discuss the

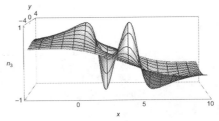

intuitive interpretation of w.) Straightforward substitution then shows that Eq. (8.16) assumes the form $\partial_{\bar{z}} w(z) = 0$ (exercise). This means that any meromorphic function

$$w = \prod_{i=1}^{W} \frac{z - a_i}{z - b_i}$$

solves the equation. For example, the figure shows the n_3-component of a $W = 2$ configuration with $(a_1, a_2, b_1, b_2) = (0.4, 4, 2, 6)$. It changes twice between ± 1, and inspection of the corresponding $n_{1,2}$ components shows that the sphere is doubly covered in the process. For more details on these types of instantons and their topological densities, see problem 8.8.2.

Once extremal configurations have been identified, we need to integrate over fluctuations. While this looks straightforward in principle, the practical execution of the program can be involved. To see why, notice that instanton field theory invokes a conflict of two principles:

> Instantons cost energy; instantons gain entropy.

More specifically, for each instanton we need to pay an action ("energy") S_0. However, in return, we gain free integration parameters ("entropy"), namely the coordinates specifying how the inhomogeneous instanton configuration is embedded in space–time. (In the info block above, a_i, b_i are such parameters.) Variations of **instanton zero modes** these **instanton zero modes** do not cost further action, and hence they counter the action cost.

On this basis, consider a sector of fixed winding number W. The option to glue fields together implies that one may combine two fields of windings W_1 and $-W_2$ to obtain one of winding $W = W_1 - W_2$. This means that the W-sector effectively contains instantons of arbitrary degree. In practice, the most important configurations are elementary $W = \pm 1$ **instantons and anti-instantons**, $\phi_\pm$, with identical action $S_0 \equiv S[\phi_+] = S[\phi_-]$. For example, for well-separated a_i, b_i, the configurations (8.4.3) are superpositions of $W = 1$ instantons. (Why would W elementary instantons contribute more to a partition sum than a single W instanton? Hint: Consider entropy.) Assuming additivity of the actions, the W-sector will thus contain W_+ instantons and $W_- = W + W_+$ anti-instantons with action $(W_+ + W_-)S_0$. **dilute instanton gas** The zero-mode center coordinates of the instantons need to be integrated over. This program is relatively straightforward, provided instantons are energetically costly and scarce – the **dilute instanton gas limit**. However, for diminishing S_0, their number increases and they start to correlate and eventually overlap in space. This defines the **instanton liquid** which is usually difficult to bring under control. For a discussion of the S^2-instanton gas problem, we refer to Ref.[28].

INFO We have already seen an example of an instanton gas, namely in the path integral treatment of the double well in section 3.3.1. Strictly speaking, the instantons considered there were not instantons in a topological sense, because no winding numbers were involved. However, to see the relevance of that discussion to the present context, 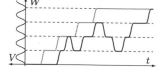 consider the particle-on-a-ring setup of section 8.2, for simplicity in the absence of flux, $A = 0$, but in the presence of a periodic potential, $\cos\phi$. Much as in section 3.3.1, (anti-)instantons are $W = \mp 1$ configurations traversing the ring once to quickly pass between

[28] A. M. Polyakov, *Gauge Fields and Strings* (Harwood, 1987).

minima of the potential. For example, the figure shows the phase profile of a configuration with total winding $W = 4$, which in the case of the bold line comprises seven forward winding instantons and three anti-instantons. For a discussion of the integration over instanton zero modes and the correlations between them, we refer to section 3.3.1.

8.4.4 Path integral of spin

In the next section, we will discuss the role played by θ-terms in the low-energy physics of antiferromagnetic spin chains. This discussion will be based on the (1+1)-dimensional field integral for a spin chain, which in turn is based on the $(0 + 1)$-dimensional path integral for individual spins. The latter is an interesting object in its own right, including from a topological perspective. However, perhaps surprisingly, we will look at the topology of the single-spin path integral only later, in section 8.5. For the moment, we just need it as a building block for the construction of the spin chain extension. A second interesting thing to notice is that the construction of the Feynman path integral of spin is not entirely straightforward – Feynman tried and failed – although a single spin is the most elementary system of quantum mechanics. In view of this situation, we offer three options: readers wishing to proceed as quickly as possible with the discussion of θ-terms are invited to take the path integral representation (8.20) with action (8.18) below as given and advance to section 8.4.6. The alternative is to put the θ-terms on hold and continue reading here about the interesting first-principle construction of the spin path integral also in the next section. Finally, there is the compromise solution to study the qualitative discussion of the path integral but skip section 8.4.4 on its technical construction.

Classical mechanics of spin

Why should the path integral of spin be challenging to construct? A cheap answer would be to reason that path integrals are integrals over exponentiated classical actions while spin-1/2 is essentially quantum and therefore does not afford a description in terms of classical objects. However, this is not correct. The classical limit of a quantum mechanical spin is an angular momentum of fixed modulus, a spinning (!) top. The configuration of a top is determined by a point on the two-sphere marking its orientation. Naïvely, one might identify the two angles required to describe this point as generalized coordinates, so that the classical mechanics of the system would be described in a four-dimensional phase space, having added to these coordinates their canonical momenta. However, this does not feel right. If these momenta existed, we would have heard of the corresponding hermitian operators in quantum mechanics, which we haven't.

To understand what is going on, it is instructive to look at the classical equations of motion for the **Larmor precession** $d_t S\mathbf{n} = SB\mathbf{n} \times \mathbf{e}_z$ of an angular momentum of modulus $\mathbf{L} \equiv S\mathbf{n}$ in the of direction $\mathbf{n}$ in the presence of a magnetic field, $B\mathbf{e}_z$ (where we have set the magnetic moment to unity). These equations are first order, and so we interpret them as Hamilton equations of motion. Let us find out the

Larmor
precession

corresponding coordinates, momenta, and Hamiltonian. To this end, we describe **n** in the usual way in terms of spherical coordinates and find (exercise) that the equations above assume the form $d_t \cos\theta = 0, d_t\phi = -B$, where we temporarily set the angular momentum, S, to unity for simplicity. These equations are structurally consistent with the Hamilton equations, $d_t q = \partial_p H, d_t p = -\partial_q H$, if we make the identifications $q = \phi, p = \cos\theta$, $H = -B\cos\theta$. The expression for $H = -\cos\theta$ makes obvious sense; it renders ϕ a cyclic variable, is periodic in θ, and states that the energy is minimized by aligning **L** with the external field. However, the more interesting observation is that the interpretation of the resulting equations of motion requires that **the two-dimensional sphere is a phase space** by itself, with $\cos\theta$ and ϕ as a coordinate and momentum pair. Also notice that these coordinates cannot be extended to cover the whole sphere (e.g., ϕ is an ill-defined variable at the north and south pole, and multi-valued at $\phi = 0$). This point, as we will see shortly, is essential to the topology underlying the present system.

Phase space

INFO The unusual nature of the pair $(\phi, \cos\theta)$ is why we do not normally hear about the Hamiltonian dynamics of angular momenta in elementary classical mechanics courses. There, the phase space of a system with f degrees of freedom is usually introduced as a $2f$-dimensional real vector space, with coordinates (q_i, p_i) and the added structure of Poisson brackets. Mathematics has a nicer definition to offer: a **phase space** is a $2f$-dimensional differentiable manifold equipped with a non-degenerate differential two-form ω, a so-called **symplectic form**. We can think of the two-form as an object very similar to a metric, the essential difference being that it is skew-symmetric, $\omega_{ij} = -\omega_{ji}$, unlike the symmetric metric tensor $g_{ij} = g_{ji}$. Manifolds with this structure are called **symplectic manifolds**, and the statement above says that the phase space of Hamiltonian dynamics is actually symplectic. Much as the local coordinate representation of a metric enters the definitions of the laws of relativity, the coordinate representation of the symplectic form gives us the Poisson brackets, the Hamiltonian structure of evolution equations, etc.[29] Specifically, in a coordinate representation, where $\omega = \sum_{i=1}^{f} dq^i \wedge dp_i$ assumes a diagonal form, we recover the standard relations for a pair of canonical variables (q^i, p_i). However, the true merit of the definition is that phase space need not be a vector space but can be a more interesting manifold, such as a sphere. The natural two-form on the sphere is the area form (see appendix section A.1.3) $\omega = d\phi \wedge d\cos\theta$ and this shows that the sphere is a two-dimensional phase space, also known as a symplectic manifold with local canonical coordinates $(\phi, \cos\theta)$. However, we do not expect to find a system of global coordinates, as it would make our sphere identical to a section of flat space.

The above discussion suggests that the **imaginary-time Hamiltonian action** describing a precessing angular momentum reads

$$S[\theta, \phi] = S \int_0^\beta d\tau \left(-i d_\tau \phi \cos\theta + B\cos\theta \right) \tag{8.18}$$

where we have reintroduced the angular momentum, S. It is straightforward to verify that the variation of these equations yields the imaginary-time variant of the

---

[29] See V. I. Arnold, *Mathematical Methods of Classical Mechanics* (Springer, 1978) for a beautiful exposition of mechanics in this language

angular precession $i\partial_\tau\phi = B$. However, the above-mentioned coordinate singularity means that the underlying variation must not be applied to configurations touching the singular north or south pole of the sphere. Fortunately, we may use the arbitrariness of coordinate choices to avoid these singularities: we always have the freedom to add to the action a full derivative $d_\tau F(\phi, \cos\theta) = \partial_\phi F d_\tau\phi + \partial_{\cos\theta} d_\tau(\cos\theta)$ without affecting the equations of motion. One may show[30] that different choices of F correspond to different choices of the singular points of the coordinate system. For instance, with $F = \pm i S d_\tau\phi$, we obtain the modified actions

$$S_{\mathrm{n/s}}[\theta, \phi] \equiv S \int_0^\beta d\tau \left(-i d_\tau\phi \left(\cos\theta \mp 1\right) + B\cos\theta\right). \tag{8.19}$$

In S_{n}, the problematic derivative term vanishes at the north pole and only the southern singularity remains. Similarly, S_{s} is singular at the north pole, etc., and so one singularity remains for the reasons mentioned above. We will address the topological interpretation of this structure in section 8.5 when we turn to the discussion of Wess–Zumino terms. However, for the time being, our main result is the classical action (8.18). Referring for a first-principle derivation to the next section, the **imaginary-time path integral describing a spin** S is given by the integral over all closed curves on the sphere weighted by the above action,

spin path integral

$$Z = \int D(\phi, \theta)\, e^{-S[\phi, \theta]}, \tag{8.20}$$

where $\int D(\phi, \theta) \equiv \prod_\tau \int \sin\theta_\tau\, d\theta_\tau d\phi_\tau$ is the integral over the invariant measure on the two-sphere at each time slice.

8.4.5 Path integral for spin (derivation)

REMARK In this section, we assume basic familiarity with $SU(2)$ and its representations at the level of a standard lecture course of advanced quantum mechanics. For a classic reference on the subject consult, e.g., Ref.[31].

As promised in the introduction to this section, here we discuss a first-principle derivation of the path integral (8.20) with action (8.18). This construction introduces methodology frequently applied in the path or field integration over target spaces with group symmetries. However, as mentioned above, it is optional where the subject of topology is concerned and can be read at any later stage.

Representation theory of $SU(2)$

The quantum mechanics of spin is intimately connected to the representation theory of $SU(2)$. Before turning to the construction of the path integral, let us summarize a few facts, most of which will be remembered from the quantum mechanics lecture

[30] H. B. Nielsen and D. Rohrloch, *A path integral to quantize spin*, Nucl. Phys. B **299**, 471 (1988).
[31] J. J. Sakurai, *Modern Quantum Mechanics* (Addison-Wesley, 1994).

courses. The **special unitary group SU**(2) is the group of two-dimensional unitary matrices with unit determinant. Counting independent components, one finds that the group has three real free parameters or, equivalently, that its **Lie algebra**, su(2), is three-dimensional. The basis vectors of the algebra – the group generators – $\hat{S}^i, i = x, y, z$, satisfy the commutation relation $[\hat{S}^i, \hat{S}^j] = i\epsilon_{ijk}\hat{S}^k$. A useful alternative basis $\{\hat{S}^+, \hat{S}^-, \hat{S}^z\}$ is defined by the **spin raising and lowering operators** $\hat{S}^{\pm} = \hat{S}^x \pm i\hat{S}^y$ with commutation relations $[\hat{S}^+, \hat{S}^-] = 2\hat{S}^z$, $[\hat{S}^z, \hat{S}^{\pm}] = \pm\hat{S}^{\pm}$.

Group elements themselves can be parameterized through the exponentiated algebra. For example, in the **Euler angle representation** they are represented as

$$g(\phi, \theta, \psi) = e^{-i\phi\hat{S}_3}e^{-i\theta\hat{S}_2}e^{-i\psi\hat{S}_3}, \qquad \phi, \psi \in (0, 2\pi), \theta \in (0, \pi). \qquad (8.21)$$

The Hilbert spaces $\mathcal{H}_S$ of quantum spin are irreducible representation spaces of SU(2). Within the spaces $\mathcal{H}_S$, SU(2) acts via matrix representations of its generators $\hat{S}_i$, and the induced representation matrices of the group elements (likewise denoted by g).[32] In each $\mathcal{H}_S$, there exists a state $|\uparrow\rangle$ – a highest-weight state in the parlance of group theory, and a z-polarized state in that of physics – defined as the normalized eigenstate of $\hat{S}^z$ with maximum eigenvalue, S.

All other states in $\mathcal{H}_S$, $|g\rangle$, can be obtained by applying group elements to this maximum-weight state, $|g\rangle = g|\uparrow\rangle$. Specifically, for the group elements represented via Euler angles, the Hilbert space is given by the set of states[33]

$$\boxed{|g(\phi, \theta)\rangle \equiv e^{-i\phi\hat{S}_3}e^{-i\theta\hat{S}_2}|\uparrow\rangle}$$

known as **spin coherent states**. Here, we have used the fact that the ψ-dependent part of Eq. (8.21) affects the maximum-weight state only via a phase $e^{-i\psi\hat{S}_3}|\uparrow\rangle = e^{-i\psi S}|\uparrow\rangle$, which we have dropped. In quantum physics, the terminology "coherent states" is reserved for states representing the best possible approximation to classical states. The above states are coherent, in that they are in one-to-one correspondence with points on a sphere representing the orientation of an angular momentum. To see how, we use the auxiliary identity $(i \neq j)$

$$e^{-i\phi\hat{S}_i}\hat{S}_j e^{i\phi\hat{S}_i} = e^{-i\phi[\hat{S}_i,]}\hat{S}_j = \hat{S}_j \cos\phi + \epsilon_{ijk}\hat{S}_k \sin\phi \qquad (8.22)$$

(which follows from expanding the commutator in odd and even contributions and using the spin commutation relations) to compute the expectation values

$$\mathbf{n} \equiv \langle g(\phi, \theta)|\mathbf{S}|g(\phi, \theta)\rangle = S\begin{pmatrix} \sin\theta\cos\phi \\ \sin\theta\sin\phi \\ \cos\theta \end{pmatrix} \equiv S\mathbf{n}. \qquad (8.23)$$

[32] To avoid confusion, keep in mind that $\mathcal{H}_S$ are vector spaces of dimension $2S + 1$, in which SU(2) – originally defined as a two-dimensional matrix group – acts via matrix generators $\hat{S}$ of equal dimension, $2S + 1$.

[33] The statement is not entirely correct since the exclusion of the angles $\theta = 0, \pi$ and $\phi = 0, 2\pi$ makes a zero-measure set of states (such as the down-polarized one $|\downarrow\rangle$) exempt from the representation. However, as we will use our representation in integrals, this is not a point of concern.

This equation is the key to understanding the terminology "spin coherent states." The expectation values of the spin operator in these states are identical to the configuration $S\mathbf{n}$ of a classical angular momentum, as described by a unit vector $\mathbf{n}$ on the sphere. In other words, the states $|g(\phi, \theta)\rangle$ define the closest approximation to an angular momentum in Hilbert space.

In the construction of the path integral, resolutions of unity realized as integrals over all group elements play a key role. These integrals have the architecture $\int dg\, f(g)$, where the integration extends over the full group (parameterized by Euler angles, or other suitable coordinates), f is a function of g, and dg is the **Haar measure**. This measure is uniquely determined by its invariance under left and right multiplication of g by fixed group elements, i.e.,

Haar measure

$$\forall h \in \mathrm{SU}(2) : \int dg\, f(gh) = \int dg\, f(hg) = \int dg\, f(g).$$

This relation states that the Haar measure is "translationally invariant" under group multiplication and treats all elements of the group on the same footing.

Construction of the path integral

With this background, we are now in a position to formulate the Feynman path integral. To be specific, let us consider a particle of spin S subject to the Hamiltonian

$$\hat{H} = \mathbf{B} \cdot \hat{\mathbf{S}},$$

where $\mathbf{B}$ is a magnetic field and $\hat{\mathbf{S}} \equiv (\hat{S}_1, \hat{S}_2, \hat{S}_3)$ is a vector of spin operators in the spin-S representation. Our aim is to calculate the imaginary-time path integral representation of the partition function $\mathcal{Z} \equiv \mathrm{tr}\, e^{-\beta \hat{H}}$. In constructing this path integral, we follow the strategy outlined at the end of section 3.2.3 and start from a trotterization of the partition sum, $\mathcal{Z} = \mathrm{tr}(e^{-\epsilon \hat{H}})^N$, where $\epsilon = \beta/N$. Next, we have the most important step in the construction – to insert a suitably chosen resolution of identity between each of the factors $e^{-\epsilon \hat{H}}$. A representation that will lead us directly to the final form of the path integral is specified by

$$\mathbb{1} = C \int dg\, |g\rangle\langle g|, \tag{8.24}$$

where $\int dg$ is a group integral over the Haar measure, C is a constant, and $|g\rangle \equiv g|\uparrow\rangle$ as discussed in the previous section.

Of course it remains to be verified that the integral (8.24) defines a proper representation of the unit operator. That this is so follows from **Schur's lemma**, which states that, if, and only if, an operator $\hat{A}$ commutes with all the representation matrices of an irreducible group representation (in our case the gs acting in the Hilbert space $\mathcal{H}_S$), $\hat{A}$ is proportional to the unit matrix. That the above group integral fulfills this commutativity criterion follows from the properties of the Haar measure, $\forall h \in \mathrm{SU}(2)$:

Schur's lemma

$$h \int dg\, |g\rangle\langle g| = \int dg\, |hg\rangle\langle g| \overset{\text{Haar}}{=} \int dg\, |hh^{-1}g\rangle\langle h^{-1}g| = \int dg\, |g\rangle\langle g|h.$$

We have thus established that the integral is proportional to the unit operator. As usual in the construction of path integrals, the constant of proportionality, C, is not of concern, as it cancels in any properly normalized expectation value.

Substituting the resolution of identity into the time-sliced partition function and computing matrix elements as

$$\langle g'|e^{-\epsilon \mathbf{B}\cdot\hat{\mathbf{S}}}|g\rangle \simeq \langle g'|g\rangle - \epsilon\langle g'|\mathbf{B}\cdot\hat{\mathbf{S}}|g\rangle \overset{\langle g|g\rangle=1}{=} 1 - \langle g|g\rangle + \langle g'|g\rangle - \epsilon\langle g'|\mathbf{B}\cdot\hat{\mathbf{S}}|g\rangle$$
$$\simeq \exp\left(\langle g'|g\rangle - \langle g|g\rangle - \epsilon\langle g'|\mathbf{B}\cdot\hat{\mathbf{S}}|g\rangle\right),$$

one obtains

$$\mathcal{Z} = \lim_{N\to\infty} \int_{g_N=g_0} \prod_{i=0}^{N} dg_i \, \exp\left[-\epsilon \sum_{i=0}^{N-1}\left(-\frac{\langle g_{i+1}|g_i\rangle - \langle g_i|g_i\rangle}{\epsilon} + \langle g_{i+1}|\mathbf{B}\cdot\hat{\mathbf{S}}|g_i\rangle\right)\right],$$

which in the limit $N\to\infty$ assumes path integral form,

$$\boxed{\mathcal{Z} = \int Dg \, \exp\left[-\int_0^\beta d\tau \left(-\langle\partial_\tau g|g\rangle + \langle g|\mathbf{B}\cdot\hat{\mathbf{S}}|g\rangle\right)\right]} \qquad (8.25)$$

with the $\mathcal{H}_S$-valued function $|g(\tau)\rangle$ as the continuum limit of $|g_i\rangle$. Equation (8.25) is a valid, if somewhat over-compact, representation of the path integral. In order to give this expression concreteness, we employ the **Euler angle representation** discussed above for the states $|g\rangle$. Substituting $|g(\phi,\theta)\rangle$ into the integral, the latter becomes an integral over the closed path on the sphere, parameterized by $(\theta(\tau),\phi(\tau))$. It can be verified that, in this representation, the Haar measure integral reduces to an integral over the canonical measure on the sphere, $dg \to \sin\theta\, d\theta d\phi$, which likewise treats all points in configuration space indiscriminately.

Let us now proceed by exploring the action of the path integral. With the auxiliary identity (8.22), it is a straightforward matter to show that

$$\Gamma[\phi,\theta] \equiv -\int_0^\beta d\tau \, \langle\partial_\tau g|g\rangle = -iS\int_0^\beta d\tau \, \partial_\tau\phi\cos\theta. \qquad (8.26)$$

Finally, using Eq. (8.25) to compute the angular representation of the **B**-dependent term, we obtain Eqs. (8.18) and Eq. (8.20) for the spin path integral.

In the next section, we will use this path integral as a building block for the construction of the *field* integral of a spin chain. However, let us mention at this point that the "canonical contribution" to the action (8.26) is a topological term in itself. It belongs to the class of **Wess–Zumino terms**, and we will discuss its topological interpretation in section 8.5.

8.4.6 Spin chains

In this section, we use the spin path integral to construct a field integral representation of an antiferromagnetic spin chain. We will find that the coupling of many spins of approximately opposite orientation – the antiferromagnetic order of the chain – leads to the emergence of a θ-term whose presence is of key importance to the physics of the chain at low temperatures.

Low-energy excitations of spin chains

In section 2.2.5, we saw that the low-energy dispersion of spin-wave excitations in antiferromagnets is approximately linear $\epsilon(\mathbf{p}) \sim v_s|\mathbf{p}|$, where v_s is the spin-wave velocity. This result was obtained by a semiclassical expansion in small values of $1/S$, where S is the magnitude of the spins. Here, we ask what happens as we push towards the quantum regime, $S = \mathcal{O}(1)$. The main conclusion will be that, via a mechanism linked to the topology discussed in section 8.4.2, the physics of a spin chain depends crucially on whether S is half-integer or integer.

For small values of S, we have no expansion parameter stabilizing the derivation of an effective action. In this situation, it is fortunate that the smallest possible choice, $S = 1/2$, is a solvable reference configuration: in problem 2.4.5, we saw that for small wave vectors, the $S = 1/2$ antiferromagnet becomes equivalent to a half-filled system of one-dimensional linearly dispersive fermions.[34] The excitations of the latter, i.e., charge density waves, are likewise linearly dispersive, consistent with the semiclassical result. Given that both $S \gg 1$ and the exactly solvable point $S = 1/2$ have similar excitations, one may speculate that the analytically inaccessible intermediate regime $S \simeq 1$ also shows the same behavior.

However, this expectation does not conform with experimental observation. While neutron scattering experiments on spin-1/2 chains have confirmed linear dispersion in the vicinity of the Néel ordering wave vector $q = \pi/a$, spin $S = 1$ chains show different behavior and do not support low-energy magnetic excitations at all (see the figure, reprinted from Ref.[35], where the main panel shows the excitation energy for wave vectors close to the value $Q_c = \pi$ corresponding to antiferromagnetic order). More generally,

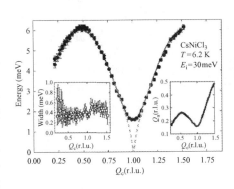

34 To be specific, the equivalent fermion system contains an interaction whose strength depends on the anisotropy Δ of the magnetic correlations. In the XY-limit, $\Delta = 0$ (vanishing coupling of the z-components), the fermion system is free and, like its equivalent spin system, supports long-range excitations. For finite Δ, bosonization techniques map the problem onto a **two-dimensional sine–Gordon model** in the universality class of the two-dimensional classical XY-model (see the discussion on page 367). From the RG flow of the latter, one infers that, for all values of the anisotropy up to the Heisenberg limit $\Delta = 1$, the system remains gapless and linearly dispersive. For further discussion of the spin-1/2 chain, we refer to A. O. Gogolin, A. A. Nersesyan, and A. M. Tsvelik, *Bosonization and Strongly Correlated Systems* (Cambridge University Press, 1998).

35 M. Kenzelmann, R. A. Cowley, W. J. L Buyers, *et al.*, *Properties of Haldane excitations and multiparticle states in the antiferromagnetic spin-1 chain compound* $CsNiCl_3$, Phys. Rev. B **66**, 24407 (2002). Copyright (2002) by The American Physical Society.

sine–
Gordon
model

experiment has shown that antiferromagnetic chains of half-integer spin are gapless and linearly dispersive, while chains of integer spins are gapped.

Physical phenomena depending on the parity of an integer quantum number (in the present case, whether $2S$ is even or odd) tend to be of topological origin. To understand why topology appears in the present context, recall that the classical configuration space of spin S is a sphere of radius S. A spin chain is thus described by a map from $(1 + 1)$-dimensional space–time into the sphere, i.e., the class of maps discussed in the previous section.

Field theory of the antiferromagnetic spin chain

To demonstrate the appearance of topology in the present setting, we start out from the quantum partition function of a field-free isolated spin (8.26),

$$\mathcal{Z}^{(1)} = \int D\mathbf{n} \, \exp\left[-iS \int_0^\beta d\tau \, \cos\theta \, \partial_\tau \phi\right],$$

where $D\mathbf{n} = D(\theta, \phi)$ is a shorthand for the measure on the sphere.

As a warm up to the following discussion of the antiferromagnetic chain, first consider the **ferromagnetic chain**. Application of Eq. (8.23) shows that, in this case, the interaction between spins at neighboring sites, i, is described by

$$-J\hat{\mathbf{S}}_i \cdot \hat{\mathbf{S}}_{i+1} \longrightarrow -JS^2 \mathbf{n}_i \cdot \mathbf{n}_{i+1} \longrightarrow \frac{JS}{2}(\mathbf{n}_i - \mathbf{n}_{i+1})^2, \qquad (8.27)$$

where J is the positive exchange constant, the first arrow points to the functional integral representation, $\hat{\mathbf{S}} \to S\mathbf{n}$, and the second holds true up to an inessential constant ($\mathbf{n}_i^2 = 1$). Adding to this "discrete derivative" the free spin action, we obtain a partition function $\mathcal{Z} = \int D\mathbf{n} \exp(-S[\mathbf{n}])$, with effective action

$$S[\mathbf{n}] = \int d\tau \sum_i \left[iS \cos\theta_i \, \partial_\tau \phi_i + \frac{JS^2}{2}(\mathbf{n}_i - \mathbf{n}_{i+1})^2\right]. \qquad (8.28)$$

Note a curiosity here: while it would seem natural to express the entire action as a functional of $\mathbf{n}$ and $\partial_\tau \mathbf{n}$, it is not possible to represent the free spin action in this way. There is a topological principle behind this, and we will discuss it in section 8.5. The positive exchange constant favors smooth configurations $\{\mathbf{n}_i\}$, which justifies the continuum limit,

$$S_{\text{ferro}}[\mathbf{n}] = a^{-1} \int d\tau \, dx \left[iS \cos\theta \, \partial_\tau \phi + \frac{JS^2 a^2}{2}(\partial\mathbf{n})^2\right], \qquad (8.29)$$

where a denotes the lattice spacing. The action (8.29) does not contain a θ-term.

EXERCISE Recapitulate what we know about the variation of the first term in the action from section 8.20 to derive the equations of motion of this action and show that the mean field dispersion of the ferromagnetic chain, $\omega \sim q^2$, is quadratic. Renormalization group analysis shows that, at large distance scales, the system flows into a strong fluctuation regime (a manifestation of the Mermin–Wagner theorem according to which magnetic order in two-dimensional systems is not possible) and builds up a gap.

In the complementary case of an **antiferromagnetic spin chain**, the negative sign of the exchange constant favors the antiparallel alignment of spins at neighboring sites. We thus start out from an *ansatz* $\mathbf{n}_i = (-)^i \mathbf{n}'_i$, with smoothly varying $\mathbf{n}'$. One should now substitute this configuration into the action and perform a gradient expansion as above. However, instead of formulating this program in detail,[36] we here determine the resulting action by semi-quantitative reasoning.

First, we know that the action supports a wave-like mode at the semiclassical level, and is rotationally invariant in $\mathbf{n}'$. The minimal action satisfying these criteria reads

$$S_0[\mathbf{n}] = \frac{S}{4} \int d\tau \, dx \left[v_{\mathrm{s}} (\partial_x \mathbf{n})^2 + \frac{1}{v_{\mathrm{s}}} (\partial_\tau \mathbf{n})^2 \right], \qquad (8.30)$$

where we have relabeled $\mathbf{n}'$ as $\mathbf{n}$ for notational simplicity, and the definition of the spin-wave velocity, $v_{\mathrm{s}} = 2aJS$, is beyond the scope of the present argument. Rescaling variables so that $\tau \to v_{\mathrm{s}}^{-1/2} \tau \equiv x_0$, $x \to v_{\mathrm{s}}^{1/2} x \equiv x_1$,

$$S_0[\mathbf{n}] \to \frac{1}{\lambda} \int d^2x \, \partial_\mu \mathbf{n} \cdot \partial_\mu \mathbf{n}, \quad \lambda = 4/S, \qquad (8.31)$$

nonlinear
σ-model

we obtain the action of the O(3) **nonlinear σ-model**.[37]

However, in addition to the O(3) action, we expect the presence of a topological term. This contribution must emanate from the – likewise topological – Wess–Zumino action. To see how this happens, let us turn back to the spatially discretized representation. In angular coordinates, a spin of orientation opposite to one described by the coordinates (θ, ϕ) has coordinates $(\pi - \theta, \phi + \pi)$. This suggests describing pairs of almost opposite spins at sites x_i and $x_{i+1} = x_i + a$, respectively, by the variables $(\theta(x_i), \phi(x_i))$ and $(\pi - \theta(x_i + a), \phi(x_i + a) + \pi)$, where the angular functions are smooth and their time dependence is left implicit. In these variables, the sum over free spin actions becomes

$$-iS \sum_i \int d\tau \left(\cos(\theta(x_i)) \, \partial_\tau \phi(x_i) + \cos(\pi - \theta(x_i + a)) \, \partial_\tau (\phi(x_i + a) + \pi) \right)$$

$$\simeq iS \sum_i \int d\tau \, \sin(\theta(x_i)) \, a (\partial_x \theta(x_i) \partial_\tau \phi(x_i) - \partial_\tau \theta(x_i) \partial_x \phi(x_i))$$

$$\simeq \frac{iS}{2} \int dx \, d\tau \, \sin \theta (\partial_x \theta \partial_\tau \phi - \partial_\tau \theta \partial_x \phi).$$

Here, we have Taylor-expanded to first order in a and integrated by parts. The integral representation in the last line contains a factor $1/2$ reflecting the fact that the "unit cell" contains two spins, so that the effective discretization interval equals $2a$. The main point here is that the final integral represents a standard coordinate

---

[36] See N. Nagaosa, *Quantum Field Theory in Condensed Matter Physics* (Springer, 1999).

[37] This common denotation is a bit of a misnomer since the target space of the theory is the two-sphere $S^2 \simeq O(3)/O(2)$ and not O(3).

representation for an area integral over the sphere. In fact, it is straightforward to verify that, in a representation of the unit vector $\mathbf{n}(\theta, \phi)$, it equals the integral (8.15) with topological angle $\theta/4\pi = S/2$, and identification $\mathbf{x} = (x, \tau)$. Notice that the $(1+1)$-dimensional θ-term obtained by staggered addition of the $(0+1)$-dimensional Wess–Zumino terms *is* representable in terms of the invariant variables $\mathbf{n}$. That, too, reflects a geometric principle whose origins we will explain in section 8.5.

Summarizing, we have found that the **field integral of the antiferromagnetic spin chain** assumes the form

$$\mathcal{Z} = \int D\mathbf{n}\, e^{-S_0[\mathbf{n}] - S_{\text{top}}[\mathbf{n}]}, \tag{8.32}$$

where S_0 is given by Eq. (8.31), and the topological action by Eq. (8.15) with angle $\theta = 2\pi S$. From the renormalization group (RG) analysis of section 6.4, we know that, at large length scales, the system described by $S_0[\mathbf{n}]$ flows into a disordered phase. How will the presence of the topological term modify this behavior? Being insensitive to small variations, it does not couple to the RG decimation of fluctuations in individual topological sectors, W. However, to see why it may nevertheless act as a game changer, let us reorganize the partition function as a sum,

$$\mathcal{Z} = \sum_{W \in \mathbb{Z}} \int D\mathbf{n}_W\, e^{2\pi i S W} e^{-S_0[\mathbf{n}_W]}, \tag{8.33}$$

where $\mathbf{n}_W$ denotes field configurations of winding number W. For integer spin, $\exp(2\pi i S W) = 1$, and the topological term just drops out. By contrast, for half-integer spin, $\exp(2\pi i S W) = (-)^W$, consecutive topological sectors are weighted by alternating signs. Notice that the topological term is susceptible to the *parity* of $2S$.

To understand the consequences heuristically, note that the RG flow to strong coupling is driven by fluctuations which at the late stages of the flow include fluctuations over topologically nontrivial, $W \neq 0$, configurations. In the integer-S case, the positivity of $\exp(-S_0)$ implies that all these fluctuations *add* up, furthering the flow into a disordered phase. However, for half-integer spin, they contribute with alternating sign, which implies a tendency to mutual cancellation. This observation is the basis of **Haldane's conjecture**:[38]

Haldane's conjecture

> Spin chains of integer S are conjectured to flow into a disordered phase with no long-range excitations, while half-integer spin chains are gapless.

This expectation is confirmed by neutron scattering measurements of the dispersion of various quasi one-dimensional magnets. However, it must be emphasized that the conjecture is not backed by an actual computation. In particular, it presumes that the topological σ-model does not change its form under renormalization.

[38] F. D. M. Haldane, *Nonlinear field theory of large-spin Heisenberg antiferromagnets: Semiclassically quantized solitons of the one-dimensional easy-axis Néel state*, Phys. Rev. Lett. **50**, 1153 (1983).

Subsequent analysis by Affleck and Haldane[39] indeed showed that this presumption was premature and that the topological $O(3)$ nonlinear σ-model flows to a different theory under renormalization. We will return to this point in section 8.5.5 after the concept of Wess–Zumino terms has been introduced in generality.

8.4.7 Integer quantum Hall effect

REMARK The integer quantum Hall effect constitues a role model showing many of the phenomena characteristic of topological condensed matter physics. Its topological origins can be addressed from different perspectives, including that of homotopy introduced in the previous section. At the same time, a discussion of the rich phenomenology of the quantum Hall state may come as a digression for readers primarily interested in a more streamlined introduction to topological field theory. To these readers, we suggest skipping this section and proceeding directly to section 8.5.

The traditional meaning of the integer quantum Hall effect (IQHE) is the quantization of the transverse, or Hall conductance of two-dimensional electron gases subject to strong perpendicular magnetic fields. When this effect was discovered in 1980,[40] it could not be foreseen that the IQHE would become the precursor of multiple conceptually related phenomena. From a modern perspective, the quantum Hall system belongs to the family of topological insulators. Referring back to section 8.1, all these materials are insulating in the bulk, with ground state wave functions carrying topological invariants. The observable manifestation of the ground state topology is the presence of gapless surface states. In the one-dimensional setting discussed in section 8.1, these assume the form of discrete zero-energy bound states sitting inside the excitation gap of the insulating bulk. In higher dimensions, they generalize to continua with gapless dispersion relations, i.e., surface *metallic* states. However, unlike conventional metals, the surface states of topological insulators are protected by topological principles against the (Anderson) localizing effects of impurity scattering.

Below we will summarize the phenomenological consequences of this topological bulk boundary correspondence for the specific case of the two-dimensional quantum Hall insulator. We will then investigate the topological origins of the IQHE, and finally introduce an effective low-energy field theory describing its experimental signatures.

Klaus von Klitzing 1943– is a German physicist renowned for his observation of the integer quantum Hall effect in 1980. For this discovery, he was awarded the 1985 Nobel Prize. The quantized unit of the transverse Hall conductance, $R_K \equiv h/e^2 = 25812.80745\cdots\,\Omega$ is named the *von Klitzing constant* in honor of his achievement.

---

[39] I. Affleck and F. D. M. Haldane, *Critical theory of quantum spin chains*, Phys. Rev. B **36**, 5291 (1987).

[40] K. v. Klitzing, G. Dorda, and M. Pepper, *New method for high-accuracy determination of the fine-structure constant based on quantized Hall resistance*, Phys. Rev. Lett. **45**, 494 (1980).

IQHE phenomenology

REMARK For the convenience of the reader, here we summarize a few constants which keep reappearing in the physics of the IQHE.

Cyclotron frequency	$\omega_c = \frac{eB}{m}$	energy spacing between Landau levels
Magnetic length	$l_0 = \sqrt{\frac{\hbar}{eB}}$	defines the extension of one flux-quantum area
Flux quantum	$\Phi_0 = \frac{h}{e}$	quantum unit of magnetic flux
von Klitzing constant	$R_K = \frac{h}{e^2}$	transverse resistance of lowest quantum Hall state

To ease the identification of these constants, we temporarily refrain from setting $\hbar$ and e to unity. Also note that the flux quantum is often defined as $\Phi_0 = h/2e$. This definition makes sense in the theory of superconductivity, where the Cooper pair charge is $2e$, but less so here.

The IQHE is observed in two-dimensional electron gases subject to a strong magnetic field (see the figure). The driving of an electric current I_x through the system is accompanied by the build up of a voltage drop V_x in the direction of the current. In a conventional conducting system, the ratio of the two, $R_{xx} = V_x/I_x$, defines the longitudinal

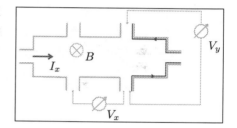

electric resistance, which in two dimensions equals the resistivity, $\rho_{xx} = R_{xx}$, up to a dimensionless geometric factor inessential to our discussion.[41] However, in a system subject to a transverse magnetic field, the driven motion in the x-diection (i.e., the current I_x) generates a Lorentz force in the y-direction, giving rise to a voltage V_y (see the exercise below). The ratio between current and voltage is now governed by a vectorial relation,

$$\begin{pmatrix} V_x \\ V_y \end{pmatrix} = \begin{pmatrix} \rho_{xx} & \rho_{xy} \\ -\rho_{xy} & \rho_{xx} \end{pmatrix} \begin{pmatrix} I_x \\ I_y \end{pmatrix}, \qquad \begin{pmatrix} I_x \\ I_y \end{pmatrix} = \begin{pmatrix} \sigma_{xx} & \sigma_{xy} \\ -\sigma_{xy} & \sigma_{xx} \end{pmatrix} \begin{pmatrix} V_x \\ V_y \end{pmatrix}, \quad (8.34)$$

conductivity tensor

where the two matrices define the resistivity tensor ρ, and the **conductivity tensor**, $\sigma = \rho^{-1}$, respectively. The two tensors are inverse to each other, $\rho = \sigma^{-1}$, and the relation $\sigma_{xy} = -\sigma_{yx}$ reflects a geometric symmetry of the system. In the physics of the quantum Hall effect, the tensorial structure of the transport coefficients is important. For example, for non-vanishing Hall conductivity, $\sigma_{xy} \neq 0$, the vanishing of the longitudinal conductivity, $\sigma_{xx} = 0$, implies the vanishing of the longitudinal resistivity, and not a divergence as one might naïvely expect.

In classical transport theory (see the exercise below), one expects a linear growth $\rho_{xy} = \mu B/\sigma_0$, where $\mu = e\tau/m$ defines the mobility of the electron gas and $\sigma_0 = ne^2\tau/m$ is the Drude conductivity. Here, m is the electron mass, τ the impurity

---

[41] Recall that the resistivity $\rho_{xx} \equiv E_x/j_x$ is defined by the ratio of electric field and current density, while the resistance $R_{xx} = V_x/I_x$ is voltage over current. In a system with homogeneous currents and electric fields, current equals current density times transverse area, $I_x = Aj_x$, and voltage equal electric field times length, $V_x = L_x E_x$. In two-dimensions, where the transverse area equals the transverse linear extension, $A = L_y$, resistivity and resistance equal each other, up to a factor L_x/L_y.

scattering time, and $n = N/A$ the carrier density defined by the total number of charges, N, in a system of area A.

Drude conductivity tensor

EXERCISE Let us investigate the **conductivity tensor within classical Drude transport theory**. To this end, consider Newton's equation

$$\dot{\mathbf{p}} = -e\left(\mathbf{E} + \frac{\mathbf{p} \times \mathbf{B}}{m}\right) - \frac{\mathbf{p}}{\tau}, \tag{8.35}$$

for the dynamics of an electron subject to the Lorentz force and to dissipative damping at a scattering rate τ. Assuming stationarity, $\dot{\mathbf{p}} = 0$, set up an equation for $\mathbf{p}$ in terms of the electric field, $\mathbf{E}$, and from there obtain the current as $\mathbf{j} = -en\mathbf{p}/m$. Show that this defines the resistivity tensor as

$$\rho = \frac{1}{\sigma_0}\begin{pmatrix} 1 & \mu B \\ -\mu B & 1 \end{pmatrix},$$

where the constants are defined above. Note that the Hall resistivity, $\rho_{xy} = B/en$, is universal in that it does not depend on scattering rates.

However, in the famous 1980 experiment[40] different behavior was observed (see the figure, courtesy of D. Leadley): instead of increasing linearly, the Hall resistivity showed a more interesting step function profile with plateau values

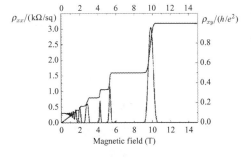

$$\rho_{xy} = \frac{1}{\nu}\frac{h}{e^2}, \tag{8.36}$$

defined by integers $\nu \in \mathbb{N}$ quantized to an accuracy of $\mathcal{O}(10^{-10})$.[42] The center of the νth plateau depends on the magnetic field strength through

$$\nu \equiv 2\pi n l_0^2, \qquad l_0 = \sqrt{\frac{\Phi_0}{2\pi B}}, \tag{8.37}$$

magnetic length

where l_0 is the **magnetic length**. A good way to way to interpret the magnetic length is the $2\pi B l_0^2 = \Phi_0/2\pi$, i.e., up to a numerical constant, l_0^2 is the geometric area containing one flux quantum $\Phi_0 = h/e$. At the center of the νth plateau, we

filling fraction

have $2\pi l_0^2 N/A = \nu$, corresponding to a **filling fraction** of ν charge carriers per flux-quantum area.

By itself, the fact that for field strength B corresponding to integer ν the resistivity is given by Eq. (8.36) is not so remarkable – at these field strengths, the classical resistivity assumes the same value. However, the truly surprising observation is that the resistivity remains constant and sharply quantized under variations

---

[42] The striking precision of the experimental data has led to the introduction of the von Klitzing constant $R_K \equiv h/e^2 = 25812.80\,\Omega$ as the unit of electrical resistance in 1990. As of 2018 the value of Planck's constant, h, and the electron charge, e, are likewise calibrated via this constant.

of B around the center value, to form extended plateaus. In the following, we will discuss how this phenomenon is born out of a conspiracy of topology and, at first unexpectedly, the presence of disorder.

INFO To understand the crucial **role of disorder in the stabilization of plateaus**, consider the idealized case of a clean (translationally invariant) electron gas subject to a magnetic field, $\mathbf{B}$, in the absence of electric fields, $\mathbf{E} = 0$, or current flow, $\mathbf{I} = 0$. Now suppose we observe the system from a frame moving with velocity v in the x-direction. An experimentalist working in that frame would measure both a current density $\mathbf{j} = -vn\mathbf{e}_x$ and, owing to the Lorentz covariance of electrodynamics, an electric field $\mathbf{E} = v\mathbf{e}_x \times \mathbf{B} = -vB\mathbf{e}_y$, where $\mathbf{B} = B\mathbf{e}_z$. With $j_x = \sigma_{xy}E_y$, we obtain $\sigma_{xy} = nB^{-1}$ for the Hall conductivity in the moving frame. Being independent of v, this result holds in all moving frames, including the static one, $v \to 0$. We conclude that, in any translationally invariant environment, the Hall conductivity is linearly related to the magnetic field.

Landau levels

As a preparation for the discussion of the IQHE, let us recapitulate the quantum mechanics of a two-dimensional system of free spinless[43] electrons in a perpendicular magnetic field B. Here, we summarize the facts essential to our discussion and refer to the info block below (or a quantum mechanics textbook) for an explicit diagonalization of the problem.

Landau levels

▷ The eigenenergies of the problem are called **Landau levels** and they are given by $\epsilon_n = \omega_c(n + 1/2)$, where $\omega_c = eB/m$ is the **cyclotron frequency**.

▷ For a system of linear extension, L, each Landau level is $(L/l_0)^2$-fold degenerate. Roughly speaking, each individual wave function in the nth level occupies an area l_0^2, accommodating one magnetic flux quantum, Φ_0. Notice that this implies a *massive* degeneracy, extensive in the system size.

▷ The form of the wave functions depends on the gauge of $\mathbf{A}$ in $\mathbf{B} = \nabla \times \mathbf{A}$, and different choices cater to different applications. For example, in the **symmetric gauge**, $\mathbf{A} = \frac{B}{2}(-y, x, 0)^T$, the wave functions display rotational symmetry relative to the center coordinate $(0, 0)$. (This gauge is often the preferred choice in theoretical studies of the quantum Hall physics in "infinite geometries.") The alternative **Landau gauge** $\mathbf{A} = (0, Bx, 0)^T$ is tailored to rectangular geometries, with wave functions extended in the y-direction and confined in the x-direction (see the info block below for an explicit construction of the wave functions in this gauge).

▷ For all electron numbers, N, different from that corresponding to precisely n filled Landau levels, $N = n(L/l_0)^2$, the degeneracy of the single-particle problem implies degeneracy of the many-body ground state. This implies high sensitivity to perturbations such as static disorder or interactions.

---

[43] For strong magnetic fields, the Zeemann splitting effectively splits different spin components and we treat them as separate.

INFO Explicit **computations of Landau level wave functions** in different gauges can be found in almost any solid state textbook. For later reference, here we discuss these states in a gauge tailored to the description of Hall bar geometries, as shown in the figure. Our starting point is the gauge $\mathbf{A} = B(0, x, 0)^T$, in which the Hamiltonian operator assumes the form,

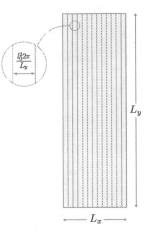

$$\hat{H} = \frac{1}{2m}\hat{p}_x^2 + \frac{1}{2m}(\hat{p}_y - eB\hat{x})^2.$$

With the ansatz $\psi(x, y) = e^{i\hbar^{-1}ky}\phi(x)$, $k = 2\pi\hbar n/L_y$, the equation $H\psi = E\psi$ reduces to

$$\left[\frac{1}{2m}\hat{p}_x^2 + \frac{1}{2m}(k_y - eB\hat{x})^2\right]\phi(x) = E\phi(x),$$

which is the equation for a harmonic oscillator centered around k_y/eB and with characteristic frequency $\omega_c = eB/m$ equal to the cyclotron frequency. The solutions are the harmonic oscillator wave functions $\phi_n(x - k_y/eB)$ and the eigenenergies $\epsilon_n = \hbar\omega_c(n+1/2)$ are the Landau levels. The number of linearly independent states in each Landau level is obtained from the condition that the center coordinates k_y/eB must lie within the transverse extension $[0, L_x]$. With $k_{y,\max} \equiv 2\pi\hbar n_{\max}/L_y = eBL_x$, we obtain the degeneracy as $eBL_xL_y/2\pi\hbar = \Phi/\Phi_0$, equal to the total flux through the system $\Phi = BL_xL_y$ in units of the flux quantum.

IQHE as a topological phenomenon

There are different approaches to understanding the IQHE as a topological phenomenon. One approach starts from an interpretation of experimental facts, another considers the IQHE as a representative of the grander scheme of topological insulators. They both provide different insights, and we discuss them in turn.

Empirical facts and the role of disorder: Let us take a closer look at the experimental data shown on page 458 and schematically reproduced in the figure. The key observations are: (i) for extended parameter regions in the applied field strength, B, the transverse Hall resistivity remains constant at values given by Eq. (8.36) with integer ν. (ii) At these plateaus, the longitudinal Hall resistivity vanishes. The plateaus are separated by transition regions, where the Hall resis-

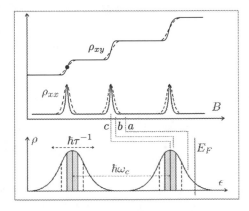

tivity changes between quantized values and the longitudinal resistivity transiently assumes a non-zero value. (iii) The transition regions get steeper as one lowers the temperature and/or increases the system size, with a common crossing point corresponding to a half-integer value of the Hall conductivity,

$$\sigma_{xy} = \frac{h}{e^2}(n + 1/2).$$

These observations are explained by a phenomenological construction that involves Landau level quantization, the effects of disorder, elements of scaling theory, and a few bold assumptions. In the disordered system, the previously singular single-particle density of states broadens and assumes the from of smooth impurity bands centered around the Landau level energies. The peak width $\sim \hbar\tau^{-1} < \hbar\omega_c$ is set by the impurity scattering rate, which we assume to be smaller than the clean Landau level spacing. According to standard localization theory, all eigenstates in a two-dimensional disordered system should be Anderson localized on the scale of the localization length, ξ. However, indiscriminate localization of *all* states would be incompatible with the experimental observation of a finite longitudinal conductivity at the magnetic field strength at which the transverse conductivity changes. In fact, we can say more: the sharpening of the steps (see the solid lines vs. the dashed lines in the left-hand panels of the figure) with increasing system size indicates the presence of a second-order **phase transition**. Tuning the magnetic field into a transition point implies the divergence of all length scales, including the localization length. In an ingenious argument[44] Halperin suggested that all these observations fall into place if we assume that:

Quantum Hall phase transition

> Static disorder Anderson localizes all eigenstates, except those in the centers of the Landau bands.

More precisely, he reasoned that the localization length $\xi(E)$ becomes a function of energy, and diverges upon approaching the center energies, $E \to \hbar\omega_c(\nu + 1/2)$.

The consequences of this proposition are best discussed in a gedanken experiment where, instead of changing the magnetic field at fixed density, the density is varied at fixed field, for a system of size L. Either way, this amounts to shifting the Landau level centers relative to the Fermi energy E_F (see the figure on the previous page). First consider a situation where $\xi(E_F) \ll L$, so that all states at the Fermi energy are localized. In this case, the conductivity $\sigma_{xx} = 0$. Let us make the additional assumption that this localization implies rigid quantization of the Hall conductivity, $\sigma_{xy} = \nu e^2/h$ (the explanation of this statement being the subject of the next section.) Inversion of the conductance tensor (8.34) then leads to $\rho_{xx} = 0$ and $\rho_{xy} = \nu^{-1}h/e^2$ (see dashed line c in the figure).

Upon approaching the Landau level center, we enter a crossover region where $\xi(E) \simeq L$ becomes comparable with the system size. At this stage, σ_{xx} begins to increase which, for finite σ_{xy}, also amounts to an increase in ρ_{xx} (dashed line b). The crossover region shrinks with increasing system size, explaining the sharpening of the step regions. Finally, at the dashed line c, we reach the Landau level center where σ_{xx} and ρ_{xx} assume a maximal value. At this point, we cannot say what value σ_{xy} assumes at these critical configurations; however, symmetry reasoning suggests *half*-integer values $\sigma_{xy} = (\nu + 1/2)e^2/h$.

---

[44] B. I. Halperin, *Quantized Hall conductance, current-carrying edge states, and the existence of extended states in a two-dimensional disordered potential*, Phys. Rev. B **25**, 2185 (1982).

Laughlin gauge argument: The above argument makes the formation of stable plateaus in the transverse conductance coefficient σ_{xy} plausible. However, it does not yet explain its integer quantization. Shortly after the experimental discovery of the IQHE, Laughlin presented a similarly ingenious **gauge argument**, showing why σ_{xy} must be quantized.[45]

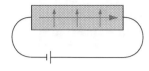

Laughlin gauge argument

Laughlin's starting point was the assumption that, if σ_{xy} is a robust ("topological") quantity, then it must be insensitive to continuous deformations of the sample geometry. Using this freedom, he considered deformation of the quantum Hall bar-geometry shown in section 8.4.7, as indicated in the figure: the bar geometry considered previously is replaced by a annular geometry, where the longitudinal voltage is generated by a time-dependent flux threading the center of the system. Laughlin then suggested considering the effect of the adiabatically slow insertion of a magnetic flux $\Phi = \Phi(t)$ through a solenoid piercing the center of the structure. This generalization is described by a vector potential $\mathbf{A}_\Phi \equiv (0, \Phi/L_y, 0)^T$ adding to the background potential, where L_y is the circumference of the ring. Recapitulating the construction of the Landau states reviewed in the info block of section 8.4.7, we note that A_Φ couples to the now azimuthal y-coordinate via a constant shift $k_y \to k_y - e\Phi/L_y$ (cf. the analogous persistent current problem of section 8.2). Specifically, the insertion of a flux quantum $\Phi = \Phi_0 = h/e$ corresponds to a shift h/L_y in momentum by one quantization step, $k_y = hn/L_y \to hn/L_y + h/L_y$, or $n \to n+1$, and hence does not affect the solution of the problem.

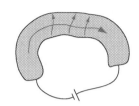

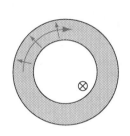

This invariance is a manifestation of the Byers and Yang theorem,[16] according to which the insertion of a flux quantum into a system with annular topology leaves the full set of wave functions and the spectrum unchanged (see the discussion in section 8.2 for a one-dimensional illustration). It does not, however, require that the occupation of individual single-particle states remains the same.

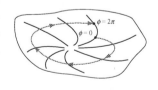

The situation is visualized in the figure, where the lines represent the full set of wave functions, and the dot tracks the fate of individual single-particle states. The insertion of flux is equivalent to a change in boundary conditions in the y-direction. For fractional fluxes, the boundary conditions are twisted, which means that the wave functions are different from the situation with no flux. After the insertion of a full flux quantum, we are back to periodic boundary conditions, and hence the original problem. However, individual single-particle states may map onto different states in the process.

---

[45] R. B. Laughlin, *Quantized Hall conductivity in two dimensions*, Phys. Rev. B **23**, 5632 (1981).

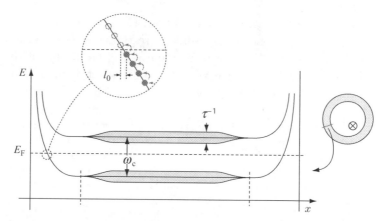

Fig. 8.4   Energy levels of a quantum Hall annulus as functions of the radial coordinate. For a discussion, see the main text.

This is what we observe in the present case, where $k_y \to k_y + h/L_y$ implies a radial shift of individual wave function centers in the x-direction by one strip, in the representation in fig. 8.4.7. The physical consequences of this shift are illustrated in fig. 8.4, which indicates the centering of the single-particle radial coordinates along a cut through the system in the x-direction, along with the single-particle energies. Deep in the system, the energies are pinned to the Landau level centers. However, upon approaching the boundaries, the inclusion of a boundary potential causes an upturn of the energies (for sufficiently soft variation, this can be described by a shift of the Landau level energies), and eventually a crossing of the Fermi energy, assumed to be located somewhere between the center energies. The insertion of a flux quantum amounts to a shift of the entire pattern by one unit. At zero temperature, where all states below the Fermi energy are initially filled, this causes the appearance of a state on the inner perimeter above the the Fermi energy and the removal of one on the outer perimeter – as if a charge had been transported from right to left. Referring to Ref.[44] for a detailed discussion, the argument survives the inclusion of disorder (indicated by the broadening of Landau levels); flux insertion adiabatically pumps a particle from one edge to the other.

This charge transport in the x-direction is caused by the time-dependent insertion of flux, which by the law of induction is equivalent to an electromotive force in the y-direction: with $\oint_y E + \partial_t \Phi = 0$, and assuming that the protocol injects one flux quantum in time t_0 at uniform rate, we have $\oint_y E \equiv V_y = \Phi_0/t_0$. The corresponding current in the x-direction is $I_x = e/t_0$, so that the transverse conductance equals $\sigma_{xy} = I_x/V_y = e/\Phi_0 = e^2/h$. The construction is topological in that it links the flux insertion protocol – equivalent to a winding number of unity "trajectory" in the circular space of boundary conditions – to an observable quantity. The structure of the quantum Hall ground state wave function enters indirectly via the stripe pattern of its single-particle orbitals. However, it would be even better if we could attach a topological index directly to the wave function itself (much as we did in the case of the SSH wave function in section 8.1.1). In the info block below, we discuss how this is done for an example which is in the same universality class as the quantum Hall setup but of simpler design.

EXERCISE In actual quantum Hall measurements (performed on systems with geometries as in the figure in section 8.4.7), charge transport occurs at the edge. To see why, consider the simplified geometry of section 8.4.7 and compute the expectation value of the current $\int dx \langle \hat{j}^y_{(x,y)} \rangle$ in the y-direction integrated over a cross section in the x-direction for individual states in the nth Landau level. With

$$\langle \psi | \hat{j}^y_{\mathbf{r}} | \psi \rangle = \frac{e}{2m} \bar{\psi}(\mathbf{r})(-i\hbar \overleftrightarrow{\partial}_y - eA_y)\psi(\mathbf{r}),$$

show that the current for the shifted harmonic oscillator states $\psi(x,y) = e^{\frac{i}{\hbar}ky}\phi_n(x - ky/eB)$ vanishes by the spatial symmetry of the wave functions ϕ_n for states deep inside the sample. For the same reason, states close to the sample boundary (how close?) do carry current. These are the **edge states** of the quantum Hall topological insulator. They are a two-dimensional analog of the edge states discussed in section 8.1.1 for the SSH chain. A voltage applied in the longitudinal direction to a system with the geometry shown on page 8.4.7 creates an imbalance in the chemical potential of individual edges and hence a net (quantized) current flow into the terminals in the transverse direction. This defines the edge interpretation of quantized Hall transport.

edge states

Note that the edge-state picture presents us with another opportunity to understand the **importance of bulk Anderson localization by disorder** in the IQHE: in the absence of localization, currents from one edge would "leak" to the other edge by diffusive transport through the bulk. This would compromise (and indeed does so in systems of narrow width) the quantization of conductance coefficients.

INFO In the theory of topological quantum matter, the quantum Hall system is classified as a two-dimensional topological insulator in symmetry class A – no symmetries of the Hamiltonian besides hermiticity. According to the periodic table, 8.1, these are systems characterized by integer-valued invariants, and our discussion above shows how these invariants define the quantized values of the Hall conductivity σ_{xy}. However, furthering the topological insulator analogy, it must be possible to express the invariant as a topological property of the quantum ground state, much as we did in the case of the SSH insulator in section 8.1.1. That this is possible was shown in the seminal paper Ref.[46]. However, the construction of this invariant is complicated by specific features of the IQH system, notably the Landau level structure and the necessary presence of disorder.

quantum anomalous Hall effect

Here, we discuss the same principles for the simpler example of the **quantum anomalous Hall (QAH) effect**.[47] The two-dimensional QAH effect is in the same symmetry class A as the conventional IQHE. However, crucially, it does not require the presence of a magnetic field, nor stabilization by disorder. In this way, it demonstrates the essential physics of two-dimensional topological insulators in a simple setting and will serve as a recurrent example of topological insulators in various sections of this text.

[46] D. J. Thouless, M. Kohmoto, M. P. Nightingale, and M. den Nijs, *Quantized Hall conductance in a two-dimensional periodic potential*, Phys. Rev. Lett. **49**, 405 (1982).

[47] F. D. M. Haldane, *Model for a quantum Hall effect without Landau levels: Condensed-matter realization of the "parity anomaly,"* Phys. Rev. Lett. **61**, 2015 (1988).

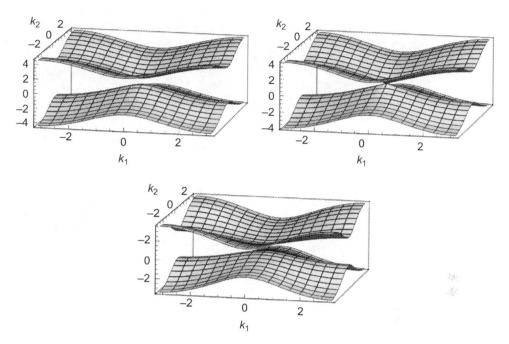

Fig. 8.5   Dispersion of the quantum anomalous Hall effect Hamiltonian (8.38), for $r = 2.5$ (left), $r = 2$ (center), and $r = 1.5$ (right).

The QAH effect is defined on a hexagonal lattice whose Hamiltonian contains (a) a staggered potential on the A and B sublattice (see the figure), (b) a nearest-neighbor hopping of strength t, and (c) next-nearest-neighbor hopping of strength t'. In dimensionless units, the Hamiltonian describing this setup can be reduced to (exercise)

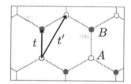

$$\hat{H}_\mathbf{k} = \sin k_1 \sigma_1 + \sin k_2 \sigma_2 + (r - \cos k_1 - \cos k_2)\sigma_3, \qquad (8.38)$$

where the Pauli matrix structure is defined in AB-space, r is the staggering potential, the sine-terms describe nearest-neighbor hopping off-diagonal in sublattice space, and the cosine-terms next-nearest-neighbor hopping diagonal in sublattice space. Finally, $k_{1,2}$ are lattice momenta aligned with the reciprocal lattice vectors (cf. discussion of hexagonal lattice structures on page 56).

The dispersion of the Hamiltonian, $\epsilon_{\pm,\mathbf{k}} = \sqrt{\sin^2 k_1 + \sin^2 k_2 + (r - \cos k1 - \cos k_2)^2}$, is shown in fig. 8.5 for the three values $r = 2.5, 2, 1.5$, respectively. At the critical value $r = 2$ the gap closes at $\mathbf{k} = (0,0)^T$. Such gap closings occur at three values, $r = -2, 0, 2$. Notice that for r close to these values, and momenta close to the gap-closing points, the dispersion is approximately linear. For these values, $\hat{H}$ may be approximated by an effective **Dirac representation**. For example, for $r \simeq 2 + m$, small "mass" m, and $|\mathbf{k}| \ll 1$,

$$\hat{H}_\mathbf{k} \simeq k_1 \sigma_1 + k_2 \sigma_2 + m \sigma_3. \qquad (8.39)$$

Much as with the SSH chain discussed in section 8.1.1, the gap closings separate insulating ground states of different topology. In simple systems such as the QAH insulator, the corresponding topological indices can be identified straightforwardly by inspection of the Hamiltonian itself. (In section 10.5, we introduce more powerful approaches shifting the focus to the quantum ground states themselves.) To see how, write the Hamiltonian (8.38) away from the critical points as $H_{\mathbf{k}} = \epsilon_{+,\mathbf{k}} \mathbf{n_k} \cdot \boldsymbol{\sigma}$, with unit vector $\mathbf{n_k}$. By way of example, consider $r \gg 2$. In this case, $\mathbf{n_k}$ points in an upward direction, with only weak dependence on $\mathbf{k}$. For $0 < r < 2$, we have the different situation that $\mathbf{n}_{(0,0)^T} = -\mathbf{e}_3$ while $\mathbf{n}_{(\pi,\pi)^T} = -\mathbf{e}_3$. Inspection of intermediate values of the momentum variable shows that, in this case, $\mathbf{n}$ covers the full unit sphere for $\mathbf{k}$ running through the Brillouin zone.

The coverage of the sphere by $\mathbf{n}$ has a status similar to the winding around the circle by a planar unit vector in the SSH context; configurations covering the sphere cannot be continuously deformed into those that do not. The corresponding winding number is given by

$$W = \frac{1}{4\pi} \int_0^{2\pi} dk_1 dk_2 \, \mathbf{n} \cdot (\partial_1 \mathbf{n} \times \partial_2 \mathbf{n}), \tag{8.40}$$

i.e., the area swept out on the sphere by $\mathbf{n}$ in units of 4π. This winding number assumes the values $0, 1, -1, 0$ depending on the value of r. It is a topological signature of the quantum ground state, encoded in the variation of $\mathbf{n}$. The technical term for this type of winding number defined in $d = 2$ (and even-dimensional systems in general) is known as the **Chern number**. In section 10.5.3, we discuss the general concept of Chern numbers and will introduce ways to compute them for general even-dimensional quantum ground states. Furthering the analogy to the IQH system, section 9.1.3 addresses the physics of edge states and shows that this system also supports gapless boundary states and quantized transport coefficients.

Chern number

Field theory of the integer quantum Hall effect

REMARK Before reading this section, we suggest that you re-familiarize yourself with the field theory approach to the disordered electron gas introduced in section 5.4. To keep the notation simple, we write the conductance coefficients σ_{ij} as dimensionless quantities measured in units of the conductance quantum e^2/h.

While Laughlin's gauge argument and other phenomenological approaches lifted much of the mystery posed by the experimental discovery of the IQHE, they were not first-principle theories. A major step towards a microscopic description of the effect was taken by Pruisken when he extended the nonlinear σ-model of disordered fermion systems to include a strong magnetic field. In the following, we will review Pruisken's field theory and apply it to a description of the critical physics of the quantum Hall insulator.

Pruisken's field theory: In section 5.4 we saw that the physics of the disordered electron gas at length scales exceeding the elastic mean free path is described by a nonlinear σ-model with action

$$S[Q] = \frac{\sigma_{11}}{8} \int d^2x \, \mathrm{tr}(\partial_i Q \partial_i Q), \tag{8.41}$$

where the fields $Q = \{Q^{as,a's'}\}$ are $2R$-dimensional matrices, $a = 1, \ldots, R$ is a **replica** index, and $s = \pm$ is a causality index. This theory was derived from a microscopic parent theory describing the disorder average of retarded ($s = +$) and advanced ($s = -$) fermion Green functions. Prior to averaging, that theory possessed invariance under spatially uniform rotations $T \in \mathrm{U}(2R)$ (reflecting the identity of the microscopic actions of retarded and advanced Green functions of different replicas). However, disorder averaging led to symmetry breaking, in that retarded and advanced Green functions then carried a "self energy," $\pm i/2\tau$. The unbroken symmetry group $\mathrm{U}(R) \times \mathrm{U}(R)$ contained only rotations between replica flavors of identical causality, $s = +$ or $-$.

The matrices $Q \in \mathrm{U}(2R)/\mathrm{U}(R) \times \mathrm{U}(R)$ are the Goldstone modes of this "causal symmetry breaking." A concrete representation is given by $Q = T\sigma_3^{\mathrm{ar}}T^{-1}$, where $T \in \mathrm{U}(2R)$ and σ_3^{ar} is a Pauli matrix in advanced/retarded (ar) space. The above Goldstone mode action is invariant under global transformations $T \to T_0 T$, where $T_0 \in \mathrm{U}(2R)$ is constant, $Q \to T_0 Q T_0^{-1}$, and under local transformations $T \to TH(\mathbf{x})$, where $H(\mathbf{x}) \in \mathrm{U}(R) \times \mathrm{U}(R)$, and Q does not change (because $[H, \sigma_3] = 0$).

INFO Note how the above symmetry breaking parallels that in a magnet: the local mean field magnetization of a Heisenberg magnet can be represented as $n^i \sigma_i \equiv T\sigma_3 T^{-1} \equiv Q$, where $T \in \mathrm{U}(2)$ describes deviations away from the 3-axis magnetization, $\mathbf{n} = \mathbf{e}_3$. The magnetization is *locally* invariant under $T \to TH$, where $H \in \mathrm{U}(1) \times \mathrm{U}(1)$ represents both, transformations by a trivial phase, and rotations around the 3-axis. The symmetry under *global* transformations $T \to T_0 T$ represents changes in the mean field magnetization axis, $n^i \to n'^i$. In this case, the Goldstone mode manifold $\mathrm{U}(2)/\mathrm{U}(1) \times \mathrm{U}(1)$ is a sphere, with two different representations in terms of $\mathbf{n}$ or in terms of Q. Similarly, the Heisenberg action affords the alternative representations $\int d^d x \, \mathrm{tr}(\partial Q \partial Q) = 2 \int d^d x \, \partial \mathbf{n} \cdot \partial \mathbf{n}$. The nonlinear σ-model of the disordered electron gas is the generalization of this theory to arbitrary R, and its Goldstone mode manifold is a higher-dimensional variant of the sphere.

The coupling constant of the action is proportional to the longitudinal conductivity, σ_{11} of the electron gas at length scales comparable with the mean free path. (At large length scales, fluctuations of the Q-matrices cause a downward renormalization of the coupling – Anderson localization.) In a series of seminal papers,[48] Pruisken investigated what happens if a strong magnetic field enters the stage. Referring for the first-principle derivation of the generalized theory to the original references, here we argue heuristically and note that we expect a contribution to the action containing a term with mixed derivatives $\partial_1 Q \partial_2 Q$, whose coupling constant is proportional to the second entry in the conductance tensor, σ_{12}. We must also require rotational invariance around the three-axis on averaging over disorder realizations. This symmetry is satisfied by a system subject to a magnetic field in the 3-direction, and must be respected by the effective theory. A term $\mathrm{tr}(\partial_1 Q \partial_2 Q)$ would not have it (why?), and hence is ruled out. The minimal rotationally invariant extension reads $\epsilon_{ij} \, \mathrm{tr}(\partial_i Q \partial_j Q)$; however this term vanishes (why?). Next in order of complexity is the integrand $\epsilon_{ij} \, \mathrm{tr}(Q \partial_i Q \partial_j Q)$, and this is indeed the term derived by Pruisken: the **field theory of the quantum Hall effect** has the action

field theory of the IQH

$$S[Q] = \frac{1}{8} \int d^2 x \, \left[\sigma_{11} \, \mathrm{tr}(\partial_i Q \partial_i Q) - \sigma_{12} \epsilon_{ij} \mathrm{tr}(Q \partial_i Q \partial_j Q) \right]. \tag{8.42}$$

[48] For a review, see A. M. M. Pruisken, *Field theory, scaling and the localization problem*, in *The Quantum Hall Effect*, eds. R. E. Prange and S. M. Girvin (Springer-Verlag, 1987).

The second term in this action is a **topological θ-term**. To understand this point, consider the case $R = 1$ in which (see the info block above) $Q = n^i \sigma_i$ becomes equivalent to a unit vector on the sphere. Substitution of this representation into the action gives

$$S_{\text{top}}[\tilde{Q}] \equiv -\frac{\sigma_{12}}{8} \int d^2x\, \epsilon_{ij} \text{tr}(Q\partial_i Q \partial_j Q) = i\frac{\sigma_{12}}{2} \int d^2x\, \mathbf{n} \cdot (\partial_1 \mathbf{n} \times \partial_2 \mathbf{n}).$$

On the left, we meet again the topological action (8.15), with $\theta = 2\pi\sigma_{12}$ as the θ-angle. For an infinitely large system with fields relaxing to a constant value $Q(\mathbf{x}) = $ const., the base space becomes topologically equivalent to a sphere, and the topological term measures the number of windings in the map $Q : S^2 \to S^2, \mathbf{x} \mapsto Q(\mathbf{x})$. Before addressing the interesting question of what happens in finite sized samples, we note that the above interpretation extends to the case of general R. Heuristically, this follows from the observation that the action S_{top} is topological – it is purely imaginary and does not change under local changes of variables – and therefore it must measure topological contents (which the $R = 1$ analysis shows is nontrivial). From mathematics we infer that

$$\pi_2 \left(\frac{\text{U}(2R)}{\text{U}(R) \times \text{U}(R)} \right) = \mathbb{Z},$$

i.e., maps from compactified real space to the field space of general Q matrices are characterized by integer winding numbers. A more detailed analysis does indeed show that these winding numbers are computed by the topological action above. Equation (8.42) demonstrates how the conspiracy of a strong magnetic field and disorder stabilizes a topological field theory with a θ-term. In the following, we will discuss the physics resulting from this description.

Quantum Hall phases: Consider a gedanken experiment where we are given a two-dimensional electron gas of linear extension $L \gtrsim l$ with "bare" values σ_{11} and σ_{12} of the conductance tensor. What will happen as the size of the system is increased?[49] We know that local fluctuations of the Q-matrices serve to decrease the value of the coupling σ_{11}. At the same time, they leave the topological action invariant. Naïvely, one might thus suspect renormalization to $\sigma_{11} = 0$ – Anderson localization – and that σ_{12} is identical to its bare value in the limit $L \to \infty$. However, an elegant argument presented in the info block below shows that a line of fixed points $(0, \sigma_{12})$ is inconsistent with fundamental principles; only configurations $(0, n)$ with integer quantized Hall conductance qualify as fixed points of the localized phase.

INFO Let us investigate **why field-theoretical localization, $\sigma_{11} \to 0$, requires integer quantization of σ_{12}.** To this end, consider the action at σ_{11}, where it has become purely topological, $S[Q] = S_{\text{top}}[Q]$. For definiteness, we consider a system with

---

[49] In a real experiment, lowering of the temperature T has a similar effect. It leads to an increase in the length scale $L(T)$ over which wave functions remain phase coherent. Where localization effects are concerned, this scale plays a role similar to the physical system size.

disk geometry, D, and assume that the boundary ∂D is parameterized by a coordinate $s \in [0, L]$. Referring to section 8.4.1 for a discussion of the underlying geometric principles, we note that the topological Lagrangian $\mathcal{L}_{\text{top}} = -\frac{\theta}{16\pi}\,\epsilon_{ij}\text{tr}(Q\partial_i Q\partial_j Q)$, $\theta = 2\pi\sigma_{12}$ affords a local representation as a full derivative. This is most easily seen in differential form notation, where $S_{\text{top}}[Q] = -\frac{\theta}{16\pi}\int_D \text{tr}(QdQ \wedge dQ)$. With the representation $Q = T\sigma_3 T^{-1}$, it is straightforward to verify[50] that

$$S_{\text{top}} = \frac{\theta}{4\pi}\int_D \text{tr}(dT \wedge \sigma_3 dT^{-1}) = \frac{\theta}{4\pi}\int_D d(\text{tr}(T\sigma_3 dT^{-1})) = \frac{\theta}{4\pi}\int_{\partial D}\text{tr}(T\sigma_3 dT^{-1}),$$

where in the last step we applied Stokes' theorem (A.17) to pass to a boundary integral. The representation is "local" in the sense that it works for a topologically trivial base space, like our disk, D. However, it is not extensible to a map $Q : S^2 \to T$ defined on compactified infinite space with its spherical topology.[51]

Represented in standard derivative form as an integral over the boundary coordinate, the topological action assumes the form

$$S_{\text{top}} = \frac{\theta}{4\pi}\int_0^L ds\,\text{tr}(T\sigma_3\partial_s T^{-1}).$$

This is a nice result. It shows that, in the deeply localized phase ($\sigma_{11} = 0$), the system is described by a linear-derivative boundary action. As one might expect (see the original reference for the details), this action describes chiral boundary current flow, i.e., the physics of the surface states discussed previously on a phenomenological basis. Also notice that the localized bulk action has completely dropped out of the picture at this stage.

However, there is something disconcerting about the result above: the required *local* symmetry under transformations $T(\mathbf{x}) \to T(\mathbf{x})H(\mathbf{x})$, $[H(\mathbf{x}), \sigma_3] = 0$, is no longer manifest in the boundary representation. While the original degrees of freedom $Q = T\sigma_3 T^{-1} \to TH\sigma_3 H^{-1}T^{-1} = T\sigma_3 T^{-1} = Q$ do not transform, substitution of the transformation into the boundary action leads to

$$S_{\text{top}}[TH] = \frac{\theta}{4\pi}\int_0^L ds\,\text{tr}(TH\sigma_3\partial_s(H^{-1}T^{-1})) = S_{\text{top}}[T] + \delta S_{\text{top}}[H],$$

where $\delta S_{\text{top}}[H] \equiv \frac{\theta}{4\pi}\int_0^L ds\,\text{tr}(H\sigma_3\partial_s H^{-1})$ and we have used the product rule and the cyclic invariance of the trace. Focusing on a single replica index and, within it, on the transformation generated by the U(1) subgroup of SU(2) rotating around σ_3, $H(s) = \exp(i\phi(s)\sigma_3)$, the H-dependent term becomes

$$\delta S_{\text{top}}[H] = \frac{\theta}{4\pi}\int_0^L ds\,\text{tr}(H\partial_s H^{-1}) = \frac{i\theta}{2\pi}\int_0^L ds\,\partial_s\phi(s) = \frac{i\theta}{2\pi}(\phi(L) - \phi(0)) = i\theta W,$$

where we note that $\phi(L) - \phi(0) \equiv 2\pi W$ may include windings. Gauge invariance requires that the path integral must not depend on these transformations, and this requires $\exp(i\theta W) = 1$, or $\theta = 2\pi n$. With $\sigma_{12} = \theta/2\pi$, this is equivalent to the condition of integer

---

[50] Readers not yet warmed up to differential-form notation may carry out the above operations in standard derivative notation with its explicit book-keeping of anti-symmetrization operations.

[51] It is a good exercise to explore this point for a $R = 1$ toy model, where Q is a two-dimensional matrix and $T \in$ SU(2) are rotation matrices. For example, with $T = \exp(-i\phi\sigma_3/2)\exp(-i\theta\sigma_2/2)$ parameterized in an Euler angle representation, one obtains (exercise) $\text{tr}(dT \wedge \sigma_3 dT^{-1}) = \sin\theta\,d\theta \wedge d\phi$. Locally, this can be written as a derivative, $d(\sin\theta\,d\phi)$. However, at the poles $d\phi$ becomes singular, demonstrating that complete extension is not possible.

quantized Hall conductance. We thus conclude that the bulk localized theory actually requires integer quantization of the Hall coefficients for its intrinsic consistency.

Quantum criticality: While the identification above of a discrete set of renormalization group fixed points, $(\sigma_{11}, \sigma_{12}) \to (0, n)$, is consistent with the phenomenological understanding of the Hall effect – bulk localization stabilizes edge states carrying quantized currents – it presents us with a conceptual problem: if perturbative renormalization lowers σ_{11} but does not change the bare value of σ_{12}, how can the renormalization group flow ever end up at the permitted target points $(0, \sigma_{12})$? This question was addressed in Pruisken's series of papers on the subject (see Ref. [48] for a review). The analysis is technically involved and here we limit ourselves to a discussion of the main ideas.

We first note that the presence of a θ-term implies that the field integration $\int DQ$ must be conceptualized as an integration over distinct topological sectors, i.e., distinct values of the topological action $S_{\text{top}}[Q]$. The subsequent renormalization program is compatible with this organization of the integral: renormalization involves successive integration over short-ranged fluctuations and does not alter the global winding of field configurations. The renormalization steps operate *within* individual topological sectors. Second, our earlier discussion of the θ-action of the spin chain showed that nontrivial topological sectors are weighted with a minimal action, the reason being that topological twists require a minimal spatial variation, for which one has to pay via the gradient term. The same argument applied to the present action shows that $S[Q] \geq 2\pi W \sigma_{11}$.

EXERCISE From the positivity of

$$\frac{1}{2} \int d^2 x \operatorname{tr} \left[(\partial_\mu Q + i\epsilon_{\mu\nu} Q \partial_\nu Q)(\partial_\mu Q + i\epsilon_{\mu\lambda} Q \partial_\lambda Q) \right] \geq 0,$$

derive the lower bound for the action mentioned in the text.

Third, we know that under a change of orientation of the coordinate system (physically, a change of direction of the applied magnetic field) both the Hall conductivity σ_{12} and the winding number W of individual field configurations change sign, while σ_{11} remains invariant. On the basis of these structures, Pruisken carried out an RG procedure, perturbatively stabilized by a large value of σ_{11} – an assumption which we know breaks down at large scales, where σ_{11} has values of $\mathcal{O}(1)$. For $\sigma_{11} \gg 1$, the exponential suppression $\sim \exp(-2\pi W \sigma_{11})$ justifies a limitation of the RG procedure to the sectors $W = 0, 1$. To this order, Pruisken obtained the flow equations

$$\beta_{11} \equiv \frac{\partial \sigma_{11}}{\partial \ln L} = -\frac{1}{2\pi^2 \sigma_{11}} - c\sigma_{11} e^{-2\pi\sigma_{11}} \cos(2\pi\sigma_{12}),$$

$$\beta_{12} \equiv \frac{\partial \sigma_{12}}{\partial \ln L} = c\sigma_{11} e^{-2\pi\sigma_{11}} \sin(2\pi\sigma_{12}),$$

(8.43)

where $c > 0$ is a numerical constant. Here, the non-exponential term in the first equation describes the logarithmically slow downward renormalization of the two-dimensional linear conductance due to disorder. The weighting of the RG equations for σ_{11} and σ_{12} with a cos or sin function reflects the above-mentioned change in the transport coefficients under a change of orientation. It implies that, for σ_{12} different from half-integer values $\mathbb{Z} + 1/2$, the Hall conductance grows or increases depending on whether the sign of the difference of σ_{12} from the nearest half-integer is positive or negative. At the same time, the longitudinal conductance keeps decreasing. The flow equations do indeed have fixed points at values $(\sigma_{11}, \sigma_{12}) = (0, n)$ that, however, are outside the highly conducting regime $\sigma_{11} \gg 1$, where the equations can be trusted. Finally, $\sigma_{12} \in \mathbb{Z} + 1/2$ defines configurations where σ_{12} does not change, while σ_{11} decreases. These are the **critical surfaces** containing the quantum Hall critical points. Since $\cos(2\pi(n + 1/2)) = -1$, the equation for σ_{11} indeed contains a fixed point at $(\sigma_{11}, \sigma_{12}) = (\sigma_0, n + 1/2)$, where $\sigma_0 = \mathcal{O}(1)$ (see the dots in the figure). This is in line with the experimental observation of a non-zero value, of $\mathcal{O}(1)$, of the longitudinal conductance at the quantum Hall transition.

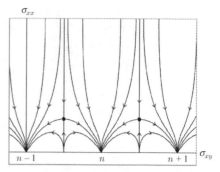

Summarizing, the above RG equations are consistent with the flow depicted in the figure. Bare configurations characterized by a pair of transport coefficients at short length scales generically flow towards the quantum Hall fixed points, $(\sigma_{11}, \sigma_{12}) \rightarrow (0, n)$, with vanishing longitudinal and quantized Hall conductance. Only along the critical lines $\sigma_{12} \in \mathbb{Z} + 1/2$ does the system end up in a quantum Hall critical state. For slight deviations away from criticality, controlled, e.g., by the strength of the external field, or deviations $\Delta E = |E_F - E^*|$ of the Fermi energy away from the centers of the Landau bands, the critical physics is controlled by a length scale $\xi(\Delta E)$, diverging at the critical point. Identifying this scale with the localization length, bulk states of energy E are localized or not depending on whether $\xi(E)$ is smaller or larger than the system size. This is consistent with the expected formation of long-range states close to the centers of the Landau levels, where ΔE is small (see the figure on page 460).

The discussion above shows that the nonlinear σ-model enriched by a topological term explains numerous features related to the interplay of topology and disorder in the physics of the quantum Hall effect. Physically backed theories are often applicable beyond the regimes where their construction is parametrically controlled. One might thus cross one's fingers and hope that even the critical physics of the quantum Hall transition is described by Pruisken's theory. However, this is where we run out of luck. As discussed above, the quantum Hall critical point is observed at values of the coupling constants deep inside the strong coupling regime of the σ-model. For these values it is out of control. Worse, we must expect that it loses its identity. Evidence that something drastic happens comes from the conceptually similar yet simpler field theory of the antiferromagnetic spin chain. In section 8.4.6, we saw that this system, too, is described by a topological nonlinear σ-model (8.32). It is identical to that of the quantum Hall insulator in the case $R = 1$ of one replica, the analog of the quantum Hall critical surfaces being half-integer spin configurations $S \in \mathbb{Z} + 1/2$. In this case, we know the critical theory, and it is different from the nonlinear σ-model!

In the next section we will introduce this theory, and along with it a new class of topological terms, the Wess–Zumino terms. However, where the IQHE is concerned, we must concede that four decades after its discovery, an effective theory describing the quantum Hall transition remains unknown.

8.5 Wess–Zumino Terms

REMARK Familiarity with the language of differential forms as introduced in appendix section A.1 will make this section easier to read.

In this section, we introduce a new class of topological terms, termed **Wess–Zumino** (WZ), **Wess–Zumino–Witten** (WZW), or **Wess–Zumino–Novikov–Witten** (WZNW) terms, honoring those who introduced them to physics. The rule of thumb is that WZ terms appear in theories in dimensions d with field target spaces T if the same theory in one dimension higher, $d + 1$, admits a θ-term. Before discussing this statement in general, let us illustrate its meaning on the teaser example of the spin path integral introduced in the previous section.

8.5.1 Spin path integral revisited

Let us take another look at the spin path integral (8.20) and specifically the "canonical" part of the action (8.26). Notice two things: first, we reasoned that this contribution to the action does not admit a representation in terms of the natural degrees of freedom, $\mathbf{n}(\tau)$, describing a time-dependent angular momentum (unlike, say, a Zeeman term in the action, which can be written as $B\cos\theta = B\mathbf{e}_z \cdot \mathbf{n}$). Second, the spin path integral exemplifies the $d \to d + 1$ principle mentioned above. In the previous section, we saw that the action (8.26) of the $(0 + 1)$-dimensional path integral seeds the θ-term (8.15) of the $(1 + 1)$-dimensional spin field integral in one dimension higher. Let us discuss the principles behind these observations and how they are related.

Spin quantization from monopole potential

We first discuss the situation as it is usually presented in physics textbooks. As explained above, the free spin action describes a canonical pair of variables $(\phi, \cos\theta)$ via a term $\int p\dot{q}\, d\tau = \int \cos\theta\, \dot{\phi} d\tau$ familiar from classical mechanics. However, the standard physics treatment takes a different view[52] and reasons that a first-order time derivative in a classical action is reminiscent of a term $\int d\tau\, \dot{\mathbf{q}} \cdot \mathbf{A}$ describing the coupling of a particle with velocity $\mathbf{v}$ to a vector potential. To identify the corresponding vector potential, we start from the canonical term of the spin path integral $\Gamma[\theta, \phi] = -iS \int d\tau\, \dot{\phi} \cos\theta$. As in Eq. (8.19), we use our freedom to add a full derivative, and consider the modification

$$\Gamma_{\mathrm{n}}[\theta, \phi] = -iS \int d\tau\, \dot{\phi}\, (\cos\theta - 1) \tag{8.44}$$

in the representation S_{n} of Eq. (8.19), where the coordinate singularity at the north pole of the sphere has been removed by the addition of a full derivative.

Wess–Zumino term — Anticipating that this functional will be a realization of a **Wess–Zumino term** we use the notation Γ, frequently found in connection with these functionals. Our particle moves on the sphere and, from the relation $\mathbf{v} = \dot{\mathbf{n}} = \dot{\theta}\,\mathbf{e}_\theta + \sin\theta\,\dot{\phi}\,\mathbf{e}_\phi$, we find that, with the choice

$$\mathbf{A}_{\mathrm{n}} = S\frac{1 - \cos\theta}{\sin\theta}\mathbf{e}_\phi, \tag{8.45}$$

we obtain the corresponding expressions $\Gamma_{\mathrm{n}} = -i \int d\tau\, \dot{\mathbf{q}} \cdot \mathbf{A}_{\mathrm{n}}$. What kind of magnetic field does $\mathbf{A}_{\mathrm{n}}$ describe? The answer is found by taking the curl: application of standard formulae of vector calculus in spherical coordinates shows that $\mathbf{B} \equiv \nabla \times \mathbf{A}_{\mathrm{n}} = S\mathbf{e}_r$. This is the uniform radial magnetic field generated by a **magnetic monopole** of strength $\int_{S^2} d\mathbf{S} \cdot \mathbf{B} = 4\pi S$ at the origin of the sphere.

INFO How can this statement be consistent with the absence of magnetic charges? First, keep in mind that we are talking about a "fictitious magnetic field." Second, the vector potential $\mathbf{A}_{\mathrm{n}}$ becomes singular at the south pole. Inspection of its behavior around the singularity (exercise) shows that is identical to the vector potential generated by a likewise fictitious thin solenoid of magnetic flux $-4\pi S$ entering the sphere through the south pole: the **Dirac string**. In this interpretation, the outward flux through the surface of the sphere enters through its south pole, and there is no net magnetic charge inside the sphere.

The spatial uniformity of the magnetic monopole field suggests computing the action via Stokes' theorem. To this end, we interpret Γ_{n} as the line integral of $\mathbf{A}_{\mathrm{n}}$ along the closed curve defined by $\mathbf{n}(\tau)$, which via Stokes' theorem equals the surface integral of $\mathbf{B}$ over an area on the sphere bounded by $\mathbf{n}$ (see the light- and dark-shaded areas in the figure). The applicability of Stokes' theorem requires that we pick the light-shaded northern area, avoiding the singularity of $\mathbf{A}_{\mathrm{n}}$ at the south pole. Since $\mathbf{B} = S\mathbf{e}_r$

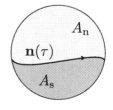

is constant, this integral yields S times the geometric area A_{n} bounded by $\mathbf{n}(\tau)$ on the sphere.

This finding immediately raises questions. First, there is the ambiguity in the choice of bounded area: northern or southern? Initially, we had to pick the northern

---

[52] Perhaps this point was not really understood at the time when the interpretation below was formulated.

option. However, this was dictated by the ad-hoc choice of $\mathbf{A}_n$, with its absence of singularities in the northern hemisphere. We might have built our analysis of the action on the alternative, $\mathbf{A}_s$, in which case the dark-shaded southern area, $-A_s$, would have been obtained, where the sign reflects the different sense of orientation of the curve $\mathbf{n}(\tau)$ relative to that bounding area. The difference between the two answers, $SA_n - S(-A_s) = 4\pi S$ equals the S times the area of the sphere. This gauge ambiguity must be inconsequential, which requires that $\exp(iS4\pi) = 1$, i.e., half-integer quantization $S \in \frac{1}{2}\mathbb{Z}$. Notice that, in the construction of the path integral via spin coherent states, we never used the quantization of S. However, the path integral requires it a posteriori to be geometrically consistent.

Finally, let us discuss the somewhat subtle issue of **coordinates** in connection with the topological action. Above, we reasoned that $S_{\text{top}}[\theta, \phi]$ cannot be expressed in a coordinate-invariant way. On the other hand, we have just argued that S_{top} equals iS times the geometric area bounded by $\mathbf{n}$. The latter clearly is a "geometric object" defined without reference to a particular coordinate system. How can these two statements be reconciled with each other? The answer lies in the observation that the geometric interpretation of the action makes reference to a higher dimension, namely the one-dimensional curve $\mathbf{n}$ interpreted as the boundary of a two-dimensional area. This extension is made more concrete by introducing a homotopic interpolation $\mathbf{n}(s, \tau)$ between the bounding curve and the north pole. Here, $s \in [0, 1]$, $n(0, \tau) = \mathbf{n}(\tau)$, and $\mathbf{n}(1, \tau) = \mathbf{e}_3$, so that the parametric interpolation in s sweeps out the bounded area. The area functional can now be represented as

$$\Gamma[\mathbf{n}] = iS \int_0^1 ds \int_0^\beta d\tau\, \mathbf{n} \cdot (\partial_s \mathbf{n} \times \partial_\tau \mathbf{n}), \qquad (8.46)$$

i.e., the standard representation of an area integral on the sphere with parameterizing coordinates (s, τ).[53] This is the **coordinate invariant representation of the WZ term** in terms of the field $\mathbf{n}$ that, however, now is upgraded to a two-dimensional field. We observe that we have two options for expressing the topological action, either in coordinates in the native dimension of the theory or in an invariant way that is one dimension higher. We will see that this reflects a feature common to all Wess–Zumino terms.

WZ term

Spin quantization from geometry

However, to understand the underlying principle, we had better abandon the monopole magnetic field interpretation. Recall that the introduction of the field became necessary as a result of a particular interpretation of the $p\dot{q}$-term. We now take another and more geometrically inspired look at this term. Recall from the info block of section 8.4.4 that a phase space comes equipped with a non-degenerate symplectic two-form $\omega \equiv \sum dp_i \wedge dq^i$, and that the two-sphere is made a phase space by the choice $\omega \equiv -d(\cos\theta) \wedge d\phi = \sin\theta\, d\theta \wedge d\phi$. Three things are important

---

[53] Notice how this expression resembles a θ-term. However, there is the crucial difference that now the variable $\mathbf{n}$ does not cover integer multiples of the full sphere, but only a fraction of it.

here: first, in this two-dimensional case, the symplectic form is **top-dimensional**, a two-form on a two-dimensional manifold. Second, the symplectic form of the sphere equals its canonical area form, $\int_{S^2} \omega = 4\pi$ (see Eq. (A.15)). Third, being top-dimensional, ω is locally, but not globally, exact, i.e., it is the derivative $\omega = d\kappa$ of a one-form. If there was a global representation of this kind, we would have $\int_{S^2} \omega = \int_{S^2} d\kappa = \int_{\partial S^2} \kappa = 0$ by Stokes' theorem, due to the absence of a boundary of the sphere. However, locally, the derivative representation can always be obtained by solution of a differential equation. For example, $\omega = \sin\theta\, d\theta \wedge d\phi = -d(\cos\theta\, d\phi)$ in spherical coordinates, which is a "local" representation in that sets of measure zero such as the poles are not reached.

The discussion thus far is formulated on the sphere, S^2. To see how it is relevant to our path integral framework, all we need to do is "pull it back" from this target space to the base space of the theory. For example, consider the one-dimensional fields integrated over in the path integral, $\varphi : S^1 \to S^2, \tau \mapsto \varphi(\tau) = \varphi(\theta(\tau), \phi(\tau))$, where $\varphi \equiv \mathbf{n}$ is an element of the sphere. The image of φ is a closed curve, γ, on the sphere. Assuming that this curve lies inside the domain of a local representation $\omega = d\kappa$, i.e., we are not accidentally running over a singularity, we may consider the integral,

$$\int_\gamma \kappa = \int_{S^1} \varphi^* \kappa = - \int_0^\beta d\tau \cos\theta\, \dot\phi\, d\tau, \tag{8.47}$$

where the first integral is on the sphere, the second computes the integral by pulling it back to the base interval $[0, \beta]$ (see Eq. (A.16)), and the third contains an explicit representation of the integral. This final representation is the canonical term in our path integral action. In fact, nothing here is special to the sphere: when we write $\int d\tau\, p_i \dot q^i$ in a Hamiltonian action, we are integrating the pullback of the local potential $p_i dq^i$ of the symplectic form $\omega = d(p_i dq^i) = dp_i \wedge dq^i$ under the phase space curve $\tau \mapsto x(\tau) = (q^1, \dots, p_f)$.

In standard classical mechanics, there is no particular reason to emphasize this view. However, with the somewhat more interesting phase space of the sphere it becomes crucial. With it, the mysteries we found earlier in our discussion of the monopole field are lifted.

First, it is evident that the canonical action does not have a global invariant representation (in terms of $\varphi = \mathbf{n}$). This is the pulled back version of the Stokes' theorem argument given above. If there were such a representation, we could demonstrate that the integral of the area form of the sphere

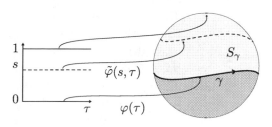

equals zero. However, we do have the option to extend our field to a two-dimensional one via a parametric construction $\tilde\varphi : I \times S^1 \to S^2, (s, \tau) \mapsto \tilde\varphi(s, \tau)$, chosen such that its image S_γ on the sphere has our curve γ as boundary: $\tilde\varphi(0, \tau) = \varphi(\tau)$. With this extension, we may consider the the pullback of the two-form ω to the two-dimensional base space $\int_{I \times S^1} \tilde{\varphi}^* \omega = \int_{S_\gamma} \omega$ in order to obtain expressions such as (8.46), which make no reference to the coordinate potential, κ, and hence is singularity free. Provided we have a local representation $\omega = d\kappa$ inside the full image, S_γ, Stokes' theorem may be applied to this integral,

$$\int_{S_\gamma} \omega = \int_{S_\gamma} d\kappa = \int_\gamma \kappa, \qquad (8.48)$$

with pullback representation Eq. (8.47).[54] Again, we have the choice between different interpolations. The invariance of the integral above under changes of coordinates implies that different interpolations to the north pole will not change the result. (If you feel uncertain about this statement, consider the definition of the integral over forms to verify it). However, an interpolation to the opposite pole, enforced by a singularity of κ on the northern hemisphere, changes it. The difference between the northern variant $S_\gamma \equiv S_{n,\gamma}$ and the southern variant $S_{s,\gamma}$ equals $\int_{S_{n,\gamma}} \omega - (-\int_{S_{s,\gamma}} \omega) = \int_{S_{n,\gamma} \cup S_{s,\gamma}} = \int_{S^2} \omega = 4\pi$, which is similar to what we had in the previous discussion. This is an alternative way of demonstrating the half-integer quantization of S required by the path integral formalism.

Equations. (8.44) and (8.46) are two alternative representations of the **Wess–Zumino action on the sphere**. While the discussion was formulated for this specific target space, the geometric construction of the action did not make reference to particular properties of the sphere. In the following, we rephrase the geometric principle of WZ terms in a more general language. On this basis, we will then discuss other realizations of WZ terms, notably the beautiful two-dimensional WZ theory with target SU(n).

Wess–Zumino action on the sphere

8.5.2 Geometry of Wess–Zumino terms

To prepare for our subsequent discussion of concrete realizations, we here summarize the principles behind the Wess–Zumino term, in direct extension of the discussion of section 8.4.1 on θ-terms. Fortunately, there is no essential work to be done: all we need to do is transcribe the discussion of the previous section on spins to generalized settings.

To this end, assume that we have a theory in $d + 1$ dimensions defined on a base manifold, which we assume to be realized as the product $N \equiv M \times I$, where I is an interval and M a d-dimensional manifold. Keep an eye on M; this is where our WZ theory will be defined.

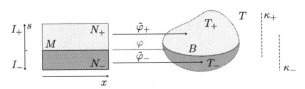

---

[54] Note that it is always possible to shift the position of singularities of κ by adding full derivatives: with $\kappa \to \kappa + df$ we obtain the same area form $\omega = d(\kappa + df) = d\kappa = \omega$, due to $dd = 0$. In our previous discussion of the action we used this freedom with $f = \pm\phi$ to work with potentials $\kappa_{n,s} = (\cos\theta \pm 1)d\phi$. However, for the reasons mentioned, it is not possible to remove all singularities.

Further assume that we have a target manifold T without boundary (think of the previous example, $T = S^2$) and that the theory with fields $\tilde{\varphi} : N \to M$ admits the construction of a θ-term. Referring back to section 8.4.1, this means the presence of a $(d+1)$ form ω on T whose pullback integrals $\int_{M \times I} \tilde{\varphi}^* \omega$ over the base produce 2π-multiples of quantized winding numbers (see Eq. (8.12), with $M \to N$ and $\omega \to 2\pi\omega$, where the 2π-scaling is introduced for later convenience). Now pick a point, say, $0 \in I$, and focus on the restriction $\varphi(x) \equiv \tilde{\varphi}(x, 0)$, defining a field in one dimension lower, $\varphi : M \to T$. The intersection at 0 splits the base manifold into two parts, $N = N_+ \cup N_-$, and φ is the common boundary configuration of both the upper and the lower extensions $\tilde{\varphi}_{+/-}$.

In the situation we want to consider, φ is our prime object of interest (for example, the one-dimensional fields of periodic time, $M = S^1$, describing a closed curve on the spin manifold $T = S^2$), and $\tilde{\varphi}$ is an extension winding once around T, so that the above integral yields 2π. On this basis, let us assume that in N_+, $\omega = d\kappa_+$ is the derivative of a d-form and, likewise in N_-, $\omega = d\kappa_-$. Since a global representation $\omega = d\kappa$ on all N cannot exist (why?), the definition of $\kappa_\pm$ necessarily makes reference to a local coordinate system. Now consider the integral over M of κ_+ pulled back by φ (corresponding to the imaginary-time integral along a closed curve on the spin manifold in the previous example):

$$\Gamma_\pm[\varphi] \equiv ik \int_M \varphi^* \kappa_\pm. \tag{8.49}$$

WZ-term (coordinate representation)

This integral defines a **coordinate representation of the WZ term** (Eq. (8.44) in the spin example). Here an arbitrary *integer* prefactor, $k \in \mathbb{Z}$, known as the **level** of the WZ theory, has been introduced.

To obtain an alternative and coordinate invariant representation, we apply Stokes' theorem to the integral over the boundary $M = \partial N_\pm$, to pass to an integral of $d\kappa_\pm = \omega$ over either N_+ or N_-:

$$\int_M \varphi^* \kappa_\pm = \int_M \tilde{\varphi}^* \kappa_\pm \overset{(A.17)}{=} \pm \int_{N_\pm} d\tilde{\varphi}^* \kappa_\pm \overset{(A.12)}{=} \pm \int_{N_\pm} \tilde{\varphi}^* d\kappa_\pm = \pm \int_{N_\pm} \tilde{\varphi}^* \omega,$$

where in the first equality we have used the identity $\varphi = \tilde{\varphi}$ on the boundary, in the second we have applied Stokes' theorem, in the third we have used the commutativity of pullback and derivative, and the sign factor again reflects the opposite orientation of M relative to N_+ and N_-. The $(d+1)$-dimensional **invariant representation of the WZ term**

WZ term (invariant representation)

$$\Gamma_\pm[\varphi] \equiv \pm ik \int_{N_\pm} \tilde{\varphi}^* \omega, \tag{8.50}$$

is formulated in terms of ω and does not rely on local coordinate representations (cf. Eq. (8.46) in the spin example). However, this comes at the cost of the dimensional extension. Which form of the WZ term is more convenient to work with depends on the application.

Finally, note that the ambiguity implied by the choices $N_\pm$ must be inconsequential. The difference between the choices equals $ik \int_{N_+} \tilde{\varphi}^* \omega - ik(-\int_{N_-} \tilde{\varphi}^* \omega -) = k \int_{N_+ \cup N_-} \tilde{\varphi}^* \omega = ik \int_N \tilde{\varphi}^* \omega = 2\pi i k$, where in the final step the unit winding property was used. In the exponentiated action, $\exp(i\Gamma)$, this difference is unobservable provided $2\pi k$ is an integer multiple of 2π, or $k \in \mathbb{Z}$. We observe that the integer quantization of the WZ-coupling constant, or its level, is required for consistency. Unlike θ-terms with their arbitrary topological angles,

> WZ terms couple to actions via integer-quantized coupling constants.
> They are called the **levels** of the theory.

The differential form representations (8.49) and (8.50) may look unfamiliar but have the advantage that they describe the structure of WZ terms in general. In applications, WZ terms assume the form of Eqs. (8.44) and Eq. (8.46), or of more complicated expressions in higher dimensions. To identify a WZ term as such – which is not always straightforward – it may be best to translate to differential form language and describe the situation in this way. We also note that in down-to-earth derivations of WZ terms via gradient expansions in effective field theories, the above quantization criterion defines an important consistency check for the validity of the computation.

8.5.3 Example: magnetic moment coupled to fermions

Let us illustrate the statements made in the last paragraph of the previous section for a simple example. Consider a single energy level ϵ of a spinful fermion system. (One may think, for example, of a discrete level of an atom.) Let us assume that the fermions inhabiting the level are coupled to a magnetic moment $\mathbf{n} = \mathbf{n}(\tau)$ with externally imposed time dependence. The coherent state action of this system is given by

$$S[\psi, \mathbf{n}] = \int_0^\beta d\tau \, \bar{\psi}(\partial_\tau + \xi + \gamma \mathbf{n} \cdot \sigma)\psi,$$

where γ is a coupling constant and, as usual, $\xi = \epsilon - \mu$. A complete specification of the problem would have to include a term $S[\mathbf{n}]$ controlling the dynamics of the uncoupled magnetic moment. However, for the purposes of the present discussion, it is sufficient to consider the moment-fermion coupling in isolation. We assume an adiabatic situation where the moment varies over large scales, $\Delta\tau$, and does not generate transitions between spin states, $\Delta\tau\gamma \gg 1$. We also assume low temperatures, $\gamma/T \gg 1$.

INFO Actions of this type appear as building blocks of extended models, with $\mathbf{n}$ representing a magnetic environment, coupling to an orbital magnetic moment, or the Hubbard–Stratonovich decoupling of an electron-electron interaction.

Integration over the fermion degrees of freedom leads to the reduced action

$$S[\mathbf{n}] = -\mathrm{tr} \ln(\partial_\tau + \xi + \gamma \mathbf{n} \cdot \sigma).$$

To proceed, we use the representation

$$\mathbf{n} \cdot \boldsymbol{\sigma} = U^{-1}\sigma_3 U, \tag{8.51}$$

where $U \in \mathrm{SU}(2)$. For example, with the standard polar representation, $\mathbf{n} = (\sin\theta\cos\phi, \sin\theta\sin\phi, \cos\theta)^T$, the choice

$$U = e^{-i\phi\sigma_3/2}e^{-i\theta\sigma_2/2}e^{-i\psi\sigma_3/2} \tag{8.52}$$

establishes this representation. Substituting Eq. (8.51) into the action, we obtain

$$\begin{aligned}
S[\mathbf{n}] \to S[U] &= -\operatorname{tr}\ln(\partial_\tau + \xi + \gamma U^{-1}\sigma_3 U) = -\operatorname{tr}\ln(U\partial_\tau U^{-1} + \xi + \gamma\sigma_3) \\
&= -\operatorname{tr}\ln(\partial_\tau + \xi + \gamma\sigma_3 + U\dot{U}^{-1}),
\end{aligned}$$

where, in the first equality we used the cyclic invariance of the trace and in the last equality we defined $\partial_\tau U \equiv \dot{U}$. Under the above adiabaticity assumptions, the effective action may be approximated by a first-order expansion of the "tr ln" operator in the time derivative, $S[U] = \operatorname{tr}(\hat{G}U\dot{U}^{-1}) + \mathcal{O}(\Delta\tau\gamma)^2$, where $\hat{G} = (-\partial_\tau - \xi - \gamma\sigma_3)^{-1}$ (how would you formally verify this statement?). Switching to a frequency representation,

$$\begin{aligned}
S[U] &= \sum_n \operatorname{tr}(G_n(U\dot{U}^{-1})_{m=0}) = -\int d\tau\,\operatorname{tr}(\Theta(-\xi - \gamma\sigma_3)\dot{U}\partial_\tau U^{-1}) \\
&= -\int d\tau\,\operatorname{tr}\left(\left[\Theta(-\xi - \gamma\sigma_3) - \frac{1}{2}\right]U\partial_\tau U^{-1}\right),
\end{aligned}$$

where we used Eq. (3.82) for the frequency summation, and approximated the low-temperature Fermi function as $n_F(x) = \Theta(-x)$. In the last line, we introduced a factor $1/2$, which is inconsequential because $\operatorname{tr}(U\partial_\tau U^{-1}) = -\partial_\tau\operatorname{tr}\ln U = -\partial_\tau\ln\det U = 0$ and U has unit determinant.

Now, if both levels are either occupied ($-\xi \mp \gamma < 0$) or unoccupied ($-\xi \mp \gamma > 0$), the action reduces to $\operatorname{tr}(\mathrm{const.}\times U\partial_\tau U^{-1}) = 0$. However, if the excited state $\xi + \gamma > 0$ is empty while $\xi - \gamma$ is occupied, we obtain

$$S[U]|_{\xi+\gamma>0>\xi-\gamma} \equiv \Gamma[U] = -\frac{1}{2}\int_0^\beta d\tau\,\operatorname{tr}(\sigma_3 U\dot{U}^{-1}). \tag{8.53}$$

Being a first derivative in time, this action is invariant under reparameterizations of the integration variable τ, indicating that it is a topological term. To identify its type, we first determine the dimensionality of the target space, T. Naïvely, one might suspect that $T = \mathrm{SU}(2)$, a three-dimensional manifold. However, this cannot be correct, as U was introduced to parameterize the two-dimensional $\mathbf{n}$ in Eq. (8.51). This equation indicates that, of the three variables in the parameterization (8.52), ψ drops out and we are left with a two-dimensional integration manifold. In the action (8.53), the redundancy of this variable manifests itself via a "gauge invariance" under transformations $U \to U\exp(i\psi\sigma_3)$. These transformations generate an additional term, $\frac{1}{2}\int d\tau\operatorname{tr}(\sigma_3(i\dot{\psi}\sigma_3)) = i(\psi(\beta) - \psi(0)) = 2\pi iW$. The phase windings W are inconsequential because $\exp(2\pi iW) = 1$. This construction demonstrates that the actual field manifold is the two-sphere, $\mathrm{SU}(2)/\mathrm{U}(1) \simeq S^2$. Its dimension is one higher than that of the one-dimensional base manifold, indicating that we have a Wess–Zumino term.

To substantiate this, we substitute the coordinate representation (8.52) into the action (exercise):

$$\Gamma[\phi, \theta] = -\frac{i}{2} \int d\tau \, (1 - \cos\theta) \, \partial_\tau \phi. \tag{8.54}$$

We identify this expression as the coordinate representation (8.44) of the $S = 1/2$ WZ term with base S^1 and target S^2. In hindsight, the rediscovery of this action is not surprising: the action of the spin path integral measures the phase associated with the traversal of a closed curve in the space of spin coherent states. Presently, we are dealing with a two-level system (a spin) enslaved to a time-dependent magnetic moment. Quantum mechanically, this is again a spin tracing out a closed curve, and the result (8.54) shows the equivalence to the previous situation.

Berry phase

REMARK Here we review the concept of Berry phases and demonstrate their relation to WZ terms for the example of the magnetic moment studied above. This discussion reveals interesting connections between topological field theory and adiabatic quantum time evolution. However, readers primarily interested in the former subject may skip it at first reading.

In the literature on time-dependent quantum magnetism, the action (8.54) is known as the **Berry phase action of a spin** $1/2$. To understand this connection, let us start with a review of Berry phases in quantum mechanics.

Berry phase action of spin

Consider a Hamiltonian $\hat{H}(x(t)) \equiv \hat{H}(t)$, where the D-component vector $x(t) \equiv \{x_i(t)\}$ parameterizes a weakly time-dependent contribution to $\hat{H}$. We assume that at each instant of time the spectrum of the operator $\hat{H}(x(t))$ is discrete and that the time variation takes place on scales much larger than the inverse spacing between consecutive energy levels. Under this **adi-**

> **Sir Michael Berry** 1941–
> is a British theoretical physicist who has made groundbreaking contributions to the field of quantum nonlinear dynamics and optics. Berry introduced the concept of the Berry phase (or geometric phase as he himself prefers to call it) and explored its manifestations in various physical contexts. (Figure courtesy of Sir Michael Berry.)

adiabatic evolution

abaticity condition, a particle prepared in the ground state of $\hat{H}(x(0))$ at time $t = 0$ will remain in the instantaneous ground state of $H(x(t))$ and, as shown by Berry, the dynamical phase acquired during its evolution assumes the form

$$\exp\left[-i\phi(t)\right] \equiv \exp\left[-i \int_0^t ds \, \epsilon_0(s) + i\gamma(t)\right], \tag{8.55}$$

where $\epsilon_0(t)$ is the instantaneous ground state energy. The first contribution to the exponent is the usual dynamical phase of quantum evolution. The second contribution, $\gamma(t)$, is of geometric origin. It depends on the path traced out by the vector x in parameter space, but not on the dynamical details. However, before discussing its connection to the phase described by the WZ action, let us review the computation of the geometric phase, γ, in the context of adiabatic quantum time evolution.

We are interested in the phase picked up in the time evolution of the ground state, $|\psi(t)\rangle$. To this end, we start from a representation $|\psi(t)\rangle = e^{-i\phi(t)}|0(t)\rangle$, where $|0(t)\rangle$ is the instantaneous ground state of the Hamiltonian at time t, $\hat{H}(t)|0(t)\rangle = \epsilon(t)|0(t)\rangle$, and $\epsilon(t)$ is its energy. Substitution of this ansatz into the time-dependent Schrödinger equation

$$\hat{H}(x(t))|\psi(t)\rangle = i\partial_t|\psi(t)\rangle,$$

and multiplication by $\langle 0(t)|$ leads to the equation $\partial_t\phi = \epsilon(t) - i\langle 0(t)|\partial_t|0(t)\rangle$. Integrating over time and comparing with our discussion above, we are led to the identification

$$\boxed{\gamma(t) = i\int_0^t ds\,\langle 0(s)|\partial_s|0(s)\rangle} \qquad (8.56)$$

Berry
phase

of the **Berry phase** (exercise: why is γ real?). Now, the instantaneous ground state inherits its time dependence from the parameters $x(t)$. We may thus write

$$\gamma(t) = i\int_0^t ds\,\langle 0(x)|\partial_{x_i}|0(x)\rangle\big|_{x(s)}\partial_s x_i(s) = i\int_c dx\,\langle 0(x)|\partial_x|0(x)\rangle = i\int_c \langle 0|d0\rangle.$$

Here, the second integral has to be interpreted as a line integral in parameter space. It is taken along a curve c which starts at $x(0)$, follows the evolution of the parameter vector, and ends at $x(t)$. Importantly, the line integral depends only on the choice of γ but not on the velocity at which this curve is traversed. In this sense, we are dealing with a phase of geometric but not of dynamic nature. (In fact, Berry himself calls γ the "geometric phase."[55]) The third integral representation above emphasizes the geometric nature of the phase even more strongly: for any value of x, we have a state $|0(x)\rangle$. We may then construct the differential (one-form) $\langle 0(x)|d0(x)\rangle$.[56] The geometric phase is obtained by evaluating the integral of this form along the curve c.

The advantage of the third representation above is that it suggests an alternative interpretation of the geometric phase in cases where a closed path in parameter space is traversed. For a closed loop γ, application of Stokes' theorem gives

$$\boxed{\gamma = i\oint_\gamma \langle 0|d0\rangle = i\int_S \langle d0| \wedge |d0\rangle} \qquad (8.57)$$

where S may be any surface in parameter space that is bounded by γ. This last representation is aesthetic, but too compact to be of real computational use. To give it a more concrete meaning, we insert a spectral decomposition in instantaneous eigenstates,

---

[55] M. V. Berry, *Quantal phase factors accompanying adiabatic changes*, Proc. R. Soc. Lond. A **392**, 45 (1984). For an earlier identification of the phase factor, see T. Kato, *On the adiabatic theorem of quantum mechanics*, J. Phys. Soc. Japan **5**, 435c (1950).

[56] A somewhat less condensed representation is obtained by expansion of the ground state in any basis $\{|\lambda\rangle\}$, $\langle 0(x)|d0(x)\rangle \equiv \langle 0(x)|\lambda\rangle d\langle\lambda|0(x)\rangle$, where $d\langle\lambda|0(x)\rangle$ is the exterior derivative of the function $\langle\lambda|0(x)\rangle$ in the parameters x.

$$\gamma = i \int_S \sum_{m \neq 0} \langle d0|m\rangle \wedge \langle m|d0\rangle.$$

EXERCISE Why does the $m = 0$ term vanish, $\langle d0|0\rangle \wedge \langle 0|d0\rangle = 0$? (Hint: Make use of the fact that $d\langle 0|0\rangle = \langle d0|0\rangle + \langle 0|d0\rangle = 0$ and of the skew-symmetry of the $\wedge$-product.)

We now evaluate the equation $0 = \langle m|d[(H - \epsilon_0)|0\rangle]$ to obtain $\langle m|d0\rangle = (\epsilon_0 - \epsilon_m)^{-1}\langle m|dH|0\rangle$ or

$$\gamma = i \int_S \sum_{m \neq 0} \frac{\langle 0|dH|m\rangle \wedge \langle m|dH|0\rangle}{(\epsilon_m - \epsilon_0)^2} \qquad (8.58)$$

As an example, we consider the **Berry phase of spin due to the adiabatic variation of an external magnetic moment**, $\mathbf{n}(t)$. In this case, the Hamiltonian reads $\hat{H} = \mu \mathbf{n} \cdot \boldsymbol{\sigma}$, where μ measures the coupling strength and the role of the parameters x_i is now assumed by the coordinates $x = (\phi, \theta)$ specifying the direction of $\mathbf{n} = \mathbf{n}(x)$ on the sphere. Using the same trick as previously, we can represent the Hamiltonian as $\hat{H} = \mu \mathbf{n} \cdot \boldsymbol{\sigma} \equiv \mu U \sigma_3 U^{-1}$, where U is the rotation matrix introduced in Eq. (8.51). The instantaneous ground state of $\hat{H}$ is given by $|0\rangle = U|\downarrow\rangle$, where $\sigma_3|\downarrow\rangle = -|\downarrow\rangle$. To compute its Berry phase along a closed loop, we can consider the first of the two representations in Eq. (8.57). Using the parameterization (8.52), we verify that $\langle 0|d0\rangle = \langle \downarrow|U^{-1}dU|\downarrow\rangle = \frac{i}{2}(1 - \cos\theta)d\phi$, and hence

$$\gamma = \frac{i}{2} \oint d\phi\,(1 - \cos\theta). \qquad (8.59)$$

This looks similar to the previously discussed WZ action of spin. Realizing the parameter dependence in terms of a periodic time variation, $(\theta, \phi) = (\theta(t), \phi(t))$, we indeed find that the exponentiated Berry phase, $\exp(i\gamma) = \exp(-i\Gamma)$, is given by the WZ action (8.54). This equivalence shows that:

> The Berry phase is a one-dimensional Wess–Zumino action.

This identification underpins the geometric nature of the Berry phase. In the one-dimensional case, the role of the higher-dimensional representation of the WZ action (8.50) is taken by the two-dimensional integral over parameter space, as in the second equation of Eq. (8.57). For the spin example, this integral assumes the form of Eq. (8.46).

INFO The Berry phase (8.59) **is experimentally observable**. The key to understanding the connection with experiment lies in the interpretation of $A = i\langle 0|d0\rangle$ as a vector potential in the space of parameters, to be discussed in section 10.2.2. This potential acts on fermions in addition to that representing extraneous magnetic fields and does contribute to the Hall conductivity. Such geometric contributions to magnetotransport have been observed in complex oxide minerals known as **pyrochlores**. The

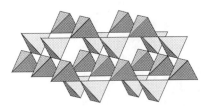

pyrochlores

frustrated lattice structure of these materials (see the figure[57]) implies spin chirality in the sense that, on average, the magnetic moments of the lattice unit cells point in an outward direction. From the point of view of itinerant electrons, this background appears like an external magnetic moment. The corresponding Berry connection provides a contribution to the Hall conductivity, which can be identified via its pronounced dependence on temperature and external magnetic field.[58]

8.5.4 Nonabelian bosonization

REMARK Familiarity with section 3.6 on bosonization of the one-dimensional Fermi gas is useful, but not essential.

In this section, we will discuss an example of WZ field theory with $U(n)$ as a target space. The focus here is on the construction of this theory; its application in quantum magnetism will be addressed below. In section 3.6, we considered the action of free fermions in one spatial and one temporal dimension, as described by the two-dimensional Euclidean Dirac theory,

$$S[\psi, \bar{\psi}] = \int dx d\tau \, \psi^\dagger (\partial_\tau - i\sigma_3 \partial_x)\psi, \tag{8.60}$$

where $\psi = (\psi_+, \psi_-)^T$ contains a left- and a right-moving field component, the Fermi velocity v_F has been set to unity and, as always with Grassmann fields, the dagger is a formal symbol, indicating that $\psi^\dagger$ and ψ are independent integration variables. We found that this theory was equivalently described by that of a free real bosonic field,

$$S[\theta] = \frac{1}{2\pi} \int d^2x \, (\partial_\mu \theta)^2,$$

where we introduced the two-component vector $(x_0, x_1)^T = (x, \tau)^T$. In a seminal paper,[59] Witten asked how we should think about bosonization if a number N of independent fermion fields ψ^a with identical actions are considered or, equivalently, if ψ in Eq. (8.60) is upgraded to an $2N$-component vector field.

Edward Witten 1951-
is a mathematical physicist and string theorist. Awarded the 1990 Fields Medal for his ground breaking work in differential geometry. Witten contributed massively to the success of string theory.

INFO Realistic quasi one-dimensional conductors often support multiple channels of left- and right-moving fermions. In general, these have different Fermi velocities. However,

[57] O. Tchernyshyov, R. Moessner, and S. L. Sondhi, *Order by distortion and string modes in pyrochlore antiferromagnets*, Phys. Rev. Lett. **88**, 067203 (2002).

[58] Y. Taguchi *et al.*, *Spin chirality, Berry phase, and anomalous Hall effect in a frustrated ferromagnet*, Science **291**, 2573 (2001).

[59] E. Witten, *Nonabelian bosonization in two dimensions*, Commun. Math. Phys. **92**, 255 (1984).

one may always rescale fields and coordinates (think how?) to achieve equivalence of the effective free fermion actions. In this way, one arrives at the setting considered by Witten, indicating that his question is relevant, including from a condensed matter perspective.

Of course, there is the obvious (and frequently applied) option to bosonize each of the theories individually, $\sum_a S[\bar{\psi}^a, \psi^a] \to \sum_a S[\theta^a]$, to arrive at an action of N independent free bosons. However, this representation misses out on an essential feature of the fermion action: the latter is invariant under uniform rotations $\psi_s \to g_s \psi_s$, $\psi_s^\dagger \to \psi_s^\dagger g_s^\dagger$ by $g_s \in \mathrm{U}(N)$, $s = \pm$. Our action thus comes with a large symmetry group $\mathrm{U}(N) \times \mathrm{U}(N)$, where the two factors represent independent transformations of the left- and right-moving fields, generalizing the abelian transformations by phases in $\mathrm{U}(1) \times \mathrm{U}(1)$ which motivated the bosonization program.

This symmetry is no longer manifest in the representation $\sum_a S[\theta^a]$. Of course, it is still present – the transformation is exact – but it is no longer visible. This is a problem in applications where, as well as the free fermion action, we have nontrivial contributions such as interactions in the problem. Symmetries are an essential part of the solution of such problems, and we would prefer to work with representations that keep them visible. This consideration motivated Witten to ask whether there is a generalization of bosonization, **nonabelian bosonization** with retained symmetry group $\mathrm{U}(N) \times \mathrm{U}(N)$.

nonabelian bosonization

Wess–Zumino action

In his paper,[59] Witten suggested an extended bosonization identity based on a symmetry construction to be reviewed in problem 8.8.4. Here, we present an abridged version of the argument, which hopefully will be convincing enough to make the result plausible. Anticipating that group structures will be center stage, let us rewrite the abelian bosonization action as

$$S_0[g] = \frac{1}{\lambda} \int dx d\tau \, \mathrm{tr}(\partial_\mu g \partial_\mu g^{-1}), \qquad \lambda = 8\pi, \qquad (8.61)$$

where $g = \exp(2i\theta) \in \mathrm{U}(1)$ and we have included the – here redundant – trace, anticipating later generalization to $\mathrm{U}(N)$. Previously, we have seen that transformations of the fermion fields $\psi \to g_s \psi$, with $g_s = e^{i\phi_s}$ affect the θ-variable as $\theta \to \theta + (\phi_+ - \phi_-)/2$. This transformation is implied if we represent the symmetry group $\mathrm{U}(1) \times \mathrm{U}(1)$ as

$$g \longrightarrow g_+ g g_-^{-1}. \qquad (8.62)$$

Upon making the generalization $\mathrm{U}(1) \to \mathrm{U}(N)$, the action (8.61) with this group representation appears to tick all the boxes where symmetries are concerned. However, there is a problem. Equation (8.61) defines a nonlinear σ-model in two dimensions. We saw in section 6.4 that, at large length scales, strong field fluctuations lead to an exponential decay of correlations, which in technical terms means a renormalization of the coupling constant λ to large values. This is at variance with the behavior of the free fermion system, which exhibits long-range fluctuations. What we need is an additional player in the action that prevents this renormalization into a disordered phase. On dimensional grounds this term must have an equal number of two-derivatives. However, the dilemma is that there is no rotationally invariant two-derivative term of the g-fields other than the one displayed in Eq. (8.61).

Nevertheless, we do know that in one dimension higher, $d = 3$, we have the U(N) winding number integral shown in the info block of page 440 in the notation $g = U$. This integral can be used for the construction of a U(N) WZ term in $d = 2$. To see how, we assume two-dimensional space–time to be compactified to a two-sphere S^2. Much as the circle, S^1, is the boundary of a northern and a southern half of S^2, the two-sphere, S^2, is the boundary of two halves of $S^3 = B \cup B'$. According to a construction known in topology, these two halves can be individually identified with three-dimensional balls, B and B', whose centers represent the north and south poles of S^3, respectively, and their surface a two-sphere, S^2. The gluing of the balls at that equatorial boundary then gives S^3.[60] We may now pick either of these balls as an integration domain, and its radial coordinate, r, to extend the field as $g(x) \to \tilde{g}(x, r)$, with $\tilde{g}(x, 1) = g(x)$. In this way we obtain the **WZ term of U(N)**,

WZ term of U(N)

$$\Gamma[g] = -\frac{i}{12\pi} \int_B d^3x \, \epsilon^{ijk} \, \mathrm{tr}(\tilde{g}^{-1}\partial_i\tilde{g}\,\tilde{g}^{-1}\partial_j\tilde{g}\,\tilde{g}^{-1}\partial_k\tilde{g}). \qquad (8.63)$$

Comparing with the winding number integral discussed on page 440, here we have set $\theta = 2\pi$ (on the basis of our discussion in section 8.5.2 – explain why). Note that $\Gamma[g]$ is a functional of g (not $\tilde{g}$), because it depends only on value of the field at the integration boundary, $r = 1$. Also notice that $\Gamma[g_+ g g_-^{-1}] = \Gamma[g]$ is invariant under the transformation (8.62), and therefore displays the full symmetry of the problem.

With S_0 given by Eq. (8.61), the sum of two terms,

$$S[g] = S_0[g] + \Gamma[g], \qquad (8.64)$$

nonabelian bosoniza-tion action

defines the **nonabelian bosonization action**. A refined version of the above symmetry analysis (see problem 8.8.4 and the next section) indicates that, at the coupling $\lambda = 8\pi$, it provides an equivalent representation of the multi-flavor chiral fermion theory. In particular, it shows the conformal invariance (see section A.3) generic for two-dimensional critical theories. Below, we provide added evidence for the equivalence to the likewise-critical free fermion theory.

bosonization dictionary

However, before going there, let us set up the **bosonization dictionary** translating different elements of the fermionic theory into the bosonic language. First, the nonabelian action (8.64) applies to fields in U(N). However, it is often convenient to split $g = e^{i\phi} g'$ into a phase and an SU(N) matrix g'. A straightforward calculation shows that $\Gamma[g] = \Gamma[g']$, i.e., the WZ term is indifferent to the phase. In

---

[60] An intuitive way to understand this construction is to go one dimension lower and imagine the northern and southern hemisphere of the two-sphere to be flattened to two disks, D. The centers of these disks represent the north or south pole, respectively, and their boundaries the circular equator.

the gradient term, we obtain $S_0[g] = S_0[g'] + \frac{1}{\lambda} \int dx d\tau \, (\partial_\mu \phi)^2$, and in this way the action splits into an SU(N) action and a phase contribution.

In abelian bosonization, we defined holomorphic and anti-holomorphic currents, $j = \psi_-^\dagger \psi_-$ and $\bar{j} = \psi_+^\dagger \psi_+$, obeying the classical conservation laws $\partial_{\bar{z}} j = \partial_z \bar{j} = 0$, where complex coordinates $z = \frac{1}{\sqrt{2}}(\tau + ix)$, $\partial_z = \frac{1}{\sqrt{2}}(\partial_\tau - i\partial_x)$ are used. The obvious generalization to nonabelian **holomorphic currents** reads

holomorphic
currents

$$j^{ab} \equiv \psi_-^{a\dagger} \psi_-^b, \qquad \bar{j}^{ab} \equiv \psi_+^{a\dagger} \psi_+^b, \tag{8.65}$$

with the same conservation laws $\partial_z \bar{j}^{ab} = \partial_{\bar{z}} j^{ab} = 0$ (see problem 8.8.4 for details). In the abelian case, these currents have a bosonic representation $j = \frac{i}{\sqrt{2\pi}} \partial_z \theta$ and $\bar{j} = -\frac{i}{\sqrt{2\pi}} \partial_{\bar{z}} \theta$. The generalization to a nonabelian **bosonic representation of the currents** reads

$$j = \frac{1}{\sqrt{8\pi}} g^{-1} \partial_z g, \qquad \bar{j} = -\frac{1}{\sqrt{8\pi}} (\partial_{\bar{z}} g) g^{-1}. \tag{8.66}$$

For $N = 1$ and $g = \exp(2i\theta)$, these expressions reduce to the abelian version. The particular ordering of the group-valued factors, which matters in the nonabelian case, is explained in problem 8.8.4.

Finally, we identify a bosonic representation for the bilinears of mixed chirality $\psi_+^{a\dagger} \psi_-^b$. In the abelian case, these were represented as (see Eq. (3.95)) $\psi_+^\dagger \psi_- = \gamma e^{2i\theta}$, with a non-universal prefactor γ. This is generalized to the nonabelian case as

$$\psi_+^{a\dagger} \psi_-^b \longrightarrow \gamma g^{ab}. \tag{8.67}$$

Note that this translation is consistent with Eq. (8.62). However, unlike the abelian case, no bosonic representation of individual fermion operators exists.

Renormalization group analysis

Previously, we reasoned that the gradient action S_0 flows to strong coupling at large length scales and hence disqualifies as a representation of the free fermion theory. Can the presence of the WZ term in Eq. (8.64) change the situation? Counting the number of derivatives in relation to the number of variable integrations, we note that the WZ term has the same engineering dimension, zero, as the gradient term. This makes it a potential game changer.

Suppose we started renormalizing the model at small values of λ such that $1/\lambda \gg 1/12\pi$ is much larger than the coupling constant of the WZ term. In this regime, the presence of the latter is irrelevant, and the RG flow will be towards increasing λ. Eventually, one reaches a regime $\lambda \simeq 12\pi$, where a competition between the two terms starts. A one-loop RG analysis (see problem 8.8.5) almost identical to that performed earlier in section 6.4 yields the scaling equation

$$\frac{d\lambda}{d \ln b} = \frac{N\lambda}{4\pi} \left[1 - \left(\frac{\lambda}{8\pi} \right)^2 \right]. \tag{8.68}$$

This result confirms our qualitative expectation: the value $\lambda^* = 8\pi$ defines an attractive fixed point at which the upwards flow of λ comes to an end. While this result is based on a perturbative RG analysis – questionable in the regime of strong coupling $1/\lambda < \mathcal{O}(1)$ – Witten[59] showed that at $\lambda = 8\pi$ the theory is conformally invariant, thus proving the fixed-point property beyond perturbation theory.

The combination of symmetry arguments, RG analysis, and conformal invariance provides compelling evidence for the equivalence of Eq. (8.64) to the multi-flavor free fermion theory. Evaluated on diagonal configurations, $g = \mathrm{diag}(e^{i\theta_1}, \ldots, e^{i\theta_N})$, it collapses to a sum of abelian bosonized actions, which we know likewise represents the free fermion theory. This raises the question of what advantages the more complicated nonabelian representation might have (beyond the aesthetic appeal of showing the full symmetry contents of the theory). It is fair to say that nonabelian bosonization has not yet become a mainstream tool in condensed matter theory, and perhaps never will. In most situations, the scalar fields of abelian bosonized theories are easier to handle than their $\mathrm{U}(N)$-valued counterparts. However, there are some applications, notably in the physics of disordered or topological electron systems, as well as quantum magnetism, where it provides results that abelian bosonization cannot give. Below, we discuss one such application as an example.

8.5.5 Spin chains: beyond the semiclassical limit

In section 8.4.6, we considered the physics of one-dimensional spin chains at strong strong inter-site coupling J. In this regime, the system is described by the partition sum of the O(3) nonlinear σ-model (8.32) with a topological θ-term. We reasoned that the role of the latter depended on whether the spin S is half-integer or integer. In the latter case, its presence was inessential and the model ended up in a gapped strong fluctuation phase at large length scales, the standard behavior of two-dimensional σ-models. However, for half-integer spin we reasoned, but could not show how, that the model will flow towards an ordered phase with long-range correlations.

In fact, the situation looks somewhat mysterious from the outset: the microscopic Hamiltonian of the antiferromagnetic spin chain has global SU(2) invariance. However, the antiferromagnetically ordered configuration around which we expanded in the derivation of the σ-model spontaneously breaks the symmetry down to U(1), and the soft modes of the model are the spherical degrees of freedom on the sphere $S^2 = \mathrm{SU}(2)/\mathrm{U}(1)$. At large length scales, we expect two things to happen. First, the symmetry must be "restored" (there is no breaking of continuous symmetries in dimension two – the Mermin–Wagner theorem), so that we expect an effective theory with SU(2)-valued degrees of freedom. In effect, the number of degrees of freedom of the theory must be dynamically enhanced from two to three, which is unusual. Second, we expect the fixed-point theory to show long-range order.

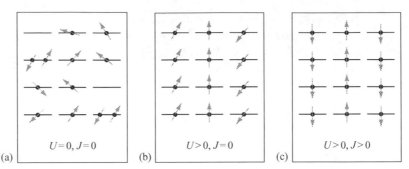

$U=0, J=0$ (a) $U>0, J=0$ (b) $U>0, J>0$ (c)

Fig. 8.6   (a) A chain of sites, each containing n_c fermions on average. (b) A strong Hund's rule coupling maximizes the spin carried by each state to $S = n_c/2$. (c) Upon the switching on of a nearest-neighbor hopping matrix element, the system becomes a spin S antiferromagnet.

Spin chain via nonabelian bosonization

As explained in a ground-breaking paper by Affleck and Haldane (AH),[61] non-abelian bosonization is the key to the understanding of this phenomenon. To see why, we first need to draw a connection between the spin chain and the bosonization machinery. Referring to the info block below for details, the idea is simple. A chain of spin-1/2 degrees of freedom is realized in terms of a chain of spinful lattice fermions, occupied by one fermion per site. Replicating $n_c = 2S$ such chains and turning on a strong on-site spin interaction aligning the individual spin-1/2, we generate a single chain of effective spin-S degrees of freedom. This approach requires a strong local interaction U, and in addition a relatively weaker inter-site hopping J, to obtain antiferromagnetic correlations (see fig. 8.6). However, following AH, we now argue boldly that nothing drastic (no phase transition) happens as we reverse the strength of the couplings to weak on-site interactions superimposed on strong inter-site coupling. The non-interacting $N = 2n_c$ flavor fermion chain at half-filling is a system of left- and right-moving fermions described by the bosonization action. Turning on interactions, we may hope that the latter will teach us something about the physics of antiferromagnetism, including in the regimes beyond the reach of the nonlinear σ-model.

INFO Consider a system of $n_c \equiv 2S$ chains of spin-1/2 lattice fermions ψ_{ia}^{α}, where i, a, α denote the site, chain, and spin index respectively. To align their spins locally at each site, we turn on a strong spin interaction

$$\hat{H}_{\text{int}} = -U \sum_i \hat{\mathbf{S}}_i \cdot \hat{\mathbf{S}}_i, \qquad \hat{\mathbf{S}}_i = \frac{1}{2} \psi_{ia}^{\alpha\dagger} \boldsymbol{\sigma}^{\alpha\beta} \psi_{ia}^{\beta}. \tag{8.69}$$

The minimization of energy in the presence of this operator requires n_c fermions at each site, i.e., half-filling of the total system.[62] On top of this, we add a small amount of inter-

[61] I. Affleck and F. D. M. Haldane, *Critical theory of quantum spin chains*, Phys. Rev. B **36**, 5291 (1987).
[62] To see this in more detail, use the relation $\sigma^{\alpha\beta} \cdot \sigma^{\gamma\delta} = 2\delta^{\alpha\delta}\delta^{\beta\gamma} - \delta^{\alpha\beta}\delta^{\gamma\delta}$ to verify that (with the site index left out for clarity) $\hat{\mathbf{S}} \cdot \hat{\mathbf{S}} = -\frac{1}{4}(\psi_a^{\alpha\dagger}\psi_\alpha^a - n_c)^2 + \cdots$, where the omitted terms

site hopping $\hat{H}_0 = -\frac{1}{2}(n_c JU)^{1/2} \sum_i \psi_{ia}^{\alpha\,\dagger} \psi_{i+1a}^{\alpha} + \text{h.c.}$ In the limit $J/U \to 0$, the half-filled system then becomes equivalent to the spin-S antiferromagnetic chain,

$$\hat{H} = \hat{H}_0 + \hat{H}_{\text{int}} \xrightarrow{\langle \hat{n}_i \rangle = n_c} H_{\text{af}} = J \sum_i \hat{\mathbf{S}}_i \cdot \hat{\mathbf{S}}_{i+1} + \mathcal{O}(J/U). \tag{8.70}$$

The easiest way to see this is to recall the situation in the standard Hubbard model (the $n_c = 1$ variant of our present model) at half-filling (cf. section 2.2.3). There, virtual deviations from half-filling led to an effective antiferromagnetic exchange coupling between the $S = 1/2$ spins carried by neighboring sites. The effective strength of this interaction was $J \sim t^2/U$, where t is the strength of the hopping term. Formally, the generalization of this mechanism to the case $n_c > 1$ can be shown, for example, by subjecting the Hamiltonian $\hat{H}$ to a canonical transformation eliminating the hopping term (all in complete analogy to the $n_c = 1$ canonical transformation discussed in section 2.2.3).

Let us then approach the problem from the perspective of $N = 2n_c$ free fermions, perturbed by the two-fermion interaction (8.69). This interaction respects certain continuous symmetries of the theory.

EXERCISE Pause reading and form an opinion as to which symmetries might be relevant to the present problem, considering which are conserved by the interaction and which are not.

Representing the fermions as ψ_{sa}^{α}, with $s = \pm$, spin index $\alpha = 1, 2$ and a "color" index[63] $a = 1, \ldots, n_c$, the interaction has U(1)-symmetry reflecting the conservation of charge, SU(2)-symmetry reflecting spin rotation invariance, and SU(n_c)-symmetry reflecting isotropy in the color indices. However, chiral symmetry is lost, i.e., these symmetry groups do not come in two copies for the $s = \pm$ sectors because the interaction mixes left- and right-moving fermions.

It makes sense to start from a fermion representation of the interactions respecting these symmetries. To this end, we focus on a subset of the current operators (8.65), namely the **charge current** $j_s^q \equiv \psi_s^{\dagger} \psi_s$, the **spin currents**, $j_s^{s,\alpha\beta} \equiv \psi_s^{\alpha\dagger} \psi_s^{\beta}$, and the **color currents**, $j_{sab}^c \equiv \psi_{sa}^{\dagger} \psi_{sb}$, where all non-appearing indices are summed over. Symmetry-compatible interaction operators then assume the form $j_+^q j_-^q$, $\text{tr}(j_+^s j_-^s)$, and $\text{tr}(j_+^c j_-^c)$. In general, momentum conservation forbids the scattering of left-moving fermions into right-moving ones. However, the presently considered case of the half-filled chain is an exception to this rule: the scattering of a pair of left-moving fermions into a pair of right-moving ones exchanges momentum $2 \times 2 \times \frac{\pi}{2a} = \frac{2\pi}{a}$, which is a reciprocal lattice vector. This type of scattering is compatible with the crystalline symmetry and is represented by the **Umklapp**

---

are not essential to the argument. The term in parentheses tells us that the extremization of the local spin in the coupled chain system requires n_c fermions per site. In passing, we note that the present way to engineer coupled spin states is not as artificial as it may seem. Large atomic spins are usually the result of the *Hund's rule* coupling of shell electrons via the above spin interaction: parallel alignment is favored because, in this case, the orbital wave functions are mutually antisymmetric, thus reducing electronic repulsion.

[63] The terminology is borrowed from particle physics, where (quark)-fermions carrying a threefold color index, red–green–blue, transform under SU(n_c) with $n_c = 3$.

scattering operator $(\psi_{+a}^{\alpha\dagger}\sigma_2^{\alpha\beta}\psi_{+b}^{\dagger\beta})(\psi_{-a}^{\gamma}\sigma_2^{\gamma\delta}\psi_{-b}^{\delta})$ + h.c. The Umklapp operator is compatible with the above symmetry groups (exercise); however, scattering left-moving fermions into right-moving fermions breaks the chiral symmetry of the problem. For the discussion of a few other interaction operators compatible with the symmetries of the problem, we refer to Ref.[61] and problem 8.8.6.

Quantum criticality

Our further course of action is now clear: we are to use the rules (8.66) and (8.67) to represent the interactions in bosonic language, apply RG reasoning to investigate their relevance, and interpret the nature of the fixed-point theory. In view of the main theme of this chapter, a detailed discussion of this program would lead us too far astray (see, however, problem 8.8.6). However, it is instructive to discuss some essential points on the example $n_c = 1$, the spin-1/2 chain. Our above modeling is designed to separate the spin from the charge dynamics, and hence it makes sense to split the U(2) matrices representing the $N = 2$ fermion systems as $g = e^{i\phi}g'$, into a phase and an SU(2) matrix g'. It is then straightforward to check that the Umklapp operator assumes the form $\cos(2\phi)$, i.e., as one would expect on the basis of its symmetries, it does not couple to the spin degrees of freedom, g'. However, it does "gap out" the chiral charge excitations: substituting the above decomposition into the perturbed WZ action, we obtain $S[\phi, g'] = S[\phi] + S[g']$, where

$$S[\phi] = \frac{1}{2\pi}\int d^2x\left[(\partial\phi)^2 + \lambda_{\mathrm{uk}}\cos(2\phi)\right] + \cdots, \tag{8.71}$$

$$S[g'] = S_0[g'] + \Gamma[g'] + \lambda_s\int d^2x\,\mathrm{tr}(\partial g'\partial g'^{-1}) + \cdots, \tag{8.72}$$

and the ellipses denote terms of lesser relevance. In this expression, charge and spin are fully separated. The Umklapp term coupling to the charge sector is strongly relevant, and gaps out ϕ-fluctuations at large distance scales. This is the bosonization way of describing the **Mott–Hubbard gap** at half-filling discussed in section 2.2.3. Turning to the spin sector, we see that the current interaction effectively modifies the coupling constant of the gradient term S_0, Eq. (8.61). However, as discussed above, the RG flow will push this deformation back to the critical value $\lambda = 1/8\pi$ at large distance scales. In other words, the theory flows towards a fixed point equivalent to a free fermion theory – the spin-1/2 chain remains gapless, including in the deep infrared. Also notice that the bosonized theory represents the critical spin-1/2 chain in terms of a field theory on SU(2), and not just on the Goldstone mode coset space $S^2 = \mathrm{SU}(2)/\mathrm{U}(1)$. In the strong-coupling regime, the mean field symmetry breaking is undone. In this way, the theory manages to reconcile the two seemingly conflicting principles "absence of a gap" and "absence of symmetry breaking."

What happens for **larger spins**? Referring to problem 8.8.6 for details, AH found that excitations in flavor space are generally gapped out. This means that, as in the spin-1/2 case, the theory is described by a perturbed WZ action for SU(2)-valued fields. However, the presence of the color sector shows itself in an integer coupling $k = n_c$ featuring as the level $k\Gamma[g']$ at the front of the WZ term. Affleck and Haldane also demonstrated that, in the case $n_c > 1$, massive perturbations $\text{tr}(g^2 + g^{-2})$ become physically allowed. A chain of beautiful arguments then leads to the conclusion that, for odd k, the chains end up in the universality class of the spin-1/2 chain, and for even k become gapped. This is the most rigorous verification of the Haldane conjecture known to date. From the perspective of topological field theory, the take-home message is that:

> Under renormalization, theories containing a θ-term may flow towards theories with a target manifold of one dimension higher and a corresponding WZ term.

8.6 Chern–Simons Terms

REMARK Chern–Simons theory is the effective theory behind many forms of strongly entangled quantum matter. This makes it a very rich theory with connections to quantum information, particle physics, and pure mathematics. Some of the mathematical and gauge-theoretical foundations of Chern–Simons theory are introduced in section 10.5, and it may be a good idea to read this section before, or in parallel, this section. Here, we will illustrate application of the theory for the important example of the fractional quantum Hall effect. However, there are numerous other applications, notably in the theory of nonabelian quantum matter and topological quantum computation. For further literature on these subjects, we refer to the reviews of Ref.[64] and Ref.[33].

In chapter 10, we will show that the topology of gauge theories in *even* dimensions is described by quantized topological actions of the field strength, which in the reading of this chapter play the role of θ-terms. These actions are closely related to the Chern–Simons (CS) actions, Eq. (10.27), in one dimension lower, which by the same rationale resemble WZ terms. In this section, we will discuss the fascinatingly rich Chern–Simons action in $2 + 1$ dimensions and its applications in the theory of two-dimensional topological matter. However, before going there, we first need to discuss why the physics of topological matter in two-dimensional real space differs from that in general dimensions.

8.6.1 Topological phases in two dimensions

REMARK Readers familiar with the concept of braiding may skip the first section. However, the second may contain new material.

Depending on whether they are fermionic or bosonic, N-body quantum wave functions stay invariant or change sign under the exchange of particle coordinates, $\Psi(x_1, x_2, \ldots, x_N) = \pm\Psi(x_2, x_1, \ldots, x_N)$. This exchange is usually introduced ax-iomatically without reference to concrete realizations of particle exchange. As late

---

[64] G. V. Dunne, in *Topological Aspects of Low-Dimensional Systems*, eds. A. Comtet *et al.* (Springer, 1999).

as 1977, Leinaas and Myrheim[65] investigated whether particle permutation can be made physical via concrete exchange protocols, and whether physical consequences beyond sign factors might ensue.

Fractional and nonabelian statistics

For a constructive realization of an exchange operation, consider an N-body ground state wave function with particles kept at positions $x_1, \ldots, x_N$ by engineered potential traps. Assuming an excitation gap, we now adiabatically rotate the position of x_2 around that of x_1 by an angle π (see the figure). In a second step, we translate as indicated, to arrive at the starting configuration, with the roles of x_2 and x_1 interchanged. Neglecting correlations with the other particles, the second operation will leave the wave function unchanged. However, the first comes with a phase, $e^{i\theta}$ associated with the π-rotation of particles. A two-fold execution of this protocol is equivalent to the rotation of x_2 around x_1 by an angle 2π and weighted with phase $e^{2i\theta}$. However, *in dimensions different from* 2, a rotation around a full circle is topologically equivalent to a null operation (the closed rotation loop can be adiabatically contracted to a point), which requires $e^{2i\theta} = 1$, i.e., $\theta = 0$ or π. The two allowed options define fermionic or bosonic exchange statistics, respectively.

However, in $d = 2$, a full rotation of a particle around another is physical, and 2θ is not constrained to take the value 2π, which leaves room for particles with fractional **anyons** transmutation statistics such as **semions**, $\exp(i\theta) = i$, or **anyons**, $\exp(i\theta) \neq \pm 1$. In cases where the ground state is degenerate, $\Psi \to \Psi_a$, with quantum numbers $a = 1, \ldots, n$, an adiabatic rotation of quasiparticles may even change the state by the nonabelian generalization of a phase, a unitary operation, $\Psi_a \to U_{ab}\Psi_b$. **nonabelian** This happens for **nonabelian anyons**, the Majorana fermions to be introduced in **anyons** section 9.1.2 being the most basic representatives.

Where the generic exchange statistics of wave functions is associated with the trivial or antisymmetric representation of the permutation group, particle transmutations in two dimensions are representations **braid** of the **braid group**. The idea of braiding becomes ap- **group** parent when we add time, τ, to two-dimensional space to keep track of the world lines of the adiabatically exchanged particles (see the figure). An exchange of two or more anyons is described by the abstract concept of a braid. Individual braids can be concatenated to form new braids, and neutralized by inverse braids. Refer- ring to Ref.[33] for an in-depth introduction of the ensu- ing group structure, here we mention a slightly differ- ent interpretation of this concept. Consider imaginary

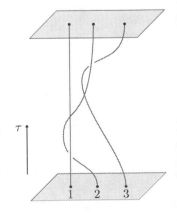

65 J. M. Leinaas and J. Myrheim, *On the theory of identical particles*, Il Nuovo Cimento B **37**, 1 (1977).

time with periodic boundary conditions, such that individual world lines become closed loops. A braiding operation now corresponds to a specific way of knotting these lines, and the bookkeeping of phases, or more generally unitary group operations, amounts to keeping track of the knot invariants associated with these structures. In section 8.8.7, we will discuss how Chern–Simons field theory describes such knots within the framework of field theory.

8.6.2 Nonabelian statistics and ground state degeneracy

In section 8.1.2, we pointed out that the concept of anyonic statistics is related to that of degenerate ground states. To understand this link, consider a spatial base manifold with periodic boundary conditions in both directions, a torus, and on it a ground state with (abelian, for simplicity) anyonic excitations with statistical angle $2\theta = 2\pi/k$. Define T_1 to be an operator that creates an anyonic quasiparticle–quasihole

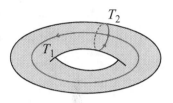

pair out of the ground state, and adiabatically drags the particle around one of the circles defining the torus to then recombine with the hole (see the figure). Let T_2 perform the same operation around the complementary circle. We now claim that these operators obey the relation

$$T_2^{-1} T_1^{-1} T_2 T_1 = e^{2\pi i/k}. \tag{8.73}$$

To understand why this is so, consider the world lines corresponding to the adiabatic motions. A good way to visualize them is to imagine the two-dimensional surface of the torus thickened into a slab whose transverse coordinate is imaginary time. On this space–time torus, T^3, the curve $T_1^{-1} T_1$ is contractible to a point and hence topologically equivalent to a closed loop. The same is true for the $T_2^{-1} T_2$ loop. However, the nesting of the operators T_1 and T_2 implies that the respective loops are linked. In other words, the action of the operators is equivalent to an elementary braid operation and hence multiplies the ground state with the corresponding phase, the right-hand side of Eq. (8.73). None of the operators leaves the ground *space* of the system.

   We next identify a space of minimal complexity realizing the commutation relation (8.73). To this end, assume the operator T_1 to be diagonalized and $e^{i\lambda}$, $0 \leq \lambda < 2\pi$, to be one of its eigenvalues. Application of the kth power, T_1^k, generates k-fold traversed loops with trivial statistical angle, $(e^{2\pi i/k})^k = 1$. This requires $\lambda k = 0$ mod 2π, which is satisfied by $\lambda = 2\pi m/k$, $l = 0, 1, \ldots, k-1$. We hence postulate that the operator T_1 assumes the diagonal form $T_1 = \mathrm{diag}(1, e^{2\pi i/k}, \ldots, e^{2\pi i(k-1)/k})$, which requires a ground space of dimension k. Equation (8.73) shows that, in the basis of states $|l\rangle$, the operator T_2 acts as a lowering operator, $T_2|l\rangle = |(l-1) \bmod k\rangle$. (In a basis where T_2 is diagonal, T_1 would assume the role of a ladder operator.)

   The bottom line is that:

> The realization of anyonic statistics with phase $2\pi/k$ on a system with periodic boundary conditions requires a k-fold degenerate ground state.

The situation gets more complicated when quasiparticles of different statistics enter the stage. For example, we know that the bonding of two electrons defines a quasiparticle of bosonic statistics (think of Cooper pairs as an example). In the same way, the **fusion** of anyons defines effective anyons whose braiding properties are induced by the compound particles. For an in-depth discussion of the corresponding **fusion rules**, the emergent types of abelian and nonabelian statistics, and the associated physical phenomena, we refer once again to the review in Ref.[33]. We also refer to section 10.4.5 for a microscopic realization of the four-fold degenerate ground state of a system with fermions, bosons, and semions in its quasiparticle spectrum.

fusion

8.6.3 Topology from Chern–Simons theory

In previous sections, the interplay of "non-topological" elements (gradient terms, etc.) and topological terms in field theories led to new physics at large distance scales. We here consider Chern–Simons terms in the more restrictive setting of **topological field theory**. In such a theory the Lagrangian is purely topological. As we will see, Chern–Simons topological field theory is the theory describing gapped topological phases in two dimensions, including their ground state degeneracies and anyonic statistics.

topological field theory

In differential form language, the **Chern–Simons action** in $2 + 1$ dimensions reads

Chern–Simons action

$$S_{\mathrm{CS}}[A] = \frac{k}{4\pi} \int_M \mathrm{tr}\left(A \wedge dA + \frac{2}{3} A \wedge A \wedge A \right) \qquad (8.74)$$

and in coordinate language

$$S_{\mathrm{CS}}[A] = \frac{k}{4\pi} \int d^3x \, \epsilon^{\mu\nu\rho} \mathrm{tr}\left(A_\mu \partial_\nu A_\rho + \frac{2}{3} A_\mu A_\nu A_\rho \right). \qquad (8.75)$$

Here $A = A_\mu dx^\mu$, where the coefficient functions $A_\mu = \{(A_\mu(x))^a{}_b\}$ are matrices taking values in the Lie algebra of a gauge group. Following standard conventions, here we work in a representation with real-valued connection forms, and real CS action. The action enters functional integrals as $\exp(iS_{\mathrm{CS}})$, i.e., via a "topological phase," as in our previous discussions of topological terms.

Nonabelian CS theory

INFO **Nonabelian Chern–Simons theory**, where A takes values in a Lie algebra of dimension higher than one, describes the properties of nonabelian topological phases of matter. In the nonabelian setting, the **level** k is integer-quantized.[66] For example,

[66] The quantization condition follows from the peculiar behavior of CS theory under nonabelian gauge transformations, addressed in problem 10.7.7. There, we show that, on manifolds M without boundary, the action changes as $S_{\mathrm{CS}} \to S_{\mathrm{CS}} + 2\pi k W$, where W is the winding number of the map $M \to G, x \mapsto g(x)$. The invariance of $\exp(iS_{\mathrm{CS}})$ under "large" gauge transformations with $W \neq 0$ requires the quantization of k. In abelian CS theory, there is no such condition (except in a theory with magnetic monopoles, where a similar condition may be derived).

$A_\mu \in \mathrm{su}(2)$ describes **Ising anyons** at level $k = 2$ and **Fibonacci anyons** at level $k = 3$, respectively. The braiding statistics of the "Fibonaccis" is rich enough to define a universal set of logical gates for quantum computation, which has made them a subject of intensive research. For an introduction into the physics of these nonabelian phases of matter and their representation in terms of CS theory, we once more refer to Ref.[33].

abelian
CS theory

Abelian CS theory is defined by the action

$$S_{\mathrm{CS},k}[A] = \frac{k}{4\pi} \int_M A \wedge dA = \frac{k}{4\pi} \int d^3 x \, \epsilon^{\mu\nu\rho} A_\mu \partial_\nu A_\rho \qquad (8.76)$$

where $A = (A_0, A_1, A_2)$ is the vector potential in $2+1$ dimensions and the coupling k may take arbitrary values. We will consider this theory in two contexts. In the first, A is an externally imposed fixed response field (which will become meaningful once matter currents are coupled to the theory). In the second, $A \equiv a$ is an intrinsic integration variable (as in quantum electrodynamics), distinguished from the external field A by the usage of a lower case letter. In cases where the level index, k, is obvious, we write $S_{\mathrm{CS},k} = S_{\mathrm{CS}}$ for simplicity.

EXERCISE To see how **magnetic monopoles may enforce level quantization in the abelian theory**, consider the case $M = S^1 \times S^2$, where S^1 is time with periodic boundary conditions and S^2 is a spatial sphere. Now assume that a represents a static magnetic monopole in the sense of section 10.5.2, such that integration of the field strength $\partial_1 a_2 - \partial_2 a_1$ over S^2 yields a factor 2π. Consider a gauge transformation $a_0 \to a_0 + \partial_t \phi$, where the U(1) phase $\phi = \phi(t)$ has a unit winding in the t-direction (but does not depend on spatial coordinates). Show that invariance under this transformation requires the integer quantization of k.

8.6.4 Universal aspects of abelian Chern–Simons theory

REMARK Throughout this section, we assume our CS theory to be defined on a space–time torus $T^3 = S^1 \times T^2$ comprising periodic time, S^1, and a spatial torus, T^2. The discussion skips over some technical details whose completion is left as an exercise.

In this section, we assume that A is an emergent degree of freedom and discuss how its integration over the action (8.76) describes three key features of gapped abelian topological matter: stability of the ground state, its degeneracy, and the fractional statistics of quasiparticle excitations.

Absence of dynamics and ground state degeneracy

The Chern–Simons action for the degrees of freedom contained in A describes a dynamical system and we can ask what is the underlying Hamiltonian. (The same question asked for the standard electromagnetic action in $3 + 1$ dimensions yields the electromagnetic field energy as an answer; see problem 1.8.2.) We eliminate gauge redundancy by fixing $a_0 = 0$, in which case the action simplifies to

$$S_{\text{CS}}[a] = \frac{k}{2\pi} \int d^3x \, a_2 \partial_0 a_1,$$

where we have integrated by parts. Interpreting this expression as a canonical action $S = \int dt (p\partial_t q - H)$, we conclude that:

> The Hamilton operator of the theory described by the
> Chern-Simons path integral vanishes.

This observation is in line with the expectation that there is no dynamics in this theory; excitations above the ground state would be represented in terms of additional contributions to the action.

A second observation is that (a_1, a_2) form a canonical pair and that the quantum operators corresponding to these variables obey the commutation relation

$$[a_1(\mathbf{x}), a_2(\mathbf{x})] = \frac{2\pi i}{k} \delta(\mathbf{x} - \mathbf{x}'), \tag{8.77}$$

where $\mathbf{x} = (x_1, x_2)^T$ is a two-dimensional coordinate vector. However, these operators are not gauge invariant: even with fixed $a_0 = 0$, we may make the change $\mathbf{a}(\mathbf{x}) \to \mathbf{a}(\mathbf{x}) + \nabla\phi(\mathbf{x})$. This disqualifies them as representatives of physical observables. Instead, we consider the gauge invariant **Wilson loop operators** (cf. section 10.1.4)

Wilson loop operators

$$T_i = \exp\left(i \int_{\gamma_i} \mathbf{a} \cdot d\mathbf{x}\right),$$

where the curves γ_i are chosen to run around the space torus in the direction of the coordinate x_i. Using the fact that the two curves intersect in one point, we have

$$\left[\int_{\gamma_i} \mathbf{a} \cdot d\mathbf{x}, \int_{\gamma_j} \mathbf{a} \cdot d\mathbf{x}\right] = \oint dx_i \oint dx_j [a_i, a_j] = \epsilon_{ij} \frac{2\pi i}{k},$$

and application of the Baker–Campbell–Hausdorff formula shows that the Wilson loop operators satisfy the commutation relations (8.73). The way to read this result is that the gauge degrees of freedom (a_1, a_2) generate Wilson loop operators whose commutation relations require a k-dimensional Hilbert space. The absence of dynamics, $H = 0$, identifies this space as the ground space of the system.

Fractional statistics

In section 8.6.1, we reasoned that the quasiparticles forming on top of the above ground states obey fractional statistics, with exchange phase $2\pi/k$. In order to describe this feature within the framework of CS theory, we need to couple to the action the quasiparticle worldlines describing the braiding of a minimum of two quasiparticles in $2 + 1$ dimensions. We do so by adding a matter gauge potential coupling as

$$S_{\text{CS}}[a] \longrightarrow S_{\text{CS}}[a] + \int d^3x \, j^\mu a_\mu, \tag{8.78}$$

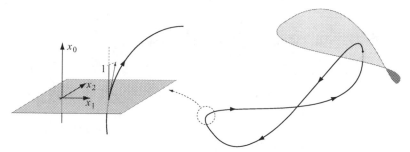

where

$$j^\mu(x) = \sum_{i=1}^{2} \partial_t x^\mu \delta(\mathbf{x} - \mathbf{x}_i(t)), \qquad x = (t, \mathbf{x}),$$

is the current density corresponding to two particles $i = 1, 2$ moving along trajectories $\mathbf{x}_i(t)$. We tentatively (for a microscopic justification, see below) interpret S_{CS} as the action of a path integral describing the adiabatic braid of two particles in the space–time continuum. As will be discussed in section 8.6.1, the braid is defined in terms of two-particle world lines $(t, \mathbf{x}(t))$, with identical coordinates $\mathbf{x}$ at the beginning and end of the protocol. We describe this feature by imposing periodic boundary conditions in time, thus converting the two worldlines to closed loops, γ_i in the space–time continuum (see fig. 8.7).

In problem 8.8.7, we show how integration over a weights each configuration of such loops with a phase $\exp(\frac{2\pi i}{k} I(\gamma_1, \gamma_2))$, where $I(\gamma_1, \gamma_2)$ is the number of times the curve γ_1 pierces the area defined by γ_2. This equals the number of exchange operations implied by the braid and hence sets the statistical phase. Referring to the problem for details, the CS functional does this counting by interpreting the loops as unit current-carrying "wires" in (3+0)-dimensional Euclidean space, time playing the role of a third coordinate. The loop γ_1 then creates a magnetic field whose curl is centered on its infinitesimal wire cross-section. After integration over a, the action assumes the form of an integral of that curl over the area spanned by the other loop, γ_2. In this way, the number of piercings gets counted.

Note how the above construction underpins the topological nature of the CS action: the set of current loops is reduced to a single piece of topological data, its degree of knotting. The computation of "knot invariants" extends to the nonabelian setting, where it plays an essential role in the description of fractional statistics in nonabelian Chern–Simons theory. For further discussion of this point, we refer to Ref.[33], or the original Ref.[67].

[67] E. Witten, *Quantum field theory and the Jones polynomial*, Commun. Math. Phys. **121**, 351 (1989).

Gauge (non-)invariance

The issue of gauge invariance in connection with CS actions is a delicate one. In problem 10.7.7, we will show that CS theory on a boundaryless space, M, is invariant under "small gauge" transformations[68] $A' = gAg^{-1} + gdg^{-1}$. However, in the physical case of **base spaces with boundaries**, both the abelian and non-abelian CS theory lack gauge invariance under even small transformations. Let us investigate this point in the abelian case. Application of Stokes' theorem (exercise) shows that, under $a \to a + df$,

$$S_{\text{CS}}[a] \longrightarrow S_{\text{CS}}[a] + \frac{k}{4\pi} \int_{\partial M} f da. \tag{8.79}$$

For example, for a base manifold $M = \mathbb{R} \times D$, where $\mathbb{R}$ is time, and D a spatial disk, $\partial M = \mathbb{R} \times S^1$ is a cylinder comprising time and a spatial boundary with circular geometry. Parameterizing that circle with a boundary coordinate x, the excess term reads $\frac{k}{4\pi} \int dt dx\, f(t, x)(\partial_t a_x(x, t) - \partial_x a_t(x, t))$. The way to read this gauge-deficit is that:

> Pure CS theory on a manifold with a boundary is not complete.

It needs to be augmented by an, as yet unspecified, theory at the boundary which is likewise gauge non-invariant, in such a way that the two f-dependent contributions cancel. The argument indicates that the boundaries of systems described by bulk CS theories are generally "alive."

Some of the dynamical phenomena generated by this principle are addressed in problem 8.8.8. In section 9.2.2, we will discuss specifically how the gauge non-invariance of CS theory is linked to the presence of **chiral anomalies** in the corresponding boundary theories. This connection is so strong that the bulk and the boundary theory largely condition each other. The combined package then is gauge invariant and anomaly free, as required by a sensible low-energy theory of a condensed matter system.

parity non-invariance

Finally, note that the CS action with its first-order derivatives also lacks **parity invariance**. Reflections of space or time, $x_\mu \to -x_\mu$, change the sign of both ∂_μ and a_μ, and hence that of the CS action. This parity breaking is again connected to a specific "anomaly," the **parity anomaly**, to be discussed in section 9.2.2.

8.6.5 Fractional quantum Hall effect (FQHE)

In this section, we discuss the fractional quantum Hall insulator as an example of a long-range entangled topological phase of matter with abelian CS theory as its effective theory. The derivation of the theory from first principles will show how anyonic quasiparticles form on the microscopic level, and how they are represented by CS gauge field fluctuations on a spatially coarse-grained mesoscopic level. We will also

---

[68] Recall that the extension of gauge invariance to large transformations requires the quantization of the level, k.

discuss how physical observables such as transport coefficients are extracted from the CS description. However, as always in this chapter, the focus is on topological principles. Many facets of the beautiful phenomenology of the FQHE will not be touched upon. For a more comprehensive coverage, we refer, e.g., to Ref.[69].

Phenomenology

In section 8.4.7, we understood the Hall conductance of the two-dimensional electron gas as a topological observable. Interpretations in terms of the number of edge channels or Chern number proved its robust integer quantization. These principles were understood soon after the discovery of the IQHE, and it therefore came as a major surprise when only slightly later Tsui, Stormer, and Gossard[70] discovered a quantized Hall plateau at $\sigma_{xy} = \frac{\nu}{2\pi}$, $\nu = 1/3$, in high-mobility samples. This spectacular observation (which earned a Nobel Prize in 1988 for Tsui

Robert B. Laughlin 1950– (left)
Horst L. Störmer 1949– (middle)
Dan C. Tsui 1939– (right)
The 1988 Nobel Prize was awarded to the experimentalists Horst Störmer and Dan Tsui for their discovery of the FQHE and to Robert Laughlin for groundbreaking contributions to its theoretical explanation.

and Stormer jointly with Laughlin) triggered an experimental effort which led to the discovery of whole families of plateaus at rational fractions of the conductance quantum (see fig. 8.8).

Salient experimental signatures of the FQHE include the following:

▷ Only relatively simple rationals n/m are observed as plateau values. The most prominent of these lie in the "principal sequence" $1/m$, where m is odd. More generally, plateaus have been observed for filling fractions

$$\nu = \frac{n}{m} = \frac{p}{2sp + 1},\tag{8.80}$$

with integer s and p.[71]

▷ At $\nu = 1/2$ (formally, the limit $p \to \infty$ for $s = 1$) the system behaves as if no external field was present. Around this filling fraction, Shubnikov–de Haas oscillations otherwise seen in weak-field Fermi liquids are observed.

---

[69] E. Fradkin, *Field Theories of Condensed Matter Systems*, 2nd edition (Cambridge University Press, 2013)

[70] D. C. Tsui, H. L. Stormer, and A. C. Gossard, *Two-dimensional magnetotransport in the extreme quantum limit*, Phys. Rev. Lett. **48**, 1559 (1982).

[71] However, this two-parameter family does not capture all experimentally observed values. A prominent exception is the plateau at $\nu = 5/2$. It is believed that this plateau reflects an exotic pairing state of the quasiparticle degrees of freedom introduced below and that it harbors nonabelian anyons as excitations. However, the discussion of the 5/2 state is beyond the scope of this text.

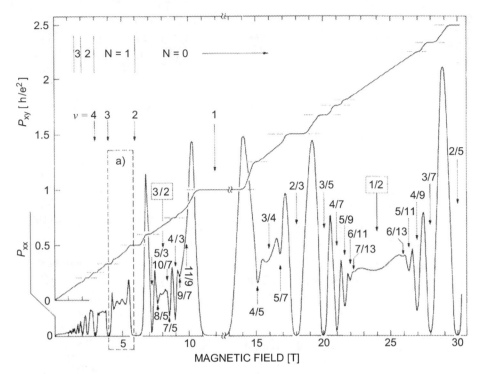

Fig. 8.8   Longitudinal and transverse resistance of a high-mobility electron gas, where the numbers indicate the filling fraction of the Landau levels. Image taken from R. Willet *et al.*, *Observation of an even-denominator quantum number in the fractional quantum Hall effect*, Phys. Rev. Lett. **59**, 1776 (1987). Copyright (1987) by the American Physical Society.

A first hint as to possible explanations of this phenomenology follows from the fact that the FQHE, unlike the IQHE, is averse to the presence of disorder. In the absence of disorder, the free electron system subject to a magnetic field is quantized in Landau levels, whose most striking feature is their macroscopic degeneracy. Perturbations such as interactions or disorder acting on the background of a massively degenerate state are expected to cause major instabilities (formally, due to the appearance of zero energy denominators in perturbation theory). We anticipate that the FQHE reflects a macroscopic reordering of the system into a novel state whose identity we aim to understand.

The resemblance of the FQHE to an ordinary QHE, albeit at different plateau values, and the observation of the Shubnikov–de Haas oscillations (which require a Fermi surface) indicates that the relevant players are fermionic quasiparticles. Early work by Jain[72] indeed suggested an ingeniously simple fermion mean field scenario wherein much of the above phenomenology affords a simple explanation. Referring to the original reference for a more substantial discussion, think of the magnetic field

[72] J. K. Jain, *Composite-fermion approach for the fractional quantum Hall effect*, Phys. Rev. Lett. **63**, 199 (1989).

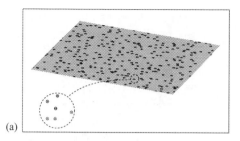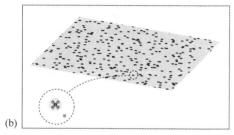

(a)   (b)

Fig. 8.9   Illustrating the idea of the composite fermion approach. Imagine that an even number of flux quanta constituting the magnetic field get tied to the electrons in the system. The new composite particles ("electron + (2n) flux quanta") continue to be fermions. They only see the remaining "free" flux quanta, i.e., a reduced external field (indicated by the lighter shading in part (b)).

as a large number N_ϕ of unit flux-quantum tubes piercing the system. Homogeneity of the field implies that the distribution of these flux tubes is macroscopically uniform. For a system with filling fraction $1 > \nu = N/N_\Phi$, the number of flux quanta exceeds that of electrons, N.[73] Now assume that, by some mechanism, each electron pairs with an even number of flux quanta to form a composite particle. For example, for $\nu = 1/3$ the composite particles would comprise one electron and two flux quanta each, leaving one third of the flux quanta unpaired. What can be said about the properties of these composites?

composite   ▷ First, the composite particles obey fermionic statistics and are hence called **com-**
fermions   **posite fermions** (CFs). To understand why, notice that the adiabatic trans-
              port of an electron around a flux-quantum tube leads to a phase 2π (cf. sec-
              tion 8.2). An exchange operation, meaning transport around a half-circle as in
              section 8.6.1, thus gives a phase π, and an even number, $2s$, of flux quanta comes
              with an unobservable phase $2s \times \pi$. Where statistics is concerned, the flux tubes
              are unobservable, and we are left with the statistics of the bare electrons.

           ▷ The CFs see an effectively reduced external field. For example, for $\nu = 1/3$,
              each electron binds to two flux quanta. Thus, the residual field seen by the
              CFs is three times lower than the original field. In other words, the number of
              remaining flux quanta, $N_\phi - 2N = N$, is equal to the number of CFs. This
              suggests a tentative identification of the $\nu = 1/3$ FQHE as an integer QHE of
              the composite fermions. For a half-filled ($\nu = 1/2$) band, we have $N_\phi - 2N = 0$,
              i.e., the CFs experience a mean field of vanishing strength. This nicely conforms
              with the experimental observation of weak field behavior close to half-filling.

Starting from a microscopic field integral representation, we will show in the next section how a specific gauge transformation introduces the CFs as effective degrees of freedom. Building on this representation we will then demonstrate that

[73] If $\nu > 1$, then $\nu - [\nu]$, where $[\nu]$ is the largest integer smaller than ν, sets the filling fraction of the highest occupied Landau level and the discussion applies to that level.

the coarse-grained representation of the problem at mesoscopic length scales is an abelian CS theory. Finally, we will discuss how this theory elegantly describes the essential physics of the FQHE.

INFO An early triumph in the understanding of the FQHE was the proposal of the **Laughlin wave function**[74]

$$\Psi(z_1,\ldots,z_N) = \mathcal{N} \prod_{1\leq i<j\leq N} (z_i - z_j)^m \prod_{k=1}^{N} \exp(-|z_k|^2), \qquad (8.81)$$

as a trial ground state wave function of the lowest Landau level at filling $\nu = 1/m$. Here, $z = (x+iy)/2l_0$ are complex dimensionless coordinates, where $l_0 = 1/\sqrt{B}$ is the magnetic length. Indeed, the harmonic oscillator-like second factor in Eq. (8.81) tells us that we are in the lowest Landau level (see the discussion in section 8.4.7). However, the first factor incorporates three essential aspects, which make the physics of the FQHE different from that of the IQHE:

▷ *Unconventional quasiparticle statistics:* Changing any coordinate z_i, we pick up a phase factor that depends on the position of all other coordinates, and the filling fraction.

▷ *Strong ground state entanglement:* The wave function is very different from a product wave function.

▷ *Particle interactions:* Whenever particle coordinates get close to each other, the wave function vanishes. In fact, it could be demonstrated that for certain types of repulsive point interactions, Eq. (8.81) gives the exact ground state.

In the rest of the section, we will demonstrate how the physics characterized by the three points above is described from first principles.

8.6.6 Chern–Simons field theory: construction

Consider the standard Hamiltonian of two-dimensional electrons subject to a perpendicular magnetic field, $\hat{H} = \hat{H}_0 + \hat{H}_{\text{int}}$, where

$$\hat{H}_0 = \int d^2x \, a^\dagger(\mathbf{x}) \frac{1}{2m} (-i\partial_{\mathbf{x}} + \mathbf{A}_0)^2 a(\mathbf{x}), \qquad (8.82)$$

$\mathbf{A}_0 = \frac{B_0}{2}(y,-x)^T$ with B_0 the external field strength, and $\hat{H}_{\text{int}}$ is the electron-electron interaction.

Singular gauge transformation

Above, we reasoned that composite fermions are natural quasiparticles in the description of the FQHE. The distinguishing feature of a CF is that charged particles moving around pick up an additional phase owing to the presence of $2s$ attached

[74] R. B. Laughlin, *Anomalous quantum Hall effect: An incompressible quantum fluid with fractionally charged excitations*, Phys. Rev. Lett. **50**, 1395 (1983).

singular
gauge
transfor-
mation
flux tubes. We now introduce this additional phase by hand via a so-called **singular gauge transformation** affecting the many-body wave function Ψ as follows:

$$\Psi(\mathbf{x}_1, \ldots, \mathbf{x}_N) \to \Psi(\mathbf{x}_1, \ldots, \mathbf{x}_N) \exp\left(-2is \sum_{i<j} \arg(\mathbf{x}_i - \mathbf{x}_j)\right), \qquad (8.83)$$

where $\arg(\mathbf{x}) = \tan^{-1}(x_2/x_1)$ is the angle enclosed between $\mathbf{x}$ and the positive real axis. This is not a true gauge transformation because the expression in the exponent is not a single-valued function of coordinates. In fact, it becomes singular whenever two coordinates $\mathbf{x}_i \to \mathbf{x}_j$ approach each other. Accordingly, the vector potential corresponding to the phase factor,

$$\mathbf{a} = -2s\partial_\mathbf{x} \sum_i \arg(\mathbf{x} - \mathbf{x}_i) = -2s \sum_i \frac{(x_1 - x_{i,1})\mathbf{e}_2 - (x_2 - x_{i,2})\mathbf{e}_1}{|\mathbf{x} - \mathbf{x}_i|^2},$$

is not a harmless gradient field, but creates the magnetic field

$$b = \epsilon^{ij}\partial_{\mathbf{x}_i} a_j = -4\pi s \sum_i \delta(\mathbf{x} - \mathbf{x}_i) \qquad (8.84)$$

of $2sN$ flux tubes centered at the coordinates of the fermions. The transformation (8.83) therefore is physical and replaces the system of N fermions by an equal number of CFs. Our rationale in the following will be the replacement N fermions $+ B_0 \to N$ CFs $+ B_0 - b$. Since the fermion density is uniform on mesoscopic scales, both representations describe the same number of fermions in an external field of net strength B_0, and hence are physically equivalent.

Derivation of the Chern–Simons action

Within the framework of second quantization, the transformation (8.83) amounts to the replacement

$$a^\dagger(\mathbf{x}) \to a^\dagger(\mathbf{x}) \exp\left[-2is \int d^2y \arg(\mathbf{x} - \mathbf{y})\hat{\rho}(\mathbf{y})\right],$$

where $\hat{\rho} = a^\dagger a$ is the density operator.

EXERCISE Expand a many-body wave function $|\Psi\rangle$ in the position basis $|\mathbf{x}_1, \ldots, \mathbf{x}_N\rangle = \prod_i a(\mathbf{x}_i)|0\rangle$ and verify that the above replacement induces the transformation (8.83).

The substitution of this transformation into the Hamiltonian (8.82) amounts to a replacement $\mathbf{A}_0 \to \mathbf{A} \equiv \mathbf{A}_0 + \hat{\mathbf{a}}$ with the gauge field operator

$$\hat{\mathbf{a}}(\mathbf{x}) = -2s \int d^2y \frac{(x_1 - y_1)\mathbf{e}_2 - (x_2 - y_2)\mathbf{e}_1}{|\mathbf{x} - \mathbf{y}|^2}\hat{\rho}(\mathbf{y}). \qquad (8.85)$$

Equivalently, this operator may be defined by the condition,

$$\hat{b} \equiv (\nabla \times \hat{\mathbf{a}})_3 = -2s\hat{\rho}, \qquad (8.86)$$

i.e., the flux of the field generated by the vector potential operator is proportional to the charge density.

We now turn to a *real-time* field integral representation $\mathcal{Z} = \int D\psi \, e^{iS[\psi]}$, with action $S = S_0 + S_{\text{int}}$ and

$$S_0[\psi] = \int d^3x \, \bar{\psi} \left(i\partial_t - \mu - \frac{1}{2m}(-i\partial_{\mathbf{x}} + \mathbf{A}[\psi])^2 \right) \psi,$$

where $x = (t, \mathbf{x})$ are the space–time variables, and the chemical potential, μ, determines the filling fraction. In this action, $\mathbf{A}[\psi] \equiv \hat{\mathbf{A}}_{a \to \psi}$ is nonlocal in ψ, reflecting the structure of Eq. (8.85). It is, thus, nonlinear and nonlocal in the integration variables, and not manageable in this form. However, we now apply an elegant trick to remove its nonlinearity by auxiliary integration variables: multiply the partition function by $1 = \int D\theta \prod_x \delta \left(\Delta\theta + 4\pi s\rho(x) \right)$, where a determinant generated by the presence of the Laplace operator under the δ-functional is absorbed in the measure. With this insertion, the functional integral assumes the form

$$\mathcal{Z} = \int D\psi \, D\theta \prod_x \delta \left(\Delta\theta + 4\pi s\rho(x) \right) \exp\left(-S[\psi] \right)$$

$$= \int D\psi \, D\theta \, D\phi \exp\left(iS[\psi] - i \int d^3x \, \phi \left(\frac{\Delta\theta}{4\pi s} + \rho \right) \right).$$

We now claim that the $\phi\Delta\theta$ combination is a CS action in disguise. To see how, define a three-component vector potential as[75]

$$a_\mu = (a_0, a_1, a_2)^T \equiv (-\phi, \partial_2\theta + \partial_1\psi, -\partial_1\theta + \partial_2\psi)^T, \qquad (8.87)$$

where θ and ψ define the transverse and longitudinal contribution to the spatial sector, $a_i = \epsilon_{ij}\theta + \partial_i\psi$. It is then straightforward to verify that

$$\frac{1}{4\pi s} \int d^3x \, \phi\Delta\theta = \frac{1}{8\pi s} \int d^3x \, \epsilon_{\mu\nu\sigma} a_\mu \partial_\nu a_\sigma.$$

EXERCISE Verify this relation. You will need to integrate by parts in space. For the time being, we assume periodic boundary conditions and will not worry about surface terms. We now understand that the $\phi\Delta\theta$ action is a CS action in the particular gauge where $\psi = 0$, or $\nabla \cdot \mathbf{a} = 0$. (This is a (2+1)-dimensional version of the Coulomb, or *transverse* gauge.)

Comparison with Eq. (8.86) and the δ-constraint shows that the spatial part of the vector potential obeys $(\nabla \times \mathbf{a}) = -2s\rho = -2s\bar{\psi}\psi$. We may thus make the substitution $\mathbf{A} \to \mathbf{A}_0 + \mathbf{a}$, and integrate over the Grassmann fields ψ, to arrive at the representation

$$\boxed{S[a] = -i \operatorname{tr} \ln \left(i\partial_t - \mu - \phi + \frac{1}{2m}(-i\partial_{\mathbf{x}} + \mathbf{A}_0 + \mathbf{a})^2 \right) + \frac{1}{8\pi s} \int d^3x \, \epsilon_{\mu\nu\sigma} a_\mu \partial_\nu a_\sigma}$$

$$(8.88)$$

---

[75] For completeness, we note that are working in a (2+1)-dimensional Lorentzian framework with (note the positioning of indices) $x^\mu = (x_0, \mathbf{x})$, $\partial_\mu = (\partial_0, \nabla)$, and $a^\mu = (+\phi, \mathbf{a})$.

where the factor $-i$ before the tr ln operator arises because we are working in a real-time ($\exp(iS)$) framework. This is the action of a gauge field a subject to a Chern–Simons action at level $k = 1/2s$ (non-integer!) and minimally coupled to the system of CFs via the first term. The field strength generated by $\mathbf{a}$ is set by Eq. (8.86) and the coupling to the effective potential $\mathbf{A} = \mathbf{A}_0 + \mathbf{a}$, indicating that the CFs are subject to a field of reduced strength. In the next section, we will discuss what can be learned from this representation about the physics of the FQHE.

8.6.7 FQHE from the Chern–Simons field integral

In view of the macroscopic occupation of such a system with (composite) fermions, we start our approach with a variational analysis, asking for stationarity of the action under variations $\delta_{a_\mu} S[\bar{a}] = 0$. Specifically, variation in $a_0 = -\phi$ yields the equation

$$\rho[\bar{a}] = -\frac{1}{4\pi s}\bar{b}, \tag{8.89}$$

where $\rho[a](x) = i(i\partial_0 + \mu - \phi + \frac{1}{2m}(-i\nabla + \mathbf{A})^2)^{-1}(x, x)$ denotes the local density of CFs. Assuming homogeneity of the density ρ at stationarity (see the info block below on this delicate point), we have $\rho = \nu B_0/2\pi$, or $\bar{b} = -2sB_0\nu$. The differentiation with respect to the space–like components $\mathbf{a}$ does not yield independent new information; all it gives us is two relations expressing the compatibility of Eq. (8.89) with the continuity equation.

EXERCISE Verify this statement. (Hint: Show that differentiation with respect to a_i generates a relation between the CF current and the electric field represented by a. Then use the Maxwell equations.)

interactions   **INFO** At this point, we can understand why **interactions are important in the physics of the FQHE**. At the beginning of this section, we formulated the expectation that the FQHE may be an IQHE of composite fermions. For that interpretation to work, we need the CFs to experience a homogeneous effective field, $B = B_0 + b$. However, in the absence of both interactions, and dispersion in the partially filled Landau band, we have to expect uncontrollable fluctuations of the background density and, by virtue of Eq. (8.89), fluctuations of the CS gauge field. The way out is to take interactions into account.

To understand their effect, consider the CFs subject to both an effective field $B = B_0 + \bar{b}$, and repulsive interactions. Let us assume that repulsive interactions render the density of CFs, and hence the induced field $\bar{b}$, homogeneous. We will have to expect Landau quantization of the CFs due to the effective field. Provided this effective quantum Hall system is at an integer filling fraction, $p \equiv 2\pi\rho_{CF}/B \in \mathbb{N}$, we have a system of fully occupied effective CF Landau levels, which will be inert against weak perturbations. Conversely, at generic filling fractions, interactions may induce all kinds of fluctuations in the degenerate effective Landau level system and may destabilize it. Note that, in this argument, interactions have the two-fold effect of homogenizing the background field strength, and favoring integer filling of the CF Landau levels.

At which "native" filling fractions $\nu = 2\pi\rho/B_0$ does this self-stabilization occur? Noting that each bare fermion defines a single composite fermion, we have $\rho = \rho_{CF}$. With the mean field result $-\bar{b} = 2\pi 2s\rho_{\mathrm{CF}}$, it is then straightforward to obtain the condition Eq. (8.80) defining the experimentally dominant filling fractions.

To conclude this brief excursion into the physics of interactions, let us investigate the effective **charge of the CF** degrees of freedom in mean field theory. To this end, consider a correlation function $\langle\psi(x)\bar{\psi}(y)\rangle$ probing the propagation amplitude of a CF in the medium. We expect this amplitude to be weighted by a factor $\sim \exp(iq_{\mathrm{CF}} \int_{\mathbf{x}}^{\mathbf{y}} d\mathbf{s} \cdot \mathbf{A}_0)$, *defining* the charge q_{CF} of the CF. Now noting that the action of the path integral resembles that of a free fermion action in the presence of an effective field $\mathbf{A}_0 + \mathbf{a}$, and that $\mathbf{a} = -2s\nu\mathbf{A}_0$ at mean field level, we conclude that the phase factor will assume the form given above, with prefactor

$$q_{\mathrm{eff}} = 1 - 2s\nu = \frac{1}{1 + 2sp}. \tag{8.90}$$

fractional
charge

We conclude that the composite fermions carry an effective **fractional charge**, lower than the unit charge of the bare fermions. We saw in section 8.4.7 that the insertion of flux into an IQHE geometry leads to an expulsion of charge in the radial direction. It is thus not surprising that the composite fermion with its attached flux tubes carries an effective screening charge. Equation (8.90) quantifies the bare electron charge minus the screening charge due to the insertion of $2s$ flux tubes.

In essence, the mean field analysis above shows that:

> The fractional QHE is an integer QHE of composite fermions.

In the following, we will include fluctuations and external response fields to see what further information can be obtained from the effective theory.

Linear response

Here, we will explore the transport physics of the FQHE on the basis of the action (8.88). Referring to the specialized literature for a more detailed discussion of the combined influence of interactions and gauge field fluctuations, our presentation will be brief and focus on the interplay of Chern–Simons gauge field fluctuations and matter currents.

The starting point is the coupling of the CF matter sector in Eq. (8.88) to an infinitesimal response field, $A \to A + A'$. Splitting the internal gauge field as $a \to \bar{a} + a$ into the mean field Eq. (8.89) plus fluctuations, the action assumes the form $S[a, A'] = S_{\mathrm{CF}}[\bar{a}+a+A'] + S_{\mathrm{CS}}[\bar{a}+a]$. Now consider the formal expansion $S[a, A'] \equiv \sum_n S^{(n)}[a, A']$, where $S^{(n)}[a, A']$ is of total order n in a and A'. The zeroth-order term $S^{(0)}$ describes the mean field CF system and was discussed in the previous section. As we are expanding around a stationary point, the first-order term $S^{(1)}$ does not contain a. We do not expect the system to support equilibrium currents, and hence no linear terms in A' either, i.e., $S^{(1)} = 0$. For the second-order action, we consider the general ansatz

$$S^{(2)}[a, A'] = \frac{1}{2} \int d^3x \, d^3y \, (a + A')^\mu(x) K_{\mu\nu}(x, y)(a + A')^\nu(y) + S_{\mathrm{CS}}[a]. \tag{8.91}$$

Here, $K_{\mu\nu}(x,y) = \frac{\delta^2 S_{\mathrm{CF}}[a]}{\delta a_\mu(x)\delta a_\nu(y)}\Big|_{a=\bar{a}}$ is the electromagnetic linear response kernel (see section 7.2) evaluated at the mean field level. On general grounds, this object is gauge invariant, $q^\mu K_{\mu\nu}(q) = K_{\mu\nu}q^\nu = 0$, where $K(q)$ is the Fourier transform in $x - y$. We also know that, in a bulk insulating material such as the FQHE system, $K_{\mu\nu}(q)$ is short-ranged, in the sense that it affords a Taylor expansion in q.[76] Since there are no gauge invariant zero-derivative actions quadratic in A, the expansion starts at first order. In a conventional (parity invariant) condensed matter system, terms of first order in derivatives are forbidden by symmetry, but not so here. The unique gauge invariant, but parity non-invariant, action that is first order in derivatives and second order in A is the CS action. We thus speculate that, to leading order in derivatives, $S^{(2)}[a, A'] = S'_{\mathrm{CS}}[a + A'] + S_{\mathrm{CS}}[a]$, where the prime indicates that the first Chern–Simons action is multiplied by an as yet undetermined coupling constant. Of course, it is possible to derive that action, including the coupling constant, by microscopic expansion of the tr ln term. However, we here take the cheaper route to obtain it by consistency reasoning.

With $S'[a] \equiv \frac{k'}{4\pi}\int d^3x\, \epsilon_{\mu\sigma\nu}a_\mu\partial_\sigma a_\nu = -\frac{ik'}{4\pi}\sum_q \epsilon_{\mu\sigma\nu}a_q^\mu q^\sigma a_{-q}^\nu$,[77] we have

$$K_{\mu\nu}(q) = -\frac{ik'}{2\pi}\epsilon_{\mu\sigma\nu}q^\sigma + \mathcal{O}(q^2). \qquad (8.92)$$

Physically, this kernel describes the linear response of the CFs subject to the effective background field $A_0 + \bar{a}$ which, as we reasoned earlier, corresponds to p filled CF Landau levels. This system has Hall conductivity $\sigma_{12}^0 = \frac{p}{2\pi}$.[78] On the other hand, we know from linear response theory that $\sigma_{12}^0 = -i\lim_{q\to 0}\omega^{-1}K_{12}(\omega, \mathbf{q})$, and comparison with Eq. (8.92) leads to the identification $k' = p$.

We have thus arrived at the result

$$S[A', a] = S_{\mathrm{CS},k}[A' + a] + S_{\mathrm{CS},k'}[a], \qquad k = p, \qquad k' = \frac{1}{2s}, \qquad (8.93)$$

for the effective low-energy action of the FQHE in the presence of an external response field, where we have made the level index explicit as in Eq. (8.76). The final step in the computation of the linear response is the integration over a.

EXERCISE **Integration over the CS action** can be carried out as in problem 8.8.7, i.e., gauge fixing followed by Gaussian integration over the non-redundant variables contained in a. However, there is a more economic procedure. We may use the fact that integration over a *Gaussian* action $S[\phi]$ amounts to finding the stationary configurations $\delta_\phi S[\bar{\phi}] = 0$, followed by substitution back into the action, $S[\phi] \to S[\bar{\phi}]$. To exponential accuracy (ignoring integration determinants), this procedure yields the result of the integral. Apply this procedure to the action Eq. (8.93). Verify that a solution of the stationary

---

[76] By contrast, in systems with a non-vanishing longitudinal conductivity, the kernel becomes non-local. For example, $K_{\mu\nu}(q) = \frac{q^\mu q^\nu - q^2}{Dq^2 + i\omega\delta_{\mu\nu}}$ in diffusive systems, where D is the diffusion constant.

[77] Be careful of the positions of the indices. We used $\epsilon_{\mu\sigma\nu}v_\mu v_\sigma v_\nu = \epsilon_{\mu\sigma\nu}\eta_{\mu\mu'}\eta_{\sigma\sigma'}\eta_{\nu\nu'}v^{\mu'}v^{\sigma'}v^{\nu'} = \epsilon_{\mu'\sigma'\nu'}v^{\mu'}v^{\sigma'}v^{\nu'}\det(\eta)$ where $\eta = \mathrm{diag}(1, -1, -1)$ is the Minkovski metric with $\det(\eta) = -1$.

[78] Of course, this is not the actual Hall conductivity of the system. The latter is obtained by inclusion of the a-field fluctuations, which will be our next step.

phase equations $\delta_a S[A', a]$ is given by $\bar{a} = -\frac{k}{k+k'} A'$. (In view of the gauge redundancy, this solution is non-unique. However, we ignore this point, trusting in the overall gauge invariance of the procedure.) Then substitute this result back into the action to obtain

$$S_{\mathrm{CS},k}[A' + a] + S_{\mathrm{CS},k'}[a] \xrightarrow{\int da} S_{\mathrm{CS},k''}[A'], \qquad k'' = \frac{1}{1/k + 1/k'}.$$

As a result of the integral obtained via the strategy outlined in the exercise, we obtain

$$S[A'] = S_{\mathrm{CS},k}[A'], \qquad k = \frac{p}{1 + 2ps}, \tag{8.94}$$

for the **linear response action of the FQHE insulator**. Above we have seen that the level of this action equals the Hall conductivity, $\sigma_{12} = k/2\pi \equiv \nu/2\pi$, so we have confirmed Eq. (8.80) from the effective theory.

The above derivation is a first-principles confirmation of the courageous phenomenological interpretation of the FQHE as an IQHE of exotic composite particles. However, there is further information to be obtained from the field theory. In section 8.6.4, we discussed the gauge non-invariance of the CS action in systems with a boundary, and the need to augment the CS description by a likewise gauge non-invariant boundary action. In the physics of the FQHE (and that of other applications of CS theory), this gauge deficit becomes an asset: in problem 8.8.8, we discuss how the boundary action partnering Eq. (8.94) describes the chiral transport of fractional charge at the FQHE boundary. These effective theories define the basis for the interpretation of FQHE experiments, which generically probe edge transport coefficients (see Refs. [79] for direct observations of the $\nu = 1/3$ quasiparticle charge by edge-current noise measurement). However, for the further discussion of the – still partially mysterious – physics of the FQHE edge and its description by effective field theories, we refer to the specialized literature.

8.7 Summary and Outlook

In the last two decades, driven by experimental breakthroughs some of which we have reviewed in this chapter, the field of topological condensed matter has shown explosive development. At this point, it has become almost impossible to keep track of all the various research directions, both theoretical and experimental. However, the availability of universal low-energy theories helps to keep oversight in a jungle of "topologically quantized" physical observables whose conceptual meaning might otherwise remain obscure. Familiarity with the three large families of effective topological field actions, θ-terms, Wess–Zumino terms, and Chern–Simons terms goes

[79] L. Saminadayar *et al.*, *Observation of the e/3 fractionally charged Laughlin quasiparticle*, Phys. Rev. Lett. **79**, 2526 (1997); R. de-Picciotto *et al.*, *Direct observation of a fractional charge*, Nature **389**, 162 (1997).

a long way towards understanding the physics of topological quantization in condensed matter physics (and in particle physics, where the concepts introduced above were originally developed).

The question whether effective field-theoretical approaches are actually *necessary* cannot be answered in general and depends on the application. For example, many of the phenomena observed in the physics of weakly entangled topological insulators or (semi-)metals afford straightforward explanations via the diagonalization of band Hamiltonians. However, the moment disorder or interactions enter the stage, effective field theories become indispensable tools, as illustrated in this chapter for various examples. Finally, where strongly entangled topological matter is concerned, we are still at the beginnings. At this point, it is not even clear whether quantum field theory will turn out to be the proper language to address this type of physics. Perhaps an entirely novel physical formalism and/or type of mathematics will be required?

8.8 Problems

8.8.1 Winding numbers

In applications of the Feynman path integral, one is often interested in phase spaces that are not simply connected. One then must integrate over topologically distinct classes of trajectories. To illustrate this situation, we consider here the application of the path integral to the single-particle problem of a particle on a ring. The problem fills in some details left out in our discussion in section 8.2.

(a) Starting with the Hamiltonian $\hat{H} = -\partial_\phi^2/2I$, where ϕ denotes an angle variable, show from first principles that the quantum partition function $\mathcal{Z} = \mathrm{tr}\, e^{-\beta \hat{H}}$ is given by

$$\mathcal{Z} = \sum_{n=-\infty}^{\infty} \exp\left[-\beta \frac{n^2}{2I}\right]. \tag{8.95}$$

(b) Formulated as a Feynman path integral, show that the quantum partition function assumes the form

$$\mathcal{Z} = \int_0^{2\pi} d\phi \sum_{m=-\infty}^{\infty} \int_{\substack{\phi(0)=\phi \\ \phi(\beta)=\phi(0)+2\pi m}} D\phi \exp\left[-\frac{I}{2}\int_0^\beta d\tau\, \dot{\phi}^2\right].$$

(c) Varying the Euclidean action with respect to ϕ, show that the path integral is minimized by the classical trajectories $\bar{\phi}(\tau) = \phi + 2\pi m\tau/\beta$. Now parameterize a general path as $\phi(\tau) = \bar{\phi}(\tau) + \eta(\tau)$, where $\eta(\tau)$ is non-winding, to obtain

$$\mathcal{Z} = \mathcal{Z}_0 \sum_{m=-\infty}^{\infty} \exp\left[-\frac{I}{2}\frac{(2\pi m)^2}{\beta}\right], \tag{8.96}$$

with $\mathcal{Z}_0$ the partition function for a free particle with open boundary conditions. Making use of the free particle propagator, show that $\mathcal{Z}_0 = \sqrt{2\pi I/\beta}$.

Poisson
summation
formula
(d) Finally, using the **Poisson summation formula**,

$$\sum_m h(m) = \sum_n \int_{-\infty}^{\infty} d\theta \, h(\phi) e^{2\pi i n\theta},$$

show that Eq. (8.96) coincides with Eq. (8.95).

Answer:

(a) Solving the Schrödinger equation, the wave functions obeying periodic boundary conditions take the form $\psi_n = e^{in\phi}/\sqrt{2\pi}$, with n integer, and the eigenvalues are given by $E_n = n^2/2I$. Cast in the eigenbasis representation, the partition function assumes the form of Eq. (8.95).

(b) Interpreted as a Feynman path integral, the quantum partition function takes the form of a propagator with

$$\mathcal{Z} = \int_0^{2\pi} d\phi \, \langle \phi | e^{-\beta \hat{H}} | \phi \rangle = \int_0^{2\pi} d\phi \int_{\phi(\beta) = \phi(0) = \phi} D\phi(\tau) \exp\left[-\int_0^\beta d\tau \, \frac{I}{2} \dot{\phi}^2 \right].$$

The trace implies that paths $\phi(\tau)$ must start and finish at the same point. However, to accommodate the invariance of ϕ under translation by 2π, we must impose the boundary conditions shown in the question.

(c) Varying the action with respect to ϕ, we obtain the classical equation of motion $I\ddot{\phi} = 0$. Solving this equation subject to the boundary conditions, we obtain the solution given in the question. Evaluating the Euclidean action, we find that $\int_0^\beta d\tau (\partial_\tau \phi)^2 = \int_0^\beta d\tau (\frac{2\pi m}{\beta} + \partial_\tau \eta)^2 = \beta (\frac{2\pi m}{\beta})^2 + \int_0^\beta d\tau (\partial_\tau \eta)^2$. Thus, we obtain (8.96), where $\mathcal{Z}_0 = \int D\eta(\tau) \exp[-\frac{I}{2} \int_0^\beta d\tau (\partial_\tau \eta)^2] = \sqrt{\frac{I}{2\pi\beta}}$ denotes the free particle partition function. This can be obtained from direct evaluation of the free particle propagator.

(d) Applying the Poisson summation formula with $h(x) = \exp[-\frac{(2\pi)^2 I}{2\beta} x^2]$,

$$\sum_{m=-\infty}^{\infty} e^{-\frac{(2\pi)^2 I m^2}{2\beta}} = \sum_{n=-\infty}^{\infty} \int_{-\infty}^{\infty} d\phi \, e^{-\frac{(2\pi)^2 I}{2\beta} \phi^2 + 2\pi i n\phi} = \sqrt{\frac{\beta}{2\pi I}} \sum_{n=-\infty}^{\infty} e^{-\frac{\beta}{2I} n^2}.$$

Multiplication by $\mathcal{Z}_0$ obtains the required result.

8.8.2 Topology of the magnetic skyrmion

In this problem, we explore the topology of the magnetic skyrmion, as described by the representation Eq. (8.4.3). As preparation for this problem, recapitulate the stereographic representation of the two-sphere.

In the info block on page 444, we showed that the meromorphic functions define stationary configurations of the two-dimensional σ-model. Qualitatively, each zero (pole) of these functions defines a zero (infinity) of $w(z)$, which is the stereographic representation of the north- (south-) pole $n_3 = \mp 1$ on the sphere. A configuration with W zero-pole pairs thus interpolates W times between the poles. Inspection of the corresponding $n_{1,2}$ coordinates shows that, in each sweep, the sphere is fully covered. In this problem, we compute the topological charge of these configurations. **(a)** Show that, in complex coordinates, $z = x_1 \pm i x_2$, the topological action assumes the form $S_{\text{top}}[\mathbf{n}] = \frac{3}{4\pi} \int dz d\bar{z} \, n_3 (\partial_z n_1 \partial_{\bar{z}} n_2 - \partial_{\bar{z}} n_1 \partial_z n_2)$. **(b)** Substitute Eq. (8.17) to obtain the representation

$$S_{\text{top}}[w] = -\frac{3i}{2\pi} \int dz d\bar{z} \frac{1 - |w|^2}{1 + |w|^2} \left(\partial_z \frac{\bar{w}}{1 + |w|^2} \partial_{\bar{z}} \frac{w}{(1 + |w|^2)} - (z \leftrightarrow \bar{z}) \right).$$

Next, show that for meromorphic functions $w(z)$,

$$S_{\text{top}}[w] = \frac{3i\theta}{2\pi} \int dz d\bar{z} \frac{(1 - |w|^2)^2}{(1 + |w|^2)^4} \partial_z w \partial_{\bar{z}} \bar{w}. \tag{8.97}$$

Naïvely, one might now change variables as $\int dz d\bar{z} \, \partial_z w \partial_{\bar{z}} \bar{w} = \int dw d\bar{w}$. However, here one has to take into account the fact that for z covering the sphere, $w(z)$ covers the sphere W times, where W is the multi-valuedness of the function $w(z)$ (think about this point). This means implies that the proper replacement rule reads $\int dz d\bar{z} \, \partial_z w \partial_{\bar{z}} \bar{w} \to W \int dw d\bar{w}$. With this replacement, show that the topological action is given by $S[w] = iW\theta$, as stated in the info block.

Answer:

(a) This involves a straightforward change of variables. **(b)** The first equation follows directly from the definition of w, and the second from the fact that, for meromorphic $w(z)$, $\partial_{\bar{z}} w = \partial_z \bar{w} = 0$. The final w-integral is most easily made in the variables $|w|^2 = s$, with $dw d\bar{w} = 2\pi ds$.

8.8.3 U(N) winding numbers

In this technical problem, we compute the winding numbers of U(N) on base spaces of dimensions one and three.

(a) Assuming periodicity in the coordinate $x \in [0,1]$, prove the quantization of the integral multiplying the topological angle θ in Eq. (8.13) in units of $i\mathbb{Z}$. (Hint: Demonstrate that the integrand is a full derivative and use the multi-valuedness of the logarithm.)

SU(2) instanton

(b) Demonstrate the integer quantization of the second integral in Eq. (8.14) for the case of SU(2), or $N = 2$. Start by constructing an **SU(2) instanton**, here understood as an SU(2)-valued field $g(x_1, x_2, x_3)$ of three coordinates, $x_i \in [0,1]$, $i = 1, 2, 3$ (for periodic boundary conditions), which cannot be continuously deformed to unity. To this end, let $\mathbf{n}(x_1, x_2)$ be a unit vector covering the two-sphere once as a function of x_1, x_2. Next define $g(x_1, x_2, x_3) \equiv \exp(i2\pi x_3 \mathbf{n}(x_1, x_2) \cdot \boldsymbol{\sigma})$. Reason why these configurations cover the group SU(2) *twice*, i.e., we have a winding number

$W = 2$ configuration. Proceed to compute the integral Eq. (8.14) and confirm this expectation. (Can you come up with a $W = 1$ configuration?)

Without proof, we mention that the computation of winding numbers in $U(N)$, $N \geq 2$, reduces to that for $SU(2) \subset U(N)$ embedded as a submanifold in those groups. The formula applied to general N therefore effectively computes $SU(2)$ winding numbers.

(c) Demonstrate that the differential forms $\mathrm{tr}(g^{-1}dg)$ and $\mathrm{tr}(g^{-1}dg \wedge g^{-1}dg \wedge g^{-1}dg)$ have vanishing exterior derivative. Arguing as in the info block of page 441, this feature safeguards the robustness of the numbers W computed above for one particular family of fields under continuous field deformations.

Answer:

(a) Using $\mathrm{tr}(U^{-1}d_xU) = d_x\mathrm{tr}\ln U$, we have $\frac{1}{2\pi}\int_0^1 dx\,\mathrm{tr}(U^{-1}d_xU) = \frac{1}{2\pi}\mathrm{tr}\ln U(x)\big|_0^1 = \frac{1}{2\pi}\ln\det(U)\big|_0^1 = iW$, where we note that $\det(U(x)) \equiv \exp(i\varphi(x))$ is a phase that may wind by integer multiples $\varphi(1) - \varphi(0) = 2\pi W$ along the base interval.

(b) The eigenvalues of an $SU(2)$ matrix must be unit-modular and multiply to unity. This implies that any such matrix can be written as $g = U\exp(i\phi\sigma_3)\,U^\dagger$ with unitary matrices U. Using the fact that $U\sigma_3 U^\dagger = \mathbf{n}\cdot\boldsymbol{\sigma}$, we obtain the representation $g = \exp(i\phi\mathbf{n}\cdot\boldsymbol{\sigma})$ and, with the identification $2\pi x_3 = \phi$, have our ansatz. Next note that for $\mathbf{n}$ covering the unit sphere, and x_3 running from 0 to 1, each configuration is visited twice: with

$$g = \cos(2\pi x_3) + i\sin(2\pi x_3)\mathbf{n}\cdot\boldsymbol{\sigma}, \tag{8.98}$$

this follows from the symmetry $g(x_3, \mathbf{n}) = g(1 - x_3, -\mathbf{n})$. To compute the winding number integral, differentiate the above representation to obtain $g^{-1}\partial_3 g = 2\pi i\mathbf{n}(x_1, x_2)\cdot\boldsymbol{\sigma}$, and $\partial_i g = -\partial_i g^{-1} = i\sin(2\pi x_3)\partial_i\mathbf{n}(x_1, x_2)\cdot\boldsymbol{\sigma}$. Using these relations, we compute the integral as

$$I \equiv \frac{\epsilon^{ijk}}{24\pi^2}\int d^3x\,\mathrm{tr}((g^{-1}\partial_i g)(g^{-1}\partial_j g)(g^{-1}\partial_k g)) = -\frac{\epsilon^{ij}}{8\pi^2}\int d^3x\,\mathrm{tr}(g^{-1}\partial_3 g\partial_i g^{-1}\partial_j g)$$

$$= -\frac{i}{4\pi}\int_0^1 dx_3\,\sin^2(2\pi x_3)\int d^2x\,\epsilon^{ij}\,\mathrm{tr}((\mathbf{n}\cdot\boldsymbol{\sigma})(\partial_i\mathbf{n}\cdot\boldsymbol{\sigma})(\partial_j\mathbf{n}\cdot\boldsymbol{\sigma}))$$

$$= \frac{1}{2\pi}\int d^2x\,\mathbf{n}\cdot(\partial_1\mathbf{n}\times\partial_2\mathbf{n}) = 2.$$

Here, the first equality follows by an elementary rearrangement of terms (note that $g^{-1}\partial_i gg^{-1} = -\partial_i g^{-1}$), and in the third we used $\mathrm{tr}((\mathbf{n}_1\cdot\boldsymbol{\sigma})(\mathbf{n}_2\cdot\boldsymbol{\sigma})(\mathbf{n}_3\cdot\boldsymbol{\sigma})) = 2i\mathbf{n}_1\cdot(\mathbf{n}_2\times\mathbf{n}_3)$. The final S^2 surface integral yields a factor of 4π and in this way we arrive at the result.

(c) The exterior derivative of the one-form yields $d\,\mathrm{tr}(g^{-1}dg) = \mathrm{tr}(dg^{-1}\wedge dg) = -\mathrm{tr}(\Phi\wedge\Phi)$, with the matrix-valued one-form $\Phi = g^{-1}dg$. In a matrix notation, this reads $-\mathrm{tr}(\Phi\wedge\Phi) = -\Phi_{ab}\wedge\Phi_{ba} = +\Phi_{ba}\wedge\Phi_{ab} = +\mathrm{tr}(\Phi\wedge\Phi)$, where the anticommutativity of the wedge product has been used. The expression equals its own negative and therefore vanishes. The moral of this computation is that traces of matrix-valued differentials satisfy the cyclic invariance property. However, sign factors arise when differential one-forms are exchanged. On the same basis, we obtain $d\operatorname{tr}(g^{-1}dg \wedge g^{-1}dg \wedge g^{-1}dg) = -3\operatorname{tr}(\Phi \wedge \Phi \wedge \Phi \wedge \Phi) = +3\operatorname{tr}(\Phi \wedge \Phi \wedge \Phi \wedge \Phi) = 0.$

8.8.4 Wess–Zumino action from symmetries

In this problem, we demonstrate how the analysis of symmetries and conservation laws of the free Dirac action suffices to fix the $U(N)$ Wess–Zumino action up to the value of its coupling constants.

(a) Writing the fermion action (8.60) as $S[\psi] = \int dx d\tau \sum_s \bar{\psi}_s(\partial_\tau - is\partial_x)\psi$, and introducing complex coordinates $z = \frac{1}{\sqrt{2}}(\tau + ix)$, $\partial_z = \frac{1}{\sqrt{2}}(\partial_\tau - i\partial_x)$ demonstrate that the two independent Noether currents associated with the $U(N) \times U(N)$ symmetry are given by Eq. (8.65), with conservation laws $\partial_z \bar{j} = \partial_{\bar{z}} j = 0$.

(b) Now consider the trial action (8.61) purportedly representing the fermion action in a bosonized language. Assuming that the plus-sector of the group symmetry is represented as $g \to g_+ g$, show that the corresponding Noether conservation law reads as $-\frac{2}{\lambda}\partial_\mu((\partial_\mu g)g^{-1}) = 0$. This does not have the holomorphic structure $\partial_z(\dots)$ of current conservation in fermionic language and therefore does not faithfully represent the symmetry.

(c) Now consider what happens if we add the WZ term (8.63) to the action. We first need to understand how the WZ term responds to the above variation. In this part of the problem, we are going to show that

$$\Gamma[(1 + W)g] - \Gamma[g] = \frac{i}{4\pi} \int d^2 x \, \epsilon_{\mu\nu} \operatorname{tr}(W \partial_\mu g \partial_\nu g^{-1}) + \mathcal{O}(W^2). \qquad (8.99)$$

Equation (8.99) may be obtained from the three-dimensional representation (8.63) in two ways. The first is a brute force verification that the variation is a full derivative, and can be represented as a two-dimensional "surface integral." The second exploits the defining properties of the closed differential three form $\omega = \operatorname{tr}(g^{-1}dg)^{\wedge 3}$ (see the info block on page 440), which defines the WZ term as a pullback integral (see Eq. (8.50)). In working with topological terms, it is often best to operate on the target manifold (here the group $SU(N)$) and go back to the base only in the end. Adopting this strategy, argue as in the info block on page 441 to demonstrate that the variation $g \to (1 + W)g$ defines a closed three-form which is only locally non-vanishing on $SU(N)$ and therefore must be exact. Then apply Stokes' theorem to obtain the above result.

(d) Add the variation of the gradient action discussed in part (b) to that of the WZ term and show that, for the particular value $\lambda = 8\pi$, it assumes the form Eq. (8.66). The construction shows that, for this choice of the coupling constant, the action $S_0 + \Gamma$ displays the same group symmetry as the fermion action. While not a mathematical proof, this provides compelling evidence for the equivalence of the two representations.

Answer:

(a) The Noether current of the group symmetry is obtained by variation of the action under a weakly space–time dependent transformation. Focusing on the $s = +$ sector for definiteness, we transform $\psi_+ \to e^W \psi_+ \simeq (1 + W)\psi_+$, with an anti-hermitian generator matrix $W = \{W^{ab}(x, \tau)\}$. Substituting this transformation, we obtain the change $\delta S = \int dx d\tau \, \psi_+^{a\dagger}(\partial_z W^{ab})\psi_+^b = -\int dx d\tau \, W^{ab}\partial_z(\psi_+^{\dagger a}\psi_+^b)$. For extremal field configurations, all first-order variations, including this one, must vanish, and this implies the conservation law $\partial_z(\psi_+^{\dagger a}\psi_+^b) = 0$ on the classical level.

(b) The result follows from a straightforward substitution of $g \to g_+ g = e^W g \simeq (1 + iW)g$ into the effective action followed by integration by parts.

(c) In differential form language, the WZ term (8.63) is the pullback integral $\Gamma[g] = -\frac{i}{12\pi}\int_B \tilde{g}^*\omega$ of the above three-form to the space–time manifold B. Substituting the variation $g^{-1}dg \to g^{-1}dWg$ into the latter, we obtain $\omega \to 3\,\mathrm{tr}(g^{-1}dWg \wedge g^{-1}dg \wedge g^{-1}dg) = -3\,\mathrm{tr}(dW \wedge dg \wedge dg^{-1}) = -3d\,\mathrm{tr}(Wdg \wedge dg^{-1})$. This is a full derivative, and so application of Stokes' theorem yields $\Gamma[(1 + W)g] - \Gamma[W] = \frac{i}{4\pi}\int_M g^*\mathrm{tr}(Wdg \wedge dg^{-1})$, where g^* is the pullback under the group-valued field $g(x, \tau)$ to the two-dimensional space–time boundary, M. Translating back to coordinates, we obtain the stated result.

(d) A straightforward computation shows that $g(\partial_{\bar{z}}j)g^{-1} = \partial_z\bar{j}$, i.e., the two conservation laws are equivalent. Demonstrating that they are equivalent to the vanishing of the first-order variation of the action at $\lambda = 8\pi$ is a matter of an equally straightforward substitution $z = \tau + ix$ in the conservation laws.

8.8.5 Renormalization of the Wess–Zumino model

In this problem, we derive the RG equation (8.68) of the $\mathrm{SU}(N)$ model. To get warmed up to this task you should recapitulate the RG analysis of the $\mathrm{SU}(N)$ nonlinear σ-model discussed in section 6.4.

Our starting point is the action (8.64) at a presumed small value of the coupling constant $\lambda \ll 1$ of the gradient action (8.61). As in our previous analysis of nonlinear σ-models, we begin by splitting the fields $g = g_s g_f$ into a slow and a fast part, and expand the latter as $g_f = \mathbb{1} + W + \frac{1}{2}W^2 + \cdots$, where $W \in \mathrm{su}(N)$ is in the Lie algebra of anti-hermitian traceless matrices. The one-loop RG equations are obtained by expansion of the action to quadratic order in W followed by the computation of all contributions to the functional integral containing one fast momentum integration and no more than two derivatives acting on a slow field.

(a) Show that the expansion of the action $S[g_s g_f]$ to second order in the generators takes the form $S[g_f g_s] = S[g_s] + S[g_s, W] + S^{(2)}[W]$, where

$$S^{(2)}[W] = -\frac{1}{\lambda}\int d^2x \, \mathrm{tr}(\partial_\mu W \partial_\mu W),$$

$$S[g_s, W] = -\frac{1}{\lambda}\int d^2x \, \mathrm{tr}\left[\left(g_s^{-1}\partial_\mu g_s - \frac{i\lambda}{8\pi}\epsilon_{\mu\nu}g_s^{-1}\partial_\nu g_s\right)[\partial_\mu W, W]\right]. \tag{8.100}$$

(b) One-loop corrections to the action are obtained by expanding the functional to second order in $S^{(2)}[g_s, W]$ and integrating over W: $S[g] \to S[g_s] - \frac{1}{2}\langle S[g_s, W]^2\rangle_W$. Use the results of section 6.4 (in particular those derived in the exercises on page 360) to confirm that only the gradient term of the action is renormalized and that the RG equation for its coupling constant is given by Eq. (8.68).

Answer:

(a) The first line of Eq. (8.100) and the first term in the second line are obtained by substituting $g_s g_f$ into the gradient term of the action (8.61) and expanding to second order in the generators W. The second term in $S[g_s, W]$, a descendant of the WZ action, is best derived in the language of differential forms: substitution of $(g_s g_f)^{-1} d(g_s g_f) = g_f^{-1}(g_s^{-1} dg_s + (dg_f)g_f^{-1})g_f$ into the pullback $(g_s g_f)^* \omega = \text{tr}(g^{-1}dg \wedge g^{-1}dg \wedge g^{-1}dg)|_{g=g_s g_f}$ gives

$$(g_s g_f)^* \omega = g_s^*(\omega) + \frac{3}{2}\text{tr}(dg_s^{-1} \wedge dg_s \wedge [W, dW]) + 3\text{tr}(dW \wedge dW \wedge g_s^{-1}dg_s) + \mathcal{O}(W^3)$$
$$= g_s^*(\omega) + \frac{3}{2}d\left[\text{tr}([W, dW] \wedge g_s^{-1}dg_s)\right] + \mathcal{O}(W^3).$$

Application of Stokes' theorem thus leads to

$$\Gamma[g_s g_f] = -\frac{i}{12\pi}\int_B (g_s g_f)^*\omega \to -\frac{i}{8\pi}\int_{B^3} d\left[\text{tr}([W, dW] \wedge g_s^{-1}dg_s)\right]$$
$$= -\frac{i}{8\pi}\int_{S^2} \text{tr}([W, dW] \wedge g_s^{-1}dg_s) = -\frac{i}{8\pi}\int_{S^2} d^2x\, \epsilon_{\mu\nu}\text{tr}([W, \partial_\mu W] \wedge g_s^{-1}\partial_\nu g_s),$$

where S^2 is symbolic notation for the two-dimensional boundary surface and the arrow stands for the isolation of terms $\mathcal{O}(W^2)$.

(b) Defining $\Phi_\mu \equiv g_s^{-1}(\partial_\mu + \frac{i\lambda}{8\pi}\epsilon_{\mu\nu}\partial_\nu)g_s$, the action $S[g_s, W]$ assumes the form $S[g_s, W] = \frac{2iL^d}{\lambda}\sum_{pq} p_\mu \text{tr}(\Phi_\mu \Phi_\mu)$. Except for the differently defined field Φ, this equals the fast–slow action of the standard SU(N) model. Using the results derived in the exercises on page 360, we thus obtain

$$-\frac{1}{2}\langle S[g_s, W]^2\rangle_W = -\frac{N\ln b}{8\pi}\int d^2x\, \text{tr}(\Phi_\mu \Phi_\mu)$$
$$= -\frac{N\ln b}{8\pi}\left(1 - \left(\frac{\lambda}{8\pi}\right)^2\right)\int d^2x\, \text{tr}(\partial g_s \partial g_s^{-1}).$$

This result confirms that only the gradient term in the action is renormalized. By proceeding in direct analogy to the discussion of section 6.4, it is a straightforward matter to derive the corresponding RG equation. The result is given by Eq. (8.68)

8.8.6 Strong-correlation physics via nonabelian bosonization

In this problem we study in more detail than in section 8.5.5 how nonabelian bosonization may reveal information on strongly interacting quantum systems that is difficult to obtain otherwise. Specifically, we will consider the bosonized representation of the higher

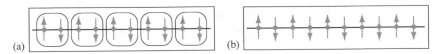

Fig. 8.10   (a) Dimerized phase of the spin chain and (b) the Néel phase.

S antiferromagnetic spin chain, perturbed by operators representing various types of lattice dimerization, to understand its physics at large distance scales.

We consider the fermion representation of an $S = n_c/2$ spin chain introduced in section 8.5.5. Affleck and Haldane[61] showed that the inclusion of all relevant interaction operators leads to a collapse $SU(2n_c) \to SU(2)$ of its symmetry, meaning that in the infrared the individual spins get locked to a single spin S with symmetry group $SU(2)$. Substituting the reduced fields $g \otimes \mathbb{1}_{n_c}$ into the unperturbed WZ action (8.64), we obtain $n_c S[g] = n_c S_0[g] + n_c \Gamma[g]$. This is the action of an $SU(2)$ field with coupling constants multiplied by n_c, due to the trace over color indices; it is called the $SU(2)$ WZ action at level $k = n_c$.

(a) To understand how this model responds to the presence of interactions, we need the bosonic representation of lattice spin operators $\mathbf{S}_l = \psi_l^{\alpha\dagger}\boldsymbol{\sigma}^{\alpha\beta}\psi_l^{\beta}$. Decompose the fermions of the half-filled system into left- and right-moving components, $\psi_l = e^{ik_F l}\psi_+(x_l) + e^{-ik_F l}\psi_-(x_l)$, where $x_l = la$, and $k_F = \pi/2a$. Then apply the bosonization identities (8.66) and (8.67) to the slowly oscillating components $\psi_\pm$ to show $\mathbf{S}_l = \mathrm{tr}([c(j_+^s + j_-^s) + c'(-)^l(g + g^{-1})]\boldsymbol{\sigma})$, where c, c' are constants inessential to our discussion.

The second term under the trace is of particular interest. It tells us that translation by one site, $l \to l + 1$, is equivalent to a sign change of the field g. This observation can be read in different ways. For example, contributions to the action that are not invariant under $g \to -g$ explicitly break translational invariance on the lattice. We do not expect translational symmetry-breaking in the action and hence exclude contritions odd in g. However, terms even in g are not *a priori* excluded. Similarly, mean field ground states $\bar{g}$ obtained by variation of a translationally invariant action might turn out to break $\bar{g} \to -\bar{g}$ symmetry, and hence *spontaneously* break translational invariance.

(b) To explore these points, we consider the dominant translationally invariant massive contribution to the action, $S_m[g] \equiv m \int \mathrm{tr}(g^2 + g^{-2})$.[80] These terms are RG relevant (why?) and we aim to understand their effect at large distance scales.

To prepare this discussion, we first note that a spin chain favoring local anti-alignment may support two competing ground states: the first is a phase defined by local dimer formation – see fig. 8.10(a). This ground state contains local singlets,

---

[80] From where might such terms come? We know that in the half-filled system, momentum conservation does not exclude the scattering of pairs of fermions between the Fermi points – Umklapp scattering. In the spin-1/2 case, there is only one such term compatible with the symmetries of the problem; the Umklapp scattering operator discussed in section 8.5.5. However, for higher spin, or $n_c > 1$, the scattering of fermions of different a-index may generate terms like S_m in the effective action.

and hence is trivially spin rotation invariant. It does not allow for gapless spin fluctuations. The second, (b), is the Néel order. This phase spontaneously breaks spin rotation invariance, and hence admits Goldstone modes. Both the dimer and the Néel phase spontaneously break translational invariance.

Identify the SU(2) ground states favored by S_m for negative and positive mass m, and show that they satisfy the criteria required by the dimer and the Néel ground state, respectively. In the Néel case, how do we describe the Goldstone mode spin fluctuations in the language of g-fields?

(c) We now understand that the ground state manifold of the Néel phase is defined by configurations $ig_0 \tau_3 g_0^{-1} \equiv i \mathbf{n} \cdot \boldsymbol{\sigma}$, where the unit vector $\mathbf{n}$ is implicitly defined by the rotation g_0 and describes the orientation of the Néel state on the two-sphere. What is the effective action of the Goldstone modes defined by spatially fluctuating orientation vectors $\mathbf{n}$? To find out, we need to substitute $g \to i \mathbf{n} \cdot \boldsymbol{\sigma}$ into the Wess–Zumino action. Show that the gradient term becomes $n_c S_0[i\mathbf{n} \cdot \boldsymbol{\sigma}] = \frac{n_c}{4\pi} \int \partial_\mu \mathbf{n} \cdot \partial_\mu \mathbf{n}$ and the Wess–Zumino term becomes $n_c \Gamma[i\mathbf{n} \cdot \boldsymbol{\sigma}] = S_{\text{top}}[\mathbf{n}]$, with topological angle $\theta = \pi n_c$. (Hint: Recycle the parameterization (8.98) and the calculation of problem 8.8.3.)

We have thus found that the extremal action configurations of the perturbed WZ models describing the $S = n_c/2$ spin chains are described by the spin non-linear σ-model at topological angle $\theta = \pi n_c$. One may object that this is no surprise; we had already arrived at the same conclusion within the semiclassical expansion of the antiferromagnetic spin chain in section 8.4.6. However, there now is the added twist that we have the connection between the σ-model and the WZ action. In this setting, the Haldane conjecture materializes as follows: starting from the WZ treatment of the $S = n_c/2$ chain, we flow to a $\theta = \pi n_c$ σ-model. Since the topological action is periodic in θ mod 2π, even/odd values of n_c are described by the action with topological coupling zero (flowing into a gapped phase), or $n_c = 1$, the action of the spin-1/2 chain. However, the latter, as we have seen in section 8.5.5 is equivalent to a level-1 WZ theory. In this way, we arrive at the conclusion that the critical theory of the half-integer spin chains is the level-1 (not n_c) WZ theory, which in turn is equivalent to a free one-dimensional fermion. On this basis, the physical properties (critical exponents, correlation functions, etc.) are now known. Note that this result cannot be obtained by standard field-theoretical methods starting from the nonlinear σ-model; the bosonization machinery is essential.

Answer:

(a) Substituting the decomposition into the definition of the spin operator, we obtain $\mathbf{S}_l = \sum_s \psi_s^\dagger(x_l) \boldsymbol{\sigma} \psi_s(x_l) + [(-)^l \psi_+(x_l)^\dagger \boldsymbol{\sigma} \psi_-(x_l) + \text{h.c.}]$, where the sign factor results from $e^{ilk_F x} = e^{i(\pi/a)la} = (-)^l$. Substitution of the bosonization rules leads to the stated idenitity.

(b) For negative m, configurations of least action (and hence maximal weight $\exp(-S_m)$) have $\text{tr}(g) = \text{tr}(g^{-2}) = 2$, the maximal value realizable for SU(2)-matrices. The eigenvalues of g thus must be unit-modular and $\det(g) = 1$ fixes $\bar{g} = \pm \mathbb{1}_2$ as extremal configurations. Notice that either choice spontaneously breaks

$g \to -g$ invariance, i.e., we have translational symmetry breaking. At the same time, the ground states are individually spin-rotation invariant, $g_0 \bar{g} g_0^{-1} = \bar{g}$, without any soft fluctuations around them; they describe the dimer phase. For positive m, we are looking for configurations with maximally negative $\operatorname{tr} g = \operatorname{tr}(g^{-2}) = -2$. This requires eigenvalues $\pm i$. A diagonal matrix satisfying this condition is $\bar{g} \equiv \exp(i\frac{\pi}{2}\sigma_3) = i\sigma_3$. Since the trace is unitarily conserved, all $ig_0\sigma_3 g_0^{-1}$ satisfy the extremal condition, i.e., we have a continuous manifold of individually spin-rotation and translational symmetry breaking ground states. These describe the Néel phase.
(c) The form of the gradient action follows immediately upon substitution of $g \to i\mathbf{n}\cdot\boldsymbol{\sigma}$ into $n_c S_0$. Concerning the WZ term, we consider the parameterization (8.98), which interpolates between $g = \mathbb{1}_2$ and $g = i\mathbf{n}\cdot\boldsymbol{\sigma}$ for $x_3 = 0 \to 1/4$. Using the fact that the winding number integral, I, studied in problem 8.8.3 equals $\Gamma/2\pi$, and reproducing the calculation of that problem, we arrive at the result.

8.8.7 Knot invariant from the Chern–Simons functional

In this problem, we derive the fractional phase of two-dimensional abelian anyon braiding from the Chern–Simons functional. The problem provides the technical details underlying the qualitative discussion of section 8.6.4.

In section 8.6.4, we described the braiding of two quasiparticles in terms of their current densities coupled to the CS functional, Eq. (8.78). We now want to integrate over a to obtain the statistical phase weighing the process. Prior to this integration, we need to fix a gauge, for otherwise the "infinite gauge freedom" would produce a divergence. In the present context, it is convenient to choose the *Lorentz gauge*, $\partial^\mu a_\mu = \partial_t a_0 - \partial_i a_i = 0$. In order to avoid positive indefinite expressions such as $q^\mu q_\mu$, we first turn to an Euclidean (imaginary-time) framework. We do so in the usual manner, by defining $i\tau = t$. This substitution entails secondary ones, $-i\partial_\tau = \partial_t$, and $-ia_0' = a_0$. (Justify this latter replacement on the basis of gauge invariance or the structure of the CS action.) In this problem, it is best to think of Euclidean space as a $(3+0)$-dimensional real space.
(a) With $x_\mu' = (\tau, \mathbf{x})$ and $a' = (a_0, \mathbf{a})$, show that the gauge condition becomes $\partial_\mu' a_\mu = 0$, the Euclidean CS action is identical to its Lorentzian form, and that the current vector potential coupling reads as $-ia_\mu' j_\mu'$ (ordinary summation convention), where $j' = j_1' + j_2'$, and $j_{i,\mu}' = \partial_\tau x_\mu' \delta(\mathbf{x} - \mathbf{x}_i(\tau))$.
(b) Next, we need to integrate over a. You may assume that the gauge fixing is implemented via addition of a contribution $\frac{\Gamma}{2}\int d^3x (\partial_\mu a_\mu)^2$ to the action, where Γ is a large positive constant and we omitted the Euclidean primes to simplify the notation. Alternatively you may assume that the integration is restricted to the components of the Fourier transform a_q satisfying the condition $q_\mu a_{\mu,q} = 0$. Apply either of these procedures to show that $\langle a_{\mu,q} a_{\nu,q'} \rangle = -i\frac{2\pi}{k}\epsilon_{\mu\sigma\nu}\frac{q_\sigma}{q^2}\delta_{q,-q'}$, and hence

$$A \equiv \left\langle \exp\left(\sum_q j_{\mu,q} a_{\mu,-q}\right) \right\rangle_{\mathrm{CS}} = \exp\left(-\frac{i\pi}{k}\sum_q \epsilon_{\mu\sigma\nu} j_{\mu,q}\frac{q_\sigma}{q^2}j_{\nu,-q}\right).$$

(c) We now turn to an interpretation of this expression inspired by magnetostatics. To this end, we interpret the densities j_i as those of two current-carrying loops $i = 1, 2$ (use fig.8.7 for guidance). Integrate over a suitably chosen surface intersecting the loops to show that they carry unit current. Verify that the current density is sourceless, $\partial_\mu j_\mu = 0$. Then verify that the expression above can be written as $A = \exp(-\frac{i\pi}{k} \int d^3x j_\mu b_\mu)$, where b is the magnetic field generated by j (for unit magnetic permeability). We are now almost there. Substituting the definition of the current vector fields, we obtain

$$A = e^{-\frac{i\pi}{k} \sum_{i,j} \oint d\tau\, \dot{x}_i(\tau) \cdot b_j(x_i(\tau))} = e^{-\frac{i\pi}{k} \sum_{i,j} \int dS_i \cdot (\nabla \times b_j)} = e^{-\frac{2i\pi}{k} I(\gamma_1, \gamma_2)}.$$

In the first expression, we collapsed the volume integral to loop integrals, over the field strengths generated by the two currents, in the second we applied Stokes' theorem, and in the third we used the fact that the integral of $\nabla \times b_i = j_i$ over loop j produces the linking number $I(\gamma_i, \gamma_j)$. The contributions from $(i, j) = (1, 2)$ and $(2, 1)$ add, which gets us to the final expression.

We have thus shown that the Chern–Simons functional produces the statistical phase weighting corresponding to the braid operation.

Answer:

(a) These features all follow from the definition of the Euclidean degrees of freedom.
(b) In the presence of the gauge-fixing term, the quadratic CS assumes the form $S_{\text{CS}}[a] = \frac{1}{2} \sum_q a_{\mu q}(-i\epsilon_{\mu\sigma\nu} q_\sigma + \Gamma q_\mu q_\nu) a_{\nu, -q}$. It is useful to think about the 3×3 matrix kernel defined by this expression in a projected representation, where $P \equiv \{q_\mu q_\nu q^{-2}\}$ projects onto the one-dimensional longitudinal subspace $\parallel q$ and $Q = 1 - P$ onto its two-dimensional transverse complement. Since the first/second contribution to the matrix is annihilated by P/Q, it assumes a block diagonal form. We now claim that $-i\frac{2\pi}{k}\epsilon_{\mu\sigma\nu}\frac{q_\sigma}{q^2}$ defines the inverse of the matrix in the transverse sector. Indeed, it is straightforward to verify that the matrix multiplication of this expression with the first/second term in the action yields the projector $Q/0$. The displayed equation then follows from the standard rules of Gaussian integration.
(c) Integrating j_i over a patch, P, in the 12 plane containing $\mathbf{x}_i(0)$, we obtain $I = \int_P dS \cdot j = 1$. The absence of current sources follows by straightforward differentiation. Taking the curl of the Ampère equation $\nabla \times b = j$ and using $\nabla \cdot b = 0$, we obtain $\Delta b = \nabla \times j$, which in momentum space assumes the form $b_{\mu, q} = -q^{-2}\epsilon_{\mu\nu\sigma}(-iq_\nu)j_\sigma$. Turning to real space, the assertion follows.

8.8.8 Fractional quantum Hall effect: physics at the edge

In section 8.4.7, we saw that the bulk physics of the quantum Hall systems is inseparably connected to that of their boundaries. Following work of Wen and Stone,[81] here we show how the edge theory of the FQHE is deduced from that of its bulk. (The discussion is limited to

---

[81] X-. G. Wen, *Theory of the edge states in fractional quantum Hall effects*, Int. J. Mod. Phys. **B6**, 1711 (1992); M. Stone, *Edge waves in the quantum Hall effect*, Ann. Phys. **207**, 38 (1991).

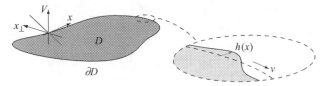

Fig. 8.11   On the formation of surface wave excitations in an FQHE droplet. For an explanation, see the main text.

filling fractions $p = 1$ in the principal sequence, $\nu = 1/(2s + 1)$. The inclusion of higher p requires theories with more than one chiral edge channel, which is beyond the scope of this problem.)

Consider an FQHE sample, D. At its boundary, ∂D, an (approximately linearly) increasing potential $V = Ex_\perp$ separates the vacuum and the filled states defining the FQHE ground state (fig. 8.11), where $x_\perp$ is a local coordinate transverse to the boundary coordinate, x.

In the first part of this problem, we will address the formation of boundary edge excitations on phenomenological grounds. The presence of a bulk excitation gap in the FQHE insulator implies $\partial\mu/\partial N \to \infty$. The thermodynamic relation $\partial\mu/\partial N \sim \kappa^{-1} \equiv -V(\partial P/\partial V)_N$ relates this divergence to a vanishing bulk compressibility, the intuitive picture being that of an incompressible fluid. The lowest-energy excitations of the system must therefore be deformations of its boundary, $h(x)$, similar to boundary distortions of a puddle of water (see fig. 8.11). With the bulk charge density, $\sigma dx dx_\perp$, $\sigma \equiv \frac{\nu B}{2\pi}$ (cf. Eq. (8.37)), this deformation is proportional to a spatially varying boundary charge density, $\rho(x)dx \equiv \sigma h(x)dx$.

(a) To derive the boundary action on phenomenological grounds, we proceed in a few steps. First, derive the energy cost of a boundary distortion ρ. Next, use the continuity equation and what you know about the Hall conductivity of the system to derive an equation of motion describing the propagation of density $\rho(x,t)$ along the boundary at a constant velocity v. Defining a new variable ϕ through $\partial_x\phi = \frac{2\pi}{\nu}\rho$, explain why

$$S[\phi] = \frac{\nu}{4\pi} \int dx\, dt \left[v \left(\partial_x\phi\right)^2 - \partial_x\phi\partial_t\phi \right] \qquad (8.101)$$

chiral Luttinger liquid   defines its effective action. This is the action of the **chiral Luttinger liquid**. The correspondence follows from the fact that it describes the propagation of a bosonic variable ϕ at constant wave velocity in one direction. (Indeed, it is straightforward to verify that the substitution of $\phi = \phi_L + \phi_R$, $\theta = \phi_L - \phi_R$ into the action (3.96) of the Luttinger liquid splits the latter into two chiral actions describing propagation in opposite directions.)

(b) In section 8.6.4, we showed that pure CS theory on a system with a boundary is not gauge invariant and that a likewise non-gauge invariant boundary action is required to define an overall gauge invariant package. We tentatively identify Eq. (8.101) as this boundary action. A first indication that this action might also lack gauge invariance follows from its identity as a chiral Luttinger liquid. Representing only one-half of a full Luttinger liquid, it defines the bosonized form of a single left- or right-moving branch of fermions.

However, we saw in section 9.2.1 that an individual chiral fermion branch displays the chiral anomaly – application of an electric field makes particles disappear into nowhere – and therefore lacks charge conservation and gauge invariance. We suspect this gauge invariance to cancel against that of the bulk CS action. In physical terms, this means that the above "nowhere" is the bulk of the system and that application of an electric field in the direction of the boundary leads to bulk boundary charge exchange.

Recalling from bosonization that the field ϕ couples to gauge transformations, $A_\mu \to A_\mu + \partial_\mu f$, as the phase of a fermion field, $\phi \to \phi + f$, find a minimal coupling of the action $S[\phi]$ to the external field A such that the gauge anomaly of the bulk action (8.79) indeed cancels.

Answer:

(a) With the boundary potential $V = Ex_\perp$, a deformation of the surface profile costs energy $H = \int dx dx_\perp \rho V = \sigma E \int dx \int_0^{h(x)} dx_\perp x_\perp = \frac{E}{2\sigma} \int dx \rho(x)^2$. In a system with Hall conductivity $\sigma_{12} = \frac{\sigma}{B}$, the confining electric field E generates a Hall current density $j = -\sigma_{12} E = -\frac{\sigma E}{B}$ tangential to the boundary. (We choose the negative sign for later convenience, as a matter of convention.) The total boundary current is obtained by integrating the current density from 0 to $h(x)$, i.e., $I(x) = -\frac{E}{B}\rho(x)$, and the continuity equation $\partial_x I(x,t) + \partial_t \rho(x,t) = 0$ implies the equation of motion $\partial_t \rho - v\partial_x \rho = 0$, with the velocity $v = E/B$. We thus obtain a uniformly propagating density profile, $\rho(x,t) = \rho(x + vt)$. Differentiating the equation of motion in x, we may rewrite it as $\partial_t \phi - v\partial_x \phi = 0$ in terms of the variable ϕ. The action (8.101) produces this equation under variation $\delta_\phi S = 0$ and contains the energy $\int dt H$ as a first term. It thus defines the (Hamiltonian) action of the boundary system.

(b) A gauge invariant coupling of the boundary action (8.101) to an external gauge field is obtained by the substitution $\partial_\mu \phi \to \partial_\mu \phi - A_\mu$. However, we now need an extra term, which cancels against the excess action $S[A, f] \equiv \frac{\nu}{4\pi} \int_{\partial M} f dA = \frac{\nu}{4\pi} \int dx dt\, f(\partial_t A_x - \partial_x A_t)$ of Eq. (8.79). In differential form language, this term is easily guessed as $\frac{\nu}{4\pi} \int d\phi \wedge A = -\frac{\nu}{4\pi} \int \phi dA$. Under a gauge transformation it changes as $-\frac{\nu}{4\pi} \int (\phi + f)d(A + df) = -\frac{\nu}{4\pi} \int \phi dA - \frac{\nu}{4\pi} \int f dA$, which contains a term countering the bulk gauge anomaly. Turning to a coordinate representation and collecting terms, we conclude that

$$S[\phi] = \frac{\nu}{4\pi} \int dx\, dt\, \left[(v(\partial_x \phi - A_x)^2 - (\partial_x \phi - A_x)(\partial_t \phi - A_t)) + \partial_t \phi A_x - \partial_x \phi A_t \right]$$

defines the gauge field coupling of the boundary theory.

9 Relativistic Field Theory

SYNOPSIS In this chapter, we discuss relativistic effects in condensed matter physics from a field-theoretical perspective. We begin with an application-oriented derivation of the Dirac action. (This means that we will avoid the early specialization to $3 + 1$ space–time dimensions that is common in particle physics texts.) Building on it, we address fundamental aspects of relativistic quantum matter including its symmetries, the appearance of anomalies, topological signatures of Dirac matter, and the coupling to gauge fields. We will discuss the realization of these principles in condensed matter contexts such as the $(1 + 1)$-dimensional Fermi gas, and different forms of topological quantum matter. Notably, the physics of topological insulators close to phase transitions is described by effective Dirac Hamiltonians, and this provides a rich spectrum of applications for concepts in relativistic quantum field theory.

We call a condensed matter fermion system *relativistic* if its Brillouin zone contains one or several singular points around which the dispersion is approximately, or even fully, linear (see the figure schematic). Before the turn of the century, such dispersion relations were known to exist in gapless superconductors and certain narrow-gap semiconductors, but were otherwise considered exotic. Two disruptive experimental developments in the first decade of this century changed that situation and triggered a surge of research on relativistic, or *Dirac*, matter. First came the discovery of graphene, a material whose fascinating properties are due to the presence of two Dirac points in its Brillouin zone (recall the info block on page 56). Shortly afterwards, the new material class of topological insulators was discovered. The surface Brillouin zones of topological insulators harbor Dirac points protected against the detrimental effects of disorder and other system imperfections by principles of topology. These surface states are experimentally observable and responsible for the majority of the unconventional physical properties of topological insulators.

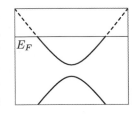

Along with these developments, the physics of Dirac quantum matter has become an experimental reality outside the realm of $(3 + 1)$-dimensional particle physics. Phenomena previously restricted to that field, such as anomalies, Klein tunneling, chiral symmetry breaking, etc., are now the subject of mainstream condensed matter research. In particular, condensed matter physics has introduced experimental platforms for realizing Dirac physics in dimensions different from four. This option has attracted the interest of particle theorists and is stimulating research at the interface between high energy and condensed matter physics.

Motivated by these developments, here we introduce some of the key concepts distinguishing relativistic quantum field theory from the non-relativistic theories discussed previously. We start out in the first section with a condensed matter-oriented introduction to the Dirac action. In the remaining sections of the chapter, we discuss the physics of Dirac field theory with an eye on applications in condensed matter physics.

9.1 Dirac Theory

REMARK This section assumes familiarity with basic notions of special relativity, notably the concept of four-vectors, Lorentz transformations, and covariant notation. For a succinct summary, see the info block below or otherwise consult a textbook to brush up your knowledge.

Derivations of the Dirac equation are a canonical part of almost every introductory text on particle physics. There, emphasis is usually placed on the historical development of Dirac's approach towards a relativistic generalization of the Schrödinger equation: requiring to replace a quadratic dispersion $\sim \hat{p}^2/2m$ by a relativistic dispersion $\sim c\hat{p}$, Dirac suggested generalizing the scalar Schrödinger equation to a four-component spinor equation. He then introduced the famous Dirac γ-matrices, engineered to satisfy certain commutation relations, and on this basis wrote down the first-order Dirac equation. The construction is ingenious and can be developed on little more than a single sheet of paper.

Paul Adrien Maurice Dirac
1902–1984
was a British physicist who made fundamental contributions to the development of quantum mechanics and its subsequent extension to relativistically invariant quantum field theory. The Dirac equation revolutionized physics in various ways: among other things, it motivated the concept of quantum *fields*, predicted the existence of antimatter, and led to a fundamentally new understanding of fermions. In 1933, Dirac shared the 1933 Nobel Prize in Physics with Erwin Schrödinger "for the discovery of new productive forms of atomic theory."

However, the price to be paid for its brevity is its limitation to $(3+1)$-dimensional space–time and that the ad hoc introduction of γ-matrices does not really elucidate the underlying physical symmetry principles.[1] On the other end of the spectrum are derivations of Dirac theory on the basis of representation theory.[2] These emphasize symmetries; they work in general dimensions, but are a little too abstract for our purpose. Here, we aim to strike a middle ground and derive the Dirac equation by extending the known quantum mechanics of the rotation group (describing the physics of spin) to a quantum mechanics of the Lorentz group. (The derivation

---

[1] It's the other way around: a plethora of symmetry principles *follows* from the equation introduced ad hoc, which led Dirac himself to comment that "the equation was more intelligent than its inventor" (a statement which may be disputed).

[2] For a succinct introduction to this approach, see the text by Göckeler and Schücker in Ref.[1]

takes six pages instead of one. However, along the way, we will introduce a lot of useful material, so even readers familiar with the equation may find the construction instructive.)

INFO For convenience, here is a quick synopsis of the **covariant notation** used in special relativity and several parts of this text. **Contravariant vectors** $v = \{v^\mu\}$, $\mu = 0, 1, 2, 3$, comprise a time-like component v^0 and three spatial components v^i, $i = 1, 2, 3$. (In general dimensions, $\mu = 0, \ldots, d + 1$.) We follow the convention of labeling space–time indices by Greek indices, $\mu, \nu, \ldots$, and spatial ones by Latin indices, $i, j, \ldots$. For example, space–time events are recorded as $x = x^\mu = (ct, \mathbf{x})$, where the inclusion of the velocity factor c (whose role in condensed matter physics will be taken by the velocity v of a linearly dispersive quasi-particle) gives all components the same physical dimension. The associated **covariant vector** is defined as $x_\mu = \eta_{\mu\nu} x^\nu$ by contraction with the

Minkowski metric,[3]

$$\eta = \{\eta_{\mu\nu}\} = \begin{pmatrix} -1 & & & \\ & 1 & & \\ & & 1 & \\ & & & 1 \end{pmatrix}. \tag{9.1}$$

The passage back to contravariant vectors is defined by $x^\mu = \eta^{\mu\nu} x_\nu$, where $\eta^{\mu\nu}$ is the inverse of the metric. (In the above representation, the Minkowski metric is trivially self-inverse. However, it need not be in other coordinate representations. For a discussion of the actual meaning of the above index-raising and index-lowering operations we refer to section A.1.3.)

Derivatives with respect to covariant and contravariant vector components are defined respectively as

$$\partial_\mu \equiv \frac{\partial}{\partial x^\mu} = (+\partial_0, \nabla), \qquad \partial^\mu \equiv \frac{\partial}{\partial x_\mu} = (-\partial_0, \nabla).$$

For example, the **d'Alembert operator** is defined as

$$\partial_\mu \partial^\mu = -\frac{1}{c^2} \partial_t + \Delta.$$

Theories following the conventions above are said to be formulated in **Lorentzian signature**. By contrast, a theory with **Euclidean signature** has a standard metric $g_{\mu\nu} = \delta_{\mu\nu}$. We can pass from one to the other by allowing Lorentzian time to take imaginary values (via a Wick rotation), $t = -i\tau$. In the τ-representation, the Minkowski metric assumes Euclidean form. The imaginary-time formalism used in many parts of this text effectively implements this change of representation.

9.1.1 Derivation of the Dirac equation

The derivation of the Dirac equation discussed in this section starts from symmetry considerations. Particular emphasis is placed on the connection between Dirac

---

[3] The metric is sometimes defined with a negative sign, $\eta \to -\eta$ (preferred in electromagnetism and particle physics). Here, we work with the choice above (preferred in gravity), or $\eta_{00} = -1$, as it naturally reflects the sign change $\eta_{00} \leftrightarrow -\eta_{00}$ when switching between real ("Lorentzian") and imaginary ("Euclidean") quantum field theory. However, the choice of convention remains physically inconsequential.

theory and the quantum mechanics of spin. For the time being, we will work in three spatial dimensions for definiteness.

Spin rotation symmetry: a warm-up

As a preparatory exercise, consider a single-particle Hamiltonian containing a coupling term $\hat{H}_{\mathrm{LS}} = \sum_i \hat{l}_i \otimes \hat{\sigma}_i$ between an orbital angular momentum and a spin-$1/2$. The tensor product indicates that the problem is defined in a Hilbert space, $\mathcal{H} = L^2(\mathbb{R}^2) \otimes \mathbb{C}^2$, comprising the space $L^2(\mathbb{R}^2)$ of square integrable wave functions in three-dimensional space and $\mathbb{C}^2$ for spin. The summation over spatial indices, $i = 1, 2, 3$, makes $\hat{H}_{\mathrm{LS}}$ rotationally invariant. However, to prepare for the discussion of the relativistic case, let us review in detail how this invariance comes about. To this end, consider a rotation of space $x_i \to R_{ij} x_j$ described by a rotation matrix $R \in \mathrm{SO}(3)$. We anticipate that the rotation (group) is represented on the state space as $|\psi\rangle \xrightarrow{R} |\psi'\rangle$, where $\psi'(x) = U\psi(R^{-1}x)$ (see the figure in section 1.6.1). Here, the as yet undetermined matrix $U \in SU(2)$ accounts for the fact that $|\psi\rangle$ is not a "scalar" field, but contains a spin structure. For a classical vector field, $\phi(x) \in \mathbb{R}^3$, the transformation would be represented as $\phi(Rx) = R\phi(R^{-1}x)$. Our task is to find out what transformation U replaces R in the case of a spinor field.

To answer this question, we pass from a "Schrödinger representation" emphasizing the transformation of states to the operator "Heisenberg picture" as follows:

$$\langle \psi' | \hat{H}_{\mathrm{LS}} | \psi' \rangle = \int d^3x \, (\bar{\psi}(R^{-1}x) U^\dagger) \hat{l}_i \sigma_i (U\psi(R^{-1}x))$$

$$= \int d^3x \, \bar{\psi}(x) \, (R\hat{l})_i (U^\dagger \sigma_i U) \, \psi(x) = \langle \psi | (R\hat{l})_i (U^\dagger \sigma_i U) | \psi \rangle, \quad (9.2)$$

where in the second equality we changed integration variables and used the fact that the operator $\hat{l}$ transforms as a vector (why?). This computation shows that (i) on the operator level, the transformation is represented as $\hat{l} \to R\hat{l}$ and $\sigma_i \to U^\dagger \sigma_i U$, and (ii) that rotational invariance $\langle \psi' | \hat{H}_{\mathrm{LS}} | \psi' \rangle = \langle \psi | \hat{H}_{\mathrm{LS}} | \psi \rangle$ requires $(R\hat{l})_i (U^\dagger \sigma_i U) = \hat{l}_i \sigma_i$, which is equivalent to the condition

$$U^\dagger \sigma_i U = R_{ij} \sigma_j. \quad (9.3)$$

We read this as an equation (almost) determining the SU(2) matrix U by the SO(3) matrix R. The key to the solution of Eq. (9.3) lies in the identity between the Lie algebras su(2) and so(3) of the two matrix groups, i.e., the generators of infinitesimal rotations. Recall that both algebras su(2) and so(3) are spanned by generators satisfying the commutation relations

$$[J_i, J_j] = i\epsilon_{ijk} J_k. \quad (9.4)$$

The difference lies in the realization of these generators as matrices:

$$\mathrm{su}(2): J_i = \frac{1}{2}\sigma_i, \qquad \mathrm{so}(3): J_i = -i\epsilon_{ijk} E^{jk}, \quad (9.5)$$

where E^{ij} is a three-dimensional matrix of zeros except for a unit entry at position (i,j). We denote these matrices by identical symbols, emphasizing that they realize the same algebra (9.4) but represented in the different vector spaces $\mathbb{C}^2$ and $\mathbb{R}^3$, respectively. The equivalence of the algebras implies a local isomorphism between the two groups: to a finite SU(2) rotation $U = \exp(i\phi_i J_i)$ it assigns the SO(3) rotation $R = \exp(i\phi_i J_i)$, with differently represented J_i. It is straightforward to verify that, with these identifications, Eq. (9.3) holds and rotational invariance is established.

spinor rep-
resentation

Summarizing, we have seen that the rotational invariance of quantum mechanics with fermionic (spin-1/2) particles relies on the combination of two different representations of the rotation group. The first is the ordinary representation in terms of SO(3) space rotations, the second is a **spinor representation** in terms of SU(2). The two representations are determined by each other via their equivalence at the algebra level, so(3) $\simeq$ su(2), and on this basis rotational invariance is established.

INFO Notice, however, that the two groups are only locally equivalent. *Globally*, the correspondence between SO(3) and SU(2) is 2 to 1: the correspondence above assigns to a rotation by 2π around, say, the three-axis the SO(3) matrix $R = \exp(2\pi i J_3) = \mathbb{1}_3$ and the SU(2) matrix $U = \exp(2\pi i J_3) = -\mathbb{1}_2$. These formulae state that a 2π rotation of Euclidean space is an empty operation. However, it changes the sign of the spinor degrees of freedom. In fact, the transformation behavior above is the defining property of

spinor

a **spinor**. Nevertheless, in Eq. (9.3) this sign factor drops out and hence does not conflict with rotational invariance.

At this point, readers may wonder how this discussion is related to the theme of this chapter, relativistic quantum field theory. In the following, we explain this point and show how a slight extension of the construction above defines a Lorentz invariant theory.

Lorentz group (and its double cover)

Lorentz
group

Recall that the **Lorentz group** is defined by the set of linear transformations, Λ, of $(3+1)$-dimensional space–time that leave the Minkowski metric invariant, $\Lambda^T \eta \Lambda = \eta$. With, $\eta = \mathrm{diag}(-1,1)$, it contains SO(3,1) as a subgroup. This is the group of unit-determinant Lorentz transformations, which can be continuously deformed to unity. It will assume the role SO(3) had in the previous discussion. In addition, there are the discrete non unit-determinant transformations **parity**, P, and **time-reversal**, T, with

$$\mathrm{P}(x^0, x^i) = (x^0, -x^i), \qquad \mathrm{T}(x^0, x^i) = (-x^0, x^i). \tag{9.6}$$

The combination of SO(3,1) transformations with P and T exhausts the contents of the Lorentz group.[4]

---

[4] The generalization of this statement to dimensions other than $3+1$ is straightforward. However, in even dimensions such as $d = 2$, we need to replace the now unit-determinant parity

The Lie algebra so$(3,1)$ is spanned by matrices satisfying $X\eta + \eta X^T = 0$. Three of its six independent generators are conveniently chosen as $J_i = -i\epsilon_{ijk}E^{jk}$, now realized as *four*-dimensional matrices. These are the space rotation generators embedded in the Lorentz group. For the remaining three we choose the symmetric matrices $K_i = i(E^{0i} + E^{i0})$, the generators of **Lorentz boosts**.

Lorentz boost

EXERCISE Consider a **Lorentz boost** in the 1-direction $\Lambda = \exp(\theta K_1)$ and convince yourself that $x'^\mu = \Lambda^\mu{}_\nu x^\nu$ describes a relativistic transformation between inertial frames.

A quick calculation shows that rotation and boost generators satisfy the **commutation relations of** so$(3,1)$,

$$[J_i, J_j] = i\epsilon_{ijk}J_k, \qquad [K_i, K_j] = -i\epsilon_{ijk}J_k, \qquad [J_i, K_j] = i\epsilon_{ijk}K_k. \qquad (9.7)$$

Guided by the construction of the previous section, we now ask whether there are spinor representations of the same abstract algebra. The answer can be found by guessing: we need a six-dimensional matrix algebra, and it must span a non-compact matrix group (the latter feature reflecting the non-compact nature of Lorentz boosts). These two features suggest sl$(2, \mathbb{C})$ as a candidate, the algebra of unit-determinant complex two-dimensional matrices. This algebra contains the hermitian rotation generators $J_i = \frac{1}{2}\sigma_i$ and three anti-hermitian generators $K_i \equiv \frac{i}{2}\sigma_i$. A quick calculation shows that, with this identification, the commutation relations (9.7) are satisfied.

To summarize, we have found

$$\text{sl}(2, \mathbb{C}) : J_i = \frac{1}{2}\sigma_i, \qquad \text{so}(3,1) : J_i = -i\epsilon_{ijk}E^{jk},$$

$$K_i = \frac{i}{2}\sigma_i, \qquad\qquad K_i = i(E^{0i} + E^{i0}), \qquad (9.8)$$

as a generalization of Eq. (9.3). Below, we discuss how this result forms the basis for a relativistically invariant generalization of quantum mechanics.

Weyl fermions

At this point, the essential groundwork towards the construction of the relativistic theory has been laid. All we need to do is generalize the above criterion of rotational invariance to Lorentz invariance. To this end, consider the linear operator, $\hat{O}_R = p_\mu \sigma^\mu$, where σ^i are the Pauli matrices as before, $\sigma^0 = \mathbb{1}_2$, and $p_\mu = -i\partial_{x^\mu}$ transforms like a covariant vector. We require the matrix elements $\langle \psi | \hat{O}_0 | \psi \rangle$ to be Lorentz invariant under transformations affecting states as $\psi'(x) = U\psi(\Lambda^{-1}x)$, with an as yet undetermined matrix U. Here, $\psi(x)$ are functions of space–time with scalar product $\langle \psi | \phi \rangle = \int d^4x\, \bar{\psi}(x)\phi(x)$. Arguing as in Eq. (9.2), the invariance condition assumes the form

$$U^\dagger \sigma^\mu U = \Lambda^\mu{}_\nu \sigma^\nu, \qquad (9.9)$$

transformation by **reflection** at a plane, for example, R$(x^0, x^1, x^2) = (x^0, -x^1, x^2)$. (Why is the choice of this plane inessential mod SO$(d, 1)$?)

generalizing Eq. (9.3). It is straightforward to verify that, for a general $SO(3,1)$ transformation, with $\Lambda = \exp(i\phi_i J_i + i\theta_i K_i)$ combining space rotations and boosts, the criterion is satisfied with the matching $SL(2,\mathbb{C})$ transformation

$$U = \exp(i\phi_i J_i + i\theta_i K_i). \tag{9.10}$$

(For all that follows, keep in mind that, in general, U is *not* unitary, $U^{-1} \neq U^\dagger$, for $\theta \neq 0$.)

EXERCISE As usual with such statements, it is sufficient to check them on the level of an infinitesimal transformation. Do so using the commutator relations (9.7).

To summarize, we have found that the equation

$$\hat{O}_R \psi_R \equiv -i\sigma^\mu \partial_\mu \psi_R = (i\partial_t - i\sigma_i \partial_i)\psi_R = 0, \tag{9.11}$$

is relativistically invariant if the solutions transform under $SO(3,1)$ transformations as $\psi_R \to \psi'_R$ with $\psi'_R(x) = U\psi_R(\Lambda^{-1}x)$. Note that the zero on the right-hand side is fixed by the invariance criterion. For example, a term like $V(x)\psi(x)$ with some scalar "potential" would not transform in the right way. (Why? Remember that $U^\dagger U \neq \mathbb{1}$ in general.) Two-component spinor fields with this property are called **Weyl fermions** (right-handed) **Weyl fermions**.

Weyl fermions

INFO Weyl fermions have recently come to prominence in the condensed matter physics of **Weyl Semimetals**.[5] These are three-dimensional materials harboring an even number of singular points in their Brillouin zone around which the dispersion is linear (see the figure). These Weyl points, or nodes, can be shifted relative to each other, both in momentum space and in energy, but cannot be individually destroyed. Eigenstates describing the physics of the system at low excitation energies relative to the Weyl nodes obey the equation $\hat{H}\psi = E\psi$, where $\hat{H} \equiv -\sigma^i \hat{p}_i$ features as the effective Hamiltonian of individual Weyl nodes. Up to a rescaling by the velocities of the band dispersion, the nodal time-dependent Schrödinger equation is equivalent to the Weyl equation Eq. (9.11).

Weyl semimetal

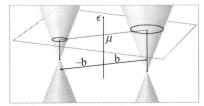

Dirac equation

At this point, we have defined an $SO(3,1)$-invariant theory. However, what about the remaining Lorentz transformations, **parity** and **time-reversal**? Following the same strategy as above, we seek a transformation $P\psi = \psi'$, where $\psi'(x) = U_P\psi(Px)$, and the action of $P = P^{-1}$ on space–time vectors is defined in Eq. (9.6). Invariance of the Weyl equation requires $P\sigma^\mu \equiv U_P^\dagger \sigma^\mu U_P = P^\mu{}_\nu \sigma^\nu \equiv \bar{\sigma}^\mu$, with $\bar{\sigma}^\mu = (\sigma^0, -\sigma^i)$.

[5] See B. Q. Lv *et al.*, *Experimental discovery of Weyl semimetal TaAs*, Phys. Rev. X **5**, 031013 (2015), for an experimental realization in TaAs and N. P. Armitage, E. J. Mele, and A. Vishwanath, *Weyl and Dirac semimetals in three-dimensional solids*, Rev. Mod. Phys. **90**, 015001 (2018) for a review.

Evaluated on the spatial indices, this means that $U_{\mathrm{P}}^{\dagger}\sigma^i U_{\mathrm{P}} = -\sigma^i$. However, there is no matrix with this property (check it!). To achieve parity invariance, we need to extend the theory.

The key lies in the observation that there exists a second representation of $\mathrm{SL}(2, \mathbb{C})$ defining an $\mathrm{SO}(3, 1)$-invariant theory. In this case, states transform as $\psi_L \to \psi_L'$ with $\psi_L'(x) = (U^{-1})^{\dagger}\psi_L(\Lambda^{-1}x)$. It is straightforward to verify that, for these states, the equation

$$\hat{O}_L\psi_L \equiv -i\bar{\sigma}^{\mu}\partial_{\mu}\,\psi_L = (i\partial_t + i\sigma_i\partial_i)\psi_L = 0 \tag{9.12}$$

is $\mathrm{SO}(3, 1)$-invariant. We distinguish between the two realizations of the theory by calling ψ_R and ψ_L **right- and left-handed Weyl spinors**.

The equation $\mathrm{P}\sigma^{\mu} = \bar{\sigma}^{\mu}$ indicates that parity couples the two representations, and the only remaining question is how the extended theory can be made P and T invariant. To this end, we combine the two Weyl fermions into a four-component **Dirac spinor**

Dirac
spinor

$$\psi = \begin{pmatrix} \psi_R \\ \psi_L \end{pmatrix}, \tag{9.13}$$

and the two Weyl equations then become

$$\begin{pmatrix} & \hat{O}_L \\ \hat{O}_R & \end{pmatrix} \begin{pmatrix} \psi_R \\ \psi_L \end{pmatrix} = 0. \tag{9.14}$$

In this extended representation, Lorentz transformations, with $\Lambda \in \mathrm{SO}(3, 1)$, are implemented as

$$\psi'(x) = U_{\Lambda}\psi(\Lambda^{-1}x), \qquad U_{\Lambda} = \begin{pmatrix} U & \\ & U^{-1\dagger} \end{pmatrix}, \tag{9.15}$$

where the subscript indicates that the $\mathrm{SL}(2, \mathbb{C})$ matrices U are functions of Λ via Eq. (9.9). However, it is now also evident how to include **parity symmetry**. Previously, we have seen that $\mathrm{P}\sigma^{\mu} = \bar{\sigma}^{\mu}$, or $\mathrm{P}\hat{O}_L = \hat{O}_R$. We thus need a unitary transformation U_{P} exchanging the role of $\hat{O}_L$ and $\hat{O}_R$ in the above equation. This transformation is easy to find: with $U_{\mathrm{P}} = \tau_1$ and the τ-matrix acting in the L/R space of Eq. (9.13), the extended equation becomes invariant under $\psi(x) \to \psi'(x) = U_{\mathrm{P}}\psi(\mathrm{P}x)$. For the slightly more tricky inclusion of time-reversal, we refer to problem 9.4.1. Summarizing, the extended representation:

> Equation (9.14), with $\hat{O}_{R,L}$ defined in (9.11) and (9.12), respectively, provides the minimal framework for a fully Lorentz invariant theory of quantum spinors.

Previously, we have seen that the Weyl equations were rigid in that they did not admit anything but zero on their right-hand side. However, perhaps the doubled representation leaves room for more flexibility? To find out, we represent the matrix operator in Eq. (9.14) as

$$-i\slashed{D} \equiv -i\partial_{\mu}\gamma^{\mu}. \tag{9.16}$$

The notation emphasizes that the operator is of first order in derivatives and contains a 4×4 matrix structure implicitly defined by the γ^μ. These – you will have guessed it – are the Dirac γ-matrices, and they will be discussed in detail in section 9.1.1. However, at this point all we need is the Lorentz invariance of the structure above, $i\partial_\mu \gamma^\mu \to U_\Lambda (i\partial_\mu \Lambda^\mu{}_\nu \gamma^\nu) U_\Lambda^{-1} = i\partial_\mu \gamma^\mu$. (Compare this expresssion with the transformation of the previously discussed Weyl equations.) There, we had a matrix structure $U(\ldots)U^\dagger$ with $UU^\dagger \neq \mathbb{1}$ in general. This is now replaced by

Dirac equation

$U_\Lambda(\ldots)U_\Lambda^{-1}$, and so we may add an invariant term $m\mathbb{1}_4 \to U_\Lambda(m\mathbb{1})U_\Lambda^{-1} = m\mathbb{1}$. Following this generalization, we arrive at the full **Dirac equation**[6]

$$\boxed{(i\slashed{\partial} - m)\psi = 0, \qquad \slashed{\partial} \equiv \partial_\mu \gamma^\mu} \tag{9.17}$$

Physically, the **mass term** defines the rest mass of the corresponding relativistic particle. It couples left and right spinors, which exist in isolation only in the ultra-relativistic limit, $m \to 0$.

Referring for an in-depth discussion of relativistic Dirac quantum theory to textbooks on particle physics, here we simply note that in **condensed matter physics** linearly dispersive fermions generally appear in even numbers containing equally many left- and right-moving branches (see section 9.1.2 for a more detailed discussion). This principle explains, e.g., the even number of linear cones in the Weyl semimetals mentioned previously. Mass terms occur when the left and the right sectors get coupled by, e.g., potential scattering. In such cases, the Weyl cones "gap out" to produce a dispersion, as in the figure at the beginning of the chapter. To see how this happens, make the plane wave ansatz $\psi \sim \psi_0 \exp(ip_\mu x^\mu)$, which reduces the Dirac equation to $(p_\mu \gamma^\mu + m)\psi_0 = 0$. With the above identification of the γ-matrices and the relation $\hat{O}_L \hat{O}_R = \hat{O}_R \hat{O}_L = p_\mu p^\mu$, it is straightforward to verify that solutions exist for $p_\mu p^\mu + m^2 = 0$. Recalling that $p_0 = E$ is the energy, we have

$$E = \pm(p_i p_i + m^2)^{1/2} \tag{9.18}$$

for the two branches of a **massive Dirac dispersion**.

γ-matrices

In his original derivation, Dirac introduced a set of γ-matrices ad hoc, in such a way that the (likewise postulated) first-order differential equation (9.17) would produce a relativistic dispersion. He found that the defining condition required

$$\boxed{[\gamma^\mu, \gamma^\nu]_+ = -2\eta^{\mu\nu}} \tag{9.19}$$

i.e., different γ-matrices anticommute and square to multiples of the unit matrix.

Clifford algebra

(These are the defining commutation relations of a **Clifford algebra**.) A beautiful aspect of Dirac's construction is that any set of matrices satisfying the Clifford algebra will define a valid representation of the theory.

[6] P. A. M. Dirac, *The quantum theory of the electron*, Proc. R. Soc. Lond. A **117**, 610 (1928).

To see how this construction is related to the present approach, we can compare the definition $-i\partial_\mu\gamma^\mu = \begin{pmatrix} & \widehat{O}_L \\ \widehat{O}_R & \end{pmatrix} = -i\partial_\mu \begin{pmatrix} & \bar\sigma^\mu \\ \sigma^\mu & \end{pmatrix}$ with the operator representation in Eq. (9.14) to obtain the explicit realization

$$\boxed{\gamma^0 = \tau_1, \qquad \gamma^i = -i\tau_2 \otimes \sigma_i} \qquad (9.20)$$

γ-matrices   These matrices satisfy the Clifford algebra relations and they define the γ-**matrices** in the so-called **Weyl (or chiral) representation**.[7] In addition to γ^μ, one often introduces a further matrix

$$\boxed{\gamma_c \equiv \gamma^5 \equiv i^{\frac{d}{2}-1}\gamma^0 \ldots \gamma^{d-1}} \qquad (9.21)$$

In $d = 4$, this defines a fifth γ matrix; and for historical reasons it is often called γ^5, including cases where $d \neq 4$. The denotation γ_c reflects that, in the Weyl representation, $\gamma_5 = \tau_3$. Comparing with Eq. (9.13), we note that the eigenstates of this operator have purely right- or left-handed chirality, hence the subscript c for chiral.

EXERCISE Note that γ^0 is hermitian, while the γ^i are anti-hermitian. Considering the Clifford algebra, why can γ^i not be hermitian? On the same basis, show why the γ^μ are traceless.

The form of the γ-matrices is fixed only up to transformations leaving the Clifford algebra invariant. For example, in the **Dirac basis representation**, $\gamma^0 = \tau_3$ and γ^i as in Eq. (9.20). Another example is the **Majorana representation**,

Majorana representation

$$\gamma^0 = \sigma_2 \otimes \tau_1, \qquad \gamma^1 = i\sigma_3, \qquad \gamma^2 = \sigma_2 \otimes (-i\tau_2), \qquad \gamma^3 = -i\sigma_1, \qquad (9.22)$$

for which the γ-matrices are purely imaginary. The physical meaning of this representation is discussed in section 9.1.2.

EXERCISE Identify the unitary transformation U mapping between the Weyl and the Majorana representations, $\gamma^\mu_{\text{Majorana}} = U\gamma^\mu_{\text{Weyl}}U^\dagger$.[8] (Hint: In the solution of such problems, two tools come in handy. First, the anticommutativity of the Pauli matrices, e.g., $\tau_1\tau_2\tau_1 = -\tau_1$, and second, the option to "rotate" them around coordinate axes, e.g., $\exp(i\frac{\alpha}{2}\sigma_1)\sigma_2\exp(-i\frac{\alpha}{2}\sigma_1) = -\sigma_3$ for $\alpha = \pi/2$.)

Spinor bilinears and currents

In Dirac theory, physical objects are usually labeled according to their transformation behavior under the Lorentz group. For example, the field $\psi(x)$ transforms under the spinor representation defined by Eq. (9.15). "Good" operators have the property that they transform spinors into spinors. For example, application of the

[7] In the literature, different yet unitarily equivalent sign conventions such as $\gamma^i \leftrightarrow -\gamma^i$ can be found.

[8] The answer is given by $U = \Pi \ \text{exp}(i\frac{\pi}{4}\sigma_2)(1 - \tau_2) \otimes \tau_2$. (printed upside down)

Dirac operator to a spinor produces $(i\slashed{D} - m)\psi$, which again is a spinor (check it) regardless of whether ψ solves the Dirac equation.

Physical observables are usually described by bilinears in the states, and here again we should classify them according to their transformation behavior. The simplest transformation in relativity is that of scalars $\phi(x) \rightarrow \phi'(x) = \phi(\Lambda^{-1}x)$. How can we build a scalar from quantum spinors? Since $U_\Lambda^{-1} \neq U_\Lambda^\dagger$ in general, the obvious guess, $\psi^\dagger\psi$, does not do it. However, with the definition

$$\bar{\psi} \equiv \psi^\dagger\gamma^0, \tag{9.23}$$

the bilinear $\bar{\psi}\psi = \psi^\dagger\gamma^0\psi$ *is* a proper **Lorentz scalar**. This follows from the identity $\gamma^0 U_\Lambda^\dagger \gamma^0 = U_\Lambda^{-1}$, which in turn follows from Eqs. (9.15) and (9.20). The same argument shows that $\bar{\psi}\phi$ is scalar even for $\phi \neq \psi$, provided both are spinors.

There are several useful generalizations to bilinears with different transformation behavior. For example, it is straightforward to verify (exercise) that the combination ^pseudo ^scalar $\bar{\psi}\gamma^5\psi$ defines a **pseudo scalar**. It is invariant under Lorentz boosts and rotations, but changes sign under parity and time-reversal. A similar computation shows that the quantity

$$j_V^\mu \equiv \bar{\psi}\gamma^\mu\psi \tag{9.24}$$

transforms like a **Lorentz vector**, $j_V^\mu(x) \rightarrow \Lambda^\mu{}_\nu j_V^\nu(\Lambda^{-1}x')$. This bilinear turns out to be the conserved current associated with the gauge invariance of the Dirac action ^vector ^current (section 9.1.2) and is hence called the **vector current**. In particular $\bar{\psi}\gamma^0\psi$ defines the **charge density** in Dirac theory. Finally, the **axial current**,

$$j_A^\mu \equiv \bar{\psi}\gamma^\mu\gamma^5\psi, \tag{9.25}$$

transforms like a vector under proper Lorentz transformations but changes sign under parity and time-reversal just like the pseudo scalar.

Although the classification above, emphasizing transformation behavior, is tailored to applications in relativistic quantum field theory, it turns out to be useful in condensed matter applications as well. We will see this in later sections when realizations of currents and scalars reappear in different contexts.

Dirac operators in lower dimensions

The symmetry-oriented derivation of the Dirac equation works in arbitrary dimensions. In the following, we illustrate this point for $(1 + 1)$-dimensions; this plays a role in numerous condensed matter applications. Another motivation for an early introduction of the lower-dimensional realization is that it is technically simpler than the $(3 + 1)$-dimensional realization, but conceptually similar; many of the intriguing phenomena observed in Dirac systems afford a technically simpler representation in two dimensions. This will allow us to understand their physical principles without being bogged down by technical complications.

The dimensionally reduced theory follows from the analysis of $SO(1, 1)$, which is now a one-dimensional group represented by matrices $\Lambda = \exp(\theta\sigma_1)$ acting on the $(1 + 1)$-dimensional space–time vectors of the theory as Lorentz boosts. The invariance condition is satisfied as before via right and left Weyl fermions acted upon by the differential operators, $i\sigma^\mu \partial_\mu$ and $i\bar\sigma^\mu \partial_\mu$ with $\sigma^\mu = (\sigma^0, \sigma^1)$ and $\bar\sigma^\mu = (\sigma^0, -\sigma^1)$. Accordingly, we end up with just two γ-matrices, $\gamma^0 = \tau_1$ and $\gamma^1 = -i\tau_2 \otimes \sigma_1$. The corresponding $\gamma_c = \gamma_0 \gamma_1 = \tau_3 \otimes \sigma_1$.

We note that there are just three γ-matrices acting in a four-dimensional spinor representation space. The "sparsity" of these matrices indicates redundancy in the representation. A more compact form follows from the observation that the Clifford relations (9.19) imply the commutation relations $[\gamma_0, \gamma_1] = 2\gamma_c$, $[\gamma_1, \gamma_c] = 2\gamma_0$, $[\gamma_c, \gamma_0] = -2\gamma_1$. Up to a sign, these are the relations satisfied by the algebra of the Pauli matrices, showing that the **two-dimensional Dirac equation** can be equivalently represented as[9]

$$(i\partial_0 \gamma^0 + i\partial_1 \gamma^1 + m)\psi = 0, \tag{9.26}$$

with

$$\gamma^0 = \tau_1, \qquad \gamma^1 = -i\tau_2, \qquad \gamma_c = \tau_3. \tag{9.27}$$

In condensed matter applications, the equation with Lorentizan signature (9.26) often describes the physics of quasi-one dimensional materials, where $x^0 = t$ is time and $x^1 = x$ is space (section 3.6.1).

There exists another way of reducing the dimensionality of Dirac equations: the **Kaluza–Klein reduction**. While the idea originated long ago in the context of gravity,[10] it affords a natural motivation from condensed matter reasoning: Dirac materials are often realized through the dimensional reduction of higher-dimensional materials. For example, a graphene sheet is a dimensional reduction of graphite, and a carbon nanotube is a dimensional reduction of graphene. Consider what happens if a d-dimensional material with a *massless* Dirac operator for its low-energy description gets "rolled up" to a cylinder of narrow circumference $L_\perp$. Choosing the d-coordinate to be the transverse coordinate, the size quantization of its eigenvalues requires $p_d = 2\pi n / L_\perp$. For sufficiently small $L_\perp$, the contribution of this momentum to the Dirac eigenvalue becomes large, and the low-energy sector is that of the transverse zero mode, $p_d = 0$. The reduced Dirac operator describing it has $d - 1$ longitudinal variables, and one γ-matrix fewer. Importantly, it also contains a higher degree of symmetry than its d-dimensional parent: the reduced operator anticommutes with γ^d, which defines a *chiral symmetry* (cf. section 8.1.1). Symmetry hierarchies defined by the successive dimensional reduction of low-energy Dirac representations provide a particularly transparent way of understanding the symmetry patterns of the periodic table of topological insulators discussed in section 8.1.1.

[9] One can show that the minimal dimension of matrix representations for d-dimensional Clifford algebras is given by $2^{[d/2]}$, where $[d/2]$ is the integer part of $d/2$, i.e., 2 in $d = 3 + 1$, 1 in $d = 2 + 1$ and $1 + 1$, etc.

[10] T. Kaluza, *On the problem of unity in physics*, Sitzungsber. Preuss. Akad. Wiss. Berlin (Math. Phys.), 966 (1921); O. Klein, *Quantum theory and five-dimensional theory of relativity*, Z. Phys. **37**, 895 (1926).

Odd-
dimensional
Dirac
operators
To summarize, there are two different strategies for realizing Dirac operators in dimensions different from $3 + 1$, dimensional reduction or "working from scratch" by constructing spinor representations of the relevant Lorentz groups. Either way, one finds that there is a key difference between Dirac operators in even and odd dimensions. **Odd-dimensional Dirac operators** have no sense of chirality, i.e., there is no notion of left- and right-handed fermions. There are different ways to see this. For example, we have defined chiral fermions as eigenstates of the matrix γ_c with eigenvalue ± 1. However, in odd space–time dimensions, γ_c *commutes* with all other γ-matrices and is therefore a multiple of the unit matrix in irreducible representations of the latter – there do not exist nontrivial eigenstates of a chirality operator. Alternatively, one may work from the fundamentals, and investigate spinor representations of, e.g., the three-dimensional group $SO(2, 1)$. This is a highly recommended exercise (play with the group $SL(2, \mathbb{R})$), which shows that the theory no longer splits into distinct L and R sectors.

9.1.2 The Dirac action

REMARK The discussion in this section glosses over a number of fundamental issues concerning the symmetries of the Dirac action, or the differences between operator and path integral representations. For an authoritative discussion of these aspects, see Ref. [11].

The derivation of the Dirac equation marked the beginning of quantum field theory. Attempts to interpret it as a relativistic generalization of the single-particle Schrödinger equation failed due to the unbounded dispersion (9.18). The suggested solution was a many-body interpretation, where all negative energy eigenstates are filled – the "Dirac sea."

Dirac sea   **INFO** The postulate of a **Dirac sea** of filled states was one of the early triumphs of Dirac theory. It showed that a consistent relativistically invariant generalization of quantum mechanics required a many-body interpretation. It also led to the prediction of **antimatter**: positively charged hole states left behind when electrons are excited out of the sea. At the same time, the Dirac sea led to conceptual problems such as the need for an infinitely large negative charge of sea states compensating the precisely equal positive charge of empty space. In modern quantum field theory, the idea of the Dirac sea has been replaced by a reinterpretation of the vacuum. (In essence, it is axiomatically declared that, for negative energy states, fermion annihilators create hole states, rather than the null vector.) However, in condensed matter physics, the Dirac sea is generally physical and represents the Fermi sea of filled valence electrons below the chemical potential.

In the field integral-oriented spirit of this text, Dirac quantum field theory is defined by a free fermion coherent state action whose variational equation is Eq. (9.17). Naïvely, one might expect a Gaussian action with the Dirac operator as a Gaussian weight. This guess is almost, but not quite, correct. Inspection of Eq. (9.17) shows that the Dirac operator is not hermitian ($\gamma_0^\dagger = \gamma_0$, $\gamma_i^\dagger = -\gamma_i$). Even though

[11] M. Stone *Gamma matrices, Majorana fermions, and discrete symmetries in Minkowski and Euclidean signature*, J. Phys. A: Math. Theor. **55**, 205401 (2022).

convergence is not an issue with free fermion actions, we require the hermiticity of one-body fermion operators.

The key to the resolution of this issue lies again in the symmetry of the problem. We have seen in section 9.1.1 that $\psi^\dagger \psi$ does not have invariant transformation behavior in a relativistic theory. However, with $\phi \equiv (i\partial_\mu \gamma^\mu - m)\psi$, the bilinear $\bar\psi\phi$ is a Lorentz scalar and thus defines a promising building block for the construction of an effective action. On this basis, we can consider

$$\boxed{S[\psi] = \int d^dx\, \bar\psi(i\slashed{D} - m)\psi} \tag{9.28}$$

as a candidate for the Dirac action (in Lorentzian, i.e., real-time representation). This action checks the two boxes of Lorentz invariance and the correct variational properties: variation in ψ trivially produces the Dirac equation (9.17). However, there is more to be learned from this representation. Using Eq. (9.20) to rewrite it as[12]

$$S[\psi] = \int d^dx\, \psi^\dagger(i\partial_t + \tau_3 \otimes \sigma_i\,(i\partial_i) - m\tau_1)\psi, \tag{9.29}$$

we observe that the operator sandwiched by the Grassmann fields is hermitian, so that one of the issues raised above is out of the way. Second, Eq. (9.29) nicely illustrates the splitting of the theory into chiral sectors of R and L **Weyl fermions**. The propagation of the right- and left-moving states is described by $i\partial_t \pm i\partial_i\sigma_i$, respectively. The two sectors are coupled by the Dirac mass operator $m\tau_1$. Notice that Eq. (9.29) is still perfectly Lorentz invariant. The special role played by γ_0 in this construction is due to the signature $(-,+,+,\dots)$ of the Minkowski metric; all it does is make a perfectly invariant theory look somewhat asymmetric.

INFO In **condensed matter applications**, Dirac theory frequently emerges in the form of Eq. (9.29). For example, the minimal Hamiltonian of a Weyl semimetal contains two Weyl nodes, individually represented by $\pm i\partial_i\sigma_i$. Impurity scattering may couple these sectors by an operator $\sim m\tau_1$.[13] The real-time action then assumes the form of Eq. (9.29). It may or may not pay to switch to the manifestly invariant representation of Eq. (9.28).

Depending on the context, one may prefer to work in an **imaginary-time framework**. Following standard protocol, the latter is defined by the analytic continuation $t \to -i\tau$. The additional factor 'i' upfront of the analytically continued time derivative would then be absorbed in the matrix $\gamma^0 \to i\gamma^0 \equiv \gamma_E$ to define an algebra of **Euclidean γ-matrices**.[14] Comparing with Eq. (9.19), the new set of matrices does indeed satisfy a Euclidean Clifford algebra, $[\gamma_E^\mu, \gamma_E^\nu] = -2\delta^{\mu\nu}$, and they are uniformly (anti-)hermitian, $\gamma_E^{\mu\dagger} = -\gamma_E^\mu$. Unfortunately, however, the definition of a Euclidean action from the Lorentzian one is not quite as straightforward. The point is that, in the construction above, the definition of

---

[12] Here, the spinor $\psi^\dagger$ and ψ are independent Grassmann fields. Alluding to the representation in terms of creation and annihilation operators, we use the (standard) †-notation to distinguish between $\bar\psi$ and $\psi^\dagger = \bar\psi\gamma_0$.

[13] In the condensed matter context, more general couplings are possible as we are not constrained by relativistic invariance.

[14] In the particle physics literature, the Euclidean γ-matrices are often labeled differently, namely $\gamma_E^{1,\cdots 4}$, with $i\gamma^0 \to \gamma_E^4$ and $\gamma^i = \gamma_E^i$.

the adjoint $\bar{\psi} = \psi^\dagger \gamma^0$ reflected the special role played by the time coordinate. The particle physics Euclidean Dirac action often (but not always) replaces the $\bar{\psi}$ adjoint by $\psi^\dagger$, so that the two variants of the action differ by more than a few factors of 'i'. The rule of thumb is as follows:

$$\text{Lorentzian (real-time) action:} \quad S[\psi] = \int d^d x\, \bar{\psi}(i\partial_\mu \gamma^\mu - m)\psi,$$

$$\text{Euclidean (imaginary-time) action:} \quad S[\psi] = \int d^d x\, \psi^\dagger(\partial_\mu \gamma^\mu_E - m)\psi. \tag{9.30}$$

However, there are no generally accepted conventions, which makes navigating the literature on Euclidean versus Lorentzian Dirac theories somewhat tedious.[15] In condensed matter physics, the situation is better in that effective Dirac theories are generally defined by an underlying microscopic theory, and in this way all factors of i and adjoints get fixed automatically.

Majorana fermions

REMARK This section discusses Majorana fermions in- and outside the context of Dirac physics. It digresses somewhat from the main theme of this chapter and can be skipped at first reading.

In section 9.1.1 we saw that the γ-matrices can be represented in purely imaginary form (9.22). In this representation, the Dirac operator $i\partial_\mu \gamma^\mu - m$ is real, and the Dirac equation (9.17) affords a solution in terms of real spinors, ψ. Within the many-body framework, such solutions give rise to neutral particles, which are their own antiparticles – the famous **Majorana fermions**.

Ettore Majorana 1906–? was an Italian theoretical physicist famed for the discovery of Majorana fermions and his work on neutrino masses. Majorana disappeared in 1938 under mysterious circumstances while on a ferry passage. His subsequent whereabouts and fate remain unknonwn.

Majorana fermions

The fundamental physics of Majorana fermions is discussed in every textbook on particle theory and will not be reviewed here. With condensed matter applications in mind, we rather focus on the **manifestations of Majorana fermions in effective Dirac theories**. To this end, consider the Dirac action (9.28) in the Majorana representation. With $\bar{\psi}(i\partial_\mu \gamma^\mu - m)\psi = \psi^\dagger(i\partial_\mu \gamma^0 \gamma^\mu - m\gamma^0)\psi$ and Eq. (9.22), we notice that the operator in parentheses is purely imaginary and hermitian. In other words, it assumes the form of i times an antisymmetric matrix. With the abbreviation $(i\partial_\mu \gamma^0 \gamma^\mu - m\gamma^0) \equiv iA$, we may rewrite the action as (the notation is somewhat symbolic[16])

$$S[\psi] = i\int \psi^\dagger A\psi = \frac{i}{2}\int (\psi^\dagger A\psi - \psi^T A^T \psi^*) = \frac{i}{4}\int (\eta^T A\eta + \nu^T A\nu) \equiv S[\eta, \nu],$$

where $\eta = \psi + \psi^*$ and $\nu = (\psi - \psi^*)/i$. This construction shows that the action can be represented as a 'real' bilinear form, in terms of the "real" and "imaginary" part

---

[15] See P. Nieuwenhuizen and A. Waldron, *On Euclidean spinors and Wick rotations*, Phys. Lett. B **389**, 29 (1996) for a discussion.

[16] In the first equality we used the fact that for arbitrary Grassmann bilinear forms, $\lambda^T X\nu = \lambda_i X_{ij} \nu_j = -\nu_j X^T_{ji} \lambda_i = -\nu^T X^T \lambda$.

of the Grassmann variables $\psi = \frac{1}{2}(\eta + i\nu)$. We are using quotes because complex conjugation for Grassmann variables is not defined, i.e., $\psi^\dagger$ and ψ are independent variables.

To make the real representation of a fermion more tangible, consider the simpler example of the **free fermion Hamiltonian**, $\hat{H} = a_i^\dagger 2h_{ij} a_j + (a_i^\dagger \Delta_{ij} a_j^\dagger + \text{h.c.})$. Here, h and Δ are arbitrary $N \times N$ matrices constrained only by hermiticity, $h = h^\dagger$, and the antisymmetry $\Delta = -\Delta^T$ required by Fermi statistics, $[a_i^\dagger, a_j^\dagger]_+ = 0$. The Hamiltonian $\hat{H}$ represents the most generic form of a free fermion Hamiltonian. It obeys no symmetries besides hermiticity, not even particle number conservation. Equivalently, we may think of $\hat{H}$ as the most general mean field superconductor Hamiltonian.

As in section 5.3, we introduce a Nambu–Gorkov representation,

$$\hat{H} = (a^\dagger, a)_i \begin{pmatrix} h & \Delta \\ -\bar{\Delta} & -h^T \end{pmatrix}_{ij} \begin{pmatrix} a \\ a^\dagger \end{pmatrix}_j \equiv \Psi^\dagger H \Psi.$$

Notice that the matrix Hamiltonian $\hat{H}$ obeys the relation $\sigma_x H^T \sigma_x = -H^T$. Except for the presence of the Pauli matrices in Nambu space, this is an antisymmetry relation similar to that we had in the Majorana representation of the Dirac operator. This suggests a transformation

$$\Psi = \begin{pmatrix} a \\ a^\dagger \end{pmatrix} = \frac{1}{2} \begin{pmatrix} 1 & i \\ 1 & -i \end{pmatrix} \begin{pmatrix} \eta \\ \nu \end{pmatrix} \equiv M\Xi \tag{9.31}$$

to the "real" **Majorana fermion operators**,

$$\eta = a + a^\dagger, \qquad \nu = \frac{1}{i}(a - a^\dagger). \tag{9.32}$$

Comparison with our previous discussion shows that this is the operator analog of the transformation of Grassmann variables $\psi \leftrightarrow a, \psi^\dagger \leftrightarrow a^\dagger$. Indeed, it is straightforward to check that, in the **Majorana representation**, the Hamiltonian assumes the form

$$\hat{H} = i\,\Xi^T X \Xi, \qquad X^T = -X, \tag{9.33}$$

with a real antisymmetric matrix.[17] The defining features of the Majorana fermion operators are their self-adjointness,

$$\Xi_a^\dagger = \Xi_a, \tag{9.34}$$

and the **Clifford commutation relations**,

$$\boxed{[\Xi_a, \Xi_b]_+ = 2\delta_{ab}} \tag{9.35}$$

where a is a container index for the $2N$ components of $\Xi = (\eta, \nu)$. Notice that we have another realization of the Clifford algebra (9.19), this time in a $2N$-dimensional

[17] As an exercise, verify that $\begin{pmatrix} (\nabla - \nabla) - ({}_L y - y) & (\nabla + \nabla)\imath - ({}_L y + y)\imath - \\ (\nabla + \nabla)\imath - ({}_L y + y)\imath & (\nabla - \nabla) + ({}_L y - y) \end{pmatrix} \frac{\imath}{\imath} = X\imath.$

Euclidean realization, $\eta^{\mu\nu} \to \delta_{ab}$. Also notice that the Majorana fermion squares to unity, $\Xi_a^2 = 1$.

We may establish contact with the previous discussion by turning to an effective Grassmann action representation. Starting from the imaginary-time action $S[\psi] = \int d\tau\, \bar{\Psi}(\frac{1}{2}\partial_\tau + H)\Psi$,[18] we then apply the above similarity transformation to obtain

$$S[\Xi] = \int d\tau\, \Xi^T \left(\frac{1}{4}\partial_\tau + iX\right) \Xi \qquad (9.36)$$

for a Euclidean variant of the Majorana action. With the identification $iA = \frac{1}{4}\partial_\tau + iX$, this resembles the previously discussed Dirac expression, $iA = i\partial_t + i\partial_i\gamma_0\gamma_i - m\gamma_0$.

INFO There exist various condensed matter realizations of free fermion Hamiltonians lacking all symmetries. Examples include semiconductors with strong spin-orbit interaction, such as InAs, tunnel-coupled to superconductors and to external magnetic fields. The corresponding Hamiltonians violate particle number conservation, time-reversal, and spin-rotation symmetry; in the classification of section 8.1.1, they belong to the class of least symmetry, D. In low dimensions, $d = 1, 2$, these systems can be pushed into "topological phases," where they harbor Majorana fermion modes ($d = 2$), or isolated Majorana bound states ($d = 1$) close to their boundaries. In the case of finite-size topological quantum wires ($d = 1$), the two boundary regions are separately described by a Majorana Hamiltonian (9.33) of *odd* dimension. (Notice that our "bulk" construction above necessarily led to *even*-dimensional Majorana Hamiltonians. This indicates that the two odd-dimensional effective edge Hamiltonian of a topological class D quantum wire cannot exist in isolation. However, the full system comprising two odd-dimensional boundary systems conforms with the dimensionality principle, even if the boundaries are separated by an arbitrarily extended insulating bulk.)

The crucial point now is that every odd-dimensional antisymmetric matrix has a zero eigenvalue (why?). The corresponding wave function is the Majorana zero-energy bound state. We may combine the left and right Majorana zero modes $\eta_{L,R}$ to form a complex fermion, $a = \frac{1}{2}(\eta_L + i\eta_R)$. The wave function of this fermion state is unusual in that it is "decentralized" with weight accumulated at the two ends of the wire. This feature makes it robust against local perturbations and gives the qubit states defined by the 0 and 1 occupation of the non-local fermion a high degree of "topological protection." In a nutshell, this is the principle behind the **Majorana qubit** and its stability. At the time of writing, various material platforms are showing indications of Majorana bound state formation (see fig. 9.1 for the example of semiconductor quantum wires). However, in spite of intensive experimental efforts, the definite proof that these states are indeed Majoranas (as opposed to conventional superconductor mid-gap resonances) remains in question.

Majorana qubit

Dirac actions describing lattice structures

Previously, we derived the Dirac action in a continuum framework. Condensed matter systems, on the other hand, are defined as lattice structures. Realizations of Dirac Hamiltonians in condensed matter physics therefore must have the status of *effective* long-range descriptions. In many cases, the passage from a lattice to

---

[18] The factor $1/2$ multiplying the time derivative follows from the rewriting of the canonical term $\int \bar{\psi}\partial_\tau\psi = \frac{1}{2}\int \bar{\Psi}\partial_\tau\Psi$ for $\bar{\Psi} = (\psi, \bar{\psi})$.

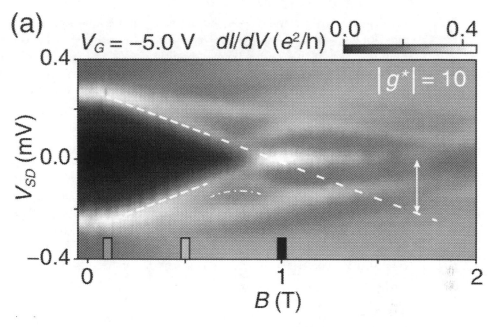

Fig. 9.1   Local density of states of an InAs quantum wire measured by tunnel spectroscopy, as a
function of energy and the strength of the external magnetic field. At a critical point, the
wire enters a topological phase and signatures of a zero-energy state appear. Figure courtesy
of C. M. Marcus, QDev Copenhagen.

a continuum is straightforward: we pick a reference point in the Brillouin zone
and expand around it in small momentum deviations. The same strategy works in
connection with Dirac fermion representations; but there is a subtlety.

The issue is best understood by going in reverse: how would a good lattice discretization of the Dirac action (9.28) look? The natural approach to answering this
question is to introduce a space–time lattice, and replace derivatives by differences,

$$\partial_\mu \psi(x) \to \frac{1}{2\Delta} x(\psi(x + \Delta x e^\mu) - \psi(x - \Delta x)).$$

In a momentum space representation, this turns into

$$\int d^d x\, \bar\psi i \partial_\mu \gamma^\mu \psi \to -\frac{1}{L^d \Delta x} \sum_p \bar\psi \gamma^\mu (\sin(\Delta x p_\mu)) \psi.$$

The issue is that the lattice representation has not just one but 2^d hot spots in its
Brillouin zone around which it is described by a Dirac representation: linearizion
around any of the points $p \in \{0, \pi\}^d$ leads back to the same continuum action,
i.e., the lattice action describes 2^d different "flavors" of Dirac fermions. In particle
fermion   physics, this is known as the **fermion doubling problem**. It presents a nuisance
doubling   in numerical approaches to Dirac problems, and a great deal of effort has been
problem   invested into minimizing its consequences.

In condensed matter physics, we generally approach the problem from a different
perspective: given a solid state system, what kind of effective Dirac low-energy

Nielsen–
Ninomiya
theorem

descriptions are possible? Possible answers to this question must conform with the **Nielsen–Ninomiya no-go theorem**.[19] In essence, this theorem is another manifestation of the conflict between multiple zeros in trigonometric functions and single linearly dispersive hot spots in Brillouin zones. More precisely, it states that in even d space–time dimensions, and under mild conditions (such as locality and translational invariance), it is impossible to realize odd numbers of Weyl fermions on a lattice. For example, in $d = 3 + 1$, the number of Weyl cones in a Weyl semimetal is necessarily even. Or, in $d = 1 + 1$, it is impossible to have a lattice with only left-moving fermions in the low-energy limit, etc.

In odd space–time dimensions (e.g., $d = 2 + 1$), there is more freedom. For example, graphene has two Dirac Hamiltonians as a low energy theory. However, the quantum anomalous Hall (QAH) insulator introduced on page 464 reduces to only one Dirac Hamiltonian at low energies. This is understood by inspection of the lattice dispersion (8.38). For $r \simeq 2$, an expansion of $\sin k_i$ shows that zero momentum, $k = (0,0)$, becomes the center of a linearly dispersive low-energy Hamiltonian

$$H_{\text{eff}} = k_1 \sigma_1 + k_2 \sigma_2 + m \sigma_3, \tag{9.37}$$

with small mass $m = r - \cos k_1 - \cos k_2$. The complementary points $k_i = \pi$ have large mass and effectively get gapped out. As an exercise, investigate how attempts to gap out only one of the two Weyl fermions in a lattice discretization of a $(3+1)$-dimensional Dirac model fail.

Symmetries of the Dirac action

In the present section, we postulated the Dirac action (9.28) associated with the Dirac equation (9.17) (for an actual derivation, see problem 9.4.2) and discussed some of its key features. It should be emphasized just how amazing is this list of features. We started from a minimal relativistically invariant extension of Schrödinger theory. That extension required a spin-1/2 spinor structure. The unboundedness of the relativistic spectrum forced on us a many-body *fermion* interpretation. (For further details, see problem 9.4.2.) This linkage between fermion statistics and odd half-integer spin is a manifestation of the **spin statistics theorem**,[20] which states that, in $(3 + 1)$-dimensional Lorentz invariant theories, particles with odd/even half-integer spin are fermions/bosons.

spin
statistics
theorem

In addition to its invariance under proper Lorentz transformations, the Dirac action possesses three discrete symmetries: **charge conjugation C, parity P, and time-reversal T**. We discussed parity in section 9.1.1; time-reversal and charge conjugation are addressed in problem 9.4.1. Specifically,

> **Julian Schwinger** 1918–1994
> an American physicist who received the 1965 Nobel Prize in Physics with Sin-Itiro Tomonaga and Richard P. Feynman, for their fundamental work in quantum electrodynamics, with far-reaching consequences for the physics of elementary particles.

[19] H. B. Nielsen and M. Ninomiya, *Absence of neutrinos on a lattice*, Nucl. Phys. B. **185**, 20 (1981).

[20] J. Schwinger, *The quantum theory of fields I*, Phys. Rev. **82**, 914 (1951). There are older references, but this one appears to provide the first watertight proof.

the symmetry under the latter implies the existence of antiparticles, particles identical to fermions but of opposite charge. From a condensed matter perspective, the Dirac antiparticles are just the holes left in the Dirac sea of filled negative energy states when one particle is excited out of it. However, to appreciate how weird the original postulate of anti-electrons (positrons) appeared to the physics community, note that Dirac himself mistakenly identified his positive-charge solutions with protons before accepting them as a new species of particle. The combined appearance of C, P, and T is a manifestation of another fundamental theorem, the **CPT-theorem**,[21] which states that Lorentz invariant theories must be invariant under the combined application of the three symmetries. As a corollary, note that, in a relativistic theory, the fulfillment of two of the three symmetries implies the third. Conversely, violation of one of them implies violation of the combination of the remaining two.

CPT-theorem

Finally, the Dirac action *conserves charge* and hence is invariant under continuous U(1) transformations $\psi \to e^{i\phi}\psi$ of the Dirac spinors. However, there is a second and less obvious U(1) symmetry. Consider the transformation

$$\psi \to e^{i\alpha\gamma^5}\psi, \qquad \bar{\psi} \to \bar{\psi}e^{i\alpha\gamma^5}. \tag{9.38}$$

Since γ^5 anticommutes with all other γ-matrices, this is a symmetry of the massless action ($m = 0$) called its **axial symmetry**.

axial symmetry

EXERCISE Show that the conserved currents associated with the two U(1) symmetries are the vector current (9.24) and axial current (9.25), respectively.

At first sight, this looks like a weird non-unitary symmetry (identical signs in the exponent of Eq. (9.38)) of the Dirac operator. However, note that the above transformation implies that $\psi^\dagger = \bar{\psi}\gamma^0$ transforms as $\psi^\dagger \to \psi^\dagger \exp(-i\alpha\gamma^5)$. The symmetry thus preserves the hermitian conjugacy between ψ and $\psi^\dagger$ in the operator interpretation of these objects. Indeed, the physical meaning of the axial symmetry becomes more obvious in the chiral representation (9.29) in terms of ψ and $\psi^\dagger$. With $\gamma^5 = \tau_3$ in the basis underlying this representation, we observe that the axial symmetry simply describes the independently conserved charges of the left- and right-handed fermions in the absence of mass. In this limit, the spinor components with τ_3-eigenvalues ± 1 may be independently transformed, with phase factors $\exp(i(\phi \pm \alpha))$.

From the perspective of the condensed matter physicist, that second continuous symmetry of the Dirac action looks suspicious. We do not expect to find independently conserved charges in lattice theories with approximately linear low-energy spectra. In the next section we will show how the inclusion of quantum fluctuations is key to reconciling these two views of the axial symmetry.

9.1.3 Massive Dirac operators and the physics of boundaries

In condensed matter physics, phase transitions between distinct topological insulator phases involve the closure of band gaps, which can be modeled in terms of

---

[21] The statement appeared first in Ref.[20] and was independently proven by Pauli, Lüders and Bell. See particle physics textbooks for further discussion and references.

low-energy Dirac Hamiltonians. Referring to Ref. [22] for a general discussion covering all dimensions and symmetry classes, the principle is illustrated by the Hamiltonian of the quantum anomalous Hall insulator (8.38). Depending on the sign of $m \equiv r - 2$, the system is either in a topological phase ($m < 0$) or in a trivial insulating phase ($m > 0$). The assignment of a topological index to the parameter r requires a complete description of the system's quantum ground state (see info block on page 464) beyond the local Dirac approximation. Nevertheless, the latter knows about the topological phase via the sign of the Dirac mass.

Now consider a spatial boundary between different topological phases realized by the change in sign of a mass function $m(x)$ smoothly changing as a function of one variable, chosen such that $m(0) = 0$ marks the position of the phase boundary. Far from that interface, our system is in an gapped insulating phase. However, at the interface, the system is in a critical configuration, and we expect a divergence of length scales symptomatic of critical points. Phenomenologically, this spatially confined criticality manifests itself in the protected surface transport properties of topological insulators. Within the Dirac description, it shows in the emergence of **surface zero modes**, zero-energy eigenstates of the Dirac operator spatially confined to the surface.

surface zero modes

Let us construct these states for the example of the low-energy Dirac representation of the QAH insulator (8.39),

$$\hat{H} = i\partial_x \sigma_x + i\partial_y \sigma_y + m(x)\sigma_z.$$

Multiplication of the equation $\hat{H}|\Psi\rangle = \epsilon|\Psi\rangle$ from the left with the matrix $i\sigma_x$ leads to $[(-\partial_x + m(x)\sigma_y) + i\sigma_x(i\partial_y \sigma_y - \epsilon)]\Psi = 0$. The separability of this differential equation, translational invariance in the y-direction, and the Pauli structure of its x-dependent part motivates an expansion $|\Psi(x, y)\rangle = \sum_{k,s=\pm} \psi_s(x) e^{iky}|s\rangle$ in y-eigenstates, $\sigma_y|s\rangle = s|s\rangle$. Substitution of this ansatz into the equation gives

$$\sum_s [(-\partial_x + m(x)s)\psi_s(x) + \psi_s(x)i\sigma_x(-ks - \epsilon)]|s\rangle = 0.$$

The x-dependent part of the equation has the formal solution

$$\psi_s(x) = \psi_s(0) \exp\left(s \int_0^x du\, m(u)\right),$$

where the actual existence of the functions $\psi_s(x)$ depends on the spatial profile of m. For m crossing from negative to positive values, $s = -1$ defines a normalizable solution exponentially decaying on a scale set by the variation of m. For m changing from positive to negative, the

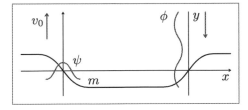

$s = 1$ solution exists. This distinction is at the origin of the definite "chirality" of the surface states. Notice that, for a finite-size geometry with, say, a $(-, +, -)$

---

[22] S. Ryu, A. P. Schnyder, A. Furusaki, and A. W. W. Ludwig, *Topological insulators and superconductors: tenfold way and dimensional hierarchy*, New Journal of Physics **12**, 065010 (2010).

sign profile of the gap in the x-direction, boundary states of opposite chirality are stabilized. Provided m varies on scales short compared with the transverse extension of the system, these states do not overlap (see the figure).

The second term in the equation above fixes the linear dispersion of the surface states as $\epsilon = -ks$. (In physical units, this would read $\epsilon = -v_0 ks$, where the characteristic velocity is determined by the low-energy dispersion of the system.) We conclude that either interface supports a single branch of linearly dispersive modes propagating with opposite velocity $\pm v_0$. These solutions define the edge modes of the QAH insulator.

While the details of **surface state identification in general dimensions** depend on the "fine print" such as the metric signature (Euclidean versus Lorentzian) or on the realization of the γ-matrices, the general principle mirrors the rationale of the computation above. To see why, consider the Lorentzian Dirac equation (9.17) with a mass parameter crossing zero as a function of, say, $x^1 = x$. Using the fact that $(\gamma^1)^2 = -1$, multiplication of the equation by $i\gamma^1$ transforms it into $(\partial_x + i\gamma^1 m(x) - \sum_{\mu \neq 1} \partial_\mu (\gamma^1 \gamma^\mu))\psi = 0$. Now, the anti-hermitian matrix γ^1 has eigenvalues $\pm i$, and a projection $\psi = \sum_s P_s \psi$, where $\gamma^1 P_s = is P_s$ reduces the x-dependent part of the equation to $(\partial_x - sm(x))P_s \psi_s = 0$. Depending on the slope of the zero crossing, this equation has a bound state solution for $s = \pm 1$. The projection of the "longitudinal" part of the equation then defines a system of linearly dispersive surface states as above. (As an exercise, fill in the details!)

In section 9.2.1, we will discuss the connection between the topological signature of the Dirac operator and its zero modes of definite chirality. In that reading, the surface modes identified above reflect a change in a topological index of the Dirac operator. The practical consequence of this topological principle is that there is a high level of protection of the surface states against, say, potential disorder or other sources of inhomogeneities.

9.2 Anomalies

In section 1.6, we introduced symmetries as transformations of fields that leave their action unchanged. Introduced in this way, the invariance criterion must hold for *all* fields. In particular, it applies to fields satisfying extremal conditions, the "classical" equations of motion, where it manifests itself in conservation laws via the Noether anomaly   theorem. A symmetry **anomaly** is realized if the inclusion of (integration over) fluctuations around the classical configuration spoils the conservation law. In view of what has just been said, this statement seems paradoxical. If all fields individually satisfy the symmetry criterion, how can it be violated as a result of an integration over fields? The only way for this to happen is that something exceptional is going on at the boundaries of the integration domain – we anticipate that anomalies are ultraviolet (UV) phenomena.

As an example, consider the action of the nonlinear σ-model (6.32) in $d = 2$. It contains as many derivatives as spatial integrations, which implies scale invariance, a symmetry under transformations $g(x) \to g(x/b)$. However, as discussed in section 6.4, fluctuations in this theory are UV singular, and their integration – i.e., renormalization group procedure – sends the theory into a disordered phase with exponentially decaying correlations and no scale invariance.

The reinterpretation of fluctuation–renormalization as an anomaly may be interesting, but is a little too vague to be useful in its own right. More concrete realizations are found in the physics of Dirac systems. These theories have linear dispersion, which implies that their propagator is inverse-linear in momenta. Loop integrals over fluctuations thus show a higher degree of UV singularity than in non-relativistic theories. In the following, we discuss the classic example of an anomaly relying on this principle – the *chiral anomaly*.

9.2.1 Chiral anomaly

chiral
anomaly Discovered by Adler,[23] and Bell and Jackiw,[24] the **chiral anomaly** refers to the non-conservation of the axial current density (9.25) in Dirac theory. The essence of this effect is explained in a $(1 + 1)$-dimensional setting.

Qualitative discussion

Consider the two-dimensional Dirac action, in the representation of Eq. (9.29), minimally coupled to a background field:

$$S[\psi] = \int d^2x\, \psi^\dagger((i\partial_t + A_0) + \sigma_3\,(i\partial_1 + A_1) - m\sigma_1)\psi. \qquad (9.39)$$

In the massless case, $m = 0$, the left- and right-moving sector decouple. The separately conserved particle numbers on the two branches define the conservation law of the axial symmetry. (In the presence of m, the two branches hybridize and only global charge conservation remains.)

EXERCISE To make the consequences of this decoupling in the effective action more explicit, make the generalization $A \to A + B\sigma_3$ to a vector potential with an "axial component," B. Write down the gauge principle for A and B tha upgrades the above two-fold symmetry under phase transformations by (ϕ, α) to a local gauge symmetry with gauge group $U(1) \times U(1)$. Show that the conserved charges of this symmetry are proportional to the particle numbers on the left- and right-moving branches. Where in this construction do you sense trouble with regard to UV regularity?

Now, let us see what happens if the system is coupled to an electric field E represented by a potential $(A_0, A_1) = (0, -Et)$. Inspection of the Dirac Hamiltonian in a first-quantized representation (see problem 9.4.2), $H = -(p + Et)\sigma_3 + m\sigma_1$, shows that the kinematic momentum $k \equiv p + Et$ changes in time. Assume that, at time

[23] S. L. Adler, *Axial-vector vertex in spinor electrodynamics*, Phys. Rev. **177**, 2426 (1969).

[24] J. S. Bell and R. Jackiw, *A PCAC puzzle: $\pi_0 \to \gamma\gamma$ in the σ-model*, Il Nuovo Cimento A **60**, 47 (1969).

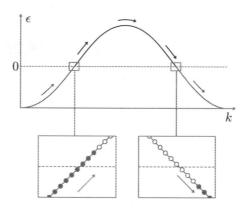

Fig. 9.2   Chiral anomaly in a $(1 + 1)$-dimensional system from a condensed matter perspective. The
            application of an electric field to a one-dimensional crystal (with cosine dispersion relation)
            shifts momentum states as indicated in the upper figure. From the perspective of the two
            low-energy left- and right-moving linearizations around the chemical potential, it looks as
            though particles are appearing or disappearing (see the two lower insets).

zero, all negative-energy states are filled. Over time the momentum increases and
the state occupation changes, as indicated in the insets in fig. 9.2. From the perspective of the right- and left-moving branches, it looks as if particles appear and
disappear, respectively, in conflict with particle number conservation. If we define
the particle density ρ_σ, $\sigma = L, R$, in terms of the number of particles outside the
Dirac sea of negative-energy states, we have (verify) $d_t \rho_\sigma = \sigma E/2\pi$, different from
zero.

Where does the conflict between a decoupling of the two branches, yet violation
of individual charge conservation, originate? The problem lies in the naïvety of our
definition of charge. To determine the total charge of the right-moving branch, say,
we would need to subtract the infinite number of all empty states from the equally
infinite number of occupied states. The alternative to this ill-defined operation is a
UV regularization, defining how the number of left- and right-moving states is limited at some large momentum cutoff Λ. This is how the problem gets resolved. Any
consistent UV regularization effectively introduces coupling between the branches
and removes the spurious conservation of axial charge.

Depending on the context, there are different options for making such UV regularization schemes explicit, and we will discuss some of them below. For the time
being, we just note that condensed matter (lattice) realizations afford a particularly
transparent interpretation. In condensed matter physics, the $(1 + 1)$-dimensional
Dirac action arises as the linearization of a cosine-lattice dispersion around two
Fermi points (see fig. 9.2). The above time-dependence of the momentum is now
understood as a sliding of the full cosine band, as indicated by the arrows. In this
way, we understand how particles get pushed (through the energetically high-lying
UV states) from the left- to the right-moving branch.

Invariant formulation

In previous sections, we emphasized the relativistically invariant formulation of Dirac theory. In the same spirit, let us formulate the above symmetry violation in an invariant form. We first note that $\partial_t(\rho_R - \rho_L)$ is the zeroth component of $\partial_\mu j_A^\mu$, where the axial current density is defined in Eq. (9.25). The non-conservation of axial current thus assumes the form

$$\partial_\mu j_A^\mu = \frac{1}{\pi}E. \tag{9.40}$$

The left-hand side of this equation is a Lorentz scalar, and so we need to represent the right-hand side as a scalar, too. What could this be?

A straightforward way to figure this out is to look at the situation in the language of differential forms.[25] In this framework, the current density in 2 dimensions is a one-form, $j_A = j_{A\nu}dx^\nu$ (see page 784), where we use a roman letter j_A to avoid confusion with the vectorial object $j_A^\mu = \epsilon^{\mu\nu}j_{A,\nu}$.[26] Its conservation is probed by the exterior derivative, $dj_A = \partial_\mu j_{A\nu}dx^\mu \wedge dx^\nu$, a two-form. The non-conservation of axial current must thus be expressed in terms of a two-form containing the electric field. Indeed, we have one such object, the two-dimensional field strength tensor $F = \frac{1}{2}F_{\mu\nu}dx^\mu \wedge dx^\nu$, whose single non-vanishing component is given by $F_{01} = \partial_0 A_1 - \partial_1 A_0 = E$. On this basis, we conclude that the invariant formulation of the non-conservation of axial current assumes the form

$$dj_A = \frac{1}{2\pi}F. \tag{9.41}$$

To translate back to a scalar (0-form) relation, we apply the Hodge star of appendix section A.1.3, which in two-dimensions maps two-forms to 0-forms. Application of Eq. (A.20) leads to

$$\partial_\mu j_A^\mu = \frac{1}{2\pi}\epsilon^{\mu\nu}F_{\mu\nu}. \tag{9.42}$$

On the right-hand side, we have the invariant representation of the electric field.

Chiral anomaly in field theory

How can we describe the chiral symmetry and its breaking within the framework of Dirac field theory? In principle, the strategy seems clear. Given that the anomaly is caused by UV singularities, we need a regulator in the action which damps out fluctuations for large momenta, $p > \Lambda$, and remains invisible at low momenta. The regulator must explicitly break the chiral symmetry (for otherwise it would remain unbroken in the regularized theory). It is also evident that we will see the manifestations of the symmetry breaking only after the UV singular fluctuations have been integrated over, i.e., at the level of "trace logarithms." If, following Noether, we then

---

[25] Readers not yet familiar with this concept may take this is as an incentive to read appendix section A.1. Differential forms will be increasingly used in what follows.

[26] The relation between the current form and vectors involves the antisymmetric tensor, i.e., it is not the usual metric index-raising operation connecting one-forms and vectors.

consider an infinitesimal axial gauge transformation (9.38), its non-commutativity with the regulator will lead to a finite contribution to the effective action. From our discussion above, we expect this contribution to include the coefficients of an external vector potential.

The expansion of the "tr ln" operator in symmetry generators and the vector potential leads to a momentum integral rendered finite by the regulators. The diagrammatic representations of these integrals are known as **triangle diagrams**,[27] and we refer interested readers to any textbook on particle physics for a discussion.

Pauli–Villars regulator

INFO There are different ways to regularize UV divergences in relativistic field theory. For example, **Pauli–Villars regularization**[28] introduces a fictitious system of heavy particles such that divergent propagators are replaced by $(p_\mu \gamma^\mu + m)^{-1} - (p_\mu \gamma^\mu + M)^{-1}$. For small momenta $|p| \ll M$, the subtracted term is irrelevant, but for large momenta, it cancels the singularities of the first. In condensed matter physics, Dirac Hamiltonians arise as low-energy approximations of band structures that define the extrapolation of the theory to larger energies. In these cases, the band structure provides UV regularization. As an example, consider the QAH insulator, Eq. (8.38). For low energies, it reduces to the two-dimensional Dirac Hamiltonian (9.37) with mass $m = r - \cos k_1 - \cos k_2 \approx r - 2$. However, at larger momenta, $m \approx r + \frac{1}{2}(k_1^2 + k_2^2)$, and the quadratic terms provide UV regularization of momentum space propagators. (The full integral over the Brillouin zone is "even more" convergent owing to the finite range of the lattice dispersion relation.)

Fujikawa approach

Below, we will compute the anomaly by a beautiful alternative strategy due to **Fujikawa**.[29] Compared with the diagrammatic formulation, the Fujikawa method is closer in spirit to the path integral approach. It makes reference to dimensionality only at late stages, and elucidates the connections between the anomaly and topology.

Fujikawa method

REMARK In this section, we work in a Euclidean framework. With $\gamma^0 \to i\gamma^0$, all γ-matrices are anti-hermitian, and the eigenvalues $\pm(p_\mu p_\mu)^{1/2} \equiv \pm p$ of the Dirac operator $i\slashed{D}$ are real. We will also make reference to the Wigner transform explained in section A.4.3. However, even readers unfamiliar with the Wigner transform will likely be able to understand the section.

The idea of Fujikawa was to pay attention to the measure of the field integral, $D\psi$. The measure controls which fluctuations are integrated over and should therefore take into account UV issues. Before investigating how axial transformations affect it, we must first define a measure suitable for later UV regularization. To this end, let

---

[27] The particle physics literature generally focuses on $(3 + 1)$-dimensions, where the anomaly manifests itself through a term quadratic in field amplitudes (see below). The corners of the diagrammatic triangle are two vertices defined by A and one defined by the infinitesimal rotation generator.

[28] W. Pauli and F. Villars, *On the Invariant Regularization in Relativistic Quantum Theory*, Rev. Mod. Phys. **21**, 434 (1949).

[29] K. Fujikawa, *Path-Integral Measure for Gauge-Invariant Fermion Theories*, Phys. Rev. Lett. **42**, 1195 (1979).

$$i\slashed{D}\phi_n = \lambda_n\phi_n \tag{9.43}$$

be the complex-valued eigenfunctions (to be distinguished from the Grassmann integration variables) of the Dirac operator $\slashed{D} \equiv (\partial_\mu - iA_\mu)\gamma^\mu$ minimally coupled to an external vector potential. In the Euclidean framework, the eigenvalues λ_n are real.

EXERCISE Consider the inner product

$$\langle f, g \rangle \equiv \int d^d x \, \bar{f}g = \int d^d x \, f^\dagger \gamma^0 g \tag{9.44}$$

and verify that the Dirac operator is hermitian[30] relative to it: $\langle f, (i\slashed{D})g \rangle = \langle (i\slashed{D})f, g \rangle$. On this basis, conclude that the set of eigenfunctions $\{\phi_n\}$ is complete. These features hold for both the Euclidean and the Lorentzian Dirac operators. However, in the latter case, $p_\mu p^\mu$ has indefinite sign and hence the eigenvalues $(p_\mu p^\mu)^{1/2}$ can be real or imaginary. How do hermiticity and the absence of a real spectrum go together? Recapitulate what linear algebra has to say about this matter to figure out why the above inner product does not exclude hermitian operators with complex eigenvalues.

Assuming these states to be orthonormalized, $\langle \phi_n, \phi_m \rangle = \delta_{nm}$, we can expand the Grassmann spinors entering the field integral as $\psi = \sum_n c_n\phi_n$, $\bar{\psi} = \sum_n \bar{c}_n\bar{\phi}_n$, with Grassmann-valued coefficients $c_n, \bar{c}_n$. The integration measure is then defined as $D\psi \equiv \prod_n d\bar{c}_n dc_n \equiv Dc$, and the massless Dirac action reads $S[\psi] = \sum_n \lambda_n \bar{c}_n c_n$.

On this basis, we now investigate the change of the measure under the infinitesimal axial transformation

$$\psi \to \psi' \equiv (1 + i\epsilon\gamma^5)\psi, \qquad \bar{\psi} \to \bar{\psi}' \equiv \bar{\psi}(1 + i\epsilon\gamma^5), \tag{9.45}$$

with real $\epsilon = \{\epsilon(x)\}$. (Since the axial transformation is not unitary, a change in the measure is to be expected.) Expanding ψ' in coefficients $\{c_n'\}$, and using the orthonormality of the eigenfunctions, it is straightforward to verify (do it) that the vector of transformed coefficients is given by $c' = Jc$, where $J = \mathbb{1} + iM$ and the matrix M has elements $M_{nm} = \langle \phi_n, \epsilon\gamma^5\phi_m \rangle$. We note that the matrix M is anti-hermitian, $M = -M^\dagger$.

EXERCISE Use the definition of the above inner product and the properties of the matrices γ^0, γ^5 to verify this statement.

The measure thus changes as $Dc = |\det(J)|^{-2} Dc'$, with a Jacobian factor

$$|\det(J)|^{-2} = \exp(-\operatorname{tr}\ln J + \text{c.c.}) = \exp(-\operatorname{tr}\ln(\mathbb{1} + iM) + \text{c.c.}) \simeq \exp(-2i\operatorname{tr}(M)),$$

where in the final step we have used the anti-hermiticity of M.

---

[30] Actually, the hermiticity (or self-adjointness) of the Dirac operator is a subtle issue. We are accustomed to "symmetry relative to a scalar product" as a criterion for the hermiticity of an operator X. However, there is another requirement, that X and $X^\dagger$ have identical domains of definition. The Dirac operator defines one of the rare cases in physics where this additional disclaimer matters, and we will touch upon this point in section 9.2.1. However, for the time being, we rely only on the exchange symmetry of the scalar product and in this sense consider it hermitian.

At this point, we have reduced the problem to a computation of the trace over $M = \epsilon\gamma^5$. However, this final step is not trivial. On the one hand, M contains the matrix γ^5, which is traceless. On the other hand, there are infinitely many terms in the n-sum. We are thus left with a $0 \times \infty$ situation, whose clarification requires a regularization of the sum.

We achieve this regularization by introduction of a "convergence generating factor,"

$$\mathrm{tr}(M) = \mathrm{tr}(\epsilon\gamma^5) \longrightarrow \mathrm{tr}\left(\epsilon\gamma^5 e^{+\delta\slashed{D}^2}\right), \tag{9.46}$$

with positive infinitesimal δ. At this point, the fact that we are using a Euclidean metric comes into play: the eigenvalues of $\slashed{D}$ are imaginary, and hence $\delta\slashed{D}^2$ defines a negative-definite damping operator. The tracelessness of M further implies that contributions to the now regularized trace must involve the participation of the exponents. At the same time, the exponent is small in δ and the smallness of that parameter must be compensated by a large summation volume, which means large momenta; we see the UV nature of the anomaly creeping in.

To make optimal use of this fact, we evaluate the trace in a Wigner transform representation (see appendix table A.1). In concrete terms, this means that we abandon the concept of "exact eigenfunctions" and represent the trace as an integral over momenta and coordinates. Products of non-commuting operators such as $\hat{p}_\mu$ and $A_\nu(\hat{x})$ are evaluated to leading order (in a Wigner expansion), $\hat{p}_\mu A_\nu(\hat{x}) \to p_\mu A_\nu(x) - \frac{i}{2}\partial_\mu A_\nu(x)$, where the expressions on the right-hand side are functions of c-number variables. In this way, the squared Dirac operator becomes

$$-\slashed{D}^2 = (\hat{p}_\mu - A_\mu)\gamma^\mu(\hat{p}_\nu - A_\nu)\gamma^\nu$$
$$\to \left(p_\mu p_\nu + \frac{i}{2}(\partial_\mu A_\nu - \partial_\nu A_\mu)\right)\gamma^\mu\gamma^\nu = p_\mu \delta^{\mu\nu} p_\nu + \frac{i}{2}F_{\mu\nu}\gamma^\mu\gamma^\nu,$$

where, in the final step, we have used the anticommutation relation (9.19) and the notation makes the presence of the (Euclidean, $\eta^{\mu\nu} \to \delta^{\mu\nu}$) metric explicit.[31] Substitution of this expansion into the trace yields

$$\mathrm{tr}\left(\epsilon\gamma^5 e^{+\delta\slashed{D}^2}\right) = \int \frac{d^dx\,d^dp}{(2\pi)^d} \mathrm{tr}\left(\epsilon(x)\gamma^5 e^{-\delta(p^2 + iF_{\mu\nu}(x)\gamma^\mu\gamma^\nu/2)}\right), \tag{9.47}$$

where we have made the coordinate dependence of ϵ and F explicit. We now see light at the end of the tunnel. The γ-matrices appearing in the exponent may counter the tracelessness of the pre-exponential γ^5. Expansion in these matrices introduces small factors of δ, which in turn are countered by the integral over momenta. Specifically, for $d = 2$, a first-order expansion and Eq. (9.27) imply that

$$\mathrm{tr}\left(\epsilon\gamma^5 e^{+\delta\slashed{D}^2}\right) = -i\delta \int d^2x\,\epsilon(x)\epsilon^{\mu\nu}F_{\mu\nu}(x) \underbrace{\int \frac{d^2p}{(2\pi)^2} e^{-\delta p^2}}_{1/4\pi\delta}.$$

---

[31] In the $\to$ step, we have ignored a term $A_\mu A^\mu$ proportional to the unit matrix, which does not contribute to the trace in the limit $\delta \to 0$.

In the final step, we turn back to the **Lorentzian formulation**, which effectively removes a factor i, through $-ix^0 \to x^0$. In this way we arrive at

$$\mathrm{tr}(\epsilon\gamma^5) \xrightarrow{d=2} \frac{1}{4\pi} \int d^2x\, \epsilon\, F_{\mu\nu}\epsilon^{\mu\nu} = \frac{1}{2\pi} \int \epsilon F \tag{9.48}$$

for the regularized trace. The upshot of the construction is that the above $0 \times \infty$ conflict is resolved via the appearance of a finite contribution $\propto \int \epsilon F$ to the regularized action.

EXERCISE Repeat the derivation above in $d = 4$ to obtain

$$\mathrm{tr}(\epsilon\gamma^5) \xrightarrow{d=4} \frac{1}{32\pi^2} \int d^2x\, \epsilon F_{\mu\nu} F_{\rho\sigma}\epsilon^{\mu\nu\rho\sigma} = \frac{1}{8\pi^2} \int \epsilon F \wedge F. \tag{9.49}$$

We may now piece everything together to compute the full change of the action under an infinitesimal axial transformation. Substitution of the transformation (9.38) into the Dirac action generates a term (verify), $\delta S = \int d^d x\, \epsilon \partial_\mu(\bar{\psi}\gamma^\mu\gamma^5\psi) = \int d^d x\, \epsilon \partial_\mu j_A^\mu$. If there was only this term we would conclude that the axial current is conserved, $\partial_\mu j_A^\mu = 0$. However, from the transformation of the measure, we pick up an additional $-2i\,\mathrm{tr}(\epsilon\gamma^5)$, whose regularized value is given by Eqs. (9.48) or (9.49). From the perspective of the path integral, the transformation is just a change of variables, and hence must remain inconsequential. Symbolically: $\int D\psi\, e^{iS} \to \int D\psi'\, e^{iS+\epsilon(\ldots)} \simeq \int D\psi\, e^{iS}(1 + \epsilon(\ldots))$. The identity of the functional in primed and unprimed variables requires the vanishing of the functional expectation value, $\int D\psi'\, e^{iS}\epsilon(\ldots) = 0$. Collecting all terms multiplied by ϵ, this condition is equivalent to the equation

$$\boxed{\langle\partial_\mu j_A^\mu\rangle = \begin{cases} \dfrac{1}{2\pi} F_{\mu\nu}\epsilon^{\mu\nu}, & d = 2 \\[2mm] \dfrac{1}{16\pi^2} F_{\mu\nu} F_{\rho\sigma}\epsilon^{\mu\nu\rho\sigma}, & d = 4 \end{cases}} \tag{9.50}$$

where the brackets stand for the functional average. (On the right-hand side, we have the externally-imposed field strengths, and the average is inessential.)

INFO The chiral anomaly is a real physical effect. In **condensed matter systems**, it describes how two chiral branches of a low-energy Dirac Hamiltonian "talk to each other" indirectly via energetically high-lying parts of the Hilbert space. We have already mentioned electric field-induced charge pumping in $1 + 1$ dimensions as an example. The $(3 + 1)$-dimensional anomaly shows in the physics of Weyl semimetals.[5] To understand how, recall the definition of the electromagnetic fields $B_i = \epsilon_{ijk}F_{jk}$ and $E_i = F_{0i}$ in terms of the strength tensor, which implies $F_{\mu\nu}F_{\rho\sigma}\epsilon^{\mu\nu\rho\sigma} = 8\mathbf{E}\cdot\mathbf{B}$. The analog of axial charge pumping in the $(1 + 1)$-dimensional case is now a transport of charge from one Weyl node to the other. With $j_A^i = 0$ and $j_A^0 = \rho_R - \rho_L$, the difference in particle density between the nodes, we obtain

$$\partial_t(\rho_R - \rho_L) = \frac{1}{2\pi^2}\mathbf{E}\cdot\mathbf{B}. \tag{9.51}$$

This equation describes the build-up of charge balance in response to an electric field and, unusually, a longitudinal magnetic field. For the (not entirely straightforward) manifestation of this equation in observable transport coefficients, we refer to Ref.[5].

Chiral anomaly and topology

REMARK The following section requires familiarity with section 10.5 and, in particular, the concept of Chern classes. It can be skipped by readers who have not yet been through chapter 8.

Consider Eqs. (9.48) and (9.49) for the regularized traces, which we saw were the essential piece in the construction of the anomaly equation. Comparison with Eqs. (10.49) and (10.50) shows that, for constant ϵ, these relations can be rewritten as[32]

$$\boxed{\operatorname{tr}(\gamma_5)_{\mathrm{reg}} = \int \mathrm{ch}_{d/2}(F) = \mathrm{Ch}_{d/2}(F)} \qquad (9.52)$$

where, on the right-hand side, we have the Chern characters and Chern numbers of the gauge theory defined by F. With $\partial_\mu j_A^\mu$ the vector representation of the invariant object dj (section 9.2.1) we formulate the anomaly equation in the language of differential geometry as

$$dj_A = 2 \begin{cases} \mathrm{ch}_1(F), & d = 2, \\ \mathrm{ch}_2(F), & d = 4, \end{cases} \qquad (9.53)$$

in terms of the first or second Chern characters.

These are remarkable identities: Eq. (9.52) establishes a connection between an "analytical" object (a trace regularized in the basis of eigenfunctions of an operator) and a topological object. On this basis, Eq. (9.53) suggests an interpretation of the **chiral anomaly as a topological phenomenon**: in section 10.5.3, we introduced Chern classes as fingerprints of gauge theories defined in topologically nontrivial contexts. To understand the consequences of this connection, consider Eq. (9.53) integrated over a space–time *without boundaries*,[33]

$$\int_M dj_A = 2 \int_M \mathrm{ch}_{d/2}(F) \equiv 2\,\mathrm{Ch}_{d/2}(F), \qquad (9.54)$$

Chern numbers

where, on the right-hand side, we have the integer-valued **Chern numbers** characterizing the gauge theory on M. At first sight, this relation looks like a triviality: on the left-hand side, we are integrating a derivative over a boundaryless domain, suggesting that $0 = 0$ is the only consistent reading. However, Eq. (9.54) must be taken with a grain of salt. For gauge theories with a nontrivial Chern class, the

---

[32] The connection A in these definitions is related to the physical connection A of the present discussion by a factor i, and the trace is irrelevant, as we are considering abelian gauge theory.

[33] This includes the case of infinitely extended space–time with vanishing boundary conditions.

derivative dj_A is defined only locally, and more than one coordinate system is required to describe the divergence. Under these circumstances, both the left- and the right-hand side evaluate to non-vanishing integers. For a concrete example, see problem 9.4.3 where we discuss a manifestation of the chiral anomaly in connection with a **topological charge pumping** protocol.

Above, we introduced the anomaly as a UV effect. Topology, on the other hand, probes large-scale structures and in this regard connects to the most extreme IR. How, then, can the chiral anomaly afford a topological interpretation? The key to understanding what is going on lies in the relation (9.52). We consider the regularized trace, so there is no reason to worry about $0 \times \infty$ pathologies. Switching back to a representation of the trace in the basis of eigenfunctions of the Dirac operator $i\slashed{D}$, we have

$$\mathrm{tr}(\gamma_5)_{\mathrm{reg}} = \sum_{n,\mathrm{reg}} \langle \phi_n | \gamma_5 | \phi_n \rangle,$$

where the sum effectively extends over finitely many terms. (Equivalently, we could damp the sum out via an exponential weight, as in (9.46).) The clue to a better understanding of this relation is the anticommutativity $[\gamma^5, i\slashed{D}]_+ = 0$. It implies that, for any eigenfunction $i\slashed{D}\phi_n = \lambda_n \phi_n$, the function $\gamma_5 \phi_n$ has negative eigenvalue $-\lambda_n$. For $\lambda_n \neq 0$, these are eigenfunctions with different eigenvalues and hence orthogonal, $\langle \phi_n | \gamma_5 \phi_n \rangle \overset{\lambda_n \neq 0}{=} 0$. We conclude that the trace reduces to one over the kernel of the Dirac operator,

$$\mathrm{tr}(\gamma_5)_{\mathrm{reg}} = \sum_{n,\lambda_n=0} \langle \phi_n | \gamma_5 | \phi_n \rangle.$$

To better understand this expression, we consider the γ-matrices in the chiral representation, Eq. (9.20). In this representation, the Dirac operator assumes the block off-diagonal form (9.14), and $\gamma_5 = -\tau_3$ relative to that block form. We note that the zero-energy eigenfunctions are of the form $\phi \equiv \phi_R \equiv \psi_R(1,0)^T$ or $\phi \equiv \phi_L \equiv \psi_L(0,1)^T$ with $\hat{O}_C \psi_C = 0$, $C = R, L$, and γ_5 eigenvalue ∓ 1, respectively. In other words:

> The zero-eigenvalue eigenstates of the Dirac operator $i\slashed{D}$ are defined by the solutions of the right and left Weyl equations (9.11) and (9.12).

Denoting the number of left or right solutions of the Weyl equation by n_C, we have

$$\mathrm{tr}(\gamma_5)_{\mathrm{reg}} = n_R - n_L \equiv \mathrm{Index}(i\slashed{D}). \tag{9.55}$$

index of the Dirac operator

The difference in the number of zero-energy eigenstates of right and left chirality appearing on the right-hand side of this equation is called the **index of the Dirac operator**. Our analysis above identifies the index with a topological invariant,

$$\boxed{\mathrm{Index}(i\slashed{D}) = \mathrm{Ch}_{d/2}(F)} \tag{9.56}$$

**Atiyah–
Singer
index
theorem**

This result connecting the analytical index to topology is known as the **Atiyah–Singer index theorem**.[34] The common sense explanation for the (topological) stability of the index is that all non-zero energy states of the Dirac operator come in left-right pairs. Nothing

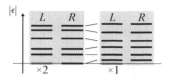

prevents a doublet defined by a left and a right zero-energy eigenfunction from gapping out and becoming such a pair: the total number $n_L + n_R$ is not protected. (See the figure for the lifting of a pair out of an $(n_L, n_R) = (2, 1)$ eigenspace.) However, by the same principle, it is not possible to shift an individual left or right mode from zero energy. In this way, we can understand how topological stability and analytical stability (i.e., the impossibility of creating L-R pairs out of L or R solutions alone) are different sides of the same coin. However, in problem 9.4.3 we demonstrate how "large gauge transformations" may alter then Chern number and at the same time change the analytical index.

INFO An important consequence of the topological interpretation is that the chiral anomaly (and other realizations of anomalies) shows a high level of **robustness in the presence of perturbations**. For example, particle interaction or translational invariance-breaking due to disorder leave the topological terms generated by the anomaly untouched. This protection mechanism implies that physical phenomena predicted on the basis of anomalies enjoy exceptional stability. For examples, see the next section.

9.2.2 Generalizations and physical manifestations of anomalies

Historically, the chiral anomaly is the oldest and most widely known anomaly. However, there are many others. The field theories of particle physics are generally linearly dispersive at large energies, with the consequence that there exists a cobweb of interrelated anomalies that only experts can oversee. Anomalies also appear in realizations not tied to linear dispersion; for an example see the discussion of the conformal anomaly in appendix section A.3.3.

Since the advent of topological quantum matter and its effective Dirac representations, anomalies have become increasingly important in condensed matter physics. While an exhaustive overview of this ongoing development would be beyond the scope of the present text, here we provide a synopsis of some of the more prominent realizations, along with applications. The context of this discussion is the physics of topological insulators and their surface states.

Anomalies and θ-terms

REMARK This section requires a survey of section 8.4 as a prequisite.

---

[34] M. F. Atiyah and I. M. Singer, *The index of elliptic operators on compact manifolds*, Bull. Amer. Math. Soc. **69** , 422 (1963).

So far, we have discussed the chiral anomaly of the massless Dirac operator. A mass term $\sim \bar{\psi} m \psi$ explicitly breaks the axial symmetry (9.38). This transformation makes the change $m \to m\cos(2\alpha) + im\sin(2\alpha)\gamma^5$. At the same

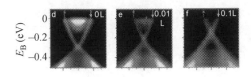

time, the transformation of the measure induces a term $2\alpha\,\mathrm{tr}(\gamma^5)_{\mathrm{reg}}$ in the effective action. While we derived this term for infinitesimal α, a repeated application of the transformation shows that the construction extends to finite parameters, α. The upshot of this argument is that we have a family of equivalent representations of the action,

$$S_\alpha \equiv \int \left(\bar{\psi}(m\cos(2\alpha) + im\sin(2\alpha)\gamma^5)\psi + 2\alpha\,\mathrm{ch}_{d/2}(F) \right), \qquad (9.57)$$

where α now is a constant parameter. From Eq. (10.51) we recognize in the second term the θ-action at topological angle $\theta = 2\alpha$. The freedom implied by the choice of α has interesting physical consequences in the physics of topological insulators (TIs). As an example, consider the three-dimensional TI in class AII (see section 8.1.1). Following the rationale of section 9.1.3, one may describe the interface of the TI and the vacuum in terms of a $(3+1)$-dimensional Dirac operator, where a spatially varying mass describes the closing and reopening of a band gap. (The figure[35] shows photoemission spectroscopy data mapping out the spectrum of the massless $(2+1)$-dimensional Dirac fermions at the surface of a TI in the $\mathrm{Bi}_{1-x}\mathrm{Sb}_x$ material class.)

However, noting that a bulk configuration with negative m is equivalent to one with positive m, subject to a rotation $\alpha = \pi/2$, the situation may alternatively be described in terms of a Dirac fermion with uniform mass signature (say, negative on both sides of the interface). In this alternative description, the "Dirac determinant" is structureless, but we have an induced $\theta = \pi$ term in the insulator. Remembering the info block section on page 550 (and Eq. (10.51)), this term represents an $\mathbf{E} \cdot \mathbf{B}$ coupling contributing to the bulk electromagnetic response. In spite of the presence of a magnetic field, it respects time-reversal invariance: the sign change of $\mathbf{B}$ under time-reversal is equivalent to $\pi \to -\pi = \pi \bmod 2\pi$. Referring to Ref.[41] for details, this term describes the unconventional bulk polarization properties of the TI and its surface conduction (recall that, on systems with boundaries, θ-terms are reducible to surface terms). A take-home message of this discussion is that the continuous symmetries of the relativistic action may be exploited to pass from "microscopic" Dirac fermions descriptions to "effective" response actions of macroscopic electromagnetic fields.

[35] Taken from D. Hsieh *et al.*, *A tunable topological insulator in the spin helical Dirac transport regime*, Nature **460**, 1101 (2009).

Parity anomaly

The chiral anomaly affects Dirac operators in even space–time dimensions. In odd dimensions, $d = 1, 3, \ldots$, the *parity* anomaly is observed instead. While this anomaly is physically different, it is intimately related to the chiral anomaly via one of the dimensional-reduction mechanisms frequently at work in this field.

Consider a Dirac action in odd space–time dimensions. Following the discussion of section 9.1.1, we may think of it as the descent of an even-dimensional action in one dimension higher, where one of the spatial fluctuation directions is gapped out by size quantization. (This reflects the reality of most condensed matter realizations in $(2 + 1)$-dimensions.) Assume the Dirac fermions are minimally coupled to an electromagnetic field. The coupling ensures that the theory is invariant under local gauge transformations. It turns out, however, that the Dirac determinant resulting from an integration over fermion degrees of freedom, i.e., the inclusion of potentially singular UV fluctuations, is no longer invariant under topologically nontrivial "large" gauge transformations. Specifically, for gauge phase configurations winding W times around the odd-dimensional space–time torus of a system with periodic boundary conditions, it changes as $(-)^W$. This is an example of a **gauge anomaly**.

gauge anomaly

Unlike physical symmetries, the breaking of "gauge symmetries" by quantum fluctuations is something that we can not tolerate. As discussed in the info block on page 589, gauge symmetries reflect the redundancy in describing physical objects in different ways, or gauges. They do not represent physical symmetries, and hence should not be broken either. It turns out that a properly regularized relativistic action does indeed respect gauge invariance, including under large transformations. However, this comes at the price of the breaking of a physical symmetry present in the classical action, *parity*, the symmetry under spatial reflection at planes.[36]

parity anomaly

More specifically, the **parity anomaly** materializes as follows.[37] Our starting point is a UV-regularized action. In realizing this regulator, we have to choose between Scylla and Charybdis: either we decide for a regulator breaking gauge invariance, or for one breaking parity symmetry via a *mass term*.[38] For the reasons outlined above, we choose option number two. Note that the explicit symmetry breaking becomes effective only in the asymptotic UV; the low-energy action continues to be parity invariant. After integration over regularized fluctuations, the action contains not just a (gauge non-invariant) Dirac determinant, but also an induced term that couples to the electromagnetic field. Without going into detail, we know that this term must be of topological nature (on a boundaryless space–time,

---

[36] In three space–time dimensions, a spatial point inversion $x^i \to -x^i$ is equivalent to a π-rotation of space. In the (-1) determinant operation, parity is thus defined as a mirror reflection, e.g., $(x^1, x^2) \to (-x^1, x^2)$.

[37] A. Redlich, *Parity violation and gauge noninvariance of the effective gauge field action in three dimensions*, Phys. Rev. D **29**, 2366 (1984).

[38] Unlike in even dimensions, a mass term in odd dimensions breaks parity (see problem 9.4.4).

it takes into account the difference between local and large gauge transformations and nothing else), and is defined in odd dimension. These two criteria single out the Chern–Simons (CS) action $S_{CS}[A]$ as the candidate for the induced term.

As discussed in section 8.6.4, the CS action is also non-invariant under large gauge transformations. It changes as $S_{CS}[A] \rightarrow S_{CS} + 2\pi W k$, where the coupling constant (or level), k, is generally chosen to be integer to safeguard the invariance of the exponentiated action $\exp(iS_{CS})$. The Chern–Simons action generated by UV fluctuations in the effective action comes with a *half-integer* level, $k = 1/2$. In this way, a large gauge transformation generates two sign factors: $(-)^W$ from the Dirac determinant, and $\exp(i\pi W)$ from the Chern–Simons action. The two factors cancel out and gauge invariance is restored. However, the Chern–Simons action does violate parity. For example, in $d = 3$, the action (8.76) contains the antisymmetric combination $\epsilon^{\mu\nu\rho}A_\mu \partial_\nu A_\rho$. For each coordinate direction, $i = 1, 2$, it contains ∂_i or A_i exactly once. If we define the parity operation as $(x^0, x^1, x^2) \rightarrow (x^0, -x^1, x^2)$ [36] the transformation $\partial_1 \rightarrow -\partial_1$ and $A_1 \rightarrow -A_1$ makes the action change sign. (On the same basis, the action breaks time-reversal invariance under $(x^0, x^1, x^2) \rightarrow (-x^0, x^1, x^2)$.) In this way, the parity and time-reversal symmetry present in the unregularized action is broken.

We finally note that:

> The parity anomaly affects only $(2n + 1)$-dimensional theories with massless fermions.

Heuristically, in theories where a mass m breaks parity,[38] there is no symmetry left to be broken, and hence there is no anomaly. The actual state of affairs is somewhat more interesting. For massive fermions, both the intrinsic mass, and the mass introduced by the regulator, individually introduce a CS action. We then end up with a CS action at level $k = (\text{sgn}(m) + \text{sgn}(M))/2$. For a physical manifestation of such contributions, see the info block below.

The question remains how the structures mentioned above are made concrete. In three-dimensions, the actual calculation generating the CS action from the regularized action is technically involved.[37] However, in problem 9.4.4, we consider a one-dimensional toy version of the parity anomaly, which has enough structure to demonstrate the workings of the regularization in a manner unburdened by technical complications. Readers interested in the derivation of the three-dimensional CS action are referred to the original Ref.[37] or to particle physics textbooks.

In **condensed matter systems**, the parity anomaly shows in systems containing gapless Dirac fermions in odd dimensions. This situation is realized, e.g., in the **QAH insulator** introduced on page 464. As another example, consider the **Weyl semimetal** defined on page 528, interpreted as a $(3 + 0)$-dimensional system. In this case, we work at fixed energy and the Hamiltonian of individual Weyl nodes, $H = \pm i\partial_i \sigma^i$, assumes the role of the massless Dirac operator. On this basis, we know that the electromagnetic response action contains a CS action. The topological protection of the anomaly implies robust manifestations in the conduction

properties of Weyl semimetals, including in the presence of strong disorder.[39] Another example of a $(2+1)$-dimensional gapless Dirac system is the **surface of the**
$(3+1)$**-dimensional topological insulator**. Naïvely, one might think that the
surface must be subject to the parity anomaly, which would imply the breaking of
P and T symmetry. However, this is contrary to physical observation. The point
is that the surface of the TI cannot be considered in isolation: it is inseparably
linked to the underlying bulk. Where anomalies are concerned, this coupling manifests itself through a mechanism called *anomaly inflow*, the cancellation of surface
anomalies against anomalies from the bulk (see below). In the case of the TI surface, the anomaly, and its cancellation from the bulk, go a long way in determining
the electromagnetic response properties.[40]

INFO In Eq. (8.39) we defined $\hat{H} = -i\partial_i\sigma^i + m\sigma_3$ as an effective low-energy Hamiltonian
describing the anomalous quantum Hall (AQH) insulator. We consider $S = \int d^3x\,(\psi^\dagger\dot{\psi} -
\psi^\dagger\hat{H}\psi)$ as the corresponding $(2+1)$-dimensional Dirac action (see problem 9.4.2 for a gen-
eral discussion of the connection between the Dirac action and Dirac Hamiltonian) and,
with $\bar{\psi} \equiv \psi^\dagger\sigma_3$, obtain $S = \int d^3x\,\bar{\psi}(i\slashed{\partial} - m)\psi$ with the realization $\gamma^\mu = (\sigma_3, i\sigma_2, -i\sigma_1)$.
Minimal coupling of the action to a vector potential defines a problem showing the par-
ity anomaly. The discussion above then leads to the prediction that the effective action
describing the system after the regulated integration over fermion fields is a CS action of
levels $\frac{1}{2} + (-\frac{1}{2}, 0, \frac{1}{2}) = (0, \frac{1}{2}, 1)$ depending on whether m is (positive, zero, negative).

In our discussion of CS linear response theory in section 8.6.7 we saw that the level of
the CS action determines the transverse conductivity as $\sigma_{12} = k/2\pi$. Our discussion thus
determines the Hall response of the QAH insulator without further calculation.

Anomaly inflow

We have seen that individual realizations of relativistic fermions display anomalies,
which in turn may represent the non-conservation of physical currents. For example,
the $(1+1)$-dimensional Dirac fermion with action (9.39) describes left- and right-
moving chiral fermions, whose individual charge is not conserved, including in the
case $m = 0$, where they are not explicitly coupled.

Now consider the particular case where such $(d + 1)$-
dimensional chiral fermions are realized as the surface states
of a $(d+2)$-dimensional bulk system. Thinking from a condensed matter perspective, where such setups describe the
low-energy physics of topological insulators, we would reason
that the system at large (including the UV degree of freedom)
must be non-anomalous and current conserving. This in turn
implies that the anomaly of individual surfaces is canceled

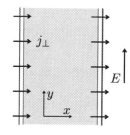

by a similarly anomalous low-energy theory of the bulk. In this way, current conservation is restored without the need to engage UV degrees of freedom (see the

[39] A. Altland and D. Bagrets, *Theory of the strongly disordered Weyl semimetal*, Phys. Rev. B
93, 75113 (2016).
[40] M. Mulligan and F. Burnell, *Topological insulators avoid the parity anomaly*, Phys. Rev. B **88**,
85104 (2013).

figure for the case where two spatially one-dimensional surfaces are coupled to a two-dimensional bulk; the horizontal arrows represent the leakage of current from one part of the system into another). This mechanism of anomaly cancellation is called **anomaly inflow**.

Anomaly inflow implies the cancellation of anomalies between theories of different dimensional "parity." (An even-dimensional surface has an odd-dimensional bulk, etc.) We have seen on multiple occasions that dimensional parity is a key factor in the manifestations of both anomalies and topology. For example, systems in even space–time dimensions display the chiral anomaly. In this case, the odd-dimensional bulk does not contain a chiral anomaly but a parity anomaly instead. The parity anomaly in turn manifests itself through a topological Chern–Simons response action. We conclude that anomaly inflow into even-dimensional surfaces happens through a channel of "communication" between CS theory and the relativistic theory of chiral fermions.

Let us illustrate the workings of this mechanism on the example of the **QAH insulator**. Earlier in this section, we reasoned that the bulk of the topological insulating phase is described by a CS response action (see Eq. (8.76))

$$S[A] = \frac{1}{4\pi} \int d^3x \, \epsilon^{\mu\nu\rho} A_\mu \partial_\nu A_\rho.$$

Now consider the current flowing in response to a static electric field E in the $y = x^2$ direction. From the variation

$$j^\mu = \frac{\delta S[A]}{\delta A_\mu} = \frac{1}{2\pi} \epsilon^{\mu\nu\rho} \partial_\nu A_\rho,$$

we obtain $j^x = \frac{1}{2\pi}(\partial_y A_0 - \partial_t A_2) = -\frac{1}{2\pi}E$. At the boundary, this violates charge conservation – there is current flow out of "nowhere" – in line with the gauge non-invariance of the CS action. The "nowhere" is the anomalous boundary theory. That boundary theory is a chiral fermion, say the left-moving branch of the Dirac action (9.39). In section 9.2.1 we saw that this action is anomalous; the axial current is non-conserved according to Eq. (9.40). Specifically, with $j^0 = \rho_L - \rho_R$, we have $\partial_t(\rho_L - \rho_R) = E/\pi$, or $\partial_t \rho_L = E/2\pi$ for a single branch. We conclude that the rate of charge leakage out of the boundary, $E/2\pi$, is what feeds the current into the bulk. Considering the whole system comprising left boundary, bulk, and right boundary, this mechanism describes the Hall current flow between the boundaries in response to a longitudinal electric field.

9.3 Summary and Outlook

We started this chapter with a symmetry-oriented construction which led to the Dirac Hamiltonian for the minimal description of spin-1/2 particles with a linear dispersion. On our way towards the Dirac Hamiltonian we came across similar but more restricted theories describing spinful particles subject to a non-relativistic linear spin-momentum coupling and to the chiral Weyl fermion Hamiltonian, respectively. We continued with a discussion of the astonishingly rich spectrum of symmetries of the Dirac Hamiltonian and derived the Dirac action describing many-body Dirac theory. On the basis of these foundations we discussed various ramifications of Dirac theory which have recently turned out to be crucially important to the physics of topological quantum matter: Dirac Hamiltonians as effective low-energy theories close to topological phase transitions, dimensional reduction and the bulk boundary-principle connecting gapless boundary phases to gapped topological bulk phases, and the breaking of continuous or discrete symmetries through anomalies. While the early sections of the chapter introduced Dirac theory as a stand-alone subject, it later became clear that its inclusion into the context of condensed matter physics is inseparably linked to two other subjects: gauge theory and topology. This brings us back to a principle mentioned in chapters 8 and 10, namely that Dirac physics, topology, and gauge theory in condensed matter physics are like a Siamese triplet, and ideally should be studied in parallel.

9.4 Problems

9.4.1 Parity, charge conjugation and time-reversal symmetry

In recent years, the fundamental symmetries charge conjugation and time-reversal have played a major role in the classification of quantum matter. In condensed matter physics, we often consider "effective" realizations of these symmetries in non-relativistic settings. However, it is important to know how these symmetries can be traced back to their fundamental origins in relativistic quantum physics.[41] Compared with the Schrödinger equation, the Dirac equation has a more authoritative standing where symmetries are concerned: it is Lorentz invariant, which includes symmetry under time-reversal and parity in their original form. The realization of these symmetries combines a transformation of space–time with a transformation of the equation's spinor degrees of freedom. In the main text, we showed how this combined transformation makes the Dirac equation parity invariant. Here, we extend the discussion to include time-reversal, which is slightly more tricky because it is an anti-linear symmetry. We also add the third fundamental symmetry of the Dirac equation, one that exchanges particles with their corresponding antiparticles, thus changing the sign of all charges. The relativistically invariant realization of charge conjugation that we discussed motivates similar transformations in non-relativistic many-body physics, testing invariance under the exchange of particle and hole degrees of freedom.

In this problem, we focus on $d = 4$, as it is relevant to the realization of symmetries in nature.

[41] S. Weinberg, *The Quantum Theory of Fields*, Cambridge University Press, 1996.

Our starting point is the Dirac action

$$S[\psi] = \int d^4x \, \bar{\psi}(i\slashed{D} - e\slashed{A} - m)\psi, \tag{9.58}$$

minimally coupled to a vector potential $\slashed{A} = A_\mu \gamma^\mu$. Since the sign of the charge will play a role in this problem, we have reintroduced e. We write the Dirac conjugate spinor as $\bar{\psi} = \psi^{\dagger T}\gamma^0$. This somewhat awkward notation emphasizes that $\psi_i \leftrightarrow a_i$ and $\psi_i^\dagger \leftrightarrow a_i^\dagger$ are Grassmann representatives of second-quantized operators. (Of course, ψ and $\psi^\dagger$ remain independent integration variables.)

When we discussed symmetries in section 9.1, the focus was on the action of Lorentz transformations on the complex valued Dirac spinors. However, since we understand Dirac theory as a many-body theory, it is more natural to switch to a discussion in Fock space or, equivalently, the language of the effective action. Our starting point is a formal representation of the symmetries $X = C, T, P$ as

$$\psi(x) \to \psi_X(x) \equiv X\psi(\Lambda_X^{-1}x)X^{-1}, \tag{9.59}$$

with the explicit realizations (the argument $\Lambda_X^{-1}x$ is omitted for clarity)

$$\begin{aligned} P\psi P^{-1} &= U_P\psi, & & & \Lambda_P &= \mathrm{diag}(1,-1,-1,-1), \\ T\psi T^{-1} &= U_T\psi, & T i T^{-1} &= -i, & \Lambda_T &= \mathrm{diag}(-1,1,1,1), \\ C\psi C^{-1} &= U_C\psi^\dagger, & & & \Lambda_C &= \mathbb{1}. \end{aligned}$$

Here, Λ_X implements the symmetry transformation of space–time coordinates, the reflection of space (P), inversion of time (T), or no transformation (C). The factors U_P are fixed and as yet undetermined unitary 4×4 matrices acting in spinor space. The second entry in the second line states that time-reversal is an antilinear transformation. Recall that the antilinearity of time-reversal is required for physical reasons, e.g., to effect an inversion of time in dynamical phases, $Te^{i\epsilon t}T^{-1} = e^{i\epsilon(-t)}$. (Consult a textbook on quantum mechanics for further discussion.) Finally, the unitary operation C exchanges creation and annihilation operators in such a way that the action of a C-transformed operator on the vacuum creates a quasiparticle of opposite charge. Note how these transformations of Grassmann variables (or of the operators represented by them) resemble those for *states* in Eq. (9.15). The difference is that the linear or antilinear operators U_X now act from both sides, as befits operators. Finally, the transformation of $\psi^\dagger$ is obtained by taking the formal adjoint of $\psi_{X,i} = U_{X,ij}\psi_j$ as $\psi_X^\dagger = \bar{U}_{X,ij}\psi_j^\dagger$, $X = P, T$, and $\psi_C^\dagger = \bar{U}_{X,ij}\psi_{C,j}$.

We now ask how these Fock space transformations become symmetries of the Dirac action, where "symmetry" means that the transformed action $XS[\psi]X^{-1} = S[\psi']$ equals the original one. What we have at our disposal in satisfying this condition are the as yet unspecified definitions of the matrices U_X.

(a) The sought-for choices for U_X must be compatible with the matrix symmetries of the γ-matrices, which are affected by the symmetry transformations. Using Eqs. (9.20) and (9.22), tabulate their symmetries under the hermitian adjoint, transposition, and complex conjugation in the Weyl and Majorana representations.

Table 9.1 Symmetry properties of γ-matrices in Weyl
and Majorana representation.

γ_{Weyl}	0	1	2	3	γ_{Majorana}	0	1	2	3
†	+	−	−	−	†	+	−	−	−
T	+	−	+	−	T	−	+	+	+
c.c.	+	+	−	+	c.c.	−	−	−	−

(b) We next discuss the symmetries of the action under discrete transformations, first setting $A = 0$ for simplicity. As a warm-up, establish the **parity** transformation previously discussed for the Dirac equation as a symmetry of the action, $S[\psi_P] = S[\psi]$. Try to define $U_P = U_P(\gamma^\mu)$, without reference to the detailed structure of the γ-matrices.

(c) Next run the same program for **charge conjugation** to identify U_C such that $S[\psi_C] = S[\psi]$. (Hint: You will need to use the fact that the action, a number, equals its own transpose. Also keep the anticommutativity of Grassmann variables in mind.) Repeat the analysis for the γ-matrices of the Majorana representation. Are the transformation matrices U_C obtained in the two representations unitarily equivalent? If not, why is this allowable?

(d) Finally, we consider **time-reversal** in the Weyl representation. The best way to proceed is to subject the entire action, considered as a (Grassmann $\leftrightarrow$ operator) bilinear form to the antiunitary operation $\mathrm{T}(\text{action})\mathrm{T}^{-1}$. Identify matrices U_T such that $\overline{S[\psi_T]} = S[\psi]$. Compared with conventional unitary symmetries, an additional complex conjugation is required to describe the time-reversal transformation of dynamical phases, $\exp(iS) \to \exp(\overline{iS})$. When writing down $\overline{S[\psi_T]}$, use the fact that ψ_T are Grassmann variables, and that complex conjugation has no effect on them. Inspect how U_T acts on the spin degrees of freedom of the Weyl fermions contained in the Dirac spinor and compare with what you know from quantum mechanical time-reversal.

(e) Finally, extend your analysis to include the coupling to the vector potential eA. On physical grounds, formulate an idea of how eA needs to be transformed in order to establish invariance under $\mathrm{P}, \mathrm{C}, \mathrm{T}$ in the presence of minimal coupling. Then repeat the steps above to verify that you have guessed correctly.

Answer:

(a) From Eq. (9.20) and the properties of the Pauli matrices, we obtain the properties summarized in table 9.1. What is unappealing about our discussion is that it makes reference to properties specific to a given basis. To remedy this issue, one would need to switch to a framework where transposition and complex conjugation are invariantly defined. However, this generalization[11] is beyond the scope of the present text. We simply have to accept that the concrete realization of the symmetries (the form of U_X) is specific to a given basis. Owing to the presence of anti-linear symmetries (T), the different representations are not unitarily equivalent.

(b) Substitution of the parity transformed field into the action yields

$$S[\psi_{\mathrm{P}}] = \int \overline{U\psi}(\Lambda^{-1}x)(i\partial_\mu\gamma^\mu - m)U\psi(\Lambda^{-1}x) = \int \overline{U\psi}(x)(i\partial_\mu\Lambda^\mu{}_\nu\gamma^\nu - m)U\psi(x)$$

$$= \int \overline{\psi}(\gamma^0 U^\dagger \gamma^0)(i\partial_\mu(\Lambda\gamma)^\mu - m)U\psi,$$

where we have made the abbreviation $\int d^4x = \int$, $U_{\mathrm{P}} = U$, $\Lambda_{\mathrm{P}} = \Lambda$ for notational simplicity; in the first equality we made the change of variables $\Lambda_{\mathrm{P}}^{-1}x \to x$, and in the second we used the definition $\bar{\psi} = \psi^{\dagger T}\gamma^0$ of the Dirac conjugate. We want this action to equal the original one, $S[\psi_{\mathrm{P}}] = S[\psi]$, and this requires that $[U, \gamma^0] = 0$ (i.e., invariance of the mass term). With $(\Lambda\gamma)^0 = \gamma^0$ and $(\Lambda\gamma)^i = -\gamma^i$, we obtain the additional conditions $[U, \gamma^i]_+ = 0$. These are simultaneously satisfied with the choice $U = U_{\mathrm{P}} = \gamma^0$. In the Weyl representation, $\psi^0 = \tau_1$, which is consistent with our discussion in section 9.1.1.

(c) We proceed as above, except that in the last line we need to exchange ψ and $\psi^\dagger$, and there is no transforming matrix Λ. This gives

$$S[\psi_{\mathrm{C}}] = \int \psi^T U^\dagger \gamma^0 (i\partial_\mu\gamma^\mu - m)U\psi^\dagger = -\int \psi^{\dagger T} U^T(-i\partial_\mu\gamma^{\mu T} - m)\gamma^{0T}\overline{U}\psi,$$

where, in the second line, we have taken the transpose of the bilinear form in the action. This operation introduces a global minus sign ($\eta^T X\nu = -\nu^T X^T\eta$ for anticommuting variables), and another one due to the antisymmetry of the derivative under partial integration. Using table 9.1, we now do our bookkeeping again to obtain $[U, \gamma^0]_+ = 0$ from the mass term and ∂_0 invariance. With $[\gamma^0, \gamma^i]_+ = 0$, the remaining conditions read $[U, \gamma^{1,3}] = 0$ and $[U, \gamma^2]_+ = 0$. These conditions are satisfied by $U = U_{\mathrm{C}} \equiv \gamma^0\gamma^2$. With this choice, we have $S[\psi_{\mathrm{C}}] = \int \bar{\psi}(i\partial_\mu\gamma^\mu - m)\psi = S[\psi]$, as required. In the Majorana representation, Eq. (9.22), the γ-matrices have different symmetries, as summarized in the second half of the table. Now we require $[U, \gamma^0] = 0$ and $[U, \gamma^i]_+ = 0$, which is satisfied for $U_{\mathrm{C}} \equiv \gamma^0$. This matrix is not related to its counterpart $U_{\mathrm{C,Weyl}}$ by the unitary transformation relating the Majorana and Weyl γ-matrices (see exercise on page 531). There is no contradiction since the derivation involves transposition, which is not a unitarily invariant concept. For example, $[U_{\mathrm{Weyl}}, \gamma^0_{\mathrm{Weyl}}]_+ = 0$ becomes $[U_{\mathrm{Majorana}}, \gamma^0_{\mathrm{Majorana}}] = 0$, while a unitary transformation would preserve the commutator structure.

(d) Consider the complex conjugate $\overline{S[\psi_{\mathrm{T}}]} = \int \bar{\psi}_{\mathrm{T}}(-i\partial_\mu\overline{\gamma^\mu} - m)\psi_{\mathrm{T}}$, where we note that complex conjugation has no effect on γ^0 in $\bar{\psi} = \psi^\dagger\gamma^0$. With $\psi_{\mathrm{T}}(x) = U\psi(\Lambda^{-1}x)$, we obtain $\overline{S[\psi_{\mathrm{T}}]} = \int \bar{\psi}\gamma^0 U^\dagger \gamma^0(-i\partial_\mu(\Lambda\overline{\gamma})^\mu - m)U\psi$. From the table and the definition of $\Lambda = \Lambda_{\mathrm{T}}$, we have $(\Lambda\overline{\gamma})^{1,3} = \gamma^{1,3}$ and $(\Lambda\overline{\gamma})^{0,2} = -\gamma^{0,2}$, leading to the invariance criteria: $[U, \gamma^{0,2}]_0$ and $[U, \gamma^{1,3}]_+ = 0$. This is satisfied by $U = U_{\mathrm{T}} \equiv \gamma^1\gamma^3$. For this choice, we have the required invariance. In the Weyl representation, $\gamma^1\gamma^3 = -i\sigma_2$. This implies that time-reversal acts on spin-$1/2$ spinor wave functions by a combination of complex conjugation and multiplication by σ_2 (consider why the factor of $-i$ is irrelevant), as is familiar from quantum mechanics.

(e) Under parity, we have $A_\mu \to A_\nu(\Lambda_{\mathrm{P}})^\nu{}_\mu$, i.e., the scalar potential is unaffected, and the vector potential changes sign. In this way, $\mathbf{E} \to -\mathbf{E}$ and $\mathbf{B} \to +\mathbf{B}$. The substitution of the parity-transformed field into the covariant derivative then establishes parity invariance of the minimally coupled action. Similarly, we have $A_\mu \to -A_\nu (\Lambda_T)^\nu{}_\mu$ corresponding to $\mathbf{E} \to \mathbf{E}$ and $\mathbf{B} \to -\mathbf{B}$ under time-reversal. Again, the symmetry analysis above goes through in this way. (Note how the extra minus sign compensates for the sign change of i under complex conjugation in $i\partial_\mu - eA_\mu$.) Finally, charge conjugation should do what its name says, i.e., exchange charge as $eA \to (-e)A$. With the global sign change in the covariant derivative, C, also holds.

9.4.2 Quantization of the Dirac action

In the main text, we leapfrogged over various conceptual steps in the passage from the Dirac equation to the many-body physics of the Dirac field. Readers not familiar with the quantization of the Dirac action from their lecture courses are advised to read about this interesting story in any textbook on particle physics. Here, we address a few aspects concerning the quantization of Dirac theory in the simplified setting of $d = 2$.

Consider the real-time two-dimensional Dirac action $S[\psi] = \int d^2x\, \bar\psi(i\partial_\mu \gamma^\mu - m)\psi$, where $\gamma^0 = \sigma_1, \gamma^1 = i\sigma_2$.

(a) Interpreting $S[\psi]$ as a classical Lagrangian action, identify the canonical momentum, π, associated with ψ, the action in Hamiltonian representation, and the Hamiltonian. How many classical degrees of freedom does the theory have? How would that number increase in $d = 4$, and what meaning do these degrees of freedom have in the quantum theory?

(b) We now want to quantize the theory. To this end, we first upgrade ψ and $\psi^\dagger$ to operators satisfying canonical commutation relations. But which ones? In principle, the two options of bosonic relations, $[\psi_i(x), \psi_j^\dagger(y)] = \delta_{ij}\delta(x - y)$ and of fermionic relations $[\psi_i(x), \psi_j^\dagger(y)]_+ = \delta_{ij}\delta(x - y)$, are on the table. In particle physics textbooks, a great deal of attention is devoted to the discussion of these two options. Energy considerations eventually invalidate quantization in terms of bosons; the Dirac field must be a fermion field. On the basis of the field integral construction, provide arguments supporting this conclusion. Why would a derivation of the Dirac field integral in terms of boson coherent states run into trouble? And why is the fermionic version safe?

Answer:

(a) Making the time-derivative explicit, we write the Dirac action as $S[\psi] = \int d^2x\, \mathcal{L}(\psi, \partial_t\psi)$, with the Lagrangian $\mathcal{L}(\psi, \partial_t\psi) = \psi^\dagger(i\partial_t - i\partial_x\sigma_3 - m\sigma_1)\psi$. Variation in ψ yields the **canonical momentum** of the Dirac field as $\pi = \partial_{\partial_t\psi}\mathcal{L} = i\psi^\dagger$.

Dirac
Hamil-
tonian From here, we define the **Hamiltonian density** as $\mathcal{H}(\psi, \pi) = \pi\dot\psi - \mathcal{L}(\psi, \pi) = \pi(\partial_x\sigma_3 - im\sigma_1)\psi = \psi^\dagger(i\partial_x\sigma_3 + m\sigma_1)\psi$. The action in the Hamiltonian representation is given by $S[\psi, \pi] = \int d^3x\,(\pi\dot\psi - \mathcal{H}(\pi, \psi))$. Being first order in derivatives, the Dirac action leads to a canonical momentum expressed in terms of the original variables, $\psi, \psi^\dagger$. (In this regard, the situation resembles that of the classical mechanics of spin; see section 8.4.4.) With 2×2 real variables contained in ψ, the theory has two degrees of freedom. In $d = 4$ that number doubles. In the quantum theory, these four degrees of freedom become the amplitude of the spin-up and spin-down components of the two Weyl fermions, respectively.

(b) Suppose we subjected the Hamiltonian above to a boson coherent state field integral construction. We would end up with an integral over the complex variables ψ with Gaussian action $S[\psi, \pi]$. What about convergence? The eigenvalues of the Dirac Hamiltonian are given by $\epsilon_p^{\pm} = \pm(p^2 + m^2)^{1/2}$. A variable transformation to eigenmodes $b_p^{\pm}$ leads to the action $S[b, b^{\dagger}] = \sum_{s=\pm} \int dt \, dp \, b_p^{s\dagger}(i\partial_t - \epsilon_p^s) b_p^s$. The problem is the instability of the negative-energy branch. This convergence issue cannot be fixed by a Wick rotation, the reason being that it is physical, not technical. The integral describes a continuum of unstable boson modes, and the divergence reflects the option to "condense" in this sea of negative-energy states. For Grassmann variables, the convergence issue does not arise. In this case, the negative-energy states are filled up (for the chemical potential-zero considered here), and excited states are interpreted as particle–hole (or particle–antiparticle) pairs.

9.4.3 Chiral anomaly and topological quantization

In section 9.2.1 we discussed the topological interpretation of the chiral anomaly on boundaryless space–time manifolds. Here, we consider the example of a charge pumping protocol in $1 + 1$ dimensions to make this abstract statement tangible. The problem requires familiarity with section 10.5.

Consider the $(1 + 1)$-dimensional Dirac action (9.39) on a spatial circle of radius L. Assume that a time-dependent flux is turned on and a single flux quantum $\phi_0 = h/e = 2\pi$ is pushed through the system during a long interval $[0, t_0]$.

(a) Describe the setting in the language of section 10.5, and especially that of the discussion of the Dirac monopole in section 10.5.2: why does the setting above define the U(1)-fiber bundle with a space–time torus as the base manifold? Describe the electric field generated by the flux in two alternative ways, first through a time-varying "magnetic vector potential" and second through a "scalar potential." Show that the two representations are equivalent up to a *large* gauge transformation. Show that two coordinate charts with nontrivial transition function are required to parameterize the situation.

(b) Translate your description to the language of differential forms, and compute Eq. (9.54). Identify the right-hand side as a non-vanishing Chern number. On the left-hand side, discuss the continuity properties of the current and obtain a statement on axial charge non-conservation. Discuss the meaning of your findings in the context of fig. 9.2.

In the second part of the problem, we want to understand the anomaly from the perspective of the Dirac index, Eqs. (9.55) and (9.56). To this end, we consider the Euclidean version of the Dirac operator defined by Eqs. (9.27): $\gamma_0 = \tau_1 \to i\tau_1$, equivalent to $\partial_t \tau_1 \to \partial_\tau i\tau_1$ with $it = \tau$. Assume this operator is minimally coupled to a vector potential $(A_0, A_1) = (A_\tau, A_x) = (0, \phi(\tau)/L)$, where ϕ is a function slowly increasing from $-\pi$ to $+\pi$ over an asymptotically large imaginary-time interval $[-\infty, \infty]$. This is the same potential as in the first part of the problem, and so we know that it has associated Chern number unity.

(c) Show that the Weyl modes of this operator obey the equation $[-\partial_\tau + C(-i\partial_x - \phi(\tau)/L)]\psi_C(x, \tau) = 0$. Consider the instantaneous eigenfunctions of the spatial operator in this equation to solve it in an adiabatic approximation. Show that the function $\phi(\tau)$ stabilizes a zero mode in the $C = 1$ sector, while there are no regular solutions for $C = -1$. In this way, the minimally coupled Dirac operator has a unit index, and the Atiyah–Singer index theorem (9.56) is satisfied.

Answer:

(a) Topologically, a 2-torus T is a product of two circles, $T = U(1) \times U(1)$. Here, the role of the first is taken by compactified space, and that of the second by time subject to periodic boundary conditions. A time-varying flux, $\phi(t)$, is represented by a vector potential with component $A_1 = \frac{\phi(t)}{L}$ along the ring, with associated field $E = \partial_t A_1 = \dot{\phi}/L$. The pushing of a single flux quantum means that $\phi(t)$ increases from, say, 0, to 2π in time t_0. *Locally*, A_1 may be removed by a

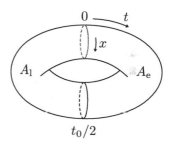

gauge transformation $\psi \to \exp(i\phi(t)\frac{x}{L})\psi$. However, this generates a 0-component $A_0 = -\dot{\phi}\frac{x}{L}$. (Of course, the field $E = -\partial_x A_0$ remains invariant.) *Globally*, the gauge transformation above is not continuously contractible to a trivial one, and hence represents a large transformation in the sense of section 10.5.1. In either representation, two coordinate charts are required to describe the system. For example, in the first, A_1 is constant in the spatial direction but would have a jump $2\pi/L$ as we pass through time $0 = t_0$. (In the other gauge, A_0 jumps by $-\dot{\phi}(t)$ at the joining point $0 = L$ of the spatial circle.) We describe the situation in terms of two charts, the first defined on $(0, t_0/2)$, and the second on $(t_0/2, t_0)$. Next define $A_{1,e} = \frac{\phi(t)}{L}$ on the earlier domain and $A_{1,l} = \frac{\phi(t)}{L}$ on the later one. In this way we have a trivial transition gauge, $A_{1,e}(t_0/2) = A_{1,l}(t_0/2)$ on the first chart boundary, and the nontrivial transition gauge $A_{1,e}(0) = A_{1,l}(t_0) - 2\pi/L$ on the second chart boundary (cf. Eq. (10.44)).

(b) The local connection forms on the two halves of the torus are given by $A_{e,l} = \frac{\phi(t)}{L}dx$. With this representation, the right-hand side of Eq. (9.54) is evaluated as

$$\frac{1}{2\pi} \int_T F = \frac{1}{2\pi} \int_{S^1} (A_e - A_l) = 2.$$

Here, S^1 is the spatial circle defining the boundary between the two charts at times $t = 0$ and t_0. At this boundary, we have the jump $A_l - A_e = \frac{2\pi}{L}dx$, and the integral leads to the stated result.

Turning to the left-hand side, the integral extends over a local derivative dj_A. Since T has no boundary, there must be a singularity involved. Decomposing the current as $j_A = j_{A,0}dt + j_{A,1}dx$, the spatial rotational symmetry of the problem implies constancy of the temporal component $j_{A,0}(t,x) = j_{A,0}(t)$ along the ring; there will not be any discontinuities in the spatial direction. However, we *do* expect the axial charge density, $j_{A,1}$, to be different at the beginning and end of the protocol. To make this quantitative, we write

$$\int_T dj_A = \int_{S^1} (j_{A,\mathrm{f}} - j_{A,\mathrm{i}}) = Q_{A,\mathrm{f}} - Q_{A,\mathrm{i}},$$

where the the subscripts f (final) and i (initial) refer to the charge density at times t_0 and 0, respectively. In this way, we arrive at the conclusion that the axial charge, i.e., the difference between the left- and the right-moving charge changes by 2 in the process. The interpretation of this number is that the insertion of a flux quantum makes individual momentum states, indicated in fig. 9.2 as dots, shift by one unit. In this way, a right-moving particle and a left-moving hole are created. The charge balance at the end of the process thus reads $1 - (-1) = 2$.

(c) With $i\slashed{D} = (i\partial_\mu + A_\mu)\gamma^\mu = -\partial_\tau \tau_1 - (\partial_x - iA_x(\tau))\tau_2$, the zero-mode equation $i\slashed{D}\psi = 0$ splits into the two stated chiral equations. The instantaneous eigenfunctions obey the equation $(-i\partial_x - A_x(\tau))\psi_n(\tau) = \lambda_n(\tau)\psi_n(\tau)$, with time-independent Fourier mode solutions $\psi_n(x,\tau) = \psi_n(x) = L^{-1/2}\exp(ik_n x)$, $k_n = 2\pi n/L$, and eigenvalues $\lambda_n(\tau) = k_n - A_x(\tau)$. With the ansatz $\psi_C(x,\tau) = f(\tau)\psi_n(x)$, the differential equation reduces to $[-d_\tau + C\lambda_n(\tau)]f(\tau) = 0$. This equation has the formal solution $f(\tau) = f(0)\exp(C\int_0^\tau ds\,\lambda_n(s))$. A solution with regular behavior for $\tau \to \pm\infty$ must satisfy $C\lambda_n(s) < 0$ for large positive times, and $C\lambda_n(s) > 0$ for negative times. In other words, the function $\lambda_n(s)$ must have a zero with positive derivative. Considering the eigenvalue equation $C\lambda_n(s) = -C(k_n + A_x(\tau))$ for $A_x(\tau) = \phi(\tau)/L$, and the given profile of $\phi(\tau)$, we realize that this condition is met for $C = R = +1$ and $n = 0$. We thus have a single R-zero mode, $n_R = 1$, and no L-zero mode, $n_L = 0$, corresponding to a unit index.

9.4.4 Parity anomaly

In this problem, we discuss the parity anomaly in $d = 1$. The one-dimensional Dirac operator, essentially a single derivative, is somewhat too simplistic to be of practical relevance. However, the merit of this toy problem is that it lets us understand the parity anomaly in a manner uncluttered by technicalities. The question relies on problem 9.4.3 as a prerequisite.

We consider a one-dimensional Dirac action $\int dx\,\bar\psi(-i\partial_x - A)\psi$ defined on a ring of circumference L. The minimal coupling to A provides invariance under local gauge transformations, but not necessarily large ones. A large gauge transformation is defined by $A = W/L$ or $\int dx\,A = 2\pi W$, corresponding to the insertion of W flux quanta through the ring. Naïvely, one would say that this is an empty operation, or that A can be removed by the inverse (large) gauge transformation $\psi \to \exp(-i2\pi Wx/L)\psi$. However, this transformation is defined for an unregularized theory and we are, in a sense, again met with a $0 \times \infty$ ambiguity.

An elegant way to make sense of the situation is to establish contact with the Dirac operator in one dimension higher, $d = 2$, with its chiral anomaly. To this end, we consider the setup of problem 9.4.3, where the additional time coordinate defines the space–time torus shown in the figure, and $A_1 \equiv A = A(\tau) = \phi(\tau)/L$ is the vector potential locally describing a large gauge transformation with Chern number $W = 1$. Choosing our coordinates such that $\phi(0) = 0$ and $\phi(t_0) = 1$, and thinking of τ as an adiabatic parameter, we notice that τ smoothly interpolates between the situations with and without an inserted flux quantum. (Intermediate values, $\phi(x, \tau)$ do not represent gauge transformations of the one-dimensional theory. They correspond to the insertion of fractional flux quanta into our ring, which do have physical effects such as changes in the spectrum, or persistent current flow. It will be useful to think about this point.)

(a) In problem (9.4.3), we showed that the above potential stabilizes a zero mode of the two-dimensional Dirac operator. Reinterpret this finding in terms of the eigenvalues of the one-dimensional Dirac operator. Specifically, show that the adiabatic time-dependent parameters $\lambda(\tau)$ appearing in the solution of the zero-mode equations assume the role of eigenvalues associated with the instantaneous eigenfunctions. On this basis, reason why the Dirac determinant $\det(-i\partial_x - A)$ must undergo a sign change as $A = A(\tau)$ interpolates between $A(0) = 0$ and $A(\tau) = 2\pi/L$.

(b) We now want to show how the regularized theory generates a Chern–Simons term after integration over the ψ-fields. With the abbreviation $D \equiv \partial_x - iA$, we regularize the action by adding a second contribution $\bar{\varphi}(-iD + iM)\varphi$, where $M > 0$ is large and φ is a bosonic field with $\bar{\varphi} = \varphi^\dagger$. Discuss how, after integration over φ, this contribution regularizes the previously UV-singular determinant; how it is inessential at low energies; and in what sense it breaks parity.

(c) The abelian Chern–Simons action in general odd dimensions is given by $S_{CS}[A] = 2\pi k \int \mathrm{ch}_{(d+1)/2}(F)$, where the Chern characters are defined in Eq. (10.49). Show that the expansion in A of the regulator action after integration over φ yields a one-dimensional Chern–Simons action at half-integer level. You may ignore all UV-singular contributions to the expansion as they will cancel against the fermion determinant. Demonstrate how this action produces a sign factor when evaluated on a large gauge transformation. (Hint: It is most economic to work in a Wigner representation; see appendix section A.4.3.)

Answer:

(a) The zero-mode equations for the $C = L, R$ sector of the two-dimensional Dirac operator discussed in the previous section were solved in terms of the instantaneous eigenfunctions $(-i\partial_x - A(\tau))\psi_n(\tau) = \lambda_n(\tau)\psi_n(\tau)$ of the one-dimensional Dirac operator. We saw that the existence of a normalizable zero mode required a solution ψ_n whose associated eigenvalue $\lambda_n(\tau)$ crosses zero in the course of adiabatic evolution. This implies that the determinant of the one-dimensional operator $\prod_n \lambda_n(\tau)$ changes sign (or, more generally, changes sign W times) as τ progresses from 0 to t_0.

(b) Integration over the added action, which is convergent owing to the contribution $\sim -M \int dx\, \bar{\varphi}\varphi$ to $\exp(iS)$, produces the inverse of the determinant of $-iD + iM$. Denoting the eigenvalues of the hermitian operator $-iD$ by λ_n, the integration thus yields a product $\prod_n \frac{\lambda_n}{\lambda_n + iM}$. For large $\lambda_n \gg M$, numerator and denominator cancel out and, for small momenta, the large mass M makes the contribution from the denominator an inessential constant, $\lambda_n + iM \simeq iM$. Finally, a parity transformation sends the spectrum to its negative, $\lambda_n \to -\lambda_n$. For finite M, this will affect the product above.

(c) Integration over φ generates the contribution $i \operatorname{tr} \ln(-iD + iM) = i \operatorname{tr} \ln(p - A + iM) \to -i \operatorname{tr}(A(P + iM)^{-1})$. In the Wigner representation, this becomes $-i \int \frac{dx\,dp}{2\pi} A(x) \frac{1}{p + iM}$. The momentum integral contains a singular real part, which we ignore for the reason mentioned. However, the imaginary part gives the finite contribution $-i\pi$, so that we obtain a term $S_{\text{CS}}[A] = -\frac{1}{2} \int dx\, A = 2\pi k \int \operatorname{ch}_1(F)$ with $k = -1/2$. For a large gauge transformation, $A = 2\pi/L$, so that $S_{\text{CS}}[A] = -\pi$, so that $\exp(iS_{\text{CS}}) = -1$.

9.4.5 Functional bosonization

In our discussion of the interacting electron gas in $(1+1)$-dimensions in section 3.6, we used symmetry arguments to guess the action of an equivalent bosonic theory. In this problem, we discuss how the action emerges directly within the functional integral approach as a manifestation of the chiral anomaly.

Consider the interacting one-dimensional electron gas as described by the relativistic action (3.86) and the interaction contribution (3.90). Throughout, it will be convenient to formulate the latter in a matrix representation,

$$S_{\text{int}}[\psi] = \frac{1}{2} \int d\tau\,dx \, \begin{pmatrix} \hat{\rho}_+ & \hat{\rho}_- \end{pmatrix} g \begin{pmatrix} \hat{\rho}_+ \\ \hat{\rho}_- \end{pmatrix}, \qquad g = \begin{pmatrix} g_4 & g_2 \\ g_2 & g_4 \end{pmatrix}.$$

To probe the response of the system to external perturbations, we add to the action a source term $S_{\text{source}}[\psi, j] \equiv \int \sum_{s=\pm} (\psi_s^\dagger j_s + j_s^\dagger \psi_s)$.

(a) Adding S_{int} to the Euclidean free action (3.86), decouple the four-fermion interaction by introducing a two-component Hubbard–Stratonovich field $\varphi^T = (\varphi_+, \varphi_-)$ and show that the result can be written as

$$S[\psi, \varphi] = \frac{1}{2} \int d^2x \, \varphi^T g^{-1} \varphi + \int d^2x \, \bar{\psi}(i\slashed{D} - \slashed{\varphi})\psi, \qquad (9.60)$$

where $\slashed{D} = \partial_\mu \gamma^\mu$, the components of the interaction field are defined by $\varphi_1 = \frac{1}{2}(\varphi_+ + \varphi_-)$ and $\varphi_2 = \frac{1}{2i}(\varphi_+ - \varphi_-)$, and we use the representation $(\gamma^1, \gamma^2) = (-i\sigma_1, -i\sigma_2)$.

The interaction field φ couples to the fermion action as a two-dimensional vector potential. As with any two-component vector, we may decompose the coefficients of φ into a longitudinal and a transverse contribution (the Hodge decomposition): $\varphi_\mu = -(\partial_\mu \xi + i\epsilon_{\mu\nu}\partial_\nu \eta)$. This is an interesting decomposition as it suggests that the vector potential φ_μ can be removed from the action by a combined vector and axial gauge transformation $\psi \to e^{i\xi + i\eta\sigma_3}\psi$, $\bar{\psi} \to \bar{\psi}e^{-i\xi i\eta\sigma_3}$ (check). However, we know that the axial transformation by η is anomalous and must be handled with care. (Think what the situation would be otherwise: we would have shown that interactions leave the one-dimensional fermion system totally unaffected!)

In principle, we might follow the Fujikawa route to track the φ-dependent change in the action caused by the axial transformation. However, as an instructive alternative, here we proceed differently and expand the action in the gauge potential. Given what we know about the anomaly, we expect the appearance of UV-problematic contributions whose regularization will yield the effective action. (This procedure parallels the historic approach to anomalies.)

(b) Expand the "tr ln" operator obtained by integration over fermions to second order in the fields $\varphi_{\pm}$. Switching to a frequency–momentum representation and approximating the Matsubara sum by an integral, one obtains an expression that is formally UV divergent. Regularize the integral by introducing a cutoff Λ in momentum space.[42] Show that the effective φ-action reads as

$$S[\varphi] = \frac{1}{2} \sum_k \varphi_k^T \left(g^{-1} + \Pi_k \right) \varphi_{-k}, \tag{9.61}$$

where $k = (\omega, q)$ and $\Pi_k \equiv \{ \frac{\delta_{ss'}}{2\pi} \frac{q}{-is\omega + q} \}$. Note that this result is independent of the cutoff. Finally, introduce the field doublet $\Gamma^T \equiv (\xi, \eta)$ to represent the action as

$$S[\Gamma] = \frac{1}{2} \sum_k{}' \Gamma_k^T D_k^T \left(g^{-1} + \Pi_k \right) D_{-k} \Gamma_{-k}, \tag{9.62}$$

where the transformation matrix $D_k \equiv \begin{pmatrix} q-i\omega & -q-i\omega \\ -q-i\omega & q+i\omega \end{pmatrix}$ mediates between the field variables Γ and φ (exercise).

We next turn our attention to the source terms. Integration over the original fermion variables generates a source contribution

$$S_{\text{source}}[\psi, j] \overset{\int D\psi}{\longrightarrow} S[j, \Gamma] = \int d^2x \, d^2x' \, \bar{j}(x) G[\Gamma](x, x') j(x')$$

$$= \int d^2x \, d^2x' \, (\bar{j} e^{-i(\xi + \eta\sigma_3)})(x) G(x, x') (e^{i(\xi + \eta\sigma_3)} j)(x'),$$

where x are space–time indices, $G[\Gamma]$ is the Green function coupled to the interaction field, and in the last step we have applied the generalized gauge transformation above to transfer the (ξ, η)-dependence to the source vectors j. The action S_{source} contains the free fermion Green function (a matrix in both space–time and the space of s-indices) as an integration kernel. To proceed, notice that its matrix elements can be obtained as correlation functions of an equivalent free bosonic theory. This connection was introduced in problem 3.8.10 on the example of a specific free fermion correlation function. Generalizing the results of that problem, one may verify that (exercise)

$$\hat{G}_{ss'}(x, x') = (2\pi a)^{-1} \langle e^{-i(\varphi + s\theta)(x)} e^{i(\varphi + s\theta)(x')} \rangle, \tag{9.63}$$

---

[42] In particle physics, it would be more natural to work with a rotationally symmetric regulator preserving the Euclidean invariance of the theory. However, in condensed matter physics we should remember that the action above is obtained by linearization of a lattice Hamiltonian. The theory thus includes an in-built momentum cutoff.

where a is the lattice spacing and the action of the bosonic doublet $\Xi^T \equiv (\phi, \theta)$ is given by

$$S_0[\Xi] = \frac{1}{2} \sum_k \Xi_k^T K_k \Xi_{-k}, \quad K_k \equiv \frac{1}{\pi}(q^2 - iq\omega\sigma_1), \tag{9.64}$$

i.e., a non-interacting variant of the Luttinger liquid action (see Eq. (3.96)).

(c) Use the Fermi–Bose correspondence to represent the generating function as a double field integral over Γ and Ξ. Next shift the integration variables Ξ to remove the field Γ from the source action, and perform the quadratic integral over Γ. Show that the final form of the action is given by (3.98).

Summarizing, we have rediscovered the action of the interacting Luttinger liquid, and the boson representation of fermion correlation functions (the latter obtained by differentiation with respect to the source parameters j). While the present **functional bosonization** approach is more involved than that of section 3.6, it has the advantage of explicitness. However, if the truth be told, the authors are not aware of cases where this aspect became relevant: usually, the standard bosonization approach is just fine and, where it is not, the formalism above would not be any better).

Answer:

(a) This is resolved by a straightforward exercise in Gaussian integration and reorganizing indices.

(b) Integrating over fermions, we obtain the effective action

$$\begin{aligned}
S[\varphi] &= \frac{1}{2} \int d^2x \, \varphi^T g^{-1} \varphi - \mathrm{tr}\ln(\slashed{\partial} - i\slashed{\varphi}) \\
&= \frac{1}{2} \int d^2x \, \varphi^T g^{-1} \varphi - \frac{1}{2}\mathrm{tr}(\slashed{\partial}^{-1}\slashed{\varphi}\,\slashed{\partial}^{-1}\slashed{\varphi}) + \mathcal{O}(\varphi^4) \\
&= \frac{1}{2} \sum_k \varphi_{k,s} \left(g_{ss'}^{-1} + \delta_{ss'}\Pi_{s,k}\right) \varphi_{s',-k} + \mathcal{O}(\varphi^4),
\end{aligned}$$

where $k = (\omega, q)$ and $\Pi_{s,k} = \int d^2p \, (\epsilon + isp)^{-1}(\epsilon + \omega + is(p+q))^{-1}$. Evidently, the structure of this integral poses a problem: while all the poles of the integrand appear to be on one side of the real axis (so that analyticity arguments might suggest a vanishing of the integral), the double integral is divergent.[43] This is a $0 \times \infty$ conflict, which we resolve by introducing a momentum cutoff:

$$\begin{aligned}
\Pi_{s,k} &= \frac{1}{\omega + isq} \int_{-\Lambda}^{\Lambda} \frac{dp}{2\pi} \int \frac{d\epsilon}{2\pi} \left[\frac{1}{\epsilon + isp} - \frac{1}{\epsilon + \omega + is(p+q)}\right] \\
&= -\frac{1}{\omega + isq} \frac{i}{2} \int_{-\Lambda}^{\Lambda} \frac{dp}{2\pi} \left[\mathrm{sgn}(sp) - \mathrm{sgn}(s(p+q))\right] = \frac{1}{2\pi} \frac{q}{-is\omega + q}.
\end{aligned}$$

Substituting this result into the action, we obtain Eq. (9.61).

---

[43] Note that the situation gets better at higher orders in the expansion: the poles remain on the same side, but now the integrals are convergent. We conclude that the expansion of the action stops at second order.

(c) Representing the fermion Green function as in Eq. (9.63), we obtain the local expression

$$\mathcal{Z} = \int D\Xi \; D\Gamma \; e^{-S_0[\Xi] - S[\Gamma]} \exp\left[-\int d^2x \left(\bar{j} e^{-i(\xi + \phi + (\eta + \theta)\sigma_3)} + e^{i(\xi + \phi + (\eta + \theta)\sigma_3)} j \right) \right],$$

where the non-universal factor $2\pi a$ has been absorbed into the definition of the source fields. (To confirm that the Ξ-integral faithfully reproduces the source action, one has to take into account the fact that $\exp(i(\phi \pm \theta)) \leftrightarrow \psi$ is a Gaussian correlated variable.) The structure of the source term suggests a shift $\phi \to \phi - \xi$, $\theta \to \theta - \eta$, or $\Xi \to \Xi - \Gamma$ for short. Denoting the now Γ-independent source contribution by $\exp(-S_{\mathrm{source}}[\Xi])$, the partition function assumes the form $\mathcal{Z} = \int D\Xi \; D\Gamma \; e^{-S_0[\Xi - \Gamma] - S[\Gamma] - S_{\mathrm{source}}[\Xi]}$. We further note that $K_k = -D_k^T \Pi_k D_{-k}$, to obtain the integral over Γ,

$$
\begin{aligned}
S_0[&\Xi - \Gamma] + S[\Gamma] \\
&= \frac{1}{2} \sum_k \left[\Xi_k^T K_k \Xi_{-k} + \Gamma_k^T (D_k^T g^{-1} D_{-k}) \Gamma_{-k} + \Xi_k^T K_k \Gamma_{-k} + \Gamma_k^T K_{\bar{q}} \Xi_{-k} \right] \\
&\xrightarrow{\int D\Gamma} \frac{1}{2} \sum_k \Xi_k^T \left[K_k - K_k (D_k^T g^{-1} D_{-k})^{-1} K_k \right] \Xi_{-k} \\
&= \frac{1}{2} \sum_k \Xi_k^T \left[K_k + 2q^2 (g_4 - g_2 \sigma_3) \right] \Xi_{-k} \\
&= \frac{1}{2\pi} \sum_k (\varphi, \theta)_k \begin{pmatrix} q^2[1 + 2\pi(g_4 - g_2)] & -iq\omega \\ -iq\omega & q^2[1 + 2\pi(g_4 + g_2)] \end{pmatrix} \begin{pmatrix} \varphi \\ \theta \end{pmatrix}_{-k},
\end{aligned}
$$

where we have used the fact that $K_k D_{-k}^{-1} = -g(\mathbf{1} - i\sigma_2)$. Transforming back to real space–time, we obtain Eq. (3.98).

10 Gauge Theory

SYNOPSIS In this chapter we introduce modern concepts and applications of gauge theory. We will start out from a geometric interpretation of gauge theory. This entry point will provide the basis for understanding the common origin of many phenomena in gauge theories. Specifically, we will discuss the general concepts of covariance, confinement, the gauge theory of topological matter, and that of strongly entangled systems with discrete gauge groups (notably $\mathbb{Z}_2$).

Reflecting the geometric origin of gauge theory, we will increasingly use differential form language as the chapter develops. Readers not familiar with this concept are encouraged to read the quick introduction in appendix section A.1 before starting this chapter, or in parallel with it.

What is gauge theory? Even a modern physics curriculum often does not provide an easy answer to this question. At an early stage, we get introduced to gauge fields as computational tools in electromagnetism. Later, we learn about the "gauge invariance" of classical and quantum mechanics, and that gauge fields have their own physical identity as mediators of forces, or as gauge particles when quantized. Later still, we get acquainted with a multitude of other gauge concepts, among them:

▷ the fact that gauge forces can be decaying or increasing with distance, the latter phenomenon being known as **confinement**;

▷ the application of gauge fields to the topological classification of solids (in that context, they are often called gauge "**connections**" – but connections of what?);

▷ the destruction of classical symmetries by quantum fluctuations (quantum **anomalies**) and its connection to topological signatures of gauge theories;

▷ gauge theories with **discrete gauge groups** featuring in the description of strongly entangled quantum matter;

▷ and parallels between **general relativity** and electromagnetism, where, e.g., the Riemann curvature tensor plays a role similar to the field strength tensor in electromagnetism.

Clearly, a concept essential to so many different physical contexts must be quite fundamental. But what is the overarching idea behind gauge theory?

To understand the principle, recall that physics describes nature by comparing (differentiating) quantities at different instances of space and time. For scalar quantities, this is simply achieved by subtraction; for example, the temperatures in a room at two different times are compared by computing their difference. However, for objects of higher structure the situation gets more involved. As an example of the relevance to **relativity**, consider a vector field, v, in the universe. How do we compare its amplitudes at two different space–time points? A naïve approach would be to pick a common basis and compute the differences between vector components, v^μ. However, the problem with this construction is the requirement of a synchronized choice of basis at different points in space–time, which is in conflict with the spirit of relativity.[1] The situation in **quantum mechanics** is similar. For example, the Bloch wave functions, ψ_k, of electrons in a solid are defined only up to a phase. One may apply this freedom to choose different phases for different Bloch momenta, $\psi_k \to e^{i\phi_k}\psi_k$. However, this arbitrariness invalidates comparisons based on the naïve subtraction of wave function components; answers would depend on the choice of phases.

> Gauge theory provides a unified framework for the unambiguous comparison of non-scalar quantities at different points of space and time.

This viewpoint actually motivates the terminology "gauge." However, the mathematician's terminology for a gauge field, *connection*, is even more to the point. Gauge fields/connections bridge between the different points of a (space–time) manifold, and in this way provide valid frames of reference.

In the first section of this chapter we introduce this way of thinking as a one-does-it-all framework covering most aspects of gauge theory relevant to condensed matter physics. This will be followed by the discussion of various applications which all have in common the feature that the geometric viewpoint facilitates the understanding of the underlying physics.

10.1 Geometric Approach to Gauge Theory

REMARK In this section, we introduce the foundations of gauge theory from a unifying perspective. To be concrete in our discussion, we will focus on the detailed discussion of two examples: the U(1) gauge theory of electromagnetism and quantum mechanics; and the GL(4) gauge theory, which is relevant to the description of space–time structures in general relativity. The second is added for illustrative purposes and may be skipped by readers who are in a hurry.[2]

In the 1930s, mathematicians invented a concept tailored to the understanding of gauge structures: *fiber bundles*. Much as linear algebra provides a perfect language

---

holographic condensed matter physics

[1] The choice of a space–time synchronized basis would require action at a distance, which is axiomatically excluded.

[2] However, we note that a basic familiarity with general relativity appears to become more important in condensed matter, notably in the context of **holographic condensed matter physics**. This emerging fields draws connections between strongly correlated quantum theories and gravitational theories in one dimension higher. For a discussion of this interesting development we refer to an increasing number of specialized textbooks.

to describe quantum mechanics, the concept of fiber bundles allows us to describe different realizations of gauge theory – electromagnetism, theories with non-abelian Lie gauge groups in particle and condensed matter physics, or theories with discrete gauge groups, etc. – in a unified setting. We begin this section with an heuristic introduction to bundles and their connection to field theory. On this basis, we will then address the problem of invariant differentiation, and its solution through the introduction of gauge fields.

10.1.1 Fiber bundles

fiber bundle

The construction of a **fiber bundle**, E, starts with a smooth manifold M. This manifold assumes the role of the base space of the theory, for example, a condensed matter Brillouin zone, or the space–time manifold of the universe. To each point $x \in M$, we attach a fiber, F_x. The fiber defines the target space of the theory. In theories with continuous targets, it is a smooth manifold in its own right; for example, $F_x = \mathbb{C}$ in the case of solid state wave functions, or $F_x = \mathbb{R}^4$ in the case of space–time vector fields defined in the universe. However, we will also discuss cases with discrete fibers, such as $F_x = \{-1, 1\}$ in the case of $\mathbb{Z}_2$ gauge theory.

gauge group

We assume that the fibers are subject to the action of a group, G. In mathematics, it is called the **structure group** of the theory, and in physics its **gauge group**. In the case of complex-valued wave functions, $G = \mathrm{U}(1)$ acts by phase multiplication and, in the case of four-dimensional vector fields, $G = \mathrm{GL}(4)$ acts as the group of basis transformations. Importantly, the actions of G at different fibers are independent of each other, which reflects the freedom to perform different phase or basis transformations at different points of the physical system.

local trivialization

The attachment of fibers to points in $x \in M$ means that, locally, E looks like a Cartesian product of M and F: there exists a neighborhood $x \in U \subset M$ such that the restriction of the bundle to the neighborhood is isomorphic to $U \times F$ (see fig. 10.1). It is important to keep in mind that such **local trivializations** of E are defined locally but, in general, not globally. Specifically, one can show that bundles whose base manifold M is not contractible to a point can not be globally trivialized. For example, a torus base space, $M = T^d$, is nontrivial and this feature is at the root of most topological structures in band insulators. We will return to this point in section 10.5, where we discuss topological aspects of gauge theory. However, for the time being, it is okay to think of the bundle as a local Cartesian product $U \times F$.

section

Finally, a **section** is a map $s : x \mapsto s(x)$ assigning values $s(x)$ in the fiber F_x to points x. Sections are the geometric analog of wave functions or fields. It is always possible to define *local* sections in trivializing neighborhoods of points in M. However, both in mathematics and physics, the existence of global sections is intimately related to the topology of the bundle, and this is again a subject to be addressed in section 10.5.[3]

principal bundle

[3] For example, one can show that **principal bundles**, i.e., bundles whose fibers are groups, admit global sections if and only if they are trivial.

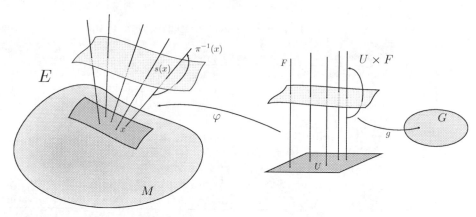

Fig. 10.1   The concept of fiber bundles: to the points of a smooth manifold, M, fibers, F, are attached. Locally, but not necessarily globally, the total space, E, defined by this construction looks like a Cartesian product $U \times F$ of subsets of $U \subset M$ and the fibers. Fields are described as local sections of the bundle, i.e., maps $x \to s(x) \in F$. Fibers, and thereby fields, are subject to the action of a (gauge) group, G.

INFO The mathematical theory of fiber bundles is a subject of great depth (see Ref. [21] for a physics-oriented introduction). Fortunately, the concept is also quite intuitive, and a heuristic understanding is sufficient for our purposes. Just for the sake of completeness, here we state the **definition of fiber bundles**, articulating what has been said above in precise terms:

    A fiber bundle (E, π, M, F, G) consists of a *total space*, E, a *base space*, M, and a *fiber*, F, which are all differentiable manifolds. The *projection*, $\pi : E \to M$ is a surjective map, whose pre-image $\pi^{-1}(x) \equiv F_x \simeq F$, $x \in M$, defines the fiber above the point x. For each point $x \in M$, we require the existence of an open neighborhood, U_i, and an isomorphism φ_i, which maps the *local trivialization* $U_i \times F \to \pi^{-1}(U_i)$, $(x, f) \mapsto \varphi_i(x, f)$ to the bundle space over U_i. Further, we require that $\pi(\varphi_i(x, f)) = x$, i.e., the trivialization does not alter base points. In cases where $x \in U_i \cap U_j$ lies in the intersection between two neighborhoods, points in the fiber $\pi^{-1}(x)$ above x are represented by two different elements of the reference fiber F. We then require that the *transition functions*, $\varphi_{ij}(x) \equiv \varphi_i^{-1}(x) \circ \varphi_j(x) : F \to F$ are elements of a transformation group of the fiber, the *structure group*, G.

For convenience, table 10.1 lists some fiber bundle vocabulary and its translation to physics. Some of the entries are not yet explained and are included for later reference.

10.1.2 Parallel transport

REMARK In this and the following sections we will interchangeably use geometry and physics parlance. Which of the terminologies is preferred depends on the context, while a mixture of languages is also widespread in the modern physics literature. For reasons of notational transparency, we frequently use subscript notation, $s_x \equiv s(x)$, for the coordinate dependence of fields.

definition of fiber bundles

Table 10.1 Fiber bundle vocabulary with its translation to physics.

Geometry	Physics
Fiber bundle, E	–
Base manifold, M	Base manifold of field theory
Fiber, F	Target manifold of field theory
(Local) section, s	Field
Structure group, G	Gauge group
Action of structure group	Gauge Transformation
(Loop) parallel transport, Γ	Wilson-loop
Connection form, A	Vector potential/Berry connection/Christoffel symbol
Curvature form, F	Field strength/Berry curvature/curvature tensor

We are now in a position to formulate the problem of invariant differentiation and its solution. To this end, consider two points $x, y \in M$ connected by a curve γ (see fig. 10.2 for a one-dimensional visualization). Let us assume that the fiber has been parameterized by coordinates, so that we may assume $F \subset \mathbb{R}^n$. The values of a section, s_x, s_y, at the points may then be compared by subtraction, $s_y - s_x$. However, the problem with this prescription is that it is not invariant under local gauge transformations, g_x. The latter change the section values as $s_x \to g_x s_x$ and $s_y \to g_y s_y$. After a transformation, the naïve comparison may yield a different result, which means that it is lacking gauge invariance.

parallel transport To define an invariant comparison scheme, additional structure is required. We define a **parallel transporter**, $\Gamma[\gamma] \in G$, which depends on the chosen curve and is an element of the structure group itself. Applied to a field value, $\Gamma[\gamma]s_x$ defines an element of the fiber F_y above the terminal point of the curve. This value is called the parallel transport of s_x to y, and it serves as a reference value

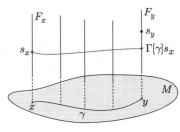

for comparison. Changes in the section are measured as $s_y - \Gamma[\gamma]s_x$, i.e., differences between the local value s_y and the parallel transported value $\Gamma[\gamma]s_x$. Specifically, a section $s_y = \Gamma[\gamma]s_x$ identical to the parallel transported value is considered to be effectively constant, even if $\Gamma[\gamma]s_x \neq s_x$.

Of course, this prescription remains formal as long as we have not defined physically, or mathematically, motivated realizations of parallel transport. However, for the moment, let us stay on a general level and implement the key condition of **gauge invariance** **gauge invariance** (see fig. 10.2). Assume that a gauge transformation has been applied such that $s_x \to g_x s_x$ and $s_y \to g_y s_y$. In the gauge-transformed setting, we work with a new realization of the parallel transporter, which we term Γ' where, to keep the notation concise, we temporarily omit the curve dependence $[\gamma]$. Applied to the gauge-transformed section value $g_x s_x$, it yields $\Gamma' g_x s_x$ in F_y. Alternatively, we may first transport to Γs_x under the old transporter and then consider the gauge transform $g_y \Gamma s_x$ at y. Gauge invariance means that the target values are the same,

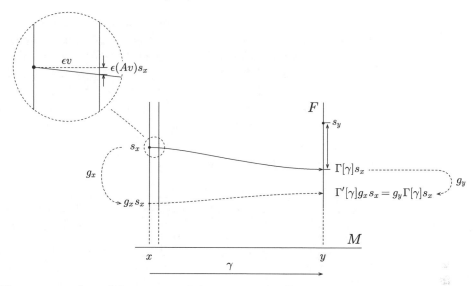

Fig. 10.2   The concept of parallel transport: elements s_x in the fiber above x get transported to the fiber above y in a manner compatible with the action of the gauge group. The inset shows an infinitesimal version of the process.

$g_y \Gamma s_x = \Gamma' g_x s_x$, irrespective of the value of s_x. Multiplication by g_x^{-1} leads to the **gauge invariance condition for parallel transport**,

$$\Gamma'[\gamma] = g_y \Gamma[\gamma] g_x^{-1}. \tag{10.1}$$

While this condition expresses the essence of gauge invariance, it is tied to the choice of a curve, γ, and therefore is not very useful in practice. As is usual in such situations, a more practical expression is obtained by considering the limiting case of *infinitesimal* parallel transport. To this end, note that an infinitesimal curve on M is characterized by a product ϵv, where ϵ is a small parameter and v a tangent vector to the manifold. For example, $v^\mu \equiv k^\mu$ might be the components of lattice momentum vectors in the case $M =$ Brillouin zone, or those of space–time vectors in the case $M =$ universe.[4] Under these conditions, parallel transport will be almost the identity operation,

$$\Gamma = \mathbb{1} - \epsilon A,$$

where $\mathbb{1}$ is the unit element of the gauge group, A, an element of its Lie algebra, and the minus sign is a matter of convention. For example, for $G = \mathrm{U}(1)$, infinitesimal transformations are described as $e^{-i\epsilon\phi}$, which identifies $A = i\phi$ as (i times) a real number. For $G = \mathrm{GL}(4)$, $A = \{A^i{}_j\}$ are real 4×4 matrices, etc. The dependence of Γ on the curve is now encoded in the dependence of A on the displacement vector. For infinitesimal arguments, this dependence is linear, i.e., $\Gamma[\gamma] \simeq \mathbb{1} - \epsilon A v$, where

---

[4] Note that, in the latter setting, the vector components v^μ appear in two different contexts: as coordinates of the fibers and as components of tangent vectors to the base manifold. This double function can be a source of confusion, especially when reading relativity textbooks.

connection
form

$Av \equiv A_\mu v^\mu$ with Lie algebra-valued coefficients A_μ. The linear functional A is the **connection form** of the theory. Its denotation in physics depends on the context, where the *vector potential* (quantum mechanics), *gauge field* (SU(n) gauge theory of particle physics), and *Christoffel symbol* (general relativity – see info block below) are specific realizations. For a d-dimensional base manifold, A has many components A_μ. Notice, however, that A is not a vector (and the denotation "vector potential" an unfortunate misnomer). Rather, it is a dual vector, or one-form, taking vectors as arguments, $v \mapsto Av = A_\mu v^\mu$. This assignment is required to be smooth on the manifold, which makes A a *Lie algebra-valued differential one-form*.

gauge the-
ory of elec-
troweak or
strong in-
teractions

EXAMPLE In the case $M = \mathbb{R}^4$ and $G = $ SU(2) or SU(3), relevant for the gauge theory of **electroweak or strong interactions**, A takes values in the Lie algebras su(n), $n = 2, 3$. It can be expanded as $A = iA^aT^a$ in the anti-hermitian generators iT^a of these algebras. In physics, it is more customary to pull out an i and work with hermitian n-dimensional matrix representations $T^a = \{(T^a)^i{}_j\}$ instead. For example, in the two-dimensional spin-1/2 representation of su(2), $T^a = \sigma^a$ are just the Pauli matrices. The coefficients $A^a = A^a_\mu dx^\mu$ are one-forms whose application to a tangent vector v^μ of M yields the generator matrix $iv^\mu A^a_\mu T^a$.

How does the **infinitesimal version of gauge invariance** look? With $y = x + \epsilon v$, the invariance equation assumes the form

$$\mathbb{1} - \epsilon A'_x v = g_{x+\epsilon v}(\mathbb{1} - \epsilon A_x v)g_x^{-1} = \mathbb{1} - \epsilon g_x A_x v g_x^{-1} + \epsilon \partial_{x^\mu} g_x v^\mu g_x^{-1} + \mathcal{O}(\epsilon^2).$$

Dropping the argument dependence for better readability, and using that $\partial_\mu g g^{-1} = -g \partial_\mu g^{-1}$, and $g \partial_\mu g^{-1} v^\mu = (g dg^{-1})v$, this assumes the form

$$A'_\mu = g A_\mu g^{-1} + g \partial_\mu g^{-1}, \qquad\qquad (10.2)$$

or

$$\boxed{A' = g A g^{-1} + g dg^{-1}} \qquad\qquad (10.3)$$

in invariant notation. In the case $G = $ U(1), relevant to electromagnetism or quantum mechanics, A is purely imaginary and, with $g = \exp(i\phi)$, we obtain the familiar gauge transformation $A'_\mu = A_\mu - i\partial_\mu \phi$. However, for theories with a non-abelian structure group, Eq. (10.2) defines the transformation of the connection form.

parallel
transport
in general
relativity

INFO In the context of **general relativity**, $F = \mathbb{R}^4$ are the local tangent spaces to the universe, M, i.e., the target spaces of space–time vector fields. One usully expands these fields in coordinate bases, defined as bases whose vectors are tangent to coordinate lines of a local coordinate system (see the figure for a two-dimensional illustration). For example, in three-dimensional space, the local basis vector $\mathbf{e}_\theta$ is tangent to the lines of varying polar angle θ in a spherical coordinate system. If $\{v^\mu\}$ are the coordinate vectors of a system of coordinates $\{x^\mu\}$,[5] and $\{w^\nu\}$ those of a system $\{y^\nu\}$, the basis transformation is mediated

covariant
notation

[5] Here, we use standard **covariant notation**, where superscripts (**contravariant indices**) label coordinate functions x^μ, or the components of vectors, u^μ, and subscripts (**covariant indices**) label partial derivatives, $\partial_\mu = \frac{\partial}{\partial x^\mu}$, or the components A_μ of forms (see info block on page 524). The index positioning specifies the behavior under coordinate transformations $x^\mu \to y^\mu$,

by the Jacobi matrix, $v^\mu = \frac{\partial x^\mu}{\partial y^\nu} w^\nu$. This means that the elements of the structure group representing transformations between coordinate bases are given by the Jacobi matrices $J \equiv \{\frac{\partial y^\nu}{\partial x^\mu}\}$, with inverse $J^{-1} = \{\frac{\partial x^\mu}{\partial y^\nu}\}$.

Following standard conventions, we denote the components of the connection form of this structure group as Γ_μ, $\mu = 0, 1, 2, 3$.[6] The individual Γ_μ are four-dimensional real matrices with matrix elements $\Gamma^\rho{}_{\sigma\mu} \equiv (\Gamma^\rho{}_\sigma)_\mu$. A change of basis amounts to the action of the structure group under which Γ changes as in (10.2), with $g = J$, such that

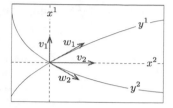

$$\Gamma' = J\Gamma J^{-1} + J d J^{-1}. \qquad (10.4)$$

In the y-representation, Γ is naturally expanded in y-coordinate differentials as $\Gamma = \Gamma_\mu dx^\mu = \Gamma_\mu \frac{\partial x^\mu}{\partial y^\nu} dy^\nu = \Gamma_\mu (J^{-1})^\mu{}_\nu dy^\nu$. This leads to the coordinate-resolved transition law, $\Gamma^\alpha{}_{\beta\gamma} \rightarrow \Gamma'^\alpha{}_{\beta\gamma} = J^\alpha{}_\mu (J^{-1})^\nu{}_\beta (J^{-1})^\rho{}_\gamma \Gamma^\mu{}_{\nu\rho} + J^\alpha{}_\rho \partial_{y^\gamma} (J^{-1})^\rho{}_\beta$. Substituting the definition of the Jacobi matrix, we arrive at the transformation identity,

$$\Gamma'^\alpha{}_{\beta\gamma} = \frac{\partial y^\alpha}{\partial x^\mu} \frac{\partial x^\nu}{\partial y^\beta} \frac{\partial x^\rho}{\partial y^\gamma} \Gamma^\mu{}_{\nu\rho} + \frac{\partial y^\alpha}{\partial x^\rho} \frac{\partial^2 x^\rho}{\partial y^\gamma \partial y^\beta}, \qquad (10.5)$$

which in this form is found in almost any textbook on relativity. Note how this equation looks much more cluttered than Eq. (10.4). The equation is not only crawling with indices, but also does not visibly distinguish between two different sectors of gauge theory, the fiber structure (encoded in the indices μ and ν) of $\Gamma^\mu{}_{\nu\rho}$ and the geometry of the base manifold, (the index ρ). Under a coordinate change, both transform but in different ways, and this leads to the mingled appearance of the equation. Equation (10.5) illustrates that it can be advisable to stay on the more transparent coordinate-invariant level, $\Gamma \rightarrow g\Gamma g^{-1} + g dg^{-1}$, for as long as possible. Of course, concrete calculations are then made in coordinates.

10.1.3 Covariant derivative

In the previous section, we argued that a gauge invariant prescription for the change in fields between different points, x, y, of the target manifold reads as $s_y - \Gamma[\gamma]s_x$, i.e., a comparison between the actual value, s_y, and the parallel transport $\Gamma[\gamma]s_x$ in the same fiber F_y. However, this expression must be taken with a grain of salt. For example, it prematurely assumes the independence of the parallel transport on the choice of the curve γ connecting x and y, a point to which we return in the next section. An unambiguous and more tractable definition is once again obtained by considering the case where the points are only infinitesimally separated, and differences become *derivatives*. Writing $y = x + \epsilon v$ as before, this leads to $s_{x+\epsilon v} -$ **covariant** $(\mathbb{1} - \epsilon A v)s_x \simeq \epsilon v^\mu (\partial_\mu + A_{x,\mu})s_x$. Division by ϵ yields the **covariant derivative**:

$$D_\mu \equiv \partial_\mu + A_\mu, \qquad (10.6)$$

or

$$\boxed{D \equiv d + A} \qquad (10.7)$$

where contravariant indices transform as $v^\mu \rightarrow \frac{\partial y^\mu}{\partial x^\nu} v^\nu$ and covariant indices as $w_\mu \rightarrow \frac{\partial x^\nu}{\partial y^\mu} w_\nu$. For further details on covariant notation, consult a textbook on relativity.
[6] Do not confuse these Γ_μ with the parallel transporter $\Gamma[\gamma]$.

in coordinate-invariant language. By construction, covariant differentiation yields $\epsilon v^\mu D_\mu s_x$ for the difference of two field values at points $y = x + \epsilon v$ and x, respectively. This expression is invariant in that, under a gauge transformation $s_x \to g_x s_x$, the covariant derivative changes as $v^\mu D_\mu s_x \to g_x(v^\mu D_\mu s_x)$.

EXAMPLE The construction above generalizes the standard **minimal coupling scheme of quantum mechanics**. With $s(x) = \psi(x)$ a wave function, and $g = e^{i\phi}$ a U(1)-gauge phase, the standard vector potential, A_μ equals i times the connection form. Its transformation under gauge transformations, $A_\mu \to A_\mu - \partial_\mu \phi$ is designed to make the covariant derivative $D_\mu = \partial_\mu - iA_\mu$ gauge covariant, $D_\mu e^{i\phi}\psi = e^{i\phi}(D_\mu \psi)$.

minimal coupling in quantum mechanics

For SU(n) gauge theory, and $A_\mu = iA_\mu^a T^a$, the covariant derivative acting on SU(n) spinors assumes the form $D_\mu \psi^i = \partial_\mu \psi^i + iA_\mu^a (T^a)^i{}_j \psi^j$, with the anti-hermitian generators T^a acting as matrices on ψ. In **general relativity** the covariant derivative of one vector field $v(x)$ in the direction of another, w, acts as $w^\nu D_\nu v^\mu = w^\nu(\partial_\nu v^\mu + \Gamma^\mu_{\nu\sigma} v^\sigma)$. In this way, it is guaranteed that, under coordinate transformations, $w^\nu D_\nu v^\mu$ transforms as a contravariant vector, i.e., covariant differentiation maps vector fields onto vector fields (verify this statement using Eq. (10.5)).

covariant derivative of general relativity

The covariant derivative can be applied to obtain a concrete expression for **parallel transport along extended curves**: by definition, a field is parallel transported from x to $x + \epsilon v$ if its covariant derivative vanishes, $\epsilon v^\mu D_\mu s = 0$. Now consider two points x, y and a connecting curve γ with coordinate representation $x(t)$, $x(0) = x$, $x(1) = y$. We transport s_x to y by a succession of infinitesimal parallel transports along the local tangent vectors $v(t) = d_t x(t) \equiv \dot{x}(t)$. All the while the parallel-transported field amplitudes obey the equation

parallel transport along curves

$$\dot{x}^\mu(t)(\partial_\mu + A_\mu)s_{x(t)} = \dot{s}_{x(t)} + \dot{x}^\mu A_{x(t),\mu} s_{x(t)} = 0. \tag{10.8}$$

This is a system of ordinary differential equations for the fiber components of the parallel transports $s(t) \equiv s_{x(t)}$ along the curve, which needs to be solved with initial condition $s(0) = s_x$.

First consider the situation where $s \equiv \psi$ is a scalar (wave function), and $A_\mu \in i\mathbb{R}$ are imaginary numbers. In this case, $d_t \psi + d_t x^\mu A_\mu \psi = 0$ is a single ordinary differential equation, which is solved by

$$\psi(t) = \exp\left(-\int_0^t du\, \dot{x}^\mu A_\mu\right)\psi(0).$$

Noting that the expression in the exponent is the line integral of the vector potential A along the curve connecting x and y, the final value $\psi_y = \psi(1)$ is obtained as

$$\psi_y = \exp\left(-\int_\gamma A\right)\psi_x. \tag{10.9}$$

This expression features frequently in quantum mechanics where, depending on the context, it is called **Berry phase**, **geometric phase**, **Aharonov–Bohm phase**, or **Peierls phase**. In each case, it describes situations where a vector potential is present and the "minimal variation" of a wave function along the base manifold is considered.

Aharonov–Bohm effect

EXAMPLE The **Aharonov–Bohm effect** of quantum mechanics is a phenomenon caused by the path dependence of parallel transport. In the experiment shown in the figure – the first observation of the effect in a condensed matter experiment[7] – the conductance of sub micron-sized gold rings threaded by a magnetic flux was measured. The conductance is sensitive to the interference of wave function amplitudes parallel transported along different paths, γ and γ' from the entry point, $\mathbf{x}$, to the terminal point, $\mathbf{x}'$, of the ring. The phase difference, $\int_\gamma A - \int_{\gamma'} A = \int_{\partial S} A = \int_S dA$ equals the flux of the magnetic field $B = dA$ through the area S bounded by $\partial S = \gamma - \gamma'$. Since this quantity appears as a phase, one expects 2π-periodic modulations of the conductance in the flux. This expectation is confirmed in the lower panel of the figure, which

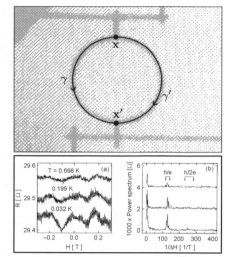

shows periodic modulations of the conductance, whose flux periodicity is revealed by the power spectrum on the right. The effect rapidly diminishes upon increasing temperature, reflecting the sensitivity of quantum mechanical wave interference to decoherence.

non-abelian parallel transport

More generally, for a gauge theory with a **non-abelian gauge group**, the differential equation for parallel transport is solved by a path-ordered exponential (in much the same way as the mathematically identical time-dependent Schrödinger equation),

$$s(t) = P \exp\left(-\int_0^t du \,\dot{x}^\mu A_\mu\right) s(0), \qquad (10.10)$$

where the path ordering $P \exp \int_0^t (\dots) \equiv \sum_n \int_0^t du_n \int_0^{u_n} du_{n-1} \cdots \int_0^{u_2} du_1 (\dots)$ accounts for the lack of commutativity of the matrices A_μ appearing in the integrand.

10.1.4 Field strength and Wilson loops

As an interesting special case, consider parallel transport along a *closed* curve, γ, parameterized as $x(t)$, $t \in [0,1]$. The transporter

$$\Gamma[\gamma] = P \exp\left(-\int_0^1 dt \,\dot{x}^\mu A_\mu\right) \qquad (10.11)$$

will now send section values s_x to $\Gamma[\gamma]s_x$ in the same fiber F_x. It is therefore a map $\Gamma[\gamma] : F_x \to F_x$, which may be identified with an element of the structure group. In cases where it is nontrivial, s_x gets parallel transported to a different value. In

holonomy

mathematics, this mechanism is called **holonomy**. While holonomy appears to be

[7] R. A. Webb, S. Washburn, C. P. Umbach, and R. B. Laibowitz, *Observation of h/e Aharonov–Bohm oscillations in normal-metal Rings*, Phys. Rev. Lett. **54**, 2696 (1985). Copyright by The American Physical Society.

of purely geometric origin, it is of defining importance to the physically observable fields of gauge theory.

To understand how this comes about, consider the example of an abelian $U(1)$ connection and a purely spatial curve, γ, parameterized as $\mathbf{r}(u)$, $u \in [0,1]$. In this case, Eq. (10.11) is a phase defined by the closed-loop line integral $\oint_\gamma du\, d_u r^i A_i = \oint_\gamma d\mathbf{s} \cdot \mathbf{A} = \int_{S_\gamma} d\mathbf{S} \cdot (\nabla \times \mathbf{A}) = \int_{S_\gamma} d\mathbf{S} \cdot \mathbf{B}$, where S_γ may be any area in M bounded by γ and $\mathbf{B} = \nabla \times \mathbf{A}$ is the gauge invariant magnetic induction. The construction demonstrates that non-trivial holonomy is equivalent to the presence of a non-zero magnetic flux.

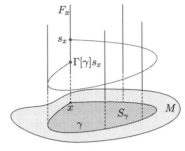

INFO In the example considered on the previous page, the difference $\gamma - \gamma'$ defines a closed loop; the area of the gold ring is a natural choice for the corresponding surface, and the probed quantity is the magnetic flux through that loop. It demonstrates that the Aharonov–Bohm effect is a manifestation of geometric holonomy.

More generally, considering curves that need no longer be space-like, application of the general Stokes theorem (see appendix section A.1.2) yields $\oint_\gamma A = \int_{S_\gamma} dA = \int_{S_\gamma} F$, where the two-form $dA = F = F_{\mu\nu} dx^\mu \wedge dx^\nu$ is the **electromagnetic field strength tensor** with components $F_{\mu\nu} = \partial_\mu A_\nu - \partial_\nu A_\mu$. We conclude that holonomy and field strength in the gauge theory of electromagnetism are different sides of the same coin!

From a physics perspective, the beauty of holonomy is that it defines physical field strengths regardless of the specific realization of a gauge theory. However, what makes our life in non-abelian theories a little more complicated is the path-ordering prescription in Eq. (10.11). A way to sidestep this problem is to consider the case of infinitesimally short loops of length ϵ. The previous example of the $U(1)$ connection established a correspondence between the line integral of the connection form (a quantity nominally $\mathcal{O}(\epsilon)$) with the integral over a surface bounded by γ (which is $\mathcal{O}(\epsilon)^2$). We thus expect the expansion of $\Gamma[\gamma]$ to start with a contribution of that order. Indeed the expansion of the parallel transporter up to second order in ϵ (see problem 10.7.1) yields $\Gamma[\gamma] = \mathbb{1} - \int_{S_\gamma} F + \cdots$ where the **general field strength tensor** is defined as

$$F_{\mu\nu} = \partial_\mu A_\nu - \partial_\nu A_\mu + [A_\mu, A_\nu]. \tag{10.12}$$

Introducing the field strength differential two-form as $F = \frac{1}{2} F_{\mu\nu} dx^\mu \wedge dx^\nu$, the invariant representation of this identity assumes the simple form,

$$\boxed{F = dA + A \wedge A} \tag{10.13}$$

In the abelian case, where $[A_\mu, A_\nu] = 0$, the second term vanishes and we are back to the formula $F = dA$ familiar from electrodynamics. More generally, the A_μ are matrices acting on the fiber degrees of freedom, where the explicit representation of the second term depends on the context.

structure constants

EXAMPLE In SU(n) gauge theory, the fields A are expanded as $A = iA^a T^a$, and their commutator is characterized by the **structure constants** of the corresponding Lie algebras, $[T^a, T^b] = f^{abc} T^c$. For example, in the case su(2), $f^{abc} = 2i\epsilon^{abc}$ is given by the antisymmetric Levi-Civita symbol. The field strength tensor (10.12) then assumes the form $F_{\mu\nu} = (i\partial_\mu A_\nu^a - i\partial_\nu A_\mu^a - f^{abc} A_\mu^b A_\nu^c) T^a$.

EXAMPLE For the connection coefficients $A_\nu \equiv \{(\Gamma^\rho{}_\sigma)_\nu\}$ of the GL(4) gauge theory, Eq. (10.12) yields

$$F \equiv R = d\Gamma + \Gamma \wedge \Gamma, \tag{10.14}$$

which becomes

$$(F^\rho{}_\sigma)_{\mu\nu} \equiv R^\rho{}_{\sigma\mu\nu} = \partial_\mu \Gamma^\rho{}_{\sigma\nu} - \partial_\nu \Gamma^\rho{}_{\sigma\mu} + \Gamma^\rho{}_{\tau\mu} \Gamma^\tau{}_{\sigma\nu} - \Gamma^\rho{}_{\tau\nu} \Gamma^\tau{}_{\sigma\mu} \tag{10.15}$$

Riemann curvature tensor

in a coordinate representation. The four-index object $R^\rho{}_{\sigma\mu\nu}$ is the **Riemann curvature tensor**. Referring for the connection with *curvature* to section 10.2.2, the above formula for R is another example of a hard-to-read coordinate representation. Quoting the differential geometer Serge Lang, "In many cases, proofs based on coordinate-free local representations are clearer than proofs that are replete with the claws of a rather unpleasant prying insect such as $R^\rho{}_{\sigma\mu\nu}$."[8] Lang's point is made by comparison of Eqs. (10.14) and (10.15).

Under a **gauge transformation**, the field strength tensor transforms as

$$F_{\mu\nu} \to F'_{\mu\nu} \equiv gF_{\mu\nu}g^{-1}, \tag{10.16}$$

or

$$\boxed{F \to F' \equiv gFg^{-1}} \tag{10.17}$$

in invariant language. This follows from the master equation (10.1), and the definition of F as an infinitesimal closed-loop parallel transporter, $\Gamma = \mathbb{1} - \int_{S_\gamma} F$. The loop transporter transforms as $\Gamma \to g\Gamma g^{-1}$, with identical factors on the left and right, and this transformation is inherited by F. However, the relation can of course also be checked by direct computation.

EXERCISE Verify Eq. (10.16) or Eq. (10.17) on the basis of the above definitions. Depending on your taste, use the differential form representation (10.3) of the gauge transformation, and the definition (10.13), or start from the component-resolved relations (10.3) and (10.12).

In abelian U(1) gauge theory, the transforming factors commute through, and the field strength tensor remains invariant. This is the principle behind the gauge invariance of the fields of electromagnetism. However, in general, the field strength tensor is subject to change.

---

[8] S. Lang, *Differential and Riemannian Manifolds*, 3rd edition (Springer, 1995).

EXERCISE Apply reasoning similar to that which led to Eq. (10.5) to verify that the **coordinate transformation of the Riemann curvature tensor** is given by

$$R'^{\alpha}_{\beta\gamma\delta} = \frac{\partial y^{\alpha}}{\partial x^{\mu}} \frac{\partial x^{\nu}}{\partial y^{\beta}} \frac{\partial x^{\sigma}}{\partial y^{\gamma}} \frac{\partial x^{\rho}}{\partial y^{\delta}} R^{\mu}_{\nu\sigma\rho}. \tag{10.18}$$

As one might have expected, it transforms as a tensor that is contravariant to first degree (the superscript index) and covariant of third degree (the three subscripts). However, while the first two of the four Jacobi matrix factors appearing in this relation assume the role of the transforming factors in Eq. (10.17), the remaining two account for the transformation of the differentials in the expansion $R = \frac{1}{2} R_{\sigma\rho} dx^{\sigma} \wedge dx^{\rho}$. The coordinate representation obscures the different roles played by fiber and manifold coordinates.

We finally note that the behavior of the field strength under gauge transformations implies an interesting type of "constancy" of the field strength: referring to problem 10.7.2 for an in-depth discussion, one can verify the **Bianchi identity**

Bianchi identitiy

$$\boxed{DF \equiv dF + A \wedge F - F \wedge A = 0} \tag{10.19}$$

where the "D" acting on F is a version of the covariant derivative appropriate to objects transforming as $X \to gXg^{-1}$ rather than as $s \to gs$. (Exercise: What is the coordinate representation of the Bianchi identity?) In electromagnetism, it reduces to $dF = 0$, which is a compact version of the homogeneous Maxwell equations. More generally, the Bianchi identity constrains the number of degrees of freedom contained in the field strength tensor, and in this way simplify the theories of gravitation, or of the strong and weak interactions.

The discussion above indicates that the majority of quantities fundamental to the theory change under gauge transformations. However, there are exceptions to the general rule. An important class of gauge invariant quantities are **Wilson loops**. For a theory whose fibers are vector spaces, and a closed curve γ in M, the Wilson loop operator is defined as $\mathrm{tr}(\Gamma[\gamma])$, i.e., the trace of the structure group element representing the parallel transporter taken over fiber space. Using Eq. (10.11),

Wilson loop

$$\boxed{W \equiv \mathrm{tr}\left[P \exp\left(-\int_0^1 dt\ \dot{x}^{\mu} A_{\mu} \right) \right]} \tag{10.20}$$

In this expression, the group elements $g(\ldots)g^{-1}$ picked up by the parallel transporter under a gauge transformation cancel out owing to the cyclic invariance of the trace. Applications of Wilson loops in the diagnostics of physical phases of gauge theories are discussed in sections 10.3 and 8.6.4. In either case, they are instrumental to the description of the degenerate ground states of topological matter.

10.2 Connections

The theory above did not make reference to a particular choice of the connection. For a given fiber bundle there are infinitely many connections, and this freedom is the essential resource of gauge theory. Depending on the application, connections are chosen according to different principles:

▷ The connection can be a **dynamical degree of freedom** in the sense that different connections are integrated subject to some action principle. This is the standard situation in the gauge theories of electromagnetism, or strong and weak interactions.

▷ The choice of a natural connection can be determined, or at least partially constrained, by **geometric structures characterizing the fiber bundle.** Important examples include adiabatic parallel transport via the Berry connection, or gravity. In the latter case, the base manifold carries a metric, which in turn defines a class of connections.

▷ We often encounter hybrid situations where integration is carried out over a connection, partially constrained by geometrical or physical principles.

In the following, we briefly consider these cases in turn.

10.2.1 Dynamical gauge theory

REMARK This section uses the language of differential forms to derive the most frequently occurring gauge field actions. Readers not yet familiar with forms may skip over this section at first reading. However, we will later refer back to the results: Eq. (10.23) for the dynamical gauge field action, and (10.25) and (10.28) for the topological actions. (Throughout, "d dimensions" means space–time dimensions, i.e., $(d-1)$-dimensional real space.)

In this section, we use the terminology "dynamical gauge theory" for theories where the connection coefficients A_μ are fields over which integration is carried out. We thus consider structures $\int DA \exp(-S[A])$ and need to address two questions: over what measure, DA, should we integrate, and what action, $S[A]$, weights the gauge field fluctuations? The answer to these questions fixes the pure gauge theory. A general theory will also contain matter fields, coupled to the gauge theory by current-vector potential terms, $\sim \int dx \, \mathrm{tr}(A_\mu j^\mu)$. However, we ignore the field–matter coupling for the moment and concentrate on the pure gauge sector.

gauge symmetries

INFO People often refer to gauge structures as **"gauge symmetries."** However, one may reason that this is a misnomer. Quoting the physicist Xiao-Gang Wen:[9] "The terms *gauge symmetry* and *gauge symmetry breaking* are two of the most misleading terms in theoretical physics." Unlike, e.g., rotation (symmetries), whose application to a magnetization vector creates physically different states of magnetization, states related by a gauge transformation represent the same physical state. The gauge transformation is not a physical transformation. Instead, it expresses the freedom to express the same state in different representations. We tend to talk of gauge symmetries when this freedom becomes a creative tool. However, one should keep in mind that these "symmetries" do not generate physical phenomena. In particular there is no such thing as gauge symmetry *breaking* (see the discussion in section 10.3).

[9] X. -G. Wen, *Quantum Field Theory of Many-Body Systems* (Oxford University Press, 2004).

gauge
fixing
condition

The proper choice of an **integration measure** must ensure that connections A and A' related to each other by a gauge transformation (10.3) should be regarded as identical and not be independently integrated over. In the abelian case, the situation can be handled by imposing a **gauge fixing** condition, e.g., by defining the integration measure as $DA = \prod_{x,\mu} dA_{x,\mu}\, \delta(\partial_\mu A^\mu)$, where the δ-function implements the Lorentz gauge, $\partial_\mu A^\mu = 0$ (and $A_\mu = g_{\mu\nu} A^\nu$ if a metric is present).[10] In non-abelian theories, more thought is required. The gauge-fixing condition must be augmented by a functional determinant securing its proper transformation behavior.

Faddeev–
Popov
ghost fields

That determinant is in turn represented by an integration over Grassmann-valued integration variables, the **Faddeev–Popov ghost fields**. However, since these complications do not arise on a daily basis in condensed matter applications, we will not discuss them here.[11]

Let us instead turn to the second aspect, realizations of **effective gauge field actions**, $S[A]$. From the theory of electromagnetism, we are accustomed to the effective action of abelian gauge theory in four space–time dimensions, Eq. (1.22). However, is this the only possible choice? And what about theories in two and three space–time dimensions? What differences arise in non-abelian theories? None of these questions is straightforward to answer by physical reasoning. However, what comes to the rescue are conditions deduced from the geometry of gauge fields. These criteria imply an almost unique fixation of gauge field actions in all dimensions, and for all realizations of the fields. The geometric analysis also tells us which actions are "topological" (in the sense of the discussion of chapter 8) and which are not. This information is also not so easy to obtain in other ways.

In the following, we will discuss these two classes of gauge field actions in turn.

Maxwell and Yang–Mills actions

Let us begin our discussion with a few general remarks. The desirable objects to integrate in a theory in d space–time dimensions are d-forms. At present, we also require gauge invariance, which means that the integrands must be independent (at least up to total derivatives) of the gauge choice. An obvious building block for the construction of integrals satisfying these criteria is the field strength tensor, F, a two-form. In differential geometry, there exists an operation, the Hodge star (see appendix section A.1.3), mapping a general p-form, ω, to a $(d-p)$-form $*\omega$.

dual field
strength
tensor

Applied to the field strength tensor, it yields the **dual field strength tensor**,

$$G \equiv - * F. \tag{10.21}$$

This is a form of degree $d - 2$, which means that the $(d = (d-2) + 2)$-degree form $F \wedge G = -F \wedge *F$ is a good object to integrate. We thus conjecture that

---

[10] As an alternative to the δ-constraint integration over all gauge sectors, one may fix a gauge first, and integrate only over one gauge sector. This is straightforward to implement in the abelian case, but less so in non-abelian theories.

[11] For an excellent introduction, see *Faddeev–Popov ghosts*, Scholarpedia **4**, 7389 (2009) by Ludwig Dmitrievich Faddeev, one of the inventors of the concept.

$$S[F] \equiv c \int_M \mathrm{tr}(F \wedge *F) \tag{10.22}$$

is a candidate for an effective gauge action. To make the representation more concrete, we note that, in a coordinate representation, $F = \frac{1}{2} F_{\mu\nu} dx^\mu \wedge dx^\nu$, the application of the Hodge star (Eq. (A.20)) yields $G = \frac{1}{2(d-2)!} F^{\rho\sigma} \epsilon_{\rho\sigma\mu_3...\mu_d} dx^{\mu_3} \wedge \cdots \wedge dx^{\mu_d}$, where $F^{\rho\sigma} \equiv g^{\rho\mu} g^{\sigma\nu} F_{\mu\nu}$ are the components of the field strength tensor with indices raised by the metric.[12] Substituting this expression into the wedge product, it is straightforward to verify that $F \wedge *F = \frac{\sqrt{|g|}}{2} F_{\mu\nu} F^{\mu\nu} dx^1 \wedge \cdots \wedge dx^d$, so that the coordinate representation of the action functional reads

$$S[F] = \frac{c}{2} \int_M d^d x \, \mathrm{tr}(F_{\mu\nu} F^{\mu\nu}). \tag{10.23}$$

Yang–Mills theory

In $d = 4$, and for non-abelian gauge groups, this is the action of **Yang–Mills theory**. It plays a key role in the formulation of the standard model, and we refer to the particle theory literature for an in-depth discussion. In the simpler abelian $U(1)$ setting, it reduces to the standard **Maxwell action** of the electromagnetic gauge field, Eq. (1.22). In this case, G, is the dual field strength tensor entering the inhomogeneous Maxwell equations. However, Eq. (10.23) equally applies to other dimensions. For example, it defines the action of $(2+1)$-dimensional electrodynamics, or the Schwinger action[13] in $(1+1)$-dimensions. In these cases, it is sometimes argued that the integrand $\sim F_{\mu\nu} F^{\mu\nu}$ is the only Lorentz-invariant expression that can be constructed from the field strength tensor (an incorrect assertion, as we will see shortly). Here, we arrive at the result via the general field strength dual, which in dimensions 3 and 2 is a one-form or an index-less zero-form, respectively.

covariant description of electromagnetism

INFO For illustrative purposes, let us review how the tensors F and G feature in the **covariant description of electromagnetism**. Starting from the definitions $F_{\mu\nu} = \partial_\mu A_\mu - \partial_\nu A_\mu$, the Minkovski metric $g = \eta = \mathrm{diag}(-1, 1, 1, 1)$, and $A^\mu = (\phi, \mathbf{A})$, a straightforward comparison shows that the field strength tensor and its dual are related to the electromagnetic fields $E_i = -\partial_i \phi - c^{-1} \partial_t A_i$ and $B_i = \epsilon_{ijk} \partial_j A_k$ as

$$\{F_{\mu\nu}\} = i \begin{pmatrix} 0 & -E_1 & -E_2 & -E_3 \\ E_1 & 0 & -B^3 & B^2 \\ E_2 & B^3 & 0 & -B^1 \\ E_3 & -B^2 & B^1 & 0 \end{pmatrix}, \qquad \{G_{\mu\nu}\} = i \begin{pmatrix} 0 & B^1 & B^2 & B^3 \\ -B^1 & 0 & -E_3 & E_2 \\ -B^2 & E_3 & 0 & -E_1 \\ -B^3 & -E_2 & E_1 & 0 \end{pmatrix}. \tag{10.24}$$

Again, we have a factor i by which the (imaginary) geometric tensor $F_{\mu\nu}$ differs from the field strength tensor of physics, $F_{\mathrm{geo}} = i F_{\mathrm{phys}}$ (see Eq.(1.21)). Expressed in terms of $F_{(\mathrm{phys})}$ and $G_{(\mathrm{phys})}$, the Maxwell equations assume the form $\partial_\mu F^{\mu\nu} = 0$ and $\partial_\mu G^{\mu\nu} = \mathrm{const.} \times j^\mu$, where $j^\mu = (c\rho, j_i)$ is the four-current vector and the constant depends on the chosen system of units (e.g., const. $= 4\pi$ in CGS units).

[12] We here assume that we are working with a metric of unit-modular determinant, such as the Minkovski metric. More generally, the coordinate representation of the field strength dual must be multiplied by a factor $\sqrt{|g|}$, where $g = \det(\{g_{\mu\nu}\})$ is the determinant of the metric tensor.

[13] J. Schwinger, *Gauge invariance and mass. II*, Phys. Rev. **128**, 2425 (1962).

In view of the importance of effectively low-dimensional systems in condensed matter physics, the study of electromagnetism in $(1 + 1)$- or $(2 + 1)$-dimensions, as described by the effective action (10.23), is becoming an increasingly important subject. The discussion above indicates that, in these dimensions, the traditional formulation of Maxwell theory in terms of vector fields is not as straightforward as in the $(3 + 1)$-dimensional case.[14] However, here we do not pursue this subject further and instead ask whether there can be effective gauge field actions different from Eq. (10.23).

Topological gauge field actions

Previously, we reasoned that a "natural" field strength action is obtained by integration of the form $\mathrm{tr}(F \wedge *F)$ over d-dimensional space–time. The presence of this $*$-operator indicates that a metric is involved; these actions are not topological. However, in even space–time dimensions, we may easily define integrands without reference to $*$: in $(1 + 1)$-dimensions consider $\mathrm{tr}\, F$ and in $(3 + 1)$-dimensions, $\mathrm{tr}(F \wedge F)$. These traces define the effective actions

$$S_{\mathrm{top}}^{(2)} \equiv c \int_M \mathrm{tr}(F), \qquad S_{\mathrm{top}}^{(4)} \equiv c \int_M \mathrm{tr}(F \wedge F), \qquad (10.25)$$

or

$$S_{\mathrm{top}}^{(2)} \equiv \frac{c}{2} \int_M d^2 x\, \epsilon^{\mu\nu} \mathrm{tr}(F_{\mu\nu}), \qquad S_{\mathrm{top}}^{(4)} \equiv \frac{c}{4} \int_M d^4 x\, \epsilon^{\mu\nu\rho\sigma} \mathrm{tr}(F_{\mu\nu} F_{\rho\sigma}), \qquad (10.26)$$

in coordinates. No reference to a metric is made, indicating that these are **topological gauge field actions**. Indeed, we will show in section 10.5.3 that the effective actions $S_{\mathrm{top}}^{(2l)}$ are topological invariants defining the so-called lth Chern class of the underlying bundle. However, the topological nature of these terms is not easy to see with the naked eye. For example, it is straightforward to show (do it!) that $\epsilon^{\mu\nu\rho\sigma}\mathrm{tr}(F_{\mu\nu} F_{\rho\sigma}) = -8\mathbf{E} \cdot \mathbf{B}$. That this term is Lorentz invariant, and yields an integer-quantized result upon integration, is not obvious (a point to which we will return in section 10.5).

Equation (10.25) defines topological actions in the even-dimensional cases $d = 2, 4$. But what about the **odd dimensions** $d = 1, 3$? To understand what is happening here, consider an odd-dimensional base manifold M as the boundary of an even-dimensional one N in one dimension higher, $M = \partial N$. Mathematically, the integrands of gauge field actions on N are differential forms of highest degree, which means that, at least locally, they can be written as exterior derivatives of something else: $\mathrm{tr}\, F = dX$ and $\mathrm{tr}(F \wedge F) = dY$ in symbolic notation. For example, in $d = 2$, $\mathrm{tr}(F) = \mathrm{tr}(dA + A \wedge A) = \mathrm{tr}(dA) = d\,\mathrm{tr}(A)$, where we noted that $\mathrm{tr}(A \wedge A) = -\mathrm{tr}(A \wedge A) = 0$ owing to the cyclicity of the trace and the anti-

---

[14] Curiously, the literature on low-dimensional electromagnetism appears to be scattered. See, however, the unpublished online resource K. T. McDonald, *Electrodynamics in one and two spatial dimensions* (physics.princeton.edu/mcdonald/examples/2dem.pdf).

commutativity of the wedge product.[15] A similar calculation in $d = 4$ shows that $\text{tr}(F \wedge F) = \text{tr}(dA \wedge dA + 2dA \wedge A \wedge A) = d\,\text{tr}(A \wedge dA + \frac{2}{3}A \wedge A \wedge A)$.

Applying Stokes' theorem to the topological action over N as $\int_N dX = \int_M X$, we obtain

$$S_{\text{CS}}^{(1)} \equiv ik \int_M \text{tr}(A), \qquad S_{\text{CS}}^{(3)} \equiv \frac{k}{4\pi} \int_M \text{tr}\left(A \wedge dA + \frac{2}{3}A \wedge A \wedge A\right) \qquad (10.27)$$

level
as candidate topological actions, where the integer-valued coupling constant k defines the **level** of the theory. (For the necessity of integer quantization, see below and section 10.5.) In coordinates, these actions read

$$S_{\text{CS}}^{(1)} \equiv ik \int_M dx\, \text{tr}(A), \qquad S_{\text{CS}}^{(3)} \equiv \frac{k}{4\pi} \int_M d^3x\, \epsilon^{\mu\nu\sigma} \text{tr}\left(A_\mu \partial_\nu A_\sigma + \frac{2}{3}A_\mu A_\nu A_\sigma\right),$$
$$(10.28)$$

Chern–Simons actions
where we note that, in one dimension, the connection has only one component. These actions are called **Chern–Simons actions** (or CS actions for brevity).[16] In the abelian case, the commutativity $[A_\mu, A_\nu] = 0$ implies that the term $A^{\wedge 3}$ vanishes, and the action reduces to the abelian form $\int A \wedge dA$. In section 10.5, we discuss the CS action from a geometric perspective and in section 8.6 from that of field theory.

gauge invariance of the CS action
Note that the odd-dimensional CS actions are formulated in terms of the gauge-connection rather than the physical field strength. This raises interesting **questions concerning gauge invariance**, whose answer depends on both the topology of space, and that of the fibers. As an example, consider the $d = 1$ case, where the fibers U(1) have the topology of a circle. A gauge transformation (10.3) leads to a shift $A \to A + e^{i\phi}de^{-i\phi}$, where ϕ is a function on M. The action then picks up a term $S_{\text{top}}^{(1)} \to S_{\text{top}}^{(1)} + ik \int_M e^{i\phi}de^{-i\phi}$. If M has a boundary, for example, M is the interval $[a, b]$, the induced term $k(\phi(b) - \phi(a))$ is non-vanishing, and the action lacks gauge invariance. The appropriate "repair mechanisms" fixing this problem at the boundary were discussed in problem 8.8.8. However, even if M is boundaryless, for example, $M \simeq S^1$, we may run into trouble. To see why, consider the choice, $\phi(x) = 2\pi x$, where $x \in [0, 1]$ is the circular coordinate parameterizing M. The gauge transformation, $g(x) = \exp(i2\pi x)$, then winds once around the structure group as x runs through M. The multi-valuedness of the exponential function, $\exp(2\pi i \times 1) = \exp(2\pi i \times 0)$, makes this a legitimate transformation; however, it is one that cannot be continuously deformed into a trivial transformation. This is the defining criterion **large gauge transformation** for a **large gauge transformation**. Under it, our topological action picks up term $2\pi k$. In this case, there is no boundary where repair mechanisms can be installed to fix the problem. The only way to safeguard gauge invariance is to require that $\exp(2\pi ik) = 1$ be integer, or $k \in \mathbb{Z}$, so that value of the induced action

---

[15] In coordinates: $\text{tr}(A \wedge A) = \text{tr}(A_\mu A_\nu)dx^\mu \wedge dx^\nu = -\text{tr}(A_\mu A_\nu)dx^\nu \wedge dx^\mu = -\text{tr}(A_\nu A_\mu)dx^\nu \wedge dx^\mu = -\text{tr}(A \wedge A)$. The same construction shows that $\text{tr}(A^{\wedge n}) = 0$ for even n, but not in general for odd n.

[16] In physics, the terminology "Chern–Simons action" often refers to the three-dimensional variant. However, the argument above shows that these actions exist in all odd-dimensional spaces.

remains inconsequential.[17] We thus see that, by a mechanism similar to that of path integral spin quantization in section 8.4.4, invariance of Chern–Simons theory under large gauge transformations requires **integer quantization of the Chern–Simons level**, k. Summarizing:

> Chern–Simons actions lack invariance under generic gauge transformations on manifolds with a boundary and under large transformations on manifolds without.

10.2.2 Connections from geometry

In this section, we discuss a second approach to the definition of physically motivated connections. Unlike the previous case, here we do not integrate over all realizations of the gauge field. Instead, we consider situations where the geometric and/or physical structure of the theory singles out a "canonical connection," or at least imposes geometric constraints on the realization of connections. This scenario is best explained using two examples.

Adiabatic transport and Berry connection

REMARK In this section, we introduce the concept of non-abelian Berry phases. For a discussion of the "standard" abelian Berry phase, see section 8.5.3.

Suppose a quantum system depends on a parameter R and that its R-dependent ground state $|0(R)\rangle$ is separated from the rest of the spectrum by an energy gap. For concreteness, let us consider a two-dimensional band insulator, where $R = \mathbf{k} = (k_1, k_2)^T \in T^2$ is a crystal momentum taking values in the two-torus $T = [0, 2\pi]^2$ (where we have set the lattice spacing to unity). The ground state is defined by n occupied Bloch eigenstates $|\psi_{1,\mathbf{k}}\rangle, \ldots, |\psi_{n,\mathbf{k}}\rangle$. These states define the valence band energetically separated from the conduction band by a global gap (fig. 10.3.) In the following, we will apply gauge theory to the description of topological properties of the system. Since topology is insensitive to the continuous deformations of system parameters, it will be sufficient to consider a fictitious variant of the system in which local transformations have been applied to collapse the spectrum to a set of degenerate and flat valence band energies, $\epsilon_{1,\mathbf{k}} = \cdots = \epsilon_{n,\mathbf{k}} \equiv \epsilon$. Without loss of generality, we set $\epsilon = 0$.

Consider a single-particle state prepared in a superposition $|\psi_{\mathbf{k}}\rangle \equiv \sum_a c_a |\psi_{a,\mathbf{k}}\rangle$ of eigenstates. We ask how this state changes in response to an adiabatically slow change of parameters $\mathbf{k}(t)$ along a curve in $\mathbb{T}^2$. (Here, "adiabaticity" means parametric variations made sufficiently slowly that excitations into the conduction band are negligible.) The evolution of the state is described by the time-dependent Schrödinger equation, $i\partial_t |\psi(t)\rangle = \hat{H}_{\mathbf{k}(t)} |\psi(t)\rangle$. At any instant, $|\psi(t)\rangle$

---

[17] We follow the general convention of coupling the Chern–Simons action as $\exp(i \times S_{\mathrm{CS}})$ to the functional integral.

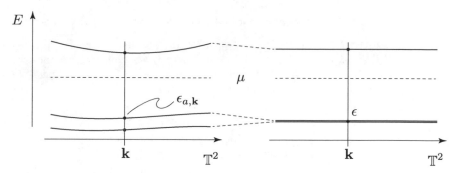

Fig. 10.3   Left: Band structure of a two-dimensional solid with two valence and one conduction band. Right: Band structure of a "deformed" Hamiltonian, topologically identical to the original one but with a degenerate band structure.

can be expanded in the states spanning the degenerate ground-space, $|\psi(t)\rangle = \sum_b c_b(t)|\psi_{b,\mathbf{k}(t)}\rangle$, with initial condition $c_b(0) = c_b$. Substituting this ansatz into the Schrödinger equation and using $\hat{H}_{\mathbf{k}(t)}|\psi_{b,\mathbf{k}(t)}\rangle = 0$, we obtain $\sum_b(\dot{c}_b(t)|\psi_{b,\mathbf{k}(t)}\rangle + c_b(t)\partial_t|\psi_{b,\mathbf{k}(t)}\rangle) = 0$. We multiply by $\langle\psi_{a,\mathbf{k}(t)}|$ and use the assumed orthogonality $\langle\psi_{a,\mathbf{k}}|\psi_{b,\mathbf{k}}\rangle = \delta_{ab}$, as well as $\partial_t|\psi_{b,\mathbf{k}(t)}\rangle = \partial_{k^\mu}|\psi_{b,\mathbf{k}(t)}\rangle\dot{k}^\mu(t)$ to obtain $\dot{c}_a + \sum_b\langle\psi_a|\partial_\mu\psi_b\rangle\dot{k}^\mu c_b = 0$. Comparing with Eq. (10.8), identifying x and $\mathbf{k}$ and the parametric section $s_{x(t)}$ with the vector $c_a(t) \equiv c_{a,\mathbf{k}(t)}$, we recognize that this equals

Berry connection   the equation for parallel transport under the **Berry connection** $A \equiv \{A_{ab}\}$, where

$$(A_{ab})_{\mathbf{k},\mu} = \langle\psi_a|\partial_{k^\mu}\psi_b\rangle, \tag{10.29}$$

or

$$\boxed{A_{ab} = \langle\psi_a|d\psi_b\rangle = (A_{ab})_{\mathbf{k},\mu}\,dk^\mu} \tag{10.30}$$

in an invariant representation. Since $0 = \partial_\mu\langle\psi_a|\psi_b\rangle = \langle\partial_\mu\psi_a|\psi_b\rangle + \langle\psi_a|\partial_\mu\psi_b\rangle = \overline{\langle\psi_b|\partial_\mu\psi_a\rangle} + \langle\psi_a|\partial_\mu\psi_b\rangle$, the Berry connection is anti-hermitian, $A_{ba} = -\overline{A_{ab}}$, and hence takes values in the Lie algebra $\mathfrak{u}(n)$ of the unitary group $U(n)$. This tells us that adiabatic parallel transport is defined on a bundle with fibers $\mathbb{C}^n \ni c_{\mathbf{k}}$ and structure group $U(n)$.

We may therefore conclude that:

> The adiabatic evolution of n-fold degenerate quantum states can be understood as parallel transport by the Berry connection (10.30) on a bundle with fibers $\simeq \mathbb{C}^n$ and base defined by the parameter manifold.

A particularly interesting situation arises when topological obstructions prohibit a continuous deformation of the Berry connection to a trivial one, $A = 0$. In this case, the parametric variation of the ground state manifold $\{\psi_{a,\mathbf{k}}\}$ contains a "twist." Referring for further discussion to section 10.5.2, we note that such twists define the fundamental principle distinguishing **topological insulators** from trivial ones.

Riemannian parallel transport and the Levi-Civita connection

REMARK   This section refers to GL(4) parallel transport, introduced earlier in various examples. Readers who skipped over these examples, should skip this section, too. Our discussion assumes familiarity with the concept of curvilinear coordinates and local coordinate bases, as introduced in section A.1.1.

Here we consider another setting in which the geometry of the bundle defines a connection. Here, the main player will be a *metric*, g, on the base manifold, M. Recall that a metric is defined by a symmetric non-degenerate bilinear form $g_{\mu\nu,x} \equiv g_x(v_{\mu,x}, v_{\nu,x})$, where $\{v_{\mu,x}\}$ is a local basis of the tangent space of M at x (see appendix section A.1.3). A metric provides the means to measure lengths and geometric orientations on a manifold.

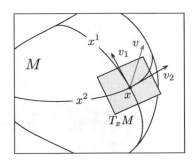

In this discussion, we show that it also provides a canonical way of comparing (covariantly differentiating) vectors on the manifold, via the *Levi-Civita connection*. This connection plays a fundamental role in general relativity.

To set the stage, consider the locally flat approximation to M at a point x, the tangent vector space, T_xM (see the figure). Importantly, we do not assume M to be embedded in a "larger space" (for all we know, the universe is not so embedded). To support the intuition, it may be helpful to think of $M = S^2$ as a unit sphere, and T_xS^2 as a two-dimensional plane locally approximating it. Assuming that the neighborhood of x is represented by a system of coordinates $\{x^\mu\}$ (such as $(x^1, x^2) = (\theta, \phi)$ for a sphere), we may define a local **coordinate basis** of vectors v_i pointing in the direction of the coordinate lines.[18] These basis vectors are not normalized in general. For example, the canonical metric on the sphere, which is defined as the restriction of the standard metric of three-dimensional space to the sphere, is given by (exercise) $g_{\theta\theta} = 1$, $g_{\phi\phi} = \sin^2\theta$, $g_{\theta\phi} = 0$. The spherical coordinate basis vectors $e_{\theta,\phi}$ about which we learn in entry level courses are the unit normalized versions of $v_{\theta,\phi}$: $e_\theta = v_\theta$, $e_\phi = \sin^{-1}\theta\, v_\phi$.

Given the metric tensor and a basis $\{v_\mu\}$, the inner product of two vectors $u, w \in T_xM$ assumes the form $g(u, w) = u^\mu w^\nu g_{\mu\nu}$, where the vectors are expanded as $u = u^\mu v_\mu$ and the subscript x is omitted for better readability.[19] We

metric connection

now define a **metric connection** by the condition that it leave the inner product of parallel-transported vectors invariant: for any curve γ connecting x and y, we require $g_x(u_x, w_x) = g_y(\Gamma[\gamma]u_x, \Gamma[\gamma]v_x)$. As usual, this condition is best analyzed in its infinitesimal form, where $y \simeq x + \epsilon v$, $\Gamma[\gamma]u = u - \epsilon A(v)u$, and $(A(v)u)^\mu = \Gamma^\mu{}_{\nu\rho}v^\rho u^\nu$, with the connection coefficients introduced in the info block on page 578. Substituting the coordinate representations of the transported vectors,

---

[18] If $x = x(x^1, \dots, x^n)$ is a coordinate representation of the point x, these vectors are obtained as tangent vectors $v_i = d_s|_{s=0}x(x^1, \dots, x^i + s, \dots, x^n)$ of the curves obtained by varying one coordinate.

[19] Pay attention to the (standard) notation: v_μ are vectors, and v^μ are the components of vectors.

we obtain the condition $u^\mu w^\nu g_{x,\mu\nu} = (u^\mu - \epsilon\Gamma^\mu{}_{\rho\sigma}u^\rho v^\sigma)(w^\nu - \epsilon\Gamma^\nu{}_{\rho\sigma}w^\rho v^\sigma)g_{x+\epsilon,\mu\nu}$. Expanding $g_{x+\epsilon v} \simeq g_x + \epsilon v^\rho \partial_\rho g$ and isolating terms of first order in ϵ, this becomes $(\Gamma^\sigma{}_{\mu\rho}g_{\sigma\nu} + g_{\mu\sigma}\Gamma^\sigma{}_{\nu\rho} - \partial_\rho g_{\mu\nu})u^\mu w^\nu v^\rho = 0$. Since this relation must hold for all tangent vectors, u, v, w, we have the condition

$$\Gamma^\sigma{}_{\mu\rho}g_{\sigma\nu} + g_{\mu\sigma}\Gamma^\sigma{}_{\nu\rho} - \partial_\rho g_{\mu\nu} = 0. \tag{10.31}$$

Christoffel symbols

In the literature, the connection coefficients $\{\Gamma^\sigma{}_{\mu\rho}\}$ satisfying this metricity condition are called **Christoffel symbols**. Defining the one-forms $dg_{\mu\nu} = \partial_\rho g_{\mu\nu}dx^\rho$, and $\Gamma^\mu{}_\nu = \Gamma^\mu{}_{\nu\rho}dx^\rho$, the equation assumes the less index-heavy form $\Gamma_{\nu\mu} + \Gamma_{\mu\nu} - dg_{\mu\nu} = 0$.

Does the metricity condition suffice to uniquely fix a connection? The answer is no. We need to fix $d \times d \times d$ Christoffel symbols $\Gamma^\mu{}_{\nu k}$. The metric tensor is symmetric, and hence contains $d(d+1)/2$ independent coefficients. We have $d^2(d+1)/2$ equations for the derivatives $\partial_\rho g_{\mu\nu}$, and thus $d^2(d-1)/2$ Christoffel symbols remain undetermined. For example, in the case $d = 4$ of relativity, 64 symbols are constraint by 40 equations, leaving 24 free parameters.

EXAMPLE As an example, consider **metric parallel transport on the two-sphere**. Using polar coordinates, the connection-form is represented by

$$\Gamma = \begin{pmatrix} \Gamma^\theta{}_\theta & \Gamma^\theta{}_\phi \\ \Gamma^\phi{}_\theta & \Gamma^\phi{}_\phi \end{pmatrix}, \tag{10.32}$$

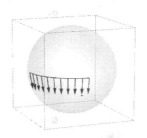

where $\Gamma^\theta{}_\phi = \Gamma^\theta{}_{\phi\theta}d\theta + \Gamma^\theta{}_{\phi\phi}d\phi$, etc. With $g = \text{diag}(1, \sin^2\theta)$ and $dg = \text{diag}(0, 2\sin\theta\cos\theta\,d\theta)$, it is straightforward to verify (exercise) that the metric condition assumes the form

$$\Gamma^\theta{}_\theta = 0, \qquad \Gamma^\phi{}_\phi = \cot\theta\,d\theta,$$
$$\sin^2\theta\,\Gamma^\phi{}_\theta + \Gamma^\theta{}_\phi = 0. \tag{10.33}$$

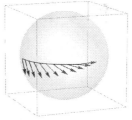

This equation fixes the diagonal elements $\Gamma^i{}_i$, $i = \theta, \phi$, but leaves $\Gamma^\phi{}_\theta$ undetermined.

To better understand the meaning of this connection, consider a south-pointing vector with components $w = (w^\theta, w^\phi) = (1, 0)$ at the point $x = (\theta, \phi) = (\pi/2, 0)$ on the equator of the sphere. Let us explore what happens as we parallel-transport the vector to the point $y = (\pi/2, \pi/2)$ separated from x by a quarter rotation around the equator. We parameterize the curve connecting the points as $x(t) = (\pi/2, t)$ with $t \in [0, \pi/2]$. With $v = (v^\theta, v^\phi) = \dot{x} = (0, 1)$, we have $\cot\theta = d\theta(v) = 0$ and $\sin\theta = 1$ along the curve. This means that the equations for parallel transport assume the form

$$d_t w^\theta = -\Gamma^\theta{}_{\phi\phi}w^\phi, \qquad d_t w^\phi = -\Gamma^\phi{}_{\theta\phi}w^\theta.$$

For $\Gamma^\theta{}_{\phi\phi} = \Gamma^\phi{}_{\theta\phi} = 0$, the solution of this system of differential equations reads $(w^\theta, w^\phi)(t) = (1, 0)$: the vector gets shifted along the curve, all the while pointing south (upper panel of figure). However, for non-vanishing off-diagonal elements of the connection form, e.g., $\Gamma^\theta{}_{\phi\phi} = c = \text{const.}$, we obtain $(w^\theta, w^\phi = (\cos(ct), \sin(ct))$. Now the vector rotates as it is transported along the curve, at constant metric length. This phenomenon is called *torsion*. We observe that the undetermined components $\Gamma^\phi{}_{\theta\phi}$ distinguish between realizations of parallel transport determined by their degree of torsion. The example also suggests that metric transport might become unique if we require the absence of torsion, in addition to metricity. As an instructive exercise, consider different choices of curves along the sphere and explore what happens under parallel transport both in the presence and absence of the torsion coefficients.

A unique metric connection is specified if the absence of **torsion** is imposed. Referring to the specialized literature for details, we note that torsionless parallel transport is defined by Christoffel symbols symmetric in the lower indices, $\Gamma^\mu{}_{\nu\rho} = \Gamma^\mu{}_{\rho\nu}$. This gives $d^2(d-1)/2$ additional equations, i.e., precisely the number of missing conditions. For example, in the case of parallel transport on the sphere, the absence of torsion requires $0 = \Gamma^\theta{}_{\theta\phi} = \Gamma^\theta{}_{\phi\theta} = 0$. We also have $-\cot\theta = \Gamma^\theta{}_{\theta\phi} = \Gamma^\theta{}_{\phi\theta}$. All connection coefficients are now fixed. (Along the equator, $\Gamma^\theta{}_{\phi\theta} = \Gamma^\theta{}_{\phi\phi} = 0$, so that the discussion of parallel transport in the example remains unaffected.)

Levi-Civita connection

The unique torsionless metric connection is called the **Levi-Civita connection**. This connection is key to the formulation of general relativity. Its curvature tensor $R^\mu{}_{\nu\rho\sigma}$ defines the **Riemann curvature tensor**. From it, two objects of lower tensorial degree, but likewise denoted by the symbol R, are built by contraction of indices: the **Ricci tensor** is obtained as $R_{\mu\nu} \equiv R^\rho{}_{\mu\rho\nu}$, and from it the **scalar curva-**

Riemann curvature tensor

> **Albert Einstein 1879–1955** was a German-born physicist awarded the Nobel Prize in Physics in 1921 "for his services to theoretical physics, and especially for his discovery of the law of the photoelectric effect." Einstein is best known for his theory of relativity and specifically mass–energy equivalence, $E = mc^2$. However, he also made breakthrough contributions to other fields, notably the theory of "old quantum mechanics," and non-equilibrium transport.

ture as $R \equiv g^{\mu\nu}R_{\mu\nu}$. Integrated over space–time, the scalar curvature defines the **Einstein–Hilbert action**,

Einstein–Hilbert action

$$S_{\text{EM}} = \frac{1}{2\kappa}\int d^4x\sqrt{-g}\,R, \tag{10.34}$$

where $\kappa \equiv 8\pi G c^{-4}$ is the **Einstein constant**, and G is the gravitational constant. The factor $\sqrt{-g}$ is the square root of the negative determinant of the metric, i.e., the canonical measure (see appendix section A.1.3) required to define a coordinate-invariant volume integral. This action plays a role analogous to that of the electromagnetic field action (10.23). The full relativistic action $S_{\text{GR}} \equiv S_{\text{EM}} + S_{\text{M}}$ is obtained by adding a matter action, similar to the $\int j_\mu A^\mu$ matter–field coupling of electromagnetism. The idea here is that matter is affected by the gravitational background g_{ij}, much as charged matter is affected by the electromagnetic field. For a simple example of the ensuing matter–field coupling, see problem 10.7.5. When these actions are considered as functionals of the space–time metric, $g_{\mu\nu}$, the (nontrivial) variation $\delta_{g_{\mu\nu}}S_{\text{GR}} = 0$ yields the **Einstein equations**

Einstein equations

$$R_{\mu\nu} - \frac{1}{2}g_{\mu\nu}R = \frac{8\pi G}{c^4}T_{\mu\nu}, \tag{10.35}$$

energy–
momentum
tensor
where the **energy–momentum tensor of gravity**, $T_{\mu\nu}$ is defined by the variation of the matter action, $T_{\mu\nu} \equiv -\frac{2}{\sqrt{-g}}\delta_{g^{\mu\nu}}S_M$. For further discussion of this equation and its physical interpretation, we refer to the extensive literature on the subject. Readers interested in exploring the connection between gauge structures and geometry on a simple example are invited to solve problem 10.7.4.

10.3 Lattice Gauge Theory

In this text, we have routinely mapped lattice models onto coarse-grained continuum field theory descriptions. Here, we take the reverse step, and put the previously discussed continuum gauge theory onto a lattice. In view of our earlier emphasis of smooth geometric structures in gauge theory, this reverse step may at first seem ill-motivated. However, it turns out that there are numerous reasons for considering gauge theory on a lattice.

▷ Theories with **discrete gauge groups** are more naturally formulated on a lattice. The most important example is $G = \mathbb{Z}_2$, a gauge group featuring in the physics of topological quantum matter, and in quantum information.

▷ The important phenomenon of **confinement**, to be discussed below, is best introduced on a lattice.

▷ Lattices are instrumental to **numerical approaches to gauge theory**, and numerical lattice gauge theory has become a research field in its own regard.

▷ Lattice systems are ideally suited to identify **dualities** between gauge theories in different physical limits.

Lattice gauge theory replaces the space–time manifold M by a d-dimensional lattice, often chosen to be a hypercubical lattice of spacing a. The gauge group acts on degrees of freedom living on the sites of the lattice. Prominent examples include $G = \mathrm{SU}(n)$ in high energy lattice gauge theory, U(1) in "compact electrodynamics" (see info block section on page 603), and $G = \mathbb{Z}_2$ in the theory of strongly entangled quantum matter. In the following, we briefly introduce the general framework of lattice gauge theory and then specialize to the particularly interesting case $G = \mathbb{Z}_2$.

10.3.1 General framework

Following Ref.[20], we denote lattice sites by n and the μth lattice unit vector emanating from n by μ. The link connecting n and $n + \mu$, is denoted by (n, μ). On each link we place a group element $U(n, \mu) \in G$. This element is interpreted as a *parallel transporter* along the microscopically short stretch $n \to n + \mu$.

---

[20] The classic reference on the subject is J. Kogut, *An introduction to lattice gauge theory and spin systems*, Rev. Mod. Phys. **51**, 659 (1979). This is a highly pedagogical paper, which continues to present a modern view of lattice gauge theory.

To connect this picture with our previous discussion, identify the lattice points n with the points of a base manifold and consider a fiber attached to each of them. For example, if $G = \mathrm{U}(m)$, we may assume these fibers to be isomorphic to $\mathbb{C}^m$, with the standard representation of group elements, U, by matrix multiplication: $U : \mathbb{C}^m \to \mathbb{C}^m$. To compare fields at different points n and m, we need parallel transporters Γ along lattice curves connecting n and m. In particular, we need parallel transporters $\Gamma_{n \to n+\mu} \equiv U(n, \mu)$ along the elementary links of the system.

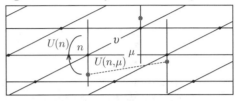

In lattice gauge theory, these link parallel transporters are center stage and assume the role of effective degrees of freedom. For a finely resolved space–time discretization of a theory with smoothly fluctuating continuum gauge fields, individual elements stay close to unity. In this case, the interpretation of parallel transport with a line integral over the connection form (see Eq. (10.9)) implies that $U(n, \mu) \simeq \mathbb{1} - aA_\mu(n) \simeq \exp(-aA_\mu(n))$, i.e., link variables in the μ-direction are determined by the corresponding components of the connection form. However, lattice gauge theory also addresses settings where this continuity assumption does not hold, and the link variables differ strongly from unity. In fact, these are the most interesting applications of lattice gauge theory.

action of lattice gauge theory

Applied to the present framework, Eq. (10.1) requires $U(n, \mu) \to U(n)U(n, \mu)U^{-1}(n + \mu)$ under local actions of the gauge group. We now need a **lattice implementation of a gauge theory action**, which takes the link variables as degrees of freedom. The form of this action is determined by two principles: it should be consistent with the above transformation condition and as local as possible. These two criteria suggest consideration of

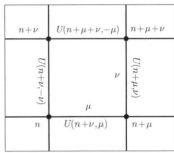

$$S[U] = -J \sum_{n,\mu,\nu} \mathrm{tr}\left[U(n, \mu)U(n + \mu, \nu)U(n + \mu + \nu, -\mu)U(n + \nu, -\nu) \right], \quad (10.36)$$

where J is a dimensionless constant. The product of link variables describes parallel transport along a minimal Wilson-loop defined by a single plaquette with corner point n and boundary vectors, μ and ν. The loops contributing to the action are gauge invariant by design. Indeed, since a gauge transformation applied at a corner n multiplies one of the U elements in the plaquette by $U(n)$ from the left and its neighbor by $U^{-1}(n)$ from the right, the factors cancel. The action (10.36) defines the partition sum of lattice gauge theory as $Z = \int DU\, e^{-S[U]}$, where the measure $DU \equiv \prod_{(n,\mu)} dU(n, \mu)$ contains the independent integration of all link variables over the group space (see below for concrete examples).

In the following, we discuss various aspects of the remarkably rich physics described by this partition sum. Specifically, we will discover *confinement transitions* separating phases with strong link fluctuations at small J from ordered phases at

Wilson
loops
large J. A remarkable feature of these transitions is that, unlike conventional phase transitions, they are not characterized by a local order parameter. Instead, spatially extended gauge invariant test observables known as **Wilson loops** assume the role of order parameters. Before turning to the physics of confinement transitions, let us introduce these loop variables and discuss their physical interpretation.

A Wilson loop is the expectation value of a parallel transporter around a rectangular surface in the lattice:

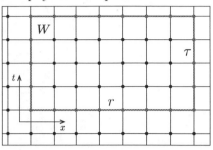

$$W \equiv \left\langle \prod_{l \in R} U(l) \right\rangle, \qquad (10.37)$$

where the product over links, l, extends over a large rectangle lying in one of the two-dimensional planes of the $(d > 1)$-dimensional lattice, and the functional average is defined as usual, $\langle \dots \rangle = Z^{-1} \int DU \, e^{-S[U]}(\dots)$. For definiteness, we consider a rectangle of extension r and τ in the direction of two coordinates denoted by x and t, respectively.

physical
meaning of
Wilson
loops
Let us discuss the physical meaning of the Wilson loops. To this end, recall that in $(3+1)$-dimensional electromagnetism, matter currents couple to the gauge field through the term $\int j^\mu A_\mu$ in the action. Specifically, a single unit-charged particle moving along a trajectory $x^i(t)$ carries a charge density $j(x) = (1, \dot{\mathbf{x}}(t))^T \delta(\mathbf{x} - \mathbf{x}(t)) = d_t x(t) \delta(\mathbf{x} - \mathbf{x}(t))$, where $x = (t, \mathbf{x})$ is the position four-vector, and the speed of light is set to unity, $c = 1$. Substitution into the gauge–matter action gives $\int A_\mu \dot{x}^\mu dt = \int_x A$, where the latter representation is a shorthand for the line-integral of the potential along the space–time curve $x(t)$ representing the particle. Exponentiating this expression, and passing to a lattice-discretized representation of the line integral, we obtain $\exp(a \sum_l A_l) = \prod_l U(l)$, where $A_l \in i\mathbb{R}$ is the imaginary generator of the link phase $U_l = \exp(ia A_l) \in \mathrm{U}(1)$.

This argument shows that products of the $\mathrm{U}(1)$ parallel transporter along lattice-curves probe the presence of charged test particles in the system. We now turn the logic upside down and postulate that the generalization to non-abelian gauge fields (10.10) *defines* the coupling of "charged" particles to the field. (For a further discussion on matter-gauge field coupling; see the info block below.)

matter–
gauge field
coupling
INFO In the abelian continuum framework, $S_\mathrm{M} \equiv \int A_\mu j^\mu$ defines the coupling of a gauge field to matter, represented by a current vector, j^μ. Under a gauge transformation, $A_\mu \to A_\mu - i\partial_\mu \phi$, this term picks up a phase contribution which, upon partial integration, assumes the form $i \int \phi \, \partial_\mu j^\mu$. This vanishes owing to continuity, $\partial_\mu j^\mu = 0$. However, a moment's thought shows that the generalization to the **coupling of matter to non-abelian gauge fields** is not straightforward. For example, assuming that the generalized currents take values in the Lie algebra of the gauge group, one might consider expressions such as $\mathrm{tr}(A_\mu j^\mu)$, where j^μ is now matrix-valued. However, this expression lacks invariance under the transformation (10.3) and therefore cannot be valid in general.

Instead of discussing this issue from a general perspective, here we proceed pragmatically and mention two manifestly gauge invariant matter–field coupling schemes employed

coupling

in practice: the standard approach in continuum field theories is **minimal coupling**. Derivatives acting on matter–fields ϕ carrying a representation of the gauge group are replaced by covariant derivatives, $\partial_\mu \phi \to (\partial_\mu + A_\mu)\phi$. This generates a matter–gauge field coupling, which need not necessarily assume the current–vector potential form above. For example, the energy–momentum tensor entering the Einstein equations (10.35), which plays a role analogous to the matter currents of electromagnetism, is defined by differentiation of a matter–gravitational field coupling action in the metric, $T_{ij} \equiv -\frac{2}{\sqrt{-g}} \delta_{g^{ij}} S_M$. This is similar to the definition of the electromagnetic current $j^\mu = \delta_{A_\mu} S_M$ as the variation of the matter–electromagnetic field action, but does not assume the linearity of S_M in fields and currents.

While the constructions above are tailored to the continuum, a matter field on a lattice is described by variables $\psi(n)$ at the nodes transforming under gauge transformations as $\psi(n) \to U(n)\psi(n)$. In fact, this transformation behavior is taken as a definition of **charged matter on a lattice**. The building blocks for currents are bilinears, $\bar{\psi}(n)\psi(n+\mu)$, where the shift μ resembles the degree of off-diagonality (represented by derivatives in a continuum theory) required to describe directional current flow, and $\bar{\psi}(n) \to U^{-1}\bar{\psi}(n)$ transforms under the inverse of the gauge transformations. These expressions are rendered gauge invariant by the addition of a parallel transporter $\bar{\psi}(n)\psi(n+\mu) \to \bar{\psi}(n+\mu)U(n,\mu)\psi(n)$. Under a gauge transformation, the latter changes to leave the overall expression invariant.

charged
matter on
a lattice

The recipe can be generalized to describe non-local insertions of matter. For example, $\bar{\psi}(n)\psi(m)$, interpreted as a particle–antiparticle pair at space–time coordinates n and m is made gauge invariant via the addition of a string variable $\bar{\psi}(n) \prod_l U(l)\psi(m)$, where the product runs over a lattice path connecting n and m. We thus realize that the coupling of particles to the gauge field naturally leads to the concept of gauge strings, as reasoned above.

Specifically, consider the product $\prod_{l \in R} U(l)$ along a rectangle in the xt-plane whose temporal extension exceeds the spatial extension by far, $\tau \gg r$. What kind of matter coupling might it represent? In the U(1) case, and on a two-dimensional lattice, we would interpret the boundaries of the rectangle in the t-direction as integrals of a current density $j_\pm = \pm\delta(x - x_\pm)(1,0)$, where $x_\pm = \pm r/2$ are the two x-coordinates of the boundaries. These are the current densities of two static particles at $\pm r/2$ with opposite charges ± 1. In the assumed limit $\tau \gg r$, the stretches in the x-direction are inessential, and we conclude that an asymptotically extended Wilson loop in the xt-plane represents the insertion of a static pair of oppositely charged particles at distance r.

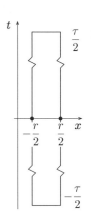

The generalization to higher-dimensional lattices does not add anything essential, only that the particles now sit at coordinates marked by the two temporal boundaries of the rectangle. On this basis we postulate that:

> space–time Wilson loops probe the presence of oppositely charged
> particles on the lattice.

In the following, we specialize to the case $G = \mathbb{Z}_2$ (the smallest of all gauge groups) to explore a confinement transition in the gauge field sector and how it is reflected in the physics of charged matter.

10.3.2 $\mathbb{Z}_2$ lattice gauge theory

Despite the simplicity of $\mathbb{Z}_2$, this gauge theory is more than a toy model: it plays an important role in the physics of topological and correlated quantum matter (see info block below), and in the the physics of "quantum codes" (section 10.4.4), where the binary-valued fields ($\rightarrow$ qubits) of the theory protect quantum information via macroscopic entanglement.

$\mathbb{Z}_2$ symmetry from fractionalization

INFO In lattice theories with strong interactions, $\mathbb{Z}_2$ gauge symmetries may emerge via the **fractionalization** of physical degrees of freedom. Referring to Ref.[21] for a detailed discussion, consider an electronic model subject to charge ($\hat{n}\hat{n}$), spin ($\mathbf{S} \cdot \mathbf{S}$), and d-wave pairing ($\Delta\Delta$) interactions. Anticipating fractionalization, we splinter the lattice fermions into structureless charge operators, or "chargons," and neutral spinful fermions, or "spinons" $c_\sigma^\dagger(n) = b^\dagger(n)f_\sigma(n)$, where $\sigma = \pm 1/2$ is spin. In this way, one introduces a local $\mathbb{Z}_2$ gauge symmetry, a transformation $b^\dagger(n) \rightarrow g(n)b^\dagger(n)$, $f_\sigma(n) \rightarrow g(n)f_\sigma(n)$, where $g(n) \in \{1, -1\}$ leaves the physical degrees of freedom unchanged. (Since a Cooper pair is represented as a double chargon, $b^\dagger b^\dagger$, this is the only symmetry consistent with the model.) In regimes where both the chargons and the spinons acquire an effective mass, one is left with an effective $\mathbb{Z}_2$ gauge theory.

In the case $G = \mathbb{Z}_2$, each link carries a binary degree of freedom, ± 1, denoted by $\sigma_3(n, \mu)$. We denote the opposite values ∓ 1 by $\sigma_1\sigma_3$.[22] The gauge group acts locally on the lattice sites via $\mathbb{1}(n)$ or $\sigma_1(n)$, where application of the nontrivial element flips all parallel transporters emanating from n as $\sigma_3(n, \mu) \rightarrow \sigma_1\sigma_3(n, \mu) = -\sigma_3(n, \mu)$.

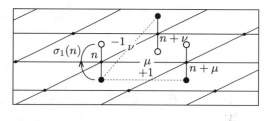

The $\mathbb{Z}_2$-variant of the general action (10.36) reads

$$S[\sigma_3] = -J \sum_{n,\mu,\nu} \sigma_3(n,\mu)\sigma_3(n+\mu,\nu)\sigma_3(n+\mu+\nu,-\mu)\sigma_3(n+\nu,-\nu), \quad (10.38)$$

where, as before, $J = \epsilon\beta$. In the following, we discuss the remarkably rich physics described by this action.

10.3.3 Confinement transition

Recall that the majority of second-order phase transitions describe the breaking of a symmetry in terms of a local order parameter. For example, in a conventional Ising model – a lattice system similar to the present one, where however, the binary lattice variables are coupled by nearest-neighbor coupling rather than by plaquette

[21] T. Senthil and M. P. A. Fisher, *$\mathbb{Z}_2$ gauge theory of electron fractionalization in strongly correlated systems*, Phys. Rev. B **62**, 7850 (2000).
[22] Denoting a binary quantity ± 1 by the symbol for the Pauli matrix σ_3, and its opposite by $\sigma_1\sigma_3$, is confusing yet standard notation. The rationale behind this convention becomes clear when we pass to the quantum theory.

exchange – symmetry is a global $\mathbb{Z}$ mapping $\sigma_3 \to -\sigma_3$ for all spins. This symmetry is broken at an Ising transition, where $\langle \sigma_3(n, \mu) \rangle \neq 0$ defines the low-temperature phase.

But what about the present system? Unlike with the single $\mathbb{Z}_2$ symmetry of the Ising model, here we have a $\mathbb{Z}_2$ freedom at each node, which seems to introduce a much higher level of symmetry. In the early literature on the subject, the symmetry-perspective of gauge theories was emphasized, and the breaking of **"gauge**

"gauge
symmetry"
breaking

symmetries" became a topic of interest. However, we have reasoned above that the gauge degrees of freedom simply reflect redundancy in the description of invariant physical objects, much as an invariant vector can be described by numerous (gauge) equivalent component representations. There is no associated physical symmetry. Specifically, the 2^N configurations on an N-site lattice related to each other by gauge transformations represent the same state. Observables "breaking the gauge symmetry" must vanish under averages over equivalent realizations. The realization that:

> "Gauge symmetries" cannot be broken

is known as *Elizur's theorem* (see info block below for further discussion of this statement).

INFO   Consider a conventional Ising system with a global $\mathbb{Z}_2$ symmetry. Symmetry-breaking expectation values of local order parameters at low temperatures $\langle \sigma_3(n, \mu) \rangle \neq 0$ are possible because the inverted expectation value $\langle \sigma_1 \sigma_3(n, \mu) \rangle = -\langle \sigma_3(n, \mu) \rangle$ is symmetry-related to $\langle \sigma_3(n, \mu) \rangle$ by the simultaneous reversal of *all* spins. In the presence of just an infinitesimal symmetry-breaking perturbation, this operation costs a divergent amount of action, and hence may be excluded.

Elitzur's
theorem

Elitzur's theorem[23] considers the different situation with a local symmetry, where

$$\langle \sigma_3(n, \mu) \rangle = \frac{1}{Z} \sum_{\{\sigma_3\}} e^{-S[\sigma_3]} \sigma_3(n, \mu) \tag{10.39}$$

includes the summation over all gauge-equivalent configurations. Specifically, $\sigma_1(n)\sigma_3(n, \mu) = -\sigma_3(n, \mu)$ differs from $\sigma_3(n, \mu)$ only by a local gauge transformation $\sigma_1(n)$, for which $S[\sigma_1(n)\sigma_3] = S[\sigma_3]$ by gauge invariance. This implies that

$$\sum e^{-S[\sigma_3]} \sigma_3(n, \mu) = \sum e^{-S[\sigma_1(n)\sigma_3]} (\sigma_1(n)\sigma_3(n, \mu)) = -\sum e^{-S[\sigma_1(n)\sigma_3]} \sigma_3(n, \mu),$$

and hence the vanishing of the expectation value. Note that the expectation value cannot be rescued by coupling the system to a small gauge "symmetry" breaking field: since the expectation value inverting transformation is local, the price in action to be paid in the presence of a symmetry non-invariant perturbations is local too, and it vanishes in the limit of zero field strength.

Elizur's theorem implies that $\langle \sigma_3(n, \mu) \rangle = 0$, for all temperatures. This raises the question of alternative order parameters detecting phase changes in the system

---

[23] S. Elitzur, *Impossibility of spontaneously breaking local symmetries*, Phys. Rev. D **12**, 3978 (1975).

following changes in temperature. In fact, this question was Wegner's motivation for inventing $\mathbb{Z}_2$ lattice gauge theory in the first place.[24]

It is clear that candidate order parameters must be gauge invariant. A single plaquette term $\langle \sigma_3 \sigma_3 \sigma_3 \sigma_3 \rangle$ satisfies this criterion, but there is no indication of a singular change in the expectation value under variations of temperature. The next obvious choice is *nonlocal* order parameters, realized through $\mathbb{Z}_2$ Wilson loops $W \equiv \langle \prod_{l \in R} \sigma_3(l) \rangle$, where R is a large rectangle embedded into the $d > 1$ dimensional lattice.

First consider the limit of small $J \ll 1$. (In the literature, the coefficient J is sometimes defined as $J = 1/2g^2$ through a coupling constant g, and $J \ll 1$ is called a **strong coupling limit**. Interpreting Z as a classical partition sum, one may also define $J \equiv \epsilon/T$, where ϵ is an energy scale, and $J \ll 1$ corresponds to the limit of high temperatures.) In this case, the partition sum can be expanded in powers of the plaquette terms, where the non-vanishing term of lowest order yields the dominant contribution to the partition sum. For the denominator, this means $\lim_{J \to 0} Z = \sum_{\{\sigma_3\}} 1 = 2^M$, where 2^M is the number of different $\mathbb{Z}_2$ configurations on a lattice with M links. In the numerator, we have the situation that the individual links of the Wilson loop rectangle vanish when averaged over spin configurations, $\sum_{\{\sigma_3\}} \sigma_3(l) = 0$. To obtain a non-vanishing contribution, we must make sure that every $\sigma_3(l)$ under the sum appears as a square, $\sigma_3(l)^2 = 1$.

strong
coupling
limit

The cheapest way to achieve this is to expand the exponentiated action to first order in each of the plaquettes inside the rectangle. Each of the links inside the rectangular perimeter then appears twice as a boundary of adjacent plaquettes. The outer perimeter of the assembly of plaquettes cancels against the links of the Wilson loops. In this way, a non-vanishing configurational average is obtained. Considering a loop of extension r and τ in the direction of the coordinates x and t as before, there are $A \equiv r\tau$ plaquettes inside the loop and this leads to the estimate $W \simeq J^A = \exp(-|\ln J|A)$. This dependence is called an **area law** for the expectation value of the loop.

area law

Now consider the opposite **weak coupling limit**, $J \gg 1$. In this case, violations of the product criterion $\sigma_3 \sigma_3 \sigma_3 \sigma_3 = 1$ around individual loops are costly. Both, the action and the loop observable are gauge invariant, and so we may organize the summation around any convenient representative of the least-action gauge-equivalence class. For convenience, we choose the uniform configuration $\sigma_3(n, \mu) = 1$. For infinite J, the Wilson loop expectation value is then trivial, $W \simeq Z^{-1} \times 1$. Single spin flip departures from this state change the sign of the $2(d-1)$ plaquettes containing the inverted link, and hence cost the action $\delta \equiv e^{-J4(d-1)}$. To lowest order in perturbation theory, we obtain $W \simeq Z^{-1}(1 + (M - 2L)\delta)$, where M is the

---

[24] F. Wegner *Duality in generalized Ising models and phase transitions without local order parameters*, J. Math. Phys. **12**, 2259 (1971). This paper is a still underrated classic. It anticipated the physics of strongly correlated topological quantum matter many years before the explosive development of the field in the first decade of this century. In view of the depth and scope of this work, it is remarkable how little credit it has received in the mainstream community (as lamented already in Kogut's 1979 review of the subject).

number of links of the lattice, $L = 2(r + \tau)$ is the circumference of the loop, and the factor $(M - 2L)$ accounts for the sign change in the product over pre-exponential σ_3 if it is one of the loop links that is flipped. Pushing the expansion to two flipped links (assumed to be separated by more than one plaquette distance), we obtain the refined estimate $W \simeq Z^{-1}(1 + (M - 2L)\delta + \frac{1}{2}(M - 2L)^2\delta^2)$. (Exercise: Do the bookkeeping of signs and explain the prefactor $(M - 2L)^2/2$.)

We recognize the beginning of an exponential series. Provided J is large enough to make contributions from close-by spin flips (for which the counting of signs becomes more complicated) statistically insignificant, resummation yields $W \simeq Z^{-1}$ $\exp[(M - 2L)\delta]$. By the same argument, $Z \simeq \exp(M\delta)$, so that

$$W \simeq e^{-2Le^{-4(d-1)J}}.$$

This expectation value contains the length of the perimeter of the loop and hence is called an **perimeter law**. Since the strong coupling area law and weak coupling perimeter law, respectively, depend in a qualitatively different ways on the loop geometry, a **phase transition** must occur at some intermediate temperature.

INFO For completeness, we mention that the construction above works only in dimensions $d = 3$ and above. The **two-dimensional Ising gauge system** is always in the disordered area law phase, no matter what the coupling. We discuss the somewhat subtle mechanism behind this phenomenon in problem 10.7.3, where we show that the two-dimensional $\mathbb{Z}_2$ gauge theory is equivalent to the one-dimensional conventional Ising chain. This is an example of various equivalences between Ising gauge systems and conventional Ising systems identified in Ref.[24] and reviewed in Ref.[20].

Phase transitions of this type occur in many lattice gauge theories and it is important to understand their **physical consequences**. To this end, we again think of t and x as a time- and a space-like coordinate, respectively. In this picture, the partition sum, $Z = \text{tr}[\exp(-\beta\hat{H})]$, becomes the imaginary-time path integral of a $(d - 1)$-dimensional quantum system, with Hamiltonian $\hat{H}$, and an inverse temperature $\beta \to \infty$ set by the extension of the system in the t-direction. (There will be more on this view in the next section.) The space–time Wilson loop describes the insertion of a particle–antiparticle pair at time $-\tau/2$ and its subsequent annihilation at time $\tau/2$. In the expectation value $W \sim \exp(-\tau c)$, the constant c is the energy associated with this process. To understand why, consider a gauge where $\sigma_3(n, t) = 1$ on all links in the t-direction (see problem 10.7.3 for the counting arguments showing that this is a possible choice). In this gauge, the Wilson loop reduces to $W = \langle \hat{X}_+(\tau/2)\hat{X}_-(-\tau/2)\rangle$, where $\hat{X}_\pm(t) = \prod_{n=(t,x)} \sigma_3(n, x)$ are the operators defined by taking the product of link operators along the spatial boundaries of the loop at times $t = \pm\tau/2$, and $\langle\ldots\rangle \equiv Z^{-1}\text{tr}[e^{-\beta\hat{H}}(\ldots)]$ is a quantum expectation value. Representing the latter through a formal insertion of eigenstates of the $(d - 1)$-dimensional system, and

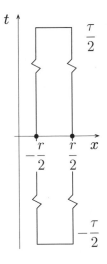

using the fact that $\beta \to \infty$ enforces a projection onto the system's ground state, $|0\rangle$,[25] we obtain $W \simeq |\langle 0|\Psi|n\rangle|^2 \exp(-E_n \tau)$, where E_n is the energy of the lowest lying state excited by application of X relative to the ground state energy itself. We identify $\Delta E_n \equiv E(r)$ with the energy required for the creation of a particle–antiparticle pair at distance r.

Comparison with the result $W \sim \exp(-2\delta L) = \exp(-4\delta(\tau + r))$ obtained in the weak coupling phase shows that, in this case, the energy $E(r) = -4\delta$ associated with the particle pair is constant and independent of the particle separation. In this case, pairs of separable particles can be created out of the vacuum at finite energy cost.

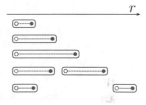

confinement

The situation in the strong coupling case is more interesting. Here, we obtained $W \sim \exp(|-\ln J|r\tau)$, indicating that $E(r) \sim r$ increases linearly in the particle separation r. This is the hallmark of **confinement**. In the confined phase, the energy associated with the separation of individually charged particles increases linearly with their separation. At some point, it becomes favorable to create a new particle pair, thus breaking the energetically expensive string separating the original pair into two. The conclusion is that it is not possible to obtain individually charged particles in isolation; particle–antiparticle pairs remain *confined* as dipoles.

quark confinement

compact electrodynamics

INFO The construction demonstrating confinement does not rely on specific properties of the gauge group $\mathbb{Z}_2$. For example, in QCD, the quarks gauge-interacting with SU(3) color gauge fields are also believed to be in a confinement phase. **Quark confinement** binds them into color-neutral configurations such as baryons or mesons. Even $U(1)$ gauge theory (which in the weak coupling phase reduces to electromagnetism) supports confinement phases. Emphasizing that this theory is described by "compact" phase factors $\exp(i\phi)$, it is also called **compact electrodynamics**. The strong coupling confinement phase of compact electrodynamics is easy to describe by adaption of the plaquette cancellation argument to $U(1)$ phases. However, a surprising amount of work[20] is required to obtain the energy of a pair of charges – the standard Coulomb potential – in the weak coupling phase! This demonstrates that the lattice language is best suited to the description of strongly fluctuating phases of matter.

10.4 Quantum Lattice Gauge Theory

REMARK Whereas, in previous sections, $\sigma_3 = \pm 1$ were the link variables, the notation σ_i now refers to an operator, acting in a space of σ_3 eigenstates $|s_3(n,\mu)\rangle$ defined on the links of the lattice.

In this section, we take a closer look at the interpretation of the d-dimensional lattice gauge partition sum as the trace $Z = \exp(-\beta \hat{H})$ of a quantum Hamiltonian. We will

---

[25] Here we assume uniqueness of the ground state. However, this assumption is not essential to the argument.

formulate this analogy in detail for a three-dimensional $\mathbb{Z}_2$ lattice gauge theory, in which case $\hat{H}$ will emerge as a two-dimensional spin-1/2-Hamiltonian.[26] The main purpose of the exercise is to identify the ground state of $\hat{H}$, and its qualitative change, when the classical parent system undergoes a confinement–deconfinement transition. An understanding of the ground state will be key to the identification of the phase transition as a topological phase transition.

10.4.1 The Hamiltonian of $\mathbb{Z}_2$ lattice gauge theory

In this text, we are biased towards thinking of thermal traces, $\operatorname{tr}\exp(-\beta\hat{H})$, as path integrals over fields. Remembering the Trotter constructions of chapter 3, these integrals are obtained by the factorization $\operatorname{tr}\exp(-\beta\hat{H}) \simeq \operatorname{tr}(e^{-\delta\hat{H}})^N$, where $\delta = \beta/N$ is chosen to be infinitesimally small such that $e^{-\delta\hat{H}}$ is close to the unit operator. We now apply this strategy to the d-dimensional lattice partition sum. The idea is to understand $Z \equiv \operatorname{tr}(\hat{T}^N)$ as a trace, where $N \to \infty$ is the extension of the lattice in the time direction and $\hat{T}$ is a **transfer matrix** yet to be identified. However, we will assume $\hat{T} \simeq \mathbb{1}$ to be close to the unit operator, and define the system's quantum Hamiltonian through $\hat{T} \equiv e^{\delta\hat{H}}$, in analogy with the general path integral construction.

transfer matrix

The condition $\hat{T} \simeq \mathbb{1}$ means that spin configurations remain almost stationary from one time slice to the next. Clearly, this assumption requires a modification of the lattice action such that the coupling constants of the spatial plaquettes, J, are chosen differently from those, J_τ, of the plaquettes in space–time planes. If we choose $J_\tau \gg J$, we make changes of spins in the time direction very costly, and in this way can implement our working assumption.[27]

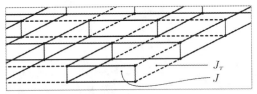

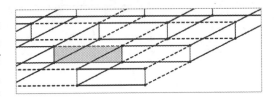

The upper panel in the figure shows a situation where spin configurations defined in a spatial plane are just copied in the time direction, and no flips occur. (In the figure, up-spin and down-spin configurations are visualized as solid and dashed lines, respectively. For better visibility, the vertical time-like bonds are shown as thinner lines.) No action $\sim J_\tau$ has to be paid in this case, and the transfer matrix acts as a unit operator.

---

[26] It is straightforward to generalize the construction to different dimensions and gauge groups.[20] However, the $(d = 3)$-dimensional $\mathbb{Z}_2$ application is particularly interesting.

[27] The introduction of this anisotropy is physically equivalent to the taking of a temporal continuum limit in the general path integral construction. It may be useful to think about this point.

More specifically, we consider the latter as a linear operator in the 2^M-dimensional space of configurations spanned by the tensor products $\otimes_{n,\mu}|s_3(n,\mu)\rangle$ of σ_3 eigenstates $|s_3(n,\mu)\rangle$ of the spins in the spatial links of the lattice. We choose a gauge where the links in the temporal direction, $|s_3(n,\tau)\rangle = |+1/2\rangle$, are all in an up-spin configuration. Under these assumptions, the zero-spin flip transfer matrix assumes a diagonal form, and comparison with the action shows that it acts as

$$\hat{T}^{(0)} \equiv e^{J\sum_\square \sigma_3\sigma_3\sigma_3\sigma_3},$$

where the sum extends over all spatial plaquettes of the lattice, and a shorthand notation for the bond operators is used. Notice that, in this case, all space–time plaquettes are locked to a $+1$ configuration. We ignore the large positive constant $\exp[J_\tau \times$ (number of space–time plaquettes)] corresponding to this configuration.

At first order in the number of flipped spins (see bottom panel of figure) a single spin flip in time direction occurs. One of the space–time plaquettes is now frustrated (the shaded one), and this comes at a price $2J_\tau$ relative to the above configuration. The transfer matrix flips one spin, and thus contains a single operator $\sigma_1(n,\mu)$:

$$\hat{T}^{(1)} \equiv e^{-2J_\tau} e^{J\sum_\square \sigma_3\sigma_3\sigma_3\sigma_3} \sum_{(n,\mu)} \sigma_1(n,\mu). \qquad (10.40)$$

We will assume $\exp(-2J_\tau)$ to be so small that higher powers of this factor, appearing in combination with multiple spin flips, are negligibly small. Under this condition, the transfer matrix can be written as

$$\hat{T} \simeq \hat{T}^{(0)} + \hat{T}^{(1)} = e^{J\sum_\square \sigma_3\sigma_3\sigma_3\sigma_3}\left(1 + e^{-2J_\tau}\sum_{(n,\mu)}\sigma_1(n,\mu)\right)$$

$$\simeq e^{J\sum_\square \sigma_3\sigma_3\sigma_3\sigma_3 + e^{-2J_\tau}\sum_{(n,\mu)}\sigma_1(n,\mu)} \stackrel{!}{=} e^{-\delta\hat{H}}.$$

Inspection of the last equation shows that the Hamiltonian of the system may be defined as

$$\hat{H} \equiv -J_{\mathrm{m}}\sum_\square \hat{B}_p - J_s\sum_l \sigma_{1,l}, \qquad (10.41)$$

where $J = \delta J_{\mathrm{m}}$, $e^{-2J_\tau} = J_s\delta$, $\sum_l$ is a shorthand for the sum over all spatial links, l, and we have introduced the standard notation $\sum_p \hat{B}_p$ for the sum over

plaquette operators all **plaquette operators** $\hat{B}_p \equiv \sigma_3\sigma_3\sigma_3\sigma_3$ around spatial plaquettes, p. Keeping the second coupling constant, J_s, fixed, we have $e^{-2J_\tau} \sim \delta$, which means that higher-order hopping terms generate higher powers of δ and can be neglected.

Our Hamiltonian has been derived descending from a gauge invariant action in one dimension higher and therefore should be gauge invariant as well. **Gauge transformations** change the sign of $\mathbb{Z}_2$ states at lattice nodes, n, and hence flip all link variables (or spin states) emanating from that node, $|s_3(n,\mu)\rangle \to |-s_3(n,\mu)\rangle$.

star operator The action of a local gauge transformation is thus described by the **star operator** $\hat{A}_s \equiv \prod_\mu \sigma_1(n,\mu)$, i.e., application of a σ_1 operation at the "star" defined by all bonds (n, μ) emanating from site n.[28] We confirm gauge invariance, $[\hat{A}_s, \hat{H}] = 0$: the σ_x in $\hat{A}_s$ trivially commute with those in $\hat{H}$, and $[\hat{A}_s, \hat{B}_p] = 0$ follows from the fact that a star and a plaquette have either two or no bonds in common, implying the commutativity of the corresponding σ_3 and σ_1 products.

10.4.2 Gauge invariant ground state

Since the gauge transformation (operators) commute with the Hamiltonian, we can seek energy eigenstates which are simultaneously eigenstates of all $\hat{A}_s$. In particular, we can decide to work in the **Hilbert space of gauge invariant states**, defined by the condition $\hat{A}_s |\psi\rangle = |\psi\rangle$ for all s. We may consider this as an additional condition imposed on top of the ground state property. Alternatively, we may define a new Hamiltonian,[29]

$$
\boxed{\hat{H}_{\mathrm{g}} \equiv -J_{\mathrm{e}} \sum_{+} \hat{A}_s - J_{\mathrm{m}} \sum_{\square} \hat{B}_p - J_s \sum_{n,\mu} \sigma_1(n, \mu)}
\tag{10.42}
$$

and seek its ground state. Since all terms in $\hat{H}_{\mathrm{g}}$ commute, this identification both reveals the ground state of $\hat{H}$ and implements the gauge invariance condition. On top of that, the study of the ground state of $\hat{H}_{\mathrm{g}}$ will reveal the topological nature of the $\mathbb{Z}_2$ confinement transition.

We begin our search for the ground state of $\hat{H}_{\mathrm{g}}$ by introducing some analogies to electrostatics: the subscripts "e" and "m" in (10.42) hint at an "electric" and "magnetic" interpretation of the plaquette and the star operator in the Hamiltonian, respectively.

EXERCISE Recall the connection between the plaquette terms of lattice gauge theory, Wilson loops, and the field strength tensor. Think what happens if we subject U(1) theory to the transfer matrix procedure (or better still, formulate it explicitly), to convince yourself that the spatial plaquette terms become the $\sim \mathbf{B}^2$ contribution of the field energy in the Hamiltonian. The denotation indicates that $J_m \sigma_3 \sigma_3 \sigma_3 \sigma_3$ is the $\mathbb{Z}_2$ analog of the magnetic energy density.

Turning to the star operator, recall from problem 1.8.2 that, in U(1) gauge theory, the components of the electric field, E_i, and the spatial components of the connection A_i form a canonical pair. Much as the momentum operator generates spatial translations, the electric field operator generates translations in the eigenstates of the vector potential (operator). In our present theory, we have no continuous structures with which to work. However, the analogs of the vector potential operator are the link variables σ_3. Their eigenstates are translated, or flipped, by the application of σ_1, and in this sense the latter assumes the role of an electric field operator.

---

[28] It is customary to label star operators by an index s, although just using the center coordinate, n, of the star would be a less redundant notation.

[29] E. Fradkin and S. H. Shenker, *Phase diagrams of lattice gauge theories with Higgs fields*, Phys. Rev. D **19**, 3682 (1979). This highly recommended paper studies an extended version of the theory, where the pure $\mathbb{Z}_2$ gauge theory is coupled to a $\mathbb{Z}_2$ matter field.

charge
density
operator

In the same spirit, the star operator $A_{s,n} = \sigma_1(n,\mu)\sigma_1(n,-\mu)\sigma_1(n,\nu)\sigma_1(n,-\nu)$ at a site n is interpreted as a **charge density operator**: this operator probes the total radial electric field emanating from site n. By virtue of Gauss's law, a negative eigenvalue of this operator is equivalent to the presence of a $\mathbb{Z}_2$ charge at the center. In this way, the gauge invariant ground state is understood as a globally charge-neutral state, $\hat{A}_s|\psi\rangle = |\psi\rangle$.

10.4.3 Confining phase

Building on these analogies, we revisit the confining phase transition, now in the quantum Hamiltonian framework. We begin with the confining phase, which corresponds to small coupling constants in the $(2+1)$-dimensional classical model. To understand what this means in the present context, let us rescale the coupling constants as $J \to J\epsilon$, $J_\tau \to J_\tau\epsilon$, and investigate what happens as $\epsilon \to 0$. From the above identifications, we have $J_m = \delta^{-1}J \sim \epsilon$ and $J_s \sim \delta^{-1}e^{-J_\tau} \sim e^{-\epsilon}$; we infer that, for diminishing ϵ, we will enter a regime where $J_s > J_m$, i.e., the electric operator in (10.42) is stronger than the magnetic operator. For simplicity, we consider a situation where $J_m = 0$ can be totally ignored.

In this case, a charge-neutral/gauge invariant ground state is easily found: simply define $|\psi\rangle = \otimes_{n,\mu}|\to\rangle_{(n,\mu)}$, where $\sigma_1|\to\rangle = |\to\rangle$, i.e., a tensor product of σ_1 eigenstates on all links. Previously, we reasoned that confinement is diagnosed by computing the energy required for the separation of two opposite charges in the system. Let us, therefore, consider a state $|\psi_{n,m}^{(2)}\rangle$ defined by the condition that it contains two charges at positions n and m and is of minimal energy. We claim that states satisfying these criteria contain strings of flipped configurations $|\leftarrow\rangle$ along the shortest path connecting n and m. To see why, note that

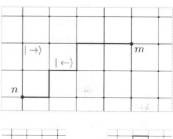

we need a single flipped link $|\leftarrow\rangle_{n,\mu}$ connected to n to satisfy the charge condition $\hat{A}_{s,n}|\psi_{n,m}^{(2)}\rangle = -|\psi_{n,m}^{(2)}\rangle$. However, we cannot end there, for a single flip would imply a charge at site $n + \mu$. The only way to avoid this is to extend the flipping to the terminal charge at m, thus creating a flip-string (see the figure).

The energy cost of this string is $|n - m|2J_s$, where $|n - m|$ is the "Manhattan-metric" distance (the minimal number of links) between the sites. The value increases linearly in the separation, and this is how the present formulation describes confinement: a separation of two test charges costs an unbounded amount of energy and is thus energetically unfeasible.

EXERCISE The proportionality of the string energy to length, $|n - m|2J_s$, implies a **string tension** (energy per unity length) $2J_s$. It is interesting to explore how the perturbative inclusion of the plaquette operator modifies this value. Apply second-order perturbation theory in J_m to show that the inclusion of plaquette terms couples the minimal string state to states containing a one-plaquette detour, as indicated in the lower panel of the figure. Show that this *lowers* the string energy by an amount $-J_m^2/(6J_s)$. (Hint: Interpret the denominator as $6 = 2 \times (4-1)$.) The first-order perturbative inclusion of plaquette operators thus diminishes the string tension, $J_s \to J_s - J_m^2/6J_s$. In the next section, we will discuss where this tendency for the strings to become more slack translates to the magnetic term becoming dominant.

10.4.4 Deconfined phase

Turning to the other side of the phase transition, we consider the opposite extreme, where $J_s = 0$ at finite J_m,

$$\hat{H}_{\rm t} \equiv -J_{\rm e}\sum_+ \hat{A}_s - J_{\rm m}\sum_\square \hat{B}_p \tag{10.43}$$

toric code   The Hamiltonian $\hat{H}_{\rm t}$ is known as the Hamiltonian of the **toric code**. It is a representative of a larger class of **stabilizer codes**, which are defined by Hamiltonians containing mutually commuting operators (all operators in Eq. (10.43) commute). The denotation *stabilizer codes* indicates that such Hamiltonians play a role in quantum information. Specifically, the toric code describes physics at the interface of information theory, topology, and gauge theory and, since its introduction in 2006,[30] has become a paradigm of many-body physics. However, before discussing why the toric code is of interest to different communities, let us compute its ground state.

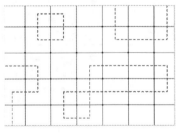

The stabilizer property implies a number of symmetries (growing extensively in system size), which indicates exact solvability. To see why, note that the commutativity of all plaquette operators means that the ground state must satisfy $\hat{B}_p|\psi\rangle = |\psi\rangle$ for all of them. It is easy to find such states. For example, the uniformly z-magnetized configuration $|\uparrow\rangle \equiv \otimes_{n,\mu}|\uparrow\rangle_{(n,\mu)}$ has this property. The problem with this ansatz is that it lacks gauge invariance. The action of any $\hat{A}_s$ flips spins, and hence violates the invariance condition $\hat{A}_s|\psi\rangle = |\psi\rangle$. However, what first looks like a problem can be turned into a construction principle for a gauge invariant ground state. Consider the state, $|\psi\rangle \equiv \mathcal{N}\sum\prod_s \hat{A}_s|\uparrow\rangle$, where $\mathcal{N}$ is a normalization factor, and $\sum\prod_s$ is a symbolic notation for the sum over all possible star configurations. This is an equal-weight superposition of all possible applications of products of star operators to the polarized state. By construction, the application of any $\hat{A}_s$ does not change this state, i.e., $\hat{A}_s|\psi\rangle = |\psi\rangle$ (think about this point), and owing to the commutativity of all operators, the condition $B_p|\psi\rangle = |\psi\rangle$ is preserved. We thus have a gauge invariant ground state at hand.

[30] A. Kitaev, *Anyons in an exactly solved model and beyond*, Ann. Phys. **321**, 2 (2006).

Application of $\hat{A}_s$ at a site n flips all spins emanating from that site. In the figure, this is indicated by a dashed square centered on the site, where a dashed line crossing a solid one means a spin flip. Application of all possible $\hat{A}_s$ combinations then amounts to a superposition of all patterns of closed dashed line loops. Alluding to the equal weight superposition of string-like objects, states of this type are called **string net condensates**.[31]

string net condensate

String net condensates are paradigms in the physics of topological phases. In section 8.1.2, we argued that **degeneracy of the ground state** is a hallmark of such phases of matter. To identify possible ground states besides the one constructed above, we impose periodic boundary conditions, i.e., we put the lattice on a torus with L sites in each direction.

EXERCISE On the torus, compare the number of constraints identifying ground states with the number of available degrees of freedom. Count the number of vertex and plaquette constraints, and the number of available link degrees of freedom to show that the latter exceeds the former by 4. This indicates a four-fold ground state degeneracy.

Repeat the exercise for a lattice on a surface containing g handles (one handle is a torus, two is a pretzel, etc.) to conclude that there are now 2^{2g} uncompensated degrees of freedom. The fact that this number depends on the surface topology suggests that the system supports a form of **topological order**.

If the degeneracy is of topological origin, the distinction between different ground states cannot depend on "details." However, it must be sensitive to the topology of the underlying surface. Guided by this principle, we consider the operators $\Sigma_3^{1,2} \equiv \prod_{h,v} \sigma_3$, where $\prod_{h,v}$ is shorthand for a product of links cutting the system in a horizontal or vertical direction, respectively, and it does not matter which line is chosen (see the figure). The ground state considered previously is an eigenstate $\Sigma_3^i |\psi\rangle = |\psi\rangle$. To understand why, note that its construction started with the up-polarized state $|\uparrow\rangle$, which trivially satisfies the eigenstate condition. Now, each of the loops defined by the application of products of A_s operators intersects the horizontal or vertical lines in the grid an *even* number of times. This means that it flips an even number of spins in the intersection and so the product operators commute with Σ_3^i.

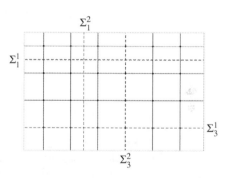

However, we might just as well have started the loop construction on a different Σ_3-polarized state, for example, one where we start with $|\uparrow\rangle$ but then flip all spins along a horizontal or vertical line cutting through the system. This operation is generated by $\Sigma_1^{1,2} = \prod_{h,v} \sigma_1$, where the product is now over a horizontal or vertical cut line; see the dashed lines in the figure. If we now let the loop construction act on these states, we generate gauge invariant ground states, which are eigenvalues of

[31] M. A. Levin and X. -G. Wen, *String-net condensation: a physical mechanism for topological phases*, Phys. Rev. B **71**, 45110 (2005).

Σ_3^i with different eigenvalues. Since these states differ from the original one by loop operators winding around great circles of the torus, they are topologically distinct. We have a totality of four such states, labeled by their eigenvalues ± 1 of $\Sigma_3^{1,2}$. It is straightforward to show (exercise) that on a genus g surface, the dimension of the ground space is given by 4^g.

It is interesting to consider the construction above from the perspective of **quantum information**. The original system is defined in terms of a macroscopically large number of "physical" spin-1/2 states, or **physical qubits** living on the links of the lattice. Embedded in the high-dimensional Hilbert space of this system, we have the ground space, which we saw is a tensor product of the much smaller number of just two (or $2g$) **logical qubits** defined by their eigenvalue of $\Sigma_3^{1,2}$. The states of these qubits are changed by application of $\Sigma_1^{1,2}$, where the operators $\Sigma_a^{1,2}$ define a Pauli matrix algebra (consider this point.) Since the logical qubits are topologically distinct, one may say that their states are "encoded" in a large number of physical qubits. This is the essential idea of **stabilizer code quantum computation**: logical information protected by being encoded in a much larger set of physical information. The advantage gained in return for the redundancy of the approach – a large number of physical qubits need to be realized for just $\mathcal{O}(1)$ information qubits – is error protection. An accidental change of physical qubits, will not in general change the logical state, unless the errors spread over the bulk of the torus (see info block in next subsection). In the context of quantum information, the ground space of a stabilizer code is sometimes called the **computational space**. Quantum information is stored and manipulated in the topologically distinct states of this space. For further discussion of the stabilizer approach to quantum computation we refer to a large body of literature, starting with the original reference[30].

INFO The degeneracy of the toric code ground states is a manifestation of their macroscopic entanglement (see section 8.1.2). Indeed, we have constructed these states as a large sum over "entangled" loops of macroscopic extension. For a spin system, this is the defining feature of a **spin liquid**, a state of correlated spins with strong fluctuations including at low temperature. For the original references on $\mathbb{Z}_2$ spin liquids, see Ref.[32]

To summarize, we now understand that the confinement–deconfinement transition in $\mathbb{Z}_2$ gauge theory is not a symmetry-breaking phase transition but a topological one. It distinguishes a confining phase from a strongly spin-fluctuating deconfined phase. The latter is long-range ordered in that the ground state is characterized by topological invariants – the eigenvalues of non-local string operators winding around the system. There is no local observable distinguishing between the different ground state sectors. The absence of local order parameters identifies the phase transition of $\mathbb{Z}_2$ Ising gauge theory as a **topological phase transition**.

---

[32] N. Read and S. Sachdev, *Large-N expansion for frustrated quantum antiferromagnets*, Phys. Rev. Lett. **66**, 1773 (1991); X. -G. Wen, *Mean field theory of spin liquid states with finite energy gaps and topological orders*, Phys. Rev. B **44**, 2664 (1991).

logical qubit

stabilizer code quantum computation

spin liquid

10.4.5 Anyonic excitations

Given the topological nature of the ground state, one may speculate that excitations also afford a topological interpretation. The elementary excitations of the system have in common that they (i) cost energy $2J_m$ or $2J_e$, respectively, (ii) can only be created in pairs, which (iii) are connected by strings. All these features follow from the schematic representation shown in the figure. Consider

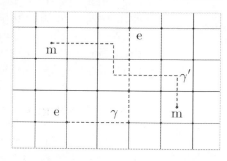

string operator the **string operator** $\Sigma_3(\gamma) \equiv \prod_{l \in \gamma} \sigma_3$, where the product is over the lattice links defining a curve, γ, on the lattice. This operator commutes with all operators in the Hamiltonian, except for the vertex operators A_s sitting at the end-points of the string, with which it anticommutes. Application of the operator to the ground state thus generates a state of energy $2 \times 2J_e$. We interpret it as a state containing two "electric" excitations, e, at the end points of the string. Similarly, the string operator $\Sigma_1(\gamma') \equiv \prod_{l \in \gamma} \sigma_1$, where the product is over all links intersecting a curve γ', commutes with everything, except for the two plaquette operators defined by the terminal points. Application of this operator costs energy $2 \times 2J_m$ and creates two "magnetic" excitations, m, at the end points.

INFO Electric and magnetic excitations of this type pose a threat to the integrity of quantum information encoded in the computational space of the code: a pair of excitations may be created by, e.g., a thermal activation. Once created, the two excitations may propagate on the lattice without further energy cost. A logical error occurs when an excitation string moves around one of the great circles of the torus before its end points recombine. The final state is again in the ground space, but in a different topological sector. Referring to the literature, we note that various passive and active schemes of **error** **correction** have been devised to counter such processes.

error correction

The end points of the string operators play the role of effective quasiparticle excitations of energy $2J_{e,m}$, respectively. Once the energy $2 \times 2J_{e,m}$ has been paid, they are free to move on the lattice. What kind of quantum statistics do these particles obey: bosonic, fermionic, or other? To answer such questions in a two-dimensional setting one needs to understand what happens if one of these particles gets dragged around another along a circle (see discussion of section 8.6.1). For two magnetic or electric quasiparticles the answer is evident: nothing happens because the motion of the end-point of a string of $\sigma_{1,3}$ operators has no consequences.

However, a more interesting situation arises if a magnetic particle gets dragged around an electric one (see the figure). Magnetic and electric string operators intersecting an odd number of times anticommute because they contain an odd number of operators $\sigma_{1,l}, \sigma_{3,l}$ on the intersecting links in common. This implies that the loop introduces a minus sign, or a phase (-1), distinguishing the wave

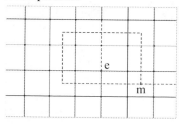

semion

functions before and after the braiding procedure. This phase is characteristic of a specific type of anyon, a **semion**. The defining feature of semions is that their exchange – which is topologically equivalent to a *half* rotation – leads to a phase factor i, half that for fermions. We have thus found that the magnetic and electric excitations of the toric code are bosons relative to themselves, and semions relative to each other.

The exchange statistics identified above is a simple example of the **braiding properties of topologically ordered ground states**. States defined by more complex string-net ground states[31] support excitations whose braiding may lead back to the ground state wave function $U|\psi\rangle$ multiplied by a nontrivial unitary matrix U. The exploration of such **non-abelian anyons** and their potential application as a computational resource is the subject of the field of **topological quantum computation**, not discussed in this text but beautifully reviewed in Ref.[33].

However, before closing this section, let us mention one entertaining fact about the simple semions of the toric code: one may consider a composite excitation defined by an electric and a magnetic excitation on a site and a neighboring plaquette, i.e., a charge with a $\mathbb{Z}_2$-gauge flux attached to it. The "composite particles" defined in this way are linked by two strings. This implies that the exchange of two of them comes with a phase $i^2 = -1$, characteristic of a *fermion*. We thus observe that the toric code – nominally a system of localized spins, is capable of generating spatially delocalized fermions as emergent quasiparticles. These fermions have the remarkable property that they always come in pairs, which means that the fermion *parity* is a conserved quantum number. They are non-local objects, in that they emerge as end-points of strings. It is interesting to see how these structures – which separately play a role in the understanding of the standard model and the fundamental structure of matter – appear as emergent signatures of a simple spin system.

10.5 Topological Gauge Theory

Up to now, we have considered gauge theories over space–time regions M equivalent to simple subsets of $\mathbb{R}^n$. In this section, we turn to situations where M has a more interesting topology. Here, the bundle is no longer equivalent to a Cartesian product, $E \neq M \times F$. This generalization can have profound consequences whose study is the subject of topological gauge theory.

10.5.1 Connections of topologically nontrivial bundles

We begin by asking how the presence of topology manifests itself in the principal building blocks of the theory: connections, states, and covariant derivatives. To this

---

[33] C. Nayak *et al.*, *Non-abelian anyons and topological quantum computation*, Rev. Mod. Phys. **80**, 1083 (2008).

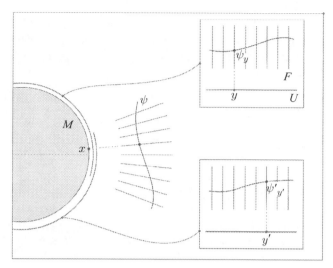

Fig. 10.4   Left: Coverage of a bundle by overlapping coordinate charts. Right: Locally the bundle is equivalent to Cartesian products $U \times F$ of coordinate domains and fibers. Depending on the chosen system, a section has different local representations, $\psi(y)$ and $\psi'(y')$, respectively, related to each other by a gauge transformation, $\psi' = g\psi$.

end, consider a region of $N \subset M$ covered by two overlapping trivializations, U and U'. In the restriction to either of these regions, the bundle is equivalent to $U \times F$ or $U' \times F'$, respectively, where U and U' may be considered subsets of $\mathbb{R}^n$ equipped with their individual coordinate representations.[34] Points $x \in N$ in the overlap region then have two different coordinate representations, $y \in U$ and $y' \in U'$ (see Fig. 10.4). The denotation of these coordinates is arbitrary and, confusingly, they may even have the same name. For example, in the case $M = S^2$, U and U' might be domains covering the northern and the southern hemisphere, with overlap in the equatorial region, both equipped with identically named spherical coordinates $y = y' = (\theta, \phi)$. The states, or sections, of the theory have independent representations, ψ_y and $\psi'_{y'}$ in $U \times F$ and $U' \times F'$, respectively. Since these are different representations of the same state, there must be a translation $g_x : F \to F, \psi_y \mapsto \psi'_{y'} \equiv g_x \psi_y$, realized through an element of the gauge group. The subscript

transition function   indicates that g_x is a **transition function**, i.e., a group-valued function in the overlap region between different trivializations. A key point now is that:

> The transition functions between different local trivializations
> encapsulate the topology of a fiber bundle.

[34] There is a potentially confusing subtlety here. In principle, one needs to distinguish between points in M and their coordinate representations – there is a difference between a geometric point on a sphere and its representation $(\theta, \phi), z, \ldots$ in spherical, stereographic, $\ldots$ coordinates. This is important, specifically in cases where M does not have a global coordinate representation. However, for brevity, one often does not distinguish between local domains $U \subset M$ and their coordinate representations $U \subset \mathbb{R}^n$. As long as one knows what one is doing, this (not very hygienic, yet convenient) practice is acceptable and we will adopt it throughout.

As a corollary we note that on a bundle characterized by nontrivial transition functions, it is not possible to describe sections as globally defined functions. Specifically, quantum wave functions become sections of a complex line bundle, with only locally defined function representations.

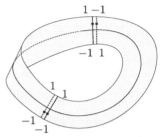

Möbius strip

INFO As an example illustrating these structures, consider the **Möbius strip**. A Möbius strip can be described as a bundle with a circle base, S^1, and a finite interval $F = [-1, 1]$ as fibers. Consider the circle covered by two overlapping coordinate charts, such that we have two overlap regions (see the figure, where the local fiber representations are indicated by solid and dashed lines, respectively). The presence of a twist is equivalent to the statement that, in one of these regions, the transition map $g : [-1, 1] \to [-1, 1], s \mapsto -s$ inverts the local representation of a section on the fibers (see the figure.) If it were not for this twist, the strip would have the topology of a cylinder.

At this point, connections have not yet entered the stage. However, this changes when we aim to monitor the local change of sections via parallel transport. In an overlap region, we then have two independent versions of a **covariant derivative**, $D\psi \equiv (d + A)\psi$ and $D' \equiv (d + A')\psi'$, and we need to find out how the coordinate representations of the connection A and A' are related to each other. To this end, we note that the covariant derivatives (evaluated along a certain direction on M) are local functions, and as such are subject to the translation protocol: $D'\psi' = g(D\psi)$, where we have omitted subscripts for brevity. Substituting $\psi' = g\psi$, this leads to $D' = gDg^{-1}$. With $gdg^{-1} = d + g(dg^{-1})$, we obtain the identification

$$\boxed{A' = gAg^{-1} + gdg^{-1}} \tag{10.44}$$

This shows that the local representations of the connection are related to each other by a gauge transformation as in Eq. (10.3), where the transition functions feature as group transformations. We may think of the gauge transformation between A and A' as a map $g : N \to G$ from the intersection of the two coordinate domains into the gauge group. The bundle is nontrivial if this map cannot be continuously deformed to unity. In the physics literature, gauge transformations with this property are called **large gauge transformations**.

large gauge transformation

10.5.2 Dirac monopole

In this section, we illustrate the principles of topological gauge theory for the example of the Dirac monopole. In the early 1930s, Dirac was interested in quantum mechanics in the presence of a magnetic monopole field.[35] Studying this problem, he ran into singularities which we now understand as a manifestation of the topological

---

[35] P. A. M. Dirac, *Quantised singularities in the electromagnetic field*, Proc. R. Soc. Lond. A **133**, 60 (1931).

structure of the gauge field representing the monopole. Here, we proceed in reverse order. We will start from a singularity-free description of a gauge field with non-trivial topology. From there, we will discover effective monopoles which, as we will see, find applications in condensed matter physics.

Monopoles as a consequence of geometry

The Dirac monopole problem is defined on a base manifold $M \simeq S^2$ of spherical geometry. We want to develop quantum mechanics on this surface, which means that the relevant fibers, $F = \mathbb{C}$, are the complex image spaces of wave functions.

complex line bundle

Mathematicians call such bundles **complex line bundles**. The sphere is different from an open subset of $\mathbb{R}^2$, meaning that more than a single coordinate chart is required to cover it. This generates the situation depicted schematically in fig. 10.4.

Let us now explore how these structures manifest themselves in the **monopole problem**. We begin by covering the northern and the southern hemisphere individually with spherical coordinates. The simplest way to glue the locally defined wave functions ψ_n and ψ_s in the equatorial overlap region is to identify $\psi_\mathrm{n}(\theta, \phi) = \psi_\mathrm{s}(\theta, \phi)$, with transition function $g(\theta, \phi) = 1$. In this case, a consistent choice of the connections in the two hemispheres reads $A_\mathrm{n} = A_\mathrm{s} = 0$. A more interesting gluing prescription is given by $\psi_\mathrm{s}(\theta, \phi) = e^{in\phi}\psi_\mathrm{n}(\theta, \phi)$, with non-vanishing integer n, where the latter condition is required to make $\psi_\mathrm{n,s}$ single valued along the equator.[36] This twist corresponds to the choice $g = e^{in\phi}$ for the U(1) structure group transformation. It is topological nontrivial in that g cannot be continuously deformed to $g = \mathbb{1}$. Conceptually, this transformation plays a role analogous to that of the coordinate sign inversion implementing the twist of the Möbius band. In the present context, it gives the complex line bundle over the sphere a nontrivial topology.

Owing to the abelian nature of the theory, the transformation between the local connections reads

$$A_\mathrm{s} = A_\mathrm{n} - in d\phi. \tag{10.45}$$

This condition cannot be satisfied by vanishing connections, $A_\mathrm{n} = A_\mathrm{s} = 0$, which conveys an important general message:

> The topological nontriviality of a bundle requires the presence of a non-vanishing connection.

In section 10.5.3, we will indeed see that the topological nontriviality of a bundle can be diagnosed via universal properties of the connections defined on them.

[36] Critical readers may object that spherical coordinates are ill-suited to cover the sphere in terms of just two charts. For example, a single coordinate ϕ cannot even parameterize the full equator. However, there do exist two-chart atlases of the sphere, a stereographic projection providing one of them. A watertight formulation would refer to ϕ and θ as the local spherical representations of a singularity-free stereographic system with different charts for the northern and southern hemisphere. However, this language would make our story intolerably complicated while not changing its conclusions.

Now consider the choice

$$A_n = \frac{in}{2}\left(+1 - \cos\theta\right)d\phi,$$

$$A_s = \frac{in}{2}\left(-1 - \cos\theta\right)d\phi.$$

The key features of these connections is that they satisfy the transformation law, and are well defined on their respective hemispheres. The first property follows via inspection of the equatorial line $\theta = \pi/2$. On it, both connections are defined and we have $A_n = \frac{in}{2}d\phi = A_s + in\,d\phi$, as required. To verify the second assertion, note that at the points of singularity of the form $d\phi$, the north- and the south-pole, the one-forms A_n and A_s, vanish, respectively. This removes the singularities in their domains of definition and makes $A_{n/s}$ valid connection forms.

However, the connections $A_{n/s}$ have further remarkable properties. We first note that they support a **non-vanishing and non-singular field strength**,

$$F = dA_{n/s} = \frac{in}{2}\sin\theta\,d\theta \wedge d\phi. \tag{10.46}$$

Mathematically, this is $in/2$ times the area form of the two sphere. Integrated over the sphere, $\int_{S^2}\sin\theta\,d\theta \wedge d\phi = 4\pi$, meaning that the total flux, or integrated field strength of the problem, equals $\int_{S^2} F = 2\pi in$. Physically, $-iA_{n/s}$ are real-valued vector potentials and $-iF$ represents its magnetic field. Translated to a vector language, $(-iA_{n/s})_\phi = (n/2)(\pm 1 + \cos\theta)$, with all other components vanishing. It is straightforward to compute the curl and confirm that the corresponding magnetic field (the vector formalism analog of F) points radially outward with uniform
magnetic
monopole strength $B_r = n/2$. This is th defining feature of a **magnetic monopole**. Further, we may observe that the strength of the induced field is integer quantized, where the index characterizes the degree of topological twisting of the bundle. (This includes the case $n = 0$ of the trivial bundle.) We observe that:

> The topological nontriviality of the complex line bundle over S^2 is equivalent to the presence of a magnetic monopole field of quantized strength.

Later in the section, we will understand this phenomenon as a manifestation of a more general structure: the topological contents of bundle structures in physics and mathematics – which are often not easy to see from the outset – may be diagnosed by computing integral invariants of their connections. This approach to classifying topology is formulated in a language reminiscent of that of electromagnetism and hence resonant with a physics way of thinking.

Finally observe that we really need a minimum of two connections $A_{n/s}$ to represent the problem. For example, upon approaching the south pole, $A_n \to in\,d\phi$ becomes ill-defined. If it were not for this singularity, we would have a global representation, $F = dA$, and by Stokes' theorem $\int_{S^2} F = \int_{\partial S^2} A = 0$, since the sphere has no boundary: the non-vanishing of the topological flux excludes the existence of a globally defined connection on M, which in turn indicates the nontriviality of the bundle.

INFO As mentioned above, Dirac approached the problem by postulating the presence
of a monopole field and exploring the consequences. Noticing that the single-valuedness
of wave functions requires magnetic charge quantization, he took an important additional
step to show[35] that *electric* charge had to be quantized as well. This led to the spectacular
conjecture that the (experimentally observed) quantization of charges might be rooted in a
(sought for but not yet observed) existence of magnetic sources in the universe. Dirac also
noticed the necessary presence of singularities in the vector potentials generating the field.
However, unlike our present discussion, he attributed physical significance to the singular
points and interpreted them as the piercing points of infinitesimally thin solenoidal flux
lines, called **Dirac strings**, between remote monopoles of opposite charge. In this way, the
physics of individual monopoles can be addressed and no conflict with the fundamental
non-existence of isolated monopoles in nature arises (see section 8.5.1 for further discussion
of the Dirac string).

Dirac
strings

Monopoles in condensed matter systems

For all we know there are no fundamental monopoles in nature. However, in condensed matter physics there are numerous realizations of *effective* gauge fields that
do not represent real magnetic fields and hence may support monopoles. In the
following, we discuss two example realizations of such fields.

QAH insulator: In the info block of section 8.4.7 we introduced the quantum anomalous Hall insulator as a two-dimensional topological insulator. We analyzed its effective Bloch-sphere Hamiltonian (8.38) with associated the integer winding number
(8.40). However, this assignment was indirect in that the *actual* object carrying a
topological index is the ground state of the system.

A more systematic approach describes the ground state as a bundle with base
Brillouin zone torus, one-dimensional fibers hosting the single-particle valence band
states, and the Berry connection, Eq. (10.29), describing the twisting of the latter.
In the exercise below, we demonstrate that, for a QAH insulator in a topological
phase, the torus contains an integer monopole charge.

EXERCISE The ground state wave function of the QAH Hamiltonian (8.38), $H_\mathbf{k} = v_\mathbf{k} \cdot \boldsymbol{\sigma}$,
$\mathbf{v}_\mathbf{k} = (\sin k_1, \sin k_2, m - \cos k_1 - \cos k_2)$, is given by $U_\mathbf{k}|\downarrow\rangle \equiv |\mathbf{n}_\mathbf{k}\rangle$, where $U_\mathbf{k}$ is the unitary
transformation diagonalizing the 2×2 matrix $H_\mathbf{k}$ and the second representation emphasizes
that these transformations rotate the spin-down state $|\downarrow\rangle$ into a direction set by the unit
vector $\mathbf{n}_\mathbf{k} \parallel \mathbf{v}_\mathbf{k}$ on the Bloch sphere. Verify that in an Euler angle representation, $U_\mathbf{k} =$
$\exp(\frac{i}{2}\phi_\mathbf{k}\sigma_3)\exp(\frac{i}{2}\theta_\mathbf{k}\sigma_2)$, the Berry connection (10.29) assumes the form $A_\mathbf{k} = \frac{i}{2}\cos\theta_\mathbf{k} d\phi_\mathbf{k}$.
In this expression, $(\phi_\mathbf{k}, \theta_\mathbf{k})$ are functions of the crystal momentum, implicitly defined by
the condition that they describe the orientation of $\mathbf{n}_\mathbf{k}$ in spherical coordinates.

In the topological phase $m \in (0, 2)$, these angles cover the full
sphere upon variation of $\mathbf{k}$. Specifically, the north pole corresponds
to the center of the Brillouin zone $\theta_{(0,0)} = 0$, and the south pole
to the boundaries $\theta_{(\pm\pi,\pm\pi)} = \pi$. The equatorial line is visited
for momentum configurations implicitly defined by the condition
$m - \cos k_1 - \cos k_2 = 0$. This line defines the boundary between two
coordinate patches on the torus (see figure for a visualization for
$|2 - m| \ll 1$, where the boundary approaches a circular geometry).
Inside (outside) that line we need to choose a gauge $A_\mathrm{n} = A + \frac{i}{2}d\phi$

($A_\mathrm{n} = A - \frac{i}{2}d\phi$) removing the singularity of A upon approaching the north-pole (south-pole), as in our previous discussion of the Dirac monopole. Show that the field strength on the torus, $F = dA = dA_\mathrm{n} = dA_\mathrm{s}$, has the coordinate representation $F = \sin\theta(\partial_1\theta\partial_2\phi - \partial_2\theta\partial_1\phi)dk_1 \wedge dk_2$ and convince yourself that it has non-vanishing flux, i.e., it describes a monopole included in the torus. Integrate the field strength to obtain its flux as

$$\frac{1}{2\pi i}\int_{T_2} F = 1. \tag{10.47}$$

(Hint: Apply Stokes' theorem for an efficient calculation of the integral.) Can you reason without explicit calculation why, in a non-topological phase, such as $m > 2$, the monopole charge vanishes?

For the simple QAH system, it does not seem to matter whether we identify an index from the Hamiltonian or its ground state. However, for more complex systems this need not be so. Specifically, the periodic table of topological insulators is built on a geometric classification of quantum ground states, taking the ground state bundle with Berry connection (section 8.1.1) as its starting point.

Weyl semimetals: In section 9.1.1 we introduced the Weyl semimetal as a gapless fermion system harboring an even number of Weyl nodes in its three-dimensional Brillouin zone. Each node is represented by the Hamiltonian $H_\mathbf{k} = k_i\sigma_i$, where the momenta are measured from the nodal center. Assuming that the system is kept at a small chemical potential, μ, the nodes are surrounded by spherical Fermi surfaces at $|\mathbf{k}| \equiv k \equiv \mu$ separating filled states, $k < \mu$, from empty ones. To this ground state, a Berry connection is assigned through $A_i = \langle\mathbf{k}|\partial_{k_i}|\mathbf{k}\rangle$, where $|\mathbf{k}\rangle$ is the ground state of the above 2×2 Weyl Hamiltonian. In the notation of the previous discussion of the Hall insulator, this translates to $\mathbf{n_k} = \mathbf{k}/k$. Representing $\mathbf{k} = \mathbf{k}(\theta, \phi, k)$ in polar coordinates, the integral of the Berry curvature around each node equals $\frac{1}{2}\int\sin\theta\,d\theta d\phi = 2\pi$, i.e., each node is the source of a Berry flux quantum.

INFO While there may be no "real" monopoles in nature, *effective* monopoles have been observed in condensed matter physics. As in Dirac's construction, they appear tied to solenoidal flux lines. Such structures have been observed in pyrochlore compounds – lattices of highly frustrated magnetic moments generating a form of matter dubbed **spin ice** (see info block on page 482). In these compounds, the local alignment of moments may generate effective solenoids with magnetic flux emerging at end-point monopoles. Such local flux sources have been observed in neutron scattering experiments on the pyrochlore compound $Dy_2Ti_2O_7$.[37]

spin ice

10.5.3 Characteristic classes

The discussion above showed that **gauge fields on topologically twisted bundles** are nontrivial. We cannot put $A = 0$ everywhere since this would be in conflict

[37] D. J. P. Morris *et al. Dirac strings and magnetic monopoles in the spin ice $Dy_2 Ti_2 O_7$*, Science **326**, 411 (2009).

with the transition rules (10.44) for the translation of vector potentials between charts; either A or A' or both must be non-vanishing.

Since the As take into account the underlying topology, one may wonder whether the logic can be turned around: can we describe the topology of bundles from the connections defined on them? This is the idea behind the concept of **characteristic** **classes**. Technically, characteristic classes are functionals of the connections of a bundle, defined to classify the underlying topology. For two reasons, they define a powerful concept in the description of topological quantum matter. First, the microscopic information describing, say, a band insulator is often delivered in the form of a vector potential, locally computed from underlying microscopic structures. A class functional may then be employed to integrate this information and obtain a number classifying the topology of the system. Second, it is irrelevant which vector potential is employed. Any vector potential permissible on the bundle structure will yield the same class information.

There are various characteristic classes, and we refer to Ref.[21] for a comprehensive discussion. Here, we focus on the realization most frequently encountered in physics applications, the *Chern classes*. In the next section, we introduce the concept of Chern classes in generality. We will then consider the example of a topological band insulator to illustrate their utility in physics.

characteristic classes (margin note)

Chern classes

REMARK The following is an informal introduction to Chern classes. For a more rigorous discussion, see Ref.[21].

Let us begin with an introduction to Chern classes formulated in the language of geometry. Consider a Lie group G, and let B be matrices in the Lie algebra $\mathfrak{g}$. For definiteness, we may think of $\mathfrak{g} = \mathbb{R}$, the Lie algebra of U(1), or su(N). We now define the matrix monomials $I_n(B) \equiv \mathrm{tr}(B^n)$. Cyclic invariance of the trace implies that these are *invariant monomials* in the sense that $I_n(gBg^{-1}) = I_n(B)$.

Now consider a setting with base manifold M of *even* dimension d and fibers on which G acts as a gauge group. Assume we have chosen a local connection A with field strength F. Expanded as $F = \frac{1}{2}F_{\mu\nu}dx^\mu \wedge dx^\nu$, the coefficients $F_{\mu\nu}$ are $\mathfrak{g}$-valued matrices, and F transforms as $F \to gFg^{-1}$. Hence it makes sense to consider $I_n(F) \equiv \mathrm{tr}(F \wedge \cdots \wedge F) \equiv \mathrm{tr}(F^{\wedge n})$, invariant polynomials evaluated on the field strength form.[38] These monomials define the so-called **Chern characters**,

Chern characters (margin note)

$$\mathrm{ch}_n(F) \equiv \frac{1}{n!}\, \mathrm{tr}\left(\frac{iF}{2\pi}\right)^{\wedge n}, \tag{10.48}$$

where the normalization factor is chosen for later convenience. Notice that Chern characters exist only up to $2n = d$ (think why?).

[38] For example, with $F = \frac{1}{2}F_{\mu\nu}dx^\mu \wedge dx^\nu$, $F_{\mu\nu}$ are matrices in $\mathfrak{g}$, and $I_2(F) = \mathrm{tr}(F \wedge F) = \frac{1}{4}\mathrm{tr}(F_{\mu\nu}F_{\rho\sigma})dx^\mu \wedge dx^\nu \wedge dx^\rho \wedge dx^\sigma$, etc.

As we will explain shortly, the importance of these objects in the topological description of bundles follows from two key properties (the **Chern–Weil theorem**): (i) they are closed, $d\,\mathrm{ch}_n(F) = 0$, and (ii), for two different field strengths on the manifold, F and F', the difference $\mathrm{ch}_n(F) - \mathrm{ch}_n(F') \equiv d\kappa_n$ is exact, where $\kappa = \kappa(A, A')$ is a differential form depending on the local vector potentials.

For an instructive proof of the Chern–Weil theorem we refer to problem 10.7.6. As useful spin-offs from this proof, we get explicit representations for κ. For example, if $F' = 0$ locally[39] then for $n = 1, 2$ (which cover most applications in condensed matter physics), we have the representations

$$
\begin{aligned}
\mathrm{ch}_1(F) &= \frac{i}{2\pi}\mathrm{tr}(F) = \frac{i}{2\pi}d\,\mathrm{tr}(A), \\
\mathrm{ch}_2(F) &= -\frac{1}{8\pi^2}\mathrm{tr}(F \wedge F) = -\frac{1}{8\pi^2}d\,\mathrm{tr}\left(A \wedge dA + \frac{2}{3}A \wedge A \wedge A\right),
\end{aligned}
\tag{10.49}
$$

which can be straightforwardly verified by computing the derivatives on the right-hand side and comparing with the definition of the field strength tensor (10.13). On the right-hand side, we encounter the differential forms defining the topological gauge field actions (10.27). Their coupling constants are obtained by multiplication of Eq. (10.49) by $2\pi k$, where k is the integer **level** of the theory. The factor 2π guarantees that ambiguities related to the global gauge structures addressed below affect actions only via unobservable phases, $2\pi i$ times an integer.

Chern numbers

After this formal construction, let us now now discuss its utility. To this end, assume we have a bundle setting comprising a (boundaryless) base manifold, M, a fiber structure, and a connection, A, as usual. For example, consider a Brillouin zone base, a Hilbert space $\mathbb{C}^n$ for n-bands as the fiber, and a Berry connection. The twisting of the bundle is encoded in the transition functions between local realizations of the connection. However, assuming that we do not know these functions yet, we now show how equivalent information is obtained by computing relatively straightforward integral invariants of the curvature tensor.

These invariants are called **Chern numbers** and they are defined as

$$
\boxed{\mathrm{Ch}_n[F] = \int_M \mathrm{ch}_n(F)}
\tag{10.50}
$$

i.e., as integrals of the above Chern characters. These intgrals are defined to produce integer-valued (Chern) numbers. Their advantage is that they produce fingerprints of the underlying topology via relatively straightforward integrals. Crucially, any field strength F compatible with the bundle structure[40] may be used to compute the

---

[39] Keep in mind that, in nontrivial cases, $F' = 0$ can hold only *locally*. The form κ_n is defined in terms of a connection form which, likewise, is a local object.

[40] Remember that on a manifold with a nontrivial twist, we cannot just set $F = 0$ globally. We need to work with the As, and the corresponding Fs, consistent with the bundle's gluing conditions.

Chern number. The reason is that the difference $\text{Ch}_n(F) - \text{Ch}_n(F') = \int_M (\text{ch}_n(F) - \text{ch}_n(F')) = \int_M d\kappa_n = 0$ vanishes by the Chern–Weil theorem and Stokes' theorem.

Let us illustrate how this works for the example of the Dirac monopole bundle. In this case, a representative of the field strength is given by Eq. (10.46), and integration over the sphere $M = S^2$ yields $\text{Ch}_1(F) = -n$. However, to make the connection to the underlying bundle twist more explicit, let us do the integral in a slightly different manner: locally the Chern characters afford the representation (10.49) in terms of patches of the connection form, i.e., $F = \frac{i}{2\pi} dA_{\text{n/s}}$ on the northern and the southern hemisphere, respectively. Splitting the integral into contributions from these two regions, we obtain

$$\text{Ch}_1(F) = \int_{\text{n}} F + \int_{\text{s}} F = \frac{i}{2\pi} \left(\int_{\text{n}} dA_{\text{n}} + \int_{\text{s}} dA_{\text{s}} \right)$$
$$= \frac{i}{2\pi} \int_{\text{equator}} (A_{\text{n}} - A_{\text{s}}) \overset{(10.45)}{=} \frac{i}{2\pi} \int_{\text{equator}} (in\, d\phi) = -n.$$

The take-home message is that the Chern number responds only to what happens in the equatorial gluing region (third equation), and there probes the twist between $A_{\text{n,s}}$. Parameterizing the equator through the azimuthal coordinate ϕ, the large gauge transformation between A_{n} and A_{s} is a map $S^1 \to \text{U}(1), \phi \mapsto e^{i\phi}$, from the equator–circle into the gauge group $\text{U}(1)$, which is topologically also a circle. The windings (homotopy classes) of that map determine the Chern numbers. Notice how this construction describes the topology of the bundle with two-dimensional base in terms of the winding numbers of a map in one dimension lower. In the info block below, we illustrate the same principle on a higher-dimensional example, likewise of relevance to topological matter.

INFO As an example of the **application of Chern numbers in condensed matter physics**, compare with the discussion of the bundle structure describing the QAH insulator in the exercise on page 617. The topological index of that system, Eq. (10.47), is the (negative of the) **first Chern number** of its Berry connection.

The **second Chern number** comes into play when we consider topological insulators of more complex structure. As an example, consider the four-dimensional band insulator described by the Hamiltonian[41]

$$\hat{H}(k) = \sum_{i=1}^{4} \sin k_i \Gamma_i + \left(m + \sum_{i=1}^{4} \cos k_\mu \right) \Gamma_0 \equiv \sum_\mu d_\mu(k) \Gamma_\mu,$$

where $k = (k_1, \cdots k_4)$ is the crystal momentum parameterizing a four-dimensional Brillouin zone, T^4, and Γ_μ, $\mu = 0, \cdots 4$ are the Euclidean Dirac γ-matrices, $[\Gamma_\mu, \Gamma_\nu]_+ = 2\delta_{\mu\nu}$ (see section 9.1.1).[42] At first sight, four-dimensional insulators seem like a purely academic concept. However, systems defined in synthetic dimensions can nowadays be engineered by techniques of quantum optics. Futhermore, insulators in dimensions > 3 play a role as parents for the construction of lower-dimensional insulators by reduction (see Refs.[41] for a discussion of this point).

[41] X. Qi, T. Hughes, S. C. Zhang, *Topological field theory of time-reversal invariant insulators*, Phys. Rev. B **78**, 195424 (2008).
[42] To be concrete, $\Gamma_0 = \gamma_0$, $\Gamma_{1,2,3,4} = -i\gamma_0\gamma_{1,2,3,5}$.

non-abelian
Berry con-
nection

At given k, $\hat{H}(k)$ is a four-dimensional matrix with two-fold degenerate eigenvalues $\pm|d(k)|$. (Verify this statement using the anticommutation relations of the Γ-matrices. Hint: Use the "ln det = tr ln" formula for the computation of the characteristic polynomial.) For generic values of the parameter $m \neq 0, \pm 2, \pm 4$, the spectrum has a gap, $|d(k)| \neq 0$, and our system is an insulator at half-filling, $\mu = 0$. Its two-fold degenerate ground state defines a **non-abelian Berry connection** $A_{ab}(k)$, $a, b = 1, 2$, Eq. (10.29), describing the variation of the ground state wave functions $\psi_a(k)$ in k.

A hint as to the topology of the system follows by inspection of the Hamiltonian, analogously to what we did in the case of the QAH insulator: without closing the energy gap, one may deform $\hat{H}(k)$ to $\sum \hat{d}_\mu(k)\Gamma_\mu$, with the four-dimensional unit vector $\hat{d} = d/|d|$. This representation suggests a classification in terms of homotopy classes (see section 8.3) of maps from the four-torus, T^4, to the four-sphere, S^4, i.e., a higher-dimensional variant of the $T^2 \to S^2$ Bloch sphere mapping describing the QAH insulator. It turns out that different classes are distinguished by an integer winding number, so that we expect a $\mathbb{Z}$-classification scheme.[43]

Referring to Ref.[41] for the conceptually straightforward yet technically lengthy details, we now outline how the situation is described via the Chern classes of the ground state bundle. The discussion again parallels that of the QAH insulator, except that the base space changes as $T^2 \to T^4$, the fiber dimension $1 \to 2$, and the gauge group $U(1) \to SU(2)$. (In dimensions > 2, the abelian factor in $U(2) = U(1) \times SU(2)$ is topologically trivial and the gauge group reduces to $SU(2)$.) For parameter values, m, corresponding to a twisted ground state, no globally non-singular connection form A_{ab} can be found. Rather, we need to segment the four-torus into two chart regions with a three-dimensional boundary. For example, if $m \simeq 4$ is close to a critical value, the "mass coefficient" of the Hamiltonian $\hat{H}(k)$ vanishes on a small three-sphere S^3 in the momentum space surrounding $k_0 = (\pi, \pi, \pi, \pi)^T$. This sphere, now plays a role analogous to that of the circle indicated in the exercise on page 617. Inside and outside this sphere, we work with different Berry connections (once again, we call them $A_{n,s}$), and on it there is a large gauge transformation connecting them, $A_n = gA_s g^{-1} + gdg^{-1}$. Here, $g = g(k)$ is a map $S^3 \to SU(2)$ which cannot be deformed to a trivial one. We now have the local representations $ch_2(F_{n,s}) = d\kappa_2(A_{n,s})$ inside and outside that sphere, respectively and, by Stokes' theorem, the topological index assumes the form of an integral over the difference of Chern–Simons three-forms $\kappa_2(A_n) - \kappa_2(A_s)$ over S^3. Assuming that, locally, $A_s = 0$, and using Eq. (10.49), it is straightforward to verify (exercise) that $\kappa(A_n) - \kappa(A_s) = \frac{1}{24\pi^2} tr[(gdg^{-1})^{\wedge 3}]$. In problem 8.8.3, we showed that the integration of this form over the boundary three-sphere yields the winding numbers of the map $S^3 \to SU(2)$ describing the twisting of the bundle.

There may be easier ways to describe the windings of the four-band insulator. However, the striking advantage of the present approach is its extensibility to different representations of the Hamiltonian, and more complicated lattice structures. Topology guarantees that the integrals defined via the structures above will always yield the same answer. This makes them valuable tools in, e.g., the numerical computation of topological indices.

SU(2)
instantons

As an aside, we mention that, in the particle physics literature, these gauge transformations are called **SU(2) instantons**. In that context, the domains introduced above assume the role of the four-dimensional universe, and S^3 is a "space–time boundary." It may be useful to consult particle theory textbooks and check how their discussion of the subject compares with the present one.

---

[43] Alert readers my wonder how this statement can be compatible with $\pi_d(T^k) = \emptyset$ for $d \geq 2$, i.e., the absence of topologically nontrivial maps from d-spheres to k-tori. The resolution is that winding maps from the torus to the sphere exists, but not the other way around. (We are grateful to A. Abanov for pointing this out.)

INFO Equations. (10.50) define homotopy classes of gauge field configurations as integer-valued integrals over fields with local representation, $F = F(A, \partial A)$. These are the defining features of the θ-terms discussed in section 8.4. In both the particle physics and the condensed matter physics literature, the coordinate representations of the integrals (10.50),

$$S_{\text{top}}[F] \equiv \int_M \text{ch}_{d/2}(F) = \text{Ch}_{d/2}[F]$$

$$= \begin{cases} \dfrac{1}{4\pi} \displaystyle\int d^2x \, \epsilon^{\mu\nu} \text{tr}(F_{\mu\nu}), & d = 2, \\[2ex] \dfrac{1}{32\pi^2} \displaystyle\int d^4x \, \epsilon^{\mu\nu\rho\sigma} \text{tr}(F_{\mu\nu} F_{\rho\sigma}), & d = 4, \end{cases} \tag{10.51}$$

θ-terms   are often referred to as θ-**terms**.[44] In particle physics, the $d = 4$ variant describes, e.g., the instanton configurations mentioned in the info block above. In condensed matter physics, four-dimensional topologically nontrivial gauge field configurations are rare. However, the $d = 4$ term may appear in effective actions nonetheless. To understand why, consider electromagnetism, where $F_{\mu\nu}$ is the familiar field strength tensor and a short calculation shows that $\epsilon^{\mu\nu\rho\sigma} \text{tr}(F_{\mu\nu} F_{\rho\sigma}) = 8\mathbf{E} \cdot \mathbf{B}$. This leads to $S_{\text{top}} = \frac{\theta}{4\pi^2} \int d^4x \, \mathbf{E} \cdot \mathbf{B}$, which in this form appears as part of the electromagnetic response action of topological insulators[41] or Weyl semimetals[5]. For further discussion, we refer to section 9.2 where these terms are discussed from the perspective of physical anomalies.

To conclude this section, Chern numbers are integer-valued invariants describing the topology of even-dimensional bundle manifolds as integrals over their gauge field strengths. Although Chern numbers do not exist for odd dimensional manifolds, Eq. (10.48) shows how the Chern characters of $(d = 2n)$-dimensional manifolds determine the topological gauge field actions in $d = 2n - 1$. The physics of these Chern–Simons actions was the subject of section 8.6.

10.6 Summary and Outlook

Much of physics involves comparing (differentiating) objects at different points in space, time, or differently defined parameter spaces. Our starting point in this chapter was the observation that naïve comparison of objects containing internal symmetries, vector fields, wave functions, etc., leaves room for ambiguities which are fixed by the introduction of connections. This simple realization defined a geometric approach to gauge theory and leads to intuitive and universally applicable definitions of all of its elements, vector potentials, field strengths, gauge "symmetries," covariant derivatives, Wilson loops, etc. By way of illustration, we considered the physics of gravity, where the view above led to a transparent representation of structures otherwise plagued by hordes of indices. We also discussed how this framework is implemented on lattice structures, and for discrete gauge groups such as $\mathbb{Z}_2$. In the final part of the chapter, we upgraded the geometric perspective to a topological

---

[44] Here, we made the replacement $iF \to F$ relative to Eq. (10.50) since the hermitian-physics A is the anti-hermitian gauge group generator iA of this section.

one, and learned how to detect topological structures by inspection of the gauge fields defined on them. This view is increasingly important in the physics of topological condensed matter physics, where gauge fields frequently appear as interfaces between the microscopic physics and the macroscopic topologies defined by it. For example, the *microscopic* band crossings of a topological insulator are described by lattice fermions minimally coupled to a Berry gauge potential. Their integration leads to a topological gauge field action, whose integration defines a Chern class *macroscopic* invariant. This example also illustrates that the contents of this chapter are inseparably linked to that of chapter 8, whose theme is the physics of non-local structures in condensed matter physics and their field-theoretical description.

10.7 Problems

10.7.1 Identifying the field strength tensor

The goal of this problem is to practice fluency in handling the mathematical objects of gauge theory. It introduces various tricks which facilitate the handling of structures containing matrix gauge fields $A_\mu = \{(A_\mu)^a{}_b\}$.

Consider the parallel transporter $\Gamma[\gamma]$, Eq. (10.11), along an infinitesimal closed loop γ of linear extension ϵ on a manifold. Our goal is to identify the contribution of leading order, ϵ^2, to $\Gamma[\gamma]$. Parameterizing the loop as $x(t) \equiv \{x^\mu(t)\}$, $t \in [0, 1]$, expand the exponential up to second order in the time integrals and show that the first order of the expansion yields the contribution $-\int_{S_\gamma} dA$, which is of second order in ϵ (think why?). In the term of second order, you will encounter products $\dot{x}^\mu(t_1) A_{x(t_1),\mu} \dot{x}^\nu(t_2) A_{x(t_2),\nu}$ of the matrix-valued gauge fields evaluated at different positions along the curve. Use the matrix commutator and anticommutator $[\,,\,]_\pm$ to decompose such products into the sum of a symmetric and an anti-symmetric matrix. Show that the symmetric contribution can be written as the square of two integrals, each contributing to $\mathcal{O}(\epsilon^2)$. Being of $\mathcal{O}(\epsilon^4)$, this term can be neglected. Turning to the anti-symmetric term, assume that the coefficients A_μ vary only weakly over the extension of the curve so that they can be pulled out of the integral. Show that one of the time integrals over the remaining parts of the integrand can be performed, and the second can be written as $\oint_\gamma dx^\mu x^\nu$. Apply Stokes' theorem and use $A = A_\mu dx^\mu$ to bring the result into the form $-\int_{S(\gamma)} A \wedge A$. Combined with the first contribution, this leads to the structure $-\int_{S(\gamma)} (dA + A \wedge A)$ for the $\mathcal{O}(\epsilon)$ contribution to the parallel transporter. Assuming a fiber-matrix structure $A = \{A^a{}_b\}$, derive the form of the integrand fully resolved in indices, μ and a.

Answer:

The expansion of the the transporter (10.11) up to second order in exponentials reads

$$\Gamma[\gamma] = \mathbb{1} - \int_0^1 dt \ \dot{x}^\mu(t) A_{t,\mu} + \int_0^1 dt \int_0^t du \ \dot{x}^\mu(t) A_{t,\mu} \dot{x}^\nu(u) A_{u,\nu} + \mathcal{O}(\epsilon^3),$$

where we have used the abbreviation $A_{t,\mu} \equiv A_{x(t),\mu}$. The first term is a line integral, which can be converted to an area integral via Stokes' theorem: $\int_0^1 dt \ \dot{x}^\mu(t) A_{t,\mu} = \oint_\gamma A = \int_{S(\gamma)} dA$, as in the abelian case. Turning to the second term, we arrange the product of matrices as $A_{t,\mu} A_{u,\nu} = \frac{1}{2}[A_{t,\mu}, A_{u,\nu}] + \frac{1}{2}[A_{t,\mu}, A_{u,\nu}]_+$, thus into a symmetric and an anti-symmetric contribution. A straightforward rearrangement of terms (never changing the order of matrices) shows that the symmetric term can be written as, $\int_0^1 dt \int_0^t du \ \dot{x}^\mu(t) \dot{x}^\nu(u) [A_{t,\mu}, A_{u,\nu}]_+ = \frac{1}{2} (\int_0^t dt \dot{x}^\mu A_\mu)^2$, i.e., a product of two line integrals. Since each of these scales as $\mathcal{O}(\epsilon^2)$, this is a sub-leading term, which we can neglect. Turning to the antisymmetric term, we assume that the dependence of $A_{x(t)}$ over the extension of the curve is weak, and so pull these factors out of the integral, to obtain $\frac{1}{2}[A_\mu, A_\nu] \int_0^1 dt \int_0^t du \ \dot{x}^\mu(t) \dot{x}^\nu(u) = \frac{1}{2}[A_\mu, A_\nu] \int_0^1 dt \ \dot{x}^\mu(t) x^\nu(t) = \frac{1}{2}[A_\mu, A_\nu] \oint_\gamma dx^\mu x^\nu$, where in the second step we performed the full-derivative integral over u (consider why the term coming from the lower integration boundary, $u = 0$, does not contribute), and in the third step turned to a parameter-invariant formulation of the integral. Rearranging indices, and once more applying Stokes' theorem, we obtain $\frac{1}{2}[A_\mu, A_\nu] \oint_\gamma dx^\mu x^\nu = \frac{1}{2} A_\mu A_\nu \oint_\gamma (dx^\mu x^\nu - dx^\nu x^\mu) = -\frac{1}{2} A_\mu A_\nu \int_{S(\gamma)} (dx^\mu \wedge dx^\nu - dx^\nu \wedge dx^\mu) = -A_\mu A_\nu \int_{S(\gamma)} dx^\mu \wedge dx^\nu = -\int_{S(\gamma)} A \wedge A$, where in the third equality we used the anti-symmetry of the $\wedge$-product, and in the final step pulled the matrices A_μ back under the integral. Combining terms, we have identified the second-order contribution to the expansion of the parallel transporter as $-\int_{S(\gamma)} (dA + A \wedge A)$. The integrand, $F \equiv dA + A \wedge A$ defines the field strength tensor of the general theory. For $A_\mu = (A_\mu)^a{}_b$, its index representation reads $F = F_{\mu\nu} dx^\mu \wedge dx^\nu$, with $(F_{\mu\nu})^a{}_b = (\partial_\mu A_\nu)^a{}_b - (\partial_\nu A_\mu)^a{}_b - (A_\mu)^a{}_c (A_\nu)^c{}_b$.

10.7.2 Bianchi identity

In this problem, we explore how the field strength tensor of a gauge theory behaves under parallel transport. This is an interesting problem both mathematically and physically. Mathematically, we consider an object that transforms under gauge transformations differently from the fiber elements discussed so far. This requires an extension of the concept of parallel transportation. Physically, the behavior of the field strength tensor of electromagnetism under parallel transport is described by the homogeneous Maxwell equations. We thus realize that these, as well as their non-abelian generalizations, are of purely geometric origin.

In section 10.1.2, we studied the parallel transport of sections/fields, which under gauge transformations change as $s_x \to g_x s_x$. If g_x acts as the matrix representing a group operation, this is called transformation under the **fundamental representation** of the group. Not all elements of gauge theory transform under the fundamental representation. An important example is the field strength tensor, which changes as $F_x \to g_x F_x g_x^{-1}$ (see Eq. (10.17)). This action is called the **adjoint representation** of the group. All that has been said above about the differentiation

adjoint
repre-
sentation

of non-scalar objects generalizes to those transforming under the adjoint. Specifically, a slightly modified variant of parallel transport defines a covariant derivative tailored to objects in the adjoint representation.

(a) Apply reasoning similar to that in section 10.1.3 to show that the covariant differentiation of the components of the field strength tensor in the direction of a tangent vector v_μ takes the form

$$v^\mu D_\mu F_{\rho\sigma} \equiv v^\mu \left(\partial_\mu F_{\rho\sigma} + A_\mu F_{\rho\sigma} - F_{\rho\sigma} A_\mu \right).$$

Verify that the differential form representation of this identity reads as

$$DF \equiv dF + A \wedge F - F \wedge A.$$

(b) Now use the definition of the field strength tensor to demonstrate that it is covariantly constant: $D_\mu F_{\rho\sigma} + D_\rho F_{\sigma\mu} + D_\sigma F_{\mu\rho}$, which in differential form notation assumes the compact form

$$\boxed{DF = 0} \qquad\qquad (10.52)$$

Bianchi identity   This is the famous (second) **Bianchi identity**. In the abelian case, where $A = A_\mu dx^\mu$ with scalar coefficients, $DF = dF$ reduces to the ordinary derivative (reflecting the gauge invariance of the field strength in the abelian case). The Bianchi identity $dF = 0$ is then equivalent to the homogeneous Maxwell equations (if you are not familiar with this statement, verify it by application of the derivative to the tensor $F = \{F_{\mu\nu}\}$ with components defined by Eq. (10.24)). The Bianchi identity plays a similarly important role in general relativity and Yang–Mills theory, where it geometrically constrains (and algebraically reduces the number of free components) of the Riemann curvature tensor and of the field strengths tensors describing the strong and weak interactions, respectively. For further discussion of this point we refer to the literature.

Answer:

(a) Applied to objects transforming under the adjoint, the covariant rule of differentiation is obtained from the difference $F_{x+\epsilon v} - (\mathbb{1} - \epsilon Av)F_x(\mathbb{1} + \epsilon Av)$. Here, the first term is the object at a displaced point $x + \epsilon v$, and the second term is the object adjointly transformed under the action of a group element infinitesimally close to unity, with $Av \equiv A_\mu v^\mu$. First-order expansion in ϵ immediately leads to the first displayed equation stated in the problem. With $F = \frac{1}{2}F_{\mu\nu}dx^\mu \wedge dx^\nu$ and $A = A_\mu dx^\nu$, this is seen to be equivalent to the invariant representation.

(b) With $F = dA + A \wedge A$, we have $dF = dA \wedge A - A \wedge dA$, where the rules of exterior multiplication (appendix section A.1.2) have been used. Further, $A \wedge F - F \wedge A = A \wedge dA - dA \wedge A$. Adding the two contributions, we obtain zero and in this way prove the covariant constancy of the field strength tensor, $DF = dF + A \wedge F - F \wedge A = 0$. The computation in coordinates $F \to (F_{\mu\nu})^i{}_j$ is likewise straightforward but more cumbersome.

10.7.3 Two-dimensional $\mathbb{Z}_2$ gauge theory

In this problem we address the physics of the two-dimensional $\mathbb{Z}_2$ gauge theory. We demonstrate its equivalence to a conventional one-dimensional Ising chain, and in this way show that it is in a strongly fluctuating phase at all temperatures.

Consider a two-dimensional variant of the $\mathbb{Z}_2$ lattice gauge theory of section 10.3. There, we argued that, at low temperatures, the reversal of an individual spin comes at an action cost of $4\beta(d-1)$. Show that the argument fails in the special case $d = 2$, where one may reverse arbitrarily large numbers of neighboring spins at the action cost of just 4β (hint: think of line defects). Why does this construction fail in $d > 2$?

Energetically cheap line defects of reversed spins must be quite effective in disordering the system: their finite energy cost competes with the large amount of entropy created by the integration over their geometric orientation. To find out which wins, energy or entropy, we may use the gauge freedom to map the system onto a simpler one. First convince yourself that there exists a gauge in which *all* links pointing in one of the directions of the lattice, the ν-direction, say, are equal: $\sigma_3(n,\nu) = 1$. Do so by counting the number of gauge-equivalent configurations in each state with definite plaquette values. Show that this number exceeds the number of links pointing in a given direction.

Finally, show that the system with uniform link variables in one direction decomposes into a stack of independent one-dimensional Ising models. This establishes the equivalence (two-dimensional $\mathbb{Z}_2$ gauge theory) $\leftrightarrow$ (one-dimensional Ising model). From our discussion of section 6.1.1, we know that the $1d$ Ising model is in a disordered phase at any non-zero temperature. The same is therefore true for its two-dimensional gauge partner system.

Answer:

Rather than graphically sketching the structure of the low-energy line defects (which would give the solution away too easily), we describe them in words: pick a link (n,μ) on the lattice and reverse it along with a stack of parallel links at positions $(n + k\nu, \mu)$, where ν and μ are perpendicular unit vectors, and k is an integer. This flips the plaquette products at the terminal points of the stack, but nowhere else. The action cost of the configuration with $k + 1$ reversed spins thus equals $2 \times 2\beta = 4\beta$. In dimensions three and above, the stack would be surrounded by a chain of plaquettes sharing just one reversed spin. Each of these would cost action, so the total action of the configuration would be extensive in its length.

On a lattice with L_μ and L_ν rows and columns, respectively, we have $N_P \equiv (L_\mu - 1)(L_\nu - 1)$ plaquettes and $N_B \equiv L_\nu(L_\mu - 1) + L_\mu(L_\nu - 1)$ bond degrees of freedom. Fixing the plaquette values, thus leaves $N_B - N_P = L_\mu L_\nu - 1$ (gauge) degrees of freedom. This is more freedom than is required to set all bonds in a given direction (μ or ν) to a specified value.

If the ν-link variables are set to $+1$, the plaquette terms in Eq. (10.38) simplify to

$$S[\sigma_3] = -J \sum_{n,\mu,\nu} \sigma_3(n, \mu)\sigma_3(n + \mu + \nu, -\mu).$$

This equals the action of a stack of independent one-dimensional Ising models in the ν-direction.

10.7.4 Geometric curvature from gauge theory

In this problem, we investigate how the Levi-Civita connection describes the geometric curvature of a simple manifold.

In the example on page 593, we have computed Christoffel symbols of the Levi-Civita connection on the two-sphere. We found that $\Gamma^\phi_{\ \phi\theta} = \Gamma^\phi_{\ \theta\phi} = \cot\theta$, $\Gamma^\theta_{\ \phi\phi} = -\sin\theta\cos\theta$, and all other symbols are zero. Show that the corresponding connection is given by the one-forms

$$\Gamma^\phi_{\ \phi} = \cot\theta\, d\theta, \qquad \Gamma^\phi_{\ \theta} = \cot\theta\, d\phi, \qquad \Gamma^\theta_{\ \phi} = -\sin\theta\cos\theta\, d\phi.$$

Next consider the Riemann curvature tensor $R^i_{\ j} = \frac{1}{2}R^i_{\ jkl}dx^k \wedge dx^l$ and verify that it is defined by the matrix of two-forms

$$R = \begin{pmatrix} R^\theta_{\ \theta} & R^\theta_{\ \phi} \\ R^\phi_{\ \theta} & R^\phi_{\ \phi} \end{pmatrix} = \begin{pmatrix} 0 & \sin^2\theta \\ -1 & 0 \end{pmatrix} d\theta \wedge d\phi.$$

Proceed to obtain the coefficients of the Ricci tensor as $R_{\phi\phi} = \sin^2\theta$, with all other matrix elements non-vanishing,

$$R = \begin{pmatrix} R_{\theta\theta} & R_{\theta\phi} \\ R_{\phi\theta} & R_{\phi\phi} \end{pmatrix} = \begin{pmatrix} 1 & \\ & \sin^2\theta \end{pmatrix}.$$

Show that this result implies a constant scalar curvature $R = 2$ of the sphere. (The scalar curvature of a two-dimensional surface equals twice its Gaussian curvature, which for a sphere equals unity.)

Answer:

The first part of the solution follows from reading the Christoffel symbols as $\Gamma^i_{\ kl} = (\Gamma^i_{\ k})_l$, where the inner indices represent the matrix structure of the connection form and the outer index defines a form as $(\Gamma^i_{\ k})_l dx^l = \Gamma^i_{\ k}$. The curvature tensor is obtained as $R = d\Gamma + \Gamma \wedge \Gamma$, or $R^i_{\ j} = d\Gamma^i_{\ j} + \Gamma^i_{\ k} \wedge \Gamma^k_{\ j}$, and the building blocks entering this construction are obtained from the connection forms as

$$\begin{aligned} d\Gamma^\theta_{\ \theta} &= 0, & \Gamma^\theta_{\ k} \wedge \Gamma^k_{\ \theta} &= 0, \\ d\Gamma^\theta_{\ \phi} &= (\sin^2\theta - \cos^2\theta)\, d\theta \wedge d\phi, & \Gamma^\theta_{\ k} \wedge \Gamma^k_{\ \phi} &= \cos^2\theta\, d\theta \wedge d\phi, \\ d\Gamma^\phi_{\ \theta} &= -\frac{1}{\sin^2\theta} d\theta \wedge d\phi, & \Gamma^\phi_{\ k} \wedge \Gamma^k_{\ \phi} &= \cot^2\theta\, d\theta \wedge d\phi, \\ d\Gamma^\phi_{\ \phi} &= 0, & \Gamma^\phi_{\ k} \wedge \Gamma^k_{\ \phi} &= 0. \end{aligned}$$

Adding these terms, we obtain the curvature tensor as a matrix of two-forms, as given above. From this result, we can read off the non-vanishing elements of the curvature tensor, $R^i_{\ j\theta\phi} = -R^i_{\ j\phi\theta}$. Specfically, the elements of the Ricci tensor are obtained as $R_{\phi\phi} = R^\theta_{\ \phi\theta\phi} = \sin^2\theta$, $R_{\theta\phi} = R^\theta_{\ \theta\theta\phi} = 0$, $R_{\phi\theta} = R^\phi_{\ \phi\phi\theta} = 0$, and $R_{\theta\theta} = R^\phi_{\ \theta\phi\theta} = 1$. The Ricci tensor $R = \mathrm{diag}(1, \sin^2\theta) = g$ thus equals the metric tensor of the sphere. This means that contraction with the *inverse* metric tensor, $R = g^{ij}R_{ij} = g^{ij}g_{ij} = 2$, as stated.

10.7.5 Scalar field coupled to gravity

In this problem we consider a scalar field as a simple example of matter–gravity coupling. Its energy–momentum tensor is derived by variation of the corresponding action in the metric.

Consider a scalar field, ϕ, defined on a four-dimensional space–time manifold with metric, $g_{\mu\nu}$. Assuming, for simplicity, that there are no potentials acting on the field, the minimal two-derivative action form-invariant under coordinate transformations reads (think why?)

$$S_\mathrm{M}[\phi, g] = -\frac{1}{2}\int d^4x\sqrt{-g}\,\partial_\mu\phi\partial^\mu\phi, \tag{10.53}$$

where the notation emphasizes that S_M is considered as a functional of the field, ϕ, and the metric, g. Vary this action in the metric to obtain the energy–momentum tensor as

$$T^{\mu\nu} = -\frac{2}{\sqrt{-g}}\frac{\partial S[\phi]}{\partial g^{\mu\nu}} = \partial_\mu\phi\,\partial_\nu\phi - \frac{1}{2}g_{\mu\nu}\partial^\rho\phi\,\partial_\rho\phi. \tag{10.54}$$

Answer:

The metric couples to the action through the determinant, $\sqrt{-g}$, and the contracted derivatives, $\partial_\mu\phi\,\partial^\nu\phi = \partial_\mu\phi\,g^{\mu\nu}\partial_\nu\phi$. The differentiation of the derivative term yields $\partial^{g_{\mu\nu}}(\partial_\mu\phi\,g^{\mu\nu}\partial_\nu\phi) = \partial_\mu\phi\,\partial_\nu\phi$. Turning to the determinant, we can write $\sqrt{-g} = \exp(\frac{1}{2}\ln(-g)) = \exp(\frac{1}{2}(-\mathrm{tr}\ln g^{-1} + i\pi))$, where we have used the identity $\ln g = -\ln\det(g^{-1}) = \mathrm{tr}\ln(g^{-1})$ and, in the last term, $\ln(g^{-1})$ is the logarithm of the matrix $g^{-1} = \{g^{\mu\nu}\}$. The partial derivative then yields $\partial_{g_{\mu\nu}}\sqrt{-g} = \partial_{g_{\mu\nu}}\exp(\frac{1}{2}(-\mathrm{tr}\ln g^{-1}+i\pi)) = \sqrt{-g}\partial_{g_{\mu\nu}}\frac{1}{2}(-\mathrm{tr}\ln g) = -\sqrt{-g}\frac{1}{2}\mathrm{tr}(gE_{\mu\nu}) = -\frac{g_{\nu\mu}}{2}\sqrt{-g}$, where $\partial_{g_{\mu\nu}}g = E_{\mu\nu}$ is the matrix containing zeros everywhere except for a unity at position (μ, ν). Combining the two terms, we obtain the stated result.

10.7.6 Chern–Weil theorem

Here, we prove the Chern–Weil theorem discussed in section 10.5.3. The problem is a technical, yet instructive, application of differential form calculus.

We aim to investigate properties of the monomials $I_n(F) = \mathrm{tr}(F^n)$. Our main allies in this problem will be the cyclic invariance of the trace (which extends to forms as $\mathrm{tr}(X \wedge Y) = \mathrm{tr}(Y \wedge X)(-)^{r \cdot s}$, where r and s are the form degrees of X and Y – think why?), and the Bianchi identity (10.52), $DF = dF + A \wedge F - F \wedge A = 0$.

(a) Apply the two relations just mentioned to prove that the invariant monomials are closed, $dI_n(F) = 0$.

(b) Now let F and F' be two connections, and F_t a one-parameter curve continuously connecting them with $F_0 = F$ and $F_1 = F'$. Use the definition of the field strength (10.13) and the relations above to identify a form κ such that $d_t F_t \equiv d\kappa_t$. Why is this sufficient to demonstrate that $I_n(F) - I_n(F')$ is exact? As an example, consider the case $F' = 0$ and choose $F_t = dA_t + A_t \wedge A_t$, with linear interpolation $A_t = tA$. Considering the case $n = 1, 2$, show that this leads to Eq. (10.49).

Answer:

(a) Using the cyclic properties of the sphere, we have $dI_n(F) = n\mathrm{tr}(dF \wedge F^{(n-1)})$. Since A and F commute as forms (but not in general as matrices), we have

$$0 = \mathrm{tr}(A \wedge F^n - F^n \wedge A) = n\,\mathrm{tr}(A \wedge F - F \wedge A)F^{n-1}). \qquad (10.55)$$

We can add the two relations to obtain $dI_n(F) = n\,\mathrm{tr}((dF + A \wedge F - F \wedge A)F^{n-1}) = n\,\mathrm{tr}(DFF^{n-1}) = 0$.

(b) We can then compute the time derivative as $d_t I_n(F_t) = nI(\dot{F}F^{n-1}) = n\,\mathrm{tr}((d\dot{A} + \dot{A} \wedge A + A \wedge \dot{A}) \wedge F^{n-1}) = n\,\mathrm{tr}(d\dot{A} \wedge F^{n-1} + \dot{A} \wedge (A \wedge F^{n-1} - F^{n-1} \wedge A))$. We now use the fact that commutators obey "chain rules," much as derivatives do: $A \wedge F^{n-1} - F^{n-1} \wedge A = \sum_{l=0}^{n-2} F^l \wedge (A \wedge F - F \wedge A) \wedge F^{n-2-l} = -\sum_{l=0}^{n-2} F^l \wedge (dF) \wedge F^{n-2-l} = -dF^{n-1}$, where the Bianchi identity is used again. Combining these formulas, $d_t I_n(F_t) = n\,\mathrm{tr}(d\dot{A} \wedge F^{n-1} - \dot{A} \wedge dF^{n-1}) = nd\,\mathrm{tr}(\dot{A} \wedge F^{n-1}) \equiv d\kappa_t$, with $\kappa_t = n\,\mathrm{tr}(\dot{A}_t \wedge F_t^{n-1})$ demonstrates the exactness on the infinitesimal level. Integration then yields $I(F) - I(F') = d\int_0^1 dt\,\kappa_t$, i.e., the difference equals the exterior derivative of another form, and therefore is exact. Under the stated conditions, we have $\kappa_t = n\,\mathrm{tr}(A \wedge (tdA + t^2 A \wedge A)^{n-1})$ and integration over t immediately leads to the stated result.

10.7.7 Gauge invariance of Chern–Simons action

This is another technical, yet instructive, problem. We demonstrate that the second Chern class of a bundle with structure group $\mathrm{SU}(2)$ is determined by the winding number of large $\mathrm{SU}(2)$ gauge transformations ("instantons") in three-dimensional space. As a by-product, we learn that the Chern–Simons action (10.27) is not gauge invariant under large gauge transformations. The problem nicely illustrates how various algebraic expressions routinely encountered in topological field theory are related to each other.

Consider the second Chern number, Eq. (10.50), of a bundle with four-dimensional base space, T, and with $\mathrm{SU}(2)$-gauge connection and Chern character (10.48). We assume that the bundle is twisted and that $T = X \cup X'$ is the union of two domains separated by a three-dimensional boundary, B. On the sub-domains, the connections generating the field strength are given by A and A', respectively, and on B the two are related by an SU(2) gauge transformation, Eq. (10.3), $A' = gAg^{-1} + gdg^{-1}$. We may now apply Stokes' theorem to compute the Chern number as

$$\mathrm{Ch}_2[F] = \frac{1}{8\pi^2} \int_T F = -\frac{1}{2\pi} \left(\int_X d\mathcal{L}_{\mathrm{CS}}(A) + \int_{X'} d\mathcal{L}_{\mathrm{CS}}(A') \right)$$

$$= -\frac{1}{2\pi} \int_B \left(\mathcal{L}_{\mathrm{CS}}(A) - \mathcal{L}_{\mathrm{CS}}(A') \right),$$

where $\mathcal{L}_{\mathrm{CS}}(A) = \frac{1}{4\pi} \mathrm{tr} \left(A \wedge dA + \frac{2}{3} A \wedge A \wedge A \right)$ is the Chern–Simons topological density (see Eq. (10.27)).

(a) Defining the matrix valued one-form $g^{-1}dg \equiv \Phi$, verify that $\mathrm{tr}(A' \wedge A' \wedge A' - A \wedge A \wedge A) = \mathrm{tr}(3A \wedge \Phi \wedge \Phi + 3A \wedge A \wedge \Phi + \Phi \wedge \Phi \wedge \Phi)$.

(b) In a similar manner verify that $\mathrm{tr}(A' \wedge dA' - A \wedge dA) = \mathrm{tr}(-2A \wedge A \wedge \Phi - 3A \wedge \Phi \wedge \Phi + \Phi \wedge dA - \Phi \wedge \Phi \wedge \Phi)$ and explain why the third term may be replaced by $\mathrm{tr}(A \wedge \Phi \wedge \Phi)$ in the action. Combine the results of (a) and (b) to demonstrate that $\mathcal{L}_{\mathrm{CS}}(A) - \mathcal{L}_{\mathrm{CS}}(A') = -\frac{1}{12\pi} \mathrm{tr}(\Phi \wedge \Phi \wedge \Phi)$ and $\mathrm{Ch}_2[F] = W$, where W is the winding number defined by the integral (8.14) at $\theta = 1$.

It is instructive to look at the result above from a slightly different perspective. Focusing on the three-dimensional boundary-less manifold B, we have found that the non-abelian Chern–Simons action $S_{\mathrm{CS}}[A] = \int_B \mathcal{L}_{\mathrm{CS}}(A)$ defined on it is *not* gauge invariant. Under an SU(2) gauge transformation $A \to A'$, it changes as $S_{\mathrm{CS}}[A] - S_{\mathrm{CS}}[A'] = 2\pi W$, where W is the winding number characterizing the transformation. The gauge invariance of the exponentiated action $\exp(ikS_{\mathrm{CS}})$ requires integer quantization of the level index k.

Answer:

(a) This follows straightforwardly from $A' = g(A + \Phi)g^{-1}$ and the cyclic invariance of the trace.

(b) The first result is obtained as in part (a). The third term in the trace appears under an integral over the three-dimensional boundary-less manifold B. (B is the boundary of X and X', and hence does not have a boundary itself.) Application of Stokes' theorem thus yields $0 = \int_B d\,\mathrm{tr}(\Phi \wedge A) = \int_B \mathrm{tr}(d\Phi \wedge A - \Phi \wedge dA)$. Now, $d\Phi = dg^{-1} \wedge dg = -g^{-1}dg \wedge g^{-1}dg = -\Phi \wedge \Phi$. Combination of these formulae shows that the substitution stated in the problem is valid.

11 Nonequilibrium (Classical)

SYNOPSIS This chapter provides an introduction to nonequilibrium statistical field theory. We start by reviewing various concepts describing many-particle systems out of statistical equilibrium (with no previous knowledge required). We will then discuss how cornerstones of this theory – notably the Langevin, master, and Fokker-Planck equation – are derived through variational principles from field integrals. Within this framework, we will discuss various signatures of non-equilibrium physics, such as metastability, the interplay between dissipation and noise, out-of-equilibrium universality classes, phase transitions, and more. Throughout the chapter, we will highlight remarkable parallels between the theory of classical nonequilibrium processes and (imaginary-time) quantum mechanics. The full scope of this parallelism will become evident in the next chapter, when we discuss quantum non-equilibrium systems.

In this chapter, we need some elements of probability theory that may be unfamiliar. The required material is reviewed in appendix section A.2.

The world around us is full of nonequilibrium phenomena: jams forming out of seemingly light traffic, charge carrier dynamics in a strongly voltage-biased electronic device, the dynamics of social or economic networks, and chemical reactions are all examples where large numbers of interacting "particles" are out of equilibrium. The ubiquity of nonequilibrium phenomena makes the understanding of physics beyond that of thermal states an important part of statistical physics. However, this statement by itself does not explain the dramatic growth of the field in the last two to three decades.

The primary driving force behind recent developments has been the progress in experimentation and device technology. While bulk solid state phases are hard to perturb out of equilibrium, the situation with cold atomic and optical systems, or miniaturized condensed matter device structures, is different. These systems are easily pushed out of equilibrium, and the ensuing phenomenology can be monitored, e.g., via imaging techniques available to atomic physics, or quantum transport in condensed matter physics.

Nonequilibrium physics is a discipline much more multi-faceted than equilibrium statistical mechanics. The multitude of phenomena addressed by this field in fact bears the risk of "over-diversification": facing a confusingly rich spectrum of physical regimes and phenomena, it is easy to get bogged down with details and over-specialized theory. The unifying field-theoretical approach introduced below counteracts this tendency. From it, the majority of theoretical tools applied in nonequilibrium physics follow by reduction. This defines a versatile and transparent top-down structure. Specifically, it reveals connections between different approaches that are not easy to see otherwise.

We begin this chapter by introducing a number of traditional approaches to nonequilibrium physics, involving the Langevin, Fokker–Planck, Boltzmann, and master equations, and their application in the physics of stochastic processes. We will then derive a unifying variational principle for these equations. This "stochastic path integral" will be the basis for the extension of the theory to higher-dimensional stochastic processes in the second half of the chapter.

INFO How does one define a state of **nonequilibrium**? "Nonequilbrium" is the opposite of **thermal equilibrium**, the latter being defined by the following two conditions:[1]

thermal equilibrium

▷ An equilibrium system is characterized by a unique set of extensive and intensive variables, which do not change over time.

▷ After isolation of the system from its environment, all the variables remain unchanged.

The second condition is necessary to distinguish an equilibrium from a **stationary nonequilibrium**. For example, the particle distribution function of a conductor subject to a voltage bias is time-independent (the first condition is fulfilled), yet different from the equilibrium Gibbs distribution. Upon removal of the "environment," defined by the attached leads, it will change and relax to the Gibbs distribution (the second condition is violated).

Note that the two conditions are quite restrictive, indicating that pristine realizations of thermal equilibrium are rare in nature.

11.1 Fundamental Concepts of Nonequilibrium Statistical Mechanics

Above, we pointed out that thermodynamic equilibrium is an exception, rather than the rule. What makes that limit so much easier to describe than the nonequilibrium limit is that its many-body distribution is known: states in thermal equilibrium are distributed according to the grand canonical ensemble $\mathcal{Z}^{-1}\exp[-\beta(H - \mu N)]$. Our task is reduced to computing observables from this distribution (which can be complicated enough, as we saw in the early chapters of the text). However, in thermal nonequilibrium, the distribution of the system is *a priori* unknown, a lack of information that bears important consequences:

▷ Identifying the many-particle distribution function becomes part of the challenge (and usually the first to be addressed).

▷ Concepts that we often take for granted – the existence of a uniquely defined temperature, homogeneity of thermodynamic variables, etc. – need to be re-examined.

---

[1] W. Ebeling and I. M. Sokolov, *Statistical Thermodynamics and Stochastic Theory of Nonequilibrium Systems* (World Scientific, 2005).

▷ There exists an unfathomable multitude of different universality classes in nonequilibrium physics. Identifying the scope of any of these can also be a difficult task.

How does one determine the statistical distribution of a system if the only known data are a Hamiltonian, the number of particles, and boundary conditions? Fundamentally, the state of a classical d-dimensional N-particle system defines a point $\mathbf{X} \in \mathbb{R}^{2Nd}$ in $2Nd$-dimensional phase space. The full information about its dynamics, and all derived physical properties, are contained in a high-dimensional Hamiltonian equation of motion.

The view above is rigorous, but largely useless in practice. However, in the late nineteenth century, **Boltzmann** introduced a much more practical approach. He suggested describing the system as a "swarm" of N points in $2d$-dimensional phase space, rather than as a single point in a high-dimensional space. This idea paved the way to a statistical formulation of the problem and may be considered as

Ludwig Boltzmann 1844–1906 was an Austrian physicist famous for his pioneering contributions to statistical mechanics. Concepts such as Maxwell–Boltzmann statistics, the Boltzmann distribution, and the logarithmic connection between entropy and probability remain foundations of this field. Boltzmann was one of the most important supporters of early atomic theory, at a time when the reality of atoms was still controversial.

the starting point of statistical mechanics. Indeed, we may now trade the excessive fine structure contained in the full coordinate dependence of the swarm for a statistical formulation. This is achieved by introducing a probability measure $f(\mathbf{x}, t)\, d^d x$, where $\mathbf{x} = (\mathbf{q}, \mathbf{p})$ is a point in phase space and the dimensionless function $f(\mathbf{x}, t)$ is the distribution of the number of particles found at time t in the volume element $d^d x = \prod_{i=1}^{d} dq_i dp_i$ (see figure). The function f is normalized as

$$\int_{\Gamma} d^d x\, f(\mathbf{x}) = N,$$

where $\int_{\Gamma}$ is an integral over classical phase space. Apart from this normalization to N (rather than unity), it is a probability distribution, i.e., Boltzmann's approach is statistical by design.

Boltzmann distribution

The **Boltzmann distribution** f is the most important object of the theory. From it, average values of physical observables X can be calculated as

$$\langle X \rangle = \int d^d x\, f(\mathbf{x}, t) X(\mathbf{x}),$$

where $X(\mathbf{x})$ is the phase space function representing the observable. For exam-

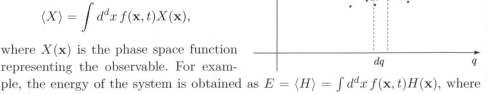

ple, the energy of the system is obtained as $E = \langle H \rangle = \int d^d x\, f(\mathbf{x}, t) H(\mathbf{x})$, where $H(\mathbf{x})$ is the Hamiltonian.

Understanding the state of a system is now equivalent to the identification of its Boltzmann distribution for a given Hamiltonian and initial state $f(.,t=0)$. As a warm-up to this general problem, we first ask how **thermal equilibrium** is described via the function f, and by which mechanisms is the equilibrium state approached if the system is allowed to relax under the influence of its own many-body dynamics. For concreteness, consider a system that is weakly interacting, or gaseous, in the sense that the dynamics is essentially of single-particle type. Particle collisions, $(\mathbf{x}_1, \mathbf{x}_2) \to (\mathbf{x}_1', \mathbf{x}_2')$, define a channel of relaxation by exchange of energy and momentum. Under these conditions, it is safe to assume that the distribution of particles will be independent in the sense that the joint probability of observing particles at $\mathbf{x}_1$ and $\mathbf{x}_2$, respectively, $p(\mathbf{x}_1, \mathbf{x}_2)$, factorizes into independent particle distributions, $p(\mathbf{x}_1, \mathbf{x}_2) = p(\mathbf{x}_1)p(\mathbf{x}_2)$. Under this assumption, the system can be described via a distribution $f(\mathbf{x}) = Np(\mathbf{x})$.

EXERCISE Think why this condition is crucially required for the definition of the Boltzmann statistical approach, and how it might be compromised in the presence of strong correlations.

The fact that coordinate configurations $(\mathbf{x}_1, \mathbf{x}_2) \leftrightarrow (\mathbf{x}_1', \mathbf{x}_2')$ are coupled by an elementary scattering event implies the conservation of probability, $f(\mathbf{x}_1, \mathbf{x}_2) = f(\mathbf{x}_1)f(\mathbf{x}_2) = f(\mathbf{x}_1')f(\mathbf{x}_2') = f(\mathbf{x}_1', \mathbf{x}_2')$. Assuming that $f(\mathbf{x}) = f(H(\mathbf{x})) \equiv f(\epsilon)$ is a function of energy, we conclude that probability conservation is compatible with energy conservation, $\epsilon_1 + \epsilon_2 = \epsilon_1' + \epsilon_2'$, if $\ln f(\epsilon) = a\epsilon + b$ is linear in energy. In this case, $f(\mathbf{x}_1)f(\mathbf{x}_2) = \exp(a(\epsilon_1 + \epsilon_2) + 2b) = \exp(a(\epsilon_1' + \epsilon_2') + 2b) = f(\mathbf{x}_1')f(\mathbf{x}_2')$ is indeed satisfied. To fix the constants a and b, we require normalization of the partition function and employ the equipartition theorem, i.e., we use the fact that, in thermodynamic equilibrium, the expectation value of the energy of each of the $2Nd$ degrees of freedom must equal $T/2$.[2] Normalization requires (Exercise) $b = \ln N - \ln \int d^d x \exp(a\epsilon)$ and from the equipartition theorem we obtain

$$NdT \overset{!}{=} dN \frac{\int_\Gamma d^d x \, e^{aH(\mathbf{x})} H(\mathbf{x})}{\int_\Gamma d^d x \, e^{aH(\mathbf{x})}} = -\frac{Nd}{a} \Rightarrow a = -\frac{1}{T} = -\beta,$$

where, again, a quadratic Hamiltonian is assumed. We thus arrive at the conclusion that

$$\boxed{f(\mathbf{x}) = N \frac{e^{-\beta H(\mathbf{x})}}{\int_\Gamma dx \, e^{-\beta H(\mathbf{x})}}} \tag{11.1}$$

Maxwell–Boltzmann distribution

which is the famous **Maxwell–Boltzmann distribution**. Exercise: In what sense is Eq. (11.1) a descendant of the many-particle Gibbs distribution?

While the construction above fixes the form of the equilibrium distribution, it does not describe the relaxational dynamics whereby equilibrium is approached. The modeling of these processes will be discussed in the next two sections.

----

[2] Here we assume that, close to thermal equilibrium, the energy of the particles will be approximately quadratic in their d coordinate and momentum degrees of freedom.

11.2 Langevin Theory

REMARK Readers familiar with the Langevin and Boltzmann equation, or those on a fast track towards field theory, may skip this and the following section and directly proceed to section 11.4.

How can we describe the dynamics of a distribution initially described by a given $f(\mathbf{x}, t = 0)$? Specifically, we wish to understand how the Maxwell–Boltzmann distribution is approached if the system is kept in isolation, and how departures from equilibrium may be caused by external forcing.

Naïvely, one might argue that phase space points $\mathbf{x} \xrightarrow{t} \mathbf{x}(t) = \exp(-t\{H, \cdot\})\mathbf{x}$ propagate according to Hamiltonian dynamics. This phase space flow implies the following time-dependence of phase space distributions (Exercise: See footnote[3])

$$(\partial_t - \{H, \cdot\})\, f(\mathbf{x}, t) = 0. \tag{11.2}$$

However, this equation does not capture the irreversible approach to equilibrium. Neglecting the interaction processes with other particles, it misses two key aspects of the dynamics.

First, interaction processes hinder the ballistic motion of individual particles. In a coarse-grained description, they *dissipation* generate **dissipation** of energy and **friction**. Second, repeated collisions *fluctuations* with other particles generate **fluctuations** in the particle trajectories. As we will see, the two mechanisms, dissipative friction and fluctuations, are not independent. Indeed, in thermal equilibrium, they completely determine each other. This mutual dependence is encapsulated by the *fluctuation–dissipation theorem* and we will discuss it extensively throughout.

Paul Langevin 1872–1946 was a French physicist who developed the concept of Langevin dynamics. He is also known for his modern interpretation of para- and diamagnetism in terms of the electron spin. Langevin was a devoted anti-fascist and lost his academic position under the Vichy regime. (Shortly before his death he was rehabilitated.)

In a seminal work,[4] the French physicist Paul Langevin described the joint influence of dissipation and fluctuation on a single test particle of mass m by a stochastic *Langevin* generalization of Newton's equation, the **Langevin equation**, *equation*

$$\boxed{m d_t \mathbf{v} + m\gamma \mathbf{v} - \mathbf{F} = \boldsymbol{\xi}(t)} \tag{11.3}$$

where $\mathbf{v}$ is the particle velocity, $\mathbf{F} = -\partial_\mathbf{q} H$ a macroscopic force acting the particle, γ a phenomenological friction coefficient, and $\boldsymbol{\xi}$ is a randomly fluctuating force describing the effect of particle collisions. Langevin modeled the components of

---

[3] Noting that $(0\,'(\jmath-)\mathbf{x})f = (\jmath\,'\mathbf{x})f$, the partial derivative of the distribution with respect to time is computed as $\{(\jmath\,'\mathbf{x})f\,'H\} = (\jmath\,'(s)\mathbf{x})f\,0{=}s|{}^s\varrho{-} = (s + \jmath\,'\mathbf{x})f\,0{=}s|{}^s\varrho = (\jmath\,'\mathbf{x})f\,\jmath{}^2\varrho.$

[4] P. Langevin, *Sur la théorie du mouvement Brownien*, CR Hebd. Acad. Sci. **146**, 530 (1908).

the stochastic force as short-range correlated Gaussian random variables with zero mean, $\langle \xi_i(t) \rangle = 0$, and variance

$$\langle \xi_i(t) \xi_{i'}(t') \rangle = A \delta_{ii'} \delta(t - t'), \tag{11.4}$$

where $A > 0$ is a constant.

INFO The development of Langevin's theory was motivated by the phenomenon of **Brownian motion**. In 1827, the Scottish botanist, Robert Brown, observed[5] erratic motion of pollen particles in aqueous immersion. The first qualitative explanations of the phenomenon in terms of random particle collisions appeared in the late nineteenth century. Langevin suggested Eq. (11.3) as an effective equation of motion governing the

Brownian motion

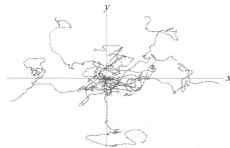

dynamics of "mesoscopic" particles, subject to collisions with microscopically small constituents of the system. The erratic nature of solutions of the Langevin equation – qualitatively consistent with Brownian motion – is illustrated in the figure, which shows the numerical solutions of two-dimensional Langevin trajectories for given realizations of the fluctuation force.

Before turning to a more detailed analysis of Eq. (11.3), let us make two general remarks in order to provide orientation for our subsequent discussion.

▷ One may ask whether the friction coefficient, γ, and the fluctuation strength, A, are independent parameters of the theory. We will show below that they are not: in thermal equilibrium, the strength of the fluctuations is related to that of the friction by temperature. This is the manifestation of the aforementioned **fluctuation–dissipation theorem** in Langevin theory, and we will discuss it in the next section.

▷ One may expect that fluctuations render the dynamics of particles partially or fully **diffusive**. Indeed, it turns out that A^{-1} is a measure of the diffusivity of a medium. The connection between Langevin theory and diffusion is addressed in full in section 11.2.3, and phenomenologically in the following section.

11.2.1 Dissipation versus fluctuations

To understand the connection between dissipation and fluctuation, we consider the Langevin equation in the absence of driving forces, $\mathbf{F} = 0$. The equation then assumes the form of a linear differential equation for $\mathbf{v}$ with inhomogeneity $\boldsymbol{\xi}$. A temporal Fourier transformation, $\mathbf{v}(\omega) = \int dt \exp(i\omega t) \mathbf{v}(t)$, yields the solution $\mathbf{v}(\omega) = \frac{1}{m(-i\omega + \gamma)} \boldsymbol{\xi}(\omega)$. Thinking of $\boldsymbol{\xi}$ as a random variable, this gives $\mathbf{v} = \mathbf{v}[\boldsymbol{\xi}]$ the

---

[5] The phenomenon had, in fact, been observed earlier by the Dutch physician Jan Ingenhousz in 1785.

status of a dependent random variable. The latter has zero mean, $\langle v_i(t)\rangle = 0$, and second moment (exercise)

$$\langle v_i^2(t)\rangle = \frac{A}{m^2} \int_{-\infty}^{\infty} \frac{d\omega}{2\pi} \frac{1}{\omega^2 + \gamma^2} = \frac{A}{2m^2\gamma}.$$

However, in thermal equilibrium, the **equipartition theorem** relates the average kinetic energy per degree of freedom to temperature, $m\langle v_i^2\rangle/2 \overset{!}{=} T/2$. Comparison with the result above then leads to the equation

$$A = 2m\gamma T, \tag{11.5}$$

stating that, in thermal equilibrium, the variance of the fluctuation is proportional to the strength of the dissipative friction forces, and to temperature. This relation is the first of several formulations of the **fluctuation–dissipation theorem**.

INFO Let us briefly review **Einstein's phenomenological derivation of Eq. (11.5)**. It starts by observing that, in a medium governed by frequent inter-particle collisions, the dynamics of individual particles will be **diffusive**. By the same principle, externally applied forces will cause **drift motion**, rather than free ballistic acceleration. The first postulate implies that a density gradient, ∂f, in the medium generates a diffusion current,

$$\boxed{\mathbf{j}_{\mathrm{d}} = -D\nabla f} \tag{11.6}$$

Fick's law i.e., a current acting to restore a uniform density profile. Equation (11.6) is **Fick's (first) law**. Application of the continuity equation $d_t f = -\nabla \cdot \mathbf{j}_{\mathrm{d}}$ shows that $\partial_t f = D\Delta f$ (Fick's second law), implying that the dynamics is indeed diffusive.

The second postulate states that $\mathbf{j}_{\mathrm{ext}} = \frac{f}{\gamma m}\mathbf{F}_{\mathrm{ext}}$, i.e., an external force generates a drift current proportional to the force, the density, and the inverse of the friction coefficient. (This formula may be obtained by dimensional analysis, or by consideration of the stationary configuration, $\partial_t \langle \mathbf{v}\rangle = 0$, obtained by averaging the Langevin equation (11.3).) In thermal equilibrium, the diffusion current and external current must compensate each other, $\mathbf{j}_{\mathrm{d}} = -\mathbf{j}_{\mathrm{ext}}$, or $D\nabla f = -f\nabla V/m\gamma$, where we assumed that the force is generated by some potential, V. This equation for the density profile is solved by $f \sim \exp(-V/Dm\gamma)$. However, compatibility with the Maxwell–Boltzmann distribution (11.1) requires that $f \sim \exp(-V/T)$, or

$$\boxed{D = \frac{T}{m\gamma}} \tag{11.7}$$

Einstein
relation
Equation (11.7) is the **Einstein relation**. Comparison with Eq. (11.5) finally leads to the identification,

$$A = \frac{2T^2}{D}, \tag{11.8}$$

i.e., the variance of the (microscopic) fluctuation forces in the medium is inversely proportional to its (macroscopic) diffusion constant.

11.2.2 Noise

REMARK This section reviews different types of noise appearing in the physics of many-body systems in and out of equilibrium. It can be skipped at first reading, and consulted later as a reference.

Our discussion above conveys the important message that dissipation and friction are linked to the presence of fluctuating forces. However, unlike the everyday phenomenon of friction, the presence of fluctuation forces is often not felt. The reason is that friction acts in a directed way (cf. the action of a brake), while the response caused by fluctuations tends to average out. Yet, there are exceptions to this rule. For example, high amplification levels render the fluctuations, or noise caused by resistive elements in electronic circuits, a noticeable (and generally unwelcome) side effect.

The figure shows three set-ups where the interplay of fluctuations and dissipation is essential: an electronic RLC-circuit, an electromagnetic cavity mode experiencing radiative losses, and a damped oscillator. In the non-dissipative limit, the dynamics of all these are harmonic. The presence of dissipation (resistor/radiative losses/mechanical damping) generates fluctuations. While these fluctuations do not affect the motion of the macroscopic mechanical oscillator, they can be noticeable in the other two systems.

Johnson–Nyquist noise

Let us take a closer look at the example of the RLC-circuit. Here, the observable of interest is the charge Q transmitted through the circuit (from which current and voltage drop across the resistive element are obtained as $I = \dot{Q}$ and $U = R\dot{Q}$, respectively). In elementary courses, we learn that the dynamics of charge is governed by the equation,

$$L d_t^2 Q + R d_t Q + \frac{Q}{C} = U_{\text{ext}}, \tag{11.9}$$

where U_{ext} is the externally applied voltage. This is the equation of motion of a dissipatively damped oscillator. However, we now know that Eq. (11.9) neglects the effect of the fluctuations accompanying dissipation. A correct way to think about Eq. (11.9) is as an equation for $\langle Q \rangle$, the coordinate Q averaged over realizations of these fluctuations. A realization-specific generalization of Eq. (11.9) compatible with the fluctuation–dissipation theorem is given by

$$L d_t^2 Q + R d_t Q + \frac{Q}{C} = U_{\text{ext}} + U_{\text{jn}}(t), \tag{11.10}$$

where the statistics of the noise is described by

$$\langle U_{\text{jn}}(t) \rangle = 0, \qquad \langle U_{\text{jn}}(t) U_{\text{jn}}(t') \rangle = 2 T R \delta(t - t'). \tag{11.11}$$

The noise term in Eq. (11.10) can be interpreted as a time-dependent fluctuating voltage, additional to the external voltage. It is important to realize that this type of noise is an inevitable consequence of the presence of resistive elements in electronic circuits; it cannot be removed by "improving the quality" of the device.

The presence of a fluctuating voltage of strength $\langle |U_{jn}(\omega)|^2 \rangle = 2TR$ was experimentally discovered by Johnson[6]. Nyquist[7] explained the phenomenon in terms of the thermal equipartition of energy of electromagnetic oscillator modes. Reflecting this background, U_{jn} is alternatively denoted as **Johnson noise**, **Nyquist noise**, **Johnson–Nyquist noise**, or just **thermal noise**.

thermal
noise

INFO   The **strength of the noise correlator** is fixed by slight modification of the arguments used in section 11.2.1: Think of a non-resistive LC-circuit as an electronic realization of the harmonic oscillator. The $R = 0$, $U = 0$ variant of Eq. (11.10) is obtained by variation of the Lagrangian action

$$S = i \int dt \left(\frac{L}{2}\dot{Q}^2 - \frac{1}{2C}Q^2 \right),$$

where $L\dot{Q}^2/2$ and $Q^2/2C$ represent the kinetic (inductive) energy and potential (capacitive) energy, respectively. We expect an interplay of fluctuation and dissipation to establish an energy balance compatible with the equipartition theorem,

$$\frac{L}{2}\langle \dot{Q}^2 \rangle = \frac{1}{2C}\langle Q^2 \rangle = \frac{T}{2},$$

i.e., $T/2$ for both kinetic and potential energy. This energy balance should hold, including in the presence of dissipation due to resistance (R). In the absence of external biasing, $U_{ext} = 0$, the temporal Fourier transform is given by

$$Q(\omega) = \frac{U_{jn}(\omega)}{i\omega R - \omega^2 L + C^{-1}}.$$

Using Eqs. (11.10) and (11.11), it is then straightforward to verify that

$$\frac{L}{2}\langle \dot{Q}(t)^2 \rangle = 2TR\frac{L}{2} \int \frac{d\omega}{2\pi} \frac{\omega^2}{\omega^2 R^2 + (\omega^2 L - C^{-1})^2} = \frac{T}{2}.$$

Similar reasoning shows that $\frac{1}{2C}\langle Q^2 \rangle = \frac{T}{2}$.

Although the voltage correlation (11.11) is consistent with the classical equipartition theorem, this very consistency generates an annoying problem: the second moment $\langle U_{jn}(t)^2 \rangle \propto \delta(0)$ diverges, i.e., our analysis makes the unphysical prediction that a voltmeter measuring noise amplitudes will detect voltage spikes of unbounded strength. Equivalently, we are making the unphysical prediction that the noise fluctuates with constant average intensity at arbitrarily high frequencies. (More formally, the **noise power**

noise
power

$$S(U_{jn}, \omega) \equiv \lim_{T \to \infty} \frac{1}{T} \left\langle \left| \int_{-T}^{T} dt\, U_{jn}(t) e^{i\omega t} \right|^2 \right\rangle = RT \tag{11.12}$$

[6] J. B. Johnson, *Thermal agitation of electricity in conductors*, Phys. Rev. **32**, 97 (1928).

[7] H. Nyquist, *Thermal agitation of electric charge in conductors*, Phys. Rev. **32**, 110 (1928).

is a measure of the fluctuation intensity at characteristic frequency ω, remains constant.)

This problem disappears if we replace the classical fluctuation energy of the harmonic oscillator, $2 \times \frac{T}{2} = T \to \omega(e^{\omega/T} - 1)^{-1}$, by the energy of a quantum oscillator. (Exercise: Explore this point.) For frequencies much larger than the temperature, $\omega \gg T$, the mode does not store energy and the problem with the diverging noise amplitude disappears.[8]

Shot noise

In electronic devices, the **discreteness of charge quanta** generates a noise source additional to the thermal noise. The current supported by a train of uncorrelated electrons moving down a wire is a random variable with Poisson distributed statistics (see section 11.4.2 for a general discussion of Poissonian stochastic processes). The average number of electrons, n, passing through the wire in a time interval Δt (large in comparison to the mean passage time) defines the mean current,

$$\langle I \rangle = \frac{\langle n \rangle}{\Delta t}.$$

However, it is a characteristic feature of Poisson processes that the variance in the number of transmitted electrons $\mathrm{var}(n) = \langle n \rangle$, which means that

$$\mathrm{var}(I) = \frac{1}{\Delta t} \langle I \rangle.$$

Although the signal-to-noise ratio $\langle I \rangle^2/\mathrm{var}(I) = \Delta t \langle I \rangle$ increases with increasing observation times and increasing mean current strength, the discreteness of charge remains a source of noise. In a continuum model of charge transport, in addition to the Johnson noise term, this **shot noise**[9] can be modeled by the further addition of a suitable noise term to the right-hand side of Eq. (11.9). (In section 0.6 of the next chapter, we explore how this addition emerges in a microscopic construction.) Unlike Johnson noise, shot noise is independent of the temperature and resistivity of the circuit. Both have in common that the noise power is largely independent of frequency.

shot noise

Other sources of noise

A variety of physical processes produce noise. Empirically one finds that these sources of noise can add to give a non-universal noise signal with pronounced frequency dependence. It is customary to denote noise signals with a power spectrum

---

[8] Notice that the expression $\omega(e^{\omega/T} - 1)^{-1}$ excludes the vacuum or zero-point energy, $\omega/2$. If the vacuum energy is kept in the energy balance, the problem with the divergence in the integrated noise power reappears. One may reason that the exclusion of the vacuum energy is justified because only physical *transitions* can participate in dissipative processes. For a more satisfactory picture, see below.

[9] Shot noise was first observed in vacuum tubes, i.e., devices where electrons are "shot" from some cathode through empty space.

$$S(\omega) \sim \omega^{-\alpha}, \qquad 0.5 < \alpha < 2,$$

as $1/f$-noise (since "f" is the commonly used variable for frequency in engineering). For further discussion of $1/f$-noise, we refer to the literature.[10]

11.2.3 Fokker–Planck equation

Individual solutions $\mathbf{q}(t) = \mathbf{q}[\boldsymbol{\xi}](t)$ of the Langevin equation depend on a given noise profile. They contain excessive microscopic information (about the details of $\boldsymbol{\xi}$), and are randomly fluctuating. However, it would be interesting to know the *probability* $p(\mathbf{v}, t)$ of observing velocity values at a given time. We expect this distribution to be a smooth function describing the interplay between drift, diffusion, dissipation, and fluctuation, which we saw is characteristic of Langevin dynamics.

Technically, $\mathbf{q}[\boldsymbol{\xi}]$ is a random variable depending on the random variable $\boldsymbol{\xi}$. The distribution $p(\mathbf{v}, t)$ is therefore an induced distribution, determined by the known (Gaussian) distribution of the noise, and by the dependence $\mathbf{q}[\boldsymbol{\xi}]$ expressed in the Langevin equation. More precisely, we wish to derive an evolution equation for the conditional probability $p(\mathbf{v}, t | \mathbf{v}_0, t_0)$ of observing a particle velocity $\mathbf{v}$ at time t given that we started with velocity $\mathbf{v}_0$ at time $t_0 < t$. The evolution of the variable $\mathbf{v}$ is controlled by the Langevin equation (11.3), a first-order differential equation in time. This means that the evolution $\mathbf{v}' \xrightarrow{\Delta t} \mathbf{v}$ from some initial configuration $\mathbf{v}'$ to $\mathbf{v}$ depends solely on the initial configuration $\mathbf{v}'$ and on the instantaneous value of $\boldsymbol{\xi}$. However, it does *not* depend on the history of the particle prior to reaching $\mathbf{v}'$. (Why is it essential here that the evolution equation has a first-order term?) This feature is expressed by the recursive relation (again, a statement to reflect upon),

$$p(\mathbf{v}, t | \mathbf{v}_0, t_0) = \int d^d v' \, p(\mathbf{v}, t | \mathbf{v}', t') p(\mathbf{v}', t' | \mathbf{v}_0, t_0), \qquad (11.13)$$

where $t' \in [t_0, t]$ is an arbitrary intermediate time. Summation over the product of probabilities $p(\mathbf{v}', t' | \mathbf{v}_0, t_0)$ to reach all possible intermediate values $\mathbf{v}'$, and over $p(\mathbf{v}, t | \mathbf{v}', t')$ to move on to the final value $\mathbf{v}$, yields the full probability $p(\mathbf{v}, t | \mathbf{v}_0, t_0)$. This convolution relation is the defining property of a **Markovian stochastic process**. (For a more systematic discussion, see section 11.4.2 below.)

Markovian
process

Evaluating Eq. (11.13) for the time arguments $t \to t + \delta t$ and $t' \to t$, and suppressing the initial data $(\mathbf{v}_0, t_0)$ for notational simplicity, we obtain an equation for the incremental evolution of probability,

$$p(\mathbf{v}, t + \delta t) = \int d^d u \, w(\mathbf{v} - \mathbf{u}, t; \mathbf{u}, \delta t) p(\mathbf{v} - \mathbf{u}, t),$$

where we define

$$w(\mathbf{v} - \mathbf{u}, t; \mathbf{u}, \delta t) \equiv p(\mathbf{v}, t + \delta t | \mathbf{v} - \mathbf{u}, t)$$

---

[10] For instance, see P. Dutta and P. M. Horn, *Low-frequency fluctuations in solids: 1/f-noise*, Rev. Mod. Phys. **53**, 497 (1981).

for the probability of short time transitions $(\mathbf{v} - \mathbf{u}, t) \to (\mathbf{v} + \mathbf{u}, t + \delta t)$. We next use that, for diffusive motion, the increment $\mathbf{u}$ during a sufficiently small time window, δt, is small in comparison to the length scales over which the probability distribution p is expected to change. This smallness invites a Taylor expansion in the argument $\mathbf{v} - \mathbf{u}$ around $\mathbf{u}$, i.e., an expansion in the underlined arguments in $w(\mathbf{v} - \underline{\mathbf{u}}, t; \mathbf{u}, \delta t) p(\mathbf{v} - \underline{\mathbf{u}}, t)$.

Kramers–
Moyal
expansion

INFO It is important to appreciate that the legitimacy of this so called **Kramers–Moyal expansion** hinges on the kinematics of the process. In cases where one-step transitions are over large distances (as is the case in, say, ballistic collisions in dilute gases), other procedures have to be applied. One of these alternative approaches, Boltzmann kinematic theory, is discussed in section (11.3) below.

Performing the expansion up to second order, we obtain the differential equation,

$$
\begin{aligned}
p(\mathbf{v}, t + \delta t) = {} & (\alpha^{(0)} p)(\mathbf{v}, t) \\
& - \frac{\partial}{\partial v_i}(\alpha_i^{(1)} p)(\mathbf{v}, t) + \frac{1}{2}\frac{\partial^2}{\partial v_i \partial v_j}(\alpha_{i,j}^{(2)} p)(\mathbf{v}, t),
\end{aligned} \tag{11.14}
$$

where

$$
\alpha^{(0)}(\mathbf{v}, t) = \int d^d u \, w(\mathbf{v}, t; \mathbf{u}, \delta t),
$$

$$
\alpha_i^{(1)}(\mathbf{v}, t) = \int d^d u \, w(\mathbf{v}, t; \mathbf{u}, \delta t) \, u_i,
$$

$$
\alpha_{i,j}^{(2)}(\mathbf{v}, t) = \int d^d u \, w(\mathbf{v}, t; \mathbf{u}, \delta t) \, u_i u_j.
$$

To compute the coefficients, w, we discretize the Langevin evolution for short times as

$$
\mathbf{v}(t + \delta t) = \mathbf{v} - \delta t \gamma \mathbf{v} + \frac{\delta t}{m}\boldsymbol{\xi}(t).
$$

This equation defines the random variable $\mathbf{v}(t + \delta t)$ in terms of the random variable $\boldsymbol{\xi}(t)$. From this information, the transition probability is obtained as[11]

$$
\begin{aligned}
w(\mathbf{v}, t; \mathbf{u}, \delta t) &= p(\mathbf{v} + \mathbf{u}, t + \delta t | \mathbf{v}, t) = \langle \delta(\mathbf{v} + \mathbf{u} - \mathbf{v}(t + \delta t)) \rangle_{\boldsymbol{\xi}} \\
&= \langle \delta(\mathbf{u} + \delta t \gamma \mathbf{v} - \delta t \boldsymbol{\xi}(t)/m) \rangle_{\boldsymbol{\xi}}.
\end{aligned}
$$

We next note that, in the time-discretized framework, the probability for the instantaneous noise variable is given by

$$
p(\boldsymbol{\xi}) = \mathcal{N}\exp\left(-\frac{\delta t}{2A}|\boldsymbol{\xi}|^2\right),
$$

where here, and in the following, $\mathcal{N}$ stands for an appropriate normalization factor.

---

[11] Here, the essential third equality uses the fact that the probability of obtaining a specific value $\mathbf{x} = \mathbf{v} + \mathbf{u}$ of a random variable $\mathbf{X} = \mathbf{v}(t + \delta t)$ that depends on another random variable $\mathbf{Y} = \boldsymbol{\xi}$ is obtained by averaging the corresponding δ-function $\langle \delta(\mathbf{x} - \mathbf{X}(\mathbf{Y})) \rangle_{\mathbf{Y}}$.

EXERCISE Write down the Gaussian functional distribution corresponding to the continuum Langevin noise Eq. (11.4); discretize it in time and verify the relation above.

From this distribution, we obtain

$$w(\mathbf{v}, t; \mathbf{u}, \delta t) = \int d^d \xi \, p(\boldsymbol{\xi}) \, \delta(\mathbf{u} + \delta t \gamma \mathbf{v} - \delta t \boldsymbol{\xi}/m) = \mathcal{N} \exp\left(-\frac{m^2}{2A\delta t}|\mathbf{u} + \gamma \delta t \mathbf{v}|^2\right),$$

where the normalization is defined through $\int d^d u \, w(\mathbf{v}, t; \mathbf{u}, \delta t) = 1$. The α-coefficients are now straightforwardly computed as

$$\alpha^{(0)}(\mathbf{v}, t) = 1, \qquad \alpha_i^{(1)}(\mathbf{v}, t) = -\gamma \delta t v_i, \qquad \alpha_{i,j}^{(2)}(\mathbf{v}, t) = \delta_{ij} \frac{2\delta t A}{m^2},$$

Fokker–Planck equation

where terms of $\mathcal{O}(\delta t^2)$ are neglected. Substitution of these results into Eq. (11.14), followed by first-order expansion in δt leads to the **Fokker–Planck equation**,[12]

$$\boxed{\left(\partial_t - \partial_{v_i} \gamma v_i - \partial^2_{v_i v_i} D_v\right) p(\mathbf{v}, t) = 0} \qquad (11.15)$$

where

$$D_v = \frac{A}{2m^2}$$

is the diffusion constant of *velocity*. In equilibrium, $D_v = T\gamma/m$, which is related to the particle diffusion constant (11.7) through the relation $D_v = \gamma^2 D$.

Equation (11.15) is a second-order partial differential equation for the probability distribution. It has to be solved with initial condition $p(\mathbf{v}, t = t_0) = \delta(\mathbf{v} - \mathbf{v}_0)$, i.e., Eq. (11.15) defines an initial-value problem (much like the time-dependent Schrödinger equation in quantum mechanics).

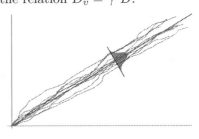

Derived from the Langevin equation with its interplay of dissipation and fluctuation, the Fokker–Planck equation describes diffusion and drift as the dominant transport mechanisms at large times. For example, the figure shows a number of simulated Langevin trajectories randomly fluctuating around a common average. We expect the corresponding distribution to diffusively expand around a drifting center, as indicated in the figure. In the following, we discuss in more detail how this evolution follows from the Fokker–Planck equation.

▷ The meaning of the **drift term** is best discussed in a framework generalized for the presence of a constant external driving force $\mathbf{F}$. Retracing the derivation of the Fokker–Planck equation, one verifies that[13] this generalization

---

[12] Anticipating later generalizations to the case of $\mathbf{v}$-dependent coefficients, v, D_v, we place the latter to the right of the differential operators ∂_{v_i}.

[13] External forcing may drive the system out of equilibrium. (Switching off the force will change the state of the system, a criterion for a nonequilibrium situation.) In this case, the fluctuation–dissipation relation $D_v = T\gamma/m$ need not hold.

amounts to a shift $\alpha_i^{(1)} \to \delta t(-\gamma v_i + m^{-1}F_i)$ of the drift coefficient. The Fokker–Planck equation thus generalizes to $\left(\partial_t - \partial_{v_i}(\gamma v_i - m^{-1}F) - \partial_{v_i v_i}^2 D_v\right) p(\mathbf{v}, t) = 0$. Multiplying this equation by v_i, and integrating over velocity, we obtain $d_t\langle \mathbf{v}\rangle + \gamma\langle \mathbf{v}\rangle - m^{-1}\mathbf{F} = 0$, a result otherwise found by averaging the Langevin equation over noise. This equation describes the relaxation of the velocity expectation value to a stationary drift configuration $\langle \mathbf{v}\rangle \to \frac{1}{m\gamma}\mathbf{F}$. Identifying that expectation value with the force-induced drift current, we obtain $\mathbf{j}_d = \mathbf{F}/m\gamma$, in agreement with the phenomenological reasoning above.

▷ The third term in Eq. (11.15) describes **diffusion**. The influence of the diffusion term on the dynamics is visualized in the figure above for the case of particles in two dimensions subject to a uniform driving force $\mathbf{F} = \text{const.} \times (2\mathbf{e}_x + \mathbf{e}_y)$. At large times, the distribution of the spread $p(\mathbf{v})$ becomes stationary. It is obtained by solution of the stationary long-time limit of the Fokker–Planck equation, $(\partial_{v_i}(\gamma v_i - m^{-1}F_i) + D_v\partial_{v_i v_i}^2)p(\mathbf{v}) = 0$, or

$$\left(\gamma v_i - m^{-1}F_i + D_v\partial_{v_i}\right)p(\mathbf{v}) = 0.$$

This is solved by $p(\mathbf{v}) \sim \exp[-\frac{\gamma}{2D_v}|\mathbf{v} - \frac{1}{m\gamma}\mathbf{F}|^2]$, which is a diffusion cloud centered around the drift trajectory. At $\mathbf{F} = 0$, the diffusion constant $D_v = T\gamma/m$ assumes its equilibrium value and $p(\mathbf{v}) = \exp[-\frac{m}{2T}|\mathbf{v}|^2]$ reduces to the **Maxwell–Boltzmann distribution**. We conclude that:

> In the absence of external driving, a combination of diffusion and drift sends the system into a state of thermal equilibrium.

EXERCISE The discussion above focused on the dynamics of the *velocity* distribution, $p(\mathbf{v}, t)$. To describe the **diffusive dynamics of the particle coordinates**, q, consider a force-free ($\mathbf{F} = 0$) Langevin equation in the overdamped limit, $m\gamma d_t\mathbf{q} = \boldsymbol{\xi}$, i.e., a limit void of external forces where the ballistic acceleration $\sim md_t^2\mathbf{q}$ is negligible. Show that the Fokker–Planck equation for $p(\mathbf{q}, t)$ takes the form

$$(\partial_t - D\partial_q^2)p(\mathbf{q}, t) = 0, \tag{11.16}$$

where D is given by Eq. (11.7). The equation confirms the expectation that, in thermal equilibrium, the coordinate diffuses, and the level of diffusivity increases with temperature.

EXERCISE Consider single-particle dynamics governed by a Hamiltonian $H(\mathbf{q}, \mathbf{p}) \equiv H(\mathbf{x})$. Describing the influence of an environment by a combination of dissipation and fluctuation as before, we consider the generalized phase space Langevin equations, $d_t\mathbf{q} = \frac{\partial H}{\partial \mathbf{p}}$, $d_t\mathbf{p} = -\frac{\partial H}{\partial \mathbf{q}} - \gamma\mathbf{p} + \boldsymbol{\xi}$. Show that the probability distribution $p(\mathbf{x}, t)$ is governed by the **phase space Fokker–Planck equation**,

$$\boxed{d_t p(\mathbf{x}, t) = -\{H, p(\mathbf{x}, t)\} + \frac{\partial}{\partial \mathbf{p}}\left(\gamma\mathbf{p} + D_p\frac{\partial}{\partial \mathbf{p}}\right)p(\mathbf{x}, t)} \tag{11.17}$$

where $D_p = m^2 D_v = T\gamma m$ is the diffusion constant of momentum. The second operator on the right-hand side describes the dissipation–fluctuation generalization of the deterministic single-particle dynamics (see Eq. (11.2)).

Finally, show that the stationary limit of the solution is given by the phase space Maxwell–Boltzmann distribution $p(\mathbf{x}) = \mathcal{N}\exp\left(-H(\mathbf{x})/T\right)$.

To summarize, we have seen how the Fokker–Planck equation describes the universal aspects of drift–diffusion dynamics, microscopically described by the Langevin equation. The derivation relied on two crucial features of the micro-evolution: its Markovianness (what happens in the future does not depend on the history of the evolution) and on the smallness of coordinate updates in short instances of time. In section 11.4, we will discuss other "stochastic processes" satisfying these two criteria and show that they too afford a description in terms of the Fokker–Planck equations. However, before turning to this generalization, let us discuss what can be done if the second of the above criteria is violated.

11.3 Boltzmann Kinetic Theory

REMARK In this section, we introduce the Boltzmann equation as an alternative to the Fokker–Planck approach in cases where stochastic variables do not locally "diffuse." The inclusion of this section pays tribute to the importance of Boltzmann kinematics in condensed matter applications. Readers interested primarily in the field-theoretical approach to nonequilibrium dynamics may skip it at first reading.

As mentioned above, a crucial condition in the derivation of the Fokker–Planck equation by the Kramers–Moyal expansion is the smallness of coordinate updates in short instances of time. (More precisely, we assumed that $\langle (\mathbf{x}(t+\delta t) - \mathbf{x}(t))^n \rangle \lesssim \delta t$, for $n = 1, 2$.) Boltzmann kinematic theory describes situations where this condition may be violated.

11.3.1 Derivation of the Boltzmann equation

By way of an example, consider a gaseous system of particles in a container. In the infinitely dilute limit (where particle collisions can be neglected), the phase space coordinates of individual particles change according to Newtonian dynamics, Eq. (11.2). Specifically, the single-particle energy ϵ is conserved, and (assuming the absence of smooth potential gradients) momentum is constrained to an energy shell $|\mathbf{p}|^2/2m = \epsilon$.

However, once interactions are taken into account, the rate at which the distribution $f(\mathbf{x}_1, t)$ changes becomes influenced by particle collisions: losses occur when particles at phase space point $\mathbf{x}_2$ scatter off particles at $\mathbf{x}_1$ into final states $\mathbf{x}_1'$ and $\mathbf{x}_2'$ (see figure). We denote the corresponding two-particle transition rate by $w(\mathbf{x}_1', \mathbf{x}_2'; \mathbf{x}_1, \mathbf{x}_2)\, d^d x_1' d^d x_2' d^d x_1 d^d x_2$, where the function w encapsulates the kinematic constraints (energy and momentum conservation) of the collision. The derivative $d_t f$ due to "out" processes then reads

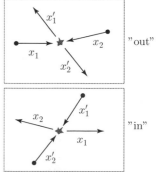

$$d_t f_1 \big|_{\text{out}} = \int d2\, d1'\, d2'\; w(1', 2'; 1, 2) f_1 f_2,$$

where we take into account that the total transition rate depends linearly on the number of available collision partners in the two initial states, and introduce a (standard) notation abbreviating coordinates by numbers, $d^d x_1 \to d1$, $f(\mathbf{x}_1) \to f_1$, etc.

Gain occurs when particles get scattered into state 1 in collisions $(1', 2') \to (1, 2)$:

$$d_t f_1 \big|_{\text{in}} = \int d2 \, d1' \, d2' \; w(1, 2; 1', 2') f_{1'} f_{2'}.$$

collision integral

The total occupation change due to collisions, $I[f] \equiv d_t f_1 \big|_{\text{in}} - d_t f_1 \big|_{\text{out}}$, is then described by the so-called **collision integral**,

$$I[f] = \int d2 \, d1' \, d2' \; \left(w(1, 2; 1', 2') f_{1'} f_{2'} - w(1', 2'; 1, 2) f_1 f_2 \right). \tag{11.18}$$

Notice that the collision integral is a nonlinear functional of the distribution functions, i.e., we are looking at a nonlinear theory.

micro-reversibility

The transition rates w are subject to symmetry relations that reflect the time-reversal invariance and unitarity of the microscopic laws of physics. For example, **microreversibility** implies that a transition $(1, 2) \to (1', 2')$ must be as probable as the time-reversed process $(1'^T, 2'^T) \to (1^T, 2^T)$, where $1^T \equiv (\mathbf{q}_1, \mathbf{p}_1)^T = (\mathbf{q}_1, -\mathbf{p}_1)$ is the time-reverse of $(\mathbf{q}_1, \mathbf{p}_1)$, i.e., $w(1', 2'; 1, 2) = w(1^T, 2^T; 1'^T, 2'^T)$. The consequences of unitarity are best exposed if we interpret the coefficients w as classical limits of quantum transition probabilities. Identifying $(1, 2) \leftrightarrow |i\rangle$ as a classical approximation to a (coherent) state in two-particle Hilbert space, and $(1', 2') \leftrightarrow |f\rangle$ as a final state, we have the identification

$$w(1', 2'; 1, 2) \leftrightarrow |S_{\text{fi}}|^2,$$

where S_{fi} is the scattering matrix. Unitarity means that $\sum_f |S_{\text{fi}}|^2 = \sum_f |S_{\text{if}}|^2 = 1$, and in particular $\sum_{f \neq i} |S_{\text{fi}}|^2 = \sum_{f \neq i} |S_{\text{if}}|^2$, where we exclude the "forward scattering" configuration $i = f$, which does not change the state. With the identification $\int d1' d2' \leftrightarrow \sum_{f \neq i}$, this implies

$$\int d1' d2' \, w(1', 2'; 1, 2) = \int d1' d2' \, w(1, 2; 1', 2'). \tag{11.19}$$

collision integral

Using this relation in the out-process, the **collision integral** can be transformed to

$$\boxed{I[f] = \int d2 \, d1' \, d2' \; w(1, 2; 1', 2') \left(f_{1'} f_{2'} - f_1 f_2 \right)} \tag{11.20}$$

Boltzmann equation

Adding $I[f] \equiv d_t f_1 \big|_{\text{in}} - d_t f_1 \big|_{\text{out}}$ to the right-hand side of Eq. (11.2), we obtain the **Boltzmann equation**[14]

$$\boxed{\left(\partial_t - \{H, \, . \} \right) f(\mathbf{x}, t) = I[f]} \tag{11.21}$$

---

[14] In the Russian literature, the Boltzmann equation is usually called the "**kinetic equation**."

11.3.2 Discussion of the Boltzmann equation

The Boltzmann equation (11.21) is one of the most important equations of statistical physics. First, it has been, and still is, a powerful tool in applied many-particle physics. Before the advent of the more modern techniques reviewed in previous chapters, it was the most important computational workhorse in many-particle physics.[15] Even today, Boltzmann equation-based approaches often represent the most economic and straightforward route to understanding the physics of interacting many-particle systems.

Second, the Boltzmann equation is of conceptual value. Various of the principal questions raised in the beginning of the chapter afford a relatively straightforward answer in terms of this equation. For example, note (cf. the related discussion before Eq. (11.1)) that the collision term vanishes in thermal equilibrium. In this case, the distribution function is given by the Maxwell–Boltzmann distribution (11.1). Energy conservation in elastic collisions, $H(\mathbf{x}_1) + H(\mathbf{x}_2) = H(\mathbf{x}'_1) + H(\mathbf{x}'_2)$, means that $f_1 f_2 = f_{1'} f_{2'}$, i.e., the vanishing of $I[f]$. In equilibrium, the losses and gains due to many-particle collisions compensate each other and $I[f]$ does not change the distribution. (Since $f(\mathbf{x}) = f(H(\mathbf{x}))$, the single-particle dynamics also conserves f, $\{H, f\} = 0$, i.e., the Maxwell–Boltzmann distribution is stationary under (11.21).)

The Boltzmann H-theorem

We expect the collision term to drive the system towards a thermal equilibrium configuration. Indeed, it does; however, it is not entirely straightforward to show this. The defining property of the thermal equilibrium is that it maximizes entropy, S. In equilibrium statistical mechanics, it is usually taken for granted (according to the second law of thermodynamics) that many-body relaxation processes increase the entropy, $d_t S \geq 0$, before the maximum of the equilibrium configuration is reached. However, the present theory should actually be able to demonstrate how a microreversible theory leads to macroscopic irreversibility and entropy increase.

H-theorem   The manner in which interactions increase entropy was demonstrated by Boltzmann in a famous construction known as the H-**theorem**. (It is called the H-theorem rather than the S-theorem because Boltzmann called entropy H.) Using the shorthand notation $\int_\Gamma d^d x \equiv \int_\Gamma \equiv \int$, consider the definition of entropy

$$S = -\int f \ln(f/e),$$

which differs from the information entropy $S = -\int p \ln p$ of the distribution $f = p$ only by the presence of the constant e. A change in S is then given by

$$d_t S = -\int \ln f \, \partial_t f = -\int \ln f \left(\{H, f\} + I[f] \right).$$

---

[15] For a detailed account of applications and solution strategies related to the Boltzmann equation, see L. D. Landau and E. M. Lifshitz, *Course of Theoretical Physics, Vol. 10 – Physical Kinetics* (Butterworth–Heinemann, 1981).

Since the full phase space integral of any function is invariant under Hamiltonian flow (exercise: why?), we have $0 = \int \{H, f \ln(f/e)\} = \int \ln f \{H, f\}$, i.e., changes in entropy are due to interactions.

To explore how interactions do the job, consider the collision integral Eq. (11.18). For an arbitrary function $\phi(\mathbf{x})$, and using the abbreviation $d\Gamma \equiv d1\, d2\, d1'\, d2'$, we have

$$
\int d1 \phi(1) I[f(1)] = \int d\Gamma\, \phi(1)\, (w(1,2;1',2') f_{1'} f_{2'} - w(1',2';1,2) f_1 f_2)
$$

$$
= \int d\Gamma\, (\phi(1) - \phi(1')) w(1,2;1',2') f_{1'} f_{2'} \qquad (11.22)
$$

$$
= \int \frac{d\Gamma}{2}\, (\phi(1) + \phi(2) - \phi(1') - \phi(2')) w(1,2;1',2') f_{1'} f_{2'},
$$

where, in the first equality, we relabeled coordinates, $(1,2) \leftrightarrow (1',2')$, and the second equation is based on the micro-reversibility principle (11.19). From this result, we can derive a few auxiliary identities. Setting $\phi = 1$, we conclude that the collision term vanishes upon integration,

$$
0 = \int d1\, I[f] = \int d\Gamma\, w(1,2;1',2')(f_{1'} f_{2'} - f_1 f_2),
$$

where the representation (11.20) was used. Introducing the abbreviations

$$
\Lambda \equiv \frac{f_{1'} f_{2'}}{f_1 f_2}, \qquad X \equiv w(1,2;1',2') f_1 f_2,
$$

this can be written as $0 = \int d\Gamma\, X(\Lambda - 1)$. Setting $\phi = \ln f$, Eq. (11.22) yields $d_t S = -\int d1\, \ln(f_1) I[f(1)] = \frac{1}{2} \int d\Gamma\, X\Lambda \ln \Lambda$. The combination of these results finally leads to

$$
d_t S = \frac{1}{2} \int d\Gamma\, X(\Lambda \ln \Lambda - \Lambda + 1).
$$

Now, the functions X and Λ are manifestly positive. For positive Λ, the combination $\Lambda \ln \Lambda - \Lambda + 1$ is also positive (exercise). This construction demonstrates that elastic particle collisions do indeed increase the entropy of the system. The asymptotic state of the system must be the thermal equilibrium state, for which we saw the collision term vanishes, and the entropy becomes stationary.

Mesoscopic evolution laws

The arguments used in the derivation of the H-theorem are useful for deriving evolution equations describing the flow of particle currents, densities, and other observables on "mesoscopic" scales larger than the collision mean free path $l = \tau v$ (where v is a typical particle velocity) yet smaller than macroscopic scales. Such equations must reflect the conservation laws of a system, notably the conservation of energy, particle number, and momentum. Interactions may change the energy or momentum of individual particles, but they will not change the total energy and momentum of a sufficiently large assembly of particles. To be more concrete, let $\delta\Gamma \equiv \delta V \oplus \mathbb{R}^3$ be a thin slice in phase space containing phase space points in the small volume element δV at coordinate $\mathbf{q}$ and of arbitrary momentum $\mathbf{p} \in \mathbb{R}^3$. The particle, energy, and momentum density averaged over the volume are given by

$$\left.\begin{array}{c} \rho(\mathbf{q},t) \\ \epsilon(\mathbf{q},t) \\ \boldsymbol{\pi}(\mathbf{q},t) \end{array}\right\} = \frac{1}{\delta V} \int_{\delta\Gamma} f(\mathbf{x},t) \left\{\begin{array}{l} 1, \\ H(\mathbf{x}), \\ \mathbf{p}. \end{array}\right.$$

Let us prove that interactions do not change these values. To this end, we assume that the transition coefficients $w(1',1';1,2)$ represent an elastic point interaction, i.e., an interaction local in space that respects energy, momentum, and particle number conservation. In this case (think why), Eq. (11.22) applies to individual phase space cells $\delta\Gamma$,

$$\int_{\delta\Gamma} d1\, \phi(1)I[f(1)] = \frac{1}{2}\int_{1\in\delta\Gamma} d1\, (\phi(1) + \phi(2) - \phi(1') - \phi(2'))w(1',2';1,2)f_{1'}f_{2'}.$$

For the specific choices $\phi = 1, \epsilon, \mathbf{p}$ (corresponding to density, energy, and momentum), the linear combinations $\phi(1) + \phi(2) - \phi(1') - \phi(2')$ on the right-hand side vanish. For example, $\epsilon(1) + \epsilon(2) = \epsilon(1') + \epsilon(2')$ for an energy-conserving interaction $(1,2) \longrightarrow (1',2')$. We thus conclude that

$$\int_{\delta\Gamma} d1\, I[f(1)] \left\{\begin{array}{l} 1 \\ H(\mathbf{x}) \\ \mathbf{p} \end{array}\right. = 0\,.$$

This identity is the key to the derivation of effective transport equations. For example, changes in the particle density are obtained as $\partial_t \rho \delta V = \partial_t \int_{\delta\Gamma} f = \int_{\delta\Gamma} \{H, f\} + \int_{\delta\Gamma} I[f]$. We see that the second term on the right-hand side vanishes, and the first evaluates to

$$\int_{\delta\Gamma} \{H, f\} = \int_{\delta\Gamma} (\partial_{q_\alpha} H \partial_{p_\alpha} f - \partial_{p_\alpha} H \partial_{q_\alpha} f)$$

$$\simeq \partial_{q_\alpha} H \int_{\delta\Gamma} \partial_{p_\alpha} f - \int_{\delta\Gamma} v_\alpha \partial_{q_\alpha} f \simeq 0 - \partial_{q_\alpha} j_\alpha,$$

current density

where we note that $0 = \int_{\delta\Gamma} \partial_{p_\alpha} f$ is a full derivative and $\mathbf{j} \equiv \delta V^{-1} \int_{\delta\Gamma} \mathbf{v} f$ defines the **current density**. We thus obtain the continuity equation,

$$\partial_t \rho + \nabla \cdot \mathbf{j} = 0.$$

Although this result may look trivial, it has been derived for arbitrary distribution functions, and under reasonably general assumptions on the microscopic interactions.

EXERCISE Show that, for a translationally invariant system, the energy and momentum density obey the conservation laws

$$\partial_t \epsilon + \nabla \cdot \mathbf{j}_\epsilon = 0, \qquad \partial_t \pi_a + \partial_{q_\beta} \Pi_{\alpha\beta} = 0, \tag{11.23}$$

where $\mathbf{j}_\epsilon \equiv \frac{1}{\delta V} \int_{\delta\Gamma} f H \mathbf{v}$ and $\Pi_{\alpha\beta} \equiv \frac{1}{\delta V} \int_{\delta\Gamma} f m v_\alpha v_\beta$ are the energy current density and the momentum current tensor respectively.

Above, we showed why the thermal distribution $f_0 \sim \exp(-\beta\epsilon)$ annihilates the collision term. Use the same reasoning to demonstrate that this vanishing extends to all distributions exponential in conserved quantities,

$$f_0 \sim \exp\left(-\beta(\epsilon(\mathbf{x}) - \mathbf{v} \cdot \mathbf{p} - \mu)\right), \tag{11.24}$$

with free coefficients $\beta, \mathbf{v}, \mu$.

To summarize, our discussion has shown how particle number, energy, and momentum conservation are hardwired into the Boltzmann theory. Other conserved quantities – for example, spin – may be included in the theory, if required. Once established, these conservation laws must be respected by approximate solutions, which can constitute a strong consistency check.

Beyond equilibrium: Boltzmann transport theory

While the Boltzmann equation predicts the thermalization of isolated systems, it is also a powerful tool for the description of nonequilibrium phenomena. Referring to Ref.[15] for a more detailed discussion, here we mention a few basics of Boltzmann theory beyond equilibrium.

We reasoned above that the collision term is annihilated by distributions (11.24) exponential in the conserved quantities. All parameters entering the distribution may vary in space and time. For example, $T = T(\mathbf{r}, t)$ may be a nontrivial temperature profile, in which case f_0 describes the *local* equilibrium, with spatio-temporal variation.

Noting that externally applied perturbations are usually weak compared with the internal forces, it is often sufficient to consider a system weakly perturbed out of equilibrium, $f = f_0 + \delta f$. We know that the collision term will act to restore equilibrium and suppress δf. This relaxation can be described phenomenologically via a simple **relaxation time approximation**,

relaxation
time
approx-
imation

$$I[f] \longrightarrow -\frac{f - f_0}{\tau}, \tag{11.25}$$

where τ is of the order of the microscopic scattering times in the system.

As an example, consider a system of charged particles subject to an **external electric field**, $\mathbf{E}(t)$, harmonically varying in time with frequency ω. Setting the particle charge to unity, the field coupling is described by the single-particle potential $-\mathbf{E}_0 \cdot \mathbf{q} \exp(-i\omega t)$. We expect this perturbation to cause local time-dependent departures from equilibrium, but no macroscopic variations in the equilibrium parameters $T, \mu, \mathbf{v}$ themselves. This suggests entering the equation with an ansatz $f(\mathbf{x}, t) = f_0(\epsilon(\mathbf{p})) + \delta f(t)$. Using the identity

$$\{H, f\} = \partial_\mathbf{q} H \cdot \partial_\mathbf{p} f - \partial_\mathbf{p} H \cdot \partial_\mathbf{q} f = -e^{-i\omega t} \mathbf{E}_0 \cdot \mathbf{v} \, \partial_\epsilon f_0,$$

the relaxation time approximation to Eq. (11.21) becomes $\partial_t \delta f + e^{-i\omega t}\mathbf{E} \cdot \mathbf{v} \, \partial_\epsilon f_0 = -\frac{\delta f}{\tau}$, which is readily solved to give

$$\delta f(\mathbf{x}, t) = \frac{\tau \mathbf{E}_0 \cdot \mathbf{v} \partial_\epsilon f_0}{1 - i\omega\tau} e^{-i\omega t}.$$

(Exercise: Interpreting $\mathbf{v} \cdot \mathbf{E}$ as the rate at which the electric field does work on a phase space volume at $(\mathbf{q}, \mathbf{p})$, discuss the physics of this result, both for high $(\omega \gg \tau^{-1})$ and low $(\omega \ll \tau^{-1})$ frequencies.) From here, the average current density is obtained as $\mathbf{j}(\mathbf{q}, t) = \int_{\delta\Gamma} \delta f(\mathbf{x}, t)\mathbf{v}$. Defining the longitudinal **AC conductivity** as $\mathbf{j}(\omega) = \sigma(\omega)\mathbf{E}(\omega)$ and, assuming a Maxwell–Boltzmann distribution $f_0(\mathbf{x}) = Z^{-1} \exp(-p^2/2mT)$, a straightforward calculation (exercise) leads to

$$\sigma(\omega) = \frac{n\tau}{m} \frac{1}{1 - i\omega\tau}.$$

One may reason that the Boltzmann machinery is a bit of an overkill for a result otherwise obtainable from simple Drude transport theory. However, unlike Drude theory, the present approach can be straightforwardly extended to the presence of external magnetic fields, disorder, the coupling to optical or phonon modes, etc. At the same time, it is simpler than fully microscopic theories. In this way, Boltzmann theory represents a favorable trade-off between microscopic explicitness and simplicity. In the next chapter we will see how elements of this theory are reflected in quantum field-theoretical approaches to nonequilibrium physics.

11.4 Stochastic Processes

In the previous sections, we have met with various examples of stochastic processes. Stochastic processes are frequently realized not just in physics but also in the life sciences, engineering, the socio–economic sciences, and others. In this section, we introduce this highly versatile concept as a generalization of the Langevin process considered previously. This discussion will provide the basis for our later field-theoretical formulation of nonequilibrium dynamics.

11.4.1 The notion of a stochastic process

A process describes the temporal evolution of a certain state. It is represented by a sequence $\{a(t_i)\}$, where $t_i, i = 1, \ldots, n$ are the discrete times at which the states are recorded and $a(t_i)$ is a (generally multi-component) state variable, for example $a(t_i) = \mathbf{q}(t_i)$ for the state vectors representing a stochastic random walk in d-dimensions. To keep the notation simple, we often write $a_i \equiv a(t_i)$ for the

stochastic process ith recording. A **stochastic process**[16] is one where the evolution $a_i \to a_{i+1}$ involves elements of randomness. The task then is to identify a probability measure $p(a_n, a_{n-1}, \ldots, a_1)$ for the realization of individual sequences $\{a_i\}$. It is useful to think of the full probability as a product

$$p(a_n, a_{n-1}, \cdots, a_1) = p(a_n|a_{n-1}, \cdots, a_1)p(a_{n-1}|a_{n-2}, \cdots, a_1) \ldots p(a_2|a_1)p(a_1),$$

---

[16] A standard reference covering theory and application of stochastic processes is N. G. van Kampen, *Stochastic Processes in Physics and Chemistry* (Elsevier, 1992).

where the individual factors measuring the probability of the next step depend conditionally on the history of previous events.

Before moving on, it is worthwhile to distinguish between a few categories of stochastic processes. A **stationary stochastic process** is one where p does not change under a uniform shift $t_i \to t_i + t_0$ of all time arguments and the theory is translationally invariant on average. The process is called **purely random** if the probabilities $p(a_j|a_{j-1}, \ldots, a_1) = p(a_j)$ are independent of the history of events. In this case,

$$p(a_n, \ldots, a_1) = \prod_{j=1}^{n} p(a_j)$$

is the product of n uncorrelated random numbers; evidently, purely random processes are not very interesting. However, next in the hierarchy of complexity are conditional probabilities, for which

$$p(a_n|a_{n-1}, \ldots, a_1) = p(a_n|a_{n-1}). \tag{11.26}$$

Markov process

Processes of this type are called **Markov process**. Markov processes have a vast spectrum of applications in statistical sciences, and deserve a separate discussion.

11.4.2 Markov processes

In a Markov process, the passage to the state a_n depends on the previous state a_{n-1}, but not on the earlier history $(a_{n-2}, \ldots)$. Put differently, a Markov process lacks memory. Few stochastic processes are rigorously Markovian. However, Markovian approximations often capture the essential contents of a stochastic evolution and at the same time are numerically or analytically tractable. The reduction of a stochastic process to a Markovian one often amounts to the choice of an update time window $\Delta t = t_{n+1} - t_n$ that is long enough to eliminate the short-time memory of a process, yet short enough not to miss relevant aspects of the dynamics (see Ref.[16] for a more detailed discussion).

Whether a process assumes a Markovian form also depends on the choice of coordinates. To see this, consider a Brownian motion represented in coordinate space as $q_1 \to q_2 \to \cdots \to q_{n-2} \to q_{n-1} \to q_n$, where the coordinates are recorded at equal time steps. If the spacing $|q_{n-1} - q_{n-2}|$ is exceptionally large, the particle will be moving at high instantaneous velocity. It is then likely that the next step $q_{n-1} \to q_n$ will also be large. The conditional probability $p(q_n|q_{n-1}, q_{n-2}, \ldots) \neq p(q_n|q_{n-1})$ thus does not reduce to a Markovian one. (Exercise: Discuss why the particle velocity *can* be assumed to be Markovian.) For a given process one will naturally try to identify coordinates making it as "Markovian" as possible.

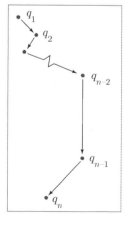

Chapman–Kolmogorov relation and master equation

The Markovian nature of a process can be expressed in a manner alternative to Eq. (11.26). Consider the joint probability of observing a at time t and a' at an earlier time t', given initial data a_i at t_i, $p(a, a'|a_i)$. From this probability, the full probability of a at t, $p(a|a_i)$ is obtained by integration over intermediate events,

$$p(a|a_i) = \int da' \, p(a, a'|a_i) = \int da' \, p(a|a', a_i)p(a'|a_i),$$

Chapman–
Kolmogorov
relation

where we have used Eq. (A.22). For a Markovian process, $p(a|a', a_i) = p(a|a')$, this convolution reduces to the **Chapman–Kolmogorov relation**

$$\boxed{p(a|a_i) = \int da' \, p_{\delta t}(a|a')p(a'|a_i)} \tag{11.27}$$

where the notation $p_{\delta t}(a|a')$ indicates the dependence of the conditional probability on the time difference δt between the observation of a at t and a' at $t' \equiv t - \delta t$.

The usefulness of Eq. (11.27) becomes evident if we consider the case $t - t' = \delta t \ll t - t_i$. Equation (11.27) then factorizes the description of the process into three blocks: (a) the probability of getting to a given an initial configuration a_i, (b) the probability of reaching an intermediate stage a' shortly before the final time, t, and (c) the short-time transition probability $a' \xrightarrow{\delta t} a$ (see the figure). This iterative, or "transfer matrix," description of the process is fundamental to most theories of Markovian dynamics.

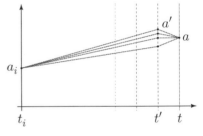

To develop the description further, assume δt to be very small. To a very good approximation, $p_{\delta t}(a|a') \simeq \delta(a - a') + \mathcal{O}(\delta t)$ will then be stationary. The term of $\mathcal{O}(\delta t)$ contains losses out of a' due to transition into another state a'' and gains due to transition from a'. Denoting the probability of transitions $a_1 \xrightarrow{\delta t} a_2$ by $W(a_2|a_1)\delta t$, we may thus write

$$p_{\delta t}(a|a') = \left(1 - \delta t \int da'' \, W(a''|a')\right) \delta(a - a') + \delta t \, W(a|a').$$

Substituting this expression into Eq. (11.27), taking the limit $\delta t \to 0$, and suppressing the initial time argument for simplicity $p(a|a_i) = p(a) \equiv p(a, t)$, we obtain the

master
equation

master equation[17] (see fig. 11.1),

---

[17] The term "master equation" was coined in A. Nordsieck, W. E. Lamb, and G. E. Uhlenbeck, *On the theory of cosmic-ray showers I: the Furry model and the fluctuation problem*, Physica **7**, 344 (1940), where a rate equation similar to Eq. (11.28) appeared as a fundamental equation of the theory.

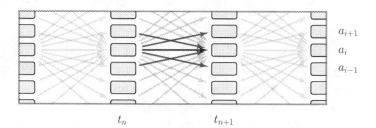

t_n $\quad\quad\quad\quad\quad$ t_{n+1}

Fig. 11.1   The evolution of Markovian probability, as described by the master equation: the probability gain of states a_i (here considered as discrete entities, while the text uses continuum notation) due to in-processes is counteracted by loss due to out-processes. The sum over all in- and out-contributions cancels, reflecting probability conservation.

$$\partial_t p(a,t) = \int da'\ [W(a|a')p(a',t) - W(a'|a)p(a,t)] \qquad (11.28)$$

Notice that the master equation conserves the total probability, i.e., $\partial_t \int da\ p(a,t) = 0$.

INFO   There is not much that can be said in generality about the master equation. However, at large time scales, solutions of the equation often relax to a stationary distribution function $p_{\mathrm{eq}}(a)$. In this limit, we have $0 = \int da'\ [W(a|a')p_{\mathrm{eq}}(a') - W(a'|a)p_{\mathrm{eq}}(a)]$, an equation that can be read as an effective sum rule obeyed by the transition matrix elements $W(a|a')$. In many cases, this balance equation holds not only in an integral sense but also locally for (a, a'), i.e., $W(a|a')p_{\mathrm{eq}}(a') = W(a'|a)p_{\mathrm{eq}}(a)$, or

$$\frac{W(a|a')}{W(a'|a)} = \frac{p_{\mathrm{eq}}(a)}{p_{\mathrm{eq}}(a')} \qquad (11.29)$$

detailed
balance
   Equation (11.29) is called a **detailed balance relation**. If detailed balance holds, knowledge of the stationary distribution contains information on the transition rates, and vice versa. It can be shown[16] that closed, isolated, classical systems obey this principle. Under more general circumstances, it is often not easy to decide whether or not a system obeys detailed balance.

Gaussian process

random
walk
   Consider a one-dimensional **random walk**, a process where a particle performs random motion on a one-dimensional lattice (see the figure) with N sites. Setting the lattice spacing to unity, the state variable $a \equiv n \in \mathbb{Z}$ takes integer values, and the goal is to compute the probability $p(n, t|n_0, t_0)$. Defining the probability density for an individual left-turn (right-turn) as $q\,dt$ $((1-q)\,dt)$, the master equation assumes the form

$$\partial_t p(n, t) = q p(n+1, t) + (1-q)p(n-1, t) - p(n, t).$$

Anticipating smoothness of the probability distribution, we convert this discrete equation into a differential one: define the scaling variable $x = n/N \in [0,1]$ and a rescaled probability distribution through $\tilde{p}(x)dx = p(Nx)dn = p(Nx)\frac{dn}{dx}dx$, which means $\tilde{p}(x) = Np(Nx)$. We may then Taylor expand as $\tilde{p}(x \pm 1/N) = \tilde{p}(x) \pm$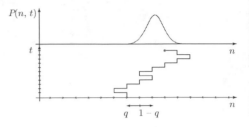
$N^{-1}\partial_x\tilde{p}(x) + (1/2)N^{-2}\partial_x^2\tilde{p}(x)$, and the master equation assumes the form of a drift–diffusion equation for the rescaled function,

$$\left(\partial_t + \gamma\partial_x - D\partial_x^2\right)\tilde{p}(x,t) = 0, \qquad \tilde{p}(x,t_0) = \delta(x - x_0),$$

with drift coefficient $\gamma = (1 - 2q)/N$, diffusion constant $D = 1/(2N^2)$, and initial point $x_0 = n_0/N$. This equation is solved by

$$\tilde{p}(x,t) = \int \frac{dk}{2\pi} e^{-(i\gamma k + Dk^2)t} e^{ik(x-x_0)} = \frac{1}{2(\pi Dt)^{1/2}} \exp\left(-\frac{(x - x_0 - \gamma t)^2}{4Dt}\right),$$

or, in terms of the original variables,

$$p(n,t) = \frac{1}{(2\pi t)^{1/2}} \exp\left(-\frac{(n - n_0 - (1-2q)t)^2}{2t}\right).$$

(Exercise: Consider how the structurally similar Schrödinger equation is solved and discuss the above solution in the light of this analogy.)

The Gaussian form of $p(n,t)$ is a manifestation of the central limit theorem: the variable n is an additive quantity obtained by summing t elementary random variables ± 1 drawn from a bimodal distribution with probability $p(1) = 1 - q$ and $p(-1) = q$. According to the central limit theorem, a large number of additions will result in a Gaussian-distributed variable n centered around the mean, $n_0 + (1 - 2q)t$, with variance $\sim \sqrt{t}$, as stated by the above result. Random processes whose distributions are Gaussian are generally called **Gaussian processes**.

Gaussian process

Example: Poisson process

Now consider a succession of sparse elementary events uncorrelated in time (apples falling off a tree, a Geiger counter exposed to weak radiation, etc.). The random variable of interest is the number, n, of events occurring in a certain time t (see the figure below). Calling the probability for an event to occur in a short time window νdt, the master equation takes the form

$$\partial_t p(n,t) = \nu p(n-1,t) - \nu p(n,t). \tag{11.30}$$

For definiteness, we impose the initial condition $p(n, t=0) = \delta_{n,n_0}$.

This equation may again be solved by assuming smoothness and using a Taylor expansion. However, it turns out that this approach is too crude to resolve the statistics of this process. (Exercise: Approximate the right-hand side of the equation by a first-order Taylor expansion, obtain the solution, and compare with the results below.) Instead, we build on the similarity of the master equation to the Schrödinger equation and employ concepts from quantum mechanics to construct a more accurate solution.

We start by rewriting the equation as $\partial_t p(n,t) = \nu(\hat{E}_{-1} - 1)p(n,t)$, where $\hat{E}_m$ is the translation operator in discrete n-space, $\hat{E}_m f(n) \equiv f(n+m)$. Thinking of n as the discrete eigenvalues of a coordinate operator $\hat{n}$, the translation operator affords a representation in terms of the conjugate operator $\hat{\phi}$, defined by the commu-

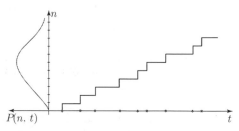

tation relation $[\hat{\phi}, \hat{n}] = -i$. In the ϕ-representation, E_{-1} is diagonal with eigenvalues $\hat{E}_{-1}(\phi) = \exp(-i\phi)$ and eigenfunctions $f_\phi(n) = (2\pi)^{-1/2}\exp(i\phi n)$. With the spectral decomposition of the initial configuration $p(n,0) = \delta_{n,n_0} = \int_0^{2\pi} \frac{d\phi}{2\pi} e^{-i\phi(n-n_0)}$, the distribution $p(n,t)$ is obtained in analogy to the solution of a time-dependent Schrödinger problem,

$$p(n,t) = \int_0^{2\pi} \frac{d\phi}{2\pi} e^{-\nu t(e^{-i\phi}-1)} e^{i\phi(n-n_0)}.$$

Using the identity $\int_0^{2\pi} \frac{dk}{2\pi} e^{i\Delta nk} = \delta_{\Delta n,0}$, a straightforward Taylor expansion of the exponent leads to the result (exercise)

$$p(n,t) = \frac{(\nu t)^{n-n_0}}{(n-n_0)!} e^{-\nu t}. \tag{11.31}$$

Comparing with Eq. (A.28), we note that the process is described by a Poisson distribution, where $\nu t = \nu dt(t/dt) \leftrightarrow np$ is interpreted as the product of $t/dt \leftrightarrow n$ attempts with individual probability $\nu dt \leftrightarrow p$. This explains the name **Poisson process**. In the present notation, the first and second cumulants of the Poisson distribution are given by $\mu_1 = \mu_2 = \nu t$: centered around $\mu_1 = \nu t$, the width of the distribution is given by $\sim (\nu t)^{1/2}$.

Poisson process

Kangaroo process

EXERCISE Kangaroo process: Consider the master equation,

$$\partial_t p(a,t) = \int da' \left[W(a|a')p(a',t) - W(a'|a)p(a,t) \right],$$

with a factorizable transition matrix $W(a|a') = u(a)v(a')$. The process is "kangaroo-like" in that the transition rates for jumps $a \to a'$ depend on a and a', but not on the distance $a - a'$. Solve the master equation with the help of the Laplace transform $p(a,z) = \int_0^\infty dt\, e^{izt} p(a,t)$. In a first step, show that the following relation holds:

$$p(a,z) = \frac{p(a,t=0) + u(a)\sigma(z)}{-iz + v(a)/\tau},$$

where $1/\tau = \int da\, u(a)$ and $\sigma(z) = \int da\, v(a)p(a,z)$. Determine $\sigma(z)$ by substituting the expression for $p(a,z)$ and solving the resulting algebraic equation.

Kubo–
Anderson
process
Kubo–Anderson process: Taking $v(a) \equiv v$, simplify the resulting equation for $p(a,z)$ and compute the inverse Laplace transform to obtain the time evolution of the distribution $p(a,t)$. Compare with the Ornstein–Uhlenbeck process (problem 11.9.2).

For readers interested in gaining a little more fluency with stochastic processes, here are some additional practice exercises.

one-step
process
EXERCISE The defining property of a **one-step process** is that the rate of change of p_n depends only on p_n and $p_{n\pm1}$. Its master equation is given by

$$\dot{p}_n = r_{n+1}p_{n+1} + s_{n-1}p_{n-1} - (r_n + s_n)p_n.$$

To solve this equation, consider the generating function $g(z,t) = \sum_{n=-\infty}^{\infty} z^n p_n(t)$, where $g(1,t) = 1$ reflects the conservation of probability. Show that, for a given g, the probability follows from the inverse Mellin transform,

$$p_n(t) = \frac{1}{2\pi i} \oint dz\, z^{-1-n} g(z,t), \tag{11.32}$$

where the integral is taken around a contour enclosing $z = 0$ (but no singularities of g).
(a) Consider the special case $r_n = s_n = 1$ defining a **symmetric random walk** with the initial condition $p_n(0) = \delta_{n,0}$. From the master equation, derive a differential equation for $g(z,t)$ and show that its solution is $g(z,t) = \exp((1/z + z - 2)t)$. Solve the integral (11.32) in the saddle-point approximation and show that the resulting behavior for large times $t \to \infty$ is diffusive, $p_n(t) \simeq e^{-n^2/4t}/\sqrt{4\pi t}$.
(b) Consider the case $r_n = 0$ and $s_n = \gamma n$ with the initial condition $p_n(0) = \delta_{n,1}$ (the **Furry process**). Determine $g(z,t)$. You can solve its partial differential equation with the help of the method of characteristics. Using Eq. (11.32), show that the probabilities for $n > 0$ are given by $p_n(t) = e^{-\gamma t}(1 - e^{-\gamma t})^{n-1}$.
(c) Consider the case $r_n = \alpha n$ and $s_n = \beta n$ relevant for the modeling of **population growth**. Imposing the initial condition $p_n(0) = \delta_{n,m}$, show that the probability that the population dies out at time t is given by $p_0(t) = \left(\frac{\alpha(1-\epsilon)}{\beta-\alpha\epsilon}\right)^m$, where $\epsilon = e^{(\alpha-\beta)t}$. Discuss the short- and long-time limits.

11.4.3 Fokker–Planck approach to stochastic dynamics

The master equation (11.28) is a linear integral equation for the probability density, p. Like the structurally similar Schrödinger equation it can be solved via a spectral decomposition of the integral kernel $\{W(a|a')\}$ (exercise: discuss this solution strategy). However, for kernels $W(a|a')$ with nontrivial dependence on the state variables a, an explicit eigenmode decomposition of p is usually out of reach.[18] On the other hand, we may use the fact that transitions between remote states are usually suppressed for short δt. In such cases, $W(a|a')$ is short range and, as in section 11.2.3, we may formulate a **Kramers–Moyal expansion** in $a - a'$. (Again, compare with quantum mechanics: for low energies, or large times, the discrete Schrödinger equation for an electron subject to short-range hopping on a lattice can be reduced to a second-order differential equation.)

Kramers–
Moyal
expansion

---

[18] However, a numerical diagonalization of W may still be the most efficient route to obtain the distribution.

Following the same strategy as in section 11.2.3, we define $W(a|a') = W_b(a')$, $b \equiv a - a'$, where $W_b(a')$ is the transition probability for $a' \to a' + b$. In this notation,

$$\partial_t p(a) = \int db \left[(W_b p)(a - b) - (W_{-b} p)(a) \right].$$

Using the relation $\int db \, (W_b(a) - W_{-b}(a)) = 0$, a Taylor expansion of $(W_b p)(a - b)$ in b around a (the subscript in W_b remains passive) yields the series

$$\partial_t p(a, t) = \sum_{n=1}^{\infty} \frac{(-)^n}{n!} \partial_a^n (\alpha_n p)(a, t), \tag{11.33}$$

where the coefficients α_n are the moments of the transition probabilities,

$$\alpha_n(a) = \int db \, W_b(a) b^n. \tag{11.34}$$

Fokker–Planck equation

Assuming that all moments beyond the second are of negligible importance, we obtain the **Fokker–Planck equation**

$$\partial_t p(a, t) = \left(-\partial_a \alpha_1(a) + \frac{1}{2} \partial_a^2 \alpha_2(a) \right) p(a, t), \tag{11.35}$$

generalizing the Fokker–Planck equation of the Langevin process (section 11.2.3).

drift term

▷ As in that case, the first derivative, or **drift term** $\sim \partial_a \alpha_1 p$, is deterministic in that it describes the evolution of the *averaged* variable $\langle a \rangle$. Indeed,

$$d_t \langle a \rangle = \int da \, a \partial_t p(a, t) = \int da \, a \left(-\partial_a \alpha_1(a) + \frac{1}{2} \partial_a^2 \alpha_2(a) \right) p(a, t)$$

$$= \int da \, \alpha_1(a) p(a, t) = \langle \alpha_1(a) \rangle,$$

where we have integrated by parts.

diffusion term

▷ In a similar manner one may verify that (exercise) the variance of the distribution grows as $d_t \mathrm{var}(a) = 2\langle a \alpha_1(a) \rangle - 2\langle a \rangle \langle \alpha_1(a) \rangle + \langle \alpha_2(a) \rangle$. In most cases, the third term on the right-hand side dominates, telling us that the second derivative, or **diffusion term** $\propto \alpha_2$, governs the diffusive spread of probability around the center of the distribution.

11.4.4 Quality of the Fokker–Planck approximation: an example

rare event

The Fokker–Planck approximation is routinely applied in the description of probability distributions. It is therefore important to identify those situations where it goes qualitatively wrong. Indeed it turns out that Fokker–Planck approaches often fail to describe the *tails* of probability distributions, i.e., regions describing large and statistically unlikely fluctuations of variables. Such **rare events** can be of profound importance: think of the probability of the emergence of a vicious viral mutant, or of a chain reaction in a nuclear power plant.

As an example, consider the Poisson process introduced in section 11.4.2. We have computed the rigorous solution of this process, Eq. (11.31), and we now compare with the results obtained by Kramers–Moyal expansion. As before, we introduce a scaled variable, $x = n/N$, with distribution $p(x,t) = p(xN,t)N$, and we subject the equation to a second-order Kramers–Moyal expansion according to the above protocol. Equivalently, we may note that the one-step translation operator appearing in the Poisson master equation enters the equation of the scaled variable as $\partial_t \tilde{p} = \nu(\hat{E}_{-1/N} - 1)\tilde{p}$, where $E_a f(x) = f(x + a)$. Second-order Taylor expansion then leads to

$$\left(\partial_t + \frac{\nu}{N}\partial_x - \frac{\nu}{2N^2}\partial_x^2\right)\tilde{p}(x,t) = 0,$$

with $p(x,0) = \delta(x)$. This equation is solved by $\tilde{p}(x,t) = N(\frac{1}{2\pi\nu t})^{1/2}\exp[-\frac{1}{2t\nu}(Nx - t\nu)^2]$ or

$$p(n,t) = \left(\frac{1}{2\pi\nu t}\right)^{1/2}\exp\left[-\frac{1}{2t\nu}(n - t\nu)^2\right], \tag{11.36}$$

in the language of the original variable.

This Gaussian approximation reproduces the two main characteristics of the Poisson distribution, the mean value νt and width $\sim (\nu t)^{1/2}$. However, in the tails of the distribution the approximation is poor. This is illustrated in (a logarithmic representation in) the figure: already at values of n about four times bigger than the mean value, the Fokker–

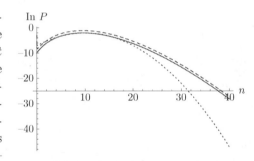

Planck prediction underestimates the probability by orders of magnitude. The origins of this error can be traced to the expansion of the operator $\hat{E}_{-N^{-1}} = e^{-\partial_x/N}$. The second-order expansion takes no account of whether the variable n is an integer. (Exercise: Why?) To obtain a better result in the tail regions, we need to keep the full identity of the hopping operator intact. This makes the direct solution of the differential equation difficult. However, we may push the quantum correspondence further, thinking of $p(n,t) = p(n,t|0,0)$ as the analog of a transition amplitude, and turn the smallness of p to our advantage: using a path integral-oriented language, and writing $p \sim \int \exp(-S)$ as the integral over an exponentiated action, we know that $S \gg 1$ in the tails. This indicates that a variational approach similar to the semiclassical approximation to quantum amplitudes might be successful. Solution schemes of this type are called **large deviation approaches** to stochastic dynamics.

large deviation approaches

Let us formulate a large-deviation analysis for the Poisson process and demonstrate that it leads to excellent results in the tail region. We first write the master equation in a form reminiscent of the imaginary-time Schrödinger equation,

$$(\partial_t + \hat{H})p(n,t) = 0, \qquad p(n,0) = \delta_{n,0},$$

governed by the non-hermitian operator $\hat{H} \equiv -\nu \left(e^{-i\hat{p}} - 1 \right)$. This equation has the formal solution

$$p(n,t) = \langle n | e^{-t\hat{H}} | 0 \rangle = \int_{q(0)=0}^{q(t)=n} D(q,p) \, e^{\int_0^t dt' \, (ip\dot{q} - H(p,q))},$$

where $\langle n | n' \rangle = \delta_{n,n'}$ is the scalar product in n-space and in the second equality we switched to a path integral representation. To avoid the mixed appearance of real and imaginary terms in the exponent, we subject the integration over the momentum variable to a Wick-rotation, $p \rightarrow -ip$. In the new variables,

$$p(n,t) = \int_{q(0)=0}^{q(t)=n} D(q,p) \, e^{\int_0^t dt' \, (p\dot{q} - H(q,p))}, \qquad H(q,p) = -\nu \left(e^{-p} - 1 \right).$$

Application of a stationary-phase approximation to this representation leads to

$$p(n,t) \sim e^{-S[q,p]}, \qquad S[q,p] = -\int_0^t dt' \, (p\dot{q} - H(q,p)),$$

where $(q,p)(t')$ are solutions of the extremal equations with the configuration-space boundary conditions $q(0) = 0, q(t) = n$, and the symbol $\sim$ indicates that fluctuations around the stationary configurations are neglected at this stage. Variation of the action leads to the Hamiltonian equations of motion,

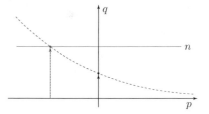

$$d_{t'} q = \partial_p H = \nu e^{-p},$$
$$d_{t'} p = -\partial_q H = 0,$$

whose solution reads $p(t') = -\ln(n/\nu t), q(t') = n(t'/t)$. The Hamiltonian form of the equation implies that energy is conserved and is given by $H(q(t'), p(t')) \equiv H = -n/t + \nu$. The figure shows a phase portrait of these solutions. Owing to the conservation of momentum, the flow lines are parallel to the q-axis with constant velocity νe^{-p}. At $p = 0$ we have $q(t') = \nu t'$, corresponding to the evolution of the mean value $\langle n(t') \rangle$ of the Poisson process. The action of this particular trajectory vanishes, $S(q, p = 0) = 0$. Final configurations $q(t) = n(t)$ different from the mean value νt can be reached at the expense of non-zero action cost. Substitution of the above solution leads to $S[q,p] = n \ln(n/\nu t) - n + \nu t$, which assumes a minimum value $S = 0$ on the mean trajectory $n = \nu t$ and is positive otherwise. We finally exponentiate this function to obtain the estimate

$$p(n,t) \simeq e^{-(n \ln(n) - n) + n \ln(\nu t) - \nu t} = e^{-(\ln(n) - n)} (\nu t)^n e^{-\nu t} \simeq \frac{(\nu t)^n}{n!} e^{-\nu t},$$

where the final expression is based on the Stirling formula, $n! \simeq \frac{1}{2} \ln(2\pi n) e^{n \ln n - n} \sim e^{n \ln n - n}$. This semiclassical approximation of the probability function is shown as a dashed curve in the figure on page 660. In the center of the distribution, where $S = 0$ and semiclassical approximations are problematic, the Fokker–Planck equation does

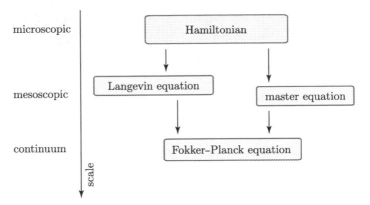

Fig. 11.2   The three levels of description of nonequilibrium systems.

a better job at approximating the Poisson distribution. However, in the tail regions, where $S \gg 1$, the semiclassical approximation works remarkably well. The quality of the result may be improved by including quadratic fluctuations around the mean field trajectory, which is equivalent to including the next to leading term in the Stirling approximation. However, in applications, it is often sufficient to work with "exponential accuracy" and just determine the action of extremal trajectories.

The take-home message here is that variational methods can be powerful in the study of rare events. The situation is similar to that in quantum mechanics, where the amplitudes of unlikely events – for example, the tunneling of particles through a large barrier – are conveniently described by WKB or instanton methods.

11.5 Field Theory I: Zero-dimensional Theories

In the previous sections, we introduced elements of classical nonequilibrium theory. We have seen that systems out of equilibrium can be described (see fig. 11.2) on:

▷ a **microscopic level**, i.e., in terms of microscopic Hamiltonians and their interactions,

▷ a **mesoscopic level**, where the microscopic transition rates are lumped into either a stochastic differential equation of Langevin type or a master equation, and

▷ a **continuum level**. In our discussion so far, the continuum description has been formulated in terms of the Fokker–Planck equation, a second-order partial differential equation in the state variables that is first order in time.

We have also seen that these formulations in terms of differential (Langevin, Fokker–Planck) or integral (master) equations show similarities to the Schrödinger equation of quantum mechanics. In the following, we will build on this analogy and formulate a path integral approach to classical non-equilibrium theory. Much of what has been said in earlier chapters to motivate path integration carries over to the present context. However, before discussing how the path integral describes nonequilibrium physics, let us first construct it.

11.5.1 Martin–Siggia–Rose (MSR) approach

As a warm-up exercise, let us construct a path integral representation for stochastic differential equations. A (first-order) **stochastic differential equation** has the structure

$$\partial_t n + f(n) = \xi, \tag{11.37}$$

where $n = n(t)$ is a time-dependent random variable, $f = f(n)$ is a function, and $\xi = \xi(t)$ is a noise term. In some applications, $\xi = \xi(n,t)$ depends on n. However, here we restrict ourselves to **linear noise**, defined through n-independent ξ.[19] For simplicity, we assume Gaussian correlations,

$$\langle \xi(t) \rangle_\xi = 0, \qquad \langle \xi(t)\xi(t') \rangle_\xi = A\delta(t - t'). \tag{11.38}$$

MSR and Onsager–Machlup functional

As usual in the construction of path integral representations, we start with a discretization of time, $n(t) \to n_i, i = 1, \ldots, N$, $t = i\Delta t$, which brings the evolution equation into the form

$$n_i - n_{i-1} + \Delta t \left[f(n_{i-1}) - \xi_{i-1} \right] = 0. \tag{11.39}$$

Denoting the solution for a given noise profile by $n[\xi]$, the expectation values of observables, $\langle \mathcal{O}[n] \rangle_\xi$, assume the form

$$\langle \mathcal{O}[n] \rangle_\xi = \int Dn\, \mathcal{O}[n] \, \langle \delta(n - n[\xi]) \rangle_\xi = \int Dn\, \mathcal{O}[n] \left\langle \left| \frac{\delta X}{\delta n} \right| \delta(X) \right\rangle_\xi,$$

where $Dn = \prod_i dn_i$ is the functional measure and $\delta(n - n[\xi]) = \prod_i \delta(n_i - n[\xi]_i)$. In the second equality, we have introduced a vector $X = \{X_i\}$, where $X_i \equiv n_i - n_{i-1} + \Delta t \left[f(n_{i-1}) - \xi_{i-1} \right]$ and $|\delta X/\delta n|$ is the (modulus) of the determinant of $\{\frac{\partial X_i}{\partial n_j}\}$. The advantage of the discretization chosen above is that $\delta X/\delta n$ is a triangular matrix with unit diagonal, i.e., the functional determinant equals unity. Thus, substituting the definition of X,

$$\langle \mathcal{O}[n] \rangle_\xi = \int Dn\, \mathcal{O}[n] \, \langle \delta(\partial_t n + f(n) - \xi) \rangle_\xi, \tag{11.40}$$

where $\partial_t n = \{n_i - n_{i-1}\}$ is shorthand for the lattice time derivative.

---

[19] For an application with nonlinear noise term, see section 11.7.1.

Ito dis-
cretization

INFO The discretization leading to a unit determinant is called **Ito discretization**. Other discretization schemes generate functional determinants, which are often represented in terms of Grassmann integrals. The resulting integrals then include both real and Grassmann variables. Interestingly, this combination gives rise to a symmetry principle, the BRS-symmetry, which plays an important role in various contexts, including particle theory. For further discussion of BRS symmetries in field theory, we refer to J. Zinn-Justin, *Quantum Field Theory and Critical Phenomena* (Oxford University Press, 1993).

Representing the δ-function in terms of a Fourier integral and switching back to a continuum notation, we arrive at the functional integral representation

$$\langle \mathcal{O}[n] \rangle_\xi = \int D(n, \tilde{n}) \, \mathcal{O}[n] \, \left\langle e^{i \int dt \, \tilde{n}(\partial_t n + f(n) - \xi)} \right\rangle_\xi,$$

where the structure of the action suggests that the two fields n and $\tilde{n}$ are canonically conjugate to each other. We finally average over the noise to arrive at the **Martin–**

MSR-
functional

Siggia–Rose–Bausch–Janssen–de Dominicis (MSR) functional integral[20]

$$\boxed{\langle \mathcal{O}[n] \rangle_\xi = \int D(n, \tilde{n}) \, \mathcal{O}[n] \exp \left(\int dt \, \left[i\tilde{n}(\partial_t n + f(n)) - \frac{A}{2}\tilde{n}^2 \right] \right)} \qquad (11.41)$$

Equivalently, we may use the path integral representation to represent the conditional probability $p(n, t|n_0, 0)$ (just write $\mathcal{O}[n] = \delta(n - n(t))$) as a path integral with boundary conditions:

$$\boxed{p(n, t|n_0, 0) = \int_{n(0)=n_0}^{n(t)=n} D(n, \tilde{n}) \exp \left(\int dt \, \left[i\tilde{n}(\partial_t n + f(n)) - \frac{A}{2}\tilde{n}^2 \right] \right)} \qquad (11.42)$$

Before proceeding, let us make a few remarks on the structure of the path integrals (11.41) and (11.42).

▷ Introducing a new set of variables (q, p) through $(n, \tilde{n}) \equiv (q, -ip)$, Eq. (11.41) assumes the form of an imaginary-time **phase space path integral**,

$$p(n, t|n_0, 0) = \int_{q(0)=n_0}^{q(t)=n} D(q, p) \, e^{\int dt \, [p\dot{q} - H(q,p)]}, \quad H(q, p) = -pf(q) - \frac{A}{2}p^2. \quad (11.43)$$

However, it is important to keep in mind that, in this representation, the integration over p extends over the imaginary axis.

▷ The variation of the action shows that we always have a solution with $p = 0$ and $\dot{q} = -f(q)$. It describes the evolution of the system in the absence of noise. A second characteristic of this solution is that its action vanishes, $S[q, p = 0] = 0$. In fact, both the action and the Hamiltonian $H(q, 0)$ vanish for all trajectories with $p = 0$, regardless of whether they are variational solutions or not.

[20] P. C. Martin, E. D. Siggia, and H. A. Rose, *Statistical dynamics of classical systems*, Phys. Rev. A **8**, 423 (1973). Although the integral has been independently derived a number of times, "MSR-functional" appears to be the most frequent denotation.

▷ Since the Hamiltonian is quadratic in momenta, we may integrate over p to obtain

$$p(n, t | n_0, 0) = \int\limits_{q(0)=n_0}^{q(t)=n} D(q) \, e^{-\frac{1}{2A} \int dt \, [\partial_t q + f(q)]^2}. \tag{11.44}$$

Onsager–
Machlup
functional

This defines the **Lagrangian variant** of the theory, the **Onsager–Machlup functional**,[21] with Lagrangian $L(q, \dot{q}) = \frac{1}{2A}\dot{q}^2 + \dot{q}\frac{f(q)}{A} + \frac{f(q)^2}{2A}$. Note that L resembles the Lagrangian of a one-dimensional point particle with mass $m = 1/A$ coupled to a velocity-dependent one-dimensional vector potential, of strength $A(q) = f(q)/A$, and a potential $V(q) = f(q)^2/2A$. Interpreting the Onsager–Machlup functional as the analog of a Feynman path integral, the corresponding **"Schrödinger equation"** reads as

$$\left(\partial_t - \frac{1}{2m}(\partial_q + A(q))^2 + V(q) \right) p(q, t) = \left(\partial_t - \frac{A}{2}\partial_q^2 - \partial_q f(q) \right) p(q, t) = 0,$$

where we note that $p(q, t) = p(n, t)$ plays the role of a wave function, subject to the initial condition $p(n, 0) = \delta(n - n_0)$ (think about this point). Comparing with our earlier discussion, we recognize the Fokker–Planck equation:

> The Fokker–Planck equation is the "Schrödinger equation" of the MSR path integral.

The discussion above is somewhat abridged in that we do not pay much attention to operator ordering, e.g., the order $\partial_q f(q)$ versus the order $f(q)\partial_q$.

Example: thermal escape

Consider a particle trapped at the bottom of a metastable potential configuration (see inset of fig. 11.3). Thermal fluctuations may assist the particle in overcoming the potential barrier and reach an energetically more favorable configuration.

thermal
escape

The corresponding **thermal escape rates** are controlled by **Arrhenius factors** whose computation is of importance in, e.g., the chemistry of thermally activated reactions. In the following, we demonstrate how the functional integral approach describes the situation in an intuitive and computationally efficient manner.

Assuming overdamped dynamics, the equation of motion of the trapped particle reads $\gamma\dot{q} + \partial_q V(q) = \xi'(t)$, where $\langle \xi'(t)\xi'(t') \rangle = 2\gamma T\delta(t - t')$, by the fluctuation–dissipation theorem. Dividing by γ, we obtain a stochastic differential equation of the type (11.37), with the identification $q \leftrightarrow n$, $f = \partial_q V$, and noise strength $A = 2T\gamma^{-1}$. We assume that T is much lower than the height ΔV of the barrier, so that thermal escape is a rare event, occurring with probability $p \sim \exp(-\Delta V/T)$. The largeness of the exponent $-\ln p \sim \Delta V/T \gg 1$ suggests a variational evaluation of the path integral (11.43), neglecting fluctuation effects.

---

[21] L. Onsager and S. Machlup, *Fluctuations and irreversible processes*, Phys. Rev. **91**, 1505 (1953).

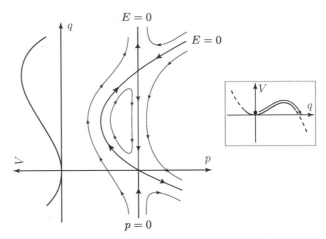

Fig. 11.3   Inset: Particle in a metastable potential well. Main figure: Phase space portrait of the corresponding MSRJD functional. For a discussion, see the main text.

With $H(q,p) = -\frac{p}{\gamma}(\partial_q V(q) + Tp)$, the Hamiltonian equations of motion of the action (11.43) read

$$\dot{q} = \partial_p H(q,p) = -\frac{1}{\gamma}\partial_q V(q) - \frac{2T}{\gamma}p, \qquad \dot{p} = -\partial_q H(q,p) = \frac{1}{\gamma}p\,\partial_q^2 V(q).$$

For the purposes of the present discussion, we do not need to solve these equations; it is sufficient to inspect the ensuing phase portrait.

As we shall see, the topology of the Hamiltonian flow in phase space is largely determined by the zero-energy contours $H(q,p) = 0$. We saw above that one of these curves is specified by $(q, p = 0)$. On the line $p = 0$, the equations of motion reduce to $\gamma\dot{q} = -\partial_q V(q)$. This is drift motion down the potential gradient towards the local minimum. We note that the point of the local potential maximum $q \equiv q^*$ defines an unstable point on that line, while $q = 0$ is stable.

Besides the obvious downward drift, one may expect the presence of a reverse trajectory $0 \to q^*$ representing thermal escape. This fluctuation-induced trajectory must emanate from the fixed points and, since H is conserved, it must also have zero energy. Indeed, the equation $H(q,p) = 0$ possesses a second solution, $p = -T^{-1}\partial_q V$. At the fixed points, $q = 0, q^*$ we have $\partial_q V = 0$, and this solution intersects the $p = 0$ line. However, as in classical mechanics, physical trajectories cannot cross it. Instead we have trajectories terminating at $(0,0)$ and $(0,q^*)$, where the time it takes to reach the fixed points, or depart from them is infinite. Further, Liouville's theorem requires the conservation of phase space volume. Since the fixed point $(0,0)$ is attractive along the $p = 0$ line, it must be repulsive along the trajectories escaping into the $p \neq 0$ continuum, in order to define an overall hyperbolic fixed point. (Recapitulate why globally attractive or repulsive fixed points are at variance with the laws of classical mechanics.) The figure shows how the interpolation between the near fixed-point regions generates a phase diagram where the $E = 0$ trajectories are *separatrices* between differently directed flows.

None of these features is specific to the present escape problem, which is one of the strong features of the present formalism. On general grounds, the fixed points of rate equations defined on the $p = 0$ lines of an MSR functional are always terminal points of $p \neq 0$ trajectories, which describe the essential physics of fluctuations in the problem. To see how this materializes in the present context, note that the fluctuation trajectories leaving or arriving at the point 0 or q^* are connected to form a single trajectory $0 \to q^*$ moving *against* the potential gradient. Its action (finite, although the energy H vanishes) is calculated as

$$S = -\int dt \, (p(q)\dot{q} - H(q,p)) = \frac{1}{T} \int_0^{q^*} dq \, \partial_q V(q) = \frac{1}{T}(V(q^*) - V(0)) = \frac{\Delta V}{T},$$

where the defining property $p = -T^{-1}\partial_q V$ has been used. Exponentiation of the action, $p \sim \exp(-S) = \exp(-\Delta V/T)$ shows that, in the present case, the escape probability is determined by the Arrhenius factor of the potential well. Of course, one might have arrived at this result with less effort, for example by reasoning similar to that applied before in connection with the Einstein relation (11.7). However, as usual with functional approaches, the advantage of the above formalism is its extensibility to the more complex problem classes addressed below.

11.5.2 Field integral representation of the master equation I

REMARK In order to avoid potential confusion with the "momentum" variable below, the symbol for probability is capitalized in this section, $p \to P$.

The MSR path integral formalism can be generalized to a general *stochastic path integral* describing the evolution of Markovian stochastic processes. For simplicity, here we consider a **one-step master equation**, i.e., an equation where a discrete state variable n changes by no more than one unit, ± 1, in each time step. Generalization to multi-step processes is straightforward. Our starting point, thus, is a general one-step master equation,

$$\partial_t P(n,t) = \left[(\hat{E}_1 - 1)f_-(n) + (\hat{E}_{-1} - 1)f_+(n) \right] P(n,t), \tag{11.45}$$

where $E_{\pm 1}f(n) \equiv f(n \pm 1)$ as before. We rewrite the equation as

$$\left(\partial_t + \hat{H}(\hat{q}, \hat{p}) \right) \hat{P}(t) = 0, \qquad \hat{P}(0) = \mathbb{1}, \tag{11.46}$$

with the "Hamiltonian operator"

$$\hat{H}(\hat{q}, \hat{p}) = -\left(e^{+i\hat{p}} - 1 \right) f_-(\hat{q}) - \left(e^{-i\hat{p}} - 1 \right) f_+(\hat{q}) \tag{11.47}$$

emphasizing the interpretation of $P(n,t;n_0,t_0) \equiv P(n,t|n_0,0) \equiv P(n,t)$ as an imaginary-time evolution operator. Throughout, we will work in a coordinate representation, where $\hat{q} \to q$ is a real variable and $\hat{p} \to -i\partial_q$ acts by differentiation. It is understood that the continuous variable q is initialized at discrete values $q \leftrightarrow n_0$. The fact that the Hamiltonian changes $q \to \pm 1$ in integer steps, then ensures that a discrete-variable stochastic dynamics is described.

We next represent the formal solution to (11.46) as an imaginary-time phase space path integral:

$$P(n, t; n_0, t_0) = \langle n | e^{-t\hat{H}} | n_0 \rangle = \int_{q(0)=n_0}^{q(t)=n} D(q, p) \, e^{\int dt \, (ip\dot{q} - H'(p,q))},$$

where $H'(q, p)$ is the classical Hamiltonian corresponding to the operator (11.47). Again, it will be convenient to remove the i-dependence of the action via a Wick rotation of the p-integration contour, $p \to -ip$. This leads to the **stochastic path integral**

stochastic
path
integral

$$P(n, t; n_0, t_0) = \int_{q(0)=n_0}^{q(t)=n} D(q, p) \, e^{\int dt \, (p\dot{q} - H(p,q))}, \tag{11.48}$$

$$H(q, p) = -\left(e^{+p} - 1\right) f_-(q) - \left(e^{-p} - 1\right) f_+(q),$$

where $H(p, q) \equiv H'(q, -ip)$ is the Hamiltonian evaluated on the new momentum variable. Let us remark on the structure of this integral:

▷ In the present variables, the integration is over **imaginary momenta**. This can become relevant when convergence issues are discussed.

▷ The stochastic path integral is a close ally of the **MSR functional** (11.43). As with the latter, its action vanishes on the axis $p = 0$. The line $p = 0$ accommodates a solution of the variational equations, $(q(t), p = 0)$, where $\dot{q} = f_+(q) - f_-(q)$. Comparison with $d_t \langle n \rangle = \langle f_+(n) \rangle - \langle f_-(n) \rangle \to f_+(\langle n \rangle) - f_-(\langle n \rangle)$ shows that this equation describes the fluctuationless dynamics, again in analogy with Langevin theory.

▷ A quadratic expansion of the Hamiltonian operator, $\hat{H} \simeq i\hat{p}(f_+(\hat{q}) - f_-(\hat{q})) + \frac{\hat{p}^2}{2}(f_-(\hat{q}) + f_+(\hat{q}))$, leads to the **Fokker–Planck approximation** of the master equation

$$\left(\partial_t + \partial_q (f_+ - f_-)(q) - \frac{1}{2}\partial_q^2 (f_- + f_+)(q)\right) P(q, t) = 0.$$

Performing this expansion on the level of the action, we obtain an MSR functional with quadratic momentum dependence and "diffusion" coefficient $A = f_- + f_+$. This observation conveys an interesting message: here, the dynamics is rendered noisy (the presence of a diffusion term) via the *discreteness* of the variable $q \leftrightarrow n$. For small q, the fact that the state variable changes in integer steps becomes essential, and this is reflected in the increased importance of the higher-order terms in the expansion (consider why). Specifically, the diffusive second-order term has a stronger influence on the drift dynamics described by the first-order term. Noise of this type is of importance, e.g., in the evolution of populations of small size, and is often called **demographic noise**.

demographic
noise

11.5.3 Doi–Peliti operator technique

Besides Eq. (11.48), there exists another path integral approach to the master equation, the **Doi–Peliti operator technique**.[22] The main feature of the Doi–Peliti formalism is that it describes the evolution of a discrete state variable with techniques borrowed from second quantization. To introduce the idea, let us consider the stochastic extinction of a population whose individuals die at a constant rate λ. It is described by a process with weight functions $f_+ = 0$, $f_- = \lambda n$, and the corresponding master equation reads

$$\partial_t P(n, t) = \lambda(\hat{E}_1 - 1)nP(n, t).$$

Again, we approach this equation from a quantum perspective; however, this time with techniques of second quantization. To this end, define a bosonic Fock space with basis states $|n\rangle$. Within this space, we introduce an algebra of Fock space operators through

$$a|n\rangle = |n-1\rangle, \qquad \bar{a}|n\rangle = (n+1)|n+1\rangle. \tag{11.49}$$

It is straightforward to verify that these operators obey canonical commutation relations, $[a, \bar{a}] = 1$, i.e., they are related to the standard ladder operator algebra of second quantization (defined by $a|n\rangle = n^{1/2}|n-1\rangle$, $a^\dagger|n\rangle = (n+1)^{1/2}|n+1\rangle$) by a simple canonical transformation. Employing these operators, the right-hand side of Eq. (11.45) assumes the form (exercise)

$$\hat{H} = -\lambda(\bar{a} - \bar{a}a), \tag{11.50}$$

and

$$P(n, t) \equiv \langle n|e^{-t\hat{H}}|n_0\rangle \tag{11.51}$$

is a Fock space representation of the distribution function.

We next apply the techniques introduced in section 0.4 to represent $P(n, t)$ as a coherent state field integral. Proceeding in the usual manner, we decompose the time interval into $N \gg 1$ steps, $e^{-t\hat{H}} = (e^{-\delta\hat{H}})^N$, $\delta = t/N$, then note that the Hamiltonian Eq. (11.50) is normal ordered, and finally insert coherent state resolutions of the identity, id. $= \int d(\bar{\psi}, \psi) \, e^{-\bar{\psi}\psi}|\psi\rangle\langle\psi|$, where the coherent states $|\psi\rangle$ obey the relations $a|\psi\rangle = \psi|\psi\rangle$, $\bar{a}|\psi\rangle = \partial_\psi|\psi\rangle$, $\langle\psi|\bar{a} = \langle\psi|\bar{\psi}$, $\langle\psi|a = \partial_{\bar{\psi}}\langle\psi|$. This leads to the representation

$$P(n, t) = \int D\psi e^{-\delta \sum_{n=0}^{N-1} \left[\delta^{-1}(\bar{\psi}_n - \bar{\psi}_{n+1})\psi_n + H(\bar{\psi}_{n+1}, \psi_n)\right] + n \ln \psi_N + n_0 \ln \bar{\psi}_0 - \ln n!}.$$

Here, $H(\bar{\psi}, \psi) = -\lambda\bar{\psi}(1-\psi)$ is the coherent state representation of the Hamiltonian and the last two terms in the exponent (action) represent the boundary matrix elements;

---

[22] M. Doi, *Second quantization representation for classical many-particle system*, J. Phys. A **9**, 1465 (1976); L. Peliti, *Path integral approach to birth-death processes on a lattice*, J. Physique **46**, 1469 (1985).

$$\langle n|\psi_N\rangle = \psi_N^n = e^{n \ln \psi_N},$$
$$\langle \psi_0|n_0\rangle = \frac{1}{n!}\bar{\psi}_0^{n_0} = e^{n_0 \ln \bar{\psi}_0 - \ln n_0!}. \tag{11.52}$$

Switching to a continuum notation, we arrive at the **Doi–Peliti functional integral**

$$\boxed{P(n,t) = \int D\psi\, e^{-\int_0^t dt'\left(-\partial_{t'}\bar{\psi}\psi + H(\bar{\psi},\psi)\right] \right) + n \ln \psi(t) + n_0 \ln \bar{\psi}(0) - \ln n_0!}} \tag{11.53}$$

Equation (11.53) defines a representation of $P(n,t)$ alternative to the phase space path integral (11.48). To understand how the two representations are related to each other, note that, in the path integral formalism of an extinction process with $f_+ = 0$ and $f_- = \lambda n$ the Hamiltonian (11.47) reduces to $\hat{H} = -\lambda(e^{\hat{p}} - 1)\hat{q}$. Comparison with Eq. (11.50) suggests the identification

$$\bar{a} = e^{\hat{p}}\hat{q}, \qquad a = e^{-\hat{p}}, \tag{11.54}$$

which, up to a constant, leads to an identical Hamiltonian. Equation (11.54) defines the **Cole–Hopf transformation**. It is straightforward to verify that it is canonical, i.e., the relation $[\hat{p},\hat{q}] = 1$ is compatible with $[a,\bar{a}] = 1$. The operator transform defines a structurally identical transformation of coherent state integration variables, whose substitution into the functional integral (11.53) transforms the bulk action, $\int dt'(-(\partial_{t'}\bar{\psi})\psi + H(\bar{\psi},\psi))$, into that of Eq. (11.48). A careful treatment of the boundary term generates the effective boundary condition $\int_{q(0)=n_0}^{q(t)=n}$.

Cole–Hopf transformation

EXERCISE Explicitly represent the coherent states above as exponentials of the operators $a, \bar{a}$ and verify the relations (11.52). Apply the Cole–Hopf transformation to the time-discretized representation of the functional integral to prove the above boundary identity in the large-n limit. In doing so, avoid taking hermitian adjoints – the operator $\bar{a}$ is different from $a^\dagger$.

Summarizing, we have constructed two alternative representations of $P(n,t)$ that are related by a Cole–Hopf transformation. The example above illustrates that the Doi–Peliti formalism often leads to simpler representations of the Hamiltonian. However, the price to be paid for this simplicity is that its variables are more abstract and less easy to interpret in physical terms. One cannot say categorically which of the two representations is better. The network of different theories constructed in the previous sections is summarized in fig. 11.4.

11.6 Field Theory II: Higher Dimensions

So far, we have been focusing on systems whose state is described by a single scalar variable – the analog of $(0 + 1)$-dimensional point quantum mechanics. However, one of the strongest features of the path integral formalism is its extensibility to

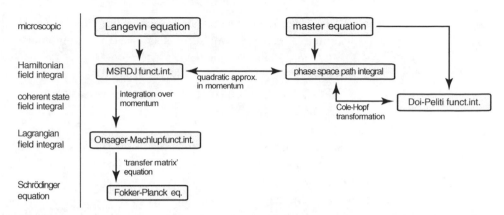

Fig. 11.4   Summary of different functional integral approaches to Langevin dynamics and stochastic processes.

continuum problems – the analog of $(d+1)$-dimensional quantum field theory. Examples in this category include the following:

▷ **interacting lattice particle systems**, whose dynamics is governed by an interplay of single-particle hopping and interactions. For example, in **exclusion processes**, a hardcore repulsion prohibits multiple occupancy of lattice sites, and in **driven diffusive lattice gases**, an external force generates nonequilibrium states.

reaction-
diffusion
systems

▷ **reaction–diffusion systems**, which are lattice systems populated by particle species $A, B, \ldots$ which may undergo reactions. They frequently appear in the modeling of chemical reaction kinetics, the description of evolutionary or ecological dynamics, traffic simulation, and related set-ups. For example, the bottleneck limiting the throughput of a fast chemical reaction $A + B \to C$ in solution may be the speed at which the agents diffuse through the liquid. In this case, a combination of diffusion and local reaction defines the minimal framework to describe the process. Adopting the language of population dynamics, elementary reaction processes include birth $\emptyset \to A$, death $A \to \emptyset$, transmutation $A \to B$, death at contact, $A + B \to A$, etc.

In such problems we typically deal with macroscopically large particle numbers. First principle solutions of stochastic evolution equations are out of the question, and besides would contain excessive information. As in Boltzmann statistical mechanics (section 11.3), one therefore starts by defining coarse-grained **effective variables**. For example, reaction–diffusion systems are often described in terms of the local concentration of agents, $n_A(\mathbf{x}), n_B(\mathbf{x}), \ldots$. Throughout, we will denote the set of effective variables of a problem by $\phi_a(\mathbf{x})$, $a = 1, \ldots, n$.

As in the Langevin theory discussed in section 11.2, the microscopic degrees of freedom influence the dynamics of coarse-grained ones in different ways: (i) combinations of dissipative and external forces may stabilize local stationary configurations.

However, (ii) such dissipative relaxation is accompanied by stochastic fluctuations. (iii) The dynamics of conserved quantities is constrained by continuity relations $\partial_t \phi + \nabla \cdot \mathbf{j} = 0$, where $\mathbf{j}$ is a current field.

11.6.1 Dynamical critical phenomena

Systems of the above kind may undergo *phase transitions* between macroscopically distinct states. As in equilibrium statistical mechanics, such phase transitions are a prime object of study. A difference is that, now, we are working under nonequilibrium conditions, implying richer forms of critical phenomena with interesting *time dependence*. Following the seminal review article[23] by Hohenberg and Halperin (HH), this section is an introduction to **dynamical critical phenomena** in general. In what follows, we will then apply this framework to the field-theoretical study of nonequilibrium phase transitions.

dynamical critical phenomena

Consider a system a system undergoing a second-order phase transition characterized by an order parameter ϕ. The distance to the transition point is measured by a relevant scaling field τ (e.g., $\tau = (T - T_c)/T_c$, where T is temperature and T_c is the critical temperature). Assume that the order parameter ϕ is conjugate to a field h, in the sense that $\langle \phi \rangle = -\partial_h F[h]$, where $F[h]$ is an effective functional describing the system. (In thermal equilibrium, this would be the thermodynamic free energy.)

In the vicinity of a critical point, the system builds up long-range correlations. For example, the static correlation function $C(\mathbf{x}) \equiv \langle \phi(\mathbf{x})\phi(0) \rangle$ is expected to show power law behavior:

$$C(\mathbf{x}) = |\mathbf{x}|^{-(d-2+\eta)} Y(|\mathbf{x}|/\xi), \qquad (11.55)$$

where Y is a dimensionless scaling function and the correlation length $\xi \sim |\tau|^{-\nu}$ defines the characteristic length scale of order parameter fluctuations. However, the order parameter may also fluctuate in time. The corresponding characteristic **order parameter time scales** are denoted by $t(\tau)$ and are expected to increase as

$$\boxed{t(\tau) \sim \xi(\tau)^z \sim |\tau|^{-\nu z}} \qquad (11.56)$$

dynamical critical exponent

where z defines the **dynamical critical exponent**. The growth of these time scales upon approaching the transition is called **critical slowing down**. Equation (11.56) can be read as a statement on the **dispersion of order parameter fluctuations**: the characteristic frequencies corresponding to Eq. (11.56), $\omega(\tau) \sim t(\tau)^{-1} \sim \tau^{\nu z}$, relate to the momenta $|\mathbf{q}| \sim \xi^{-1} \sim |\tau|^{\nu}$ according to

$$\omega \sim |\mathbf{q}|^z f(|\mathbf{q}|\xi),$$

where f is a dimensionless function.

[23] P. C. Hohenberg and B. I. Halperin, *Theory of dynamic critical phenomena*, Rev. Mod. Phys. **49**, 435 (1977).

To make these statements more concrete, consider a situation where ϕ has been prepared in an initial state and is subsequently allowed to relax. It is reasonable to describe the relaxation profile by the scaling ansatz

$$\phi(t, \tau) = |\tau|^\beta g(t/t(\tau)),$$

where β is the standard order parameter exponent and g is a dimensionless scaling function. For large arguments, $g(x \to \infty) \to$ const., reflecting that, for $t \gg t(\tau)$, the order parameter scales as in the static theory: $\phi \sim |\tau|^\beta$ below the transition point, and $\phi = 0$ above. For short times, $t \to 0$, the relaxation process does not feel the full extension of the correlation volume, and $\phi(t, \tau)$ must be independent of τ. This leads to the condition $g(t/t(\tau)) \overset{t \to 0}{\sim} |\tau|^{-\beta}$, or

$$\phi(t) \sim t^{-\alpha}, \qquad \alpha = \beta/\nu z.$$

dynamic suscep-tibility We next define two correlation functions characterizing the fluctuation/correlation behavior of the order parameter: The **dynamic susceptibility**, χ, is defined through the linear response relation

$$\langle \phi(q) \rangle \overset{h \to 0}{=} \chi(q) h(q), \tag{11.57}$$

where $q = (\mathbf{q}, \omega)$. Since $\phi(t)$ can depend only on $h(t' < t)$ (by causality), $\chi(\omega)$ is an analytic function in the upper complex ω-plane (recapitulate why). The spatio-dynamic correlation function temporal fluctuation behavior of the order parameter is described by the **dynamic correlation function**

$$C(x) \equiv \langle \phi(x)\phi(0) \rangle - \langle \phi(0) \rangle^2, \tag{11.58}$$

where $x = (\mathbf{x}, t)$. While (11.57) probes the response of the average order parameter to changes in h, and is therefore affected by dissipation, Eq. (11.58) is a measure of the fluctuations. In section 11.6.4, we will see that, in thermal equilibrium, the two quantities satisfy a variant of the **fluctuation–dissipation theorem**, $C(\mathbf{q}, \omega) = -\frac{2T}{\omega} \mathrm{Im}\, \chi(\mathbf{q}, \omega)$. Violations of this relation are an indication of nonequilibrium conditions.

11.6.2 Field theories of finite-dimensional Langevin systems

With theses definitions in place, we now formulate an effective field theory approach to dynamical critical phenomena. Let us assume that the energy stored in an order parameter configuration is described by an energy functional $H[\phi]$. Following HH,[23] we anticipate that fluctuations in H cause temporal changes in ϕ, where the specifics depend on the nature of the order parameter field.

▷ In a model for a system with **non-conserved order parameter** (model A, in the terminology of HH), ϕ will relax to the minima of the energy functional.

Defining $x = (\mathbf{x}, t)$ as usual, this is described by a variational equation of the form[24]

$$\partial_t \phi(x) = -\frac{\delta H[\phi]}{\delta \phi(x)}.$$

For example, for a Hamiltonian with quadratic order parameter dependence, $H[\phi] = \frac{\gamma}{2} \int d^2x \, \phi^2(\mathbf{x})$, the equation above leads to exponential relaxation, $\partial_t \phi = -\gamma \phi$.

▷ However, in a system with conserved order parameter (model B), the continuity equation $\partial_t \phi + \nabla \cdot \mathbf{j} = 0$ has to be obeyed. Assuming drift current away from regions of increased energy, $\mathbf{j}(x) = -\nabla \frac{\delta H[\phi]}{\delta \phi(x)}$, it assumes the form

$$\partial_t \phi(x) = \Delta \frac{\delta H[\phi]}{\delta \phi(x)}.$$

With the choice $D = \gamma$, the above quadratic Hamiltonian now leads to a variant of Fick's law, $\mathbf{j} = -D\nabla\phi$, with diffusion $\partial_t \phi = D\Delta\phi$.

The two cases above can be represented in one equation as $\partial_t \phi(x) = \Delta^n \frac{\delta H[\phi]}{\delta \phi(x)}$, where $n = 0, 1$ for models A and B, respectively. This equation describes the dissipative relaxation of the order parameter, and hence a noise term must be added to account for the presence of fluctuations. This defines the **generalized Langevin equation**,

generalized Langevin equation

$$\boxed{\partial_t \phi(x) = -(-\Delta)^n \frac{\delta H[\phi]}{\delta \phi(x)} + \xi_n(x)} \tag{11.59}$$

where, for model A, the noise is correlated as

$$\langle \xi_0(x)\xi_0(\mathbf{x}', t') \rangle = 2T\delta(t - t')\delta(\mathbf{x} - \mathbf{x}') = 2T\delta(x - x'). \tag{11.60}$$

This ansatz assumes a time resolution sufficiently coarse-grained that the noise is effectively Markovian and short-range in space. The noise strength defines a parameter T, playing the role of an effective temperature. For model B, we assume that $\xi_1 = \partial_i \eta_i$ is the divergence of a noisy current field whose components η_i are independently Gaussian correlated as $\langle \eta_i(x)\eta_j(x') \rangle = 2T\delta_{ij}\delta(x - x')$. In this way the conservation of the field ϕ is guaranteed, $\partial_t \phi = -\partial_i(j_i - \eta_i)$.

Functional averages $\langle \mathcal{O}[\phi] \rangle_\xi$ may now be computed by straightforward generalization of the $(0 + 1)$-dimensional functional (11.40):

$$\langle \mathcal{O}[\phi] \rangle_\xi = \int D\phi \, \mathcal{O}[\phi] \left\langle \prod_x \delta\left(\partial_t \phi(x) + (-\Delta)^n \frac{\delta H[\phi]}{\delta \phi(x)} - \xi_n(x)\right) \right\rangle_\xi.$$

Again, we represent the δ-functional as a integral over a field ψ and integrate over the noise to arrive at

$$\langle \mathcal{O}[\phi] \rangle_\xi = \int D(\phi, \psi) \, \mathcal{O}[\phi] \, e^{i \int dx \, \psi(x)\left(\partial_t \phi(x) + (-\Delta)^n \frac{\delta H[\phi]}{\delta \phi(x)}\right) - T \int dx \psi(x)(-\Delta)^n \psi(x)},$$

---

[24] The Hamiltonian is a function of $\phi(\mathbf{x})$. In $\delta_{\delta\phi(x)} H = \delta_{\phi(x)} H|_{\phi(\mathbf{x},t)}$, we evaluate the variational derivative at a specific time-dependent configuration.

where $\int dx \equiv \int d^d x dt$, and the derivative present in the noise correlation of model B leads to the appearance of the Laplacian $(-\Delta)$ in the second term in the action. A Wick rotation $\psi \to -i\psi$ finally brings us to the finite-dimensional generalization of the **MSR-functional**

MSR-
functional

$$\langle \mathcal{O}[\phi] \rangle_\xi = \int D(\phi, \psi) \, \mathcal{O}[\phi] \, e^{\int dx \, \psi \left(\partial_t \phi + (-\Delta)^n \frac{\delta H[\phi]}{\delta \phi} \right) + T \int dx \, \psi(-\Delta)^n \psi} \qquad (11.61)$$

where we omit the argument x of the fields for notational simplicity. Before discussing the physics described by this functional, we extend the formalism to include more general Markovian stochastic processes.

11.6.3 Field theory of finite-dimensional stochastic processes

Problems in reaction kinetics or population dynamics are frequently described by **master equations defined on a lattice**,

$$\partial_t p_t[n] = \sum_{n'} \left(W[n, n'] p_t[n'] - W[n', n] p_t[n] \right). \qquad (11.62)$$

The notation $P_t[n]$ indicates that the distribution depends on the discrete field $\{n_i\}$, where i is a lattice point and n_i is a generally multi-component variable describing the local state of the system (the concentrations $A_i, B_i, \ldots$ of chemicals, etc.). In **reaction–diffusion systems**, the transition rates W describe the competition of diffusive spreading (a random walk on a lattice) and particle reactions. Field theory representations of such systems are obtained by straightforward generalization of the $(0 + 1)$-dimensional techniques introduced in sections 11.5.1 and Eq. (11.50).

As an example, consider the simple process $A \xrightarrow{\lambda} \emptyset$ describing the extinction of an agent A at constant rate λ. Defining $n_i \equiv$ as the number of A-agents at lattice site i, we may ask how the probability $p_t[n]$ evolves under the joint influence of annihilation and diffusive spreading. A field theory representation of this process can be obtained by generalization of the Doi–Peliti operator algebra,

$$\begin{aligned} a_i | \ldots, n_i, \ldots \rangle &= | \ldots, n_i - 1, \ldots \rangle, \\ \bar{a}_i | \ldots, n_i, \ldots \rangle &= (n_i + 1) | \ldots, n_i + 1, \ldots \rangle, \end{aligned} \qquad (11.63)$$

where the population $\{n_i\}$ is described as a Fock space state. A straightforward generalization of the discussion of section 11.5.3 shows that the Hamiltonian generating the process is given by

$$\hat{H} = -D \sum_{ij} \bar{a}_i \Delta_{ij} a_j - \lambda \sum_i \bar{a}_i (1 - a_i), \qquad (11.64)$$

where Δ_{ij} is the lattice Laplacian and D a diffusion constant. The corresponding coherent state field integral action then reads

$$S[\bar{\psi}, \psi] = \int d^d x dt \left[\bar{\psi}(\partial_t - D\Delta)\psi - \lambda \bar{\psi}(1 - \psi) \right], \qquad (11.65)$$

where we have switched to a continuum notation. The variational equations of this action,

$$(-\partial_t + D\Delta)\psi = \lambda(\psi - 1),$$
$$(\partial_t + D\Delta)\bar{\psi} = \lambda\bar{\psi},$$

have the fixed point $(\psi, \bar{\psi}) = (1, 0)$. Remembering the Cole–Hopf transformation (11.54), $(\psi, \bar{\psi}) = (e^{-p}, e^p q)$, this is identified as the lifeless and noiseless configuration $(q, p) = (0, 0)$. In agreement with the general principle, $p = 0$, or $\psi = 1$, while the conjugate variable describes a combination of extinction at rate $-\lambda\bar{\psi}$ and diffusive spreading.

11.6.4 Fluctuation–dissipation theorem

How can one tell whether a system is in thermal equilibrium? One way to answer the question is to compute the stationary limit of the probability distribution and check whether it is of Maxwell–Boltzmann form. However, this procedure is often too elaborate to be practical. In general, it is more economical to check whether a system satisfies the fluctuation–dissipation theorem (FDT) – a hallmark of thermal equilibrium. We have already encountered various manifestations of the FDT. In this section, we discuss the FDT from a general perspective, first in the terminology of response functions, and then from a field theory perspective.

Equilibrium linear response

Linear response was introduced in section 0 as a means to compute the expectation value of observables $\hat{X}(\mathbf{x}, t) \equiv \hat{X}(x)$ in response to the application of a generalized force, F, conjugate to $\hat{X}$. Assuming that, in the absence of force, $X(x) \equiv \langle \hat{X}(x) \rangle = 0$, we obtained the linear relation

$$X(\omega) = \chi_q(\omega)F(\omega) + \mathcal{O}(F^2), \tag{11.66}$$

where $\chi_q(\omega)$ is a **generalized quantum susceptibility**. Equation (11.66) defines a connection between the two macroscopic quantities F and X in an environment containing a large number of microscopic degrees of freedom. In general, the latter have a damping, or dissipative, effect on external perturbations and, in this sense, χ_q probes the dissipative response of the system. In section 0.3 we showed that $\chi_q(\omega) = C_q^+(\omega)$ is the Fourier transform of the **retarded response function**

$$C_q^+(t) = -i\Theta(t)\langle [\hat{X}(t), \hat{X}(0)] \rangle.$$

We also considered the **time-ordered correlation function**

$$C_q^T(t) \equiv -i\langle T_t \hat{X}(t)\hat{X}(0) \rangle,$$

which describes the average second moment, or *fluctuations*, of the observable X. It turned out that the dissipative C_q^+ and the fluctuation C_q^T are related by (cf. Eq. (0.19))

$$\operatorname{Im} C_{\mathrm{q}}^{T}(\omega) = \operatorname{Im} C_{\mathrm{q}}^{+}(\omega) \coth(\beta\omega/2). \qquad (11.67)$$

FDT in
linear
response

This is the **FDT of (quantum) linear response**.

We want to understand the **classical limit** of this relation. Referring for a more rigorous discussion to appendix section A.4, we note that in this limit $\hat{X} \to X(\mathbf{x})$ becomes a function in phase space, and $\hat{X}(t) \to X(\mathbf{x}(t))$ is the evolution of that function according to classical mechanics. The time-ordering in C_{q}^{T} becomes irrelevant, so that the object replacing the time-ordered expectation value reads

$$C(t) \equiv \langle X(t)X(0) \rangle,$$

where the phase space average $\langle \dots \rangle \equiv \int_{\Gamma} \rho(\mathbf{x}) \dots$ assumes the role of the quantum thermal trace and we do not include the factor $-i$ present in the definition of C_{q}^{T}. Noting that $C(t) = C(-t)$, the left-hand side of the FDT is replaced as $\operatorname{Im} C_{\mathrm{q}}^{T}(\omega) \to C(\omega)$. Turning to the retarded response function, the commutator $[\hat{A}, \hat{B}] \to \hbar i \{A, B\}$ becomes a Poisson bracket, so that $C_{\mathrm{q}}^{+} \to \hbar C^{+}$ with

classical
suscep-
tibility

$$C^{+}(t) = \chi(t) \equiv \Theta(t)\langle \{X(t), X(0)\} \rangle, \qquad (11.68)$$

where χ defines the **classical susceptibility**. Finally, we note that the dimension $[\beta\omega] = \text{action}^{-1}$ of the argument of the coth-factor requires another $\hbar$, so that $\coth(\hbar\beta\omega/2) \to 2/\hbar\beta\omega$. This $\hbar$ in the denominator cancels that multiplying the Poisson bracket, and the substitution of all factors leads to the **classical fluctuation–dissipation theorem**

classical
fluctuation–
dissipation
theorem

$$\boxed{C(\omega) = -\frac{2T}{\omega}\operatorname{Im}\chi(\omega)} \qquad (11.69)$$

We obtained this expression by taking the $\hbar \to 0$ limit of its quantum analog. However, it is equally possible to derive the FDT entirely within the framework of classical linear response theory. Interested readers may consult the original reference on the subject,[25] or consult section 11.9.3.

EXERCISE   Consider a classical harmonic oscillator $\hat{H}(\mathbf{x}) = \frac{1}{2}(p^2 + q^2)$ (with the mass and spring constant set to unity for simplicity) in a state of thermal equilibrium, $p(\mathbf{x}) = \mathcal{N}\exp(-H(\mathbf{x})/T)$. Compute the two response functions and verify that the FDT is satisfied. (Hint: It may be more economical to evaluate the Poisson brackets in a pair of canonical variables different from $\mathbf{x} = (q, p)$.) For more details and the solution, see section 11.9.3. Below, we will approach this problem from a somewhat different perspective.

It is sometimes useful to consider a variant of the FDT integrated over frequencies. To this end, we note that the integrated fluctuation function

$$C(t = 0) = \int \frac{d\omega}{2\pi} C(\omega) = \langle X(0)X(0) \rangle \equiv \langle X^2 \rangle$$

[25] R. Kubo, *The fluctuation–dissipation theorem*, Rep. Prog. Phys. **29**, 255 (1966). Note that Kubo uses a different convention for the temporal Fourier transform. This leads to a few sign differences.

just probes the second moment of the variable X in the thermal distribution function. Turning to the right-hand side, $\chi(\omega)$ is analytic in the upper half-plane. Application of the theorem of residues to an integration contour shifted infinitesimally into the lower half plane then yields $\int \frac{d\omega}{2\pi} \frac{\mathrm{Im}\chi(\omega)}{\omega} = \frac{\chi(0)}{2}$. This leads to the **static version of the FDT**,

$$\boxed{C(t = 0) = T\chi(\omega = 0)} \tag{11.70}$$

where the notation highlights the non-local character of the statement: fluctuations local in time are related to the zero-frequency susceptibility, which is the susceptibility integrated over time.

Field theory

Here, we discuss how the FDT (11.69) is implied by the path integral approach. For definiteness, let us consider the quadratic Hamiltonian

$$H[\phi] = \int d^d x \left(\frac{D}{2}(\partial\phi)^2 + \frac{r}{2}\phi^2 - h\phi \right),$$

describing the relaxation of an effective variable, ϕ. Assuming noisy dynamics, the corresponding MSR functional (11.61) reads

$$\mathcal{Z} \equiv \int D(\phi, \psi) \, e^{\int dx \, [\psi(\partial_t \phi + (-\Delta)^n(-D\Delta\phi + r\phi - h)) + T\psi(-\Delta)^n \psi]}.$$

Switching to $(d + 1)$-dimensional momentum space, $q = (\omega, \mathbf{k})$, we represent the action as a bilinear form, $S[\phi, \psi] = S_0[\phi, \psi] + \int dq \, \psi_q k^{2n} h_{-q}$, where

$$S_0[\phi, \psi] = -\frac{1}{2} \int dq \, (\phi, \psi)_{-q} \begin{pmatrix} 0 & (g_q^-)^{-1} \\ (g_q^+)^{-1} & 2Tk^{2n} \end{pmatrix} \begin{pmatrix} \phi \\ \psi \end{pmatrix}_q,$$

$$g_q^{\pm} \equiv \left(\mp i\omega + k^{2n}(Dk^2 + r) \right)^{-1}.$$

Here, $\int dq \equiv \int \frac{d^d k \, d\omega}{(2\pi)^{d+1}}$, and we use an index notation g_q, instead of $g(q)$, for improved notational clarity.

We next employ this functional to compute the **dynamic susceptibility**. Owing to the evenness of S_0 in the fields, $\langle\phi\rangle_0 = 0$, where $\langle\ldots\rangle_0 \equiv \int D(\phi, \psi) \, e^{-S_0[\phi,\psi]}$. To leading order in the driving field, we thus obtain $\langle\phi_q\rangle \simeq \int dq' \, \langle\phi_q k'^{2n} \psi_{-q'}\rangle h_{q'}$, giving

$$\chi(q) = \int dq' \, \langle\phi_q k'^{2n} \psi_{-q'}\rangle = g_q^+ k^{2n}. \tag{11.71}$$

This leads to the identification $\chi(q) = q^{2n} g_q^+$. On the other hand, the **fluctuation correlation function** is given by

$$C(q) = \int dq' \, \langle\phi_q \phi_{q'}\rangle_0 = 2Tk^{2n} g_q^+ g_q^- = \frac{2T}{\omega} \mathrm{Im}\chi_q, \tag{11.72}$$

where from the last equality we note $\mathrm{Im}g_q^+ = \omega g_q^+ g_q^-$. This calculation demonstrates the consistency of the *free* (quadratic) equilibrium field theory with the

FDT. However, it can be much more challenging to show that this compatibility extends to the interacting case. A common strategy is first to consider limiting cases, where a system of interest is in equilibrium. Compatibility with the FDT is then demonstrated, if necessary by approximate or perturbative methods. Building on this reference point, departures from equilibrium can then be diagnosed by FDT violation.

dynamic
structure
factor

INFO The function $C(q) \equiv C(\mathbf{q}, \omega)$ describes spatio-temporal fluctuations at the characteristic length scale $|\mathbf{q}|^{-1}$ and time scale ω^{-1}, and for this reason is sometimes called the **dynamic structure factor**. Another frequently used notation is $S(q) \equiv C(q)$. The FDT establishes a connection between the dynamic structure factor and the dynamic susceptibility. Likewise, the static version of the FDT (11.70) relates the **static structure factor** $C(\mathbf{q}) \equiv S(\mathbf{q})$ to the static susceptibility.

It may be worth recapitulating that, in systems kept at some predetermined noise level (set, e.g., by the coupling to an external environment), the noise strength *determines* the effective temperature via the FDT condition.

11.7 Field Theory III: Applications

In this section, we apply the formalism developed thus far to explore one of the most important universality classes of nonequilibrium physics, *directed percolation*. Within the framework of this discussion, we will address the following questions:

▷ How can the effective Hamiltonians $H[\phi]$ introduced in section 11.5.2 be constructed for concrete physical systems?

▷ What can the theory say about the different phases realized in and out of equilibrium?

▷ And what can be said about the transitions between these phases? To what degree can we generalize concepts developed for equilibrium systems – critical dimensions, mean field theory, renormalized theory of fluctuations, etc. – to the present setting?

11.7.1 Directed percolation

In equilibrium statistical mechanics, the ϕ^4- or Ising-class defines perhaps the most fundamental universality class. For systems defined by a single scalar order parameter ϕ, it describes the spontaneous breaking of reflection symmetry $\phi \leftrightarrow -\phi$ in a phase transition. In nonequilibrium statistical mechanics, the directed percolation class plays a similarly fundamental role. In this section, we will introduce the concept of directed percolation, demonstrate its importance to non-equilibrium physics, and describe its critical phenomena.

Directed percolation: phenomenology

The term **directed percolation**[26] (DP) refers to a class of models describing the directed spreading of a substance. Examples include liquids dispersing through porous rock under the influence of a gravitational force, the spreading of an epidemic in time, and many other systems. DP models are often defined on d-dimensional hypercubic lattices. As in 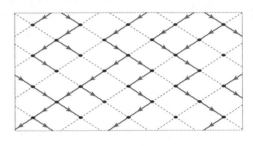 non-directed percolation models, the links of the lattice are chosen to be open or closed with probability p and $1-p$, respectively. Percolation becomes directed when a specific main diagonal of the lattice is singled out to give the bonds a sense of preferred orientation (see the figure).

At a critical value, $p = p_c$, the model undergoes a **directed percolation phase transition** (see the figure). The order parameter of this transition is the probability P_∞ that a randomly chosen site of the lattice is at the origin of an infinitely large connected cluster.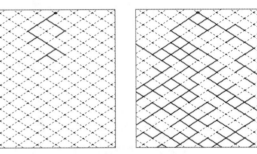

At $p < p_c$, the system is in a **dry phase** void of extensive clusters, and at $p > p_c$ it is **wet**. It is evident that the critical value $p_c = p_c(d)$ depends on the dimensionality. In one dimension, any broken link truncates a cluster, such that $p_c(1) = 1$. In infinite dimensions, the unlimited number of options for departing from individual lattice sites implies $p_c(d \to \infty) \searrow 0$. However beyond such qualitative statements, the critical properties of the DP-transition are not fully understood. Numerical studies indicate that the critical exponents characterizing it assume complicated, and likely irrational, values.

It turns out that many **reaction–diffusion processes** fall into the DP universality class. To understand the connection, note that a reaction–diffusion system can be defined on a hypercubical lattice, where one main diagonal represents the direction of time. In this formulation, covered bonds are segments of the world lines of particles, whose state may change at the nodes of the network. This includes a diffusive change of direction and reactions (see fig. 11.5). A cut through the network at a fixed instance of time defines the instantaneous state of the system. The rules describing the generation update at the next plane of nodes can now be chosen to describe even very complicated reaction–diffusion processes; some basic examples are shown in fig. 11.5. However, it turns out that many of these choices imply a

---

[26] See H. Hinrichsen, *Nonequilibrium critical phenomena and phase-transitions into absorbing states*, Adv. Phys. **49**, 815 (2000) for an excellent review of the subject.

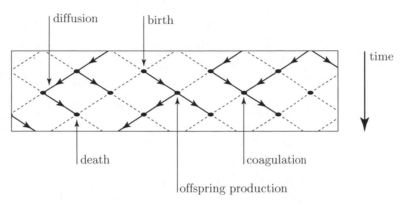

diffusion   birth   time

death   coagulation

offspring production

Fig. 11.5   Dynamical interpretation of a DP network. Each row of sites represents a generation of particles. The generation dynamical update then involves diffusion, birth and death, coagulation
and the generation of offspring.

phase transition between two principal alternatives, a dry and a wet phase characterized by full and absent coverage of the lattice, respectively. It is a challenge to
theory to describe the universal properties of such transitions, and their connection
to the basic DP universality class. However, before turning to this point, we first
discuss the physics of the DP as initially defined.

Elements of scaling theory

The physics of the DP transition is conveniently described in terms of scaling laws.
Introducing $\tau = p - p_c$ as a parameter measuring the distance from the transition
point, correlations in the spatial directions of the system are characterized by a
correlation length $\xi \sim \tau^{-\nu}$ and those in the time direction by a **correlation
time** $t \sim \xi^z \sim \tau^{-\nu z}$. The transition as such can be diagnosed via two candidate
order parameters. The first is the **average number of active sites**,

$$\rho(t) \equiv N^{-1} \left\langle \sum_i s_i(t) \right\rangle,$$

where N is the total number of sites within one time layer, and where $s_i = 1$ if site
i has occupied bonds emanating from it, and zero otherwise. In the vicinity of the
transition, ρ is expected to exhibit power law scaling, $\rho(t \to \infty) \equiv \rho \sim \tau^\delta, \tau < 0$.
More generally, the critical time dependence of order parameter fluctuations is
described by the ansatz

$$\rho(t) \sim |\tau|^\beta g(t/t(\tau)),$$

with short-time asymptotics (see the discussion in section 11.6.1),

$$\rho(t) \sim t^{-\alpha}, \qquad \alpha = \beta/\nu z.$$

Another possible order parameter is the **infinite-cluster size probability** P_∞
introduced above. It can be generalized to a time-dependent quantity by defining

$P(t)$ as the probability that a cluster that started at $t = 0$ is still active at time t. Since the spatial size of the cluster grows indefinitely with time, $P_\infty = P(t \to \infty)$. The scaling behavior of P is described by $P_\infty = \tau^{\beta'}$, $\tau > 0$, and

$$P(t) \sim |\tau|^{\beta'} g'(t/t(\tau)),$$

with short-time limit $P(t) \sim t^{-\alpha'}$, $\alpha' = \beta'/\nu z$. The four exponents ν, z, β, β' describe the basic scaling behavior of the system. (For the elementary DP class, $\beta = \beta'$, however, this is not a general identity.)

Another important quantity is the **pair connectedness function** $C(\mathbf{x}, t)$, defined as the probability that an active site at $(0, 0)$ is connected to the site at $(\mathbf{x}, t)$ by an open path. The scaling hypothesis implies that

$$C(\mathbf{x}, t) = t^{\theta - d/z} F(|\mathbf{x}|/t^{1/z}, \tau t^{1/\nu z}), \tag{11.73}$$

where it is understood that $|\mathbf{x}|$ is small enough that the final point is within reach of generic clusters spreading out from 0. The dimension θ can be determined via a few consistency checks: the probability of finding connectivity along a fixed coordinate in space, $(\mathbf{x} = 0, 0) \to (0, t)$, is proportional to both the probability of finding a cluster of temporal extension $P(t)$, and the density of active sites $\rho(t)$: $C(0, t) \overset{t \to \infty}{\sim} \tau^{\beta + \beta'}$. On the other hand, the expected stationarity of $C(0, t \to \infty)$ requires that $F(0, \tau t^{1/\nu z}) \overset{!}{\sim} t^{-(\theta - d/z)}$, i.e., $F(0, \tau t^{1/\nu z}) \sim \tau^{-\nu z(\theta - d/z)}$. Comparing these two conditions, we obtain the identification $\tau^{-\nu z(\theta - d/z)} \overset{!}{\sim} \tau^{\beta + \beta'}$, or

$$\theta = \frac{1}{z}\left(-\frac{\beta + \beta'}{\nu} + d\right). \tag{11.74}$$

Equations (11.73) and (11.74) describe the correlation behavior of the cluster.[27] In the next section, we discuss how these quantities can be obtained by field-theoretical methods.

Field theory

Above, we described cluster formulation in a language whose translation to a field-theoretical description is not entirely obvious. We begin by formulating a dynamical mean field equation implementing the nodal update rules. This equation will then be used as input for the construction of a field theory, along the lines of section 11.6.3.

The mean field equation describes the evolution of the mean density of active sites, $\rho(t)$. This quantity increases at a rate proportional to the product of the probability of open bonds, p, the density, ρ, of active sites prior to the update, and the density, $1 - \rho$, of sites that may be converted to active sites. Conversely, it decreases at a rate proportional to product of the density of active sites prior to the update and the probability of closed bonds, $1 - p$. This defines the evolution

---

[27] Notice, however, that the construction assumes that the center of the cluster remains stationary. In high dimensions, this condition may be violated, and clusters may branch out along different directions in space. Closer analysis shows that the validity of Eq. (11.74) is limited to dimensions below the upper critical dimension of the universality class.

equation $\partial_t \rho = c_1 p\rho(1 - \rho) - c_2(1 - p)\rho$, where $c_{1,2}$ are non-universal constants. We may rescale density and time in such a way that the equation assumes the form,

$$\partial_t \rho = (\lambda - 1)\rho - \lambda\rho^2. \tag{11.75}$$

A key feature of this equation is that it possesses a stationary configuration $\rho = 0$, and a non-empty stationary state $\rho = \frac{\lambda-1}{\lambda}$, provided $\lambda > 1$. The critical value $\lambda_c = 1$ at which the non-empty mean field begins to emerge marks the position of the continuous DP transition. Notice the simplicity of the DP mean field equation, which explains the ubiquity of the DP class: one may conjecture that generic processes described by a rate equation free of constant terms exhibit a DP transition, provided that the first two terms $\sim \rho$ and $\sim \rho^2$ admit a non-zero fixed-point solution.

mean field critical exponents Equation (11.75) fixes the **mean field critical exponents** of the system. Using the notation $\tau = \lambda - \lambda_c = \lambda - 1$ for the difference from the critical point, near criticality, $\rho \simeq \tau$, i.e., $\beta_{\mathrm{mf}} = 1$. In the dry phase, $\tau < 1$, the asymptotic temporal decay of density is given by $\rho \sim \exp(-|\tau|t)$, which means that $-\nu z = 1$. However, lacking spatial structure, mean field theory cannot say anything about ν (nor z) individually.

Equation (11.75) can be generalized to a spatially resolved rate equation by upgrading $\rho(t) \to \rho(\mathbf{x}, t) \equiv \rho(x)$ to a local density profile and adding a diffusion term:

$$\partial_t \rho = D\partial^2\rho + \tau\rho - \lambda\rho^2. \tag{11.76}$$

Dimensional analysis shows that $\nu_{\mathrm{mf}} = 1/2$. In combination with the results above (which survive generalization to a local equation – exercise) we then have the mean field, or tree level, prediction

$$\beta_{\mathrm{mf}} = 1, \qquad \nu_{\mathrm{mf}} = 1/2, \qquad z_{\mathrm{mf}} = 2. \tag{11.77}$$

A field theory can now be constructed by interpreting Eq. (11.77) as a model A rate equation along the lines of our discussion in section 11.6.2. To this end, we generalize Eq. (11.76) to a stochastic equation,

$$\partial_t \rho = D\partial^2\rho + \tau\rho - \lambda\rho^2 + \xi,$$

where the noise is correlated as

$$\langle \xi(x) \rangle = 0, \qquad \langle \xi(x)\xi(x') \rangle = 2A\rho(x)\delta(x - x'). \tag{11.78}$$

The crucial feature here is the scaling $\xi \sim \sqrt{\rho}$. It reflects the assumption that the system self-generates noise through its active sites. Since we are to interpret ξ as a coarse-grained variable sampling fluctuations in a large number of active sites, the central limit theorem requires that the fluctuations scale as $\sqrt{\rho}$. We are thus multi-plicative noise dealing with a Langevin equation governed by **multiplicative noise**, i.e., noise that scales in some power of the Langevin variable itself.

In the **MSR field theory representation**, we have $\mathcal{Z} = \int D(\phi,\psi)\, e^{-S[\phi,\psi]}$, where the action (cf. Eq. (11.61)) corresponding to Eq. (11.77) reads[28]

$$S[\phi,\psi] = \int dx\, \left[\psi(\partial_t - D\partial^2 - \tau)\phi + \kappa\psi\phi(\phi - \psi)\right] \tag{11.79}$$

with $\kappa \equiv (A\lambda)^{1/2}$. (In deriving Eq. (11.79), we rescaled the fields so as to make the coefficients of the two nonlinear terms $\sim \phi\psi^2$ and $\sim \phi^2\psi$ equal.) We next aim to understand the importance of the various terms in the action at large length scales. To this end, we consider what happens under the rescaling transformation

$$
\begin{aligned}
\mathbf{x} &\to \mathbf{x}/b, & \phi &\to b^{d_\phi}\phi, \\
t &\to t/b^z, & \psi &\to b^{d_\psi}\psi.
\end{aligned}
\tag{11.80}
$$

A quick calculation shows that the individual terms scale as

$$
\int dx\,
\begin{cases}
\psi\partial_t\phi \\
\psi\partial^2\phi \\
\psi\phi \\
\psi\phi^2 \\
\psi^2\phi
\end{cases}
\to b^{d+z} \int dx\,
\begin{cases}
b^{-(d_\phi+d_\psi)-z} & \psi\partial_t\phi, \\
b^{-(d_\phi+d_\psi)-2} & \psi\partial^2\phi, \\
b^{-(d_\phi+d_\psi)} & \psi\phi, \\
b^{-(2d_\phi+d_\psi)} & \psi\phi^2, \\
b^{-(d_\phi+2d_\psi)} & \psi^2\phi,
\end{cases}
\tag{11.81}
$$

upper crit-
ical di-
mension

Choosing $z = 2$ and $d_\phi = d_\psi = d/2$, the first two contributions to the action become scale invariant. The operator $\sim \tau\psi\phi$ measuring the distance from criticality is strongly relevant, with dimension 2. Finally, the nonlinear operators $\sim \lambda\psi\phi^2$ and $\sim A\psi^2\phi$ carry dimension $4 - d$. This identifies the **upper critical dimension** of directed percolation as $d = 4$.

Before discussing the role of fluctuations below $d = 4$, let us see how the mean field limit materializes in the field theory (11.79). As with the zero-dimensional problems discussed in section 11.4, the phase structure of the system is determined by the pattern of intersecting zero-energy lines of the mean field Hamiltonian, $H = \psi\phi(\phi - \psi - \tau)$. Depending on the sign of τ, three distinct phases need to be distinguished (see fig. 11.6). At $\tau > 0$, the system is in its **active phase**: on the fluctuationless manifold $\psi = 0$, the concentration variable ϕ is driven towards the stable configuration $\phi = \tau$. (Notice that only positive values of ϕ are physically meaningful.) At $\tau < 0$, the empty state $\phi = 0$ is stable, the **inactive phase**. The mean field **phase transition** between the two phases happens at $\tau = 0$. The phase space picture of directed percolation was introduced by Kamenev[29] as the basis for a classification of phase transitions in reaction–diffusion systems. For further

[28] Curiously, this theory has a background in particle physics, where it is known as **Reggeon field theory**; see M. Moshe, *Recent developments in Reggeon field theory*, Phys. Rep. **37**, 255 (1978) for a review.

[29] A. Kamenev, *Classification of phase transitions in reaction–diffusion models*, Phys. Rev. E **74**, 41101 (2006).

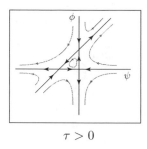 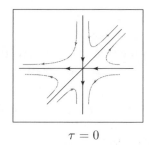 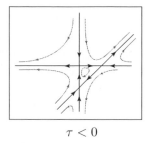

$$\tau > 0 \qquad\qquad\qquad \tau = 0 \qquad\qquad\qquad \tau < 0$$

Fig. 11.6 Three distinct phase portraits of directed percolation. Left: phase supporting a non-vanishing concentration of active sites, center: phase transition, right: empty state.

aspects of the utility of the Hamiltonian approach in this field, we refer to the original reference.

Perturbative renormalization group

How will **fluctuations** alter this picture? Close to the upper critical dimension, in $d = 4 - \epsilon$, renormalization group (RG) techniques can be applied to answer this question. This analysis shows how the two coupling constants (τ, κ) evolve if short-distance fluctuations of the field variables are successively integrated out, and the other coefficients in the action are kept constant using the freedom to rescale space, time, and fields. Referring for the actual computation to problem 11.9.5, this leads to the RG flow equations

$$\frac{d\tau}{d\ln b} = -DC\kappa^2 + \tau\left(2 - \frac{3}{4}C\kappa^2\right),$$

$$\frac{d\kappa}{d\ln b} = \kappa\left(\frac{\epsilon}{2} - \frac{3}{2}C\kappa^2\right),$$
(11.82)

where the constant $C = \frac{S_4}{(2\pi)^4 D^2}$ and S_4 is the area of the four-dimensional unit-sphere. The exponents used to rescale space–time and fields as in Eq. (11.80) are given by

$$z = 2 - \frac{C\kappa^2}{4}, \qquad \chi = \frac{C\kappa^2}{4}.$$

These equations possess a nontrivial fixed point at

$$\tau^* = \frac{D\epsilon}{6} + \mathcal{O}(\epsilon^2), \qquad \kappa^* = \left(\frac{\epsilon}{3C}\right)^{1/2},$$

which means that fluctuations shift the position of the DP transition to a non-vanishing value of τ. At the fixed point, $z = 2 - \epsilon/12$ and $\chi = \epsilon/12$. Linearization of the RG equations around the fixed point leads to

$$\frac{d\tilde\tau}{d\ln b} = \left(2 - \frac{\epsilon}{4}\right)\tilde\tau - \tilde\kappa CD\kappa^*\left(2 + \frac{\epsilon}{4}\right),$$

$$\frac{d\tilde\kappa}{d\ln b} = -\epsilon\tilde\kappa,$$

one-loop
critical
exponents

where $\tau = \tau^* + \tilde{\tau}$ and $\kappa = \kappa^* + \tilde{\kappa}$. The linearized equations imply that the coupling constant $\tilde{\tau}$ is a relevant scaling field with $\tilde{\tau} \sim b^{2-\epsilon/4}$. This result can in turn be used to determine the **one-loop critical exponents**; the scaling $b^{-1} \sim \xi \sim \tilde{\tau}^{-\nu} \sim b^{-\nu(2-\epsilon/4)}$ leads to the identification $\nu = 1/2 + \epsilon/16 + \mathcal{O}(\epsilon^2)$. The critical exponent β follows from the scaling relation $\tau^\beta \overset{!}{\sim} \langle \phi \rangle = b^{-(4-\epsilon)/2 + \chi}\phi(\tau b^{2-\epsilon/4})$. Setting $b \sim \tau^{-\frac{1}{2-\epsilon/4}}$, we obtain the identification $\beta = 1 - \epsilon/6 + \mathcal{O}(\epsilon^2)$. Summarizing, the one-loop RG analysis generates the list of exponents

$$z = 2 - \frac{\epsilon}{12} + \mathcal{O}(\epsilon^2), \qquad \nu = \frac{1}{2} + \frac{\epsilon}{16} + \mathcal{O}(\epsilon^2), \qquad \beta = 1 - \frac{\epsilon}{6} + \mathcal{O}(\epsilon^2).$$

How do these values compare with reality? Numerical simulations for $(3+1)$-dimensional clusters[30] obtain $(z, \nu, \beta) = (1.90(1), 0.581(5), 0.81(1))$, which compares reasonably well with the $\epsilon = 1$ extrapolation of the exponents above, $(z, \nu, \beta) = (1.92, 0.56, 0.83)$. A two-loop extension of the RG[31] leads to excellent agreement. However, in lower dimensions, the situation is worse. For example, in the extreme case of $(1+1)$-dimensional percolation clusters, the RG exponents differ from the results of simulations by around 40%. Even so, the field theory analysis sheds light on the physical mechanisms generating critical behavior, and this is information that cannot be obtained from direct simulations.

11.8 Summary and Outlook

In this chapter, we discussed the foundations of nonequilibrium systems and introduced three pathways to their description: the Boltzmann approach, Langevin dynamics, and probabilistic formulations via master equations. We saw that the latter two afford a unified description, the MSR formalism, in terms of functional integrals. Interpreting the latter as a Hamiltonian functional integral, we constructed its Lagrangian partner – the Onsager–Machlup functional – and recovered the Fokker–Planck equation as the functional variational equation. Noting similarities to quantum mechanics, we saw that the MSR path integral provides a framework for the unified description of nonequilibrium phenomena. Examples of functional-based methods included the variational approach to large fluctuation, and RG approaches to the understanding of fluctuations in higher-dimensional systems. In the final chapter, we will discover another facet of the MSR functional: its interpretation as the classical limit of a nonequilibrium *quantum* theory.

[30] I. Jensen, *Critical behavior of the three-dimensional contact process*, Phys. Rev. E **45**, R563 (1992).

[31] J. B. Bronzan and J. W. Dash, *Higher order epsilon terms in the renormalization group approach to Reggeon field theory*, Phys. Lett. B **51**, 496 (1974).

11.9 Problems

11.9.1 Wigner surmise

At the lowest energy scales, the description of quantum problems subject to randomness can often be reduced to just a pair of states governed by a Hamiltonian with statistically distributed matrix elements. The distribution of the spacing between these levels is then induced by the distribution of matrix elements. Specifically, for Gaussian statistics of the Hamiltonian, the spacing distribution assumes a form known as the "Wigner surmise." What makes this finding important is that Wigner spacing distributions are frequently observed in realistic quantum systems. For example, the spacings of consecutive scattering resonances in complex nuclei often show this form of statistics.

Technically, the derivation of the Wigner surmise is an exercise in identifying the distribution of a dependent variable (the spacing) from that of a primary variable (the matrix elements).

Consider a real Hamiltonian $\hat{H}$ having the form of a 2×2 matrix parameterized as

$$\hat{H} = \begin{pmatrix} H_1 & \Delta \\ \Delta & H_2 \end{pmatrix}.$$

Suppose that the three real numbers H_1, H_2, and Δ are drawn from a Gaussian distribution $P(H_1, H_2, \Delta) \propto e^{-\frac{1}{2}\text{tr}\,\hat{H}^2}$. Derive the joint distribution $P(E_1, E_2)$ of the two energy levels E_i, $i = 1, 2$, of the Hamiltonian $\hat{H}$. Use this result to derive the distribution of the energy splitting $s = |E_1 - E_2|$, and show that it has the form

Wigner surmise

of the **Wigner surmise**

$$P(s) = c_1\,s\,e^{-c_2 s^2},$$

with constants c_1 and c_2. As a further exercise, show that if $\hat{H}$ is complex hermitian, i.e., both real and imaginary parts of Δ are drawn from a Gaussian distribution, the probability distribution vanishes as $P(s) \sim s^2$ for small s.

Answer:

The joint eigenvalue distribution is given by

$$P(E_+, E_-) \sim \int dH_1\,dH_2\,d\Delta\,\delta(E_+ - \lambda_+)\delta(E_- - \lambda_-)e^{-(E_+^2 + E_-^2)/2},$$

where $\lambda_{\pm} = \frac{H_1 + H_2}{2} \pm (\Delta^2 + (\frac{H_1 - H_2}{2})^2)^{1/2}$ denote the eigenvalues of the matrix $\hat{H}$ and we assumed $E_+ > E_-$ (in the opposite case, one has to interchange $\lambda_+ \leftrightarrow \lambda_-$). Integrating over $(H_1 + H_2)/2$ and setting $\delta H = H_1 - H_2$, one obtains $P(E_+, E_-) \sim \int d\delta H\,d\Delta\,\delta(E_+ - E_- - 2(\Delta^2 + \delta H^2)^{1/2})e^{-(E_+^2 + E_-^2)/2}$. Finally, setting $r = (\Delta^2 + \delta H^2)^{1/2}$, and integrating over the angular variable, one obtains $P(E_+, E_-) \sim \int_0^\infty dr\,r\,\delta(E_+ - E_- - 2r)e^{-(E_+^2 + E_-^2)/2} \sim (E_+ - E_-)e^{-(E_+^2 + E_-^2)/2}$. From this result, we obtain the distribution

$$P(s) \sim \int_{E_+ > E_-} dE_+\, dE_-\, P(E_+, E_-)\delta(E_+ - E_- - s) \sim s e^{-s^2/8} \,.$$

We therefore find that $c_2 = 1/8$, and $c_1 = 1/4$ is obtained from the normalization condition. This result shows that the probability of finding a degeneracy is vanishing i.e., the levels repel each other. The generalization to a complex Hamiltonian along the same lines is straightforward.

11.9.2 Ornstein–Uhlenbeck process

The Ornstein–Uhlenbeck process[32] is the "harmonic oscillator" of nonequilibrium dynamics. It describes stochastic evolution in cases where the averaged dynamics of a Langevin particle is described by a linear first-order differential equation. Realizations include the dynamics of a particle subject only to friction (but no external forces), or that of an overdamped particle in a harmonic external potential. The dissipative dynamics close to extremal potential points is often described by variants of this process.

(a) Consider a Brownian particle in one dimension whose velocity is governed by the stochastic differential equation

$$\dot{v} + \gamma v = \frac{f(t)}{m}, \tag{11.83}$$

where f is a Langevin force, and $\langle f(t)f(t')\rangle = A\delta(t - t')$. Determine the general solution of Eq. (11.83) for the velocity with the initial condition $v(0) = v_0$, and calculate $\langle v(t)\rangle$ and $\langle v^2(t)\rangle$. Using the long-time limit of the latter, identify the coefficient A with the help of the equipartition theorem.
(b) Consider the time-dependent velocity distribution $p(v, t)$. It is governed by the Fokker–Planck equation

$$\left[\frac{\partial}{\partial t} + \frac{\partial}{\partial v} a_1(v) - \frac{1}{2} \frac{\partial^2}{\partial v^2} a_2(v) \right] p(v, t) = 0. \tag{11.84}$$

With the help of Eq. (11.84), derive equations for $\partial_t \langle v \rangle$ and $\partial_t \langle v^2 \rangle$ and identify the coefficients a_1 and a_2, e.g., by using the results of part (a). Show that the corresponding Fokker–Planck equation for the generating function $g(k, t) = \int dv\, e^{-ikv} p(v, t)$ is given by

$$\left[\frac{\partial}{\partial t} + \gamma k \frac{\partial}{\partial k} + \frac{\gamma T}{m} k^2 \right] g(k, t) = 0. \tag{11.85}$$

(c) The first-order partial differential equation (11.85) can be solved, for example with the method of characteristics. Show that a general solution is of the form

$$g(k, t) = e^{-\frac{Tk^2}{2m}} \phi(ke^{-\gamma t}).$$

[32] G. E. Uhlenbeck and L. S. Ornstein, *On the theory of Brownian motion*, Phys. Rev. **36**, 823 (1930).

Determine the function ϕ from the initial condition $p(v, t = 0) = \delta(v - v_0)$ and derive the distribution $p(v, t)$. Discuss the short- and long-time limits of your result.

Computing the inverse transform, we obtain the **probability distribution function of the Ornstein–Uhlenbeck process,**

$$p(v, t) = \left(\frac{2\pi T}{m} (1 - e^{-2\gamma t}) \right)^{-1/2} \exp\left[-\frac{(v - v_0 e^{-\gamma t})^2}{2\frac{T}{m}(1 - e^{-2\gamma t})} \right].$$

For short times, the solution $p(v, t) \simeq \sqrt{\frac{1}{2\pi D t}} \exp[-\frac{(v-v_0)^2}{4Dt}]$, with diffusion constant $D = \frac{T}{m}\gamma$, spreads around the initial configuration, and in the long time limit it becomes thermal, $p(v, t) \simeq \sqrt{\frac{m}{2\pi T}} \exp[-\frac{mv^2}{2T}]$.

Answer:

(a) Integrating the equation, one obtains the solution,

$$v(t) = e^{-\gamma t} \left[v_0 + \int_0^t dt' \frac{f(t')}{m} e^{\gamma t'} \right],$$

where $v_0 \equiv v(0)$. Averaging over the distribution for the Langevin force and noting that $\langle f \rangle = 0$, we obtain $\langle v(t) \rangle = v_0 e^{-\gamma t}$. Similarly, making use of the expression for the correlator, we have

$$\langle v^2(t) \rangle = e^{-2\gamma t} \left[v_0^2 + \int_0^t dt' \, dt'' \frac{\langle f(t')f(t'') \rangle}{m^2} e^{\gamma(t'+t'')} \right]$$

$$= e^{-2\gamma t} \left[v_0^2 + \frac{A}{2\gamma m^2} (e^{2\gamma t} - 1) \right].$$

In the long-time limit, $\langle v^2(t) \rangle = \frac{A}{2\gamma m^2} \overset{\text{FDT}}{=} \frac{T}{m}$, i.e., $A = 2\gamma m T$. Setting $\delta v(t) = v(t) - \langle v(t) \rangle$, one may further show that $\langle \delta v(t)\delta v(t') \rangle = \frac{T}{m}(e^{-\gamma|t-t'|} - e^{-\gamma(t+t')})$.

(b) Using the Fokker–Planck equation, we have

$$\partial_t \langle v \rangle = -\int dv \, v \partial_v (a_1 P) + \int dv \frac{1}{2} v \partial_v^2 (a_2 P)$$

$$= \int dv \, a_1 P - \frac{1}{2} \int dv \, \partial_v (a_2 P) = \langle a_1 \rangle.$$

From this result, we obtain $a_1(v) = -\gamma v$. Similarly,

$$\partial_t \langle v^2 \rangle = -\int dv \, v^2 \partial_v (a_1 P) + \int dv \frac{1}{2} v^2 \partial_v^2 (a_2 P)$$

$$= \int dv \, 2v a_1 P - \int dv \, v \partial_v (a_2 P) = \langle 2va_1 \rangle + \langle a_2 \rangle.$$

From part (a), we have $\partial_t \langle v^2 \rangle = -2\gamma v_0^2 e^{-2\gamma t} + \frac{A}{m^2} e^{-2\gamma t}$. A little bit of algebra then leads to the result $\langle a_2 \rangle = \frac{2\gamma T}{m}$. The derivation of Eq. (11.85) amounts to a straightforward substitution of the Fokker–Planck equation into the Fourier integral defining $g(k)$.

(c) Using the method of characteristics, the left-hand side of the equation in (b) gives $-\gamma k \, dt + dk = 0$, i.e., $\frac{dk}{dt} = \gamma k$ and $k(t) = c e^{\gamma t}$. From the variation of g along the characteristic, $\frac{dg}{g} = -\frac{T}{m} k \, dk$, we obtain $g(k,t) = \text{const.} \times e^{-k_B T k^2/2m}$, i.e.,

$$g(k,t) = e^{-k_B T k^2/2m} \phi(k e^{-\gamma t}).$$

With the initial condition $g(k,0) = e^{-ikv_0}$, we obtain $\phi(k) = e^{-ikv_0} e^{k_B T k^2/2m}$, and this leads to

$$g(k,t) = \exp\left[-ikv_0 e^{-\gamma t} - \frac{T}{2m} k^2 (1 - e^{-2\gamma t})\right].$$

11.9.3 Classical linear response theory

In this problem, we derive the correlation functions describing the linear response of an observable to a force in classical dynamics.

Consider a Hamiltonian, $H(\mathbf{x},t) = H_0(\mathbf{x}) + F(t)X(\mathbf{x})$, where F is a generalized time-dependent force conjugate to an observable X. For example, for F a mechanical force, $X(\mathbf{x}) = q$ is a real space coordinate. For simplicity, we assume that, in the absence of the force, $\langle X \rangle = 0$, where $\langle \ldots \rangle = \int_\Gamma \rho_0(\ldots)$, and $\rho_0 = \mathcal{N} \exp(-H/2T)$ is the equilibrium thermal average of the unperturbed system. To leading order in F, we must then have a relation $\langle X(t) \rangle = \int dt' \chi(t - t') F(t')$, where the kernel χ defines the susceptibility. We aim to represent χ as a correlation function of the observable X.

(a) As a warm-up, show that for arbitrary phase space functions f, g, h, we have the algebraic relation, $\int_\Gamma \{f, g\} h = \int_\Gamma f\{g, h\}$. Defining the inner product $\langle f, g \rangle = \int_\Gamma f(\mathbf{x}) g(\mathbf{x})$, discuss in what sense $\{H, \}$ is an anti-hermitian operator and $e^{t\{H, \}}$ is a unitary operator in the space of functions defined on Γ.

(b) Assuming that $F(t \to -\infty) = 0$ and $\rho(t \to -\infty) = \rho_0$, consider the Hamiltonian equations of motion describing the evolution of the phase space function ρ. Solve them to first order in F to verify that the distribution changes as $\rho \to \rho_0 + \delta\rho$, where

$$\delta\rho(t) = -\int_{-\infty}^t dt' \, e^{-(t-t')\{H_0, \}} F(t')\{X, \rho_0\}.$$

(Hint: Let yourself be guided by the structural similarity of the problem to time-dependent quantum mechanical perturbation theory.)

(c) Use this result and the unitarity of the classical time evolution operator derived in (a) to obtain the classical susceptibility (11.68). In doing so, note that if a phase space function $X(\mathbf{x}) \equiv X(\mathbf{x},0) = X(\mathbf{x}(0))$ is interpreted as the initial value along a Hamiltonian flow, then $e^{t\{H\}} X(\mathbf{x},0) = X(\mathbf{x}(t)) = X(t)$ describes the evolution of that function.

(d) Show that, in thermal equilibrium, $\langle \{f, g\} \rangle = \frac{1}{T} \langle (\partial_t f) g \rangle$.

(e) Combine the results of (c) and (d) to derive the classical FDT (11.69).

(f) Test the theorem on the harmonic oscillator, i.e., do the problem stated in more detailed terms in the exercise in section 11.6.4. Choosing $X = q$, it is best to represent the Hamiltonian $H = \frac{1}{2}(q^2 + p^2) \equiv \omega_0 I$ in action–angle variables (I, ϕ), with $q = \sqrt{I}\cos(\phi)$, $p = \sqrt{I}\sin(\phi)$, and $\omega_0 = 1/2$. This set of variables is the classical analog of the number–phase variables frequently used in the description of the quantum harmonic oscillator. Check that the transformation $(q, p) \to (I, \phi)$ is canonical, and use the simplicity of the equations of motion in the latter variables to compute the building blocks of the FDT.

Answer:

(a) With $\{f, g\} = \partial_{q_i} f \partial_{p_i} g - \partial_{p_i} f \partial_{q_i} g$, the first identity follows immediately from an integration by parts. The Poisson bracket $\{H, \ \} : f \mapsto \{H, f\}$ acts as a linear operator in the vector space of phase space functions. Integrating by parts, one finds that $\langle \{H, f\}, g \rangle = \int_\Gamma \{H, f\} g = -\int_\Gamma f \{H, g\} = -\langle f, \{H, g\} \rangle$, showing its anti-hermiticity. The unitarity of its exponential, $\langle e^{t\{H, \ \}} f, g \rangle = \langle f, e^{-t\{H, \ \}} g \rangle$, then follows on general grounds, or more explicitly by Taylor expansion in t and iterative application of the anti-hermiticity relation.

(b) The classical equation of motion describing the evolution of the phase space function ρ reads $\partial_t \rho = -\{H, \rho\}$. Writing $\rho = \rho_0 + \delta\rho$, an expansion to first order in F leads to $\partial_t \delta\rho = -\{H_0, \delta\rho\} - F(t)\{X, \rho_0\}$. This is an inhomogeneous first-order linear differential equation for $\delta\rho$. Noting that the homogeneous equation is solved by $\delta\rho(t) = e^{(t-t')\{H, \ \}} \delta\rho(t')$, we immediately obtain $\delta\rho$ as stated in the problem.

(c) Computing the average of the observable X at time t, we obtain

$$
\begin{aligned}
\langle X(t) \rangle &= \int_\Gamma \delta\rho(\mathbf{x}, t) X(\mathbf{x}) = -\int_{-\infty}^t dt' \int_\Gamma \left(e^{-(t-t')\{H_0, \ \}} F(t')\{X, \rho_0\} \right) X \\
&= -\int_{-\infty}^t dt' F(t') \int_\Gamma \{X, \rho_0\} e^{(t-t')\{H_0, \ \}} X \\
&= -\int_{-\infty}^t dt' F(t') \int_\Gamma \{X(0), \rho_0\} X(t' - t) \\
&= \int_{-\infty}^t dt' F(t') \int_\Gamma \rho_0 \{X(0), X(t' - t)\} = \int_{-\infty}^t dt' F(t') \langle \{X(0), X(t' - t)\} \rangle \\
&= \int_{-\infty}^\infty dt' \Theta(t - t') \langle \{X(t - t'), X(0)\} \rangle F(t') \equiv \int_{-\infty}^\infty dt' \chi(t - t') F(t'),
\end{aligned}
$$

where in the fourth line we used the properties of the Poisson bracket derived in (a), and in the last line the temporal translation invariance of phase space averages, $\langle f(0) f(t) \rangle = \langle f(s) f(t + s) \rangle$. Comparison with (11.68) proves the statement.

(d) This is proven as $\langle \{f, g\} \rangle = \int_\Gamma \rho_0 \{f, g\} = \int_\Gamma \{\rho_0, f\} g = -\frac{1}{T} \int_\Gamma \rho_0 \{H, f\} g = \frac{1}{T} \int_\Gamma \rho_0 (\partial_t f) g = \frac{1}{T} \langle (\partial_t f) g \rangle$, where we used $\rho_0 = \mathcal{N} \exp(-H/T)$, the chain rule for the Poisson bracket, and the equation of motion.

(e) The FDT makes reference to the imaginary part of the Fourier transformed susceptibility, which we represent as

$$\text{Im}\chi(\omega) = \frac{1}{2i} \int_0^\infty dt\, e^{i\omega t} \langle \{X(t), X(0)\} \rangle - \text{c.c.} = \frac{1}{2i} \int_{-\infty}^\infty dt\, e^{i\omega t} \langle \{X(t), X(0)\} \rangle.$$

Here, the second equality is based on the fact that the phase space expectation value is real, and on its time translational invariance, already used in (c). Using the identity proven in (d), this becomes

$$\text{Im}\chi(\omega) = \frac{1}{2iT} \int dt\, e^{i\omega t} \langle \partial_t X(t) X(0) \rangle = -\frac{\omega}{2T} \int dt\, e^{i\omega t} \langle X(t) X(0) \rangle = -\frac{\omega}{2T} C(\omega).$$

(f) The canonicity of the action–angle variables follows from the quick check $1 = \{q, p\} = (\partial_I \sqrt{I} \cos(\phi))(\partial_\phi \sqrt{I} \sin(\phi)) - (\partial_\phi \sqrt{I} \cos(\phi))(\partial_I \sqrt{I} \sin(\phi))$, proving the invariance of the Poisson bracket. This shows that, in action–angle variables, the equations of motion assume the form $\dot{I} = 0$ and $\dot{\phi} = -\partial_I H = -\omega_0$. We thus obtain $q(t) = \sqrt{I} \cos(\phi - \omega_0 t)$, where (I, ϕ) define the initial coordinates of the trajectory. From here, the Poisson bracket entering the susceptibility is obtained as $\{q(t), q(0)\} = (\partial_I \sqrt{I} \cos(\phi - \omega_0 t))(\partial_\phi \sqrt{I} \cos(\phi)) - (\partial_\phi \sqrt{I} \cos(\phi - \omega_0 t))(\partial_I \sqrt{I} \cos(\phi)) = -\sin(\omega_0 t)$. This function does not depend on the phase space variables, and therefore equals its average. The Fourier transformation of $\chi(t) = \Theta(t)\langle\{q(t), q(0)\}\rangle = -\Theta(t)\sin(t\omega_0)$ leads to $\text{Im}\chi(\omega) = \frac{\pi}{2}(\delta(\omega + \omega_0) - \delta(\omega - \omega_0))$. (What is the physical meaning of this result?) Noting that $\langle \ldots \rangle = \mathcal{N} \int_0^\infty dI \int_0^{2\pi} d\phi\, e^{-\omega_0 I/T}(\ldots)$, a quick calculation shows that $\langle q(t) q(0) \rangle = \langle I \cos(\phi - \omega_0 t) \cos(\phi) \rangle = \frac{T}{2\omega_0} \cos(\omega_0 t)$. Computing the Fourier transform as above, this gives $C(\omega) = \frac{\pi T}{2\omega_0}(\delta(\omega + \omega_0) + \delta(\omega - \omega_0)) = -\frac{\pi T}{2\omega}(\delta(\omega + \omega_0) - \delta(\omega - \omega_0))$. Comparison of the two formulae shows that the FDT is satisfied.

11.9.4 Ornstein–Uhlenbeck process revisited

Here, we approach the Ornstein–Uhlenbeck process from the point of view of path integration. As with the harmonic oscillator in quantum mechanics, the application of the path integral to this simple problem may seem like shooting sparrows with canons. However, what makes this integral useful is that it frequently appears in the representation of more complex problems. (In this regard, it has a status similar to that of the path integral of the quantum harmonic oscillator.)

Once more, we write P instead of p to distinguish probability distributions from momenta p in phase space.

Consider the stochastic equation describing a particle in a harmonic potential with overdamped dynamics, $\dot{q} + \frac{\omega_0^2}{\gamma} q = \xi$ with $\langle \xi(t)\xi(t') \rangle = A\delta(t - t')$, where $A = 2T\frac{\omega_0^2}{\gamma^2}$. In the path integral representation, the probability distribution reads

$$P(q_f, t | q_i, 0) = \int_{q(t) = q_f, q(0) = q_i} D[q, p] e^{-S[q,p]},$$

where $S = -\int_0^t dt'\,(p\dot{q} - H)$, and $H = -\alpha pq - \frac{A}{2}p^2$ with $\alpha = \frac{\omega_0^2}{\gamma}$.

(a) Solve the Hamiltonian equations of motion and determine the corresponding classical action S_{cl}. **(b)** To go beyond the leading semiclassical approximation

$P(q_f, t|q_i, 0) \sim e^{-S[q,p]}$, we need to determine the fluctuation determinant. In section 0.2 we saw that the fluctuation determinant of a path integral assumes the form

$$\frac{1}{\sqrt{2\pi}} \sqrt{-\frac{\partial^2 S_{\rm cl}}{\partial q_i \partial q_f}}, \tag{11.86}$$

where the action under the square root is multiplied by a real factor $i \to -1$, since we are doing an imaginary-time integral. However, the direct substitution of expression (11.86) for the fluctuation determinant would be premature: (11.86) applies to theories whose Hamiltonian is of conventional type, $\hat{H} = \hat{T} + \hat{U}$, where $\hat{T}$ and $\hat{U}$ are the kinetic and potential energy, respectively.

In order to bring the Hamiltonian into this form, interpret q and $p \to \partial_q$ as operators with commutation relation $[p, q] = 1$, and $H(q, p) \to \hat{H}(\hat{q}, \hat{p})$ as the operator governing a Fokker–Planck equation (recall the positioning of derivatives in the Fokker–Planck operator). Next rearrange terms to bring this operator into a $\hat{T} + \hat{U}$ representation, where the kinetic energy contains a suitably defined vector potential. This expression defines the Hamiltonian function appearing in an imaginary-time path integral. Show that the action has picked up a constant piece, $S \to S - \frac{\alpha}{2}t$ relative to the formula given above. At first sight, this seems to indicate the presence of a diverging factor $e^{-S} \to e^{-S}e^{\alpha t}$ in the long-time evolution. However: **(c)** Evaluate the determinant (11.86) to obtain the result

$$P(q_f, t|q_i, 0) = \left(\frac{2\alpha}{A}\right)^{1/2} \frac{e^{-\frac{\alpha}{A}\frac{(q_i - q_f e^{\alpha t})^2}{e^{2\alpha t}-1}}}{(1 - e^{-2\alpha t})^{1/2}}.$$

Notice that the factor $\exp(\alpha t)$ obtained in (b) cancels a factor coming from the fluctuation determinant.

Answer:

(a) The classical equations of motion read

$$\dot{q} = \partial_p H = -\alpha q - Ap, \qquad \dot{p} = -\partial_q H = \alpha p,$$

with solution $p(t) = p_0 e^{\alpha t}$ and $q(t) = ce^{-\alpha t} - \frac{Ap_0}{2\alpha}e^{\alpha t}$. From the boundary conditions, one finds $-\frac{Ap_0}{2\alpha} = \frac{q_i - q_f e^{\alpha t}}{1 - e^{2\alpha t}}$ and $c = \frac{q_i - q_f e^{-\alpha t}}{1 - e^{-2\alpha t}}$. When substituted into the classical action, after some algebra one obtains

$$S_{\rm cl} = -\int_0^t dt' \left(p\dot{q} + \alpha pq + \frac{A}{2}p^2\right) = \frac{\alpha}{A}\frac{(q_i - q_f e^{\alpha t})^2}{e^{2\alpha t} - 1},$$

with $\alpha/A = \gamma/2T$. (You may check for consistency that $\partial S/\partial q_i = p_i$.) **(b)** The Fokker–Planck differential operator reads $\hat{H} = -\alpha\partial_q q - \frac{A}{2}\partial_q^2$. This can be equivalently represented as $\hat{H} = -\frac{A}{2}\left(\partial_q + \frac{\alpha}{A}q\right)^2 + \frac{\alpha^2}{2A}q^2 - \frac{\alpha}{2}$. Substituting the function corresponding to this representation ($\partial_q \to p$) into the path integral, and rearranging terms, we obtain $H = -\alpha pq - \frac{A}{2}p^2 - \frac{\alpha}{2}$. Integration over time yields a shift $-\frac{\alpha t}{2}$ relative to the na"ive action. **(c)** The solution involves a straightforward exercise in differentiation.

11.9.5 Directed percolation

In this problem, we apply the field theory (11.79) to explore the role of fluctuations in directed percolation slightly below the upper critical dimension, $d = 4 - \epsilon$. The analysis yields a set of RG equations, whose physical significance is discussed in the main text.

Phase transitions in the **directed perco-lation** universality class are described by the field theory (11.79). In this problem, we apply elements of diagrammatic perturbation theory to explore the impact of fluctuations on the critical theory. To this, end, we first note that contractions of the fields ϕ, ψ with respect to the free action of the theory, $\kappa = 0$, generate a Green function,

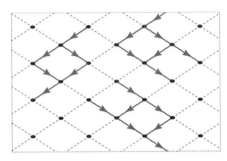

$$\langle \phi_q \psi_{q'} \rangle = g_q^+ \delta_{q,-q'}, \qquad (g_q^+)^{-1} = -i\omega + Dk^2 - \tau, \qquad (11.87)$$

where the corresponding advanced Green function is given by $g_q^- = g_{-q}^+$, and $q \equiv (\omega, \mathbf{k})$ is the four-momentum.

(a) Perturbation theory: To familiarize ourselves with the structure of the perturbation theory in κ, we first consider perturbative corrections to the Green function and the vertex. Show that the Green function generalizes as $g^{-1} \to g^{-1} - \Sigma$, where, to lowest order in κ, the self-energy correction reads

$$\Sigma_q = -2\kappa^2 \int \frac{d^d k'}{(2\pi)^d} \frac{1}{-i\omega + Dk'^2 + D(\mathbf{k} - \mathbf{k}')^2 - 2\tau}. \qquad (11.88)$$

Show that the first correction to the coupling constant, $\kappa \to \kappa + \delta\kappa$, appears at third order and has the form,

$$\delta\kappa \simeq -2\kappa^3 \int \frac{d^d k'}{(2\pi)^d} \frac{1}{(Dk^2 - \tau)^2}. \qquad (11.89)$$

Convince yourself that the diagrams contributing to the renormalization of κ and τ correspond to the configuration-space processes indicated in the figure above.

(b) Renormalization group: Consider the theory regularized with a hard cutoff Λ for the (spatial) momentum integral. Rescale (spatial) momentum by this cutoff, $\mathbf{k} \to \mathbf{k}/\Lambda$, which means that the momentum integrals now extend over the support $|\mathbf{k}| \leq 1$ and all coupling constants are measured in units Λ^{d_x}, where d_x is the relevant engineering dimension.

Integrate out perturbatively the spatial fast modes within the momentum shell $(1, 1/b)$ with $0 < \ln b \ll 1$. Show that, in spatial dimension $d = 4 - \epsilon$, this modifies the coupling constants in the following way:

$$\omega \to \omega \left(1 - \frac{x}{2} \ln b\right), \qquad D \to D \left(1 - \frac{x}{4} \ln b\right),$$

$$\tau \to \tau \left(1 - \left(\frac{D}{\tau} + 1\right) x \ln b\right), \qquad \kappa \to \kappa \left(1 - 2x \ln b\right),$$

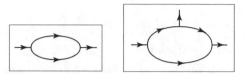

Fig. 11.7   Skeleton structure of the two diagrams contributing to the renormalization of the propagator
(left) and interaction vertex (right).

where we have introduced

$$x \equiv \frac{S_4}{(2\pi)^4} \left(\frac{\kappa}{D}\right)^2, \tag{11.90}$$

and S_4 is the area of the four-dimensional unit-sphere. Next, rescale coordinates
and fields as $q \to q/b$, $\omega \to \omega/b^z$, $\phi \to \phi b^{-\frac{4-\epsilon}{2}+\chi}$. Choose the dynamical exponent
z and the field renormalization exponent χ so as to make the diffusion constant
D and the coefficient of frequency invariant under renormalization. Show that this
condition generates the equations

$$z = 2 - \frac{x}{4}, \qquad \chi = \frac{x}{4}.$$

The mass term τ and the interaction constant κ flow according to (11.82). The
meaning of these equations is discussed in the main text.

Answer:

(a) In the Fourier basis, the interaction contribution to the action $S = S_0 + S_{\text{int}}$
is given by $S_{\text{int}}[\phi, \psi] = \kappa \int dq\, dq'\, \psi_q \phi_{q'} (\phi_{-q-q'} - \psi_{-q-q'})$, where the measure $dq = d\omega\, d^d k/(2\pi)^{d+1}$. Expanding the action to second order in κ, and applying the contraction rule (11.87), we obtain the self-energy

$$\Sigma_q = 2\kappa^2 \int dq'\, (g_{q-q'} g_{q'} + g_{q+q'} g_{q'}).$$

The frequency integration $\int d\omega'$ over $g_{q+q'} g_{q'}$ vanishes because the integrand falls off
as $\sim 1/\omega'^2$ and has no poles in the upper complex plane. Performing the contour
integration over the remaining contribution, we readily obtain Eq. (11.88). The
diagrammatic representation of this term is shown in fig. 11.7, where the external
field vertices (which do not contribute to the self-energy) have been included for
the sake of notational clarity.

To obtain the renormalization of the coupling constant of the interaction, κ, it
is necessary to expand to third order in perturbation theory. The Wick contraction
of these terms (up to the external field vertices entering the interaction operator,
$\sim \psi \phi^2$ and $\sim \psi^2 \phi$) yields two one-loop diagrams. In one of these, all Green functions
have their poles on one side of the real axis, which implies that they vanish upon
frequency integration. The surviving diagram, whose graphical representation is
shown in fig. 11.7 has the analytic representation

$$\frac{(-1)^3}{3!}\langle S_{\text{int}}^3 \rangle \to -\frac{2^4 3}{3!}\kappa^3 \int dq\, dq'\, dq''\, \phi_{q'}\psi_{q''}(-\phi_{q-q'-q''} + \psi_{q-q'-q''}) \int dp\, g_p^2 g_{-p},$$

where we have neglected the "small" momenta q, q', q'' in the arguments of the "fast" Green functions g_p. The required renormalization of $\delta\kappa$ thus reads $\delta\kappa = -(2\kappa)^3 \int dq\, g_q^2 g_{-q}$. Performing the integral over frequency, we obtain Eq. (11.89).

(b) Consider the contribution $\delta\Sigma$ to Eq. (11.88) due to the integration over the fast momentum layer, $\int_f d^d k' \equiv \int_{1/b < |\mathbf{k}'| \leq 1} d^d k'$. An expansion of the self-energy in small corrections to the fast momenta gives

$$\delta\Sigma_q \simeq 2\kappa^2 \left(\frac{I_{-2}}{2D} + \frac{(i\omega + 2\tau)I_{-4}}{(2D)^2} + \frac{k^2}{4D}\left(\frac{2}{d} - 1\right)I_{-4} + \cdots \right),$$

where we have introduced $I_n \equiv \int_f \frac{d^d k'}{(2\pi)^d} k'^n$. Likewise, the fast momentum contribution to the coupling constant correction (11.89) reads

$$\delta\kappa \simeq -\frac{2\kappa^3}{D^2}I_{-4}.$$

In dimensions $4 - \epsilon$, we have $I_{-2} \simeq I_{-4} = \frac{S_4}{(2\pi)^4}\ln b$. Substituting this result into the fast-fluctuation induced change in the Green function, $g^{-1} \to g^{-1} - \delta\Sigma$, we obtain the required renormalization of the coupling constants.

The rescaling of coordinates and fields modifies the renormalization of coupling constants according to

$$\omega \to \omega\left(1 + \left(2\chi - \frac{x}{2}\right)\ln b\right), \qquad D \to D\left(1 + \left(2\chi + z - 2 - \frac{x}{4}\right)\ln b\right),$$

$$\tau \to \tau\left(1 + \left(2\chi + z - \frac{Dx}{\tau} - x\right)\ln b\right),$$

$$\kappa \to \kappa\left(1 + \left(\frac{\epsilon}{2} + 3\chi + z - 2 - 2x\right)\ln b\right).$$

The invariance of the first two terms generates the required conditions on χ and z. We finally substitute χ and z into the remaining two equations to obtain the RG equations (11.82).

12 Nonequilibrium (Quantum)

SYNOPSIS In the previous chapter, we saw that the description of physics away from equilibrium requires a whole new conceptual framework. We will now extend this theory to the quantum world. Specifically, we will derive and discuss quantum master equations, and construct a nonequilibrium quantum field theory, which reduces to the MSR functional integral in the classical limit. After introducing the required theoretical foundations in the first sections of the chapter, we will apply them to one extended case study: out-of-equilibrium transport through mesoscopic quantum devices. This application will illustrate how the various concepts introduced earlier are required in order to address realistic problems in nonequilibrium physics.

Departures from equilibrium require theoretical frameworks different from the Matsubara formalism developed in earlier chapters. The latter builds on the grand canonical density operator, and hence becomes inapplicable outside thermal equilibrium. This begs the question as to how quantum many-body systems can be described if their state is not known *a priori*. One approach would be to quantize the individual approaches introduced in the previous chapter to derive quantum–Langevin, or quantum master, equations. This can and has been done. However, from the perspective of the present text, there exists a more powerful strategy. We will build our approach on a formalism introduced by Leonid Keldysh[1] to describe nonequilibrium systems under general conditions. Originally formulated in the language of second quantization, Keldysh theory is tailored to a functional integral representation. Its classical limit turns out to be the MSR integral introduced in the previous chapter to describe a classical nonequilibrium. In this sense, the Keldysh functional is a "theory of everything" describing both classical and quantum aspects of nonequilibrium physics within one coherent framework.

Unlike the previous chapter, here we will follow a top-down approach and begin with an introduction of the general Keldysh functional integral representation. It will then be used as a platform from which various more specialized tools can be derived by reduction. While the present exposition will be introductory and self-contained, readers interested in a more expansive coverage of the subject are referred to the excellent textbook of Ref.[2].

---

[1] L. V. Keldysh, *Diagram technique for nonequilibrium processes*, Sov. Phys. JETP **20**, 1018 (1965).

[2] A. Kamenev *Field Theory of Non-Equilibrium Systems* (Cambridge University Press, 2011).

INFO The probabilistic nature of classical nonequilibrium theory originates in the effectively random influence of a large number of microscopic degrees of freedom on a smaller number of mesoscopic ones. Quantum mechanics, on the other hand, is *intrinsically* probabilistic. Even for small systems, with deterministic classical dynamics, **quantum uncertainty** makes the measurement of observables probabilistic. We will thus be confronted with two "layers" of stochasticity: extrinsic- and intrinsic-quantum.

To understand the connection between these two, consider a classically deterministic system, prepared in a definite quantum state $|\Psi\rangle$. The quantum expectation value of an observable $\hat{X}$ is then given by $(\hat{X}) \equiv \langle\Psi|\hat{X}|\Psi\rangle = \sum_n X_n|\Psi_n|^2$, where X_n is the nth eigenvalue of $\hat{X}$, $\Psi_n \equiv \langle n|\Psi\rangle$, $|n\rangle$ is the nth eigenstate, and we denote the quantum expectation value by parentheses to distinguish it from the average over fluctuations below. The formal analogy to a distribution with probabilities $P_n = |\Psi_n|^2$ makes the probabilistic nature of quantum mechanics manifest.

However, for our present purposes, it will be more useful to represent the expectation value as $(\hat{X}) = \mathrm{tr}\,(\hat{\rho}_{|\Psi\rangle}\hat{X})$, where $\hat{\rho}_{|\Psi\rangle} \equiv |\Psi\rangle\langle\Psi|$ is the density operator of a pure quantum state. This representation immediately generalizes to mixed states, $\rho_{|\Psi\rangle} \to \hat{\rho} \equiv \sum_a p_a|\Psi_a\rangle\langle\Psi_a|$, where the system is realized with probability p_a in the states of an orthonormal basis $\{|\Psi_a\rangle\}$. In a many-particle system, the inevitable presence of fluctuations– cf. the principles discussed in the previous chapter – may render the coefficients p_a effectively random. These fluctuations may be externally imposed, or caused by integration over microscopic degrees of freedom of the system. Summarily denoting the average over fluctuations by $\langle\ldots\rangle_\mathrm{f}$, we then have

$$\langle\hat{X}\rangle \equiv \langle(\hat{X})\rangle_\mathrm{f} = \left\langle\mathrm{tr}\,(\hat{\rho}\hat{X})\right\rangle_\mathrm{f} = \mathrm{tr}\,(\langle\hat{\rho}\rangle_\mathrm{f}\hat{X}) = \sum_a \langle p_a\rangle_\mathrm{f}\,\langle\Psi_a|\hat{X}|\Psi_a\rangle.$$

This representation makes the distinction between classical and extrinsic fluctuations manifest: the latter make the quantum probabilities random variables subject to external fluctuations. In a canonical (basis-invariant) manner, all aspects of quantum mechanics – interference, wave coherence, etc. – are encapsulated in the mathematical properties of the density operator and the trace operation. The statistics of fluctuations (both thermal, and externally imposed) is contained in the average over coefficients, $\langle\ldots\rangle_\mathrm{f}$. This average may effectively be generated by a taking a quantum average (i.e., a trace operation) over many microscopic degrees of freedom, leaving only a smaller number of mesoscopic quantities in a "reduced density matrix" (we will discuss this principle in the following section). For a summary of the above discussion, see the following table:

Classical	Quantum
variable X	hermitian operator $\hat{X}$
values of X, x	eigenvalues of $\hat{X}$, x
probability distribution p	density operator $\hat{\rho}$
moments $\langle X^n\rangle = \int dx\, p(x)x^n$	moments $\langle\hat{X}^n\rangle = \mathrm{tr}(\hat{\rho}\hat{X}^n)$

12.1 Prelude: Quantum Master Equation

REMARK This section introduces several quick approaches to the description of nonequilibrium systems and, at the same time, motivates the subsequent construction of a field theory. Readers wishing to proceed in a streamlined manner may proceed directly to section 12.2.1.

Consider the standard set-up of a "system" coupled to a "bath." The bath may represent external degrees of freedom (e.g., electromagnetic modes coupled to a quantum dot), or internal microscopic degrees of freedom coupled to fewer collective quantum variables (e.g., quasiparticles affecting the phase of a superconducting Josephson junction). As in the previous chapter, our strategy will be to integrate, or trace

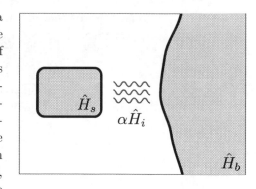

over, the environmental degrees of freedom to generate an effective description of the system's degrees of freedom.

12.1.1 Nakajima–Zwanzig equation

Consider the density operator $\hat{\rho}$ describing the state of the total system in the product Hilbert space $\mathcal{H} \equiv \mathcal{H}_s \otimes \mathcal{H}_b$ of system (s) and bath (b). The information in which we are interested is contained in the **reduced density matrix**, $\hat{\rho}_s \equiv \mathrm{tr}_b(\hat{\rho})$, where tr_b denotes the trace over the Hilbert space of the bath. Note that $\hat{\rho}_s$ is a linear operator in $\mathcal{H}_s$ and that it has the hermiticity, positivity, and unit-trace properties of a density matrix. The dynamics of the full system is controlled by the Hamiltonian operator $\hat{H} \equiv \hat{H}_s + \hat{H}_b + \alpha \hat{H}_i$, where the coupling between system and bath, $\hat{H}_i$, is multiplied by a dimensionless coupling constant α. We assume that, at some initial time, $t = 0$, the system and bath were decoupled, and that the latter is in a state of thermal equilibrium, $\hat{\rho}(0) = \hat{\rho}_s(0) \otimes \hat{\rho}_b^{\mathrm{eq}}$.

reduced density matrix

An efficient strategy to obtain an effective evolution equation controlling the dynamics of the system has been introduced by Nakajima and Zwanzig.[3] Its starting point is an exact representation of the von Neumann equation

$$(\partial_t - \hat{L})\hat{\rho} = 0,$$

in terms of the "quantum Liouville operator" $\hat{L} = \hat{L}_s + \hat{L}_b + \alpha \hat{L}_i$, where $\hat{L}_{s,b,i} \equiv -i[\hat{H}_{s,b,i}, \ ]$. We next introduce the projector

$$\mathcal{P} \equiv \hat{\rho}_b^{\mathrm{eq}} \, \mathrm{tr}_b(\,.\,),$$

where the trace operation acts on everything to the right, and unit normalization of $\hat{\rho}_b^{\mathrm{eq}}$ is assumed (why is $\mathcal{P}$ a projector?). The idea behind introducing this projector is that we assume the bath stays close to a thermal state. In this sense, $\mathcal{P}$ projects onto the preferred state of the system, and the complementary projector $\mathcal{Q} \equiv \mathrm{id} - \mathcal{P}$ projects onto transient states. Without loss of generality,[4] we assume that the

---

[3] S. Nakajima, *On quantum theory of transport phenomena: Steady diffusion*, Prog. Theory. Phys. **20**, 948 (1958); R. Zwanzig, *Ensemble method in the theory of irreversibility*, J. Chem. Phys. **33**, 1338 (1960).

[4] Should this condition not be met by the interaction, the term $\mathrm{tr}_b(\hat{\rho}_b^{\mathrm{eq}} \hat{H}_i) = 0$ can be interpreted as part of the *system* Hamiltonian operator.

thermal trace of the interaction operator over $\mathcal{H}_b$ vanishes, $\mathrm{tr}_b(\hat{\rho}_b^{\mathrm{eq}}\hat{H}_i) = 0$. The projector then obeys the equations,

$$\mathcal{P}\hat{L}_b = \hat{L}_b\mathcal{P} = 0, \qquad [\hat{L}_s, \mathcal{P}] = 0, \qquad \mathcal{P}\hat{L}_i\mathcal{P} = 0, \tag{12.1}$$

where the first equation follows from the cyclic invariance of the trace and the commutativity $[\hat{H}_b, \hat{\rho}_b^{\mathrm{eq}}] = 0$, the second should be obvious, and the third expresses the presumed vanishing of the interaction under the bath-trace. Introducing the shorthand notation $\hat{\rho}_P \equiv \mathcal{P}\hat{\rho}$, $\hat{\rho}_Q \equiv \mathcal{Q}\hat{\rho}$, the Liouville equation may now be split into two,

$$\begin{aligned}\partial_t\hat{\rho}_P &= \hat{L}_s\hat{\rho}_P + \alpha\mathcal{P}\hat{L}_i\hat{\rho}_Q, \\ \partial_t\hat{\rho}_Q &= \mathcal{Q}\hat{L}\hat{\rho}_Q + \alpha\mathcal{Q}\hat{L}_i\hat{\rho}_P.\end{aligned} \tag{12.2}$$

The second of these is an inhomogeneous linear equation solved by

$$\hat{\rho}_Q(t) = \alpha\int_0^t dt'\, e^{t'\mathcal{Q}\hat{L}\mathcal{Q}}\hat{L}_i\hat{\rho}_P(t - t').$$

Substitution of this result into the first equation yields

$$\partial_t\hat{\rho}_P = \hat{L}_s\hat{\rho}_P + \alpha^2\int_0^t dt'\,\mathcal{P}\hat{L}_i e^{t'\mathcal{Q}\hat{L}\mathcal{Q}}\hat{L}_i\hat{\rho}_P(t - t'). \tag{12.3}$$

Finally, using the definition of the projection operation $\hat{\rho}_P = \mathcal{P}\hat{\rho} = \hat{\rho}_b^{\mathrm{eq}}\otimes\hat{\rho}_s$, this can be rewritten as

$$\boxed{\partial_t\hat{\rho}_s = \hat{L}_s\hat{\rho}_s + \alpha^2\int_0^t dt'\,\left\langle\hat{L}_i e^{t'\mathcal{Q}\hat{L}\mathcal{Q}}\hat{L}_i\right\rangle_b\hat{\rho}_s(t - t')} \tag{12.4}$$

where $\langle\ldots\rangle_b \equiv \mathrm{tr}_b((\ldots)\hat{\rho}_b^{\mathrm{eq}})$.

Nakajima–
Zwanzig
equation

Equation (12.4) is called the **Nakajima–Zwanzig equation**[3] or **quantum master equation**. The latter denotation is misleading in that this equation (together with a complementary equation for $\hat{\rho}_Q$) contains the full information on the quantum evolution of the density operator. Specifically, the information on the environmental coupling is stored in the time-nonlocal memory kernel on the right hand side. By contrast, master equations describe memoryless Markovian processes.

Markovian approximation and Lindblad equation

To reduce Eq. (12.4) to the simpler form of a master equation, we apply a **Markovian approximation**, which assumes that the time-scales over which ρ_s changes are large in comparison to the relaxation times t' of the integral kernel (for the discussion of a setting where such approximations fail, see problem 12.9.1). In this case, the equation reduces to the time-local form $\partial_t\hat{\rho}_s = (\hat{L}_s + \alpha^2\hat{X})\hat{\rho}_s(t)$, where $\hat{X} \equiv \int_0^\infty dt'\,\langle\hat{L}_i e^{t'\mathcal{Q}\hat{L}\mathcal{Q}}\hat{L}_i\rangle_b$.

The equation can be simplified further if we assume weak coupling α in order to approximate the generator in the exponent by $\hat{L} \simeq \hat{L}_0 \equiv \hat{L}_s + \hat{L}_b$ (a variant of a "Born approximation"). The Markovian approximation $\hat{\rho}_P(t - t') \simeq \hat{\rho}_P(t)$ to the integral kernel in Eq (12.3) then reduces to

$$\mathcal{P}\hat{L}_i e^{t'\mathcal{Q}\hat{L}_0}\mathcal{Q}\hat{L}_i\,\hat{\rho}_P(t) = \mathcal{P}\hat{L}_i e^{t'\hat{L}_0}\hat{L}_i\,\hat{\rho}_P(t).$$

EXERCISE Use Eq. (12.1) above to verify this relation.

Further, defining $\hat{H}_0 = \hat{H}_s + \hat{H}_b$ and $\hat{O}(t) \equiv e^{it\hat{H}_0}\hat{O}e^{-it\hat{H}_0}$ for the interaction-picture time-representation of an operator $\hat{O}$, we have

$$\mathcal{P}\hat{L}_i e^{t'\hat{L}_0}\hat{L}_i\,\hat{\rho}_P(t) \simeq \mathcal{P}\hat{L}_i\hat{L}_i(-t')\,\hat{\rho}_P(t).$$

EXERCISE For an arbitrary operator $\hat{O}$, show that $e^{t\hat{L}_0}\hat{O} = \hat{O}(t)e^{t\hat{L}_0}$. The equality above is approximate since, once again, we neglect the action of $e^{t'\hat{L}_0}$ on the (slow) operator $\hat{\rho}_P$.

Markovian quantum master equation

Using this result, we find that Eq. (12.4) reduces to the **Markovian quantum master equation**

$$\boxed{\partial_t \hat{\rho}_s = \left(\hat{L}_s + \alpha^2 \int_0^t dt' \left\langle \hat{L}_i(0)\hat{L}_i(-t')\right\rangle_b\right)\hat{\rho}_s} \qquad (12.5)$$

Under a mild set of further assumptions, this equation can be reduced to a yet more manageable form due to Lindblad.[5] We first represent the system–bath coupling as

$$\hat{H}_i = \sum_n \left(c_n \hat{\Gamma}_n \otimes \hat{X}_n + \bar{c}_n \hat{\Gamma}_n^\dagger \otimes \hat{X}_n^\dagger\right),$$

Lindblad operators

where the system operators $\hat{X}_n$ are now called **Lindblad operators** or **jump operators**, and the c_n are coupling constants defined to absorb the global constant α above. We now assume that correlations between different bath operators, Γ_n, vanish, $\langle \Gamma_n^\dagger \Gamma_m \rangle_b = 0$ for $n \neq m$, and that the bath memory is instantly lost,

$$\langle \Gamma_n(0)\Gamma_n^\dagger(t)\rangle_b = \kappa_n^+ \delta(t),$$
$$\langle \Gamma_n^\dagger(0)\Gamma_n(t)\rangle_b = \kappa_n^- \delta(t).$$

Both assumptions rely on a separation of energy scales, i.e., that the dynamical phase factors in $\langle \Gamma_n^\dagger(0)\Gamma_m(t)\rangle_b$ oscillate so rapidly that the correlation can be neglected for $n \neq m$ and that they decay nearly instantly on time-scales relevant to the evolution of $\hat{\rho}_s$. Under these assumptions, substitution of the interaction into Eq. (12.5) leads to the **Lindblad equation**

Lindblad equation

$$\begin{aligned}\partial_t \hat{\rho}_s = \hat{L}_s\hat{\rho}_s + \sum_n \Big[&\gamma_n^+ \left(2\hat{X}_n^\dagger\hat{\rho}_s\hat{X}_n - \{\hat{X}_n\hat{X}_n^\dagger, \hat{\rho}_s\}\right) \\ &+\gamma_n^- \left(2\hat{X}_n\hat{\rho}_s\hat{X}_n^\dagger - \{\hat{X}_n^\dagger\hat{X}_n, \hat{\rho}_s\}\right)\Big],\end{aligned} \qquad (12.6)$$

where $\gamma_n^\pm \equiv \kappa_n^\pm|c_n|^2$.

[5] G. Lindblad, *On the generators of quantum dynamical semigroups*, Commun. Math. Phys. **48** (2), 119 (1976).

EXERCISE Derive this equation from Eq. (12.5).

The crude assumptions on the bath correlation functions and the dynamical phases of the system evolution which led to the Lindblad equation can all be relaxed. Such refinements are important in **atomic molecular and optical (AMO) physics**, where the Lindblad equation is extensively applied to the description of externally driven or open quantum systems. The resulting equations still have the algebraic structure of (12.6), while the details of the physics sit in the dependence of the couplings $\gamma_n^{\pm}$ on the system parameters.

AMO physics

Combining all jump operators $\hat{X}_n, \hat{X}_n^{\dagger}$ into a single set of (generally non-hermitian) operators $\{\hat{\Xi}_i\} = \{\hat{X}_1, \hat{X}_1^{\dagger}, \hat{X}_2, \dots\}$, the Lindblad equation assumes the form

$$\partial_t \hat{\rho}_s = \hat{L}_s \hat{\rho}_s + \sum_i \gamma_i \left(2\hat{\Xi}_i^{\dagger} \hat{\rho}_s \hat{\Xi}_i - \{\hat{\Xi}_i \hat{\Xi}_i^{\dagger}, \hat{\rho}_s\} \right) \equiv \mathcal{L} \hat{\rho}_s \qquad (12.7)$$

Lindbladian

The linear operator on the right is called the **Lindbladian**. It describes the irreversible system dynamics under the influence of the bath and obeys a number of important properties (see Ref.[5] for a more complete discussion):

▷ Lindbladian dynamics is **trace preserving**, $\partial_t \hat{\rho}_s = 0$, implying that the general condition $\hat{\rho}_s = 1$ is not violated. This follows by taking the trace of the right-hand side and using cyclic invariance.

▷ Lindbladian evolution defines a completely **positive map**. Here, positivity means that the formal solution $\hat{\rho}_s(t) = e^{t\mathcal{L}} \rho_s(0)$ maps positive initial configurations $\hat{\rho}_s(0)$ (all eigenvalues positive) to positive configurations. For a discussion of the stronger attribute *completely* positive, we refer to the original reference.

▷ The evolution defines a **dynamical semigroup** in the sense that $e^{t\mathcal{L}} e^{s\mathcal{L}} = e^{(t+s)\mathcal{L}}$. It is not a group, because the inverse of the Lindbladian evolution operator does not in general exist (irreversibility).

12.1.2 Example: oscillator coupled to a bath

The Lindblad equation is an important tool in various areas of nonequilibrium quantum physics. For example, it provides an efficient description of the decoherence processes affecting quantum systems coupled to environments (see problem 12.9.3). In this example, we focus on a different aspect, namely, the emission, absorption and thermalization processes caused by the coupling of a system to a bath.

We model the system as a single harmonic oscillator, and the bath as an assembly of oscillators: $\hat{H} = \hat{H}_s + \hat{H}_b + \alpha \hat{H}_i$, where

$$\hat{H}_s = \epsilon(a^\dagger a + 1/2), \quad \hat{H}_b = \sum_k \omega_k(a_k^\dagger a_k + 1/2), \quad \hat{H}_i = \sum_k c_k a_k^\dagger a + \text{h.c.} \quad (12.8)$$

Using $a(t) = e^{-i\epsilon t}a$ and $a_k(t) = e^{-i\omega_k t}a_k$, it is straightforward to verify that

$$\int_0^t dt' \left\langle \hat{L}_i(0)\hat{L}_i(-t') \right\rangle_b \hat{\rho}_s$$

$$= -\sum_k |c_k|^2 \int_0^\infty dt'$$

$$\times \left(e^{-i(\omega_k - \epsilon)t'} \left(\langle\hat{n}_k\rangle \left(\hat{\rho}_s a a^\dagger - a^\dagger \hat{\rho}_s a\right) + \langle\hat{n}_k + 1\rangle \left(a^\dagger a \hat{\rho}_s - a\hat{\rho}_s a^\dagger\right) \right) \right.$$

$$\left. + e^{+i(\omega_k - \epsilon)t'} \left(\langle\hat{n}_k\rangle \left(a a^\dagger \hat{\rho}_s - a^\dagger \hat{\rho}_s a\right) + \langle\hat{n}_k + 1\rangle \left(\hat{\rho}_s a^\dagger a - a\hat{\rho}_s a^\dagger\right) \right) \right)$$

$$= \pi \sum_k |c_k|^2 \delta(\epsilon - \omega_k)$$

$$\times \left(\langle\hat{n}_k\rangle \left(2a^\dagger \hat{\rho}_s a - \hat{\rho}_s a a^\dagger - a a^\dagger \hat{\rho}_s\right) + \langle\hat{n}_k + 1\rangle \left(2a\hat{\rho}_s a^\dagger - \hat{\rho}_s a^\dagger a - a^\dagger a\hat{\rho}_s\right) \right)$$

$$\simeq \pi |c_\epsilon|^2 \rho(\epsilon)$$

$$\times \left(\langle\hat{n}_\epsilon\rangle \left(2a^\dagger \hat{\rho}_s a - \hat{\rho}_s a a^\dagger - a a^\dagger \hat{\rho}_s\right) + \langle\hat{n}_\epsilon + 1\rangle \left(2a\hat{\rho}_s a^\dagger - \hat{\rho}_s a^\dagger a - a^\dagger a\hat{\rho}_s\right) \right).$$

Here, $\hat{n}_k = a_k^\dagger a_k$ is the number operator of the bath, and we have made the abbreviation $\langle\ldots\rangle_b \equiv \langle\ldots\rangle$. In the fourth equality we introduced the spectral density of the bath, $\rho(\omega) \equiv \sum_k \delta(\omega - \omega_k)$, and we assumed that $c_k \equiv c_{\omega_k}$ and $\langle\hat{n}_k\rangle \equiv \langle\hat{n}_{\omega_k}\rangle$ depend only on the energy of the reference state.

EXERCISE In the second equality above we assumed that $[\hat{\rho}_s, a^\dagger a] = 0$. Verify that a relaxation of that assumption leads to terms with structure $\sim P \int d\omega \frac{|c_\omega|^2 \rho(\omega)}{\omega - \epsilon} \langle\hat{n}_\omega\rangle [a^\dagger a, \hat{\rho}_s]$, where $P \int$ is the principal value integral. Interpret these expressions in terms of an **energy shift** of the oscillator energy due to virtual transitions into the bath.

Assuming again the commutativity of $\hat{\rho}_s$ with $\hat{n} \equiv a^\dagger a$, we have $\hat{L}_s\hat{\rho}_s = 0$ and thus

$$\partial_t \rho_s = \pi |c_\epsilon|^2 \rho(\epsilon) \left[\langle\hat{n}_\epsilon\rangle \left(2a^\dagger \hat{\rho}_s a - \hat{\rho}_s a a^\dagger - a a^\dagger \hat{\rho}_s\right) \right.$$
$$\left. + \langle\hat{n}_\epsilon + 1\rangle \left(2a\hat{\rho}_s a^\dagger - \hat{\rho}_s a^\dagger a - a^\dagger a\hat{\rho}_s\right) \right], \quad (12.9)$$

which has the form of a **Lindblad equation**. It is instructive to look at (12.9) from a number of different perspectives. The commutativity $[\hat{\rho}_s, \hat{n}] = 0$ implies that $\hat{\rho}_s = \hat{\rho}_s(\hat{n})$. Using commutator relations such as $a^\dagger \hat{\rho}_s(\hat{n}) = \hat{\rho}_s(\hat{n} - 1)a^\dagger$, it is then straightforward to bring Eq. (12.9) into the form

$$\partial_t \rho_s(\hat{n}) = 2\pi |c_\epsilon|^2 \rho(\epsilon) \left[\underbrace{\langle\hat{n}_\epsilon\rangle \left(\hat{n}\hat{\rho}_s(\hat{n} - 1) - (\hat{n} + 1)\hat{\rho}(\hat{n})\right)}_{\text{absorption}} \right.$$

$$\left. + \underbrace{\langle\hat{n}_\epsilon + 1\rangle \left((\hat{n} + 1)\hat{\rho}_s(\hat{n} + 1) - \hat{n}\hat{\rho}_s(\hat{n})\right)}_{\text{emission}} \right].$$

The right-hand side of this equation has a straightforward interpretation in terms of the absorption and emission of bath bosons (see the figure). The absorption term is proportional to the number of available bosons that are resonant with the oscillator frequency, $\langle \hat{n}_\epsilon \rangle$, the golden rule transition rate, $\sim |c_\epsilon|^2 \rho(\epsilon)$, and the number of absorbing bosons (plus one). Absorp-

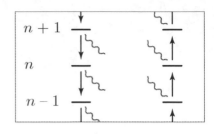

tion appears on the right-hand side of the equation in the characteristic rate-balance form of a master equation: interpreting $\hat{\rho}_s(n)$ as the probability of having n system oscillator quanta, this quantity increases due to an in-process $\propto \hat{\rho}_s(\hat{n}-1)$ and it diminishes due to an out-process $\propto \hat{\rho}_s(\hat{n})$. The equation also describes the opposite emission process, where bath bosons are resonantly created from oscillator bosons. Emission, too, enters the master equation as a sum of an in- and an out-process. Notice the proportionality $\langle \hat{n}_\epsilon + 1 \rangle$, where the $+1$ is interpreted as **spontaneous emission**, i.e., the emission of energy by the system independently of the occupation of the bath. This process survives in the limit of zero bath occupancy, $\langle \hat{n}_\epsilon \rangle = 0$, where no bath quanta are present. By contrast, the emission/absorption rates $\propto \langle \hat{n}_\epsilon \rangle$ are stimulated in that they are resonantly triggered by bath quanta.

spontaneous
emission

A few more comments can be made on the structure of the master equation:

▷ The competition of in- and out-terms implies the **conservation of probability**: $\partial_t \mathrm{tr}(\hat{\rho}_s) = 0$ (Exercise).

▷ Assume the bath to be in equilibrium, $\langle \hat{n}_\epsilon \rangle = (\exp(\beta\epsilon)-1)^{-1}$. It is then straightforward to verify that the distribution $\hat{\rho}_s$ becomes stationary (the right-hand side of the master equation vanishes) at the equilibrium configuration

$$\hat{\rho}_s(\hat{n}) = \mathcal{Z}^{-1} e^{-\beta\epsilon\hat{n}},$$

where $\mathcal{Z} = (1 - e^{-\beta\epsilon})^{-1}$ is the normalizing partition function. Interaction thus leads to **thermalization** of the system at the bath temperature. Notice that this process is based on the interplay of emission and absorption.

▷ The in- and out-terms in the master equation are obtained from contributions of the form $\sim a^\dagger \hat{\rho}_s a$ and $\sim \hat{\rho}_s a^\dagger a$ in (12.9), respectively. It is instructive to interpret this structure from a perspective emphasizing the underlying quantum dynamics. Recall that the time-evolution of the density operator is given by (in symbolic notation) $\hat{\rho}_s(t) = \hat{U}(t)\hat{\rho}_s(0)\hat{U}^\dagger(t)$, where $\hat{U}$ is the time-evolution operator. We may visualize the time-evolution described by these operators by two time-lines, one directed forward and one backward (see fig. 12.1).

If we now compare $\hat{\rho}_s(t + \Delta t)$ with $\hat{\rho}_s(t)$ to probe the incremental change of $\hat{\rho}_s$ in time, we need to expand the tensor product $\hat{U} \otimes \hat{U}^\dagger$ to first order in the system–bath coupling. This produces the sum of two contributions $\sim \alpha(\hat{H}_i(t)\hat{U} \otimes \hat{U}^\dagger + \hat{U} \otimes \hat{U}^\dagger\hat{H}_i(t))$. However, to obtain a non-vanishing result, we need to expand to one more order. This gives a factor $\sim \alpha \int dt' \hat{H}_i(t')$ which, again, may act to the left or the right of the density operator, or on the forward

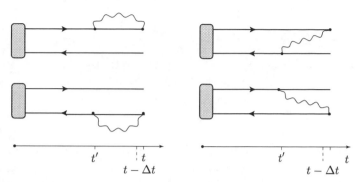

Fig. 12.1   Evolution of the density operator (shaded box) along the double time-contour. Left: Out-processes (self-energy corrections). Right: In-processes (vertex corrections).

or backward contour. This leads to four combinations as visualized in the figure. The contributions where $\hat{H}_i$ acts on different time-contours correspond to terms where the density operator is sandwiched between creation and annihilation operators. The presence of these operators changes the state of the system in a manner in which the forward and backward time-evolution are correlated via a bath correlation function. In the diagrammatic representation of the figure, they represent **vertex corrections** to a two-particle propagator. These are the in-terms. Conversely, the terms with two $\hat{H}_i$ on the same contour do not change the state of the system (it is first changed at t' and then changed back at t.) These out-terms resemble **self-energy corrections** to the two-particle propagator. Indeed, we have seen on various occasions that the imaginary part of a self-energy correction represents a decay rate, here corresponding to the rate of the out-process.

For further discussion of the quantum master equation, we refer to the textbooks by Weiss,[6] van Kampen,[7] and Haake.[8]

12.2 Keldysh Field Theory: Basics

In this section, we take the emergence of the double contour picture as a guiding principle for the construction of a field integral approach to quantum nonequilibrium: the Keldysh formalism. We start by introducing the idea behind this formalism and then make it concrete for the simple example of a free bosonic theory. This construction will then be the starting point for the discussion of various nontrivial applications.

[6] U. Weiss, *Quantum Dissipative Systems* (World Scientific Publishing, 1993).
[7] N. G. van Kampen, *Stochastic Processes in Physics and Chemistry* (Elsevier, 1992).
[8] F. Haake, *Quantum Signatures of Chaos* (Springer-Verlag, 2001).

12.2.1 The idea

Consider a general quantum system initialized in a known state $\hat{\rho}$ at time $t = 0$. For concreteness, we may think of a (system + bath) configuration initialized as $\hat{\rho}_s \otimes \hat{\rho}_b^{eq}$, or a single system initialized in a thermal state. The state then evolves as $\hat{\rho} \rightarrow \hat{\rho}(t) = \hat{U}(t)\hat{\rho}\hat{U}^\dagger(t)$, where $\hat{U}(t) = T_t \exp(-i \int_0^t dt' \hat{H}(t'))$ with T_t the time-ordering operator, and we allow for explicit time-dependence of the Hamiltonian. This time-dependence serves to implement a protocol often used in nonequilibrium problems: we assume that, at the initial time, the evolution is trivial in that the system–bath coupling or many-body correlations are absent, $\hat{H}_i = 0$. These complications are then gradually turned on and eventually switched off again at $t_0 \rightarrow \infty$. (Although these assumptions are not strictly needed, they lead to a formally cleaner description of the dynamics.)

Consider the following trace identity $1 = \text{tr}(\hat{U}(t_0)\hat{\rho}\hat{U}^\dagger(t_0))$. At first sight, considering a complicated representation of unity may seem like a waste of time. However, Keldysh's brilliant idea[1] was to interpret this expression as a unit-normalized path integral, $\mathcal{Z}$, from which useful information can be extracted in a second step, via the introduction of suitable source terms. To see this picture emerging, let $\{\psi_a\}$ be a complete set of states, chosen to di-

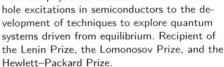

Leonid. V. Keldysh 1931–2016 was a Russian physicist and former Director of the Lebedev Physical Institute, Moscow. Keldysh made seminal contributions to solid state theory, from the physics of electron–hole excitations in semiconductors to the development of techniques to explore quantum systems driven from equilibrium. Recipient of the Lenin Prize, the Lomonosov Prize, and the Hewlett–Packard Prize.

agonalize the initial density operator, $\hat{\rho}$. We may then write the unit trace as $1 = \mathcal{Z} = \sum_a \langle \psi_a | \hat{U}\hat{\rho}\hat{U}^\dagger | \psi_a \rangle = \text{tr}(\hat{U}^\dagger | \psi_a \rangle\langle \psi_a | \hat{U}\hat{\rho}) = \langle \psi_b | \hat{U}^\dagger | \psi_a \rangle\langle \psi_a | \hat{U} | \psi_b \rangle \rho_b$, where summation over a and b is implied. Now consider what happens if we "trotterize" the time-evolution operator in order to pass to a path integral representation. Starting from a configuration $|\psi_b\rangle$, we would obtain a path integral describing the evolution under $\hat{U}$ to $|\psi_a\rangle$, in time $0 \rightarrow t_0$. The subsequent application of $\hat{U}(t) = \tilde{T}_t \exp(-i \int_{+t_0}^0 dt\hat{H}(t))$ describes time-evolution in reversed chronological order, where $\tilde{T}_t$ is an anti time-ordering operator, arranging factors in $\hat{H}_{\text{int}}(t)$ such that later times appear to the right of earlier times. The trotterization of this factor leads to the same path integral, except that time is integrated in reversed order. The structure of the formula above suggests combining $\hat{U}$ and $\hat{U}^\dagger$ into one path integral, where $|\psi_a\rangle\langle\psi_a|$ appears as a central trotterization step in the middle at $t = t_0$. This defines a path integral where time is integrated along a closed double contour, as shown at the bottom of fig. 12.2, where the fat dot indicates the presence of the initial state $\hat{\rho}$.

The reversal of chronological orders involved in the construction of the double **Keldysh contour** path integral can sometimes be inconvenient. In such cases, one may replace the double contour parameterization $t : 0 \rightarrow t_0 \rightarrow 0$ by a single

Keldysh
contour

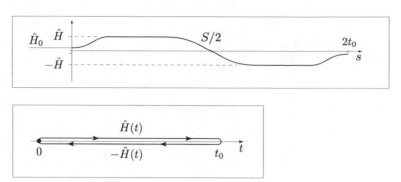

Fig. 12.2   Upper panel: Keldysh contour parameterized in a forward-running parameter s. Lower panel: Keldysh contour parameterized in a time variable.

one, $s : 0 \to S$, where $S = 2t_0$, and the new parameter variable runs in the forward direction. In this case, the sign change previously implemented by a chronological reversal is described by a change in the integrand: $\hat{H}(s) = \hat{H}(t)$, with $s = t$ on the first half of the contour, $0 \leq s \leq t_0$, and $\hat{H}(s) = -\hat{H}(t)$, with $t = 2t_0 - s$ on the second half, $t_0 \leq s \leq 2t_0$ (see the upper panel in fig. 12.2), where we also indicate the adiabatic switching on and off in the terminal regions. Within this formulation, the dynamics is described by the operator $\hat{U} = T_s \exp(-i \int_0^S ds\, \hat{H}(s))$, where T_s orders along ascending s.[9] Note that this operator equals unity by design, and that $\mathcal{Z} = \mathrm{tr}(\hat{U}\hat{\rho}) = 1$.

To get a first idea of the usefulness of this construction, consider what happens if we add a perturbation breaking the perfect $\pm\hat{H}$ symmetry on the s-contour: $\hat{H}(s) \to \hat{H}(s) + a\delta(s - t)\hat{X}$, where $t < t_0$ is in the first half of the contour, $\hat{X}$ is some operator, and a is a source parameter. Let $\hat{U}_{\mathrm{int}}(a)$ be the time-evolution operator in the interaction representation relative to this perturbation. The full evolution operator then assumes the form $\hat{U}(a) = \hat{U}\hat{U}_{\mathrm{int}}(a) = \hat{U}_{\mathrm{int}}(a)$, where $\hat{U} = \mathbb{1}$ is the unperturbed evolution operator, and $\hat{U}_{\mathrm{int}}(a) = \mathbb{1} + ia \int_0^S ds\delta(s - t)\hat{X}(s) + \mathcal{O}(a^2) = \mathbb{1} + ia\hat{X}(t) + \mathcal{O}(a^2)$, with $\hat{X}(t) = \hat{U}(t)\hat{X}\hat{U}^\dagger(t)$. Note that, for $s = t$, $\hat{U}(s) = \hat{U}(t)$ evolves along the forward directed contour and is equal to the time-evolution operator $\hat{U}(t) = \hat{U}(t, 0)$. Insertion of this result into the trace leads to the result

$$-i\partial_a\Big|_{a=0} \mathrm{tr}(\hat{U}(a)\hat{\rho}) = \mathrm{tr}(\hat{U}_0^\dagger(t,0)\hat{X}\hat{U}(t,0)\hat{\rho}) = \mathrm{tr}(\hat{X}\hat{\rho}(t)),$$

where $\hat{\rho}(t) = \hat{U}(t, 0)\hat{\rho}\hat{U}^\dagger(t, 0)$ is the density operator evolved to t. Thus, differentiation of the density operator with respect to the source generates expectation values of operators in the dynamically evolved state of the system. This shows that $\mathcal{Z}(a) \equiv \mathrm{tr}(\hat{\rho}\hat{U}(a))$ is a normalized ($\mathcal{Z}(0) = 1$) **generating function** for dynamically evolved quantum expectation values of the operator $\hat{X}$. Crucially, the construction does not assume a thermal state, which makes it superior to the equilibrium Matsubara approach. However, before demonstrating how Keldysh theory

---

[9] To be precise, $T_s \hat{A}(s_1)\hat{A}(s_2) = \hat{A}(s_1)\hat{A}(s_2)$, if $s_1 > s_2$, and $\hat{A}(s_2)\hat{A}(s_1)$ otherwise.

is applied to describe out-of-equilibrium physics, we need to make the path integral representations more concrete.

12.2.2 Case study

To prepare for the functional representation of the Keldysh partition function, we first consider a miniature quantum system containing only a single bosonic state of energy ω. This leads to a simple path integral whose subsequent generalization to more complex systems will be straightforward. Consider the partition function

$$\mathcal{Z} \equiv \mathrm{tr}\left(T_s e^{-i\int_0^S ds\,\hat{H}(s)}\hat{\rho}_0\right),\qquad(12.10)$$

where

$$\hat{\rho}_0 = \mathcal{Z}_0^{-1}\,e^{-\beta(\hat{H}-\mu\hat{N})},\qquad(12.11)$$

with equilibrium partition sum $\mathcal{Z}_0 = (1 - e^{-\beta(\omega-\mu)})^{-1}$, is the initial density operator, $\hat{H} = \omega a^\dagger a$ is the Hamiltonian operator, and $\hat{N} = a^\dagger a$. Since the theory is free, there is no need for an adiabatic switching-on procedure, and we assume constancy of $\hat{H}$ along the contour.

EXERCISE Show that

$$\langle\bar{\psi}|e^{ca^\dagger a}|\psi\rangle = e^{e^c\bar{\psi}\psi},\qquad(12.12)$$

where $\langle\bar{\psi}|$ and $|\psi\rangle$ are boson coherent states. (Hint: Differentiate by c to derive a first-order differential equation and use the uniqueness of its solution.)

Let us now apply the standard recipe – insertion of a large number $(2N)$ of coherent state resolutions of unity into a time-slice dissection of $\mathcal{Z}$ – to construct a path integral. This leads to the representation

$$\mathcal{Z} = \langle\psi_1^-|\mathbb{1}|\psi_N^+\rangle \times (\langle\psi_N^+|\hat{U}_\epsilon|\psi_{N-1}^+\rangle\langle\psi_{N-1}^+|\hat{U}_\epsilon\cdots|\psi_1^+\rangle) \times \langle\psi_1^+|\hat{\rho}_0|\psi_N^-\rangle$$
$$\times (\langle\psi_N^-|\hat{U}_{-\epsilon}|\psi_{N-1}^-\rangle\langle\psi_{N-1}^-|\hat{U}_{-\epsilon}\cdots|\psi_1^-\rangle),$$

where $\epsilon = t_0/N$, $\hat{U}_\epsilon = e^{-i\epsilon\hat{H}}$, and the factors in parentheses are the trotterized representations of the propagation along the two contours. The functional integral representation then becomes

$$\mathcal{Z} = \mathcal{Z}_0^{-1}\int D\psi\, e^{\sum_{j=2}^N \left(\bar{\psi}_j^+[\psi_{j-1}^+ - \psi_j^+ - i\epsilon\omega\psi_{j-1}^+] + \bar{\psi}_j^-[\psi_{j-1}^- - \psi_j^- + i\epsilon\omega\psi_{j-1}^-]\right)}$$
$$\times e^{-\bar{\psi}_1^+\psi_1^+ - \bar{\psi}_1^-\psi_1^- + \kappa\bar{\psi}_1^+\psi_N^- + \bar{\psi}_1^-\psi_N^+},\qquad(12.13)$$

where $\kappa \equiv \exp(-\beta(\omega-\mu))$. Before turning to a continuum representation, let us interpret the exponent of the functional integral as a bilinear form $i\bar{\psi}G^{-1}\psi$, with the $(2N\times 2N)$-dimensional matrix kernel

$$
G^{-1} = i
\left[
\begin{array}{cccc|cccc}
1 & & & & & & & -\kappa \\
-a_+ & 1 & & & & & & \\
& \ddots & \ddots & & & & & \\
& & -a_+ & 1 & & & & \\
\hline
& & & -1 & 1 & & & \\
& & & & -a_- & \ddots & & \\
& & & & & \ddots & 1 & \\
& & & & & & -a_- & 1
\end{array}
\right],
\qquad (12.14)
$$

where $a_\pm \equiv 1 \mp i\epsilon\omega$. This matrix has a number of interesting properties. First, it is straightforward to verify that

$$
\det(-iG^{-1}) = 1 - \kappa(a_+ a_-)^{N-1}.
$$

EXERCISE   Verify this result. (Hint: Use the identity "det = exp tr ln.") Expand the logarithm in powers of the difference from the unit matrix, $(-iG^{-1} - 1)$ (essentially the side-diagonal containing the coefficients $a_\pm$, plus the corner element κ). Take the trace and re-sum the series into another logarithm. Exponentiation leads to the result.

Taking the limit $N \to \infty$, we obtain

$$
\det(-iG^{-1}) = 1 - \kappa(1 + (\epsilon\omega)^2)^{N-1} = 1 - \kappa\left(1 + \left(\frac{T\omega}{N}\right)^2\right)^{N-1} \overset{N\to\infty}{\longrightarrow} 1 - \kappa = \mathcal{Z}_0^{-1}.
$$

Since $\mathcal{Z} = \mathcal{Z}_0^{-1} \det(-iG^{-1})^{-1}$, this result proves the **unit normalization** of the functional Keldysh partition function. Later, we will want to compute expectation values $\langle \bar{\psi}_i \psi_j \cdots \psi_k \rangle$ by Wick's theorem. For this purpose, we need to know the elementary contractions

$$
G_{ii'}^{CC'} = -i \langle \psi_i^C \bar{\psi}_{i'}^{C'} \rangle,
$$

where $C, C' = \pm$. To compute these elements, we introduce the block decomposition $G^{-1} = \begin{pmatrix} M^{++} & M^{+-} \\ M^{-+} & M^{--} \end{pmatrix}$. It is a straightforward exercise to show that

$$
\left[M^{CC}\right]_{ij}^{-1} = -i\Theta(i - j) a_\pm^{i-j},
$$

where $\Theta(n) = 1$ if $n \geq 0$ and zero otherwise.

EXERCISE   To verify this result, write $M = \mathbb{1} - H$, where H contains the a-dependent terms next to the main diagonal. Consider $M^{-1} = (\mathbb{1} - H)^{-1}$, expand in H, and interpret the powers H^k as kth-order directed hopping processes in time.

We may now use the general formulae for the inversion of 2×2 block matrices to obtain $G^{++} = (M^{++} - \Xi^{++})^{-1}$, where $\Xi^{++} = M^{+-}(M^{--})^{-1}M^{-+}$. Substitution of $\Xi_{ij}^{++} = i\delta_{i1}\delta_{jN}a_-^{N-1}\kappa$ into this formula followed by a straightforward series expansion in Ξ leads to the result

$$G_{ij}^{++} = -ia_+^{i-j}\left(\Theta(i-j) + \frac{\kappa(a_+a_-)^{N-1}}{1-\kappa(a_+a_-)^{N-1}}\right). \qquad (12.15)$$

INFO This formula affords an **intuitive interpretation** (see the figure above). The matrix G^{++} is the amplitude for propagation between two discrete points j and i on the $+$ segment of our closed time-contour. To get from j to i we may either go directly, which is possible if $i > j$ (because of the time-ordering). In this case, we pick up $i - j$ hopping amplitudes a_+. This is the first term in the equation. Alternatively, we may go via round-trips through the $-$ segment of the contour. In this case, and no matter what the chronological ordering is between i and j, we first go from j to N ($N - j$ amplitudes a_+), then from 1 to N on the bottom contour ($N - 1$ amplitudes a_-), back to the upper contour (giving a factor κ), and finally from 1 to i ($(i-1)$ amplitudes a_+). This gives the contribution of first order in κ to the second term. Now, orbiting around the contour can be continued, where each additional revolution contributes a factor $\kappa a_-^{N-1}a_+^{N-1}$. Summation over all these processes generates the denominator of the equation.

The remaining three matrix elements are computed in the same manner, with the result

$$G_{ij}^{--} = -ia_-^{i-j}\left(\Theta(i-j) + \frac{\kappa(a_-a_+)^{N-1}}{1-\kappa(a_-a_+)^{N-1}}\right),$$
$$G_{ij}^{+-} = -i\frac{\kappa a_-^{N-j}a_+^{i-1}}{1-\kappa(a_-a_+)^{N-1}}, \qquad (12.16)$$
$$G_{ij}^{-+} = -i\frac{a_+^{N-j}a_-^{i-1}}{1-\kappa(a_+a_-)^{N-1}}.$$

Now, let us take the continuum limit. To the discrete time-steps on the upper/lower contour we assign a time variable (see fig. 12.2)

$$\begin{aligned} +, &\quad t = i\epsilon; \\ -, &\quad t = (N-i)\epsilon. \end{aligned} \qquad (12.17)$$

Noting that $\lim_{N\to\infty}(a_+a_-)^N = 1$, and $a_+^i \to e^{-i\omega t}, a_-^i \to e^{i\omega(T-t)}$, we obtain the **propagators in a continuous-time representation**, Eq. (12.36), where $n(\omega) = \kappa/(1-\kappa) = (e^{\beta(\omega-\mu)} - 1)^{-1}$ is the Bose distribution function.

INFO In the literature, a few **different conventions for the Green functions** of Keldysh field theory are in use. For later referencing purposes, we summarize all these in the info block section of page 716. Note that Green functions on the Keldysh contour are often written as $G^{++} \equiv G^T$, $G^{--} \equiv G^{\bar{T}}$, $G^{+-} \equiv G^<$, $G^{-+} \equiv G^>$, a convention that we will not adopt in this text.

12.2.3 Continuum field theory

Keldysh
functional
integral

Having derived the continuum propagators, we now represent the action (12.13) in a continuum form. Using Eq. (12.17), we immediately obtain the **Keldysh functional integral**

$$
\mathcal{Z} = \mathcal{Z}_0^{-1} \int D\psi \, e^{i \int_0^{t_0} dt \, \bar{\psi}\sigma_3(i\partial_t - \omega)\psi}
\tag{12.18}
$$

where $\psi \equiv (\psi_+, \psi_-)^T$ is a two-component field in contour space and σ_i are Pauli matrices in that space. Notice that, in the continuum notation, the terms gluing the boundaries at $t = t_0$ in Eq. (12.13) and the presence of the density operator at $t = 0$ are suppressed. However, as we saw above, these terms play a crucial role as initial conditions for the computation of the propagators $G_{tt'}^{CC'}$ of the theory. Before discussing how the correlation between the contours is described in the continuum formalism, let us represent the path integral in a new set of integration variables.

One more thing to notice is that the Green function G contains a degree of **redundancy**. Inspection of Eqs. (12.15) and (12.16) shows, for example, that $G^{++} + G^{--} = G^{+-} + G^{-+}$. Defining the linear transformation

$$
U \equiv \frac{1}{\sqrt{2}} \begin{pmatrix} 1 & 1 \\ 1 & -1 \end{pmatrix},
\tag{12.19}
$$

we obtain an alternative representation of the block Green function:

$$
G_{tt'} \to G'_{tt'} \equiv UG_{tt'}U^\dagger = \frac{1}{2} \begin{pmatrix} 1 & 1 \\ 1 & -1 \end{pmatrix} \begin{pmatrix} G^{++} & G^{+-} \\ G^{-+} & G^{--} \end{pmatrix}_{tt'} \begin{pmatrix} 1 & 1 \\ 1 & -1 \end{pmatrix}
$$
$$
= -ie^{-i\omega(t-t')} \begin{pmatrix} 1 + 2n(\omega) & \Theta(t - t') \\ -\Theta(t' - t) & 0 \end{pmatrix} \equiv \begin{pmatrix} G^K & G^+ \\ G^- & 0 \end{pmatrix}_{tt'},
\tag{12.20}
$$

where the three block Green functions are defined by the last equation,[10] or the explicit representation of Eq. (12.43) below. Here, the functions $G^\pm$ are identical to the **retarded and advanced Green function** of quantum mechanics, i.e.,

$$
G_{t-t'}^+ = -i\Theta(t - t')\langle[a(t), a^\dagger(t')]\rangle,
$$
$$
G_{t-t'}^- = +i\Theta(t' - t)\langle[a(t), a^\dagger(t')]\rangle,
$$

Keldysh
Green
function

where $a(t) = e^{i\hat{H}t}ae^{-i\hat{H}t} = e^{-i\omega t}a$ are the Heisenberg-evolved boson operators of the theory. The **Keldysh Green function**, G^K, is a new acquaintance. Notice that G^K, and only G^K, contains the initial distribution. We will obtain a better understanding of this function as we go along.

The correspondence $G_{tt'}^{CC'} = -i\langle\psi_t^C \bar{\psi}_{t'}^{C'}\rangle$ implies that $G_{tt'}'^{CC'} \equiv (UGU^\dagger)_{tt'}^{CC'} = -i\langle(U\psi)_t^C (\bar{\psi}U^\dagger)_{t'}^{C'}\rangle \equiv -i\langle\psi_t'^C \bar{\psi}_{t'}'^{C'}\rangle$. It is customary to denote the transformed fields as

$$
\psi' = U\psi = \frac{1}{\sqrt{2}} \begin{pmatrix} \psi^+ + \psi^- \\ \psi^+ - \psi^- \end{pmatrix} \equiv \begin{pmatrix} \psi_c \\ \psi_q, \end{pmatrix},
\tag{12.21}
$$

---

[10] or, expressed in terms of the block Green functions, $G^\pm = G^{++} - G^{\pm\mp}$ and $G^K = G^{++} + G^{--}$.

classical
and quan-
tum fields
where ψ_c and ψ_q define the **classical and quantum fields**, of the theory. Formally, ψ_c and ψ_q are the center of mass and the difference of the field amplitudes $\psi^\pm$, respectively. To understand the origin of the terminology classical/quantum, notice that in quantum mechanics dynamical evolution is governed by the product of two independent wave function amplitudes, $\psi^\pm$. It collapses to the classical limit if these two amplitudes become identical, $\psi^+ = \psi^- = \psi_c/\sqrt{2}$, and $\psi_q = 0$.

With $\psi = U^\dagger \psi'$ and $\bar{\psi} = \bar{\psi}' U$, and noting that $U\sigma_3 U^\dagger = \sigma_1$, the action in the new variables assumes the form

$$S[\bar{\psi}, \psi] \equiv \int dt \, (\bar{\psi}_c, \bar{\psi}_q) \begin{pmatrix} & i\partial_t - \omega \\ i\partial_t - \omega & \end{pmatrix} \begin{pmatrix} \psi_c \\ \psi_q \end{pmatrix}, \tag{12.22}$$

where now $\psi = (\psi_c, \psi_q)^T$ and we have omitted the primes for clarity.

On this basis, let us now return to the question of how to represent the information on the **initial distribution in the continuum theory**. We first notice that the temporal exponents in Eq. (12.20) must be understood as infinitesimally damped, $\exp(-it\omega) \to \exp(-it\omega - \delta|t|)$, $\delta > 0$. Fourier transformation of the Green functions then leads to

$$G^{\pm,K}(\epsilon) = \int_{-\infty}^{\infty} dt \, e^{i\epsilon t} G_{t,0}^{\pm,K}, \tag{12.23}$$

with

$$\begin{aligned} G^\pm(\epsilon) &= \frac{1}{\epsilon^\pm - \omega}, \\ G^K(\epsilon) &= -2\pi i F(\epsilon)\delta(\epsilon - \omega) = F(\epsilon)\left[G^+(\epsilon) - G^-(\epsilon)\right]. \end{aligned} \tag{12.24}$$

Here, we have defined the function

$$\boxed{F(\epsilon) \equiv 1 + 2n(\epsilon) = \coth\left(\frac{\epsilon - \mu}{2T}\right)} \tag{12.25}$$

Note that the δ-function appearing in combination with F locks the argument as $F(\epsilon)\delta(\epsilon - \omega) = F(\omega)\delta(\epsilon - \omega)$. As discussed below, this locking is important and ensures the convergence of the integral. Also notice that the integration has been extended to infinity, in spite of the fact that the theory was introduced on a finite interval $[0, t_0]$. The rationale behind this procedure is that, in applications of the theory, t_0 must be chosen sufficiently large that the correlation functions of the theory decay on scales $\ll t_0$. Formally, this decay is enforced by the order of limits $\lim_{\delta \to 0} \lim_{t_0 \to \infty}$. However, as we will see shortly, actual theories generate finite decay times such that the formal limiting procedure is not necessary.

Anticipating that for arbitrary Hamiltonians, $\hat{H}$, anti-hermiticity will be a general feature of the Keldysh Green function, $\hat{G}^{K\dagger} = -\hat{G}^K$ and, recalling that $\hat{G}^{+\dagger} = \hat{G}^-$, the structures above suggest the representation

$$\hat{G}^K \equiv \hat{G}^+ \hat{F} - \hat{F}\hat{G}^-, \tag{12.26}$$

where $\hat{F}$ is an hermitian operator and the energy dependence $\hat{X} = \hat{X}(\epsilon) \equiv \hat{X}_\epsilon$ is suppressed. With this we obtain

$$\hat{G} = \begin{pmatrix} \hat{G}^K & \hat{G}^+ \\ \hat{G}^- & 0 \end{pmatrix} = \begin{pmatrix} \hat{G}^+\hat{F} - \hat{F}\hat{G}^- & \hat{G}^+ \\ \hat{G}^- & 0 \end{pmatrix}, \tag{12.27}$$

$$\hat{G}^{-1} = \begin{pmatrix} 0 & (\hat{G}^-)^{-1} \\ (\hat{G}^+)^{-1} & -(\hat{G}^+)^{-1}\hat{G}_K(\hat{G}^-)^{-1} \end{pmatrix} = \begin{pmatrix} 0 & (\hat{G}^-)^{-1} \\ (\hat{G}^+)^{-1} & (\hat{G}^+)^{-1}\hat{F} - \hat{F}(\hat{G}^{-1})^{-1} \end{pmatrix}.$$

(Note that these formulas do not assume the commutativity of $\hat{G}^\pm$ and $\hat{F}$.) Finally, we substitute this representation into the energy–Fourier representation of the action, $S[\psi] = \int \frac{d\epsilon}{2\pi} \bar{\psi}_\epsilon \hat{G}_\epsilon^{-1} \psi_\epsilon$ to obtain the structure

$$\boxed{S[\psi] = \int \frac{d\epsilon}{2\pi} (\bar{\psi}_c, \bar{\psi}_q)_\epsilon \begin{pmatrix} 0 & (\hat{G}^-)^{-1} \\ (\hat{G}^+)^{-1} & (\hat{G}^+)^{-1}\hat{F} - \hat{F}(\hat{G}^{-1})^{-1} \end{pmatrix}_\epsilon \begin{pmatrix} \psi_c \\ \psi_q \end{pmatrix}_\epsilon} \tag{12.28}$$

In the specific context of our toy model, where $\hat{F}$ commutes with $\hat{G}^\pm$, this reads

$$S[\psi] = \int \frac{d\epsilon}{2\pi} (\bar{\psi}_c, \bar{\psi}_q)_\epsilon \begin{pmatrix} & \epsilon^- - \omega \\ \epsilon^+ - \omega & 2i\delta \coth\left(\frac{\epsilon-\mu}{2T}\right)\delta(\epsilon - \omega) \end{pmatrix} \begin{pmatrix} \psi_c \\ \psi_q \end{pmatrix}_\epsilon. \tag{12.29}$$

This formula suggests a new **interpretation of the δ-broadening** used in the Fourier transform. First, notice that in applications the parameter δ is physical. It appears whenever the imaginary part of the self-energy of a propagator, $(\hat{G}^+)^{-1} \to (\hat{G}^+)^{-1} - \Sigma$ is non-zero, in which case $-\text{Im}\Sigma \leftrightarrow \delta$ plays the role of a damping rate. This damping makes the otherwise purely oscillatory integral convergent: on physical grounds, $\omega > \mu$ ensures that the single-boson level is stable and has a finite occupation. The effective real damping rate $2\delta \coth(\frac{\epsilon-\mu}{2T})$ then safeguards the convergence. At the same time, it injects the information on the distribution into the continuum representation, via the factor $\hat{F}$. The construction above shows that even if that coupling is infinitesimally small, $\propto \delta$, it affects the propagator $\hat{G}$ as an $\mathcal{O}(1)$ effect. In less trivial applications, $\hat{F}$, will *change* in the course of time, and in this way will describe the evolution from an initially trivial distribution to different distributions.

In the next section, we generalize the construction above to settings rich enough to show changes in the state of a system via interaction and thermalization processes.

12.2.4 Generalization

Retracing the construction steps leading to the field integral (12.18), one notices that the simple structure of the one-level free boson Hamiltonian was nowhere essential. For a higher-dimensional interacting system governed by the sum of a free Hamiltonian $\int d^d r\, a^\dagger(\mathbf{r})\hat{H}a(\mathbf{r})$ and an interaction $\frac{1}{2}\int d^d r d^d r'(a^\dagger a)(\mathbf{r})V(\mathbf{r} - \mathbf{r}')(a^\dagger a)(\mathbf{r}')$, trotterization in terms of boson coherent states $\psi(\mathbf{r})$ leads to the generalization $\mathcal{Z} = \mathcal{Z}_0^{-1}\int D\psi\, e^{iS_0[\psi]+iS_{\text{int}}[\psi]}$, where

$$S_0[\psi] \equiv \int dx \sum_{C=\pm} C\bar{\psi}^C(x)(i\partial_t - \hat{H})\psi^C(x),$$

$$S_{\text{int}}[\psi] \equiv -\frac{1}{2}\int dt d^d r d^d r' \sum_C C\,(\bar{\psi}^C\psi^C)(x)\,V(\mathbf{r} - \mathbf{r}')\,(\bar{\psi}^C\psi^C)(x'), \tag{12.30}$$

$x = (t, \mathbf{r})$, and $dx = dt\, d^d r$. Switching to classical and quantum fields through the transformation (12.21), and assuming a contact interaction $V(\mathbf{r}) = 2g\delta(\mathbf{r})$ for notational simplicity, these relations assume the form

$$S_0[\psi] \equiv \int dx\, (\bar{\psi}_c, \bar{\psi}_q) \begin{pmatrix} 0 & (\hat{G}^-)^{-1} \\ (\hat{G}^+)^{-1} & (\hat{G}^{-1})^K \end{pmatrix} \begin{pmatrix} \psi_c \\ \psi_q \end{pmatrix},$$

$$S_{\text{int}}[\psi] \equiv -g \int dx\, \left(\bar{\psi}_c \bar{\psi}_q (\psi_c^2 + \psi_q^2) + \text{c.c.}\right),$$

(12.31)

where

$$(\hat{G}^\pm)^{-1} = i\partial_t \pm i\delta - \hat{H},$$

$$(\hat{G}^{-1})^K = (\hat{G}^+)^{-1}\hat{F} - \hat{F}(\hat{G}^-)^{-1}$$

(12.32)

are the retarded and advanced Green functions, and the Keldysh sector of the inverse of the Green function is given by Eq. (12.27). Building on this representation, let us now turn to a more thorough discussion of the different elements of Keldysh field theory.

Retarded and advanced Green function

The infinitesimal increments entering the definition of the retarded and advanced Green function imply the causality condition $G^\pm(t, t') \propto \Theta(\pm(t - t'))$. For a time-independent Hamiltonian,

$$G^\pm(t, t') = G^\pm(t - t') = \mp i\Theta(\pm(t - t'))e^{-i\hat{H}(t-t')},$$

$$G_\epsilon^\pm = \frac{1}{\epsilon \pm i\delta - \hat{H}}.$$

(12.33)

If $\hat{H}$ contains explicit time-dependence, we need to include a time-ordering procedure,

$$G^\pm(t, t') = \mp i\Theta(\pm(t - t'))T_\pm e^{-i\int_t^{t'} d\tilde{t}\, \hat{H}(\tilde{t})},$$

where $T_\pm$ time-orders in the chronological/anti-chronological direction. Finally, the Green functions obey the composition law $G^\pm(t, t') = G^\pm(t, t'')G^\pm(t'', t')$ and the limit behavior $G^+(t, t) + G^-(0, 0) = 0$.

Keldysh Green function

As discussed above, the representation (12.32) for the Keldysh block of the inverse Green function, $(\hat{G}^{-1})^K$,[11] is motivated by the anti-hermiticity of $\hat{G}^K$ and $(\hat{G}^{-1})^K$. The operator $\hat{F}$ appearing in this construction holds the information on the state of the system, and determining its dynamical evolution is an important part of the theory. This task is greatly simplified by the ansatz $\hat{F} = \{F(t, t')\}$, depending on

---

[11] Carefully distinguish between this object and the inverse of the Keldysh Green function, $(\hat{G}^K)^{-1} = (\hat{G}^+\hat{F} - \hat{F}\hat{G}^-)^{-1}$.

time arguments, but independent of the Hilbert space indices of the problem ($\mathbf{r}$ in the current continuum field theory).

INFO To **motivate this form for** $\hat{F}$, we recall its origins in the non-interacting sector of the theory (see Eq. (12.25)) and interpret $\hat{F}$ as a generalized (quasi-)particle distribution function. In a non-interacting theory, it is natural to assume that $\hat{F}(\epsilon) = \hat{F}(\epsilon, \hat{H})$ depends on the Hamiltonian of the system. The straightforward generalization of Eq. (12.24) to a higher-dimensional Hilbert space, $\hat{G}^K(\epsilon) = -2\pi i \hat{F}(\hat{H}, \epsilon)\delta(\epsilon - \hat{H})$, then suggests the eigenfunction representation $\hat{G}^K(\epsilon, \omega_a) = -2\pi i \hat{F}(\epsilon, \omega_a)\delta(\epsilon - \omega_a)$, where the spectral δ-function locks the eigenenergies ω_a to the energy argument. Since $\hat{F}$ always appears in combination with the spectral function, dropping its Hilbert space index, $\hat{F}(\epsilon, \omega_a) = \hat{F}(\epsilon)$, does not imply a loss of information.

For problems without explicit time-dependence, $F(t_1, t_2) = F(t_1 - t_2)$ and the Fourier transform $F(\epsilon)$ determine the occupation of quasi-particles at energy ϵ. For example, in thermal equilibrium $F(\epsilon) = \coth(\epsilon T/2) = 1 + 2n(\epsilon)$, with the Bose distribution function $n(\epsilon)$. However, nonequilibrium problems often contain external time-dependence. In such cases, $F(t_1, t_2) = F(t, \Delta t)$ depends on the center time $t = (t_1 + t_2)/t$ and on the difference $\Delta t = t_1 - t_2$ and it is mostly a good idea to Fourier transform in the latter, and keep the former as a parameter:

$$F(\epsilon, t) \equiv \int d\Delta t \, e^{i\epsilon t'} F(t + \Delta t/2, t - \Delta t/2). \tag{12.34}$$

The function $F(\,.\,, t)$ then represents the instantaneous particle distribution function at time t.

Interaction

In many ways, the treatment of particle interactions in Keldysh theory parallels that in equilibrium theories, the repertoire including diagrammatic perturbative methods, Hubbard–Stratonovich transformations, mean field treatments, renormalization group methods, etc. One often starts by considering the propagator of an interacting theory,

$$\boxed{-i\langle \psi_\alpha(\mathbf{r}, t)\bar{\psi}_{\alpha'}(\mathbf{r}', t')\rangle = \begin{pmatrix} G^K & G^+ \\ G^- & 0 \end{pmatrix}(\mathbf{r}, t; \mathbf{r}', t')_{\alpha\alpha'}} \tag{12.35}$$

Once this function is known, the particle distribution function, $\hat{F}$ is defined through Eq. (12.27).

In perturbative approaches, elements of the matrix propagator are usually represented by a diagrammatic code, as in fig. 12.3. In the same spirit, the interaction vertex is represented as in the lowest row of that figure. Notice that the interaction vertex connects to at least one Keldysh (dashed) field component, the reason being that $S_{\text{int}}[\psi] = 0$ for $\psi_q = 0$ (think why). Other than that, not much can be said in general terms. However, in the remainder of the chapter, we will discuss various applications illustrating how interacting theories are handled in practice.

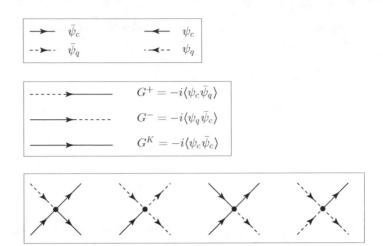

Fig. 12.3 Building blocks of bosonic Keldysh diagrammatic perturbation theory. Top, field vertices; middle, propagators; bottom, interaction vertices.

To conclude the discussion of general structures, the info block below provides a reference to standard conventions in bosonic Keldysh theory. While the different forms of Keldysh matrix Green functions can be confusing at first, the conventions listed below are standard and are used in this form with little or no alteration in the literature. Over time, one just gets used to them.

INFO For the convenience of the reader, we here provide a **reference chart summarizing conventions** for bosonic quantum fields and propagators. To simplify the notation, we list formulas for a single-level Hamiltonian, $\hat{H} = \omega a^\dagger a$, with equilibrium distribution $n(\omega) = (e^{\beta(\omega-\mu)} - 1)^{-1}$. We consider the free theory with infinitesimal damping constant $i\delta$. A straightforward generalization to more complex Hamiltonians and particle distributions was discussed earlier in this section.

Contour representation:

$$S[\phi] = \sum_{C=\pm} C \int dt \, \bar{\psi}_t^C (i\partial_t - \omega)\psi_t^C = \sum_{C=\pm} C \int \frac{d\epsilon}{2\pi} \, \bar{\psi}_\epsilon^C (\epsilon - \omega)\psi_\epsilon^C.$$

Coupling between contours (mediated via boundary terms) is often left implicit in this representation. Green functions in the contour representation are represented as

$$-i\langle \psi_t^C \bar{\psi}_{t'}^{C'} \rangle = \begin{pmatrix} G^{++} & G^{+-} \\ G^{-+} & G^{--} \end{pmatrix}_{tt'} \equiv \begin{pmatrix} G^T & G^< \\ G^> & G^{\tilde{T}} \end{pmatrix}_{tt'}, \tag{12.36}$$

where

$$G_{tt'}^{+-} \equiv G_{tt'}^< = -ie^{-i\omega(t-t')} n(\omega),$$
$$G_{tt'}^{-+} \equiv G_{tt'}^> = -ie^{-i\omega(t-t')}(1 + n(\omega)),$$
$$G_{tt'}^{++} \equiv G_{tt'}^T = G_{tt'}^> \Theta(t - t') + G_{tt'}^< \Theta(t' - t),$$
$$G_{tt'}^{--} \equiv G_{tt'}^{\tilde{T}} = G_{tt'}^> \Theta(t' - t) + G_{tt'}^< \Theta(t - t'),$$

with the constraint $G^{++} + G^{--} = G^{+-} + G^{-+}$.

Keldysh representation: Fields in the Keldysh basis are defined as (cf. Eq. (12.21))

$$
\begin{pmatrix} \psi_c \\ \psi_q \end{pmatrix} \equiv \frac{1}{\sqrt{2}} \begin{pmatrix} \psi^+ + \psi^- \\ \psi^+ - \psi^- \end{pmatrix}, \qquad \begin{pmatrix} \psi^+ \\ \psi^- \end{pmatrix} \equiv \frac{1}{\sqrt{2}} \begin{pmatrix} \psi_c + \psi_q \\ \psi_c - \psi_q \end{pmatrix}. \tag{12.37}
$$

The transformation $(\psi^+, \psi^-) \to (\psi_c, \psi_q)$ is unitary. Depending on the interpretation of the fields, the alternative definitions $\psi^\pm = \psi_c \pm \frac{1}{2}\psi_q$, $\psi_c = \frac{1}{2}(\psi^+ + \psi^-)$, $\psi_q = \psi^+ - \psi^-$ are sometimes used. The Keldysh action (see Eqs. (12.27) and (12.28) for a generalization not assuming the commutativity of $\hat{F}$ and $\hat{G}^\pm$) is defined by

$$
S[\bar{\psi}, \psi] \equiv \int dt\, \bar{\psi}_t \left(G^{-1} \right)^{-1}_t \psi_t = \int \frac{d\epsilon}{2\pi} \bar{\psi}_\epsilon \left(G^{-1} \right)^{-1}_\epsilon \psi_\epsilon, \tag{12.38}
$$

where

$$
\bar{\psi} G^{-1} \psi = (\bar{\psi}_c, \bar{\psi}_q) \begin{pmatrix} & (G^-)^{-1} \\ (G^+)^{-1} & (G^{-1})^K \end{pmatrix} \begin{pmatrix} \psi_c \\ \psi_q \end{pmatrix}, \tag{12.39}
$$

with

$$
\begin{aligned}
(G^\pm)^{-1}_t &= i\partial_t \pm i\delta - \omega, & (G^\pm)^{-1}_\epsilon &= \epsilon \pm i\delta - \omega, \\
(G^{-1})^K_t &= 2i\delta F(\omega), & (G^{-1})^K_\epsilon &= 2i\delta F(\omega).
\end{aligned} \tag{12.40}
$$

Here, the subscripts serve to distinguish between the time and energy representations. Note that, in either case, the Keldysh blocks $(G^{-1})^K$ are constants defined by the distribution

$$
F(\omega) = 1 + 2n(\omega) = \coth\left(\frac{\omega - \mu}{2T} \right) \tag{12.41}
$$

and depend on the energy eigenvalue. The Keldysh Green function is given by

$$
-i \begin{pmatrix} \langle \psi_{c,t} \quad \bar{\psi}_{c,t'} \rangle & \langle \psi_{c,t} \quad \bar{\psi}_{q,t'} \rangle \\ \langle \psi_{q,t} \quad \bar{\psi}_{c,t'} \rangle & \langle \psi_{q,t} \quad \bar{\psi}_{q,t'} \rangle \end{pmatrix} = \begin{pmatrix} G^K & G^+ \\ G^- & 0 \end{pmatrix}_{tt'},
$$
$$
\tag{12.42}
$$
$$
-i \begin{pmatrix} \langle \psi_{c,\epsilon} \quad \bar{\psi}_{c,\epsilon} \rangle & \langle \psi_{c,\epsilon} \quad \bar{\psi}_{q,\epsilon} \rangle \\ \langle \psi_{q,\epsilon} \quad \bar{\psi}_{c,\epsilon} \rangle & \langle \psi_{q,\epsilon} \quad \bar{\psi}_{q,\epsilon} \rangle \end{pmatrix} = \begin{pmatrix} G^K & G^+ \\ G^- & 0 \end{pmatrix}_\epsilon,
$$

with Keldysh factors defined by

$$
\begin{aligned}
G^+_{tt'} &\equiv -i\Theta(t - t')e^{-i\omega(t-t')}, & G^+_\epsilon &= \frac{1}{\epsilon^+ - \omega}, \\
G^-_{tt'} &\equiv +i\Theta(t' - t)e^{-i\omega(t-t')}, & G^-_\epsilon &= \frac{1}{\epsilon^- - \omega}, \\
G^K_{tt'} &\equiv -iF(\omega)e^{-i\omega(t-t')}, & G^K_\epsilon &= -2\pi i F(\omega)\delta(\epsilon - \omega).
\end{aligned} \tag{12.43}
$$

In the equilibrium situation considered here, we have the FDT relation

$$
G^K_\epsilon = F(\omega)\left(G^+_\epsilon - G^-_\epsilon \right). \tag{12.44}
$$

The relation to the Green functions of the contour representation is given by

$$
G^+ = G^T - G^<, \qquad G^- = G^T - G^>, \qquad G^K = G^T + G^{\tilde{T}}. \tag{12.45}
$$

Also note the relations

$$
(G^\pm)^\dagger = G^\mp, \qquad (G^K)^\dagger = -G^K. \tag{12.46}
$$

12.2.5 Fluctuation–dissipation theorem

In thermal equilibrium, the Keldysh Green function is fully determined by the retarded and advanced Green functions: with $\hat{F} = 1 + 2n = \coth(\epsilon/2T)$ and $\hat{G}^K = \hat{G}^+\hat{F} - \hat{F}\hat{G}^{-1}$, we have

$$\hat{G}^K(\epsilon) = \coth\left(\frac{\epsilon}{2T}\right)\left(\hat{G}^+(\epsilon) - \hat{G}^-(\epsilon)\right). \tag{12.47}$$

fluctuation–
dissipation
relation This relation resembles the **fluctuation–dissipation relation** reviewed in section 11.6.4. In the following, we will make this connection more concrete and at the same time explain the physical meaning of $\hat{G}^K$. Turning back to a representation in terms of contour fields, the Keldysh Green function is defined as

$$\hat{G}^K(t,t') = -i\langle\psi_c(t)\bar{\psi}_c(t')\rangle = \frac{1}{2}\langle(\psi_+(t) + \psi_-(t))(\bar{\psi}_+(t') + \bar{\psi}_-(t'))\rangle,$$

where Hilbert space indices are omitted for clarity. These correlation functions constitute the field integral representation of the contour-ordered expectation values, $\langle\psi_C(t)\bar{\psi}_{C'}(t')\rangle \leftrightarrow \langle T_\gamma a_C(t)a^\dagger_{C'}(t')\rangle$, where the angular brackets on the right-hand side give the thermal expectation value $\langle\ldots\rangle = \text{tr}(\ldots\hat{\rho}_0)$, and the combination (C,t) fixes the position of the operator on the closed time-contour. Employing an exact spectral decomposition (see info block below), it is a straightforward exercise to demonstrate that

$$\hat{G}^K(\epsilon) = 2i\,\text{Im}\,\hat{G}^T(\epsilon). \tag{12.48}$$

In words:

> The Keldysh Green function $\hat{G}^K$ equals twice the imaginary part of the time-ordered correlation function $\hat{G}^T$.

(The latter is defined in Eq. (7.11) for $X_1 = a$ and $X_2 = a^\dagger$.) Substitution of this identification into Eq. (12.47) then gives the standard representation of the fluctuation–dissipation relation:

$$\text{Im}\,G^T(\epsilon) = \coth\left(\frac{\epsilon}{2T}\right)\text{Im}\,\hat{G}^+(\epsilon).$$

INFO   To prove Eq. (12.48), we employ the Heisenberg representation of the operators, as $a_C(t) = U^{-1}(C,t)aU(C,t)$, where $U(C,t) = T_\gamma \exp\left(i\int_0^{(C,t)} dt\,\hat{H}(t)\right)$ describes the evolution along the contour up to the point specified by (C,t). For example, re-

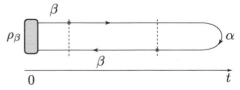

calling that times on the backward contour are always Keldysh-later than those on the forward contour, $\langle T_\gamma a^\dagger_-(t')a_+(t)\rangle = \text{tr}(e^{i\hat{H}t'}a^\dagger e^{i\hat{H}(t-t')}ae^{-i\hat{H}t}\rho_0) = \langle a^\dagger(t')a(t)\rangle$, where the time-dependence in the last expression refers to the standard Heisenberg representation. Inserting a formal decomposition in system eigenstates, this yields the Lehmann representation $\langle T_\gamma a^\dagger_-(t')a_+(t)\rangle = \rho_\beta|a_{\alpha\beta}|^2 e^{i\Xi_{\alpha\beta}(t-t')}$, where $a_{\alpha\beta} \equiv \langle\alpha|a|\beta\rangle$, $\Xi_\alpha = E_\alpha - \mu N_\alpha$,

$\Xi_{\alpha\beta} = \Xi_\alpha - \Xi_\beta$, $\rho_\beta = \mathcal{Z}_0^{-1}e^{-\beta\Xi_\beta}$, and a summation over α, β is implied. Proceeding in the same way with the remaining contributions, one obtains $G^K(t,0) = -i\mathcal{Z}_0^{-1}|X_{\alpha\beta}|^2 e^{i\Xi_{\alpha\beta}t}(\rho_\alpha + \rho_\beta)$, which Fourier transforms to

$$G^K(\epsilon) = -2\pi i \mathcal{Z}_0^{-1}|X_{\alpha\beta}|^2 \delta(\epsilon + \Xi_{\alpha\beta})(\rho_\alpha + \rho_\beta).$$

Comparison with Eq. (7.16) then leads immediately to Eq. (12.48).

12.2.6 Classical limit I

Readers who went through the previous chapter may have noticed apparent similarities between the Keldysh theory outlined in the previous section and the **MSR functional** of section 11.5.1. Specifically, the triangular form of the quadratic action, containing retarded/advanced Green functions and the distribution functions as building blocks, resembles that of the MSR theory. In this section, we indicate how classical structures emerge in interacting Keldysh theories as a particular class of stationary configurations. A more comprehensive discussion of the ensuing structures and connections with MSR theory will then be the subject of section 12.3.

Prior to taking the semiclassical limit of a theory, we need to let $\hbar$ re-enter the description. Presently, the Keldysh action represents a quantum phase, which requires the scaling $iS \to \frac{i}{\hbar}S$. In the limit $\hbar \to 0$, the functional is governed by stationary configurations obeying

$$\frac{\partial S[\psi_c, \psi_q]}{\partial \psi_q} = \frac{\partial S[\psi_c, \psi_q]}{\partial \psi_c} = 0. \qquad (12.49)$$

Among the solutions of these equations, there is always one that reflects the classical physics of the problem. Recalling that an expansion of the action in ψ_q starts at linear order (the purely classical action $S[\psi_c, 0]$ vanishes), we identify this **classical saddle-point** by the equations,

classical
saddle-
point

$$\psi_q = 0,$$
$$\left.\frac{\partial S[\psi_c, \psi_q]}{\partial \psi_q}\right|_{\psi_q=0} = 0. \qquad (12.50)$$

The first line, $\psi_q = \psi_+ - \psi_- = 0$, underpins the classical nature of this solution.

In the case of the interacting Bose action (12.30), the equation assumes the form

$$\left(i\partial_t - \hat{H} - V|\psi_c|^2\right)\psi_c = 0. \qquad (12.51)$$

Gross–
Pitaevskii
equation

This is the **Gross–Pitaevskii equation**, a time-dependent generalization of the static mean field equation (5.10) discussed in chapter 5 in connection with superfluidity. The Gross–Pitaevskii equation describes the non-linear evolution of a bosonic order parameter amplitude ψ_c self-interacting with its own density $\sim |\psi_c|^2$. Referring for a discussion of this particular equation to the info block below, here we note that the approach above exemplifies a general principle: typical many-body actions are large in comparison with $\hbar$. One therefore often applies stationary phase analysis: identify stationary configurations first, then include fluctuations. In the sections below, we will illustrate on more examples how this program works in practice.

INFO We may think of the **Gross–Pitaevskii equation (GPE)** as the time-dependent Ginzburg–Landau equation of the transition into a superfluid state. Alternatively, it may be interpreted as a nonlinear Schrödinger equation describing the wave function of a Bose–Einstein condensate. Inhomogeneous solutions of this equation describe collective excitations of the condensate field.

As an example illustrating this mechanism, consider a one-dimensional spatially homogeneous system where $\hat{H} = -\frac{1}{2m}(\hbar\partial_x)^2 - \mu$ includes a chemical potential. Passing to the Fourier energy representation, $i\hbar\partial_t \to E$, the equation assumes the form

$$\left(E + \frac{(\hbar\partial_x)^2}{2m} + \mu - V|\psi_c|^2\right)\psi_c = 0.$$

The solution of lowest energy, $E = 0$, is the spatially homogeneous condensate amplitude $|\psi_c|^2 = \frac{\mu}{V}$ (see the discussion in section 5.2.1). However, the GPE also contains information on spatially inhomogeneous excitations of the condensate. For example, it is straightforward to verify that, for any x_0, the function

$$\psi_c \equiv \psi_0 \tanh\left(\frac{(x - x_0)}{\xi}\right) \tag{12.52}$$

dark
soliton

solves the GPE. Here, ψ_0 is a complex amplitude with $|\psi_0|^2 = \mu/g$, and $\xi \equiv \hbar/\sqrt{4m\mu}$ is the coherence length. The wave function (12.52) is called a **dark soliton**, the attribute *dark* alluding to the vanishing of the condensate amplitude $|\psi|$ at the soliton center. The sign change in ψ cannot be removed by a continuous deformation, indicating that the soliton is a topological excitation. As an exercise, try to identify a bright soliton for a system with attractive interactions, $V < 0$.

12.3 Particle Coupled to an Environment

In the previous section we reasoned that, starting from an initial distribution at negative infinity, a system may generate an effective distribution via interaction processes. We also saw that the theory may be analyzed starting from its classical stationary phase configurations. In this section, we illustrate these principles on the example of a quantum particle coupled to a bath.

Keldysh theory of a quantum particle

Consider the quantum mechanics of a point particle with Hamiltonian $\hat{H} = \frac{\hat{p}^2}{2m} + V(\hat{q})$, initially prepared in a state $\hat{\rho}_0$. In this case, the Keldysh theory is best formulated in the language of a Feynman path integral, rather than as a coherent state field integral:

$$\mathcal{Z}_s = \mathcal{Z}_0^{-1} \int D(q,p) \, e^{i \sum_C sc \int dt \, (p_C \partial_t q_C - H(q_C, p_C))}$$

$$= \mathcal{Z}_0^{-1} \int Dq \, e^{i \sum_C sc \int dt \, \left(\frac{m}{2} (\partial_t q_C)^2 - V(q_C) \right)}$$

$$= \mathcal{Z}_0^{-1} \int Dq \, e^{i \int dt \, \left(m \partial_t q_q \partial_t q_c - V\left(\frac{1}{\sqrt{2}} (q_c + q_q) \right) + V\left(\frac{1}{\sqrt{2}} (q_c - q_q) \right) \right)}, \qquad (12.53)$$

where, in the second equality, we integrated over momenta to pass to a Lagrangian representation. To explore the **classical limit** of this expression, we scale $\hbar$ as $iS \to i\hbar^{-1}S$. Recalling the quantum nature of configuration differences $q_+ - q_- \sim q_q$ above, it is also natural to make the scaling $q_c \to \hbar q_c$. A first-order expansion of the action in $q_c \hbar$ then identifies the $\hbar$-independent (classical) sector of the action:

$$S_c[q_c, q_q] \equiv \int dt \, q_q \left[-m \partial_t^2 q_c - \sqrt{2} \, V'\left(\frac{1}{\sqrt{2}} q_c \right) \right].$$

Integrating over the quantum component, we obtain the constraint $-m\partial_t^2 q_c - \sqrt{2} \, \partial_q V(q_c/\sqrt{2}) = 0$ or, upon rescaling coordinates so that $q_c/\sqrt{2} \equiv q$, **Newton's equations of motion**

$$m \partial_t^2 q = -\partial_q V(q). \qquad (12.54)$$

This equation determines the classical stationary phase configurations of the system. We next ask what happens if the coupling to a bath environment is switched on.

Coupling to an oscillator bath

The bath itself is described by a partition function such as Eq. (12.31). Assume the bath degrees of freedom to be oscillator-like, $\bar{\psi} \hat{H} \psi \equiv \sum_k \bar{\psi}_k \omega_k \psi_k$, with wave-like dispersion $\omega_k \equiv c|k|$, we get

$$S_b[\psi] \equiv \int (d\omega) \sum_k \bar{\psi}_{k,\omega} \, \hat{G}_{k,\omega}^{-1} \, \psi_{k,\omega}, \qquad (12.55)$$

with $\psi = (\psi_c, \psi_q)^T$, $(d\omega) = d\omega/2\pi$, and the oscillator Green function,

$$\hat{G}_{k,\omega}^{-1} = \begin{pmatrix} 0 & \omega - i\delta - \omega_k \\ \omega + i\delta - \omega_k & 2i\delta \coth(\frac{\omega_k}{2T}) \delta(\omega - \omega_k) \end{pmatrix}.$$

The coupling between system and bath is now modeled by the action

$$S_{sb}[\psi, q] \equiv \int dt \sum_{k,C} s_C q_C(t) \left(\gamma_k \psi_{C,k}(t) + \text{c.c.} \right) = \int dt \sum_k q(t)^T \sigma_1 \left(\gamma \psi_k(t) + \text{c.c.} \right)$$

$$= \int (d\omega) \sum_k \left(\gamma_k q_\omega^T \sigma_1 \psi_{k,-\omega} + \bar{\gamma}_k \bar{\psi}_{k,\omega}^T \sigma_1 q_{-\omega} \right), \qquad (12.56)$$

where in the second line we switched to the Keldysh representation $q^T = (q_c, q_q)$, and $\gamma(k)$ are complex coupling constants, assumed to scale as $|\gamma(k)| = \lambda|k|^\alpha$ for definiteness.

Integration over oscillator modes

At this stage, the oscillators may be integrated out. Using the fact that $-i\langle \psi \bar{\psi}^T \rangle = \hat{G}$ (see. Eq. (12.35)), we obtain the effective action, $S_{\mathrm{eff}} = S_s + S_{\mathrm{diss}}$, where the **dissipative action** is given by

$$S_{\mathrm{diss}}[q] = -\int (d\omega) q_\omega^T \sigma_1 \left(\sum_k |\gamma_k|^2 \hat{G}_{k,\omega} \right) \sigma_1 q_{-\omega},$$

and the Green function has the form

$$\hat{G}_{k,\omega} = \begin{pmatrix} -2\pi i \coth\left(\frac{\omega_k}{2T}\right) \delta(\omega - \omega_k) & (\omega + i\delta - \omega_k)^{-1} \\ (\omega - i\delta - \omega_k)^{-1} & 0 \end{pmatrix}.$$

Using the relation $|\gamma_k|^2 = \lambda^2 k^{2\alpha}$, and noting that the Green function enters in a frequency-symmetrized form, $\sum_\omega q_\omega^T \hat{G}_\omega q_{-\omega} = \frac{1}{2} \sum_\omega q_\omega^T (\hat{G}_\omega + \hat{G}_{-\omega}) q_{-\omega}$, summation over the k-modes produces the expression (exercise)

$$S_{\mathrm{diss}}[q] = \int (d\omega)\, q_\omega^T \hat{K}_\omega q_\omega, \qquad (12.57)$$

dissipation kernel

where the **dissipation kernel** $\hat{K}$ is given by

$$\hat{K}_\omega = \pi \nu \lambda^2 \left(\frac{\omega}{c}\right)^{2\alpha} \begin{pmatrix} 0 & -i \\ i & 2i \coth\left(\frac{\omega}{2T}\right) \end{pmatrix}. \qquad (12.58)$$

Here we have introduced the density of modes, $\nu \equiv \sum_k \delta(\omega - \omega_k)$, and omitted a frequency-independent contribution to K (which can be absorbed into the chemical potential). Physically, K_ω may be interpreted as the self-energy of the particle due to its coupling to the oscillator bath. The frequency dependence of K_ω signals that S_{diss} is a time-nonlocal contribution to the action. The temporal profile of the dissipation kernel depends on the coupling γ. In the following, we concentrate on the particularly important case of **Ohmic dissipation** where $\alpha = 1/2$ and

Ohmic dissipation

$$\hat{K}_\omega = g \begin{pmatrix} 0 & -i\omega \\ i\omega & 2i\omega \coth\left(\frac{\omega}{2T}\right) \end{pmatrix}, \qquad (12.59)$$

with $g \equiv \frac{\pi \nu \lambda^2}{c}$. Notice that the addition of S_{diss} to the action renders the qq-sector of the quadratic action non-zero. In the present context, this contribution plays the role of an effective self-energy (see the comments after Eq. (12.29)). We next discuss how this coupling affects the dynamics of the system coordinate.

INFO The power law $\alpha = 1/2$ is realized for the **coupling to an elastic medium** and therefore is physically relevant. To see how this comes about, define $\gamma = i\lambda \,\mathrm{sgn}(k) \sqrt{|k|}$ and recall the representation of the coherent state amplitudes in terms of oscillator coordinates and momenta, $\psi_k = \frac{1}{\sqrt{2}}(\sqrt{\omega_k} q_k + \frac{i}{\sqrt{\omega_k}} p_k)$. Substitution of these expressions into the coupling term gives

$$\sum_k (\gamma_k \psi_k + \mathrm{c.c.}) \sim \sum_k \mathrm{sgn}(k) \sqrt{|k|} \left(i\sqrt{\omega_k} q_k - \frac{1}{\sqrt{\omega_k}} p_k + \mathrm{c.c.} \right) \sim \sum_k ikq_k \sim \partial_x q(x)\big|_{x=0},$$

where, noting that $q(x)$ and $p(x)$ are real, we used $\bar{q}_k = q_{-k}$ and $\bar{p}_k = p_{-k}$. The last term describes the local coupling of the system coordinate to the stress $\partial_x q(x)$ acting on an elastic string. Macroscopic degrees of freedom are frequently coupled to harmonic microscopic modes in this way. One example is the coupling of an electric field to the microscopic modes defining a resistive medium. This coupling mechanism explains the denotation "ohmic" dissipation.

Langevin equation

Assuming that the dynamics of the system coordinate is almost classical, let us explore the role of the system–bath coupling, and indicate how quantum effects resolve problems plaguing the purely classical theory. To this end, we reintroduce $\hbar$ as follows

$$(i) : iS \to \frac{i}{\hbar}S,$$

$$(ii) : \coth(\omega/2T) \to \coth(\omega\hbar/2T) \to \frac{2T}{\hbar\omega}, \qquad (12.60)$$

$$(iii) : \psi_q \to \hbar\psi_q,$$

where, in the second line, we used the fact that, in the semiclassical limit and at fixed frequency ω, the coth can be linearized. The extra power of $\hbar$ appearing in the denominator of this expression compensates for the additional power in $\hbar$ due to two factors q_q in the quantum–quantum sector of the action. Summarizing, the effective classical action is given by

$$S_c[q_c, q_q] = \int dt \left[q_q \left(-m\partial_t^2 q_c - \sqrt{2}\,V'\left(\frac{1}{\sqrt{2}}q_c\right) - 2g\partial_t q_c \right) + 4iTg q_q^2 \right],$$

which has the characteristic structure of an MSR action. From here, we may pass to a Langevin-type differential equation for q by Hubbard–Stratonovich decoupling of the q_q term:

$$S_c[q_c, q_q] \to S_c[q_c, q_q, \xi] = \int dt \left[q_q \left(-m\partial_t^2 q_c - \sqrt{2}\,V'\left(\frac{q_c}{\sqrt{2}}\right) - 2g\partial_t q_c + \sqrt{2}\xi \right) \right]$$
$$+ \frac{i}{8gT} \int dt\,\xi^2.$$

Introducing the rescaled variable $q = q_c/\sqrt{2}$ and integrating over q_q, this leads to

$$m\partial_t^2 q + V'(q) + 2g\partial_t q = \xi, \qquad (12.61)$$

where the noise correlator

$$\langle \xi(t)\xi(t') \rangle = 4Tg\delta(t - t') \qquad (12.62)$$

is consistent with the fluctuation–dissipation theorem. In this way, we see that the coupling of the system to a bath renders the classical limit noisy. Recalling the discussion of the previous chapter, the combination of damping and noise will push the system towards a thermal distribution, at a temperature set by that of the bath.

INFO **The semiclassical theory solves a problem** that plagued the purely classical theory discussed in the previous chapter. Discussing the dissipative damping of voltage fluctuations in a resistor network, we saw that Johnson noise is problematic inasmuch as the white noise correlator $\langle \xi(t)\xi(t')\rangle = 2RT\delta(t-t')$ predicts singularly strong noise levels $\langle \xi(t)^2 \rangle$. The coupling to a quantum bath regularizes this infinity at Planck scales. Comparison with Eq. (12.59) and the subsequent steps of the derivation show that the frequency representation of the actual noise correlaton reads

$$\langle \xi_\omega \xi_{\omega'} \rangle = \frac{g}{\pi}\delta(\omega+\omega')\omega \coth\left(\frac{\hbar\omega}{2T}\right).$$

In the classical limit $\hbar \to 0$, this reduces to the white noise limit $\frac{2Tg}{\pi\hbar}\delta(\omega+\omega')$, which Fourier transforms to Eq. (12.62). However, at large frequencies $\hbar\omega \gg T$, we obtain

$$\langle \xi_\omega \xi_{\omega'} \rangle \xrightarrow{\hbar\omega \gg T} \frac{g}{\pi}\delta(\omega+\omega')|\omega|.$$

It turns out that the replacement $T \to \hbar|\omega|/2$ in this expression removes the ultraviolet singularity of the classical theory. However, the actual derivation of the regularized noise statistics requires additional discussion beyond the scope of the present text, and the interested reader is referred to the literature.[12]

The discussion above exemplifies how Keldysh theory describes the coupling of a quantum system to an environment, and its subsequent thermalization. In the remaining sections of the chapter, we will go beyond this level and address problems out of thermal equilibrium. However, since most of these involve the coupled appearance of bosonic and fermionic degrees of freedom, we first need to construct a fermionic extension of Keldysh theory.

12.4 Fermion Keldysh Theory

REMARK The construction of a fermionic Keldysh theory parallels that for the bosonic one. However, as might be expected, the anticommutation of fermion operators introduces a few sign changes. Impatient readers may skip this discussion and consult the reference chart on page 727 for a summary of the essential changes.

In the following, we construct the fermion Keldysh functional by retracing the construction steps of the bosonic one, indicating the changes required by fermion statistics.

12.4.1 Single level

As in the bosonic case, let us begin with a single-level Hamiltonian $\hat{H} = \omega a^\dagger a$, where a and $a^\dagger$ are fermion operators. The definitions of the Keldysh partition function (12.10), and of the equilibrium density operator (12.11), remain unchanged.

---

[12] See, e.g., G. B. Lesovik and R. Loosen, *On the detection of finite-frequency current fluctuations*, JETP Lett. **65**, 295 (1997).

However, the normalizing factor is now given by $\mathcal{Z}_0 = 1 + e^{-\beta(\omega-\mu)}$. We again represent the partition function in terms of (now fermionic) coherent states.

It is straightforward to verify that the auxiliary identity (12.12) generalizes to the fermionic case. This implies that the discrete functional (12.13) remains as it is, except for a sign change $\kappa \to -\kappa$. As in the construction of the Matsubara equilibrium functional, this is the sign factor picked up when the final one of the $2N$ coherent state amplitudes required to trotterize the time-evolution along the Keldysh contour is commuted through all others to complete the construction of the discrete coherent state integral; see the discussion below Eq. (3.67) (think about this point). Likewise, the inverse of the discrete fermionic Green function is given by Eq. (12.14), again up to a sign change $\kappa \to -\kappa$. Otherwise, all the formulae of section 12.2.2 generalize to the fermionic case. Specifically, $\det(-iG^{-1}) \overset{N\to\infty}{\longrightarrow} 1 + \kappa = 1 + e^{-\beta(\omega-\mu)} = \mathcal{Z}_0$ defines the normalization of the functional integral. Taking the limit $N \to \infty$, and switching to a continuum representation, we obtain the **fermionic Green functions** (12.71).

We may now pass to a continuum representation of the functional integral whose unregularized form (with no accounting for infinitesimal damping and boundary conditions) is given by the Grassmann version of Eq. (12.18). A glance at Eq. (12.71) shows that the redundancy in the theory, which motivated the "Keldysh rotation," now assumes the form $G^{++} + G^{--} = G^{+-} + G^{-+}$. In the case of Grassmann variables, it is customary to remove this redundancy by a field transformation slightly different from the bosonic Keldysh rotation:[13] using the independence of the Grassmann variables ψ and $\bar{\psi}$, we introduce new fields by

$$\psi_1 \equiv \frac{1}{\sqrt{2}}(\psi_+ + \psi_-), \qquad\qquad \psi_2 \equiv \frac{1}{\sqrt{2}}(\psi_+ - \psi_-),$$

$$\bar{\psi}_1 \equiv \frac{1}{\sqrt{2}}(\bar{\psi}_+ - \bar{\psi}_-), \qquad\qquad \bar{\psi}_2 \equiv \frac{1}{\sqrt{2}}(\bar{\psi}_+ + \bar{\psi}_-). \qquad (12.63)$$

With the notation $\psi \equiv \left(\begin{smallmatrix}\psi_+ \\ \psi_-\end{smallmatrix}\right)$, $\psi' \equiv \left(\begin{smallmatrix}\psi_1 \\ \psi_2\end{smallmatrix}\right)$, $\bar{\psi} = (\bar{\psi}_+, \bar{\psi}_-)$, $\bar{\psi}' = (\bar{\psi}_1, \bar{\psi}_2)$, this assumes the form of a non-unitary transformation

$$\psi' \equiv U\psi, \qquad \bar{\psi}' \equiv \bar{\psi}\sigma_3 U^T,$$

where U was defined in Eq. (12.19). The form of the above transformation is motivated by the transformed contraction rule

$$-i\langle\psi'\bar{\psi}'\rangle = U\langle\psi\bar{\psi}\rangle\sigma_3 U^T = U\hat{G}U^T\sigma_3 \equiv \hat{G}',$$

---

[13] Owing to the non-unitarity of the transformation, and the fact that Grassmann fields are never "classical," these fields are labeled 1 and 2 instead of classical and quantum. The motivation for sticking to the particular field decomposition (12.63) is probably historical. (It was introduced in A. I. Larkin and Yu. N. Ovchinnikov, *Vortex motion in superconductors*, in *Nonequilibrium Superconductivity*, eds. D. N. Langenberg and A. I. Larkin (Elsevier, 1986), to give the Green function the form (12.64), considered convenient for perturbation theory.) However, in the recent literature, various authors have been adopting an alternative "center of mass," previously used in the bosonic case. The transformation to those coordinates leads to a different, but no more complicated, structure for the Green function; and it is conveniently unitary.

where $\hat{G} = \begin{pmatrix} \hat{G}^{++} & \hat{G}^{+-} \\ \hat{G}^{-+} & \hat{G}^{--} \end{pmatrix}$; the transformed Green function assumes the form

$$\hat{G}' = \begin{pmatrix} \hat{G}^+ & \hat{G}^K \\ & \hat{G}^- \end{pmatrix}, \qquad (12.64)$$

and the retarded, advanced, and Keldysh Green functions are defined by analogy to Eq. (12.42) as

$$
\begin{aligned}
G_{tt'}^+ &\equiv -i\Theta(t - t')e^{-i\omega(t-t')}, \\
G_{tt'}^- &\equiv +i\Theta(t' - t)e^{-i\omega(t-t')}, \\
G_{tt'}^K &\equiv -i(1 - 2n_F(\omega))e^{-i\omega(t-t')}.
\end{aligned}
\qquad (12.65)
$$

Since we will mostly work with the transformed Green function $\hat{G}'$, we omit the prime throughout, $\hat{G}' \to \hat{G}$.

12.4.2 Generalization

As with the bosonic theory, the form of the Green function (12.65) suggests the introduction of a fermionic distribution operator $\hat{F} = \{F(\epsilon)\}$, with

$$F(\epsilon) \equiv 1 - 2n_F(\epsilon). \qquad (12.66)$$

Expressed as a Fourier transform, the Keldysh Green function then assumes the form

$$\hat{G}^K = \hat{G}^+\hat{F} - \hat{F}\hat{G}^-, \qquad (12.67)$$

as in the bosonic theory. The generalization of the toy model above to an interacting theory is described by the Grassmann version of the functional, Eq. (12.30). (The option to add spin-indices should be straightforward.) In contrast with the bosonic case, where mean field approaches to the interacting theory lead to immediate results, most fermionic theories call for a Hubbard–Stratonovich decoupling of the interaction term. Prior to passing to the Keldysh-rotated theory, we therefore introduce an auxiliary bosonic field to decouple the interaction:

$$
e^{iS_{\mathrm{int}}[\psi]} = \int D\phi \, e^{-\frac{i}{2}\int dt\, d^d r d^d r' \sum_C s_C \phi_C(\mathbf{r},t)V^{-1}(\mathbf{r}-\mathbf{r}')\phi_C(\mathbf{r}',t)}
$$
$$
\times e^{+i\int dt\, d^d r \sum_C s_C \bar{\psi}_C(\mathbf{r},t)\phi_C(\mathbf{r},t)\psi_C(\mathbf{r},t)}.
$$

We may now implement the Keldysh rotation to arrive at the effective action $S[\psi, \phi] = S[\phi] + S[\psi] + S_{\mathrm{int}}[\psi, \phi]$, where (omitting obvious space–time coordinates)

$$
\begin{aligned}
S[\phi] &= -i\int d^d r \, dt \, \phi_c(\mathbf{r}, t)V^{-1}(\mathbf{r} - \mathbf{r}')\phi_q(\mathbf{r}', t), \\
S[\psi] &= \int d^d r \, dt \, \bar{\psi}\hat{G}_0^{-1}\psi, \\
S_{\mathrm{int}}[\psi, \phi] &= \frac{1}{\sqrt{2}}\int d^d r \, dt \, \bar{\psi}(\phi_c\sigma_0 + \phi_q\sigma_1)\psi.
\end{aligned}
\qquad (12.68)
$$

Here

$$\hat{G}_0^{-1} = \begin{pmatrix} (G_0^+)^{-1} & (\hat{G}_0^{-1})^K \\ & (\hat{G}_0^-)^{-1} \end{pmatrix} \qquad (12.69)$$

is the inverse of the free Green function[14] and $(\hat{G}_0^{\pm})^{-1} = (i\partial_t \pm i0 - \hat{H})$. We finally note that elements of the Green function are generated by the fermionic variant of Eq. (12.35)

$$\boxed{-i\langle \psi_\alpha(\mathbf{r},t)\bar{\psi}_{\alpha'}(\mathbf{r}',t')\rangle = \begin{pmatrix} G^+ & G^K \\ 0 & G^- \end{pmatrix}(\mathbf{r},t;\mathbf{r}',t')_{\alpha\alpha'}} \qquad (12.70)$$

INFO Here, we complement the bosonic reference chart on page 716 with a chart listing the **conventions of fermionic Keldysh theory**. Again, we assume a single-level Hamiltonian, $\hat{H} = \omega a^\dagger a$, now with fermionic distribution $n_F(\omega) = (e^{\beta(\omega-\mu)} + 1)^{-1}$.

Contour representation:

$$S[\phi] = \sum_{C=\pm} C \int dt\, \bar{\psi}_t^C (i\partial_t - \omega)\psi_t^C = \sum_{C=\pm} C \int \frac{d\epsilon}{2\pi}\, \bar{\psi}_\epsilon^C(\epsilon - \omega)\psi_\epsilon^C.$$

Again, the correlation between contours – via a coupling term at the temporal boundaries now involving the fermionic distribution function – is left implicit in this representation. The Green functions in the contour representation are given by

$$-i\langle \psi_t^C \bar{\psi}_{t'}^{C'}\rangle = \begin{pmatrix} G^{++} & G^{+-} \\ G^{-+} & G^{--} \end{pmatrix}_{tt'} \equiv \begin{pmatrix} G^T & G^< \\ G^> & G^{\tilde{T}} \end{pmatrix}_{tt'},$$

$$G_{tt'}^{+-} \equiv G_{tt'}^< = +ie^{-i\omega(t-t')}n_F(\omega),$$

$$G_{tt'}^{-+} \equiv G_{tt'}^> = -ie^{-i\omega(t-t')}(1 - n_F(\omega)), \qquad (12.71)$$

$$G_{tt'}^{++} \equiv G_{tt'}^T = G_{tt'}^> \Theta(t - t') + G_{tt'}^< \Theta(t' - t),$$

$$G_{tt'}^{--} \equiv G_{tt'}^{\tilde{T}} = G_{tt'}^> \Theta(t' - t) + G_{tt'}^< \Theta(t - t'),$$

where $n_F(\omega) = \frac{1}{1+e^{\beta(\omega-\mu)}}$ is the Fermi distribution. This differs from the bosonic variant, Eq. (12.36), in the replacement $n(\omega) \to -n_F(\omega)$, where the sign change reflects the fermionic statistics on the contour. Note the constraint $G^{++} + G^{--} = G^{+-} + G^{-+}$.

Keldysh representation: Fields in the Keldysh basis are defined as (see Eq. (12.63))

$$\begin{pmatrix} \psi_1 \\ \psi_2 \end{pmatrix} \equiv \frac{1}{\sqrt{2}}\begin{pmatrix} \psi^+ + \psi^- \\ \psi^+ - \psi^- \end{pmatrix}, \qquad \begin{pmatrix} \bar{\psi}_1 \\ \bar{\psi}_2 \end{pmatrix} \equiv \frac{1}{\sqrt{2}}\begin{pmatrix} \bar{\psi}^+ - \bar{\psi}^- \\ \bar{\psi}^+ + \bar{\psi}^- \end{pmatrix},$$

$$\begin{pmatrix} \psi^+ \\ \psi^- \end{pmatrix} \equiv \frac{1}{\sqrt{2}}\begin{pmatrix} \psi_1 + \psi_2 \\ \psi_1 - \psi_2 \end{pmatrix}, \qquad \begin{pmatrix} \bar{\psi}^+ \\ \bar{\psi}^- \end{pmatrix} \equiv \frac{1}{\sqrt{2}}\begin{pmatrix} \bar{\psi}_1 + \bar{\psi}_2 \\ \bar{\psi}_2 - \bar{\psi}_1 \end{pmatrix}. \qquad (12.72)$$

Note the signs in the definition of $\bar{\psi}_i$, which make this transformation non-unitary. The Keldysh action retains its form (12.38); however, the block representation of the Green function changes to

$$\bar{\psi}G^{-1}\psi = (\bar{\psi}_1, \bar{\psi}_2)\begin{pmatrix} (G^+)^{-1} & (G^{-1})^K \\ & (G^-)^{-1} \end{pmatrix}\begin{pmatrix} \psi_1 \\ \psi_2 \end{pmatrix}, \qquad (12.73)$$

---

[14] The off-diagonal block $(\hat{G}^{-1})^K$ is given by $(\hat{G}^{-1})^K = -(\hat{G}^+)^{-1}\hat{G}^K(\hat{G}^-)^{-1} = (\hat{G}^+)^{-1}\hat{F} - \hat{F}(\hat{G}^-)^{-1} = 2i\delta\hat{F}$, where the last identity holds if $\hat{F}$ commutes with $(\hat{G}^\pm)^{-1}$.

where the entries are given by Eq. (12.40) with the fermionic equilibrium distribution

$$F(\omega) = 1 - 2n_F(\omega) = \tanh\left(\frac{\omega - \mu}{2T}\right). \tag{12.74}$$

The Green functions are defined by

$$
\begin{aligned}
-i\begin{pmatrix} \langle \psi_{1,t}\bar{\psi}_{1,t'} \rangle & \langle \psi_{1,t}\bar{\psi}_{2,t'} \rangle \\ \langle \psi_{2,t}\bar{\psi}_{1,t'} \rangle & \langle \psi_{2,t}\bar{\psi}_{2,t'} \rangle \end{pmatrix} &= \begin{pmatrix} G^+ & G^K \\ 0 & G^- \end{pmatrix}_{tt'}, \\
-i\begin{pmatrix} \langle \psi_{1,\epsilon}\bar{\psi}_{1,\epsilon} \rangle & \langle \psi_{1,\epsilon}\bar{\psi}_{2,\epsilon} \rangle \\ \langle \psi_{2,\epsilon}\bar{\psi}_{1,\epsilon} \rangle & \langle \psi_{2,\epsilon}\bar{\psi}_{2,\epsilon} \rangle \end{pmatrix} &= \begin{pmatrix} G^+ & G^K \\ 0 & G^- \end{pmatrix}_{\epsilon},
\end{aligned}
\tag{12.75}
$$

with entries given by Eq. (12.43) (now with the fermionic distribution Eq. (12.74)). The equilibrium FDT (12.44) and the relations (12.45) and (12.46) remain valid.

12.5 Kinetic Equation

The action (12.68) defines an exact representation of an interacting Fermi system and is the starting point for numerous approximation schemes. As an example, here we discuss the kinetic equation formalism, which is tailored to the description of **effective particle distributions** forming as a result of interactions. The methods introduced in this section are powerful and are routinely applied in the analysis of out-of-equilibrium quantum systems.

Let us interpret the Hubbard–Stratonovic field employed to decouple the four-fermion interaction as an effective bosonic degree of freedom. Seen in this way, the nonlinearity described by S_{int} represents a boson–fermion interaction vertex. We aim to describe the ensuing boson and fermion self-energies $\hat{\Sigma}_f$ and $\hat{\Sigma}_b$ dressing the respective propagators. To this end, we apply an RPA-type approximation, which 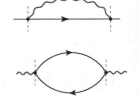 discards intersecting propagator lines. The topology of $\hat{\Sigma}_f$ is then given by the upper diagram shown in the figure, with external legs removed, and all propagators representing full propagators, with a self-consistent account of the self-energies. In a similar manner, $\hat{\Sigma}_b$ is obtained from the lower diagram. Inspection of the diagrams shows that the two self-energies contain the propagators as $\Sigma_f \sim GD$, $\Sigma_b \sim GG$, where G and D are the fermion and boson Green function, respectively. However, what this simple observation is not able to describe is the matrix structure of the self-energies in Keldysh space.

To understand this point recall that, in a functional formulation, the fermion self-energy is obtained by the contraction of interaction operators at fixed external field indices. Specifically, the fermion self-energy is defined as

$$\int dx dx'\, \bar{\psi}'(x)\Sigma_f(x,x')\psi'(x') = -\frac{i}{2}\int dx dx'\, \bar{\psi}'(x)\left\langle \hat{\phi}(x)\psi(x)\bar{\psi}(x')\hat{\phi}(x')\right\rangle \psi'(x'),$$

where $\hat{\phi} \equiv \phi_c \sigma_0 + \phi_q \sigma_1$ and we have switched to the space–time abbreviation $x \equiv (\mathbf{r}, t)$. The self-consistency of the RPA scheme requires that the contractions must be interpreted as the full Green functions of the theory. Comparison of the two sides of the equation and Eqs. (12.35) and (12.70) then leads to the identification,

$$\Sigma_f(x, x') = -\frac{i}{2} \left\langle \hat{\phi}(x)\psi(x)\bar{\psi}(x')\hat{\phi}(x') \right\rangle = -\frac{1}{2} \left\langle \hat{\phi}(x)\hat{G}(x, x')\hat{\phi}(x') \right\rangle$$

$$= -\frac{i}{2} \left(\hat{G}(x, x')D^K(x, x') + \hat{G}(x, x')\sigma_1 D^+(x, x') + \sigma_1 \hat{G}(x, x')D^-(x, x') \right),$$

where we follow the standard convention of denoting the bosonic Green functions of a real field by $D^{+,-,K}$. Using the above expression, the components of the self-energy operator

$$\hat{\Sigma}_f = \begin{pmatrix} \hat{\Sigma}_f^+ & \hat{\Sigma}_f^K \\ 0 & \hat{\Sigma}_f^- \end{pmatrix},$$

are identified as

$$\Sigma_f^\pm(x, x') = -\frac{i}{2} \left(G^\pm(x, x')D^K(x, x') + G^K(x, x')D^\pm(x, x') \right) \tag{12.76}$$

$$\Sigma_f^K(x, x') = -\frac{i}{2} \left(G^K(x, x')D^K(x, x') + G^+(x, x')D^+(x, x') + G^-(x, x')D^-(x, x') \right)$$

$$= -\frac{i}{2} \left(G^K(x, x')D^K(x, x') + (G^+ - G^-)(x, x')(D^+ - D^-)(x, x') \right),$$

where we have made use of the fact that causality implies $G^+(x, x')D^-(x, x') = 0$, etc. On the same basis, we have $D^\pm(x, x')G^\mp(x, x') = 0$, which implies the vanishing of the lower right-hand block of the self-energy operator.

Turning to the **propagator of the effective bosonic field**, comparison with the first line in Eq. (12.68) shows that it has the structure

$$\hat{D}^{-1} = \hat{V}^{-1}\sigma_1 - \hat{\Sigma}_b = \begin{pmatrix} 0 & \hat{V}^{-1} - \hat{\Sigma}_b^- \\ \hat{V}^{-1} - \hat{\Sigma}_b^+ & -\hat{\Sigma}_b^K \end{pmatrix}. \tag{12.77}$$

EXERCISE Apply a construction as above for $\hat{\Sigma}_f$ to show that the bosonic self-energy is given by

$$\Sigma_b^\pm(x, x') = -\frac{i}{2} \left(G^\mp(x', x)G^K(x, x') + G^\pm(x, x')G^K(x', x) \right), \tag{12.78}$$

$$\Sigma_b^K(x, x') = -\frac{i}{2} \left(G^K(x', x)G^K(x, x') + G^+(x', x)G^-(x, x') + G^-(x', x)G^+(x, x') \right)$$

$$= -\frac{i}{2} \left(G^K(x', x)G^K(x, x') - (G^+ - G^-)(x', x)(G^+ - G^-)(x, x') \right).$$

In the present context, where the boson field actually represents fluctuations in a scalar potential, we may think of the self-energy as an effective **polarization operator** screening field fluctuations.

12.5.1 Quasiclassical theory

REMARK This section uses the Wigner transform. Consult appendix section A.4.3 for a discussion of this representation.

We now have everything we need to understand the effective Fermi and Bose distributions stabilized by particle interactions. The understanding of the underlying relaxation mechanisms is the basis for the description of the out-of-equilibrium distributions forming, e.g., if a system is exposed to strong external fields.

Derivation of the kinetic equation

In this section, we will derive the kinetic equation, a differential equation governing the evolution of the distribution $\hat{F}$ in an interacting environment. In view of the spatio-temporal scale separation of the difference and center of mass coordinates discussed above, it will be convenient to formulate this discussion in a Wigner transform representation. Note that this makes $F(x, p) \equiv F(\epsilon, t, \mathbf{r}, \mathbf{p})$ a function of phase space coordinates describing the distribution of particle densities. In view of this interpretation, we may anticipate similarities between the kinetic equation and the Boltzmann equation discussed in section 11.3 of the previous chapter.

Our starting point is the formal operator equation

$$\begin{pmatrix} \hat{\epsilon}^+ - \hat{H} - \Sigma_f^+ & -\Sigma_f^K \\ 0 & \hat{\epsilon}^- - \hat{H} - \Sigma_f^- \end{pmatrix} \begin{pmatrix} \hat{G}^+ & \hat{G}^K \\ 0 & \hat{G}^- \end{pmatrix} = \begin{pmatrix} \mathbb{1} & 0 \\ 0 & \mathbb{1} \end{pmatrix}.$$

Writing $\hat{G}^K = \hat{G}^+ \hat{F} - \hat{F}\hat{G}^-$, this matrix equation is equivalent to

$$(\hat{\epsilon} - \hat{H} - \hat{\Sigma}^{\pm})\hat{G}^{\pm} = \mathbb{1},$$
$$(\hat{\epsilon} - \hat{H} - \hat{\Sigma}^+)(\hat{G}^+ \hat{F} - \hat{F}\hat{G}^-) - \Sigma^K \hat{G}^- = 0.$$

Multiplying the second equation by $(\hat{G}^-)^{-1}$ and using the first, we obtain

$$\hat{F}(\hat{\epsilon} - \hat{H} - \hat{\Sigma}^-) - (\hat{\epsilon} - \hat{H} - \hat{\Sigma}^+)\hat{F} - \hat{\Sigma}^K = 0.$$

At this stage, it is convenient to pass to the Wigner transform. Making use of the identities summarized in table A.1, we obtain the **kinetic equation**

$$\boxed{(\partial_t + \{H, \ \}) F = i(\Sigma_f^K - (\Sigma_f^+ F - F\Sigma_f^-)) \equiv I_{\text{col}}[F]} \qquad (12.79)$$

The approximation underlying this equation is the first-order Moyal expansion, i.e., the semiclassical assumption that additional derivatives acting on the constituents yield contributions small in comparison to inverse powers of $\hbar$. The left-hand side describes the evolution of the distribution function $F(\epsilon, t, \mathbf{r}, \mathbf{p})$ generated by the classical Hamiltonian function corresponding to the single-particle Hamilton operator. On the right-hand side the **collision term** describes the effect of interactions and plays a role similar to the collision term of the Boltzmann equation. Its influence on the distribution function is discussed below.

(margin note: kinetic equation)

(margin note: collision term)

Collision term

in/out-
term

To lowest order in the Moyal expansion, the operator products appearing in the collision term are just products of Wigner functions, $\Sigma_f^+ F \equiv \Sigma_f^+(\epsilon, t, \mathbf{r}, \mathbf{p}) F(\epsilon, t, \mathbf{r}, \mathbf{p})$, etc. Within the framework of this approximation, we will now show that the term Σ^K represents an **in-term**, i.e., a gain in particle occupancy driven by interactions. Conversely, $\Sigma^+ F - F\Sigma^-$ is an **out-term**. The physical principles behind this assignment are best understood using a concrete example, such as a system of interacting electrons.

Recall the discussion earlier on, where we reasoned that the Hubbard–Stratonovich transformation of the electron–electron interaction introduces an effectively bosonic degree of freedom with propagator $\hat{D}$, Eq. (12.77). The inversion of this equation leads to the identification

$$\hat{D}^\pm = (\hat{V}^{-1} - \hat{\Sigma}_b^\pm)^{-1} = \hat{D}^+(\hat{V}^{-1} - \Sigma_b^\mp)\hat{D}^-,$$
$$\hat{D}^K = \hat{D}^+\hat{\Sigma}_b^K\hat{D}^-$$

of the building blocks of $\hat{D}$. From here, it is a straightforward if somewhat technical exercise to determine the quasiclassical representation of the collision term. We first need an auxiliary identity for the Wigner transform of the composites $X(x, x')Y(x', x)$ appearing in the self-energies, Eqs. (12.76) and (12.78).

EXERCISE Consider an operator, $\hat{Z}$, with space–time argument dependence $Z(x, x') = X(x, x')Y(x, x')$. Assuming that $X(x, x') = X(x - x')$ depends only on coordinate differences, show that the Wigner transform $X(p)$ is a function of the space–time momentum only. With the same assumption for $\hat{Y}$, show that the Wigner transform of the composite is given by

$$X(x, x')Y(x, x') \to \int dp' X(p')Y(p - p'). \tag{12.80}$$

Similarly, show that

$$X(x, x')Y(x', x) \to \int dp' X(p')Y(p' - p). \tag{12.81}$$

We now start from the definition of the collision term in Eq. (12.79), substitute the self-energies (12.76) and (12.78), use the fact that translational invariance implies $G(x, x') = G(x - x')$, etc., and pass to the Wigner representation to obtain

$$I_{\text{col}}[F](p) \overset{(12.76),(12.80)}{=} \frac{1}{2}\int dq\Big[G^K(p+q)D^K(q) + (G^+ - G^-)(p+q)(D^+ - D^-)(q)$$
$$- \big((G^+ - G^-)(p+q)D^K(q) + G^K(p+q)(D^+ - D^-)(q)\big)F(p)\Big]$$
$$\overset{(12.67)}{=} -\frac{i}{2}\int dq\, D^+(q)D^-(q)A(p+q)$$
$$\times \Big[(F(p+q) - F(p))\Sigma_b^K(q) + (\Sigma_b^+ - \Sigma_b^-)(q)(1 - F(p+q)F(p))\Big]$$
$$\overset{(12.78),(12.81)}{=} \frac{1}{4}\int dqdp'\, D^+(q)D^-(q)A(p+q)A(p'+q)A(p')$$

$$\times \Big[(F(p+q) - F(p))(1 - F(p'+q)F(p'))$$

$$- (F(p'+q) - F(p'))(1 - F(p+q)F(p)) \Big],$$

where we have introduced the **spectral function** $A \equiv -2\,\mathrm{Im}\,G^+$ as a measure of
the fermion spectral density (see Eq. (7.28)). To understand the meaning of this
expression, note that the distribution functions, $F(p) = F(\epsilon, \mathbf{p}) = 1 - 2n_\epsilon$ depend
only on energy. This motivates the split of integrations over $p = (\epsilon, \mathbf{p})$ and the
definition of the **transition probability**

$$T(\epsilon, \epsilon', \omega) \equiv \int \frac{d^d p'}{(2\pi)^d} \frac{d^d q}{(2\pi)^d} D^+(q) D^-(q) A(p+q) A(p'+q) A(p').$$

Adding and subtracting a term $n_\epsilon n_{\epsilon+\omega} n_{\epsilon'} n_{\epsilon'+\omega}$, it is then straightforward to show
that the collision integral takes the final form

$$I_{\mathrm{col}}[F](p) = \int \frac{d\omega}{2\pi} \frac{d\epsilon'}{2\pi} T(\epsilon, \epsilon', \omega)$$

$$\times (n_\epsilon(1 - n_{\epsilon+\omega})n_{\epsilon'+\omega}(1 - n_{\epsilon'}) - (1 - n_\epsilon)n_{\epsilon+\omega}(1 - n_{\epsilon'+\omega})n_{\epsilon'}).$$

$$(12.82)$$

We **interpret this expres-**
sion as the interaction-induced
rate change of the distribution
of fermions with energy ϵ. The
skeleton process responsible for
the out-rate is shown in the
left part of the lower panel of
the figure: **Pauli blocking** requires an empty state at the
target energy $\epsilon + \omega$ (with distribution factor $1 - n_{\epsilon+\omega}$). The

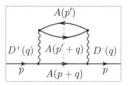

required energy for this transition must be provided by the conversion of an oc-
cupied state at energy $\epsilon' + \omega$ (with distribution factor $n_{\epsilon'+\omega}$) into an empty state
at ϵ' (with distribution factor $1 - n_{\epsilon'}$). The in-process shown at the right is interpreted analogously. For a fixed configuration $(\epsilon, \epsilon', \omega)$, this process is weighted by
the transition amplitude, $T(\epsilon, \epsilon', \omega)$, which in turn is given by the absolute square
of the quantum amplitudes indicated in the lower part of the figure. The rate samples all momentum arguments, $(\mathbf{p}', \mathbf{q})$, of the quantum states participating in the
process, and the propagators encode the underlying dynamics, as indicated in the
upper part where the dashed line indicates the interpretation of the diagram as the
absolute square of a transition amplitude.

A few more comments are due on the the interpretation of the quantum collision
term:

▷ The structure of I_{col} resembles that of the classical Boltzmann collision term
discussed in section 11.3 (see Eq. (11.18)). Quantum mechanics enters through

the **Pauli blocking factors** $1 - n_\epsilon$, requiring emptiness of the final states of the scattering process. In the classical theory, these factors are absent.

▷ Just as the collision term of Boltzmann theory was annihilated by the Maxwell–Boltzmann distribution, the collision term Eq. (12.82) vanishes in the **Fermi–Dirac distribution**: substitution of $n_\epsilon = (1 + e^{(\epsilon-\mu)/T})^{-1}$ makes the in- and the out-part balance and $I_{\rm col}$ vanish. This happens regardless of the parameters (μ, T), i.e., interactions drive equilibration into a Fermi–Dirac distribution at arbitrary temperature and chemical potential. For example, in situations where the constituents of the theory vary in space and/or time (meaning that the Wigner transforms pick up a dependence on the center coordinate), Fermi–Dirac distributions with varying parameters, $T(t, \mathbf{r})$ and $\mu(\mathbf{r}, t)$, often define a good ansatz for the solution of a theory.

12.6 Non-equilibrium Quantum Transport

In the previous sections, we discussed how interaction effects drive a system towards local equilibrium configurations and how departures from that configuration can be described. However, we have not yet investigated much of the fascinating phenomenology of out-of-equilibrium phases. Instead of attempting a complete overview – an impossible task – we focus here on one selected application: quantum transport beyond the regime of linear response studied in chapter 7.

The smaller a quantum system, the easier it is to drive it out of equilibrium. Ongoing progress in miniaturization implies that quantum devices are more frequently operated out of equilibrium and that the concepts discussed in this section are increasingly important in fields such as quantum information devices, quantum optics, and the general physics of driven and open quantum systems. We begin by introducing an exemplary setup illustrating this physics: a small metallic island sandwiched between macroscopic electrodes kept at a relative voltage difference. We will then demonstrate how almost all the concepts introduced earlier in this and the foregoing chapter are relevant to this system. Specifically, we will address the formation of non-thermal distribution functions, the generation of various forms of noise, their influence on quantum observables, and the statistics of transport coefficients in regimes beyond linear response.

12.6.1 Out-of-equilibrium quantum dot

REMARK Recapitulate problems in section 5.6.5 on the physics of a small metallic island tunnel-coupled to external leads. Our discussion below will generalize this setting to an out-of-equilibrium situation.

Consider a metallic or semiconducting island of mesoscopic proportions.[15] The island is connected by tunnel electrodes to
two leads, kept at a voltage difference V.
We also assume capacitive coupling to a
gate electrode (indicated by the horizontal plate in the figure), which determines
the electrostatically preferred number of
electrons on the dot.

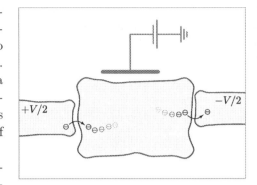

We describe this system by the Hamil-
tonian $\hat{H} = \sum_{\sigma=L,R}(\hat{H}_\sigma + \hat{H}_{t,\sigma}) + \hat{H}_d + \hat{H}_c$, where

$$\hat{H}_\sigma = \sum_{ab} c_{\sigma,a}^\dagger H_{\sigma,ab} c_{\sigma,b}, \qquad \hat{H}_{t,\sigma} = \sum_{a\mu} c_{\sigma,a}^\dagger T_{\sigma,a\mu} d_\mu + \text{h.c.},$$

$$\hat{H}_d = \sum_{\mu\nu} d_\mu^\dagger H_{d,\mu\nu} d_\nu, \qquad \hat{H}_c = \frac{E_C}{2}\left(\sum_\mu d_\mu^\dagger d_\mu - N_0\right)^2. \qquad (12.83)$$

Here the operators $c_{\sigma=L/R,a}$ create fermions in state $|a\rangle$ of the left/right lead,
d_μ creates fermions in state $|\mu\rangle$ of the dot, and $\hat{H}_\sigma$ and $\hat{H}_d$ are the corresponding
single-particle Hamiltonians. The operators $\hat{T}_{\sigma=L/R}$ describe tunneling between the
leads and the dot. Finally, E_C is the electrostatic charging energy on the dot,
and N_0 defines its electrostatically-preferred charge (notice that N_0 need not be
an integer). The voltage bias between the leads will be introduced momentarily
through a chemical potential difference.

The fermion Keldysh action describing this system is given by

$$S[\psi, V] \equiv \sum_{\sigma=L,R}(S_\sigma[\psi_\sigma] + S_{t,\sigma}[\psi_\sigma, \psi_d]) + S_d[\psi_d, V] + S_c[V], \qquad (12.84)$$

where

$$S_\sigma[\psi_\sigma] = \int dt\, \bar{\psi}_\sigma \begin{pmatrix} i\partial_t + E_F^+ - \hat{H}_\sigma & 2i\delta F_\sigma \\ 0 & i\partial_t + E_F^- - \hat{H}_\sigma \end{pmatrix} \psi_\sigma,$$

$$S_{t,\sigma}[\psi_\sigma, \psi_d] = \int dt \left[\bar{\psi}_\sigma \begin{pmatrix} \hat{T}_\sigma & 0 \\ 0 & \hat{T}_\sigma \end{pmatrix} \psi_d + \bar{\psi}_d \begin{pmatrix} \hat{T}_\sigma^\dagger & 0 \\ 0 & \hat{T}_\sigma^\dagger \end{pmatrix} \psi_\sigma \right],$$

$$S_d[\psi_d, V] = \int dt\, \bar{\psi}_d \begin{pmatrix} i\partial_t + E_F^+ - \hat{H}_d - V_c & -V_q/2 + 2i\delta F_d \\ -V_q/2 & i\partial_t + E_F^- - \hat{H}_d - V_c \end{pmatrix} \psi_d,$$

$$S_c[V] = \int dt \left(\frac{1}{2E_C} V_c V_q + N_0 V_q \right).$$

---

[15] The island is "mesoscopic" in the sense that it contains a macroscopically large number of electronic degrees of freedom, yet is sufficiently small that the quantum coherence of single-particle
wave functions is maintained over the full system size. At low temperatures of $\mathcal{O}(10\,\mathrm{mK})$, quantum phase coherence is only weakly disrupted by environmental noise, and so systems satisfying
this condition can be as large as several μm, scales that are within easy reach of modern device
technology.

Here, $E_F^{\pm} \equiv E_F \pm i\delta$ is the common Fermi energy of leads and dot, infinitesimally shifted by the regulator, $i\delta$. The information on the biasing of the leads is introduced via distribution functions

$$F_{\text{L,R}} - \coth\left(\frac{\epsilon \pm V/2}{2T}\right). \tag{12.85}$$

Finally, $V_{c,d}(t)$ are the classical and quantum components of a Hubbard–Stratonovich field decoupling the interaction.

EXERCISE Adapt the **Hubbard–Stratonovich transformation** introduced in section 12.4.2 to the charging interaction in Eq. (12.83), to reproduce the action above. In Eq. (12.84), the Hubbard–Stratonovich classical and quantum fields V_c and V_d are defined as $V_c \equiv \frac{1}{2}(V_+ + V_-)$, $V_q \equiv V_+ - V_-$, in terms of the fields $V_{\pm}$ decoupling the interaction on the contours. This differs by a factor $\sqrt{2}$ from earlier conventions and facilitates the interpretation of V_c as a classical degree of freedom.

Physically, $V(t)$ is a fluctuating potential, or voltage, conjugate to the charge on the island. As with the equilibrium situation studied in problem 5.6.5, it will be convenient to remove V by a gauge transformation on the dot by introducing the change of variables, $\psi_d \to e^{-i\hat{\phi}}\psi_d$, $\bar{\psi}_d \to \bar{\psi}_d e^{i\hat{\phi}}$, where $\hat{\phi} = \phi_c + \phi_q \sigma_1/2$ and the phases are defined by $\partial_t \phi_{c,q} = V_{c,d}$, V disappears from the bulk action while the tunneling matrix becomes dynamical,[16]

$$\hat{T}_\sigma \to \hat{T}_\sigma \, e^{-i\hat{\phi}} \equiv \tilde{T}_\sigma. \tag{12.86}$$

At this point, the fermions can be integrated out to generate the familiar "tr ln" term. The rest of the derivation then proceeds as in problem 5.6.5:

EXERCISE Adapt the derivation of problem 5.6.5 to the present setting. Expand the "tr ln" operator to second order in the tunneling matrix elements to **generate the tunneling action**

$$S_{\text{tun}}[\phi] = i\sum_\sigma \text{tr}(\hat{G}_\sigma \tilde{T}_\sigma \hat{G}_d \tilde{T}_\sigma^\dagger).$$

Evaluate this expression using the fact that the Hamiltonian operators $\hat{H}_\sigma$ and $\hat{H}_d$ are diagonal in the bases $|a\rangle$ and $|\mu\rangle$, respectively, and assuming that, in the range of relevant energies, the modulus $|T_{\sigma,a\mu}|^2 \equiv |T_\sigma|^2$ does not depend significantly on Hilbert space indices. Represent the Green functions as $\hat{G}^{\pm}(\epsilon) = \int(d\epsilon')\frac{\hat{A}(\epsilon')}{\epsilon^{\pm} - \epsilon'}$, via the spectral functions $\hat{A}_{\sigma,d}(\epsilon) = -2\,\text{Im}\,\hat{G}_{\sigma,d}^+(\epsilon) \equiv 2\pi\nu_{\sigma,d}(\epsilon)$. Assuming the energy independence of the single-particle density of states, $\nu_{\sigma,d}$, of the leads and dot, respectively, show that the tunneling action assumes the form (12.87).

The derivation sketched in the formulation of the exercise above then leads to the **tunneling action**

---

[16] The gauge transformation also alters the distribution function F_d (Exercise: How?). However, this change will not be of importance to our discussion.

$$S_{\text{tun}}[\phi] = -\frac{i}{4} \sum_\sigma g_\sigma \text{tr} \left(\Lambda_\sigma e^{-i\hat{\phi}} \Lambda_d e^{i\hat{\phi}} \right) \tag{12.87}$$

where

$$g_\sigma \equiv 4\pi^2 \nu_d \nu_\sigma |T_\sigma|^2 \tag{12.88}$$

defines the **tunneling conductance** between the dot and lead, $\sigma = \text{L/R}$, and $\Lambda_{\sigma,d}$ is a matrix with energy dependence,

$$\Lambda_x \equiv 2i \begin{pmatrix} g^+ & g_x^K \\ 0 & g^- \end{pmatrix}_\epsilon, \qquad g_x^K \equiv (g^+ - g^-)F_x, \qquad x = d, \sigma. \tag{12.89}$$

Here, $g^{\pm}$[17] are auxiliary Green functions defined by for a unit spectral density

$$g_\epsilon^{\pm} \equiv \int (d\epsilon') \frac{1}{\epsilon^{\pm} - \epsilon'}. \tag{12.90}$$

These objects have two key properties that will be of importance in the following: first, the imaginary part $\text{Im}\, g^{\pm} = \mp i/2$, owing to the Dirac identity. Second, causality implies that $g^+(t, t') = 0$ for $t < t'$, and the other way around for g^-. Specifically, for time-local functions X, Y, integrals such as $\text{tr}(g^+ X g^- Y) = \int dt\, dt'\, g^+(t, t') X(t') g^-(t', t) Y(t)$ vanish because the integration extends over acausal time configurations.

The representation (12.87) is a little too compact to be useful for practical calculations. To bring it into a more practical form, think of the phase ϕ as a bosonic field coupled to the fermion system of the dot and leads via tunneling. This should be reflected in the emergence of a "self-energy" of the ϕ-action very similar to that of the boson–fermion coupling (12.77). To see this structure emerging, we define the abbreviations,

$$c \equiv e^{i\phi_c} \cos(\phi_q/2), \qquad s \equiv e^{i\phi_c} \sin(\phi_q/2),$$

and substitute $e^{i\hat{\phi}} = c + is\sigma_1$ and $e^{-i\hat{\phi}} = \bar{c} - i\bar{s}\sigma_1$ into Eq. (12.87) to obtain

$$S_{\text{tun}}[\phi] = 2g \int dt_1\, dt_2\, (\bar{c}\ - i\bar{s})_{t_1} \begin{pmatrix} 0 & \Sigma^- \\ \Sigma^+ & \Sigma^K \end{pmatrix}_{t_1-t_2} \begin{pmatrix} c \\ is \end{pmatrix}_{t_2},$$

where (notice the similarity to the self-energy of a boson coupled to fermions (12.78))

$$\Sigma_t^{\pm} = \frac{i}{2} \sum_\sigma \left(g_t^{\pm} g_{\sigma,-t}^K + g_{d,t}^K g_{-t}^{\mp} \right),$$

$$\Sigma_t^K = \frac{i}{2} \sum_\sigma \left(g_{\sigma,-t}^K g_{d,t}^K - (g^+ - g^-)_t (g^+ - g^-)_{-t} \right).$$

Here, the absence of a $\bar{c}c$ coupling is due to the above causality principle, $g_t^+ g_{-t}^+ = 0$. On the same basis, we added $g^+ g^+$ and $g^- g^-$ terms to Σ^K to obtain the

---

[17] It is customary to denote both the Green functions, $g^{\pm}$, and the tunneling conductance, g_σ, by the same symbol. However, which is which follows from the index structure, and the context.

products of $(g^+ - g^-)$ factors in the second line. The motivation for this extension is that, as usual with dissipative actions, the most important contribution to the self-energies $\Sigma^\pm$ is given by the imaginary parts. Discarding the real parts, and recalling Eq. (12.89), we note that the Green functions appear solely in the combination $g^+ - g^- = -i$. Using this replacement and passing to the Fourier representation, we obtain

$$
S_{\text{tun}}[\phi] = \int (d\omega)\, (\bar{c}, -i\bar{s})_\omega \begin{pmatrix} 0 & \Sigma^- \\ \Sigma^+ & \Sigma^K \end{pmatrix}_\omega \begin{pmatrix} c \\ is \end{pmatrix}_\omega,
$$

$$
\Sigma_\omega^+ \simeq -\Sigma_\omega^- = \frac{i}{2} \sum_\sigma g_\sigma \int (d\epsilon)\, (F_{d,\epsilon} - F_{\sigma,\epsilon-\omega}), \qquad (12.91)
$$

$$
\Sigma_\omega^K = i \sum_\sigma g_\sigma \int (d\epsilon)\, (1 - F_{d,\epsilon} F_{\sigma,\epsilon-\omega}).
$$

The tunneling action (12.91) describes the coupling of the dot to the outside world via the effective self-energy of the phase field. Adding to this the ϕ-representation of the action S_c in Eq. (12.84),

$$
S_c[\phi] = \int dt \left[\frac{1}{2E_C} \partial_t \phi_c \partial_t \phi_q + N_0 \partial_t \phi_q \right], \qquad (12.92)
$$

we obtain the full **effective action in the phase representation**, $S[\phi] = S_c[\phi] + S_{\text{tun}}[\phi]$.

In the case of strong interactions, $E_C \gg T$, the charge action does not suppress phase fluctuations. (What is the physical meaning of this statement?) If, in addition, the dot is nearly insulated, $g_\sigma \ll 1$, the phases fluctuate wildly (also interpret this statement), and no longer represent suitable degrees of freedom. In this case, the system is more appropriately described in terms of the canonically conjugate **charge degree of freedom**, n. (Indeed, the dot charge will fluctuate only moderately in the limit of small g_x.)

Technically, the charge variable is introduced by a Hubbard–Stratonovich decoupling of the charging term. This leads to the **action in the phase–charge representation**,

$$
\boxed{S[n, \phi] \equiv \int dt\, (n_c \partial_t \phi_q + n_q \partial_t \phi_c - 2E_c(n_c + N_0)n_q) + S_{\text{tun}}[\phi]} \qquad (12.93)
$$

Conceptually, the introduction of n amounts to passing from a Lagrangian formulation – one variable, ϕ, governed by an action with "velocity squared" term $\dot{\phi}^2$ – to a Hamiltonian action $\int (n\dot{\phi} - H(n, \phi))$ defined in the phase space of variables (n, ϕ). Notice that the classical component n_c is conjugate to the quantum field ϕ_q, and vice versa.

INFO The representation (12.93) is a good starting point to explore the phenomenon of mesoscopic **charge quantization**. For the isolated dot, $g_\sigma \to 0$, we expect quantization of charge in integer units. However, the action above does not manifestly show this constraint. Rather, n appears to have the status of a continuous Gaussian variable. The key to its quantization lies in the **boundary conditions** of the conjugate variable ϕ. (Think of the quantum mechanics of a particle on a ring, where the quantization of momentum is a consequence of the periodic boundary conditions of its conjugate real space coordinate.)

The **boundary conditions of the phase field**, are best discussed in the closed Keldysh contour $s \in [0, 2t_0]$ (see fig. 12.2). As a phase field, $\phi(s)$ is defined only up to multiples of 2π. This implies the boundary conditions $\phi(2t_0) = \phi(0) + 2\pi W$, where summation is required over the integer W (see the discussion of functional integrals over S^1-valued fields in chapter 3). We implement the condition by the ansatz, $\phi(s) = \tilde{\phi}(s) + \frac{2\pi W}{2t_0}(s - t_0)$, where $\tilde{\phi}$ obeys periodic boundary conditions. Defining the classical and quantum fields through the field on the s-contour as $\phi_c(t) = \frac{1}{2}(\phi(t) + \phi(2t_0 - t))$ and $\phi_q(t) = \phi(t) - \phi(2t_0 - t)$, it is straightforward to check that this condition translates to an unconstrained $\phi_c = \tilde{\phi}_c$, and to

$$\phi_q(t) = \tilde{\phi}_q(t) + \frac{2\pi W}{t_0}(t - t_0), \tag{12.94}$$

with Dirichlet boundary conditions $\tilde{\phi}_q(0) = \tilde{\phi}_q(t_0) = 0$. The quantization of the field n_c may now understood as follows: to begin, assume that the W-dependence of the tunneling action is negligible. The winding number then enters the action through the contribution $2\pi W \frac{1}{t_0} \int_0^{t_0} dt\, n_c$. Summation over W thus generates the condition $\frac{1}{t_0} \int_0^{t_0} dt\, n_c \in \mathbb{Z}$: the temporal average of the classical charge on the dot is integer. However, the operators $\sim \exp(\pm i\hat{\phi}_q/2)$ in the action change the charge only in half-integer units, $n_c \to n_c \pm 1/2$ (think why a tunneling event on the upper or lower Keldysh contour changes the classical charge by a half-integer), and the numbers of positive and negative jumps are equal. This implies that the average charge equals the initial charge, $\frac{1}{t_0} \int_0^{t_0} dt\, n_c = n_c(0)$. The winding number summation thus enforces the quantization $n_c(0) \in \mathbb{Z}$, and the subsequent tunneling processes change this integer in quantized steps. As an instructive exercise, consider the W-dependence of the tunneling action and show that the conclusion remains unaltered.

12.6.2 Dot distribution function

The action (12.93) contains the distribution function of the dot via the tunneling term (12.91) as an unknown element. Its first-principles calculation is a complicated recursive problem: the distribution function is the result of the interaction of particles whose state depends in turn on the distribution function. It is generally more practical to start the analysis of the distribution with an educated guess and see how far one can get from there.

In the present context, we will identify a dot distribution function F_d ignoring interactions. This ansatz reflects the assumption that the coupling to the leads defines an effective distribution on the dot at time-scales $\sim \nu|T_\sigma|^2$, that are faster than the relaxation times due to interactions. Its validity must of course be self-consistently checked.

In the absence of interactions, all single-particle states below the energy $-V/2$ (relative to E_F) are occupied. At energies in the window $[-V/2, V/2]$ states in the left lead are occupied and in the right lead are empty. It is natural to assume that empty and occupied states hybridize with dot states at coupling strengths $g_{L,R}$, respectively. This motivates the ansatz,

$$F_d = \frac{g_L F_L + g_R F_R}{g_L + g_R}. \qquad (12.95)$$

The corresponding distribution function $n_d = (-F_d + 1)/2$ assumes the form of a **double-step distribution** built by the superposition of two single-step Fermi–Dirac distributions.

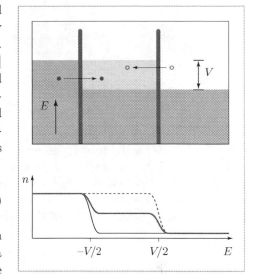

EXERCISE It is instructive to back up the guess (12.95) by a **variational calculation**. To this end, we ask for which distribution F_d does the action become stationary in the limit of weak voltage fluctuations, $\hat{\phi} \simeq 0$. The most economic way to answer this question is to start from Eq. (12.87) and verify that the vanishing of the first-order expansion in $\hat{\phi}$ requires that $\sum_\sigma g_\sigma \mathrm{tr}([\Lambda_d, \Lambda_\sigma]\hat{\phi}) = 0$. Using Eq. (12.89), compute the commutator and convince yourself that, with Eq. (12.95), the condition is satisfied. Recalling the connection between phase and voltage, argue why the stationarity condition is equivalent to the absence of stationary current flow onto the dot.

EXERCISE Let us find out how the double-step distribution affects the dissipative action of the phase field. Using the auxiliary relations

$$\int \frac{d\epsilon}{2\pi} \left(F(\epsilon + \omega) - F(\epsilon) \right) = \frac{\omega}{\pi}, \qquad \int \frac{d\epsilon}{2\pi} \left(1 - F(\epsilon - \omega) F(\epsilon) \right) = \frac{\omega}{\pi} F_b(\omega),$$

with $F(\epsilon) = \tanh(\epsilon/2T)$ and $F_b(\omega) = \coth(\omega/2T)$, and assuming equal barrier transparencies $g \equiv g_L = g_R$ for simplicity, show that the substitution of Eq. (12.95) into Eq. (12.91) leads to

$$\Sigma^+ = -\Sigma^- = \frac{ig\omega}{\pi}, \qquad \Sigma^K = \frac{2ig}{\pi} K(\omega),$$

$$K(\omega) \equiv \frac{1}{2}\omega F_b(\omega) + \frac{1}{4} \sum_{s=\pm 1} (\omega + sV) F_b(\omega + sV). \qquad (12.96)$$

In section 12.6.4, we will use this result to explore the impact of dissipation on the quantum mechanics of phase fluctuations on the dot.

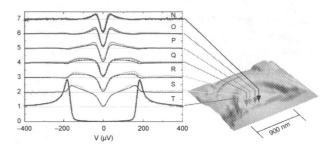

Fig. 12.4   Tunneling density of states as a function of voltage (energy) in a mesoscopic SNS structure. The arrows mark the position where, in the normal wire, the TDoS was recorded. Figure taken from H. le Sueur, P. Joyez, H. Pothier, C. Urbina, and D. Esteve, *Phase controlled superconducting proximity effect probed by tunneling spectroscopy*, Phys. Rev. Lett. **100**, 197002 (2008). Copyright (2008) from the American Physical Society.

12.6.3 Tunneling density of states

The action (12.93) with the double-step distribution (12.95) completes our description of the dot, and we may now turn to the discussion of physical phenomena. Specifically, in this section, we take a look at the **tunneling density of states (TDoS)** of the dot. Assuming no external time-dependence, the TDoS is defined as

$$\nu(\epsilon) \equiv -\frac{1}{\pi}\mathrm{Im}\,\mathrm{tr}\,(\hat{G}^{+}(\epsilon)) = -\frac{1}{\pi}\mathrm{Im}\int dt\,e^{it\epsilon}\mathrm{tr}\,(\hat{G}^{+}(t)). \qquad (12.97)$$

This observable is physically interesting from various perspectives:

▷ The **TDoS can be measured** by tunneling spectroscopy. See fig. 12.4 for an example of local probes of the TDoS in a mesoscopic superconductor–normal–superconductor (SNS) wire. In the figure, the "external wire" shows as a canyon ending in the central region. In this system, the TDoS is suppressed owing to the superconductor proximity effect: the superconductor induces pair correlations in the adjacent normal metal region, thus suppressing the spectral density. The size of the spectral gap, a function of voltage (energy), diminishes with the distance from the superconductor.

▷ As follows from the definition, the function $\nu(\epsilon)$ probes the amplitude of retarded quasiparticle propagation at time-scales $t \sim \epsilon^{-1}$. This propagation amplitude comes with a quantum dynamical phase. Out of thermal equilibrium, we expect this quantum phase to be affected by **nonequilibrium fluctuations**.

orthogonality
catastrophe

Specifically, the TDoS is affected by a mechanism known as the **orthogonality catastrophe**: the quantum state, "bare quasiparticle plus unperturbed system," realized right after the tunneling event is generally very different from (i.e., "orthogonal to") the stationary state approached in the long-time limit. The formation of that final state involves the readjustment of a large number of particles and generally takes much more time than the tunneling process itself.

The corresponding signatures in the energy dependence of the spectral density are called **zero-bias anomalies**.

INFO The **Coulomb blockade** discussed in problem 5.6.4 is an extreme manifestation of the zero-bias anomaly: tunneling onto an isolated quantum dot creates a state that is energetically forbidden, i.e., a "sub-barrier," by an amount E_c. This means that the TDoS vanishes from $\epsilon = 0$ to $\epsilon = E_C$. (Recall that energy is measured from the Fermi energy.) Equivalently, we may say that tunneling processes onto the dot are limited to short duration $t \sim E_c^{-1}$. **Tunneling into a metallic system** leads to more complex behavior. The tunneling process generates an energetically unfavorable (sub-barrier) initial charge distribution, which subsequently relaxes via diffusive spreading. The corresponding action weights the quasiparticle propagator and hence the TDoS. For example, in two dimensions, the relaxation action diverges,[18] which implies a singular zero-bias anomaly. Below, we will investigate how the zero-bias anomaly manifests itself in the quantum dot.

We start by expressing $\nu(\epsilon)$ in terms of the degrees of freedom entering the effective action. In a non-interacting setting, the formulae of section 12.4.1 imply that

$$\nu(\epsilon) = -\frac{1}{2\pi i} \int dt\, e^{i\epsilon t} \left(G^{-+}(t) - G^{+-}(t) \right).$$

We continue to build in interactions via the Hubbard–Stratonovich field, ϕ. The latter affects the Green functions as $G_{tt'}^{-+} = -i\langle \psi_t^- \bar{\psi}_{t'}^+ \rangle \to -i e^{i(\phi_-(t)-\phi_+(t'))}\langle \psi_t^- \bar{\psi}_{t'}^+ \rangle = e^{i(\phi_-(t)-\phi_+(t'))} G_{0,tt'}^{-+}$, where G_0 is the non-interacting Green function. For the multi-level dot, the single-level representation of G_0 given in Eq. (12.71) generalizes to

$$G_{0,tt'}^{-+} = -i \sum_\mu e^{-i\epsilon_\mu(t-t')}(1 - n_d(\epsilon_\mu)) = -i \int d\epsilon\, \nu(\epsilon) e^{-i\epsilon(t-t')}(1 - n_d(\epsilon))$$

$$\simeq -2\pi i \nu_d (1 - n_d)(t - t'),$$

where $n_d(t)$ is the Fourier transform of the function $n(\epsilon)$ and we assumed constant non-interacting density of states ν_d as before.[19] Adding the analogous construction for $G^\pm$, we obtain $\nu(\epsilon) = \nu_e(\epsilon) + \nu_h(\epsilon)$, where

$$\nu_e(\epsilon) = \nu_d \operatorname{Re} \int dt\, e^{i\epsilon t} e^{i(\phi_-(t+\bar{t})-\phi_+(\bar{t}))}(1 - n_d)(t),$$

$$\nu_h(\epsilon) = \nu_d \operatorname{Re} \int dt\, e^{i\epsilon t} e^{i(\phi_+(t+\bar{t})-\phi_-(\bar{t}))} n_d(t). \tag{12.98}$$

Here $\nu_{e,h}$ are to be interpreted as the contributions of electrons (holes) tunneling onto the dot, and $\bar{t}$ is an arbitrary reference time marking the beginning of the tunneling process. The hole contribution, ν_h, differs from ν_e in the replacement $(1 - n_d) \to n_d$, and in a sign change in the phase of the oscillatory exponential.

These formulae represent the TDoS in terms of the effective variable ϕ of the tunneling action. In the following, we explore the physics of this expression for a

[18] L. S. Levitov and A. V. Shytov, *Semiclassical theory of the Coulomb anomaly*, Pisma Zh. Eksp. Teor. Fiz. **66**, 200 (1997) [JETP Lett. **66**, 214 (1997)].

[19] Here, $(1 - n_d)(t) = \delta(t) - n_d(t)$ is the Fourier transform of $1 - n_d(\epsilon)$.

dot well connected to the external leads. The complementary case of an almost closed dot is addressed in problem 12.9.4.

12.6.4 Open quantum dot

REMARK To simplify the notation we assume equal transparencies $g \equiv g_{\mathrm{L}} = g_{\mathrm{R}}$ throughout this section.

At large values of the tunneling conductance, $g \gg 1$, fluctuations of the phase ϕ are largely quenched, while the conjugate charge degree of freedom, n, fluctuates wildly. Specifically, the smallness of fluctuations in ϕ_q implies that the dot stays close to the classical limit and behaves similarly to an RC-resistor unit kept at voltage $\sim \dot\phi_c$. However, Eq. (12.98) shows that, even in this semiclassical limit, the dynamics of ϕ will affect the quantum observable ν.

To see this in more concrete terms, we expand the tunneling action (12.96) to second order in ϕ around its stationary configuration $\phi = 0$. With $c \simeq 1 + i\phi_c$ and $s \simeq \phi_q/2$, this gives (exercise)

$$S^{(2)}[\phi] = \frac{1}{2E_C} \int dt\, \partial_t \phi_c \partial_t \phi_q + \frac{g}{2\pi} \int \frac{d\epsilon}{2\pi} (\phi_c, \phi_q)_\omega \begin{pmatrix} 0 & -i\omega \\ i\omega & iK(\omega) \end{pmatrix} \begin{pmatrix} \phi_c \\ \phi_q \end{pmatrix}_\omega , \quad (12.99)$$

where the kernel K is defined in Eq. (12.96) and we neglected the N_0-contribution to the charge action (12.92).[20]

EXERCISE Compute the ϕ-Green function corresponding to the above action and discuss how in the absence of biasing, $V = 0$, it conforms with the fluctuation–dissipation theorem. Conversely, departures from equilibrium, $V \neq 0$, lead to FDT violation.

Classical resistor circuit

The action (12.99) resembles that of a particle with coordinate ϕ coupled to an environment (see section 12.3). To better understand the physics of this connection, we decouple the $\phi_q\phi_q$-term to obtain a Langevin-type system:

$$S[\phi, \xi] = \int dt\, \phi_q \left[\left(-\frac{1}{2E_C}\partial_t^2 - \frac{2g}{2\pi}\partial_t \right) \phi_c + \xi \right] + i \int \frac{d\omega}{2\pi} \xi_\omega \frac{\pi}{2gK(\omega)} \xi_{-\omega}. \quad (12.100)$$

Integration over ϕ_q in Eq. (12.100) produces the classical field configurations solving the equation

$$\left(\frac{1}{2E_C}\partial_t^2 + \frac{2g}{2\pi}\partial_t \right) \phi_c = \xi,$$

---

[20] The formal justification for this is that, for large g, topologically excited configurations (i.e., with winding number W) of the phase field are strongly suppressed (why?). However, for $W = 0$, $N_0 \int dt\, \partial_t \phi_q = 0$. (Consider the physical reason for the irrelevancy of the N_0-term in the wide open quantum dot.)

where the noise field ξ is correlated according to

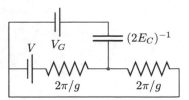

$$\langle \xi(t)\xi(t') \rangle = \frac{g}{\pi} K(t - t'),$$

and $K(t) = \int \frac{d\omega}{2\pi} e^{-i\omega t} K(\omega)$ is the Fourier transform of the kernel K. Physically, this Langevin equation describes the dynamics of voltage fluctuations in an effective resistor circuit equivalent to the quantum dot. The circuit is depicted in the figure where the node represents the quantum dot, coupled by two resistors of resistance $R = g^{-1}$ to a bias voltage source. The dot is also coupled to a capacitor of capacitance $C = 1/2E_C$. That capacitor is kept at a gate voltage $V_G = N_0/C$ (which, however, as we have seen is irrelevant to the physics of the open dot).

To make the connection to a near-classical circuit more explicit, we change notation as[21]

$$g = \frac{2\pi}{R}, \qquad E_C = \frac{1}{2C}, \qquad U = \partial_t \phi_c, \qquad \xi = C\eta. \qquad (12.101)$$

The **Langevin equation describing voltage fluctuations** away from $V/2$ then assumes the form

$$\partial_t U + \gamma U = \eta, \qquad (12.102)$$

RC-time
scale

where $\langle \eta(t)\eta(t') \rangle = \frac{\gamma}{C} K(t - t')$ and $\gamma \equiv \frac{2}{RC}$ is set by the inverse of the *RC*-**time scale** of the circuit. The meaning of this equation is not difficult to understand: in the absence of fluctuations, $\eta = 0$, the voltage on the dot relaxes to its stationary value, $V/2$, at the relaxation scale γ.[22]

However, in contrast with our previous studies of dissipative systems, the coupling to two distinct leads ("baths") implies that the voltage distribution on the dot cannot relax to an equilibrium configuration. To explore this point, let us take a closer look at the **fluctuation kernel**, K:

▷ In the **absence of biasing**, $V = 0$, the fluctuation kernel reduces to $K(\omega) = \omega \coth(\omega/2T)$, equivalent to that induced by an ohmic bath of oscillators, discussed in section 12.3. In the present context, these oscillators are realized via the bosonic particle–hole excitations of the Fermi liquids contained in the dot and leads. However, the observable effect is the same: **Nyquist–Johnson noise**, regularized by the quantization of the oscillator ground state energy at zero temperature, through $K(\omega) = |\omega|$.

---

[21] The factor 2π in the definition of R is explained as follows: g is the dimensionless conductance of the tunnel barriers. The physical conductance $G \equiv R^{-1} = ge^2/h = (g/2\pi)/\hbar$. In our units, $e^2 = \hbar = 1$ and this reduces to $g = 2\pi/R$.

[22] The factor of 2 in $\gamma = 2/RC$ reflects the coupling of the dot via two resistances R, with parallel resistance $R/2$.

shot noise

▷ A **non-zero bias**, $V > 0$, hinders the dot in reaching thermal equilibrium. Focusing on $T = 0$ for simplicity, inspection of Eq. (12.91) reveals a crossover,

$$K(\omega) = \left\{ \begin{array}{ll} |\omega|, & |\omega| \gg |V|, \\ \frac{1}{2}|V|, & |\omega| \ll |V|. \end{array} \right. \qquad (12.103)$$

This equation states that, at large frequencies, $|\omega| \gg |V|$, the noise is dominated by the equilibrium fluctuations of oscillator modes. However, at lower frequencies, the biasing of the dot generates a mean current flow through the system. This in turn induces noise with the signatures of **shot noise**.

Recall from section 11.2.2 that n charges flowing through an elementary resistor in a time-window $[\bar{t}, \bar{t} + \Delta t]$ define a current $I_{\Delta t} = n/\Delta t$. Assuming the lack of correlations between these charges, $\mathrm{var}(I_{\Delta t}) = \bar{I}/\Delta t$, where $\bar{I} = \langle I_{\Delta t} \rangle$ is the average current. We may write the instantaneous current through the resistor as $I = \bar{I} + \delta I$, where $\langle \delta I(t)\delta I(t') \rangle = f\delta(t - t')$ describes short-range correlated current fluctuations. Computing the variance,

$$\mathrm{var}(I_{\Delta t}) = \frac{1}{(\Delta t)^2} \int_{\bar{t}}^{\bar{t}+\Delta t} dt\, dt'\, \langle \delta I(t)\delta I(t') \rangle = \frac{f}{\Delta t}, \qquad (12.104)$$

we find that $f \equiv \bar{I}$ establishes compatibility with Poissonian shot noise. Turning to the slightly more complex two-resistor setting, we expect that the observable noise is due to the superposition of the shot noises of the individual contacts. To see this explicitly, we write the Langevin equation as

$$Cd_t U = -\frac{2}{R}U + C\eta, \qquad (12.105)$$

i.e., an equation that relates the rate of change of the charge on the dot (left-hand side) to current flow (right-hand side). Specifically, $C\eta \equiv \delta I_{\mathrm{L}} + \delta I_{\mathrm{R}}$ is to be interpreted as the sum of current fluctuations through the left and right tunnel resistor. Each of the two contributions $I_\sigma, \sigma = \mathrm{L}, \mathrm{R}$ is expected to express Poissonian shot noise statistics, $\langle I_\sigma(t)I_\sigma(t') \rangle = \bar{I}\delta(t - t') = V/(2R)\delta(t - t')$. Compatibility with our considerations above then requires

$$\langle \eta(t)\eta(t') \rangle = \frac{1}{C^2}\left(\langle \delta I_{\mathrm{L}}(t)\delta I_{\mathrm{L}}(t') \rangle + \langle \delta I_{\mathrm{R}}(t)\delta I_{\mathrm{R}}(t') \rangle \right) = \frac{V}{RC^2}\delta(t - t'),$$

in agreement with Eq. (12.102) and $K(t - t') = (V/2)\delta(t - t')$, or $K(\omega) = V/2$. This shows that Eq. (12.103) correctly models the double-barrier shot noise statistics at low frequencies.

INFO The analysis of noise levels in systems with several independent Poissonian noise sources plays an important role in **nonequilibrium mesoscopic physics**. It is customary to quantify the noise in a composite system by the **Fano factor** $F \equiv \frac{S}{S_0}$. This factor compares the **DC-noise power** in the system

$$S \equiv 2 \int dt\, \langle \delta I(0)\delta I(t) \rangle \qquad (12.106)$$

Fano factor

with that of an elementary resistor unit with Poissonian statistics, S_0. For an elementary Poissonian resistor, Eq. (12.104) implies that $\int dt \langle \delta I(t) \delta I(0) \rangle = \bar{I} = V/R$ and $S_0 = 2V/R$. The analysis above implies that (exercise) $S = V/R$ for the double dot system, corresponding to a Fano factor $F = 1/2$. The relatively lower noise power reflects a partial averaging of the individual noise sources $\sigma = L, R$.

Zero-bias anomaly

Here, we explore how nonequilibrium fluctuations of the collective variable ϕ affect the quantum mechanics of the TDoS, and in particular the development of the **zero-bias anomaly** (see the discussion at the beginning of the section). Being sensitive to the overlap between two many-body wave functions, the latter represents a "deep quantum" observable (as opposed to an observable with stable semiclassical limit). We therefore expect that the disruption of quantum coherence due to nonequilibrium noise will show in the manifestation of the anomaly. We will discuss this phenomenon using the example of the TDoS of the quantum dot at low frequencies $\omega < \gamma \sim 2/RC$, smaller than the inverse of the RC-time.

EXERCISE Consider the double-step distribution function $n_d(\epsilon)$ in the limit of zero temperature, where the Fermi functions reduce to step functions. Show that its Fourier transform is given by

$$n_d(t) = i \frac{\cos(Vt/2)}{2\pi t^+}, \qquad (1 - n_d)(t) = -i \frac{\cos(Vt/2)}{2\pi t^-}, \qquad (12.107)$$

where $t^\pm = t \pm i\delta$.

First consider the particle contribution ν_e to the density of states in Eq. (12.98). Choosing $\bar{t} = -t/2$ to symmetrize the notation and using that $\phi_c(t + \bar{t}) - \phi_c(\bar{t}) = \int_{-t/2}^{t/2} dt' \, U(t')$, we obtain

$$\nu_e(\epsilon) = \nu_d \, \mathrm{Re} \int dt \, e^{i\epsilon t} \left\langle e^{i \int_{-t/2}^{t/2} dt' \, U(t')} e^{\frac{i}{2}(\phi_q(-t/2) + \phi_q(t/2))} \right\rangle_\phi (1 - n_d)(t)$$

$$= \nu_d \, \mathrm{Re} \int dt \, e^{i\epsilon t} \left\langle e^{i \int_{-t/2}^{t/2} dt' \, U_d(t')} \right\rangle_\eta (1 - n_d)(t), \qquad (12.108)$$

where U_d is the actual time-dependent classical voltage on the dot and is the solution to the Langevin equation

$$(\partial_{t'} + \gamma) \, U_d(t') = \eta + \frac{1}{2C} \sum_{s=\pm} \delta(t' - st/2).$$

EXERCISE Verify Eq. (12.108). To this end, represent the $\langle \ldots \rangle_\phi$-functional average via Eq. (12.100). Next redefine variables as in Eq. (12.101) and integrate over ϕ_q to generate a constraint equivalent to the Langevin equation above.

Physically, this equation is easy to interpret. The two δ-functions on the right-hand side generalize the Langevin equation (12.102) to the presence of voltage peaks caused by the tunneling of charges at $\pm t/2$.[23] The voltage on the dot is thus driven by the competition of an initial δ-inhomogeneity, noise, and RC-relaxation.

The effect of this dynamics on the TDoS is obtained by averaging the "propagator" $\exp(i \int_{-t/2}^{t/2} dt' \, U_d(t'))$ over the fluctuating voltage U_d. We start by solving the Langevin equation for U_d in its Fourier representation,

$$U_d(\omega) \simeq \frac{\eta(\omega)}{\gamma} + \frac{R}{4} \sum_s e^{is\omega t/2t} \equiv \frac{\eta(\omega)}{\gamma} + U_0(\omega),$$

where we have neglected $i\omega$ in comparison with γ, since we are interested in low frequencies $\omega \ll \gamma$. The integral over the noiseless part of the voltage, U_0 gives $\exp(i \int dt' \int \frac{d\omega}{2\pi} e^{-i\omega t} U_0(\omega)) = \exp(\frac{iR}{4} \mathrm{sgn}(t)) \simeq (1 + \frac{iR}{4} \mathrm{sgn}(t))$, and the average over the noisy part of the voltage with correlator $\langle \eta(\omega)\eta(\omega') \rangle = 2\pi\delta(\omega + \omega') \frac{\gamma}{C} K(\omega)$ is obtained as

$$\left\langle e^{\frac{i}{\gamma} \int_{-\frac{t}{2}}^{\frac{t}{2}} dt' \int \frac{d\omega}{2\pi} e^{-i\omega t'} \eta(\omega)} \right\rangle_\eta = e^{-\frac{R}{4} \int_{-\frac{t}{2}}^{\frac{t}{2}} dt' dt'' \int \frac{d\omega}{2\pi} e^{-i\omega(t'-t'')} K(\omega)} \equiv e^{-S(t)},$$

with effective **tunneling action**

$$S(t) \equiv R \int \frac{d\omega}{2\pi} \sin^2\left(\frac{\omega t}{2}\right) \frac{K(\omega)}{\omega^2}. \tag{12.109}$$

Combining terms, we arrive at

$$\nu_e(\epsilon) = \nu_d \, \mathrm{Re} \int dt \, e^{i\epsilon t} \left(1 + \frac{iR}{4} \mathrm{sgn}(t)\right) e^{-S(t)} (1 - n_d)(t).$$

Inspection of the hole contribution in Eq. (12.98) shows that the latter differs from ν_e by a replacement $(1 - n_d) \to n_d$, and by a sign change of the signum function. Adding the two contributions and using Eqs. (12.107), the full TDoS is thus obtained as

$$\nu(\epsilon) = \nu_d \left(1 - \frac{R}{4\pi} \int_0^\infty \frac{dt}{t} \cos(\epsilon t) \cos\left(\frac{Vt}{2}\right) e^{-S(t)}\right). \tag{12.110}$$

In this expression, the cosine factors encode the information on the double-step distribution governing the TDoS in the non-interacting limit. The essential factor $\exp(-S(t))$ contains the tunneling action associated with the combination of interactions, noise, and external biasing.

We first consider its impact on the TDoS in the case of zero temperature, and **no external bias**, $V = 0$, where the dot is in thermal equilibrium. In this limit, Eq. (12.103) states that $K(\omega) = |\omega|$. In Eq. (12.109), this implies an IR singularity with cutoff at low frequencies $\omega \sim t^{-1}$. Remembering that we are working under the assumption $\omega \lesssim \gamma$ and, replacing $\sin^2 \simeq 1/2$ by its average, we obtain $S(t) \simeq \frac{R}{2} \int_{t^{-1}}^{\gamma} \frac{d\omega}{2\pi|\omega|} = \frac{R}{4\pi} \ln(\gamma t)$. The logarithmic increase indicates a weak suppression of

---

[23] There is no reason to be puzzled by the presence of two δ-functions. The Langevin equation describes the retarded evolution of U_η in the time-window $[-t/2, t/2]$ (or $[t/2, -t/2]$ for negative t). For $t > 0$, only $\delta(t' + t/2)$ affects the evolution in the time-window $[-t/2, t/2]$. For negative t, the other δ-function is effective, but never both.

coherence at large times. The cause of this suppression is that the system acts as its own bath, creating a residual noise background even at zero temperature.

To explore the consequences, we substitute $\exp(-S) = (\gamma t)^{-\frac{R}{4\pi}}$ into Eq. (12.110), note that at large times $\cos(\epsilon t)$ cuts the integral at $t \sim \epsilon^{-1}$, and again remember that the derivation holds only for $t \gtrsim \gamma^{-1}$. Using these two scales as limits for the integral, we arrive at the estimate

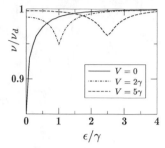

$$\nu(\epsilon) \simeq \nu_d \left(1 - \frac{R}{4\pi} \int_{\gamma^{-1}}^{|\epsilon|^{-1}} \frac{dt}{t} (\gamma t)^{-\frac{R}{4\pi}} \right) = \frac{4\pi}{R} \left(\frac{|\epsilon|}{\gamma} \right)^{\frac{R}{4\pi}}.$$

This is a radical manifestation of the **orthogonality catastrophe**: the orthogonality of the state (one tunneling particle + the dot ground state) to the true dot ground state leads to a complete suppression of the TDoS at zero energy. Remarkably, this holds even in the presently considered case where the dot is well connected to the environment. In our units $e = \hbar = 1$, the conductance quantum $g_0 = 2e^2/h = 1/\pi$, which means that the exponent can be expressed in terms of the dimensionless conductance $g = G/g_0$ as $R/4\pi = 1/(4\pi G) \equiv 1/(4\pi g g_0) = 1/(4g)$. For large conductance, $g \gg 1$, the suppression range gets narrower, but the TDoS is still vanishing at zero energy. This effect depends sensitively on the long-time quantum phase coherence being maintained under zero-temperature equilibrium conditions. For example, even at zero temperature, it gets destroyed by the shot noise accompanying a non-zero voltage bias, as demonstrated in the following exercise.

EXERCISE Explore what happens to the zero-bias anomaly in the presence of **voltage bias**. First show that, for $V \neq 0$, the tunneling action (12.87) increases linearly in time $S(t) \propto V|t|$. Verify that, in this case, the suppression of the TDoS is no longer complete. Also show that the minimum of the TDoS splits into two at $\pm V/2$ (see figure). Interpret this splitting as a consequence of the double step distribution.

12.7 Full Counting Statistics

Above, we noted that miniaturized quantum devices are easily pushed out of equilibrium. By the same principle, statistical fluctuations of quantum observables play an increasingly important role, the smaller the system. We have already seen how deviations from Poissonian statistics contain information on the architecture of a system, or how changes in the fluctuations of some observables may have a qualitative impact on the behavior of others. In fields where small complex quantum systems are center stage – atomic, molecular, and optical physics, and increasingly condensed matter physics – the full statistical distribution of observables is therefore an important object of study. Such analysis is usually applied in connection with quantized particle transport, such as photon transmission in optics, or electric current in quantum electronics, and in condensed matter physics is referred to as **full counting statistics (FCS)**. In this section, we introduce the basic concepts of FCS and apply them to electric transport through the quantum dot introduced in section 12.6. For more in-depth reviews, we refer to Refs.[24]

INFO Historically, the concept of full (photon) counting statistics was introduced in **quantum optics** (see L. Mandel, *Fluctuations of photon beams: The distributions of the photoelectrons*, Proc. Phys. Soc. **74**, 233 (1959) for an early reference and L. Mandel and E. Wolf, *Optical Coherence and Quantum Optics* (Cambridge University Press, 1995) for a review). The fruitfulness of these concepts in condensed matter physics was recognized in the beginning of the 1990s, when micron-sized mesoscopic quantum devices became a new subject of study (see L. S. Levitov and G. B. Lesovik, *Charge distribution in quantum shot noise*, Pis'ma Zh. Eksp. Teor. Fiz. **58**, 225 (1993) [JETP Lett. **58**, 230 (1993)]). This statistical approach defined a new field, giving unprecedented insight into the dynamical processes governing complex electronic quantum systems.

12.7.1 Generalities

A fundamental observable describing transport through a quantum device is the number of particles transmitted in a time-interval Δt,

$$\hat{N} \equiv \int_{-\Delta t/2}^{\Delta t/2} dt\, \hat{I}(t), \tag{12.111}$$

where $\hat{I}$ is the current operator, and the center of the observation time-window has been set to zero. In the definition of $\hat{N}$, the choice of Δt must be optimized to be sufficiently large to acquire enough statistics, and small enough not to average over physically interesting fluctuations.

Within a field-theoretical framework, information on the statistics of $\hat{N}$ is obtained by coupling the observable to a source variable, $\hat{\chi}$. The moments of $\hat{N}$, and in fact the entire distribution, can then be obtained by evaluation of the sourced functional integral in the presence of χ.

To start the construction of a suitable source, recall that current densities, $\mathbf{j}(x)$, are obtained by differentiation of effective actions in the vector potential, $\mathbf{A}(x) \equiv \mathbf{A}(t, \mathbf{x})$. Specifically, in Keldysh field theory,

$$\mathbf{j}(x) = -i \frac{\delta}{\delta \mathbf{A}(x)} \bigg|_{\mathbf{A}=0} \mathcal{Z}\left[\mathbf{A} \otimes \sigma_3/2\right],$$

where the notation $\mathcal{Z}[\mathbf{A} \otimes \sigma_3/2]$ indicates that the action is minimally coupled to a purely quantum vector potential (with opposite signs on the two Keldysh contours).

---

[24] W. Belzig, *Full counting statistics in quantum contacts*, Proceedings of the Summer School on Functional Nanostructures, Karlsruhe, 2003; L. S. Levitov, *The statistical theory of mesoscopic noise*, Quantum Noise in Mesoscopic Physics: Proceedings of the NATO Advanced Research Workshop, Delft, the Netherlands, 2002, and M. Kindermann and Yu. V. Nazarov, *Full counting statistics in electric circuits*, in: *Quantum Noise in Mesoscopic Physics*, ed. Yu. V. Nazarov, (Kluwer (Dordrecht), 403 (2002)).

According to our general discussion of the Keldysh contour in section 12.2.1, the coupling of $\mathbf{A}$ to σ_3 makes it a suitable source variable.[25] Building on this definition, we may introduce a **source variable for currents** by defining the vector potential,

$$\mathbf{A}(\mathbf{x},t) = \chi(t)\mathbf{e}_\perp \int_S d^{d-1}x'\,\delta(\mathbf{x}-\mathbf{x}'), \qquad (12.112)$$

where the surface integral confines the support of $\mathbf{A}$ to a planar section S of the system, $\mathbf{e}_\perp$ is normal to S (the generalization to curved sections is straightforward),

counting field

and the definition of the **counting field**,

$$\chi(t) \equiv \chi\Theta(t-\bar{t})\Theta(\bar{t}+\Delta t - t),$$

implies a projection onto the counting time-interval. Differentiation of $\mathcal{Z}[\mathbf{A}\otimes\sigma_3/2] \equiv \mathcal{Z}(\chi)$ with respect to χ yields the average of the transmitted particle number,

$$-i\frac{\partial}{\partial\chi}\Big|_{\chi=0}\mathcal{Z}(\chi) = \int_{\bar{t}}^{\bar{t}+\Delta t}\int_S dS\cdot\langle\mathbf{j}(\mathbf{x},t)\rangle = \langle\hat{N}\rangle.$$

Repeated differentiation – and this is the prime advantage of the above construction – generates **moments of the counting variable**:

$$\langle\hat{N}^n\rangle = (-i)^n\frac{\delta^n}{\delta\chi^n}\Big|_{\chi=0}\mathcal{Z}(\chi). \qquad (12.113)$$

moment-generating function

This identifies $\mathcal{Z}(\chi)$ as the **moment-generating function** and its logarithm

$$\ln g(\chi) \equiv \ln\mathcal{Z}(\chi),$$

as the **cumulant-generating function**. From it, cumulants characterizing the distribution can be obtained by differentiation in χ. For a discussion of the meaning of the first few of these cumulants, we refer to appendix section A.2.1.

12.7.2 Types of current noise

In this section, we introduce two frequently occurring types of noisy current distributions. We then compare these distributions with the transport statistics of the quantum dot introduced earlier in the chapter.

Quantum point contact: Consider an isolated scatterer embedded into a single-channel quantum conductor. In modern device technology, such quantum point contacts are realized as artificial imperfections or tunneling bridges (picture courtesy of Nanocenter Basel) in a few channel quantum wires. Charge carriers incident upon the point contact get transmitted with probability T and reflected

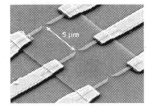

---

[25] In section 12.2.1, sources were introduced on the upper Keldysh contour. A sign inverted source on the lower contour generates the same observable, and the choice above is symmetrized over the two representations.

with probability $1 - T$. This means that the probability of transmitting n charges in N events is given by the **binomial distribution** (cf. section A.2.2),

$$p(n) = \binom{N}{n} T^n (1 - T)^{N-n}.$$

The cumulant-generating function of this distribution reads

$$g(\chi) \equiv \ln \sum_n e^{i\chi n} p(n) = N \ln(1 + T(e^{i\chi} - 1)).$$

Assuming spinless electrons, a perfect single-channel conductor has dimensionless conductance $g = e^2/h = 1/2\pi$ (in our standard units $e = \hbar = 1$), i.e., half the conductance quantum. If the system is biased, $N = I\Delta t = V\Delta t/2\pi$ charges will be incoming, and the **statistics of current in the quantum point contact** (at zero temperature[26]) is described by

$$\ln g(\chi) = \frac{V\Delta t}{2\pi} \ln(1 + T(e^{i\chi} - 1)). \tag{12.114}$$

Ohmic resistor: In an ohmic resistor, the situation is different. The transmission of charge is no longer the result of a single scattering event. However, different transmission events continue to be uncorrelated. Under these circumstances, the transmitted charge may be assumed to be Poisson distributed (appendix section A.2.2),

$$p(n) = e^{-\nu} \frac{\nu^n}{n!} \longrightarrow \ln g(\chi) = \nu \left(e^{i\chi} - 1 \right). \tag{12.115}$$

Here, the rate is determined by the applied voltage bias, barrier transparencies, temperature, and possibly other system parameters.

Bidirectional distribution: The current through a conductor connected to two terminals (for the generalization to multi-terminal geometries, see Refs.[24]) is obtained by the additive superposition of two counter-propagating current flows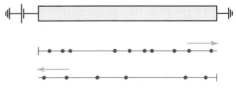
(see figure). For simplicity, let us assume the two distributions $p_i(n_i)$, $i = 1, 2$ of the counter-propagating charges to be statistically independent. The distribution $p(n)$ of the total number of transmitted charges, $n = n_1 - n_2$ can then be computed as follows:

$$p(n) = \sum_{n_1 n_2} \delta_{n, n_1 - n_2} p_1(n_1) p_2(n_2)$$

$$= \sum_{n_1 n_2} \int \frac{d\chi}{2\pi} e^{-i\chi(n - n_1 + n_2)} \int \frac{d\chi_1}{2\pi} e^{-i\chi_1 n_1} g(\chi_1) \int \frac{d\chi_2}{2\pi} e^{-i\chi_2 n_2} g_2(\chi_2)$$

$$= \int \frac{d\chi}{2\pi} e^{-i\chi n} g_1(\chi) g_2(-\chi).$$

---

[26] For the generalization to non-zero temperatures, see Refs.[24].

This means that the generating function of the transmitted charge is the product $g(\chi) \equiv g_1(\chi) g_2(-\chi)$ of the partial distributions. For example, in the case where the two distributions are individually Poissonian, we find

$$\ln g(\chi) = \nu_1 \left(e^{i\chi} - 1 \right) + \nu_2 \left(e^{-i\chi} - 1 \right), \qquad (12.116)$$

where ν_1 and ν_2 determine the average rates.

In the following, we consider the biased double-barrier quantum dot and explore to what extent its current statistics reflects the limiting cases discussed above.

12.7.3 Full counting statistics of the double-barrier quantum dot

In this section, we consider the double-barrier dot introduced in section 12.6.1. We will monitor the current through the tunnel barriers connecting the dot to the left lead. To define a suitable counting field note that, in the presence of a vector potential $A e_\perp$ parallel to the lead axis, the tunneling matrix elements $T_{\sigma,a\mu}$ defined in section 12.6.1 generalize as $T_{\sigma,a\mu} \to T_{\sigma,a\mu} \exp(i \int dx A_\sigma)$, where the integral runs over the transverse extension of the barrier $\sigma = \mathrm{L}, \mathrm{R}$. The definition (12.112) then implies that the fields χ_σ probing charge transport through the barriers couple to the tunneling matrix elements on the Keldysh contour as

$$T_{\sigma,a\mu} \to T_{\sigma,a\mu} \exp(i\chi_\sigma(t)\sigma_1/2).$$

Notice that the coupling of the counting field to the tunneling matrix elements is identical to that of the quantum component of the dynamical phase field (cf. Eq. (12.86)), and can be described by the shift $\phi_q \to \phi_q + \chi_\sigma/2$. This observation fixes the coupling of the counting field to the effective action without further calculation: the tunneling action (12.87) generalized for the presence of counting fields, reads as

$$S_{\mathrm{tun}}[\chi] = -\frac{i}{4} \sum_{\sigma = L, R} g_\sigma \mathrm{tr} \left(\Lambda_\sigma(\chi_\sigma) e^{-i\hat{\phi}} \Lambda_d e^{i\hat{\phi}} \right), \qquad \Lambda_\sigma(\chi_\sigma) \equiv e^{i\frac{\chi_\sigma}{2}\sigma_1} \Lambda_\sigma e^{-i\frac{\chi_\sigma}{2}\sigma_1},$$

where we have generalized to barriers with different tunnel conductances, g_σ. Following the logic of section 12.6.2, our next step will be to determine the matrix Λ_d. As before (see exercise on page 739), we do this by requiring stationarity under variations of ϕ. Variation of the action in the limit of vanishingly weak interaction, $\phi \simeq 0$, generates the condition

$$\left[\Lambda_d, \sum_\sigma g_\sigma \Lambda_\sigma(\chi_\sigma) \right] \overset{!}{=} 0. \qquad (12.117)$$

Now, the transformations $\Lambda \to \exp(i\chi\sigma_1/2)\Lambda \exp(-i\chi\sigma_1/2)$ render the matrices $\Lambda_\sigma(\chi_\sigma)$ non-triangular. This means that in general the stationary configurations Λ_d will also no longer be of the form (12.89). However, the transformation by the counting field does not alter the nonlinear equation $\Lambda^2 = \sigma_0$ satisfied by the matrices of Eq. (12.89). It is straightforward to verify that the solution of Eq. (12.117) obeying this normalization is given by[27]

$$\Lambda_d = \frac{g_L \Lambda_L(\chi_L) + g_R \Lambda_R(\chi_R)}{\left(g_L^2 + g_R^2 + g_L g_R [\Lambda_L(\chi_L), \Lambda_R(\chi_R)]_+\right)^{1/2}}.$$

(Notice that the the anti-commutator matrix in the denominator commutes with the numerator, i.e., the relative ordering of numerator and denominator immaterial. Also notice that, in the limit $\chi = 0$, the solution reduces to that discussed in section 12.6.2.) Substitution of this configuration into the action leads to

$$S_{\text{tun}}[\chi] = -\frac{i}{4} \, \text{tr} \left(g_L^2 + g_R^2 + g_L g_R [\Lambda_L(\chi_L), \Lambda_R(\chi_R)]_+\right)^{1/2}.$$

Assuming that the inverse of the counting time-window $(\Delta t)^{-1}$ is large in comparison with the energy scales relevant to the distribution functions, we will evaluate the trace over energy/time indices within the leading-order Wigner approximation (see appendix section A.4.3), $\text{tr}(\dots) \to \int \frac{d\epsilon dt}{2\pi}(\dots)$, $F(\hat{\epsilon}) \to F(\epsilon)$, $\chi(\hat{t}) \to \chi(t)$, where $F(\epsilon)$ and $\chi(t)$ are ordinary functions of energy and time, respectively. As a result of a straightforward calculation (exercise) one finds that the matrix $[\Lambda_L(\chi_L), \Lambda_R(\chi_R)]$ is proportional to the unit matrix, and that the trace evaluates to

$$S_{\text{tun}}[\chi] = -\frac{i}{2} \int \frac{dt d\epsilon}{2\pi} X(\chi(t)) = -\frac{i\Delta t}{2} \int \frac{d\epsilon}{2\pi} X(\chi) + C, \qquad (12.118)$$

where

$$X(\chi) \equiv \left((g_L + g_R)^2 + 4 g_L g_R \left((e^{i\chi} - 1)n_L(1 - n_R) + (e^{-i\chi} - 1)n_R(1 - n_L)\right)\right)^{1/2}.$$

Here, in the first equality, we defined $\chi \equiv \chi_L - \chi_R$, and in the second specialized to the time-dependence $\chi(t) \equiv \chi \Theta(t - \bar{t}) \Theta(\bar{t} + \Delta t - t)$, with χ a constant differentiation parameter and C an inessential constant. From this result we find the cumulant-generating function $\ln g(\chi) \simeq \ln \exp(iS[\chi])$, i.e.,

$$\ln g(\chi) = \frac{\Delta t}{2} \int \frac{d\epsilon}{2\pi} X(\chi). \qquad (12.119)$$

Let us try to make sense of this expression. Comparison with Eq. (12.116) shows that $\ln g$ contains the generating functions of two Poisson distributions as building blocks. To explore the meaning of this result, we first consider the limit of zero temperature and voltage bias V, $n_{L/R}(\epsilon) = \Theta(\pm V/2 - \epsilon)$. The generating function then reduces to

$$\ln g(\chi) = \frac{\Delta t V}{4\pi} \left((g_L + g_R)^2 + 4 g_L g_R \left(e^{i\chi} - 1\right)\right)^{1/2} + C.$$

---

[27] More precisely, the above result is obtained in an approximation where the Green functions appearing in the matrices Λ of Eq. (12.89) are replaced by their imaginary parts, $\text{Im}\, g^{\pm} = \mp 1/2$. A more careful analysis shows that the real parts regularize the superficially divergent constant C in Eq. (12.118).

According to Eq. (12.113), the **first moment** of the transmitted charge through the system is given by

$$\langle \hat{N} \rangle = -i\partial_\chi\big|_{\chi=0} \ln g(\chi) = \frac{\Delta t \, V}{2\pi} \frac{g_L g_R}{g_L + g_R}.$$

Comparison with the definition of the conductance, $G = \langle \hat{I} \rangle / V = \langle \hat{N} \rangle / V \Delta t$, gives $G = \frac{1}{2\pi} \frac{g_L g_R}{(g_L + g_R)}$, which we recognize as the mean conductance of two tunnel barriers shunted in series (in units of the conductance quantum $e^2/h = 1/2\pi$). Turning to the **current statistics**, let us consider the limit $g_L \ll g_R$. We may then expand the square root to first order in the ratio g_L/g_R to obtain a bi-directional Poisson distribution (12.116), with (time-integrated) characteristic rates identified as

$$\nu_1 = \Delta t \, g_L \int \frac{d\epsilon}{2\pi} \, n_L (1 - n_R), \qquad \nu_2 = \Delta t \, g_{rmL} \int \frac{d\epsilon}{2\pi} \, n_R (1 - n_L).$$

These coefficients may be interpreted as the integrated rate at which filled states in the right lead scatter into empty states in the right lead, and vice versa. As one may expect, the statistics of the current is caused predominantly by the bottleneck in the system, i.e., the conductance of the weaker tunnel barrier, g_L. At non-zero temperature we have $\nu_1 \simeq \Delta t g_L V / 2\pi$, while $\nu_2 \simeq \exp(-V/T) \, \Delta t \, T / 2\pi$ shows that thermal activation is necessary to push charges against the voltage gradient.

The second (cumulative) moment defines the **noise power** (cf. Eq. (12.106)),

$$S_0 = \frac{2}{\Delta t} \text{var}\,(\hat{N}) = -\frac{2}{\Delta t} \partial_\chi^2\big|_{\chi=0} \ln g[\chi] = \frac{2}{\Delta t}(\nu_1 + \nu_2) = 2\frac{g_L}{2\pi} V \coth(V/2T).$$

This shows how the noise interpolates between the equilibrium noise, $\langle \delta I(t) \delta I(t') \rangle \overset{V \to 0}{\sim} g_L T$ and the shot noise limit $\langle \delta I(t) \delta I(t') \rangle \overset{V \gg T}{\sim} g_L |V|$. In the more general case of barriers of comparable transparency, $g_L \simeq g_R$, the current statistics is more complicated and is described by the full expression (12.119).

In principle, we could now include phase fluctuations to explore the interesting question of how interaction effects influence full counting statistics (FCS). However, this topic[28] lies beyond the scope of the present text.

12.8 Summary and Outlook

In this final chapter, we have introduced the Keldysh formalism as a versatile and powerful tool in quantum nonequilibrium theory. Admittedly, the Keldysh framework comes with a steep learning curve. However, after a while, one begins to realize that it is actually more intuitive than the technically more straightforward Matsubara formalism. It is also a unifying framework in that other approaches to nonequilibrium are included in the Keldysh functional. Specifically, we discussed the

[28] see D. A. Bagrets and Y. V. Nazarov, *Full counting statistics of charge transfer in Coulomb blockade system*, Phys. Rev. B **67**, 085316 (2003).

derivation of quantum master equations and quantum kinetic equations, the connection to equilibrium field theory, the classical limit in terms of the MSR formalism of the previous chapter, nonequilibrium variants of diagrammatic perturbation theory, and more. At the time of writing of this book, physics beyond thermal equilibrium is becoming more and more important, with key driving factors including device miniaturization in condensed matter physics and experimental advances in cold atomic and optical many-body systems. It stands to reason that familiarity with the Keldysh framework will be an indispensable element of the repertoire required by future generations of researchers. The material introduced in this chapter is far from complete. However, it should be sufficient for readers to engage in their own research in the fascinating field of nonequilibrium physics.

12.9 Problems

12.9.1 Atom–field Hamiltonian

The **atom–field Hamiltonian** is a simple model Hamiltonian reducing the interaction of atoms with the electromagnetic field to a two-level system ("the atom") coupled to an assembly of oscillator modes. In spite of its simplistic nature, the model gives rise to rich phenomenology, and is often employed in quantum optics. In this problem, we study the simplest variant of the system, the exactly solvable limit of a single field mode. In this limit, the model shows fully coherent quantum dynamics. We use it as a benchmark system to compare with incoherent approximations underlying the quantum master equation of section 12.1. In the follow-up problem 12.9.2, we then explore the generalization to multi-mode coupling.

Consider an atom exposed to electromagnetic radiation. Assuming that the field modes predominantly couple two atomic states $|a\rangle$ and $|b\rangle$ (see the figure), and forgetting about the complications introduced by the polarization of the electromagnetic field, we describe this setup by the model Hamiltonian

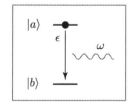

$$\hat{H} = \frac{\epsilon}{2}\sigma_3 + \sum_k \omega_k a_k^\dagger a_k + \sum_k \left(g_k \sigma_+ a_k + \bar{g}_k \sigma_- a_k^\dagger \right), \qquad (12.120)$$

where ϵ is the energy difference between the excited state, $|a\rangle$, and the lower state, $|b\rangle$, the Pauli matrices σ_i act in the two-dimensional Hilbert space defined by these states, and $\sigma_\pm = (\sigma_1 \pm i\sigma_2)/2$ as usual. This **atom–field Hamiltonian** describes excitation processes $|b\rangle \to |a\rangle$ by field quantum absorption (at coupling constants g_k), and the corresponding relaxation processes by quantum emission.

We may simplify the problem further by assuming that only a single mode of the electromagnetic field satisfies the resonance condition $\omega \simeq \epsilon$ required for significant field–state coupling. This, defines the **single-mode Hamiltonian**

$$\hat{H} = \frac{\epsilon}{2}\sigma_3 + \omega a^\dagger a + g\left(\sigma_+ a + \sigma_- a^\dagger\right), \tag{12.121}$$

where we have omitted the mode index, k, and gauged the coupling constant g to become real.

(a) Using the parlance of section 12.1, consider the atom as the "system," prepared at time $t = 0$ in a mixed state,

$$\hat{\rho}_s = \rho_a P_+ + \rho_b P_-,$$

where $P_\pm$ are projectors onto the upper state a or lower state b and $\rho_{a,b}$, $\rho_a + \rho_b = 1$, are the probabilities of occupation of these states. Apply the Markovian approximation of section 12.1 to derive an equation of motion for the reduced system density matrix $\hat{\rho}_s$ coupled to the "bath" defined by the mode. (For the moment, do not worry about the appearance of singular couplings $\sim \delta(\epsilon - \omega)$ and treat them as formal coefficients.) Derive a closed expression for the diagonal elements of the reduced density operator $\rho_x \equiv \langle x|\hat{\rho}_s|x\rangle$, $x = a, b$, and verify that the stationary limit $\rho_{x,\infty} \equiv \rho_x(t \to \infty)$ obeys the **detailed balance** relation

$$\frac{\rho_{a,\infty}}{\rho_{b,\infty}} = \frac{\langle n\rangle}{\langle n+1\rangle}, \tag{12.122}$$

where n is the number of mode quanta and the expectation value is over the thermal distribution of the bath. Accordingly, the **population imbalance** between the levels approaches the limit

$$\Delta\rho \equiv \rho_{a,\infty} - \rho_{b,\infty} = -\frac{1}{2\langle n\rangle + 1}. \tag{12.123}$$

Discuss the meaning of this expression. Is it physical? If not, where do you think the derivation failed?

(b) Let us compare the predictions of the Markovian approximation with reality. To this end, solve the time-dependent Schrödinger equation defined by the Hamiltonian (12.121). Assume that the system is prepared in the excited atomic state $|a\rangle\langle a| \otimes |n\rangle\langle n|\rho_n$, where ρ_n is the nth eigenvalue of the thermal-mode density operator, $\rho_n = \mathcal{Z}^{-1}\exp(-\beta\omega(n+1/2))$. Using the fact that the Schrödinger equation couples only the states $|a, n\rangle$ and $|b, n+1\rangle$, solve the time-dependent problem with an initial condition corresponding to the above density operator.

With $\rho_x \equiv \sum_n \langle x, n|x, n\rangle$ denoting the probability for the system to be in state $|x\rangle$, compute the population imbalance and compare with the predictions of the Markovian approach (12.123). Assuming the mode population to be thermal, it is instructive to plot the imbalance as a function of dimensionless time, gt, for different values of temperature and frequency mismatch $\Delta = \epsilon - \omega$ (see figure on page 757.)

(c) Discuss qualitatively the origin of the discrepancies between the two approaches. Why is the Markovian approximation not appropriate under the present circumstances, and why does the exact solution not predict the relaxation of an initially occupied state $|a\rangle$ to the ground state, even at zero temperature?

Answer:

(a) In the interaction representation, $a(t) = e^{-i\omega t}a$, $\sigma_\pm(t) = e^{\pm i\epsilon t}\sigma_\pm$, the interaction Hamiltonian $\hat{H}_i \equiv g\left(\sigma_+ a + \sigma_- a^\dagger\right)$ reads

$$\hat{H}_i(t) = g\left(e^{i\Delta t}\sigma_+ a + e^{-i\Delta t}\sigma_- a^\dagger\right),$$

where $\Delta \equiv \epsilon - \omega$ is the energy mismatch between the level splitting and mode frequency. Defining $\hat{L}_i = -i[\hat{H}_i, \ ]$, it is then straightforward to verify that Eq. (12.4) assumes the form

$$\partial_t \hat{\rho}_s = -g^2 \int^t dt' \, e^{+i\Delta t'} \left(+\langle n+1\rangle[\sigma_+, \sigma_- \hat{\rho}_s(t-t')] - \langle n\rangle[\sigma_+, \hat{\rho}_s(t-t')\sigma_-]\right)$$

$$- g^2 \int^t dt' \, e^{-i\Delta t'} \left(-\langle n+1\rangle[\sigma_-, \hat{\rho}_s(t-t')\sigma_+] + \langle n\rangle[\sigma_-, \sigma_+ \hat{\rho}_s(t-t')]\right),$$

$$(12.124)$$

where the expectation value is over the mode distribution. The initial state defined above is diagonal as a matrix in the two-state basis, and the evolution equation preserves this form. The Markovian approximation neglects the time-dependence of $\hat{\rho}(s)$ under the integral. Under this assumption, the evolution assumes the form of a **master equation**,

$$\partial_t \begin{pmatrix} \rho_a \\ \rho_b \end{pmatrix} = \Gamma \begin{pmatrix} -\langle n+1\rangle & \langle n\rangle \\ \langle n+1\rangle & -\langle n\rangle \end{pmatrix} \begin{pmatrix} \rho_a \\ \rho_b \end{pmatrix}, \tag{12.125}$$

where the rate $\Gamma = 2\pi g \delta(\Delta)$ is singular at resonance. This equation predicts an exponential approach to a stationary limit satisfying the detailed balance relation (12.122).

(b) For given $n \geq 0$ the Hamiltonian acts in the two-dimensional space spanned by the states $|a, n\rangle$ and $|b, n+1\rangle$. Specifically, $\hat{H}_i(t)|a, n\rangle = ge^{-i\Delta t}(n+1)^{1/2}|b, n+1\rangle$ and $\hat{H}_i(t)|b, n+1\rangle = ge^{+i\Delta t}(n+1)^{1/2}|a, n\rangle$. Introducing wave functions by $|\psi(t)\rangle = \psi_{a,n}(t)|a, n\rangle + \psi_{b,n+1}(t)|b, n+1\rangle$, the time-dependent Schrödinger equation $i\partial_t \psi = \hat{H}_i(t)\psi$ assumes the form

$$i\partial_t \psi_{a,n} = g_{\text{eff}} e^{+i\Delta t}\psi_{b,n+1}, \qquad i\partial_t \psi_{b,n+1} = g_{\text{eff}} e^{-i\Delta t}\psi_{a,n}, \tag{12.126}$$

where we have introduced the effective coupling constant $g_{\text{eff}} = g(n+1)^{1/2}$. These equations are solved by

$$\psi_{a,n}(t) = e^{+i\frac{\Delta}{2}t}\left(\psi_{a,n}(0)\cos(\Omega_n t) - i\frac{g_{\text{eff}}\psi_{b,n+1}(0) + \frac{\Delta}{2}\psi_{a,n}(0)}{\Omega_n}\sin(\Omega t)\right),$$

$$\psi_{b,n+1}(t) = e^{-i\frac{\Delta}{2}t}\left(\psi_{b,n+1}(0)\cos(\Omega_n t) - i\frac{g_{\text{eff}}\psi_{a,n}(0) - \frac{\Delta}{2}\psi_{b,n+1}(0)}{\Omega_n}\sin(\Omega t)\right),$$

with $\Omega_n \equiv (g_{\mathrm{eff}}^2 + (\Delta/2)^2)^{1/2}$. The wave function $\rho(0) = \sum_n |\psi_{n,a}(0)\rangle\langle\psi_{n,a}(0)|$ has initial value $\psi_{a,n}(0) = \sqrt{\rho_n}$. Substitution of this value leads to an exact result for the population imbalance:

$$\Delta\rho = \sum_n \left(|\psi_{a,n}|^2 - |\psi_{b,n}|^2\right) = \sum_n \rho_n \left(1 - \frac{2g^2(n+1)}{\Omega_n^2}\sin^2(\Omega_n t)\right).$$

This result is very different from the one obtained within the Markovian approach, Eq. (12.123): no stationary limit is approached. A short period of decay of the initial value $\Delta\rho(0) = 1$ merges into a pattern of irregular fluctuations – the result of a superposition of contributions of incommensurate frequencies

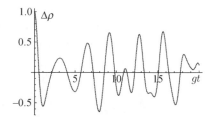

(see the figure). In quantum optics, the phenomenon of transient near-recoveries of the initial value is known as **collapse and revival** and the fluctuations of the two-state atom are **Rabi oscillations** caused by the field mode. Their oscillatory pattern is the result of maintained quantum coherence and reversibility of the dynamics. Notice that, even at zero temperature, the atom does not relax by emission of field quanta: at $T = 0$, only the $n = 0$ term (zero field quanta) contributes to the sum above. This leads to oscillatory behavior of the density operator in which the initial state is recovered at regular intervals $t \sim g$, but no relaxation. For further discussion of the fluctuation pattern, we refer to Ref.[29]. Here, the take-home message is that the prediction of irreversible dynamics derived in (a) is incorrect.

(c) The Markovian approximation fails because a single quantum oscillator mode is different from a bath. Indeed, the mode–atom coupling is strongest at resonance, $\Delta = 0$, when "system" and "bath" fluctuate at comparable time-scales. In this case, the memory of the latter is comparable to that of a system, and a Markovian approximation is not justified.

12.9.2 Weisskopf–Wigner theory of spontaneous emission

In the previous problem, we saw that the coupling of an atom to a single electromagnetic mode does not lead to radiative relaxation. Here, we study how irreversibility emerges when a large number of modes are coupled. The modeling introduced in the present problem plays an important role in, e.g., the theory of lasing. (This problem should be attempted only after problem 12.9.1.)

(a) Consider the Hamiltonian (12.120) of a two-level atom coupled to a multi-mode field. Assuming zero temperature, so that all mode occupation numbers are vanishing, start from the ansatz $\psi \equiv \psi_0|a\rangle \otimes |0\rangle + \sum_k \psi_k|b\rangle \otimes a_k^\dagger|0\rangle$, where $|0\rangle$ is the zero-temperature photon vacuum, to generalize the Schrödinger equation (12.126)

[29] M. O. Scully and M. S. Zubairy, *Quantum Optics* (Cambridge University Press, 1997).

for the initial configuration $|a\rangle \otimes |0\rangle$ to the presence of many modes of frequency ω_k coupled to the system with coupling constants g_k. Formally integrate the second equation to convert the system of two equations to a single integro-differential equation. Assume that $|g_k|^2 = g^2(\omega_k)$ depends only on the mode energy and formulate the equation in terms of the **density of bath modes** $\rho(\omega) = \sum_k \delta(\omega - \omega_k)$. Assuming that $\rho(\omega) \equiv \rho$ and the coupling strength $g^2(\omega) \equiv g^2$ vary only negligibly on the energy scales relevant to the states ψ – the **Weisskopf–Wigner approximation** – derive an approximate solution of this equation.

Weisskopf–Wigner approximation

Compute the population imbalance between the two atomic levels (hint: consider the unit normalization of the wave function), $\Delta\rho \equiv |\psi_0|^2 - \sum_k |\psi_k|^2$, and show that it relaxes as $\Delta\rho(t) = 2e^{-\Gamma t} - 1$ at the **golden rule decay rate**

$$\Gamma \equiv 2\pi\rho g^2. \tag{12.127}$$

(b) For arbitrary temperature and initial population, attack the multi-mode problem by generalization of the projector formalism applied in problem 12.9.1 (a) to the single-mode case and compare with the results of the Weisskopf–Wigner theory. Convince yourself that the approximation used there is equivalent to a Markovian approximation.

Answer:

(a) The multi-mode generalization of the Schrödinger equation (12.126) reads

$$i\partial_t\psi_0 = \sum_k g_k e^{i\Delta_k t}\psi_k, \qquad i\partial_t\psi_k = g_k e^{-i\Delta_k t}\psi_0,$$

where $\Delta_k = \epsilon - \omega_k$. We integrate the second equation and substitute the result into the first equation to obtain

$$\partial_t\psi_0(t) = -\sum_k g_k^2 \int_0^t dt'\, e^{i\Delta_k(t-t')}\psi_0(t')$$

$$= -\int d\omega\, \rho(\omega)g(\omega) \int_0^t dt'\, e^{i(\epsilon-\omega)(t-t')}\psi_0(t') \simeq -\pi\rho g^2\psi_0(t),$$

where in the second equation we applied the Weisskopf–Wigner approximation and evaluated the frequency integral as $\int d\omega\, e^{-i\omega(t-t')} = 2\pi\delta(t-t')$.[30] The (irreversible) effective equation for ψ_0 is now trivially solved as $\psi_0(t) = e^{-\pi\rho g^2 t}$.

With the normalization condition $1 = |\psi|^2 = |\psi_0|^2 + \sum_k |\psi_k|^2$, the population imbalance is obtained as $\Delta\rho = 2|\psi_0|^2 - 1$, and substitution of $\psi_0(t)$ leads to the stated result.

(b) Comparing with the discussion in the previous problem, we verify that the multi-mode generalization of Eq. (12.124) reads

---

[30] We count $\int_0^t dt'\delta(t - t') = 1/2$ since the δ-function lies at the boundaries of the integration domain.

$$\partial_t \hat{\rho}_s = - \int d\omega \, \rho(\omega) g^2(\omega) \int^t dt'$$
$$\times \left[e^{+i(\epsilon-\omega)t'} \left(+\langle n(\omega)+1\rangle [\sigma_+, \sigma_- \hat{\rho}_s(t-t')] - \langle n(\omega)\rangle [\sigma_+, \hat{\rho}_s(t-t')\sigma_-] \right) \right.$$
$$\left. - e^{-i(\epsilon-\omega)t'} \left(-\langle n(\omega)+1\rangle [\sigma_-, \hat{\rho}_s(t-t')\sigma_+] + \langle n(\omega)\rangle [\sigma_-, \sigma_+\hat{\rho}_s(t-t')] \right) \right],$$

where $n(\omega)$ is the boson distribution function. The integral now contains the superposition of contributions fluctuating at different time-scales and decays rapidly. Unlike the previous problem, this justifies the assumption of near constancy of the density operator. (Think why this is equivalent to the Weisskopf–Wigner assumption of the approximate frequency independence of ρg^2.) Doing the integral, we obtain the master equation (12.125), where $\langle n \rangle = \langle n(\epsilon) \rangle$ is the mean number of bath quanta at the resonance energy and the decay rate is given by Eq. (12.127). Solution of this equation yields the population imbalance

$$\Delta \rho(t) = \left(\Delta \rho(0) + \frac{1}{2\langle n \rangle + 1} \right) e^{-\Gamma(2n+1)t} - \frac{1}{2\langle n \rangle + 1}.$$

With the initial condition $\Delta \rho(0) = 1$ and at zero temperature, $\langle n(\epsilon) \rangle = 0$, this reduces to the results obtained in (a).

12.9.3 Qubit decoherence from the Lindblad equation

One of the most important and problematic aspects of system–bath couplings is the decoherence of quantum information. Here, we study this phenomenon for the example of a single qubit coupled to a bath.

qubit   A **qubit** is a quantum mechanical two-level system. As an example, consider a qubit subject to the Hamiltonian $\hat{H} = \epsilon \sigma_3 + \sum_k \omega_k a_k^\dagger a_k \sigma_3$, where the second term describes the fluctuations of the qubit energy 2ω due to the coupling of the qubit to a system of bath modes with the Hamiltonian $\hat{H}_b = \sum_k \omega_k (a_k^\dagger a_k + 1/2)$. We aim to explore how these fluctuations compromise phase coherence.

(a) We will be interested in the evolution of the density operator representing the qubit. As a first step, show that any density operator of a two dimensional Hilbert space can be represented as $\hat{\rho} = \frac{1}{2} + m_i \sigma_i$, where **m** is a real vector of norm $|\mathbf{m}| \leq 1$. Now assume that the qubit has been initialized in a **pure state**, $\hat{\rho} = |\psi\rangle\langle\psi|$, where $|\psi\rangle = \alpha| \uparrow \rangle + \beta| \downarrow \rangle$ is a linear superposition. Find an efficient way to obtain the corresponding expansion coefficients as $\mathbf{m} = (\mathrm{Re}(\bar{\alpha}\beta), \mathrm{Im}(\bar{\alpha}\beta), \frac{1}{2}(|\alpha|^2 - |\beta|^2))$.

(b) Assuming a Lindblad form, write down the evolution equation for the density operator. For the purposes of this exercise, it is not important to work out the value of the Lindbladian coupling strength in detail; just call it γ. From there, show that the parameterizing vector evolves as

$$\partial_t \begin{pmatrix} m_1 \\ m_2 \\ m_3 \end{pmatrix} = \begin{pmatrix} -2\epsilon m_2 - 4\gamma m_2 \\ 2\epsilon m_1 - 4\gamma m_2 \\ 0 \end{pmatrix}.$$

(c) Discuss the meaning of this equation and describe what happens for a qubit initialized in a pure state.

Answer:

(a) Any 2×2 matrix can be represented as a linear combination of the unit- and Pauli-matrices. The condition $\mathrm{tr}(\hat{\rho}) = 1$ fixes the coefficient of the unit-matrix to $1/2$. Hermiticity requires the reality of the expansion coefficients, and the positivity of the eigenvalues $1 \pm |\mathbf{m}|$ constraints the norm of the expansion vector.

Considering the ansatz $|\psi\rangle\langle\psi| = \frac{1}{2} + m_i\sigma_i$, we take the trace $\mathrm{tr}(|\psi\rangle\langle\psi|\sigma_j) = \mathrm{tr}(1 + m_i\sigma_i)\sigma_j$ to obtain $m_j = \frac{1}{2}\mathrm{tr}(|\psi\rangle\langle\psi|\sigma_j) = \frac{1}{2}\langle\psi|\sigma_j|\psi\rangle$. From here we immediately obtain the result.

(b) With the hermitian Lindblad operator $\hat{X} = \sigma_3$, the Lindblad equation assumes the form

$$\partial_t\hat{\rho} = -i\epsilon[\sigma_3, \hat{\rho}] + 2\gamma\left(\sigma_3\hat{\rho}\sigma_3 - \hat{\rho}\right), \tag{12.128}$$

where the positive constant γ measures the coupling strength. Substitution of the expansion then leads to the equation stated in the problem.

(c) The evolution equation describes the precession of the transverse components (m_1, m_2) at a frequency $\propto \epsilon$. However, it also shows that they diminish at a rate 4γ. In the parlance of decoherence theory, this time-scale is called the **transverse**

relaxation
time

relaxation time, T_1. The transverse relaxation drives the density matrix towards a coherence-free mixed state $\hat{\rho} = c_\uparrow|\uparrow\rangle\langle\uparrow| + c_\downarrow|\downarrow\rangle\langle\downarrow|$, $c_{\uparrow\downarrow} = \frac{1}{2}(1 \pm (|\alpha|^2 - |\beta|^2))$, without off-diagonal elements.

A more complete model of decoherence would also include bath couplings with jump operators $\sigma_{1,2}$ inducing qubit flips. The corresponding Lindblad equation (whose derivation and discussion along the lines of this problem is an instructive exercise) describes the diminishing of the longitudinal component at a rate set by the **longitudinal relaxation time**, T_2. The two time-scales $T_{1,2}$ are perhaps the most important parameters characterizing the quality of a qubit.

12.9.4 Keldysh theory of single-electron transport

(Recapitulate section 12.6 before turning to this problem. If not stated otherwise, the notation of that section will be used throughout.) Here, we consider the quantum dot introduced in section 12.6 in a regime of near isolation from its environment. Under these conditions, the charge on the dot is almost perfectly quantized. Here, we show how Keldysh theory can describe transport through the system at a level where the dynamics of individual electrons is resolved.

In problems 5.6.4 and 5.6.5 we considered an equilibrium quantum dot in a state of perfect or near isolation from its environment. At low temperatures, the dot admits only the integer quantum of charges that minimizes its capacitive energy.

Coulomb
blockade

This **Coulomb blockade** manifests itself in the partition function of the isolated quantum dot,

$$\mathcal{Z} = \sum_n \exp\left[-\frac{E_c}{T}(n - N_0)^2\right], \tag{12.129}$$

where the optimal charge number is determined by the parameter $N_0 \in \mathbb{R}$ (which may be set by changing the external gate voltages) and E_c is the charging energy.

As usual, the first challenge we meet in the Keldysh approach to the problem is the identification of its distribution function. Keep in mind that Keldysh theory is constructed assuming a non-interacting theory at large negative times, where the dot distribution is assumed to be thermal (or double-step, if an external voltage is applied). If we then adiabatically switch on a charge interaction, the energies of sectors of fixed n shift, but the distribution functions remain unchanged. Equilibration towards an effective distribution as in Eq. (12.129) can happen only if the Hamiltonian contains a contribution capable of changing n and hence the occupation of states of fixed n. It turns out that the switching on of a weak dot–lead coupling achieves just this. Indeed, the dot-coupling term reads (cf. section 12.6.1) $\bar{\psi}_d T e^{i\phi} \psi_\sigma$, where ψ_d are the fermion fields of the dot, $\psi_{\sigma=L,R}$ the fields of the leads, and $\dot{\phi}$ is the Hubbard–Stratonovich field of the interaction. This term both creates an excitation in the field ϕ (physically, a voltage fluctuation) and changes the occupation of the dot. In this way it is capable of altering the distribution, as we will show below. To keep the discussion simple, we assume a vanishing gate voltage, $N_0 = 0$, and zero temperature, $T = 0$, throughout. (As an instructive exercise, generalize the discussion to non-zero gate voltage and temperatures.)

(a) Our first step is purely technical: observing that at weak coupling, $g_T \ll 1$, fluctuations in the field ϕ are strong, we will trade the integration over ϕ for an integration over different charge states n of the dot. To this end, expand the action (12.93) in the tunneling action (12.91) and integrate over the phase degrees of freedom (hint: keep the quantization condition $n_c(0) \equiv n \in \mathbb{Z}$ in mind; see info block on page 738, and do the integration in the contour representation $\phi_\pm$) to obtain the representation

$$\mathcal{Z} = \sum_{m=0}^{\infty} \frac{1}{m!} \left(-\frac{g}{2} \right)^m \sum_{\{\sigma_k\}} \int Dt \, e^{-iE_c \int dt \, (n_+^2(t) - n_-^2(t))} \prod_{k=1}^{m} L_{\sigma_{2k-1}, \sigma_{2k}}(t_{2k-1} - t_{2k}),$$
(12.130)

where $\sum_{\{\sigma_k\}}$ is a sum over all sign configurations $\sigma_k \in \{-1, 1\}^{2m}$, the integration measure $Dt = \prod_{k=1}^{2m} dt_k$, and the charge profiles are given by

$$n_\sigma(t) = n + \sigma \sum_{k=1}^{2m} (-)^k \Theta(t - t_k) \delta_{\sigma_k, \sigma}.$$
(12.131)

Finally, the matrix elements of the kernel L are defined as

$$L_{\sigma\sigma'} = \frac{1}{4} \left(\sigma\sigma' \Sigma^K + \sigma\Sigma^+ + \sigma'\Sigma^- \right),$$
(12.132)

where the self-energies $\Sigma^{K,\pm}$ were introduced in Eq. (12.91).

This representation expresses the partition function as a sum over quasiparticle in- and out-tunneling events, connected by elements of the kernel L. We next need to make physical sense of this expansion.

(b) The temporal entanglement of tunneling events makes a closed computation of the partition function impossible. However, for weak tunneling processes containing intersecting or nested propagator  lines, $L(t, t')$, is negligibly small. Estimate the temporal range of the propagator L to derive a criterion for the applicability of the **non-interacting blip approximation (NIBA)**, wherein tunneling events occur in a sequential manner (see figure.) For simplicity, you may assume an unbiased situation, $V = 0$. However, it is worth checking that, for $V \neq 0$, the quality of the approximation improves. (Hint: Note the frequency dependence of the self-energy in Eq. (12.96).)

(c) The lack of entanglement of tunneling events in the NIBA makes the computation of the Keldysh partition function a lot easier. The basic picture now is that occasional charge tunneling events (**blips**[31]) are interspersed in long periods of time wherein the charge contour profile stays in a diagonal state, $n_+ = n_- = n_{cl} \equiv n$ (**sojourns**). During a blip, the quantum component $n_+ - n_- = n_q \equiv \xi \in \{-1, 0, 1\}$ jumps to a value ± 1, depending on the configuration $(\sigma_{2k-1}, \sigma_{2k})$ of the tunneling event, and the sign of the time-difference $t_{2k-1} - t_{2k}$ (see figure).

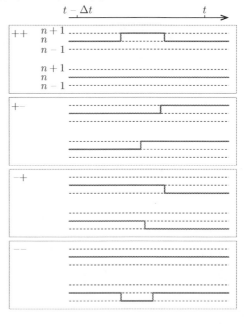

Building on this structure, and assuming zero biasing, $V = 0$, derive a master equation for the quantity $P(n, t) \equiv P(n, t|n_0, 0)$, i.e., the probability that the system evolves into a charge state $(n_+, n_-) = (n, n)$ at time t, give that it started in (n_0, n_0) at $t = 0$. To this end, interpret $P(n, t)$ as the Keldysh field integral (12.130), subject to the constraints (see Eq. (12.131)) $n_\sigma(t) = n$ and $n_\sigma(0) = 0$. Relate $P(n, t)$ to $P(n', t - \delta t)$, where $\delta t \gg E_c^{-1}$ is much larger than the typical duration of blips, yet smaller than the average spacing between blips, $\Delta t \ll (gE_c)^{-1}$.

Apply a continuum approximation $(\Delta t)^{-1}(P(n, t) - P(n, t - \Delta t)) \simeq \partial_t P(n, t)$ to obtain the **master equation**

$$\partial_t P(n, t) = \left[(\hat{E}_1 - 1)W_{n,n-1} + (\hat{E}_{-1} - 1)W_{n,n+1} \right] P(n, t), \qquad (12.133)$$

where $\hat{E}_{\pm 1}f(n) \equiv f(n \pm 1)$ are charge raising and lowering operators, the transition rates are

$$W_{n,n\pm 1} = \frac{g}{\pi} E_c(n, n \pm 1)\Theta(E_c(n, n \pm 1)),$$

---

[31] U. Weiss, *Quantum Dissipative Systems* (World Scientific Publishing, 1993).

and

$$E_c(n, n') \equiv E_c(n^2 - n'^2) \qquad (12.134)$$

is the relative energy of different charge states.

(d) Assuming that only the charge states $n = -1, 0, 1$ energetically closest to the ground state $n = 0$ are significantly occupied, solve the master equation (12.133) and show that, at large times, the system relaxes to the ground state, $P(n, t \to \infty) = \delta_{n,0}$. Show that the relaxation rate governing the approach is given by $\Gamma \equiv g E_c / \pi = 1/RC$, the RC time of the circuit.

(e) Generalize to the case of non-zero bias voltage V. Will the ground state occupancy change?

Answer:

(a) Expressed in terms of the contour representation $\phi_c = (\phi_+ + \phi_-)/2$, $\phi_q = \phi_+ - \phi_-$, the charging contribution to the action $S_c = S|_{g=0}$ reads as

$$S_c[n, \phi] = \int dt \left(n_+ \partial_t \phi_+ - n_- \partial_t \phi_- - E_c(n_+^2 - n_-^2) \right),$$

where $n_\pm = n_c \pm \frac{n_q}{2}$. The quantization condition on n_c translates to $n_+(0) = n_-(0) \in \mathbb{Z}$. The relative sign change between the first two terms tells us that the operator $e^{i\phi_+}$ raises the charge on the upper contour by one, $n_+ \to n_+ + 1$, while $e^{i\phi_-}$ lowers the charge on the lower contour by one, $n_- \to n_- - 1$. To make this action explicit, we first transform the tunneling action (12.91) to contour fields,

$$S_{\text{tun}}[\phi] = \frac{g}{2} \int dt dt' \left(e^{-i\phi_+(t)}, e^{-i\phi_-(t)} \right) L(t - t') \begin{pmatrix} e^{i\phi_+(t')} \\ e^{i\phi_-(t')} \end{pmatrix},$$

where the matrix kernel $L = \{L_{\sigma\sigma'}\}$ is defined in Eq. (12.132). We may now expand $\exp(iS_{\text{tun}})$ in powers of the coupling constant to obtain the series

$$e^{iS_{\text{tun}}[\phi]} = \sum_{m=0}^{\infty} \frac{1}{m!} \left(\frac{ig}{2} \right)^m \sum_{\{\sigma_k\}} \int Dt \, e^{i \sum_{k=1}^{2m} (-)^k \phi_{\sigma_k}(t_k)} \prod_{k=1}^{m} L_{\sigma_{2k-1}, \sigma_{2k}}(t_{2k-1} - t_{2k}),$$

where $Dt = \prod_{k=1}^{2m} dt_k$. The expansion weights in- and out-tunneling events at times t_{2k-1} and t_{2k}, respectively, with elements of the kernel L (see figure.) We now pass to a charge representation by integrating this expression against the

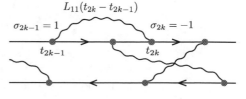

charging action S_c. This generates the constraint $\partial_t n_\sigma = \sigma \sum_{k=1}^{2m} (-)^k \delta(t - t_k) \delta_{\sigma_k, \sigma}$. This equation is solved by Eq. (12.131), and substitution of this result into the n-dependent part of the charging action yields the representation (12.130).

(b) The scaling $L(\omega) \sim \omega$ implied by Eq. (12.96) is equivalent to power law temporal decay, $L(t) \sim t^{-2}$. This means that a charge tunneling event carries the statistical weight $\sim \int dt\, e^{\pm iE_c t}t^{-2} \sim E_c$. The characteristic temporal range of this integral is $\delta t \sim E_c^{-1}$. With this, we estimate the statistical weight of a tunneling event occurring somewhere in a time-window of duration t_0 as $\sim gE_c t_0$. This means that the average number of tunneling events in t_0 is

$$\langle m \rangle \simeq \frac{\sum_m \frac{m}{m!}(gE_c t_0)^m}{\sum_m \frac{1}{m!}(gE_c t_0)^m} = gE_c t_0.$$

The temporal spacing between events follows as $t_0/\langle m \rangle \sim (gE_c)^{-1}$, and this relates to the duration of the event as $t_0/\delta t \sim g^{-1}$: at low tunneling, $g \ll 1$, the spacing between events exceeds their duration, and a sequential approximation becomes justified.

(c) For $E_c^{-1} \ll \Delta t \ll (gE_c)^{-1}$, the increment of P in the time-window $[t - \Delta t, t]$ is determined by zero-blip or one-blip processes:

$$P(n,t) \simeq P(n,-\Delta t) + \sum_{n'=n-1}^{n+1} C_{nn'} P(n',t-\Delta t),$$

where the coefficients $C_{nn'}$ are one-blip transition probabilities. An individual blip is characterized by its center time, $t_0 \in [t-\Delta t, t]$, its duration, s, and the sign structure (σ,σ'). Specifically, inspection of the figure shows that the connection between the signs and the post-blip increment in classical charge is given by $(+,+),(-,-)$: $n \to n$, $(+,-) : n \to n-1$, $(-,+) : n \to n+1$. Comparison with Eq. (12.130) shows that, e.g.,

$$C_{n,n-1} = \frac{ig\Delta t}{2} \int ds\, e^{iE_c(n-1,n)s} L_{-+}(s) = \frac{ig\Delta t}{2} L_{-+}(E_c(n-1,n))$$

$$= \frac{ig\Delta t}{2}(-\Sigma^K - \Sigma^+ + \Sigma^-)(E_c(n-1,n))$$

$$= \frac{g\Delta t}{\pi} E_c(n-1,n)\Theta(E_c(n-1,n)),$$

where $E_c(n,n')$ is the relative charging energy of Eq. (12.134), the prefactor Δt results from the integration over the center time, and the Θ-function results from the zero-temperature distribution function. In an analogous manner, we obtain

$$C_{n,n+1} = \frac{g\Delta t}{\pi} E_c(n+1,n)\Theta(E_c(n+1,n)),$$

$$C_{nn} = \frac{g\Delta t}{\pi}\left(-E_c(n,n+1)\Theta(E_c(n,n+1)) - E_c(n,n-1)\Theta(E_c(n,n-1))\right),$$

where, in computing C_{nn}, it is important to keep in mind that $\Sigma^\pm(t) \propto \Theta(\pm t)$ carry retarded and advanced causality. Substituting this result into the evolution equation above, dividing by Δt, and taking the continuum limit, we obtain the master equation (12.133).

(d) The restriction of the master equation to the sub-system $n = 0, \pm 1$ reads

$$\partial_t P(0,t) = \Gamma(P(1,t) + P(-1,t)),$$

$$\partial_t P(\pm 1, t) = -\Gamma P(\pm 1, t).$$

This is solved by $P(\pm 1, t) = e^{-\Gamma t} P(\pm 1, 0)$ and

$$P(0, t) = 1 - P(1, 0) - P(-1, 0) + \Gamma \int_0^t dt' \, e^{-\Gamma t'} (P(1, 0) + P(-1, 0)).$$

This solution describes the relaxation towards the state $P(n, t) \overset{t \to \infty}{\longrightarrow} \delta_{n,0}$ at the rate Γ, as stated in the problem.

(e) Non-zero voltages affect the theory through a redefined Keldysh self-energy Σ^K. Comparison with Eq. (12.96) shows that

$$\Sigma^K(V) - \Sigma^K(0) = \frac{i}{2\pi}(V - |\omega|)\Theta(V - |\omega|).$$

The transition rates thus change as follows:

$$W_{n,n\pm 1} \to W_{n,n\pm 1} + \frac{g}{4\pi}(V - |E_c(n \pm 1, n)|)\Theta(V \pm |E_c(n - 1, n)|).$$

This expression is easy to interpret: for voltages $|V| < |E_c(n\pm 1, n)|$ smaller than the charging energies, the transition rates of the unbiased problem remain unaltered. This means that the excess energy of external charge carriers $\sim V$ needs to exceed the charging energy in order to lift the exponential suppression of charged dot states. At large voltages, $|V| \gg |E_c(n \pm 1, n)|$, the tunneling rates cross over to values $\sim gV/4\pi$. For these rates, different charging states become equally populated at a rate $\sim gV$ set by the average current through the dot interfaces.

Appendix

A.1 Differential Geometry and Differential Forms

SYNOPSIS This appendix is a quick introduction to basic differential geometry and the language of differential forms. We aim to introduce computational tools useful in quantum field theory as succinctly as possible. While our presentation of the subject cannot substitute for a thorough introduction, it will perhaps motivate readers to dig deeper into the beautiful mathematics of differential geometry.

In chapter 1, we introduced fields as maps from a base *manifold* to a target manifold. Most of the time, physicists work with these maps in a language of "coordinates," where they are represented as smooth functions $\phi^i(\mathbf{x})$ of an argument vector $\mathbf{x} = \{x^i\}$. However, in many instances, such representations work only locally. Think of a phase field taking values in the group U(1) as an example. It can be locally described by a variable ϕ. However, at the boundaries $0, 2\pi$ this is no longer a function. The local variable ϕ jumps and hence is defined only on the open interval $(0, 2\pi)$, "locally" covering the field manifold up to a point. For the circle, this locality constraint is easy to work with, for instance via the introduction of the winding numbers frequently employed in earlier chapters. However, for more complex fields (taking values in group manifolds as $SU(N)$, for example) things get more involved, and an efficient framework for working with local coordinates as descriptors of global structures is required.

Besides these technical aspects, an overly strong emphasis of local views tends to obscure the physically important presence of global structures. (A phase tells us only indirectly, via its winding numbers, that a circle is under consideration.) In areas where global geometric structures are key – for example in topological quantum field theory, areas of gauge field theory, or relativity – it is often preferable to stay for as long as possible on a global level, and postpone the introduction of coordinates to the final stages when "concrete calculations of observables" are performed. Differential geometry and the language of differential forms provide us with a framework for such invariant representations. For concrete evidence and motivation, consider the numerous formulae in chapters 9, 10, or 8, which looked hostile in coordinate representation but took a friendly form when expressed in differential geometry language.

This appendix is an introduction to concepts of differential geometry in field theory. We emphasize that the presentation is intentionally terse. It should be sufficient to give a first overview, or serve as a reference manual. Either way, it cannot replace a – highly recommended – deeper study of the subject.[1]

A.1.1 Differentiable manifolds

differentiable manifold

A **differentiable manifold** is a set M that locally, but not necessarily globally, looks like an open subset of $\mathbb{R}^n$. With few exceptions, all fields discussed in this text take values in differentiable manifolds, hence the denotation field manifold. Similarly, the domains of definitions of fields, the base manifolds, are also locally identical to open subsets of $\mathbb{R}^d$.

local coordinate

In more concrete terms, local identification means that, for all $p \in M$, we have a subset $r \ni N \subset M$ and a **local coordinate** map $x : U \to N, x \mapsto r(x)$, where $U \subset \mathbb{R}^n$ is an open subset. Here, $x \in \mathbb{R}^n$ is the coordinate vector of the point $r \in M$. (In differential geometry, one generally does not represent vectors in a boldface notation. To keep the notation slim, the coordinate map $x : U \to N$,

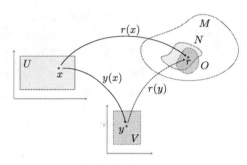

and the coordinates themselves, x, as in $r = r(x)$, are often denoted by the same symbol.) A point r may be included in the domain of two different coordinate representations, $r \in O$, where $y : V \to O, y \mapsto r(y)$. In this case, the same point r is described as $r = r(x) = r(y)$ in different ways, and the map $y : U \to O, x \mapsto y(x)$ describes the coordinate change. The defining feature of a differentiable manifold is that these **coordinate transformations** are smooth diffeomorphisms, i.e., infinitely often differentiable, and one-to-one.

coordinate transformation

chart

INFO In the parlance of differential geometry, coordinate maps are called **charts** and a minimal collection of charts large enough that each point on the manifold is covered by an **atlas**. The complete coverage of topologically nontrivial manifolds such as circles, spheres, or tori, requires multi-chart atlases.

A few more comments on the concept of manifolds:

▷   The definition above does not assume the embedding of the manifold in some larger space. For example, we often visualize the differentiable manifold "two-sphere" as embedded in three-dimensional space. However, in differential geometry, it is good practice to avoid this view. The two-sphere might just as well be

---

[1]   See, for example, M. Nakahara, *Geometry, Topology and Physics* (IOP Publishing, 2003), which has a focus on topology, is fairly rigorous, and comprehensive; M. Göckeler and T. Schücker, *Differential Geometry, Gauge Theories, and Gravity* (Cambridge University Press, 2011), which provides a miraculously concise, readable, and modern particle physics-oriented perspective; and A. Altland and J. von Delft *Mathematics for Physicists* (Cambridge University Press, 2019), which has an emphasis on pedagogy, and is far more basic than the texts above.

embedded in the group SU(2), or appear as a stand-alone object, without any embedding.

▷ Computations with differential manifolds are exclusively performed in their coordinate domains, U. We never differentiate or integrate in M (as this would require the embedding of M into a vector space). This is more than a formality. Getting used to the mindset "geometry in M, calculus in U" avoids confusion.

▷ The coordinate domains $U, V, \ldots$ are differentiable manifolds in their own regard. (Being subsets of $\mathbb{R}^n$, they are trivially identical to subsets of $\mathbb{R}^n$.) This view is frequently useful in practice. Unlike the generic case, operations of calculus *can* be performed in the coordinate manifolds.

covariant
notation

▷ In **covariant notation**, coordinate indices as in x^i are generally written as superscripts, and called **contravariant indices**. In this way, they are distinguished from symbols (see below) v_i with subscripts, called **covariant indices**.

▷ In practical computations, one often identifies functions defined on $f : M \to \mathbb{R}, r \mapsto f(r)$, and their local coordinate representations, $f : U \to \mathbb{R}, x \mapsto f(x) = f(r(x))$, where the "$f$" in $f(x)$ is a function of coordinates, and in $f(r(x))$ is a function on M. However, occasionally one must be careful not to take this identification too far. For example, a circle has a coordinate angle function ϕ, which is defined in the coordinate domain $(0, 2\pi)$, but not on the full circle manifold.

EXAMPLE   The **two-sphere**, S^2, is a differentiable manifold, and the standard spherical coordinates $(\theta, \phi)^T$ are a coordinate system on it. In this case, $U = (0, \pi) \times (0, 2\pi)$, and the map $r(\theta, \varphi)$ covers almost all, but importantly not all, of the sphere. For an alternative coordinate system, consider the **stereographic coordinates**, de-

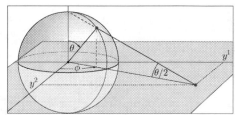

fined by the projection of points on the sphere onto a plane, as indicated in the figure. These coordinates also do not cover the entire sphere. (What is the covered subset $N \subset S^2$ in this case?) They are defined by the coordinate transformation (exercise)

$$\begin{pmatrix} y^1(\theta, \phi) \\ y^2(\theta, \phi) \end{pmatrix} = \cot(\theta/2) \begin{pmatrix} \cos\phi \\ \sin\phi \end{pmatrix}. \tag{A.1}$$

Tangent space

Intuitively, differentiable manifolds are smooth objects, which look flat from a close-up perspective. For example, a sphere looks locally like a two-dimensional plane. However, a problem with this picture is that the tangent plane lies "outside" the sphere and therefore requires embedding into a larger space. Mathematics offers an alternative description of tangency that, after a bit of getting used to it, is intuitive and does not require an embedding space.

This alternative approach starts from the observation that all local characterizations of manifolds ask questions about the variation of functions, f, defined in a neighborhood, N, of points $p \in M$. For example, how do the angles defining a local coordinate system on a sphere change under small departures from p? Or how does a potential $V(p)$ change?

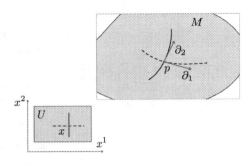

By virtue of a local coordinate embedding, we may consider f as a function $f : U \to \mathbb{R}, x \mapsto f(x)$, of local coordinates. Infinitesimal variations are now probed by directional derivatives of f at x. This line of thought suggests *defining* the **tangent space**, $T_p M$, to M at p as the space of linear operators ∂_v acting on functions by directional derivatives. For a given coordinate system, this space has a natural basis: the set of operators $\partial_{x^i} \equiv \partial_i$, acting by partial differentiation, $\partial_i f(x)$ in the coordinate direction. Defined in this way, the tangent vector ∂_i probes how f varies in the direction of the coordinate lines on M defined by the variation of the coordinate x^i with all other coordinates fixed (see the figure). Generic tangent vectors may then be defined by the linear combination

$$\partial_v = \sum_i v^i \partial_i, \qquad (A.2)$$

with coefficients v^i. Defined in this way, the vector acts as a common directional derivative: $d_t f(x + vt) = v^i \partial_i f = \partial_v f$. Also note that, for a given tangent vector, ∂_v, its expansion coefficients in the basis $\{\partial_{x^i}\}$ are obtained by the action of ∂_v on the coordinate functions themselves: $(\partial_v)^i \equiv \partial_v x^i = v^j \partial_j x^i = v^i$, where we have used $\partial_i x^j = \delta^j{}_i$.

While the definition above has been formulated in a specific coordinate system, it is easy to switch to a different one. Consider the basis tangent vectors ∂_{y^j} defined by a competing coordinate system. By definition, the components of ∂_v in the y-representation are obtained as $\frac{\partial y^j}{\partial v} = v^i \frac{\partial y^j}{\partial x^i}$ such that

$$\partial_v = v^i \frac{\partial y^j}{\partial x^i} \partial_{y^j}. \qquad (A.3)$$

This formula can be read as a basis change for tangent vectors, or as the chain rule for partial derivatives. Either way, it is compatible with the action $\partial_v f(x) = v^i \partial_{x^i} f(x) = v^i \frac{\partial y^j}{\partial x^i} \partial_{y^j} f(y)$ on functions.

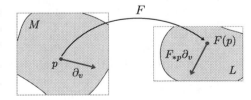

We frequently consider maps $F : M \to L, p \mapsto F(p)$ between manifolds. Any such map defines a **pushforward map** between tangent spaces, $F_{*p} : T_p M \to T_{F(p)} L$. Intuitively, this is the "infinitesimal version" of the map, describing how small variations (vectors) on the argument manifold map onto the image. In our

(margin notes) tangent space   pushforward

current way of describing tangent spaces, the action of the pushed vector $F_{*p}\partial_v$ on functions g defined in the vicinity of $F(p)$ is defined as $(F_{*p}\partial_v)g = \partial_v(g \circ F)$. In words: we probe how the composite map $g \circ F$ responds to variations in v. Specifically, for $g = y^j$, a coordinate function and $F^j = y^j \circ F$ the coordinates of the function values in L, we obtain the components of the pushforward vector as $(F_{*p}\partial_v)^j \equiv (F_{*p}\partial_v)y^j = \partial_v F^j = v^i \frac{\partial F^j}{\partial x^i}$, or

$$\boxed{\partial_v = v^i \partial_{x^i} \quad \Rightarrow \quad F_*\partial_v = v^i \frac{\partial F^j}{\partial x^i} \partial_{y^j}} \tag{A.4}$$

This is the differential geometric reading of what in coordinate language is a chain rule.

EXAMPLE Consider the local basis vectors $(\partial_\theta, \partial_\phi)$ probing how functions on the sphere change in the polar and azimuthal directions, respectively. We can expand them in the basis of the stereographic system as

$$\partial_\theta = \frac{\partial y^1}{\partial \theta} \partial_{y^1} + \frac{\partial y^2}{\partial \theta} \partial_{y^2} = -\frac{1}{2\sin^2(\theta/2)}(\cos\phi\,\partial_{y^1} + \sin\phi\,\partial_{y^2}).$$

Technically, the coordinate change $y : U \to V, x \mapsto y(x)$, is a map between the coordinate manifolds U and V. With the identifications $y = F$, and $\partial_v = \partial_{x^i}$, the equation above shows that the coordinate-transformed vector is the pushforward of ∂_{x^i} under $y(x)$.

Finally, it follows directly from the definition that repeated pushforward obeys the composition rule $(G \circ F)_* = G_* \circ F_*$.

A.1.2 Differential forms

In this section, we introduce the concept of differential forms. We begin with one-forms, which are the differential geometric variant of the dual vectors perhaps remembered from linear algebra. Next we will introduce forms of higher degree and learn how to build them from one-forms by a product operation known as the wedge product, or by (exterior) differentiation. Finally, we will discuss how differential forms are key to the definition of integrals over manifolds.

Differential one-forms

dual space   Recall that the **dual space**, V^*, of a vector space, V, is the space of all linear functions, w, of vectors $w(av + a'v') = aw(v) + a'w(v')$. Elements of the dual vector space are called dual vectors or one-forms. For a basis $\{e_i\}$ of V, the dual basis $\{e^j\}$ of V^* is defined through $e^j(e_i) = \delta^j{}_i$. Dual vectors are expanded as $w = w_i e^i$, with covariant coefficients w_i. The latter are obtained as $w_i = w(e_i)$ by the action of w on the basis vectors.

differential one-form   A **differential one-form** is a field of one-forms, ϕ_p, where $\phi_p \in (T_pM)^*$ is a one-form on the tangent vector space T_pM, and the dependence of ϕ_p on p is smooth.

cotangent
space

differential
of function

The dual space $(T_pM)^* \equiv T_p^*M$ of the tangent space T_pM is called the **cotangent space** to M at p, and $T^*M \equiv \bigcup_{p \in M} T_p^*M$ is the **cotangent bundle** of M.

In practice, differential one-forms are mostly realized as **differentials of functions**. A function $f : M \to \mathbb{R}$ defines a one-form df whose action on vectors is given by $df(\partial_v) \equiv \partial_v(f)$ (why is this a one-form?). From the expansion $\partial_v = v^i \partial_i$, we see that $df(\partial_v) = v^i \partial_i f$, showing that $df(\partial_v)$ is just the derivative of f in the direction of the vector defined by the components v^i. Specifically, for coordinate functions x^i, we have $dx^i(\partial_j) = \delta^i{}_j$, from which we learn two things: $\{dx^i\}$ is the dual basis of T_p^*M corresponding to the basis $\{\partial_j\}$ of T_pM, and arbitrary forms afford the expansion

$$\omega = \omega_i dx^i,$$

where the covariant coefficients are obtained as $\omega_i = \omega(\partial_i)$. Specifically, the form df is expanded as

$$df = \partial_i f \, dx^i, \tag{A.5}$$

which is consistent with $df(\partial_i) = \partial_i f$. Equation (A.5) shows that the differential of a function familiar from calculus affords an interpretation as a differential one-form. Indeed, the differential wants to "eat" vectors to produce the corresponding directional derivatives as numbers, and this assignment is linear, i.e., it is a differential form in the sense of the above definition.

The expansion of a form in a different basis dy^i is obtained from $\omega(\partial_{y^j}) = \omega_i dx^i(\partial_{y^j}) = \omega_i \partial_{y^j} x^i$ as

$$\omega = \omega_i \frac{\partial x^i}{\partial y^j} dy^j, \tag{A.6}$$

which is the form analog of the vector identity (A.3).

EXAMPLE The forms $d\theta$ and $d\phi$ on the sphere are locally defined as differentials of the coordinate functions (θ, ϕ). The differential form $\omega = \cos\theta \, d\phi$ is not realized as the differential of a function. What is the representation of the basis forms dz^1, dz^2 of the stereographic representation in the basis $(d\theta, d\phi)$?

Much as vectors get pushed forward under maps $F : M \to L$, differential forms, ω, defined on $T_{F(p)}^*L$ get "pulled back" to forms in the pre-image tangent space T_pM.

pullback

The **pullback** map $F_p^* : T_{F(p)}^*L \to T_p^*M$, $\omega \mapsto F^*\omega$ is defined by $(F^*\omega)(\partial_v) = \omega(F_*\partial_v)$, i.e., the action of the argument form in $T_{F(p)}^*$ on the pushforward $F_*\partial_v$. Pullback and its partner operation pushforward simply lift the apparatus of differential geometry from one domain to another, $\omega(F_*\partial_v) = (F^*\omega)(\partial_v)$, i.e., a form evaluated on a pushed vector equals the pulled form evaluated on the original vector (see figure).

For a form represented on L as $\omega = \omega_j dy^j$, the coefficients of $F^*\omega$ are obtained as $(F^*\omega)_i = (F^*\omega)(\partial_i) = \omega(F_*\partial_i) = \omega((\partial_i F^j)\partial_j) = \omega_j \partial_i F^j$, or

$$\boxed{\omega = \omega_j dy^j \quad \Rightarrow \quad F^*\omega = (\omega_j \circ F)\frac{\partial F^j}{\partial x^i} dx^i} \tag{A.7}$$

This formula, which is the partner relation of Eq. (A.4), shows that pullback is
the differential form representation of the
"chain rule for differentials." For example, for a curve $x = x(t)$, the transfor-
mation $dx = \frac{\partial x}{\partial t} dt$ is interpreted as the
pullback of the coordinate form dx under
the map $t \mapsto x(t)$ to the differential form
$\frac{\partial x}{\partial t} dt$ in the time domain. The advantage
of the differential-form representation is

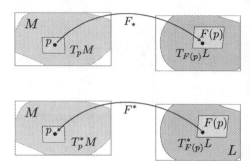

its simplicity. For example, $F^* df$ is more transparent than $(\partial_{y^j} f|_{F(x)}) \partial_{x^i} F^j$, etc.

Finally, note that for repeated pullback operations we have the composition rule
$(G \circ F)^* = F^* \circ G^*$: a form is pulled back by first pulling it under the final map,
then under the initial one (a result that follows directly from the definition).

INFO Many objects sold as vectors in standard physics teaching actually are differential
forms. One-forms and vectors are easily confused because they have equally many components and transform similarly (but not identically!) under coordinate transformations.
Examples of forms in vector disguise include **mechanical forces, vector potentials,
electric fields**, and many others besides. To understand why, say, mechanical force, F,
is a form and not a vector, recall that forces are *measured* by recording the work required
for the displacement of test objects, $W = F_i v^i$, where v^i are the coefficients of a small
displacement vector, W is work, and F_i are the components of force. The latter enters as a
linear map (a one-form) acting on a vector to produce a number. This construction demonstrates that the conceptualization of force as a form is more physical (and at the same
time no more abstract) than the standard vectorial formulation. The situation with the
other examples mentioned above is similar. Unfortunately, the "all-are-vectors" paradigm
is so deeply rooted into the social system of physics that it appears to be impossible to
eradicate.

Forms of higher degree and wedge products

Above we introduced differential one-forms as objects measuring the effects of linear
displacements on manifolds. In a similar manner, differential forms of higher degree
probe surfaces, volumes, etc.

differential
r-form

Technically, a **differential r-form** is a multilinear and antisymmetric map ω_p :
$\otimes^r T_p M \to \mathbb{R}, (\partial_{v^1}, \ldots, \partial_{v^r}) \to \omega_p(\partial_{v^1}, \ldots, \partial_{v^r})$, assigning to r tangent vectors
a number, where the dependence on p is smooth. Antisymmetry means that ω_p
changes sign under the permutation of any of its argument vectors. Owing to linearity and antisymmetry, a differential r-form is fully determined by its action on
ordered r-tuples of basis vectors $\omega(\partial_{i_1}, \ldots, \partial_{i_r}) \equiv \omega_{i_1, \ldots, i_r}$, where $i_1 < \cdots < i_r$.
Antisymmetry further implies that $\omega_{\ldots, i, \ldots, j, \ldots} = -\omega_{\ldots, j, \ldots, i, \ldots}$, resulting in $\binom{n}{r}$ independent coefficients. It follows that the space of r-forms at $p \in M$, often denoted as
$\Lambda^r(T_p M)$, is $\binom{n}{r}$-dimensional and that forms of degree higher than n do not exist.
Forms of degree n are called **top-dimensional**, or just top-forms. At the other end

of the spectrum, we have the space of 0-forms, $\Lambda^0 M$, defined as the set of smooth functions on M.

**wedge
product**

Forms of degree $r > 1$ may be built hierarchically as **wedge products** of forms of lesser degree. The wedge product of two one-forms ϕ and ψ is a two-form defined as

$$(\phi \wedge \psi)(\partial_v, \partial_w) = \phi(\partial_v)\psi(\partial_w) - \phi(\partial_w)\psi(\partial_v).$$

The generalization of this operation to forms of higher degree $\omega \in \Lambda^r M$ and $\kappa \in \Lambda^s M$ is a form of degree $r + s$ defined as

$$(\omega \wedge \kappa)(\partial_{v_1}, \ldots, \partial_{v_{r+s}})$$
$$= \frac{1}{r!s!} \sum_{P \in S_{r+s}} \mathrm{sgn}(P) \omega(\partial_{v_{P1}}, \ldots, \partial_{v_{Pr}}) \kappa(\partial_{v_{P(r+1)}}, \ldots, \partial_{v_{P(r+s)}}),$$

where the sum over the permutation group implements antisymmetry. Note that the anti-symmetrization implies the commutation relation

$$\boxed{\omega \wedge \kappa = \kappa \wedge \omega \, (-)^{rs}} \tag{A.8}$$

e.g., $d\omega \wedge d\eta = -d\eta \wedge d\omega$ for the product of two one-forms, or $dA \wedge dF = dF \wedge dA$ for that of a one-form and two-form, etc.

Specifically, the wedge products of coordinate basis one-forms $dy_{i_1} \wedge \cdots \wedge dy_{i_n}$ define $\binom{d}{n}$ independent n-forms, which serve as a natural basis of $\Lambda^n M$. In this basis, generic forms are expanded as

$$\boxed{\omega = \frac{1}{n!} \omega_{j_1 \ldots j_n} dy^{j_1} \wedge \cdots \wedge dy^{j_n}} \tag{A.9}$$

area form

EXERCISE The **area form of the two-sphere** is defined as $\omega = \sin\theta \, d\theta \wedge d\phi$. Expand this form in the basis of stereographic coordinates.

The **pullback** of higher degree forms is defined by an obvious generalization of the one-form operation $F^* \omega(\partial_{v_1} \ldots \partial_{v_n}) = \omega(F_* v_1, \ldots, F_* v_n)$. Verify that application of the pullback to forms expanded as Eq. (A.9) is conveniently obtained as the product of the individually pulled one-forms:

$$F^* \omega = \frac{1}{n!} (\omega_{j_1 \ldots j_n} \circ F) \frac{\partial F^{j_1}}{\partial x^{i_1}} \cdots \frac{\partial F^{j_n}}{\partial x^{i_n}} dx^{i_1} \wedge \cdots \wedge dx^{i_n}. \tag{A.10}$$

INFO The space of forms on three-dimensional manifolds has dimensionality $\binom{3}{2} = 3$, the same as the space of vectors, and one-forms. It should therefore not be surprising that numerous three-component objects treated as vectors in physics teaching are two-forms. For example, the magnetic induction $\mathbf{B}$ or current density, $\mathbf{j}$, belong to this category. These objects describe (magnetic or current) fluxes through surface elements. One should think of them as a locally defined function of two vectors producing the flux flowing through the area spanned by the vectors as an output. For example, in differential-form language, a magnetic field of strength B through the 12-plane is described by the form

$B = B_0 dx^1 \wedge dx^2$, and the flux through the area spanned by two tangent vectors $a^1 \partial_1$ and $a^2 \partial_2$ is given by $B(a^1 \partial_1, a^2 \partial_2) = B_0 a^1 a^2$. This definition reflects how fluxes are actually measured. However, in physics, we describe the setting by a vector $\mathbf{B} = B_0 \mathbf{e}_3$ standing perpendicular to the direction of the flux, via a construction that requires a scalar product. This vector construction is less physical: it requires the excess baggage of a scalar product, its connection to the protocols by which fluxes are measured is indirect, and it leads to the infamous transformation of magnetic field vectors under, say, spatial inversion. (The magnetic field is a "pseudo" or "axial" vector in that it remains invariant under space-inversion.)

Exterior derivative

Besides the wedge product, there is a second operation increasing the degree of differential forms: the exterior derivative. To start with, consider a function ϕ on M. This is a zero-form, and we have seen above that it defines a one-form as $d\phi$. Building on this definition, we introduce the **exterior derivative** as a linear operation, $d : \Lambda^n M \to \Lambda^{n+1}(M)$, increasing the degree of general n-forms by one. It is defined by

exterior derivative

$$
\begin{aligned}
d(\omega_{i_1 \ldots i_n} dx^{i_1} \wedge \cdots \wedge dx^{i_n}) &\equiv (d\omega_{i_1 \ldots i_n}) \wedge dx^{i_1} \wedge \cdots \wedge dx^{i_n} \\
&= (\partial_j \omega_{i_1 \ldots i_n}) dx^j \wedge dx^{i_1} \wedge \cdots \wedge dx^{i_n}.
\end{aligned}
\tag{A.11}
$$

EXERCISE Prove that the definition above is independent of the choice of coordinates. For example, $d(\omega_i dx^i)$ equals $d(\tilde{\omega}_j dy^j)$ with $\tilde{\omega}_j = \omega_i \frac{\partial x^i}{\partial y^j}$, where the second exterior derivative is carried out in y-coordinates. (Hint: Keep in mind that contractions of symmetric tensors such as $\partial_{ij} F$ with antisymmetric tensors such as $dx^i \wedge dx^j$ vanish.)

EXAMPLE The area two-form on the sphere in polar coordinates $\omega = \sin\theta d\theta \wedge d\phi$ is the exterior derivative of $\kappa \equiv -\cos\theta \, d\phi$. For the potential one-form of electrodynamics, $A = A_\mu dx^\mu$, the exterior derivative $F = dA = \partial_\nu A_\mu dx^\nu \wedge dx^\mu = \frac{1}{2}(\partial_\nu A_\mu - \partial_\mu A_\nu) dx^\nu \wedge dx^\mu$ yields the electromagnetic field strength tensor with components $F_{\mu\nu} = \partial_\nu A_\mu - \partial_\mu A_\nu$.

Finally, we mention without proof that **pullback** and **exterior differentiation commute**,

$$
\boxed{dF^*\omega = F^*d\omega}
\tag{A.12}
$$

In applications, it may be either more convenient to first differentiate and then pull back, or to proceed the other way around. The rule above tells us that we need not worry about the order. As an instructive exercise, prove this relation using Eq. (A.7) for one-forms, or for forms of higher degree using Eqs. (A.10) and (A.11).

Integration

In elementary vector calculus, we distinguish between line integrals, various types of surface integrals, volume integrals, etc. Differential geometry subsumes all these species under one umbrella concept, the integral of forms of top-degree, d, over d-dimensional manifolds.[2]

For a chart domain $N \subset M$, we define the integral of a top-form $\omega = \omega_{i_1 \cdots i_n} dx^{i_1} \wedge \ldots dx^{i_n}$ as

$$\int_N \omega = \int_N \omega_{i_1 \cdots i_n} dx^{i_1} \wedge \cdots \wedge dx^{i_n} \equiv \int_U \omega_{i_1 \cdots i_n} dx^{i_1} \cdots dx^{i_n} \qquad (A.13)$$

where the final form is the ordinary multi-dimensional Riemann integral (no wedge products) of the coefficient functions over the coordinate domain. In a slightly more general representation, for $r(x)$ a coordinate representation, and ω a top-form, the integral is defined as

$$\int_N \omega = \int_U r^* \omega, \qquad (A.14)$$

i.e., the integral of the pullback of the form to the coordinate domain. The last equality in Eq. (A.13) defines this integral in $U \subset \mathbb{R}^n$ as a conventional integral. For example, the **integral of the area form on the sphere** is defined as

Sir George Gabriel Stokes
1819–1903
As Lucasian Professor of Mathematics at Cambridge, Stokes established the science of hydrodynamics with his law of viscosity (1851), describing the velocity of a small sphere through a viscous fluid. Furthermore, he investigated the wave theory of light, named and explained the phenomenon of fluorescence, and theorized an explanation of the Fraunhofer lines in the solar spectrum.

$$\int_{S^2} \sin\theta \, d\theta \wedge d\phi = \int_U \sin\theta \, d\theta \wedge d\phi \equiv \int_0^\pi \sin\theta \, d\theta \int_0^{2\pi} d\phi = 4\pi. \qquad (A.15)$$

These definitions entail a particularly transparent way of **understanding integral transforms**: for $F : N \to L$ a diffeomorphism, we have

$$\int_L \omega = \int_N F^* \omega \qquad (A.16)$$

i.e., the integral of a form over L equals that of the pullback $F^* \omega$ of the form over N. To understand this equality, note that $F \circ r : x \mapsto F(r(x))$ defines a coordinate coverage of L. Thus, using that $(F \circ r)^* = r^* \circ F^*$ (why?), we have $\int_L \omega = \int_U (F \circ r)^* \omega = \int_U r^* (F^* \omega) = \int_N F^* \omega$. Notice how these identities are proven in full generality (for all dimensions, etc.) without the need for explicit variable transformations or the computation of Jacobians.

EXAMPLE Let $r : I \to \mathbb{R}^3, t \mapsto r(t)$, describe a curve. In $\mathbb{R}^3$, consider a differential one-form $A = A_i dx^i$. Denoting the image of the map r (i.e., the geometric curve) by γ, we have

---

[2] Actually, there exists a slightly more general concept, that of the integration of *densities* over manifolds. Closely related to top-forms, densities include the book-keeping over the orientation of coordinate systems (left-handed versus right-handed). The latter becomes relevant on non-orientable manifolds that we will not discuss here.

$$\int_{\gamma} A = \int_{I} r^* A = \int_{I} A_i(r(t)) r^* dx^i = \int_{I} A_i(r(t)) \frac{dr^i}{dt} dt,$$

where in the final step we have used the single-component version of the pullback formula Eq. (A.7). On the final right-hand side we recognize the familiar line integral, except that the integrated object is defined by the covariant components of a form, and not those of a contravariant vector: the natural objects appearing in physically motivated line integrals (electric fields, vector potentials, forces, etc.) are all one-forms.

EXERCISE Consider the Cartesian coordinates $x^i(y) = x^i(r(y))$, where $r(y) \in S_2$ is an arbitrary coordinate representation of the sphere, e.g., $x^1(\theta, \phi) = \sin\theta\cos\phi$ for spherical coordinates. Given a two-form $B \equiv \frac{1}{2} B_{ij} dx^i \wedge dx^j$ and assuming that the coordinate map $r : U \to S_2$ covers the sphere up to domains of measure zero (which do not matter in an integral), show that

$$\int_{S^2} B = \frac{1}{2} \int_{U} B_{ij} \left(\frac{\partial x^i}{\partial y^1} \frac{\partial x^j}{\partial y^2} - \frac{\partial x^i}{\partial y^2} \frac{\partial x^j}{\partial y^1} \right) dy^1 dy^2.$$

How does one have to relate the three independent components B_{ij} to a vector to obtain the familiar parameter representation of spherical surface integrals? For $y = (\theta, \phi)^T$, compute the term in parentheses and relate the integral to familiar representations of surface integrals.

Stokes'
theorem

A powerful tool in the integration of forms is the **generalized Stokes' theorem**:

$$\boxed{\int_{M} d\kappa = \int_{\partial M} \kappa} \tag{A.17}$$

– one of the most beautiful formulae of mathematics. Here, κ is an $(n-1)$-form, and ∂M is the boundary of the integration domain M. Stokes' theorem generalizes various identities of (vector) calculus equating the integral of the derivative of a quantity to a boundary integral of that quantity.

EXAMPLE Consider the one-form $\kappa = (1 - \cos\theta) d\phi$ defined on the northern hemisphere S^+ of the two-sphere. (Why is κ not extensible to the full sphere?) With $d\kappa \equiv \omega = \sin\theta d\theta \wedge d\phi$ we have $\int_{S^+} d\kappa = 2\pi$. On the other hand, ∂S^+ is the equator where $\cos\theta = 0$ and $\kappa = d\phi$. Integration gives $\int_{\partial S^+} d\phi = 2\pi$, in agreement with the area integral.

A.1.3 Metric

The structures introduced thus far make no reference to the actual "geometry" of a manifold. Geometric structures such as length, angles, areas, volumes, etc. are described by a metric. A **metric** is a scalar product on a tangent space: a bilinear and symmetric form g_p assigning to two vectors $\partial_v, \partial_w \in T_p M$ the number $g_p(\partial_v, \partial_w)$, where, as always, smooth dependence on the base point, p, is required.

metric

positive
definite
Riemannian
manifold

Writing $g_p = g$ for brevity, symmetry means that $g(\partial_v, \partial_w) = g(\partial_w, \partial_v)$, and non-degeneracy means that if $g(\partial_v, \partial_w) = 0$ for all ∂_w, then this implies that $\partial_v = 0$. A metric is **positive definite** if $g(v, v) > 0$ for all non-vanishing v. A manifold equipped with a positive metric is called a **Riemannian manifold** and one with a positive indefinite metric is called pseudo-Riemannian.

Metric tensor

metric
tensor

Given a (coordinate) basis, $\{\partial_i\}$, the metric is represented by the **metric tensor**,

$$g_{ij} = g(\partial_i, \partial_j), \tag{A.18}$$

and general scalar products are represented as $g(\partial_v, \partial_w) = v^i g_{ij} w^j$. The symmetric tensor of second degree, g, is sometimes represented as $g = g_{ij} dx^i \otimes dx^j$, where the tensor product acts on vectors as $dx^i \otimes dx^j(\partial_v, \partial_w) = v^i w^j$.

EXAMPLE The natural **metric on the sphere** is given by $g(\partial_\theta, \partial_\theta) = 1$, $g_{\phi\phi} = \sin^2 \theta$, $g_{\theta,\phi} = g_{\phi,\theta} = 0$. The **Minkovski metric** $g = \eta$ is the canonical metric of flat four-dimensional space–time, $\eta_{00} = -1$, $\eta_{ii} = 1$, for $i = 1, 2, 3$ and $\eta_{\mu\nu} = 0$, $\mu \neq \nu$. Here, ∂_0 is the tangent vector in the time-like direction of $x^0 = ct$. In the literature, the symbol η is often used for metrics in orthonormal systems $|\eta(\partial_i, \partial_j)| = \delta_{ij}$. Depending on the field, η or $-\eta$ is called the Minkowski metric; the global sign change is physically and mathematically of no significance.

A metric provides us with a **canonical map between vectors and one-forms**. Given a vector, ∂_v, we define a one-form $J\partial_v$ by the condition that, for all ∂_w, $g(\partial_v, \partial_w) = (J\partial_v)(\partial_w)$, i.e., the application of the newly-defined form $J\partial_v$ is equal to the scalar product $g(\partial_v, \partial_w)$. Notice that this is a canonical (basis invariant) passage between TM and $(TM)^* = \Lambda^1 M$.[3] With the expansion $J\partial_v = (J\partial_v)_i dx^i$, the equalities $g(v, w) = v^i g_{ij} w^j$ and $(J\partial_v)(\partial_w) = (J\partial_v)_i w^i$ lead to the identification $(J\partial_v)_i = g_{ij} v^j$.

The inverse passage $\Lambda^1 M \to TM$ between forms and vectors is defined in an analogous manner. To each form ω, we assign a vector $J^{-1}\omega$ through the condition $\omega(\partial_w) = g(J^{-1}\omega, \partial_w)$. With $\omega = \omega_i dx^i$, and $w = \partial_j$, this defines a vector with components $\omega^i \equiv (J^{-1}\omega)^i = g^{ij} \omega_j$, where $g^{ij} g_{jk} = \delta^i{}_k$ defines the **inverse metric tensor**. For example, the **gradient of a function**, $\nabla \phi = J d\phi$, is the vector conjugate to its differential one-form. It has components $(\nabla \phi)^i = g^{ij} \partial_j \phi$.

Summarizing:

> The one-form $J\partial_v$ canonically assigned to ∂_v has components $v_i \equiv g_{ij} v^j$, and the vector $J^{-1}\omega$ assigned to ω has components $\omega^i \equiv g^{ij} \omega_i$.

[3] The previously discussed definition of a dual basis is specific to the choice of a basis in TM and is therefore not canonical.

In physics, the **raising and lowering of indices** as in $\omega^i \equiv g^{ij}\omega_j$ is often treated as a purely formal, or notational, operation. However, the present construction shows that it is conceptual. It implements the passage between vectors and differential forms via a scalar product.

EXAMPLE In physics, **mechanical work** is defined as the scalar product $\mathbf{F} \cdot \mathbf{v} \equiv g(F, v) = F^i g_{ij} v^j$ between a vector and a spatial increment. Previously, we have seen that a more natural definition understands work as $F_i v^i$, i.e., the application of a differential form with covariant components F_i to the vector. The discussion above shows that the physics definition refers to the vector $F^i = g^{ij}F_j$ canonically assigned to that form. In this way, it requires the excess baggage of a scalar product, which does not feature in the natural definition.

INFO The above passage between co- and contravariant objects generalizes to tensors of higher degree. For example, the components $B_{\mu\nu}$ of a differential two-form $B = \frac{1}{2}B_{\mu\nu}dx^\mu \wedge dx^\nu$ define an alternating bilinear form acting on vectors as $B_{\mu\nu}v^\mu w^\mu$. One may raise one of its indices to define the *matrix* $B^\rho{}_\nu = g^{\rho\mu}B_{\mu\nu}$. This matrix acts on a single vector to produce another vector, $B^\rho{}_\nu v^\nu$.

Canonical volume form

volume
form

On a general manifold, the objects to integrate are n-forms, as in Eq. (A.13). For Riemannian manifolds, there exists a special n-form, the **canonical volume form**, ω, whose integral $\int_M \omega = \mathrm{Vol}(M)$ defines the generalized[4] volume of the manifold. It is defined as

$$\boxed{\omega = \sqrt{|g|}dx^1 \wedge \cdots \wedge dx^n, \qquad g = \det\{g_{ij}\}} \tag{A.19}$$

The connection with our intuitive understanding of volume is best seen by inspection of specific cases: for example, with $n = 2$, the application of a volume form to two basis vectors yields $\omega(\partial_1, \partial_2) = \sqrt{|g|}$, where the determinant of the two-dimensional metric tensor $g_{11}g_{22} - g_{12}^2$ equals the parallelogram area spanned by the basis vectors (think why). Integration then yields the geometric area of M as the sum over infinitely many of these area elements.

Although the definition above is formulated in a basis, it is canonical. To see why, change variables to a different basis system dy^i. The product of differentials changes as

$$dy^1 \wedge \cdots \wedge dy^n \rightarrow \frac{\partial y^1}{\partial x^{i_1}} \cdots \frac{\partial y^n}{\partial x^{i_n}}dx^{i_1} \wedge \cdots \wedge dx^{i_n}$$

$$= \frac{\partial y^1}{\partial x^{i_1}} \cdots \frac{\partial y^n}{\partial x^{i_n}}\varepsilon_{i_1,\cdots,i_n}dx^1 \wedge \cdots \wedge dx^n = Jdx^1 \wedge \cdots \wedge dx^n,$$

with Jacobian $J = \det(\frac{\partial y}{\partial x})$. At the same time, the metric tensor in y-coordinates is given by

---

[4] That is, the length of a curve in $n = 1$, the geometric area in $n = 2$, volume in $n = 3$, etc.

$$g^y_{ij} \equiv g(\partial_{y^i}, \partial_{y^j}) = \frac{\partial x^{i'}}{\partial y^i}\frac{\partial x^{j'}}{\partial y^j}g(\partial_{x^{i'}}, \partial_{x^{j'}}) = \frac{\partial x^{i'}}{\partial y^i}\frac{\partial x^{j'}}{\partial y^j}g^x_{i'j'},$$

with determinant $g^y = J^{-2}g^x$. Substituting these expressions into Eq. (A.19), we see that the Jacobians cancel out, proving the invariance of the volume form.

EXAMPLE For the sphere, we have $\det(g) = g_{\theta\theta}g_{\phi\phi} - g^2_{\theta\phi} = \sin^2\theta$, and hence $\sqrt{|g|} = \sin\theta$. This gives the canonical volume (or, better to say, area) form $\sin\theta\, d\theta \wedge d\phi$.

EXERCISE Compute the spherical coordinate basis of three-dimensional space, $M = \mathbb{R}^3$. Show that the standard Cartesian metric $g = \delta_{ij}dx^i \otimes dx^j$ assumes the form $g = dr \otimes dr + r^2 d\theta \otimes d\theta + r^2 \sin^2\theta d\phi \otimes d\phi$. From there, we obtain the familiar volume element, $\omega = r^2 \sin\theta dr \wedge d\theta \wedge d\phi$.

Hodge star

The Hodge star is an operation $* : \Lambda^r M \to \Lambda^{d-r}M, \omega \mapsto *\omega$ assigning $(d-r)$-forms to r-forms. In field theory, it is frequently used to engineer actions (objects obtained by integration) from the natural building blocks (forms) describing the theory. For example, in three-dimensional space with Cartesian coordinates, $dx^3 = *(dx^1 \wedge dx^2)$ is the $(1 = 3-2)$-form assigned to the two-form $dx^1 \wedge dx^2$ such that $dx^1 \wedge dx^2 \wedge *(dx^1 \wedge dx^2) = dx^1 \wedge dx^2 \wedge dx^3$ is a top-form, which may feature in an integral. However, this assignment is not canonical.

The Hodge star employs the metric to define a canonical map. We here define it in (an arbitrary system of) coordinates, and leave it to the reader as an instructive exercise to demonstrate the coordinate independence of the definition: the **Hodge**
Hodge star   **star** of an r-form, $\phi = \phi_{i_1,\cdots,i_r}dx^{i_1} \wedge \cdots \wedge dx^{i_r}$, is defined as

$$(*\phi)_{i_{r+1},\cdots,i_n} = \frac{\sqrt{|g|}}{r!(n-r)!}\phi^{i_1,\cdots,i_r}\epsilon_{i_1,\cdots,i_n}, \tag{A.20}$$

where $\phi^{i_1,\cdots,i_r} = g^{i_1 j_1}\cdots g^{i_r j_r}\phi_{j_1,\ldots,j_r}$ are the contravariant components of the form.

EXERCISE Consider three-dimensional space parameterized by spherical coordinates. Verify that the inverse of the metric determined in the previous exercise reads $g^{rr} = 1$, $g^{\theta\theta} = r^{-2}$, $g^{\phi\phi} = (r\sin\theta)^{-2}$. Use this result to show that

$$*dr = r^2 \sin\theta\, d\theta \wedge d\phi, \qquad *d\theta = \sin\theta\, d\phi \wedge dr, \qquad *d\phi = (\sin\theta)^{-1}\, dr \wedge d\theta.$$

EXAMPLE The Hodge star of the field strength tensor $F = \frac{1}{2}F_{\mu\nu}dx^\mu \wedge dx^\nu$ defines the **dual field strength tensor** $G \equiv -*F = -\frac{1}{4}\epsilon_{\mu\nu\rho\sigma}F^{\mu\nu}dx^\rho \wedge dx^\sigma$. In $(3+1)$-dimensions, this gives us two natural forms to integrate: $\int F \wedge F$ and $\int F \wedge *F$, respectively. The absence of a metric in the first one hints at a topological origin of this "action." The second is metric and non-topological. For a physical discussion, see section 10.2.1. Note that in $(2+1)$ dimensions, $\int F \wedge F$ is not defined. However, $\int F \wedge *F$ still is, and it defines the canonical action of $(2+1)$-dimensional electrodynamics. (As an instructive exercise, compute the Cartesian coordinate representation of this action.)

current density

EXAMPLE In differential-form language, current flow in d-dimensional systems is described by a **current density form** of degree $d-1$. With

$$j = \frac{1}{(d-1)!} j_{i_1,\cdots,i_{d-1}} dx^{i_1} \wedge \cdots \wedge dx^{i_{d-1}}$$

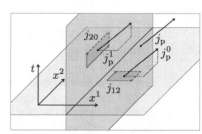

it is defined such that its application to $d-1$ vectors in a hyperplane yields the current through the generalized parallelepiped spanned by these vectors. For example, in $d=3$ and $\partial_i \equiv \Delta x^i \partial_{x^i}$, $i=1,2$, two vectors in the 12-plane, $j(\partial_1, \partial_2) = j_{12}\Delta x^1 \Delta x^2$ is equal to the coefficient j_{12} times the area spanned by the two vectors. In theories with d *space–time* dimensions, and coordinates $x = (x^0, \ldots, x^{d-1})$, the component $j_{1,\ldots,d-1}$ defines the charge density, and the application of j to the spatial vectors, $\partial_1, \ldots, \partial_d$, gives the charge $\Delta Q = j_{1,\cdots,d}\Delta x^1 \cdots \Delta x^d$ contained in the volume spanned by them. When applied to temporal and spatial vectors, $j(\partial^0, j^1, \cdots j^{d-1}) = j_{0,1,\cdots,d-1}\Delta t \Delta x^1 \cdots \Delta x^{d-1}$ equals the charge in time Δt flowing through a surface area $\Delta x^1 \cdots \Delta x^{d-1}$ in the $12 \cdots (d-1)$ hyperplane (see figure for the current flow through the 1-line in $2+1$ dimensions). Of course, the application of j generalizes to argument vectors that are not perpendicular to each other.

Application of the Hodge star yields a one-form $*j$ whose d components are given by

$$(*j)_i = \frac{1}{d-1} j^{i_1,\ldots,i_{d-1}} \epsilon_{i_1,\ldots,i_{d-1},i}.$$

The current vector commonly used in physics has components $j_{\mathrm{p}}^i = g^{ik} j_k$. If the metric is trivial, $g_{ij} = \delta_{ij}$, the current vector and the current differential form are related to each other by a straightforward relabeling of components. For example, in $d=3$, $j_{12} = j^3$, the physics reading is that current flow through the 12-plane is described by a current density vector with components *normal* to that plane (see figure for an illustration).

The advantage of the differential-form approach is that it is based on a measurement protocol. For example, a water current is measured by recording the flow of water through a geometric area per unit time. The translation to a vector normal to that plane requires the excess baggage of a scalar product. In theories where a trivial scalar product is implicit, this is no problem. However, in cases where there is no natural scalar product (topological cases), or the scalar product is center stage (as in the case of gravity), the differential-form approach to current description is more natural.

A.2 Elements of Probability Theory

SYNOPSIS This section contains a review of the elements of probability theory required in this text, especially in chapters 11 and 12.

Assume that the probability for a variable X to assume the value $x_i, i = 1, \ldots, N$ (or continuous values $x \in [a, b]$) in a large number of measurements is known. We represent this knowledge in terms of a **probability distribution**,

probability distri- bution

$$\text{discrete:} \quad p_i \equiv P(X = x_i), \qquad p_i \geq 0, \qquad \sum_{i=1}^{N} p_i = 1,$$

$$\text{continuous:} \quad p(x)dx \equiv P(x < X < x + dx), \qquad p(x) \geq 0, \qquad \int_a^b p(x)\,dx = 1.$$

random variable

Variables whose properties are defined via distributions are called **random variables**. To avoid discrete/continuous case distinctions, we will express the distribution of a discrete random variable in continuum form as $p(x)dx \equiv \sum_i \delta(x - x_i)dx$.

multivariate distri- bution

conditional probability

An n-component vector of random variables $(X_n, \ldots, X_1)$ is described by a **multivariate distribution** $p(x_n, \ldots, x_1)$. In applications, one often aims to relate bivariate probabilities $p(x_2, x_1)$ to known information on a monovariate probability $p(x_1)$. The information bridging between $p(x_1)$ and $p(x_2, x_1)$ is the **conditional probability** $p(x_2|x_1)$, which answers the question, "what is the probability of obtaining x_2 provided x_1 has been observed?" It is defined implicitly by

$$p(x_2, x_1) = p(x_2|x_1)p(x_1). \tag{A.21}$$

Summing over all possible realizations of the random variable X_1, we get the monovariate distribution of X_2, $\int dx_1\, p(x_2, x_1) = p(x_2)$, or

$$p(x_2) = \int dx_1\, p(x_2|x_1)p_1(x_1).$$

Considering the case where x_1 is a composite variable in its own right, $x_2 \to x_n$ and $x_1 \to (x_{n-1}, \ldots, x_1)$, we may iterate Eq. (A.21) to obtain

$$p(x_n, \ldots, x_1) = p(x_n|x_{n-1}, \ldots, x_1)p(x_{n-1}|x_{n-2}, \ldots, x_1) \cdots p(x_2|x_1)p_1(x_1). \tag{A.22}$$

This relation plays an important role, e.g., in the description of discrete time series of mutually dependent events.

A.2.1 Expectation values

Expectation values of functions $f(X)$ depending on a random variable are defined as

$$\mathrm{E}(f(X)) \equiv \mathrm{E}f(X) \equiv \langle f(X)\rangle \equiv \int dx\, p(x)f(x).$$

The E-notation is prevalent in the mathematical and quantum information literature. However, in this text, we will stick to $\langle \ldots \rangle$. Important examples of expectation values include:

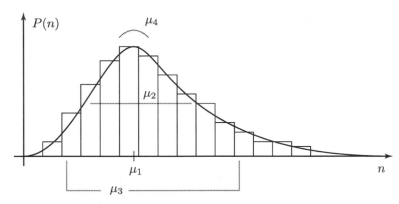

Fig. A.1   Schematic of a probability distribution defined via a histogram containing the count fre-
            quency of measurements in a fictitious experiment. To a first approximation, the shape of
            the distribution is described via its mean, width, skewness, and kurtosis.

moments   ▷ the **mean value** $\langle X \rangle \equiv \int dx\, p(x) x$ of the distribution and its **moments** $X_n \equiv$
            $\langle X^n \rangle \equiv \int dx\, p(x) x^n$. Notice that the moments of distributions need not neces-
            sarily exist. For example, the Lorentzian distribution $p(x) = \frac{1}{\pi} \frac{a}{a^2 + x^2}$ does not
            have moments at all.

cumulants   ▷ The **cumulants**

$$
\begin{aligned}
\mu_1 &\equiv X_1, \\
\mu_2 &\equiv X_2 - X_1^2, \\
\mu_3 &\equiv X_3 - 3X_1 X_2 + 2X_1^3, \\
\mu_4 &\equiv \dots,
\end{aligned}
\tag{A.23}
$$

describe how moments differ from the products of moments of lesser degree: the
first cumulant gives the **average** value of the distribution, and the second its
width. The third cumulant measures the **skewness** of the distribution relative
to its center, and the fourth its **kurtosis**. High kurtois means a sharply peaked
distribution with fat tails; low kurtosis a softer distribution, with broader shoul-
ders. The central limit theorem implies a tendency for the distributions $p(y)$ of
variables $Y = \sum_a X_a$ obtained by the superposition of a large number of "microscopic" variables (a macroscopic current obtained by superposition of microscopic single-electron currents, etc.) to become Gaussian (see the next section).
This means that cumulants beyond the second are usually small, but all the
more interesting.

cumulant   A systematic way to define cumulants is by expansion of the **cumulant gen-**
generating   **erating function**:
function

$$
\ln \langle \exp(itX) \rangle = \sum_{i=1}^{\infty} \frac{(it)^n}{n!} \mu_n.
$$

▷ The probability distribution itself may be represented as the expectation value
of a δ-distribution,

$$
p(x) = \langle \delta(x - X) \rangle.
$$

This representation is quite useful in practice. It often defines the most efficient way to obtain the distribution of variables $Y \equiv F(X)$ depending on a fundamental random variable, X. Using the δ-representation, the probability distribution $p(y)$ is obtained from $p(x)$ as

$$p(y) = \langle \delta(y - Y) \rangle = \int dx\, p(x)\delta(y - F(x)) = p(x(y)) \left| \frac{\partial x}{\partial y} \right|_{x=x(y)},$$

where we assume a unique functional relation $y = F(x)$. Alternatively, this relation may be obtained by direct transformation of the probability measure, $p(x)dx = p(x(y))|\frac{\partial x}{\partial y}|dy \equiv p(y)dy$. However, the δ-representation is often more convenient to use, especially when x is not just a simple scalar variable but a vector, or even a field with random components.

▷ The full set of information on a random variable is contained in the expectation value

$$g(t) \equiv \langle \exp(itX) \rangle. \tag{A.24}$$

moment generating function

From this **moment generating function** (or **characteristic function**) of a distribution, all moments can be obtained by power series expansion and all cumulants by expansion of the **cumulant generating function**, $\ln g(t)$:

$$g(t) = \sum_{n=0}^{\infty} \frac{(it)^n}{n!} X_n, \qquad \ln(g(t)) = \sum_{n=0}^{\infty} \frac{(it)^n}{n!} \mu_n.$$

The full probability distribution is obtained (exercise) from the generating function $g(t)$ by Fourier transformation,

$$p(x) = \int \frac{dt}{2\pi}\, e^{-itx} g(t).$$

▷ An important hallmark of a probability distribution is its **(information) entropy**

information entropy

$$S = - \int dx\, p(x) \ln(p(x)) = -\langle \ln(p) \rangle. \tag{A.25}$$

The rationale behind this expression is best understood in a discrete setting, $S = -\sum_i p_i \ln(p_i)$. An event of low probability $p_i \ll 1$ (such as a lottery win) conveys essential information. No information is gained from $p_i = 1$ events ("sun will set tonight"). The information entropy provides a logarithmic measure for this difference via $-\ln(p_i) \gg 1$ versus $-\ln(1) = 0$. In this way, $\ln(p_i)$ becomes a random variable whose average, S, describes the information stored in the distribution. Don't be confused that "high information entropy" means "broad" or "unstructured." Also note that the information entropy of the thermal distribution $P = \rho = \exp(-\beta H)/Z$ coincides with the thermodynamic entropy, S, defined via the free energy $F = -(T/\beta) \ln Z = E - TS$.

A.2.2 Examples of distributions

For later reference, we introduce here a few distributions frequently occurring in
practice. The **Gaussian distribution** is defined by

Gaussian

$$p(x)dx = \frac{1}{\sqrt{2\pi\mu_2}} \exp\left(-\frac{(x-\mu_1)^2}{2\mu_2}\right) dx. \tag{A.26}$$

It often realized as the probability of large sums $X = \sum_{n=1}^{N} X_n$ where the distribution of the 'microscopic" random variables X_n has as first two cumulants $\tilde{\mu}_1$ and
$\tilde{\mu}_2$. In this case, the **central limit theorem** states that the distribution of X is
given by Eq. (A.26), with first and second cumulant $\mu_1 = N\tilde{\mu}_1$ and $\mu_2 = N\tilde{\mu}_2$,
respectively. (All higher cumulants of the Gaussian distribution vanish. Exercise:
Verify this statement by showing that the generating function of the Gaussian distribution is again a Gaussian.) The ubiquity of additive random variables in nature
explains the importance of the Gaussian distribution in science.

*central
limit
theorem*

INFO The **central limit theorem is proven** by a variant of the large-N expansion
methods introduced in section 4.3.2:

$$p(x) = \langle\delta(x - \sum_{n=1}^{n} x_n)\rangle = \prod_{n=1}^{N} \int dx_n \, p(x_n) \int dk \, e^{ik(x_n - x)}$$

$$= \int dk \exp\left(N\ln\int dx\, p(x)e^{ikx} - ixk\right) = \int dk \exp\left(N\ln g(k) - ixk\right)$$

$$= \int dk \exp\left(N\left[ik(\mu_1 - x/N) - k^2\mu_2/2 + \mathcal{O}(k^3)\right]\right) \simeq \frac{1}{\sqrt{2\pi N\mu_2}} \exp\left(\frac{(x - N\mu_1)^2}{2N\mu_2}\right),$$

where in the first line we used the Fourier representation of the δ-function, g in the second
line is the generating function of the distribution p, and in the crucial last equality we
noted that anharmonic corrections to the quadratic k-exponent vanish in the large-N
limit.

*binomial
distri-
bution*

The **binomial distribution** $P(m)$ describes the probability of observing n events
in $n > m$ trials when the events are mutually uncorrelated[5] and occur with individual probability p (think of a coin toss, where $p = 1/2$). Straightforward combinatorics yields

$$P(m) = \binom{n}{m} p^m (1-p)^{n-m}. \tag{A.27}$$

The binomial distribution has average $\mu_1 = np$ and variance $\mu_2 = np(1-p)$.

In the case where the probability of individual events becomes small, $p \ll 1$,
and we try many times, $n \gg m$, the binomial distribution asymptotes towards the
Poisson distribution,

*Poisson
distri-
bution*

$$P(m) = \frac{(np)^m}{m!} \exp(-np). \tag{A.28}$$

---

[5] Two random variables X and Y are **uncorrelated** if $\langle XY\rangle = \langle X\rangle\langle Y\rangle$. The correlation obviously vanishes if X and Y are **independent random variables**, i.e., if $p(x,y) = p_x(x)p_y(y)$.
(The opposite statement "lack of correlation $\to$ independence" is not true in general.)

The Poisson distribution has identically equal cumulants $\mu_l = np$.

EXERCISE Obtain the Poisson distribution from the binomial distribution. Show that, in the limit $n \to \infty$ and $p \to 0$ at fixed np, the derivation becomes exact. Show that, for $n \to \infty$ at fixed p, the variable m becomes Gaussian distributed instead.

Lorentzian distribution

The **Lorentzian distribution** or **Breit–Wigner** or **Cauchy distribution** is defined by

$$p(x) = \frac{1}{\pi} \frac{a}{(x - x_0)^2 + a^2}. \tag{A.29}$$

In physics, the Lorentzian distribution describes the energy dependence of scattering resonances, the broadening of many-particle spectral functions by interactions, the line-shape distribution of damped electromagnetic modes, and many other phenomena governed by an interplay of driven, oscillatory and damping behaviour. As mentioned above, its moments are undefined. The distribution is centered around x_0, has width a, and is exceptionally broad.

EXERCISE Compute the characteristic functions (A.24) for the Gaussian, binomial, Poisson, and Lorentzian distribution. From these results, verify the above statements about the cumulants of the respective distributions.

A.3 Conformal Field Theory Essentials

SYNOPSIS Conformal field theory (CFT), and especially two-dimensional conformal field theory, is a key concept of modern theoretical physics. Two-dimensional conformal field theories possess an infinite number of symmetries giving them a much higher degree of structure and solvability than generic theories. At the same time, *critical* field theories are almost always conformal, which makes CFT crucially important in understanding the physics of two-dimensional phase transitions. CFTs are usually described in languages different from those of the functional field theories central to this text. They possess so many symmetries that it becomes more rewarding to study directly the constraints imposed by these on correlation functions. This methodological orthogonality is one reason why we have not included an extended discussion of CFT in this introduction to field theory. The other is that CFT is a deep subject and that any halfway-complete introduction would require another textbook – adding to the list of excellent treatises that are already out there.[6]

---

[6] CFT can be looked at from different perspectives, all represented via excellent texts in the literature. Here is an incomplete list of suggested references. The "bible" of CFT, called the "yellow pages" for its volume and the color of the book cover, P. Di Francesco, P. Mathieu, and D. Sénéchal, *Conformal Field Theory* (Springer-Verlag, 1997) covers everything; G. Mussardo, *Statistical Field Theory: An Introduction to Exactly Solved Models in Statistical Physics* (Oxford University Press, 2020) has a friendly statistical mechanics-oriented introduction in part III; and J. Cardy, *Scaling and Renormalization in Statistical Physics* (Cambridge University Press, 1996) is a concise and authoritative introduction by one of the architects of the field. Orig-

At the same time, CFT is too important to be left out completely. This appendix contains a synopsis of CFT concepts that the authors believe are so essential that they should be common knowledge. We will skip over technical derivations and details in favor of brevity. Ideally, this mini-introduction will make readers curious and motivate them to delve deeper into this exceptionally beautiful theory.

From the self-similarity of critical theories follows scale invariance, which has been a major theme in this text. However, self-similarity actually implies[7] invariance under the larger set of conformal transformations introduced in section 1.6.2. Transformations of coordinates $x^\mu = y^\mu(x)$ are conformal if they are angle preserving: curves intersecting at a certain angle are mapped onto image curves intersecting at the same angle. A more compact and equivalent statement is that the local metric $g_{\mu\nu}(y(x)) = \Omega(x)g_{\mu\nu}(x)$ may change by a local stretching factor at most. Intuitively, such **conformal maps** leave geometric structures *locally* form invariant. (For example, the figure on page 34 shows the conformal image of a rectangular tiling of the plane.) Conformal maps define a group, the **conformal group**, which in generic dimensions is finite-dimensional. (Besides dilatations and rotations, it contains the so-called special conformal transformations, roughly angle-preserving inversions of space; see the discussion in section 1.6.2.) Why aren't these transformations more prominently addressed in field theory texts?[8] The reason is that scale transformations ($x^\mu = by^\mu$, $g_{\mu\nu}(x) = b^{-2}g_{\mu\nu}(y(x))$) suffice to assess the operator content of a field theory, and this in most cases provides sufficient information to describe the behavior of physical observables.

A.3.1 Conformal invariance in two dimensions

However, conformal symmetry in two dimensions is special. First, $d = 2$ is the lower critical dimension of many theories in statistical and condensed matter theory and is therefore of interest *per se*. Second, the group of two-dimensional conformal transformations is infinite dimensional, and therefore provides critical theories with an enormous reservoir of symmetries. In the rest of this section, we introduce the representation of these symmetries in complex coordinates – the standard language of the field – and define what is meant by the conformal symmetry of a theory.

---

inal papers are often surprisingly pedagogical, and CFT is no exception in this regard: see, e.g., A. A. Belavin, A. M. Polyakov, and A. B. Zamolodchikov, *Infinite conformal symmetry in two-diemensional quantum field theory*, Nucl. Phys. B. **241** (1984). A quite pedagogical introduction with a barely noticeable tilt towards string theory applications is provided by D. Tong, *Introducing conformal field theory*, online lecture notes damtp.cam.ac.uk/user/tong/string/four.pdf.

[7] While there seem to be only a few exceptions, the question under which conditions scale invariance implies conformal invariance is a subject of ongoing research.

[8] A notable exception is the emerging field of holographic principles, which establishes connections between bulk gravitational theories in $d + 1$ dimensions and conformal field theories at their d-dimensional boundaries.

Conformal group in $d = 2$

Consider the two-dimensional plane (x, y) parameterized in complex coordinates, $z = x + iy$. The striking feature of $d = 2$ is that any *any holomorphic function* $z \mapsto w(z) \equiv u(z) + iv(z)$ defines a conformal transformation. The infinite dimensionality of the conformal group follows from the existence of infinitely many holomorphic functions.

EXERCISE To understand the statement above, recall the Cauchy–Riemann differential equations $\partial_x u = \partial_y v$, $\partial_x v = -i\partial_y u$ and discuss why they define an angle-preserving map $(x, y) \mapsto (u, v)$ of the two-dimensional plane. (Hint: Interpret $\partial_x u$, etc. as components of tangent vectors.)

EXAMPLE One of the most important conformal transformations is the **logarithmic transformation mapping the plane onto a cylinder**, $z = e^w$. (The figure on page 34 shows contour lines, $(\mathrm{Re}(w), \mathrm{Im}(w))$, of this transformation. They intersect at 90 degree angles locally respecting the rectangular structure of the argument lines $z = x + iy$.)

-------- space   -------- time

For $z = r\exp(i\phi)$ spanning the plane, $w = i\phi + \ln r \equiv ix + t$ covers a cylinder (see the figure). Conformal field theories of finite-size systems are often quantized in the w-language. In that context, w is a space-time cylinder; the circumferential coordinate $\phi \in [0, 2\pi]$ parameterizes real space (with periodic boundary conditions) and $t \in [-\infty, \infty]$ parameterizes time. In the z-language, time progresses from the distance past at the origin to the future at radial infinity, and the circles represent equal-time spatial contours.

We also note that a cylinder of finite circumference 2π is easier to implement on a computer, which makes the logarithmic transformation an asset in the numerical approach to CFTs.

complex
coordinates

INFO As indicated above, CFT is is usually described in **complex coordinates**. For the convenience of the reader, the translation to complex language from a language with real coordinates $x^\mu = (x, y)$ and Euclidean signature[9] $g_{\mu\nu} = \delta_{\mu\nu}$ is summarized here:

$$z = x + iy, \quad \bar{z} = x - iy, \quad x = \frac{1}{2}(z + \bar{z}), \quad y = \frac{1}{2i}(z - \bar{z}),$$

$$\partial_x = \partial_z + \partial_{\bar{z}}, \quad \partial_y = i\partial_z - i\partial_{\bar{z}}, \quad \partial_z \equiv \partial = \frac{1}{2}(\partial_x - i\partial_y), \quad \partial_{\bar{z}} \equiv \bar{\partial} = \frac{1}{2}(\partial_x + i\partial_y),$$

$$g = dx\, dx + dy\, dy, \quad g = dz\, d\bar{z}. \tag{A.30}$$

A word of caution: in CFT, we need to be careful about the difference between generic functions of the complex coordinate, such as $f(z, \bar{z}) = |z|^2$, holomorphic functions such as $g(z) = z^2$, and anti-holomorphic functions such as $\bar{g}(\bar{z}) = \bar{z}^2$. Following standard conventions, we distinguish between these alternatives through the function arguments indicated, although z and $\bar{z}$ in $f(z, \bar{z})$ remain dependent variables.

---

[9] In condensed matter applications, $x = x$ is usually real space and $y = \tau$ imaginary time.

Conformal invariance

Intuitively, a system has conformal invariance if changes of length scale, including local changes, are irrelevant as long as angular proportions are left intact. However, how can this feature be defined in more concrete terms? The question is not easy to answer in general. For a given quantum field theory, establishing conformal invariance requires knowledge of the behavior of all its constituents – fields, action, measure – under arbitrary conformal transformations. There are different ways to sidestep this complication. One is to shift the focus from the field-theoretical partition sum to its correlation functions. In fact, one may *define* a CFT axiomatically as a set of all possible correlation functions constrained by the infinitely many conditions following from conformal invariance. A related approach focuses on the fields themselves and on their representation as operators in the Hilbert space of the theory. Either way, the focus in CFT shifts from functional integrals to fields and their correlations functions.

fields of CFT

INFO Consider a theory with local degrees of freedom $\phi = \{\phi^a\}$, where $\phi = \phi^a(x, y) = \phi^a(z, \bar{z})$. In the parlance of CFT, **fields** are arbitrary local expressions built from ϕ. This includes functions such as ϕ^n or $\exp(i\phi)$, and derivatives $\partial\phi, \bar{\partial}\phi, \partial\bar{\partial}\phi, \ldots$.

free boson

Instead of addressing conformal invariance in general, let us consider the example of the **free boson** with action $S[\phi] = \frac{1}{4\pi} \int d^2x (\partial_\mu \phi)^2$. (This theory will be our guinea pig for the exemplification of various features of CFT throughout the section.) In complex coordinates, this assumes the form (verify)

$$S[\phi] = \frac{1}{2\pi} \int dz d\bar{z} \, \partial\phi\bar{\partial}\phi. \tag{A.31}$$

This theory is critical (why?) and we expect conformal invariance. To see how this comes about, consider the action as an ordinary two-dimensional integral over a function $\phi(z, \bar{z})$ and apply a holomorphic change of variables $z \mapsto w(z)$, $\bar{w} \mapsto \bar{w}(\bar{z})$ in the sense of calculus. With the substitution

$$\phi'(w, \bar{w}) \equiv \phi(z(w), \bar{z}(\bar{w})), \tag{A.32}$$

we have $\partial_z \phi(z, \bar{z}) = \frac{dw}{dz} \partial_w \phi'(w, \bar{w})$, where the prefactor cancels the change in the measure $dz = \frac{dz}{dw} dw$. The same cancellation happens in the $\bar{z}$ sector, so that the action looks the same in the old and the new variables. The way to read this observation as a symmetry is to say that $S[\phi] = S[\phi']$, where the field ϕ' is a map defined by Eq. (A.32).

In this way, the conformal symmetry of the free boson is established. Notice that the construction relies on the holomorphy $w = w(z)$ and will not work for a general function $w = w(z, \bar{z})$. The symmetry also goes lost in the presence of a mass term, $m\phi^2$. In essence, it is based on a cancellation of the scaling dimension -1 of the derivative operators against the dimension $+1$ of the measure factors. Finally notice that inspection of the action suffices to establish conformal invariance only in the case of simple free field theories. For example, we know that, in interacting theories, the integration over UV fluctuations may change the naïve dimension of operators by anomalous dimensions. Theories may also show anomalies, where the "classical" conformal symmetry of an action is spoiled by quantum fluctuations.

INFO These comments indicate that in CFT one must distinguish carefully between **the classical and the quantum theory**. As usual, "classical" means structures applying to variational solutions of effective actions, while "quantum" includes fluctuations. As an example, consider the variational equation of the free boson theory, $\partial\bar{\partial}\phi(z,\bar{z}) = 0$, with solutions

$$\phi(z,\bar{z}) \equiv \phi(z) + \bar{\phi}(\bar{z}) \tag{A.33}$$

(i.e., the complex representation of left- and right-propagating solutions of a wave equation). The classical field thus splits into a holomorphic and an anti-holomorphic component. However, this feature does not generalize to the quantum level. Texts on CFT do not always emphasize this distinction and one must be careful not to miss it.

A.3.2 Stress tensor, operator product expansion, and primary fields

As mentioned above, CFT places emphasis on the correlation functions of its operators. The most important operator in the study of CFTs is the stress-energy tensor. In the next section, we define this operator and then discuss how it is employed to describe the general operator contents of a CFT.

Stress–energy tensor

In section 1.6.2, we defined the stress–energy tensor as the Noether current associated with the translational invariance of a theory. Translations are conformal transformations and it is therefore not surprising that this operator features in CFTs. However, what may not be evident is just how important it is.

A priori, the **energy–momentum tensor**, **stress–energy tensor**, or just stress tensor, $T_{\mu\nu}$, of a theory with coordinates x^μ is a matrix field whose invariance is expressed under translations $x^\mu \to x^\mu + a^\mu$. In problem 1.8.5, we showed that it is conveniently obtained from its action as $T_{\mu\nu} = -4\pi\delta_{g_{\mu\nu}}S$ by variation in the elements of the metric, where the factor -4π is a matter of convention and we assume a unit determinant $\sqrt{g} = 1$. In the classical theory, the tensor $T_{\mu\nu}$ is evaluated on solutions of the variational equations of the theory. Translational invariance implies the conservation laws $\partial^\mu T_{\mu\nu} = 0$. However, the stronger condition of conformal invariance imposes additional conditions. In complex coordinates, where $T_{\alpha,\beta}(z,\bar{z})$, $\alpha,\beta = z,\bar{z}$, is a two-dimensional matrix, it can be shown to be diagonal, $T_{z\bar{z}} = 0$, with holomorphic entries $T_{zz}(z,\bar{z}) \equiv T(z)$ and anti-holomorphic entries $T_{\bar{z}\bar{z}}(z,\bar{z}) \equiv \bar{T}(\bar{z})$.

EXERCISE As an example illustrating the above features, consider the free boson. Derive the **stress tensor of the free boson** by variation of the metric, $g \to g + \delta g$. Realize this variation in either one of two representations. The first involves performing the variation

stress–
energy
tensor

of the Euclidean metric, $g = \delta_{\mu\nu}$, and transcribing the result to complex coordinates. The second involves working in complex coordinates from the beginning, where $g_{zz} = g_{\bar{z}\bar{z}} = 0$ and $g_{z\bar{z}} = g_{\bar{z}z} = 1/2$, and performing the variation in that language. Show that either procedure leads to the result

$$T(z) = -\partial\phi(z)\partial\phi(z), \qquad \bar{T}(\bar{z}) = -\bar{\partial}\bar{\phi}(\bar{z})\bar{\partial}\bar{\phi}(\bar{z}), \qquad (A.34)$$

where on the right-hand sides we have the holomorphic and the anti-holomorphic components of the classical field.

Recall that the component T_{00} defines the energy density associated with translational invariance in time. Interpreting $y = t$ as time, and using Eq.(A.30) to transform to the complex language, we find that $-(T(z) + \bar{T}(\bar{z}))$ defines the **energy density** of the theory. Also recall that the "conserved charge" associated with a symmetry is obtained by integrating the zero-component of its conserved current over space. Presently, this means that the **energy of a system with conformal invariance** is given by the integral of the above sum over a space-like surface $y = t = \text{const.}$ We will come back to this point in section A.3.2.

Operator product expansion

As indicated above, CFT places emphasis on the study of field correlation functions. Conformal symmetry implies scale invariance, which in turn implies power law singularities $\langle \phi(z,\bar{z})\phi(w,\bar{w}) \rangle \sim |z - w|^{-\alpha}$ as fields come close to each other. This observation motivates one of the most important concepts of CFT, the **operator**

operator
product
expansion

product expansion (OPE). Let $\{O_i\}$ be a container symbol for all the different fields appearing in a given CFT (we will soon be more concrete). When two of these fields appear in close proximity in a correlation function $\langle \cdots O_i(z,\bar{z})O_j(w,\bar{w}) \cdots \rangle$, where $|z - w|$ is much smaller than the separation of any other field coordinate, we expect two things to happen: power law singularities in $|z - w|$, and the option to represent the "regular contents" $O_i O_j$ as a linear combination in other operators O_j. This anticipation defines the OPE of two operators as

$$\boxed{O_i(z,\bar{z})O_j(w,\bar{w}) = \sum_k C_{ij}^k(z - w, \bar{z} - \bar{w})O_k(z,\bar{z}) + \cdots} \qquad (A.35)$$

where C_{ij}^k are singular functions of their arguments and the ellipses denote non-singular contributions disregarded in the expansion. As it stands, this is a symbolic representation and some fine print is required to make it concrete: as with many expressions in CFT, Equation (A.35) makes sense only as an insertion in a correlation function. Think of it as being plugged into $\langle \cdots \rangle$, where the angle-brackets stand for functional averaging over an action and O_i, O_j are fields in the sense defined above. This reading indicates that the OPE is a concept of the *quantum* theory, fluctuations around stationary configurations

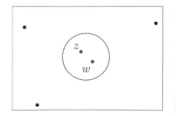

included. Thinking of O_i as functional representatives of quantum operators, the correlation functions are evaluated subject to a time ordering prescription. However, this aspect will not play a crucial role in our introductory discussion and we refer to the literature for precise definitions.

In practice, the OPE of two operators O_i, O_j is computed by representing them in terms of "elementary operators," which are then contracted by Wick's theorem. In doing so, we must be careful to avoid hard divergences. For example, a product of two fields $\phi(z)\phi(z)$ has divergent Wick contraction $\langle \phi(z)\phi(z)\rangle \sim \frac{1}{0}$. To avoid these contractions at coinciding points, operators in CFT are understood as **normal ordered**, which in practical terms means that self-contractions of the operators O_i do not enter the Wick protocol.

EXAMPLE In problem 3.8.10, we considered the regularized correlation function of the free boson, which in complex notation assumes the form

$$\langle \phi(z, \bar{z})\phi(w, \bar{w})\rangle = -\frac{1}{2}\ln(|z - w|^2) + \text{const.}, \qquad (A.36)$$

with a formally divergent constant. Splitting the logarithm as $\ln(|z-w|^2) = \ln(z - w) + \ln(\bar{z} - \bar{w})$, we find $\langle \partial_z\phi(z, \bar{z})\partial_w\phi(w, \bar{w})\rangle = -\frac{1}{2}\frac{1}{(z-w)^2}$. The holomorphic form of this correlation function motivates the following (somewhat symbolic) representation of the OPE:

$$\langle \partial_z\phi(z)\partial_w\phi(w)\rangle = -\frac{1}{2}\frac{1}{(z - w)^2} \qquad (A.37)$$

for the holomorphic field $\partial_z\phi(z, \bar{z}) \equiv \partial_z\phi(z)$.

Most OPEs in CFT involve the **stress tensor**. As an example, consider the OPE of the holomorphic stress tensor $T(z) = -\partial\phi(z)\partial\phi(z)$ of the free boson with $\partial_w\phi(w)$. Wick contraction using Eq. (A.37) leads to

$$T(z)\partial_w\phi(w) = \frac{\partial_z\phi(z)}{(z - w)^2} = \frac{\partial_w\phi(w)}{(z - w)^2} + \frac{\partial_w^2\phi(w)}{z - w} + \dots, \qquad (A.38)$$

where we have Taylor expanded, disregarding all non-singular terms. We emphasize again that all these expressions must be understood as building blocks in correlation functions.

Primary fields

primary field The structure of the OPE (A.38) gives rise to the definition of a **primary field**, or just "primary," as a field $\theta(z)$ whose OPE with the stress tensor assumes the form

$$\boxed{\begin{aligned} T(z)\theta(w, \bar{w}) &= h\frac{\theta(w, \bar{w})}{(z - w)^2} + \frac{\partial_w\theta(w, \bar{w})}{z - w} \\ \bar{T}(\bar{z})\theta(w, \bar{w}) &= \bar{h}\frac{\theta(w, \bar{w})}{(\bar{z} - \bar{w})^2} + \frac{\partial_{\bar{w}}\theta(w, \bar{w})}{\bar{z} - \bar{w}} \end{aligned}} \qquad (A.39)$$

where $(h, \bar{h})$ define the **dimensions** of the field. For example, from (A.38), we infer that $\partial_z\phi(z)$ is a primary of the free boson theory with dimensions $(1, 0)$.[10]

---

[10] The statement must be taken with a grain of salt. The vanishing of the dimension $\bar{h}$ means that the OPE of the holomorphic $\partial\phi$ with the anti-holomorphic $\bar{T}$ is empty, including the second (nominally $\bar{h}$-independent) term.

As written, the definition (A.39) is quite formal and does not explain the meaning of primary fields. While an appropriately detailed explanation is beyond the scope of this synopsis, let us summarize the main ideas:

▷ We first note that Eq. (A.39) is a *symmetry relation* in disguise. To understand why, ask how the functional representation of a correlation function $\langle \cdots \hat{O} \cdots \rangle = \int D\phi \exp(-S[\Phi])(\cdots O[\phi] \cdots)$ changes under a small change of variables $\phi \to \phi + \epsilon \delta \phi$. Of course, the answer is that it remains unaltered; a transformation of variables does not change an integral. However, both the action, $S \to S + \epsilon \delta S$, and operator representations $O \to O + \epsilon \delta O$ may change individually,[11] and the resulting total change $0 = \epsilon \langle \cdots (\delta S O + \delta O) \cdots \rangle = 0$ can lead to interesting representations of zero known as **Ward identities**.

Ward identity

In the present context, the natural "small changes of variables" are infinitesimal holomorphic and anti-holomorphic transformations, $z \to w(z) = z + \epsilon(z)$, and $\bar{z} \to \bar{w}(\bar{z}) = \bar{z} + \bar{\epsilon}(\bar{z})$, where the notation $\epsilon(z)$ indicates holomorphy through the argument (z) and smallness through ϵ. Primary fields are distinguished by their change under such transformations, $\theta \to \theta + \delta \theta + \delta \bar{\theta}$, with

$$
\begin{aligned}
\delta\theta(w, \bar{w}) &\equiv -\left(h \frac{d\epsilon(w)}{dw} + \epsilon(w)\partial_w \right) \theta(w, \bar{w}), \\
\delta\bar{\theta}(w, \bar{w}) &\equiv -\left(\bar{h} \frac{d\bar{\epsilon}(\bar{w})}{d\bar{w}} + \bar{\epsilon}(\bar{w})\partial_{\bar{w}} \right) \theta(\bar{w}, \bar{w}),
\end{aligned}
\tag{A.40}
$$

where the notation emphasizes that, after the change of variables, we are working in the w-language. Note that, to first order in ϵ, we have $z(w) = w - \epsilon(w)$ and $\epsilon(w) = \epsilon(z)$. This representation reveals the meaning of the parameters h as "dimensions." To see why, consider the particular case $\epsilon(z) = \epsilon z$ of a scale transformation, $z \to z/b$, $b = 1 - \epsilon$. The second term in Eq. (A.40) represents the change of arguments, $O(z, \bar{z}) \to O(z(w), \bar{z}(\bar{w})) = O(w - \epsilon w, \bar{w} - \epsilon \bar{w}) = O(w, \bar{w}) - (\epsilon w \partial_w + \bar{\epsilon} \bar{w} \partial_{\bar{w}}) O(w, \bar{w})$.

However, we know from our discussion of renormalization that the actual change in an operator under a scale transformation contains a factor reflecting its physical dimension. For example, in the free boson theory $\int d^2 x (\partial_\mu \phi)^2$, the field ϕ has engineering dimension zero and $\partial_\mu \phi$ dimension -1 owing to the derivative. In a scaling transformation, this dimension shows as a factor $\partial_\mu \phi \to b \partial_\mu \phi = (1 - \epsilon) \partial_\mu b$. The first term in Eq. (A.40) is the complex coordinate representation of this factor and, in combination with the second, it describes the full change in the field.

▷ We have just seen that primaries are generalizations of fields with definite scaling behavior to fields subject to conformal transformations. This interpretation

---

[11] Depending on the realization of the transformation, the measure $D\phi \to D\phi + \epsilon \delta D\phi$ may be affected too.

becomes even more evident when we generalize Eq. (A.40) to finite transformations $z \to w(z)$. The primary field then transforms as

$$O(z, \bar{z}) \to \tilde{O}(w, \bar{w}) = \left(\frac{dw(z)}{dz} \right)^{-h} \left(\frac{d\bar{w}(\bar{z})}{d\bar{z}} \right)^{-\bar{h}} O(z(w), \bar{z}(\bar{w})). \qquad \text{(A.41)}$$

As an exercise, think of the transformation as segmented into many infinitesimal steps and in this way convince yourself that this is the proper finite generalization of Eq. (A.40). Consistency with this relation implies strong constraints on the **correlation functions of primary fields**. Specifically, the two-point function of two primaries has the form

$$\boxed{\langle O(z, \bar{z}) O'(z', \bar{z}') \rangle \sim \frac{\delta_{h,h'} \delta_{\bar{h},\bar{h}'}}{|z - z'|^{2h} |\bar{z} - \bar{z}'|^{2\bar{h}}}} \qquad \text{(A.42)}$$

once more underpinning the meaning of h as a dimension. Equation (A.37) illustrates this behavior for the example of the $(h, h') = (1, 0)$ primary $\partial_z \phi$ of the free boson. Two remarks on the general formula (A.42): first, it states that correlation functions of fields with non-integer dimensions, $2h \notin \mathbb{Z}$, have branch cuts in the complex plane, indicating a connection to the anyon statistics discussed in section 8.6.1. Second, as with the OPE, Eq. (A.42) describes the leading short-distance singularities of the correlation function.

EXERCISE Consider the free boson primary in the w and z representations of the logarithmic transformation $w = \ln z$, using Eq. (A.41). In either language, the correlations are locally described by Eq. (A.42). However, that short-distance asymptotic does not know about the periodicity of the theory in $\text{Im}(w)$. It is instructive to play with Eqs. (A.41) and (A.42) and explore this point in detail.

operator–
state corre-
spondence

▷ Think of CFT as the field integral representation of a $(1+1)$-dimensional quantum theory. In this reading, fields represent quantum states created by the action of operators out of the vacuum. Within the framework of this **operator–state correspondence**, primary fields correspond to a distinguished class of states, roughly equivalent to the highest z-angular momentum states in a representation of spin. All other states in the representation can be generated from the primaries via the action of "lowering operators." We will return to this point in section A.3.3.

Identifying primaries and their associated dimensions is one of the most important steps in the characterization of a CFT. The info block section below illustrates that this identification can be non-obvious, even in simple field theories.

EXERCISE We have already identified $\partial \phi$ and $\bar{\partial} \phi$ as **primaries of the free boson theory**. Are there more? First, the field ϕ itself is *not* a primary. This follows from its OPE with the stress–energy tensor, which violates the structure Eq. (A.39) (think why).

For the same reason, higher derivatives $\partial_z^n \partial_{\bar{z}}^m \phi$ are not primary either. One might think that this exhausts the field contents of the theory, but this is not so; consider the operator

$$\Psi(z, \bar{z}) = \exp[i\alpha\phi(z, \bar{z})]. \tag{A.43}$$

Expand in ϕ to show that the contraction with the stress tensor $T(z) = -\partial_z\phi(z,\bar{z})\partial_z\phi(z,\bar{z})$ gives $T(z)\Psi(w,\bar{w}) = \frac{i\alpha\partial_z\phi(z,\bar{z})\Psi(w,\bar{w})}{z-w} + \frac{\alpha^2\Psi(w,\bar{w})}{4(z-w)^2}$. Rearranging terms and noting that, to leading order, $i\alpha\partial_z\phi\Psi = \partial_w\Psi$, we obtain the OPE

$$T(z)\Psi(w,\bar{w}) = \frac{\alpha^2}{4}\frac{\Psi(w,\bar{w})}{(z-w)^2} + \frac{\partial_w\Psi(w,\bar{w})}{(z-w)}$$

and an analogous expression for $\bar{T}(\bar{z})$. This demonstrates that Ψ is a primary with dimension $h = \bar{h} = \alpha^2/4$. In hindsight, the discovery of this transcendental primary may be not too surprising: we know from the discussion in section 3.6.2 that the free boson is dual to the free fermion, and that the latter is represented through an exponential construction as Eq. (A.43). Our finding reflects the hidden presence of fermions as basic fields in the theory.

Central charge

Perhaps surprisingly, the most important field of the theory, its stress tensor, is not a primary field. Rather the OPE with itself assumes the form

$$\boxed{T(z)T(w) = \frac{c}{2(z-w)^4} + \frac{2T(w)}{(z-w)^2} + \frac{\partial_w T(w)}{z-w}} \tag{A.44}$$

central
charge

with an analogous formula for $\bar{T}(\bar{z})$, where the real constant c is the **central charge** of the theory. How can we understand the structure of this expansion? First, the stress tensor has physical dimension $[z]^{-2}$ (think why). Its OPE with itself can therefore include terms up to $\mathcal{O}(z-w)^{-4}$, and the formula shows that, for non-vanishing c, this is what happens. The remaining terms have the same dimension, are of lesser singularity, and respect the exchange symmetry $z \leftrightarrow w$.

The central charge is the single most important number characterizing a CFT. It affords different interpretations, only one of which we address in this synopsis: the central charge is a **measure of the number of degrees of freedom of a CFT**. Beware that this number need not be integer. For example, the free field theory of the Majorana fermion – essentially half a fermion – has $c = 1/2$.

If the stress tensor is not primary, then how does it transform under conformal transformations? The answer is found from Eq. (A.44) by methods similar to those determining the transformation of primaries from their OPE with T: under a conformal transformation $z \to w(z)$, the stress tensor changes as

$$T(z) \to \tilde{T}(w) \equiv \left(\frac{dw(z)}{dz}\right)^{-2}\left(T(z(w)) - \frac{c}{12}\{w, z\}\right), \tag{A.45}$$

Schwarzian
derivative

where the **Schwarzian derivative** is defined as

$$\{w, z\} \equiv \frac{w'''}{w'} - \frac{3}{2}\frac{w''^2}{w'^2}, \qquad w' \equiv \frac{dw(z)}{dz}. \tag{A.46}$$

EXERCISE Consider the mapping $w = \ln z$ from the z-plane to the w-cylinder. Compute the Schwarzian derivative to show that

$$T_{\mathrm{cyl}}(w) = z^2 T_{\mathrm{pl}}(z) - \frac{c}{24}. \tag{A.47}$$

The constant appearing on the right affords an interpretation as a **Casimir energy** of the cylindrical geometry. Thinking of the quantum states of a CFT as extended and wave-like, a transformation of these states from an infinite planar space–time geometry to a that of a cylinder confines these states and should invoke Casimir pressure (see page 29). For free field theories, the corresponding contribution to the ground state energy can be worked out explicitly and is compatible with the constant appearing in the above transformation. Consult the literature for further discussion of this point.

A.3.3 Quantum theory

In this section, we briefly discuss the structure of the $(1 + 1)$-dimensional quantum theories described by functional integrals with conformal symmetry. The main result will be the identification of a powerful mathematical object, the **Virasoro group**. The latter is to the quantum mechanics of conformal symmetry what SU(2) is to spin rotation symmetry.

Recall the physics of spin rotation invariance: it starts with the identification of the relevant symmetry group, SU(2), and that of its irreducible representations in Hilbert space, i.e., spaces of definite angular momentum, l. In each space, we then identify states of maximal angular momentum component in one direction, $m = l$. From these "maximum weight states," a complete set of basis states with angular momenta $l, l - 1, \ldots, -l$ is obtained by application of lowering operators. The algebraic approach to conformal symmetry is essentially similar, except that three-dimensional rotation symmetry is replaced by infinite-dimensional conformal symmetry.

The first thing that we need in defining a quantum theory from the field integral formulation is an identification between the fields of the latter and the states of the former (i.e., the path integral construction in reverse). This is best carried out in the radial representation of space–time introduced in the example on page 787. There, we mapped a base space comprising a compact circular space coordinate and an infinite time coordinate to a representation where time is ordered along the radial coordinate. Specifically, the infinite past corresponds to the origin $z = \bar{z} = 0$ of the plane, and a time-ordered correlation function $\lim_{z, \bar{z} \to 0} \langle \ldots \phi(z, \bar{z}) \rangle$ should be interpreted as one involving the action of the quantum operator represented by ϕ on the vacuum state of the theory. In this sense, one may think of fields as operators. Commutators between operators are computed by comparing fields at slightly different radial coordinates. (The presence of non-vanishing commutators follows from the short-distance singularities of the OPE expansion.)

Virasoro algebra

Let us investigate the commutation relations characterizing the most important operator of the theory, the stress tensor. In the radial framework, time is measured by distance to the origin, and circular contours are equal-time contours. This is conveniently described in a representation where the stress tensor is expanded in a Laurent series,

$$T(z) = \sum_{n=-\infty}^{\infty} \frac{L_n}{z^{n+2}}, \qquad L_n = \frac{1}{2\pi i} \oint dz\, T(z) z^{n+1}. \tag{A.48}$$

In these expressions, the Laurent coefficients are interpreted as operators of the quantum theory. Recall that CFT formulae are to be understood as insertions in correlation functions. The contour integral on the right extracting the operator L_n from T must then be performed around contours that do not contain dangerous (non-holomorphic) operator insertions.

On this basis, we can ask, what are the commutator relations $[L_n, L_m]$? The strategy of this computation is determined by Eq. (A.48) and the OPE (A.44): we represent both L_n and L_m as in Eq. (A.48). Here, the radii of the contours is set by the time at which the operators act. Assuming time ordering, building the difference

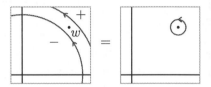

$[L_n, L_m] = L_n L_m - L_m L_n$ amounts to a comparison of the L_n contour as slightly larger and smaller than the L_m contour. Denoting the integration variable entering the computation of L_n and L_m by z and w, respectively, and considering the double integral at fixed w, we have the situation indicated in the figure. The difference of the two z-contours for a z-integrand that is holomorphic except for potential singularities at $z = w$ equals an integral performed over a small circle around w. At this stage, the factors $T(z)$ and $T(w)$ enter the integral at close-by coordinates, and the principles of the OPE apply. Inserting Eq. (A.44) and performing the integral by the theorem of residues (exercise), we find

$$\boxed{[L_n, L_m] = (n - m)L_{n+m} + \frac{c}{12}n(n^2 - 1)\delta_{n+m,0}} \tag{A.49}$$

Virasoro algebra

Equation (A.49) defines the famous **Virasoro algebra**. To understand the meaning of this expression, let us first ignore the c-dependent piece. The commutation relations then assume a form resembling that of the su(2) algebra ($[L_i, L_j] = 2\epsilon_{ijk}L_k$)), indicating that in the quantum physics of conformal symmetry, the Virasoro algebra plays a role analogous to that of the angular momentum algebra in rotationally symmetric quantum mechanics.

To understand the interpretation of the L_n as generators of a symmetry algebra, recall that our story started with the study of the $z \mapsto w(z)$ as elements of the conformal group, and of their infinitesimal version $z \mapsto z + \epsilon(z)$. Laurent-expanding $\epsilon(z) \equiv \sum \epsilon_n z^n$, such transformations act on functions as $f(z) \to f(z + \epsilon(z)) \simeq f(z) + \sum \epsilon_z^n \partial f(z)$, suggesting an interpretation of the $L_n = z^{n+1} \partial$ as generators of the conformal symmetry transformation. A straightforward computation shows that $[L_n, L_m] = (m - n) L_{n+m}$, in agreement with the first term in Eq. (A.49). In mathematics, the L_n with these commutator relations are known as the generators of the **Witt algebra**.

At this point, we have identified the (Witt) algebra of the conformal group. This group describes the conformal symmetry on the "classical level," i.e., its representation in the action functional. However, our actual theory is quantum; it involves an action functional integrated over fluctuations. We know that such fluctuations may interfere with the representation of symmetries via anomalies (see section 9.2) generated by UV singularities. The strongest UV singularity we encountered in the present context was the $(z - w)^{-4}$ contribution to the OPE of the stress tensor (A.44). At the same time, the stress tensor probes the conformal invariance of the theory. These observations suggest an interpretation of the second term in Eq. (A.49) as an anomaly making the Virasoro symmetry algebra on the quantum level different from the classical Witt algebra.[12]

INFO The low-lying Virasoro generators $\{L, L_{\pm 1}\}$ have individual physical meaning. As an example, consider the conserved energy of a conformal quantum system on the spacetime cylinder. It is defined as the energy density (see section A.3.2) $E = -i \oint dw \, (T(w) + \bar{T}(\bar{w}))$ integrated over the spatial coordinate in $w = ix + t$ at arbitrary t. Let us transform this expression to the plane to let the Virasoro generators enter the stage. Recalling Eq. (A.47), we find that

$$E = i \oint dz \, \frac{dw}{dz} \left(z^2 T_{\mathrm{pl}}(z) + \bar{z}^2 \bar{T}_{\mathrm{pl}}(\bar{z}) - \tfrac{c}{12} \right),$$

where the integral is over a circle whose radius is set by the time coordinate. With Eq. (A.48) and $dw/dz = 1/z$, we find $E = 2\pi (L_0 + \bar{L}_0) + \frac{\pi c}{6}$, which is independent of the radius, as expected. This result shows that, in the radial framework, the **Hamiltonian operator** of the theory is given by

$$H = L_0 + \bar{L}_0,$$

where we have absorbed the factor 2π in a rescaling of energy units, and ignored the c-number Casimir energy. Notice that, in the classical theory, the generators act as $L_0 z \partial_z$. With Eq. (A.30), we find $H = x\partial_x + y\partial_y = r\partial_r = \partial_{\ln r} = \partial_t$, where we have used the identification of time as $\ln r = t$. In the spirit of Noether's theorem, we read this finding as: energy, H, is the conserved charge of translational invariance in time. This charge acts as the infinitesimal generator of the symmetry transformation, ∂_t. (Think about this interpretation.)

Hamiltonian operator *(margin note)*

---

[12] In mathematical terms, the Virasoro algebra is a *central extension* of the Witt algebra. Roughly, a central extension is constructed by including a c-number, c, as an element of the operator algebra, $\{L_n\} \to \{L_n, c\}$. It commutes with all other elements, but appears on the right-hand side of the commutator expansion via the "structure constants" specified by Eq. (A.44).

EXERCISE Consider the Laurent coefficients L_n and the primaries $\phi(z, \bar{z})$ as operators of the theory. Following the same logic as in the computation of the commutators $[L_n, L_m]$, we have the representation

$$[L_n, \phi(w, \bar{w})] = \oint_w dz\, T(z) z^{n+1} \phi(w, \bar{w}).$$

Insert the OPE (A.39) to obtain

$$[L_n, \phi(w, \bar{w})] = h(n+1) w^n \phi(w, \bar{w}) + w^{n+1} \partial_w \phi(w, \bar{w}), \qquad (A.50)$$

and an analogous relation for $\bar{L}_n$.

Hilbert space

Assume we have a CFT characterized in terms of a stress tensor, and its primary fields. On the basis of this data, we would like to construct a Hilbert space of states with definite behavior under the symmetry generators L_n. Note that this program implies a passage from the correlation function representation emphasized so far to a more algebraic representation of CFT. Also note the similarity to SU(2)-symmetric quantum mechanics, where Hilbert space is organized in representation spaces of the angular momentum symmetry operators, $\hat{L}_i$.

vacuum state   As an additional assumption, we require the presence of a **vacuum state**, $|0\rangle$, implicitly defined as the state to the right in correlation functions $\cdots O_1 O_2 \cdots |0\rangle$. We work in the radial framework, where operator ordering is determined by the proximity to the origin. Of the vacuum state, we will not require much, only that it behaves regularly under application of the stress tensor and the primaries. (However, in view of the partially singular structure of OPEs and Laurent expansions, this assumption contains more information than one might think.)

Picking up on the last remark, consider the state $\lim_{z \to 0} T(z)|0\rangle$. Inserting the Laurent expansion (A.48) and requiring regularity, we find

$$L_n|0\rangle = \bar{L}_n|0\rangle = 0, \qquad n \geq -1. \qquad (A.51)$$

primary states   Note in particular that the vacuum is a zero-energy state in the sense that $(L_0 + \bar{L}_0)|0\rangle = 0$. Next in line in our study of quantum states are the **primary states** defined as

$$|h, \bar{h}\rangle \equiv \lim_{z, \bar{z} \to 0} \phi(z, \bar{z})|0\rangle, \qquad (A.52)$$

where the notation anticipates that the properties of these states are fixed by their conformal dimensions. We now combine Eqs. (A.50) and (A.51), and use the regularity of the vacuum state, $\lim_{w, \bar{w} \to 0} w^m \phi(w, \bar{w})|0\rangle = \lim_{w, \bar{w} \to 0} w^m \partial_w \phi(w, \bar{w})|0\rangle = 0$, $n > 0$, to obtain a number of operator relations characterizing the primary states:

$$L_0|h, \bar{h}\rangle = h|h, \bar{h}\rangle, \qquad \bar{L}_0|h, \bar{h}\rangle = \bar{h}|h, \bar{h}\rangle,$$
$$L_n|h, \bar{h}\rangle = 0, \qquad \bar{L}_n|h, \bar{h}\rangle = 0, \qquad n > 0.$$

The first line shows that the primary states are eigenstates of the Hamiltonian $-(h + \bar{h})$. Interpreting the operators $L_n, n > 0$, as the analog of the angular momentum lowering operators $L_- = L_x - iL_y$, the second line indicates an analogy to the minimum weight states $|l, m = -l\rangle$ in theories with SU(2) symmetry. The same analogy suggests considering sequences of states (corresponding to $|l, m\rangle$) defined by the action of raising operators on those states. In the present context, the role of the raising operators is taken by the Virasoro generators L_n, $n < 0$. The states $L_{-1}|h, \bar{h}\rangle, L_{-1}^2|h, \bar{h}\rangle, L_{-2}|h, \bar{h}\rangle, \ldots$ constructed in this way are called the **descen-** **descendants** **dants** of the primary state. Reflecting the infinite dimensionality of the Virasoro algebra, the **conformal tower** of states obtained in this way is of course more complicated than the $2l + 1$ angular momentum states in a definite representation sector of SU(2). However, as in that case, key properties of the descendants are fixed by the algebra commutation relations. For example, application of Eq. (A.49) to the case $m = 0$ shows that the descendants are energy eigenstates with

$$L_0 L_n|h, \bar{h}\rangle = (h - n) L_n|h, \bar{h}\rangle, \qquad n < 0.$$

We thus get the whole sequence of eigenstates for free!

However, the situation is not quite as simple as the discussion above might suggest: not all states in the conformal tower need to be linearly independent, indicating that there can be finite- and infinite-dimensional representation spaces of the symmetry. How can they be classified? What concrete form do the primary states and their descendants assume for a specific CFTs? Are the operators introduced so far on a purely algebraic basis necessarily unitary? For the discussion of these and many other fascinating questions revolving around the quantum physics of CFT, we refer to the literature.

A.4 Fourier and Wigner Transforms

For the convenience of the reader, here we summarize the Fourier transform conventions favored in this text. We also discuss the Wigner transform frequently applied as a partial Fourier transform to quantities depending on two space–time arguments.

A.4.1 Spatial transform

For **finite systems with a spatially continuous variable**, we favor a convention in which a volume factor stands in front of the momentum sum:

$$f(\mathbf{x}) = \frac{1}{L^d} \sum_{\mathbf{k}} e^{i\mathbf{k}\cdot\mathbf{x}} f_{\mathbf{k}}, \qquad f_{\mathbf{k}} = \int d^d x\, e^{-i\mathbf{k}\cdot\mathbf{x}} f(\mathbf{x}),$$

where momentum components k_i are summed over multiples of $2\pi/L$, and we assumed a cubic geometry ($L_i = L$) for simplicity. In this way, the limit of **infinite systems** assumes the form

$$f(\mathbf{x}) = \int \frac{d^d k}{(2\pi)^d} e^{i\mathbf{k}\cdot\mathbf{x}} f(\mathbf{k}), \qquad f(\mathbf{k}) = \int d^d x\, e^{-i\mathbf{k}\cdot\mathbf{x}} f(\mathbf{x}).$$

We write $f_{\mathbf{k}}$ for functions of a discrete Fourier variable, and $f(\mathbf{k})$ in the continuum. In cases, where momentum integrals appear in clusters, we often use the compactified notation

$$\int \frac{d^d k}{(2\pi)^d} \equiv \int (dk).$$

For **lattice systems** with spacing a and $f(\mathbf{x}) \to f_{\mathbf{v}}$ with $v_i \in a\mathbb{Z}$, we use

$$f_{\mathbf{v}} = \frac{1}{N} \sum_{\mathbf{k}} e^{i\mathbf{k}\cdot\mathbf{v}} f_{\mathbf{k}}, \qquad f_{\mathbf{k}} = \sum_{\mathbf{v}} e^{-i\mathbf{k}\cdot\mathbf{x}} f_{\mathbf{v}},$$

where the momentum sum now runs over $k_i \in [-\pi/a, \pi/a)$, and $N = (L/a)^d$ is the total number of sites. (For an infinite lattice system, the sum again turns into an integral, $N^{-1}\sum_{\mathbf{k}} \to a^d \int (dk)$.)

While these are our preferred conventions, they sometimes need to be modified. For example, for a finite-size system, the unit normalized plane wave states $\langle \mathbf{k}|\mathbf{k}'\rangle = \delta_{\mathbf{k}\mathbf{k}'}$ have the real space representation $\langle \mathbf{x}|\mathbf{k}\rangle = L^{-d/2}\exp(i\mathbf{k}\cdot\mathbf{x})$. In this case, the representation change $\psi(\mathbf{x}) \equiv \langle \mathbf{x}|\psi\rangle = \sum_{\mathbf{k}}\langle \mathbf{x}|\mathbf{k}\rangle\langle \mathbf{k}|\psi\rangle = L^{-d/2}\sum_{\mathbf{k}}\exp(i\mathbf{k}\cdot\mathbf{x})\psi_{\mathbf{k}}$ leads to a Fourier transform with $L^{-d/2}$-normalization.

A.4.2 Temporal transform

For the transform to **a real temporal variable**, we use[13]

$$f(t) = \int \frac{d\omega}{2\pi} e^{-i\omega t} f(\omega), \qquad f(\omega) = \int dt\, e^{i\omega t} f(t).$$

Again, we may use the compact abbreviation $\int \frac{d\omega}{2\pi} \to \int (d\omega)$.

In **imaginary-time** field theory we have a temporal variable $\tau \in [0, T^{-1}]$ and

$$f(\tau) = T \sum_{n} e^{-i\omega_n \tau} f_{\omega_n}, \qquad f_{\omega_n} = \int d\tau\, e^{i\omega_n \tau} f(\tau),$$

with bosonic ($\omega_n = 2n\pi T$) or fermionic ($\omega_n = 2(n+1)\pi T$) Matsubara frequencies. This is the convention prevalent in condensed matter physics, and we will stick to it in this text.[14]

A.4.3 Wigner transform

Often in this text, we work with operators (or functions) $\hat{X} = \hat{X}(x_1, x_2)$ depending on two coordinates in such a way that the dependence on the center coordinate

---

[13] Note the sign difference relative to the spatial transform. In spatio-temporal transforms, we then encounter combinations $i(-\omega t + \mathbf{k}\cdot\mathbf{x}) = ik_\mu x^\mu$, consistent with the covariant contraction of a four-momentum $k^\mu = (\omega, \mathbf{k})$ with the coordinate $x^\mu = (t, \mathbf{x})$ under the Minkowski metric Eq. (9.1).

[14] However, it would be more natural to use a sign convention opposite to that of the real-time formalism. In that case, $i(\omega_n \tau + \mathbf{k}\cdot\mathbf{x}) = k_\mu x_\mu$ would turn into the Euclidean scalar product of four vectors. Thermal field theory in particle physics sometimes uses this convention. It appears not to be standard in condensed matter.

$x = (x_1 + x_2)/2$ is smooth, but that on the difference coordinate $\Delta x = x_1 - x_2$ is rapid. (Consider a fermion Green function in the presence of a smoothly varying external field as an example.) In such cases, it is often convenient to pass from the coordinates (x_1, x_2) to (x, p), where the Fourier argument, p, is conjugate to **Wigner** Δx. In the following, we discuss the implementation of this **Wigner transform** **transform** separately for the temporal and the spatial coordinate dependences of the theory. **Temporal Wigner transform:** Focusing on the time-like arguments, $X(t_1, t_2)$, we define $t \equiv (t_1 + t_2)/2$, and define the **temporal Wigner transform** as

$$X(\epsilon, t) \equiv \int d\Delta t \, e^{i\Delta t \epsilon} X(t + \Delta t/2, t - \Delta t/2),$$
$$X(t_1, t_2) = \int \frac{d\epsilon}{2\pi} \, e^{-i(t_1 - t_2)\epsilon} X(\epsilon, t). \tag{A.53}$$

In applications, we often need to know the Wigner representation of the product of two operators, $\hat{X}\hat{Y}$. Using the definition above, it is straightforward to derive the **Moyal** **Moyal product** identity **product**

$$(\hat{X}\hat{Y})(\epsilon, t) = e^{-\frac{i\hbar}{2}(\partial_{t_1}\partial_{\epsilon_2} - \partial_{\epsilon_1}\partial_{t_2})}\Big|_{\substack{\epsilon_1 = \epsilon_2 = \epsilon \\ t_1 = t_2 = t}} X(\epsilon_1, t_1)Y(\epsilon_2, t_2)$$
$$= X(\epsilon, t)Y(\epsilon, t) - \frac{i\hbar}{2}(\partial_t X(\epsilon, t)\partial_\epsilon Y(\epsilon, t) - \partial_\epsilon X(\epsilon, t)\partial_t Y(\epsilon, t)) + \mathcal{O}(\hbar^2 \partial_{\epsilon, t}^2)$$
$$= \left(XY - \frac{i}{2}\{X, Y\} + \mathcal{O}(\hbar^2 \partial_{\epsilon, t}^2)\right)(\epsilon, t), \tag{A.54}$$

where in the third line we have introduced the Poisson bracket in temporal variables

$$\{X, Y\} = \partial_t X \partial_\epsilon Y - \partial_\epsilon X \partial_t Y. \tag{A.55}$$

In Eq. (A.54) we have temporarily reintroduced $\hbar$ to indicate that higher-order terms contain Planck's constant in comparison with higher-derivative combinations. **Spatial Wigner transform:** Similar reasoning applied to the space-like coordinate dependence defines the **spatial Wigner transform**

$$X(\mathbf{r}, \mathbf{p}) \equiv \int d\Delta \mathbf{r} \, e^{-i\Delta \mathbf{r} \cdot \mathbf{p}} X(\mathbf{r} + \Delta \mathbf{r}/2, \mathbf{r} - \Delta \mathbf{r}/2),$$
$$X(\mathbf{r}_1, \mathbf{r}_2) = \int \frac{d^d p}{(2\pi)^d} \, e^{i(\mathbf{r}_1 - \mathbf{r}_2) \cdot \mathbf{p}} X(\mathbf{r}, \mathbf{p}), \tag{A.56}$$

where the choice of signs in the exponent reflects the conventions standard in spatial and temporal Fourier transforms, respectively.

For later reference, a number of Wigner transform identities are summarized in table A.1, where the spatial analog of Eq. (A.55) is defined as $\{f, g\} = \partial_\mathbf{r} f \partial_\mathbf{p} g - \partial_\mathbf{p} f \partial_\mathbf{r} g$.

Table A.1 Wigner transform identities.

Temporal		Spatial	
Operator	Wigner representation	Operator	Wigner representation
$\hat{X}(\hat{t})$	$X(t)$	$\hat{X}(\hat{\mathbf{r}})$	$X(\mathbf{r})$
$\hat{X}(\hat{\epsilon}) = \hat{X}(i\partial_t)$	$X(\epsilon)$	$\hat{X}(\hat{\mathbf{p}}) = \hat{X}(-i\partial_{\mathbf{r}})$	$X(\mathbf{p})$
$[\hat{Y}, \hat{X}]$	$-i\{Y, X\} + \dots$	$[\hat{Y}, \hat{X}]$	$i\{Y, X\} + \dots$
$\mathrm{tr}(\hat{X})$	$\int \frac{d\epsilon dt}{2\pi} X(\epsilon, t)$	$\mathrm{tr}(\hat{X})$	$\int \frac{d^d r d^d p}{(2\pi)^d} X(\mathbf{r}, \mathbf{p})$

Finally, it is often convenient to combine temporal and spatial transforms into a **spatio-temporal Wigner transform**,

$$
X(x, p) \equiv \int d\Delta x \, e^{i\Delta x^T \eta p} X(x + \Delta x/2, x - \Delta x/2),
$$
$$
X(x_1, x_2) = \int dp \, e^{-i(x_1 - x_2)^T \eta p} X(x, p),
$$
(A.57)

where $x = (t, \mathbf{x})^T = (x_1 + x_2)/2$, $p = (\epsilon, \mathbf{p})^T$, the Minkowski metric $\eta = \mathrm{diag}(1, -\mathbf{1}_d)$, and $dx \equiv dt d^d x$, $dp \equiv d\epsilon d^d p/(2\pi)^{d+1}$.

Notice that the Wigner transform represents operators $\hat{X}$ as functions of a single phase space coordinate $X(\mathbf{r}, \mathbf{p})$, rather than through matrix elements $X(\mathbf{r}_1, \mathbf{r}_2)$. This makes it an optimal representation for semiclassical approximation schemes.

Index